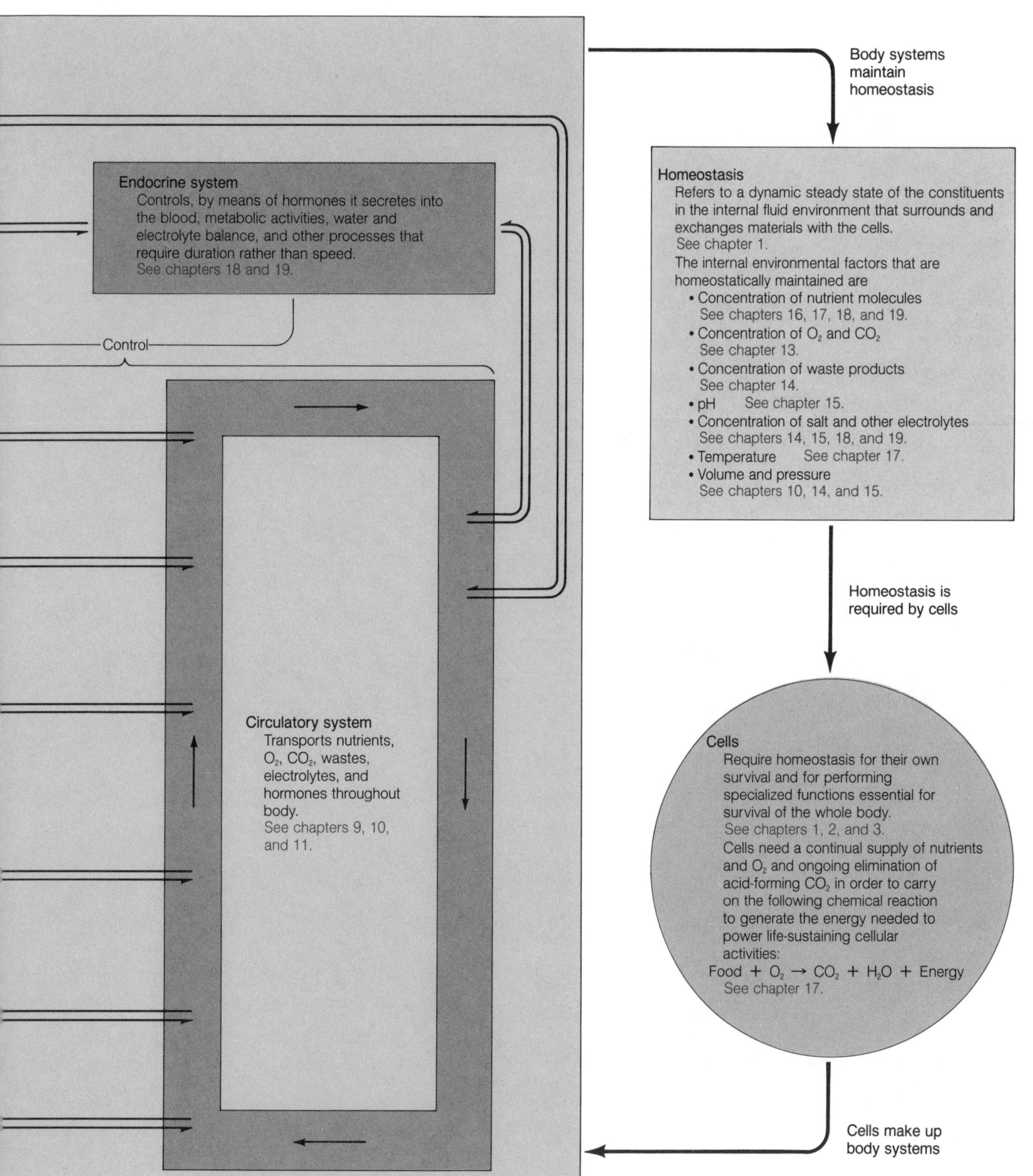

Body systems maintain homeostasis
Endocrine system
Controls, by means of hormones it secretes into the blood, metabolic activities, water and electrolyte balance, and other processes that require duration rather than speed.
See chapters 18 and 19.
Control
Circulatory system
Transports nutrients, O_2, CO_2, wastes, electrolytes, and hormones throughout body.
See chapters 9, 10, and 11.
Homeostasis
Refers to a dynamic steady state of the constituents in the internal fluid environment that surrounds and exchanges materials with the cells.
See chapter 1.
The internal environmental factors that are homeostatically maintained are
• Concentration of nutrient molecules
See chapters 16, 17, 18, and 19.
• Concentration of O_2 and CO_2
See chapter 13.
• Concentration of waste products
See chapter 14.
• pH See chapter 15.
• Concentration of salt and other electrolytes
See chapters 14, 15, 18, and 19.
• Temperature See chapter 17.
• Volume and pressure
See chapters 10, 14, and 15.
Homeostasis is required by cells
Cells
Require homeostasis for their own survival and for performing specialized functions essential for survival of the whole body.
See chapters 1, 2, and 3.
Cells need a continual supply of nutrients and O_2 and ongoing elimination of acid-forming CO_2 in order to carry on the following chemical reaction to generate the energy needed to power life-sustaining cellular activities:
Food + O_2 → CO_2 + H_2O + Energy
See chapter 17.
Cells make up body systems

HUMAN PHYSIOLOGY

FROM CELLS TO SYSTEMS

HUMAN PHYSIOLOGY

FROM CELLS TO SYSTEMS

Lauralee Sherwood

DEPARTMENT OF PHYSIOLOGY
SCHOOL OF MEDICINE
WEST VIRGINIA UNIVERSITY

WEST PUBLISHING COMPANY

St. Paul ■ *New York* ■ *Los Angeles* ■ *San Francisco*

Text copyediting Judith Lary, Naples Editing Service
Art copyediting Jo-Anne Naples, Naples Editing Service
Composition Rolin Graphics
Artwork Wayne Clark, Cyndie Clark-Huegel, Darwen and Vally Hennings, Sandra McMahon, Rolin Graphics, and John and Judy Waller. Individual credits follow Index.
Dummy artist David J. Farr/IMAGESMYTHE Inc.
Cover design David J. Farr/IMAGESMYTHE Inc.
Cover/title page artwork *Men Working* by Leonardo da Vinci. Reproduced with permission from Windsor Castle, Royal Library. Copyright ©1988 Her Majesty Queen Elizabeth II. This work of art was selected for the cover because the figures illustrate human activities that require the functional integration of a variety of body systems.

Printed in the United States of America

96 95 94 93 92 91 90 89 8 7 6 5 4 3 2 1 0

Library of Congress Cataloging-in-Publication Data

Sherwood, Lauralee.
Human physiology.

Includes bibliographies and index.
1. Human physiology. I. Title. [DNLM: 1. Anatomy.
2. Physiology. QT 104 S554h]
QP34.5.S48 1989 612 88-37830
ISBN 0-314-47230-4

TO *my husband, Peter Marshall, who supported me in every possible way;*

my parents, Larry and Lee Sherwood, who gave me the confidence, training, and encouragement to pursue my dreams;

and my daughters, Melinda and Allison Marshall, for sacrifices made in family life.

BRIEF CONTENTS

C O N T E N T S

CHAPTER 6

Peripheral Nervous System: Afferent Division; Special Senses 158

CHAPTER 7

Peripheral Nervous System: Efferent Division 202

CHAPTER 8

Muscle Physiology 216

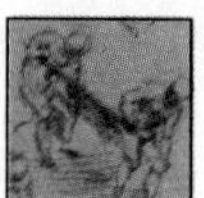

CHAPTER 9

Cardiac Physiology 258

CHAPTER 10

The Vasculature and Blood Pressure 298

CHAPTER 11

Blood 342

CHAPTER 12

Defense Mechanisms of the Body 362

C H A P T E R 13

Respiratory System 406

C H A P T E R 14

Urinary System 462

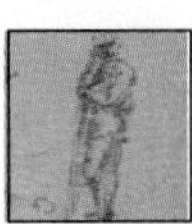

C H A P T E R 15

Fluid Balance and Acid-Base Balance 510

C H A P T E R 16

Digestive System 546

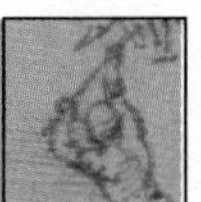

C H A P T E R 17

Energy Balance and Temperature Regulation 602

C H A P T E R 18

Principles of Endocrinology; Central Endocrine Organs 622

C H A P T E R 19

Peripheral Endocrine Organs 662

CHAPTER 20

Reproductive Physiology 710

PREFACE

PHILOSOPHY, GOALS, AND THEME

A baby first discovering that it can control its own hands will spend many hours, fascinated, as it manipulates them in front of its face. Most of us, even infants, have a natural curiosity about how our bodies work. Our bodies are quite miraculous. No machine has been constructed that can take over even a portion of a natural body function as effectively. By capitalizing on students' natural curiosity about themselves, I have strived to make physiology a subject that they can enjoy learning.

Even the most tantalizing subject matter, however, can become drudgery to study and difficult to comprehend if not effectively presented. Therefore, this book has a logical, understandable format that is unencumbered by unnecessary details and that emphasizes how each concept is an integral part of the whole subject matter. Too often, students view isolated sections of a physiology course as separate entities; by understanding how each component of the body depends on other components, a student can appreciate the integrated functioning of the human body. The text focuses on the mechanisms of body function from cells to systems, organized around the central theme of homeostasis—how the body meets changing demands while maintaining the internal constancy necessary for all cells and organs to function.

The text is written with its primary target audience—undergraduate students preparing for health-related careers—in mind. Its approach and depth are appropriate, however, for other undergraduate student populations. Because it is intended to serve as an introductory text and, for most students, may be their only exposure to a formal physiology text, there is broad coverage of all aspects of physiology, yet depth, where needed, is not sacrificed. The scope of this text has been limited by judicious selection of pertinent content that a student can reasonably be expected to assimilate in a one-semester physiology course.

Because of the diverse background of students for whom the text is intended, I do not presume that they have had extensive math and science prerequisite courses. Quantitative approaches are minimized, although biochemical and biophysical concepts are introduced when necessary. A particular effort has been made to clarify all discussions of math, physics, and chemistry. Likewise, because anatomy is not a prerequisite course, enough relevant anatomy is integrated within the text to make the inseparable relation between structure and function meaningful.

Considering the clinical orientation of the target audience, research methodologies and data are not emphasized, although the material presented is based on up-to-date evidence. Some controversial ideas and hypotheses and acknowledgment of gaps in knowledge are presented to illustrate that physiology is a dynamic, changing discipline.

To keep pace with the rapid advances in the health sciences, today's health professional students must be able to draw on their conceptual understanding of physiology instead of merely recalling isolated facts that soon may be outdated. Therefore, a didactic writing style is used with emphasis on understanding the basic principles and concepts of physiology. The text is written in simple, straightforward language, and every effort is made to assure smooth reading through good transitions, logical reasoning, and integration of ideas throughout the text.

FEATURES; PEDAGOGICAL AIDS

Topical outline; opening vignette

Each chapter begins with a brief outline, **Chapter Contents at a Glance**, followed by an attention-getting vignette relevant to the chapter. This feature is designed to capture the students' interest and draw them into the chapter.

Homeostatic model

A unique homeostatic model depicting the relations among cells, systems, and homeostasis is developed in the introductory chapter and featured on the book's inside front cover.

Reference to the full model accompanies a brief chapter summary that emphasizes how the body part discussed in the chapter fits in with the whole body. This pedagogical feature is designed to facilitate the students' comprehension of the interactions and interdependency of body systems, even though each system is discussed separately.

Analogies

A notable feature of the text is the use of many analogies and frequent references to common life experiences as a way of helping students relate to the concepts. Students who have limited science and math background may find this alternative approach of relating physiology concepts to something familiar particularly helpful.

Pathophysiology

An important aspect of keeping the students' interest is to help them realize they are learning worthwhile and applicable material. Because most students using this text will have health-related careers, frequent references to pathophysiology and clinical physiology are interspersed throughout the text to demonstrate the content's relevance as a foundation for future professional goals.

Full-color visual aids

This text is the first in the field to employ a full-color art program throughout the entire book as a functional tool to learning. Flow diagrams, anatomical illustrations, schematic representations, photographs, tables, and graphs are designed to complement and reinforce the written material. Color is used to enhance understanding, not merely for aesthetics. For example, inhibitory pathways are shown in red consistently throughout the text. Unique to this book, the people depicted in the various illustrations are realistic representatives of a cross section of humanity.

Cellular approach

Even though the text is primarily organized according to body systems, it provides extensive coverage of cellular physiology in the beginning and incorporates explanations of function at the cellular and molecular level throughout as a basis for understanding organ function. The cellular/molecular approach is slighted in most general undergraduate physiology textbooks, yet this approach is at the cutting edge of much of physiology research.

Boxed feature on exercise physiology

A relevant box entitled **A Closer Look at Exercise Physiology** is featured in each chapter. Current concepts related to exercise physiology are included for three reasons: increasing national awareness of the importance of physical fitness; increasing recognition of the value of prescribed therapeutic exercise programs for a variety of conditions; and growing career opportunities related to fitness and exercise.

Feed-forward statements as subsection titles

Instead of using traditional topic titles for each subsection (for example, **Heart valves**), the title is a feed-forward statement that alerts the student to the main point of the subsection to come (for example, **Heart valves ensure the proper direction of blood flow through the heart**).

Cross-references

Cross-references to related material in earlier chapters enable students to quickly refresh their memories and allow instructors more flexibility in organizing the presentation of materials.

Key terms; glossary index

Key terms are boldfaced and defined as they appear in the text. Word derivations are provided as necessary to enhance understanding of new words. A glossary index in the back of the book is available for reference to the pages at which key terms are defined.

End-of-chapter overviews and review exercises

A brief end-of-chapter overview, **Chapter in Perspective**, points out how the body part just discussed fits in with the homeostatic theme. An accompanying logo of the homeostatic model and reference to the full model on the inside front cover serve to remind the student of the integrated functioning of the whole body.

Review exercises pertaining to the main points are provided at each chapter's end for use as self-study aids, homework assignments, or a pool of essay-exam questions. One thought-provoking question not specifically covered in the text, **A Point to Ponder**, is included. If desired, these questions could be used as a basis for class discussions. Explanations of **A Point to Ponder** are provided in the Instructor's Manual.

Chemistry appendix

Most undergraduate physiology texts have a chapter on chemistry, yet physiology instructors rarely teach basic chemistry concepts. The decision was made, therefore, to reserve valuable text space for physiological concepts and to provide instead a chemistry appendix as a handy reference for students who need an introduction to or review of basic chemistry concepts that are essential to understanding physiology.

Organization

There is no ideal organization of physiological processes into a logical sequence. With the sequence I have chosen, most chapters build on material presented in immediately preceding chapters, yet each chapter is designed to stand on its own to allow the instructor flexibility in curriculum design. The general flow is from introductory background information to cells to excitable tissue to organ systems. Every attempt has been made to provide logical transitions from one chapter to the next. For example, Chapter 8, *Muscle Physiology*, ends with a discussion of cardiac muscle, which is carried forward into Chapter 9, *Cardiac Physiology*. Even topics that seem unrelated in sequence, such as Chapter 12, *Defense Mechanisms of the Body*, and Chapter 13, *Respiratory System*, are linked together, in this case by ending Chapter 12 with a discussion of respiratory defense mechanisms.

Several organizational features warrant specific mention. The most difficult decision in organizing this text was choosing the placement of the chapters on the endocrine system. Intermediary metabolism of absorbed nutrient molecules is largely under endocrine control, providing a link from digestion (Chapter 16) and energy balance (Chapter 17) to the endocrine system (Chapters 18 and 19). There is merit in placing the chapters on the nervous and endocrine systems in close proximity because of these systems' roles as the body's major control systems. Placing the endocrine system chapters earlier, immediately after the discussion of the nervous system (Chapters 4 through 7), however, would have created two problems. First, it would have disrupted the logical flow of material related to excitable tissue. Second, I could not have covered the endocrine system at the level of depth its importance merits if it had been discussed before the students were provided the background essential to understanding this system's roles in maintaining homeostasis. Placing the endocrine system chapters late in the book does not mean, however, that students are not exposed to endocrine function or hormones until near the book's completion. Endocrine control and hormones are defined in Chapter 1 and are compared in Chapter 5 with nervous control. Specific hormones are introduced in appropriate chapters, such as vassopressin and aldosterone in the chapters on kidney and fluid balance. Chapters 18 and 19 explore the basic characteristics of endocrine glands and hormones as well as the control and functions of specific endocrine secretions.

Unique to this book, the skin is covered in the chapter on defense mechanisms of the body in consideration of the skin's newly recognized immune functions. Bone is also covered more extensively in the endocrine chapters than in most undergraduate physiology texts, especially with regard to hormonal control of bone growth and bone's dynamic role in calcium metabolism.

Departure from traditional groupings of material in several important instances has permitted more independent and more extensive coverage of topics that are frequently omitted or buried within chapters concerned with other subject matter. For example, a separate chapter is devoted to fluid balance and acid-base regulation, topics often tucked within the kidney chapter. The grouping of the autonomic nervous system, alpha motor neurons, and the neuromuscular junction in an independent chapter on the efferent division of the peripheral nervous system, which serves as a link between the nervous system chapters and the muscle chapter, is another example.

Although there is a rationale for covering the various aspects of physiology in the order given here, it is by no means the only logical way of presenting the topics. Each chapter is able to stand on its own, especially with the cross references provided, so that the sequence of presentation can be varied at the instructor's discretion. Some chapters may even be omitted, depending on the students' needs and interests and the time constraints of the course. For example, a cursory explanation of the defense role of the leukocytes is covered in the chapter on blood, so an instructor could choose to omit the more detailed explanations of immune defense in Chapter 12. Similarly, the in-depth coverage of topics in Chapters 2, 6, 15, 17, and 19 could selectively be omitted without sacrificing a student's general appreciation of systems-approach physiology.

As an alternative to total omission of certain chapters, the Learning Resource Manual could be used as a supplement for less-comprehensive coverage of these topics.

Ancillaries

Learning Resource Manual

The Learning Resource Manual incorporates two types of supplemental learning aids. First, concise narrative *Section*

Synopses are provided to serve a two-fold purpose: they are a useful vehicle for student review; and they can be assigned in lieu of the full chapter for topics that need a condensed overview in a particular course. This permits an instructor maximum flexibility.

Second, the Learning Resource Manual includes *Learning Checks* to accompany each section of all chapters. The Learning Checks consist of various objective-type questions that invite the student to answer right in the manual. An answer key, provided on a separate page, includes the reasons that any false statements of a true/false question are not true. These self-check exercises appear after each section instead of only at the end of the chapter to help students absorb information and concepts before continuing on to new subject matter.

In addition, the Learning Resource Manual offers a *Supplemental Reading List* for students interested in pursuing particular subject matter in greater depth.

Instructor's Manual

This teaching aid offers lecture suggestions and a list of pertinent films and software that may be ordered to complement the lectures. Explanations of **A Point to Ponder** questions in the textbook's review exercises are provided. For the instructor's convenience, an extensive assortment of author-generated test questions of various objective formats and degrees of difficulty are also included.

Computerized test service

Computerized tests composed of questions of the instructor's choice from the Instructor's Manual are available from West Publishing Company.

Colored transparency acetates and transparency masters

Ready-to-use colored transparency acetates plus masters for instructor preparation of transparencies are available for selected illustrations in the text. A complete list of those available is included by figure number and title in the Instructor's Manual.

ACKNOWLEDGMENTS

I gratefully acknowledge the many people who helped bring this textbook to fruition over a four-year period. A special thank-you goes to two individuals who contributed substantially to the content of the book: Rachel Yeater (professor, Sports Exercise Program, and director, Human Performance Laboratory, School of Physical Education, West Virginia University), who contributed the material for the boxed features **A Closer Look at Exercise Physiology**; and Spencer Seager (chairman, Chemistry Department, Weber State College) who prepared Appendix A, **A Review of Chemical Principles.**

During the book's creation, many colleagues provided assistance. George Hedge and Robert Goodman, Department of Physiology, West Virginia University, deserve a special note of gratitude for their willingness to share materials used for their recent publication, **Clinical Endocrine Physiology** (Saunders, 1987), and for their thoughtful reviews of this book's chapters on endocrinology and reproduction. Appreciation is also extended to Elizabeth Walker and Dennis Overman, Department of Anatomy, West Virginia University, who provided many custom-made light and electron micrographs for the book.

Others at West Virginia University deserving of recognition for providing resource materials or countless clarifications include James Culberson and William Beresford, Department of Anatomy; Ronald Gaskins, Department of Medicine; Sidney Schochet, Jr. and Val Vallyathan, Department of Pathology; and Paul Brown, John Connors, Gunter Franz, Wilbert Gladfelter, Michael Johnson, Ping Lee, Philip Miles, Ronald Millechia, and William Stauber, Department of Physiology.

I express sincere appreciation to the following reviewers for their conscientious reading and assessment of the manuscript at its various stages of development and for offering valuable advice on ways to improve the text.

Christopher C. Barney
Hope College

Milo C. Barone
Fairfield University

Richare Beil
Chicago State University

C.H. Bennett
Kentucky State University

Craig Black
University of Toledo

Alan Brush
University of Connecticut at Storrs

Cynthia Carey
University of Colorado—Boulder

William A. Cooper
West Texas State University

Dwayne H. Curtis
California State University at Chico

Milton Fingerman
Tulane University

Bruce Gladden
University of Louisville

Lewis Greenwald
Ohio State University

John Harley
Eastern Kentucky University

Richard W. Heninger
Brigham Young University

Narinder Kapoor
Concordia University

T. Daniel Kimbrough
Virginia Commonwealth University

Charles Leavell
Fullerton College

Florence Ledwitz-Rigby
Northern Illinois University

Sheldon Lustick
Ohio State University

Fred McCorkle
Central Michigan University

Katie Mechlin
Wright State University

Esmail Meisami
University of Illinois at Urbana-Champaign

A. Kenneth Moore
Seattle Pacific University

Richard Moss
University of Wisconsin—Madison

W. Brian O'Connor
University of Massachusetts

Carmello Privitera
State University of New York—Buffalo

Dell Redding
Evergreen Valley College

Jerome B. Senturia
Cleveland State University

Gregory A. Stephens
University of Delaware

M.H. Stetson
University of Delaware

Gary Whitson
University of Tennessee at Knoxville

Robert Williams
East Texas State University

Jack Wood
Western Michigan University

Gerald Yochim
University of Kansas

It has been a personal and professional pleasure to work with two outstanding editors from West Publishing Company: Jerry Westby, acquisitions editor, a wonderful facilitator who guided me from prospectus to production with unfailing support, helpful suggestions, and gentle prodding to meet important deadlines; and Barbara Fuller, production editor, a master of organizational skills and diplomacy who pulled this complex project together on schedule with meticulous attention to detail, unflagging dedication, and sustained good cheer. Also at West Publishing Company are Liz Lee, who oversaw development of the ancillary materials; Maureen Rosener, promotion manager for the textbook; and Kristen Weber, interior designer.

The quality of the text's artwork exceeds my expectations, thanks to the following artists whose creativity and concern for both accuracy and aesthetics are greatly appreciated: John and Judy Waller, Darwen and Vally Hennings, Cyndie Clark-Huegel, Wayne Clark, Rolin Graphics, and Sandy McMahon.

My heartfelt appreciation is extended to my dear friend, Avonell Painter, for supporting me in numerous ways in this endeavor, including typing a large share of the manuscript. My daughter, Melinda Marshall, contributed to the typing effort, as did Carolyn Wisman, Diane Kinney, and Lisa Forbes. The final revisions were prepared on the word processor by my husband, Peter Marshall. I also wish to thank my family and friends for their tolerance of my preoccupation with the textbook during an extended period of time. I could not have completed the project without their understanding and support.

Human Physiology

From Cells to Systems

CHAPTER 1

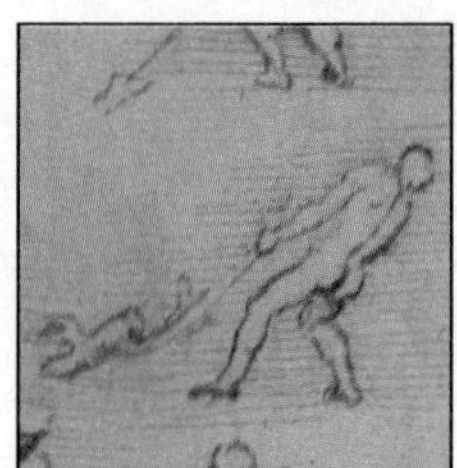

HOMEOSTASIS: THE FOUNDATION OF PHYSIOLOGY

CHAPTER CONTENTS AT A GLANCE

INTRODUCTION *Human physiology is the study of the functions of the human body, or how the body works. Physiologists view the body as a machine whose mechanisms of action can be explained in terms of the same physical and chemical principles that apply to other components of the universe. It is important to distinguish between the mechanistic and teleological approaches to explaining the various events that occur in the body. With the* **mechanistic approach,** *which is employed by physiologists, the* mechanisms *of action are emphasized. The events that occur in the body are explained in terms of a cause-and-effect sequence of physical and chemical processes. With a* **teleological approach,** *phenomena that occur in the body are explained in terms of their particular purpose in fulfilling a bodily* need, *with no consideration being given to how this outcome is accomplished.*

A simple example will help you to distinguish between these approaches. A teleological explanation of why you shiver when you are cold is "to keep warm," because shivering generates heat. A physiologist's mechanistic explanation of why you shiver is that when temperature-sensitive nerve cells detect a fall in body temperature, they signal the hypothalamus, the part of the brain responsible for temperature regulation. In response, the hypothalamus activates nerve pathways that ultimately bring about involuntary, oscillating muscle contractions (that is, shivering).

Because those mechanisms of action that are most beneficial to survival have prevailed throughout evolutionary history, it is helpful when studying physiology to predict what mechanism would be useful to the body under a particular circumstance. Chances are such a mechanism will exist. This does not mean that the discipline of physiology is an elaborate guessing game. Predictions must be verified by facts ascer-

tained by careful scientific investigation. However, the fact that most bodily mechanisms do serve a useful purpose (having been naturally selected throughout evolutionary time) allows you to apply a certain amount of logical reasoning to each new situation you encounter in your study of physiology. If you always try to find the thread of logic in what you are studying, you can avoid a good deal of pure memorization, and, more importantly, you will better understand and appreciate the concepts being presented.

Physiology is closely interrelated with ***anatomy,*** *the study of the structure of the body. Just as the functioning of an automobile depends on the shapes, organization, and interactions of its various parts, the same is true of the human body. Structure and function are inseparable. Therefore, as we tell the story of how the body works, we will provide sufficient anatomical background for you to understand the function of the body part being discussed.*

Levels of Organization in the Body

Cells are the basic units of life.

The basic unit of both structure and function is the **cell,** the foundation of all living organisms. The cell is the smallest unit capable of carrying out the processes associated with life. In fact, simple life forms include **unicellular** (single-celled) **organisms,** such as bacteria and amoebae. Humans are **multicellular** (many-celled) **organisms,** with the adult human body being an aggregate of more than 75 trillion cells.

All cells, whether they exist as solitary cells or are part of a multicellular organism, perform certain basic functions essential for survival of the cell and, in turn, survival of the organism. These basic cell functions include:

1. obtaining nutrients and oxygen (O_2) from the environment surrounding the cell;
2. performing various chemical reactions that use food and O_2 to acquire energy for the cell;
3. eliminating to the cell's surrounding environment carbon dioxide (CO_2) and other wastes produced during these chemical reactions;
4. synthesizing proteins and other components needed for cellular structure, for growth, and for carrying out particular cell functions;
5. being sensitive and responsive to changes in the environment surrounding the cell;
6. controlling to a large extent the exchange of materials between the cell and its surrounding environment;
7. moving materials from one part of the cell to another in carrying out cellular activities, with some cells even being able to move in entirety through their surrounding environment; and
8. in the case of most cells, reproducing. (Some body cells have lost the ability to reproduce, such as nerve cells and muscle cells. When these cells are destroyed through trauma or disease processes, they cannot be replaced. With other body cells, repair is possible.)

Cells are remarkable in the similarity with which they carry out these functions. Thus, all cells share many common characteristics. In multicellular organisms, each cell also performs a specialized function, which is usually a modification or elaboration of one of the basic cell functions. The following are a few examples:

□ By taking special advantage of their protein-synthesizing ability, the gland cells of the digestive system secrete digestive enzymes, which are all proteins.

□ Capitalizing on the basic ability of cells to respond to changes in their surrounding environment, nerve cells generate and transmit to other regions of the body electrical impulses that relay information about changes to which the nerve cells are responsive.

□ The ability of kidney cells to selectively retain the substances needed by the body while eliminating unwanted substances in the urine depends on these cell's highly specialized ability to control exchange of materials between the cell and its environment.

□ Muscle contraction, which involves selective movement of internal structures to bring about shortening of muscle cells, is an elaboration of the inherent capability of these cells to produce intracellular ("within the cell") movement.

It is important to recognize that each cell performs these specialized activities in addition to carrying on the unceasing, fundamental activities required of all cells. The fundamental cellular activities are essential for survival of each individual cell, whereas the specialized contributions and interactions among the cells of which a multicellular organism is composed are essential for the survival of the whole organism.

Cells are progressively organized into tissues, organs, systems, and, finally, the whole body.

Just as a machine does not function unless its various parts are properly assembled, the cells of the body must be specifically organized to carry out the life-sustaining processes of the body as a whole, such as digestion, respiration, and circulation.

There are four levels of organization in the body: cells, tissues, organs, and systems.

Cells of similar structure and function are organized into **tissues,** of which there are four primary types: muscle, nervous, epithelial, and connective tissue. Each tissue consists of cells of a single specialized type, along with varying amounts of extracellular ("outside of the cell") material.

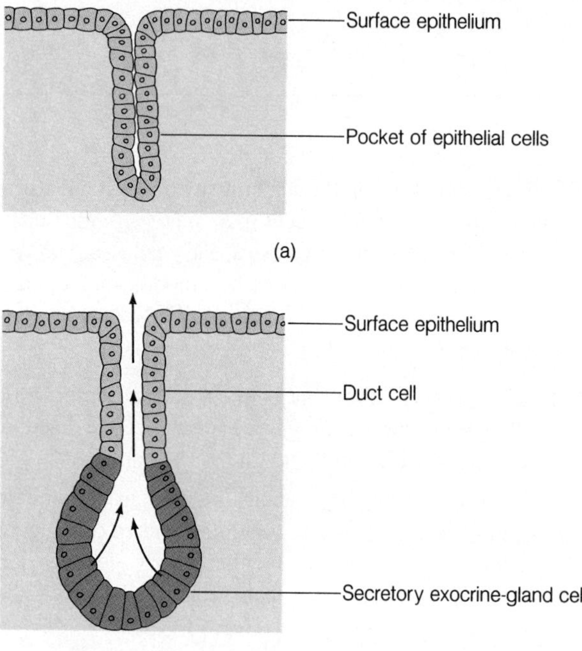

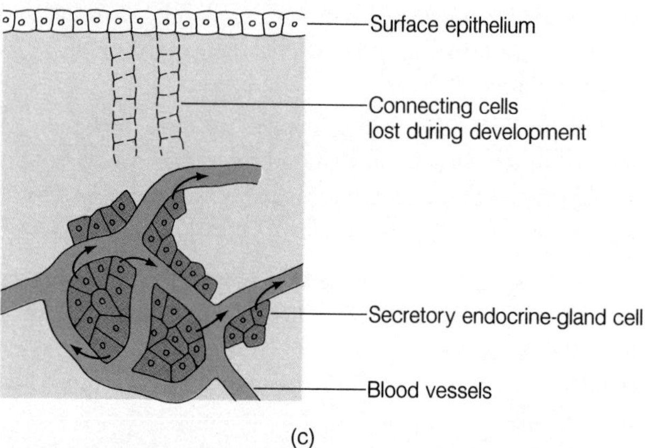

Figure 1–1 Exocrine and Endocrine Gland Formation *(a) Glands arise during development from the formation of pocketlike invaginations of surface epithelial cells. (b) If the cells at the deepest part of the invagination become secretory and release their product through the connecting duct to the surface, an exocrine gland is formed. (c) If the connecting cells are lost and the deepest secretory cells release their product into the blood, an endocrine gland is formed.*

Muscle tissue is composed of cells specialized for contraction and force generation. There are three types of muscle tissue: skeletal muscle, which accomplishes movement of the skeleton; cardiac muscle, which is responsible for pumping blood out of the heart; and smooth muscle, which encloses and controls movement of contents through hollow tubes and organs, such as the movement of food through the digestive tract.

Nervous tissue consists of cells specialized for initiation and transmission of electrical impulses, sometimes over long distances. These electrical impulses serve as signals for relaying information from one part of the body to another. Nervous tissue is found in: (1) the brain; (2) the spinal cord; (3) the nerves that signal information about the environment and about the status of various factors in the body that are subject to regulation, such as blood pressure; and (4) the nerves that influence muscle contraction or gland secretion.

Epithelial tissue is made up of cells specialized in the exchange of materials between the cell and its environment. This tissue is organized into two general types of structures: epithelial sheets and secretory glands. Epithelial cells are joined together very tightly to form sheets of tissue that cover and line various parts of the body. For example, the outer layer of the skin is epithelial tissue, as is the lining of the digestive tract. In general, these epithelial sheets serve as boundaries that separate the body from the external environment and from the contents within cavities that communicate with the external environment, such as the digestive-tract lumen. (A **lumen** is the cavity within the interior of hollow organs and tubes.) Only selected transfer of materials is permitted between the regions separated by an epithelial barrier. The type and extent of controlled exchange varies, depending on the location and function of the epithelial tissue. For example, very little can be exchanged between the body and external environment across the skin, whereas the epithelial cells lining the digestive tract are specialized for absorption of nutrients, and the cells lining the kidney tubules are specialized in exchanging materials with the blood during the formation of urine.

Glands are epithelial-tissue derivatives that are specialized for secretion. **Secretion** refers to the release from the cell, in response to appropriate stimulation, of specific products that have in large part been synthesized by the cell. Glands are formed during embryonic development by pockets of epithelial tissue that dip inward from the surface. There are two categories of glands—exocrine and endocrine (Fig. 1–1). If during development the connecting cells between the epithelial surface and the secretory-gland cells within the depths of the invagination remain intact as a duct between the gland and the surface, an exocrine gland is formed. **Exocrine glands** (*exo* means "external"; *crine* means "secretion") secrete through ducts to the outside of the body (or into a cavity that communicates with the outside). Examples are sweat glands and glands that secrete digestive juices. If, on the other hand, the connecting cells disappear during development and the secretory-gland cells are isolated from the surface, an endocrine gland is formed. **Endocrine glands** (*endo* means "internal") lack ducts and release their secretory products, known as *hormones*, internally into the blood within the body. An example is the parathyroid gland, which secretes parathyroid hormone into the blood. By means of the blood, this hormone reaches its sites of action at the bones and kidneys.

Connective tissue is distinguished by relatively few cells dispersed within an abundance of extracellular material. As implied by its name, connective tissue serves to connect, support, and anchor various body parts. It includes such diverse structures as the loose connective tissue that attaches epithelial tissue to underlying structures; tendons that attach skeletal muscles to bones; bone, which gives the body shape, support, and protection; and blood, which transports materials from one part of the body to another. Except for blood, the cells within connective tissue produce specific molecules that they release into the extracellular spaces between the cells. An example is the rubber-band–like protein fiber **elastin,** whose presence facilitates a structure's stretching and recoiling, such as in the air sacs of the lung, which alternately inflate and deflate during breathing.

Muscle, nervous, epithelial, and connective tissue are the primary tissues in a classical sense; that is, each is formed by the joining together of groups of cells of the same specialized structure and function. The term *tissue* is also frequently used in another sense to refer to the aggregate of various cellular and extracellular components that make up a particular organ (for example, lung tissue or liver tissue).

Organs are composed of two or more types of primary tissue organized to perform a particular function or functions. The stomach is an example of an organ made up of all four primary tissue types. The tissues that compose the stomach function collectively to store the ingested food and move it forward into the remainder of the digestive tract as well as to begin the digestion of protein. The stomach is lined with epithelial tissue that restricts the transfer of harsh digestive chemicals and undigested food from the stomach lumen into the blood. Epithelially derived gland cells in the stomach include exocrine cells, which secrete protein-digesting juices into the lumen, and endocrine cells, which secrete a hormone that helps regulate the stomach's exocrine secretion and muscle contraction. The walls of the stomach contain smooth-muscle tissue whose contraction mixes ingested food with the digestive juices and propels the mixture forward into the intestine. Also within the walls is nervous tissue, which, along with hormones, controls muscle contraction and gland secretion. These various tissues are all bound together by connective tissue.

Organs are further organized into **body systems,** each of which is a collection of organs that perform related functions and interact to accomplish a common activity that is essential for survival of the whole body. For example, the digestive system consists (in addition to the stomach) of the mouth, pharynx, throat, esophagus, small intestine, large intestine, salivary glands, pancreas, liver, and gall bladder. These digestive organs cooperate to accomplish the breakdown of dietary food into small nutrient molecules that can be absorbed into the blood.

The total body—a single living individual—is composed of the various organ systems structurally and functionally linked together as an entity that is separate from the external environment. Thus, the body is made up of living cells organized into life-sustaining systems.

CONCEPT OF HOMEOSTASIS

Body cells are in contact with a privately maintained internal environment instead of with the external environment that surrounds the body.

If each cell possesses basic survival skills, why can't the body cells live without performing specialized tasks and being organized according to specialization into systems that accomplish functions essential for the whole body's survival? The reason that the cells in a multicellular organism must contribute to the survival of the organism as a whole and cannot live and function without contributions from the other body cells is that the vast majority of the cells are not in direct contact with the external environment in which the organism lives. A unicellular organism such as an amoeba can directly obtain nutrients and O_2 from and eliminate wastes into its immediate external surroundings. A muscle cell, or any other cell, in a multicellular organism has the same need for life-supporting nutrient and O_2 uptake and waste elimination, yet the muscle cell cannot directly make these exchanges with the environment surrounding the body because of the cell's isolation from this external environment.

How is it possible for a muscle cell to make vital exchanges with the external environment with which it has no contact? The key is the presence of an aqueous **internal environment** with which the body cells are in direct contact. Various body systems accomplish exchanges between the external environment and internal environment. The body cells, in turn, make life-sustaining exchanges with the internal environment. The internal environment consists of the extracellular fluid, which is made up of **plasma,** the fluid portion of the blood, and **interstitial fluid,** which surrounds and bathes all the cells.

The digestive system transfers from the external environment into the plasma the nutrients required by all body cells. Likewise, the respiratory system transfers O_2 from the external environment into the plasma. The circulatory system distributes these nutrients and O_2 throughout the body. Thorough mixing and exchange of materials takes place between the plasma and the interstitial fluid across the thin, pore-lined walls of the capillaries, the smallest of the blood vessels. As a result, the nutrients and O_2 originally obtained from the external environment are delivered to the interstitial fluid that surrounds the cells. No matter how remote a cell is from the body surface, it is able to take in from this internal environment the nutrients and O_2 needed to support its own existence. Similarly, wastes extruded from the cells into the interstitial fluid are picked up by the plasma and transported to organs specialized in eliminating these wastes from the internal environment to the external environment. The lungs remove CO_2 from the plasma, and other wastes are removed by the kidneys for elimination in the urine.

Thus, a body cell takes in essential nutrients from and eliminates wastes into its watery surroundings, just as an amoeba does. The major difference is that the internal environment must constantly be maintained at the composition required to support the existence of all the cells by the same cells that it supports. In contrast, an amoeba does nothing to regulate its surroundings.

Each cell requires homeostasis, and each cell, as part of an organized system, contributes to homeostasis.

The body cells can live and function only when they are bathed by extracellular fluid that is compatible with their survival. This means that the chemical composition and physical state of the internal environment can be allowed to deviate only within narrow limits. As cells remove nutrients and O_2 from the internal environment, these essential materials must constantly be replenished in order for the cells' ongoing maintenance of life processes to continue. Likewise, wastes must constantly be removed from the internal environment so that they do not reach toxic levels. Other elements within the internal environment that are important for the maintenance of life also must be kept relatively constant. Maintenance of a relatively stable internal environment is termed **homeostasis** (*homeo* means "the same"; *stasis* means "to stand or stay").

The functions performed by each body system are aimed at maintaining homeostasis, thereby maintaining within the body the environment required for the survival and function of all the cells of which the body is composed. This is the central theme of physiology and of this book: ***each cell requires homeostasis for its own survival, and, in turn, each cell, through its specialized activities, contributes as part of a body system to the maintenance of the internal environment shared in common by all cells.***

The fact that the internal environment must be kept relatively stable does not mean that it is of absolutely unchanging composition, temperature, and so on. External and internal factors continuously threaten to disrupt homeostasis. For example, exposure to a cold environmental temperature tends to reduce the body's internal temperature. Likewise, addition of CO_2 into the internal environment tends to raise the concentration of this gas within the body. When any factor starts to move the internal environment away from optimal conditions, appropriate counterreactions are initiated to restore these conditions. For example, when the body temperature starts to fall on a cold day, compensatory shivering is initiated, which internally generates heat that restores the body temperature to normal. Similarly, a rise in the CO_2 levels within the internal environment triggers an increase in respiratory activity. The extra CO_2 is blown off to the external environment, restoring the CO_2 concentration in the extracellular fluid to normal. Thus, homeostasis should be viewed not as a fixed state but as a dynamic steady state in which the changes that do occur are minimized by compensatory physiological responses. The small fluctuations around the optimal level for each factor in the internal environment are normally kept within the narrow limits compatible with life by carefully regulated mechanisms. The various systems' activities must be regulated and coordinated to maintain a relatively steady state in the internal environment despite changes that continually threaten to disrupt the conditions essential for sustaining life. Furthermore, some changes in regulated factors that occur during exercise are considered normal under those circumstances, yet would be abnormal in a resting individual (see the accompanying boxed feature, A Closer Look at Exercise Physiology).

Figure 1–2 sums up the tripod that serves as the foundation for modern-day physiology: *cells require homeostasis, body systems maintain homeostasis, and cells make up body systems*. We have already described how cells are organized according to specialization into body systems. Why cells require homeostasis and how body systems maintain this internal constancy are the topics of the remainder of this book. For now we will provide a brief overview so that it will be easier for you to keep each chapter in perspective.

Among the factors of the internal environment that must be homeostatically maintained are the following (Fig. 1–3):

A CLOSER LOOK AT EXERCISE PHYSIOLOGY

What is Exercise Physiology?

Exercise physiology is the study of both the functional changes that occur in response to a single bout of exercise and the adaptations that occur as a result of regular, repeated bouts of exercise. Exercise disrupts homeostasis. The changes that occur in response to exercise are the body's attempt to reduce the stress that has been placed on the entire organism.

Changes that are normal during exercise would be considered abnormal if they occurred in a nonexercising individual. For example, the level to which the body temperature rises during exercise would be considered a fever if the person were not exercising.

Heart rate is one of the easiest factors to monitor that shows both an immediate response to exercise and long-term adaptation to a regular exercise program. When a person begins to exercise, more O_2 is used by the active muscle cells to support their increased energy demands. Heart rate increases to deliver more oxygenated blood to the exercising muscles. The heart adapts to regular exercise of sufficient intensity and duration by increasing its strength and efficiency so that it pumps more blood per beat. Because of increased pumping ability, the heart does not have to beat as rapidly during exercise in order to pump a given quantity of blood to perform a particular activity as it did before physical training.

Exercise physiologists study the mechanisms responsible for the changes that occur as a result of exercise. Much of the knowledge gained from the study of exercise is used to develop appropriate exercise programs to increase the functional capacities of people, ranging from athletes to the infirm.

1. *Its concentration of nutrient molecules*. Cells need a constant supply of nutrient molecules to serve as a metabolic fuel for energy production. Energy, in turn, is needed to support life-sustaining and specialized cellular activities.
2. *Its concentration of O_2 and CO_2*. Cells need O_2 to perform chemical reactions that extract from nutrient molecules the maximum energy yield for use by the cell. The CO_2 produced during these chemical reactions must be balanced by CO_2 removal from the lungs so that acid-forming CO_2 does not increase the acidity of the internal environment.
3. *Its concentration of waste products*. Various chemical reactions produce undesirable end-products that exert a toxic effect on the body's cells if these wastes are allowed to accumulate beyond a certain limit.
4. *Its pH*. Among the most pronounced effects of changes in the acidity of the internal fluid environment are alterations in the electrical signaling mechanism of nerve cells and alterations in enzyme activity of all cells.
5. *Its concentration of salt and other electrolytes*. The concentration of salt (NaCl) in the internal environment is important in maintaining the proper volume of the cells. Cells do not function normally when they are swollen or shrunken. Other electrolytes perform a variety of vital functions. For example, the rhythmic beating of the heart depends on a relatively constant concentration of potassium (K^+) in the extracellular fluid.
6. *Its temperature*. Body cells function optimally within a narrow temperature range. The cells slow down too much

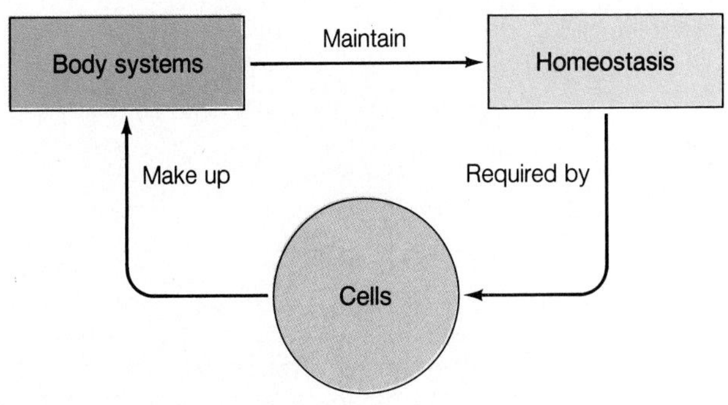

Figure 1–2 Relationship of Cells, Body Systems, and Homeostasis

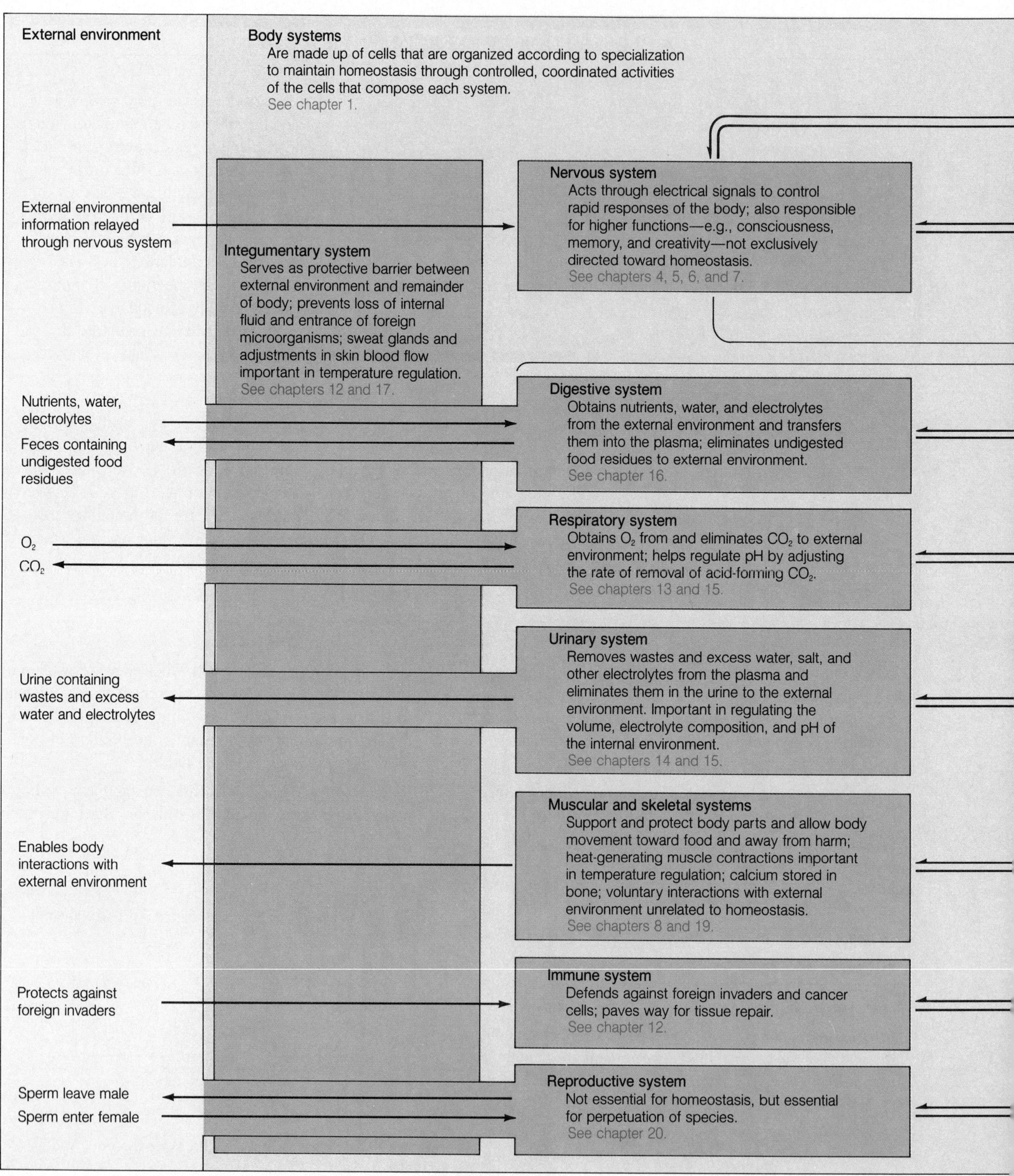

External environment
Body systems
Are made up of cells that are organized according to specialization to maintain homeostasis through controlled, coordinated activities of the cells that compose each system.
See chapter 1.
External environmental information relayed through nervous system
Nervous system
Acts through electrical signals to control rapid responses of the body; also responsible for higher functions—e.g., consciousness, memory, and creativity—not exclusively directed toward homeostasis.
See chapters 4, 5, 6, and 7.
Integumentary system
Serves as protective barrier between external environment and remainder of body; prevents loss of internal fluid and entrance of foreign microorganisms; sweat glands and adjustments in skin blood flow important in temperature regulation.
See chapters 12 and 17.
Nutrients, water, electrolytes
Feces containing undigested food residues
Digestive system
Obtains nutrients, water, and electrolytes from the external environment and transfers them into the plasma; eliminates undigested food residues to external environment.
See chapter 16.
O_2
CO_2
Respiratory system
Obtains O_2 from and eliminates CO_2 to external environment; helps regulate pH by adjusting the rate of removal of acid-forming CO_2.
See chapters 13 and 15.
Urine containing wastes and excess water and electrolytes
Urinary system
Removes wastes and excess water, salt, and other electrolytes from the plasma and eliminates them in the urine to the external environment. Important in regulating the volume, electrolyte composition, and pH of the internal environment.
See chapters 14 and 15.
Enables body interactions with external environment
Muscular and skeletal systems
Support and protect body parts and allow body movement toward food and away from harm; heat-generating muscle contractions important in temperature regulation; calcium stored in bone; voluntary interactions with external environment unrelated to homeostasis.
See chapters 8 and 19.
Protects against foreign invaders
Immune system
Defends against foreign invaders and cancer cells; paves way for tissue repair.
See chapter 12.
Sperm leave male
Sperm enter female
Reproductive system
Not essential for homeostasis, but essential for perpetuation of species.
See chapter 20.

Figure 1-3 Role of the Body Systems in Maintaining Homeostasis

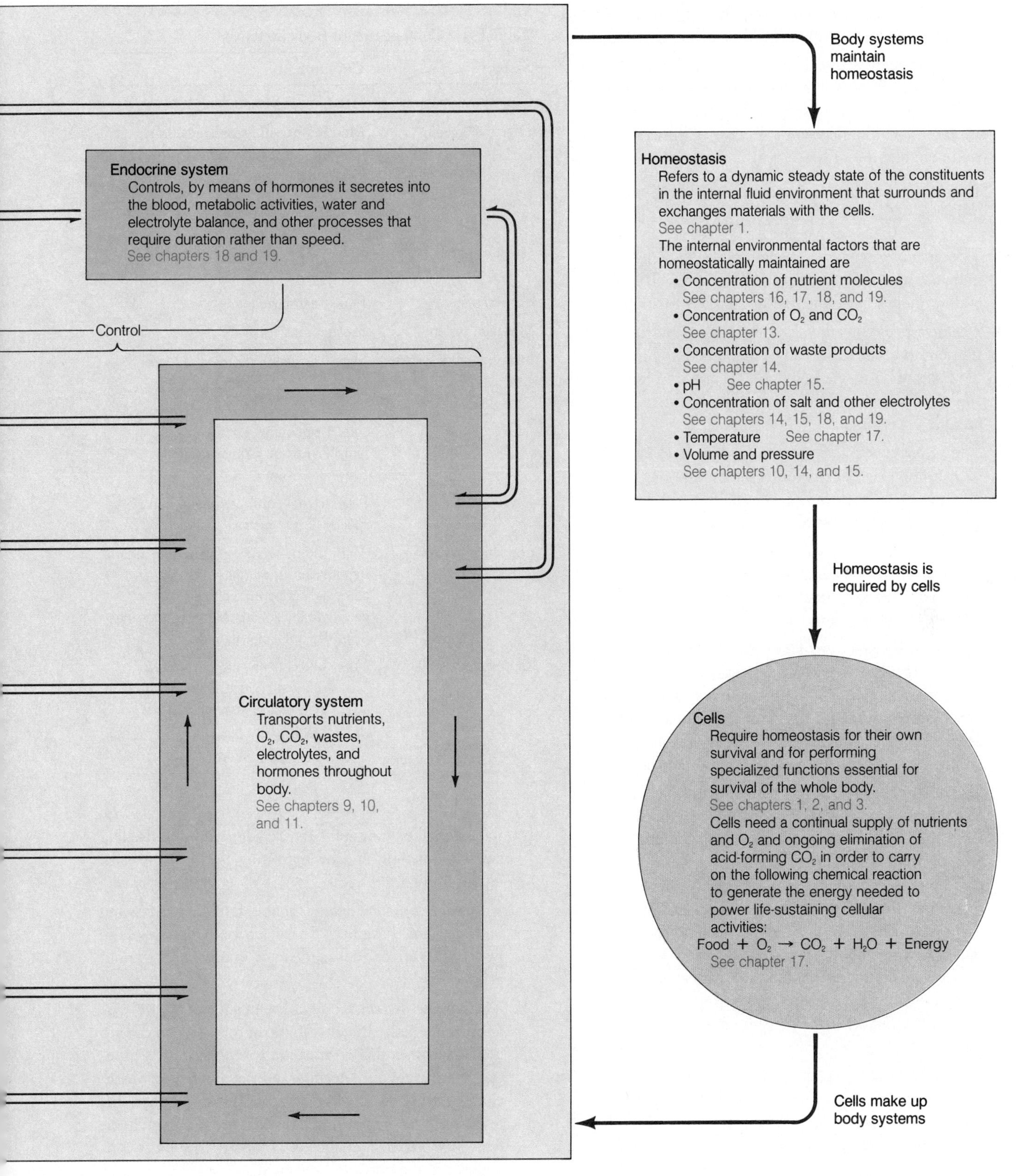

Body systems maintain homeostasis
Endocrine system
Controls, by means of hormones it secretes into the blood, metabolic activities, water and electrolyte balance, and other processes that require duration rather than speed.
See chapters 18 and 19.
Control
Circulatory system
Transports nutrients, O_2, CO_2, wastes, electrolytes, and hormones throughout body.
See chapters 9, 10, and 11.
Homeostasis
Refers to a dynamic steady state of the constituents in the internal fluid environment that surrounds and exchanges materials with the cells.
See chapter 1.
The internal environmental factors that are homeostatically maintained are
• Concentration of nutrient molecules
See chapters 16, 17, 18, and 19.
• Concentration of O_2 and CO_2
See chapter 13.
• Concentration of waste products
See chapter 14.
• pH See chapter 15.
• Concentration of salt and other electrolytes
See chapters 14, 15, 18, and 19.
• Temperature See chapter 17.
• Volume and pressure
See chapters 10, 14, and 15.
Homeostasis is required by cells
Cells
Require homeostasis for their own survival and for performing specialized functions essential for survival of the whole body.
See chapters 1, 2, and 3.
Cells need a continual supply of nutrients and O_2 and ongoing elimination of acid-forming CO_2 in order to carry on the following chemical reaction to generate the energy needed to power life-sustaining cellular activities:
Food + O_2 → CO_2 + H_2O + Energy
See chapter 17.
Cells make up body systems

if they are too cold and, worse yet, their structural and enzymatic proteins are impaired if they get too hot.

7. *Its volume and pressure*. The circulating component of the internal environment, the plasma, must be maintained at adequate volume and blood pressure to assure body-wide distribution of this important link between the external environment and the cells.

There are eleven major body systems (Table 1–1) that contribute to homeostasis in the following major ways (Fig. 1–3):

1. The *circulatory system* is the transport system that carries materials such as nutrients, O_2, CO_2, wastes, electrolytes, and hormones from one part of the body to another.
2. The *digestive system* breaks down dietary food into small nutrient molecules that can be absorbed into the plasma for distribution to the body cells. It also transfers water and electrolytes from the external environment into the internal environment.
3. The *respiratory system* obtains O_2 from and eliminates CO_2 to the external environment. By adjusting the rate of removal of acid-forming CO_2, it is also important in maintaining the proper pH of the internal environment.
4. The *urinary system* eliminates waste products other than CO_2 and plays a key role in regulating the volume, electrolyte composition, and acidity of the extracellular fluid.
5. The *skeletal system* provides support and protection for the soft tissues and organs. It also serves as a storage reservoir for calcium (Ca^{++}), an electrolyte whose plasma concentration must be maintained within very narrow limits. Together with the muscular system, the skeletal system also enables movement of the body and its parts.
6. The *muscular system* moves the bones to which the skeletal muscles are attached. From a purely homeostatic view, this system enables an individual to move toward food or away from harm. Furthermore, the heat generated by muscle contraction is important in temperature regulation. In addition, because skeletal muscles are under voluntary control, a person is able to use them to accomplish myriad other movements of his or her own choice that are not directed toward maintaining homeostasis, ranging from the fine motor skills required to do delicate needlework to the powerful movements required for weight lifting.
7. The *integumentary system* serves as an outer protective barrier to prevent loss of internal fluid from the body and entrance of foreign microorganisms into the body. This system is also important in the regulation of body temperature. The amount of heat lost from the body surface to the external environment can be adjusted by controlling sweat production and by regulating the flow of warm blood through the skin.
8. The *immune system* defends against foreign invaders and body cells that have become cancerous. It also paves the way for repair or replacement of injured or worn-out cells.
9. The *nervous system* is one of the two major control systems of the body. In general, it controls and coordinates bodily activities that require swift responses. It is especially important in detecting and initiating reactions to changes in the external environment. Furthermore, it is responsible for higher functions that are not entirely directed toward maintenance of homeostasis, such as consciousness, memory, and creativity.

Table 1–1 Components of Body Systems

System	Components
Circulatory system	Heart, blood vessels, blood
Digestive system	Mouth, pharynx, esophagus, small intestine, large intestine, salivary glands, exocrine pancreas, liver, gall bladder
Respiratory system	Nose, pharynx, larynx, trachea, bronchi, lungs
Urinary system	Kidneys, ureters, urinary bladder, urethra
Skeletal system	Bones, cartilage, joints
Muscular system	Skeletal muscles
Integumentary system	Skin, hair, nails
Immune system	White blood cells, thymus, bone marrow, tonsils, adenoids, lymph nodes, spleen, appendix, gut-associated lymphoid tissue, skin-associated lymphoid tissue
Nervous system	Brain, spinal cord, peripheral nerves, special sense organs
Endocrine system	All hormone-secreting tissues, including hypothalamus, pituitary, thyroid, adrenals, endocrine pancreas, parathyroids, gonads, kidneys, intestine, heart, thymus, pineal, and skin
Reproductive system	Male: testes, penis, prostate gland, seminal vesicles, bulbourethral glands, and associated ducts Female: ovaries, oviducts, uterus, vagina, breasts

10. The *endocrine system* is the other major control system. In general, the hormone-secreting glands of the endocrine system regulate activities that require duration rather than speed. This system is especially important in controlling the concentration of nutrients and, by adjusting kidney function, controlling the internal environment's volume and electrolyte composition.
11. The *reproductive system* is not essential for homeostasis and therefore is not essential for survival of the individual. It is essential, however, for perpetuation of the species.

As we examine each of these systems in greater detail, always keep in mind that the body is a coordinated whole even though each system provides its own special contributions. It is easy to forget that all of the body parts actually fit together into a functioning, interdependent whole body. A "Chapter in Perspective" is provided at the end of each chapter to help you focus on how the body part just described fits in homeostatically. As a further tool to help you keep track of how all the pieces fit together, Figure 1–3 is duplicated on the inside front cover as a handy reference.

Keep another point in mind as you read through the book from cells to systems: The functioning whole is greater than the sum of its separate parts. In other words, through specialization, cooperation, and interdependence, cells combine to form a coordinated, unique single living organism with more diverse and complex capabilities than is possessed by any one of the cells that make it up. For humans, these capabilities go far beyond the processes needed to maintain life. Obviously a cell, or even a random combination of cells, cannot create an artistic masterpiece or design a spacecraft, but body cells working together permit those capabilities in an individual.

Negative feedback is a common regulatory mechanism for maintaining homeostasis.

To maintain homeostasis, the body must be able to detect deviations in the internal environmental factors that need to be held within narrow limits, and it must be able to control the various body systems responsible for adjusting these factors. For example, to maintain the concentration of CO_2 in the extracellular fluid at an optimal value, there must be a way of detecting a change in CO_2 concentration and then of appropriately altering respiratory activity so that CO_2 concentration is returned to the desirable level.

There are two general categories of control systems that operate to maintain homeostasis—intrinsic and extrinsic controls. **Intrinsic controls** (*intrinsic* means "within") are those that are built in or inherent to an organ. For example, as an exercising muscle rapidly uses up O_2 and produces CO_2 to generate energy to support its contractile activity, the O_2 concentration falls and the CO_2 concentration increases within the muscle. By acting directly on the smooth muscle in the walls of the blood vessels that supply the exercising muscle, these local chemical changes cause the smooth muscle to relax and the vessels to open widely to accommodate increased blood flow into the exercising muscle. This local mechanism contributes to the maintenance of an optimal level of O_2 and CO_2 in the internal fluid environment surrounding the exercising muscle's cells.

Most factors in the internal environment are maintained, however, by **extrinsic controls** (*extrinsic* means "outside of"), which are regulatory mechanisms initiated outside of an organ that alter the activity of the organ. Extrinsic control of the various organs and systems is accomplished by the nervous and endocrine systems, the two major control systems of the body. Extrinsic control permits coordinated regulation of several organs toward a common goal, unlike intrinsic controls, which are self-serving for the organ in which they occur. Coordinated, overall regulatory mechanisms are critical for maintaining the dynamic steady state in the internal environment as a whole. For example, to restore blood pressure to the proper level when it falls too low, the nervous system simultaneously acts on the heart and the blood vessels throughout the body to increase the blood pressure to normal.

The body's homeostatic control mechanisms operate on the principle of negative feedback. **Negative feedback** exists when a change in a regulated variable triggers a response that opposes the change, driving the variable in the opposite direction of the initial change. A common example of negative feedback is a thermostatically controlled furnace (Fig. 1–4a). The room temperature is determined by the activity of the furnace, a heat source that can be turned on or off. When the temperature-sensitive thermostat detects that the room temperature has fallen below a set-point level, it activates the furnace, which produces heat to increase the room temperature. Once the room temperature reaches the set point, the thermostat and furnace are switched off. Thus, the heat produced by the furnace counteracts, or is "negative" to, the original fall in temperature. If heat production were to continue unabated, the room temperature would be increased above the set point. This overshooting beyond the set point does not occur because the heat "feeds back" to shut off the thermostat that triggered its output, thus limiting its own production by controlling the input to itself.

A homeostatic negative-feedback system operates in the same way to maintain the controlled factor in a relatively steady state. For example, when the pressure-monitoring nerve cells detect a *decrease* in blood pressure below the desired level, they bring about a sequence of events that culminate in nerve-controlled changes in the opposite direction within the circulatory system to *increase* blood pressure to the

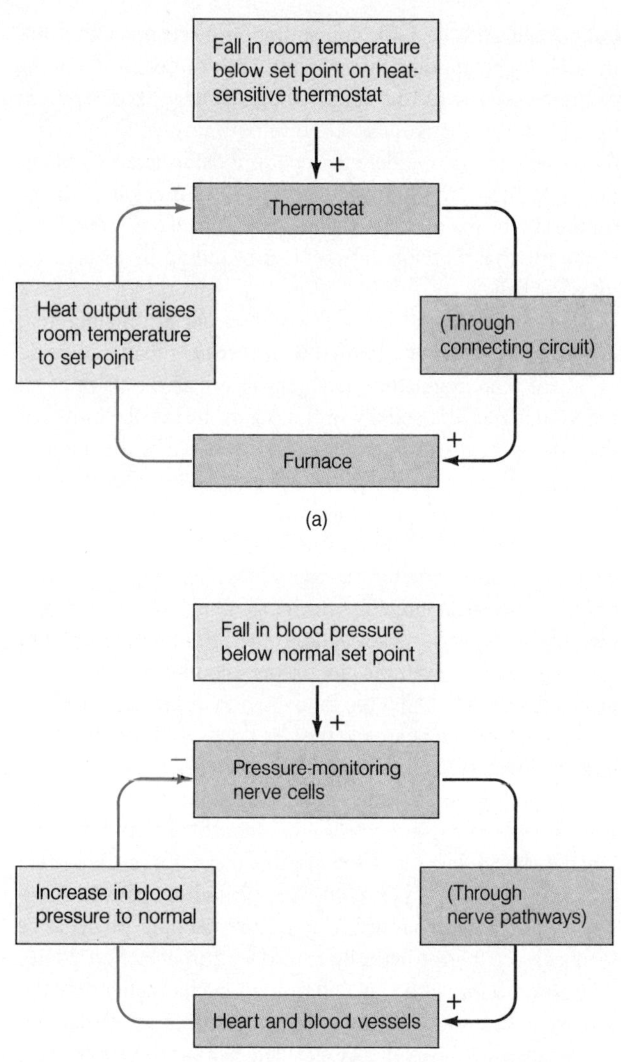

Figure 1–4 Negative Feedback *(a) Negative-feedback control of home heating system. (b) Negative-feedback control of blood pressure.*

proper level (Fig. 1–4b). When the blood pressure increases to the set point, the stimulatory input to the heart and blood vessels arising from the pressure-monitoring nerve cells is turned off. As a result, the blood pressure does not continue to increase above the set point. The opposite events occur when the original change is an elevation in blood pressure above normal. The pressure-monitoring nerve cells bring about a reduction in blood pressure to normal by triggering compensatory responses in the circulatory system. The blood pressure does not fall too low, because the pressure-monitoring nerve cells cease triggering the pressure-reducing responses when the blood pressure reaches the right level.

Positive feedback occurs less frequently in the body. In negative feedback, a control system's input and output act in reciprocal fashion to resist change and maintain a relatively steady set point for a regulated factor. With **positive feedback,** however, the input and output continue to enhance each other. Instead of bringing about a response that counteracts an initial change, positive feedback reinforces the change in the same direction. This would be comparable to the heat generated by a furnace triggering the thermostat to call for even more heat output from the furnace so that the room temperature would continuously rise.

Because positive feedback moves the controlled variable even farther from a steady state, it does not happen very often in the body, in which the major goal is maintenance of stable, homeostatic conditions. In certain instances, however, positive feedback does occur, a prime example being during birth of a baby. The hormone oxytocin causes powerful contractions of the uterus. As uterine contractions push the baby against the cervix (the exit from the uterus), a sequence of events is triggered to bring about the release of even more oxytocin, which causes even stronger uterine contractions, which triggers the release of more oxytocin, and so on. This positive-feedback cycle does not cease until the baby is finally expelled.

Disruptions in homeostasis can lead to illness and death.

When one or more of the body's systems fails to function properly, homeostasis is disrupted and all of the cells suffer because they no longer have an optimal environment in which to live and function. Various pathophysiological states ensue, depending on the type and extent of homeostatic disruption. **Pathophysiology** refers to the abnormal functioning of the body (altered physiology) associated with disease. When a homeostatic disruption becomes so severe that it is no longer compatible with survival, death results.

Chapter in Perspective

The complex society of cells that compose the body is able to survive and function only because of the cells' ability to interact and cooperate to maintain the conditions within the body that are essential for the life of each cell and thus of the whole body. Such maintenance of a stable internal fluid environment—that is, relative constancy of the chemical composition and physical state of the extracellular fluid that bathes and accomplishes exchanges with all of the body's cells—is known as homeostasis. Every cell requires homeostasis for its own survival, and every cell contributes to homeostasis for the whole body's survival.

All cells perform basic cellular functions plus a specialized function that contributes to homeostasis. Cells of similar structure and specialized function are organized into tissues, of which there are four primary types—muscle, nervous, epithelial, and connective tissues. Tissues are organized into functional aggregates known as organs, which are further organized into body systems. A body system is a collection of organs that perform related functions to accomplish a common activity essential for the survival of the whole body. Eleven major systems collectively compose the body.

Maintenance of homeostasis necessitates control. Negative feedback, whereby a change in a regulated variable triggers a response that opposes the change, is the most common regulatory mechanism for maintaining homeostasis.

Review Exercises

1. Define physiology.
2. Compare the mechanistic and teleological approaches to explaining the various events that occur in the body.
3. Describe the levels of organization in the body.
4. What are the basic cell functions?
5. What are the four primary types of tissue?
6. Define homeostasis. What is the internal environment?
7. What factors of the internal environment must be homeostatically maintained?
8. What are the components and contributions to homeostasis of each of the eleven major body systems?
9. Compare negative and positive feedback.
10. Distinguish between intrinsic and extrinsic control.
11. **A point to ponder:** Considering the nature of negative-feedback control and the function of the respiratory system, what effect do you predict that a decrease in CO_2 in the internal environment would have on how rapidly and deeply a person breathes?

CHAPTER 2

CELLULAR STRUCTURE AND FUNCTIONS

CHAPTER CONTENTS AT A GLANCE

INTRODUCTION *The setting was seventeenth-century England at a meeting of the prestigious Royal Society of London. Asked to prepare a demonstration for this meeting, newly appointed Curator of Experiments, Robert Hooke, wanted to impress his elite scientific audience. He chose to treat them to a view of the previously unseen miniscule world made visible with the new tool he had invented, the microscope. For his demonstration he selected cork because of its mysterious properties—it was solid, yet floated. The thin sliver of cork Hooke and the Royal Society viewed appeared to consist of little empty boxes, presumably filled with air, which would account for the cork's ability to float. Reminding Hooke of the rows of small rooms (or cells) found in a monastery, he named the little boxes "cells." Thus began the exciting exploration into the microscopic cellular world of which all living organisms are composed.*

Cells are the bridge between molecules and humans.

All bodily functions ultimately depend on the activities of the individual cells that compose the body. Even though the chemicals of which cells are made have been analyzed, it has not been possible in a laboratory to organize these chemicals into a living cell. It is not merely the presence of various molecules but rather the complex organization and interaction of these molecules within the cell that confers the unique characteristics of life. Inanimate chemical molecules are organized within each cell into a living entity. Cells, in turn, serve as the living building blocks for the immensely complicated whole body. Modern physiologists are unraveling many of the broader mysteries of how the body works by probing deeper into the molecular structure and organization of the cells that make up the body.

Increasingly better tools are revealing the complexity of cells.

The cells that compose the human body are so small that they cannot be seen by the unaided eye. The smallest visible particle is about five to ten times larger than a typical human cell, which averages about 10 to 20 micrometers (μm) in diameter (1 μm is equal to 1/1,000,000 meter, with a meter being 39.87 inches). Refer to Table 2–1 for a size reference and the inside back cover for a comparison of metric units and their English equivalents.

Not until Hooke invented the microscope in the middle of the seventeenth century was the existence of cells revealed. Actually, when Hooke first described a microscopic specimen of cork as consisting of little air-filled boxes or cells, he was not observing cells at all but the remaining walls of dead and disintegrated plant cells. Not until the early part of the nineteenth century, with the development of better light microscopes, was it learned that all plant and animal tissues are composed of individual cells. Another important discovery was that cells are not filled with air but that they are instead filled with a fluid, which, with the microscopic capabilities of the time, appeared to be a rather homogeneous, soupy mixture believed to be the elusive "stuff of life."

Cells are not only small but also are generally colorless and transparent. Only when staining techniques became available late in the nineteenth century were cells observed to contain numerous granular, oval, and filamentous structures. Not until the 1940s, when the technique of electron microscopy was first employed to observe living matter, did an understanding of the great diversity and complexity of the internal structure of cells begin to emerge. (Electron microscopes are about one-hundred times more powerful than light microscopes.) Now with the availability of even more sophisticated microscopic tools, biochemical techniques, and genetic engineering, the concept of the cell as a microscopic bag of amorphous fluid has given way to our present-day knowledge of the cell as a complex, highly organized, compartmentalized structure.

A cell is subdivided into the plasma membrane, nucleus, and cytoplasm.

Even though there is no such thing as a "typical" cell because of diverse structural and functional specializations, all cells share many common features. All cells have three major subdivisions: the plasma membrane, the nucleus, and the cytoplasm (Table 2–2). The **plasma membrane** or **cell mem-**

Table 2–1 The Size of a Typical Human Cell Compared to Other Structures

Structure	Average Size	Size Compared to Size of an Atom	Size Compared to Smallest Point Visible to Naked Eye
Atom	0.1 nm	Equal in size	1,000,000 × smaller
Sugar molecule	1 nm	10 × larger	100,000 × smaller
Globular protein	10 nm	100 × larger	10,000 × smaller
Bacterium	1,000 nm (1 μm)	10,000 × larger	100 × smaller
Human cell	10,000 nm (10 μm)	100,000 × larger	10 × smaller
Smallest point visible to naked eye	100,000 nm (0.1 mm, about 0.004 inch)	1,000,000 × larger	Equal in size

Table 2–2 Summary of Cell Structures and Functions

Cell Part	Number per Cell	Structure	Function
Nucleus	1	DNA and specialized proteins enclosed by a double-walled nuclear envelope	Control center of the cell, providing storage of genetic information Provision of codes for the synthesis of structural and enzymatic proteins that determine the specific nature of each cell Blueprint for cell replication
Plasma Membrane	1	Lipid bilayer studded with proteins and small amounts of carbohydrate	Selective barrier between cellular contents and extracellular fluid; control of traffic in and out of the cell
Cytoplasm			
Organelles			
Endoplasmic reticulum	1	Extensive, continuous membranous network of fluid-filled tubules and flattened sacs, partially studded with ribosomes	Formation of new cell membrane and other cell components and manufacture of products for secretion
Golgi complex	1 to several hundred	Sets of stacked, flattened membranous sacs	Modification, packaging, and distribution center for newly synthesized proteins
Lysosomes	300	Membranous sacs containing hydrolytic enzymes	Digestive system of the cell, destroying unwanted material such as foreign substances and cellular debris
Peroxisomes	200	Membranous sacs containing oxidative enzymes, hydrogen peroxide, and catalase	Detoxification activities
Mitochondria	100–2,000	Rod- or oval-shaped bodies enclosed by two membranes, with the inner membrane folded into cristae that project into the interior matrix	Energy organelles; major site of ATP production; contain enzymes for citric-acid cycle and electron-transport chain.

brane is a very thin membranous structure that encloses each cell, separating the cell's contents from its surroundings. The fluid contained within all of the cells of the body is known collectively as **intracellular fluid** (ICF), and the fluid outside of the cells is referred to as **extracellular fluid** (ECF). The plasma membrane does not merely serve as a mechanical barrier to hold in the contents of the cells; it has the ability to selectively control movement of molecules between the ICF and ECF. Plant cells, unlike animal cells, are further covered by a thicker protective coat known as the cell wall. It was the cell walls of dead cork cells that Hooke first observed under his microscope.

The two major parts of the cell's interior are the nucleus and the cytoplasm. The **nucleus,** which is typically the largest single organized cellular component, can be seen as a distinct spherical or oval structure, usually located near the center of the cell. It is surrounded by a double-layered membranous structure, the **nuclear envelope,** which separates the nucleus from the remainder of the cell. Sequestered within the nucleus is the cell's genetic material, deoxyribonucleic acid (DNA), which has two important functions. First, DNA provides codes or "instructions" for directing synthesis of specific structural and enzymatic proteins within the cell. By directing the kinds and amounts of various enzymes and other proteins that are produced, the nucleus indirectly governs most cellular activities and serves as the cell's control center. Second, DNA serves as a genetic blueprint during cell replication to insure that the cell produces additional cells just like itself, thus continuing the identical type of cell line within the body. Furthermore, in the reproductive cells, the DNA blueprint serves to

Table 2–2 Summary of Cell Structures and Functions (continued)

Cell Part	Number per Cell	Structure	Function
Cytosol			
Intermediary metabolism enzymes	Many	Sequential arrangement within the cytoskeleton	Intracellular chemical reactions involving the degradation, synthesis, and transformation of small organic molecules
Ribosomes	Many	Granules of RNA and proteins—some attached to rough endoplasmic reticulum; some free in the cytoplasm, suspended by the microtrabecular lattice	Protein synthesis
Secretory vesicles	Varies	Membrane-bound packages of secretory products	Concentration and storage of secretory products until signaled to empty contents to the outside
Inclusions	Varies	Glycogen granules, fat droplets	Storage of excess nutrients
Cytoskeleton (see Table 2–3 for detail)	Many cytoskeletal elements	Microtubules, microfilaments, intermediate filaments, and microtrabecular lattice forming a complex and dynamic structural network	Bone and muscle of the cell, giving it shape, providing for internal organization, and regulating various cellular movements
Centrioles	2 located near nucleus	Cylindrical bodies composed of nine sets of microtubule triplets	Formation of mitotic spindle during cell division
Cilia	Many on cells lining respiratory airways and oviducts	Small, motile, hairlike projections containing 9 + 2 microtubule arrangement	Moving material across the surface of the cell, clearing airways of debris, guiding egg
Flagellum	1 per sperm cell	Elongated, motile, whiplike projection containing 9 + 2 microtubule arrangement	Sperm propulsion
Microvilli	Many on cells lining small intestine and kidney tubules; also on hair cells in inner ear	Tiny, stiff, hairlike projections containing bundles of interlinked actin filaments	Increasing absorptive surface area; important for hearing and equilibrium

pass on genetic characteristics to future generations. These two roles of the nucleus—controlling protein synthesis and serving as a genetic blueprint—will be described in more detail later in the chapter.

The **cytoplasm** refers to that portion of the cell interior not occupied by the nucleus. It is not a semiuniform fluid, as was once believed, but instead contains a number of distinct, highly organized, membrane-bound structures—the **organelles**—dispersed within a complex, gel-like mass called the **cytosol.** Nearly all cells contain five main types of organelles—the endoplasmic reticulum, Golgi complex, lysosomes, peroxisomes, and mitochondria (Fig. 2–1). Each of these organelles are similar in all cells, although there are some variations depending on the specialized capabilities of each cell type. Organelles are like intracellular "specialty shops," each being a separate internal compartment that contains a specific set of chemicals for carrying out a particular cellular function. One advantage of this compartmentalization is that it permits chemical activities that would not be compatible with each other to occur simultaneously within the cell. For example, the enzymes that destroy unwanted proteins in the cell do so within the protective confines of the lysosomes without the risk of destroying essential cellular proteins. An equally important advantage is the organelles' ability to have a localized concentration of particular enzymes and chemical reactants to permit an increased rate of certain enzyme-mediated reactions. For example, enzymes and reactants important in extracting energy from nutrient molecules are concentrated within the mitochondria. About half of the total cell volume is occupied by organelles.

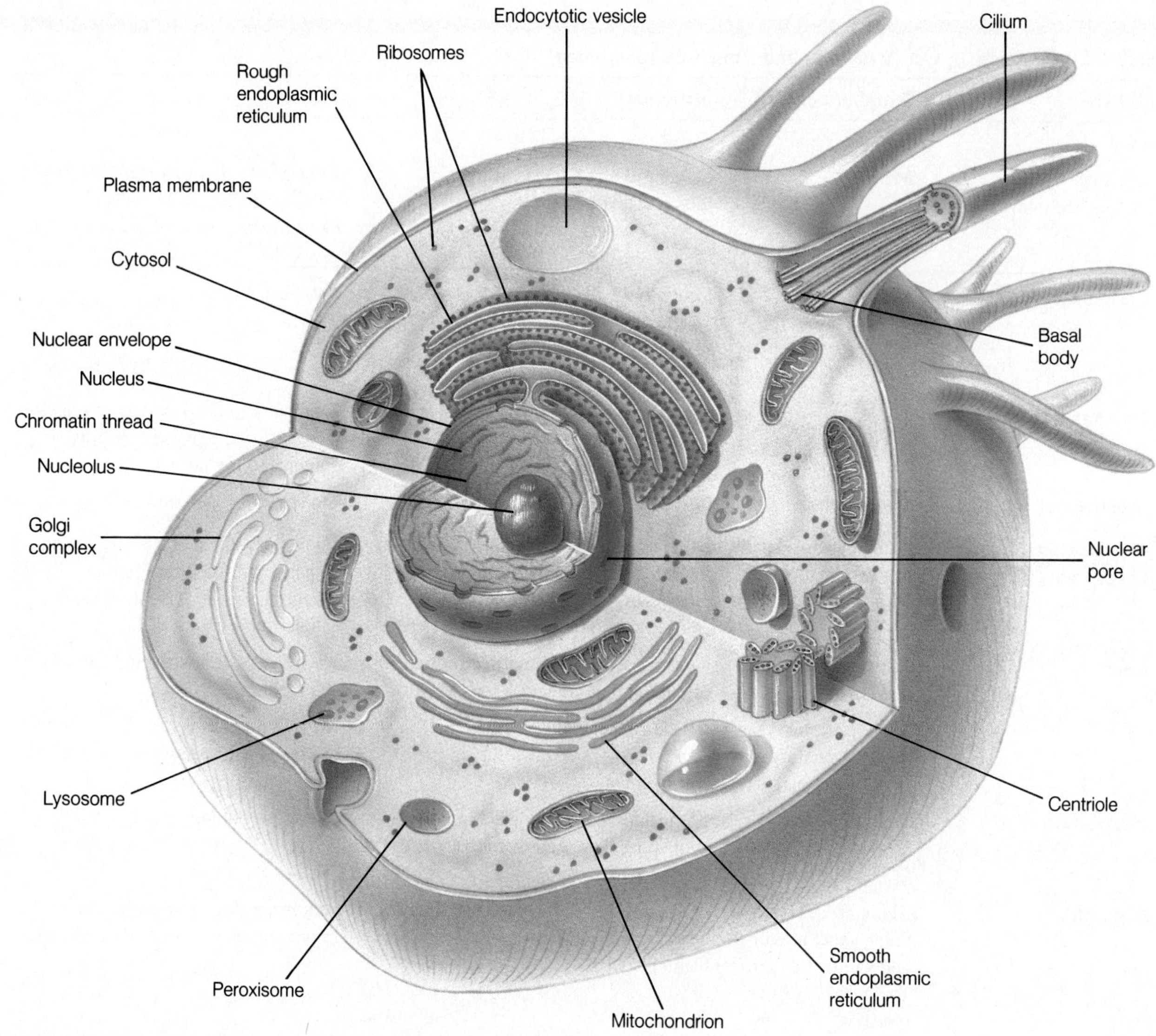

Figure 2–1 Schematic Three-Dimensional Illustration of Cell Structures Visible under Electron Microscope

The remainder of the cytoplasm that is not occupied by organelles consists of cytosol, a semiliquid mass laced with an elaborate protein network that constitutes the cytoskeleton. Many of the chemical reactions that are compatible with each other are carried on in the cytosol. The cytoskeletal network gives the cell its shape, provides for its internal organization, and regulates its various movements. In this chapter we will examine each of the intracellular components in more detail, concentrating first on the organelles.

Organelles

The endoplasmic reticulum is a synthesizing factory.

The **endoplasmic reticulum** (**ER**) is an elaborate membranous system that consists of a network of fluid-filled tubules and flattened sacs distributed extensively throughout the cyto-

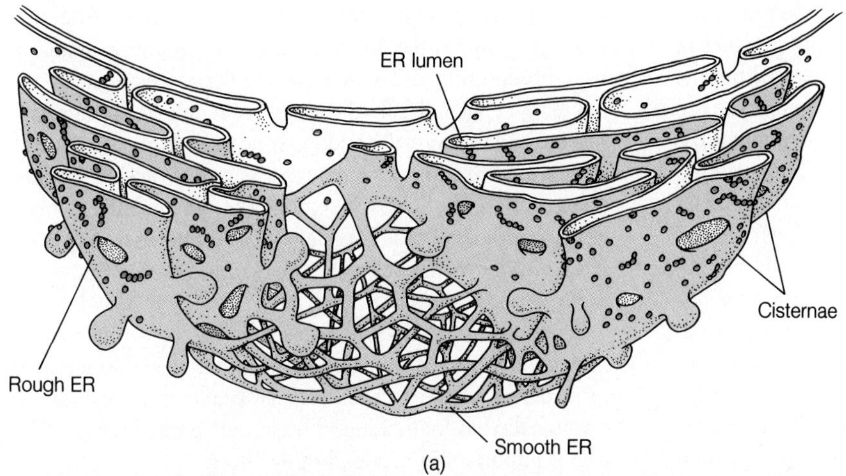

Figure 2–2 Endoplasmic Reticulum (ER) *(a) Schematic three-dimensional representation of the relationship between the rough and smooth endoplasmic reticulum. (b) Electron micrograph of cisternae of rough endoplasmic reticulum.*

SOURCE: Part (b) courtesy of Elizabeth R. Walker, Associate Professor, Department of Anatomy, School of Medicine, West Virginia University.

plasm. Two distinct types of endoplasmic reticulum—the smooth ER and the rough ER—can be distinguished. The **smooth ER** is a meshwork of tiny interconnected tubules, whereas the **rough ER** projects outward as stacks of relatively flattened sacs called **cisternae** (Fig. 2–2). Even though these two regions differ considerably in appearance and function, they are thought to be continuous with each other. In other words, the endoplasmic reticulum is one continuous organelle with many interconnected channels. It basically consists of a highly convoluted, continuous sheet of membrane that completely encloses a single fluid-filled space, or lumen. The outer surface of the rough-ER membrane is studded with small, dark-staining particles that give it a "rough" or granular appearance. These particles are **ribosomes,** which are special ribonucleic acid (RNA)–protein complexes that synthesize proteins under the direction of nuclear DNA. Not all ribosomes in the cell are attached to the rough ER. Unattached or "free" ribosomes are dispersed throughout the cytosol. The relative amount of smooth and rough ER varies between cells, depending on the activity of the cell.

ROUGH ENDOPLASMIC RETICULUM. The rough ER, in association with its ribosomes, synthesizes and releases into the lumen a variety of new proteins, which serve one of two purposes. Some proteins are destined for export to the cell's exterior as secretory products, such as proteinaceous enzymes or hormones. Other proteins are transported to sites within the cell for use in the construction of new cellular membrane (either new plasma membrane or new organelle membrane) as well as other protein components of organelles. The membranous wall of the endoplasmic reticulum also contains enzymes essential for the synthesis of nearly all the lipids (fats) needed for the production of new membranes. These newly synthesized lipids enter the cisternal lumen along with the proteins. Cellular membranes consist predominantly of these lipids and proteins. Predictably, the rough ER is most abundant in cells specialized for protein secretion (for example, cells that secrete digestive enzymes) or in cells that require extensive membrane synthesis (for example, rapidly growing cells such as immature egg cells).

Once assembled, each ribosome participates in the synthesis of only one type of protein. There is no difference between the free and ER-bound ribosomes except for the different proteins that they help to synthesize. In contrast to the rough-ER ribosomes, the free ribosomes synthesize enzymatic proteins that are used intracellularly within the cytosol. All ribosomes are produced in the cell's nucleus under the direction of DNA, with each being "programmed" at any given time to facilitate the synthesis of only one specific protein needed by that particular cell.

Currently it is thought that newly formed ribosomes destined to become attached to the rough ER start to synthesize a **leader sequence** that acts as a signal, much like an address on a letter (Fig. 2–3). Another protein present in the cytosol, the **signal-recognition protein,** binds to one of these ribosomes upon recognizing the newly synthesized leader sequence. Subsequently, the signal-recognition protein, acting much like a mail carrier, delivers the ribosome to the proper "address" on the ER membrane. How does the signal-recognition protein know the proper address? Special proteins called **ribophorins** found exclusively in the rough regions of the ER membrane act as the "house address" on this membrane. The ribophorins serve as binding sites for preferential ribosomal attachment. The signal-recognition protein (the "mail carrier") can recognize both the leader-sequence signal on the ribosome (the "address on the envelope") and the ribophorin binding site on the ER (the "house address"), thereby delivering the proper ribosome to the proper site on the rough ER for binding.

Once a ribosome is attached to the ER membrane and starts directing synthesis of a specific protein, the developing protein threads its way across the ER membrane into the lumen. Here the leader sequence is removed and the protein chain is folded into its final conformation. The protein may also be modified in other ways, such as being "pruned" (Fig. 2–4) or having sugar molecules attached to it. After this processing within the ER lumen, a new protein is unable to pass through the ER membrane and therefore becomes permanently separated from the cytosol as soon as it has been synthesized. In this way, the endoplasmic reticulum provides cells with a mechanism for separating the newly produced molecules that are destined for synthesis of new cellular components or for export out of the cell (those synthesized within the ER) from those that belong in the cytosol (those produced by the free ribosomes).

How do the newly synthesized molecules within the ER lumen get to their destinations at other intracellular sites or to the exterior of the cell if these molecules cannot pass out through the ER membrane? The smooth endoplasmic reticulum is important in accomplishing this feat.

SMOOTH ENDOPLASMIC RETICULUM. The smooth endoplasmic reticulum does not contain ribosomes; hence it is "smooth." Lacking ribosomes, it is not involved in protein synthesis. Instead it serves a variety of other purposes that vary in different cell types.

In the majority of cells, the smooth ER is rather sparse and serves primarily as a central packaging and discharge site for molecules that are to be transported from the ER. Newly synthesized proteins and lipids pass from the rough ER to gather in the smooth ER. Portions of the smooth ER then "bud off" (that is, are pinched off), giving rise to **transport vesicles** that contain the new molecules enclosed in a membrane layer derived from the smooth-ER membrane (Fig.2–5). Newly synthesized membrane components are rapidly incorporated into the ER membrane itself to replace the membrane that was used to "wrap" the transport vesicle.

Compared to the sparseness of smooth ER seen in most cells, some specialized types of cells have an extensive smooth ER, which has additional responsibilities as follows:

□ The smooth ER is abundant in cells that specialize in lipid metabolism—for example, cells that secrete steroid hormones. (A steroid is a special type of lipid derived from cholesterol.) The lipid-producing enzymes in the membranous wall of the rough ER alone are insufficient to carry out the extensive lipid synthesis necessary to maintain adequate steroid-hormone secretion levels. These cells have an expanded smooth ER compartment to house the additional enzymes necessary to keep pace with demands for hormone secretion.

□ In liver cells, the smooth ER has a special capability. It contains enzymes that are involved in detoxifying harmful substances produced within the body by metabolism or substances that enter the body from the outside in the form of drugs or other foreign compounds. These detoxification enzymes alter

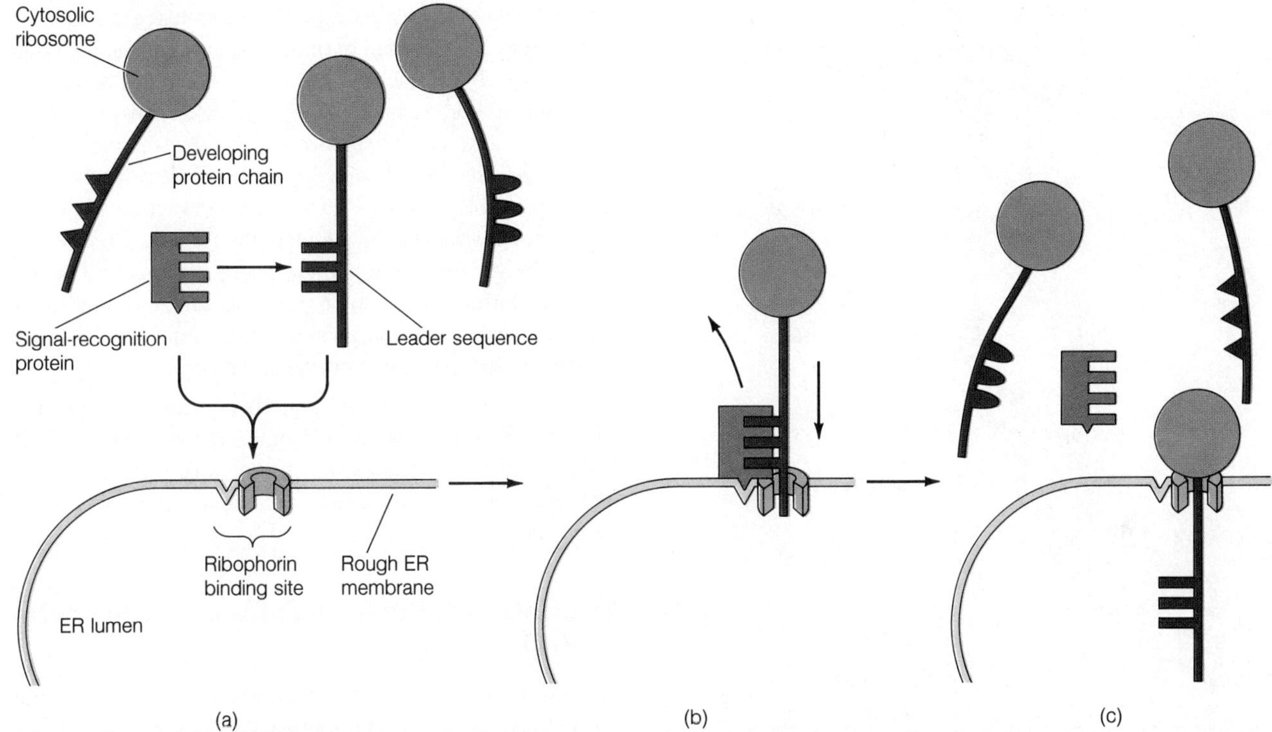

Figure 2–3 Delivery of Ribosomes to Rough Endoplasmic Reticulum *(a) The signal-recognition protein "recognizes" the correct leader sequence on the developing protein chain of a cytosolic ribosome destined to become attached to the rough ER. (b) The signal-recognition protein carries the ribosome to the correct ribophorin "address" on the ER membrane. (c) The signal-recognition protein departs, and the protein chain threads through the ER membrane into the lumen; the proper ribosome is now attached. Other ribosomes synthesizing cytosolic proteins are not recognized by the signal-recognition protein.*

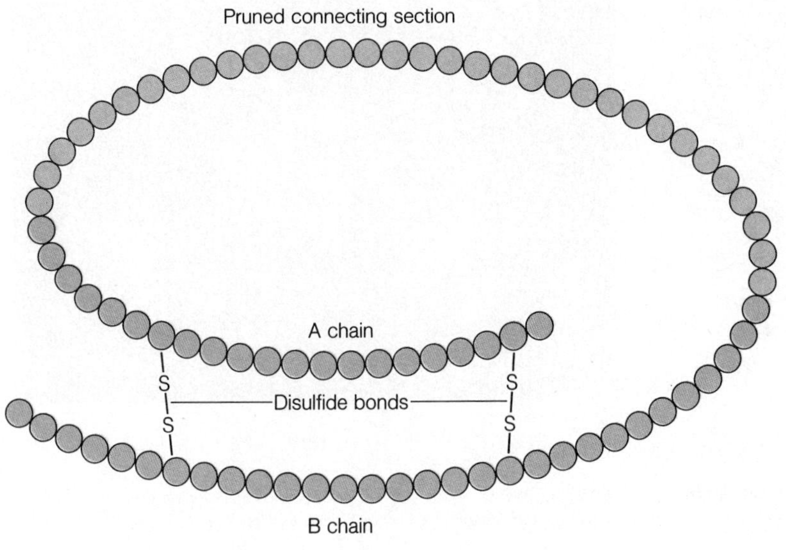

Each circle represents a specific amino acid.

Figure 2–4 Formation of Insulin Molecule within Endoplasmic Reticulum *Insulin is originally formed as a single long polypeptide chain that is folded back on itself, with the two overlapping ends being joined by two disulfide (sulfur-to-sulfur) bonds. The connecting piece (in gray) between the two overlapping ends is then pruned away, leaving the insulin molecule (in aqua) consisting of an A chain and a B chain connected by the disulfide bonds.*

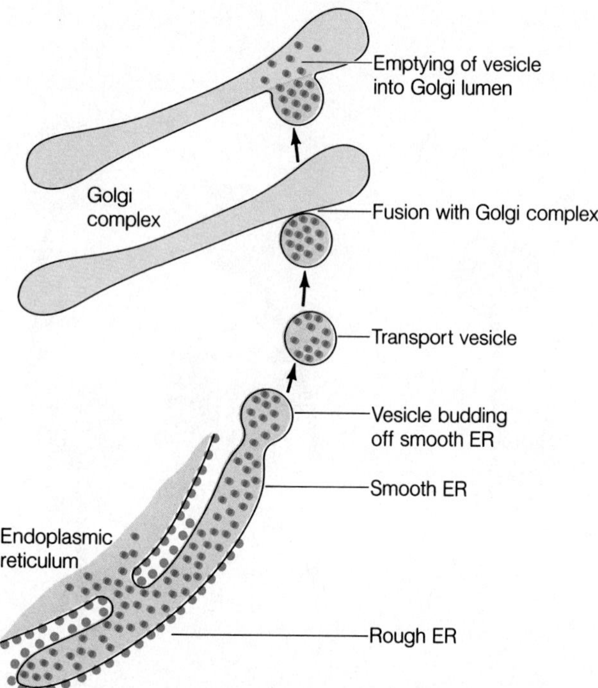

Figure 2–5 Vesicular Transport between the Endoplasmic Reticulum and the Golgi Complex

toxic substances so that the latter can be eliminated more readily in the urine. The amount of smooth ER available in liver cells for the task of detoxification can vary dramatically, depending on the need. For example, if phenobarbital (a barbiturate drug used as a sedative) is administered in large quantities, the amount of smooth ER with its associated detoxification enzymes is doubled within a few days, only to return to normal within five days after drug administration ceases. The mechanisms involved in regulating these changes are not well understood. Unfortunately, in some cases the same enzymes transform otherwise harmless substances into carcinogenic substances that play a role in cancer development.

□ Muscle cells have developed another specialized use for the smooth ER. They have an elaborate, modified smooth ER known as the sarcoplasmic reticulum, which stores calcium and plays an important role in the process of muscle contraction.

The Golgi complex is a refining plant and directs molecular traffic.

Closely associated with the endoplasmic reticulum is the **Golgi complex** or **Golgi apparatus.** It consists of sets of flattened, slightly curved, membrane-bounded sacs stacked in layers (Fig. 2–6). Note that the flattened sacs, or cisternae, are

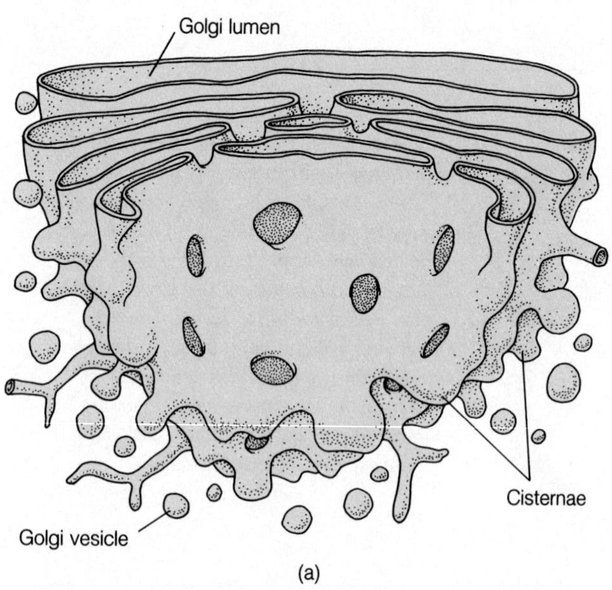

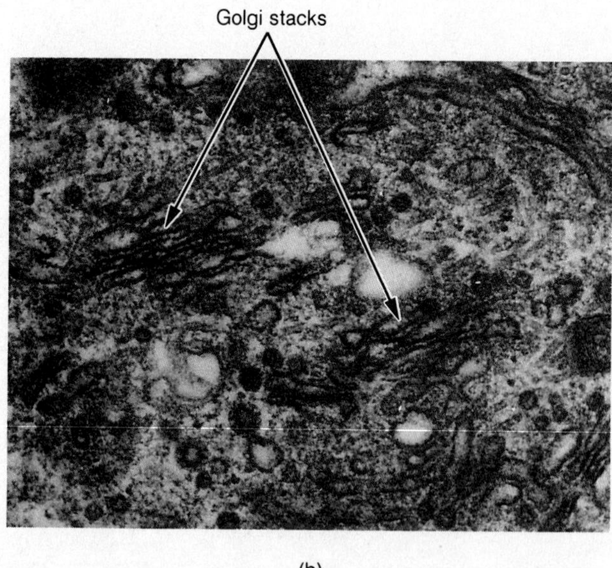

Figure 2–6 Golgi Complex *(a) Schematic three-dimensional representation of a Golgi complex. (b) Electron micrograph of Golgi stacks.*

SOURCE: Part (b) courtesy of Elizabeth R. Walker, Associate Professor, Department of Anatomy, School of Medicine, West Virginia University.

thin in the middle but have dilated edges. The number of Golgi stacks varies, depending on the cell type. Some cells have only one stack, whereas cells highly specialized for protein secretion may have hundreds of stacks.

The majority of the newly synthesized molecules that have just budded off from the smooth endoplasmic reticulum enter a Golgi stack. When a transport vesicle carrying its newly synthesized cargo reaches the Golgi complex, the vesicle membrane fuses with the membrane of a stack's innermost cisterna. The vesicle membrane opens up and becomes a new part of the Golgi membrane, and the contents of the vesicle are released to the interior of the sac (Fig. 2–5).

By unknown means, these newly synthesized raw materials from the ER travel through the layers of the Golgi stack, where two important interrelated functions take place:

1. *Processing of the raw materials into finished products*—Within the Golgi complex, the proteins from the ER are further modified into their final form, largely by adjustments being made in the sugars attached to the protein. The biochemical pathways that the protein molecules undergo during their passage through the Golgi complex are elaborate, precisely programmed, and specific for each final product.
2. *Sorting and directing the finished products to their final destinations*—The Golgi complex is responsible for sorting and segregating different types of products according to their function and destination, namely, molecules that are destined for secretion to the exterior, molecules that will become part of the plasma membrane, and molecules that will become incorporated into other organelles.

The means by which the Golgi complex directs this molecular traffic is complex and still is not clearly understood, although it is known that products destined for intracellular transport are packaged differently than those to be extruded from the cell. Those products targeted for intracellular sites are packaged in coated vesicles, whereas those to be secreted are packaged in secretory vesicles.

Coated vesicles originate from the dilated rims of the Golgi sacs, which pinch off to form clusters of small vesicles that can be seen along the edge of a Golgi stack. These vesicles are "coated" or enclosed in a cage-like protein framework consisting of **clathrin,** a structural protein that forms the latticework of the coat, plus accessory proteins derived from the Golgi membrane (Fig. 2–7). The accessory proteins serve two important functions. First, specific accessory proteins on the interior surface of the coat act as "recognition markers" for the recognition and attachment of specific molecules that have been processed in the Golgi lumen, thereby ensuring that the proper cargo is captured and packaged for the appropriate destination. Second, other specific accessory proteins on the outer surface of each vesicle act as "docking markers" that are able to fuse only with the membrane of the designated intracellular site. These docking markers assure that the coated transport vesicle docks and unloads its cargo only at the appropriate intracellular "address." Thus coated vesicles appear to act as miniature sorting machines; specific molecules are selectively trapped in a vesicle before it buds off, with its destiny being fusion with only one designated type of intracellular structure. There are different classes of coated vesicles, each of which contains specific captured cargo that is bound to a specified intracellular compartment.

After delivery of its contents to the appropriate intracellular organelle, the membrane components of the vesicle that fused with the organelle membrane may be returned to the Golgi complex for recycling if the recipient organelle does not need additional membrane. The fused membrane components are retrieved and subsequently bud and pinch off from the organelle, this time without cargo, to return to and fuse again with the Golgi complex.

Cells specialized for secretion handle molecules destined for export out of the cell somewhat differently than intracellular cargo is handled. In secretory cells, numerous large **secretory vesicles** (about two-hundred times larger than coated vesicles) can be seen in association with a Golgi stack. The molecules to be secreted are trapped within these large secretory vesicles, which pinch off the dilated rims of the outermost sac closest to the plasma membrane. The secretory proteins trapped within these structures are in a dilute solution, but the contents become concentrated over time as the secretory vesicle gradually loses water. The concentrated secretory proteins remain stored within the secretory vesicles until the cell is stimulated by a specific signal that indicates a need for release of that particular secretory product. On appropriate stimulation, the vesicles move to the cell's periphery. Vesicular contents are quickly released to the cell's exterior by fusion of the

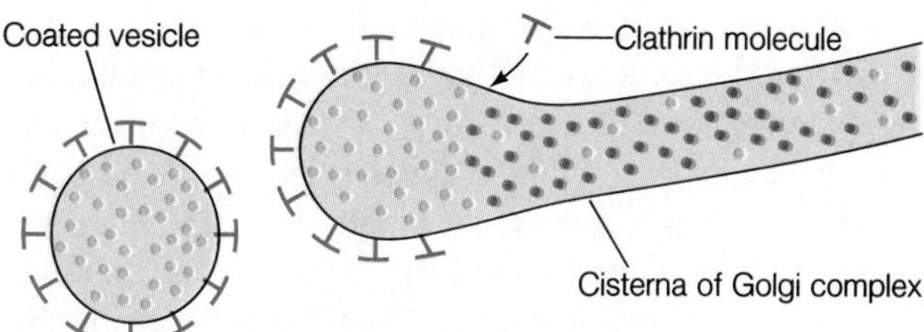

Figure 2–7 Formation of a Coated Vesicle *A coated vesicle, coated with clathrin molecules and specific accessory proteins, captures specific cargo from the lumen of the Golgi complex and then buds off the dilated rim of a Golgi cisterna.*

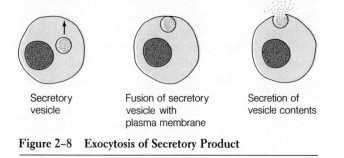

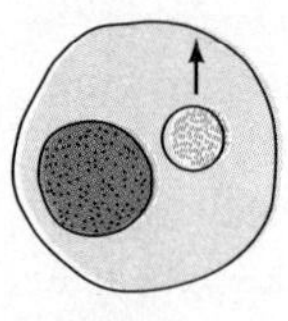
Secretory vesicle

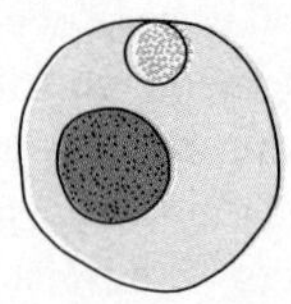
Fusion of secretory vesicle with plasma membrane

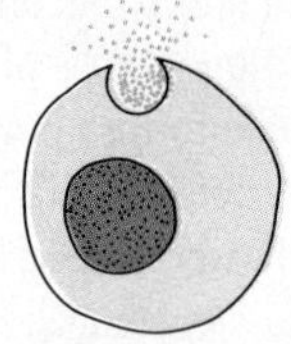
Secretion of vesicle contents

Figure 2–8 Exocytosis of Secretory Product

vesicle with the plasma membrane, followed by an opening of the vesicle and the emptying of its contents to the outside (Fig. 2–8). Release of the contents of the secretory vesicle constitutes the process of **secretion.** This mechanism of extrusion to the exterior of substances originating within the cell is referred to as **exocytosis** (*exo* means "out of"; *cyto* means "cell"). Secretory vesicles fuse only with the plasma membrane and not with any of the internal membranes that bound the organelles, which thereby prevents fruitless or even dangerous discharge of secretory products into the organelles. As with intracellular transport, new membrane added to the plasma membrane during the process of secretion can be recycled as needed.

Lysosomes serve as the intracellular digestive system.

Lysosomes are membrane-bound sacs containing powerful hydrolytic enzymes (see hydrolysis p. A–13) capable of digesting and thereby removing various unwanted cellular debris and foreign material, such as bacteria that have been internalized within the cell. Thus lysosomes serve as the intracellular "digestive system." Forty different hydrolytic enzymes have been isolated from lysosomes.

On the average, there are about three-hundred lysosomes per cell. Instead of having a uniform structure as is characteristic of all other organelles, lysosomes vary in size and shape, depending on the contents they are digesting. Most commonly, lysosomes are small (0.2–0.5 μm in diameter) oval or spherical bodies that appear granular during inactivity. These granules are protein aggregates of the powerful digestive enzymes contained within. The surrounding membrane that confines these enzymes normally prevents them from destroying the cell that houses them. Both the membrane and the enzymes are derived from the Golgi complex. A specific collection of newly synthesized hydrolytic enzymes is captured in a coated vesicle that is pinched off from the Golgi complex to become a new lysosome.

Extracellular material to be attacked by lysosomal enzymes is brought into the interior of the cell through the process of **endocytosis** (*endo* means "within"). Endocytosis can be accomplished in one of two ways (Fig. 2–9). In most cases, the plasma membrane invaginates (dips inward), forming a pouch that contains a small bit of extracellular fluid, usually with a specific particle that has bound to a surface receptor. The plasma membrane then seals at the surface of the pouch, forming a small, intracellular, membrane-bound vesicle with the contents of the pouch trapped inside. A few cell types, most notably white blood cells, perform a special form of endocytosis known as phagocytosis. When a white blood cell encounters a large multimolecular particle, such as a bacterium or tissue debris, it extends surface projections that completely surround or engulf the particle, forming an internalized vesicle that traps the particle within. A lysosome fuses with the membrane of the internalized vesicle and releases its hydrolytic enzymes into the vesicle. These potent enzymes safely attack the bacterium or other trapped material within the enclosed confines of the vesicle without damaging the remainder of the cell.

Lysosomes that have completed their digestive activities are known as **residual bodies.** The hydrolytic enzymes largely break down the engulfed material into products such as amino acids, glucose, and fatty acids that can be used by the cell. These small products can readily pass through the lysosomal membrane to enter the cytoplasm for future use. Usually, undigestible substances left within the residual body are eventually eliminated from the cell by exocytosis. Less commonly they are retained and accumulate within the cell as it ages.

Lysosomes also can fuse with aged or damaged organelles to remove these useless parts of the cell. This selective self-digestion makes way for new replacement parts. All organelles are renewable. If the whole cell is severely damaged or dies, the lysosomes rupture and release their destructive enzymes into the cytosol so that the cell digests itself entirely. In most tissues, elimination of a nonfunctional cell clears the way for its replacement with a healthy new one through cell division. However, in tissues in which cell reproduction is impossible, such as the heart and brain, scar tissue replaces the self-destructed dead cells.

In specific instances, lysosomes cause intentional self-destruction of healthy cells. This happens as a normal part of embryonic development when certain unwanted tissues that form are programmed for destruction. For example, embryonic ducts capable of forming a male reproductive tract are deliberately destroyed during the development of a female fetus. Lysosomes also play an important role in tissue regression, examples of which are the normal reduction in the uterine lining

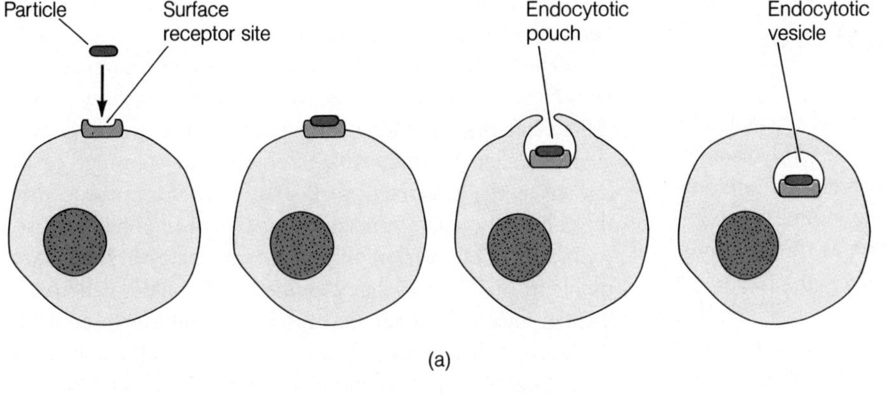

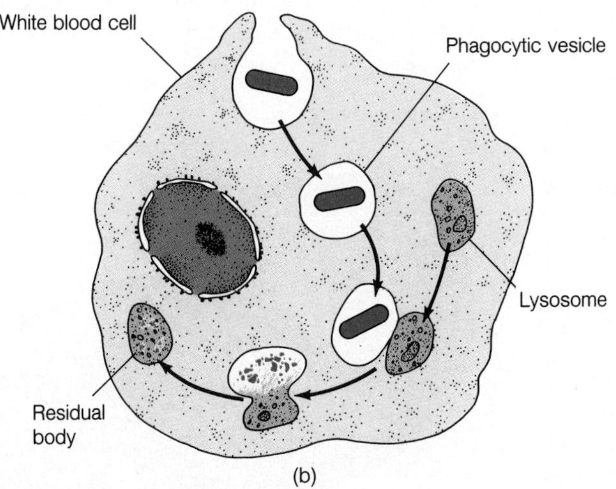

Figure 2–9 Endocytosis *(a) On attachment of a particle to a specific surface receptor site, the membrane dips inward to form a pouch, then seals the surface to internalize the particle within an intracellular vesicle. (b) White blood cells perform a special form of endocytosis known as phagocytosis. They internalize multimolecular particles such as bacteria by extending surface projections that seal in the targeted material. A lysosome fuses with the internalized vesicle, releasing enzymes that attack the engulfed material within the confines of the vesicle.*

following pregnancy and the diminution of mammary-gland (breast) tissue upon cessation of lactation (milk production). The mechanisms that control lysosomal activity under such circumstances are still unknown.

There is an inherent danger in a healthy, intact cell of the inadvertent rupture of lysosomal membranes. Because of their potential ability to self-destruct the cell, the discoverer of these organelles named them "suicide bags." However, two factors make this threat less severe than imagined. First, the lysosomal hydrolytic enzymes function best in an acid environment. The lysosomal membrane transports hydrogen ions (acid formers) into the lysosome, making it considerably more acidic than the remainder of the cell. Should the enzymes accidentally leak into the cytosol, they would be less potent than they are in their acidic home. Second, in most instances cells could tolerate the limited damage that would occur on the inadvertent rupture of only one or two lysosomes because most parts of the cell are renewable. The biggest danger is accidental digestion of part of the irreplaceable DNA molecule within the nucleus. Such nuclear damage would alter the cell's genetic properties, with the defect being perpetuated to all of the cell's progeny.

Occasionally individuals lack the ability to synthesize one or more of the lysosomal enzymes. The result is massive accumulation within the lysosomes of the specific compound that is normally digested by the missing enzyme. Clinical manifestations often accompany such disorders because the engorged lysosomes interfere with normal cell activity. The nature and severity of the symptoms depend on the type of substance that is accumulating, which in turn depends on what lysosomal enzyme is missing. Among these so-called storage diseases is **Tay-Sachs disease.** It is characterized by abnormal accumulation of gangliosides, which are complex molecules found in nerve cells. Profound symptoms of progressive nervous-system degeneration result as the accumulation process continues.

Peroxisomes house oxidative enzymes that detoxify various wastes, producing H_2O_2 in the process.

Peroxisomes are the most recently discovered organelles. A few cell types have peroxisomes similar in size to lysosomes, but most cells have much smaller peroxisomes that are only one-third to one-half the average size of lysosomes. Accordingly, peroxisomes are alternatively known as **microbodies.** Typically, several hundred small peroxisomes are present in a cell.

Peroxisomes are similar to lysosomes in that they are membrane-bound sacs containing enzymes, but unlike the hydrolytic enzymes found in lysosomes, peroxisomes house several powerful oxidative enzymes and contain most of the cell's catalase. **Oxidative enzymes,** as the name implies, use oxygen (O_2), in this case to strip hydrogen from specific substrates. Such a reaction is important in detoxifying various wastes produced within the cell or foreign compounds that have entered the cell, such as ethanol consumed in alcoholic beverages. The major product generated in the peroxisome is **hydrogen peroxide** (**H_2O_2**), which is formed by molecular oxygen and the hydrogen atoms stripped from the waste.

Hydrogen peroxide, itself a powerful oxidant, is potentially destructive if it is allowed to accumulate or escape from the confines of the peroxisome. However, peroxisomes also contain an abundance of **catalase,** an enzyme that decomposes potent H_2O_2 into harmless H_2O and O_2. This latter reaction is an important safety mechanism that destroys the potentially deadly peroxide at its site of production, thereby preventing its possible devastating escape into the cytosol.

Mitochondria are the energy organelles.

Mitochondria are the energy organelles or "powerhouses" of the cell; they extract energy from the nutrients in food and transform it into a usable form to energize cellular activities. The number of mitochondria per cell varies greatly, depending on the energy needs of each particular cell type. There may be as few as one hundred or as many as several thousand mitochondria in a single cell. In some cell types, the mitochondria are densely compacted in cellular regions that use most of the cell's energy. For example, mitochondria are packed between the contractile units in the muscle cells of the heart and are wrapped tightly around the origin of the propulsive tail in a sperm.

Mitochondria are rod- or oval-shaped structures about the size of bacteria. In fact, there is considerable evidence that mitochondria are descendants of a type of bacteria that invaded or was engulfed by primitive cells early in evolutionary history. It is generally believed that a symbiotic ("living together") relationship subsequently evolved, with the bacteria eventually becoming permanent organelles.

Each mitochondrion is enclosed by a double membrane—a smooth outer membrane that surrounds the mitochondrion itself and an inner membrane that forms a series of infoldings or shelves called **cristae,** which project into an inner cavity known as the **matrix** (Fig. 2–10). Granules embedded in these cristae contain crucial enzymes that ultimately are responsible for conversion of much of the energy in food into a usable form. The fact that the inner membrane is generously thrown into folds greatly increases the surface area available for housing these important enzymes. The matrix contains a concentrated, gel-like mixture of hundreds of different dissolved enzymes that are important in preparing nutrient molecules for final extraction of usable energy by the cristae enzymes.

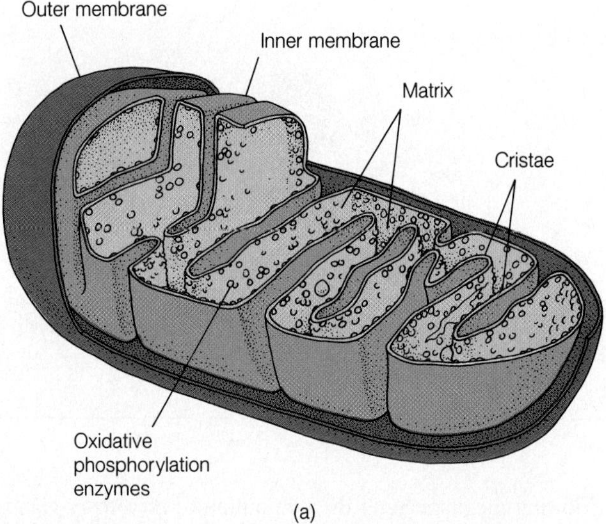

Figure 2–10 Mitochondrion *(a) Schematic representation of a mitochondrion. (b) Electron micrograph of a mitochondrion.*

SOURCE: Part (b) courtesy of Elizabeth R. Walker, Associate Professor, Department of Anatomy, School of Medicine, West Virginia University.

Energy derived from food is stored in ATP.

The source of energy for the body is the chemical energy stored in the carbon-hydrogen bonds within ingested food. Body cells, however, are not equipped to directly use this energy. Instead they must extract energy from food nutrients and convert it into an energy form that they can use—namely, the high-energy phosphate bonds of **adenosine triphosphate (ATP)**, which consists of adenosine with three phosphate groups attached (see p. A–14). When a high-energy bond such as that which binds the terminal phosphate to adenosine is split, a substantial amount of energy is released. Adenosine triphosphate is the universal energy carrier, or common energy "currency," of the body. Cells can "cash in" ATP to pay the energy "price" for running the cellular machinery. To obtain immediately usable energy, cells split the terminal phosphate bond of ATP, which yields **adenosine diphosphate (ADP**; adenosine with two phosphate groups attached) plus inorganic phosphate (P_i) plus energy:

$$\text{ATP} \xrightarrow{\text{splitting}} \text{ADP} + P_i + \text{Energy for use by the cell}$$

In this energy scheme, food might be thought of as the "crude fuel" whereas ATP is the "refined fuel" for operating the body's machinery. Let us elaborate on this fuel conversion process. Dietary food is digested or broken down by the digestive system into smaller absorbable units that can be transferred from the digestive-tract lumen into the circulatory system. For example, dietary carbohydrates are broken down primarily into glucose, which can be absorbed into the blood. No usable energy is released during the digestion of food. On being delivered by the blood to the cells, the nutrient molecules are transported across the plasma membrane into the cytosol. Among the thousands of enzymes within the cytosol are those responsible for **glycolysis,** a chemical process involving nine separate sequential reactions that break down the simple six-carbon sugar molecule, glucose, into two pyruvic-acid molecules, each of which contains three carbons. During this process, some of the energy stored in the chemical bonds of glucose are used to convert ADP into ATP (Fig. 2–11). However, this process is not very efficient in terms of energy extraction. Through glycolysis, one molecule of glucose has a net yield of only two molecules of ATP. Much of the energy originally contained within the glucose molecule is still locked within the chemical bonds of the pyruvic-acid molecules. This low-energy yield of glycolysis is insufficient to support the body's demand for ATP. This is where the mitochondria come into play.

The pyruvic acid produced by glycolysis in the cytosol can be selectively transported into the mitochondrial matrix. Here it is further broken down into a two-carbon molecule, acetic acid, by enzymatic removal of one of the carbons in the form of carbon dioxide (CO_2), which eventually is eliminated from the body as a waste (Fig. 2–12). During this process, a carbon-hydrogen bond is disrupted so that hydrogen is also released. This hydrogen atom is held by a hydrogen carrier molecule, the function of which will be discussed shortly. The acetic acid thus formed combines with coenzyme A, a derivative of pantothenic acid (a B vitamin), producing the compound acetyl coenzyme A (acetyl CoA).

Acetyl CoA then enters the **citric-acid cycle,** which consists of a cyclical series of eight separate biochemical reactions that are directed by the mitochondrial-matrix enzymes. This cycle of reactions can be compared to one revolution around a ferris wheel. On the top of the ferris wheel, acetyl CoA, a two-carbon molecule, enters a seat already occupied by oxaloacetic acid, a four-carbon molecule. These two link together to form a six-carbon citric-acid molecule, and the trip around the citric-acid cycle begins. (This cycle is alternatively known as the **Krebs cycle** in honor of its principal discoverer, Sir Hans Krebs, or the **tricarboxylic-acid cycle,** because citric acid contains three carboxylic acid groups.) With each new position as the seat moves around the cycle, matrix enzymes modify the passenger molecule, forming a slightly different molecule. The important consequences of these molecular alterations are as follows:

1. Two carbons are sequentially "kicked off the ride" as they are removed from the six-carbon citric-acid molecule, converting it back into the four-carbon oxaloacetic acid, which

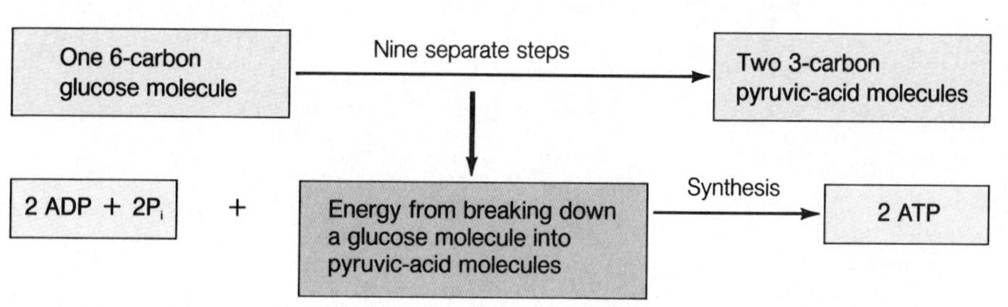

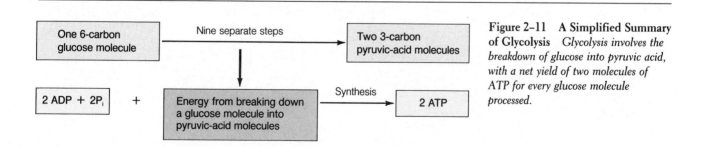

Figure 2–11 A Simplified Summary of Glycolysis *Glycolysis involves the breakdown of glucose into pyruvic acid, with a net yield of two molecules of ATP for every glucose molecule processed.*

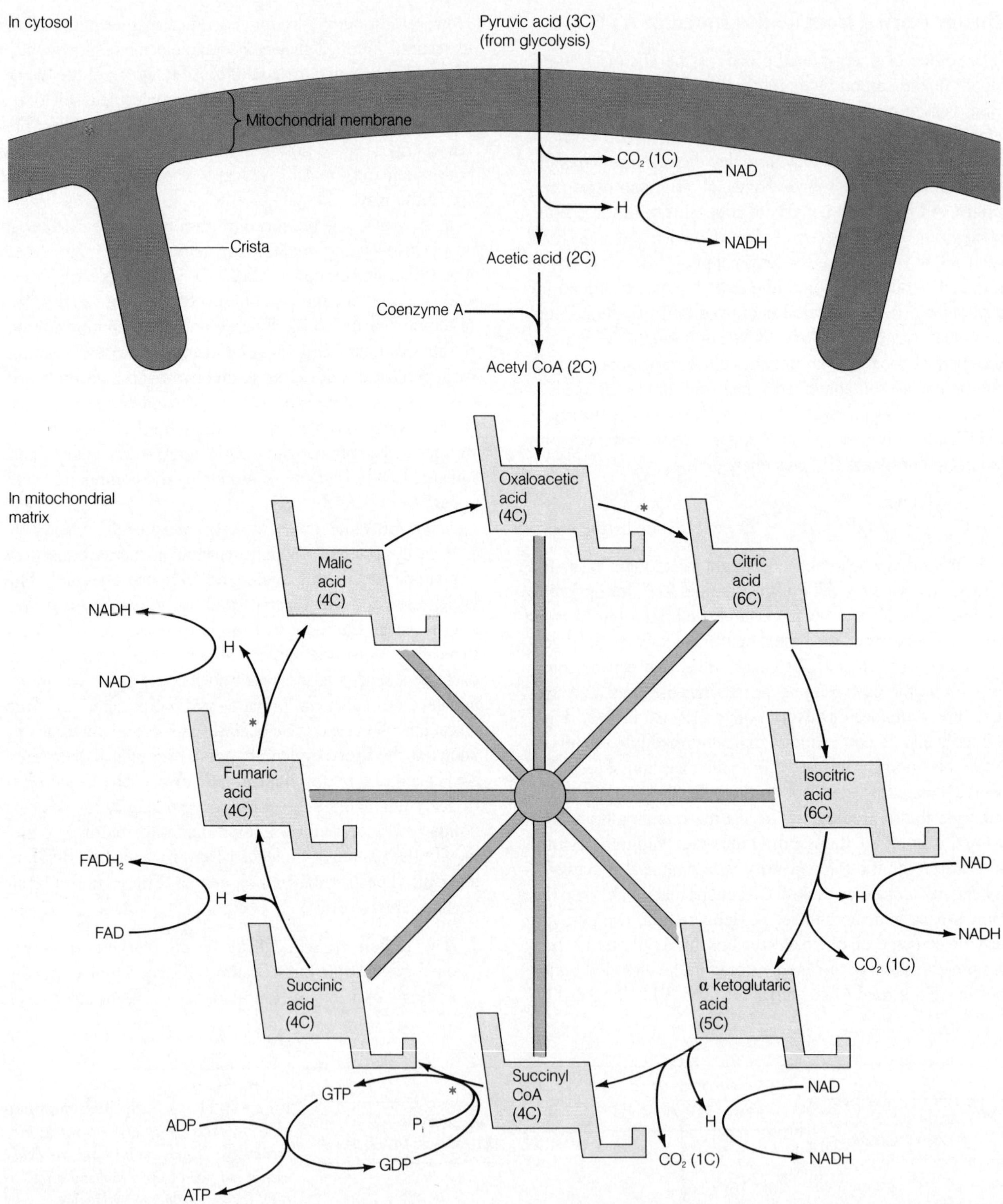

C = carbon atom.
H_2O enters the cycle at the steps marked with an asterisk.

Figure 2–12 Citric-Acid Cycle *A simplified version of the citric-acid cycle, showing how the two carbons entering the cycle by means of acetyl CoA are eventually converted to CO_2, with oxaloacetic acid, which accepts the acetyl CoA, being regenerated at the end of the cyclic pathway. Also denoted is the release of hydrogen atoms at specific points along the pathway, with these hydrogens binding to the hydrogen carrier molecules NAD and FAD for further processing. One molecule of ATP is generated for each molecule of acetyl CoA that enters the citric-acid cycle, for a total of two molecules of ATP for each molecule of processed glucose.*

is now available at the top of the cycle to pick up another acetyl CoA for another revolution through the cycle.

2. The released carbon atoms, which were originally present in the acetyl CoA that entered the cycle, are converted into two molecules of CO_2. This CO_2, as well as the CO_2 produced during the formation of acetic acid from pyruvic acid, passes out of the mitochondrial matrix and subsequently out of the cell to enter the blood. In turn, the blood carries it to the lungs, where it is finally eliminated into the atmosphere through the process of breathing. The oxygen used to make CO_2 from these released carbon atoms is derived from the molecules that were involved in the reactions, not from free molecular oxygen supplied by breathing.
3. Hydrogen atoms are also "bumped off" during the cycle at four of the chemical conversion steps. These hydrogens are "caught" by two other compounds that act as hydrogen carrier molecules—**nicotinamide adenine dinucleotide (NAD)** and **flavine adenine dinucleotide (FAD)**. These compounds are converted by the transfer of hydrogen to NADH and $FADH_2$, respectively.
4. One more molecule of ATP is produced for each molecule of acetyl CoA processed. Actually, ATP is not directly produced by the citric-acid cycle. The released energy is used to directly link inorganic phosphate to **guanosine diphosphate (GDP)** to form **guanosine triphosphate (GTP)**, a high-energy molecule similar to ATP. The energy from GTP can then be transferred to ATP as follows:

$$ADP + GTP \rightleftharpoons ATP + GDP$$

Because each glucose molecule is converted into two acetic-acid molecules, thus permitting two turns of the citric-acid cycle, two more ATP molecules are produced from each glucose molecule.

These few additional ATPs are still not much of an energy profit. However, the citric-acid cycle is important in preparing the hydrogen carrier molecules for their entry into the **electron-transport chain,** which produces far more energy than the sparse amount of ATP produced by the cycle itself. Considerable untapped energy is still stored in the released hydrogen atoms, which contain electrons that exist at high energy levels. The "big pay-off" comes when NADH and $FADH_2$ enter into the electron-transport chain located in the inner mitochondrial membrane lining the cristae (Fig. 2–13). Here the high-energy electrons are extracted from the hydrogens being held within NADH and $FADH_2$ and are transferred sequentially to the electron carrier molecules that compose the electron-transport system of the cristae, freeing NAD and FAD to pick up more hydrogen atoms. The electron-transport molecules are arranged in a specifically ordered fashion on the inner membrane so that the high-energy electrons are progressively transferred through a chain of reactions, with the electrons falling to successively lower energy levels with each step.

This electron-transport chain is also called the **respiratory chain** because it is critical to cellular respiration, which refers to the intracellular oxidation of nutrient derivatives. Ultimately, the electrons are passed to molecular oxygen (O_2) derived from the air we breathe. Electrons bound to O_2 are in their lowest energy state. Oxygen breathed in from the atmosphere is picked up by the blood, which delivers it to the cells. Within the cells, O_2 enters the mitochondria to serve as the final electron acceptor of the electron transport chain. This negatively charged oxygen (negative from having acquired additional electrons) then combines with the positively charged hydrogen ions (positive from having donated the electrons at the beginning of the electron transport chain) to form water. Energy released by the electrons as they move through this chain of reactions to ever lower energy levels is harnessed by **ATP synthetase.** This enzyme, which is present in the granules of the cristae, converts ADP plus P_i to ATP, providing a rich yield of thirty-two more ATP molecules for each glucose molecule thus processed. The harnessing of energy into a useful form as the electrons tumble from a high energy state to a low energy state can be likened to a power plant that converts the energy of water tumbling down a waterfall into electricity. Because O_2 is used in these final steps of energy conversion when a phosphate is added to form ATP, this process is known as **oxidative phosphorylation.**

The series of steps that lead to oxidative phosphorylation might at first seem like an unnecessary complication. Why not just directly oxidize, or "burn" food molecules to release their energy? This can be done outside of the body, but all of the energy stored in the food molecule is released explosively in the form of heat (Fig. 2–14). In the body, oxidation of food molecules occurs instead in many small, controlled steps so that the food molecule's chemical energy is gradually made available for convenient packaging in a storage form that is useful to the cell. The cell, by means of its mitochondria, can more efficiently capture the energy from the food molecules

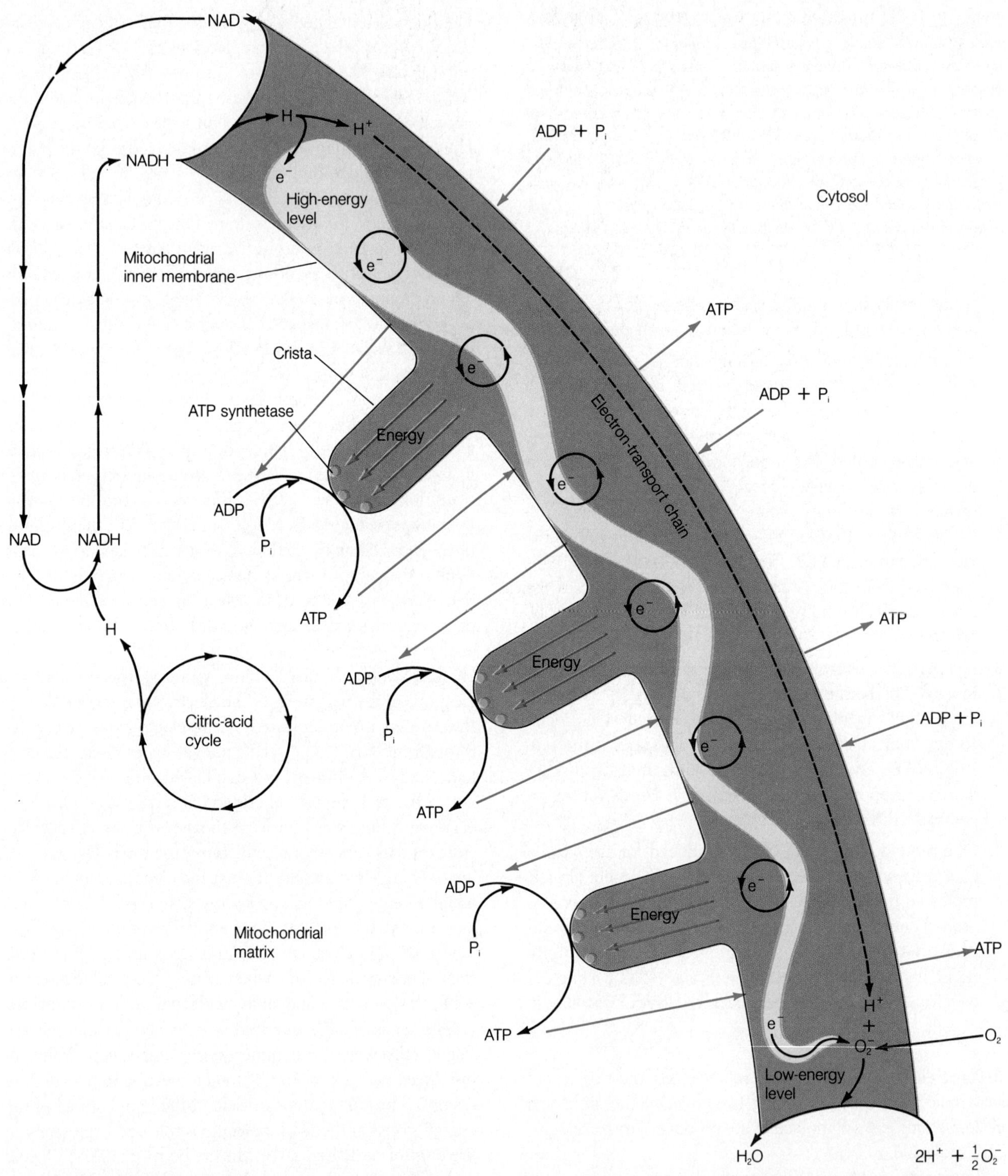

Figure 2–13 Passage of High-Energy Electrons through Electron-Transport Chain *High-energy electrons extracted from hydrogen that is released during the degradation of carbon-containing food molecules are passed through the electron-transport chain located on the mitochondrial inner membrane. Energy is gradually released as the electrons fall to successively lower energy levels by moving through the electron-transport chain of reactions. The released energy is harnessed by ATP synthetase within the mitochondrial inner-membrane granules to synthesize ATP from ADP and* P_i*. Molecular oxygen, which is essential to the process as the final electron acceptor, combines with the generated hydrogen ions to produce water.*

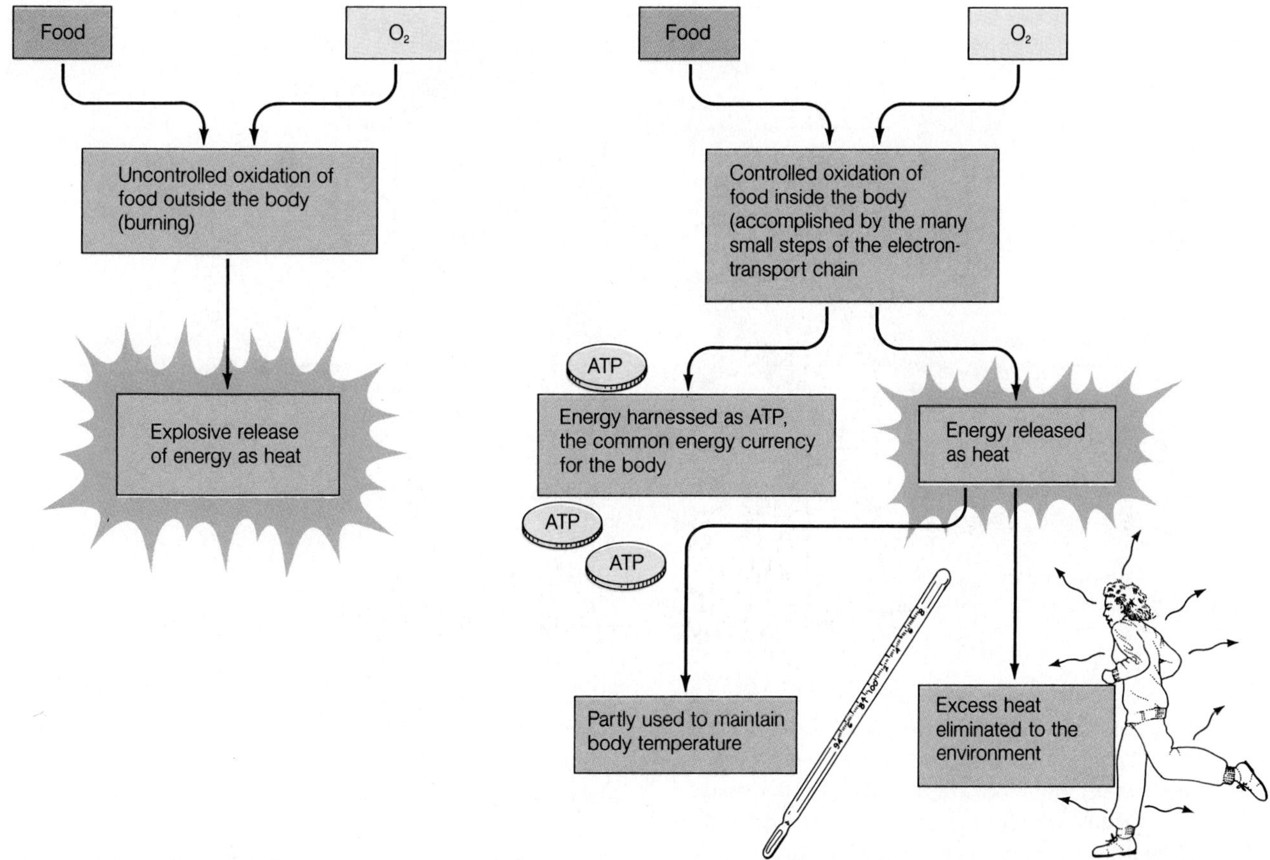

Figure 2–14 Uncontrolled versus Controlled Oxidation of Food *Schematic illustration of how, through the controlled oxidation of food in the body, much of the energy that is released as heat when the same food undergoes uncontrolled oxidation (burning) outside the body is instead harnessed and stored in useful form.*

within ATP bonds when it is released in small quantities, with much less of the energy being converted to heat. The heat that is produced is not completely wasted energy; it is used to help maintain body temperature, with any excess heat being eliminated to the environment.

The cell is a much more efficient energy converter when oxygen is available (Fig 2–15). In an **anaerobic** ("lack of air," specifically lack of O_2) condition, the degradation of glucose cannot proceed beyond glycolysis. Recall that glycolysis takes place in the cytosol and involves the breakdown of glucose into pyruvic acid, producing a low yield of two molecules of ATP per molecule of glucose. The untapped energy of the glucose molecule remains locked within the bonds in the pyruvic-acid molecules, which are eventually converted to lactic acid if they do not enter the pathway that ultimately leads to oxidative phosphorylation. When sufficient O_2 is present—an **aerobic** ("with air or with O_2") condition—mitochondrial processing (that is, the citric-acid cycle in the matrix and the electron-transport chain on the cristae) harnesses sufficient energy to generate thirty-four more molecules of ATP, for a total net yield of thirty-six ATPs per molecule of glucose processed. The overall reaction involving the oxidation of food molecules to yield energy is as follows:

Food +	O_2	→	CO_2	+	H_2O	+	ATP
	(necessary for oxidative phosphorylation)		(produced by the citric-acid cycle)		(produced by the electron-transport chain)		(produced by the electron-transport chain)

Molecular oxygen is consumed, whereas carbon dioxide and water are produced as by-products during the oxidation of foodstuff for the purpose of transferring food energy to the common energy carrier, ATP. (For a description of aerobic exercise, see accompanying boxed feature, A Closer Look at Exercise Physiology.)

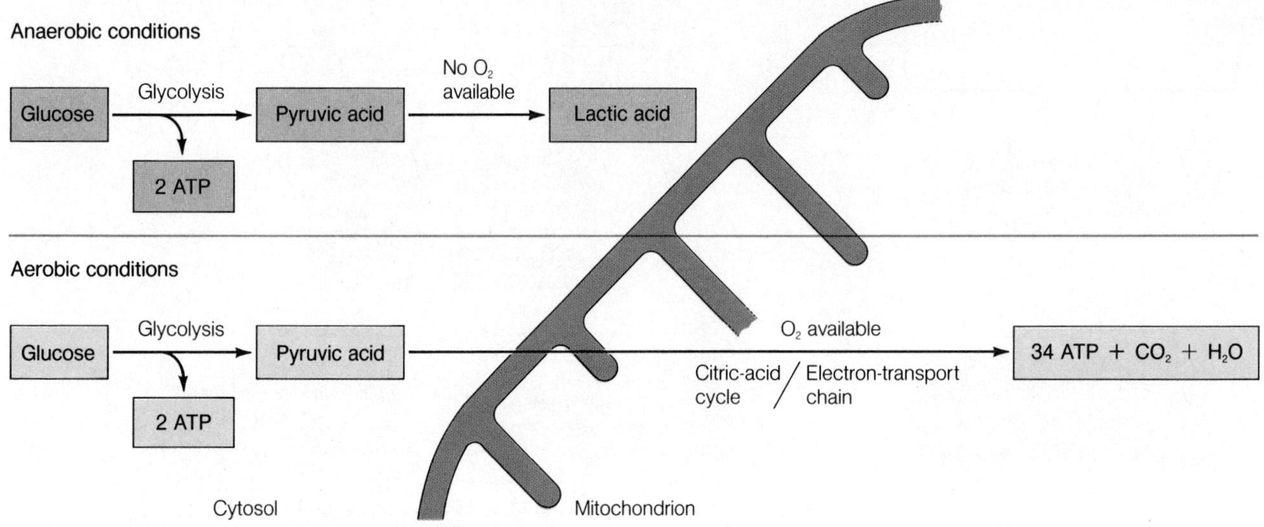

Figure 2–15 Comparison of Energy Yield and Products under Anaerobic and Aerobic Conditions

Glucose, the principal nutrient derived from dietary carbohydrates, is the fuel preference of most cells. However, nutrient molecules derived from fats (fatty acids) and, if necessary, from protein (amino acids) can also enter the citric-acid cycle at specific points to eventually produce energy. Amino acids are usually used for protein synthesis instead of energy production, but they can be used as fuel if there is insufficient glucose and fat available.

Note that the oxidative reactions within the mitochondria generate energy, unlike the oxidative reactions controlled by the peroxisome enzymes. Both organelles use O_2 but for different purposes.

The energy stored within ATP is used for synthesis, transport, and mechanical work.

Once formed, ATP is transported out of the mitochondria and is then available as an energy source as needed within the cell. Cellular activities that require energy expenditure fall into three main categories:

1. *synthesis of new chemical compounds*, such as protein synthesis by the endoplasmic reticulum. Some cells use up to 75% of the ATP they generate just to synthesize new chemical compounds. This is especially true for cells with a high rate of secretion and cells in the growth phase.
2. *membrane transport*, such as the selective transport of molecules across the kidney tubules during the process of urine formation. Kidney cells can expend as much as 80% of their ATP currency to operate their selective membrane-transport mechanisms.
3. *mechanical work*, such as the tremendous quantities of ATP required to power contraction of the heart muscle to pump blood or contraction of skeletal muscles to lift an object.

As a result of cellular energy expenditure to support these various activities, large quantities of ADP are produced. These energy-depleted ADP molecules enter the mitochondria for "recharging" and then recycle back into the cytosol once again as energy-rich ATP molecules after having participated in oxidative phosphorylation. A single ADP/ATP molecule may shuttle back and forth between the mitochondria and cytosol for this recharging-expenditure cycle thousands of times per day.

The high demands for ATP render glycolysis alone an insufficient as well as inefficient supplier of power for most cells. If it were not for the mitochondria, which house the metabolic machinery for oxidative phosphorylation, our energy capability would be very limited. However, glycolysis does provide cells with a sustenance mechanism to produce at least some ATP under anaerobic conditions. Skeletal-muscle cells in particular take advantage of this ability during short bursts of strenuous exercise, when energy demands for contractile activity outstrip the body's ability to bring adequate O_2 to the exercising muscles to support oxidative phosphorylation. Also, red blood cells, which are the only cells that do not contain any mitochondria, rely solely on glycolysis for their limited energy

A CLOSER LOOK AT EXERCISE PHYSIOLOGY

Aerobic Exercise: What For and How Much?

Aerobic ("with O_2") exercise involves large muscle groups and is performed at a low-enough intensity and for a long-enough period of time that fuel sources can be converted to ATP by using the citric-acid cycle as the predominant metabolic pathway. Aerobic exercise can be sustained for from fifteen to twenty minutes to several hours at a time. Short-duration, high-intensity activities, such as weight training and the 100 meter dash, which last for a matter of seconds and rely solely on energy stored in the muscles and on glycolysis, are forms of ***anaerobic*** ("without O_2") exercise.

Inactivity is associated with increased risk of developing both hypertension (high blood pressure) and coronary artery disease (blockage of the arteries that supply the heart). To reduce the risk of hypertension and coronary artery disease and to improve physical work capacity, the American College of Sports Medicine recommends that an individual participate in aerobic exercise a minimum of three times per week for twenty to sixty minutes. The intensity of the exercise should be based on a percentage of the individual's maximal capacity to work. The easiest way to establish the proper intensity of exercise and to monitor intensity levels is by checking heart rate. The estimated maximal heart rate is determined by subtracting the person's age from 220. Significant benefits can be derived from aerobic exercise performed between 70% and 80% of maximal heart rate. For example, the estimated maximal heart rate for a twenty year old is 200 beats per minute. If this person exercised three times per week for twenty to sixty minutes at an intensity that increased the heart rate to 140 to 160 beats per minute, the participant should significantly improve his or her aerobic work capacity and reduce the risk of cardiovascular disease.

production. However, the energy needs of red blood cells are low because they also lack a nucleus, so they do not have the capability of synthesizing new substances, the biggest energy expenditure for most noncontractile cells.

Cytosol and Cytoskeleton

Occupying about 55% of the total cell volume, the cytosol is the semiliquid portion of the cytoplasm that surrounds the organelles. Its amorphous appearance under an electron microscope belies the fact that the cytosol is not a uniform liquid mixture but is actually more like a highly organized, gelatinous mass with differences in composition and consistency existing between various regions of the cell.

The cytosol is important in intermediary metabolism, ribosomal protein synthesis, and storage of fat and glycogen.

Four general categories of activities are associated with the cytosol: (1) enzymatic regulation of intermediary metabolism; (2) ribosomal protein synthesis; (3) storage of fat and carbohydrate; and (4) the establishment of a cytoskeleton that gives shape to the cell, provides an intracellular organizational framework, and is responsible for various cell movements.

ENZYMATIC REGULATION OF INTERMEDIARY METABOLISM. **Intermediary metabolism** refers collectively to the large set of intracellular chemical reactions that involve the degradation, synthesis, and transformation of small organic molecules such as simple sugars, amino acids, fatty acids, and nucleotides. These reactions are critical for ultimately capturing energy to be used for cellular activities and for providing the raw materials needed for maintenance of the cell's structure and function and for the cell's growth. All intermediary metabolism occurs in the cytoplasm, with most of it being accomplished in the cytosol. Thousands of enzymes involved in glycolysis and other intermediary biochemical reactions are found in the cytosol.

RIBOSOME PROTEIN SYNTHESIS. Also dispersed throughout the cytosol are the free ribosomes, which synthesize proteins for use in the cytosol itself, in contrast to rough-ER ribosomes, which synthesize proteins for secretion and for construction of new cellular components. Among the proteins synthesized by cytosolic ribosomes are the enzymes involved in the chemical reactions that make up intermediary metabolism. Often, cytosolic ribosomes that are synthesizing identical proteins are clustered together in "assembly lines" known as **polyribosomes.**

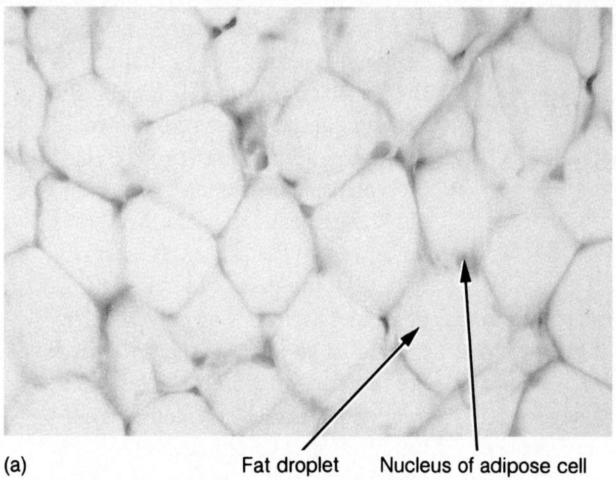

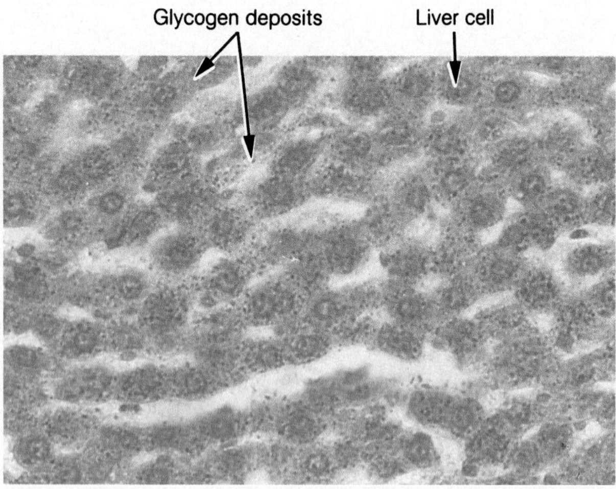

Figure 2–16 Inclusions *(a) Light micrograph depicting fat storage in an adipose cell. Note that the fat droplet occupies almost the entire cytosol. (b) Light micrograph depicting glycogen storage in a liver cell. The red-staining granules throughout the liver cell's cytosol are glycogen deposits.*

SOURCE: Photos courtesy of Elizabeth R. Walker, Associate Professor, and Dennis O. Overman, Associate Professor, Department of Anatomy, School of Medicine, West Virginia University.

STORAGE OF FAT AND GLYCOGEN. Excess nutrients not immediately used for ATP production are converted in the cytosol into storage forms that are readily visible even under a light microscope. Such nonpermanent masses of material are known as **inclusions.** The largest and most important storage product is fat. Small fat droplets can be seen within the cytosol in various cells. In **adipose tissue,** the tissue specialized for fat storage, the stored fat molecules can occupy almost the entire cytosol, coalescing to form one large fat droplet (Fig. 2–16a). The other visible storage product is **glycogen,** the storage form of glucose, which appears as aggregates or clusters dispersed throughout the cell (Fig. 2–16b). Cells vary in their ability to store glycogen, with liver and muscle cells having the greatest stores. When food is not available to provide fuel for the citric-acid cycle and electron-transport chain, stored glycogen and fat are broken down to release glucose and fatty acids, respectively, which can feed the mitochondrial energy-producing machinery. These energy stores are tapped even during a normal overnight fast. An average adult has enough glycogen stored to provide sufficient energy for about a day of normal activities, and typically enough fat is stored to provide energy for two months.

CYTOSKELETON. Permeating the cytosol is the **cytoskeleton,** a complex protein network that acts as the "bone and muscle" of the cell. The distinct shape, size, complexity, and intracellular specialization of the various body cells necessitate intracellular scaffolding to support and organize the cellular components into an appropriate arrangement and to control their movements. This is the role of the cytoskeleton. There are at least four distinct elements of this elaborate and dynamic network: (1) microtubules, (2) microfilaments, (3) intermediate filaments, and (4) the microtrabecular lattice (Table 2–3). In addition, numerous accessory proteins are involved in interactions between these structures and other cell components. The different parts of the cytoskeleton are structurally linked together and functionally coordinated to provide certain integrated functions for the cell. Because of the complexity of this network and the variety of functions it serves, its elements will be addressed separately.

Microtubules are essential for maintaining asymmetrical cell shapes and are important in complex cell movements.

The **microtubules** are the largest of the cytoskeletal elements. They are very slender [22 nanometer (nm) diameter; 1 nm = 1 billionth of a meter], long, hollow, unbranched tubes composed primarily of tubulin, a small, globular protein molecule (6 nm diameter) (Fig. 2–17). Microtubules are essential for maintaining an asymmetrical cell shape. An example of an asymmetrical shape is the elongated axon of a nerve cell, which may extend up to a meter in length from the origin of the cell body in the spinal cord to the termination of the axon at a muscle (Fig. 2–18). Microtubules, along with specialized intermediate filaments, stabilize this asymmetrical axonal extension. In cells whose mature shape and organization do not depend on the presence of microtubules, these cytoskeletal structures are present at an earlier developmental stage of the cell, functioning as temporary scaffolding to create a shape and organization that is later maintained by other cellular elements. For example, spiral rings of microtubules wrap around the base of the tail of a developing sperm and subsequently disappear. Before their disappearance, the microtubules serve as a transient "mold" for arranging the mitochondria spirally around the tail, where these energy organelles will be readily available to meet the cell's energy needs for motility.

Microtubules also play an important role in coordinating numerous complex cell movements, including: (1) transport of secretory vesicles from one region of the cell to another, (2) movement of specialized cell projections such as cilia and flagella, and (3) distribution of chromosomes during cell division.

TRANSPORT OF SECRETORY VESICLES. Axonal transport provides a good example of the importance of an organized system for moving secretory vesicles. In a nerve cell, specific chemicals are released from the terminal end of the elongated axon to influence a muscle or another structure that the nerve cell controls. These chemicals are largely produced within the cell body, where the nuclear DNA blueprint, endoplasmic reticular factory, and Golgi packaging and distribution outlet are located. Yet these chemicals ultimately function at the end of the axon, which may be a meter away. If these chemicals had to diffuse on their own from the cell body to a distant axon terminal, it would take them about fifty years to get there—obviously an impractical solution. The microtubules provide a "highway" for vesicular traffic along the axon, with the driving force depending on ATP (Fig. 2–18). Reverse vesicular traffic also occurs along these microtubular highways. Vesicles that contain debris are transported from the axon terminal to the cell body for degradation by lysosomes, which are confined within the cell body.

MOVEMENT OF CILIA AND FLAGELLA. Microtubules are also the dominant structural and functional components of cilia and flagella. These specialized protrusions from the cell surface allow a cell to move materials across its surface (in the case of a stationary cell) or to propel itself through its environment (in the case of a motile cell). **Cilia** are numerous tiny, hairlike protrusions, whereas a **flagellum** is a single, long, whiplike appendage. Even though they project from the sur-

Table 2–3 Functions of the Cytoskeleton

Integrated Functions of Entire Cytoskeleton

Responsible for the shape, rigidity, and spatial geometry of each type of cell

Responsible for directing intracellular transport and for regulating cellular movements

Appears to play a role in regulating growth and division of cells

Specific Functions of Cytoskeletal Elements

Microtubules

Maintain asymmetrical cell shapes

Serve as temporary scaffolding to create shape and organization within a cell that will later be maintained by other cellular elements

Coordinate complex cell movements

- Facilitate transport of secretory vesicles, such as axonal transport
- Serve as dominant structural and functional component of cilia and flagella
- Form mitotic spindle during cell division

Microfilaments

Play a vital role in various cellular contractile systems

- Play a dominant role in muscle contraction
- Form nonmuscle contractile assemblies, such as the contractile ring that divides the cell in half during cell division
- Important in amoeboid movement of motile cells

Serve as a mechanical stiffener for microvilli

- Increase the surface area available for absorption in intestines and kidneys
- Are specialized to detect sound and positional changes in ear

Intermediate Filaments

Play a structural role in parts of the cell subject to mechanical stress

Microtrabecular Lattice

Suspends, arranges, and reorients larger cytoskeletal elements and various organelles

Contributes to gel-like consistency of cytosol

Links components of cytoplasm into a functional unit

- Organizes cytosolic enzymes
- Plays an important role in the dispensation of newly synthesized proteins

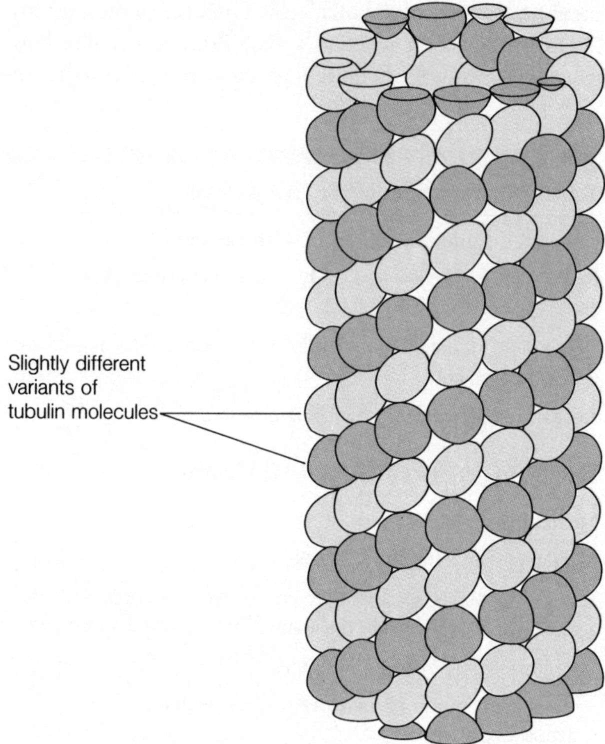

Figure 2–17 Arrangement of Tubulin Molecules into a Microtubule

face of the cell, cilia and flagella are both intracellular structures covered by the plasma membrane.

Cilia beat or stroke in unison, much like the coordinated efforts of the crew of a rowing team. Each cilium exerts a rapid active stroke, which moves material on the cell surface forward, followed by a recovery phase in which the cilium more slowly returns to its original position with a kind of unrolling, backward movement that does not exert much force. In this way, the material just pushed forward is not futilely pushed backward but instead is moved only in the direction of the forward stroke.

In humans, ciliated cells are found in the stationary cells that line the respiratory tract and the oviduct of the female reproductive tract. Respiratory cilia help keep foreign particles out of the lungs. The thousands of cilia lining the respiratory airways project into a layer of sticky mucus that traps dust and other inspired particles. The coordinated stroking action of these cilia sweeps this dust-laden mucus up to the throat, where it can be expectorated (spit out) or swallowed and eventually eliminated in the feces. In the female reproductive tract, the sweeping action of the cilia that line the oviduct draws the egg (ovum) released from the ovary during ovulation into the oviduct and then guides it toward the uterus (womb).

The only human cells that bear flagella are sperm. The whiplike motion of the flagellum or "tail" enables a sperm to move through its environment. This is particularly useful when the sperm maneuvers for final penetration of the ovum during fertilization.

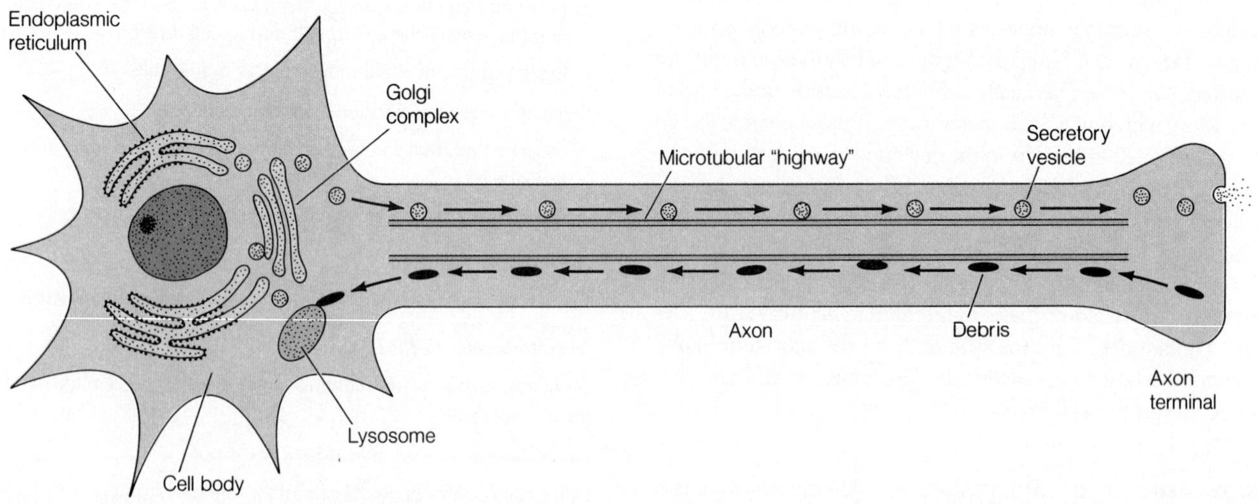

Figure 2–18 Two-Way Vesicular Axonal Transport Facilitated by the Microtubular "Highway"

Cilia and flagella have the same basic internal structure. Both consist of nine fused pairs of microtubules (doublets) arranged in an outer ring around two single unfused microtubules in the center (Fig. 2–19). This characteristic "nine plus two" array of microtubules, which extends throughout the length of the motile appendage, originates from a specialized cytoplasmic structure, the basal body, which is located inside the main part of the cell. Each basal body is a short cylinder composed of a parallel microtubular symmetry similar to that of the cilium or flagellum, but in this case the outer ring consists of nine fused triplets of microtubules (three microtubules in each peripheral bundle), with no microtubules in the center. The basal body provides a base from which the microtubules grow to form a cilium or flagellum.

Associated with these microtubules are accessory proteins that maintain the microtubules' organization and play an essential part in the microtubular movement that causes bending of the entire structure. The most important of these accessory proteins is **dynein,** which forms a set of armlike projections from each doublet of microtubules (Fig. 2–19).

The bending movements of cilia and flagella are produced by the sliding of adjacent microtubule doublets past each other. There is good evidence that the sliding is accomplished by the dynein arms, which have the capability of splitting ATP and then somehow using the released energy to "crawl" along the neighboring microtubule doublet to cause relative displacement of the doublets (Fig. 2–20). Groups of cilia working together are oriented to beat in the same direction and contract in a synchronized manner through poorly understood controlling mechanisms. These mechanisms appear to involve the single microtubules at the cilium's center and their surrounding accessory proteins, the **inner sheath** (Fig. 2–19).

Defects in flagellate and ciliary function have been observed in humans. Some hereditary forms of male sterility involving nonmotile sperm have been traced to defects in the dynein arms or inner sheath. These same individuals usually also have long histories of recurrent respiratory-tract disease. This is because the same types of defects are present in their respiratory cilia, which are unable to clear mucus and inhaled particles from the respiratory system.

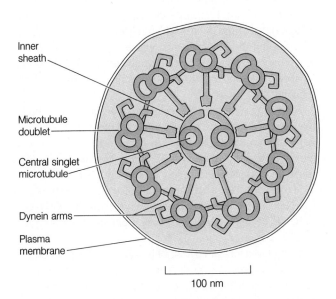

Figure 2–19 Schematic Diagram of a Cilium Showing the Dynein Arms and Other Accessory Proteins

SOURCE: Adapted and reproduced with permission from Bruce Alberts, Dennis Bray, Julian Lewis, Martin Raff, Keith Roberts, and James D. Watson: *Molecular Biology of the Cell* (New York: Garland Publishing), Figure 10–27, p. 565.

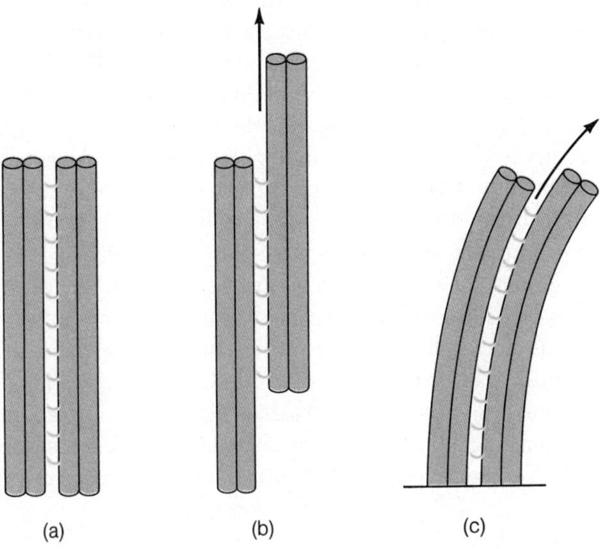

Figure 2–20 Bending of Cilium Accomplished by Relative Sliding of Anchored Microtubule Doublets *(a) Dynein arms extend between the microtubule doublets within a cilium. Relative sliding of adjacent microtubule doublets is believed to be accomplished by ATP-powered dynein "crawling." (b) If the microtubule doublets were not anchored at their base, they would freely slide past each other as a result of this dynein activity. (c) Because the doublets are anchored at their base, bending occurs when the doublets slide with respect to each other as a result of dynein activity.*

SOURCE: Adapted and reproduced with permission from Bruce Alberts, Dennis Bray, Julian Lewis, Martin Raff, Keith Roberts, and James D. Watson: *Molecular Biology of the Cell* (New York: Garland Publishing), Figure 10–30, p. 567.

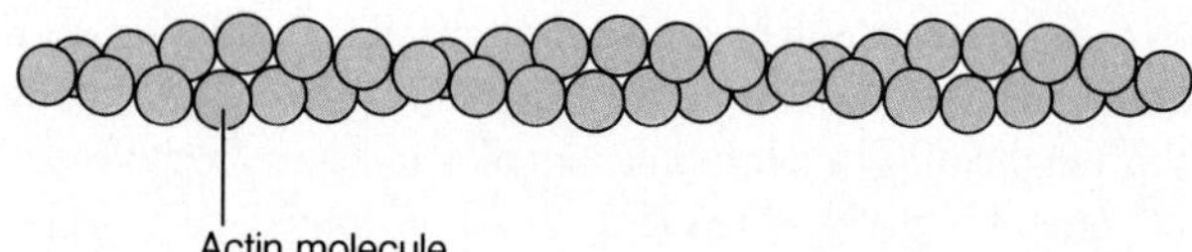

Figure 2–21 Arrangement of Actin Molecules in a Microfilament

FORMATION OF THE MITOTIC SPINDLE. Cell division involves two discrete but related activities: mitosis (nuclear division) and cytokinesis (cytoplasmic division). **Mitosis** involves replication of the DNA-containing chromosomes, which are then evenly distributed in the two halves of the cell. **Cytokinesis** involves constriction of the plasma membrane in the middle of the cell, which leads to separation of the two halves into two new daughter cells, each with a full complement of chromosomes.

Microtubules are transiently assembled to form a **mitotic spindle,** which organizes and directs the movement of the replicated chromosomes away from each other toward opposite ends of the cell so that the genetic material is evenly distributed when the cell divides. The mitotic spindle is formed by the **centrioles,** a pair of short cylindrical structures that lie at right angles to each other near the nucleus (Fig. 2–1). The centrioles also duplicate during cell division. After self-replication, the centriole pairs move toward opposite ends of the cell and form the spindle apparatus between them through a precisely organized assemblage of microtubules.

Besides their role in mitotic-spindle formation, the centrioles and surrounding complex of densely staining proteinaceous material together assemble the many microtubules that normally radiate throughout the cytoskeleton. The centrioles are identical in structure to basal bodies. In fact, under some circumstances the centrioles and basal bodies are interconvertible. During development of ciliated human cells, the centriole pair migrates to the region of the cell where the cilia will be formed and duplicates itself to produce the many basal bodies that will form the cilia.

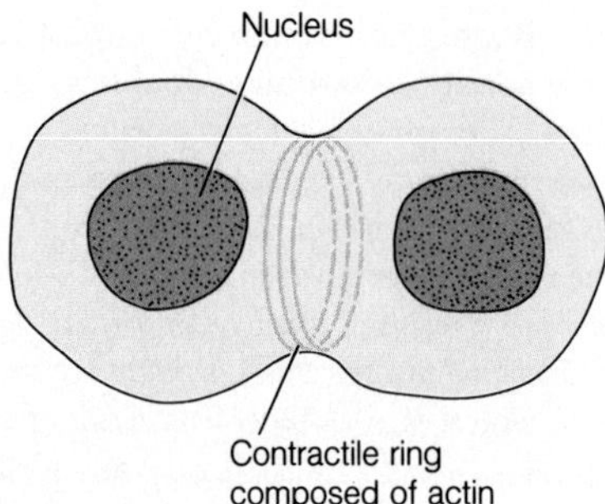

Figure 2–22 Dividing Cell

Microfilaments are important in cellular contractile systems and as mechanical stiffeners.

The **microfilaments** are the smallest (6 nm diameter) elements of the cytoskeleton visible with a conventional electron microscope. The most obvious microfilaments in most cells are those composed of **actin,** a protein molecule that has a globular shape similar to tubulin. Unlike tubulin, which forms a hollow tube, actin is assembled into two twisted strands, much like two strings of pearls twisted into a helix (spiral) to form a microfilament (Fig. 2–21). In muscle cells, another protein called **myosin** forms a distinct, different kind of microfilament. In most cells myosin is not as abundant and does not form such distinct filaments.

Microfilaments serve at least two different functions. First, they play a vital role in various cellular contractile systems, and second, they act as mechanical stiffeners for several specific cellular projections.

MICROFILAMENTS IN CELLULAR CONTRACTILE SYSTEMS. Actin-based assemblies are involved in muscle contraction, cell division, and cell locomotion. The most obvious, best organized, and most clearly understood cellular contractile system is that found in muscle. Muscle contains an abundance of actin and myosin filaments, which are organized so that when they are triggered by electrically induced ATP splitting to slide past each other, they generate a contractile force (chapter 8).

Surprisingly, nonmuscle cells may also contain "muscle-like" assemblies. Some of these microfilament contractile systems are transiently assembled to perform a specific function when needed. A good example is the contractile ring that forms during cytokinesis to split apart the duplicate cell halves. The ring consists of a beltlike bundle of actin filaments located just beneath the plasma membrane in the middle of the cell. When this ring of fibers contracts, it pinches the cell in two (Fig. 2–22).

Complex actin-based assemblies are also responsible for most cell locomotion. Four types of human cells are capable of moving on their own—sperm, white blood cells, fibroblasts, and skin cells. Many other cell types migrate to specific permanent sites during embryonic development, but these four are the only cells that retain locomotor ability. Sperm move by the flagellar mechanism already described. Motility for the other cells is accomplished by **amoeboid movement,** a process that depends on actin filaments and that is similar to the

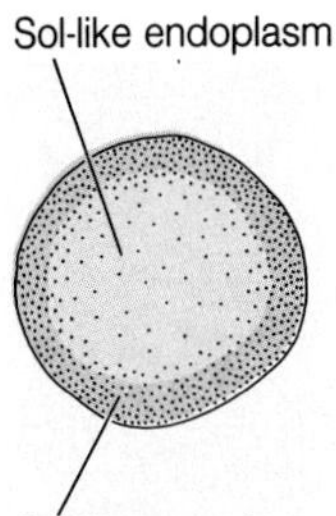

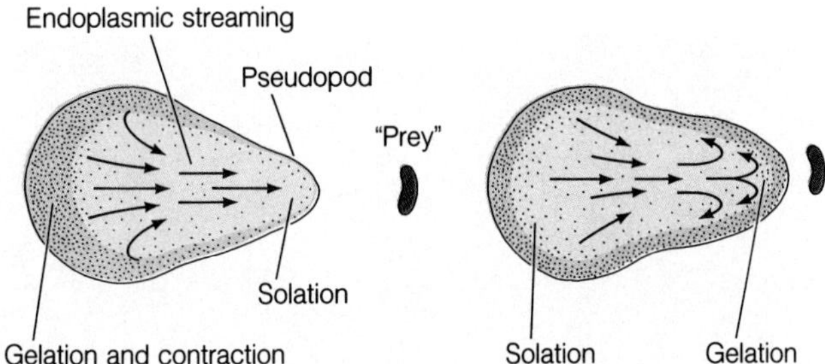

Figure 2–23 Amoeboid Movement *See text for an explanation of how amoeboid movement is accomplished by transitions of portions of the cytosol between a gel and a sol state.*

mechanism used by amoeba to maneuver through their environment. White blood cells leave the circulatory system and travel by amoeboid movement to areas of infection or inflammation, where they engulf and destroy microorganisms and cellular debris. Fibroblasts ("fiber formers") move into a wound from adjacent connective tissue to help repair the damage, being responsible for scar formation. Skin cells, which are ordinarily stationary, can become modestly mobile by means of amoeboid motion to move toward a cut area to restore the skin surface.

Two regions of cytosol can be distinguished in a cell undergoing amoeboid movement: an outer layer of cytosol that is quite gelatinous, the **plasma gel** or **ectoplasm** ("outer cytoplasm"), and an inner core of a more fluid cytosol, the **plasma sol** or **endoplasm** ("inner cytoplasm"). The more gelatinous nature of the ectoplasm appears to be caused by a greater abundance of actin filaments just beneath the plasma membrane compared to the remainder of the cell.

Although the explanation is still controversial, amoeboid movement apparently involves local transitions between the gel and sol states within the cell as a result of alternate assembly and disassembly of actin filaments (Fig. 2–23). The trigger for initiating amoeboid movement might be the proximity of a food particle (in the case of an amoeba) or the presence of a bacterium (in the case of a white blood cell). Movement is accomplished by *solation* at the end of the cell closest to the prey, causing thinning of the ectoplasm at the "front," accompanied by simultaneous *gelation* of endoplasm at the opposite end, causing thickening of the ectoplasm in the "rear." The thickened ectoplasm in the rear (thicker presumably because of the assembly of more actin filaments during gelation) contracts, squeezing out the more-fluid endoplasm, which streams toward the leading end of the cell. This streaming endoplasm causes the cell to bulge out into a finger-like extension, a **pseudopod** ("false foot"), at the thinned front end of the cell (thinned because of the disassembly of actin filaments during solation). The thickened ectoplasm at the trailing end of the cell then dissolves and flows forward into the extending pseudopod. This streaming of endoplasm from rear to front into the pseudopod moves the cell forward. After reaching the tip of the pseudopod, the streaming endoplasm turns towards the sides of the cell like a fountain spray and gels to stop the flow and restore ectoplasm in this region. The cycle then repeats itself so that the cell progressively moves forward.

MICROFILAMENTS AS MECHANICAL STIFFENERS. Besides their role in cellular contractile systems, the actin filaments' second major function is that of serving as mechanical supports or stiffeners for several cellular extensions, the most common of which is **microvilli.** The latter are microscopic, nonmotile, hairlike projections from the surface of epithelial cells lining the small intestine and kidney tubules. A single small intestinal cell may have several thousand of these microvilli, which are packed together like the bristles of a brush, projecting from its free surface. This bristly appearance gives rise to the alternative name of **brush border** for these microvilli (Fig. 2–24). Their presence greatly increases the surface area available for transferring material across the plasma membrane. In the case of the small intestine, the microvilli increase the area available for absorbing digested nutrients. In the kidney tubules, brush borders enlarge the absorptive surface that salvages useful substances passing through the kidney so that these materials are saved for the body instead of being eliminated in the urine. Within each microvillus, a core consisting of parallel actin filaments linked together forms a rigid mechanical stiffener that keeps these valuable surface projections intact.

A remarkable specialization of microvilli is found in the hair cells of the inner ear. In that portion of the inner ear responsible for hearing, the actin-stiffened projections on the surface of the hair cells are exquisitely sensitive to vibrations produced by incoming sound. In the part of the inner ear that plays a key role in equilibrium and balance, the specialized microvilli are responsive to changes in head movement and position.

Both microtubules and microfilaments form *stable* structures, such as cilia, flagella, muscle contractile units, and microvilli, and also form *labile* structures, such as mitotic spindles and contractile rings, as the need arises. Pools of unas-

Figure 2–24 Scanning Electron Micrograph of Intestinal Microvilli

SOURCE: From *Tissues and Organs: A Text-Atlas of Scanning Electron Microscopy* by Richard G. Kessel and Randy H. Kardon. Copyright © 1979 W.H. Freeman and Company. Reprinted with permission.

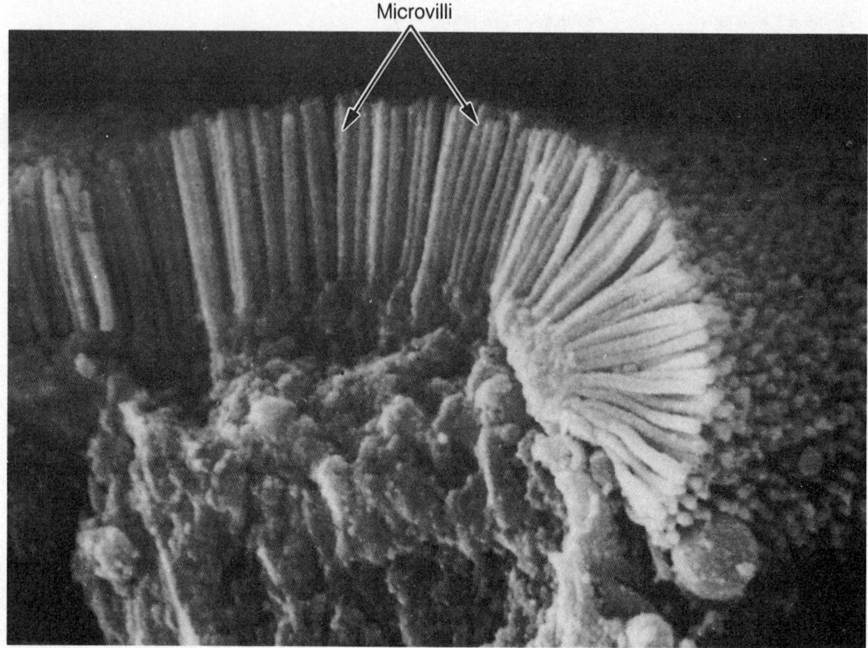

sembled tubulin and actin subunits in the cytosol can be rapidly assembled into organized structures to perform specific activities and then can be disassembled when they are no longer needed.

Intermediate filaments are important in regions of the cell subject to mechanical stress.

In contrast to the other cytoskeletal elements, the **intermediate filaments** are highly stable structures. There is no evidence for a reversible pool between unassembled and assembled intermediate-filament proteins. The intermediate filaments are intermediate in size between the microtubules and the microfilaments (7–11 nm diameter)—hence their name. The proteins that compose the intermediate filaments vary between cell types, but in general they appear as irregular, threadlike molecules. These proteins form tough, durable fibers that are structurally important in the parts of the cell subject to mechanical stress. There are different types of intermediate filaments, depending on their structural or tension-bearing role in specific cell types. In general, only one class of intermediate filament is found in a particular cell type. Several important examples follow:

□ Neurofilaments are intermediate filaments found in nerve-cell axons. Together with microtubules, neurofilaments confer strength and stability to these elongated cellular extensions.

□ Intermediate filaments in skeletal-muscle cells hold the actin-myosin contractile units in proper alignment.

□ Skin cells contain irregular networks of keratin-containing intermediate filaments. These intracellular filaments interconnect with extracellular filaments that tie adjacent cells together, thereby creating a continuous filamentous network that extends throughout the skin and gives it strength. When the surface skin cells die, their tough keratin skeletons persist to form a protective, waterproof outer layer. Hair and nails are also keratin structures.

The microtrabecular lattice provides a dynamic structural framework.

The **microtrabecular lattice** is the most recently discovered element of the cytoskeleton, having first been made visible by high-voltage electron microscopy in the early 1970s. Using this technique, which provides a three-dimensional view of the internal organization of the cell, the microtrabecular lattice is visible as a meshwork of exceedingly fine ($<$ 2nm diameter), interlinked filaments that pervade the cytoplasm and that are connected to the inner layer of the plasma membrane. Some cell biologists believe that the microtrabecular lattice is not a separate entity but that it constitutes intricate interconnections between the three other types of cytoskeletal structures. This latticework appears to suspend the microtubules and microfilaments, as well as various organelles. The free ri-

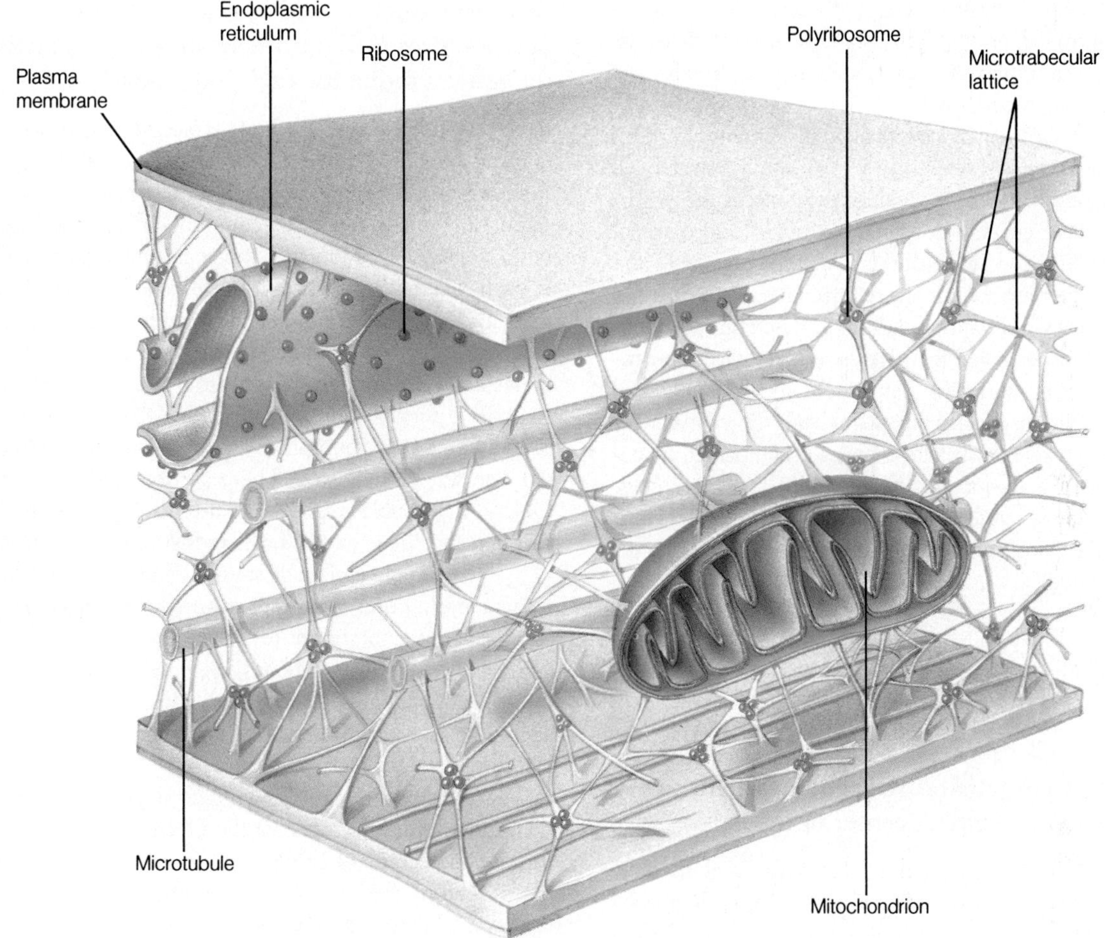

Figure 2–25 Microtrabecular Lattice in Relation to Other Cytoskeletal Structures and Organelles

bosomes are not freely floating in the cytosol as originally thought; they are entrapped as polyribosomal clusters at junctions of the microtrabecular lattice and thus have the appearance of flies caught in a spider web (Fig. 2–25).

By attaching to the inner surface of the plasma membrane and forming an internal scaffolding, the microtrabecular lattice, in association with the other cytoskeletal elements, supports the plasma membrane and is responsible for the particular shape, rigidity, and spatial geometry of each different cell type. Thus, this internal framework acts as the cell's "skeleton." In addition, by lacing throughout the entire cell, the microtrabecular lattice appears to link the different components of the cytoplasm into a functional unit. For example, the coordinated action of the cytoskeletal elements is responsible for directing intracellular transport and for regulating numerous cellular movements, thereby serving also as the cell's "musculature."

The lattice apparently also plays a role in organizing the cytosolic enzymes. Reactions that take place in the cytosol, such as glycolysis, are too well choreographed and too rapid to occur by random contacts between enzymes and their substrates (the substances acted on by the enzymes). There is convincing evidence that these enzymes are somehow incorporated into the lattice, probably in some sort of sequential alignment that guides glucose through the steps of the glycolytic pathway. Furthermore, the lattice, by sequestering the free ribosomes at its intersections, plays an important role in governing the dispensation of newly synthesized cytosolic proteins. Once proteins have been synthesized, they are available in a controlled, nonrandom fashion for enzymatic activities and for the assembly of microtubules, microfilaments, and other cytosolic components.

The microtrabecular lattice is not a rigid, static structure. Its structure has been observed to vary reversibly when the cell changes shape or when intracellular movements are occurring. The lattice changes constantly through local contractions, expansions, and deformations so that the organelles and cytoskeletal fibers are continually redistributed and reoriented as the cell carries on its various activities. Based on recent observations, it is speculated that signals received via an organized cytoskeleton normally play a role in regulating cell growth and division, a factor that might be disrupted in cancer cells. Conspicuous cytoskeletal changes accompanied by abnormal growth behavior are often present in cancer cells. Very little is known about the interactions between the components of the cytoskeleton in carrying out these and other integrated activities.

NUCLEUS

The nucleus perpetuates the genetic blueprint and serves as the control center of the cell.

The nucleus of the cell houses **deoxyribonucleic acid (DNA)**, the genetic blueprint that is unique for each individual. This genetic material serves two essential functions. First, DNA contains "instructions" for assembling the structural and enzymatic proteins of the cell. Cellular enzymes, in turn, control the formation of other cellular structures and also determine the functional activity of the cell by regulating the rate at which metabolic reactions proceed. The nucleus serves as the cell's control center by directly or indirectly controlling almost all cell activities through the role its DNA plays in governing protein synthesis. Since cells make up the body, the DNA code determines the structure and function of the body as a whole. The DNA an organism possesses not only dictates whether the organism is a human, a toad, or a pea but also determines the unique physical and functional characteristics of that individual, all of which ultimately depend on the proteins produced under DNA control. Second, by replicating (making copies of itself), DNA perpetuates the genetic blueprint within all new cells formed within the body and is responsible for passing on genetic information from parents to children. We will first examine the coding mechanism used by DNA before turning our attention to the means by which DNA replicates itself and controls protein synthesis.

Deoxyribonucleic acid is a double helix composed of nucleotides arranged in a particular sequence unique for each individual.

Deoxyribonucleic acid is a huge molecule, composed in humans of millions of nucleotides arranged into two long, paired strands that spiral around each other to form a double helix. Each **nucleotide** has three components: (1) a *nitrogenous base*, a ring-shaped organic molecule containing nitrogen; (2) a five-carbon ring-shaped sugar molecule, which in the case of DNA is *deoxyribose;* and (3) a phosphate group. Nucleotides are joined end to end by linkages between the sugar of one nucleotide and the phosphate group of the adjacent nucleotide to form a long polynucleotide ("many nucleotide") strand with a sugar-phosphate backbone and bases projecting out one side (Fig. 2–26). There are four different bases in DNA—the double-ringed bases **adenine (A)** and **guanine (G)** and the single-ringed bases **cytosine (C)** and **thymine (T)**. The two polynucleotide strands within a DNA molecule are wrapped around each other and oriented so that their bases all project to the interior of the helix. The strands are held together by weak hydrogen bonds (see p. A–5) formed between the bases of adjoining strands. Base pairing is highly specific: adenine only pairs with thymine and guanine only pairs with cytosine (Fig. 2–27).

The composition of the repetitive sugar-phosphate backbones that form the "sides" of the DNA "ladder" is identical for every molecule of DNA, but the sequence of the linked bases that form the "rungs" varies among different DNA molecules. The particular sequence of bases in a DNA molecule serves as "instructions" or a "code" that dictates the assembly of amino acids into a given order for the synthesis of specific **polypeptides** (chains of amino acids linked by peptide bonds; see p. A–12). A **gene** is a stretch of DNA that codes for the synthesis of a particular polypeptide. Polypeptides, in turn, are folded into a three-dimensional configuration to form a functional protein. Not all portions of a DNA molecule code for structural or enzymatic proteins. Some stretches of DNA code for proteins that regulate genes. Other segments appear to be important in organizing and packaging DNA within the nucleus. Still other regions are "nonsense" base sequences that have no apparent significance.

Deoxyribonucleic acid is precisely packaged within the nucleus.

The DNA molecules within each human cell, if lined up end to end, would extend more than 2 meters (2,000,000 μm), yet these molecules are packed into a nucleus that is only 5 μm in diameter. These molecules are not randomly crammed into the nucleus but are precisely organized into **chromosomes.** Each chromosome consists of a different DNA molecule and contains a unique set of genes. **Somatic** (body) **cells** contain forty-six chromosomes (the **diploid number**), which can be sorted into twenty-three pairs on the basis of various distinguishing features. Chromosomes composing a matched pair are termed **homologous chromosomes,** one member of each pair having been derived from the individual's maternal parent and the other member from the paternal parent. **Germ** (reproductive) **cells** (that is, sperm and eggs) contain only one

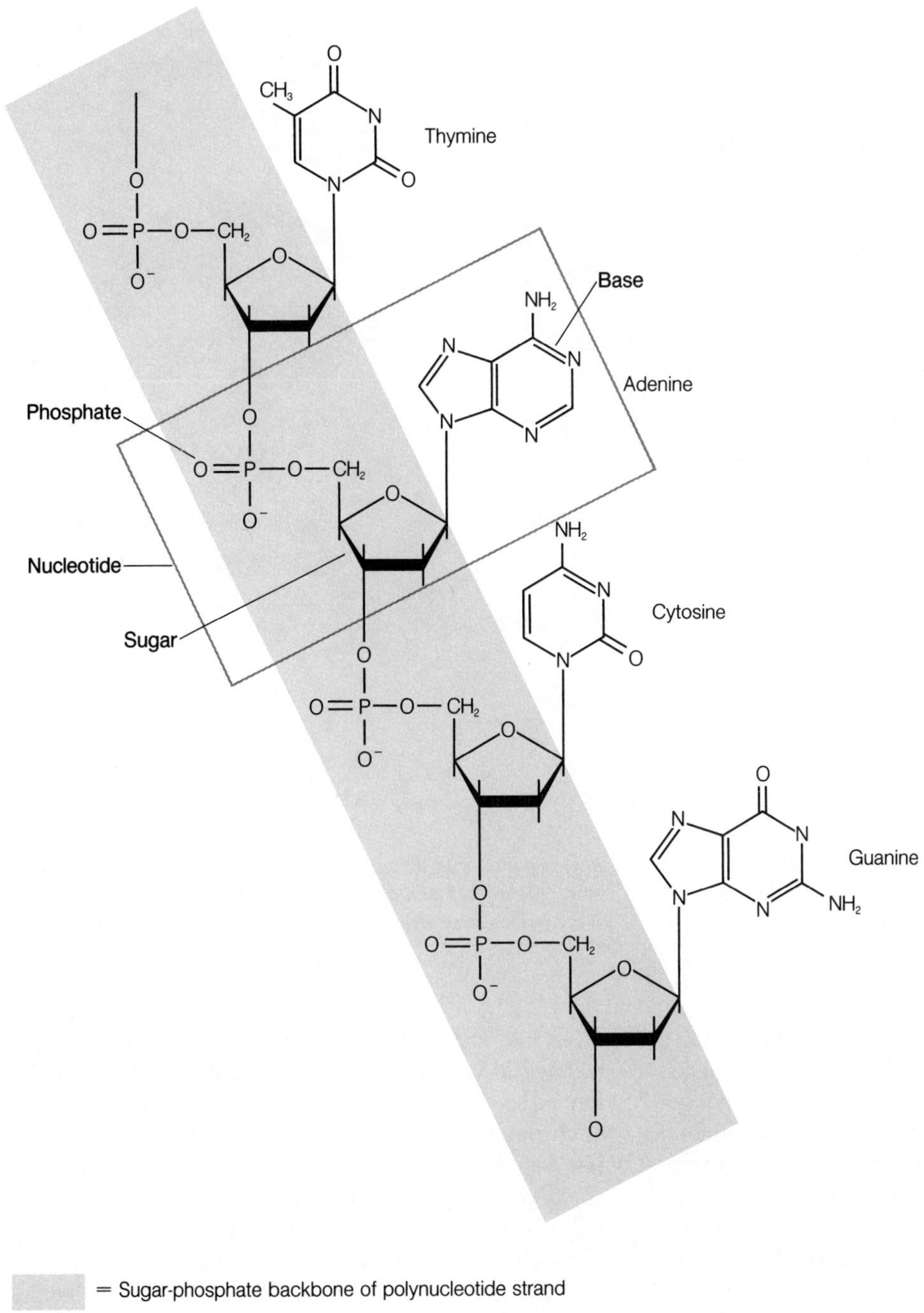

Figure 2–26 Polynucleotide Strand *Sugar-phosphate bonds link adjacent nucleotides together to form a polynucleotide strand with bases projecting to one side of the sugar-phosphate backbone. The sugar-phosphate backbone is identical in all polynucleotides, but the sequence of the bases varies.*

= Sugar-phosphate backbone of polynucleotide strand

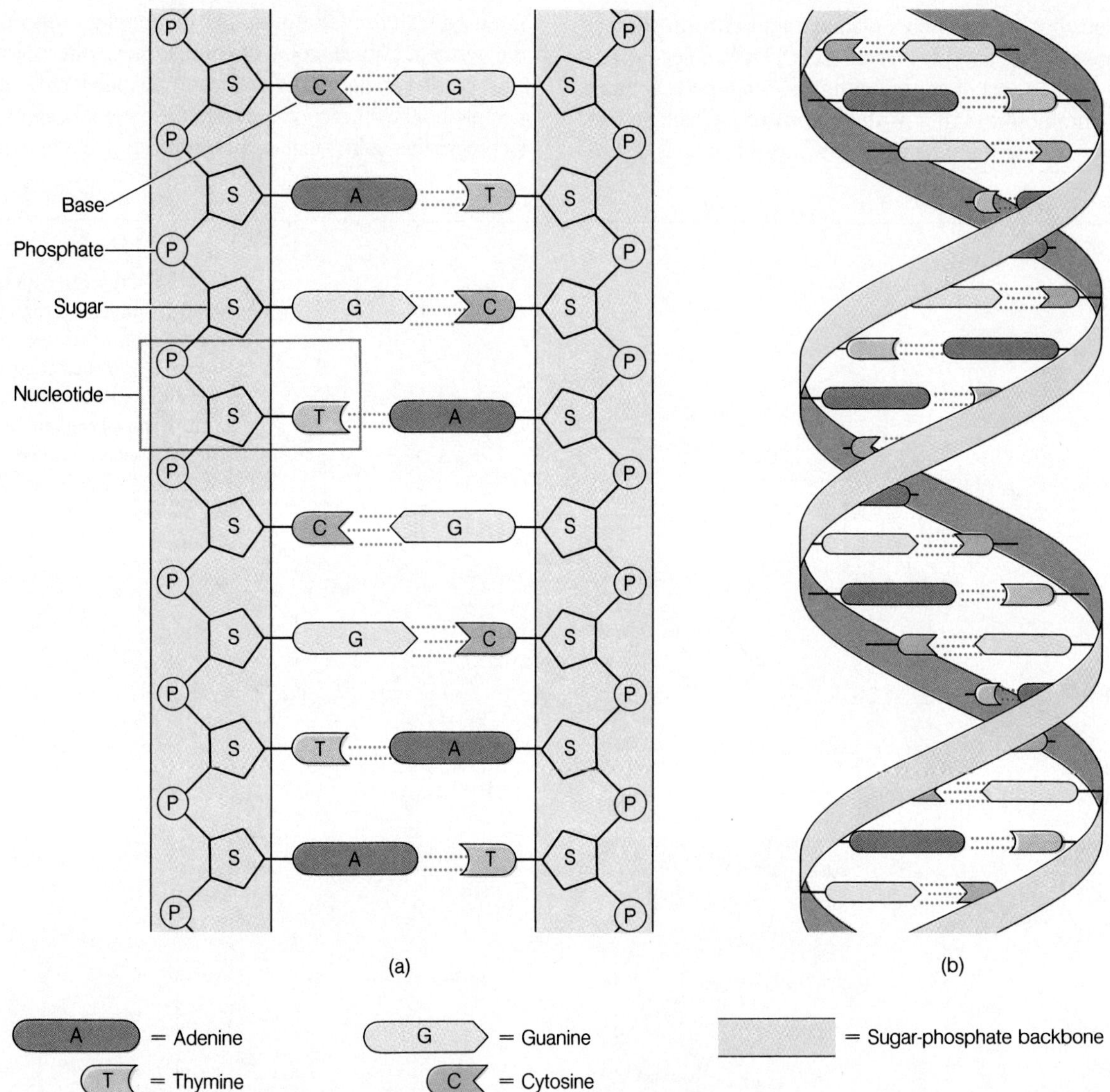

Figure 2–27 Complementary Base Pairing in DNA
(a) Two polynucleotide strands held together by weak hydrogen bonds formed between the bases of adjoining strands—adenine always paired with thymine and guanine always paired with cytosine. (b) Arrangement of the two bonded polynucleotide strands of a deoxyribonucleic acid (DNA) molecule into a double helix.

member of each homologous pair for a total of twenty-three chromosomes (the **haploid number**). Union of a sperm and an egg results in a new diploid cell with forty-six chromosomes, consisting of a set of twenty-three chromosomes from the mother and another set of twenty-three from the father.

The packaging and compression of DNA molecules into discrete chromosomal units is accomplished in part by nuclear proteins associated with DNA. Two classes of proteins—histone and nonhistone proteins—bind with DNA. **Histones** form bead-shaped bodies that play a key role in packaging DNA into its chromosomal structure. The **nonhistones** are believed to be important in gene regulation. The complex formed between the DNA and its associated proteins is known as **chromatin.** The long threads of DNA within a chromosome are wound around histones at regular intervals, thus compressing a given DNA molecule to about one-sixth its fully extended length. This "beads on a string" structure is further folded and supercoiled into higher and higher levels of organization to further condense DNA into rodlike chromosomes that are readily visible by means of a light microscope

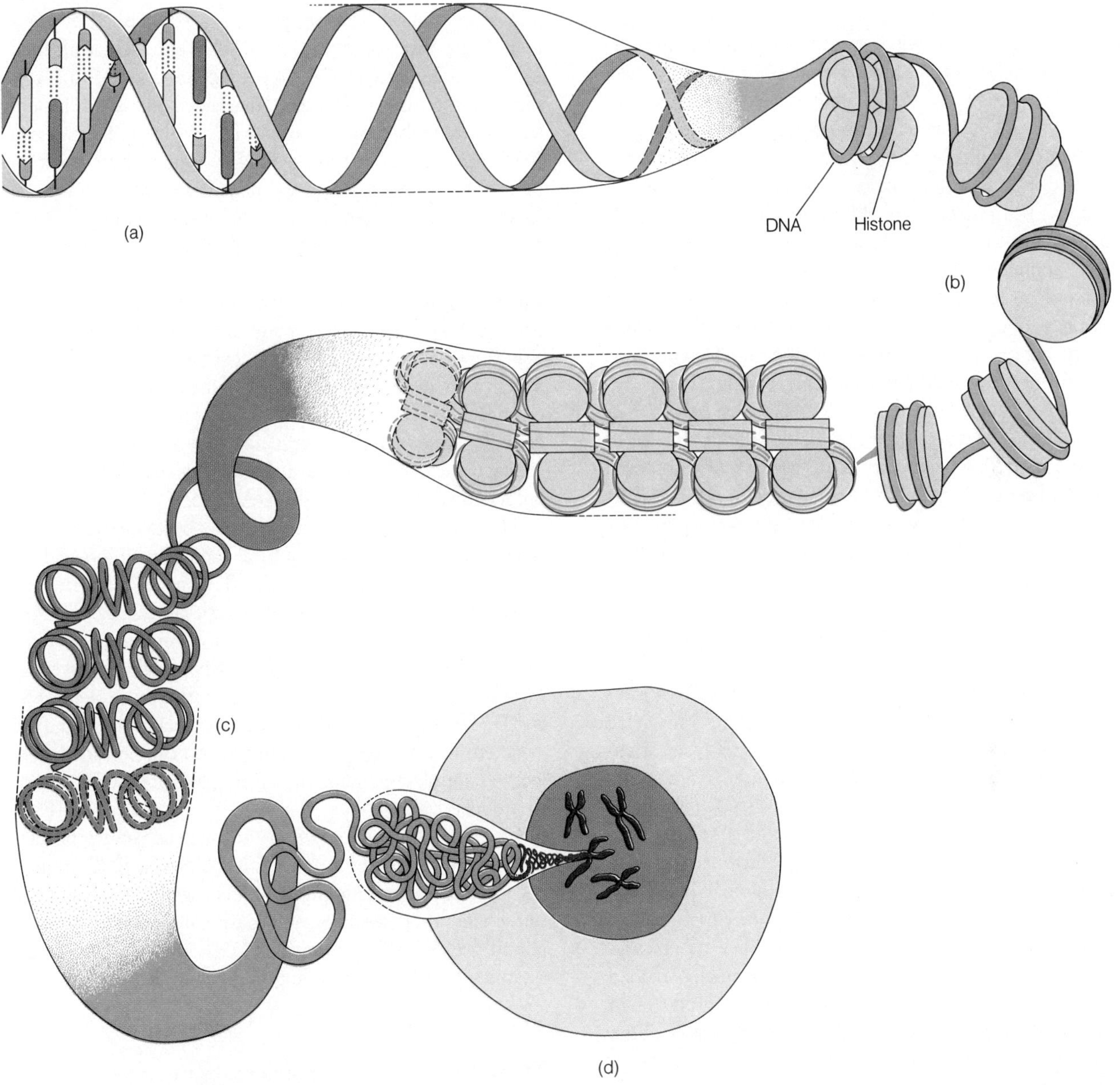

Figure 2–28 Levels of Organization of DNA *(a) Double helix of DNA molecule. (b) DNA molecule wound around histone proteins, forming a "beads-on-a-string" structure. (c) Further folding and supercoiling of DNA-histone complex. (d) Rodlike chromosome, the most condensed form of DNA, which is visible in the cell's nucleus during cell division.*

during cell division (Fig. 2–28). When the cell is not dividing, the chromosomes partially "unravel" or decondense to a less compact form of chromatin that is indistinct under a light microscope but that appears as thin strands and clumps with an electron microscope. The decondensed form of DNA is its working form; that is, it is the form used as a template for protein assembly.

Complementary base pairing serves as the foundation for both DNA replication and the initial step of protein synthesis.

During replication, the two DNA strands "unzip" as the weak bonds between the paired bases are enzymatically broken. New nucleotides present within the nucleus pair with the exposed bases from each strand (Fig. 2–29). New adenine-bearing nucleotides pair with exposed thymine-bearing nucleotides in an old strand, and new guanine-bearing nucleotides pair with exposed cytosine-bearing nucleotides in an old strand. This complementary base pairing is initiated at one end of the old strands and proceeds in an orderly fashion to the other end. The new nucleotides attracted to and thus aligned in a prescribed order by the old nucleotides are sequentially joined by sugar-phosphate linkages to form two new strands that are complementary to each of the old strands. This replication process results in two complete double-stranded DNA molecules, one strand within each molecule having come from the original DNA molecule and one strand having been newly formed by complementary base pairing. These two DNA molecules are both identical to the original DNA molecule, with the "missing" strand in each of the original separated strands having been produced as a result of the imposed pattern of base pairing. This replication process, which occurs only during cell division, is essential for assuring the perpetuation of the genetic code in both of the new daughter cells. The duplicate copies of DNA are separated and evenly distributed to the two halves of the cell before it divides.

At other times when DNA is not replicating in preparation for cell division, it serves as a blueprint for dictating cellular protein synthesis. How is this accomplished when DNA is sequestered within the nucleus and protein synthesis is carried out by ribosomes within the cytoplasm? Several types of another nucleic acid, **ribonucleic acid (RNA)**, serve as the "go-between." Ribonucleic acid differs structurally from DNA in three regards: (1) the five-carbon sugar in RNA is *ribose* instead of deoxyribose, the only difference between them being the presence of a single oxygen atom in ribose that is absent in deoxyribose; (2) RNA contains the closely-related base **uracil** instead of thymine, with the three other bases being the same as in DNA; and (3) RNA is single-stranded and not self-replicating. All RNA molecules are produced in the nucleus using DNA as a template or mold, then exit the nucleus through the **nuclear pores** (Fig. 2–1). These pores are large enough for passage of RNA molecules but preclude passage of the much larger DNA molecules.

The DNA instructions for assembling a particular protein coded in the base sequence of a given gene are "transcribed" into a molecule of **messenger RNA (mRNA)**. The segment of the DNA molecule to be copied uncoils, and the base pairs separate to expose the particular sequence of bases in the gene. In any given gene, only one of the DNA strands is used as a template for transcribing RNA, with the copied strand varying for different genes along the same DNA molecule. The beginning and end of a gene within a DNA strand are designated by particular base sequences that serve as "start" and "stop" signals. **Transcription** is accomplished by complementary base pairing of free RNA nucleotides with their DNA counterparts in the exposed gene (Fig. 2–30). The same pairing rules apply except that uracil, the RNA nucleotide substitute for thymine, pairs with adenine in the exposed DNA nucleotides. As soon as the RNA nucleotides pair with their DNA counterparts, sugar-phosphate bonds are formed to join the nucleotides together into a single-stranded RNA molecule that is released from DNA once transcription is complete. The original conformation of DNA is then restored. The RNA strand is much shorter than a DNA strand, because only a one-gene segment of DNA is transcribed into a single RNA molecule. The length of the finished RNA transcript varies, depending on the size of the gene. Within its nucleotide base sequence, this RNA transcript contains instructions for assembling a particular protein. Note that the message is coded in a base sequence that is ***complementary to, not identical to***, the original DNA code.

Messenger RNA delivers the final coded message to the ribosomes for **translation** into a particular amino-acid sequence to form a given protein. This genetic information flows from DNA (which can replicate itself) through RNA to protein. This is accomplished first by ***transcription*** of the DNA code into a complementary RNA code, followed by ***translation*** of the RNA code into a specific protein (Fig. 2–31). The structural and functional characteristics of the cell as determined by its protein composition can be varied, subject to control, depending on which genes are "switched on" to produce mRNA.

Free nucleotides present in the nucleus cannot be randomly joined together to form either DNA or RNA strands, because the enzymes required to link together the sugar and phosphate components of nucleotides are active only when bound to DNA. This assures that DNA, mRNA, and protein assembly occur only according to genetic plan.

Three forms of RNA participate in protein synthesis.

Besides messenger RNA, two other forms of RNA are required for translation of the genetic message into cellular protein: these are ribosomal RNA and transfer RNA. Messenger RNA carries the coded message from nuclear DNA to a cytoplasmic ribosome, where it directs the synthesis of a particular protein. **Ribosomal RNA (rRNA)** is an essential component of ribosomes, the "workbenches" for protein synthesis. Ribosomal RNA "reads" the base-sequence code of mRNA and translates it into the appropriate amino-acid sequence during protein synthesis. **Transfer RNA (tRNA)** transfers the ap-

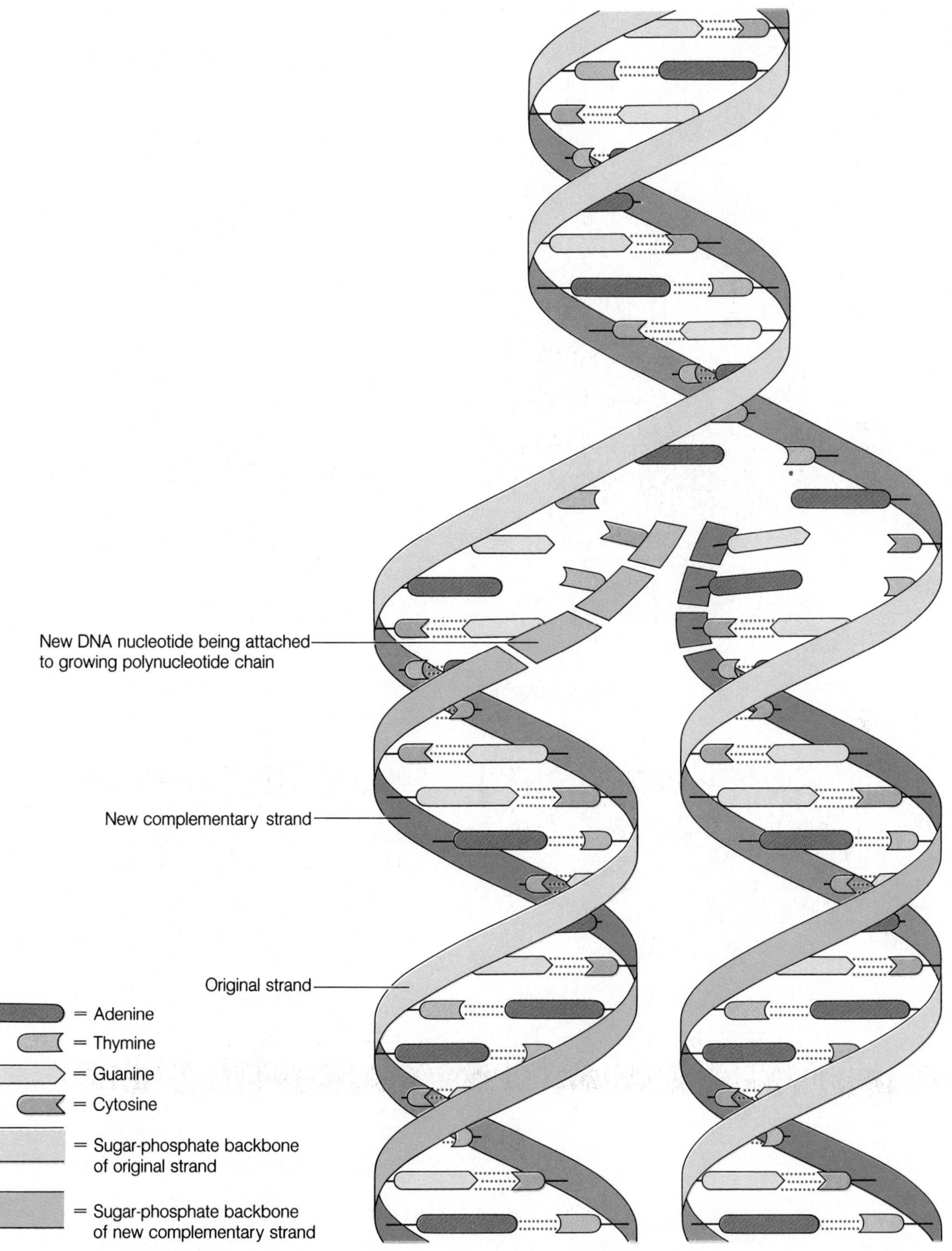

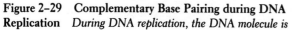

Figure 2–29 Complementary Base Pairing during DNA Replication *During DNA replication, the DNA molecule is unzipped and each old strand directs the formation of a new strand; the result is two identical double-helix DNA molecules.*

Figure 2-30 Complementary Base Pairing during DNA Transcription *During DNA transcription, a messenger RNA molecule is formed as RNA nucleotides are assembled by complementary base pairing at a given segment of one strand of an unzipped DNA molecule (that is, a gene).*

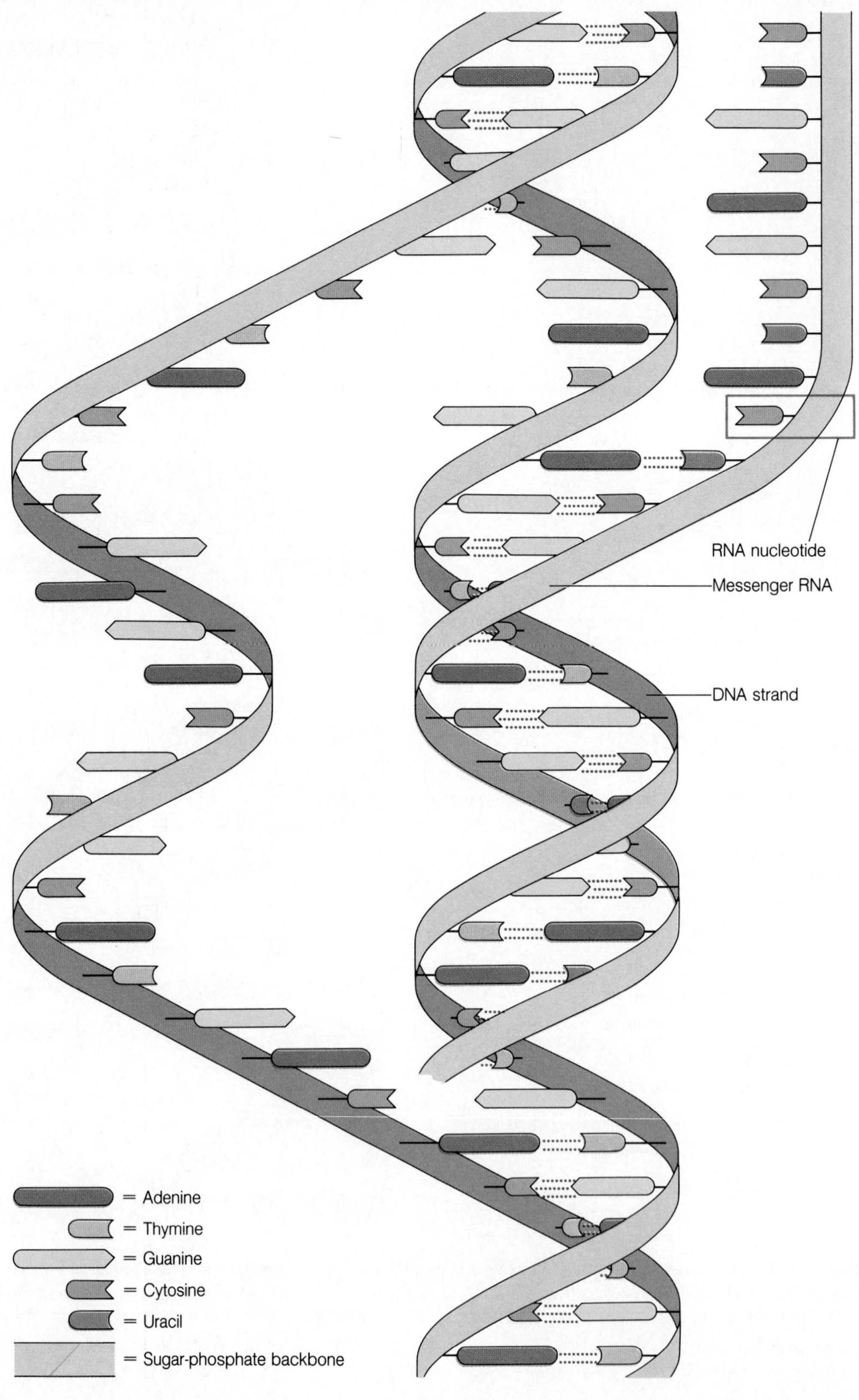

propriate amino acids in the cytosol to their designated site in the amino-acid sequence of the protein under construction.

Twenty different amino acids are used to construct proteins, yet only four different nucleotide bases are used to code for these twenty amino acids. In the "genetic dictionary," each different amino acid is specified by a **triplet code** that consists of a specific sequence of three bases in the DNA nucleotide chain. For example, the DNA sequence ACA (adenine, cytosine, adenine) specifies the amino acid cysteine, whereas the sequence ATA specifies the amino acid tyrosine. Each DNA triplet code is transcribed into mRNA as a complementary code word, or **codon,** consisting of a sequenced order of the three bases that pair with the DNA triplet. For example, the DNA triplet code ATA is transcribed as UAU (uracil, adenine, uracil) in mRNA.

Sixty-four different DNA triplet combinations (and, accordingly, sixty-four different mRNA codon combinations) are possible using the four different nucleotide bases (4^3). Of these possible combinations, sixty-one code for a specific amino acid and the remaining three serve as "stop signals." A stop signal acts as a "period" at the end of a "sentence" that specifies the amino-acid sequence in a particular protein; that is, ribosomal RNA releases the finished polypeptide product when it reaches a stop codon. Because sixty-one triplet codes each specify a particular amino acid and there are twenty different amino acids, a given amino acid may be specified by more than one base-triplet combination. For example, tyrosine is specified by the DNA sequence ATG as well as by ATA. In addition, one DNA triplet code, TAC (mRNA codon sequence AUG) functions as a "start signal" in addition to specifying the amino acid methionine. This code marks the place on mRNA where translation is to begin so that the message is started at the correct end and thus reads in the right direction. Interestingly, the same genetic dictionary is used universally; a given three-base code stands for the same amino acid in all living things, including microorganisms, plants, and animals.

The three steps of protein synthesis are initiation, elongation, and termination.

A ribosome brings together all components that participate in protein synthesis—mRNA, tRNA, and amino acids—and provides the enzymes and energy required for linking the amino acids together. The nature of the protein synthesized by a given ribosome is determined by the mRNA message that is being translated. Each mRNA serves as a code for only one particular polypeptide.

A ribosome is an rRNA-protein structure organized into two subunits of unequal size. Only when a protein is being synthesized are these subunits brought together (Fig. 2–32a).

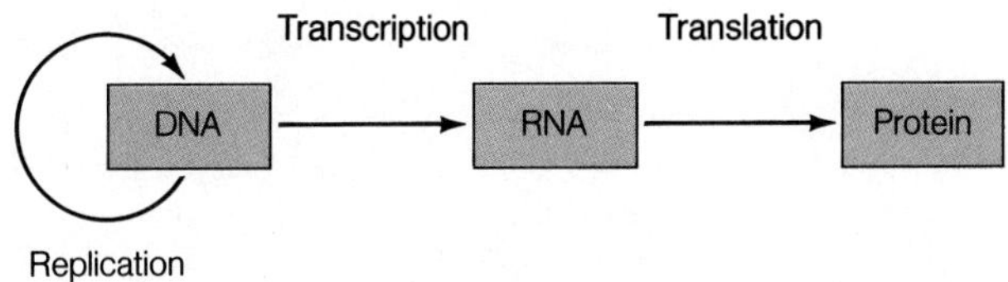

Figure 2–31 Flow of Genetic Information from DNA through RNA to Protein by Transcription and Translation

During assembly of a ribosome, a mRNA molecule attaches to the smaller of the ribosomal subunits by means of a *leader sequence*, a section of mRNA that precedes the start codon. The small subunit with mRNA attached then binds to a large subunit to form a complete, functional ribosome. When the two subunits unite, a groove is formed that accommodates the mRNA molecule as it is being translated.

Free amino acids in the cytoplasm are not able to "recognize" and bind directly with their specific codons in mRNA. Transfer RNA is required to bring the appropriate amino acid to its proper codon. Even though tRNA is single-stranded, as are all RNA molecules, it is folded back onto itself into a t-shape with looped ends (Fig. 2–33). The open-ended stem portion recognizes and binds to a specific amino acid. There are at least twenty different varieties of tRNA, each able to bind with only one of the twenty different kinds of amino acids. A tRNA is said to be "charged" when it is carrying its passenger amino acid. The loop end of a tRNA opposite the amino-acid binding site contains a sequence of three exposed bases, known as the **anticodon,** which is complementary to the mRNA codon that specifies the amino acid being carried. Through complementary base pairing, a tRNA can bind with mRNA and insert its amino acid into the protein under construction only at the site designated by the codon for the amino acid. For example, the tRNA molecule that binds with tyrosine bears the anticodon AUA, which can pair only with the mRNA codon UAU, which specifies tyrosine. This dual binding function of tRNA molecules assures that the correct amino acids are delivered to mRNA for assembly in the order specified by the genetic code. Transfer RNA can only bind with mRNA at a ribosome, so protein assembly does not occur except in the confines of a ribosome.

Protein synthesis is initiated when a charged tRNA molecule bearing the anticodon specific for the start codon binds at this site on mRNA. A second charged tRNA bearing the anticodon specific for the next codon in the mRNA sequence then occupies the site next to the first tRNA (Fig. 2–32). At any given time, a ribosome can accommodate only two tRNA molecules bound to adjacent codons. Through enzymatic action, a peptide bond is formed between the two amino acids that are linked to the stems of the adjacent tRNA molecules.

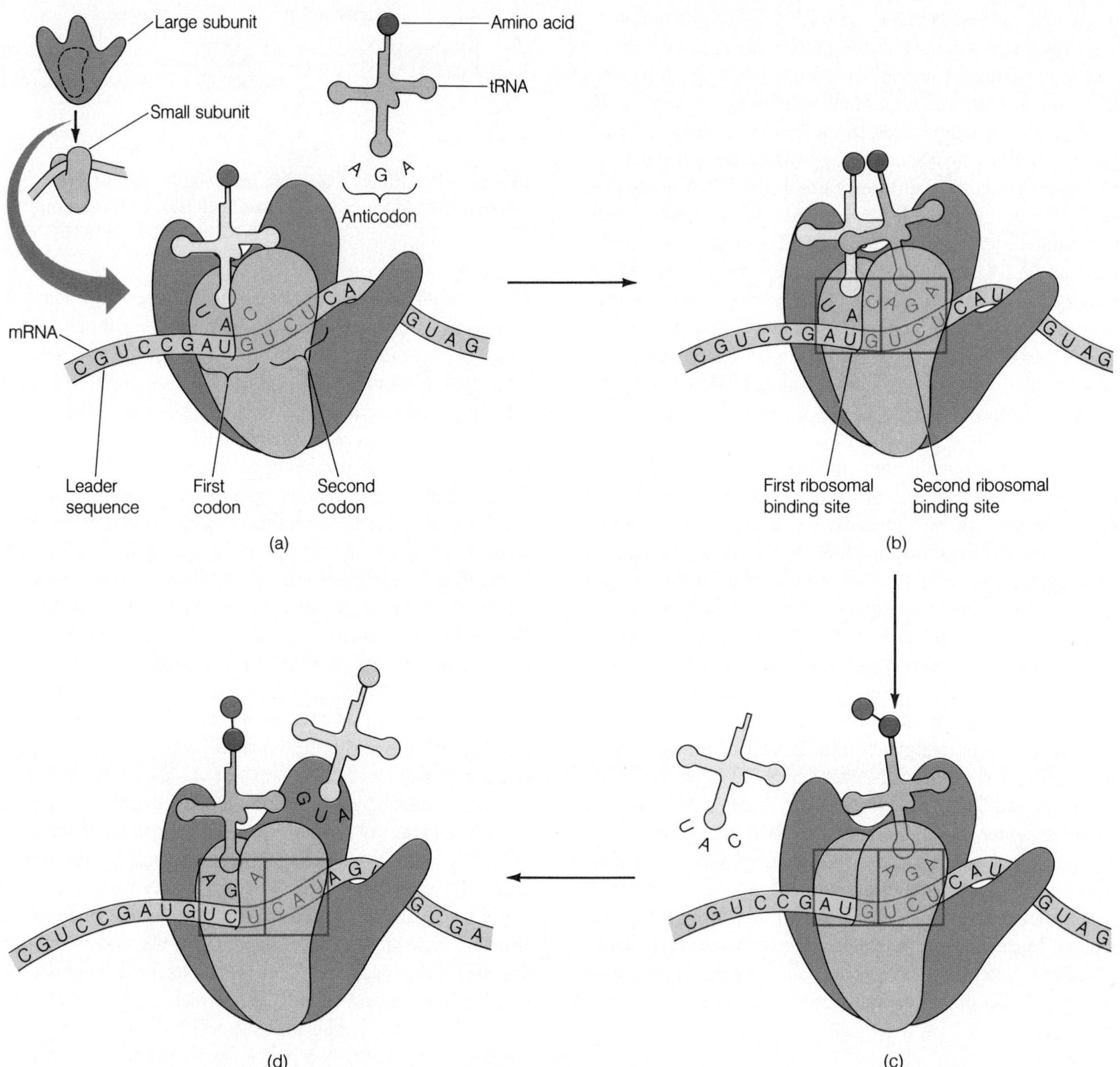

Figure 2–32 Ribosomal Assembly and Protein Translation *(a) On binding with a messenger RNA (mRNA) molecule, the small ribosomal subunit joins with the large subunit to form a functional ribosome. A transfer RNA (tRNA), charged with its specific amino-acid passenger, binds to mRNA by means of complementary base pairing between the transfer RNA anticodon and the first mRNA codon positioned in the first ribosomal binding site. (b) Another tRNA molecule attaches to the next codon on mRNA positioned in the second ribosomal binding site. (c) The amino acid from the first tRNA is linked to the amino acid on the second tRNA. The first tRNA detaches. (d) The mRNA molecule shifts forward one codon (a distance of a three-base sequence). Another charged tRNA moves in to attach with the next codon on mRNA, which has now moved into the second ribosomal binding site. The amino acids from the tRNA in the first ribosomal site are linked with the amino acid in the second site. This process continues, with the polypeptide chain continuing to grow, until a stop codon is reached and the polypeptide chain is released.*

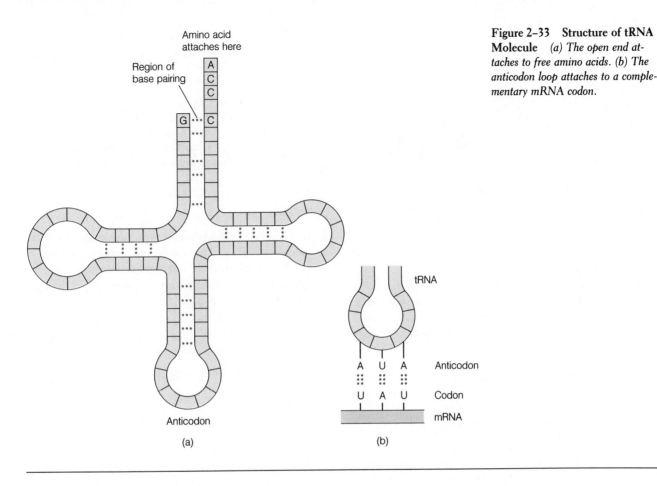

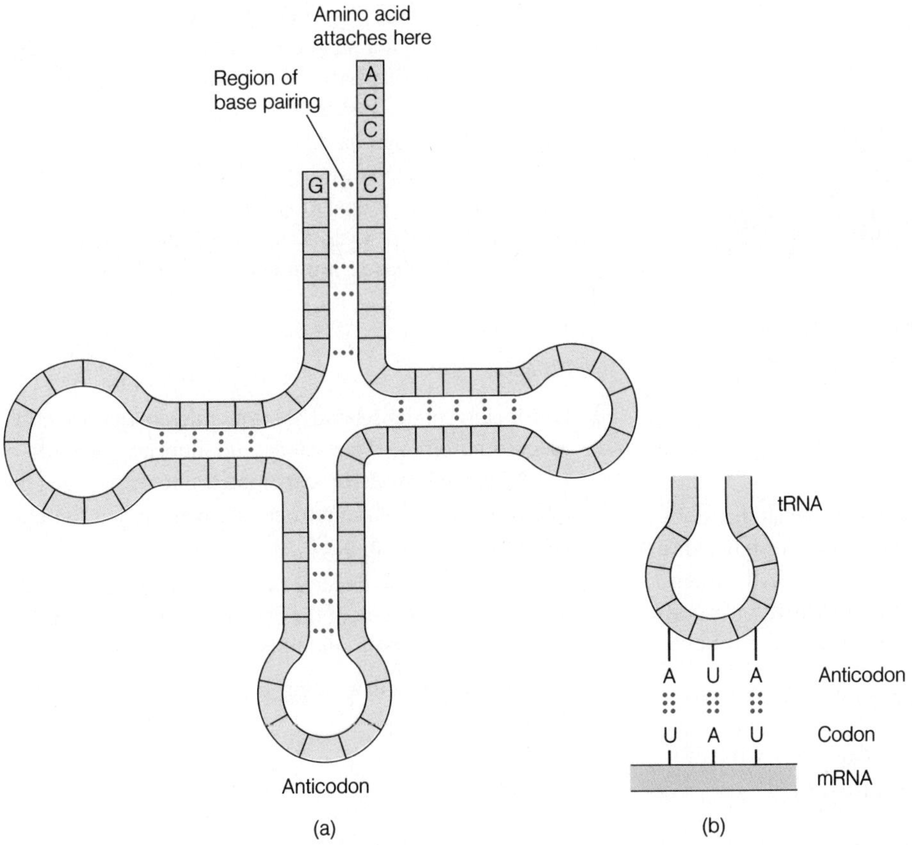

Figure 2–33 Structure of tRNA Molecule *(a) The open end attaches to free amino acids. (b) The anticodon loop attaches to a complementary mRNA codon.*

The linkage is subsequently broken between the first tRNA and its amino-acid passenger, leaving the second tRNA with a chain of two amino acids. The uncharged tRNA molecule (that is, the one minus its amino-acid passenger) is released from mRNA. The ribosome then moves along the mRNA molecule by precisely three bases, a distance of one codon, so that the tRNA bearing the two–amino-acid chain is moved into the number one ribosomal site for tRNA. Then, an incoming charged tRNA with a complementary anticodon for the third codon in the mRNA sequence occupies the number two ribosomal site that was vacated by the second tRNA. The two–amino-acid chain subsequently binds with and is transferred to the third tRNA to form a three–amino-acid chain. Through repetition of this process, amino acids are subsequently added one at a time to a growing polypeptide chain in the order designated by the mRNA codon sequence as the ribosomal translation machinery moves stepwise along the mRNA molecule one codon at at time. This process is rapid. Up to ten to fifteen amino acids can be added per second. Elongation of the polypeptide chain continues until the ribosome reaches a stop codon in the mRNA molecule, at which time the polypeptide is released. The polypeptide is then folded and modified into a full-fledged protein. The ribosomal subunits dissociate and are free to reassemble into another ribosome for translation of other mRNA molecules.

Protein synthesis is energetically expensive. Attachment of each new amino acid to the growing polypeptide chain requires a total investment of splitting four high-energy phosphate bonds— two to charge tRNA with its amino acid, one to bind tRNA to the ribosomal-mRNA complex, and one to move the ribosome forward one codon.

A number of copies of a given protein can be produced from a single mRNA molecule before the latter is chemically degraded. As one ribosome moves forward along the mRNA molecule, a new ribosome attaches at the starting point on mRNA and also starts translating the message. Attachment of many ribosomes to a single mRNA molecule results in a polyribosome. Multiple copies of the identical protein are produced as each ribosome moves along and translates the same message (Fig. 2–34). The released proteins are used within

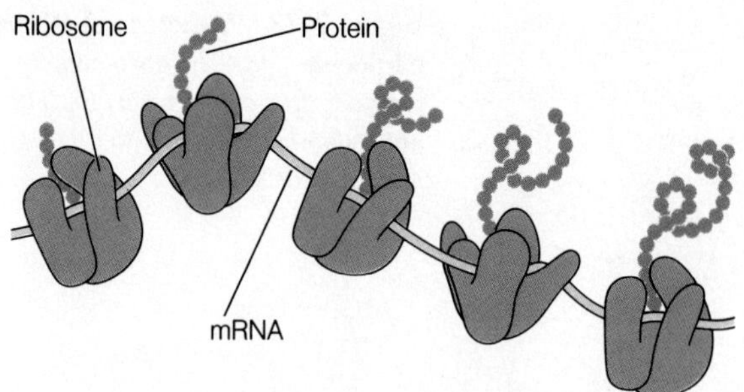

Figure 2–34 A Polyribosome *A polyribosome is formed by numerous ribosomes simultaneously translating mRNA.*

the cytosol, except for the few that move into the nucleus through the nuclear pores.

In contrast to the cytosolic polyribosomes, recall that ribosomes directed to bind with the rough endoplasmic reticulum feed their growing polypeptide chains into the ER lumen. The resultant proteins are subsequently packaged for export out of the cell or for replacement of membrane components within the cell.

Mitosis is essential for cell reproduction, whereas meiosis is essential for the formation of reproductive cells.

Most cells in the human body have the ability to reproduce themselves, a process important in growth, replacement, and repair of tissues. The rate at which cells divide is highly variable. Cells within the deeper layers of the intestinal lining divide every few days to replace cells that are continually sloughed off the surface of the lining into the lumen of the digestive tract. In this way, the entire intestinal lining is replaced about every three days. At the other extreme are nerve cells, which permanently lose the ability to divide beyond a certain period of growth and development. Consequently, when nerve cells are lost through trauma or disease, they cannot be replaced. In between these two extremes are cells that divide infrequently except when needed to replace damaged or destroyed tissue. The factors that control the rate of cell division remain obscure.

Recall that cell division involves two components: nuclear division and cytoplasmic division (cytokinesis). Nuclear division in somatic cells is accomplished by mitosis, in which a complete set of genetic information (that is, a diploid number of chromosomes) is distributed to each of two new daughter cells. Nuclear division in the specialized case of germ cells is accomplished by **meiosis,** in which only a half set of genetic information (that is, a haploid number of chromosomes) is distributed to each daughter cell.

A cell capable of dividing alternates between periods of mitosis and nondivision. The interval of time between cell division is known as **interphase.** Since mitosis takes less than an hour to complete, the vast majority of cells in the body at any given time are in interphase.

Replication of DNA and growth of the cell take place during interphase in preparation for mitosis. Although mitosis is a continuous process, it displays four distinct phases: **prophase, metaphase, anaphase,** and **telophase** (Fig. 2–35a).

PROPHASE.

1. Chromatin condenses and becomes microscopically visible as chromosomes. The condensed duplicate strands of DNA, known as **sister chromatids,** remain joined together within the chromosome at a point called the **centromere** (Fig. 2–36, p. 56).
2. The centriole pair divides, and the daughter centrioles move to opposite ends of the cell, where they assemble between them a mitotic spindle made up of microtubules.
3. The membrane surrounding the nucleus starts to break down.

METAPHASE.

1. The nuclear membrane completely disappears.
2. The forty-six chromosomes, each consisting of a pair of sister chromatids, align themselves at the midline, or equator, of the cell. Each chromosome becomes attached to the spindle by means of several spindle fibers that extend from the centriole to the centromere of the chromosome.

ANAPHASE.

1. The centromeres split, converting each pair of sister chromatids into two identical chromosomes, which separate and move toward opposite poles of the spindle. The spindle fibers are responsible for pulling the chromosomes toward the poles by an unknown mechanism.
2. At the end of anaphase, an identical set of forty-six chromosomes is present at each of the poles, for a transient total of ninety-two chromosomes in the soon-to-be-divided cell.

TELOPHASE.

1. The cytoplasm divides through formation and gradual tightening of an actin contractile ring at the midline of the cell, thus forming two separate daughter cells, each with a full diploid set of chromosomes.
2. The spindle fibers disintegrate.

3. The chromosomes uncoil to their decondensed chromatin form.
4. A nuclear membrane reforms in each new cell.

Cell division is complete with the end of telophase. Each of the new cells now enters interphase.

Meiosis differs from mitosis in several important regards (Fig. 2–35b). Specialized diploid germ cells undergo one chromosome replication followed by two nuclear divisions to produce four haploid germ cells.

MEIOSIS I.

1. During prophase of the first meiotic division, the members of each homologous pair of chromosomes line up side by side to form a **tetrad,** which is a group of four sister chromatids with two identical chromatids within each member of the pair.
2. The process of crossing over occurs during this period, when the maternal and paternal copy of each chromosome are paired. **Crossing over** involves a physical exchange of chromosome material between nonsister chromatids within a tetrad (Fig. 2–37, p. 56). This process yields new chromosome combinations, thus contributing to genetic diversity.
3. During metaphase, the twenty-three tetrads line up at the equator.
4. At anaphase, homologous chromosomes, each consisting of a pair of sister chromatids joined at the centromere, separate and move toward opposite poles. Maternally and paternally derived chromosomes migrate to opposite poles in random assortments of one member of each chromosome pair without regard for its original derivation. This genetic mixing provides novel new combinations of chromosomes.
5. During the first telophase, the cell divides into two cells. Each cell contains twenty-three chromosomes consisting of two sister chromatids.

MEIOSIS II.

1. Following a brief interphase in which no further replication occurs, the twenty-three unpaired chromosomes line up at the equator, the centromeres split, and the sister chromatids separate for the first time into independent chromosomes that move to opposite poles.
2. During cytokinesis, each of the daughter cells derived from the first meiotic division forms two new daughter cells. The end result is four daughter cells, each containing a haploid set of chromosomes.

Control of gene activity and protein transcription are incompletely understood.

Union of a haploid sperm and haploid egg results in a **zygote** (fertilized egg) that contains the diploid number of chromosomes. Development of a new multicellular individual from the zygote is accomplished by mitosis and cell differentiation. Since DNA is normally faithfully replicated in its entirety during each mitotic division, all cells in the body possess an identical aggregate of DNA molecules. How, then, are structural and functional variations between different cell types possible? If each cell has the identical DNA blueprint, you might assume that they would all produce the same proteins. This is not the case, however, because different cell types are able to transcribe different sets of genes and thus synthesize different sets of structural and enzymatic proteins. For example, only red blood cells are able to synthesize hemoglobin, even though all body cells carry the DNA instructions for hemoglobin synthesis. Only about 7% of the DNA sequences in a typical cell are ever transcribed into mRNA for ultimate expression as specific proteins.

Control of gene expression is believed to involve gene regulatory proteins that activate ("switch on") or repress ("switch off") the genes that code for specific proteins within a given cell. Various DNA segments that do not code for structural and enzymatic proteins code for synthesis of these regulatory proteins. The molecular mechanisms by which these regulatory genes, in turn, are controlled in human cells are only beginning to be understood. In some instances, regulatory proteins are controlled by **gene-signaling factors** that bring about differential gene activity among various cells to accomplish specialized tasks. The largest group of known gene-signaling factors in humans is the hormones. Some hormones exert their homeostatic effect by selectively altering the transcription rate of the genes that code for enzymes that are in turn responsible for catalyzing the reaction(s) regulated by the hormone. For example, the hormone cortisol promotes the breakdown of fat stores by stimulating synthesis of the enzyme that catalyzes the conversion of stored fat into its component fatty acids. In other cases, gene action appears to be time-specific; that is, certain genes are expressed only at a certain developmental stage in the individual. This is especially important during embryonic development.

Mutations can be harmless, deleterious, fatal, or beneficial.

It is estimated that in the vicinity of 10^{16} cell divisions take place in the body during the course of a person's lifetime to accomplish growth, repair, and normal cell turnover. Because more than 3 billion nucleotides must be replicated during

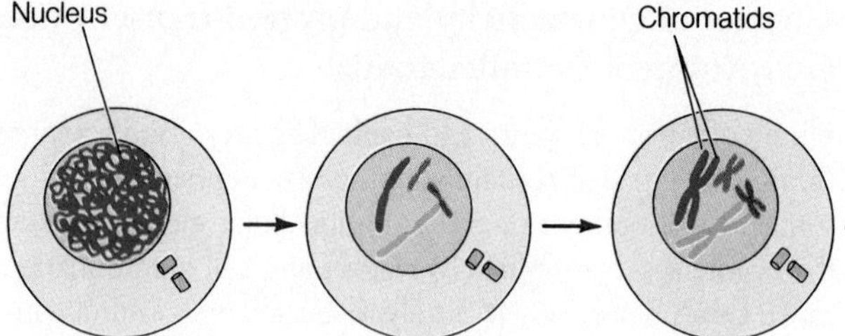

(a)

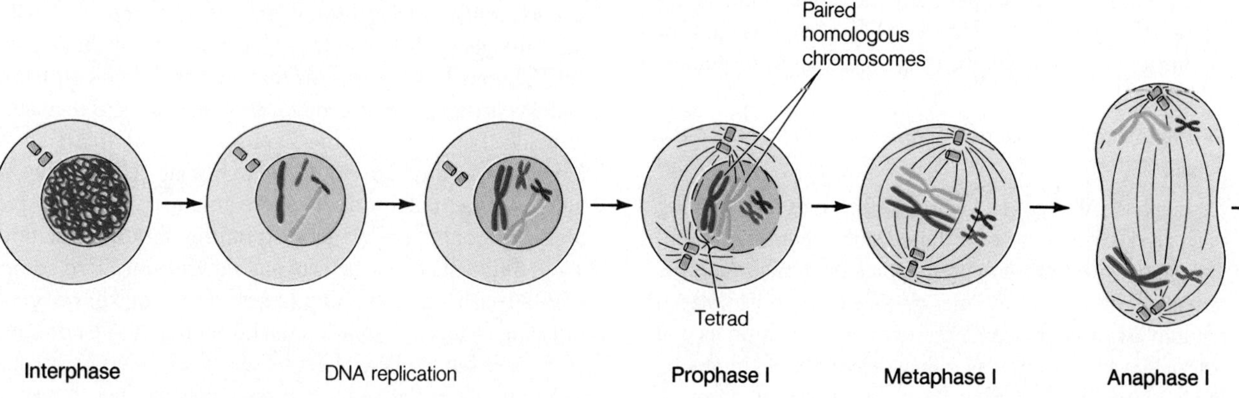

(b)

Figure 2–35 A Comparison of Events in Mitosis and Meiosis *(a) Mitosis. (b) Meiosis.*

each cell division, it is no wonder that "copying errors" occasionally occur. Any change in the DNA sequence is known as a **point** (or **gene**) **mutation.** A point mutation arises when a base is inadvertently substituted, added, or deleted during the replication process.

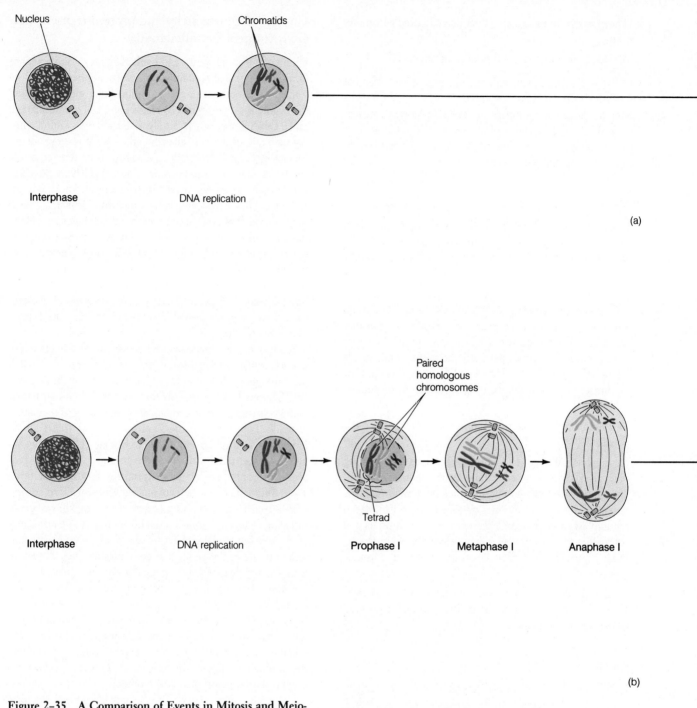

When a base is inserted in the wrong position during DNA replication, the mistake can often be corrected by a built-in "proofreading" system. Repair enzymes remove the newly replicated strand back to the defective segment, at which time normal base pairing resumes to resynthesize

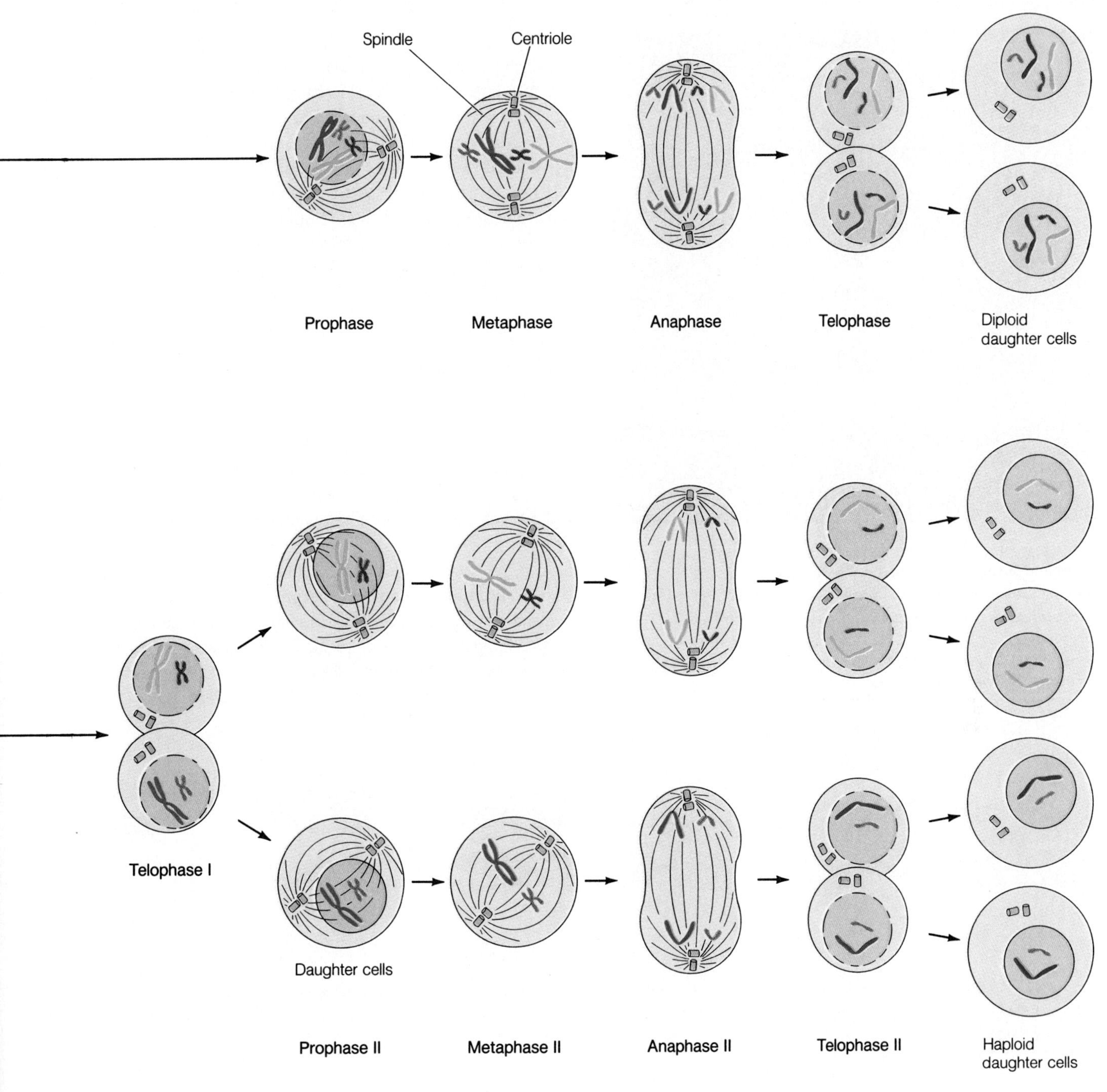

a corrected strand. Not all mistakes can be corrected, however.

Mutations can arise spontaneously by chance alone or they can be induced by **mutagens,** which are factors that increase the rate at which mutations take place. Mutagens include various chemical agents as well as ionizing radiation such as x-rays and atomic radiation. Mutagens promote mutations either by chemically altering the DNA base code through a variety of mechanisms or by interfering with the repair enzymes so that abnormal base segments cannot be cut out.

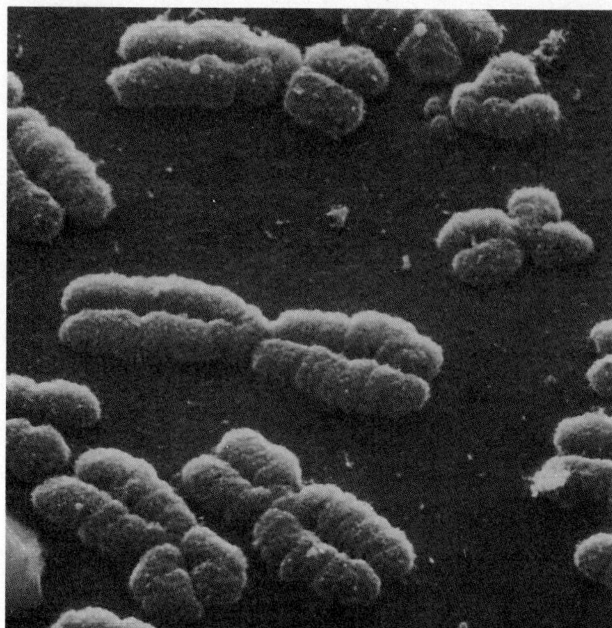

Figure 2–36 A Scanning Electron Micrograph of Human Chromosomes from a Dividing Cell
The replicated chromosomes appear as double structures, with identical sister chromatids joined at a common centromere.

SOURCE: From Christine J. Harrison et al.: "Cytogenetics," Cell Genetics 35: 21–27 (1983), Figure 3B. Reprinted with permission from Dr. Christine J. Harrison and S. Karger AG, Basel.

Depending on the location and nature of a change in the genetic code, a given mutation may: (1) have no noticeable effect if it does not alter a critical region of a cellular protein; (2) adversely alter cell function if it impairs the function of a crucial protein; (3) be incompatible with the life of the cell, in which case the cell dies and the mutation is lost with it; or (4) in rare cases, prove to be beneficial if a more efficient structural or enzymatic protein results. If a mutation occurs in a body cell (a **somatic mutation**), the outcome will be reflected as an alteration in all future copies of the cell in the affected individual, but it will not be perpetuated beyond the life of the individual. If, on the other hand, a mutation occurs in a sperm- or egg-producing cell (**germ-cell mutation**), the genetic alteration may be passed on to succeeding generations.

In most instances, cancer results from multiple somatic mutations that occur over a course of time within DNA segments known as **proto-oncogenes.** Proto-oncogenes are normal genes whose coded products are important in the regulation of cell growth and division. These genes have the potential of becoming overzealous **oncogenes** ("cancer genes"), which induce the uncontrolled cell proliferation characteristic of cancer. Proto-oncogenes can become cancer-producing as a result of several sequential mutations in the gene itself or by changes in adjacent regions that regulate the proto-oncogenes. Less frequently, tumor viruses become incorporated in the DNA blueprint and act as oncogenes.

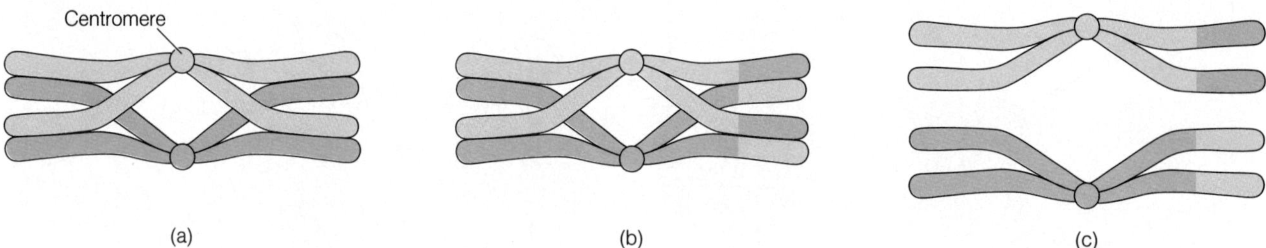

Figure 2–37 Phenomenon of Crossing Over *(a) During prophase I of meiosis, each homologous pair of chromosomes lines up side by side to form a tetrad. (b) Physical exchange of chromosome material occurs between nonsister chromatids. (c) The result of this crossing over is new combinations of genetic material within the chromosomes.*

CHAPTER IN PERSPECTIVE

Cells are the body's living building blocks. A cell is made up of three major parts—a plasma membrane, which encloses the cell and controls movement of materials between the intracellular fluid and extracellular fluid; the nucleus, which houses the cell's genetic material; and the cytoplasm, which is organized into discrete organelles dispersed throughout the semiliquid cytosol that is pervaded by a cytoskeletal network. Organelles are specialized intracellular compartments, each of which contains a specific set of enzymes for performing a particular cellular function that must be accomplished in seclusion from the remainder of the cell. Other chemical activities that are compatible with each other are carried on in the cytosol. The cytoskeleton gives the cell its shape, provides for its internal organization, and regulates its various movements.

Through the coordinated action of each of these cellular components, every cell is capable of performing certain basic functions essential to its own survival and of performing a specialized task, usually an elaboration of one of the basic cell functions, that contributes to the maintenance of homeostasis. Every body function ultimately depends on the functioning, specialization, and cooperation among the cells that compose the body.

All of the structural and functional components of a cell are produced according to genetic instructions that are coded in the sequence of bases within the deoxyribonucleic acid (DNA) molecules sequestered in the cell's nucleus. These molecules, which are packaged into chromosomes and which serve as blueprints for directing all cellular protein synthesis, are replicated during cellular division for perpetuation of the genetic information.

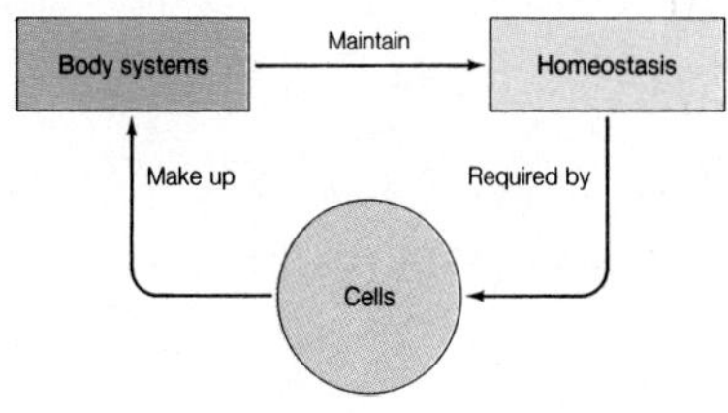

See inside front cover for an expanded version of this model.

REVIEW EXERCISES

1. What are a cell's three major subdivisions?
2. Describe the structure and functions of each of the organelles.
3. Compare exocytosis and endocytosis.
4. What is the body's universal energy carrier?
5. Briefly describe and compare the energy yield of glycolysis, the citric-acid cycle, and the entry of hydrogen carrier molecules into the electron-transport chain.
6. What three categories of cellular activities require energy expenditure?
7. List and describe the functions of each of the components of the cytoskeleton.
8. Describe the structure of DNA. What are DNA's two main functions? What are the relationships among DNA, chromosomes, and genes?
9. Distinguish between the diploid and haploid number of chromosomes.
10. Describe the process of DNA replication.
11. Describe the processes of transcription and translation. What are the roles of the three forms of RNA? How are the instructions for protein synthesis coded in DNA and RNA?
12. What are the two components of cell division? Compare mitosis and meiosis.
13. **A point to ponder:** Let's consider how much ATP you synthesize in a day. Assume that you consume one mole of O_2 per hour or 24 moles/day (a mole is the number of grams of a chemical equal to its molecular weight). About 6 moles of ATP are produced per mole of O_2 consumed. The molecular weight of ATP is 507.

$$24 \text{ moles } O_2\text{/day} \times 6 \text{ moles ATP/mole } O_2 = 144 \text{ moles ATP/day}$$

$$144 \text{ moles ATP/day} \times 507 \text{ gm ATP/mole} = 73{,}000 \text{ gm ATP/day}$$

Given that 1,000 gm equals 2.2 pounds, how many pounds of ATP do you produce per day at this rate? (This is under relatively inactive conditions!)

C H A P T E R 3

Plasma Membrane and Membrane Potential

Chapter Contents at a Glance

Introduction

Membrane Structure and Composition
Trilaminar appearance
Membrane composition; fluid mosaic model
Functions of membrane components

Membrane Receptors and Postreceptors Events
Receptor-activated channels
Second messenger systems

Cell-to-Cell Adhesions
Extracellular matrix
Specialized cell junctions

Membrane Transport
Influences of lipid solubility and size of molecule
Passive versus active forces
Passive diffusion down electrochemical gradient
Osmosis
Carrier-mediated transport
Endocytosis; exocytosis

Membrane Potential
Definition of potential
Ionic basis of resting membrane potential

INTRODUCTION *Specialists in membrane physiology frequently spend a lot of time with ghosts – specifically ghosts of red blood cells. When red blood cells are placed in a solution more dilute than their intracellular contents, water enters the cells by osmosis. This excess water causes the red blood cells to swell and burst open so that their intracellular fluid escapes. Left behind are their plasma membranes or "ghosts," the thin membranous layers that had enclosed the cells. It is a fairly simple procedure to wash the membranes free of any remaining intracellular fluid, leaving pure plasma membranes to investigate. In fact, the plasma membranes of human red blood cells have probably been studied more thoroughly than any other membrane for three reasons: (1) the ease with which red blood cells can be obtained in large numbers; (2) the ease with which their ghosts can be isolated; and (3) the fact that red blood cells do not contain any membrane-bound organelles so that membrane preparations from these cells are not contaminated by internal membranes.*

This chapter will be devoted primarily to our emerging understanding of the structure and function of the plasma membrane, which is derived from studies of these red blood cell ghosts as well as from investigations of other cells.

Membrane Structure and Composition

The plasma membrane separates the intracellular and extracellular fluid.

The survival of every cell depends on the maintenance of intracellular contents unique for that cell type despite the remarkably different composition of the extracellular fluid surrounding it. This difference in fluid composition inside and outside of a cell is maintained by the **plasma membrane,** an extremely thin layer of lipids and proteins that forms the outer boundary of every cell and encloses the intracellular contents. In addition to serving as a mechanical barrier that traps needed molecules within the cell, the plasma membrane plays an active role in determining the composition of the cell by selectively permitting passage of specific substances between the cell and its environment. Many of the functional differences between cell types are due to subtle variations in the composition of their plasma membranes, which in turn enable the cells to interact in unique ways with essentially the same extracellular fluid environment.

The plasma membrane is a fluid lipid bilayer embedded with proteins.

The plasma membrane is too thin to be seen under an ordinary light microscope, but it is visible with an electron microscope as a **trilaminar** (three-layered) **structure,** appearing as two dark layers separated by a light middle layer (Fig. 3–1). The specific arrangement of the molecules that make up the plasma membrane is believed to be responsible for this three-layered "sandwich" appearance.

All plasma membranes consist mostly of lipids (fats) and proteins plus small amounts of carbohydrate. The most abundant membrane lipids are phospholipids, with lesser amounts of cholesterol. **Phospholipids** have a polar (electrically charged) head containing a negatively charged phosphate group and two nonpolar (electrically neutral) fatty acid tails (Fig. 3–2a). The polar end is hydrophilic (water-loving) because it can interact with water molecules, which are also polar; the nonpolar end is hydrophobic (water-fearing) and will not mix with water. Such two-sided molecules self-assemble into a double layer of lipid molecules when in contact with water (Fig. 3–2b). The hydrophobic tails bury themselves in the center away from the water, while the hydrophilic heads line up on both sides, where they are in contact with the water. The outer surface of the layer is exposed to extracellular fluid (ECF) whereas the inner surface is in contact with the intracellular fluid (ICF).

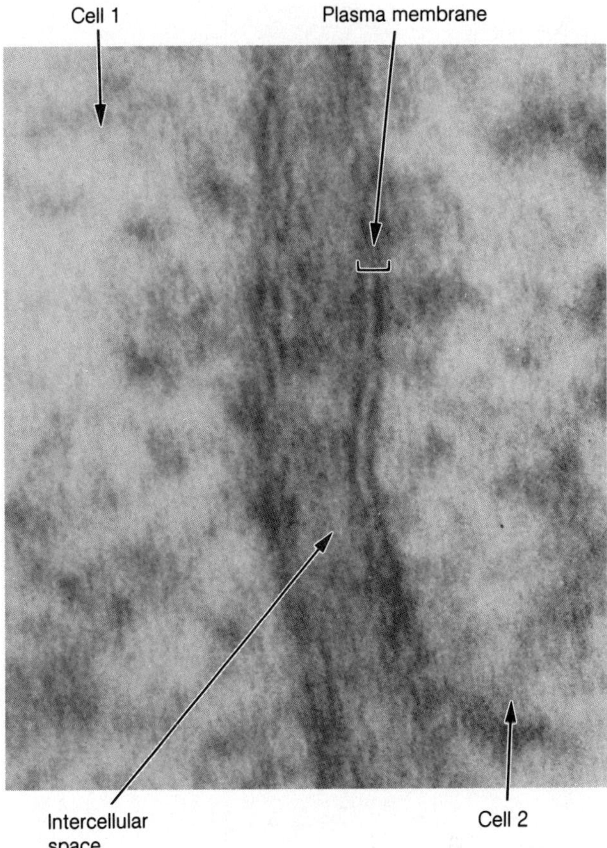

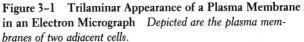

Figure 3–1 Trilaminar Appearance of a Plasma Membrane in an Electron Micrograph *Depicted are the plasma membranes of two adjacent cells.*

SOURCE: Photo courtesy of Mathew Nadakavukaren, Professor of Botany and Electron Microscopy, Illinois State University.

This lipid bilayer is not a rigid structure but instead is fluid in nature, having a consistency more like liquid cooking oil than solid shortening. The phospholipids, not being held together by chemical bonds, are able to twirl around rapidly as well as move about within their own half of the layer, much like skaters on a crowded ice pond.

Also contributing to the fluidity as well as the stability of the membrane is **cholesterol.** By being tucked in between the phospholipid molecules, the cholesterol molecules prevent the fatty acid chains from packing together and crystallizing, a process that would drastically reduce membrane fluidity.

The fluid nature of the membrane permits it to be flexible so that the cell can change its shape. Red blood cells, for example, must change shape considerably as they squeeze their way single-file through the capillaries, the tiniest of blood vessels. It is suspected that other essential membrane functions, such

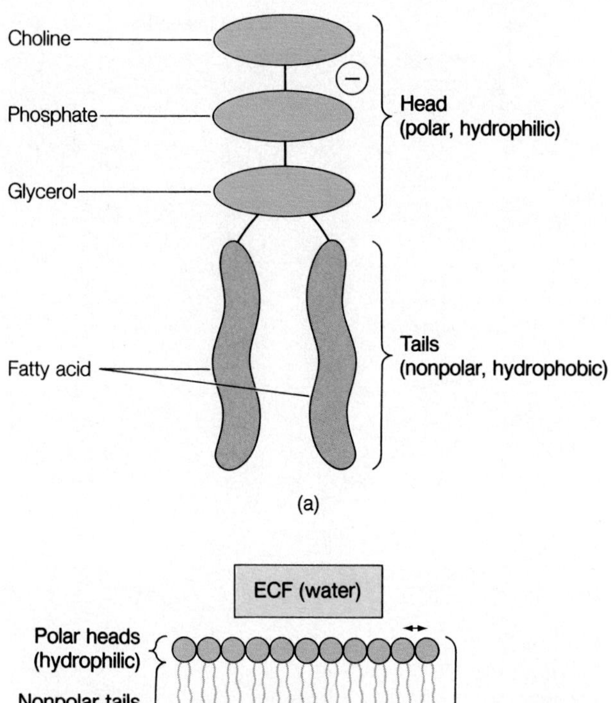

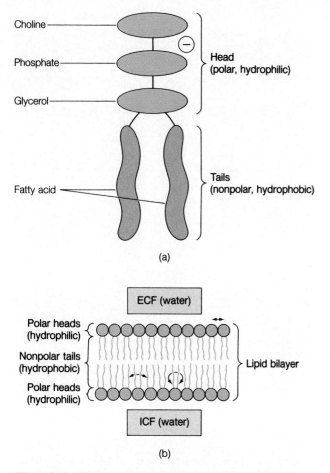

Figure 3–2 Structure and Organization of Phospholipid Molecules in a Lipid Bilayer *(a) Phospholipid molecule. (b) When in contact with water, lipid molecules organize themselves into a lipid bilayer with the polar heads interacting with the polar water molecules at each surface and the nonpolar tails all facing the interior of the bilayer.*

as transport processes, are also dependent on the fluidity of the lipid bilayer.

Attached to or inserted within the lipid bilayer are the **membrane proteins** (Fig. 3–3). Some of these proteins, having polar regions at both ends joined by a nonpolar central portion, extend entirely through the thickness of the membrane. Other proteins stud only the outer or inner surface, being anchored by interactions with a protein that spans the membrane or by attachment to a fatty acid chain that is inserted into the lipid bilayer. The fluidity of the lipid bilayer enables the membrane proteins to float freely for the most part, like "icebergs" in a moving "sea" of lipid, although proteins that perform a specialized function in a specific area of the cell are restricted in their mobility. This view of membrane structure is known as the **fluid mosaic model,** in reference to the membrane fluidity and to the ever-changing mosaic pattern of the proteins embedded within the lipid bilayer.

The small amount of membrane **carbohydrate** is located only at the outer surface. Short-chain carbohydrates protrude from the outer surface, bound primarily to membrane proteins and to a lesser extent to lipids, forming glycoproteins and glycolipids respectively (Fig. 3–3).

This proposed structure can account for the trilaminar appearance of the plasma membrane. The two dark lines are believed to be caused by the preferential staining of the hydrophilic polar regions of the lipid and protein molecules, whereas the light space between corresponds to the hydrophobic core formed by the nonpolar regions of these molecules.

Even though the outer and inner layers have the same appearance when viewed with an electron microscope, it is obvious from our description that the plasma membrane actually is asymmetrical; that is, the surface of the membrane facing the extracellular fluid is strikingly different from the surface facing the cytoplasm. Carbohydrate is located only on the outer surface; the outer and inner surfaces bear different amounts and types of protein; and even the lipid composition of the outer half of the bilayer varies somewhat from the inner half. This distinct sidedness of the membrane is related to the different functions carried out at the outer and inner surfaces.

The lipid bilayer forms the primary barrier to diffusion, whereas proteins perform most of the specific membrane functions.

The various components of the plasma membrane are responsible for carrying out different functions as follows:

LIPID BILAYER. The lipid bilayer serves at least three important functions:

1. It forms the basic structure of the membrane (the "fence" around the cell).
2. Its hydrophobic interior serves as a barrier to passage of water-soluble substances between the ICF and ECF. Water-soluble substances cannot dissolve in and pass through the lipid bilayer.
3. It is responsible for the fluidity of the membrane.

MEMBRANE PROTEINS. A variety of different proteins within the plasma membrane serve the following specialized functions :

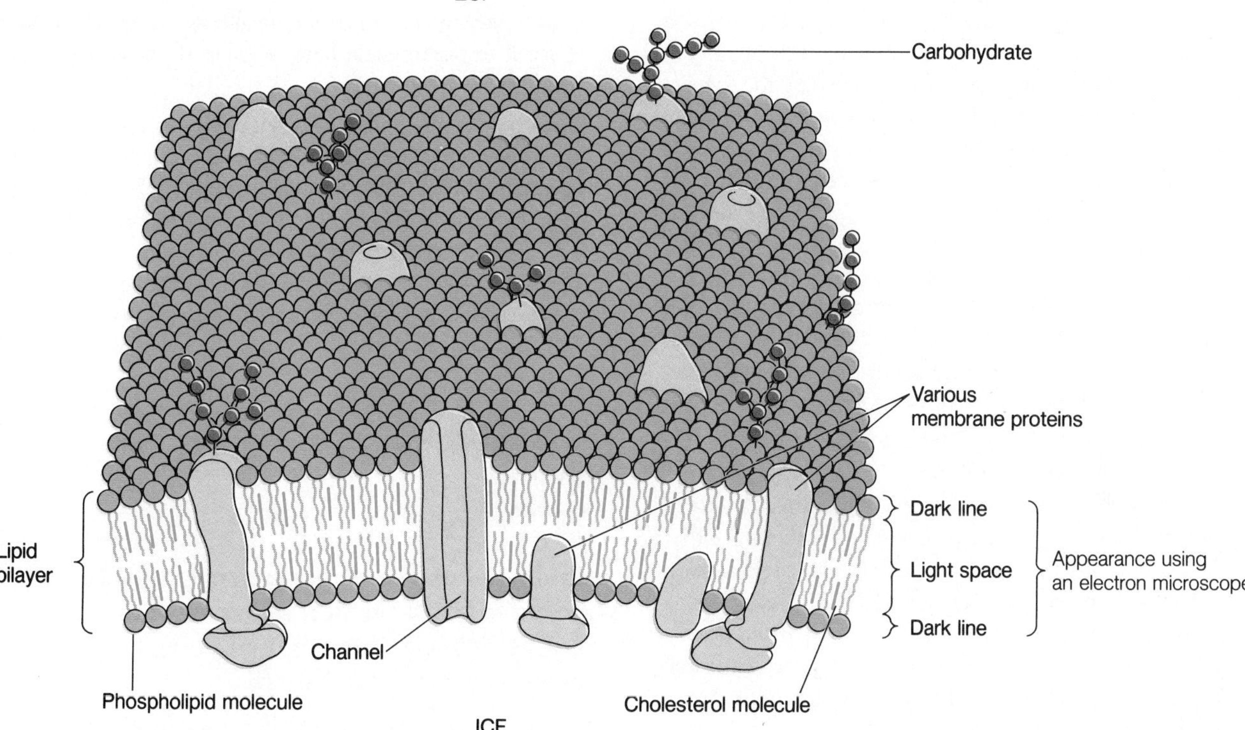

Figure 3–3 Fluid Mosaic Model of Plasma Membrane Structure *The plasma membrane is composed of a lipid bilayer embedded with proteins that penetrate the thickness of the membrane, are partially submerged in the membrane, or are loosely attached to the surface of the membrane. Short carbohydrate chains are attached to proteins or lipids on the outer surface only.*

1. Some proteins that span the membrane form water-filled pathways or **channels** across the lipid bilayer. Their presence enables water-soluble substances that are small enough to enter a channel, such as ions, to pass through the membrane without coming into direct contact with the hydrophobic lipid interior (Fig. 3–3). The channels are highly selective. Not only does their small diameter preclude passage of particles greater than 0.8 nanometer (nm) in diameter (a nm is a billionth of a meter or 0.00000004 in), but a given channel can also selectively attract or repel particular ions. For example, sodium (Na^+) channels and potassium (K^+) channels can accommodate the passage of only Na^+ or K^+ respectively. This selectivity is believed to be due to specific arrangements of charges on the interior surfaces of the proteins that form the channel walls. Cells vary in the number, kind, and activity of channels they possess. It is even possible for a given channel to vary in its ability to permit passage of a particular ion by changing its spatial configuration (shape) in response to a controlling mechanism; that is, a channel can be *open* or *closed* to its specific ion.

2. Other proteins serve as **carrier molecules** that transfer substances unable to cross the membrane on their own. Each carrier can transport only a particular molecule or closely related molecules. Variability in the kinds of carriers different cells possess permits them to selectively transport different substances across their membranes. For example, the thyroid gland requires iodide for the synthesis of thyroid hormone. Accordingly, the plasma membranes of thyroid gland cells uniquely possess carriers for iodide, enabling this essential element to be transported from the blood into thyroid gland cells, a capability not present in other body cells.

3. Many of the proteins on the outer surface serve as **receptor sites** that "recognize" and bind with specific molecules in the environment of the cell. This binding initiates a series of membrane and intracellular events that alter the ac-

tivity of the particular cell. (The postreceptor pathways involved in altering cell function are discussed in the next section.) In this way, chemical messengers in the blood, such as hormones, are able to influence only the specific cells that possess receptors for the messenger while having no effect on other cells, even though every cell is exposed to the same messenger via its widespread distribution by the blood. To illustrate, the anterior pituitary gland secretes into the blood thyroid-stimulating hormone (TSH), which can attach only to the surface of thyroid gland cells to stimulate secretion of thyroid hormone. No other cells have receptor sites for TSH, so only thyroid cells are influenced by TSH despite its ubiquitous distribution. Every cell type has a unique array of these molecular signal-detecting "antennae" for receiving special messages that can alter and regulate its activity.

4. Another group of proteins function as **membrane-bound enzymes** that control specific chemical reactions at either the inner or outer cell surface. Cells display specialization in the types of enzymes embedded within their plasma membranes. For example, the outer layer of the plasma membrane of skeletal muscle cells contains an enzyme that destroys the chemical messenger that triggers muscle contraction, thus enabling the muscle to relax.
5. Some proteins are arranged in a **filamentous meshwork** on the inner surface of the membrane and are secured to certain internal protein elements of the cytoskeleton. These membrane proteins appear to be structurally important in the maintenance of cell shape and probably participate in surface changes accompanying cell movements.
6. Finally, still other proteins, especially in conjunction with carbohydrates, are important in the cells' ability to recognize "self" and in cell-to-cell interactions.

MEMBRANE CARBOHYDRATES. The function of carbohydrates on the outer membrane surface remains obscure. Following are among the leading suggestions for possible roles of these short sugar chains:

1. They may orient, anchor, and stabilize membrane proteins.
2. The complexity and diversity of these carbohydrate chains as well as their location on the external surface suggest that they play an important role in recognition of "self" and in cell-to-cell interactions. Cells are able to recognize other cells of the same type and join together to form tissues. If cultures of embryonic cells of two different types are mixed together, such as nerve cells and muscle cells, the cells will sort themselves into separate aggregates of nerve cells and muscle cells. Apparently there are surface markers, presumably a unique combination of sugar chains projecting from the surface of membrane glycoproteins, which allow a cell to discriminate between cells of its own kind and others.
3. Carbohydrate-containing surface markers also appear to be involved in tissue growth, which is normally held within certain limits of cell density. Cells do not "trespass" across the boundaries of neighboring tissues; that is, they do not overgrow their own territory. Abnormal surface glycosphingoproteins, one class of these carbohydrate markers, have been identified in certain tumor cells, suggestive that this might underlie the uncontrolled growth of tumor cells.

Membrane Receptors and Postreceptor Events

Binding of chemical messengers to membrane receptors brings about a wide range of responses in different cells through use of only a few remarkably similar pathways.

Dispersed within the outer surface of the plasma membrane are specialized protein receptors that bind with the selected chemical messengers that come into contact with the cell — for example, hormones delivered by the blood or chemicals released from nerve endings. This combination of messenger with receptor triggers a sequence of cellular events that ultimately controls a particular cellular activity important in the maintenance of homeostasis, such as membrane transport, secretion, contraction, or metabolism.

In spite of the wide range of possible responses, there are only two general means by which binding of the receptor with the extracellular chemical messenger (the **first messenger**) brings about the desired intracellular response: (1) by opening or closing specific channels in the membrane to regulate the movement of particular ions into or out of the cell; or (2) by transferring the signal to an intracellular chemical messenger (the **second messenger**), which in turn triggers a preprogrammed series of biochemical events within the cell. Because of the universal nature of these postreceptor events, let us examine each more closely.

CHANNEL REGULATION MECHANISM. The first mechanism, that of altering channels, regulates the flow of specific ions across the membrane. This ionic movement can be responsible for two different cellular events:

1. A small, short-lived movement of Na^+, K^+ or both across the membrane (1 on Fig. 3-4) alters the electrical activity

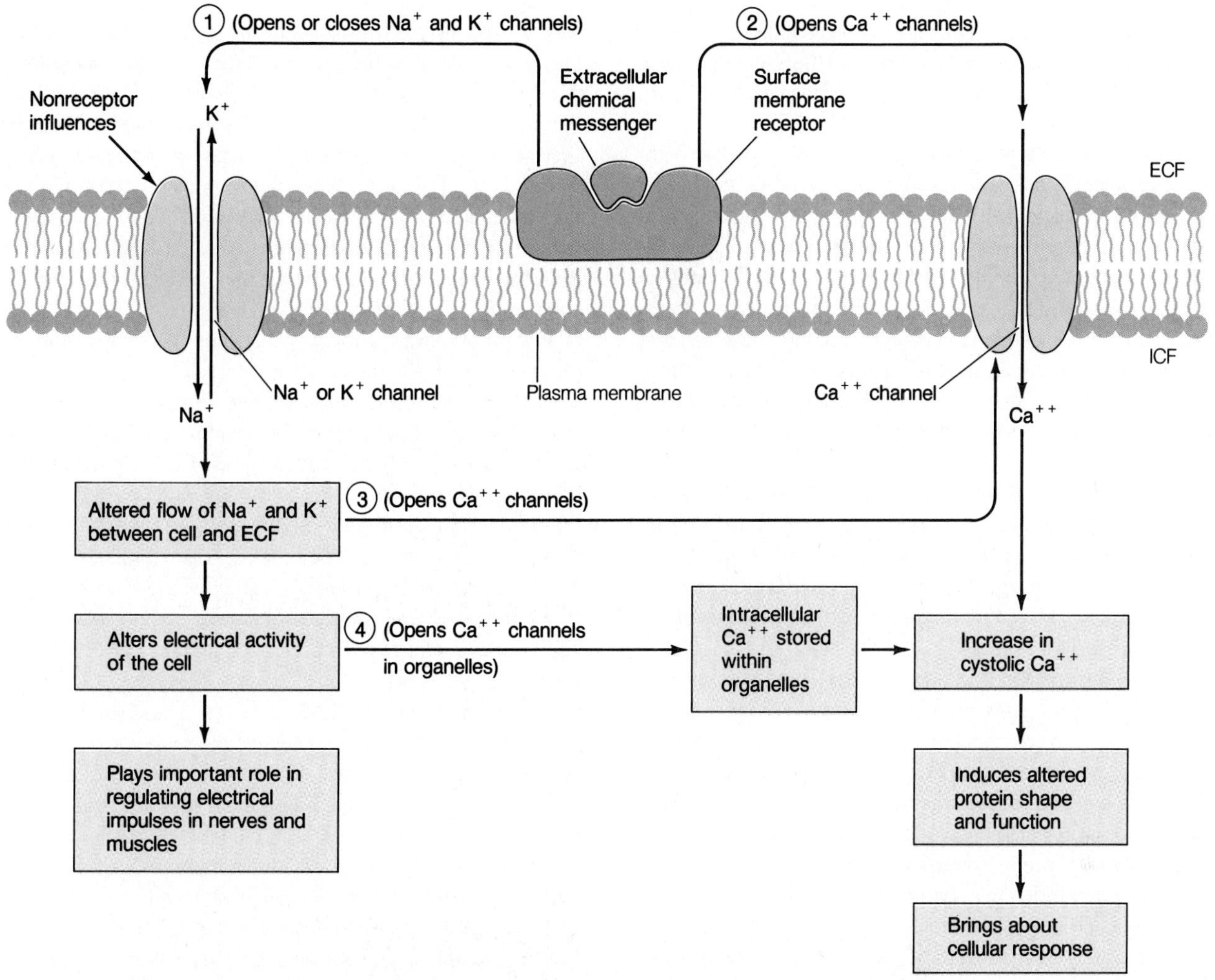

Figure 3–4 Postreceptor Event: Channel Regulation *See text for an explanation of the numbered items in the illustration.*

of cells that are capable of generating electrical signals (or impulses), such as nerve and muscle cells.

2. A transient flow of calcium (Ca^{++}) into the cell through opened Ca^{++} channels (2 on Fig. 3–4) triggers an alteration in shape and function of specific intracellular proteins, which leads to the cell's response. Illustrative is the increase in cytosolic Ca^{++} responsible for triggering the release of secretory product from many gland cells.

Upon completion of the response, the ions that moved across the membrane through opened channels to trigger the response are quickly returned to their original location by special carrier mechanisms in the membrane.

In some instances, the chemical messenger (or another stimulus) acts indirectly to open the Ca^{++} channels by altering Na^+ and K^+ channels to induce an electrical impulse in the cell. The electrical impulse, in turn, is directly responsible for opening the Ca^{++} channels (3 on Fig. 3–4). Release of chemicals from nerve cells in response to a nerve impulse is one such example.

In some cells, a rise in cytosolic Ca^{++} can be brought about by release of Ca^{++} from intracellular stores instead of Ca^{++} entry through membrane channels. For example, large amounts of Ca^{++} are stored within a modified endoplasmic reticulum (the sarcoplasmic reticulum) in skeletal muscle cells. An electrical impulse in these cells triggers the release of Ca^{++} from this organelle into the cytosol (4 on Fig. 3–4). This increased cytosolic Ca^{++} then alters a specific protein within the skeletal muscle cell to initiate the events leading to contraction.

Cytosolic Ca^{++} may be increased in yet another way; intracellular Ca^{++} can be released in a second messenger pathway independent of any electrical events, as will be described next.

SECOND MESENGER MECHANISM. The intracellular pathways activated by a second messenger in response to binding of the first messenger to a surface receptor are remarkably similar among different cells despite the diversity of ultimate responses to that signal. The variability in response depends on the specialization of the cell, not on the mechanism utilized. For example, activation of the identical second messenger system brings about modification of heart rate in the heart, stimulation of the formation of female sex hormones in the ovaries, and control of water conservation during urine formation in the kidneys.

Two major second messenger pathways are now known, one utilizing **cyclic adenosine monophosphate** (**cyclic AMP** or **cAMP**) as a second messenger and the other employing Ca^{++} in this role. Both pathways share much in common. The initial stages are similar in that binding of a chemical messenger to a receptor on the outer surface of the membrane brings about almost identical events within the membrane. In both cases this binding leads, via a series of biochemical steps, to activation of an enzyme on the cytoplasmic side of the membrane. This enzyme, in turn, activates intracellular second messengers that diffuse throughout the cell to trigger the appropriate cellular response. In both pathways, the cellular response is accomplished by altering the structure and subsequent function of particular cell proteins. For example, a particular enzymatic protein regulating a specific metabolic event may be modified so that its activity is increased or decreased.

In the cyclic AMP pathway, binding of an appropriate extracellular messenger to a special surface receptor eventually activates (or in some instances inhibits) the enzyme **adenylate cyclase** (1 on Fig. 3–5) on the inner surface of the membrane. Adenylate cyclase then induces the conversion of intracellular ATP to cyclic adenosine monophosphate by cleaving off two of the phosphates (2 on Fig. 3–5). (This is the same ATP used as the common energy currency in the body.) The extracellular messenger cannot gain entry into the cell to "personally" deliver its message to the proteins that carry out the desired response. Instead, it initiates membrane events that "arouse" an intracellular messenger, cAMP, which is given the responsibility of seeing that the response dictated by the extracellular messenger is brought about. To fulfill this function, cAMP activates one or more specific intracellular enzymes known collectively as **cAMP-dependent protein kinases** (3 on Fig. 3–5). Each protein kinase in turn phosphorylates (4 on Fig. 3–5) a specific intracellular protein, such as an enzyme important in a particular metabolic pathway. **Phosphorylation** refers to the transfer of a phosphate group from ATP to the protein at the expense of degrading ATP to ADP. Attachment of a phosphate group to the protein induces the protein to change its shape and function (either activating or inhibiting it) to bring about the desired response (5 on Fig. 3–5). After the response is accomplished, the participating chemicals are inactivated by intracellular enzymes so that the response can be terminated. Otherwise, once triggered, the response would go on indefinitely until the cell ran out of necessary supplies.

It is important to recognize that there are many different protein kinases, varying according to cell type. Therefore, *a common second messenger, cAMP, can induce widely differing responses in different cells*, depending on the differences between the activated protein kinases and the proteins that they phosphorylate and thereby modify. Cyclic AMP can be thought of as a commonly used molecular "switch" that can "turn on" (or "turn off") different cellular events depending on the unique specialization of a particular cell type. The variable responsiveness once the switch is turned on is due to the genetically programmed differences in the sets of proteins within different cells.

Instead of cAMP, some cells use Ca^{++} as a second messenger. In such cases, binding of the first messenger to the surface receptor eventually leads to activation of the enzyme **phospholipase C,** which is bound to the inner side of the membrane (1 on Fig. 3–6). This enzyme breaks down a special type of phospholipid within the membrane itself (2 on Fig. 3–6). The products of this breakdown mobilize intracellular Ca^{++} stores to increase cytosolic Ca^{++} (3 on Fig. 3–6). Many of the Ca^{++}-dependent cellular events are triggered by activation of **calmodulin,** an intracellular Ca^{++}-binding protein (4 on Fig. 3–6). Activation of calmodulin by Ca^{++} is similar to activation of a protein kinase by cAMP. From here the patterns of the two pathways are similar. Once bound to Ca^{++}, activated calmodulin alters other cellular proteins (5 on Fig. 3–6), either activating or inhibiting them, to bring about the ultimate desired cellular response.

The cAMP and Ca^{++} pathways frequently overlap in bringing about a particular cellular activity. For example, cAMP and Ca^{++} can influence each other. Calcium-activated calmodulin can regulate adenylate cyclase and thus influence cAMP, whereas cAMP-dependent kinases may phosphorylate and thereby change the activity of Ca^{++} channels or carriers. Furthermore, in some instances both Ca^{++} and cAMP regulate the same intracellular protein.

Even though the Ca^{++} and cAMP effects can become complexly intertwined, it is still notable that a great many diverse cellular events can be traced to a surprisingly small number of pathways: channel effects, one of the two known second messenger pathways, or some combination of these. It would

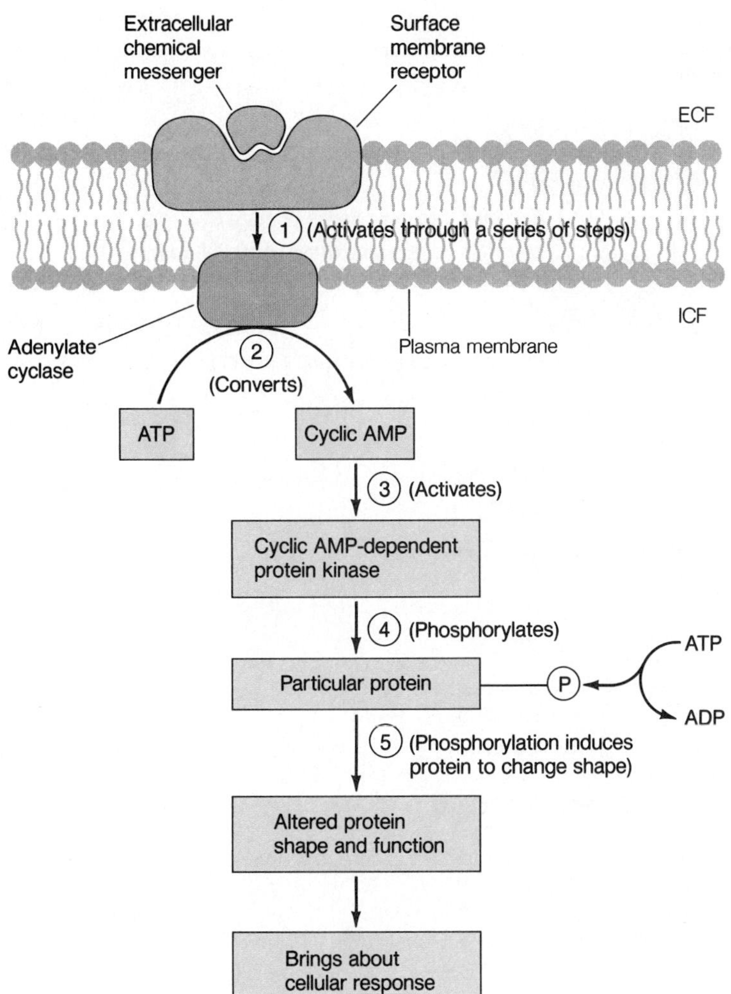

Figure 3–5 Postreceptor Event: Cyclic AMP Second Messenger System *See text for an explanation of the numbered items in the illustration.*

not be correct, however, to leave the impression that these are the only possible pathways. For example, **cyclic guanosine monophosphate (cyclic GMP)** is suspected of playing a role in postreceptor events in some cells.

Several remaining points about receptor activation and the ensuing events merit attention. First, a multiplying (cascading) effect of these pathways greatly amplifies the initial signal (Fig. 3–7). Binding of one chemical messenger molecule to a receptor activates a number of adenylate cyclase molecules (let us arbitrarily say ten), each of which activates many (in our example one hundred) cAMP molecules. Each cAMP molecule then acts on a single protein kinase, which phosphorylates and thereby influences many (again, let us say one hundred) specific proteins, such as enzymes. Each enzyme, in turn, is responsible for producing many (perhaps one hundred) molecules of a particular product, such as a secretory product. The result of this cascade of events, with one event triggering the next event in sequence, is a tremendous amplification of the initial signal; in other words, the magnitude of the output is much greater than the input. In the hypothetical example in Figure 3–7, one chemical messenger molecule has been responsible for inducing a yield of ten million molecules of a secretory product. In this way, very low concentrations of hormones and other chemical messengers can trigger pronounced cellular responses.

While membrane receptors serve as links between extracellular first messengers and intracellular second messengers in the regulation of specific cellular activities, the receptors themselves are also frequently subject to regulation. In many instances, the number and affinity (attraction of a receptor for its chemical messenger) can be altered, depending on the circumstances.

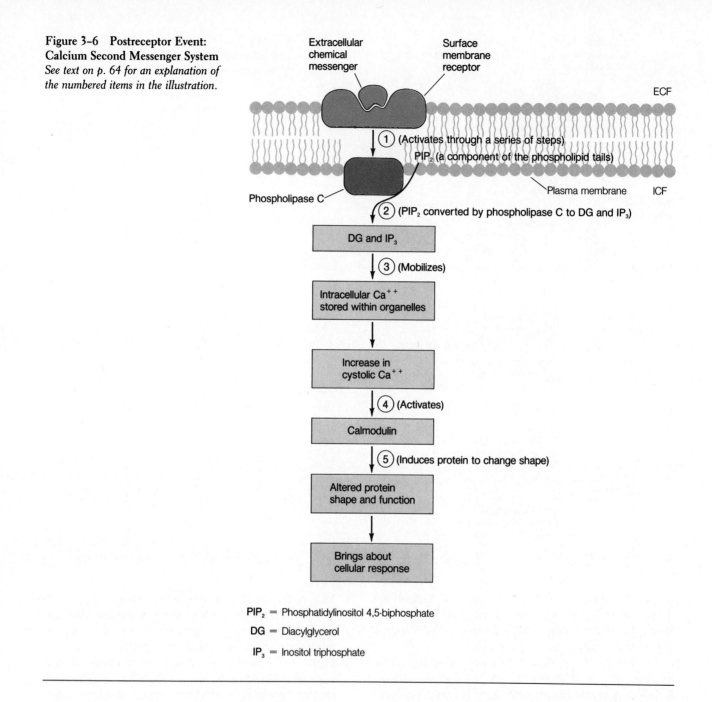

Figure 3-6 Postreceptor Event: Calcium Second Messenger System
See text on p. 64 for an explanation of the numbered items in the illustration.

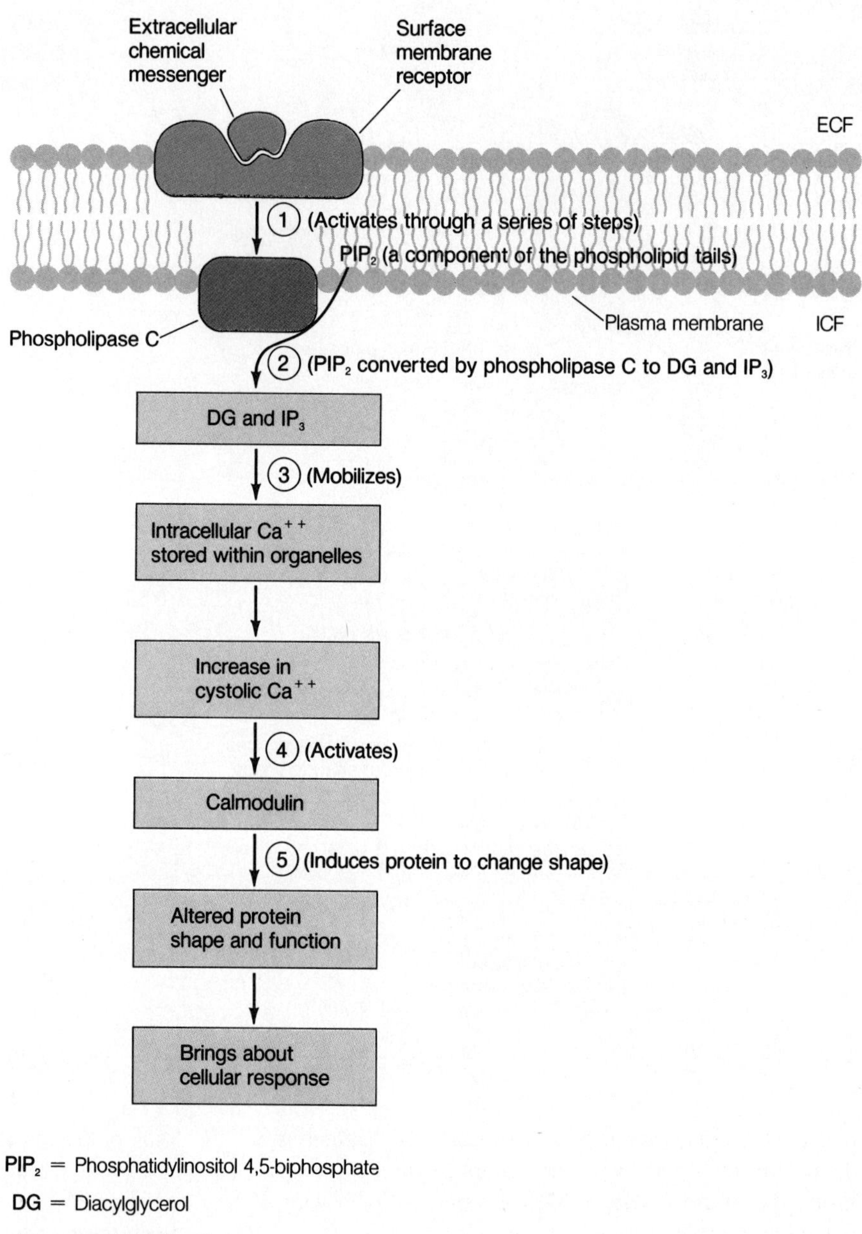

Many disease processes can be linked to malfunctioning receptors or defects in one of the components of the ensuing pathways. As an example, defective receptors are responsible for the extreme muscular weakness that characterizes myasthenia gravis. With this disease, affected skeletal muscle receptors are unable to respond to the chemical messenger released by nerves that normally triggers muscle contraction. On the other hand, in the intestinal disease cholera, the receptors are all functional but the toxin produced by the cholera pathogen prevents the normal inactivation of cAMP in intestinal cells. The role of cAMP in these cells is to induce fluid secretion into the lumen. The severe diarrhea characteristic of cholera is caused by the continued secretion of this fluid into the gut, triggered by the continual presence of cAMP.

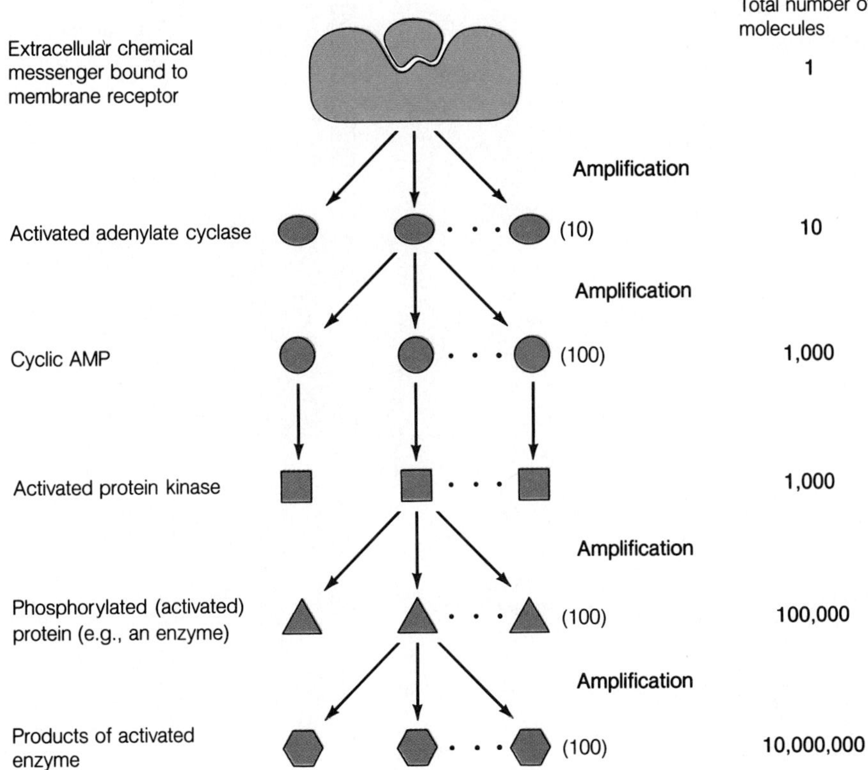

Figure 3-7 Amplification of Initial Signal by Second Messenger System *Through amplification, very low concentrations of extracellular chemical messengers, such as hormones, can trigger pronounced cellular responses.*

Cell-to-Cell Adhesions

In multicellular organisms such as humans, plasma membranes not only serve as the outer boundaries of all cells but also participate in cell-to-cell adhesions that allow groups of cells to bind together into tissues and to be packaged further into organs. Organization of cells into appropriate groupings may be at least partially attributable to the carbohydrate chains on the membrane surface. Once arranged, there are two different means by which cells are held together: (1) by the extracellular matrix and (2) by specialized cell junctions.

The extracellular matrix serves as the biological glue.

The cells within a tissue for the most part are not in direct physical contact with neighboring cells. Instead, they are held together by the **extracellular matrix,** an intricate meshwork of fibrous proteins embedded in a watery, gel-like ground substance composed of complex carbohydrates. The watery gel provides a pathway for diffusion of nutrients, wastes, and other water-soluble traffic between the blood and tissue cells. Interwoven within this gel are three major types of protein fibers: collagen, elastin, and fibronectin.

1. **Collagen** forms cablelike fibers or sheets that provide tensile strength (resistance to longitudinal stress). In scurvy, a condition caused by Vitamin C deficiency, these fibers are not properly formed. As a result, the tissues become very fragile, especially those of the skin and blood vessels. This leads to bleeding in the skin and mucous membranes, which is especially noticeable in the gums.
2. **Elastin** is a rubberlike protein fiber most abundant in tissues that must be capable of easily stretching and then recoiling after the stretching force is removed. It is found, for example, in the lungs, which stretch and recoil as air moves in and out, and in the large blood vessels, which stretch and recoil as blood is pumped into them from the heart and then is drained off into smaller vessels.
3. **Fibronectin** promotes cell adhesion and holds cells in position. Reduced amounts of this protein have been found within certain types of cancerous tissue, which could account for the fact that cancerous tissue cells do not adhere

well to each other. Accordingly, they tend to break loose and metastasize (spread elsewhere in the body).

The extracellular matrix is secreted by local cells, most commonly **fibroblasts** ("fiber formers") present in the matrix. Often the matrix plus the cells within it are known collectively as *connective tissue* because it connects cells together into tissues and tissues into organs. In some tissues the matrix becomes highly specialized to form such structures as cartilage or tendons, or, upon appropriate calcification, the hardened structures of bones and teeth.

Some cells are directly linked together by specialized cell junctions.

In addition to the tissue cohesion provided by the extracellular matrix, some cells are directly linked together by one of three types of specialized cell junctions: (1) desmosomes (adhering junctions); (2) tight junctions (impermeable junctions); and (3) gap junctions (communicating junctions).

At a **desmosome,** filaments of unknown composition extend between the plasma membranes of two closely adjacent but not touching cells, acting as "spot rivets" to anchor the cells together (Fig. 3–8). Desmosomes are distributed widely throughout the body, being most abundant in tissues that are subject to considerable stretching such as the skin, heart, muscle, and uterus. In certain sheets of epithelial tissue such as the skin, the extracellular desmosome fibers are also linked to intracellular filaments (keratin filaments, see p. 40) that extend across the interior of the cell. This forms a strong, continuous fibrous network throughout the entire epithelial sheet.

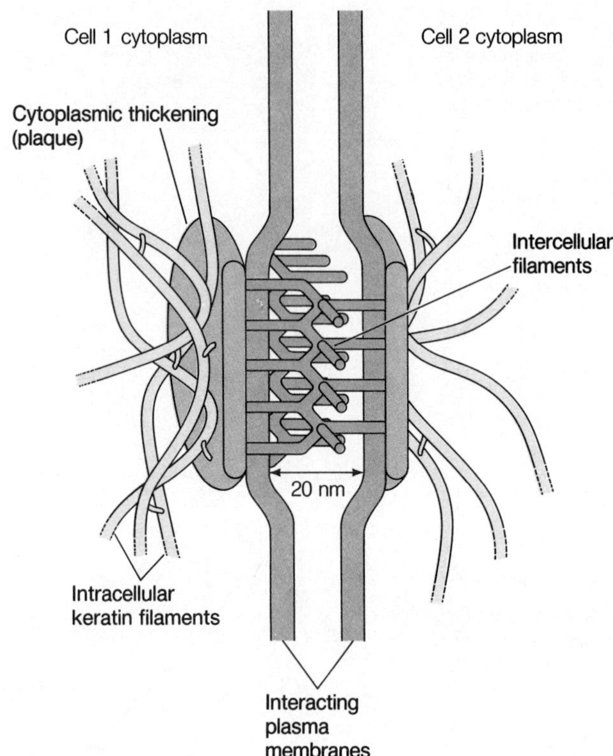

Figure 3–8 Spot Desmosome

Sheets of epithelial tissue are also joined by **tight junctions.** Epithelial tissue covers the surface of the body and lines all of its internal cavities. All of these epithelial sheets serve as highly selective barriers between two compartments that have considerably different chemical compositions. For example, the epithelial sheet lining the digestive tract separates the food and potent digestive juices within the inner cavity (lumen) from the blood vessels that lie on the other side. It is important that only completely digested food particles and not undigested food particles or digestive juices move across the epithelial sheet from the lumen to the blood. Accordingly, the lateral (side) edges of adjacent cells within the epithelial sheet are joined together in a tight seal near their luminal border by direct fusion of proteins on the outer surfaces of the two interacting plasma membranes (Fig. 3–9). These tight junctions are impermeable, thus preventing passage of materials between the cells. Passage across the epithelial barrier, therefore, must take place *through* the cells, *not between* them. This transcellular (across the cell) traffic is regulated by means of the channels and carriers present. Thus, the epithelial sheet separates the two compartments and can regulate passage of materials between them. If the cells were not joined by tight junctions, uncontrolled exchange of molecules could take place between the compartments by unpoliced traffic through the spaces between adjacent cells.

Tight junctions also perform another function besides preventing undesirable leaks within epithelial sheets. They restrict the movement of membrane proteins to one of two cell surfaces: the luminal border (facing the lumen) or basolateral border (facing the blood) (Fig. 3–9). Proteins are free to wander within their own surface but cannot move to the other surface because of the presence of the tight junction. The separation of proteins between the two parts of the surface membrane is functionally important. These proteins include carriers specific for transporting materials either from the lumen into the epithelial cell (those located on the luminal border) or from the epithelial cell into the blood (basolateral carriers). The tight junctions keep these different carriers restricted to the regions of the membrane where they can properly carry out their specific job.

The third type of junction between cells is a communicat-

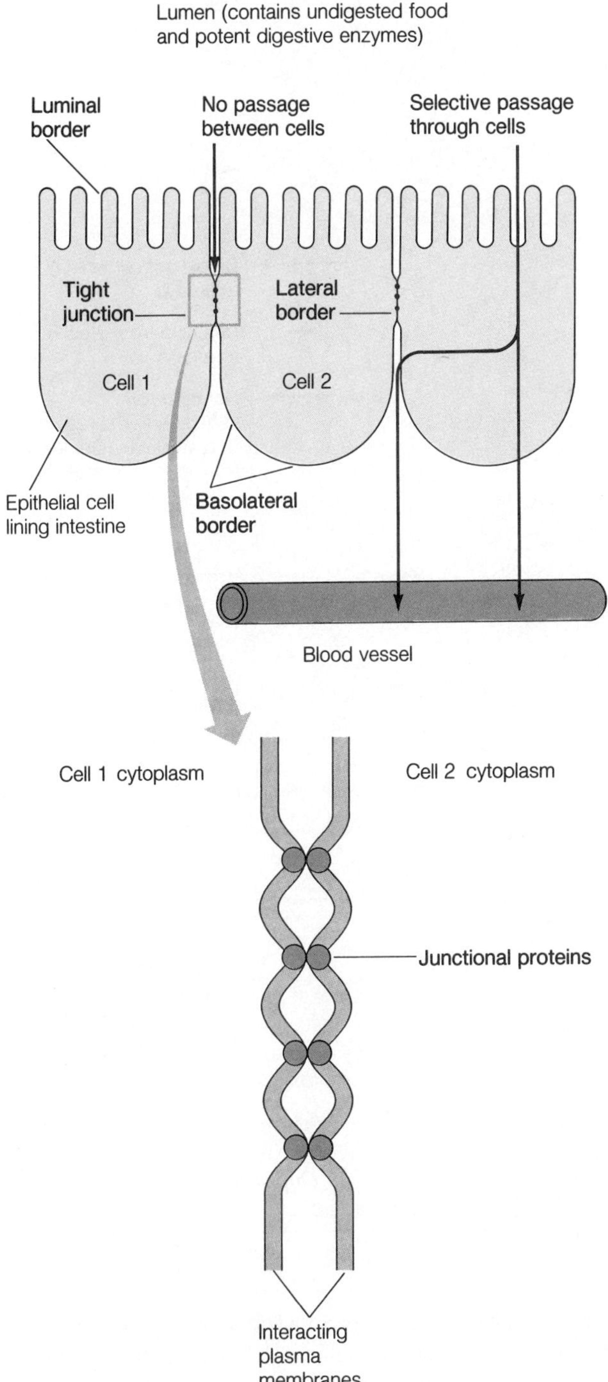

Figure 3–9 Tight Junction

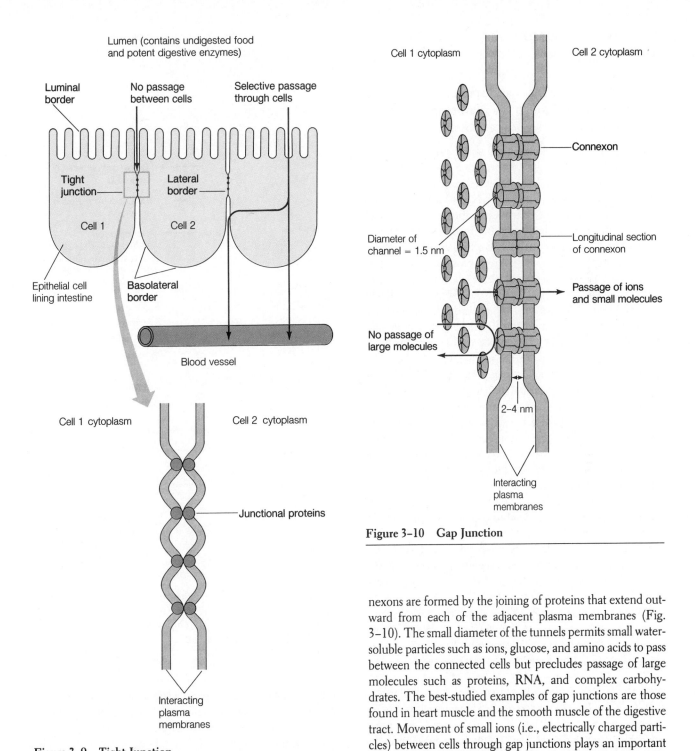

ing junction known as a **gap junction.** As implied by the name, a gap exists between two adjacent cells that are linked by small connecting tunnels known as **connexons.** Connexons are formed by the joining of proteins that extend outward from each of the adjacent plasma membranes (Fig. 3–10). The small diameter of the tunnels permits small water-soluble particles such as ions, glucose, and amino acids to pass between the connected cells but precludes passage of large molecules such as proteins, RNA, and complex carbohydrates. The best-studied examples of gap junctions are those found in heart muscle and the smooth muscle of the digestive tract. Movement of small ions (i.e., electrically charged particles) between cells through gap junctions plays an important role in transmission of electrical activity throughout an entire muscle mass. Because this electrical activity brings about contraction, the presence of gap junctions enables synchronized contraction of a whole muscle mass, such as the heart. Many tissues that are not electrically active also are connected by gap junctions, the significance of which is still unclear.

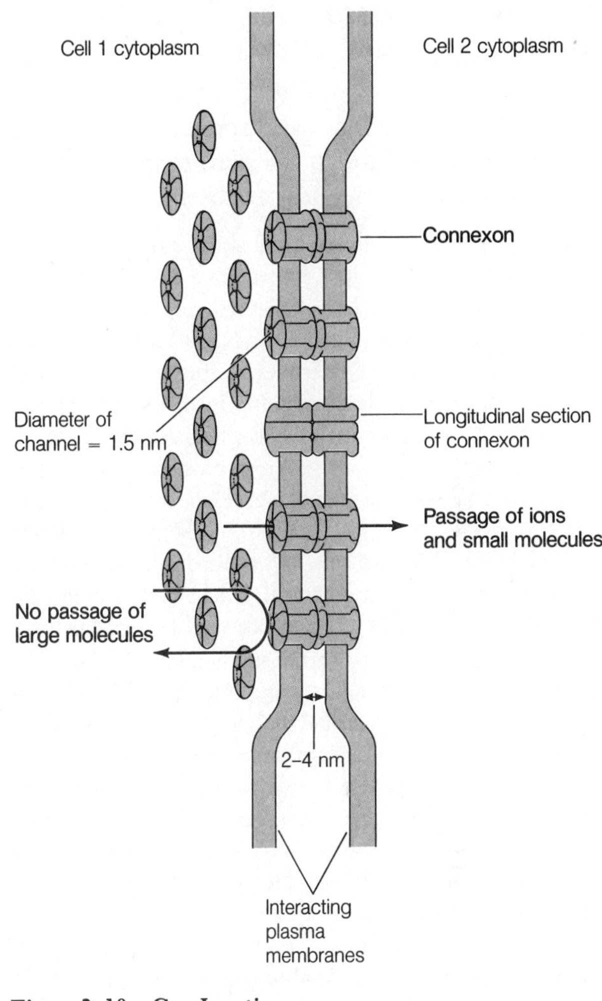

Figure 3–10 Gap Junction

Membrane Transport

Lipid-soluble substances and small ions can passively diffuse through the plasma membrane down their electrochemical gradient.

Anything that passes between a cell and the surrounding extracellular fluid must be able to penetrate the plasma membrane. If a substance can cross the membrane, the membrane is said to be **permeable** to that substance; if a substance is unable to pass, the membrane is **impermeable** to it. The plasma membrane is **selectively permeable** in that some particles are permitted to pass through while others are excluded.

Two properties of particles influence whether or not they can permeate the plasma membrane: (1) the relative solubility of the particle in lipid and (2) the size of the particle. Highly lipid-soluble particles are able to dissolve in the lipid bilayer and pass through the membrane. Uncharged or nonpolar molecules (such as O_2, CO_2, and fatty acids) are highly lipid-soluble and readily permeate the membrane. On the other hand, charged particles (ions such as Na^+ and K^+) and polar molecules (such as glucose and proteins) have low lipid solubility, being very soluble in water instead. The lipid bilayer serves as an impermeable barrier to particles poorly soluble in lipid. For the water-soluble ions less than 0.8 nm in diameter, the protein channels serve as an alternate route for passage across the membrane. Only ions for which specific channels are available can permeate the membrane.

Even though a particle might be capable of permeating the membrane by virtue of its lipid solubility or its ability to fit through a channel, there must be some force to produce its movement. A particle does not just decide it wants to be on the other side. There are two general types of forces involved: (1) those that do not require the cell to expend energy to produce movement (**passive forces**); and (2) those requiring expenditure of cellular energy (ATP) in the transport of a substance across the membrane (**active forces**). We will now examine the various methods of membrane transport, indicating whether each uses a passive or active mechanism.

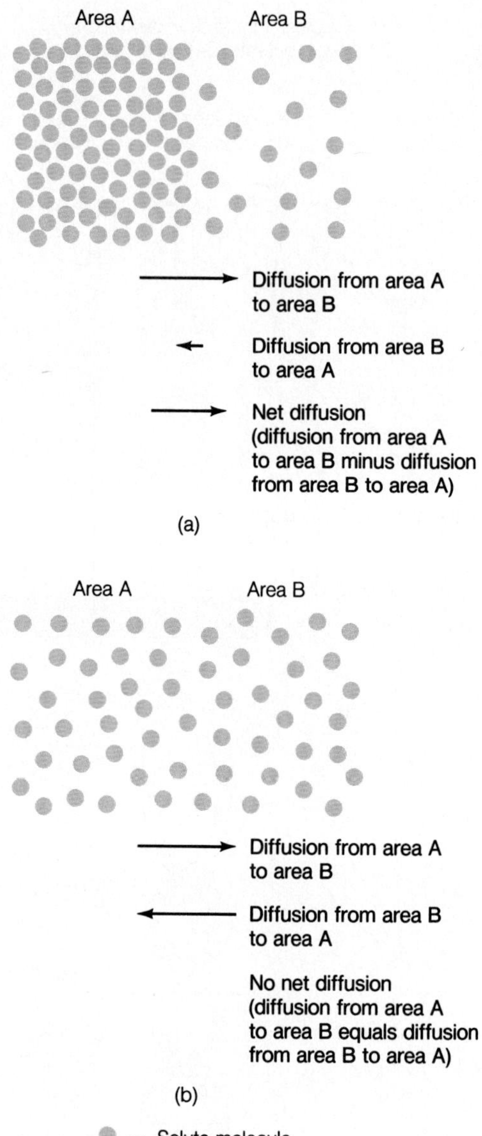

Figure 3-11 Diffusion *(a) Diffusion down concentration gradient. (b) State of equilibrium.*

DIFFUSION DOWN A CONCENTRATION GRADIENT. All molecules are in continuous random motion at temperatures above absolute zero as a result of heat (thermal) energy. This is most evident in liquids and gases, in which the individual molecules have more room to move. Each molecule moves separately and randomly in any direction. As a consequence of this haphazard movement, frequent collisions occur between the molecules, causing them to bounce off each other in different directions like billiard balls striking each other. Such random intermingling is known as **diffusion.** Obviously, the greater the molecular concentration of a substance in a solution, the greater the likelihood of collisions. For example, in Figure 3-11a, a difference in concentration exists between area A and area B in a solution. Such a difference in concentration between two adjacent areas is referred to as a **concentration gradient** (or **chemical gradient**). More frequent molecular collisions will randomly occur in area A because of its greater concentration of molecules. As a result, more molecules will bounce from area A over into area B than in the opposite direction. In both areas the individual molecules will move

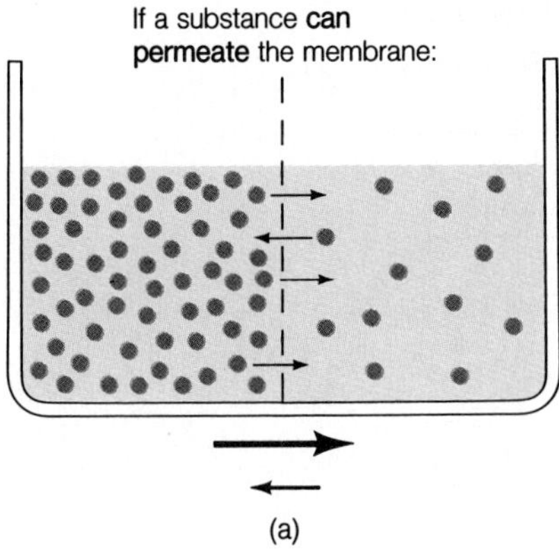

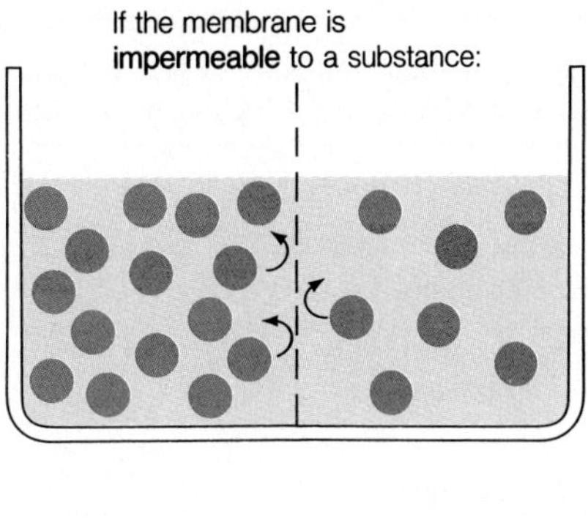

Figure 3–12 Diffusion through a Membrane *(a) Net diffusion across the membrane down a concentration gradient. (b) No diffusion through the membrane despite the presence of a concentration gradient.*

randomly and in both directions, but the net movement of molecules by diffusion will be from the area of higher concentration to the area of lower concentration.

Net diffusion refers to the difference between two opposing movements. If ten molecules move from A to B while two molecules simultaneously move from B to A, the net diffusion is eight molecules moving from A to B. Molecules will spread in this way until the substance is uniformly distributed between the two areas and there is no longer a concentration gradient (Fig. 3–11b). At this point there will be no net diffusion. Movement of molecules from A to B will be exactly matched by movement of molecules from B to A. This is known as a **state of equilibrium** or **steady state.** Even though movement is still taking place, no net movement is occurring.

What if varying concentrations of a substance are separated by a plasma membrane? If the substance can permeate the membrane, net diffusion of the substance will occur through the membrane down its concentration gradient from the area of high concentration to the area of low concentration (Fig. 3–12a). No energy is required for this movement, so it is a passive mechanism of membrane transport. The transfer of O_2 across the lung membrane occurs by this means. The blood carried to the lungs is low in O_2, having given up O_2 to the body tissues for cellular metabolism. The air in the lungs, on the other hand, is high in O_2 because it is continuously exchanged with fresh air by the process of breathing. Because of this concentration gradient, net diffusion of O_2 occurs from the lungs into the blood as blood flows through the lungs. Thus, as blood leaves the lungs for delivery to the tissues, it is high in O_2.

Several factors influence the rate of net diffusion across a membrane (**Fick's law of diffusion**) (Table 3–1):

1. *The magnitude* (or *steepness*) *of the concentration gradient*. The greater the difference in concentration, the faster the rate of net diffusion. For example, if the tissues are using O_2 more rapidly as a result of exercise, the blood delivered to the lungs will be lower in O_2 than normal, creating a larger than normal O_2 difference between the air in the lungs and the blood entering the lungs. Accordingly, more O_2 will diffuse from the lungs into the blood before equilibrium is achieved. This greater transfer of O_2 provides more O_2 for exercise.
2. *The permeability of the membrane to the substance*. The more permeable the membrane is to a substance, the more rapidly the substance can diffuse down its concentration gradient. Note that if the membrane is impermeable to the substance, no diffusion can take place across the membrane, even though a concentration gradient may exist (Fig. 3–12b). For example, because the plasma membrane is impermeable to the vital intracellular proteins, they are unable to escape from the cell, even though they are in much greater concentration in the ICF than in the ECF.
3. *The surface area of the membrane across which diffusion is taking place*. Obviously, the larger the surface area available, the greater the rate of diffusion it can accommodate. For example, absorption of nutrients across the membranes in the small intestine is enhanced by surface foldings and projections that greatly increase the available absorptive surface.
4. *The molecular weight of the substance*. Lighter molecules such as O_2 and CO_2 bounce farther upon collision than do heavier molecules. Consequently, O_2 and CO_2 diffuse rapidly, permitting rapid exchanges of these gases across the lung membranes.
5. *The distance through which diffusion must take place*. The greater the distance, the slower the rate of diffusion. Accordingly, membranes across which diffusing particles must travel are normally relatively thin, such as the mem-

Table 3–1 Factors Influencing Rate of Net Diffusion of a Substance across a Membrane (Fick's Law of Diffusion)

Factor	Effect on Rate of Net Diffusion
↑ Concentration gradient of substance	↑
↑ Permeability of membrane to substance	↑
↑ Surface area of membrane	↑
↑ Molecular weight of substance	↓
↑ Distance (thickness)	↓

Modified Fick's equation:

$$\text{Net rate of diffusion} = \frac{\text{Concentration gradient} \times \text{Permeability} \times \text{Surface area}}{\text{Distance} \times \text{Molecular weight}}$$

$$\left(\frac{\text{Permeability}}{\sqrt{\text{Molecular weight}}} = \text{Diffusion coefficient} \right)$$

$$\text{Restated: Net rate of diffusion} \propto \frac{\text{Concentration gradient} \times \text{Surface area} \times \text{Diffusion Coefficient}}{\text{Distance}}$$

branes separating the air and blood in the lungs. Thickening of this air-blood interface (as in pneumonia, for example) slows down exchange of O_2 and CO_2. Furthermore, diffusion is efficient only for short distances between cells and their surroundings. It becomes an inappropriately slow process for distances of more than a few centimeters. To illustrate, it would take months to years for O_2 to diffuse from the surface of the body to the cells in the interior. This makes it imperative that the circulatory system provide a network of tiny vessels that deliver and pick up materials at every "block" of a few cells, with diffusion accomplishing short local exchanges between the blood and surrounding cells.

MOVEMENT ALONG AN ELECTRICAL GRADIENT. Movement of ions (electrically charged particles that have either lost or gained an electron) is also affected by their electrical charge. Like charges (those with the same kind of charge) repel each other whereas opposite charges attract each other. If a relative difference in charge exists between two adjacent areas (Fig. 3–13), the positively charged ions (**cations**) tend to move toward the more negatively charged area, whereas the negatively charged ions (**anions**) tend to move toward the more positively charged area. A difference in charge between two adjacent areas thus produces an **electrical gradient** that passively induces the movement of ions. When an electrical gradient exists between the ICF and ECF, only ions that can permeate the plasma membrane are able to move down this gradient. The electrical gradient and concentration (chemical) gradient existing simultaneously for a particular ion are collectively known as the **electrochemical gradient.** Later in this chapter you will learn how electrochemical gradients are important in the determination of the electrical properties of the plasma membrane.

Osmosis is the net diffusion of water down its own concentration gradient.

Water can readily permeate the plasma membrane. The driving force for diffusion of water across the membrane is the same as for any other diffusing molecule, namely, its concentration gradient. Usually the term *concentration* refers to the density of the solute (dissolved substance) in a given volume of water. It is important to recognize, however, that the addition of a solute to pure water in essence decreases the water concentration. As a generalization, one molecule of a solute displaces one molecule of water.

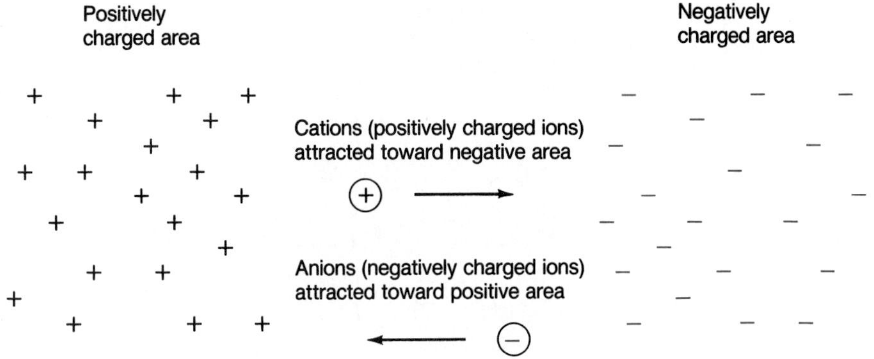

Figure 3-13 Movement Down Electrical Gradient

Compare the water and solute concentrations in the two containers in Figure 3-14. Container A is full of pure water, so the water concentration is 100% while solute concentration is 0%. In container B, 10% of the water molecules have been replaced by solute. The water concentration is now 90% (lower water concentration than A) and the solute concentration is 10% (higher solute concentration compared to A). Note that as the solute concentration increases, the water concentration decreases correspondingly.

If solutions of varying solute concentration (and hence varying water concentration) are separated by a membrane that permits passage of water, such as the plasma membrane (Fig. 3-15), water will diffuse down its own concentration gradient from the area of higher water concentration (lower solute concentration) to the area of lower water concentration (higher solute concentration). This net diffusion of water is known as **osmosis.** Because solutions are always referred to in terms of concentration of solute, *water moves by osmosis to the area of higher solute concentration*. Very loosely, then, the solute can be thought of as "drawing" or attracting water, but in reality, osmosis is nothing more than the diffusion of water down its own concentration gradient.

Thus far in our discussion we have ignored the solute. Let us compare the final outcome, depending on whether or not the solute can permeate the membrane.

1. If the membrane is permeable to the solute as well as to water, the solute is able to move down its own concentration gradient in the opposite direction of net water movement (Fig. 3-16). This movement continues until both

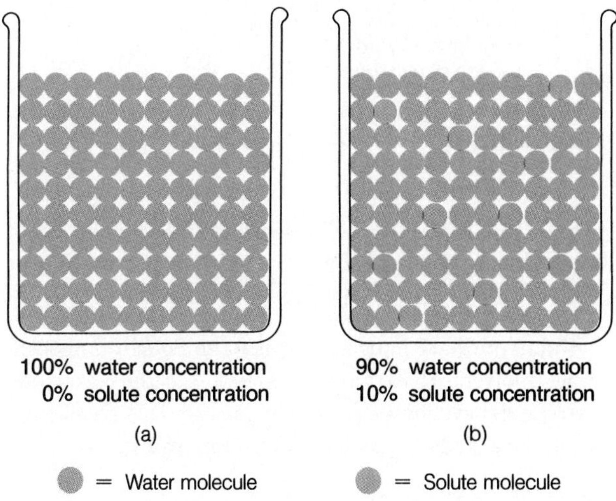

Figure 3-14 Relationship between Solute and Water Concentration in a Solution *(a) Pure water. (b) Solution.*

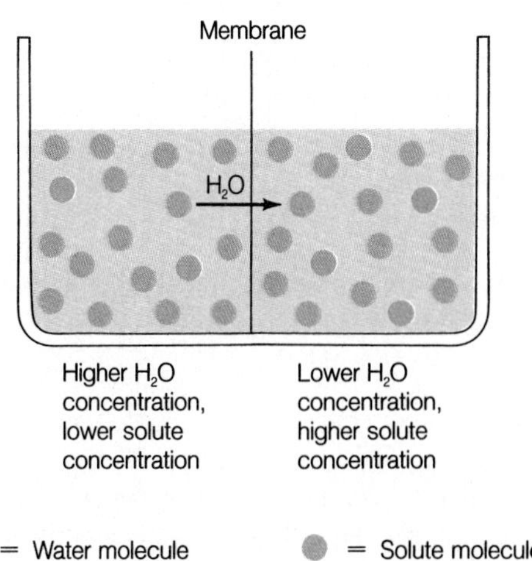

Figure 3-15 Osmosis *Osmosis is the net diffusion of water down its own concentration gradient (to the area of higher solute concentration).*

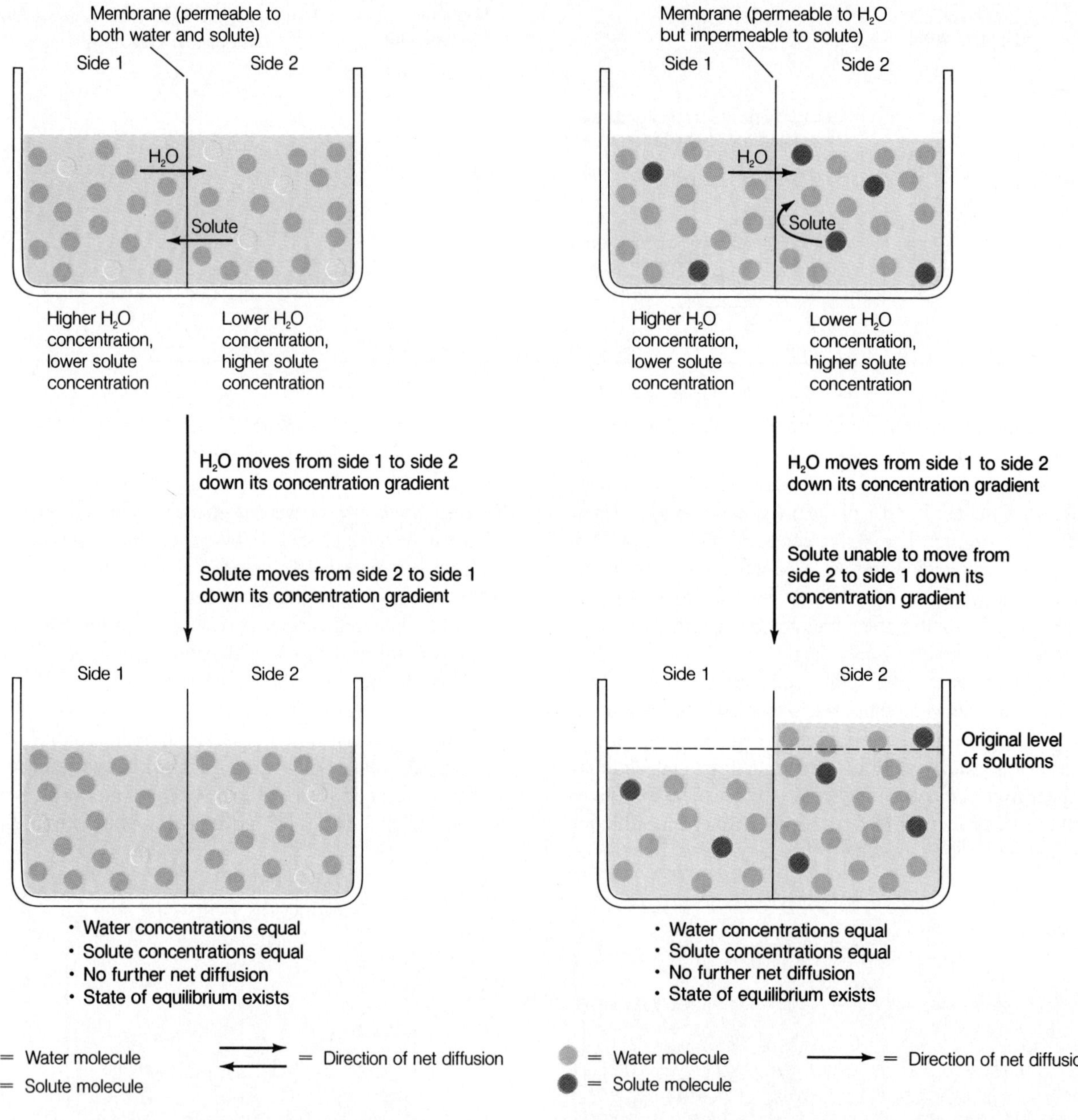

Figure 3–16 Movement of Water and a Permeable Solute Unequally Distributed across a Membrane

Figure 3–17 Osmosis in the Presence of an Unequally Distributed Nonpenetrating Solute

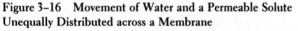

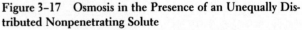

the solute and water are evenly distributed across the membrane. Even though some transitory differences in volume between the two compartments may occur because of differences in the speed with which water and the solute diffuse through the membrane, the final volume of the compartments when equilibrium is achieved is the same as at the onset. Water and solute molecules have merely exchanged places between the two compartments until their distributions have equalized; that is, an equal number of water molecules have moved from side 1 to side 2 as solute molecules have moved from side 2 to side 1. With all concentration gradients abolished, osmosis ceases.

2. If the membrane is impermeable to the solute, the solute is not able to cross the membrane down its concentration gradient (Fig. 3–17). At the onset, the concentration gra-

dients are similar to the previous example. However, even though net diffusion of water takes place from side 1 to side 2, the solute cannot move. As a result of water movement alone, side 2's volume increases while the volume of side 1 correspondingly decreases. Loss of water from side 1 increases the concentration of solutes on this side of the membrane, whereas addition of water to side 2 reduces solute concentration on that side. Eventually the concentrations of water and solute on both sides of the membrane become equal, and net diffusion of water ceases. Unlike the situation in which the solute can also permeate, diffusion of water alone has resulted in a change in final volume compared to the onset. The side originally containing the greater solute concentration is larger, having gained water.

What if a nonpenetrating solute is present on side 2 whereas only pure water is present on side 1 (Fig. 3–18)? Osmosis occurs from side 1 to side 2, but the concentrations between the two compartments can never become equal. No matter how dilute side 2 becomes because of water diffusing into it, it can never become pure water, nor can side 1 ever acquire any concentration of solute. With equilibrium of concentrations impossible to achieve, does net diffusion of water (osmosis) continue unabated until all of the water has left side 1? No. As the volume expands in compartment 2, a difference in **fluid pressure** (**hydrostatic pressure**) between the two compartments is created that opposes osmosis. The magnitude of opposing pressure necessary to completely stop osmosis is equal to the **osmotic pressure** of the solution on side 2. The osmotic pressure can be related directly to the concentration of nonpenetrating solute. The greater the concentration of nonpenetrating solute ⟶ the lower the concentration of water ⟶ the greater the drive for water to move by osmosis from pure water into the solution ⟶ the greater the opposing pressure required to stop the osmotic flow ⟶ the greater the osmotic pressure of the solution. Therefore, a solution with a high solute concentration exerts greater osmotic pressure than does a solution with a lower solute concentration.

Approximately one hundred times the volume of water in a cell crosses the plasma membrane every second. Even though water is rapidly entering and leaving cells, they normally do not experience any net gain (swelling) or loss (shrinking) of volume. This is because the concentration of solutes in the ECF is normally carefully regulated (primarily by the kidneys) to maintain the same level of osmotic activity as is present within the cells; thus there is no net diffusion of water. For example, the plasma in which the red blood cells are suspended normally has the same osmotic activity as the fluid inside these cells, enabling the cells to maintain a constant volume. If red blood cells are placed in a solution with a lower concentration of solutes (and, therefore, a higher concentration of water), water enters the cells by osmosis, causing them to swell, perhaps to the point of rupturing or *lysing*. In fact, this is how red blood cell ghosts are prepared (see p. 58).

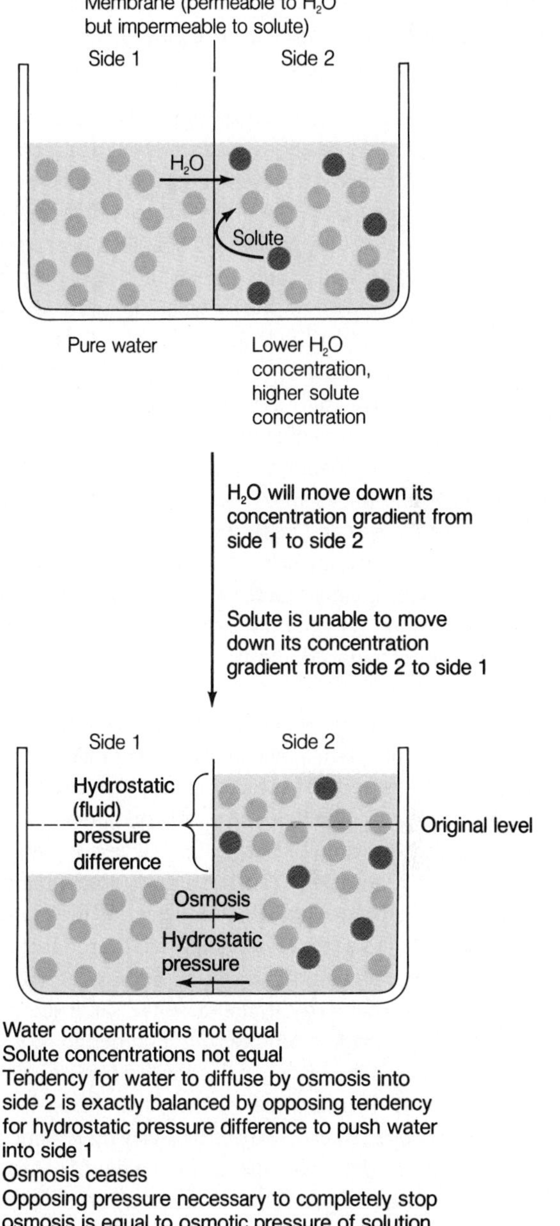
Membrane (permeable to H_2O but impermeable to solute)
Side 1
Side 2
H_2O
Solute
Pure water
Lower H_2O concentration, higher solute concentration
H_2O will move down its concentration gradient from side 1 to side 2
Solute is unable to move down its concentration gradient from side 2 to side 1
Side 1
Side 2
Hydrostatic (fluid) pressure difference
Original level
Osmosis
Hydrostatic pressure
• Water concentrations not equal
• Solute concentrations not equal
• Tendency for water to diffuse by osmosis into side 2 is exactly balanced by opposing tendency for hydrostatic pressure difference to push water into side 1
• Osmosis ceases
• Opposing pressure necessary to completely stop osmosis is equal to osmotic pressure of solution
= Water molecule
= Solute molecule
= Direction of net diffusion

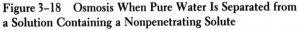

Figure 3–18 Osmosis When Pure Water Is Separated from a Solution Containing a Nonpenetrating Solute

Special mechanisms are used to transport selected molecules unable to cross the plasma membrane on their own.

All of the forces we have discussed thus far—diffusion down concentration gradients, diffusion down electrical gradients, and osmosis—produce net movement of molecules capable of permeating the plasma membrane by virtue of their lipid solubility or small size. Large, poorly lipid-soluble molecules such as proteins, glucose, and amino acids cannot cross the plasma membrane on their own no matter what forces are acting on them. This ensures that the large polar intracellular proteins cannot escape from the cell. It also means, however, that the cell must provide mechanisms for transporting into the cell essential nutrients, such as glucose for energy and amino acids for the synthesis of proteins, as well as for transporting out of the cell metabolic wastes and secretory products, such as proteinaceous hormones and enzymes. Furthermore, the movement of small ions cannot always be accounted for on the basis of passive diffusion alone. Cells utilize two different mechanisms to accomplish these selective transport processes: carrier-mediated transport and vesicular transport.

CARRIER-MEDIATED TRANSPORT. All carrier proteins span the thickness of the plasma membrane and are able to undergo reversible changes in shape so that specific binding sites can alternately be exposed at either side of the membrane. The details of the conformational changes that carriers undergo are unknown, but Figure 3–19 is a schematic representation of how this transport process might take place. As the molecule to be transported attaches to a binding site on the carrier on one side of the membrane (Step 1), it presumably triggers a change in the carrier's shape to expose the same site to the other side of the membrane (Step 2). The bound molecule then detaches from the carrier, in this way having been moved from one side of the membrane to the other (Step 3).

Carrier-mediated transport systems display three important characteristics that determine the kind and amount of material that can be transferred across the membrane: specificity, saturation, and competition.

1. **Specificity.** Each carrier protein is specialized to transport a specific substance, or at most a few closely related chemical compounds. For example, amino acids cannot bind to glucose carriers, although several similar amino acids may be able to utilize the same carrier. Cells vary in the types of carriers they possess, thus permitting transport selectivity among cells. There are a number of inherited diseases involving defects in transport systems for a particular substance. Cysteinuria (cysteine in the urine) is such a disease involving defective cysteine carriers in the kidney membranes. This transport system normally removes cysteine from the fluid destined to become urine and returns this essential amino acid to the blood. When this carrier is malfunctional, large quantities of cysteine remain in the urine, where it is relatively insoluble and tends to precipitate. This is one cause of urinary stones.

2. **Saturation.** There is a limit to the amount of a substance that can be transported across the membrane via a carrier in a given time; that is, a limited number of carrier binding sites are available within a particular plasma membrane for a specific substance. This limit is known as the **transport maximum (T_m)**. Until the T_m is reached, the number of carrier binding sites occupied by a substance, and, accordingly, the substance's rate of transport across the membrane, is directly related to its concentration. The more of a substance available to be transported, the more will be transported. When the T_m is reached, the carrier is saturated (all binding sites are occupied), and the rate of the substance's transport across the membrane is maximal. Further increases in concentration of the substance are not accompanied by corresponding increases in rate of transport (Fig. 3–20).

   As an analogy, consider a ferry boat that can maximally carry one hundred people across a river in an hour. If twenty-five people board the ferry, twenty-five will be transported that hour. Doubling the number of people boarding to fifty will double the rate of transport to fifty people per hour. Such a direct relationship will exist between the number of people waiting to board (the concentration) and the rate of transport until the ferry is fully occupied (its T_m is reached). The ferry can maximally transport one hundred people per hour. Even if one hundred and fifty people are waiting to board, still only one hundred will be transported per hour.

   Saturation of carriers is a critical rate-limiting factor in the transport of selected substances across the kidney membranes during urine formation and across the intestinal membranes during absorption of digested foods. Furthermore, in some important instances it is possible to regulate (for example, by hormones) the rate of carrier-mediated flux by varying the affinity (attraction) of the binding site for its passenger or by varying the number of binding sites. For example, the hormone insulin greatly increases the carrier-mediated transport of glucose into most cells of the body. Deficiency of insulin (diabetes mellitus) drastically impairs the body's ability to utilize glucose as the primary energy source.

3. **Competition.** Several closely related compounds may compete for a ride across the membrane on the same carrier. If a given binding site can be occupied by more than one type of molecule, the rate of transport of each sub-

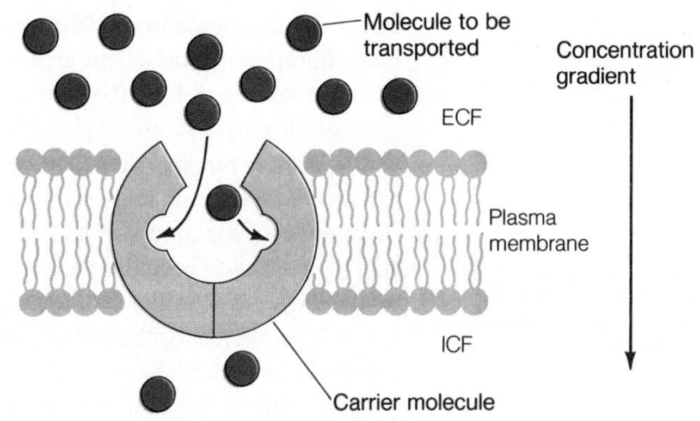

Step 1
Conformation X of carrier (binding sites exposed to outside of membrane)
Molecule to be transported binds to carrier

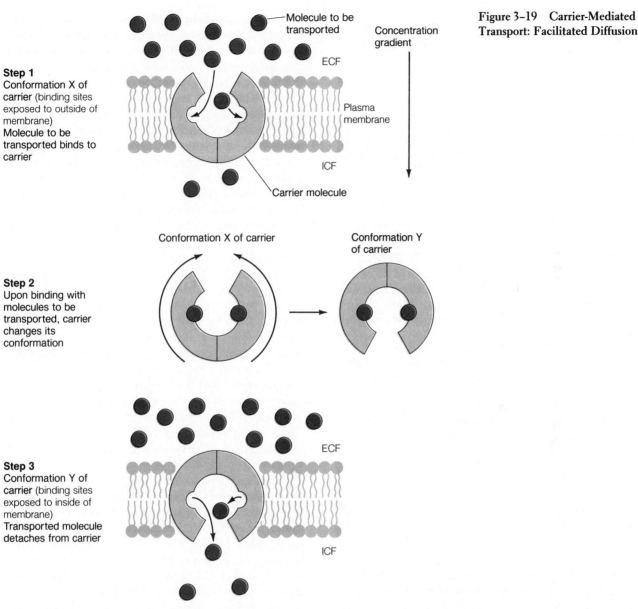

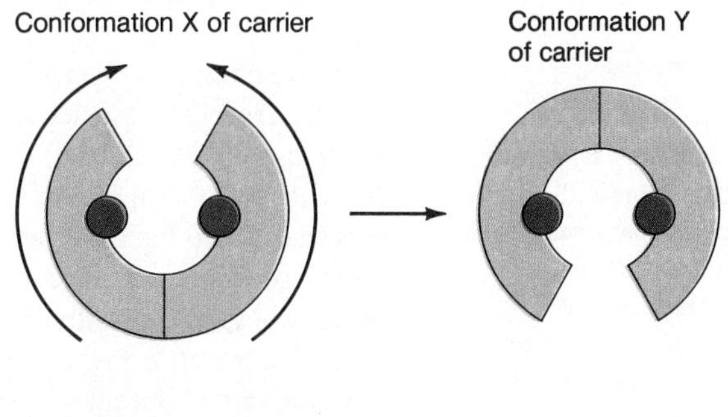

Step 2
Upon binding with molecules to be transported, carrier changes its conformation

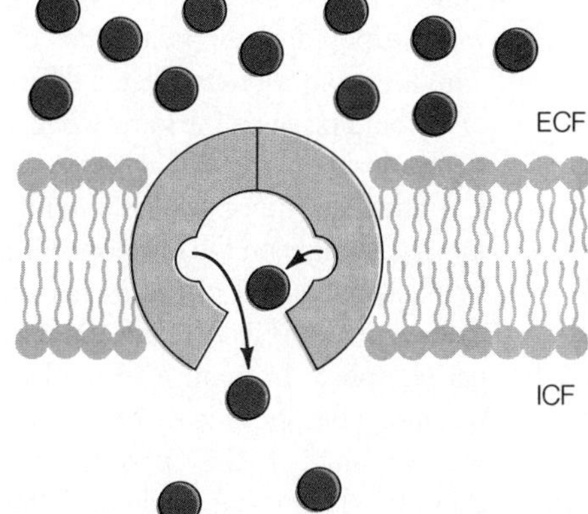

Step 3
Conformation Y of carrier (binding sites exposed to inside of membrane)
Transported molecule detaches from carrier

Figure 3-19 Carrier-Mediated Transport: Facilitated Diffusion

stance is less when both molecules are present than when either is present by itself. To illustrate, assume the ferry has one hundred seats (binding sites) that can be occupied by either men or women. If only men are waiting to board, up to one hundred men can be transported during each trip, the same holding true if only women are waiting to board. If, however, both men and women are waiting to board, they will compete for the available seats so that fewer men and fewer women will be transported than when either group is present alone. Fifty of each might make the trip, although the total number of people transported will still be the same, one hundred people. In other words, when a carrier is able to transport two closely related substances, such as the amino acids glycine and alanine, the presence of both diminishes the rate of transfer of either.

There are two types of carrier-mediated transport, depending on whether or not energy must be supplied to complete the process: facilitated diffusion and active transport. **Facilitated diffusion** utilizes a carrier to facilitate (assist) the trans-

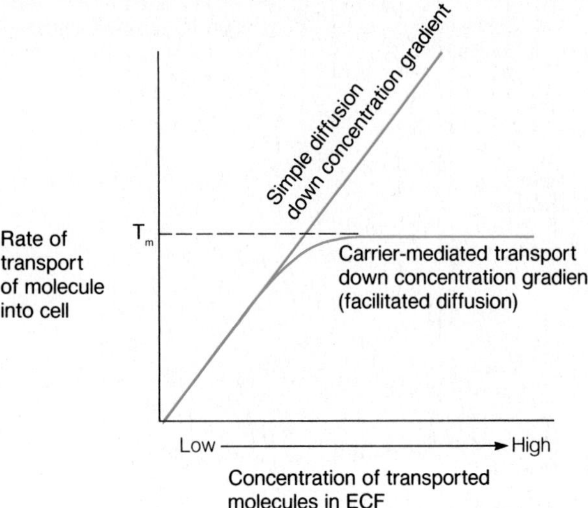

Figure 3–20 Comparison of Carrier-Mediated Transport and Simple Diffusion Down Concentration Gradient *With simple diffusion of a molecule down its concentration gradient, the rate of transport of the molecule into the cell is directly proportional to the extracellular concentration of the molecule. With carrier-mediated transport of a molecule down its concentration gradient, the rate of transport of the molecule into the cell is directly proportional to the extracellular concentration of the molecule until the carrier is saturated, at which time the rate of transport reaches a maximal value (transport maximum, or* T_m*). The rate of transport does not increase with further increases in the ECF concentration of the molecule.*

fer of a particular substance across the membrane "downhill" from high to low concentration. This is a passive process because it does not require energy, the driving force being the concentration gradient. **Active transport,** on the other hand, requires energy expenditure by the carrier to transfer its passenger "uphill" against a concentration gradient, from an area of lower concentration to an area of higher concentration. An analogous situation is a car on a hill. To move the car downhill requires no energy; it will coast from the top down. Driving the car uphill, however, requires the utilization of energy (gasoline).

The most notable example of facilitated diffusion is the transport of glucose into cells. Glucose is in higher concentration in the blood than in the tissues. Fresh supplies of this nutrient are regularly added to the blood by eating and from reserve energy stores in the body. Simultaneously, the cells metabolize glucose almost as rapidly as it enters the cells from the blood. As a result, there is a continuous gradient for net diffusion of glucose into the cells. Being a polar molecule, however, glucose cannot cross most cell membranes on its own. Without the glucose carrier molecules to facilitate membrane transport of glucose, the cells would be deprived of their preferred source of fuel. (See the accompanying boxed feature, A Closer Look at Exercise Physiology.)

The carrier binding sites involved in facilitated diffusion can bind with their passenger molecules when exposed to either side of the membrane (Fig. 3–19). Because more passengers are waiting to board on the high-concentration side than on the low-concentration side, the net movement always proceeds down the concentration gradient from higher to lower concentration. As is characteristic of mediated transport, the rate of facilitated diffusion is limited by saturation of the carrier binding sites, unlike the rate of simple diffusion, which is always directly proportional to the concentration gradient (Fig. 3–20).

Active transport also involves the use of a protein carrier to transfer a specific substance across the membrane, but the carrier transports the substance against its concentration gradient. For example, the uptake of iodide by thyroid gland cells necessitates active transport because 99% of the iodide in the body is concentrated in the thyroid. To move iodide from the blood, where its concentration is low, into the thyroid, where its concentration is high, requires expenditure of energy to drive the carrier. Specifically, energy in the form of ATP is required in active transport to vary the affinity of the binding site when exposed on opposite sides of the plasma membrane. This is in contrast to facilitated diffusion, in which the affinity of the binding site is the same when exposed to either the outside or inside of the cell.

With active transport, the binding site has a greater affinity for its passenger on the low-concentration side, brought about by phosphorylation of the carrier on this side (Fig. 3–21a). The carrier exhibits ATPase activity in that it splits the terminal phosphate from an ATP molecule to yield ADP plus a free inorganic phosphate (see p. 27). Phosphorylation involves the binding of this phosphate group to the carrier. Phosphorylation and binding of the passenger on the low-concentration side induces a conformational change in the carrier protein so that the passenger is now exposed to the high-concentration side of the membrane (Fig. 3–21b). The change in carrier shape is accompanied by dephosphorylation; that is, the phosphate group detaches from the carrier and remains in the ICF to be utilized in the synthesis of more ATP. Removal of phosphate alters the affinity of the binding site so that the passenger is released on the high-concentration side. The carrier then returns to its original conformation. Thus, ATP energy is used in the phosphorylation-dephosphorylation cycle of the carrier. It alters the affinity of the carrier's binding sites on opposite sides of the membrane so that transported particles are moved uphill from an area of low concentration to an area of higher concentration. These active trans-

A CLOSER LOOK AT EXERCISE PHYSIOLOGY

Exercising Muscles Have a "Sweet Tooth"

During exercise, muscle cells use more glucose and other nutrient fuels than usual to power their increased contractile activity. The rate of glucose transport into exercising muscle may increase more than ten-fold during moderate or intense physical activity. The mechanisms responsible for increased glucose uptake by exercising muscle are still unclear. In many cells, including resting muscle cells, facilitated diffusion of glucose into the cells depends on the hormone insulin. Because plasma-insulin levels fall during exercise, however, insulin is probably not responsible for the increased transport of glucose into exercising muscles. One possibility presently being studied is that exercise increases the availability of glucose carriers in the plasma membrane of muscle cells. This has been demonstrated in rats that have undergone physical training.

Exercise influences glucose transport into cells in yet another way. Regular aerobic exercise (see p. 33) has been shown to increase both the affinity (degree of attraction) and number of plasma-membrane receptor sites that bind specifically with insulin. This adaptation results in an increase in insulin sensitivity; that is, the cells are more responsive than normal to a given level of circulating insulin.

Because insulin enhances the facilitated diffusion of glucose into most cells, an exercise-induced increase in insulin sensitivity is one of the factors that makes exercise a beneficial therapy for controlling diabetes mellitus, a disorder characterized by insulin deficiency (see chapter 19). As a result of inadequate insulin action, glucose entry into most cells is impaired. Plasma levels of glucose become elevated because glucose remains in the plasma instead of being transported into the cells. In the type II form of the disease, insulin is being produced, but not in sufficient quantities to meet the body's need for glucose uptake. By increasing the cells' responsiveness to the limited amount of insulin available, regular aerobic exercise helps drive glucose into the cells, where it can be used for energy production, instead of remaining in the plasma, where it leads to detrimental consequences for the body.

port mechanisms are frequently called **pumps,** analogous to water pumps that require energy to raise water up against the downward pull of gravity.

The simplest active transport systems pump a single type of passenger. An example is the **hydrogen ion (H^+) pump** used by specialized stomach cells to transport H^+ into the stomach lumen in association with the secretion of hydrochloric acid during digestion of a meal. This pump moves H^+ against a gradient three to four million times more concentrated.

Other more complicated active transport mechanisms involve the transfer of two different passengers, either simultaneously in the same direction or sequentially in opposite directions. As an example, the plasma membrane of all cells contains a sequentially active **Na^+-K^+ ATPase pump (Na^+-K^+ pump** for short). This carrier transports Na^+ out of the cell, concentrating it in the ECF, and picks up K^+ from the outside, concentrating it in the ICF (Fig. 3–22). Splitting of ATP via ATPase activity and the subsequent phosphorylation of the carrier on the intracellular side increases the carrier's affinity for Na^+ and induces a change in carrier shape, leading to deposition of Na^+ to the exterior. The subsequent dephosphorylation of the carrier increases its affinity for K^+ on the extracellular side and restores the original carrier conformation, thereby transferring K^+ into the cytoplasm. There is not a direct exchange of Na^+ for K^+, however. The Na^+-K^+ pump moves three Na^+ out of the cell for every two K^+ it pumps in. (To appreciate the magnitude of active Na^+-K^+ pumping that takes place, it has been estimated that a single nerve cell membrane contains perhaps one million Na^+-K^+ pumps capable of transporting about 200 million ions/second.) The Na^+-K^+ pump plays three important roles:

1. It establishes Na^+ and K^+ concentration gradients across the plasma membrane of all cells, which is critically impor-

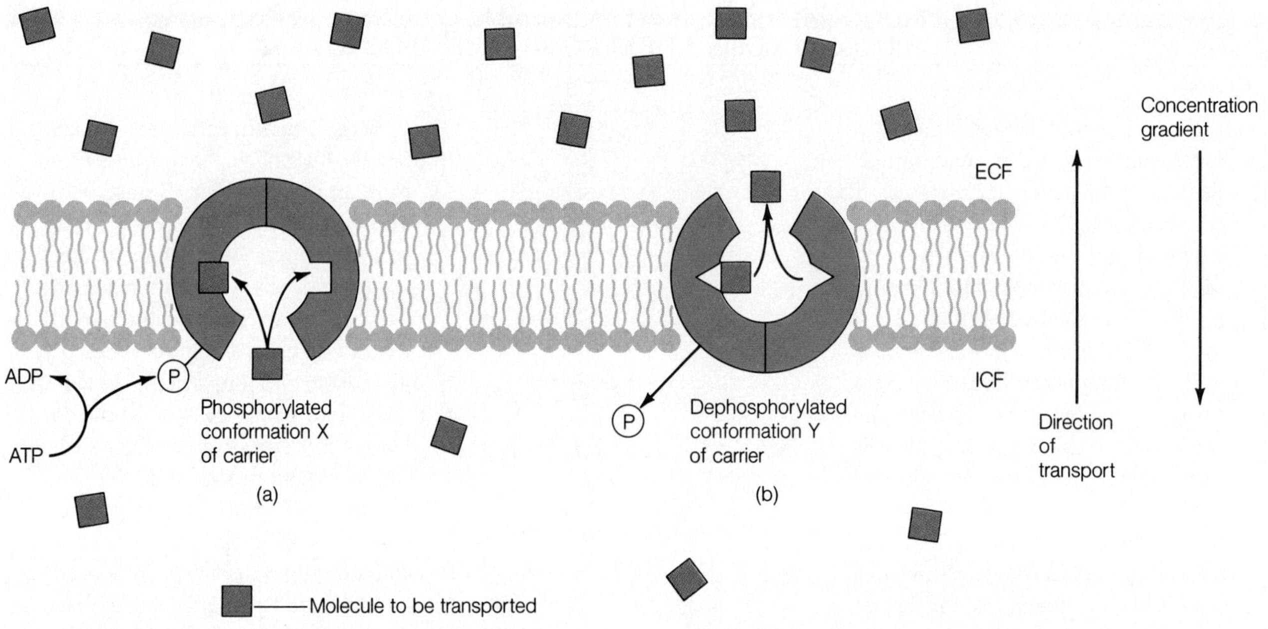

Figure 3–21 Active Transport *The energy of ATP is required in the phosphorylation-dephosphorylation cycle of the carrier to transport the molecule uphill from a region of low concentration to a region of high concentration. (a) The phosphorylated conformation of the carrier has high-affinity binding sites for the molecule to be transported so the molecule binds to the carrier. (b) The dephosphorylated conformation of the carrier has low-affinity binding sites for the molecule being transported so the molecule is released from the carrier.*

tant in the ability of nerve and muscle cells to generate electrical impulses essential to their functioning (a topic that will soon be discussed more thoroughly).

2. It helps regulate cell volume by controlling the concentrations of solutes inside the cell to minimize osmotic effects that would induce swelling or shrinking of the cell.
3. The energy used to run the Na^+-K^+ pump also indirectly serves as the energy source for the co-transport of glucose and amino acids across intestinal and kidney cells.

In the latter regard, unlike most cells of the body, the intestinal and kidney cells actively transport glucose and amino acids by moving them uphill from low to high concentration. The intestine transports these nutrients from the lumen into the blood, concentrating them in the blood until none of these molecules are left in the lumen to be lost in the feces. The kidneys save these nutrient molecules for the body by transporting them out of the fluid that is to become urine, moving them against a concentration gradient into the blood. However, energy is not directly supplied to the carrier in these instances. The luminal carriers are co-transport carriers in that they have two binding sites, one for Na^+ and one for the nutrient molecule. Since more Na^+ is present outside the cell than inside the cells because of the energy-requiring Na^+-K^+ pump that transports Na^+ out of the cell, more Na^+ binds to the luminal co-transport carrier when it is exposed to the outside (Fig. 3–23). Binding of Na^+ to the co-transport carrier increases the carrier's affinity for its other passenger (for example, glucose), so the carrier has a high affinity for glucose when exposed to the outside. When both Na^+ and glucose are bound to the carrier, it undergoes a conformational change and opens to the inside of the cell. Both Na^+ and glucose are released to the interior, Na^+ because of the lower intracellular Na^+ concentration, and glucose because of the reduced affinity of the binding site upon release of Na^+.

The movement of Na^+ into the cell by this co-transport carrier is downhill because the intracellular Na^+ concentration is low, but the movement of glucose is uphill because glucose becomes concentrated in the cell. The released Na^+ is quickly pumped out by the active Na^+-K^+ transport mechanism, keeping the level of intracellular Na^+ low. The glucose passes out of the cell to enter the blood down its concentration gradient by facilitated diffusion, mediated by another passive carrier in the plasma membrane. The energy expended in this process is not directly utilized to run the co-transport carrier, because phosphorylation is not required to alter the affinity of

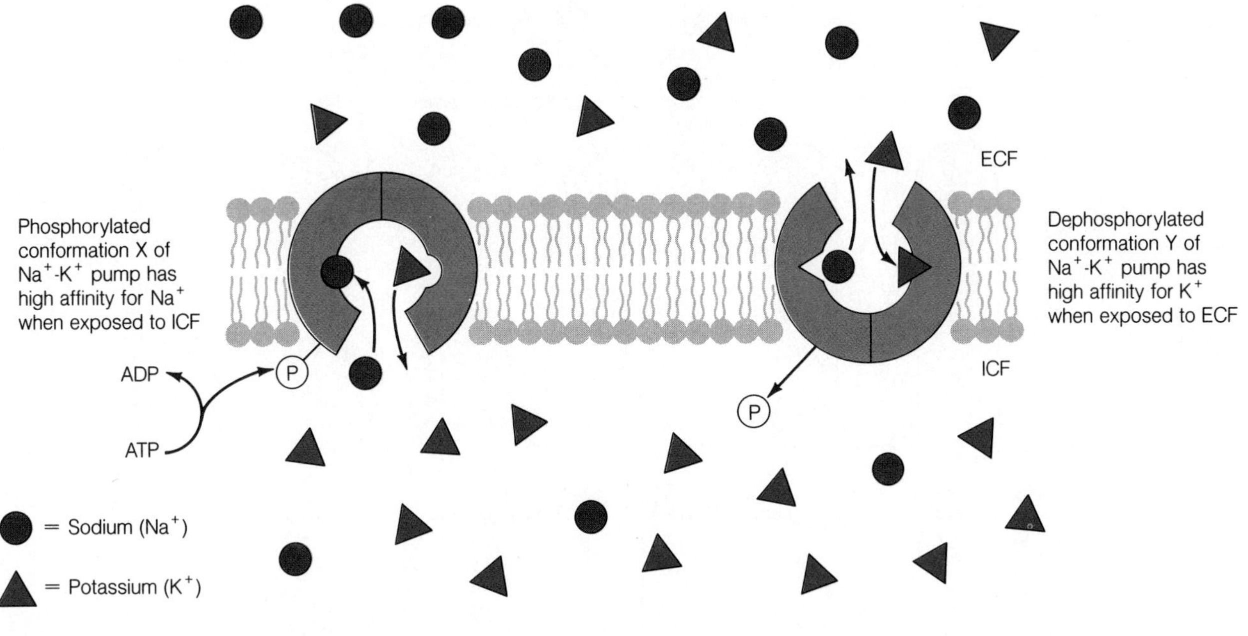

ECF
Phosphorylated conformation X of Na^+-K^+ pump has high affinity for Na^+ when exposed to ICF
Dephosphorylated conformation Y of Na^+-K^+ pump has high affinity for K^+ when exposed to ECF
ADP
P
ATP
ICF
P
= Sodium (Na^+)
= Potassium (K^+)

Figure 3–22 Na^+-K^+ ATP-ase Pump

the binding site to glucose. Instead, it is the establishment of a Na^+ concentration gradient by a primary active transport mechanism (the Na^+-K^+ pump) that drives this secondary active transport mechanism (Na^+-glucose co-transport carrier) to move glucose against its concentration gradient. With **primary active transport,** energy is directly required to move a substance uphill. With **secondary active transport,** energy is required in the entire process, but it is not directly required to run the pump. Rather it uses "second-hand" energy stored in the form of an **ion concentration gradient** (for example, a Na^+ gradient) to move the co-transported molecule uphill. This is a very efficient interaction. The co-transported molecule is essentially getting a free ride, because Na^+ must be pumped out anyway to maintain the electrical and osmotic integrity of the cell.

VESICULAR TRANSPORT: ENDOCYTOSIS/EXOCYTOSIS. The special carrier-mediated transport systems embedded within the plasma membrane can selectively transport ions and small polar molecules. But what about large polar molecules or even multimolecular materials that must leave or enter the cell, such as during secretion of protein hormones by endocrine cells or during ingestion of invading bacteria by white blood cells? To be transferred between the ICF and ECF, these large particles are enclosed in a membrane-bound vesicle. Transport into the cell in this manner is termed **endocytosis,** whereas transport out of the cell is called **exocytosis** (see p. 24).

In the case of endocytosis, the plasma membrane surrounds the substance to be ingested, then fuses over the surface, pinching off a membrane-bound vesicle that encloses the engulfed material within the cell. The transported material has not actually passed through the surface membrane but has gained entrance to the interior of the cell by being wrapped in a piece of the membrane. If fluid is internalized by endocytosis, the process is termed **pinocytosis** (cell drinking). If large multimolecular particles are engulfed, such as bacteria or cellular debris, the process is called **phagocytosis** (cell eating). Most cells perform pinocytosis, but only a few specialized cells are capable of phagocytosis. The latter are the "professional" phagocytes, the most notable of which are certain types of white blood cells that play an important role in the body's defense mechanisms.

Once inside the cell, there are two possible destinies for an engulfed vesicle:

1. In most instances, lysosomes fuse with the vesicle to degrade and release its contents into the intracellular fluid (see pp. 24 and 25).
2. In some cells, the endocytotic vesicle bypasses the lysosomes and travels to the opposite side of the cell, where it releases its contents by exocytosis. This provides a pathway to shuttle intact particles through the cell. Such vesic-

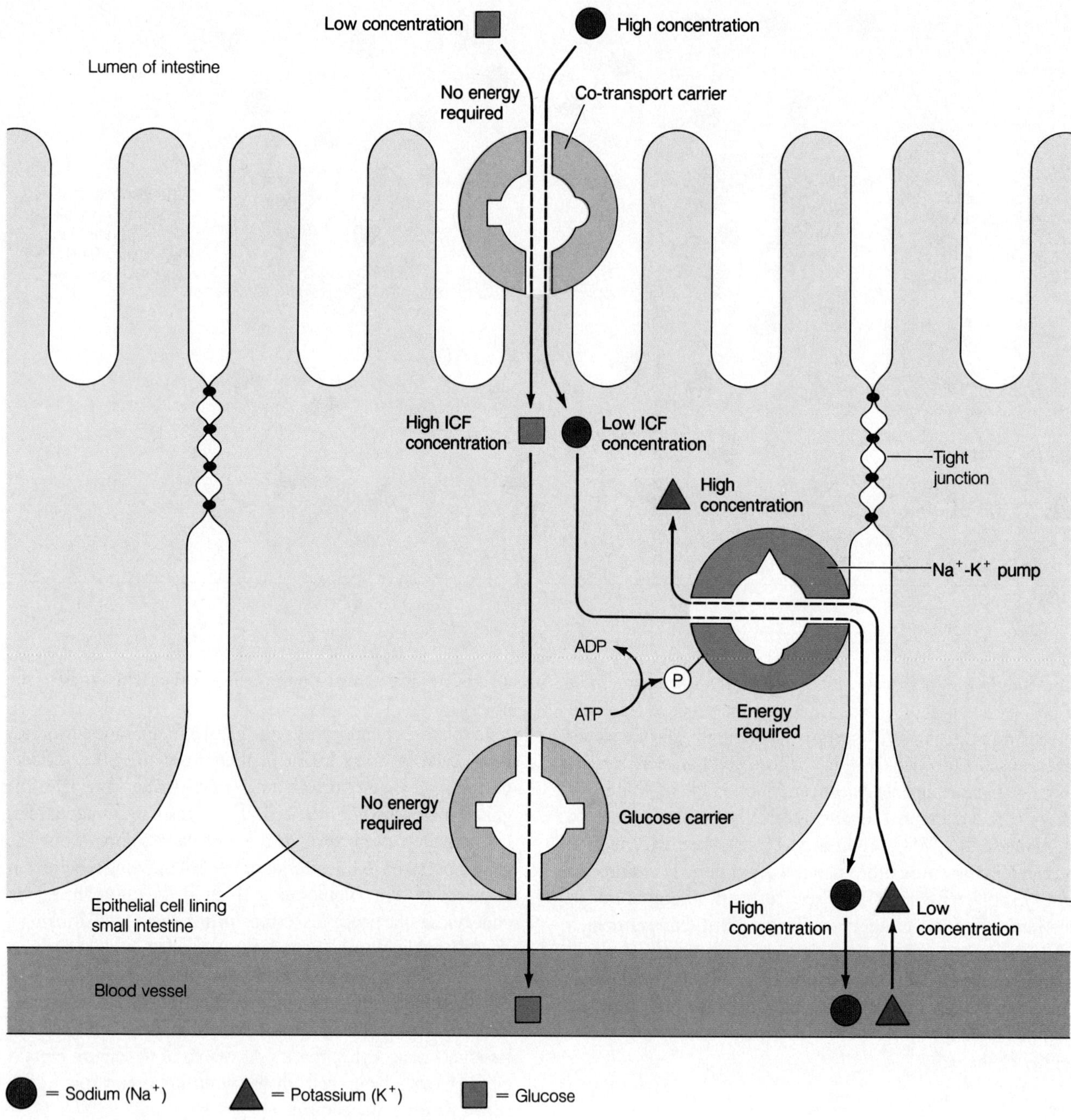

Figure 3–23 Secondary Active Transport *See text on p. 80 for an explanation of the roles of these various carriers in secondary active transport.*

ular traffic is one means by which materials are transferred through the thin cells lining the capillaries, across which exchanges are made between the blood and surrounding tissues.

In exocytosis, almost the reverse of endocytosis occurs. A membrane-bound vesicle formed within the cell fuses with the plasma membrane, then opens up and releases its contents to the exterior. Only materials packaged for export by the endoplasmic reticulum/Golgi complex can be externalized by exocytosis (see p. 23).

Two different purposes are served by exocytosis:

1. It provides a mechanism for secreting large polar molecules, such as proteinaceous hormones and enzymes, that are unable to cross the plasma membrane. In this case, the vesicular contents are highly specific and are released only upon receipt of appropriate signals.
2. It enables the cell to add specific components to the membrane, such as selected carriers, channels, or receptors, depending on the cell's needs. In such cases, the composition of the membrane surrounding the vesicle is important and the contents may be merely a sampling of ICF.

Furthermore, the rates of endocytosis and exocytosis must be kept in balance to maintain a constant membrane surface area and cell volume. More than 100% of the plasma membrane may be used in an hour to wrap internalized vesicles in a cell actively involved in endocytosis, necessitating rapid replacement of surface membrane by exocytosis. On the other hand, when a secretory cell is stimulated to secrete, it may temporarily insert up to thirty times its surface membrane through exocytosis. This added membrane must be specifically retrieved by an equivalent level of endocytotic activity. Thus, through exocytosis and endocytosis, portions of the membrane are constantly being restored, retrieved, and generally recycled.

The controlling mechanisms guiding endocytosis, exocytosis, and the associated membrane recycling have not been thoroughly elucidated. Both processes are known to require energy and are considered to be active mechanisms. In some instances of endocytosis, receptor sites on the surface membrane recognize and bind specific molecules in the environment of the cell. This combination triggers a localized invagination process that selectively traps the bound material. In fact, one of the primary ways in which antibodies help defend us against invading organisms is by attaching to the bacterial surface to form a coat that can be recognized by specific receptor sites in the plasma membrane of the phagocytic white blood cells. Such "marked" bacteria are quickly engulfed and destroyed.

Exocytosis of secretory products also appears to be a triggered event. In most instances, a specific nervous or hormonal stimulus brings about opening of Ca^{++} channels in the plasma membrane of the secretory cell. As Ca^{++} enters the cell down its concentration gradient, the resultant rise in cytosolic Ca^{++} triggers fusion of the exocytotic vesicle with the plasma membrane and the subsequent release of its secretory product. For example, nerve cells controlling skeletal muscles release a specific chemical via Ca^{++}-induced exocytosis in response to a nerve impulse. This chemical, in turn, initiates a series of events to bring about contraction of the muscle.

This completes our discussion of membrane transport. Table 3-2 summarizes the pathways by which materials can pass between the ECF and ICF.

Membrane Potential

Membrane potential refers to a separation of opposite charges across the plasma membrane.

The unequal distribution of a few key ions between the ICF and ECF and their selective movement through the plasma membrane are responsible for the electrical properties of the membrane. All plasma membranes have a membrane potential or are polarized electrically. **Membrane potential** refers to a separation of charges across the membrane, or to a difference in the relative number of cations and anions in the ICF and ECF. Recall that opposite charges tend to attract each other and like charges tend to repel each other. Work must be performed (energy expended) to separate opposite charges after they have come together. Conversely, when oppositely charged particles have been separated, the electrical force of attraction between them can be harnessed to perform work when the charges are permitted to come together again. This is the basic principle underlying electrically powered devices. Because separated charges have the "potential" to do work, a separation of charges across the membrane is referred to as a *membrane potential.*

Since the concept of potential is fundamental to understanding nerve and muscle physiology, it is important to understand what is meant by the term. The membrane in Figure 3-24a is electrically neutral. The same number of positive (+) and negative (−) charges are on each side of the membrane, so no membrane potential exists. In Figure 3-24b, some of the + charges from the right side have moved to the left. Excess + charges are now on the left, leaving behind an excess of − charges unbalanced on the right. In other words, there is a separation of charges across the membrane, or there is a difference in the relative number of + and − charges between the two sides (i.e., a membrane potential exists). The attractive force between these separated charges will cause them to accumulate in a thin layer along the outer and inner surfaces of the plasma membrane (Fig. 3-24c). These separated charges represent only a small fraction of the total number of charged particles (ions) present in the ICF and ECF. The vast majority of the fluid inside and outside the cells is electrically neutral (Fig. 3-24d). The electrically balanced ions can be ignored, because they do not contribute to membrane potential. Thus, an almost insignificant fraction of the

Table 3–2 Characteristics of the Methods of Membrane Transport

Methods of Transport	Substances Involved	Energy Requirements and Force Producing Movement	Limit to Transport
Diffusion			
Through lipid bilayer	Large or small nonpolar molecules (e.g., O_2, CO_2, fatty acids)	Passive; concentration gradient (from high to low concentration)	Continues until the gradient is abolished (state of equilibrium with no net diffusion)
Through protein channel	Specific small ions (e.g., Na^+, K^+, Ca^{++}, Cl^-)	Passive; electrochemical gradient (from high to low concentration and attraction of ion to area of opposite charge)	Continues until there is no net movement and a state of equilibrium is established
Special case of osmosis	Water only	Passive; water concentration gradient (water moves to area of lower water concentration—i.e., higher solute concentration)	Continues until concentration difference is abolished or until stopped by an opposing hydrostatic pressure or until cell is destroyed
Carrier-Mediated Transport			
Facilitated diffusion	Specific polar molecules for which a carrier is available (e.g., glucose)	Passive; concentration gradient (from high to low concentration)	Displays a transport maximum (T_m); carrier can become saturated
Primary active transport	Specific ions for which carriers are available (e.g., Na^+, K^+, H^+)	Active; moves against concentration gradient (from low to high concentration); requires ATP	Displays a transport maximum; carrier can become saturated
Secondary active transport	Specific polar molecules and ions for which carriers are available (e.g., glucose, amino acids, some ions, including Ca^{++})	Active; moves against concentration gradient (from low to high concentration); driven directly by ion gradient (usually Na^+) established by ATP-requiring primary pump	Displays a transport maximum; carrier can become saturated
Vesicular Transport			
Endocytosis			
Pinocytosis	Small volume of ECF fluid, perhaps with specific bound molecules (e.g., protein); also important in membrane recycling	ATP required	Control poorly understood; in instances of transport of a specific protein, necessitates binding to receptor site on membrane surface
Phagocytosis	Multimolecular particles (e.g., bacteria and cellular debris)	ATP required	Necessitates binding to specific receptor site on membrane surface
Exocytosis	Secretory products (e.g., hormones and enzymes) as well as large molecules passing through cell intact; also important in membrane recycling	ATP required; increase in cytosolic Ca^{++} induces fusion of vesicle with plasma membrane	Secretion triggered by specific neural or hormonal stimuli; other controls involved in transcellular traffic and membrane recycling not known

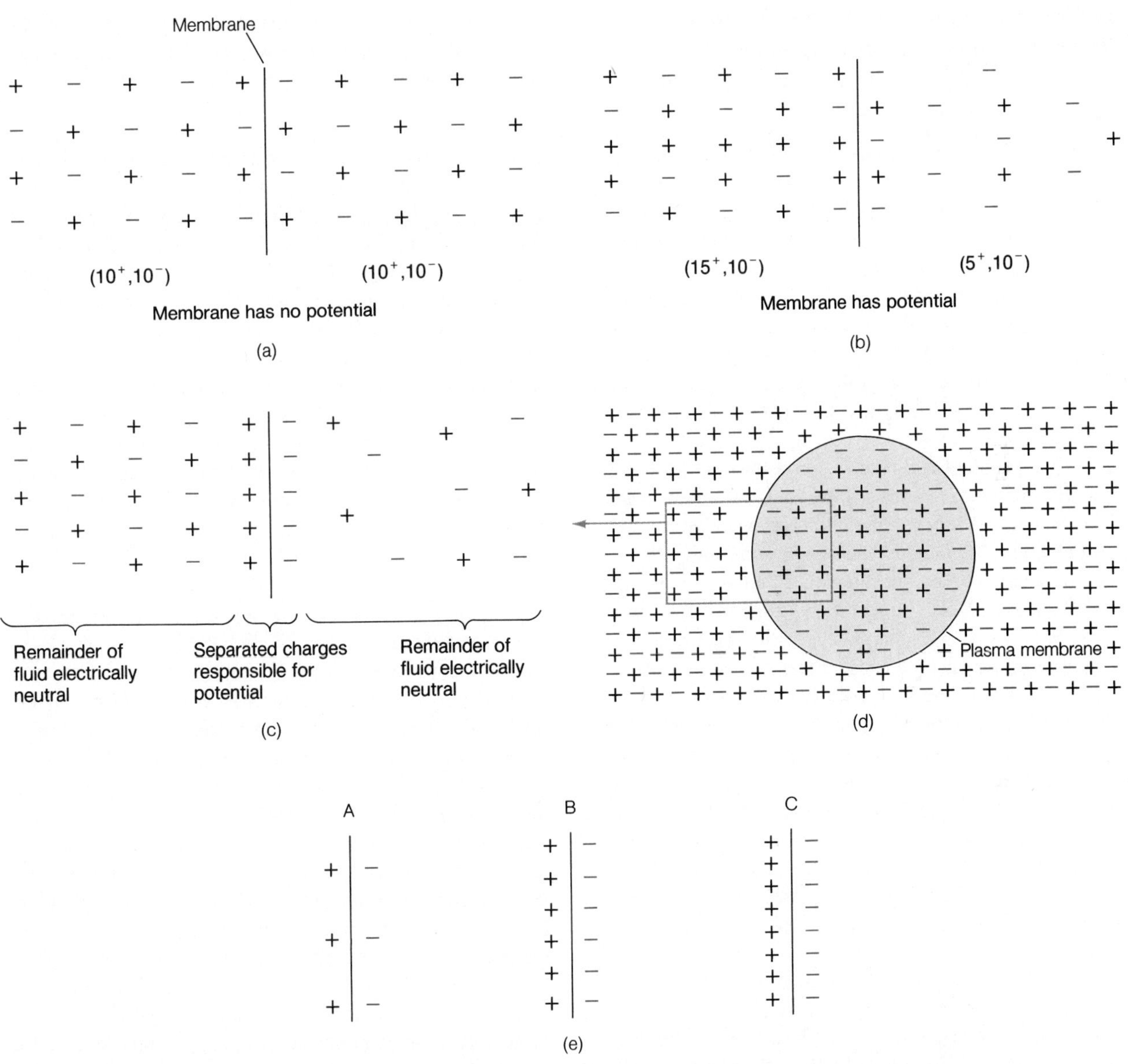

Figure 3-24 Determination of Membrane Potential by Unequal Distribution of Positive and Negative Charges across the Membrane *(a) When the positive and negative charges are equally balanced on each side of the membrane, no membrane potential exists. (b) When there is a separation of charges across the membrane, membrane potential exists. (c) The unbalanced charges responsible for the potential accumulate in a thin layer along opposite surfaces of the membrane. (d) The vast majority of fluid in the ECF and ICF is electrically neutral. The unbalanced charges accumulate in a thin layer along the plasma membrane. (e) Membrane B has more potential than membrane A and less potential than membrane C.*

total number of charged particles present in the body fluids is responsible for the membrane potential.

Potential is measured in units of volts (the same as the voltage in electrical devices), but in the body, because the potential is relatively low, the unit used is **millivolts** (mV) (1 mv = 1/1,000 volt). The magnitude of the potential depends on the degree of separation of opposite charges; the greater the number of charges separated, the larger the potential. Therefore, in Figure 3-24e, membrane B has more potential than A and less potential than C.

Membrane potential is primarily due to differences in distribution and membrane permeability of sodium, potassium, and large intracellular anions.

All living cells have a membrane potential characterized by a slight excess of positive charges outside and a correspondingly slight excess of negative charges on the inside. In the body, electric charges are carried by ions. The ions primarily responsible for the generation of membrane potential are Na^+, K^+, and A^-. The latter refers to the large, negatively charged intracellular proteins. Other ions (calcium, magnesium, bicarbonate, and phosphate, to name a few) do not make a direct contribution to the electrical properties of the plasma membrane in most cells, even though they play other important roles in the body.

The concentrations and relative permeabilities of the ions critical to membrane electrical activity are compared in Table 3–3. Note that *Na^+ is in greater concentration in the extracellular fluid* and *K^+ is in much higher concentration in the intracellular fluid.* These concentration differences are maintained by the Na^+-K^+ pump at the expense of energy. Because the plasma membrane is virtually impermeable to A^-, these large, negatively charged proteins are found *only inside* the cells. After they have been synthesized from amino acids transported into the cell, they remain trapped within the cell. In addition to the active carrier mechanism, Na^+ and K^+ can passively cross the membrane through protein channels specific for them. It is usually much easier for K^+ than Na^+ to get through the membrane because typically more K^+ channels than Na^+ channels are open. In a nerve cell at rest (i.e., when it is not conducting a nerve impulse), the membrane is about fifty to seventy-five times more permeable to K^+ than to Na^+.

Armed with a knowledge of the relative concentrations and permeabilities of these ions, we can now analyze the forces acting across the plasma membrane. This analysis will be broken down as follows: first, we will consider the direct contribution of the Na^+-K^+ pump on membrane potential; second, the effect the movement of K^+ alone would have on membrane potential; third, the effect of Na^+ alone; and finally the situation that exists in the cells when both K^+ and Na^+ effects are taking place concurrently.

EFFECT OF THE SODIUM-POTASSIUM PUMP ON MEMBRANE POTENTIAL. About 20% of the membrane potential is directly generated by the Na^+-K^+ pump. Recall that this active transport mechanism pumps three Na^+ out for every two K^+ it transports in. Because Na^+ and K^+ are both positive ions, this unequal transport generates a membrane potential, with the outside becoming relatively more positive than the inside as more positive ions are transported out than in. However, most of the membrane potential, the remaining 80%, is caused by the passive diffusion of K^+ and Na^+ down concentration gradients. Thus, most of the Na^+-K^+ pump's role in producing membrane potential is indirect via its critical contribution in maintaining the concentration gradients directly responsible for the ion movements that generate most of the potential.

EFFECT OF THE MOVEMENT OF POTASSIUM ALONE ON MEMBRANE POTENTIAL. Given a hypothetical situation of: (1) the concentrations that exist for K^+ and A^- across the plasma membrane, (2) free permeability of the membrane to K^+ only, and (3) no potential as yet present, the concentration gradient for K^+ would tend to move this ion out of the cell (Fig. 3–25). Because the membrane is permeable to K^+, this ion would readily pass through. As potassium ions moved to the outside, they would carry their positive charge with them, so more positive charges would be on the outside whereas negative charges in the form of A^- would be left behind on the inside, similar to the situation in Figure 3–24b. (Remember that the large protein anions cannot diffuse out, despite a tremendous concentration gradient.) A membrane potential would now exist. Because an electrical gradient also would be present, K^+, being a positively charged ion, would be attracted toward the negatively charged interior and repelled by the positively charged exterior. Thus, two opposing forces would now be acting upon K^+: the concentration gradient tending to

Table 3–3 Concentration and Permeability of Ions Responsible for Membrane Potential in a Resting Nerve Cell

	Concentration (millimoles/liter)		
Ion	*Extracellular*	*Intracellular*	**Relative Permeability**
Na^+	150	15	1
K^+	5	150	50–75
A^-	0	65	0

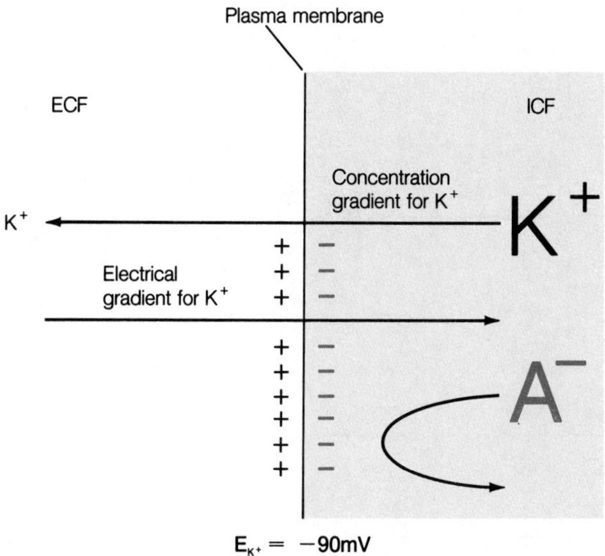

Figure 3–25 Equilibrium Potential for K^+ *At the equilibrium potential for K^+, the outward concentration gradient is exactly counterbalanced by the inward electrical gradient. The membrane potential at this point is $-90mV$.*

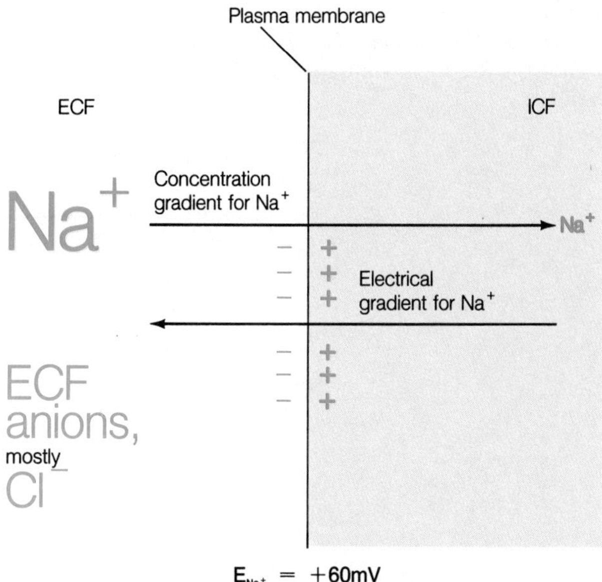

Figure 3–26 Equilibrium Potential for Na^+ *At the equilibrium potential for Na^+, the inward concentration gradient is exactly counterbalanced by the outward electrical gradient. The membrane potential at this point is $+60mV$.*

move K^+ out of the cell; and the electrical gradient tending to move these same ions into the cell.

Initially, the concentration gradient would be stronger than the electrical gradient, so net diffusion of K^+ out of the cell would continue and the membrane potential would increase. As more and more K^+ moved out down its concentration gradient, however, the opposing electrical gradient would also become greater as the outside became increasingly more positive and the inside more negative. Net outward diffusion would gradually be reduced as the strength of the electrical gradient approached that of the concentration gradient. Finally, when these two forces exactly balanced each other, no further net movement of K^+ would occur. The potential that would exist at this equilibrium is known as the **equilibrium potential for K^+ (E_{K^+})**. At this point a large concentration gradient for K^+ would still exist, but no more K^+ would move out down this concentration gradient because of the exactly equal opposing electrical gradient.

The membrane potential at E_{K^+} is -90 mV. This is not really a negative potential. By convention, *the sign always designates the polarity of the excess charge on the inside of the membrane.* A membrane potential of -90 mV means that the potential is of a magnitude of 90 mV, with the inside being negative relative to the outside. A potential of $+90$ mV would have the same strength, but in this case the inside would be more positive than the outside.

EFFECT OF THE MOVEMENT OF SODIUM ALONE ON MEMBRANE POTENTIAL. A similar hypothetical situation could be developed for Na^+ alone (Fig. 3–26). The concentration gradient for Na^+ would move this ion into the cell, producing a build-up of positive charges on the interior of the membrane and leaving negative charges unbalanced outside [these being primarily in the form of chloride (Cl^-); Na^+ and Cl^- are the predominant ECF ions]. Net diffusion inward would continue until equilibrium was established by the development of an opposing electrical gradient that exactly counterbalanced the concentration gradient. At this point, given the concentrations that exist for Na^+, the **Na^+ equilibrium potential (E_{Na^+})** would be $+60$ mV. In this case the inside of the cell would be positive, in contrast to the equilibrium potential for K^+. The magnitude of E_{Na^+} is somewhat less than for E_{K^+} (60 mV compared to 90 mV) because the concentration gradient for Na^+ is not as large (Table 3–3); thus the opposing electrical gradient (membrane potential) is not as great at equilibrium.

CONCURRENT POTASSIUM AND SODIUM EFFECTS ON MEMBRANE POTENTIAL. In the cell, the effects of both K^+ and Na^+ must be taken into account. Because the membrane at rest is fifty to seventy-five times more permeable to K^+ than to Na^+, K^+ passes through more readily than Na^+ so that K^+ influences the resting membrane potential to a much greater

Figure 3–27 Effect of Concurrent K^+ and Na^+ Movement on Establishing the Resting Membrane Potential

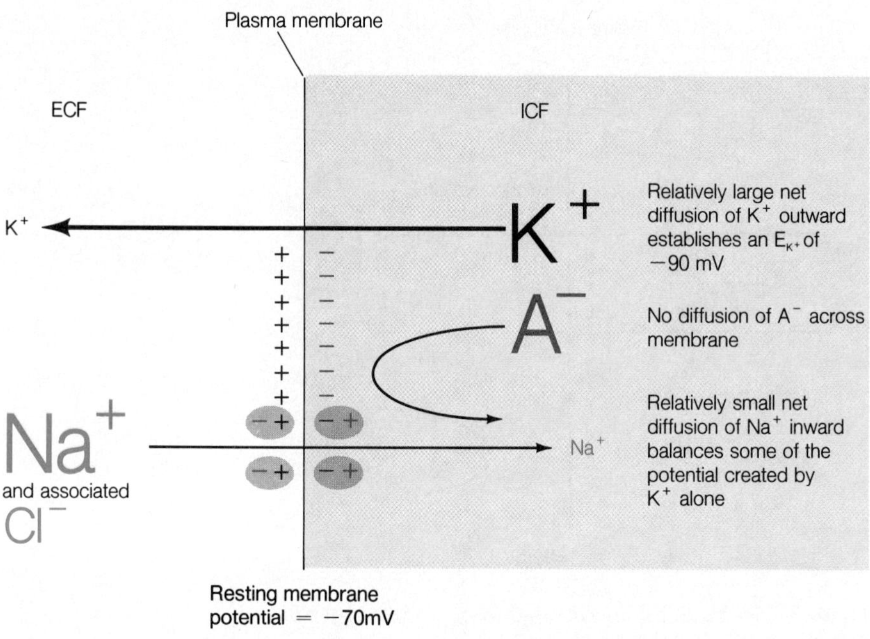

Plasma membrane
ECF
ICF
K+
K+
Relatively large net diffusion of K+ outward establishes an E_K+ of −90 mV
A−
No diffusion of A− across membrane
Na+
and associated
Cl−
Na+
Relatively small net diffusion of Na+ inward balances some of the potential created by K+ alone
Resting membrane potential = −70mV

extent than does Na^+. Recall that K^+ acting alone would establish an equilibrium potential of -90 mV. The membrane is somewhat permeable to Na^+, however, so some Na^+ enters the cell in a limited attempt to reach its equilibrium potential. This Na^+ influx neutralizes or cancels some of the potential produced by K^+ alone.

To facilitate an understanding of this concept, assume that each separated pair of charges in Figure 3–27 represents 10 mV of potential. (This is not technically correct because in reality, many separated charges must be present to account for a potential of 10mV.) In this simplified example, nine separated pluses and minuses, with the minuses on the inside, would represent the E_{K^+} of -90 mV. Superimposing the slight influence of Na^+ on this K^+-dominated membrane, assume that two sodium ions enter the cell down the Na^+ concentration and electrical gradients. (Because more Na^+ is outside the cell and the inside is negative, both gradients favor the inward movement of Na^+.) The inward movement of these two positively charged sodium ions neutralizes some of the potential established by K^+, so now only seven pairs of charges are separated, and the potential is -70 mV. This is the **resting membrane potential** of a typical nerve cell. The resting potential is much closer to E_{K^+} than to E_{Na^+} because of the greater permeability of the membrane to K^+, but it is slightly less than E_{K^+} (-70 mV is a lower potential than -90 mV) because of the weak influence of Na^+.

At resting potential, neither K^+ nor Na^+ are at equilibrium. A potential of -70 mV does not exactly counterbalance the concentration gradient for K^+; it takes a potential of -90 mV to do that. Thus, there is a continual tendency for K^+ to ***leak*** out through its channels. In the case of Na^+, the concentration and electrical gradients do not even oppose each other; they both favor the inward movement of Na^+. Therefore, Na^+ continually leaks inward down its electrochemical gradient, but only slowly because of its low permeability.

Since such leaking goes on all the time, why doesn't the intracellular concentration of K^+ continue to fall and the concentration of Na^+ inside the cell progressively increase? This does not happen because of the Na^+-K^+ pump. This active transport mechanism counterbalances the rate of leakage (Fig. 3–28). At resting potential, the pump transports back into the cell essentially the same number of potassium ions that have leaked out and simultaneously transports to the outside the sodium ions that have leaked in. Through this balance between leak and pump, the concentration gradients for K^+ and Na^+ remain constant across the membrane. Thus, not only is the Na^+-K^+ pump initially responsible for the Na^+ and K^+ concentration differences across the membrane, but it also maintains these differences.

As just discussed, it is the presence of these concentration gradients, together with the difference in permeability of the membrane to these ions, that accounts for the resting membrane potential. In this resting state, the potential remains constant. There is no net movement of any ions. All passive forces

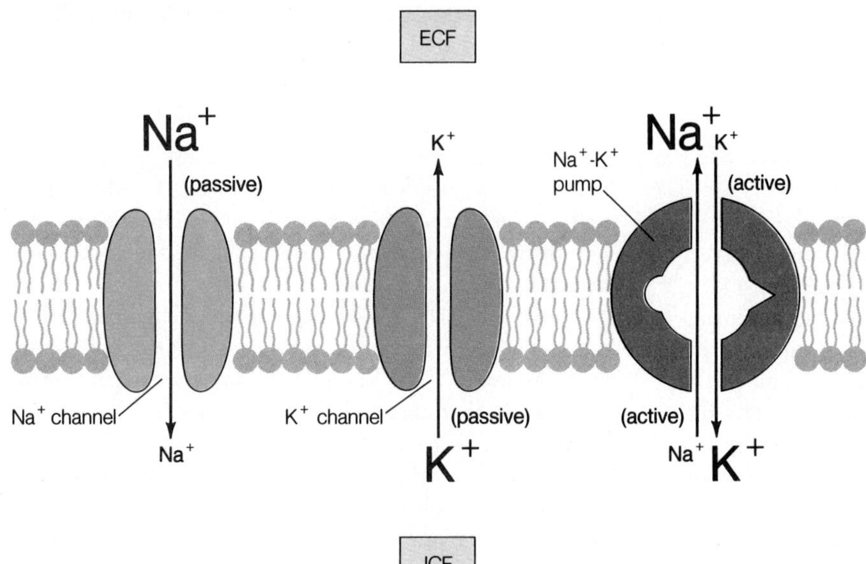

Figure 3–28 Counterbalance between Passive Na^+ and K^+ Leaks and the Active Na^+-K^+ Pump *At resting membrane potential, the passive leaks of Na^+ and K^+ down their electrochemical gradients are exactly counterbalanced by the active Na^+-K^+ pump, so that there is no net movement of Na^+ and K^+ and the membrane potential remains constant.*

are exactly balanced by active forces. A steady state exists, even though there is still a strong concentration gradient for both K^+ and Na^+ in opposite directions, and there is a slight excess of positive charges in the ECF accompanied by a corresponding slight excess of negative charges in the ICF (enough to account for a potential of the magnitude of 70 mV). At this point, although movement across the membrane is taking place via passive leaks and active pumping, the exchange of charges between the ICF and ECF are exactly balanced, with the potential that has been established by these forces remaining constant.

There is one other ion, Cl^-, present in high concentration in the ECF that we have largely ignored thus far. Chloride is the principal ECF anion. Its equilibrium potential is -70 mV, exactly the same as the resting membrane potential. Movement alone of negatively charged Cl^- into the cell down its concentration gradient would produce an opposing electrical gradient, with the inside negative compared to the outside. When physiologists were first examining the ionic effects that could account for the membrane potential, it was tempting to think that Cl^- movements and establishment of the Cl^- equilibrium potential could be solely responsible for producing the identical resting membrane potential. Actually, the reverse is the case. The membrane potential is responsible for driving the distribution of Cl^- across the membrane.

Most cells have high permeability to Cl^- but have no active transport mechanisms for Cl^-. With no active forces acting on it, Cl^- passively distributes itself to achieve an individual state of equilibrium. In this case, the established electrical gradient (the membrane potential, which is negative on the inside) drives the negative Cl^- out of the cell until an exactly counterbalanced opposing concentration gradient is established, with more Cl^- outside than inside the cells. Thus, the concentration difference for Cl^- between the ECF and ICF is passively brought about by the presence of the membrane potential, rather than being maintained by an active pump, as is the case for K^+ and Na^+. Therefore, in most cells Cl^- does not influence membrane potential; instead, membrane potential passively influences Cl^- distribution. (Some specialized cells have an active Cl^- pump, with subsesquent movement of Cl^- accounting for part of the potential.)

Nerve and muscle cells have developed a specialized use for membrane potential. They are able to rapidly and transiently alter their membrane permeabilities to the involved ions in response to appropriate stimulation, thereby bringing about fluctuations in membrane potential. The rapid fluctuations in potential are responsible for producing nerve impulses in nerve cells and for triggering contraction in muscle cells. These activities will be the focus of the next five chapters. Even though all cells display a membrane potential, its significance in other cells is uncertain, although the potential of some secretory cells may somehow be linked to their level of secretory activity.

Chapter in Perspective

All cells are enveloped by a plasma membrane composed of a thin, flexible lipid bilayer that separates the contents of the cell from its surrounding environment. To carry on life-sustaining and specialized cellular activities, each cell must exchange materials with the internal fluid environment that surrounds it. This internal environment must be homeostatically maintained in order to support the cells' ongoing needs. All exchanges between a cell and the internal fluid environment are accomplished across the cell's plasma membrane. This discriminating barrier is penetrated and studded by specific proteins, some of which enable selective passage of materials through the membrane. Variability between the types and numbers of protein channels and carriers that different cells possess allows them to be differentially selective in what is permitted to enter and leave. The transport selectivity of each cell type, in turn, is related to the specialization of the cell.

Other proteins in the plasma membrane serve as receptor sites for interaction with specific chemical messengers in the cell's environment. Control of specialized cellular activities, which is critical to the maintenance of homeostasis, depends in large part on the ability of the plasma membrane to respond by means of its receptor sites to these regulatory messengers.

Projecting from the outer surface of the membrane proteins and lipids are chains of complex sugars, which apparently are the "trademark" of a particular cell. These surface carbohydrates enable the cell to recognize others of its own kind in tissue formation. Cells are organized into tissues and organs that are bound together within the body by the extracellular matrix, a biological glue secreted by connective-tissue cells. This matrix further provides tensile strength and elasticity to tissues.

Cells are characterized by the presence of a membrane potential. They have a slight excess of negative charges lined up along the inside of the plasma membrane separated from a slight excess of positive charges on the outside. This separation of charges is generated and maintained by a balanced interplay of active and passive forces, as well as by differential permeabilities involving K^+, Na^+, and the intracellular protein anions. The importance of this potential in most cells is not known. However, the specialization of nerve and muscle cells depends on membrane potential and the ability of these cells to alter their potential on appropriate stimulation.

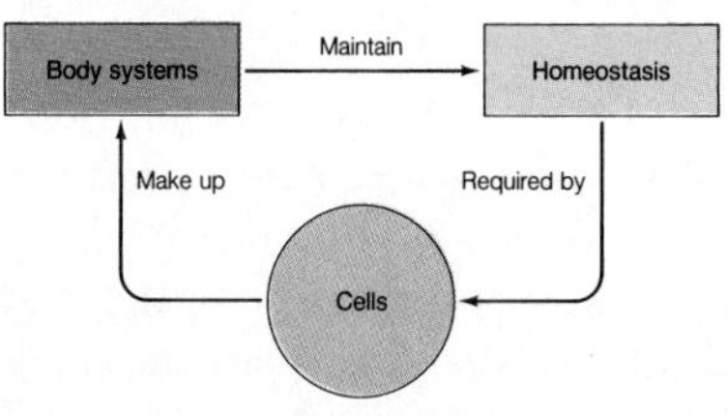

See inside front cover for an expanded version of this model.

Review Exercises

1. Describe the fluid mosaic model of membrane structure.
2. List the functions of each of the three major membrane components.
3. Compare the cyclic-AMP and calcium second-messenger pathways.
4. What are the functions of the three major types of protein fibers in the extracellular matrix?
5. Describe the structure and function of desmosomes, tight junctions, and gap junctions.
6. What are the two properties of a particle that influence whether or not it can permeate the plasma membrane?
7. List and describe the methods of membrane transport. Indicate what types of substances are transported by each method, and state whether each is a passive or active means of transport.
8. As stated by Fick's law of diffusion, what factors influence the rate of net diffusion across a membrane?
9. State three important roles of the Na^+-K^+ pump.
10. Describe the contribution of each of the following to the establishment and maintenance of membrane potential: (a) the Na^+-K^+ pump; (b) passive movement of K^+ across the membrane; (c) passive movement of Na^+ across the membrane; (d) presence of large intracel-

lular anions; (e) passive movement of Cl^- across the membrane.

11. **A point to ponder:** Which of the following methods of transport is being utilized for transfer of the substance into the cell in the accompanying graph?
    a. diffusion down a concentration gradient
    b. osmosis
    c. facilitated diffusion
    d. active transport
    e. vesicular transport
    f. It is impossible to tell with the information provided.

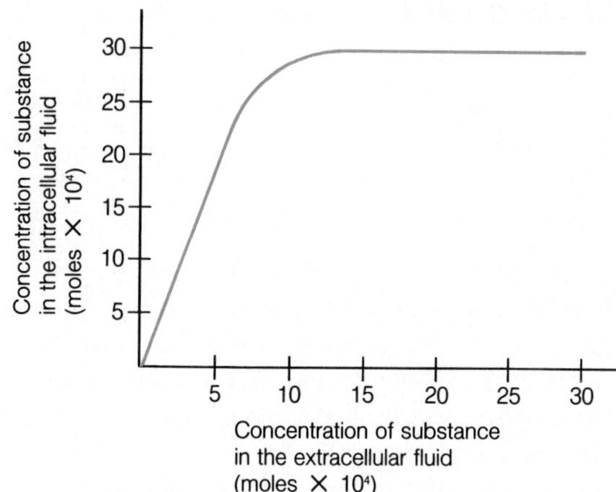

C H A P T E R 4

Neuronal Physiology

Chapter Contents at a Glance

Introduction
Electrical Signals: Graded Potentials and Action Potentials
Graded potentials—short-distance signals
Action potentials—long-distance signals
Permeability changes; ionic fluxes
Potential changes
Propagation of action potentials; role of myelin
Refractory period
All-or-none law
Synapses and Neuronal Integration
Chemical transmission of electrical signal
Excitatory and inhibitory synapses
Neurotransmitters
Grand postsynaptic potential; summation
Action potential initiation at axon hillock
Modification of synaptic effectiveness
Convergence; divergence

INTRODUCTION *What interest did Benjamin Franklin and Luigi Galvani have in common? Electricity. Franklin was the first to prove that lightning was electricity with his famous "kite in the thunderstorm" experiment, whereas Galvani was the first to record experiments that gave evidence of animal electricity.*

Galvani, an Italian anatomist, pioneered in electrophysiology, the branch of physiology relating to electricity. Actually, he stumbled on the concept by accident. Galvani was fascinated with electricity and had acquired a device for generating it. Being an anatomist, he frequently had frog legs in his laboratory. While playing around with his electricity machine one day, he accidentally discovered that applying an electrical current to a frog leg would cause it to contract. Galvani's interpretation was that "animal electricity," similar to the kind his machine generated, must be responsible for inducing muscle contraction in a living animal. His further electrical studies on frogs led to his being ridiculed as the "frog's dancing master," but, more importantly, they provided the foundation for our present-day knowledge of the role of electricity in generating nerve impulses and causing muscle contraction.

Electrical Signals: Graded Potentials and Action Potentials

Nerve and muscle are excitable tissues.

All cells of the body possess a membrane potential related to the nonuniform distribution and differential permeability to Na^+, K^+, and large intracellular anions. Two types of cells, *nerve cells* and *muscle cells,* have developed a specialized use for this membrane potential. Specifically, these cells are able to undergo transient, rapid changes in their membrane potentials. These fluctuations in potential serve as electrical signals, of which there are two basic forms: (1) graded potentials, which serve as short-distance signals, and (2) action potentials, which signal over long distances.

Nerve and muscle are considered to be **excitable tissues** because they are capable of producing electrical signals when excited. The constant membrane potential that exists when an excitable tissue cell is not displaying rapid changes in potential is referred to as the **resting membrane potential** (although the membrane is far from "resting" because of the balanced leak-pump activity constantly going on). Let us examine these types of signals in more detail.

Graded potentials die out over short distances.

Graded potentials are local changes in membrane potential that occur in varying grades or degrees of magnitude or strength. The magnitude of a graded potential is related to the magnitude of the triggering event that brings about the potential change; that is, *the stronger the triggering event, the larger the graded potential.* Depending on the location or function of the graded potential, a triggering event might be: (1) a stimulus, such as light stimulating specialized nerve cells in the eye; (2) an interaction of a chemical messenger with a surface receptor on a nerve or muscle cell membrane; or (3) a spontaneous change of potential caused by imbalances in the leak-pump cycle.

When a graded potential occurs locally in a nerve or muscle cell membrane, a different potential exists in this area than in the remainder of the membrane that is still at resting potential. Because opposite charges attract each other, current (movement of charges) passively flows between the involved area and the adjacent resting regions on both the inside and outside of the membrane. For example, assume that a triggering event has temporarily reversed the charges in a particular region of an excitable tissue membrane, a region now called an *active area* (Fig. 4–1). Current will flow on both sides of the membrane between the active and neighboring *inactive* (still at resting potential) *areas.* By convention, the direction of current flow is always designated by the movement of the positive charges, but keep in mind that negative charges are moving simultaneously in the opposite direction.

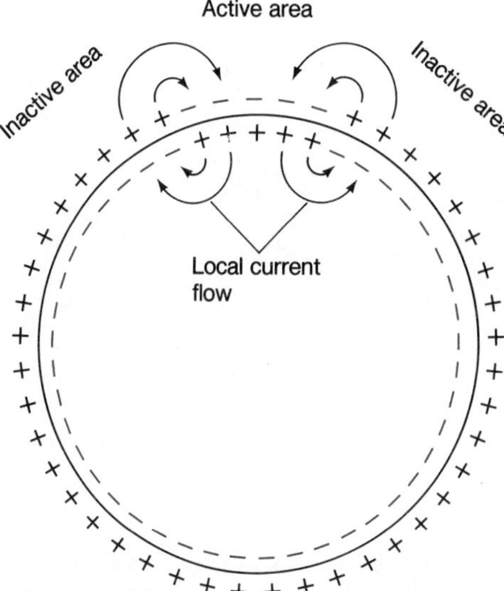

Figure 4–1 Local Current Flow between Active and Adjacent Inactive Areas of a Membrane

There does not have to be an actual reversal of charges in the active area for local current flow to occur. A reduction in potential in the active area compared with that in the remainder of the membrane, which usually is the case with graded potentials, will also initiate current flow between the active and neighboring inactive areas. Current flow is easier to visualize, however, if we assume a reversal of charges. This local flow of current alters the potential in the previously inactive area, so this area's potential now differs from that of the region immediately next to it on the other side, inducing further current flow at this next site, and so on.

This passive current flow is similar to the means by which current is carried through electrical wires. We know from experience that current leaks out of an electrical wire unless the wire is covered with an insulating material such as rubber to prevent this leakage. The regions on excitable tissues at which graded potentials take place are not wrapped in insulating material. Therefore, current does leak through the membrane (via movement of charged particles through membrane channels). As a result of this leaking, the magnitude of the graded potential is *decremental;* that is, it continues to decrease the farther it moves away from the initial active area (Fig. 4–2). In fact, these local currents die out within a few millimeters from the initial site of potential change and consequently can function only as signals for very short distances. This is similar to two people speaking directly to each other. Over short dis-

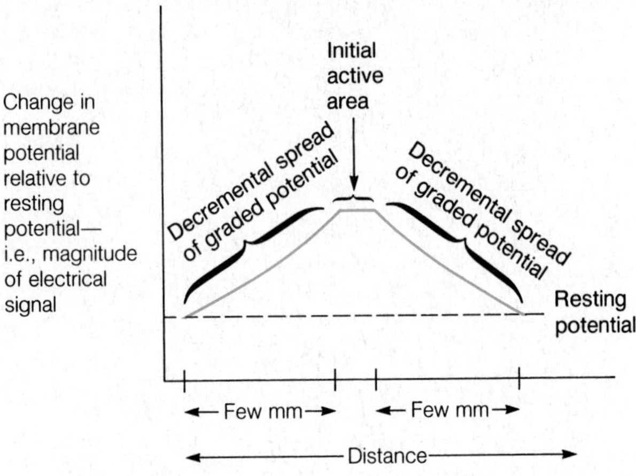

Figure 4–2 Decremental Spread of Graded Potentials

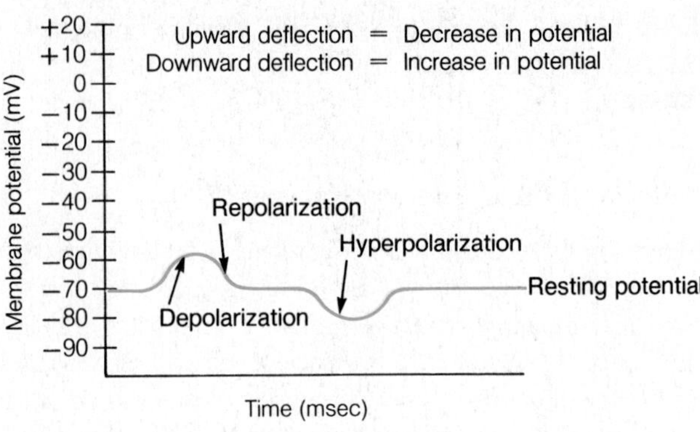

Figure 4–3 Types of Changes in Membrane Potential

tances the voices are clearly understood, but as the two people move further apart, the sounds progressively diminish until further communication by this means becomes impossible. Just as there are methods, such as telephones and telegraphs, for long-distance communication between people, action potentials provide a means of transmitting an electrical signal long distances through the body. The limited signaling distance of graded potentials does not mean that they are of no value, however. The following graded potentials are critically important to the body: (1) postsynaptic potentials; (2) receptor potentials; (3) end-plate potentials; and (4) pacemaker potentials. These are unfamiliar terms to you now, but you will become well acquainted with them as we continue discussing nerve and muscle in the next several chapters.

Action potentials are brief reversals of membrane potential brought about by rapid changes in membrane permeability.

Because the passive current flow accompanying a graded potential fades very quickly as it moves away from its site of initiation, there must be another mechanism by which an electrical signal may be transmitted over long distances, with the strength of the signal being maintained as it travels away from its site of initiation. When appropriately triggered, nerve and muscle cell membranes undergo brief, rapid reversals of membrane potential known as **action potentials,** which are able to spread throughout the membrane in nondecremental fashion. To understand the processes that occur during an action potential, it is necessary to be familiar with the following terms (Fig. 4–3):

1. **polarization** — the membrane has potential; there is a separation of opposite charges.
2. **depolarization** — the membrane potential is reduced from resting potential; it has decreased or moved toward O mV; fewer charges are separated than at resting potential.
3. **hyperpolarization** — the potential is greater than resting potential; it has increased or become even more negative; more charges are separated than at resting potential.
4. **repolarization** — return to resting potential after having been depolarized.

A possibly confusing point is worth clarifying. On the device used for recording rapid changes in potential, a ***decrease*** in potential is represented as an ***upward*** deflection whereas an ***increase*** in potential is represented as a ***downward*** deflection.

With these terms in mind, we will consider the changes that occur in the membrane potential during an action potential (Fig. 4–4). Recall that the resting potential of a typical nerve cell is −70 mV. To initiate an action potential, a triggering event causes the membrane to depolarize. Depolarization proceeds slowly at first until it reaches a critical level known as **threshold potential,** typically between −50 to −55 mV. At threshold potential, an explosive depolarization takes place. A recording of the potential at this time shows a sharp upward deflection to +30 mV as the potential rapidly decreases toward 0 mV, then reverses itself so that the inside of the cell becomes positive compared to the outside. Just as rapidly, the potential drops back to resting potential as the membrane repolarizes. Often the forces responsible for driving the membrane back to resting potential push it too far, causing a transient hyperpolarization (the **after hyperpolarization**), during which the inside of the membrane becomes even more negative than normal (for example −80 mV). The entire

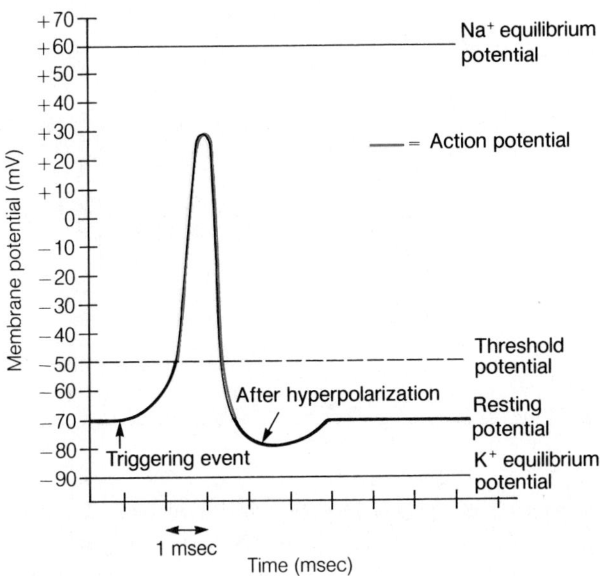

Figure 4–4 Changes in Membrane Potential during an Action Potential

rapid change in potential from threshold to peak reversal and then back to resting is called the ***action potential.*** In a nerve cell, an action potential lasts for only 1 msec (0.001 sec). It lasts longer in muscle, with the duration varying depending on the muscle type. Often an action potential is referred to as a **spike** because of its spike-like recorded appearance. Alternatively, when an excitable membrane is triggered to undergo an action potential, it is said to **fire.** Thus the terms *action potential, spike,* and *firing* all refer to the same phenomenon of rapid potential reversal.

How is the resting potential, which is usually maintained at a constant level by the counterbalancing mechanisms described in the last chapter, thrown out of balance to such an extent as to produce an action potential? Recall that K^+ makes the greatest contribution to the resting potential because the membrane when at rest is considerably more permeable to K^+ than to Na^+. During an action potential, marked changes in membrane permeability to Na^+ and K^+ take place to permit rapid fluxes of these ions down their electrochemical gradients. These ion movements carry the current responsible for the potential changes that occur during an action potential.

The membrane channels behave as if they have "gates" that can be open or closed, depending on the circumstances. The channels are formed by proteins that span the thickness of the membrane. Changes in conformation (shape) of these proteins are believed to alternatively block the channel (gates closed) or permit passage through it (gates open). There are two kinds of channels, depending on the factor that induces the change in channel conformation: (1) **voltage-gated channels,** which open or close in response to changes in membrane potential; and (2) **chemical messenger-gated channels,** which change conformation in response to binding of a specific messenger with a membrane receptor that is in close association with the channel. Voltage-gated channels are the ones involved in action potentials. (The roles of receptor-activated channels will be described later.)

Because proteins that compose the voltage-gated channels contain a number of charged groups, the electric field (potential) surrounding them can exert a distorting force on the channel structure; that is, various charged portions of the channel protein are electrically attracted or repelled by an unequal distribution of cations and anions in the fluids surrounding the membrane. Unlike the majority of membrane proteins that remain stable in spite of fluctuations in membrane potential, the voltage-gated channel proteins are especially sensitive to voltage changes. Small distortions in shape induced by potential changes can cause them to flip to another conformation. There are apparently at least three different channel conformations (Fig. 4–5): (1) gates closed but capable of opening; (2) gates open (activated); and (3) gates closed and not capable of opening (inactivated).

At resting potential (−70 mV), many K^+ channels are open but most of the Na^+ channels are closed; thus the resting membrane is fifty to seventy-five times more permeable to K^+. When a membrane starts to depolarize toward threshold as a result of a triggering event, some of its voltage-gated Na^+ channels open. Since both the concentration and electrical gradients for Na^+ favor its movement into the cell, Na^+ starts to move in, carrying its positive charge with it. This depolarizes the membrane further, thereby opening more voltage-gated Na^+ channels and allowing more Na^+ to enter. As a result, still further depolarization occurs, which opens more Na^+ channels, and so on in a positive feedback cycle (Fig. 4–6).

At threshold potential, there is an explosive increase in Na^+ permeability (**P Na^+**) as the membrane becomes six hundred times more permeable to Na^+ than to K^+. Each individual channel is either closed or open and cannot be partially open. However, the delicately poised gating mechanisms of the various Na^+ channels are jolted open by slightly different voltage changes. During the early depolarizing phase, Na^+ channels are gradually opened as the potential progressively decreases. By threshold the gates of all the Na^+ channels have swung open so that Na^+ permeability now dominates the membrane, in contrast to the K^+ domination of the resting potential. Thus, at threshold Na^+ rushes into the cell, rapidly eliminating the internal negativity and even making the inside of the cell more positive than the outside. The potential reaches

Figure 4–5 Conformations of Voltage-Gated Na$^+$ Channel

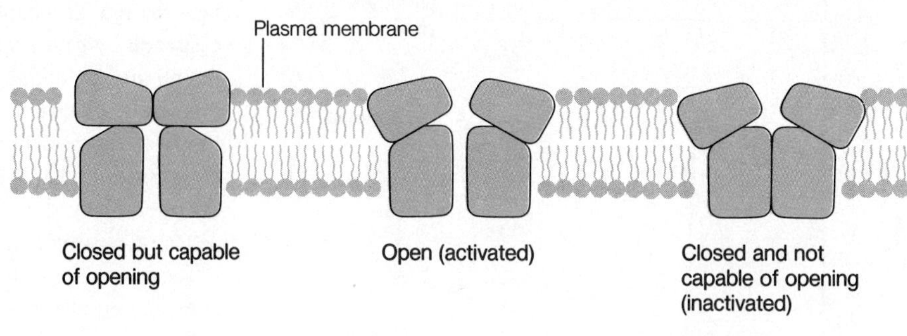

+30 mV, close to the Na$^+$ equilibrium potential. The potential does not become any more positive because, at the peak of the action potential, the Na$^+$ channels rapidly close to the inactive state because of the positive charge inside the cell, with P Na$^+$ falling to its low resting value. The Na$^+$ gates open for less than 0.5 msec, only to "slam" closed again and remain shut in an inactivated state until the membrane potential has been restored to its negative resting value.

Simultaneous to inactivation of the Na$^+$ channels, K$^+$ permeability (**P K$^+$**) greatly increases to about three hundred times the resting P Na$^+$. This opening of even more K$^+$ channels is a delayed voltage-gated response triggered by the initial depolarization to threshold. The marked increase in P K$^+$ causes K$^+$ to rush out of the cell down its concentration and electrical gradients, carrying positive charges back to the outside. Note that at the peak of the action potential, the internal positivity of the cell tends to repel the positive K$^+$ ions, so the electrical gradient for K$^+$ is outward, unlike at resting potential. The outward movement of K$^+$ rapidly restores the internal negativity and returns the potential to resting.

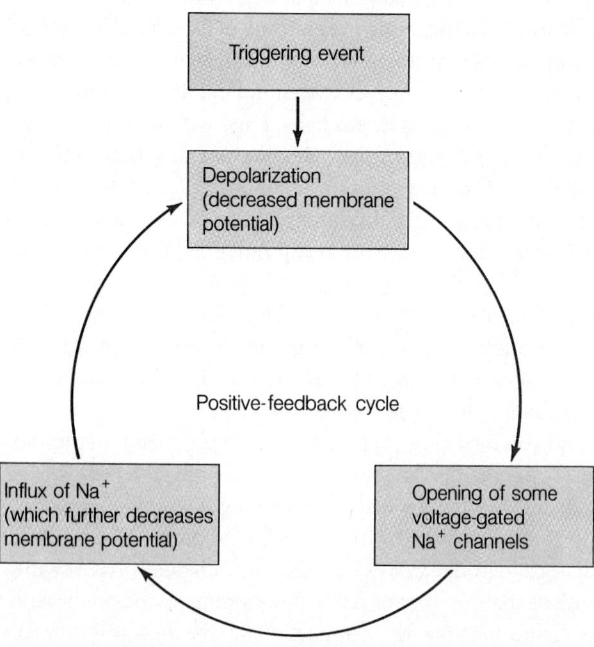

Figure 4–6 Positive-Feedback Cycle Responsible for Opening Sodium Channels at Threshold

To review (Fig. 4–7), *the rising phase of the action potential (depolarization) is due to Na$^+$ influx* (Na$^+$ entering the cell) induced by an explosive increase in P Na$^+$ at threshold. *The falling phase (repolarization) is brought about by K$^+$ efflux* (K$^+$ leaving the cell) caused by the marked increase in P K$^+$ occurring simultaneously with the slamming shut of the Na$^+$ gates at the peak of the action potential. As the potential returns to resting, the changing voltage shifts the Na$^+$ channels to their "closed but able to be opened" conformation. Meanwhile, the newly opened K$^+$ channels also close, so the membrane returns to the resting number of open K$^+$ channels. Sometimes more K$^+$ leaves than is necessary to bring the potential to resting because the K$^+$ channels do not close quickly enough. This slight excessive K$^+$ efflux makes the interior of the cell even more negative than resting potential, causing the after hyperpolarization.

The Na$^+$-K$^+$ pump restores the concentration gradient disrupted by action potentials.

At the completion of an action potential, the membrane potential has been restored to its resting condition but the ion distribution has been altered slightly. Sodium has entered the cell during the rising phase and a comparable amount of K$^+$ has

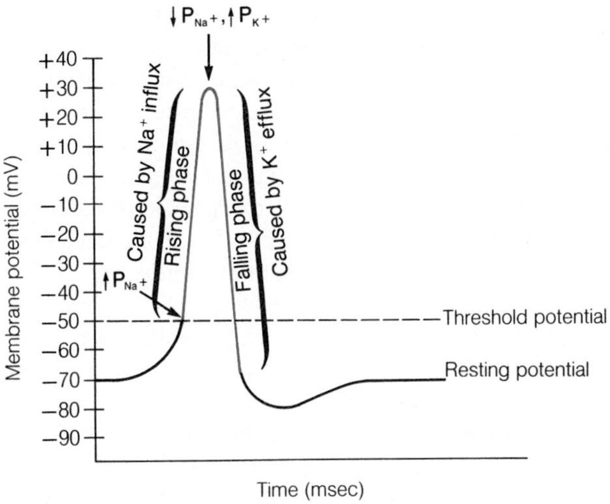

Figure 4–7 Permeability Changes and Ionic Fluxes during an Action Potential

left during the falling phase. It is the task of the Na^+-K^+ pump to restore these ions to their original locations.

The active pumping process takes much longer to restore Na^+ and K^+ to their original locations than it takes for the passive fluxes of these ions during an action potential. This does not mean, however, that the membrane must wait until the Na^+-K^+ pump slowly restores the concentration gradients before it can undergo another action potential. Actually, the movement of only relatively few of the total number of Na^+ and K^+ ions present is responsible for the dramatic swings in potential that occur during an action potential. Only about 1 out of 100,000 K^+ ions present in the cell leaves during an action potential while a comparable number of Na^+ ions enter from the ECF. Shifting locations of these extremely small percentages of the total Na^+ and K^+ during a single action potential produces only infinitesimal changes in their ICF and ECF concentrations. There is still much more K^+ inside the cell than outside, and Na^+ is still predominantly an extracellular cation. Consequently, the Na^+ and K^+ concentration gradients still exist to permit repeated action potentials without the necessity of the pump keeping pace to restore the gradients. Of course, were it not for the pump, even these tiny fluxes accompanying repeated action potentials would eventually "run down" the concentration gradients so that further action potentials would be impossible.

If the concentrations of Na^+ and K^+ were equal between the ECF and ICF, changes in permeability to these ions would not bring about ionic fluxes, so no change in potential would occur. Thus, the Na^+ - K^+ pump is critical in the long run to maintaining the concentration gradients. However, it does not have to perform its role between every action potential nor is it directly involved in the ion fluxes or potential changes that occur during an action potential.

Once initiated, action potentials are propagated throughout an excitable cell.

A single action potential involves only a small patch of the total surface membrane of an excitable cell. If action potentials are to serve as long-distance signals, obviously they cannot be merely isolated events occurring in a limited area of a nerve or muscle cell membrane. Mechanisms must exist to conduct or spread the action potential throughout the entire cell membrane. Furthermore, the signal must be transmitted from one cell to the next cell (for example, along specific nerve pathways). Let us first examine how an action potential (nerve impulse) is conducted throughout a nerve cell before turning our attention to how the impulse is passed to another cell.

A single nerve cell or **neuron** typically consists of three basic parts: the cell body, the dendrites, and the axon (although there are variations, depending on the location and function of the neuron). The nucleus and organelles are housed in the **cell body** (Fig. 4–8), from which numerous extensions known as **dendrites** project like antennae to increase the surface area available for reception of signals from other nerve cells. In most neurons the plasma membrane of the cell body and dendrites contains protein receptors for binding chemical messengers from other neurons. The **axon** or **nerve fiber** is a single, elongated tubular process that conducts action potentials away from the cell body and eventually terminates at other cells. The first portion of the axon plus the region of the cell body from which the axon leaves is known as the **axon hillock.** This is the site where action potentials are initiated in a neuron (with the exception of neurons specialized to carry sensory information, a topic described in a later chapter). The impulses are then propagated along the axon to its typically highly branched ending, the **axon terminals.** These terminals release chemical messengers to simultaneously influence numerous other cells with which they come into close association. The axon frequently gives off side branches or **collaterals** along its course.

Axons vary in length from less than a millimeter in those neurons that communicate only with neighboring cells to longer than a meter in neurons that communicate with other distant parts of the nervous system or with peripheral organs. For example, the axon of the nerve cell innervating your big toe must traverse the distance between the origin of its cell body within the spinal cord in the lower region of your back all the way down your leg to your toe. This asymmetrical cell structure is maintained by special cytoskeletal structures that also serve as highways for axonal transport of materials be-

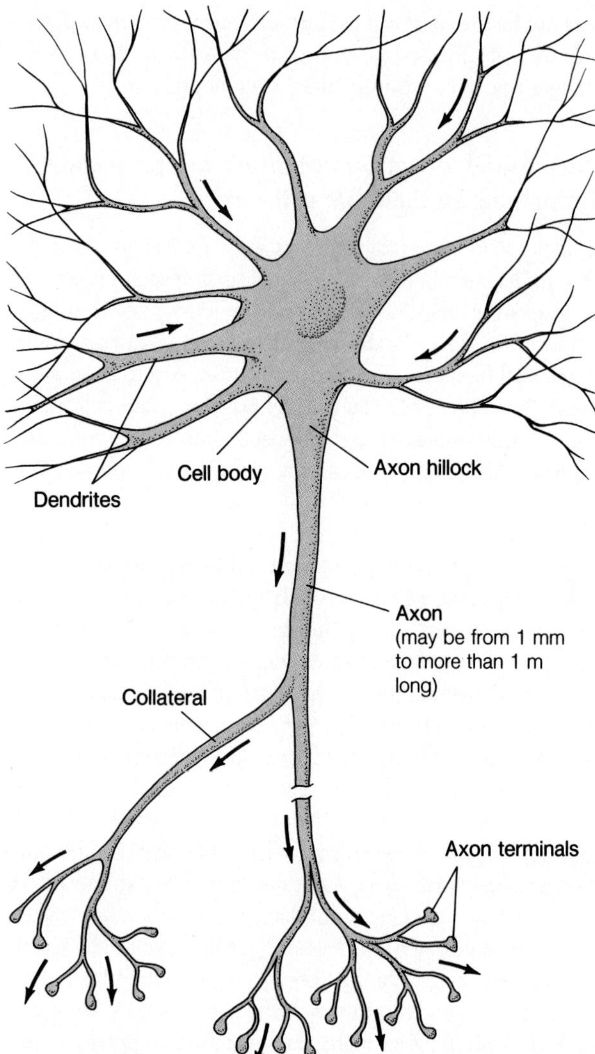

Figure 4–8 Anatomy of a Neuron

tween the cell body and the distant axon terminals (see p. 36). Coincidentally, these cytoskeletal highways may also serve as pathways for the movement of such infectious agents as Herpes virus, poliomyelitis virus, and rabies virus, which travel retrograde (backwards) along nerves from their surface site of contamination to the central nervous system.

Once an action potential is initiated at the axon hillock, no further triggering event is necessary to activate the remainder of the nerve fiber. The impulse is automatically conducted throughout the neuron without further stimulation by one of two methods of propagation: conduction by local current flow or saltatory conduction.

Figure 4–9 illustrates **conduction by local current flow.** You are viewing a schematic representation of a longitudinal section of the axon hillock and the portion of the axon immediately beyond it. The axon hillock is at the peak of an action potential. The inside of the cell is positive in this active area because Na^+ has already entered the nerve cell at this point. The remainder of the axon, still at resting potential and negative inside, is considered to be inactive. For the action potential to spread from the active to the inactive areas, the inactive areas must somehow be depolarized to threshold before they can undergo an action potential. This depolarization is accomplished by local current flow between the area already undergoing an action potential (positive inside, negative outside) and the adjacent inactive area (negative inside, positive outside), similar to the current flow responsible for the spread of graded potentials. Because opposite charges attract, current is able to flow locally between the active area and the neighboring inactive area both on the inside and outside of the membrane. This local current flow in effect neutralizes or eliminates some of the unbalanced charges in the inactive area; that is, it reduces the number of opposite charges separated across the membrane or reduces the potential in this area. This depolarizing effect quickly brings the involved inactive area to threshold, at which time the voltage-gated Na^+ channels in this region of the membrane are all thrown open, leading to an action potential in this previously inactive area.

Meanwhile, the original active area returns to resting potential as a result of K^+ efflux. (Local current flow does *not* make a significant contribution to returning the original active area to resting. Repolarization is accomplished primarily by a concurrent drop in P Na^+ and rise in P K^+, as described earlier.) In turn, beyond the new active area is another inactive area, so the same thing happens again. Local current flow brings this next inactive area to threshold, causing it to fire and become a new active area. This cycle repeats itself until the action potential has spread to the end of the axon. ***Once an action potential is initiated in one part of a nerve cell membrane, a self-perpetuating cycle is initiated so that the action potential is propagated throughout the rest of the fiber automatically.*** This is similar to a firecracker fuse that needs to be lit at only one end. Once ignited, the fire spreads down the fuse; it is not necessary to hold a match to every separate section of the fuse.

Note that the original action potential does not travel along the membrane. Instead, it triggers an identical new action potential in the adjacent area of the membrane, with this process being repeated along the axon's length. Since each new action potential in the conduction process is a fresh event dependent on the induced permeability changes and electrochemical gradients, which are virtually identical down the length of the axon, the last action potential at the end of the axon is identical to the original one, no matter how long the axon. Thus, an ac-

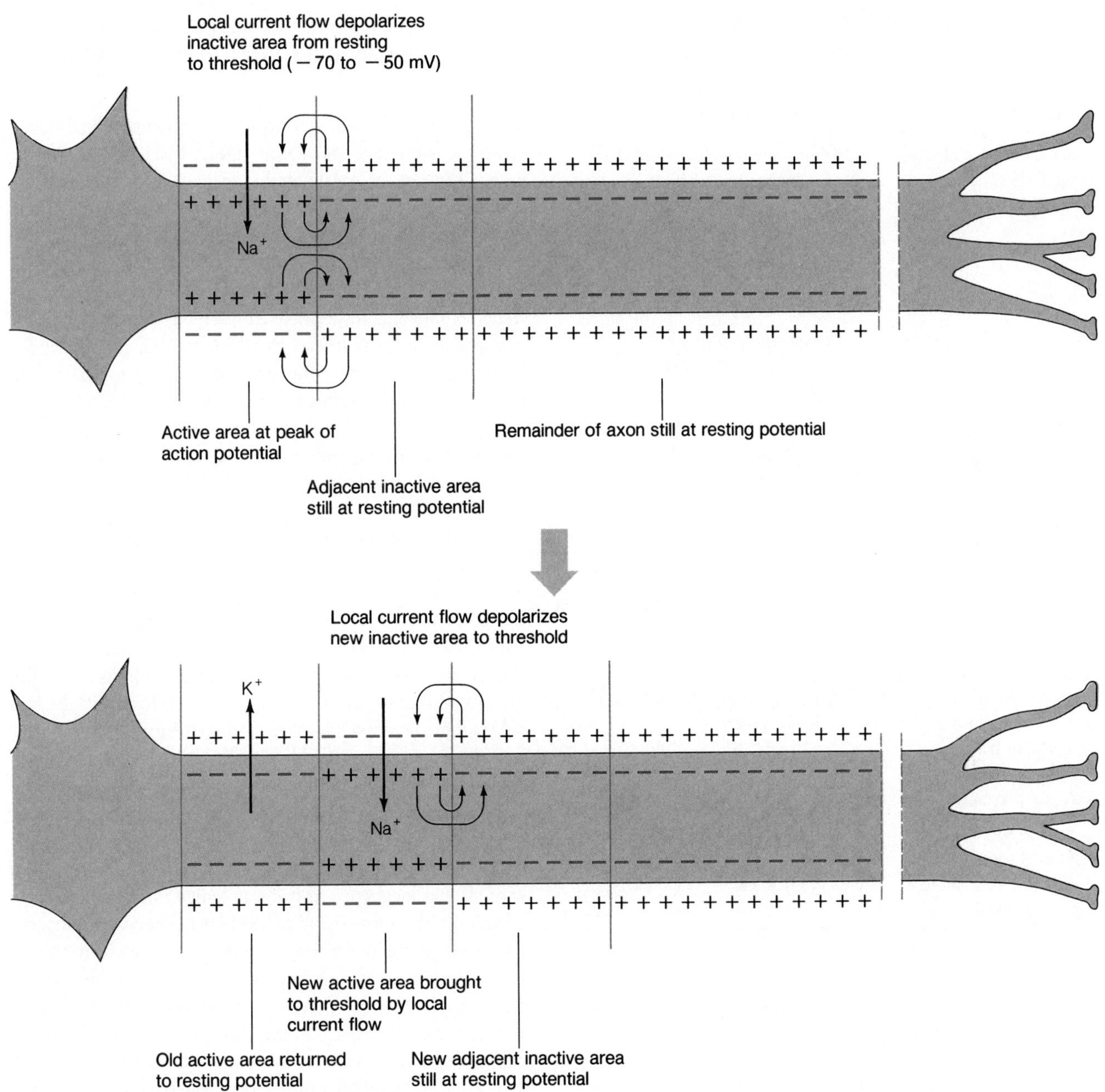
Local current flow depolarizes inactive area from resting to threshold (−70 to −50 mV)
Na+
Active area at peak of action potential
Adjacent inactive area still at resting potential
Remainder of axon still at resting potential
Local current flow depolarizes new inactive area to threshold
K+
Na+
New active area brought to threshold by local current flow
Old active area returned to resting potential
New adjacent inactive area still at resting potential

Figure 4-9 Conduction by Local Current Flow

tion potential is spread throughout the axon in undiminished fashion. It always goes to maximal amplitude, rather than getting progressively smaller as it moves down the axon. In this way, action potentials can serve as faithful long-distance signals without attenuation or distortion.

This is in contrast to the decremental spread of a graded potential, which dies out over a very short distance because it is not able to regenerate itself. Typically, regions of excitable cells where graded potentials take place do not have an achievable threshold for undergoing action potentials because of a sparsity of voltage-gated Na^+ channels. There is no threshold phenomenon for the occurrence of graded potentials themselves. Therefore, sites specialized for graded potentials do not undergo action potentials, even though they might be depo-

Table 4–1 Comparison of Graded Potentials and Action Potentials

Graded Potentials	Action Potentials
Graded potential change; magnitude varies with magnitude of triggering event	All-or-none membrane response; once initiated, magnitude of response independent of triggering event; magnitude of triggering event coded in frequency rather than amplitude of action potentials
Decremental conduction; magnitude diminishes with distance from initial site	Propagated throughout membrane in undiminishing fashion
Passive spread to neighboring inactive areas of membrane	Self-regeneration in neighboring inactive areas of membrane
No refractory period	Refractory period
Can be summed	Summation impossible
No threshold	Threshold
Can be depolarization or hyperpolarization	Always depolarization and reversal of charges
Triggered by stimulus, by combination of neurotransmitter with receptor, or by spontaneous shifts in leak-pump cycle	Triggered by depolarization to threshold, usually through spread of graded potential
Occurs in specialized regions of membrane designed to respond to triggering event	Occurs in regions of membrane with abundance of voltage-gated Na^+ channels

larized considerably. However, as will be elaborated on later, graded potentials can, before dying out, trigger action potentials in adjacent portions of the membrane by bringing these more sensitive regions to threshold through local current flow spreading from the site of the graded potential. Table 4–1 summarizes the differences between graded potentials and action potentials, some of which are yet to be discussed.

Local anesthetics function by blocking Na^+ channels, thus precluding the ionic fluxes responsible for the generation and propagation of action potentials. When administered in the vicinity of pain fibers (which conduct action potentials to the central nervous system to cause the sensation of pain), local anesthetics prevent these fibers from signaling the brain that local tissue injury is taking place.

Myelination increases the speed of conduction of action potentials and conserves energy in the process.

The velocity, or speed, with which an action potential travels down the axon depends on two different factors: (1) whether or not the fiber is myelinated; and (2) the diameter of the fiber.

Myelinated fibers, as the name implies, are covered with **myelin** at regular intervals along the length of the axon (Fig. 4–10a). Myelin is composed primarily of lipids. Because the water soluble ions responsible for carrying current across the membrane cannot permeate this thick lipid barrier, the myelin coating acts as an insulator, like rubber, to prevent current leakage across the myelinated portion of the membrane. Myelin is not actually a part of the nerve cell but consists of separate myelin-forming cells that wrap themselves around the axon in jelly-roll fashion (Fig. 4–10b). These myelin-forming cells are **oligodendrocytes** in the central nervous system (the brain and spinal cord) and **Schwann cells** in the peripheral nervous system (the nerves running between the central nervous system and the various regions of the body). The lipid composition of myelin is due to the presence of layer upon layer of the lipid bilayer that composes the plasma membrane of these myelin-forming cells. Between the chunks of myelin, the axonal membrane is bare and exposed to the ECF. It is only at these bare spaces, called **nodes of Ranvier,** that membrane potential can exist and current can flow across the membrane (Fig. 4–10c). Sodium channels are concentrated at the nodal areas; the myelin-covered regions are almost devoid of these special passageways.

The nodes are usually about 1 mm apart, a distance short enough that local current from an active node can reach an adjacent node before dying off. When an action potential occurs at one of the nodes, opposite charges attract from the adjacent inactive node, reducing its potential to threshold so that it undergoes an action potential, and so on. Consequently, in a myelinated fiber, the impulse "jumps" from node to node, skipping over the myelinated sections of the axon (Fig. 4–11). This is called **saltatory conduction** (*saltere* means "to jump or leap"). Saltatory conduction propagates action potentials more rapidly than does conduction by local current flow, because the action potential leaps over myelinated sections but must be regenerated within every section of an unmyelinated axonal membrane from beginning to end. Myelinated fibers conduct impulses about fifty times faster than nonmyelinated fibers of comparable size. As a general rule, the most urgent types of information are transmitted via myelinated fibers, whereas the nervous pathways carrying less urgent information are unmyelinated.

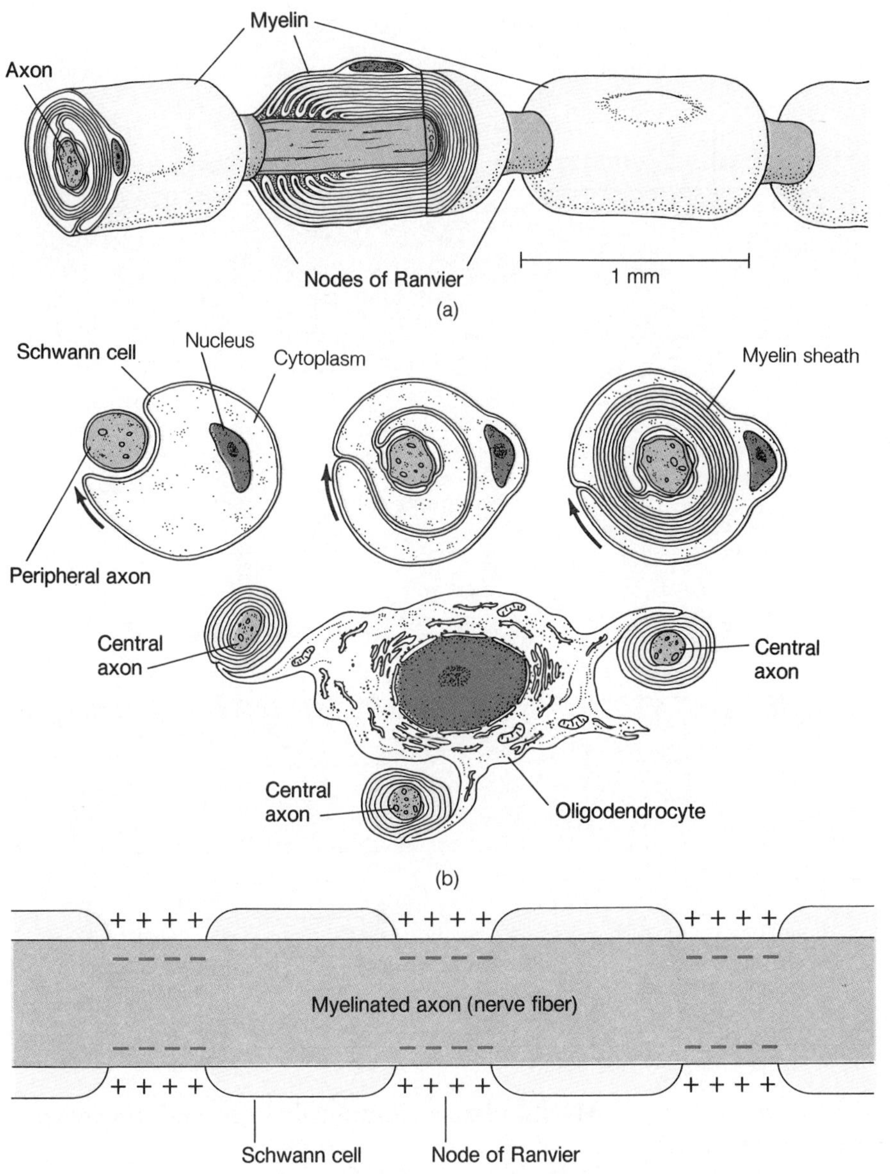

Figure 4–10 Myelinated Fibers *(a) A myelinated fiber is surrounded by myelin at regular intervals. The intervening unmyelinated regions are known as nodes of Ranvier. (b) In the peripheral nervous system, each patch of myelin is formed by a separate Schwann cell that wraps itself jelly-roll fashion around the nerve fiber. In the central nervous system, each of the several processes of a myelin-forming oligodendrocyte forms a patch of myelin around a separate nerve fiber. (c) Membrane potential exists only at the nodes of Ranvier where the bare axon is exposed to the ECF. No charges exist across the insulated myelinated regions.*

In addition to permitting action potentials to travel faster, a second advantage of myelination is that it conserves energy. Since the ion fluxes associated with action potentials are confined to only the nodal regions, the extent to which the energy-consuming Na^+-K^+ pump must restore these ions to their respective sides of the membrane following propagation of an action potential is decreased.

Fiber diameter also influences the velocity of action potential propagation.

Besides the effect of myelination, the speed with which an axon can conduct action potentials is also influenced by the *diameter* of the fiber. The larger the diameter of the nerve fiber, the faster it can propagate action potentials. Large myelinated

Figure 4–11 Saltatory Conduction
The impulse "jumps" from node to node in a myelinated fiber.

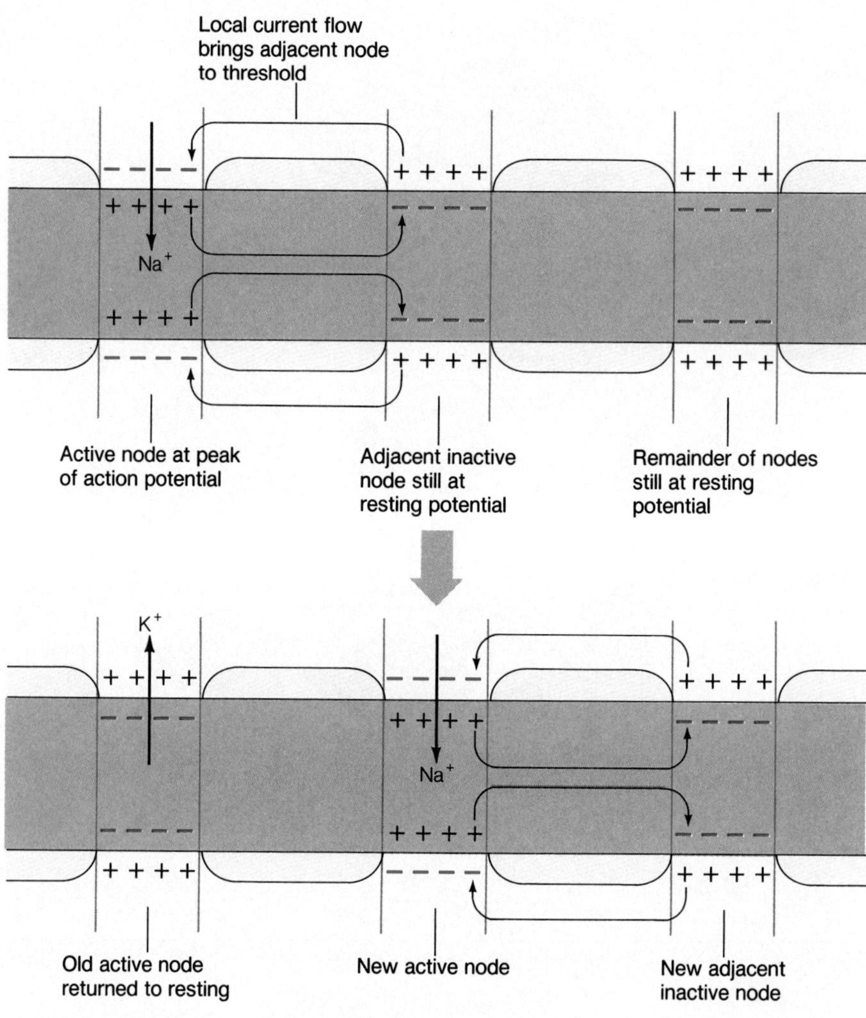

fibers, such as those supplying skeletal muscles, can conduct action potentials at a speed of up to 120 meters (m)/sec (360 miles/hr), compared with a conduction velocity of 0.7 m/sec (2 miles/hr) in small unmyelinated fibers such as those supplying the digestive tract. This variability in speed of propagation of action potentials is appropriate when considering the urgency of the information being conveyed. Rapidity of transmission of a signal to skeletal muscles for execution of a particular movement (for example, to prevent you from falling as you trip on something) is more important than modification of a slow-acting digestive process. Were it not for myelination, axon diameters within urgent nerve pathways would have to be very large and cumbersome to achieve the necessary conduction velocities. Indeed, this is the case in many invertebrates. Vertebrates have escaped the necessity of very large fibers by wrapping the axons in myelin to permit economic, rapid long-distance signaling.

Myelin plays a central role in several disorders.

Multiple sclerosis (**MS**) is a pathophysiological condition in which demyelination of nerve fibers occurs in various locations throughout the central nervous system. The cause of MS is uncertain. Its symptoms vary considerably, depending on the extent and location of the myelin damage. Loss of myelin slows transmission of impulses in the affected neurons. Also, scarring associated with myelin damage can injure the underlying axons, further interfering with action potential propagation.

Myelin defects may be the culprit in MS, but the presence of myelinating cells can be of tremendous benefit when an axon is cut. In the case of a cut axon in a peripheral nerve, the portion of axon farthest from the cell body degenerates and the surrounding Schwann cells phagocytize the debris. The Schwann cells themselves remain and form a **regeneration**

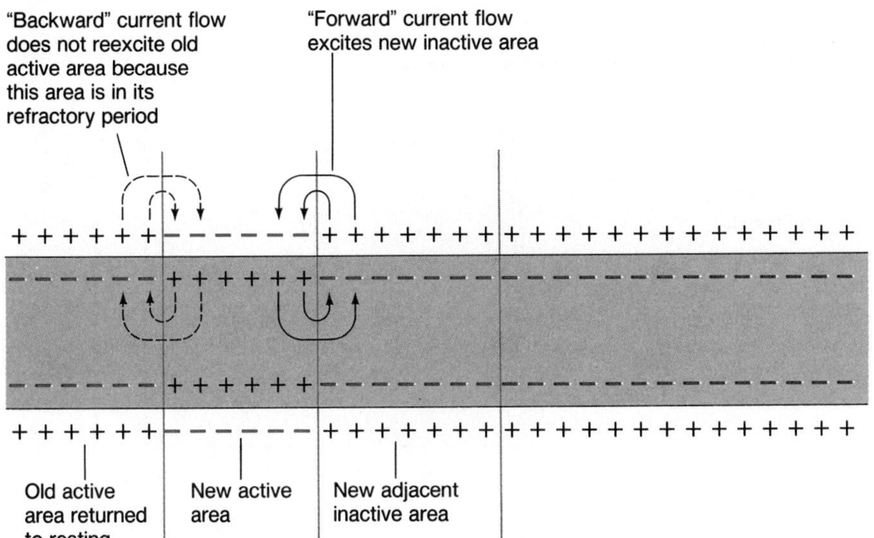
"Backward" current flow does not reexcite old active area because this area is in its refractory period
"Forward" current flow excites new inactive area
Old active area returned to resting
New active area
New adjacent inactive area

Figure 4–12 Value of Refractory Period *"Backward" current flow is prevented by the refractory period.*

tube to guide the regenerating nerve fiber to its proper destination. The remaining portion of the axon connected to the cell starts to grow and move forward within the Schwann cell column by ameoboid movement (see p. 39). It is believed that the growing axon tip "sniffs" its way forward in the proper direction, guided by a chemical secreted into the regeneration tube by the Schwann cells. Successful fiber regeneration is responsible for the return of sensation and movement after a period of time following traumatic peripheral nerve injuries, although regeneration is not always successful. Fibers in the central nervous system, which are myelinated by oligodendrocytes rather than Schwann cells, do not have this same regenerative ability. Therefore, damaged neuronal fibers in the brain and spinal cord never regenerate.

The refractory period assures unidirectional propagation of the action potential and limits the frequency of action potentials.

What assures the one-way propagation of an action potential away from the initial site of activation? Note in Figure 4–12 that once the action potential has been regenerated at a new neighboring site (now positive inside) and the original active area has returned to resting (once again negative inside), the close proximity of opposite charges between these two areas is conducive to local current flow taking place in the backward direction as well as in the forward direction (into as-yet unexcited portions of the membrane). If such backward current flow were able to bring the just-inactivated area to threshold, another action potential would be initiated here, which would spread both forward and backward, initiating another action potential, and so forth. This would create a chaotic situation of numerous action potentials bouncing back and forth along the axon until the nerve cell eventually fatigued. Fortunately, neurons are saved from this fate of oscillating action potentials by the existence of the *refractory period*. During the time that a particular patch of axonal membrane is undergoing an action potential, it is incapable of initiating another action potential, no matter how strongly it is stimulated (Fig. 4–13). This is known as the **absolute refractory period.** It corresponds to the time period during which the Na^+ gates are first opened and then closed and inactivated. Not until the potential has returned to resting and the voltage-dependent Na^+ channels are restored to their "closed but capable of opening" conformation can they respond to another depolarization with an explosive increase in P Na^+ to initiate another action potential.

Following the absolute refractory period is a **relative refractory period** during which a second action potential can be produced only by a stimulus considerably stronger than is usually necessary. During this time, the K^+ gates (those which had opened at the peak of the action potential to bring about repolarization) are still closing. Only when all channels have been restored to their resting conformation is the patch of membrane that has just undergone an action potential ready to respond again in a normal fashion. Meanwhile, the impulse has continued to be rapidly propagated in the forward direction only. By the time the original site has recovered from its refractory period and is capable of being restimulated, the action potential is so far away that it can no longer influence the original site. Thus, *the refractory period assures the unidirec-*

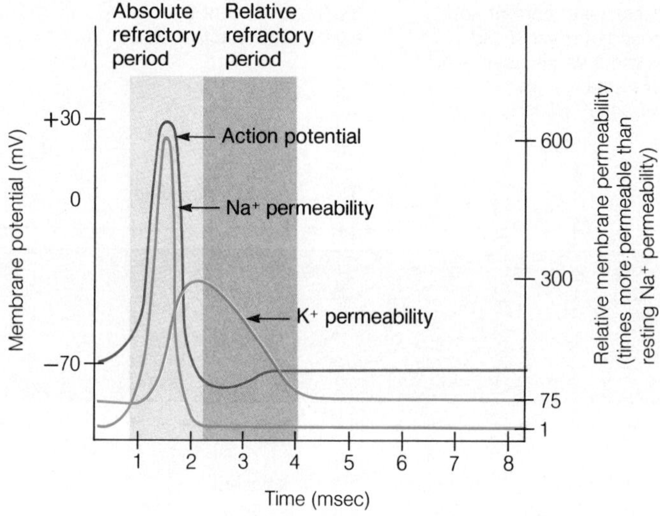

Figure 4–13 Refractory Period *During the absolute refractory period, the portion of the membrane that has just undergone an action potential cannot be restimulated. It corresponds to the time during which the Na^+ gates are not in their resting conformation. During the relative refractory period, the membrane can be restimulated only by a stronger stimulus than is usually necessary. It corresponds to the time during which the K^+ gates that were opened during the action potential are still closing.*

tional propagation of the action potential down the axon away from the initial site of activation.

The refractory period is also responsible for setting an upper limit on the frequency of action potentials; that is, it determines the maximum number of new action potentials that can be initiated and propagated along the fiber in a given period of time. The original site must recover from its refractory period before a new impulse can be triggered to follow the first impulse. The length of the refractory period varies for different types of neurons. The longer the refractory period, the greater the delay before a new action potential can be initiated and the lower the frequency with which a nerve cell can respond to repeated or ongoing stimulation.

Action potentials occur in all-or-none fashion.

If any portion of the neuronal membrane, usually the axon hillock, is depolarized to threshold, an action potential is initiated and relayed throughout the membrane in undiminished fashion. Futhermore, once threshold has been reached, the resultant action potential always goes to maximal height, because the changes in voltage during an action potential are due to ion movements down concentration and electrical gradients, which are not affected by stimulus strength. A stimulus stronger than one necessary to bring the membrane to threshold does not produce a larger action potential. On the other hand, a stimulus that fails to depolarize the membrane to threshold does not trigger an action potential at all. Thus, *an excitable membrane either responds to a stimulus with a maximal action potential that spreads nondecrementally throughout the membrane, or it does not respond with an action potential at all.* This is called the **all-or-none law.**

This all-or-none concept is analogous to firing a gun. Either the trigger is not pulled sufficiently to fire the bullet at all (threshold is not reached), or it is pulled hard enough to elicit the full firing response of the gun (threshold is reached). Squeezing the trigger harder does not produce a greater explosion. Just as it is not possible to fire a gun halfway, it is not possible to have a halfway action potential.

The threshold phenomenon is a means by which some discrimination can take place regarding important versus unimportant stimuli. Stimuli too weak to bring the membrane to threshold do not initiate an action potential and therefore do not clutter up the nervous system with transmission of insignificant signals. How is it possible, however, to differentiate between two stimuli of varying strengths if both bring the membrane to threshold and generate action potentials of the same magnitude? For example, how can one distinguish between touching a warm object or a very hot object if both trigger identical action potentials in the nerve fiber relaying information about skin temperature to the central nervous system? The answer lies in the *frequency* with which the action potentials are generated. A stronger stimulus does not produce a larger action potential, but it does trigger a greater number of action potentials per second to be propagated along the fiber.

Synapses

A neurotransmitter carries the signal across a synapse.

What happens once an action potential reaches the end of an axon? A neuron may terminate at one of three structures: a

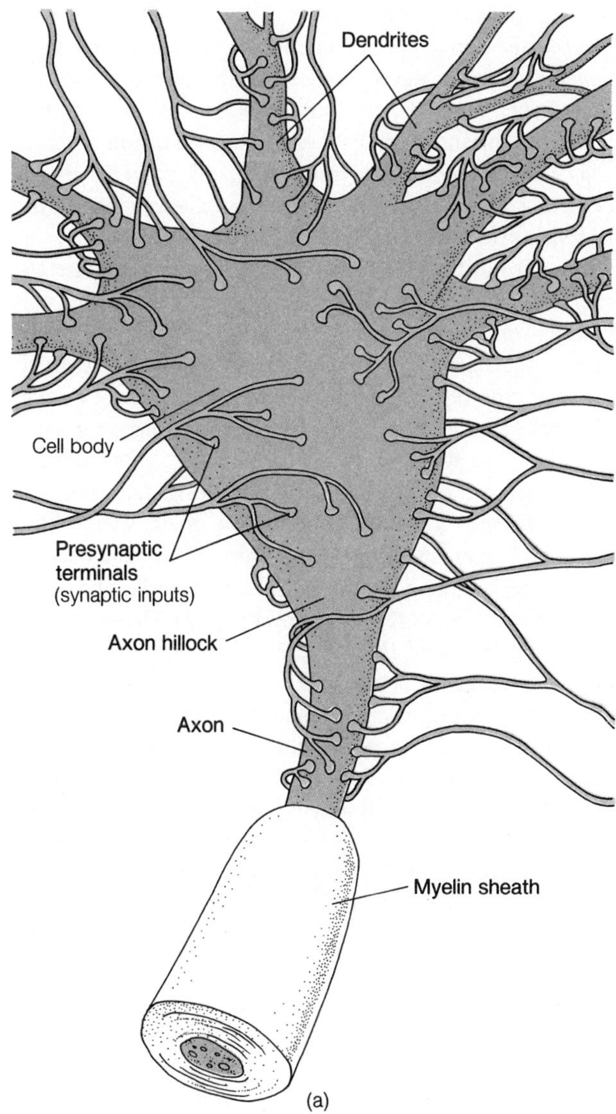

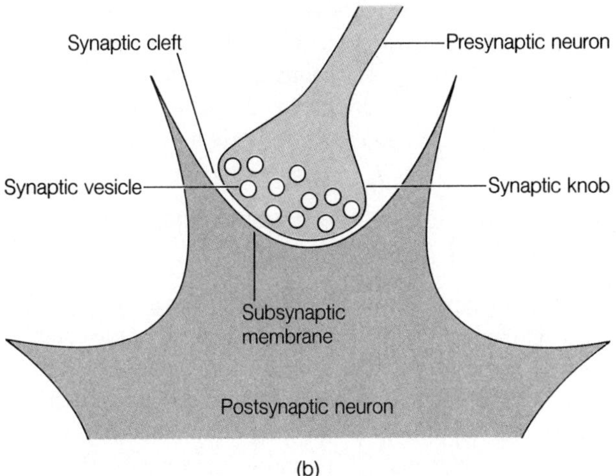

Figure 4–14 Synaptic Anatomy *(a) Presynaptic input to single neuron. (b) Schematic representation of single synapse.*

muscle, a gland, or another neuron. The junctions between nerves and the muscles and glands that they innervate will be described later. For now we will concentrate on the junction between two neurons—a **synapse.**

Typically a neuron-to-neuron synapse involves a junction between an axon terminal of one neuron and the dendrites or cell body of a second neuron. Less frequently, there are axon-to-axon connections and dendrite-to-dendrite connections. Most neuronal cell bodies and associated dendrites receive thousands of synaptic inputs, which are axon terminals from many other neurons. It has been estimated that some neurons within the central nervous system receive as many as 100,000 synaptic inputs (Fig. 4–14a).

The anatomy of one of these thousands of synapses is shown in Figure 4–14b. The axon terminal of the **presynaptic neuron,** which conducts its action potentials *toward* the synapse, ends in a slight swelling, the **synaptic knob.** The synaptic knob contains **synaptic vesicles,** which store a specific chemical messenger, a **neurotransmitter,** that has been synthesized and packaged by the presynaptic neuron. The synaptic knob comes into close proximity to, but does not actually directly contact, the **postsynaptic neuron,** the neuron whose action potentials are propagated *away from* the synapse. The space between the presynaptic and postsynaptic neurons, the **synaptic cleft,** is too wide for the direct spread of current from one cell to the other. This prevents action potentials from electrically passing between the neurons. The portion of the postsynaptic membrane immediately underlying the synaptic knob is referred to as the **subsynaptic membrane.**

Synapses operate in one direction only; that is, the presynaptic neuron influences the postsynaptic neuron, but the postsynaptic neuron does not influence the presynaptic neuron. The reason for this becomes readily apparent when one examines the events that occur at a synapse.

When an action potential in a presynaptic neuron has been propagated to the axon terminal, this change in potential triggers the opening of voltage-gated Ca^{++} channels in the synaptic knob. Because Ca^{++} is in much higher concentration in the ECF, this ion flows into the synaptic knob. Here it induces the release by exocytosis (see p. 24) of a neurotransmitter from some of the synaptic vesicles into the synaptic cleft. The released neurotransmitter diffuses across the cleft and combines with specific protein receptor sites on the subsynaptic membrane. This binding triggers the opening of specific ion channels in the subsynaptic membrane, thereby altering the permeability of the postsynaptic neuron. This is an example of chemical messenger-gated channels, in contrast to the voltage-gated channels responsible for the action potential and for the Ca^{++} influx into the synaptic knob.

Some synapses excite the postsynaptic neuron whereas others inhibit it.

There are two types of synapses, depending on the permeability changes induced in the postsynaptic neuron by the combination of transmitter substance with receptor sites: these are excitatory synapses and inhibitory synapses. At an **excitatory synapse,** the response to the receptor-neurotransmitter combination is an opening of Na^+ and K^+ channels within the subsynaptic membrane, thus increasing permeability to both of these ions. Both the concentration and electrical gradients for Na^+ favor its movement into the postsynaptic neuron at resting potential, whereas only the concentration gradient for K^+ favors its movement outward. Therefore, the permeability change induced at an excitatory synapse results in the simultaneous movement of a few K^+ ions out of the postsynaptic neuron while a relatively larger number of Na^+ ions enter this neuron, resulting in a net movement of positive ions into the cell. This makes the inside of the membrane less negative than at resting potential, producing a *small depolarization* of the postsynaptic neuron. Activation of one excitatory synapse can rarely depolarize the postsynaptic membrane sufficiently to bring it to threshold. There are just too few channels involved at a single subsynaptic membrane to permit adequate depolarizing fluxes to reduce the potential to threshold. This slight depolarization, however, does bring the membrane of the postsynaptic neuron closer to threshold, increasing the likelihood that threshold will be reached and an action potential will occur. Accordingly, such a postsynaptic potential change occurring at an excitatory synapse is called an **excitatory postsynaptic potential,** or **EPSP** (Fig. 4–15a).

At an **inhibitory synapse,** the combination of the released chemical messenger with its receptor sites increases the permeability of the subsynaptic membrane to either K^+ or Cl^- by altering these ions' respective channel conformations. In either case, the resulting ion movements bring about a *small hyperpolarization* of the postsynaptic neuron (greater internal negativity). In the case of increased P K^+, more positive charges leave the cell via K^+ efflux, leaving more negative charges behind on the inside; in the case of increased P Cl^-, negative charges enter the cell in the form of Cl^- ions because Cl^- concentration is highest outside of the cell. This slight hyperpolarization moves the membrane potential even farther away from threshold (Fig. 4–15b), lessening the likelihood that the postsynaptic neuron will reach threshold and undergo an action potential. The membrane is said to be inhibited under these circumstances, and the small hyperpolarization of the postsynaptic cell is called an **inhibitory postsynaptic potential** or **IPSP.** (In cells in which the equilibrium potential for Cl^- exactly equals the resting potential, an increased P Cl^- does not result in a hyperpolarization because there is no driving force to produce Cl^- movement. Opening of Cl^-

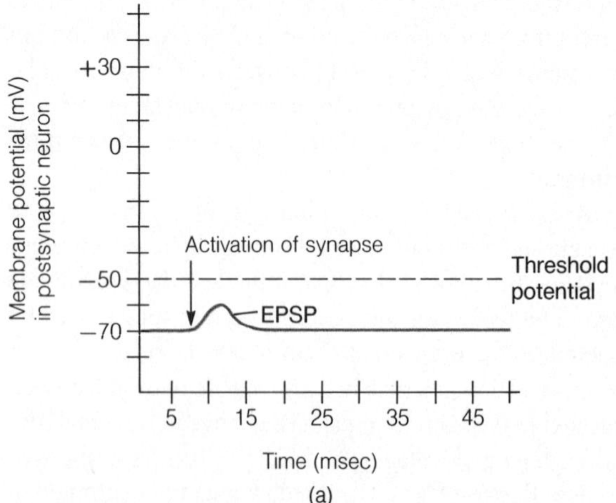

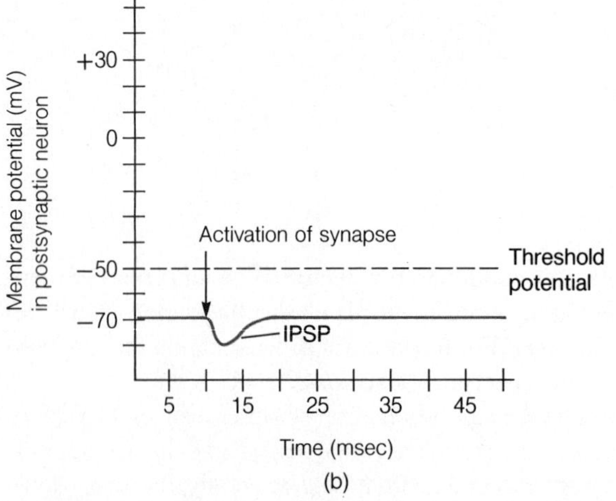

Figure 4–15 Postsynaptic Potentials *(a) Excitatory synapse. An excitatory postsynaptic potential (EPSP) brought about by activation of an excitatory presynaptic input brings the postsynaptic neuron closer to threshold potential. (b) Inhibitory synapse. An inhibitory postsynaptic potential (IPSP) brought about by activation of an inhibitory presynaptic input moves the postsynaptic neuron farther from threshold potential.*

channels in these cells tends to hold the membrane at resting potential, thus reducing the likelihood that threshold will be reached.)

This conversion of the electrical signal in the presynaptic neuron (an action potential) by chemical means (via the neurotransmitter-receptor combination) to an electrical signal in the postsynaptic neuron (either an EPSP or IPSP) takes time. This **synaptic delay** is usually about 0.5 to 1 msec. Chains of neurons often must be traversed along a specific neural pathway. The more complex the pathway, the more synaptic delays, and the longer the ***total reaction time*** (the time required to respond to a particular event).

More than thirty different chemicals are known or suspected to serve as neurotransmitters. Even though transmitter substances vary from synapse to synapse, the same transmitter is always released at a particular synapse. Furthermore, a given presynaptic neuron will always release the same neurotransmitter from all of its synaptic endings because each neuron is genetically coded to synthesize a particular neurotransmitter. The importance of neurotransmitters lies not so much in their names or chemical nature but in the particular responsiveness of the postsynaptic membrane upon their binding with subsynaptic receptors. One particular neurotransmitter will always induce EPSPs whereas another will always induce IPSPs. Yet another may even produce an EPSP at one synapse and an IPSP at another synapse. What is important, though, is that the response of a given transmitter-receptor combination is always constant. A ***given synapse is either always excitatory or always inhibitory***. It does not give rise to an EPSP under one circumstance and produce an IPSP at another time.

Neurotransmitters are quickly removed from the synaptic cleft to wipe the postsynaptic slate clean.

As long as the neurotransmitter remains in combination with the receptor sites, the alteration in membrane permeability responsible for the EPSP or IPSP continues. It is desirable to have the neurotransmitter inactivated or removed after it has produced the appropriate response in the postsynaptic neuron so that the postsynaptic "slate" is "wiped clean," leaving it ready to receive additional messages from the same or other presynaptic inputs. Thus, after combining with the postsynaptic receptor, chemical transmitters are removed and the response is terminated in one of three ways:

1. inactivation of the transmitter by specific enzymes within the subsynaptic membrane;
2. passive diffusion of the transmitter away from the synaptic cleft; or
3. active reuptake of the transmitter into the axon terminal by transport mechanisms in the presynaptic membrane. Once within the synaptic knob, the transmitter can be: (a) stored and released another time (recycled) in response to a subsequent action potential; or (b) destroyed by enzymes within the synaptic knob.

The method employed depends on the particular synapse.

Some neurotransmitters function through intracellular second messenger systems rather than by directly altering membrane permeability.

Most, but not all, neurotransmitters function by changing the conformation of chemical messenger–gated channels, thereby altering membrane permeability and ionic fluxes across the postsynaptic membrane. Another mode of synaptic transmission utilized by some neurotransmitters involves the activation of second messengers (see p. 64) within the postsynaptic neuron. Activation of the intracellular messenger, cyclic AMP, can induce both short- and long-term effects. In the short term, cAMP can lead to opening of ionic gates, a task that the other neurotransmitter-receptor combinations do directly without involvement of the intracellular messenger. The gating effects can be either excitatory or inhibitory. In addition, cAMP may trigger more long-term changes in the postsynaptic neuron cell, even to the extent of possibly altering the cell's genetic expression. Such long-term cellular changes may play a role in learning and memory.

The grand postsynaptic potential depends on the sum of the activities of all presynaptic inputs.

The events that occur at a single synapse result in either an EPSP or IPSP at the postsynaptic neuron. If a single EPSP is inadequate to bring the postsynaptic neuron to threshold and an IPSP moves it even farther from threshold, how is it possible to initiate an action potential in the postsynaptic neuron? Recall that a typical neuronal cell body receives thousands of presynaptic inputs from many other neurons. Some of these presynaptic inputs may be carrying sensory information brought from the environment; some may be signaling internal changes in homeostatic balance; others may be transmitting signals from control centers in the brain; and still others may arrive carrying other bits of information. At any given time, any number of these presynaptic neurons (probably hundreds) may be firing and thus influencing the postsynaptic neuron's level of activity. The total potential in the postsynaptic neuron, the **grand postsynaptic potential (GPSP)**, is a composite of all EPSPs and IPSPs occurring at approximately the same time.

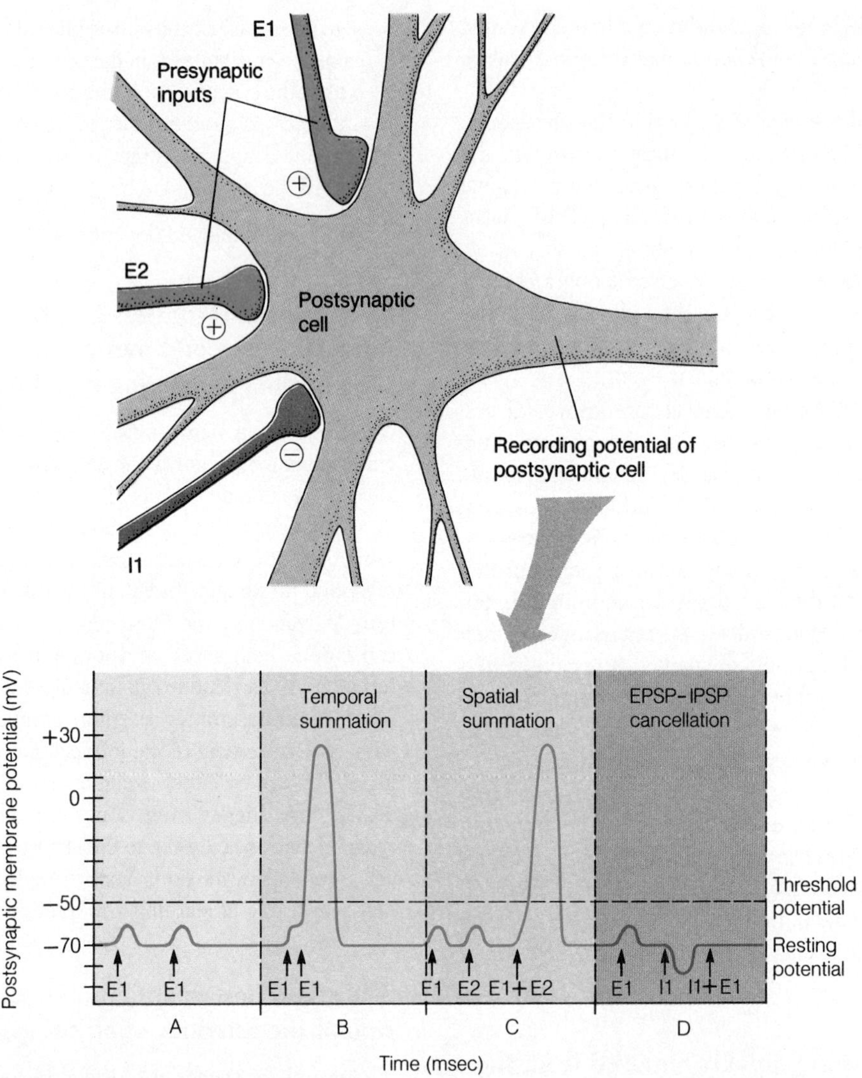

Figure 4–16 Determination of Grand Postsynaptic Potential by Sum of Activity in Presynaptic Inputs *See text for an explanation of the changes in postsynaptic potential that occur in response to variable patterns of stimulation of presynaptic inputs as depicted in panels* A, B, C, *and* D.

There are two ways in which the postsynaptic neuron can be brought to threshold: (1) temporal summation and (2) spatial summation. To illustrate these methods of summation, we will examine the possible interactions of three presynaptic inputs—two excitatory (E1 and E2) and one inhibitory (I1)—on a hypothetical postsynaptic neuron (Fig. 4–16). The recording represents the potential in the postsynaptic cell. Bear in mind during our discussion of this simplified version that many thousands of synapses are actually interacting in the same way on a single cell body.

Now suppose that E1 has an action potential that causes the release of neurotransmitter from a few of its synaptic vesicles to bring about an EPSP in the postsynaptic neuron. Because EPSPs (as well as IPSPs) are graded potentials, the EPSP spreads only a short distance before dying off. If another action potential subsequently occurs in E1, an EPSP of the same magnitude takes place (Panel A on Fig. 4–16). Next assume that E1 has two action potentials in close succession to each other (Panel B). The first action potential in E1 produces an EPSP in the postsynaptic neuron. While the postsynaptic membrane is still partially depolarized from this first EPSP (before it has returned to resting), the second presynaptic action potential produces a second EPSP in the postsynaptic neuron. The second EPSP will add on to the first EPSP, bringing the membrane to threshold so that an action potential can occur in the postsynaptic neuron. Graded potentials do not have a refractory period, so this additive effect is possible. Summing of several EPSPs occurring very close together in time because of successive firing of a single presynaptic neuron is known as **temporal summation** (*tempus* means "time"). The actual situation is much more complex than just described. Up to fifty EPSPs might have to sum to bring the postsynaptic membrane to threshold. Each action potential in a presynaptic neuron triggers the emptying of a certain number of synaptic vesicles. The amount of neurotransmitter released, and the resultant magnitude of the change in postsynaptic potential, is thus directly related to the frequency of presynaptic action potentials. One way, then, in which the postsynaptic membrane can be brought to threshold is through rapid, repetitive excitation from a single persistent input.

Let us now see what will happen in the postsynaptic neuron if both excitatory inputs are stimulated simultaneously (Panel C). An action potential in either E1 or E2 will produce an EPSP in the postsynaptic neuron; however, neither of these alone will bring the membrane to threshold to elicit a postsynaptic action potential. Simultaneous action potentials in E1 and E2, however, will each produce EPSPs that add to each other, bringing the postsynaptic membrane to threshold so that an action potential occurs. Such summation of EPSPs originating simultaneously from several different presynaptic inputs (i.e., from different points in "space") is known as **spatial summation.** A second way, therefore, to elicit an action potential in a postsynaptic cell is through concurrent activation of several excitatory inputs. Again, in reality up to fifty EPSPs arriving simultaneously on the postsynaptic membrane are required to bring it to threshold. Similarly, IPSPs can undergo temporal and spatial summation. As IPSPs add together, however, they progressively move the potential farther from threshold.

If an excitatory and an inhibitory input are simultaneously activated, the concurrent EPSP and IPSP more or less cancel each other out, the extent of cancellation depending on their respective magnitudes. In most cases, the postsynaptic membrane potential remains close to resting (Panel D).

Thus, the grand postsynaptic potential depends on the sum of activity in all presynaptic inputs. There are four possible outcomes of the GPSP:

1. If the active presynaptic inputs are predominantly excitatory, either from a single persistent excitatory input (temporal summation) or from several simultaneously activated excitatory inputs (spatial summation), the postsynaptic neuron may reach threshold and have an action potential. This action potential will cause the postsynaptic neuron to release a neurotransmitter that will influence all of the cells innervated by the neuron.
2. If excitatory input exceeds inhibitory input but the membrane still is not sufficiently depolarized to reach threshold, the membrane is said to be **facilitated.** It is more likely than usual to have an action potential.
3. If, on the other hand, predominantly inhibitory input is activated, the postsynaptic cell will be driven farther from threshold, precluding it from having an action potential and thus preventing it from influencing the cells it innervates.
4. If there is a balance of activated excitatory and inhibitory inputs, these effects will negate each other, and the postsynaptic neuron will essentially be unaffected.

The following is an oversimplified real-life example to which you can relate that demonstrates the benefits of this neuronal integration. The explanation is not completely accurate technically, but the principles of summation are accurate. Assume for simplicity's sake that urination is controlled by a postsynaptic neuron supplying the urinary bladder. (Actually, voluntary control of urination is accomplished by postsynaptic integration at the neuron controlling the external urethral sphincter rather than the bladder itself.) As the bladder starts to fill with urine and becomes stretched, a reflex is initiated that ultimately produces EPSPs in the postsynaptic neuron responsible for causing bladder contraction. Partial filling of the bladder does not cause sufficient excitation to bring the neuron to threshold, so urination does not take place (Panel A of Fig. 4–16). As the bladder becomes progressively filled, the frequency of action potentials is progressively increased in the presynaptic neuron that signals the postsynaptic neuron of the extent of bladder filling (E1 in Panel B of Fig. 4–16). When the frequency becomes great enough that the EPSPs are temporally summed to threshold, the postsynaptic neuron has an action potential that stimulates bladder contraction.

What if the time is inopportune for urination to take place? IPSPs can be produced at the bladder postsynaptic neuron by presynaptic inputs originating in higher levels of the brain responsible for voluntary control (I1 in Panel D of Fig. 4–16). These "voluntary" IPSPs in effect cancel out the "reflex" EPSPs triggered by stretching of the bladder. This keeps the postsynaptic neuron at resting potential so that it does not have an action potential, thereby keeping the bladder from contracting and emptying even though it is full.

What if the bladder is only partially filled so that the presynaptic input originating from this source is insufficient to bring the postsynaptic neuron to threshold to cause bladder contraction, yet the person needs to supply a urine specimen for laboratory analysis? The person can voluntarily activate an excitatory presynaptic neuron (E2 in Panel C of Fig. 4–16), which spatially summates with the reflex-activated presynaptic neuron (E1) to bring the postsynaptic neuron to threshold. This achieves the action potential necessary to stimulate bladder contraction, even though the bladder is not full.

This example illustrates the importance of postsynaptic neuronal integration. Each postsynaptic neuron in a sense "computes" all the input it receives and makes a "decision" about whether or not to pass the information on (i.e., whether or not to reach threshold). Each postsynaptic neuron filters out and does not pass on information it receives that is not significant enough to bring it to threshold. If every action potential in every presynaptic neuron that impinges upon a particular postsynaptic neuron were to cause an action potential in the postsynaptic neuron, the neuronal pathways would be overwhelmed with trivia. Only if an excitatory presynaptic signal is reinforced by other supporting signals through summation will the information be passed on. Furthermore, interaction of postsynaptic potentials provides a way for one set of signals to offset another set (IPSPs negating EPSPs.) This allows a fine degree of discrimination and control in determining what information will be passed on. (See the accompanying boxed feature, A Closer Look at Exercise Physiology, on p. 113.)

Action potentials are initiated at the axon hillock because it has the lowest threshold.

Threshold potential is not uniform throughout the postsynaptic neuron. The lowest threshold is present at the axon hillock, because this region has an abundance of voltage-gated Na^+ channels, making it considerably more sensitive to changes in potential than the remainder of the cell body and dendrites. The latter regions have a significantly higher threshold than the axon hillock. Because of local current flow, changes in membrane potential occurring anywhere on the cell body or dendrites (EPSPs or IPSPs) spread throughout the cell body, dendrites, and axon hillock. When summation of EPSPs takes place, the lower threshold of the axon hillock is reached first, whereas the cell body and dendrites at the same potential are still considerably below their own much higher thresholds. Therefore, the action potential originates in the axon hillock and is propagated from there throughout the rest of the neuron.

The effectiveness of synaptic transmission can be modified by a number of factors.

Numerous factors are able to modify synaptic effectiveness: (1) built-in mechanisms for fine-tuning neural transmission; (2) drug therapy aimed at deliberately manipulating synaptic transmission; and (3) various disease processes that undesirably interfere with neurotransmission. We will briefly explore each of these possible synaptic modifiers.

PRESYNAPTIC INHIBITION OR FACILITATION. Sometimes a third neuron influences activity between a presynaptic ending and a postsynaptic neuron. A presynaptic axon terminal (labeled A in Fig. 4–17) may itself be innervated by another axon terminal (labeled B) that can modify the amount of transmitter released from presynaptic terminal A. When the neurotransmitter released from terminal B binds with receptor sites on terminal A, the amount of transmitter released from terminal A in response to action potentials is altered. If the amount of transmitter released from A is reduced, the phenomenon is known as **presynaptic inhibition.** If the effect is to enhance release of transmitter, it is called **presynaptic facilitation.** The mechanism for these effects is unclear, but it most likely involves Ca^{++} because the amount of cytosolic Ca^{++} in the synaptic knob determines the extent to which synaptic vesicles release neurotransmitter.

What is the importance of this presynaptic modulation? The amount of transmitter released from presynaptic terminal A influences the potential in the postsynaptic neuron with which it is interacting (labeled C on Fig. 4–17). For example, if A, an excitatory input to C, is prevented from releasing its transmitter by presynaptic inhibition via B, the formation of EPSPs on postsynaptic membrane C will specifically be prevented from input A. As a result, no change in the potential of the postsynaptic neuron will occur in spite of action potentials in A. Could the same thing be accomplished by a simultaneous production of an IPSP through activation of an inhibitory input to negate an EPSP produced by activation of A? Not quite. The entire postsynaptic membrane is hyperpolarized by IPSPs, thereby negating excitatory information fed into any part of the cell from *any* presynaptic input. Presynaptic inhibition (or presynaptic facilitation), on the other hand, provides a means by which certain inputs to the postsynaptic neuron can be selectively altered without affecting the contributions of any other inputs. For example, firing of B does not have any influence on another excitatory presynaptic input (labeled D on Fig. 4–17). This type of neuronal integration is another means by which electrical signaling between nerve cells can be carefully fine-tuned.

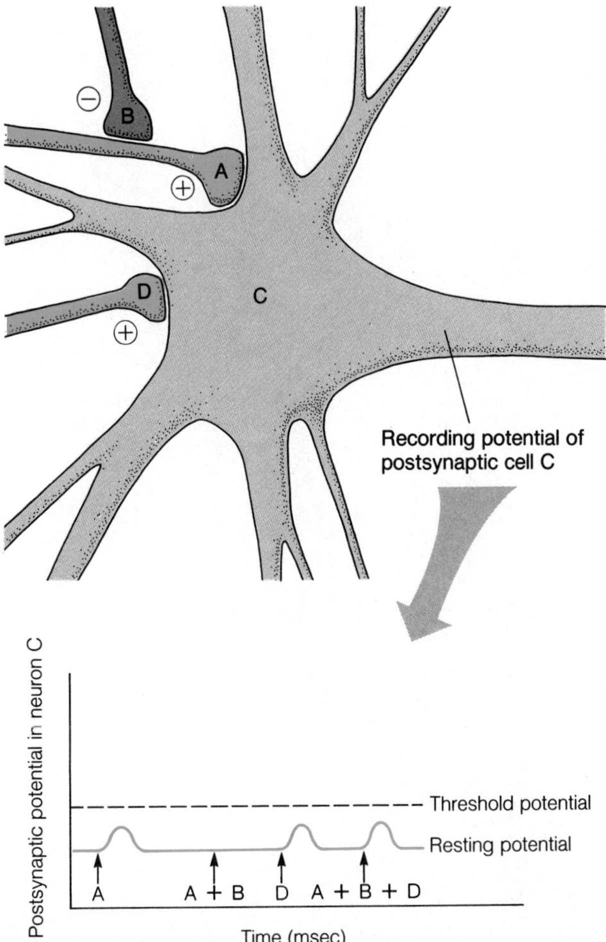

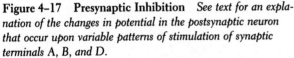

Figure 4–17 Presynaptic Inhibition *See text for an explanation of the changes in potential in the postsynaptic neuron that occur upon variable patterns of stimulation of synaptic terminals* A, B, *and* D.

NEUROMODULATION. Another means by which synaptic effectiveness can be subtly depressed or enhanced is through **neuromodulation.** Neuromodulators are chemical messengers that bind to neuronal receptors at nonsynaptic sites. In so doing, they activate second messenger systems that produce long-term intracellular biochemical effects that alter the effectiveness of ongoing synaptic activity. Neuromodulators may act at either presynaptic or postsynaptic sites. For example, a neuromodulator may influence the level of an enzyme critical in the synthesis of a specific neurotransmitter by a presynaptic neuron, or it may alter the sensitivity of postsynaptic receptors to a particular neurotransmitter. Thus, neuromodulation is another way to delicately fine-tune the synaptic response.

Interestingly, the known or presumed neuromodulators include many substances that also have distinctly different roles as hormones or neurotransmitters. It appears that a number of chemical messengers are quite versatile in the different roles they can assume, depending on their source, their distribution, and their interaction with a particular type of cell.

DRUGS AND DISEASES. The opportunities for influencing synaptic transmission by drugs are numerous. In fact, the vast majority of drugs that influence the nervous system perform their function by altering synaptic mechanisms. Synaptic drugs may act to block an undesirable effect or to enhance a desirable effect. Possible drug actions include the following: (1) altering the synthesis, axonal transport, storage, or release of a neurotransmitter; (2) modifying neurotransmitter interaction with the postsynaptic receptor; (3) influencing neurotransmitter reuptake or destruction; or (4) replacing a deficient neurotransmitter with a substitute transmitter.

As an example of the latter, Parkinson's disease is attributable to a deficiency of a particular neurotransmitter, dopamine. When patients with this disease are given levodopa (L-dopa), a compound closely related to dopamine, the L-dopa can gain access to the brain and be taken up by the dopamine-deficient synaptic knobs, thereby substituting for the lacking "homegrown variety" of this neurotransmitter. This greatly alleviates the symptoms associated with the deficit in most patients. Dopamine itself cannot be administered because it is unable to cross the blood-brain barrier (see p. 124), whereas L-dopa can enter the brain from the blood.

Synaptic transmission is also vulnerable to a number of disease processes, including defects at both presynaptic and postsynaptic sites. For example, two different neural poisons, strychnine and tetanus toxin, act at different synaptic sites to block inhibitory impulses while leaving the excitatory inputs unchecked. *Strychnine* competes with one of the inhibitory neurotransmitters (glycine) at the postsynaptic receptor site. This poison combines with the receptor but does not directly alter the potential of the postsynaptic cell in any way. However, strychnine blocks the receptor so that it is not available for interaction with glycine when the latter is released from the inhibitory presynaptic ending. Thus, postsynaptic inhibition (formation of IPSPs) is abolished in nerve pathways that use glycine as an inhibitory transmitter. Unchecked excitatory pathways lead to convulsions, muscle spasticity, and death. *Tetanus toxin,* on the other hand, prevents the release of another inhibitory transmitter, gamma-aminobutyric acid (GABA) from presynaptic inputs terminating on neurons that supply skeletal muscles. Unchecked excitatory inputs to these neurons result in uncontrolled muscle spasms. These spasms occur especially in the jaw muscles early in the disease, giving rise to the common name of lockjaw for this condition. Later they progress to the muscles responsible for breathing, at which point the disease proves to be fatal. The outcomes of both strychnine and tetanus toxin poisoning are similar, but one (strychnine) blocks specific postsynaptic inhibitory receptors, whereas the other (tetanus toxin) prevents the presynaptic release of a specific inhibitory neurotransmitter. There are other drugs and diseases too numerous to mention, but these examples should give you an idea that any site along the synaptic pathway is vulnerable to interference, either pharmacological (drug-induced) or pathological (disease-induced).

Neurons are linked to each other through convergence and divergence to form vast and complex nerve pathways.

Two important relationships exist between neurons: convergence and divergence. Any given neuron may have many other neurons synapsing upon it. Such a relationship is known as **convergence** (Fig. 4–18). Through this converging input, a single cell is influenced by thousands of other cells. This single cell, in turn, influences the level of activity in many other cells by divergence of output. **Divergence** refers to the branching of axon terminals so that a single cell synapses with many other cells.

Note that a particular neuron is postsynaptic to the neurons converging upon it but is presynaptic to the other cells upon which it terminates. Thus, the terms ***presynaptic*** and ***postsynaptic*** refer only to a single synapse. Most neurons are presynaptic to one group of neurons and postsynaptic to another group.

There are an estimated 100 billion neurons in the brain alone. When you consider the vast and intricate interconnections possible between these neurons through converging and diverging pathways, you can begin to imagine how complex the wiring mechanism of our nervous system really is. Even the most sophisticated computers are far less complex than the human brain.

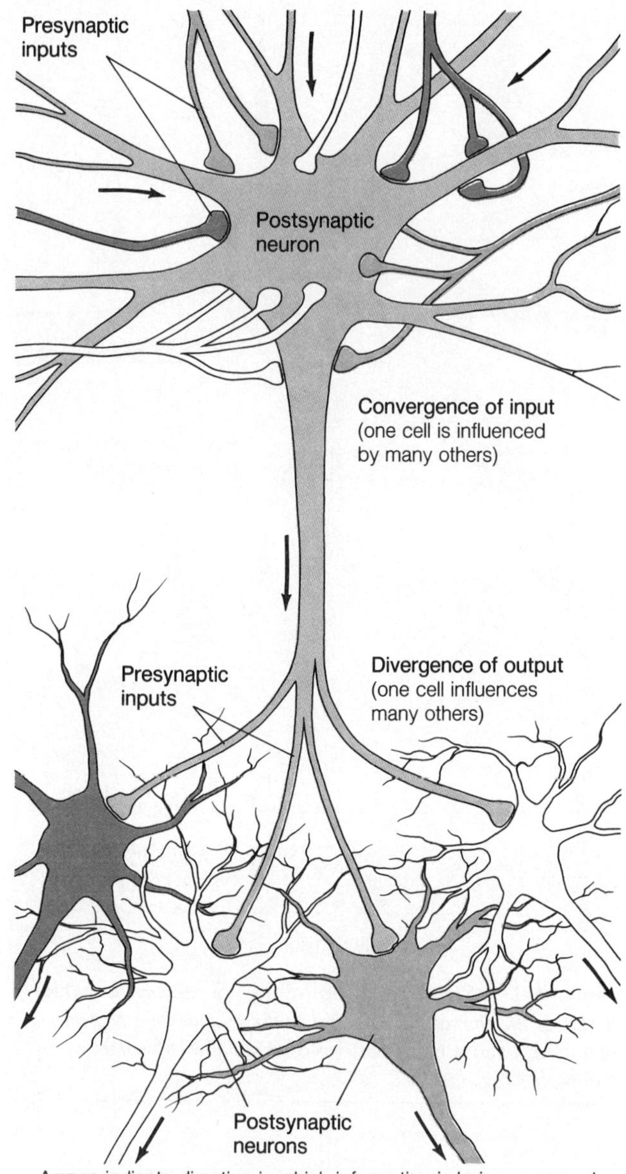

Figure 4–18 Convergence and Divergence

A CLOSER LOOK AT EXERCISE PHYSIOLOGY

The Yells of a Weight Lifter May Serve a Physiological Purpose

Skeletal muscles are innervated by a special type of neuron called an alpha motor neuron (see chapter 7). Action potentials in alpha motor neurons initiate the contractile process in the muscle on which they terminate. Both excitatory and inhibitory presynaptic inputs from a variety of sources impinge on the cell body and dendrites of the alpha motor neurons. These neurons fire, thereby initiating muscle contraction, when their membrane is brought to threshold by summation of EPSPs. A person can voluntarily contract a given skeletal muscle by consciously activating nerve pathways that generate EPSPs at the appropriate alpha motor neurons.

Among the inhibitory inputs to alpha motor neurons are those arising from the Golgi tendon organ, a sensory device located within the tendon that attaches a muscle to the skeleton (see chapter 8). The Golgi tendon organ is sensitive to stretch imposed on it by the muscle's contraction. When stretched, the Golgi tendon organ initiates a nerve pathway that generates IPSPs on the alpha motor neurons that innervate the very same contracting muscle in which the sensory device is located. If the grand postsynaptic potential of the involved alpha motor neurons is reduced below threshold by the inhibitory input from the Golgi tendon organ, the muscle is no longer stimulated and it relaxes. This reflex relaxation is protective in nature; its purpose is to prevent the muscle from contracting hard enough to rupture the muscle or tendon or break a bone.

Weight training requires the voluntary control of this natural muscular inhibitory reflex. When persons begin to weight train, their ability to lift dramatically increases in the first two weeks even though no physiologic change has yet occurred in the muscle. This initial improvement in strength is thought to be brought about by the body learning to voluntarily produce more EPSPs, thus voluntarily overriding the muscles' inhibition by their stretched Golgi tendon organs. As weight training progresses, muscles get physiologically stronger as the muscle cells enlarge, the tendons get thicker, and the Golgi tendon organs decrease their sensitivity.

Power lifters sometimes yell as they lift to try to further increase the excitatory stimuli during heavy lifting. In sports such as arm wrestling, the inhibitory reflex is overridden by voluntary control to the point that participants have ruptured muscles or tendons and have even broken the bones in the arm during the excitement of competition.

Chapter in Perspective

Neurons make up the nervous system, one of the two major control systems of the body. The nervous system exerts control over the body's muscular and glandular activities, most of which are directed toward maintaining homeostasis.

Neurons are specialized for rapid electrical and chemical signalling. They are able to process, initiate, code, and conduct changes in their membrane potential as a means of rapidly transmitting a message throughout their length. Furthermore, neurons have developed chemical means of passing this information through intricate nerve pathways from neuron to neuron as well as to muscles and glands. These pathways may be incredibly complex, but the nature of the electrical impulse (action potential) and chemical-converting mechanism (neurotransmitter-receptor combination) is always the same.

The cell body and dendrites of most neurons have an abundance of receptor proteins that bind with neurotransmitters from the thousands of presynaptic inputs converging on this area of the cell. Each neurotransmitter-receptor combination transiently changes the permeability of this region of the cell in a predictable fashion by altering either Na^+, K^+, or Cl^- channels to induce ionic fluxes. These ionic fluxes, in turn, bring about excitatory or inhibitory graded potential changes in the postsynaptic cell. The grand postsynaptic potential is a composite of the effects of all the excitatory and inhibitory presynaptic signals. If the combined effects of these inputs sufficiently excites the postsynaptic membrane to bring it to a critical threshold level, voltage-gated Na^+ channels concentrated in the axon hillock region of the neuron are electrically jolted into their "open" conformation. The subsequent influx of Na^+ brings about a rapid reversal of membrane potential, which is quickly restored to its resting state by K^+ efflux as voltage-dependent K^+ gates swing open simultaneous to closure of the Na^+ gates. This explosive change in membrane potential is known as an action potential. Current flowing between an area undergoing an action potential and its neighboring resting patch of membrane brings the adjacent area to threshold, so that the action potential is regenerated at this site, repeating the process until the end of the axon is reached.

Each action potential is an all-or-none response. The electrical signal initiated at the axon hillock is faithfully transmitted down to the axon terminals in undiminished fashion, thereby enabling long-distance transmission of the signal. The presence of an action potential at the axon terminal triggers the release of neurotransmitter, which in turn chemically carries the message to all of the cells with which this particular neuron communicates by means of divergence of output.

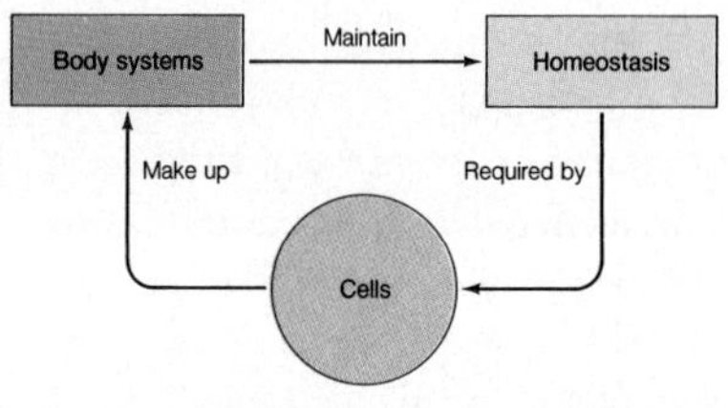

See inside front cover for an expanded version of this model.

Review Exercises

1. What are the two types of excitable tissue?
2. Compare graded potentials and action potentials.
3. Define the following terms: polarization, depolarization, hyperpolarization, repolarization, resting membrane potential, threshold potential, action potential, absolute refractory period, relative refractory period, all-or-none law, convergence, divergence.
4. Compare the two kinds of channels, depending on the factor that induces the change in channel conformation.
5. Describe the permeability changes and ionic fluxes that occur during an action potential.
6. Explain the role of the Na^+-K^+ pump with regards to action potentials.
7. Compare conduction by local current flow and saltatory conduction.
8. Compare the events that occur at excitatory and inhibitory synapses.
9. Discuss the four possible outcomes of the grand postsynaptic potential brought about by interactions between EPSPs and IPSPs.
10. Distinguish between presynaptic inhibition and an inhibitory postsynaptic potential.
11. **A point to ponder:** If a neuron were experimentally stimulated simultaneously at both ends,
    a. the action potentials would pass in the middle and travel to the opposite ends.
    b. the action potentials would meet in the middle and then be propagated back to their starting positions.
    c. the action potentials would stop as they met in the middle.
    d. the strongest action potential would override the weaker action potential.
    e. summation would occur when the action potentials met in the middle, resulting in a larger action potential.

C H A P T E R 5

Central Nervous System

Chapter Contents at a Glance

INTRODUCTION *The neurosurgeon carefully examines the portion of the patient's brain that has been exposed, looking for appropriate landmarks on the neural map. Before cutting into this precious, nonregenerative tissue, the surgeon explores with a tiny stimulating electrode, asking the patient to describe what happens with each stimulation—the flick of a finger. . . a prickly feeling on the bottom of the foot. . . nothing? Because the brain itself is insensitive to pain, all of this is done while the patient is awake with only local anesthesia used along the cut scalp, allowing the surgeon to be absolutely certain of location before penetrating the brain.*

Specific areas of the brain are known to be responsible for controlling particular activities or for receiving sensory input from designated regions of the body. These areas have been mapped in general, but there are subtle variations between individuals. Just as each of us has two eyes, a nose, and a mouth and yet no two faces have these features arranged in exactly the same way, so it is with brains. Thus, the surgeon must explore the region carefully before an incision is made to ensure that an important area is not damaged unintentionally.

Our understanding of the brain is rudimentary because the brain is so complex and because there is no good model for study of its most sophisticated functions.

Although some of its areas have been mapped, the responsibilities and mechanisms of function of much of the human brain (Fig.5–1) are still very poorly understood. Part of our lack of understanding derives from the complexity of the brain. It is the most complicated, mysterious, awesome organ on earth. Considering the myriad interconnections possible between the estimated 100 billion neurons in the brain, it is a wonder that scientists have been able to unravel even the bits of information that they have.

Furthermore, unlike other organs of the body, the human brain is so unique that no experimental animal models are available for study of its most complex and sophisticated functions, such as language and creativity. By comparison, because human hearts are essentially the same as dog hearts, investigations probing canine cardiac function provide valuable insights into human heart function. In addition, many aspects of brain function involve subjective feelings that cannot be measured or communicated by nonverbal animals. We do not know for sure to what extent, if any, animals experience what we know of as happiness, sadness, love, fear, jealousy, and so on. We can only observe in them behavioral aspects that we associate with certain emotional states (e.g., a dog wagging its tail when it is "happy"). Therefore, studies on experimental animals have been limited to examining less advanced, objectively measurable brain activities, such as control of movement, shared in common between humans and other species.

Knowledge about the brain in the past was gleaned from: (1) stimulation during brain surgery; (2) analysis of changes in electrical activity reaching the surface of the skull when a person is engaged in various activities; (3) observation of deficits in function associated with diseases or destruction of particular areas of the brain; (4) detailed anatomical studies in cadavers; and (5) inferences from animal studies. The latter, of course, are limited to characteristics shared in common between these species and humans, leaving most of the very characteristics that confer upon us our uniqueness as humans largely unexplored. However, major developments in experimental methodology are rapidly expanding the boundaries of neural science. Several exciting new noninvasive techniques of imaging the living human brain, as well as new methods of probing cellular and molecular functions of neurons, are dramatically adding new pieces of knowledge needed to ultimately bridge the gap between our understanding of the biology of nerve cells and the behavior and intellect of the human brain.

The brain is modified in response to environmental influences.

The way humans act and react depends on complex, organized, discrete neuronal processing. Many of the basic life-supporting neuronal patterns, such as those controlling respiration and circulation, are similar in all individuals. However,

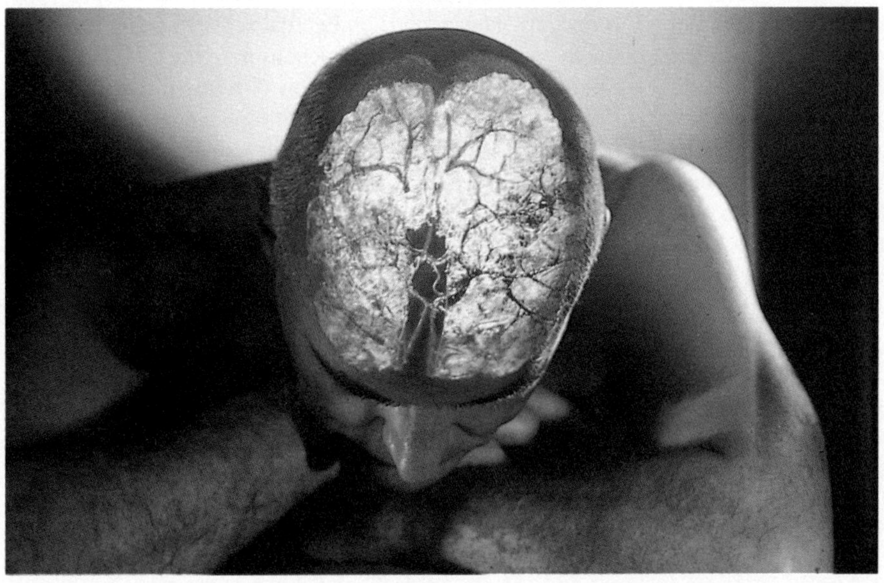

Figure 5–1
Top View of Human Brain

there must be subtle differences in neuronal integration between someone who is a talented composer versus someone who cannot carry a tune, or someone who is a math wizard versus someone who struggles with long division. Some differences in the nervous systems between individuals are genetically endowed. The rest, however, are due to environmental encounters and experiences. When the immature nervous system develops according to its genetic plan, an overabundance of neurons and synapses are formed. Depending on external stimuli and the extent of use of these pathways, some are retained, firmly established, and even enhanced, whereas others are eliminated. A case in point is **amblyopia (lazy eye)** in which the weaker of the two eyes in not used for vision. A lazy eye that does not get appropriate visual stimulation during a critical developmental period will almost completely and permanently lose the power of vision. The blind eye itself is completely normal; the defect lies in the lost neuronal connections in the brain visual pathways. If, however, the weak eye is forced to work by covering the stronger eye with a patch during the sensitive developmental period, its vision will be retained. There truly are instances of "use it or lose it" regarding maturation of the nervous system. Once matured, ongoing modifications still occur as we continue to learn from our unique set of experiences. For example, the act of reading this page is somehow altering the neuronal activity of your brain as you (it is hoped) tuck the information away in your memory.

The nervous and endocrine systems have different regulatory responsibilities but share much in common.

The nervous system is one of the two main control systems of the body, the other being the endocrine system. The **nervous system** via its swift transmission of impulses generally coordinates the rapid activities of the body, such as muscle movements. The **endocrine system** primarily controls metabolic and other activities that require duration rather than speed, such as maintenance of blood glucose levels. The endocrine glands secrete hormones into the blood, which carries these chemical messengers to their sites of action (i.e., their **target cells**).

Even though the two major control systems are distinctly different, they share much in common. They both ultimately alter their target cells by release of chemical messengers that interact in particular ways with specific receptors of the target cells. The main differences are: (1) the distance traveled by the messenger (only across a synaptic cleft in the case of neurotransmitters; long distances through the blood in the case of hormones) and (2) the signal for the release of the messenger (an action potential in the case of neurons; numerous specific signals, which may include action potentials, in the case of endocrine cells.) A given messenger, (for example norepinephrine), may even be a neurotransmitter when released from a nerve ending or may be a hormone when secreted by an endocrine cell. Besides sharing similar or identical messengers, the nervous and endocrine systems are also intricately interrelated in their control activities. The nervous system has important control functions over the secretion of many hormones. At the same time, many hormones act as neuromodulators to alter synaptic effectiveness (see p. 111). The presence of certain key hormones is even essential for the proper development and maturation of brain tissue.

For now we will concentrate on the nervous system and will examine the endocrine system in more detail in later chapters. Throughout the text we will continue to point out the numerous ways in which these two control systems interact so that the body is a coordinated whole, even though each system has its own "realm of authority."

The nervous system is organized into the central nervous system and the peripheral nervous system.

The nervous system is organized into the **central nervous system (CNS)**, consisting of the **brain** and **spinal cord,** and the **peripheral nervous system (PNS)**, consisting of nerve fibers that carry information between the CNS and other parts of the body (the periphery) (Fig. 5–2). The PNS is further subdivided into afferent and efferent divisions. The **afferent division** (*afferent* means "carrying toward") carries information *to* the CNS, apprising it of the external environment and providing status reports on internal activities being regulated by the nervous system. Instructions *from* the CNS are transmitted via the **efferent division** (*efferent* means "carrying from") to **effector organs** — the muscles or glands that carry out the orders. The efferent nervous system is divided into the **somatic nervous system,** which consists of **motor neurons** that supply the skeletal muscles, and the **autonomic nervous system,** which innervates smooth muscle, cardiac muscle, and glands. The latter system is further subdivided into the **sympathetic nervous system** and the **parasympathetic nervous system,** both of which innervate most of the organs supplied by the autonomic system.

It is important to recognize that all of these "nervous systems" are really subdivisions of a single, integrated nervous system. They are arbitrary divisions based on differences in the structure, location, and functions of the various diverse parts of the whole nervous system.

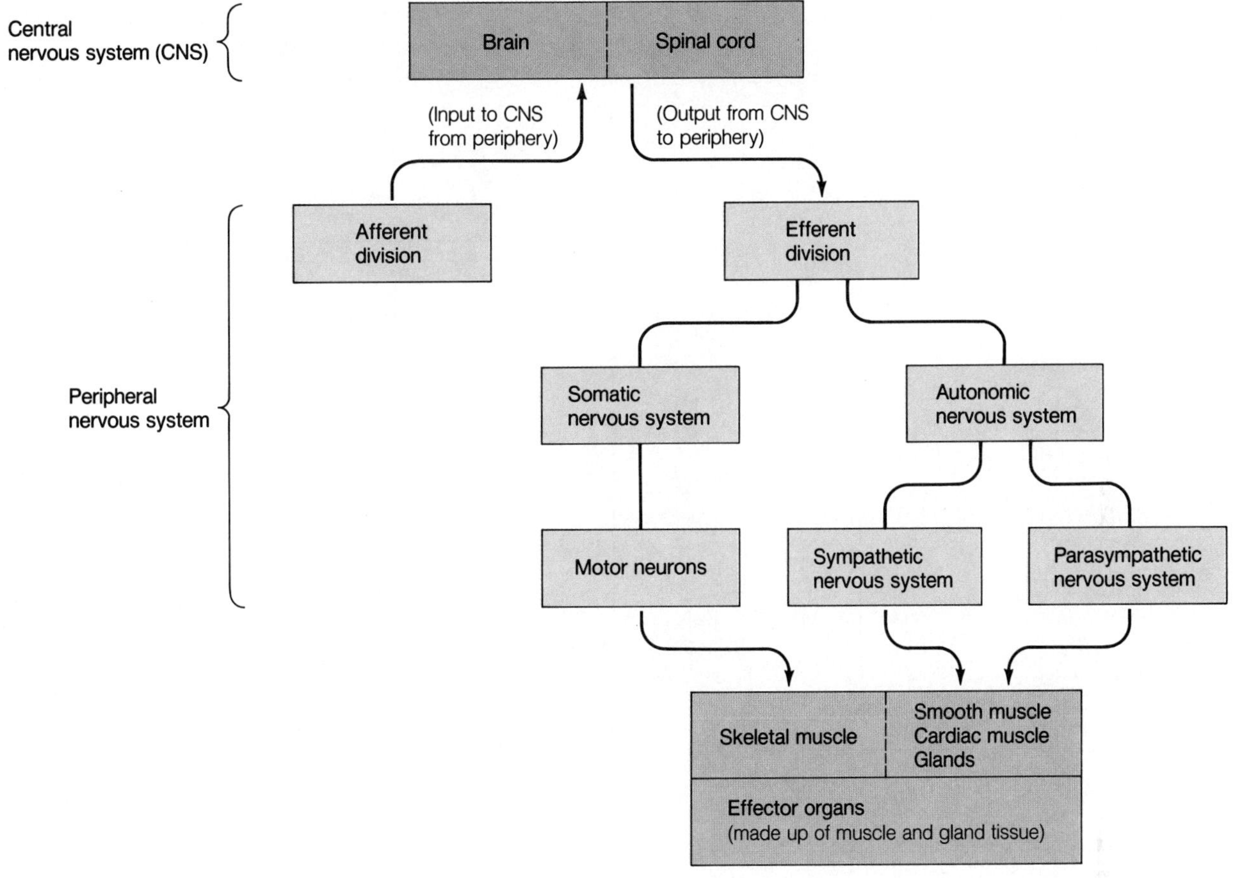

Figure 5–2 Organization of Nervous System

There are three classes of neurons.

Three classes of neurons make up the nervous system: afferent neurons, efferent neurons, and interneurons. The afferent nervous system is composed of **afferent neurons,** which are shaped differently than efferent neurons and interneurons (Fig. 5–3). At its peripheral ending, an afferent neuron has a **sensory receptor** that generates action potentials in response to a particular type of stimulus. (This stimulus-sensitive receptor should not be confused with the special protein receptors of the plasma membrane that bind chemical messengers.) The afferent neuron cell body, which is devoid of dendrites and presynaptic inputs, is located adjacent to the spinal cord. A long peripheral axon extends from the receptor to the cell body, and a short central axon passes from the cell body into the spinal cord. The terminals of the central axon diverge and synapse with other neurons within the spinal cord, in this way disseminating information about the stimulus. Afferent neurons thus lie primarily within the peripheral nervous system, with only a small portion of their central axon endings projecting into the spinal cord to relay peripheral signals.

Efferent neurons also lie primarily in the peripheral nervous system (Fig. 5–3). The cell bodies of efferent neurons originate in the CNS, where many centrally located presynaptic inputs converge upon them to influence their outputs to the effector organs. Efferent axons leave the CNS to course their way to the muscles or glands they innervate, conveying their integrated output for the effector organs to put into effect. (An autonomic nerve pathway actually consists of a two-neuron chain between the CNS and effector organ.)

Interneurons lie entirely within the CNS. About 99% of all neurons belong to this category. They serve two main roles. First, as their name implies, they lie between (*inter* means "between") the afferent and efferent neurons and are important in the integration of peripheral responses to peripheral information. For example, upon receiving information through afferent neurons that you are touching a hot object, appropriate interneurons signal efferent neurons that transmit to hand

Figure 5–3 Structure and Location of the Three Classes of Neurons

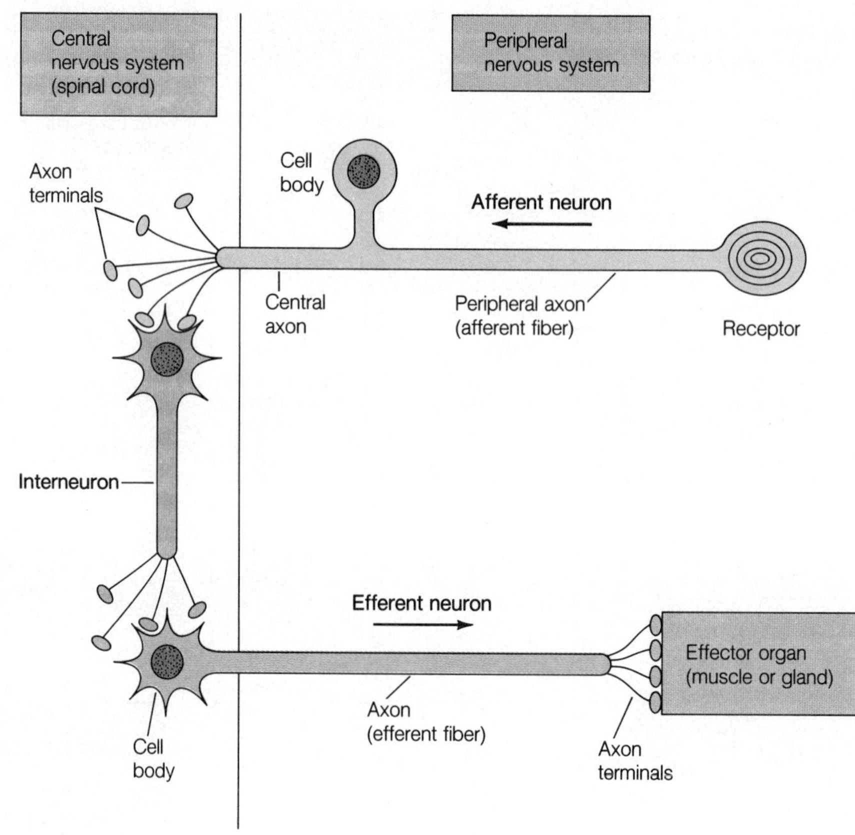

and arm muscles the message, "Pull the hand away from the hot object!" The more complex the required action, the greater the number of interneurons interposed between the afferent message and efferent response. Second, interconnections between interneurons themselves are responsible for the abstract phenomena associated with the "mind," such as thoughts, emotions, memory, creativity, intellect, and motivation. These activities are the least well understood of nervous system function.

With this brief introduction to the types of neurons and how they are organized into various divisions of the nervous system, we will next turn our attention to the central nervous system, followed in later chapters by a discussion of the peripheral nervous system.

Protection and Nourishment of the Brain

Neuroglia physically support the interneurons and help sustain them metabolically.

About 90% of the cells within the CNS are not neurons but are **glial cells** or **neuroglia.** Despite their large numbers, the glial cells occupy only about half of the volume of the brain because they do not branch as extensively as the neurons do. Unlike neurons, glial cells do not initiate or conduct nerve impulses. They are important, however, in the viability of the CNS. They serve as the connective tissue of the CNS and as such help support the neurons both physically and metabolically. The four major types of glial cells in the CNS are astrocytes, oligodendrocytes, ependymal cells, and microglia.

Astrocytes provide a number of critical functions. First, as the main "glue" of the CNS, they hold the neurons together in proper spatial relationships. Second, they are important in the repair of brain injuries and in neural scar formation. Third, they help support the neurons metabolically. Finally, they take up excess K^+ from the brain ECF when high action potential activity outpaces the ability of the Na^+-K^+ pump to return the effluxed K^+ to the neurons. If brain ECF K^+ levels were allowed to rise, the resultant lower K^+ concentration gradient between the neuronal ICF and surrounding ECF would reduce the neural membrane potential closer to threshold, even at rest. This would increase the excitability of the brain. In fact an elevation in brain ECF K^+ concentration may be one of the factors responsible for the brain cells' explosive convulsive discharge that occurs during epileptic seizures.

Oligodendrocytes form the insulative myelin sheaths around axons in the CNS. An oligodendrocyte has several elongated projections, each of which is wrapped jelly-roll fashion around a section of an interneuronal axon to form a patch of myelin (see Fig. 4–10).

Ependymal cells line the internal cavities of the CNS. As the nervous system develops embryonically from a hollow neural tube, the original central cavity of this tube is maintained and modified to form the **ventricles** of the brain and the **central canal** of the spinal cord. The ependymal cell lining of the ventricles contributes to the formation of cerebrospinal fluid .

The **microglia** are the scavengers of the CNS. They are phagocytic cells delivered by the blood to the central nervous tissue, where they remain stationary until activated by an infection or injury. They then migrate to the affected area to remove any foreign invaders or tissue debris.

Unlike neurons, glial cells do not lose the ability to undergo cell division, so most brain tumors of neural origin consist of glial cells (**gliomas**). Neurons themselves do not form tumors because of their inability to divide and multiply. Brain tumors of nonneural origin are of two types: (1) those that metastasize (spread) to the brain from other sites and (2) **meningiomas,** which originate from the meninges, the protective membranes covering the central nervous system.

The delicate central nervous tissue is well-protected.

Central nervous tissue is very delicate. This, coupled with the fact that damaged nerve cells cannot be replaced because of the inability of neurons to divide, makes it imperative that this fragile, irreplaceable tissue be well-protected. Four major features help protect the CNS from injury:

1. It is enclosed by hard, bony structures. The **cranium (skull)** encases the brain and the **vertebral column** surrounds the spinal cord.
2. Three protective and nourishing membranes, the **meninges,** lie between the bony covering and the nervous tissue.
3. The brain "floats" in a special cushioning fluid, the **cerebrospinal fluid** (**CSF**).
4. A highly selective **blood-brain barrier** limits access of potentially harmful materials into the vulnerable brain tissue.

The role of the first of these protective devices, the bony covering, is self-evident. However, the latter three protective mechanisms warrant further discussion.

Three meningeal "mothers" provide protection and nourishment for the central nervous system.

The meninges, the three membranes that wrap the central nervous system are, from the outermost to the innermost layer, the dura mater, the arachnoid mater, and the pia mater (Fig. 5–4). (*Mater* means "mother," indicative of the protective and supportive role played by these membranes.)

The **dura mater** (*dura* means "tough") is a tough, inelastic covering consisting of two layers. Usually these layers are closely adherent, but in some regions they are separated to form blood-filled cavities, **dural sinuses,** or in the case of the larger cavities, **venous sinuses.** Venous blood draining from the brain empties into these sinuses to be returned to the heart. Cerebrospinal fluid also reenters the blood at these sinus sites.

The **arachnoid mater** (*arachnoid* means "spiderlike") is a delicate, richly vascularized layer that is "cobwebby" in its appearance. The space between the arachnoid layer and the underlying pia mater, the **subarachnoid space,** is filled with CSF. Protrusions of arachnoid tissue, the **arachnoid villi,** penetrate through gaps in the overlying dura and project into the dural sinuses. It is across the surfaces of these villi that CSF is reabsorbed into the blood circulating within the sinuses.

The innermost meningeal layer, the **pia mater** (*pia* means "gentle"), is the most fragile. It is highly vascular and closely adheres to the surfaces of the brain and spinal cord, following every ridge and valley. In certain areas it dips deeply into the brain to bring a rich blood supply into close contact with the ependymal cells lining the ventricles. This relationship is important in the formation of CSF, a topic to which we now turn our attention.

The central nervous system is suspended in its own special cerebrospinal fluid.

Cerebrospinal fluid is formed primarily by the **choroid plexuses** found in particular regions of the ventricle cavities of the brain. Choroid plexuses consist of richly vascularized, cauliflower-like masses of pia mater tissue that dip into pockets formed by ependymal cells. Once CSF is formed, it flows through the four interconnected ventricles within the interior of the brain and through the spinal cord's narrow central canal, which is continuous with the last ventricle. Cerebrospinal fluid escapes through small openings from the fourth ventricle at the base of the brain to enter the subarachnoid space and subsequently flows between the meningeal layers over the entire surface of the brain and spinal cord (Fig 5–4). When the CSF reaches the upper regions of the brain, it is reabsorbed from the subarachnoid space into the venous blood through the arachnoid villi.

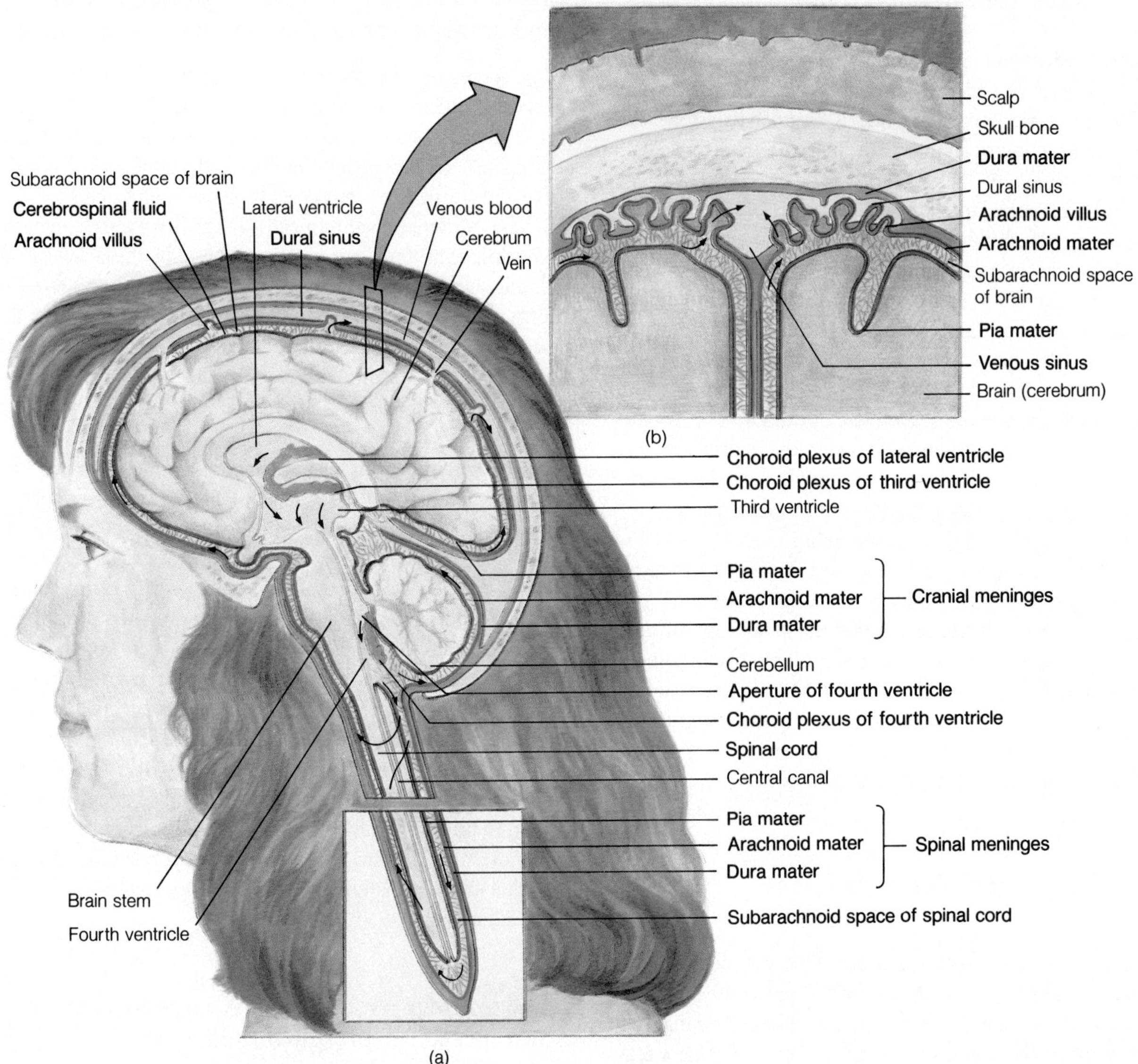

Figure 5-4 Relationship of Meninges and Cerebrospinal Fluid to Brain and Spinal Cord *(a) Brain, spinal cord, and meninges in sagittal section. The arrows depict the direction of flow of cerebrospinal fluid (in yellow), which (1) is produced by the choroid plexuses, (2) circulates throughout the ventricles, (3) exits the fourth ventricle at the base of the brain, (4) flows in the subarachnoid space between the meningeal layers, and (5) is finally reabsorbed from the subarachnoid space into the venous blood across the arachnoid villi. (b) Frontal section in the region between the two cerebral hemispheres of the brain, depicting the meninges in greater detail.*

Flow of CSF through this system is facilitated by circulatory and postural factors that result in a CSF pressure of about 10 mm Hg. Reduction of this pressure by removal of even a few milliliters (ml) of CSF during a spinal tap for laboratory analysis may produce severe headaches.

Through the ongoing processes of formation, circulation, and reabsorption of CSF, its entire volume of about 125 to 150 ml is replaced more than three times a day. **Hydrocephalus** ("water on the brain") occurs if any one of these processes is defective so that excess CSF accumulates. The resulting in-

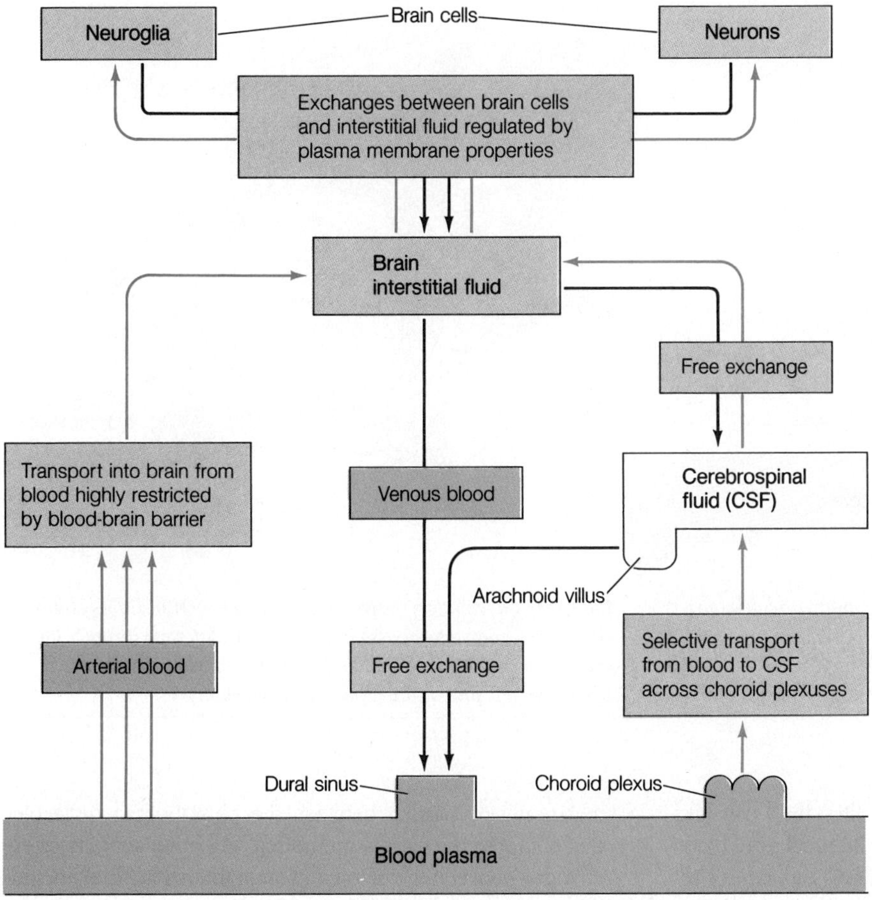

Figure 5–5 Exchanges between Blood and Brain Cells

crease in CSF pressure can lead to brain damage and mental retardation if untreated. Treatment consists of surgically shunting the excess CSF to veins elsewhere in the body.

The major function of CSF is to serve as a shock-absorbing (cushioning) fluid to prevent the brain from bumping against the interior of the hard skull when the head is subjected to sudden, jarring movements. The density of the CSF is about the same as that of the brain itself, so the brain is essentially suspended in its special fluid environment.

In addition to protecting the delicate brain from mechanical trauma, the CSF performs an important role related to the exchange of materials between the body fluids and the brain. In all body tissues, the spaces between the cells are filled with a type of extracellular fluid known as interstitial fluid or tissue fluid. It is the interstitial fluid of the brain and not the blood plasma or the CSF that comes into direct contact with the neuronal and glial cells (Fig. 5–5). Since the brain interstitial fluid directly bathes the neural cells, its composition is critical. The composition of brain interstitial fluid is influenced more by changes in the composition of CSF than by alterations in blood plasma, because there is fairly free exchange of materials between the CSF and brain interstitial fluid but only limited exchange between the blood and brain interstitial fluid. Accordingly, it is essential that the composition of the CSF be regulated within very narrow limits.

Cerebrospinal fluid is formed as a result of selective transport mechanisms across the membranes of the choroid plexuses. The composition of CSF is different than that of plasma. For example, CSF is lower in K^+ and higher in Na^+, making it an ideal environment for the movement of these ions down concentration gradients, a process essential for conduction of nerve impulses (see pp. 95–99).

A highly selective blood-brain barrier carefully regulates exchanges between the blood and brain.

The brain is carefully shielded from harmful changes in the blood by the blood-brain barrier. Exchange of materials be-

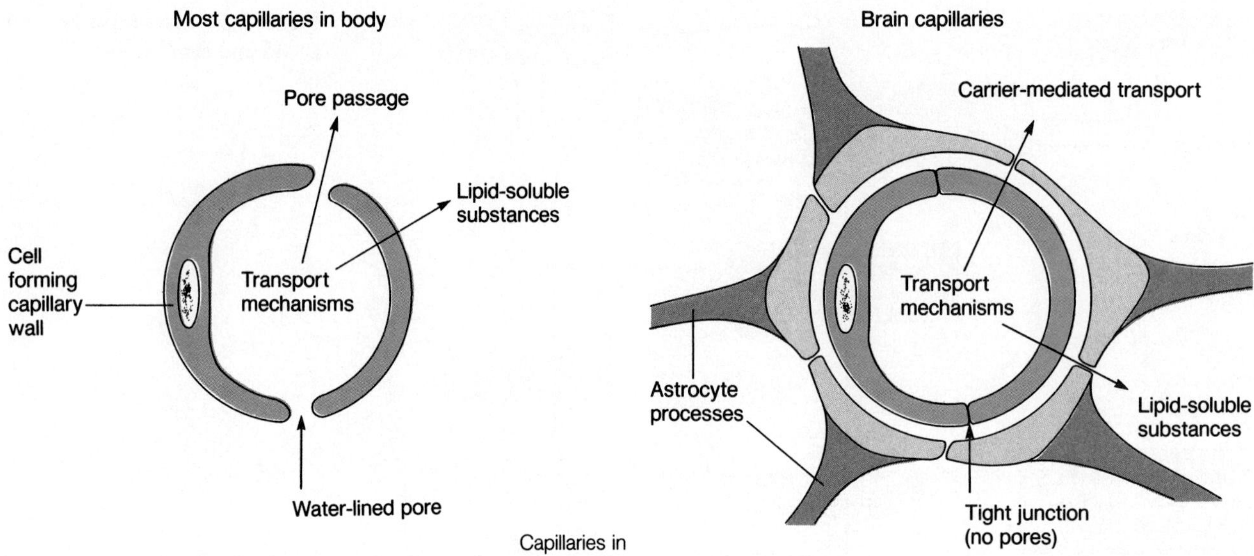

Figure 5–6 Blood-Brain Barrier *Unlike most capillaries in the body, the cells forming the walls of brain capillaries are joined by tight junctions that prevent passage of materials between the cells. The only passage across brain capillaries is through the cells that form the capillary walls. With the exception of lipid-soluble substances and water, passage of all other materials through these cells is physiologically regulated by carrier-mediated systems, which are not present in capillaries elsewhere.*

tween the blood and surrounding interstitial fluid can take place only across the walls of capillaries, the smallest of blood vessels. Unlike the rather free exchange across capillaries elsewhere in the body, there are strict limitations on permissible exchanges across brain capillaries. Changes in most plasma constituents do not easily influence the composition of brain interstitial fluid because only selected exchanges can be made. For example, even if the K^+ level in the blood is doubled, little change occurs in the K^+ concentration of the fluid bathing the central neurons. This is beneficial, because alterations in interstitial fluid K^+ would be detrimental to neuronal function.

The blood-brain barrier consists of both anatomical and physiological factors. Capillary walls throughout the body are formed by a single layer of cells. Usually there are holes or pores between the cells making up the capillary wall that permit free exchange of all plasma components with the surrounding interstitial fluid, with the exception of the large plasma proteins. In the brain capillaries, however, the cells are joined by **tight junctions** that completely seal the capillary wall so that nothing can be exchanged across the wall by passing *between* the cells (Fig. 5–6). The only possible exchanges are *through* the capillary cells themselves. Lipid-soluble substances such as O_2, CO_2, alcohol, and steroid hormones penetrate these cells easily by dissolving in their lipid plasma membrane. Small water molecules also diffuse through readily, apparently by passing between the phospholipid molecules that compose the plasma membrane. All other substances exchanged between the blood and brain interstitial fluid, including such essential materials as glucose, amino acids, and ions, are transported by highly selective membrane-bound carriers. Accordingly, transport across the capillary walls between the cells is anatomically prevented and transport through the cells is physiologically restricted.

The brain capillaries are surrounded by astrocyte processes, which at one time were thought to be physically responsible for the blood-brain barrier. Evidence now indicates that the astrocytes have three roles: (1) They appear to signal the cells forming the brain capillaries to "get tight." Capillary cells do not have an inherent ability to form tight junctions; they do so only at the command of a signal within their neural environment. (2) Astrocytes are believed to participate in the cross-cellular transport of some substances, such as K^+. (3) They also serve as support frameworks to maintain structural relationships between the capillaries and surrounding tissues.

The blood-brain barrier protects the delicate brain and spinal cord from chemical fluctuations in the blood and minimizes the possibility that potentially harmful blood-borne substances might reach the central neural tissue. It further prevents certain circulating hormones that could also act as neurotransmitters from reaching the brain, where they could produce uncontrolled nervous activity. On the negative side, the blood-brain barrier limits the use of drugs for the treatment of brain and spinal cord disorders because many drugs are unable to penetrate this barrier.

Certain areas of the brain are not subject to the blood-brain barrier, most notably a portion of the hypothalamus. Functioning of the hypothalamus depends on its "sampling" the blood and adjusting its controlling output accordingly in order to maintain homeostasis. Part of this output is in the form of hormones that must enter hypothalamic capillaries to be transported to their sites of action. Appropriately, these hypothalamic capillaries are not sealed by tight junctions.

The brain depends on constant delivery of oxygen and glucose by the blood.

Even though many substances in the blood never actually come in contact with most of the brain tissue, the brain is highly dependent on a constant blood supply, more so than any other tissue. Unlike most tissues that can resort to anaerobic metabolism to produce ATP in the absence of O_2 for at least short periods (see p. 32), the brain cannot produce ATP in the absence of O_2. Furthermore, in contrast to most tissues that can use other sources of fuel for energy production in lieu of glucose, the brain normally uses only glucose but does not store any of this nutrient. Therefore, the brain is absolutely dependent on a continuous, adequate blood supply of O_2 and glucose. Accordingly, brain damage results if this organ is deprived of its critical O_2 supply for more than four to five minutes or if its glucose supply is cut off for more than ten to fifteen minutes.

Brain damage may occur in spite of protective mechanisms.

Even though the brain is carefully protected in its cushiony vault, traumatic head injuries may damage the delicate brain tissue. Direct brain damage occurs if the brain is violently shaken or jarred by a forceful impact, such as a fall or a blow, or if crushed cranial bones are pushed against the underlying neural tissue. Further indirect brain damage may occur following a head injury as a consequence of swelling or hemorrhaging within the enclosed confines of the cranium. The resultant increase in intracranial pressure may cause compression and damage of brain tissue.

The most common cause of brain damage is not traumatic head injuries, however, but **cerebrovascular accidents (strokes)**. Rupture or blockage of a brain (cerebral) blood vessel deprives the brain tissue being supplied by the involved vessel of its vital O_2 and glucose supply. This leads to damage and usually death of the deprived tissue. Brain damage also may occur as a result of infectious or degenerative neural disorders or brain tumors, all of which can lead to destruction of brain tissue.

No matter what the cause, the nature of the ensuing loss of neurological function depends on the area of the brain involved and the extent of permanent damage. The following exemplify the range of possible outcomes: (1) death if an area responsible for maintenance of a vital function is destroyed (for example, the brain center controlling respiration); (2) severe or mild loss of specific sensory awareness; (3) motor (movement) disorders of varying severity; (4) impaired mental abilities; or (5) full functional recovery.

The brain displays a degree of **plasticity,** that is, an ability to change or be functionally remolded in response to the demands placed on it. This ability is more pronounced in the early developmental years, but even adults retain some plasticity. When an area of the brain associated with a particular activity is destroyed, in some instances other areas of the brain gradually assume some or all of the responsibilities of the damaged region. How this is accomplished is unclear.

Destroyed CNS neurons cannot themselves be replaced through cell division. Furthermore, damaged axons cannot regenerate within the CNS as they can in the peripheral nervous system, perhaps at least in part because of differences in the myelin-forming cells. The Schwann cells of the peripheral nervous system appear to release a growth-promoting factor that encourages the regeneration of a cut axon, whereas the oligodendrocytes of the CNS apparently are unable to perform this same role. Damaged CNS axons can regenerate experimentally in the presence of a Schwann cell graft, whereas further regeneration of damaged peripheral axons ceases upon exposure to CNS glial cells. A promising future prospect is the isolation, identification, and therapeutic use of this neural growth-promoting factor to permit regeneration of damaged CNS axons. There is even hope that entire damaged regions of the brain might be replaced by brain grafts.

Headaches are seldom due to brain damage.

Headaches are the most common form of pain. Almost everyone experiences headaches at least occasionally. Fortunately, most headaches are not associated with brain damage. They usually result from tension or increased pressure within pain-sensitive structures inside the cranium. The following are among the causes of headaches:

1. *tension* associated with sustained tightening of muscles in the neck, scalp, and forehead in conjunction with anxiety, stress, or fatigue;
2. *swelling of the mucous membranes* lining the sinuses in response to respiratory infections or allergies;

3. *eye disorders* accompanied by straining of eye muscles;
4. *dilation of cerebral blood vessels* in association with high blood pressure, hangovers, or migraine headaches.
5. *increased intracranial pressure* accompanying brain tumors or intracranial hemorrhaging; or
6. *inflammation and swelling* in association with meningeal infections (**meningitis**) or infection of the brain itself (**encephalitis**).

Cerebral Cortex

Newer, more sophisticated regions of the brain are piled on top of older, more primitive regions.

Although the brain is a functional whole, it is organized into several different regions. There are various ways to arbitrarily group the parts of the brain based on anatomical distinctions,

Table 5–1 Overview of Structures and Functions of the Major Components of the Brain

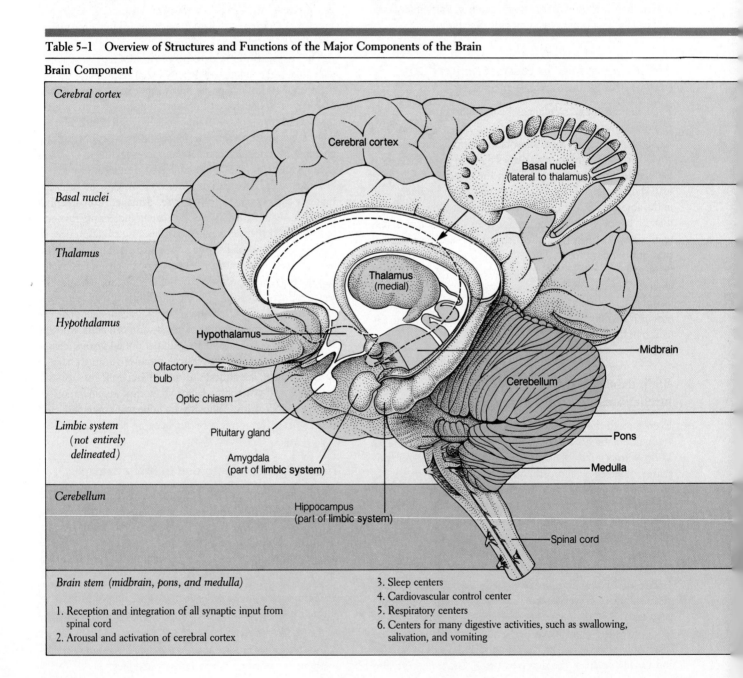

Brain Component	
Cerebral cortex	
Basal nuclei	
Thalamus	
Hypothalamus	
Limbic system (not entirely delineated)	
Cerebellum	
Brain stem (midbrain, pons, and medulla) 1. Reception and integration of all synaptic input from spinal cord 2. Arousal and activation of cerebral cortex	3. Sleep centers 4. Cardiovascular control center 5. Respiratory centers 6. Centers for many digestive activities, such as swallowing, salivation, and vomiting

functional specialization, and evolutionary development. We will use the following grouping (Table 5–1):

1. Brain stem
2. Cerebellum
3. Forebrain
   a. Diencephalon
      1. Hypothalamus
      2. Thalamus
   b. Cerebrum
      1. Basal nuclei
      2. Cerebral cortex

The order in which these components are listed generally represents both their anatomical location and their complexity and sophistication of function from the least specialized, oldest level to the newest, most specialized level.

A primitive nervous system consists of comparatively few interneurons interspersed between afferent and efferent neurons. During evolutionary development, the interneuronal component progressively expanded, formed more complex interconnections, and became localized at the head end of the nervous system, forming the brain. Newer, more sophisticated layers of the brain were added on to the older, more primitive layers, reaching its present peak of development in the human brain.

The ***brain stem***, the oldest and smallest region of the brain, is continuous with the spinal cord. It controls many of the life-sustaining processes, such as breathing, circulation, and digestion, that are shared in common with many of the lower vertebrate forms. These are often referred to as "vegetative" functions because, with the loss of higher brain functions, these lower brain levels, in accompaniment with appropriate supportive therapy such as intravenous feeding, can still sustain the functions essential for survival. However, because the person has no awareness or control of that life, the condition is sometimes referred to as "being a vegetable."

Attached at the top rear portion of the brain stem is the ***cerebellum***, which is concerned with maintaining proper position of the body in space and subconscious coordination of motor activity (movement). On top of the brain stem, tucked within the interior of the cerebrum is the ***diencephalon***. It houses two brain components, the ***hypothalamus***, which controls many homeostatic functions important in maintaining stability of the internal environment, and the ***thalamus***, which performs some primitive sensory processing. On top of this "cone" of lower brain regions is the ***cerebrum***, the "scoop" of which gets progressively larger and more highly convoluted the more advanced the vertebrate species is. The cerebrum is most highly developed in humans, constituting about 80% of the total brain weight. The outer layer of the cerebrum is the highly convoluted ***cerebral cortex***, which caps an inner core that houses the ***basal nuclei***. The cerebral cortex plays a key role in the most sophisticated neural functions, such as voluntary initiation of movement, final sensory perception, conscious thought, language, personality traits, and other factors we associate with the mind or intellect. It is the highest, most complex integrating area of the brain. Each of these regions of the brain will be discussed in turn, commencing with the cerebral cortex.

Major Functions
1. Voluntary control of movement 2. Sensory perception 3. Language 4. Personality traits 5. Sophisticated mental events, such as thinking, memory, decision making, creativity, and self-consciousness
1. Coordination of slow, sustained movements 2. Inhibition of muscle tone 3. Suppression of useless patterns of movement
1. Relay station for all synaptic input 2. Crude awareness of sensation 3. Some degree of consciousness 4. Role in motor control
1. Regulation of many homeostatic functions, such as temperature control, thirst, urine output, and food intake 2. Extensive involvement with emotion and basic behavioral patterns 3. Important link between nervous and endocrine systems
1. Basic emotional expression 2. Learning 3. Motivation 4. Sociosexual behavioral patterns
1. Regulation of muscle tone 2. Maintenance of balance and posture 3. Coordination and planning of skilled voluntary muscle activity 4. Control of eye movements
7. Origin of majority of peripheral cranial nerves 8. Regulation of muscle reflexes involved with equilibrium and posture 9. Modulation of pain sensation

The cerebral cortex is an outer shell of gray matter covering an inner core of white matter.

The **cerebrum,** by far the largest portion of the human brain, is divided into two halves, the right and left **cerebral hemispheres.** They are connected to each other by the **corpus callosum,** a thick band consisting of an estimated 300 million neuronal axons traversing between the two hemispheres.

Each hemisphere is composed of a thin outer shell of **gray matter,** the **cerebral cortex,** covering a thick central core of **white matter.** Located deep within the white matter is another region of gray matter, the basal nuclei. Throughout the entire CNS, gray matter consists predominantly of densely packaged cell bodies and dendrites. Bundles or tracts of myelinated nerve fibers (axons) constitute the white matter; the lipid (fat) composition of the myelin is responsible for its white appearance. The fiber tracts in the white matter transmit signals from one part of the cerebral cortex to another or between the cortex and other regions of the CNS. Such communication enables integration between different areas of the cortex and elsewhere. This is essential for even a relatively simple task such as picking a flower. Vision of the flower is received by one area of the cortex, reception of its fragrance takes place in another area, and movement is initiated by still another area. More subtle neuronal responses, such as appreciation of the flower's beauty and the urge to pick it, are poorly understood but undoubtedly extensively involve interconnecting fibers between different cortical regions.

The cerebral cortex is organized into layers and functional columns.

The cerebral cortex is organized into six well-defined layers based on varying distributions of the cell bodies and locally associated fibers of several distinctive cell types. These layers are organized into functional vertical columns that extend perpendicularly from the surface down through the depths of the cortex to the underlying white matter. The neurons within a given column are believed to function as a "team," with each cell being involved in different aspects of the same specific activity— for example, perceptual processing of the same stimulus from the same location.

The functional differences between various areas of the cortex result from different layering patterns within the columns and from different input-output connections, not from the presence of unique cell types or different neuronal mechanisms. For example, those regions of the cortex responsible for perception of senses have an expanded layer 4, a layer rich in **stellate cells,** which are responsible for initial processing of sensory input to the cortex. In contrast, the cortical areas that control output to skeletal muscles have a thickened layer 5, which contains an abundance of large **pyramidal cells.** These cells send fibers down the spinal cord from the cortex to terminate on the efferent motor neurons that innervate the skeletal muscles.

The four pairs of lobes in the cerebral cortex are specialized for different activities.

It is important to recognize that even though a discrete activity is ultimately attributed to a particular region of the brain, no part of the brain functions in isolation. Each part depends on complex interplay among numerous other regions, both for incoming and outgoing messages. With this in mind, let us now consider the locations of the major functional areas of the brain.

The anatomical landmarks used in cortical mapping are certain deep folds that divide each half of the cortex into four major lobes: the occipital, temporal, parietal, and frontal lobes (Fig. 5–7). Refer to the basic functional map of the cortex in Figure 5–8 during the following discussion of the major activities attributed to various regions of these lobes.

OCCIPITAL AND TEMPORAL LOBES. The **occipital lobes,** which are located posteriorly (at the back of the head), are responsible for initially processing visual input. Sound sensation, on the other hand, is initially received by the **temporal lobes,** located laterally (on the sides of the head).

PARIETAL LOBES. The parietal lobes and frontal lobes, located on the top of the head, are separated by a deep infolding, the **central sulcus,** which runs roughly down the middle of the lateral surface of each hemisphere. The parietal lobes lie to the rear of the central sulcus on each side, and the frontal lobes lie in front of it.

The **parietal lobes** are primarily responsible for receiving and processing sensory input from the surface of the body, such as touch, pressure, heat, cold, and pain. These sensations are collectively known as **somesthetic sensations** (*somesthetic* means "body feelings"). The parietal lobes also perceive awareness of body position, a phenomenon referred to as **proprioception.** The **somatosensory cortex,** the site for initial cortical processing of this somesthetic and proprioceptive input, is located at the front of each parietal lobe immediately behind the central sulcus (Fig. 5–9a). Each region within the somatosensory cortex receives sensory input from a specific area of the body. This distribution of cortical sensory processing is depicted in Figure 5–9b. Note on this so-called **sensory homunculus** (*homunculus* means "little man") that the body is represented upside down on the somatosensory cortex and, more importantly, *that different parts of the body are not*

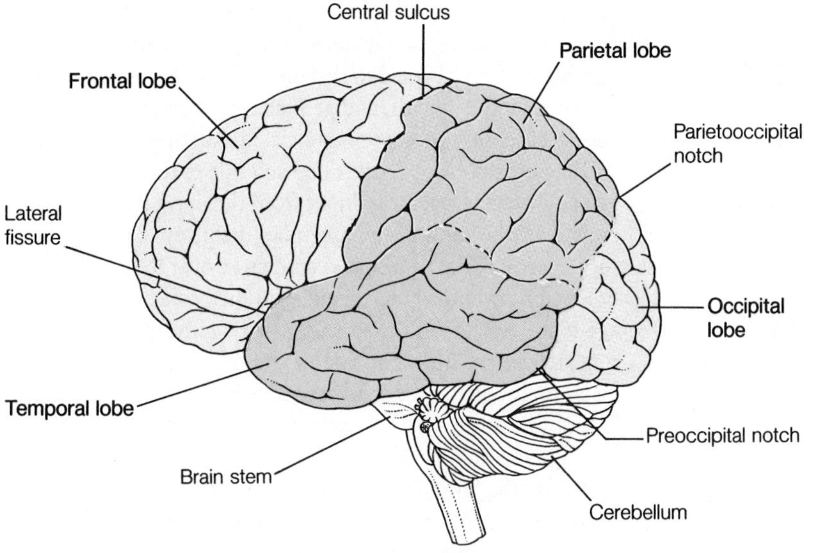

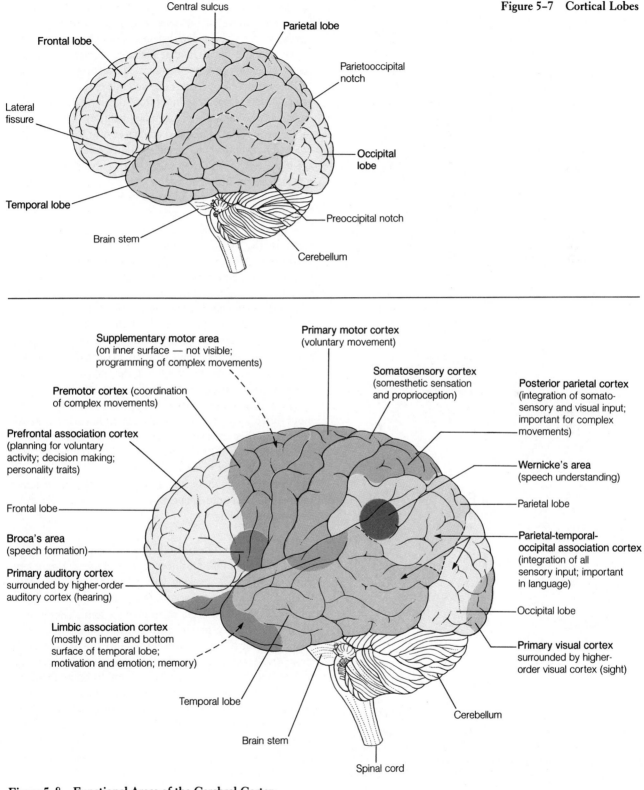

Figure 5–7 Cortical Lobes

Primary motor cortex (voluntary movement)

Supplementary motor area (on inner surface — not visible; programming of complex movements)

Somatosensory cortex (somesthetic sensation and proprioception)

Premotor cortex (coordination of complex movements)

Posterior parietal cortex (integration of somatosensory and visual input; important for complex movements)

Prefrontal association cortex (planning for voluntary activity; decision making; personality traits)

Wernicke's area (speech understanding)

Frontal lobe

Parietal lobe

Broca's area (speech formation)

Parietal-temporal-occipital association cortex (integration of all sensory input; important in language)

Primary auditory cortex surrounded by higher-order auditory cortex (hearing)

Occipital lobe

Limbic association cortex (mostly on inner and bottom surface of temporal lobe; motivation and emotion; memory)

Primary visual cortex surrounded by higher-order visual cortex (sight)

Temporal lobe

Cerebellum

Brain stem

Spinal cord

Figure 5–8 Functional Areas of the Cerebral Cortex

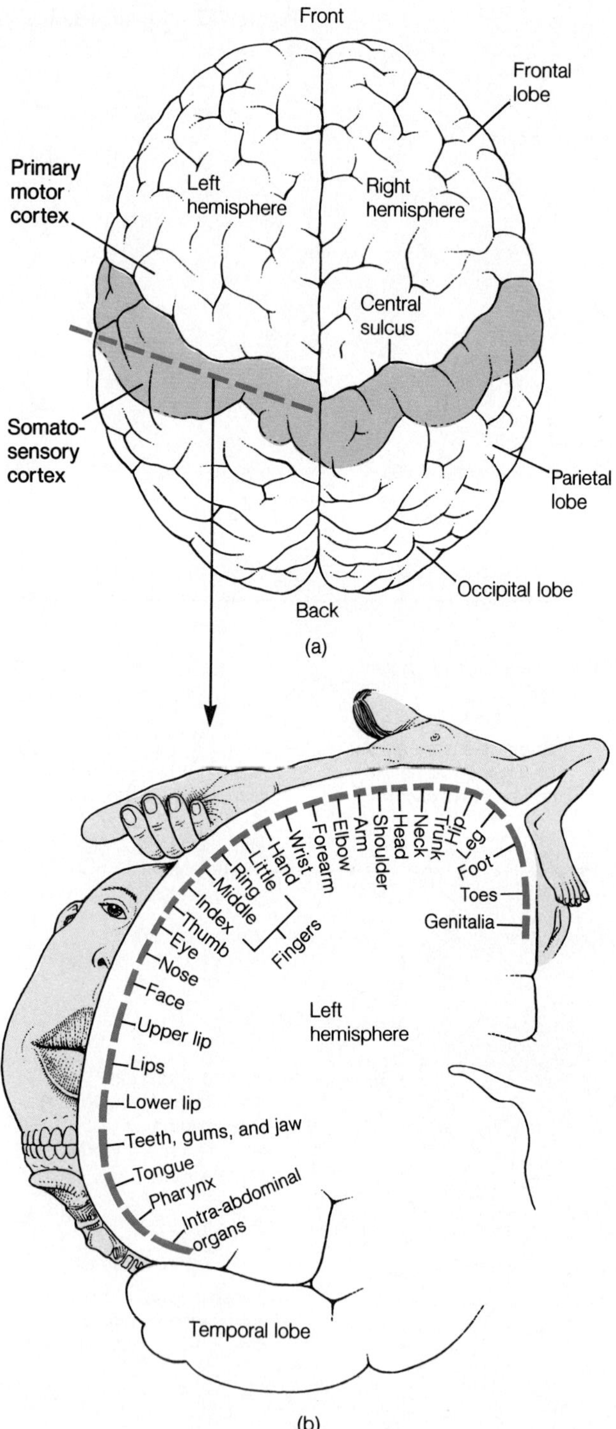

Figure 5-9 Somatotopic Map of Somatosensory Cortex *(a) Top view of cerebral hemispheres. (b) Sensory homunculus—showing the distribution of sensory input to the somatosensory cortex from different parts of the body. The distorted graphic representation of the body parts is indicative of the relative proportion of the somatosensory cortex devoted to reception of sensory input from each area.*

equally represented. The size of each body part in this homunculus is indicative of the relative proportion of the somatosensory cortex devoted to that area. The exaggerated size of the face, tongue, hands, and genitalia is indicative of the high degree of sensory perception associated with these body parts.

The somatosensory cortex on each side of the brain for the most part receives sensory input from the opposite side of the body, because most of the ascending pathways carrying sensory information up the spinal cord cross over to the opposite side before eventually terminating in the cortex. Thus, damage to the left half of the somatosensory cortex produces sensory deficits on the right side of the body, whereas sensory losses on the left side are associated with damage to the right half of the cortex.

Simple awareness of touch, pressure, or temperature is detected by the thalamus, a lower level of the brain, but the somatosensory cortex goes beyond pure recognition of sensations to fuller sensory perception. The thalamus makes you aware that something hot versus something cold is touching your body but it does not tell you where or of what intensity. The somatosensory cortex localizes the source of sensory input and perceives the level of intensity of the stimulus. It also is capable of spatial discrimination, so it can discern shapes of objects being held and can distinguish subtle differences in similar objects that come into contact with the skin.

The somatosensory cortex, in turn, projects this sensory input via white matter fibers to adjacent **higher sensory areas** for even further elaboration, analysis, and integration of sensory information. These higher areas are important in the perception of complex patterns of somatosensory stimulation—for example, simultaneous appreciation of the texture, firmness, temperature, shape, position, and location of an object you are holding.

FRONTAL LOBES. The **frontal lobes,** lying at the front of the cortex, are responsible for three main functions: (1) voluntary motor activity, (2) speaking ability, and (3) elaboration of thought. The area at the rear of the frontal lobe immediately in front of the central sulcus adjacent to the somatosensory cortex is the **primary motor cortex** (Fig. 5-10a). It confers voluntary control over movement produced by skeletal muscles. As with sensory processing, the motor cortex on each side of the brain primarily controls muscles on the opposite side of the body. Neuronal tracts originating in the motor cortex of the left hemisphere cross over before passing down the spinal cord to terminate on efferent motor neurons that trigger skeletal muscle contraction on the right side of the body. Accordingly, damage to the motor cortex on the left side of the brain produces paralysis on the right side of the body. The converse is also true.

Stimulation of different areas of the primary motor cortex brings about movement in different regions of the body. As with the somatosensory cortex, the **motor homunculus,** which depicts the location and relative amount of motor cortex devoted to output to the muscles of each body part, is upside down and distorted (Fig. 5–10b). The fingers and thumbs as well as the muscles important in speech, especially those of the lips and tongue, are grossly exaggerated, indicative of the fine degree of motor control with which these body parts are endowed. Compare this to how little brain tissue is devoted to the trunk, arms, and lower extremities, which are not capable of such complex movements. Thus, the extent of representation in the motor cortex is proportional to the fineness and complexity of motor skills required of the respective part.

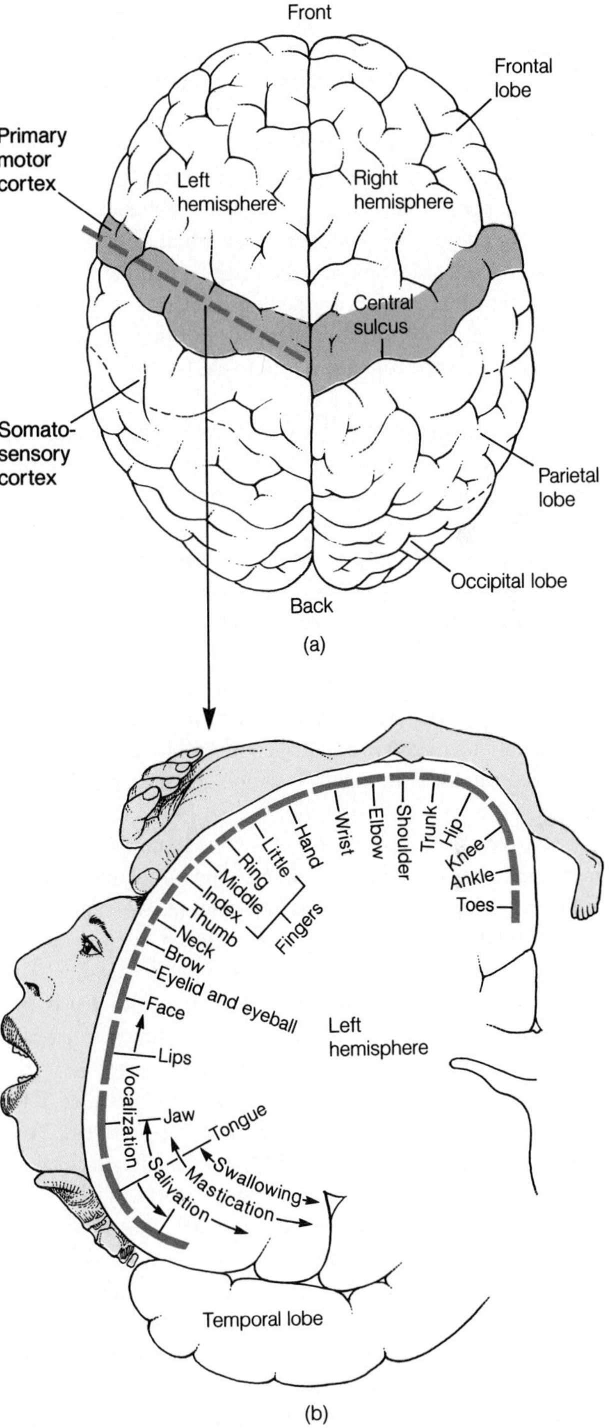

Figure 5–10 Somatotopic Map of Primary Motor Cortex *(a) Top view of cerebral hemispheres. (b) Motor homunculus—showing the distribution of motor output from the primary motor cortex to different parts of the body. The distorted graphic representation of the body parts is indicative of the relative proportion of the primary motor cortex devoted to controlling skeletal muscles in each area.*

Other regions of the nervous system besides the primary motor cortex are important in motor control.

Even though signals from the primary motor cortex terminate on the efferent neurons that trigger voluntary skeletal muscle contraction, the motor cortex is not the only region of the brain involved with motor control. First, lower brain regions and the spinal cord control involuntary skeletal muscle activity, such as the maintenance of posture. Some of these same regions also play an important role in monitoring and coordinating voluntary motor activity that has been set in motion by the primary motor cortex. Second, although fibers originating from the motor cortex can activate motor neurons to bring about muscle contraction, the motor cortex itself does not initiate voluntary movement. The motor cortex is activated by a widespread pattern of neuronal discharge, the **readiness potential,** which occurs about 750 msec before specific electrical activity is detectable in the motor cortex. The higher motor areas of the brain believed to be involved in this voluntary decision-making period include the ***supplementary motor area***, the ***premotor cortex***, and the ***posterior parietal cortex*** (Fig. 5–8). These higher areas all command the primary motor cortex. Furthermore, a subcortical region of the brain, the ***cerebellum***, appears to play an important role in the planning, initiation, and timing of movement by sending input to the motor areas of the cortex.

These four regions of the brain are important in programming and coordinating complex movements involving simultaneous contraction of many muscles. Even though electrical stimulation of the primary motor cortex brings about contraction of particular muscles, no purposeful coordinated movement can be elicited, just as pulling on isolated strings of a puppet does not produce any meaningful movement. A puppet displays purposeful movements only through coordinated manipulation of the strings by a skilled puppeteer. In the same way, these four regions (and perhaps other areas as yet undetermined) develop a **motor program** for the specific voluntary task and then "pull" the appropriate pattern of "strings" in the primary motor cortex to bring about the sequenced contraction of appropriate muscles to accomplish the desired complex movement.

The **supplementary motor area** lies on the medial (inner) surface of each hemisphere anterior to (in front of) the primary motor cortex. It appears to play a preparatory role in programming complex sequences of movement. Stimulation of various regions of this motor area brings about complex patterns of movement, such as opening or closing of the hand. Lesions here do not result in paralysis but they do interfere with performance of more complex, useful integrated movements.

The **premotor cortex,** located on the lateral surface of each hemisphere in front of the primary motor cortex, is believed to be important in orienting the body and arms toward a specific target. The premotor cortex is guided by sensory input processed by the **posterior parietal cortex,** a region that lies posterior to (in back of) the primary somatosensory cortex. These two higher motor areas have many anatomical interconnections and appear to be closely interrelated functionally. With damage to either of these areas, the individual cannot process complex sensory information to accomplish purposeful movement in a spatial context. These patients, for example, cannot successfully manipulate silverware when eating.

Even though these higher motor areas command the primary motor cortex and are important in preparing for the execution of deliberate, meaningful movement, we cannot say that voluntary movement is actually *initiated* by these areas. This pushes one step further the question of how and where voluntary activity is initiated. Probably no single area is responsible; undoubtedly there are numerous pathways that can ultimately bring about deliberate movement.

For example, think about the neural systems called into play during the simple act of picking up an apple to eat. You know that the fruit is located in a bowl on the kitchen counter because of your memory. Sensory systems, coupled with your knowledge based on past experience, enable you to identify the apple from the other varieties of fruit in the bowl. Motor systems, upon receiving this integrated sensory information, issue commands to the exact muscles of the body in the proper timed sequence to enable you to move to the fruit bowl and pick up the targeted apple. During execution of this act, minor adjustments in the motor command are made as needed, based on continual updating provided by sensory input about the position of the body relative to the goal. Then there is the issue of motivation and behavior. Why are you reaching for the apple in the first place? Is it because you are hungry (detected by a neural system in the hypothalamus) or because of a more complex behavioral scenario unrelated to a basic hunger drive, such as the fact that you started to think about food because you just saw someone eating on television? Why did you select an apple rather than a banana when you have both in the fruit bowl and you like the taste of both, and so on? Thus, initiation and execution of purposeful voluntary movement actually includes a complex neuronal interplay involving output from the motor regions guided by integrated sensory information and ultimately dependent on motivational systems and elaboration of thought, played against a background of memory stores from which meaningful decisions about desirable movements can be made.

Somatotopic maps vary slightly between individuals and are dynamic, not static.

Although the general organizational pattern of sensory and motor somatotopic ("body representation") maps of the cortex is similar in all people, the precise distribution varies uniquely for each individual. Furthermore, an individual's somatotopic mapping is not "carved in stone" but is subject to constant subtle modifications based on use. The general pattern is governed by genetic and developmental processes, but the individual cortical architecture appears capable of being influenced by **use-dependent competition** for cortical space. For example, when monkeys were encouraged to use their middle fingers instead of their other fingers to press a bar for food, after only several thousand bar presses the "middle finger area" of the motor cortex was greatly expanded and encroached upon territory previously devoted to the other fingers. This competition for cortical territory based on use probably accounts for the fact that some individuals born without arms have been able to accomplish amazing "dexterity" with their feet. They are able to use their feet almost like hands to feed themselves and perform other skills requiring a much finer degree of control over toe muscles than most people have.

Thus, the particular architecture of your own rather plastic brain has been and continues to be influenced by your unique set of experiences. It is important to realize, however, that what you do and do not do cannot totally shape the organization of your cortex. Some limits are genetically established. Also, there are developmental limits as to the extent to which modeling can be influenced by patterns of use. Whereas some cortical activities maintain their plasticity throughout life, especially the ability to add new memory stores and learn, other cortical regions can be modified by use for only a specified time after birth before becoming permanently fixed. The length of this critical developmental period varies for different cortical regions.

Language ability has several discrete components controlled by different regions of the cortex.

Language ability is an excellent example of early cortical plasticity coupled with later permanence. Unlike the sensory and motor regions of the cortex, which are present in both hemispheres, the areas of the brain responsible for language ability are found in only one hemisphere—the left hemisphere in the vast majority of the population. However, if a child under the age of two accidentally suffers damage to the left hemisphere, language functions are transferred to the right hemisphere with no delay in language development but at the expense of less obvious nonverbal abilities for which the right hemisphere is normally responsible. Up to about the age of ten, language ability can usually be reestablished in the right hemisphere after damage to the left hemisphere following a temporary period of loss. If damage occurs beyond the early teens, however, language ability is permanently impaired, even though some limited restoration may be possible. The regions of the brain involved in comprehending and expressing language apparently are permanently assigned before adolescence.

Even in normal individuals, there is evidence for early plasticity and later permanence regarding language development. Infants are able to distinguish between and articulate the entire range of speech sounds, but each language utilizes only a percentage of these sounds. As children mature, they often lose the ability to distinguish between or express speech sounds that are not important in their native language. For example, Japanese children can distinguish between the sounds of "r" and "l," but many Japanese adults cannot perceive the difference between them.

Language is a complex form of communication in which written or spoken words symbolize objects and convey ideas. It involves the integration of two distinct capabilities—namely, *expression* and *comprehension*—each of which is related to a specific area of the cortex. The primary areas of cortical specialization for language are Broca's area and Wernicke's area, which are usually developed only in the left hemisphere. **Broca's area,** which is responsible for speaking ability, is located in the left frontal lobe in close association with the motor areas of the cortex that control the muscles necessary for articulation (Figs. 5–8 and 5–11). **Wernicke's area,** located in the parietotemporal region, is concerned with language comprehension. It plays a critical role in understanding both spoken and written messages. Furthermore, it is responsible for formulating coherent patterns of speech that are transferred via a bundle of fibers to Broca's area, which in turn controls articulation of this speech. Wernicke's area also receives input from the visual cortex in the occipital lobe, a pathway important in reading comprehension and in describing objects seen, as well as from the auditory cortex in the temporal lobe, a pathway essential for understanding spoken words. According to the leading model of language, precise interconnecting pathways between these localized cortical areas are involved in the various aspects of speech (Fig. 5–11).

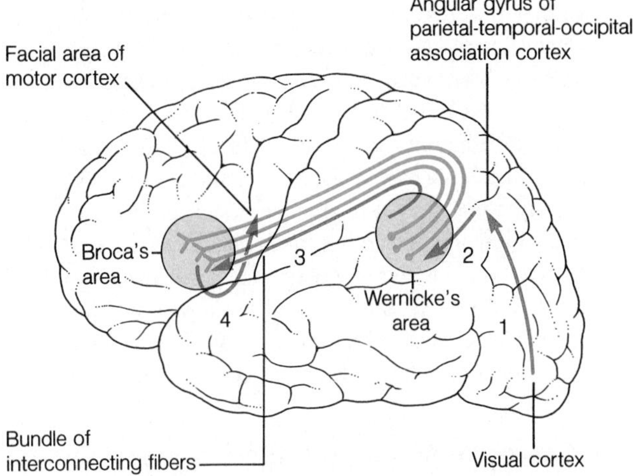

Figure 5–11 Cortical Pathway for Speaking a Written Word or Naming a Visual Object *To describe an object seen, the brain first transfers the visual information from the visual cortex to a specific area (the angular gyrus) of the parieto-temporal-occipital association cortex, a region concerned with integrating such sensory inputs as sight, sound, and touch (path 1). From here the information is transferred to Wernicke's area (path 2), where the choice and sequence of words to be spoken are formulated, and then is transmitted to Broca's area (path 3). Broca's area, in turn, translates the message from Wernicke's area into a programmed sound pattern that is conveyed to the precise areas of the primary motor cortex (path 4) to activate the appropriate facial muscles that will cause the desired words to be spoken. Similarly, appropriate muscles of the hand can be commanded to write the desired words.*

Because various aspects of language are localized in different regions of the cortex, selective disturbances of language can result from damage to specific regions of the brain. Damage to Broca's area results in a failure of word formation, although the patient can still understand the spoken and written word. Such individuals know what they want to say but are unable to express themselves. Even though they can move their lips and tongue, they cannot establish the proper motor command to articulate the desired words. In contrast, patients with a lesion in Wernicke's area cannot understand words they see or hear. They are able to speak fluently even though their perfectly articulated words make no sense. They cannot attach meaning to words or choose appropriate words to convey their thoughts. Such language disorders caused by damage to specific cortical areas are known as **aphasias,** most of which result from strokes. Aphasias should not be confused with speech impediments caused by a defect in the mechanical aspect of speech, such as weakness or incoordination of the muscles controlling the vocal apparatus. It is probable that **dyslexia,** a difficulty in learning to read because of inappropriate interpretation of letters or words as a reverse image (for example, *bad* is "seen" as *dab*), arises from developmental abnormalities in connections between the visual and language areas of the cortex or within the language areas themselves.

The association areas of the cerebral cortex are involved in many higher functions.

The motor, sensory, and language areas account for only about half of the total cerebral cortex. The remaining areas, called **association areas,** are involved in higher functions. At one time these regions were called "silent" areas because stimulation does not produce any observable motor response or sensory perception. There are three association areas (Fig. 5–8): (1) the prefrontal association cortex, (2) the parieto-temporal-occipital association cortex, and (3) the limbic association cortex.

The **prefrontal association cortex** is the front portion of the frontal lobe just anterior to the premotor cortex. The roles attributed to this region are: (1) planning for voluntary activity; (2) weighing consequences of future actions and choosing between different options for various social or physical situations; and (3) personality traits. Stimulation of this area does not produce any observable effects, but deficits in this area result in changes in personality and social behavior.

The **parieto-temporal-occipital association cortex** is found at the interface of the three lobes for which it is named. In this strategic location, it pools and integrates somatic, auditory, and visual sensations projected from these three lobes for complex perceptual processing. It enables us to "get the complete picture" of the relationship of various parts of our bodies with the external world. For example, it integrates visual information with proprioceptive input to enable you to place what you are seeing in proper perspective, such as realizing that a bottle is in an upright position in spite of the angle from which you view it (that is, whether you are standing up, lying down, or hanging upside down from a tree branch). This region is also involved in the language pathway connecting Wernicke's area to the visual and auditory cortices.

The **limbic association cortex** is located mostly on the bottom and adjoining inner portion of each temporal lobe. This area is concerned primarily with motivation and emotion and is extensively involved in memory.

Indeed, all association areas appear to be involved in sophisticated mental events such as memory, thinking, decision-making, creativity, and self-consciousness. None of these higher brain functions are controlled by a specific cortical region. All are believed to depend on complex interrelated pathways involving several different regions. The cortical association areas are all interconnected by bundles of fibers within the cerebral white matter. Collectively, the association areas integrate diverse information for purposeful action. An oversimplified basic sequence of linkage between the various funtional areas of the cortex is schematically represented in Figure 5–12.

The cerebral hemispheres have some degree of specialization.

The cortical areas described thus far appear to be equally distributed in both the right and left hemispheres, except for the language areas found only on one side, usually the left. The left side is also most commonly the dominant hemisphere for fine motor control. Thus, most people are right-handed, because the left side of the brain controls the right side of the body. Furthermore, each hemisphere is somewhat specialized in the types of mental activities it best carries out. The **left hemisphere** excels in the performance of logical, analytical, sequential, and verbal tasks, such as math, language forms, and philosophy. In contrast, the **right hemisphere** excels in nonlanguage skills, especially in spatial perception and artistic and musical endeavors. Whereas the left hemisphere tends to process information in a fragmentary way, the right hemisphere views the world holistically. Normally there is much sharing of information between the two hemispheres so that they complement each other, but in many individuals the skills associated with one hemisphere appear to be more strongly developed. Left cerebral hemisphere dominance tends to be associated with "thinkers" whereas the right hemispheric skills dominate in "creators."

An electroencephalogram is a record of postsynaptic activity in cortical neurons.

Extracellular current flow arising from electrical activity within the cerebral cortex can be detected via placement of recording electrodes on the scalp to produce a graphic record known as an **electroencephalogram** or **EEG.** These "brain waves" for the most part are not due to action potentials but instead represent the momentary collective postsynaptic potential activity (that is, EPSPs and IPSPs) in the cell bodies and dendrites located in the cortical layers under the recording electrode.

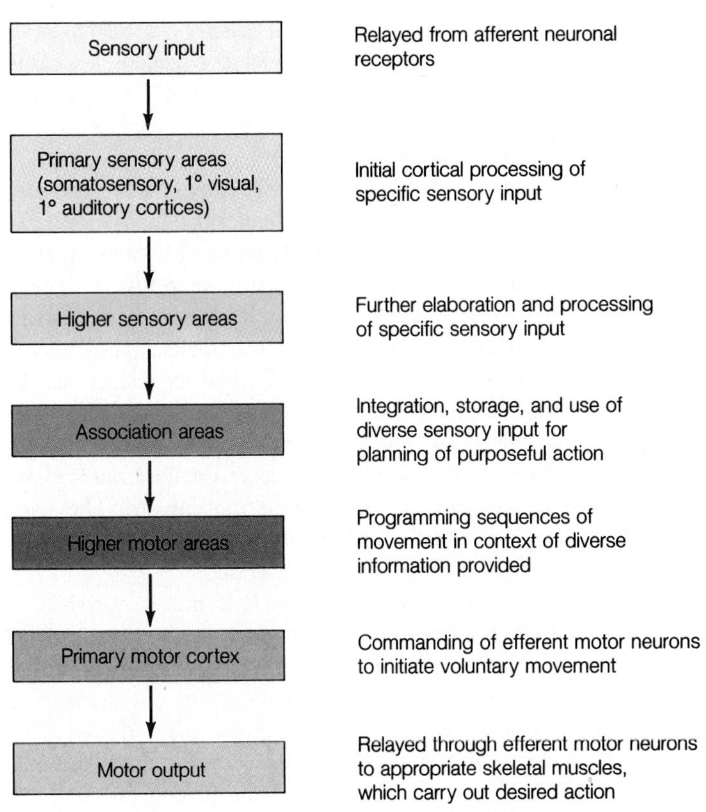
Sensory input
Relayed from afferent neuronal receptors
Primary sensory areas (somatosensory, 1° visual, 1° auditory cortices)
Initial cortical processing of specific sensory input
Higher sensory areas
Further elaboration and processing of specific sensory input
Association areas
Integration, storage, and use of diverse sensory input for planning of purposeful action
Higher motor areas
Programming sequences of movement in context of diverse information provided
Primary motor cortex
Commanding of efferent motor neurons to initiate voluntary movement
Motor output
Relayed through efferent motor neurons to appropriate skeletal muscles, which carry out desired action
For simplicity, a number of interconnections have been omitted.

Figure 5-12 Schematic Linking of Various Regions of Cortex

Electrical activity can always be recorded from the living brain, even during sleep and unconscious states, but the wave forms vary depending on the degree of activity of the cerebral cortex. Often the wave forms appear irregular, but sometimes distinct patterns can be observed on the basis of the wave's amplitude and frequency. A dramatic example of this is illustrated in Figure 5-13, in which the EEG wave form recorded over the occipital (visual) cortex changes markedly in response to simply opening and closing the eyes.

The EEG has three major uses:

1. It is often used as a ***clinical tool in the diagnosis of cerebral dysfunction***. Diseased or damaged cortical tissue often gives rise to altered EEG patterns. One of the most common neurological diseases accompanied by a distinctively abnormal EEG is **epilepsy.** Epileptic seizures occur when a large collection of neurons abnormally undergo synchronous action potentials that produce stereotypical, involuntary spasms and alterations in behavior. The seizures may be partial or generalized, depending on the location and extent of the abnormal neuronal discharge. Each type of seizure displays different EEG features.
2. The EEG is also used to ***distinguish various stages of sleep***, as will be described later in this chapter.
3. The EEG finds further use in the ***legal determination of brain death***. Even though a person may have stopped

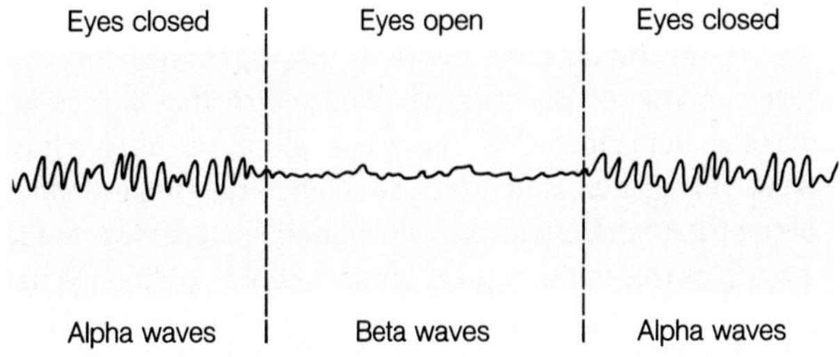
Eyes closed
Eyes open
Eyes closed
Alpha waves
Beta waves
Alpha waves

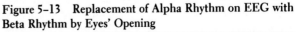

Figure 5-13 Replacement of Alpha Rhythm on EEG with Beta Rhythm by Eyes' Opening

breathing and the heart may have stopped pumping blood, it is often possible to restore and maintain circulatory and respiratory activity if resuscitative measures are instituted soon enough. Yet, because of the susceptibility of the brain to O_2 deprivation, irreversible brain damage may have already occurred before heart and lung function have been reestablished, resulting in the paradoxical situation of a dead brain in a living body. There are important medical, legal, and social implications in determining whether a comatose patient being maintained by artificial respiration and other supportive measures is alive or dead. The need for viable organs for modern transplant surgery has made the timeliness of such life/death determinations of utmost importance. Physicians, lawyers, and the public in general have accepted the notion of brain death—that is, a brain that is not functioning, with no possibility of recovery—as the determinant of death under such circumstances. The most widely accepted indication of brain death is *electrocerebral silence*—an essentially flat EEG. This, however, must be coupled with other stringent criteria, such as absence of eye reflexes, to guard against a false terminal diagnosis in individuals with a flat EEG that is due to causes that can be reversed, as in certain kinds of drug intoxication.

Subcortical Structures and Their Relationship with the Cortex in Higher Brain Functions

The **subcortical** ("under the cortex") **regions** of the brain interact extensively with the cortex in the performance of their functions. These regions include the basal nuclei located in the cerebrum and the thalamus and hypothalamus located in the diencephalon.

The basal nuclei play an important inhibitory role in motor control.

The **basal nuclei** (also known as **basal ganglia**) consist of several masses of gray matter located deep within the cerebral white matter (Fig. 5–14). A nucleus (plural, nuclei) is a functional aggregation of neuronal cell bodies. Our understanding of the basal nuclei is sparse, although they are known to play a complex role in the control of movement in addition to having nonmotor functions that are less well understood. They appear to be particularly important in: (1) inhibiting muscle tone throughout the body (proper muscle tone is normally maintained by a balance of excitatory and inhibitory inputs to the neurons that innervate skeletal muscles); (2) helping monitor and coordinate slow sustained contractions, especially those related to posture and support; and (3) selecting and maintaining purposeful motor activity while suppressing useless or unwanted patterns of movement.

To accomplish these complex integrative roles, the basal nuclei receive and send out much information, as is indicated by the tremendous number of fibers linking them to other regions of the brain. One important pathway consists of strategic interconnections that form a complex feedback loop linking the cerebral cortex (especially its motor regions), the basal nuclei, and the thalamus (Fig. 5–15). It is speculated that the thalamus positively reinforces voluntary motor behavior initiated by the cortex, whereas the basal nuclei modulate this activity by exerting an inhibitory effect on the thalamus to eliminate antagonistic or unnecessary movements. The basal nuclei also exert an inhibitory effect on motor activity by acting through neurons in the brain stem.

The importance of the basal nuclei in motor control is evident in diseases involving this region, the most common of which is **Parkinson's disease.** This condition is associated with a deficiency of dopamine, an important neurotransmitter in the basal nuclei (see p. 111). This was the first disease of the CNS shown to be due to a deficiency of a chemical transmitter. With the basal nuclei lacking sufficient dopamine to exert their normal inhibitory role, the positive feedback loop between the thalamus and cortex runs unchecked. Three types of motor disturbances characterize Parkinson's disease: (1) involuntary, useless, or unwanted movements, such as **resting tremors** (for example, hands rhythmically shaking, making it difficult or impossible to hold a cup of coffee); (2) increased muscle tone or rigidity; and (3) slowness in initiating and carrying out different motor behaviors. It is difficult for those who suffer from Parkinson's disease to stop ongoing activities. If sitting down, they tend to remain seated, and if they get up, they do so very slowly.

The thalamus is a sensory relay station and is important in motor control.

Deep within the brain near the basal nuclei is the **diencephalon,** a midline structure that forms the walls of the third ventricular cavity. The diencephalon consists of two main parts, the thalamus and the hypothalamus (Figs. 5–14 and 5–16).

The **thalamus** serves as a "relay station" and synaptic integrating center for preliminary processing of all sensory input on its way to the cortex. It screens out insignificant signals and routes the important sensory impulses to appropriate areas of the somatosensory cortex, as well as to other regions of the brain. The thalamus, along with the brain stem and cortical association areas, is important in our ability to direct attention to

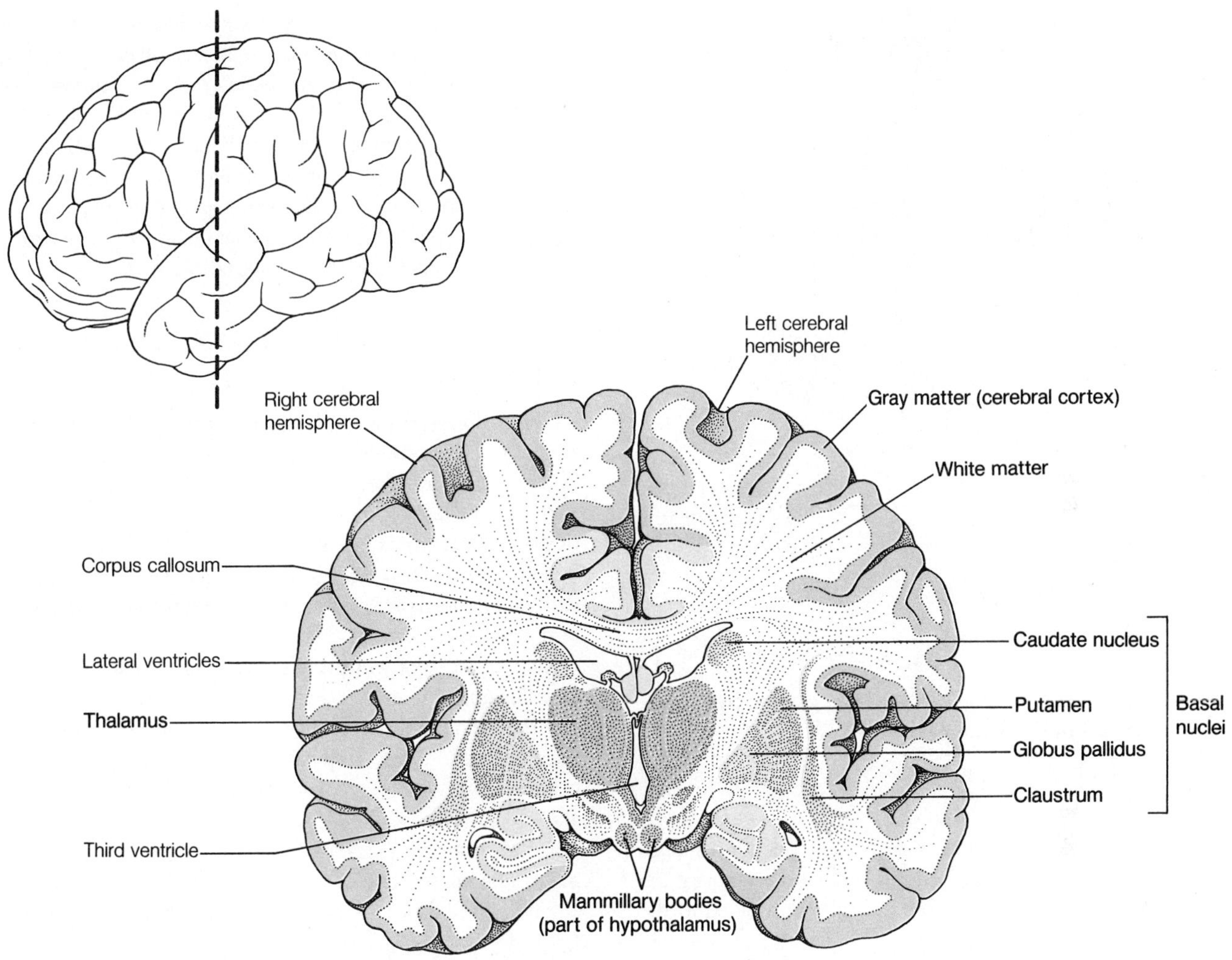

Figure 5-14 Frontal Section of Brain, Showing Cerebral Cortex (Gray Matter) Surrounding Core of White Matter, with Basal Nuclei Deep within White Matter

stimuli of interest. For example, parents can sleep soundly through the noise of outdoor traffic but be instantly aware of their baby's slightest whimper. The thalamus is also capable of crude awareness of various types of sensation but cannot distinguish their location or intensity. Some degree of consciousness resides here as well. As described in the preceding section, the thalamus also plays an important role in motor control.

The hypothalamus regulates many homeostatic functions.

The **hypothalamus** is a collection of specific nuclei and associated fibers that lie beneath the thalamus. It is an integrating center for many important homeostatic functions and serves as an important link between the autonomic nervous system and the endocrine system. Specifically, the hypothalamus: (1) controls body temperature; (2) controls thirst and urine output; (3) controls food intake; (4) controls anterior pituitary hormone secretion; (5) produces posterior pituitary hormones; (6) controls uterine contractions and milk ejection; (7) serves as a major autonomic nervous system coordinating center, which in turn affects all smooth muscle, cardiac muscle, and exocrine glands; and (8) plays a role in emotional and behavorial patterns, including psychosomatic events.

The hypothalamus is the area of the brain most notably involved in the direct regulation of the internal environment. For example, when the body is cold, the hypothalamus initi-

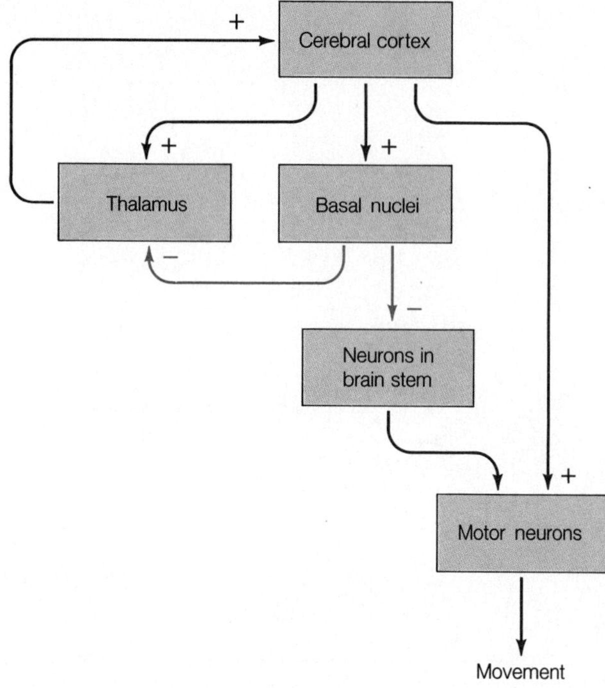

Figure 5–15 Relationship of Thalamus, Basal Nuclei, and Cortex in Control of Movement

ates internal responses to increase heat production (such as shivering) and to decrease heat loss (such as constricting the skin blood vessels to reduce the flow of warm blood to the body surface, where heat could be lost to the external environment). Other areas of the brain, such as the cerebral cortex, act more indirectly to regulate the internal environment. For example, a person who feels cold is motivated to voluntarily put on warmer clothing, close the window, turn up the thermostat, and so on. Even these voluntary behavioral activities are strongly influenced by the hypothalamus, which, as a part of the limbic system, functions in conjunction with the cortex in controlling emotions and motivated behavior.

The limbic system plays a key role in emotion and behavior.

The **limbic system** is not a separate structure but refers to a ring of forebrain structures that surround the brain stem and that are interconnected by intricate neuronal pathways (Fig. 5–17). It includes portions of each of the following: the lobes of the cerebral cortex, the basal nuclei, the thalamus, and the hypothalamus. This complex interacting network is associated with emotions, basic survival and sociosexual behavioral patterns, motivation, and learning.

The concept of **emotion** encompasses subjective emotional feelings and moods (such as anger, fear, and happiness) plus the overt physical responses that occur in association with these feelings. These responses include specific behavioral patterns (for example, preparation for attack or defense when angered by an adversary) and observable emotional expressions (for example, laughing, crying, or blushing). Evidence points to a central role of the limbic system in all aspects of emotion. Stimulation of specific regions within the limbic sys-

Figure 5–16 Location of Thalamus and Hypothalamus in Sagittal Section

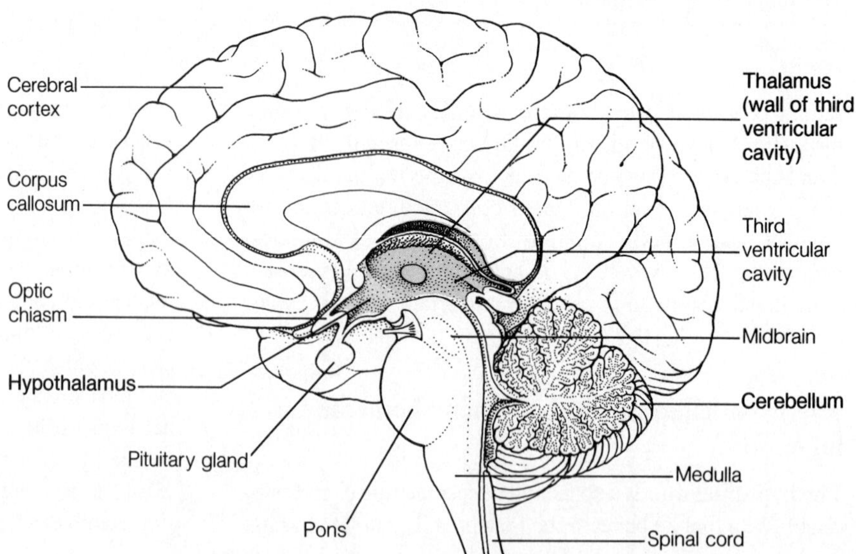

tem of humans during brain surgery produces various vague subjective sensations described by the patient as joy, satisfaction, or pleasure in one region and discouragement, fear, or anxiety in another.

Behavioral patterns controlled at least in part by the limbic system include those aimed at survival of the individual (attack, searching for food) and those directed toward perpetuation of the species (sociosexual behaviors conducive to mating). In experimental animals, stimulation of and lesions in the limbic system bring about complex and even bizarre behaviors, as evidenced by the following examples. Stimulation in one area can elicit responses of anger and rage in a normally docile animal, whereas stimulation in another area results in placidity and tameness, even in an otherwise vicious animal. Stimulation in yet another limbic area, the **amygdala,** can induce sexual behaviors such as copulatory movements. Experimental lesions in the amygdala can alter sexual behavior so that the animal even attempts to mate with members of different species.

The relationships among the hypothalamus, limbic system, and higher cortical regions regarding emotions and behavior are still not well understood. It appears that the extensive involvement of the hypothalamus in the limbic system is responsible for the involuntary internal responses of various body systems in preparation for appropriate action to accompany a particular emotional state. For example, the increased heart rate and respiratory rate, elevation of blood pressure, and diversion of blood to skeletal muscles that occur in anticipation of attack when angered are controlled by the hypothalamus. These preparatory changes in internal state require no conscious control.

In executing complex behavioral activities such as attack, flight, or mating, the individual (animal or human) must interact with the external environment. Higher cortical mechanisms are called into play to connect the limbic system and hypothalamus with the outer world so that appropriate overt behaviors are manifested. At the simplest level, the cortex provides the neural mechanisms necessary for implementation of the appropriate skeletal muscle activity required to approach or avoid an adversary, participate in sexual activity, or display emotional expression. For example, the stereotypical sequence of movement for the universal human emotional expression of smiling is apparently preprogrammed in the cortex and can be called forth by the limbic system. One can also voluntarily call forth the smile program, as when posing for a picture. Even individuals blind from birth have normal facial expressions; that is, they do not learn to smile by observation. Smiling means the same thing in every culture, despite widely differing environmental experiences. Such behavior patterns shared in common among all members of a species are believed to be more abundant in lower animals.

In humans and to an undetermined extent in other species, the cortex is additionally crucial for conscious awareness of emotional feelings. It further is capable of reinforcing, modifying, or suppressing basic behavioral responses so that actions can be guided by planning, strategy, and judgment based on an understanding of the situation. For example, even if you were angry at someone and your body was internally prepar-

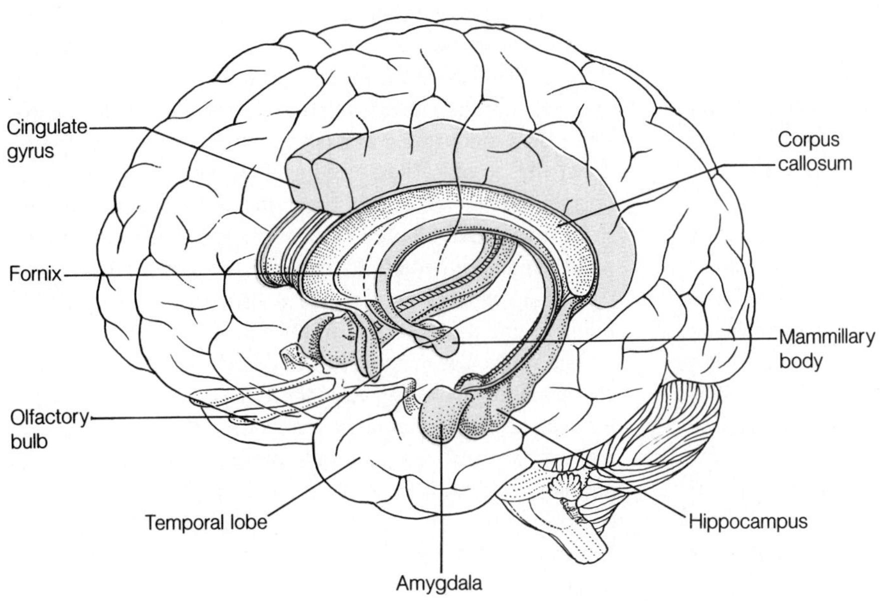

Figure 5–17 Limbic System *This partially transparent view of the brain reveals the structures composing the limbic system.*

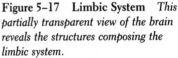

ing for attack, you probably would judge that an attack would be inappropriate and could consciously suppress the external manifestation of this basic emotional behavior. Thus, the cortex, particularly the prefrontal and limbic association areas, is important in conscious learned control of innate behavior patterns.

Motivated behavior and learning are believed to be influenced by rewards and punishments.

The most important factor responsible for the desire and ability to alter inborn behavior is learning from experience. An individual tends to reinforce behaviors that have proved to be gratifying and suppress behaviors that have been associated with unpleasant experiences. Certain regions of the limbic system have been designated as **"reward"** and **"punishment" centers** because stimulation in these respective areas gives rise to pleasant or unpleasant sensations. An experimental animal with an implanted self-stimulating device will self-deliver up to 5,000 stimulations per hour when the device is implanted in a reward center, even shunning food when starving in preference for the pleasure derived from self-stimulation. On the other hand, animals will avoid stimulation at all costs when the device is implanted in a punishment center. They will even work to prevent stimulation if the device is wired so that it continuously delivers current unless the animal does something to stop it, such as pushing a lever. Reward centers are found most abundantly in regions involved in mediating the highly motivated behavioral activities of eating, drinking, and sexual activity.

Motivation is the ability to direct behavior toward specific goals. Some **goal-directed behaviors** are aimed at satisfying specific identifiable physical **needs** related to homeostasis. **Homeostatic drives** represent the subjective urges associated with specific bodily needs that motivate appropriate behavior to satisfy those needs. As an example, the sensation of thirst accompanying a water deficit in the body drives an individual to drink to satisfy the homeostatic need for water. However, whether water, a soft drink, or another beverage is chosen as the thirst-quencher is unrelated to homeostasis.

Much of human behavior is not dependent on purely homeostatic drives related to simple tissue deficits such as thirst, but it is influenced by experience, learning, and habit, shaped in a complex framework of unique personal gratifications blended with cultural expectations. To what extent, if any, motivational drives unrelated to homeostasis, such as the drive to pursue a particular career or to win a certain race, are involved with the reinforcing effects of the reward and punishment centers is unknown. Indeed, some individuals motivated toward a particular goal may even deliberately "punish" themselves in the short-term to achieve their long-range gratification (for example, the temporary pain of training in preparation for winning a competitive athletic event).

Norepinephrine, dopamine, and serotonin serve as neurotransmitters in motivated behavioral and emotional pathways.

The underlying neurophysiological mechanisms responsible for the psychological observations of motivated behavior and emotions largely remain a mystery, although the neurotransmitters **norepinephrine, dopamine,** and **serotonin** all have been implicated. Norepinephrine and dopamine, both chemically classified as **catecholamines,** are known to be transmitters in the regions that elicit the highest rates of self-stimulation in animals equipped with do-it-yourself devices. A number of drugs are available that affect moods in humans, some of them being drugs that have also been shown to influence self-stimulation in experimental animals. For example, increased self-stimulation is observed when drugs that increase catecholamine synaptic activity are administered, such as amphetamine, an "upper" drug. Although most of these drugs are used therapeutically to treat various mental disorders, others, unfortunately, are abused.

In some cases, the effectiveness of various drugs in treating a specific disorder has provided an important clue to the underlying biochemical defect responsible for the condition. For example, two different lines of biochemical evidence suggest that **schizophrenia,** a mental disorder characterized by delusions and hallucinations, probably results from excess dopamine transmission. First, all drugs effective in the treatment of schizophrenia interfere with dopamine transmission in one way or another. Second, drugs that enhance dopamine activity can induce symptoms resembling those of schizophrenia. An unanswered question has been the anatomic site of the abnormal dopamine transmission. The situation is similar to learning that there is a gas leak in your home but not knowing where it is. Studies identifying the regions of the brain that extensively use dopamine as a neurotransmitter have narrowed down the possible sites. Among these are neural networks linked to the limbic system. Since the limbic system is extensively involved in emotions and in triggering behavior appropriate to environmental circumstances—functions that are abnormal in schizophrenic patients—it is speculated that this might be the defective site in schizophrenia.

The molecular mechanisms of other mental disorders are also beginning to be unraveled through similar pharmacological evidence. For example, a functional deficiency of serotonin or norepinephrine or both is implicated in **depression,** a disorder characterized by a pervasive unpleasant mood accompanied by a generalized loss of interests and inability to experience pleasure. All effective antidepressant drugs increase the available concentration of these neurotransmitters in the CNS. The fact that serotonin and norepinephrine are synaptic messengers in the regions of the brain involved in pleasure and motivation suggests that the pervasive sadness and lack of interest (no motivation) in depressed patients is related at least in part to disruption of these regions due to deficiencies or decreased effectiveness of these neurotransmitters. As our understanding of the molecular mechanisms of these and other mental disorders is expanded in the future, there is optimism that many psychiatric problems can be corrected or managed through drug intervention, a hope of great medical significance.

Learning is the acquisition of knowledge as a result of experiences.

Learning is the acquisition of knowledge or skills as a consequence of experience, instruction, or both. It is widely believed that rewards and punishments are integral parts of many types of learning. If, upon responding in a particular way to a stimulus, an animal is rewarded, there is increased likelihood that it will respond in the same way again to the same stimulus as a consequence of this experience. Conversely, if a particular response is accompanied by punishment, the animal is less likely to repeat the same response to the same stimulus. When behavioral responses that give rise to pleasure are reinforced or those accompanied by punishment are avoided, learning has taken place. Housebreaking a puppy is an example. If the puppy is praised when it urinates outdoors but is scolded when it wets the carpet, it will soon learn the acceptable place to empty its bladder. Wild animals, in contrast, have no such training experience, so they do not learn to confine bladder voiding to particular locations. Thus, learning is a change in behavior that occurs as a result of experiences. It is highly dependent on the organism's interaction with its environment. The only limits to the effects that environmental influences can have on learning are the biological constraints imposed by species-specific and individual genetic endowments.

Memory is laid down in stages.

Memory is the storage of acquired knowledge for later recall. Learning and memory form the basis by which individuals adapt their behavior to their particular external circumstances. Without these mechanisms, planning for successful interactions and intentional avoidance of predictably disagreeable circumstances would be impossible.

The neural change responsible for retention or storage of knowledge is known as the **memory trace** or **engram.** Concepts, not verbatim information, are generally stored. As you read this page, you are storing the concept discussed, not the specific words. Later, when you retrieve the concept from memory, you will convert it into your own words. It is possible, however, to memorize bits of information word by word.

Storage of acquired information is believed to be accomplished in at least two stages: short-term memory and long-term memory (Fig. 5–18). **Short-term memory** lasts for

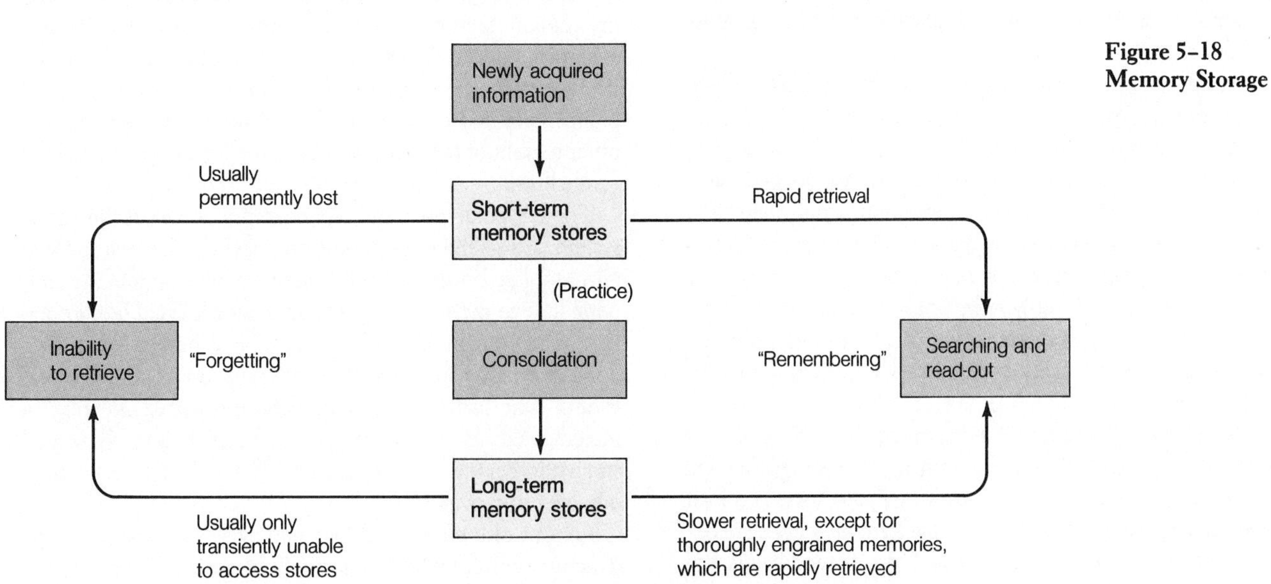

Figure 5–18 Memory Storage

seconds to hours, whereas **long-term memory** is retained for days to years. The transfer and fixation of short-term memory traces into long-term memory stores is known as **consolidation.**

Newly acquired information is initially deposited in short-term memory, which has a limited capacity for storage. Information in short-term memory has one of two eventual fates. Either it is soon forgotten (for example, forgetting a telephone number after you have looked it up and finished dialing) or it is transferred into the more permanent long-term memory mode through *active practice* or *rehearsal*. Recycling of newly acquired information through short-term memory increases the likelihood of long-term memory consolidation. (Therefore, when you cram for an exam, your long-term retention of the information is poor!) This relationship can be likened to developing photographic film. The originally developed image (short-term memory) will rapidly fade unless chemically fixed (consolidation) to provide a more enduring image (long-term memory). Sometimes only parts of memories are fixed while others fade away. Information of interest or importance to the individual is more likely to be recycled and fixed in long-term stores, whereas less important information is quickly erased.

The long-term memory bank has a much larger storage capacity than that for short-term memory. Different informational aspects of long-term memory traces seem to be processed and codified, then stored in conjunction with other memories of the same type; for example, visual memories are stored separately from auditory memories. This facilitates future searching of memory stores at a later date for retrieval of desired information. For example, in remembering a woman you once met, you may use various recall cues from different storage pools, such as her name, what she looked like, the fragrance she had on, what song was playing in the background, and so on.

Because long-term memory stores are larger, it takes longer to retrieve information from long-term memory than from short-term memory. **Remembering** is the process of retrieving specific information from memory stores; **forgetting** is the inability to retrieve stored information. Information lost from short-term memory is permanently forgotten, but information in long-term storage is frequently forgotten only transiently. Often you are only temporarily unable to access the information, as, for example, being unable to remember an aquaintance's name, then having it suddenly "come to you" later.

Some forms of long-term memory involving information or skills used on a daily basis are essentially never forgotten and are rapidly accessible, such as knowing your own name or being able to write. Even though long-term memories are relatively stable, there may be a gradual loss or modification of stored information over time unless it is thoroughly ingrained as a result of years of practice.

Occasionally individuals suffer from a lack of memory that involves whole portions of time rather than isolated bits of information. This condition, known as **amnesia,** occurs in two forms. The most common form, **retrograde** ("going backwards") **amnesia,** is the inability to recall recent past events. It usually follows a traumatic event that interferes with electrical activity of the brain, such as a concussion or stroke. If a person is knocked unconscious, the content of short-term memory is essentially erased, resulting in loss of memory of activities that occurred within about the last half hour before the event. Severe trauma may interfere with access to recently acquired information in long-term stores as well.

Anterograde ("going forward") **amnesia,** on the other hand, is the inability to store memory in long-term storage for later retrieval. It is usually associated with lesions of the medial portions of the temporal lobes, which are generally considered to be critical regions for memory consolidation. Individuals suffering from this condition may be able to recall things they learned before the onset of their problem, but they are unable to establish new permanent memories. New information is lost as quickly as it fades from short-term memory. In one case study, the individual could not remember where the bathroom was in his new home but still had total recall of his old home.

Memory traces are present in multiple regions of the brain.

This raises the question, "What parts of the brain are responsible for memory?" Apparently the neurons involved in memory traces are widely distributed throughout the subcortical and cortical regions of the brain, because some traces of memory remain even after extensive damage. However, there is evidence that certain learning tasks are profoundly affected by damage to specific regions of the brain. The regions of the brain implicated in memory include the temporal lobes and other regions of the cerebral cortex, the limbic system and the cerebellum.

The temporal lobes and hippocampus, a part of the limbic system, are essential for transferring new memories into long-term storage. Further evidence of the involvement of the temporal lobe in memory comes from clues provided by stimulation of this region during brain surgery. When patient's temporal lobes are stimulated, they often vividly experience past events. One patient, upon stimulation, reported "hearing" a melody that she believed she had heard before. Repeated stimulation of the same point caused her to hear the same melody over and over again.

Lesions of the temporal lobe and closely associated limbic structures result in deficits in **declarative memories.** These are memories of facts that often result after only one experience and that can be declared in a statement such as, "I saw the Statue of Liberty last summer." In contrast, the cerebellum seems to play an essential role in **procedural memories** involving motor skills gained via repetitive training, such as memorizing a particular dance routine. The distinct localization of these two types of memory is apparent in individuals who have temporal/limbic lesions. They are able to perform a skill, such as playing a piano, but the next day they have no recollection that they played it.

As surprising as it may seem, the cerebral cortex is not believed to be directly essential for many types of learning and memory except to help focus attention. It appears to play a direct role only in memories that involve language, complex spatial information, and other skills that are finally integrated at the cortical level.

Short-term and long-term memory appear to involve different molecular mechanisms.

Another question besides the "where" of memory is the "how" of memory. Despite a vast amount of psychological data, only a few tantalizing scraps of physiological evidence concerning the cellular basis of memory traces are available. Obviously, some change must take place within the neural circuitry of the brain to account for the altered behavior that follows learning. Different mechanisms appear to be responsible for short-term and long-term memory. Evidence suggests that short-term memory involves transient modifications in the function of preexisting synapses. Long-term memory, in contrast, is believed to involve relatively permanent functional or structural changes between existing neurons in the brain.

SHORT-TERM MEMORY. In ingenious experiments in the sea snail, *Aplysia*, two forms of short-term memory—habituation and sensitization—have been shown to be due to modification of different channel proteins in presynaptic terminals of specific afferent neurons. **Habituation** is a decreased responsiveness to repetitive presentations of an indifferent stimulus—that is, one that is neither rewarding nor punishing. **Sensitization** refers to increased responsiveness to mild stimuli following a strong or noxious stimulus. *Aplysia* reflexly withdraws its gill when its siphon is touched. Afferent neurons responding to touch of the siphon (presynaptic neurons) directly synapse on efferent motor neurons (postsynaptic neurons) controlling gill withdrawal. The snail becomes habituated when its siphon is repeatedly touched; that is, it learns to ignore the stimulus and no longer withdraws its gill in response. Sensitization, a more complex form of learning, takes place in *Aplysia* when it is given a hard bang on the siphon. Subsequently, the snail withdraws its gill more vigorously in response to even mild touch. Interestingly, these different forms of learning affect the same site—the synapse between a siphon afferent and a gill efferent—in opposite ways. Habituation depresses this synaptic activity, whereas sensitization enhances it (Fig. 5–19).

In habituation, Ca^{++} entry into the presynaptic terminal is reduced because of closing of the Ca^{++} channels, which leads to a decrease in neurotransmitter release. As a consequence, the postsynaptic potential is reduced compared to normal, resulting in a decrease or absence of the behavioral response controlled by the postsynaptic efferent neuron (gill withdrawal). Thus, the memory for habituation in *Aplysia* is stored in the form of modification of specific Ca^{++} channels. With no further training, this reduced responsiveness lasts for several hours. A similiar process is responsible for short-term habituation in other invertebrates studied. This suggests that Ca^{++} channel modification might be a general mechanism of habituation, although the process is somewhat more complicated by action of intervening interneurons in higher species. Habituation is probably the most common form of learning and is believed to be the first learning process to take place in human infants. In learning to ignore indifferent stimuli, the animal or person is free to attend to other more important stimuli.

Sensitization in *Aplysia* has also been shown to involve channel modification, but a different channel and mechanism are involved. In contrast to habituation, Ca^{++} entry into the presynaptic terminal is enhanced in sensitization. The subsequent increase in neurotransmitter release produces a larger postsynaptic potential, resulting in a more vigorous gill withdrawal response. Sensitization does not have a direct effect on the presynaptic Ca^{++} channels. Instead, it indirectly enhances Ca^{++} entry via presynaptic facilitation (see p. 110). The neurotransmitter serotonin is released from a facilitating interneuron that synapses on the presynaptic terminal. Serotonin alters activity in the presynaptic terminal to bring about increased release of presynaptic neurotransmitter in response to an action potential. It does so by triggering activation of a cyclic AMP second messenger (see p. 64) within the presynaptic terminal, which ultimately brings about blocking of K^+ channels. This blockage prolongs the action potential in the presynaptic terminal since K^+ efflux through opened K^+ channels is required for the action potential to return to resting (repolarization). Since the presence of a local action potential is responsible for opening of Ca^{++} channels in the terminal, a prolonged action potential permits the greater Ca^{++} influx associated with sensitization.

Thus, existing synaptic pathways may be functionally interrupted (habituated) or enhanced (sensitization) during simple learning. It is speculated that much of short-term memory is similarly a temporary modification of already existing processes. Several lines of research suggest that the cyclic AMP cascade plays an important role, at least in elementary forms of learning and memory.

LONG-TERM MEMORY. Long-term memory storage appears to involve rather permanent physical changes in the brain, changes that endure the ongoing turnover of cellular materials over the years. No new neurons are formed, but permanent structural or functional alterations between existing neurons can take place. Examples of such alterations include the formation of new synaptic connections, permanent changes in pre- or postsynaptic membranes, or an increase or decrease in transmitter synthesis.

Studies comparing the brains of experimental animals reared in a sensory-deprived environment with those raised in a sensory-rich environment demonstrate readily observable microscopic differences. The animals afforded more environ-

Figure 5–19 Habituation and Sensitization in *Aplysia*

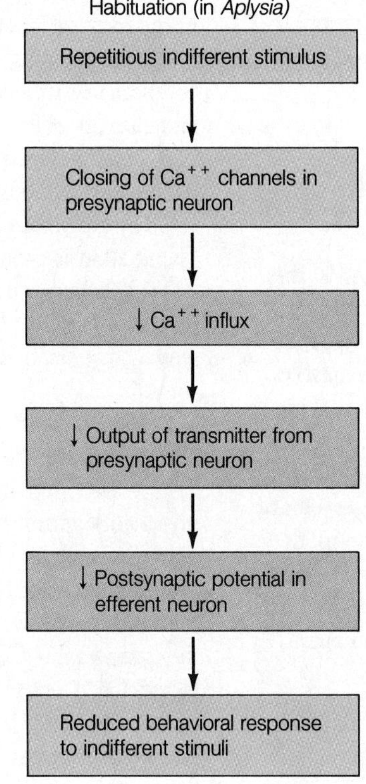

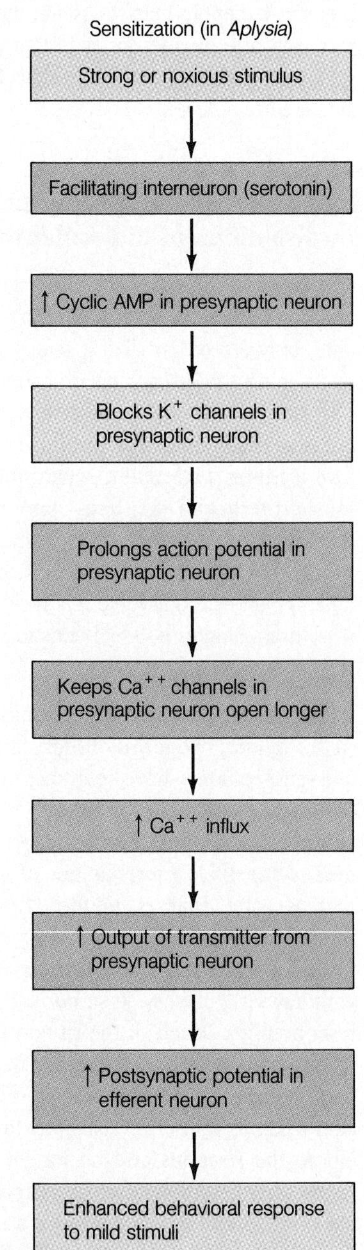

mental interactions—and supposedly, therefore, more opportunity to learn—displayed greater branching and elongation of dendrites in nerve cells in regions of the brain suspected to be involved with memory storage. Greater dendritic surface area presumably provides more sites for synapses.

Other studies suggest that protein synthesis plays a critical role in long-term memory storage. Drugs that block protein synthesis have been shown to interfere with long-term retention of learned behavior. Whether the critical protein is part of a structural change or is necessary for the synthesis of neurotransmitters remains speculative.

Recent clues suggest that different long-term changes may be responsible for declarative and procedural memories. Researchers believe that declarative memories involve a breakdown of certain structural components of the dendrites of affected neurons. This structural breakdown presumably enables the neuron to change shape or to form new connections, thus encoding the declarative memory in this modified synaptic structure. In contrast, studies suggest that laying down of procedural memories somehow requires protein synthesis.

CEREBELLUM

The cerebellum is important in balance as well as in planning and execution of voluntary movement.

The **cerebellum** is attached to the back of the upper portion of the brain stem, lying underneath the occipital lobe of the cortex (Fig. 5–16). It is concerned primarily with motor activity, yet like the basal nuclei, it does not have any direct influence on the efferent motor neurons. It functions indirectly by modifying the output of major motor systems of the brain.

The cerebellum consists of three functionally distinct parts, which are believed to have developed successively during evolution. These parts have different sets of inputs and outputs, and, accordingly, each has different functions (Fig. 5–20).

1. The **vestibulocerebellum** is important for the maintenance of balance and controls eye movement.
2. The **spinocerebellum** regulates muscle tone and coordinates skilled, voluntary movements. When cortical motor

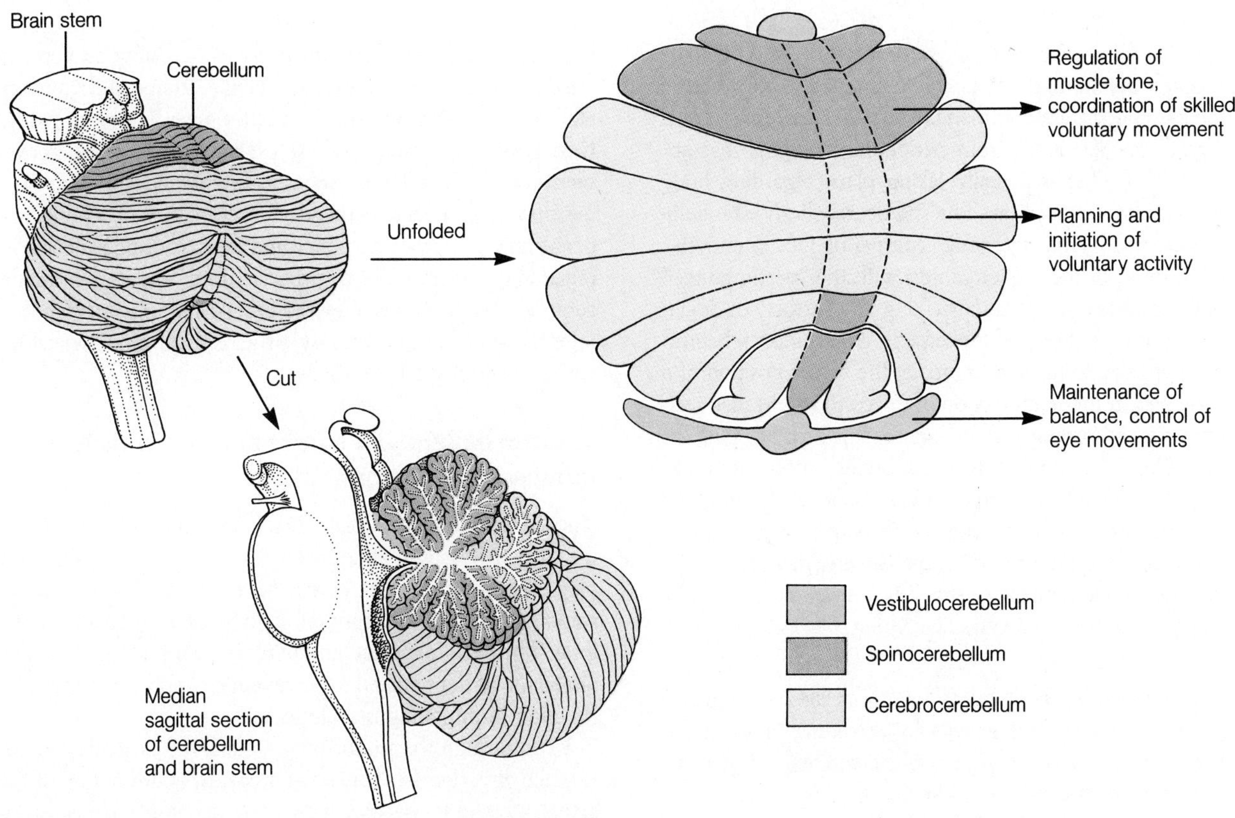

Figure 5–20 Cerebellum *(a) Gross structure of cerebellum. (b) Unfolded cerebellum, revealing its three functionally distinct parts. (c) Internal structure of cerebellum.*

Figure 5–21 Role of the Spinocerebellum

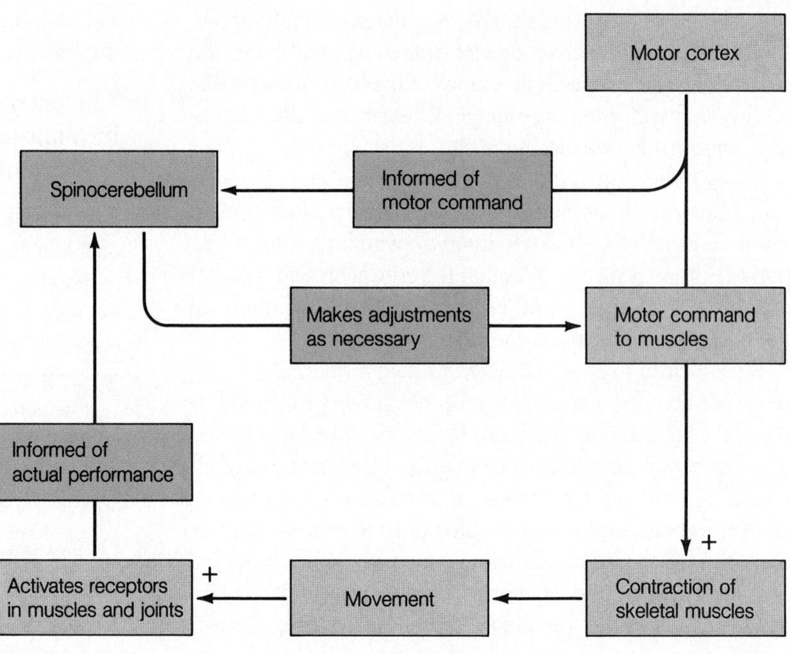

areas send messages to muscles for the execution of a particular movement, the spinocerebellum is also informed of the intended motor command (Fig.5–21). In addition, this region receives input from peripheral receptors that apprise it of what is actually taking place regarding body movement and position. The spinocerebellum essentially acts as "middle management," comparing the "intentions" or "orders" of the higher centers with the "performance" of the muscles and then correcting any "errors" or deviations from the intended movement. The spinocerebellum even appears to be able to predict the future position of a body part in the next fraction of a second and to make adjustments accordingly. If you are reaching for a pencil, for example, this region "puts on the brakes" soon enough to stop the forward movement of your hand at the intended location rather than allowing you to overshoot your target. These ongoing adjustments, which assure smooth, precise, directed movement, are especially important for rapidly changing (phasic) activities like typing, playing the piano, or running.

3. The **cerebrocerebellum** plays a role in the planning and initiation of voluntary activity by providing input to the cortical motor areas. This is also the cerebellar region involved in procedural memories.

The following range of symptoms that characterize cerebellar disease are all referable to a loss of these functions: poor balance; nystagmus (rhythmical, oscillating eye movements); reduced muscle tone but no paralysis; inability to perform rapid movements smoothly; and inability to stop and start skeletal muscle action quickly. The latter gives rise to an **intention tremor** characterized by oscillating to-and-fro movements of a limb as it approaches its intended destination. As a person with cerebellar damage attempts to pick up a pencil, he or she may overshoot the pencil and then rebound excessively, repeating this to-and-fro process until success is finally achieved. No tremor is observed except in the performance of intentional activity, in contrast to the resting tremor associated with disease of the basal nuclei.

The cerebellum and basal nuclei exert different influences on motor activity.

The cerebellum and basal nuclei both monitor and adjust motor activity commanded from the motor cortex, but they play different roles. The cerebellum helps maintain balance; smooths out fast, phasic motor activity; and enhances muscle tone. The basal nuclei are important in coordinating slow, sustained movement related to posture and support, and they also function in inhibiting muscle tone.

Even though the motor command for a particular voluntary activity arises from the motor cortex, coordination of the actual execution of that activity is accomplished subconsciously by subcortical regions, especially the cerebellum and basal nuclei. To illustrate, you can voluntarily decide that you want to walk, but you do not have to consciously think about the specific sequence of movements that will have to be performed to accomplish this intentional act. Accordingly, much of voluntary activity is actually involuntarily regulated.

Brain Stem

The brain stem is a vital link between the spinal cord and higher brain regions.

The **brain stem,** which consists of the **medulla, pons,** and **midbrain,** is a critical connecting link between the remainder of the brain and the spinal cord. All fibers traversing between the periphery and higher brain centers must pass through the brain stem, some relaying sensory information to the brain, others carrying command signals from the brain for efferent output. A few fibers merely pass through, but most stop or send side branches within the brain stem for important synaptic processing. The functions of the brain stem include the following:

1. The majority of the twelve pairs of cranial nerves arise from the brain stem. With one major exception, these nerves contain sensory and motor fibers that supply structures in the head and neck. They are important in the functions of sight, hearing, taste, smell, sensation of the face and scalp, eye movement, chewing, swallowing, facial expressions, and salivation. The major exception is cranial nerve X, the **vagus nerve.** Instead of innervating regions in the head, most of the branches of the vagus nerve supply organs in the thoracic and abdominal cavities. The vagus is the major nerve of the parasympathetic nervous system.
2. Collected within the brain stem are neuronal clusters or "centers" that control heart and blood vessel function, respiration, and many digestive activities.
3. This region further plays a role in modulating the sense of pain (see chap. 6).
4. Running throughout the entire brain stem and into the thalamus is an incredibly widespread network of interconnected neurons called the **reticular formation.** This network receives and integrates all synaptic input. Ascending fibers originating in the reticular formation carry signals upward to arouse and activate the cerebral cortex. These fibers compose the **reticular activating system (RAS)**, which controls the overall degree of cortical alertness and is important in the ability to direct attention toward specific events.
5. The reticular formation is also important in controlling motor activity, particularly the regulation of muscle reflexes involved with equilibrium and support of the body against gravity.
6. Finally, intricately involved with the reticular activating system are the centers responsible for sleep, which are also housed within the brain stem.

Sleep is an active process consisting of alternating periods of slow-wave and paradoxical sleep.

Consciousness refers to subjective awareness of the external world and self, including awareness of the private inner world of one's own mind, that is, awareness of thoughts, perceptions, feelings, dreams, and so on. No one has any idea of the neural mechanisms that give rise to conscious awareness. Even though the final level of awareness resides in the cerebral cortex and a crude sense of awareness is detected by the thalamus, conscious experience depends on the integrated functioning of many parts of the nervous system.

The following various states of consciousness are listed based on the extent of interaction between peripheral stimuli and the brain, in decreasing order of levels of arousal:

- ☐ maximum alertness
- ☐ wakefulness
- ☐ sleep (several different types)
- ☐ coma

Maximum alertness is dependent on attention-getting sensory input that "energizes" the RAS and subsequently the level of activity of the CNS as a whole. At the other extreme, **coma** refers to total unresponsiveness of a living person to external stimuli, caused either by brain stem damage that interferes with the RAS or by widespread depression of the cerebral cortex as a whole, such as that accompanying O_2 deprivation.

The **sleep-wake cycle** is a normal cyclical variation in awareness of surroundings. In contrast to being awake, sleeping individuals are not consciously aware of the external world, but they do have inward conscious experiences such as dreams. Furthermore, they can be aroused by external stimuli, such as an alarm going off.

Sleep is an *active* process, not just the absence of wakefulness. Contrary to what one might suspect, the brain's overall level of activity is not reduced during sleep. During certain stages of sleep, O_2 uptake by the brain is even increased above normal waking levels.

There are two types of sleep, characterized by different EEG patterns and different behaviors: **slow-wave sleep** and **paradoxical** or **REM** sleep (Table 5–2). Slow-wave sleep occurs in four stages, each displaying progressively slower EEG waves of higher amplitude (hence "slow wave" sleep). At the

Table 5–2 Comparison of Slow-Wave and Paradoxical Sleep

	Type of Sleep	
Characteristic	*Slow-Wave Sleep*	*Paradoxical Sleep*
EEG	Displays slow waves	Similar to EEG of alert, awake person
Motor activity	Considerable muscle tone; frequent shifting	Abrupt inhibition of muscle tone; no movement
Heart rate, respiratory rate, blood pressure	Minor reductions	Irregular
Dreaming	Rare (mentation is extension of waking-time thoughts)	Common
Arousal	Sleeper easily awakened	Sleeper hard to arouse but apt to wake up spontaneously
Percent of sleeping time	80%	20%
Other important characteristics	Has 4 stages; sleeper must pass through this type of sleep first	Rapid eye movements

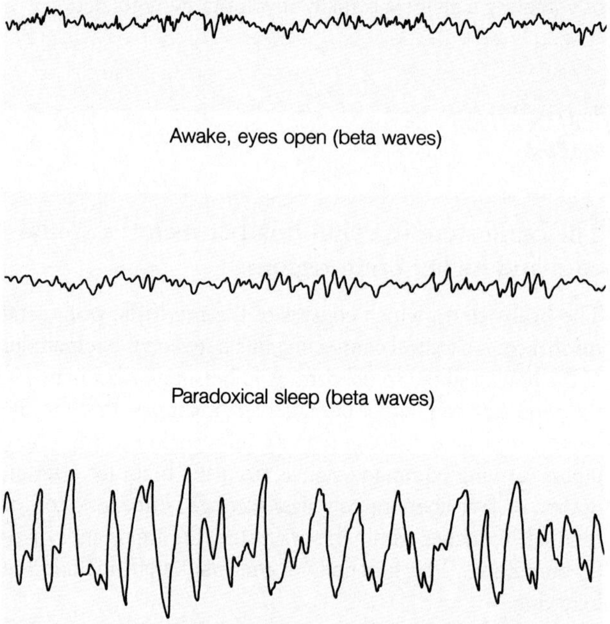

Awake, eyes open (beta waves)

Paradoxical sleep (beta waves)

Slow-wave sleep, stage 4 (delta waves)

Figure 5–22 EEG Patterns during Different Types of Sleep

onset of sleep, an individual moves from the light sleep of stage 1 to the deep sleep of stage 4 during a period of thirty to forty-five minutes, then reverses through the same stages in the same amount of time. A ten to fifteen-minute episode of paradoxical or REM sleep punctuates the end of each slow-wave sleep cycle. Paradoxically, the EEG pattern during this time abruptly becomes similar to that of a wide-awake, alert individual, even though the person is still asleep (Fig. 5–22). Following the paradoxical episode, the stages of slow-wave sleep are repeated once again. A person cyclically alternates between the two types of sleep throughout the night. Toward morning, the episodes of paradoxical sleep occur more frequently and last longer whereas the deeper stages of slow-wave sleep disappear. On the average, paradoxical sleep occupies 20% of total sleeping time throughout adolescence and most of adulthood. Infants spend considerably more time in paradoxical sleep. In contrast, paradoxical as well as stage 4 slow-wave sleep decline in the elderly.

Paradoxical sleep can either be considered the deepest sleep, since it is hardest to arouse sleepers from this stage, or the lightest sleep, since sleepers are most apt to awaken on their own during this stage. Several other observations support the contention that it is the deepest sleep. First, in a normal sleep cycle, a person always passes through slow-wave sleep before entering paradoxical sleep. Second, individuals who require less total sleeping time than normal spend proportionately more time in paradoxical and stage 4 sleep and less time in the lighter stages of slow-wave sleep.

In addition to distinctive EEG patterns, the two types of sleep are distinguished by behavioral differences. It is difficult to pinpoint exactly when an individual drifts from drowsiness into slow-wave sleep. In this type of sleep, the person still has considerable muscle tone and frequently shifts body position. Only minor reductions in respiratory rate, heart rate, and blood pressure occur. During this time the sleeper can be easily awakened and rarely dreams. The mental activity associated with slow-wave sleep is less visual than dreaming. It is more conceptual and plausible—like an extension of waking-time thoughts concerned with everyday events—and it is less likely to be recalled. The major exception is nightmares, which occur during stages 3 and 4. People who walk and talk in their sleep do so during slow-wave sleep.

The behavioral pattern accompanying paradoxical sleep is marked by abrupt inhibition of muscle tone throughout the body. The muscles are completely relaxed with no movement taking place. Paradoxical sleep is further characterized by **rapid eye movements;** this is the reason it is also called REM sleep. Heart rate and respiratory rate become irregular and blood pressure may fluctuate. The sleeper is temporarily unable to regulate body temperature, which begins to change in the direction of the room temperature. (If you wake up at this time, you may adjust the covers because you feel too hot or too cold.) Another characteristic of REM sleep is **dreaming.** There is little evidence that the rapid eye movements are related to "watching" the dream imagery. The eye movements seem to be driven in a locked oscillating pattern uninfluenced by dream content.

The sleep-wake cycle is probably controlled by interactions among three brain stem regions.

The sleep-wake cycle as well as the various stages of sleep are believed to be due to the cyclical interplay of three different neural systems in the brain stem: (1) an arousal system, which is part of the reticular activating system; (2) a slow wave sleep center; and (3) a paradoxical sleep center. The pattern of interaction among the three identified neural regions, which brings about the fairly predictable cyclical sequence between being awake and passing alternately between the two stages of sleep, is the subject of intense investigation.

The normal cycle can easily be interrupted, with the arousal system more readily overriding the sleep systems than vice versa; that is, it is easier to stay awake when you are sleepy than to fall asleep when you are wide-awake. The arousal system can be activated by afferent sensory input (for example, a person has difficulty falling asleep when it is noisy) or by input descending to the brain stem from higher brain regions. Intense concentration or strong emotional states, such as anxiety or excitement, can keep a person from falling asleep, just as motor activity, such as getting up and walking around, can arouse a drowsy person.

An unusual sleep disturbance is **narcolepsy.** It is characterized by brief (five-to thirty-minute) irresistible **sleep attacks** during the day. A person suffering from this condition suddenly falls asleep during any ongoing activity, often without warning. Narcolepsy is suspected of being linked to abnormal activation of neurons in the paradoxical sleep center during the waking state. Narcoleptic patients, in fact, can enter into paradoxical sleep directly without the normal prerequisite passage through slow-wave sleep.

Even though humans spend about a third of their lives sleeping, the reason sleep is needed largely remains a mystery. Sleep is not accompanied, as once was suspected, by a *reduction* in neural activity (that is, the brain cells are not "resting"), but rather by a profound *change* in activity. It is speculated that sleep is necessary to allow the brain to "shift gears" in order to accomplish the long-term structural and chemical adjustments necessary for learning and memory. This might explain why infants require so much sleep. Their highly plastic brains are rapidly undergoing profound synaptic modifications in response to environmental stimulation. In contrast, mature individuals in whom neural changes are less dramatic sleep less. Not much is known about the brain's need for the two types of sleep, although a specified amount of paradoxical sleep appears to be required. Individuals experimentally deprived of paradoxical sleep for a night or two by being aroused every time the paradoxical EEG pattern appeared suffered hallucinations and spent proportionally more time in paradoxical sleep during subsequent undisturbed nights, as if to "make up for lost time."

Spinal Cord

The spinal cord extends through the vertebral canal and is connected to the spinal nerves.

Extending from the brain stem is a long slender cylinder of nerve tissue, the **spinal cord.** It is about 45 cm long (18 inches) and 2 cm in diameter (about the size of your little finger). Exiting through a large hole in the base of the skull, the spinal cord descends through the vertebral canal, enclosed by the surrounding protective vertebral column. Paired spinal nerves emerge from the spinal cord through spaces formed between the bony, wing-like arches of adjacent vertebrae. The spinal nerves are named according to the region of the spinal cord from which they arise (Fig. 5–23a and b): there are eight pairs of *cervical* nerves (that is, C1–C8); twelve *thoracic* nerves, five *lumbar* nerves; five *sacral* nerves; and one *coccygeal* nerve.

During development, the vertebral column grows about 25 cm longer than the spinal cord. Because of this differential growth, segments of the spinal cord that give rise to the various spinal nerves are not aligned with the corresponding intervertebral spaces. Most of the spinal nerve roots must descend along the cord before emerging from the vertebral column at the corresponding space. The spinal cord itself extends only to the level of the first or second lumbar vertebra, so the nerve roots of the remaining lumbar, sacral, and coccygeal nerves are greatly elongated in order to exit the vertebral column at their appropriate space. The thick bundle of elongated nerve roots within the lower vertebral canal is known as the **cauda equina** ("horse's tail") because of its appearance (Fig.

Figure 5–23 Spinal Nerves *There are thirty-one pairs of spinal nerves named according to the region from which they arise. Because the spinal cord is shorter than the vertebral column, spinal nerve roots must descend along the cord before emerging from the vertebral column at the corresponding intervertebral space, especially those beyond the level of the first lumbar vertebra (L1). Collectively these rootlets are called the cauda equina, literally "horse's tail." (a) Posterior view of brain and spinal cord. (b) Lateral view of spinal cord and emergence of spinal nerves from vertebral column.*

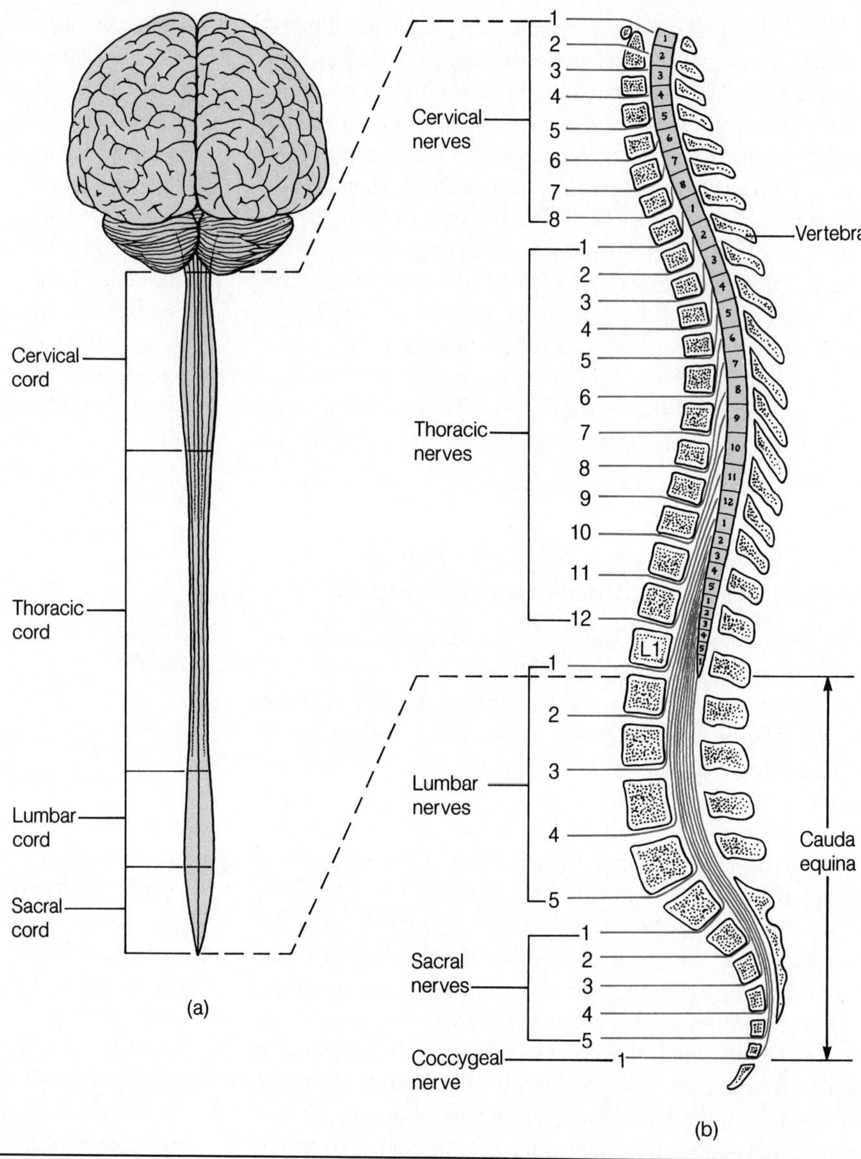

5–23b). This region is the site for performing "spinal taps" to obtain a sample of CSF. Insertion of a needle into the canal below the level of the second lumbar vertebra does not run the risk of penetrating the spinal cord. The needle pushes aside the nerve roots of the cauda equina so that the sample of the surrounding fluid can be withdrawn safely.

Although there are some slight regional variations, the cross-sectional anatomy of the spinal cord is generally the same throughout its length (Fig. 5–24). Unlike in the brain, the gray matter in the spinal cord forms a butterfly-shaped region on the inside that is surrounded by the outer white matter. As in the brain, the cord gray matter consists primarily of neuronal cell bodies and their dendrites, short interneurons, and glial cells. The white matter is organized into **tracts,** which are bundles of nerve fibers (axons of long interneurons) with a similar function. The bundles are grouped into columns that extend the length of the cord. Each of these tracts begins or ends within a particular area of the brain, and each is specific in the type of information that it transmits. Some are **ascending** (cord to brain) **tracts** that transmit to the brain signals derived from afferent input. Others are **descending** (brain to cord) **tracts** that relay messages from the brain to efferent neurons (Figs. 5–25).

The tracts are generally named for their origin and termination. For example, the **corticospinal tract** is a descending pathway; its cell bodies originate primarily in the motor region of the cerebral cortex and its axons travel down to terminate in the spinal cord on the cell bodies of efferent motor neurons

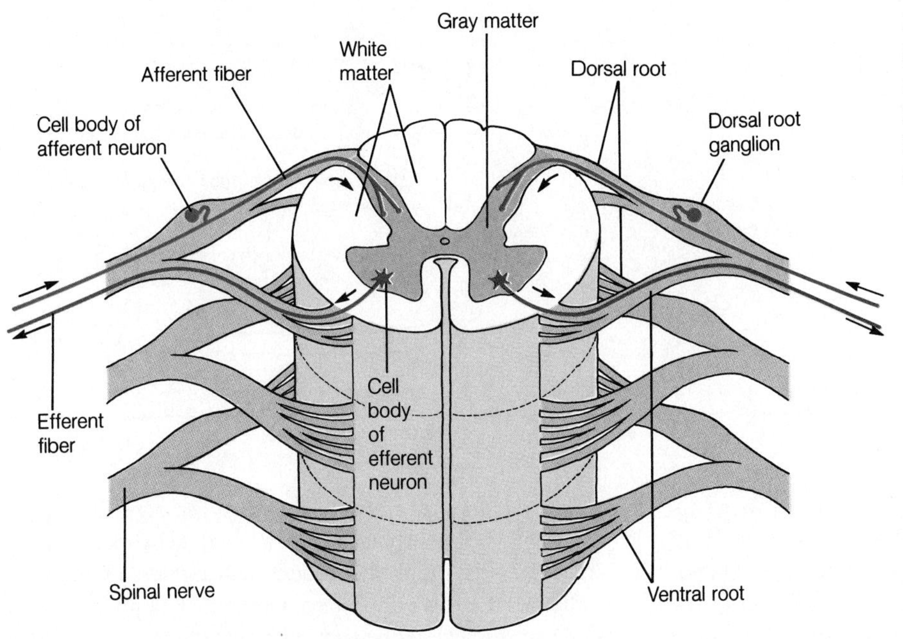

Figure 5–24 Spinal Cord in Cross Section *The afferent fibers enter through the dorsal root, and the efferent fibers exit through the ventral root. Afferent and efferent fibers are enclosed together within a spinal nerve.*

supplying skeletal muscles. In contrast, the **lateral spinothalamic tract** is an ascending pathway that originates in the spinal cord and runs laterally through the cord until it synapses in the thalamus. It carries sensory information about pain and temperature that has been delivered from various regions of the body through the spinal cord to the thalamus, which in turn sorts and relays the information to the somatosensory cortex. It is important to appreciate that various types of signals are segregated within the spinal cord and, accordingly, that damage to particular areas of the cord can interfere with some functions while others remain intact.

The centrally located gray matter is also functionally organized (Fig. 5–26). The central canal, which is filled with CSF, lies in the center of the gray matter. Each half of the gray mat-

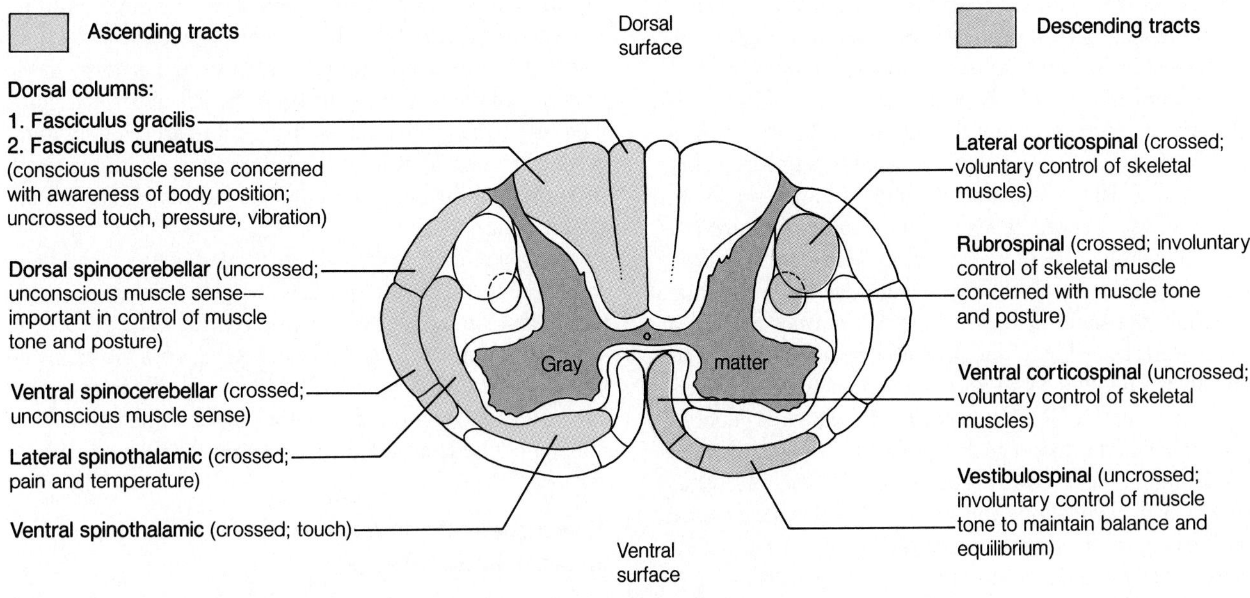

Figure 5–25 Major Ascending and Descending Tracts in the White Matter

Figure 5-26 Regions of the Gray Matter

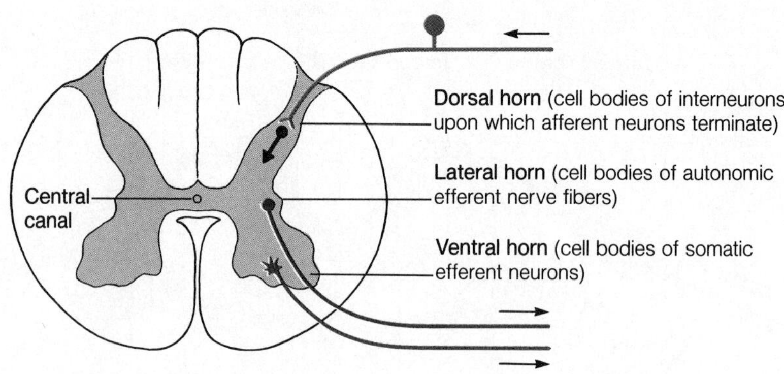

ter is arbitrarily divided into a **dorsal (posterior) horn,** a **ventral (anterior) horn,** and a **lateral horn.** The dorsal horn contains cell bodies of interneurons upon which afferent neurons terminate. The ventral horn contains cell bodies of the efferent motor neurons supplying skeletal muscles. Autonomic nerve fibers supplying cardiac and smooth muscle and exocrine glands originate at cell bodies found in the lateral horn.

Spinal nerves attach to each side of the spinal cord by a **dorsal root** and a **ventral root** (Fig. 5-24). Afferent fibers enter the spinal cord through the dorsal root; efferent fibers leave through the ventral root. The cell bodies for the afferent neurons at each level are clustered together in a **dorsal root ganglion.** (A collection of neuronal cell bodies located outside of the CNS is called a *ganglion*, whereas a functional collection of cell bodies within the CNS is referred to as a *center* or a *nucleus*). The cell bodies for the efferent neurons originate in the gray matter and send axons out through the ventral root.

The dorsal and ventral roots at each level join to form a **spinal nerve** that emerges from the vertebral column. A spinal nerve contains both afferent and efferent fibers traversing between a particular region of the body and the spinal cord. Note the relationship between a *nerve* and a *neuron*. A **nerve** is a bundle of peripheral neuronal axons, some afferent and some efferent, which are enclosed by a connective tissue covering and follow the same pathway. A nerve does not contain a complete nerve cell, only the axonal portions of many neurons. (By this definition, there are no nerves in the CNS! Bundles of axons in the CNS are called tracts.) The individual fibers within a nerve do not generally have any direct influence on each other. They travel together for convenience, just as many individual telephone lines are carried within a telephone cable, yet any particular connection can be private without interference or influence from other lines in the cable.

The thirty-one pairs of spinal nerves, along with the twelve pairs of cranial nerves that arise from the brain, constitute the *peripheral nervous system*. After they emerge, the spinal nerves progressively branch to form a vast network of peripheral nerves that supply the tissues. Each segment of the spinal cord gives rise to a pair of spinal nerves that ultimately supply a particular region of the body with both afferent and efferent fibers. Thus, the location and extent of sensory and motor deficits associated with spinal cord injuries can be clinically important in determining the level and extent of the cord injury.

With reference to sensory input, each specific region of the body surface supplied by a particular spinal nerve is called a **dermatome.** These same spinal nerves also carry fibers that branch off to supply internal organs, and sometimes pain originating from one of these organs is "referred" to the corresponding dermatome supplied by the same spinal nerve. **Referred pain** originating in the heart, for example, may appear to come from the left shoulder and arm. The mechanism responsible for referred pain is not completely understood. Presumably, inputs arising from the heart share a pathway to the brain in common with inputs from the left upper extremity. The higher perception levels, being more accustomed to receiving sensory input from the left arm than from the heart, may interpret the input from the heart as having arisen from the left arm.

The spinal cord is strategically located between the brain and afferent and efferent fibers of the peripheral nervous system so that it may fulfill its two primary functions: (1) serving as a link for transmission of information between the brain and remainder of the body; and (2) integrating reflex activity between afferent input and efferent output without involving the brain. The latter is known as a **spinal reflex.**

The spinal cord is responsible for the integration of many basic reflexes.

A **reflex** is any response that occurs automatically without conscious effort. There are two types of reflexes: **simple** or

A CLOSER LOOK AT EXERCISE PHYSIOLOGY

Swan Dive or Belly Flop: It's a Matter of CNS Control

Sport skills must be learned. Much of the time, strong basic reflexes must be overridden in order to perform the skill. Learning to dive into water, for example, is very difficult initially. Strong head-righting reflexes controlled by sensory organs in the neck and ears initiate a straightening of the neck and head before the beginning diver enters the water, causing what is commonly known as a "belly flop." In a backward dive, the head-righting reflex causes the beginner to land on his or her back or even in a sitting position. To perform any motor skill that involves body inversions, somersaults, back flips, or other abnormal postural movements, the person must learn to consciously inhibit basic postural reflexes. This is accomplished by having the person concentrate on specific body positions during the movement. For example, in the somersault, the person must concentrate on keeping the chin tucked and grabbing the knees in order to perform the skill. After the skill is performed repeatedly, new synaptic patterns are formed in the CNS, and the new or conditioned response substitutes for the natural reflex responses. Sport skills must be practiced until the movement becomes automatic; then the athlete is free during competition to think about strategy or the next move to be performed in a routine.

basic reflexes, which are built-in, unlearned responses, such as closing the eyes when an object moves toward them; and **acquired,** or **conditioned reflexes,** which are a result of practice and learning, such as turning the steering wheel of a car to follow a curve. You do this automatically, but only after considerable conscious training effort. (See the accompanying boxed feature, A Closer Look at Exercise Physiology.)

The neural pathway involved in accomplishing reflex activity is known as a **reflex arc,** which typically includes the five following basic components:

1. receptor
2. afferent pathway
3. integrating center
4. efferent pathway
5. effector

The **receptor** responds to a **stimulus,** which is a detectable physical or chemical change in the environment of the receptor. In response to the stimulus, the receptor produces an action potential that is relayed by the **afferent pathway** to the integrating center for processing. Usually the **integrating center** is the CNS. The spinal cord and brain stem are responsible for the integration of the basic reflexes, whereas higher brain levels usually process acquired reflexes. The integrating center processes all information available to it from this receptor as well as from all other inputs, then "makes a decision" about the appropriate response. The instructions from the integrating center are transmitted via the **efferent pathway** to the **effector**—a muscle or gland—which carries out the desired response. Unlike conscious behavior, in which any one of a number of responses is possible, a reflex response is predictable because the pathway between the receptor and effector is always the same.

A basic spinal reflex is one integrated by the spinal cord; that is, all components necessary for linking afferent input to efferent response are present within the spinal cord. The **withdrawal reflex** can serve to illustrate a basic spinal reflex (Fig. 5-27). When a person touches a hot stove (or receives another painful stimulus), a reflex is initiated to pull the hand away from the stove (to withdraw from the painful stimulus). The skin has different receptors for warmth, cold, light touch, pressure, and pain. Even though all information is sent to the CNS by way of action potentials, the CNS can distinguish between various stimuli because different receptors and consequently different afferent pathways are activated by different stimuli. When a receptor is stimulated sufficiently to reach threshold, an action potential is generated in the afferent neuron. The stronger the stimulus, the greater the frequency of action potentials generated and propagated to the CNS. Once the afferent neuron enters the spinal cord, it diverges to syn-

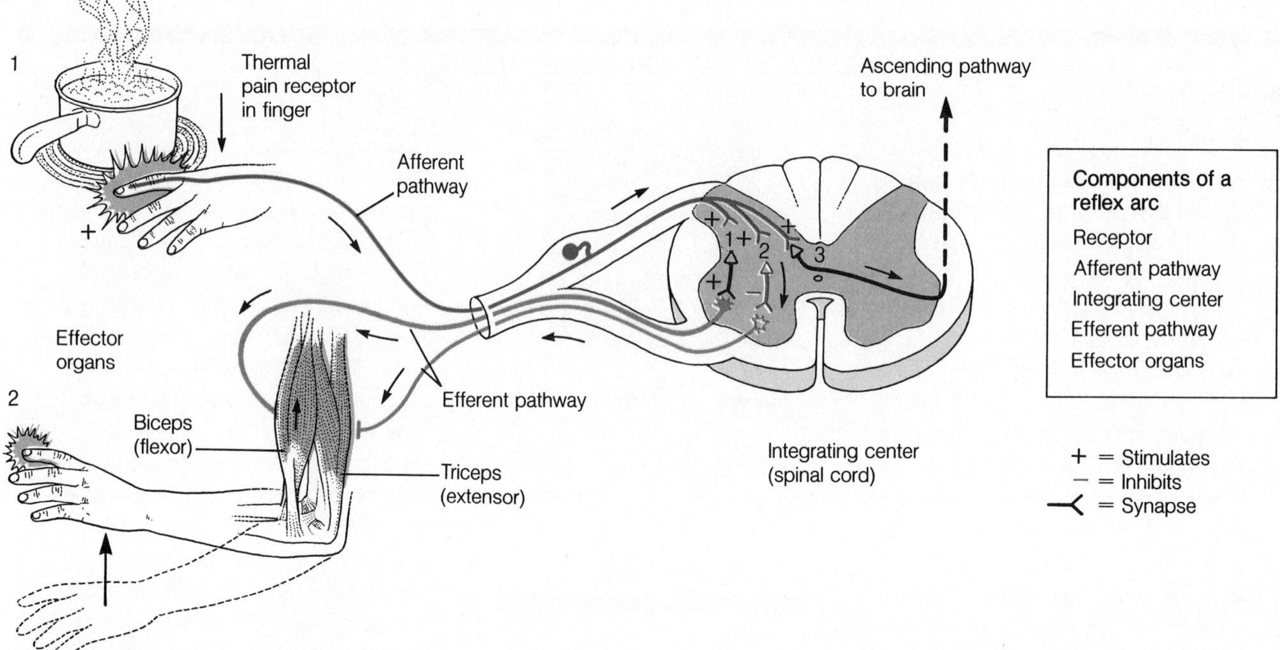

Figure 5–27 Withdrawal Reflex *See text for discussion of the various numbered components of the withdrawal reflex.*

apse with the following different interneurons (the numbers correspond to those on Fig. 5–27):

1. It stimulates some interneurons that in turn stimulate the efferent motor neurons supplying the biceps, the muscle in the arm that flexes (bends) the elbow joint. The resultant contraction of the biceps pulls the hand away from the hot stove.
2. It stimulates other interneurons that in turn inhibit the efferent neurons supplying the triceps to prevent it from contracting. The triceps is the muscle in the arm that extends (straightens out) the elbow joint. When the biceps is contracting to flex the elbow, it would be counterproductive for the triceps to be contracting. Therefore, built into the withdrawal reflex is inhibition of the muscle that antagonizes (opposes) the desired response. This type of neuronal connection involving stimulation of the nerve supply to one muscle and simultaneous inhibition of the nerves to its antagonistic muscle is known as **reciprocal innervation.**
3. It stimulates still other interneurons that carry the signal up the spinal cord to the brain via an ascending pathway. Only when the impulse reaches the sensory area of the cortex is the person aware of the pain, its location, and the type of stimulus. Also, when the impulse reaches the brain, the information can be stored as memory, and the person can start thinking about the situation—how it happened, what to do about it, how to prevent it from happening again, and so on. All this activity at the conscious level is above and beyond the basic reflex.

As is characteristic of all spinal reflexes, the brain can influence the withdrawal reflex. Impulses may be sent down descending pathways to the efferent neurons supplying the involved muscles to override the input from the receptors, actually preventing the biceps from contracting in spite of the painful stimulus. When your finger is being pricked to obtain a blood sample, pain receptors are stimulated that initiate the withdrawal reflex. However, knowing that you must be brave and not pull your hand away, you can consciously override the reflex by sending IPSPs via descending pathways to the motor neurons supplying the biceps and EPSPs to those supplying the triceps. The activity in these efferent neurons depends on the sum of activity of all their synaptic inputs. Because the neurons supplying the biceps are now receiving more IPSPs from the brain (voluntary) than EPSPs from the afferent pain pathway (reflex), these neurons are inhibited and do not reach threshold. Therefore, the biceps is not stimulated to contract and withdraw the hand. Simultaneously, the neurons to the triceps are receiving more EPSPs from the brain than IPSPs via the reflex arc, so they reach threshold, fire, and consequently stimulate the triceps to contract, thus keeping the arm

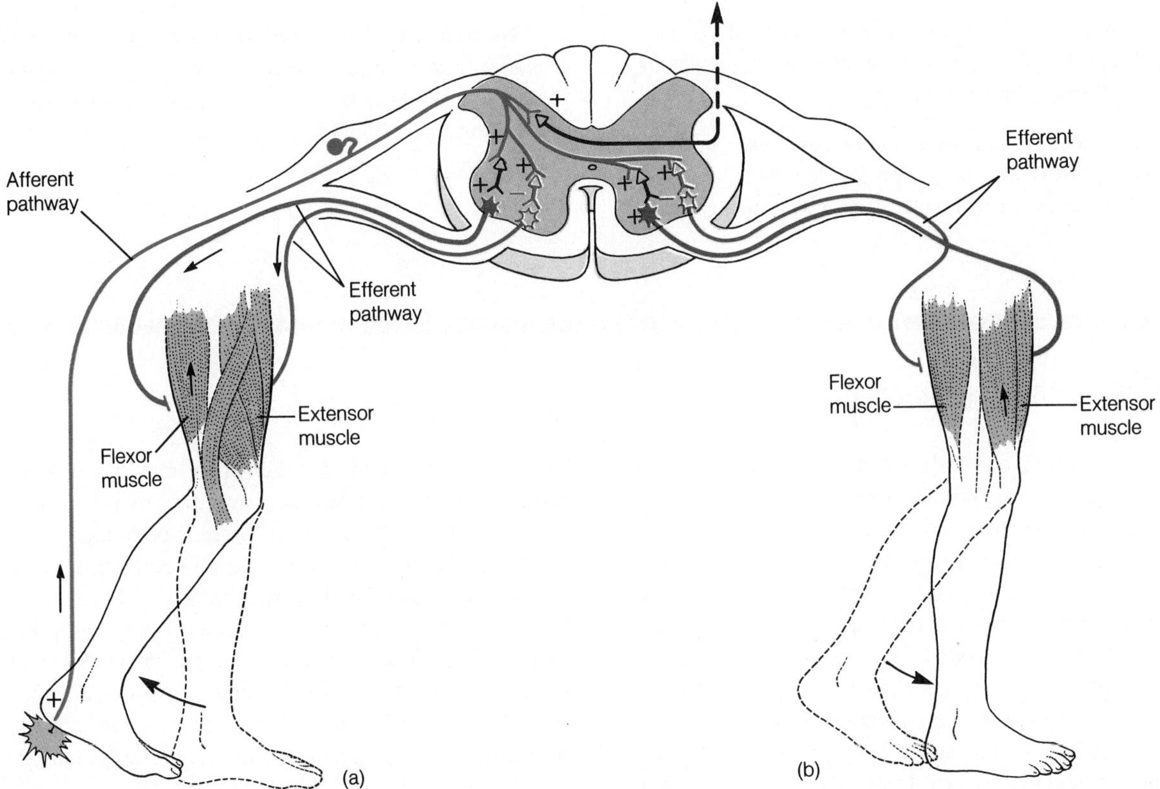

Figure 5–28 Crossed Extensor Reflex Coupled with Withdrawal Reflex *(a) Withdrawal reflex, which causes flexion of injured extremity to withdraw from painful stimulus. (b) Crossed extensor reflex, which extends opposite limb to support full weight of body.*

extended in spite of the painful stimulus. In this way, the withdrawal reflex has been voluntarily overridden.

Only one reflex is simpler than the withdrawal reflex; this is the **stretch reflex,** in which an afferent neuron originating at a stretch-detecting receptor in a skeletal muscle terminates directly on the efferent neuron supplying the same skeletal muscle to cause it to contract and counteract the stretch. This is a **monosynaptic** ("one synapse") **reflex,** because the only synapse in the reflex arc is the one between the afferent and efferent neuron. The withdrawal reflex and all other reflexes are **polysynaptic** ("many synapses"), because interneurons are interposed in the reflex pathway and, therefore, a number of synapses are involved.

Spinal reflex action is not necessarily limited to motor responses on the side of the body to which the stimulus is applied. Assume that a person steps on a tack instead of burning a finger. A reflex arc is initiated to withdraw the injured foot from the painful stimulus, while the opposite leg simultaneously prepares to suddenly bear all of the weight so that the person does not lose balance or fall (Fig. 5–28). Unimpeded bending of the injured extremity's knee is accomplished by concurrent reflex stimulation of the muscles that flex the knee and inhibition of the muscles that extend the knee. At the same time, unimpeded extension of the opposite limb's knee is accomplished by activation of pathways that cross over to the opposite side of the spinal cord to reflexly stimulate this knee's extensors and inhibit its flexors. This **crossed extensor reflex** assures that the opposite limb will be in a position to bear the weight of the body as the injured limb is withdrawn from the stimulus.

Besides protective reflexes such as the withdrawal reflex and simple postural reflexes such as the crossed extensor reflex, basic spinal reflexes also mediate emptying of pelvic organs (for example, urination, defecation, and expulsion of semen). All spinal reflexes can be voluntarily overridden at least temporarily by higher brain centers.

Not all reflex activity involves a clear-cut reflex arc, although the basic principles of an automatic response to a detectable change are still present. The two general ways in which pathways for unconscious responsiveness digress from the typical reflex arc are the following:

1. *Responses mediated at least in part by hormones*. A particular reflex activity may be mediated solely by either neurons or hormones or may involve a pathway utilizing both.
2. *Local responses that do not involve either nerves or hormones*. For example, the blood vessels in an exercising muscle dilate because of local metabolic changes, thereby increasing blood flow to match the active muscle's metabolic needs.

Chapter in Perspective

The nervous system is one of the two major control systems of the body, the other being the endocrine system. A complex interactive network of three basic types of neurons—afferent neurons, efferent neurons, and interneurons—constitutes the nervous system. The central nervous system (CNS) is composed of the brain and spinal cord, whereas the peripheral nervous system (PNS) consists of the cranial and spinal nerves. The peripheral nerves contain both afferent and efferent fibers. Receptors at the peripheral endings of afferent fibers detect and transfer information about the external and internal environment to the CNS. The CNS sorts and processes this input, then initiates appropriate directions in efferent neurons, which carry the instructions from the CNS to effector organs to bring about the desired muscular contraction or glandular secretion. Many of these neurally controlled activities are aimed toward maintaining homeostasis. At the simplest level, the spinal cord integrates many basic protective and evacuation reflexes that do not require conscious participation. The brain, in addition to serving as a more complex integrating link between afferent input and efferent output, is also responsible for abstract neural phenomena such as thinking, learning, remembering, consciousness, and emotions.

The nervous system has become progressively more complex during evolutionary development. Newer, more complicated, and more sophisticated layers of the brain have been piled on top of older, more primitive regions. Many of the basic activities necessary for survival are built into the older parts of the brain. The newer, higher levels progressively modify, enhance, or nullify actions coordinated by lower levels in a hierarchy of command, and they also add new dimensions of capabilities. The human brain may be arbitrarily divided into: (1) the brain stem, the oldest region, which primarily regulates vital life-sustaining functions and which is essential for arousal of the entire brain and for the sleep-wake cycle; (2) the cerebellum, which is concerned with maintaining body position and coordination of rapid skeletal-muscle movements; (3) the subcortical structures, which coordinate many homeostatic functions, accomplish preliminary processing of sensory input, refine certain aspects of motor activity, and are extensively involved with motivated behavior, emotions, and learning; and (4) the cerebral cortex, the evolutionary newcomer. The peak of neural capability is reached in the cerebral cortex, which is the dominant structure of the human brain. It is responsible for initiating all voluntary activity, for language, for complex perceptual awareness of environment and self, and for the individual's personality.

Our knowledge of the human brain is rudimentary because of its uniqueness and complexity. All neural activity—ranging from the most private thoughts to commands for motor activity to retrieval of stored memories from the distant past—is ultimately attributable to propagation of action potentials along individual nerve cells and chemical transmission between cells. Although the exact pathways, neurotransmitters, and cellular changes involved are far from being decoded, progress is being made. In fact, it is the human cortex that is developing techniques for unraveling the secrets of its own complex nature.

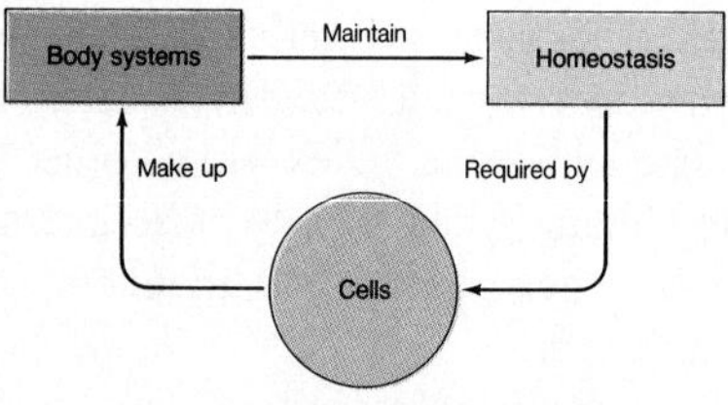

See inside front cover for an expanded version of this model.

Review Exercises

1. Outline the various subdivisions of the nervous system.
2. Compare the three classes of neurons.
3. Discuss the function of each of the following: glial cells, the meninges, the cerebrospinal fluid, and the blood-brain barrier.
4. To what phenomenon associated with the brain does *plasticity* refer?
5. Compare the composition of the white and the gray matter.
6. Draw and label the major functional areas of the cerebral cortex, indicating the function or functions attributable to each area.
7. Describe the roles played by various regions of the brain in motor control.
8. List the functions performed by each of the following noncortical structures of the brain: basal nuclei, thalamus, hypothalamus, limbic system, vestibulocerebellum, spinocerebellum, cerebrocerebellum, and brain stem.
9. Explain the role of rewards and punishments in learning.
10. Compare short-term and long-term memory.
11. Compare slow-wave and paradoxical sleep.
12. Draw and label a cross-section of the spinal cord.
13. List the components of a reflex arc; describe the components of the withdrawal reflex as an example. Distinguish between a monosynaptic and polysynaptic reflex.
14. **A point to ponder:** Special studies designed to assess the specialized capacities of each cerebral hemisphere have been performed on "split-brain" patients. These are individuals in whom the corpus callosum—the bundle of fibers that links the two halves of the brain together—has been surgically cut to prevent the spread of epileptic seizures from one hemisphere to the other. Even though no overt changes in behavior, intellect, or personality occur in these patients because both hemispheres individually receive the same information, deficits are observable with tests designed to restrict information to one brain hemisphere at a time. One such test involves limiting a visual stimulus to only one half of the brain. Because of a crossover in the nerve pathways from the eyes to the occipital cortex, the visual information to the right of a midline point is transmitted to only the left half of the brain, whereas visual information to the left half of this point is received by only the right half of the brain. A split-brain patient presented with a visual stimulus that reaches only the left hemisphere accurately describes the object seen, but when a visual stimulus is presented to only the right hemisphere, the patient denies having seen anything. The right hemisphere does receive the visual input, however, as demonstrated by nonverbal tests. Even though a split-brain patient denies having seen anything after having an object presented to the right hemisphere, he or she can correctly match the object by picking it out from among a number of objects, usually to the patient's surprise. What is your explanation of this finding?

C H A P T E R 6

Peripheral Nervous System: Afferent Division; Special Senses

Chapter Contents at a Glance

INTRODUCTION *Is the world as we perceive it reality? The answer is a resounding no.* ***Perception*** *is our conscious interpretation of the external world as created by the brain from a pattern of nerve impulses delivered to it from sensory receptors. This perception is different from what is really "out there" for several reasons. First, humans only have receptors to detect a limited number of existing energy forms. We perceive sounds, colors, shapes, textures, smells, tastes, and temperature but are not informed of magnetic forces, polarized light waves, radio waves, or x rays because we do not have receptors to respond to the latter energy forms. A limited response range exists even for the energy forms for which we do have receptors. This is exemplified by a dog whistle; dogs can hear it, but its pitch is above our level of detection. Second, the information channels to our brains are not high-fidelity recorders. During precortical processing of sensory input, some features of stimuli are accentuated and others are suppressed or ignored. Third, the cerebral cortex further manipulates the data, comparing it with other incoming information as well as with memories of past experiences to extract the significant features—for example, sifting out a friend's words from the hubbub of sound in a school cafeteria. In the process, the cortex often fills in or distorts the information to abstract a logical perception; that is, it "completes the picture." As a simple example, you "see" a white square in Figure 6–1 even though there is no white square but merely right-angle wedges taken out of four purple circles. Optical illusions also illustrate how the brain interprets reality according to its own rules. Do you see two faces in profile or a wineglass in Figure 6–2? You can alternately see one or the other out of identical visual input. Thus, our perceptions do not replicate reality. Other species equipped with different receptor types and sensitivities and*

Figure 6–1 Do You "See" a White Square That Is Not Really There?

Figure 6–2 Variable Perceptions from the Same Visual Input *Do you see two faces in profile or a wineglass?*

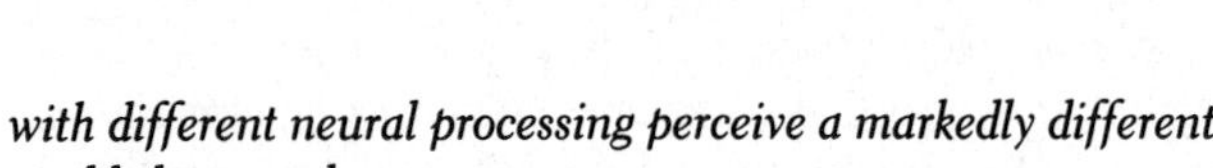

with different neural processing perceive a markedly different world than we do.

Receptors; Sensory Pathways

Receptors have differential sensitivities to various stimuli.

Afferent neurons have **receptors** at their peripheral endings that apprise the CNS of detectable changes, or **stimuli,** in both the external world and internal environment by generating action potentials in response to the stimuli. These action potentials are transmitted via the afferent fibers to the CNS. Stimuli exist in a variety of energy forms, or **modalities,** such as heat, light, sound, pressure, and chemical changes. Because the only way the body can transmit information to the CNS is via action potential propagation, receptors must convert these other forms of energy into electrical energy (action potentials). This energy conversion process is known as **transduction.**

Each type of receptor is specialized to respond more readily to one type of stimulus, its **adequate stimulus,** than to any other stimuli. For example, receptors in the eye are most sensitive to light, receptors in the ear to sound waves, and warmth receptors in the skin to heat energy. We cannot "see" with our ears or "hear" with our eyes because of this differential sensitivity of receptors, a principle known as the **law of specific nerve energies.** Some receptors can respond weakly to stimuli other than their adequate stimulus, but even when activated by a different stimulus, a receptor still gives rise to the sensation usually detected by that receptor type. As an example, the adequate stimulus for eye receptors (photoreceptors) is light, to which they are exquisitely sensitive, but these receptors can also be activated to a lesser degree by mechanical stimulation. When hit in the eye, a person often "sees stars" because the photoreceptors are stimulated by the mechanical pressure. Thus, the sensation perceived depends on the type of receptor stimulated rather than on the type of stimulus. However, because receptors typically are activated by their adequate stimulus, the sensation usually corresponds to the stimulus modality.

Depending on the type of energy to which they can respond, receptors are categorized as follows:

- □ **Photoreceptors** are responsive to light.
- □ **Mechanoreceptors** are sensitive to mechanical energy. Examples include the skeletal muscle receptors sensitive to stretch, the receptors in the ear containing fine hair cells that are bent as a result of sound waves, and blood pressure-monitoring baroreceptors.
- □ **Thermoreceptors** are sensitive to heat and cold.
- □ **Chemoreceptors** are sensitive to specific chemicals. Included here are the receptors for smell and taste, as well as those located deeper within the body that detect O_2 and CO_2 concentrations in the blood or the chemical content of the digestive tract.
- □ **Nociceptors** or **pain receptors** are sensitive to tissue damage such as pinching or burning or to distortion of tissue. Intense stimulation of any receptor is also perceived as painful.

Some sensations are compound sensations in that their perception arises from central integration of several simultaneously activated primary sensory inputs. For example, the perception of wetness comes from touch, pressure, and thermal receptor input; there is no such thing as a "wet receptor."

The information detected by receptors is conveyed via afferent neurons to the CNS, where it is used for a variety of purposes. First, afferent input is essential for the control of efferent output, both for regulation of motor behavior in accordance with external circumstances and for coordination of internal activities directed toward maintenance of homeostasis. At the most basic level, afferent input provides information (of which the individual may or may not be consciously aware) for the CNS to use in directing activities necessary for survival. Second, processing of sensory input by the reticular activating system in the brain stem is critical for cortical arousal and consciousness. Third, central processing of sensory information gives rise to our perceptions of the world around us. Finally, selected information delivered to the CNS may be stored for future reference.

Altered membrane permeability of receptors in response to a stimulus produces a graded receptor potential.

A receptor may be either a specialized ending of the afferent neuron or a separate cell closely associated with the peripheral ending of the neuron. Stimulation of a receptor alters its membrane permeability, usually via a nonselective opening of all small ion channels. The means by which this permeability change takes place is individualized for each receptor type. Because the electrochemical driving force is greater for Na^+ than for other small ions at resting potential, the predominant effect is an inward flux of Na^+, which depolarizes the receptor membrane. (There are exceptions; for example, photoreceptors are hyperpolarized upon stimulation.)

This local depolarizing change in potential is known as a **receptor potential** in the case of a separate receptor or as a **generator potential** if the receptor is a specialized ending of an afferent neuron. The receptor (or generator) potential is a graded potential (see p. 100) that can be varied in amplitude and duration, depending on stimulus strength and the rate of application or removal of the stimulus. The stronger the stimulus, the greater the permeability change and the larger the receptor potential. As is true of all graded potentials, receptor potentials have no refractory period, so summation in response to rapidly successive stimuli is possible. Because the receptor region has a very high threshold, action potentials do not take place at the receptor itself. For long-distance transmission, the receptor potential must be converted into action potentials that can be propagated along the afferent fiber.

A receptor potential triggers the release of a chemical messenger that diffuses across the small space separating the receptor from the ending of the afferent neuron, similar to a synapse (Fig. 6–3a). Binding of the chemical messenger with specific protein receptor sites on the afferent neuron opens chemical messenger-gated Na^+ channels (see p. 95). The resulting ionic flux initiates a self-propagating action potential in the afferent neuron. In the case of a generator potential (Fig. 6–3b), local current flow between the activated receptor ending and the cell membrane adjacent to the receptor brings about opening of voltage-gated Na^+ channels in this adjacent region. Once threshold is reached, an action potential is initiated that is propagated along the afferent fiber. (For convenience, from here on we will refer to both receptor potentials and generator potentials as receptor potentials.)

The intensity of the stimulus is reflected by the magnitude of the receptor potential, which in turn is encoded by the fre-

Stimulus
Chemical messenger-gated channels
Na^+
Diffusion of chemical messenger
Afferent neuron fiber
Na^+
Receptor (separate cell)
(a)

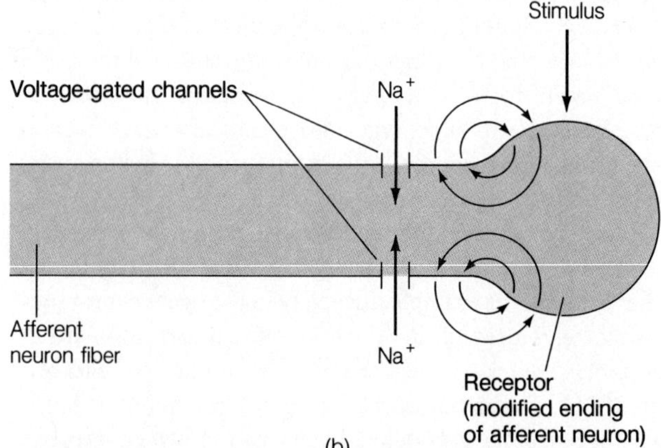

Figure 6–3 Conversion of Receptor and Generator Potentials into Action Potentials *(a) Receptor potential. The chemical messenger released from a separate receptor initiates an action potential in the fiber by opening chemical messenger-gated Na^+ channels. (b) Generator potential. The local current flow between the depolarized receptor ending and the afferent fiber initiates an action potential in the fiber by opening voltage-gated Na^+ channels.*

quency of action potentials generated in the afferent neuron. A larger receptor potential cannot bring about a larger action potential (all-or-none law), but it can induce more rapid firing of action potentials. This is one way stimulus strength is coded; the stronger the stimulus, the greater the frequency of action potentials. Stimulus strength is also reflected by the size of the area stimulated. Stronger stimuli usually affect larger areas, so a correspondingly greater population of receptors respond. For example, a light touch does not activate as many pressure receptors in the skin as a more forceful touch applied to the same area. Stimulus intensity is therefore distinguished both by the frequency of action potentials generated in the afferent neuron (**frequency code**) and by the number of receptors activated within the area (**population code**).

Receptors may slowly or rapidly adapt to sustained stimulation.

Stimuli of the same intensity do not always elicit receptor potentials of the same magnitude from the same receptor because of **adaptation.** Some receptors have the ability to diminish their extent of depolarization in spite of a sustained stimulus strength, subsequently decreasing the frequency of action potentials generated in the afferent neuron. The receptor "adapts" to the stimulus by no longer responding to it to the same degree.

There are two types of receptors—tonic receptors and phasic receptors—based on the speed of adaptation. **Tonic receptors** do not adapt at all or adapt slowly. This type of receptor is important in situations in which maintained information about a stimulus is valuable. Examples are muscle stretch receptors, which monitor muscle length, and joint proprioceptors, which measure the degree of joint flexion. The CNS must be continually apprised of the degree of muscle length and joint position to maintain posture and balance. It is important, therefore, that these receptors do not adapt to a stimulus but continue to generate action potentials to relay this information to the CNS (Fig 6–4a).

Phasic receptors, on the other hand, are rapidly adapting receptors. These receptors often exhibit an **off response** (Fig. 6–4b). The receptor rapidly adapts by no longer responding to a maintained stimulus, but when the stimulus is removed, the receptor responds with a slight depolarization. Phasic receptors are useful in situations in which it is important to signal a change in stimulus intensity rather than to relay status quo information. An example of a rapidly adapting receptor is the **Pacinian corpuscle,** a receptor in the skin that signals changes in pressure. Because these receptors adapt rapidly, you are not continually conscious of wearing your watch, rings, and clothing. When you put something on, you soon become accustomed to it because of the Pacinian corpuscles' rapid adaptation. When you take the item off, you are aware of its removal because of the off response.

The mechanism by which adaptation is accomplished varies for different receptors and has not been fully elucidated for all receptor types. The mechanism of adaptation for a Pacinian corpuscle is known to be due to its physical properties. A Pacinian corpuscle is a specialized receptor ending that consists of concentric layers of connective tissue resembling layers of an onion wrapped around the peripheral terminal of an afferent neuron (Fig. 6–5). When pressure is first applied to the Pacinian corpuscle, the underlying terminal responds with a receptor potential of a magnitude that reflects the intensity of the stimulus. As the stimulus continues, the pressure energy is dissipated because it causes slippage between the receptor layers (just as steady pressure on a peeled onion causes its layers to slip). Because this physical effect filters out the steady component of the applied pressure, the underlying neuronal ending no longer responds with a receptor potential; that is, adaptation has occurred. Adaptation in this case is due entirely to the physical features of the receptor. Experimentally stripping a Pacinian corpuscle of its connective tissue lay-

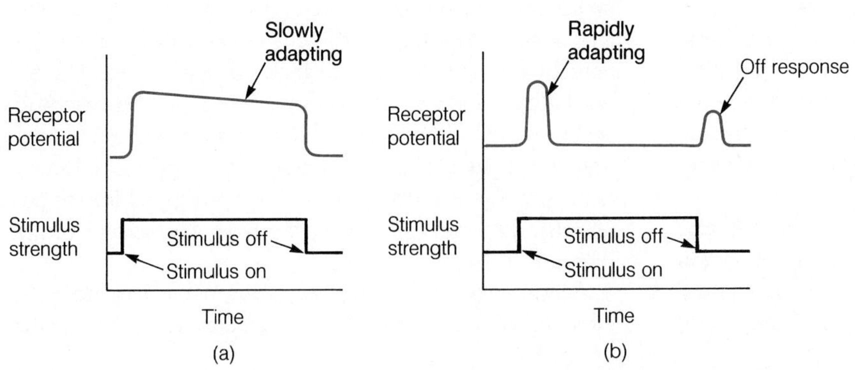

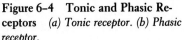

Figure 6–4 Tonic and Phasic Receptors *(a) Tonic receptor. (b) Phasic receptor.*

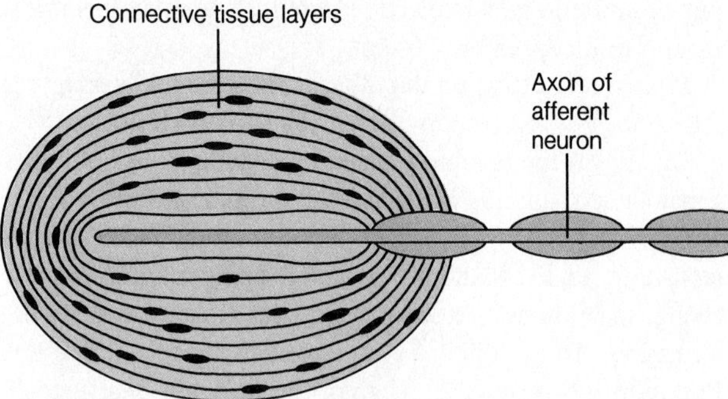

Figure 6–5 Pacinian Corpuscle *The Pacinian corpuscle, a skin receptor that signals a change in pressure, consists of concentric layers of connective tissue wrapped around the peripheral terminal of an afferent neuron.*

ers converts the naked terminal into a slowly adapting receptor.

Adaptation should not be confused with habituation (see p. 143). Although both of these phenomena involve decreased neural responsiveness to repetitive stimuli, they operate at different points in the neural pathway. Adaptation is a receptor adjustment in the PNS, whereas habituation involves a modification in synaptic effectiveness in the CNS.

Each somatosensory pathway is "labeled" according to modality and location.

On reaching the spinal cord, afferent information has two possible destinies: (1) it may become part of a reflex arc, bringing about an appropriate effector response; or (2) it may be relayed upward to the brain via ascending pathways for further processing and possible conscious awareness. Although there is afferent information that concerns the status of the internal environment, such as blood pressure and concentration of CO_2 in the body fluids, one is not made aware of this information, so it is not considered to be sensory information. The incoming pathway for subconscious information derived from the internal viscera is called a **visceral afferent.** If the information detected by the receptors is propagated to the conscious levels of the brain, it is called **sensory information,** and the incoming pathway is considered to be a **sensory afferent.**

Sensory information is categorized as either: (1) **somatic sensation** arising from the body surface, including *somesthetic sensation* arising from the skin and *proprioception* from muscles, joints, skin, and inner ear, or (2) **special sensation,** including *seeing, hearing, tasting,* and *smelling.* (See the accompanying boxed feature, A Closer Look at Exercise Physiology.)

Pathways conveying somatic sensation, the **somatosensory pathways,** consist of discrete chains of neurons synaptically interconnected in a particular sequence to accomplish progressively more sophisticated processing of the sensory information. The afferent neuron with its peripheral receptor that first detects the stimulus is known as a **first-order sensory neuron.** It synapses on a **second-order sensory neuron,** either in the spinal cord or the medulla, depending on which sensory pathway is involved. This neuron then synapses on a **third-order sensory neuron** in the thalamus, and so on. With each step, further processing of the input occurs. A particular sensory modality detected by a specialized receptor type is sent over a specific afferent and ascending pathway (a neural pathway committed to that modality) to excite a defined area in the somatosensory cortex. Thus information is kept separated within specific **labeled lines** between the periphery and the cortex. In this way, even though all information is propagated to the CNS via the same signal (action potentials), the brain can decode the type and location of the stimulus (Table 6–1).

Activation of a sensory pathway at any point gives rise to the same sensation that would be produced by stimulation of the

Table 6–1 Coding of Sensory Information

Stimulus Property	Mechanism of Coding
Type of stimulus (stimulus modality)	Distinguished by the type of receptor activated and the specific pathway over which this information is transmitted to a particular area of the cerebral cortex
Location of stimulus	Distinguished by the location of the activated receptor field and the pathway that is subsequently activated to transmit this information to the area of the somatosensory cortex representing that particular location; enhanced by lateral inhibition
Intensity of stimulus (stimulus strength)	Distinguished by the frequency of action potentials initiated in an activated afferent neuron and the number of receptors (and afferent neurons) activated

A CLOSER LOOK AT EXERCISE PHYSIOLOGY

Back Swings and Pre-Jump Crouches: What Do They Share in Common?

Proprioception, the sense of the body's position in space, is critical to any movement and is especially important in athletic performance, whether it be a figure skater performing triple jumps on ice, a gymnast performing a difficult floor routine, or the football quarterback throwing perfectly to a spot 60 yards down field. In order to control skeletal-muscle contraction to achieve the desired movement, the CNS must be continuously apprised of the results of its actions by means of sensory feedback information.

A number of receptors provide proprioceptive input. Muscle proprioceptors provide feedback information on muscle tension and length. Joint proprioceptors provide feedback on joint acceleration, angle, and direction of movement. Skin proprioceptors inform the CNS of weight-bearing pressure on the skin. Proprioceptors in the inner ear, along with those in neck muscles, provide information about head and neck position so that the CNS can orient the head correctly. For example, neck reflexes facilitate essential trunk and limb movements during somersaults, and divers and tumblers use strong movements of the head to maintain spins.

The most complex and probably one of the most important proprioceptors is the muscle spindle (see p. 246). Muscle spindles are found throughout a muscle but tend to be concentrated in its center. Each spindle lies parallel to the muscle fibers within the muscle. The spindle is sensitive to both the muscle's rate of change in length and the final length achieved. If a muscle is stretched, each muscle spindle within the muscle also is stretched and the afferent neuron whose peripheral axon terminates on the muscle spindle is stimulated. The afferent fiber passes into the spinal cord and synapses directly on the motor neurons that supply the same muscle. Stimulation of the stretched muscle as a result of this stretch reflex causes the muscle to contract sufficiently to relieve the stretch.

Older persons or those with weak quadriceps (thigh) muscles unknowingly take advantage of the muscle spindle by pushing on the center of the thighs when they get up from a sitting position. Contraction of the quadriceps muscle extends the knee joint, thus straightening the leg. The act of pushing on the center of the thighs when getting up slightly stretches the quadriceps muscle in both limbs, stimulating their muscle spindles. The resultant stretch reflex aids in contraction of the quadriceps muscles and helps the person to assume a standing position.

In sports, people use the muscle spindle to advantage all the time. To jump high, as in a basketball jumpball, an athlete starts by crouching down. This action stretches the quadriceps muscles and increases the firing rate of their spindles, thus triggering the stretch reflex that reinforces the quadriceps muscles' contractile response so that these extensor muscles of the legs gain additional power. The same is true for crouch starts in running events. The back swing in tennis, golf, and baseball similarly provides increased muscular excitation through reflex activity initiated by stretched muscle spindles.

---

receptors in the body part itself. This is the basis for **phantom pain** (for example, pain perceived as originating in the foot by a person whose leg has been amputated at the knee). Irritation of the severed endings of the afferent pathways in the stump triggers action potentials that, upon reaching the foot region of the somatosensory cortex, are interpreted as pain in the missing foot.

Acuity is influenced by receptive field size and lateral inhibition.

Each sensory neuron responds to stimulus information only within a circumscribed region of the skin surface surrounding it, which is known as its **receptive field.** The size of receptive fields varies inversely with the density of receptors in the region;

the more closely receptors of a particular type are spaced, the smaller the area of skin each monitors. The smaller the receptive field in a region, the greater its **acuity** or **discriminative ability.** Compare the **tactile** (touch) **discrimination** in your fingertips with that in your elbow by "feeling" the same object with both. You are able to discern more precise information about the object with your richly innervated fingertips because the receptive fields are small; as a result, each neuron signals information about small, discrete portions of the object's surface. In contrast, the skin over the elbow is served by relatively few sensory endings with larger receptive fields. Subtle differences within each large receptive field cannot be detected (Fig. 6–6). The distorted cortical representation of various body parts in the sensory homunculus corresponds precisely with the innervation density; more cortical space is allotted for sensory reception from areas with smaller receptive fields and, accordingly, greater tactile discriminative ability.

Besides receptor density, a second factor influencing acuity is **lateral inhibition.** You can appreciate the importance of this phenomenon by slightly indenting the surface of your skin with the point of a pencil (Fig. 6–7a). The receptive field is excited immediately under the center of the pencil point where the stimulus is most intense, but the surrounding receptive fields are also stimulated, only to a lesser extent because they are less distorted. If information from these marginally excited afferent fibers in the fringe of the stimulus area were to reach the cortex, localization of the pencil point would be blurred. To facilitate localization and sharpen contrast, lateral inhibition occurs within the CNS (Fig. 6–7b). The most strongly activated signal pathway originating from the center of the stimulus area inhibits the less excited pathways from the fringe areas. This occurs via inhibitory interneurons that pass laterally between ascending fibers serving neighboring receptive fields. Blockage of further transmission in the weaker inputs increases the contrast between wanted and unwanted information so that the pencil point can be precisely localized. The extent of lateral inhibitory connections within sensory pathways varies for different modalities. Those with the most lateral inhibition— touch and vision—bring about the most accurate localization.

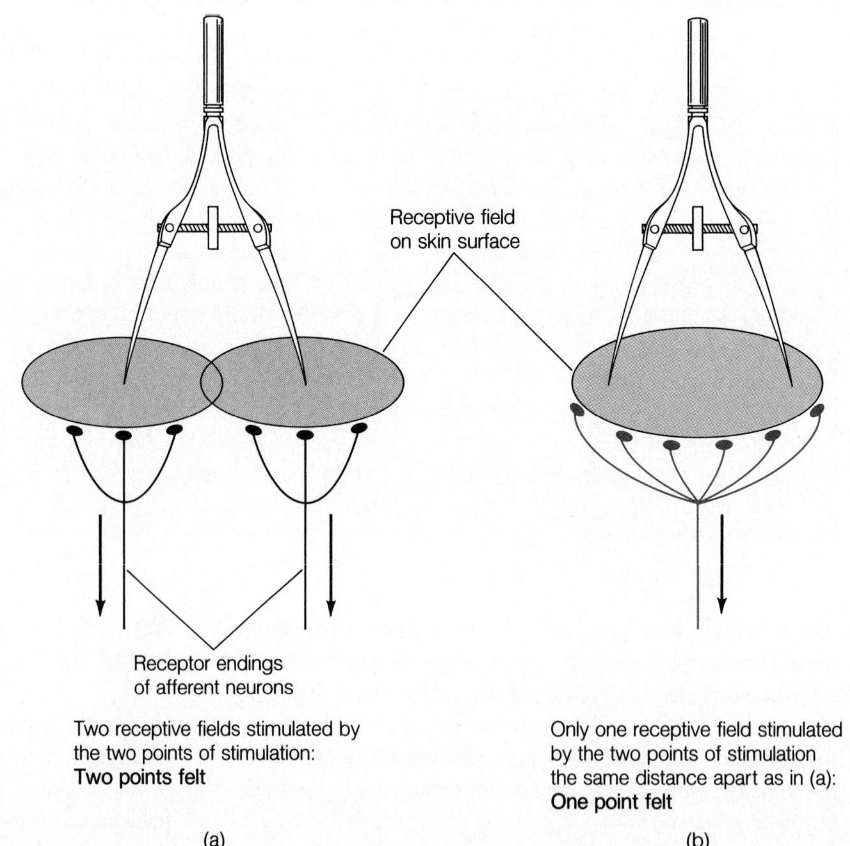

Figure 6–6 Comparison of Discriminative Ability of Regions with Small versus Large Receptive Fields *The relative tactile acuity of a given region can be determined by use of the* ***two-point threshold of discrimination test.*** *If the two points of a pair of calipers applied to the surface of the skin stimulate two different receptive fields, two separate points will be felt. If the two points touch the same receptive field, they will be perceived as only one point. By adjusting the distance between the caliper points, one can determine the minimal distance at which the two points can be recognized as two rather than one, which is a reflection of the size of the receptive fields in the region. With this technique, it is possible to plot the discriminative ability of the body surface. The two-point threshold ranges from 2 mm in the fingertip (enabling one to read Braille, the raised dots of which are spaced 2.5 mm apart) to 48 mm in the poorly discriminative skin of the calf. (a) Region with small receptive fields. (b) Region with large receptive fields.*

Pain

Stimulation of nociceptors elicits the perception of pain plus motivational and emotional responses.

Pain is primarily a protective mechanism meant to bring to conscious awareness the fact that tissue damage is occurring or is about to occur. Unlike other somatosensory modalities, it is accompanied by motivated behavioral responses (such as withdrawal or defense) as well as emotional reactions (such as crying or fear). Also, unlike other sensations, the subjective perception of pain can be modulated by other past or present experiences (for example, heightened pain perception accompanying fear of the dentist or lowered pain perception in an injured athlete during a competitive event).

There are three categories of pain receptors: **mechanical nociceptors** respond to mechanical damage such as cutting, crushing, or pinching; **thermal nociceptors** respond to temperature extremes, especially heat; and **polymodal nociceptors** respond equally to all kinds of damaging stimuli, including irritating chemicals released from injured tissues. None of the nociceptors have specialized receptor structures; they are all naked nerve endings. Because of their value to survival, nociceptors do not adapt to sustained or repetitive stimulation. On the other hand, all nociceptors can be sensitized by the presence of *prostaglandins*, which greatly enhance the receptor response to noxious stimuli (that is, it hurts more when prostaglandins are present). Prostaglandins are a special group of fatty acid derivatives that act locally on being released. Aspirin-like drugs inhibit the synthesis of prostaglandins, accounting at least in part for the drugs' **analgesic** (pain-relieving) properties.

Pain impulses originating at nociceptors are transmitted to the CNS via one of two types of afferent fibers (Table 6–2). Signals arising from mechanical and thermal nociceptors are transmitted over large myelinated **A-delta fibers** at rates of up to 30 meters/sec (the **fast pain pathway**). Impulses from polymodal nociceptors are carried by small unmyelinated **C fibers** at a much slower rate of 1 meter/sec (the **slow pain pathway**). Think about the last time you cut or burned your finger. You undoubtedly felt a sharp twinge of pain at first, with a more diffuse, disagreeable pain commencing shortly thereafter. Pain typically is perceived initially as a brief, sharp, prickling sensation that is easily localized (fast pain pathway originating from specific mechanical or heat nociceptors). This is followed by a dull, aching, poorly localized sensation that persists for a longer time and is more unpleasant (slow pain pathway activated by chemicals, especially **bradykinin,** a

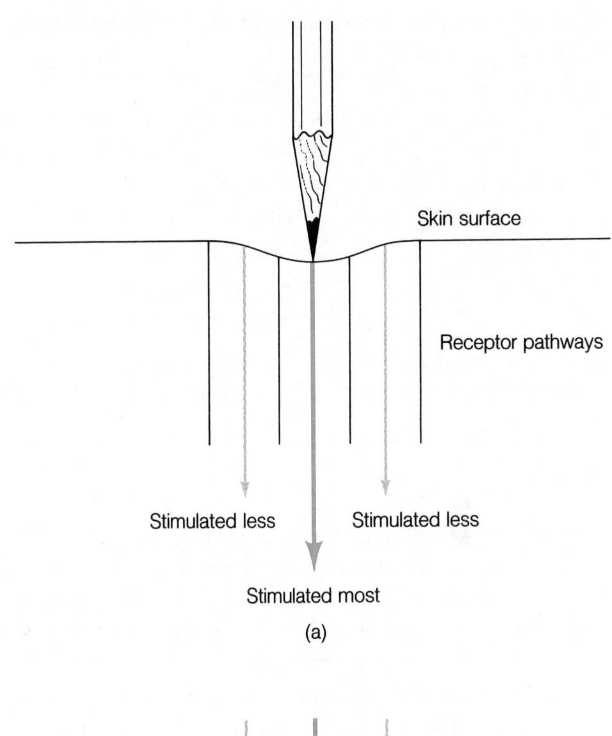

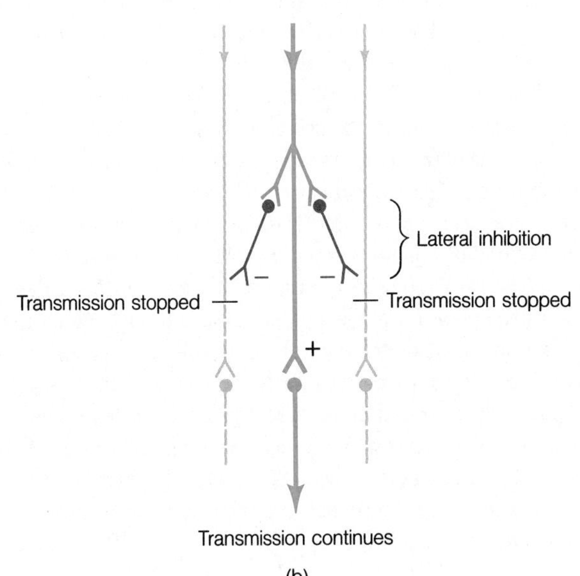

Figure 6–7 Lateral Inhibition *(a) The receptor at the site of most intense stimulation is activated to the greatest extent. Surrounding receptors are also stimulated but to a lesser degree. (b) The most intensely activated receptor pathway halts transmission of impulses in the less intensely stimulated pathways through lateral inhibition. This process facilitates localization of the site of stimulation.*

Table 6–2 Characteristics of Pain

Fast Pain	Slow Pain
Carried by A-delta fibers	Carried by C fibers
Sharp, prickling sensation	Dull, aching, burning sensation
Easily localized	Poorly localized
Occurs first	Occurs second; persists for longer time; more unpleasant
Occurs upon stimulation of mechanical and thermal nociceptors	Occurs upon stimulation of polymodal nociceptors

normally inactive substance that is activated by enzymes released into the ECF from damaged tissue). Bradykinin and related compounds not only provoke pain, presumably by stimulating the polymodal nociceptors, but they also contribute to the inflammatory response to tissue injury (chap. 12). The persistence of these chemicals might explain the long-lasting, aching pain that continues after removal of the mechanical or thermal stimulus that caused the tissue damage.

The primary afferent fibers synapse with specific second-order interneurons in the dorsal horn of the spinal cord (Fig. 6–8a). One of the neurotransmitters released from these afferent pain terminals is **substance P,** which is believed to be unique to pain fibers. Ascending pain pathways have poorly understood destinations in the *somatosensory cortex*, the *thalamus*, and the *reticular formation*. The role of the cortex in pain perception is not clear, although it is probably important at least in localizing the pain. Pain can still be perceived in the absence of the cortex, presumably at the level of the thalamus. The reticular formation increases the level of alertness associated with the noxious encounter. Interconnections from the thalamus and reticular formation to the hypothalamus and limbic system elicit the behavioral and emotional responses accompanying the painful experience.

The brain has a built-in analgesic system.

In addition to the chain of neurons connecting peripheral nociceptors with higher CNS structures for pain perception, the CNS also contains a neuronal system that suppresses pain. Our knowledge about this built-in **analgesic system** is still fragmentary. It appears that neural mechanisms exist to suppress transmission in the pain pathways as they enter the spinal cord. Electrical stimulation of the **periaqueductal gray matter** (gray matter surrounding the cerebral aqueduct, a narrow canal that connects the third and fourth ventricular cavities) results in profound analgesia, as does stimulation of the reticular formation within the brain stem. These two regions are thought to be part of a descending analgesic pathway (Fig. 6–8b) that blocks by presynaptic inhibition (see p. 110) the release of substance P from afferent pain fiber terminals.

The built-in analgesic system is dependent on the presence of **opiate receptors.** It has long been known that **morphine,** a derivative of the opium poppy, is a powerful analgesic. It seemed highly unlikely that the body would be endowed with opiate receptors only to interact with chemicals derived from a flower! A search was therefore undertaken to discover the substances that normally bind with these opiate receptors. The result was the discovery of endogenous morphine-like substances, the **endorphins** and **enkephalins,** which are important in the body's natural analgesic system. According to a proposed model for the analgesic system, an endogenous opiate neurotransmitter, enkephalin, is released from the descending pathway and binds with opiate receptors on the afferent presynaptic terminal. This binding suppresses the release of substance P, thereby blocking further transmission of the pain signal. Morphine binds to these same opiate receptors, accounting for its analgesic properties.

It is unclear how the natural pain-suppressing mechanisms are activated. Factors known to modulate pain include exercise (endorphins are believed to be released during prolonged exercise and presumably are responsible for the "runner's high"); acupuncture and hypnosis, which suppress pain by poorly understood mechanisms; and stress. There is evidence that some types of stress induce analgesia via the opiate pathway and other less well-understood non-opiate mechanisms. It is sometimes disadvantageous for a stressed organism to display the normal reaction to pain. For example, when two males are fighting for dominance of the herd, it would mean certain defeat to withdraw, escape, or rest when injured.

Pain can frequently be managed by drugs that suppress transmitter activity at some point along the pain pathway. Surgical intervention is occasionally necessary for relief of intractable pain, as may occur in terminally ill cancer patients. Surgical relief from pain entails either interrupting the ascending pain pathways within the spinal cord or severing certain pathways in the brain to modify emotional response to the pain. Recall that a painful experience includes the sensation of pain plus an emotional and behavioral reaction to it. These two components can be dissociated. The person still feels pain but does not mind as much.

EYE: VISION

Somatic sensation is detected by widely distributed receptors that provide information about the body's interactions with

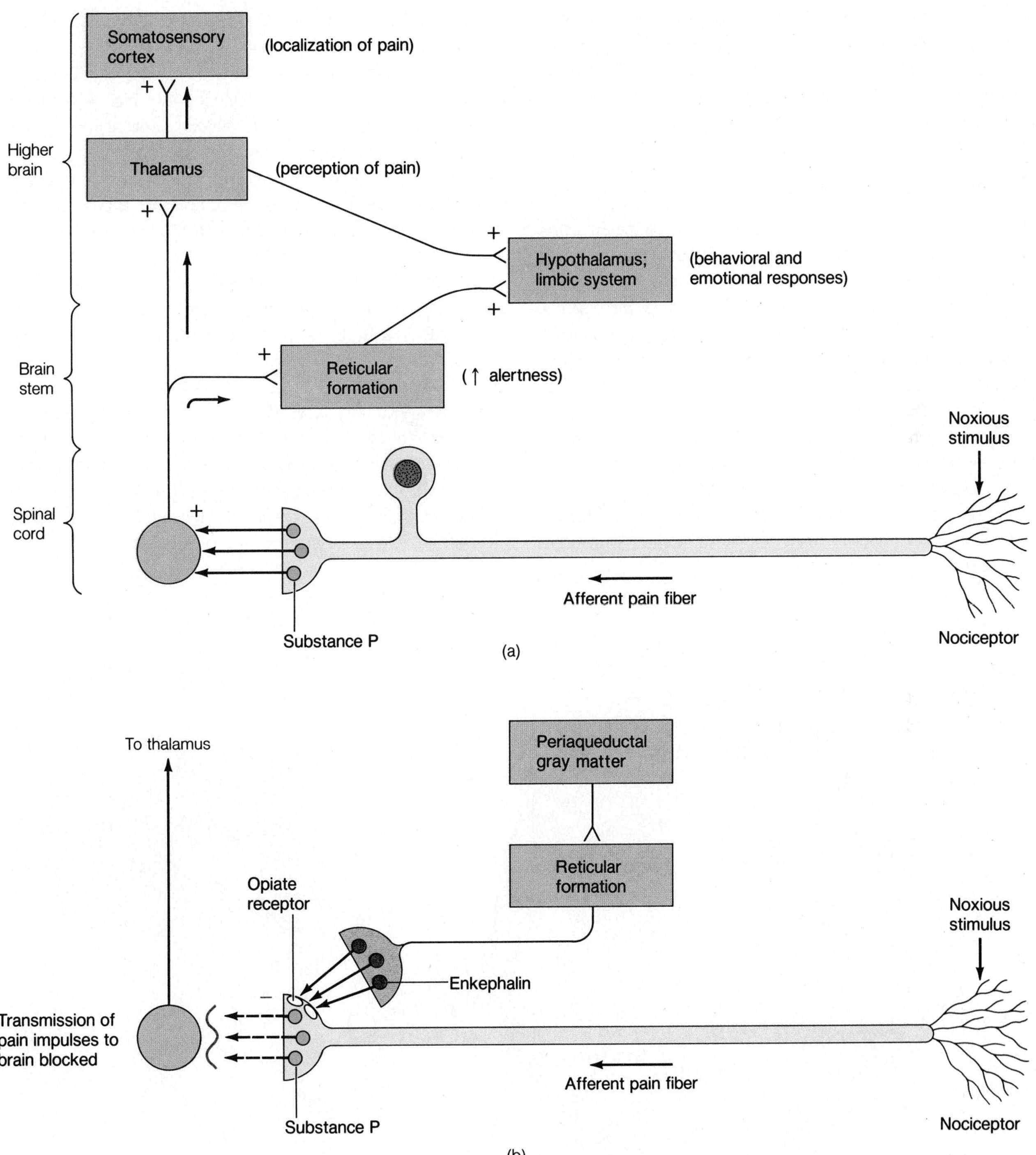

Figure 6–8 Pain Pathway and Proposed Analgesic Pathway *(a) Pain pathway. (b) Proposed analgesic pathway.*

the environment in general. In contrast, each of the special senses has highly localized, extensively specialized receptors that respond to unique environmental stimuli. The special senses include **vision, hearing, taste** and **smell,** to which we now turn our attention, commencing with vision.

The eye is like a camera.

The eyes capture the patterns of illumination in the environment as an "optical picture" on a layer of light-sensitive cells, the retina, much like a camera captures an image on film. Just as film can be developed into a visual likeness of the original image, the coded image on the retina is transmitted through a series of progressively more complex steps of visual processing until it is finally consciously perceived as a visual likeness of the original image.

The **eye** is a spherical, fluid-filled structure enclosed by three layers. From outermost to innermost these are: (1) the sclera/cornea; (2) the choroid/ciliary body/iris; and (3) the retina (Fig. 6–9a). Most of the eyeball is covered by a tough outer layer of connective tissue, the **sclera,** which forms the visible white part of the eye (Fig. 6–9b). Anteriorly (toward the front), the outer layer consists of the transparent **cornea** through which light rays pass into the interior of the eye. The middle layer underneath the sclera is the highly pigmented **choroid,** which contains many blood vessels that nourish the retina. The choroid layer becomes specialized anteriorly to form the ***ciliary body*** and ***iris***. The innermost coat under the choroid is the **retina** consisting of an outer pigmented layer and an inner nervous tissue layer. The latter contains the **rods** and **cones,** the photoreceptors that convert light energy into nerve impulses. Like the black walls of a photographic studio, the pigment in the choroid and retina absorbs light after it strikes the retina to prevent reflection or scattering of light within the eye.

The interior of the eye consists of two fluid-filled cavities, separated by a **lens,** all of which are transparent to permit passage of light through the eye from the cornea to the retina. The anterior (front) cavity between the cornea and lens contains a clear watery fluid, the **aqueous humor,** and the larger posterior (rear) cavity between the lens and retina contains a semifluid, jellylike substance, the **vitreous humor.**

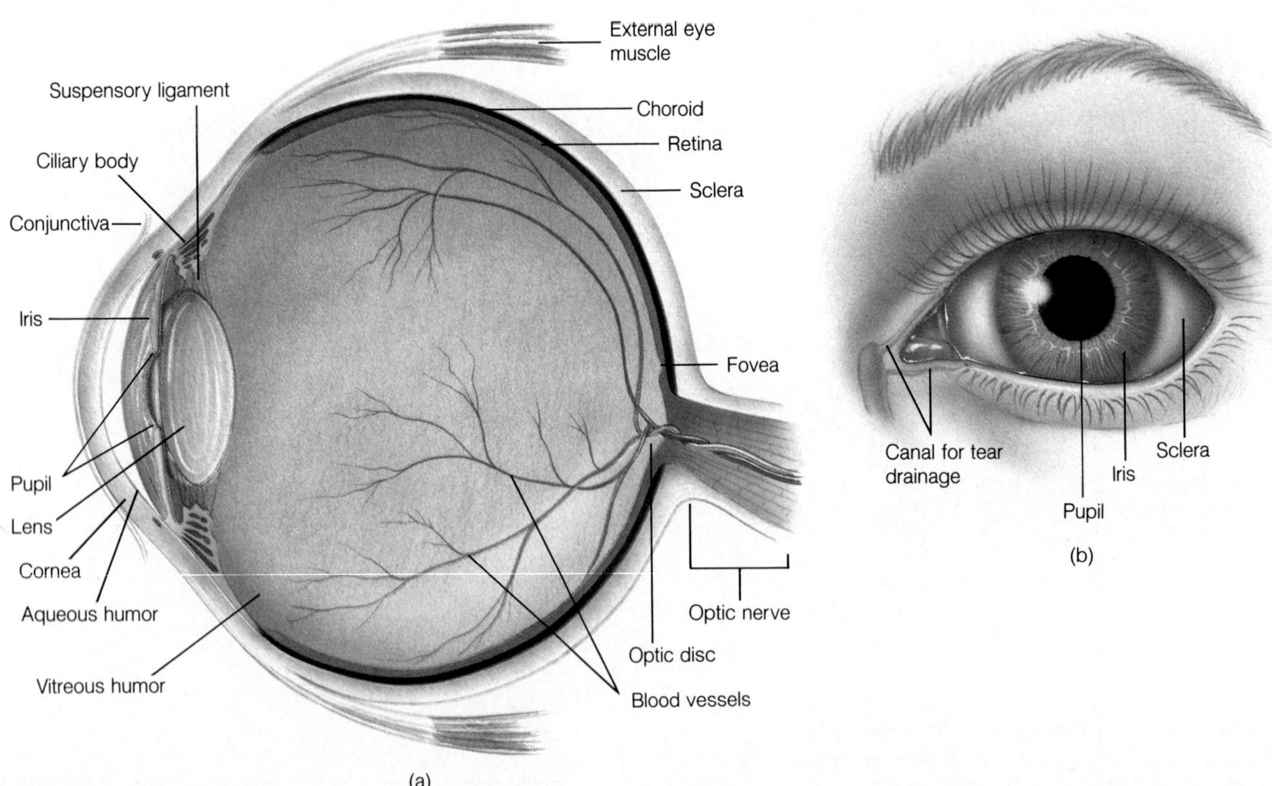

Figure 6–9 Structure of the Eye *(a) Internal sagittal view. (b) External front view.*

The vitreous humor is important in maintaining the spherical shape of the eyeball. The aqueous humor carries nutrients for the cornea and lens, both of which lack a blood supply because blood vessels in these structures would impede the passage of light to the photoreceptors. Aqueous humor is produced at a rate of about 5 ml/day by a capillary network within the **ciliary body,** a specialized anterior derivative of the choroid layer. This fluid drains into a canal at the edge of the cornea and eventually enters the blood. If aqueous humor is not drained as rapidly as it is formed (for example, because of a blockage in the drainage canal), excess aqueous humor will accumulate in the anterior cavity, causing the intraocular ("within the eye") pressure to rise. This condition is known as **glaucoma.** The excess aqueous humor pushes the lens backward into the vitreous humor, which in turn is pushed against the inner neural layer of the retina. This compression causes retinal damage that can lead to blindness if it is not treated.

The amount of light entering the eye is controlled by the iris.

Not all of the light passing through the cornea reaches the light-sensitive photoreceptors because of the presence of the **iris,** a thin, pigmented smooth muscle that forms a visible ring-like structure within the aqueous humor (Fig. 6–9a and b). The pigment in the iris is responsible for eye color. The round opening in the center of the iris through which light enters the interior portions of the eye is the **pupil.** The size of this opening can be adjusted by variable contraction of the iris muscles to admit more or less light, much like the shutter in a camera. The iris contains two sets of smooth muscle networks, one *circular* (the muscle fibers of which run in a ring-like fashion within the iris) and the other *radial* (the fibers of which project outward from the pupillary margin like bicycle spokes) (Fig. 6–10). Because muscle fibers shorten when they contract, the pupil gets smaller when the **circular** (or **constrictor**) **muscle** contracts and forms a smaller ring. This reflex pupillary constriction occurs in bright light to decrease the amount of light entering the eye. When the **radial** (or **dilator**) **muscle** shortens, the size of the pupil increases. Such pupillary dilation occurs in dim light to allow the entrance of more light.

Iris muscles are controlled by the autonomic nervous system. Parasympathetic nerve fibers innervate the circular muscle and sympathetic fibers supply the radial muscle. Acting via the autonomic nervous system, conditions other than light can induce changes in pupillary size. For example, dilation of the pupils accompanies generalized discharge of the sympathetic nervous system in response to actual or impending danger.

The eye refracts the entering light to focus the image on the retina.

Light is a form of electromagnetic radiation composed of particle-like individual packets of energy called **photons** that travel in wave-like fashion. The distance between two wave peaks is known as the **wavelength** (Fig. 6–11). The photoreceptors in the eye are only sensitive to wavelengths between 400 and 700 nm (billionths of a meter between peaks). This **visible light** is only a small portion of the total electromagnetic spectrum (Fig. 6–12). Light of different wavelengths in this visible band is perceived as different color sensations.

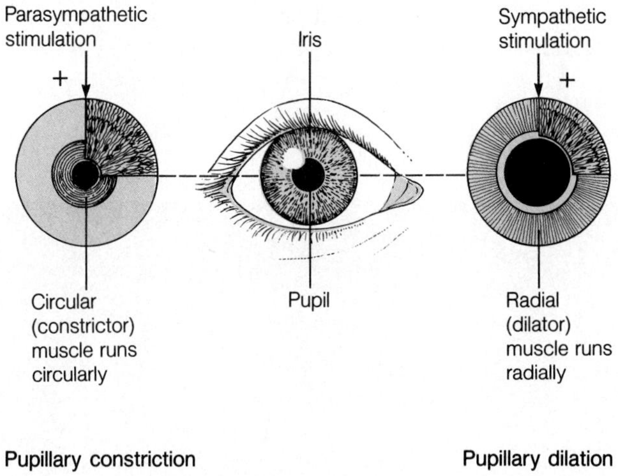

Parasympathetic stimulation
+
Iris
Sympathetic stimulation
+
Circular (constrictor) muscle runs circularly
Pupil
Radial (dilator) muscle runs radially
Pupillary constriction
Pupillary dilation

Figure 6–10 Control of Pupillary Size

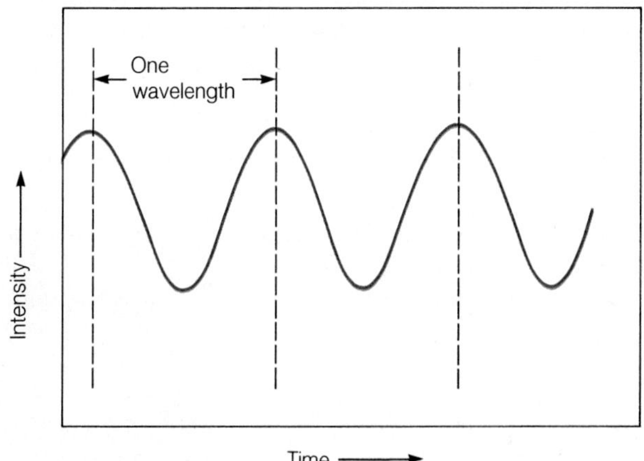

Figure 6–11 Properties of an Electromagnetic Wave *A wavelength is the distance between two wave peaks. Intensity refers to the amplitude of the wave.*

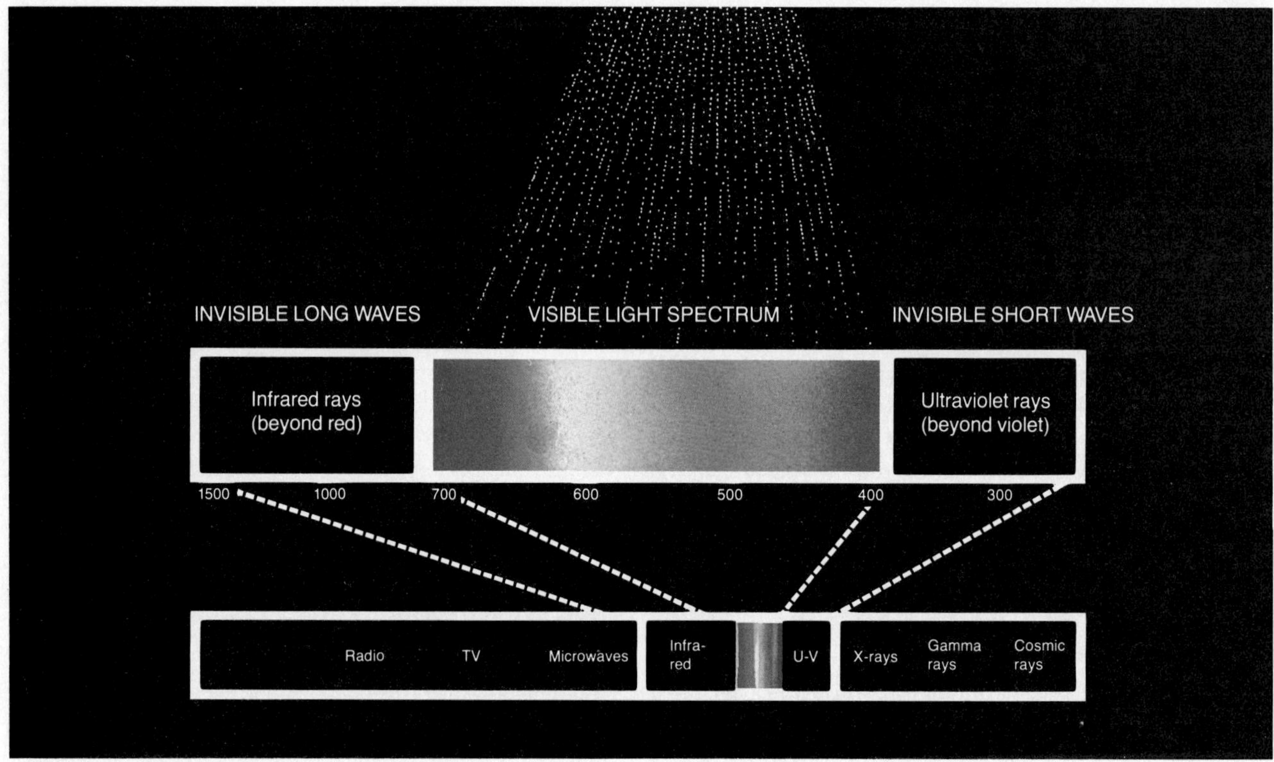

Figure 6–12 Electromagnetic Spectrum *The wavelengths in the electromagnetic spectrum range from 10^4 m (10 km—e.g., long radio waves) to less than 10^{-14} m (quadrillionths of a meter—e.g., gamma and cosmic rays). The visible spectrum includes wavelengths ranging from 400 to 700 nm (billionths of a meter).*

Short wavelengths are sensed as violet and blue; long wavelengths are interpreted as orange and red.

In addition to having variable wavelengths, light energy also varies in **intensity,** in reference to the amplitude or height of the wave (Fig. 6–11). Dimming of bright red lights does not change their color; they just become less intense or less bright.

Light waves ***diverge*** or radiate outward, in all directions from every point of a light source. The forward movement of a light wave in a particular direction is known as a **light ray.** Divergent light rays reaching the eye must be bent inward to be focused back into a point on the light-sensitive retina to provide an accurate image of the light source (Fig. 6–13).

Figure 6–13 Focusing of Divergent Light Rays *Diverging light rays must be bent inward to be focused.*

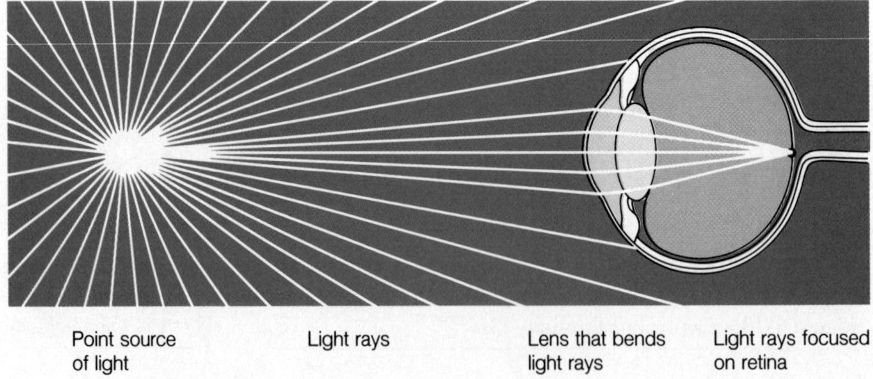

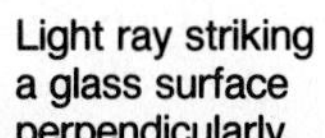

Figure 6–14 Refraction

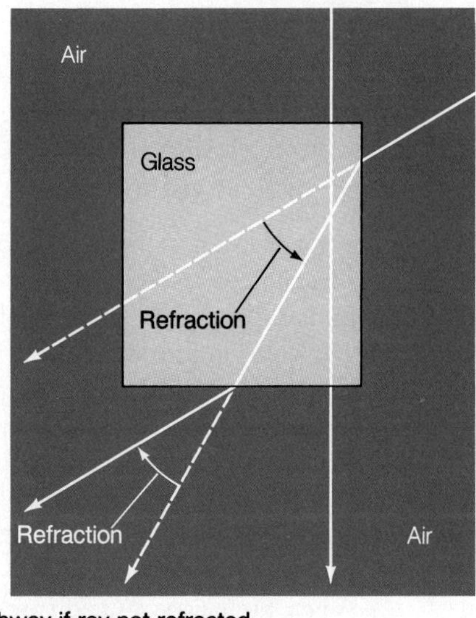

The bending of a light ray (**refraction**) occurs when the ray passes from a medium of one density into a medium of a different density (Fig. 6–14). Light travels faster through air than through other transparent media, such as water and glass. When a light ray enters a medium of greater density, it is slowed down. The converse is also true. The ray changes its course of direction if it strikes the surface of the new medium at any angle other than perpendicular. (This is why trees appear bent when you view their reflection in a lake).

Two factors contribute to the degree of refraction: the comparative densities of the two media (the greater the difference in density, the greater the degree of bending) and the angle at which the light strikes the second medium (the greater the angle, the greater the refraction).

With curved surfaces such as lens, the greater the curvature, the greater the degree of bending and the stronger the lens. When a light ray strikes the curved surface of any object of greater density, the direction of refraction depends on the angle of the curvature (Fig. 6–15). A lens with **convex** surfaces converges light rays, bringing them closer together, a requirement

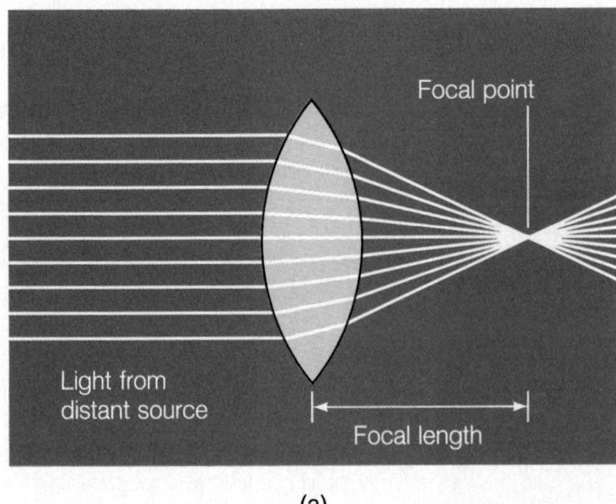

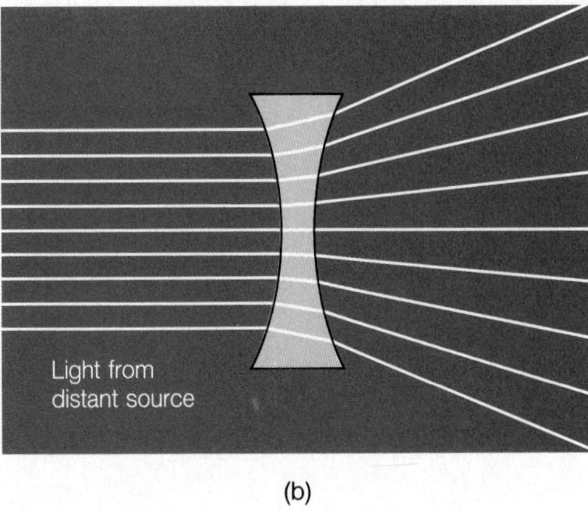

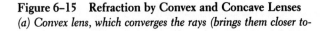

Figure 6–15 Refraction by Convex and Concave Lenses *(a) Convex lens, which converges the rays (brings them closer together). (b) Concave lens, which diverges the rays (spreads them farther apart).*

for bringing an image to a focal point. Refractive surfaces of the eye are therefore convex. A lens with **concave** surfaces diverges light rays, or spreads them farther apart, which is useful for correcting certain refractive errors of the eye.

The two structures most important in the eye's refractive ability are the *cornea* and the *lens*. The curved corneal surface, the first structure light passes through as it enters the eye, contributes most extensively to the eye's total refractive ability. This is because the difference in density at the air/corneal interface is much greater than the differences in density between the lens and the fluids surrounding it. In **astigmatism,** the curvature of the cornea is uneven, so light rays are unequally refracted and cannot come to a common focal point. Special cylindrical lens can correct this refractive error. The refractive ability of the cornea remains constant because the curvature of the cornea never changes. In contrast, the refractive ability of the lens can be adjusted by changing its curvature as needed to accommodate for near or far vision.

The refractive structures of the eye must bring light images into focus on the retina for clear vision. If an image is focused before it reaches the retina or is not yet focused when it reaches the retina, it will be blurred (Fig. 6–16). Light rays originating from near objects are more divergent when they reach the eye than are those from distant sources. Rays from light sources greater than twenty feet are considered to be parallel by the time they reach the eye. With a given refractive ability of the eye, a near source of light requires a greater distance behind the lens to be focused than a far source does, because the near source rays are still diverging when they reach the eye (Fig. 6–17a and b).

In a particular eye, the distance between the lens and the retina always remains the same. To bring both near and far light sources into focus on the retina (that is, in the same distance), a stronger lens must be used for the near source (Fig. 6-17c).

Accommodation increases the strength of the lens for near vision.

The ability to adjust the strength of the lens so that both near and far sources can be focused on the retina is known as **accommodation.** The strength of the lens depends on its shape, which in turn is regulated by the ciliary muscle.

The **ciliary muscle** is part of the cilary body, an anterior specialization of the choroid layer. The ciliary body has two major components: the ciliary muscle and the capillary network that produces the aqueous humor. The ciliary muscle is a circular ring of smooth muscle attached to the lens by means of **suspensory ligaments** (Fig. 6–18a and b).

When the ciliary muscle is relaxed, the suspensory ligaments are taut. This pulls the lens into a flattened, weakly refractive shape (Fig. 6–18b). As the muscle contracts, its circumference decreases, which slackens the tension within the suspensory ligaments (Fig. 6–18c). When the lens is subjected to less tension by the suspensory ligaments, it assumes a more spherical shape because of its inherent elasticity. The greater curvature of the rounded-up lens increases its strength, providing additional bending of light rays.

In the normal eye, the ciliary muscle is relaxed and the lens is flat for far vision, but the muscle contracts to allow the lens

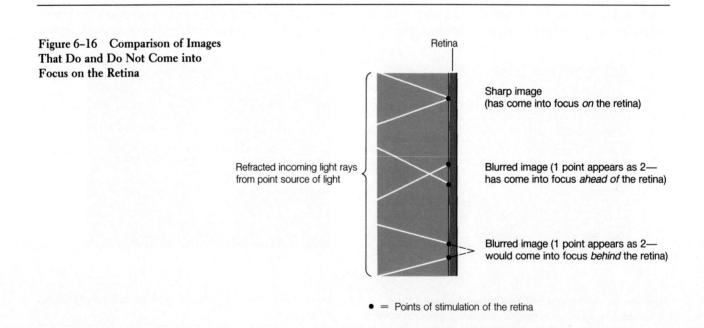

Figure 6–16 Comparison of Images That Do and Do Not Come into Focus on the Retina

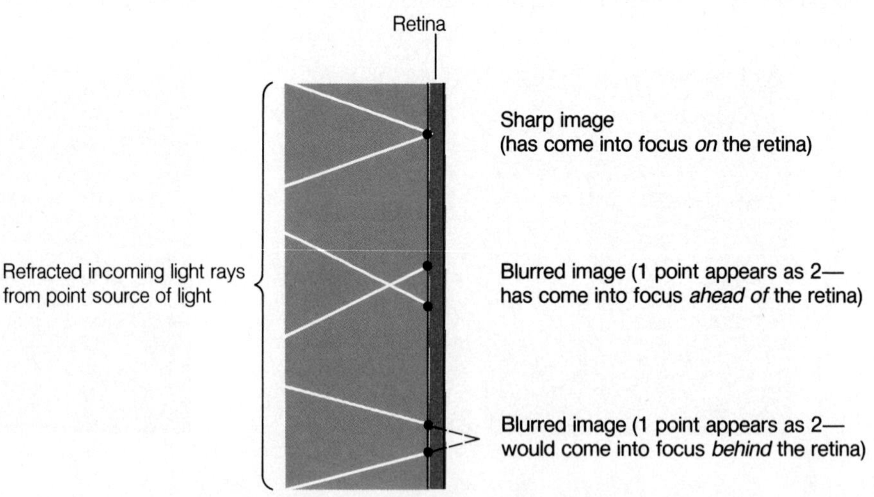

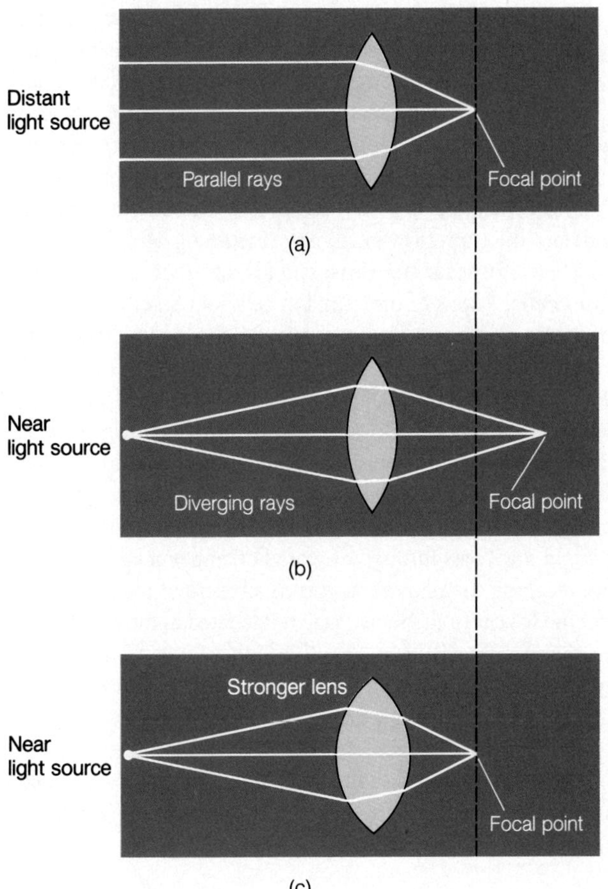

Figure 6–17 Focusing of Distant and Near Sources of Light *(a) A distant (far) light source is a light source greater than 20 feet from the eye. By the time the rays reach the eye from a distant source, they are considered to be parallel. (b) The rays from a near light source (a light source less than 20 feet from the eye) are still diverging by the time they reach the eye. A longer distance is required for a lens of a given strength to bend the diverging rays from a near light source into focus than to bend the parallel rays from a distant light source into focus. (c) To focus both a distant and near light source in the same distance (the distance between the lens and retina), a stronger lens must be used for the near source. A stronger lens is able to focus a near image in the same distance as a weaker lens does a distant image.*

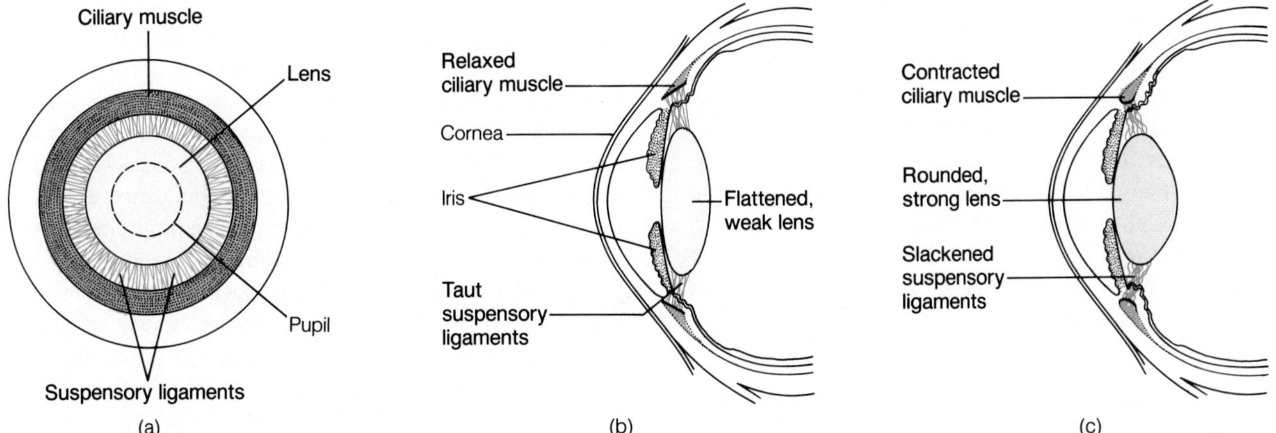

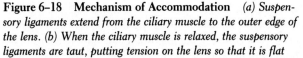

Figure 6–18 Mechanism of Accommodation *(a) Suspensory ligaments extend from the ciliary muscle to the outer edge of the lens. (b) When the ciliary muscle is relaxed, the suspensory ligaments are taut, putting tension on the lens so that it is flat and weak. (c) When the ciliary muscle is contracted, the suspensory ligaments become slack, reducing the tension on the lens. This enables the lens to assume a stronger, rounder shape because of its elasticity.*

to become more convex and stronger for near vision. The ciliary muscle is controlled by the autonomic nervous system. Sympathetic nerve fibers induce relaxation of the ciliary muscle for far vision, whereas the parasympathetic nervous system causes the muscle's contraction for near vision.

The lens is an elastic structure consisting of transparent fibers. Occasionally these fibers become opaque so that light rays cannot pass through, a condition known as a **cataract.** The defective lens can usually be surgically removed and vision restored by an implanted artificial lens or compensating eyeglasses.

Throughout life only cells at the outer edges of the lens are replaced. Cells within the center of the lens are in double jeopardy. Not only are they the oldest but they also are the farthest away from the aqueous humor, the lens' nutrient source. With advancing age, these nonrenewable central cells die and become stiff. With loss of elasticity, the lens is no longer able to assume the spherical shape required to accommodate for near vision. This age-related reduction in accommodative ability, **presbyopia,** affects most people by middle age (45 to 50), requiring them to resort to corrective lens for near vision (reading).

Other common vision disorders are near-sightedness (myopia) and far-sightedness (hyperopia). In a normal eye (Fig. 6–19a), a far light source is focused on the retina without accommodation, whereas the strength of the lens is increased by accommodation to bring a near source into focus. In **myopia** (Fig. 6–19b1), because the eyeball is too long or the lens is too strong, a near light source is brought into focus on the retina without accommodation (even though accommodation is normally used for near vision), whereas a far light source is focused in front of the retina and is blurry. Thus, a myopic individual has better near vision than far vision, a condition that can be corrected by a concave lens (Fig. 6-19b2). With **hyperopia** (Fig. 6-19c1), either the eyeball is too short or the lens is too weak. Far objects are focused on the retina only with accommodation, whereas near objects are focused behind the retina even with accommodation and, accordingly, are blurry. Thus, a hyperopic individual has better far vision than near vision, a condition that can be corrected by convex lens (Fig. 6–19c2). Such vision tends to get worse as the person gets older because of loss of accommodative ability with the onset of presbyopia.

Light must pass through several retinal layers before reaching the photoreceptors.

The major function of the eye is to focus light rays from the environment on the photoreceptor cells of the retina. The retina then transforms the light energy into electrical signals for transmission to the CNS.

The receptor-containing portion of the retina is actually an extension of the CNS and not a separate peripheral organ. During embryonic development, the retinal cells "back out" of the nervous system, so the retinal layers, surprisingly, are facing backward! The neural portion of the retina consists of three layers (Fig. 6–20): (1) the outermost layer closest to the choroid containing the *rods* and *cones*, whose light-sensitive ends face the choroid away from the incoming light; (2) a middle layer of **bipolar neurons;** and (3) an inner layer of **ganglion cells.** Axons of the ganglion cells join together to form the **optic nerve,** which leaves the retina slightly off-center. The point on the retina at which the optic nerve leaves and blood vessels pass through is the **optic disc** (Fig. 6–9a). This region is often referred to as the **blind spot;** no image can be detected in this area because it is devoid of rods and cones. We are normally not aware of the blind spot because what is missed by the blind spot in one eye is seen by the other eye.

Light must pass through the ganglion and bipolar layers before reaching the photoreceptors in all areas of the retina except the **fovea.** In the fovea, which is located in the exact center of the retina (Fig. 6–9a), the bipolar and ganglion cell layers are pulled aside so that light directly strikes the photoreceptors. This, coupled with the fact that the cones (which have greater acuity or discriminative ability than the rods) are concentrated here, makes the fovea the point of most distinct vision. Thus, we turn our eyes so that the object at which we are looking is focused on the fovea.

Phototransduction by retinal cells converts light stimuli into neural signals.

Photoreceptors consist of three parts (Fig. 6–21): (1) an outer segment facing the choroid that detects the light stimulus; (2) an inner segment that contains the metabolic machinery of the cell; and (3) a synaptic terminal that transmits the signal to the bipolar cells. The outer segment, which is rod-shaped in rods and cone-shaped in cones, is composed of stacked, flattened membranous disks containing an abundance of photopigment molecules, which undergo chemical alterations when activated by light. A **photopigment** consists of an enzymatic protein called **opsin** combined with **retinal,** a derivative of Vitamin A. There are four different photopigments, one in the rods and one in each of three types of cones. Retinal is identical in all four photopigments, but the photoreceptors' opsins vary slightly, enabling them to differentially absorb various wavelengths of light. **Rhodopsin,** the rod photopigment, cannot discriminate between various wavelengths in the visible spectrum; therefore, rods provide vision only in shades of gray by detecting varying intensities, not varying colors. The photopigments in the three types of cones—**red, green,** and

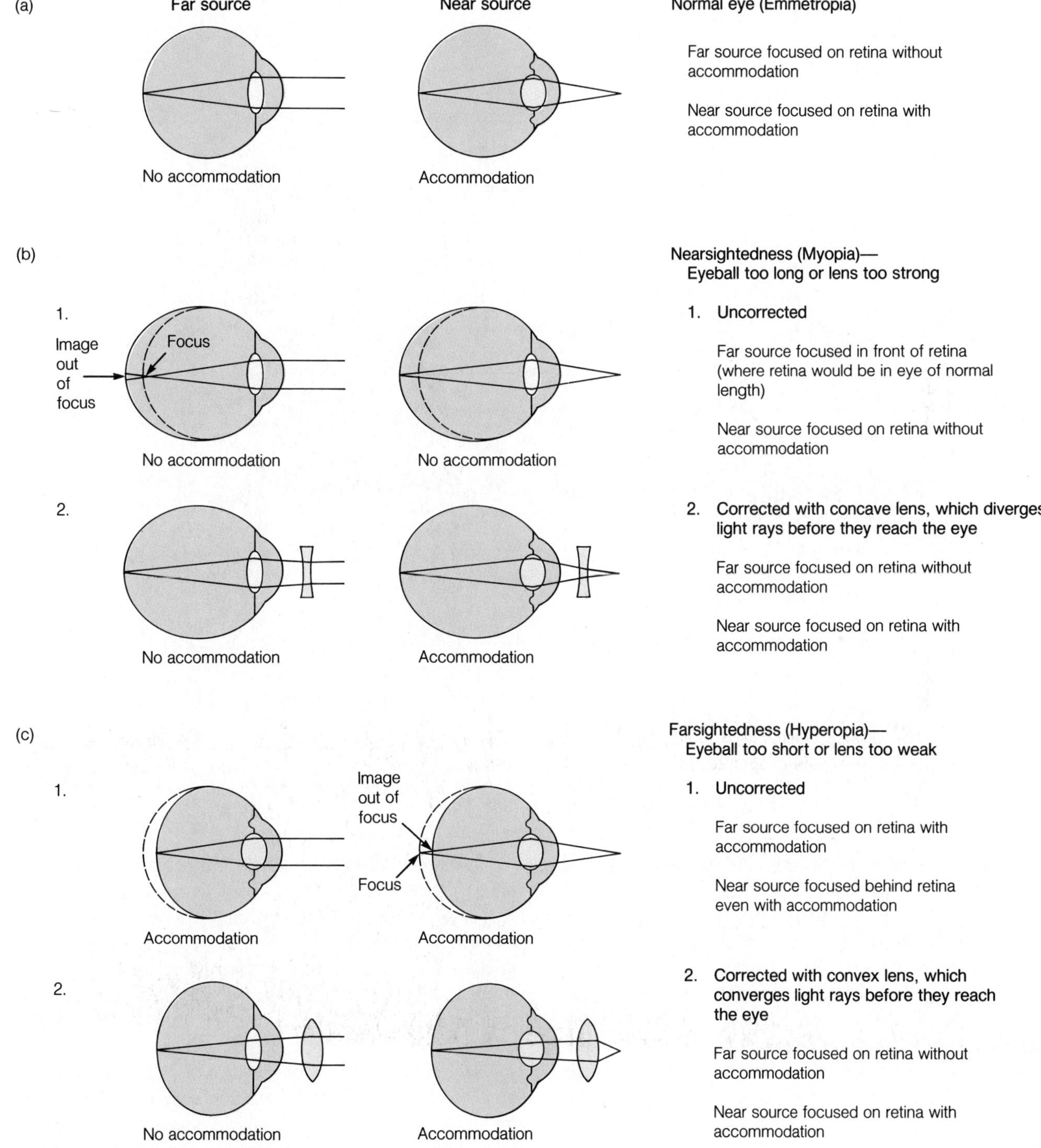
(a)
Far source
Near source
No accommodation
Accommodation
Normal eye (Emmetropia)
Far source focused on retina without accommodation
Near source focused on retina with accommodation
(b)
Nearsightedness (Myopia)—
Eyeball too long or lens too strong
1.
Image out of focus
Focus
No accommodation
No accommodation
1. Uncorrected
Far source focused in front of retina (where retina would be in eye of normal length)
Near source focused on retina without accommodation
2.
No accommodation
Accommodation
2. Corrected with concave lens, which diverges light rays before they reach the eye
Far source focused on retina without accommodation
Near source focused on retina with accommodation
(c)
Farsightedness (Hyperopia)—
Eyeball too short or lens too weak
1.
Image out of focus
Focus
Accommodation
Accommodation
1. Uncorrected
Far source focused on retina with accommodation
Near source focused behind retina even with accommodation
2.
No accommodation
Accommodation
2. Corrected with convex lens, which converges light rays before they reach the eye
Far source focused on retina without accommodation
Near source focused on retina with accommodation

Figure 6–19 Myopia and Hyperopia

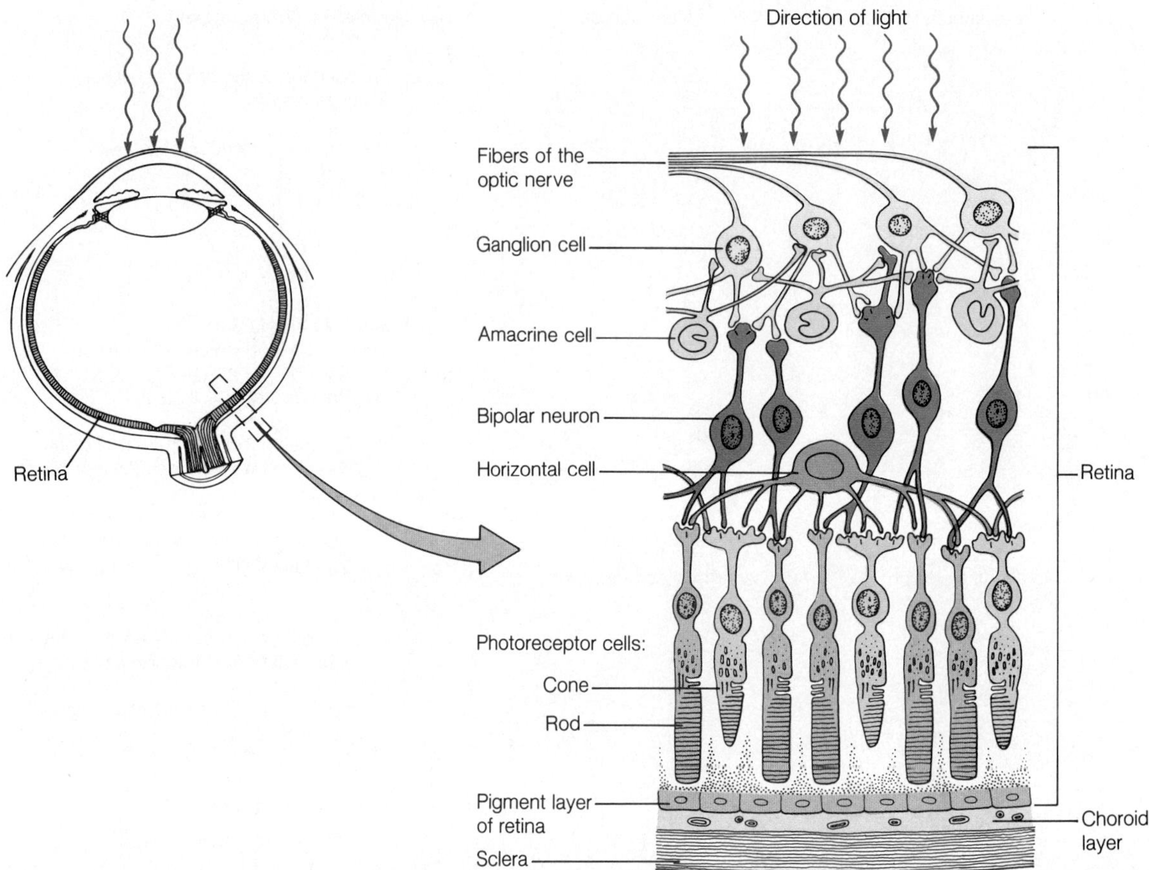

Figure 6–20 Retinal Layers *The retinal visual pathway is from the photoreceptor cells (rods and cones) to the bipolar cells to the ganglion cells. The horizontal and amacrine cells act locally for retinal processing of visual input, such as lateral inhibition.*

blue cones—do respond selectively to various wavelengths of light, making color vision possible (Fig. 6–22).

The mechanism of excitation is basically the same for all photopigments (Fig. 6–23). When a photopigment molecule absorbs light, its retinal portion changes shape. This light-induced biochemical change in the photopigment leads to closing of chemical messenger-gated Na^+ channels in the outer segment's membrane. Unlike other chemical-gated channels that respond to external chemical messengers, these channels respond to an internal second messenger, **cyclic GMP** (cyclic guanosine monophosphate), which links photopigment light absorption with Na^+ channel closing. Cyclic GMP is activated by a series of enzymatic reactions initiated by light excitation of the photopigment, similar to activation of the cyclic AMP second messenger system by an extracellular first messenger (see p. 64). The altered photopigments are gradually restored to their original conformation in the dark.

Unlike most receptors, the Na^+ channels of a photoreceptor are open in the absence of stimulation, that is, in the dark (Fig. 6–24). The resultant passive inward Na^+ leak depolarizes the photoreceptor and keeps the Ca^{++} channels in its synaptic terminal open, thereby permitting release of transmitter while in the dark. Closure of Na^+ channels during the light-excitation process stops the depolarizing Na^+ leak, bringing about membrane hyperpolarization, which passively spreads from the outer segment to the synaptic terminal of the photoreceptor. This hyperpolarization leads to a reduction in transmitter release from the synaptic terminal. Thus, photoreceptors are inhibited by their adequate stimulus (hyperpolarized by light) and excited in the absence of stimulation (depolarized by darkness). The hyperpolarizing potential and subsequent decrease in transmitter release is graded according to the intensity of light. The brighter the light, the greater the hyperpolarizing response and the greater the reduction in transmitter release.

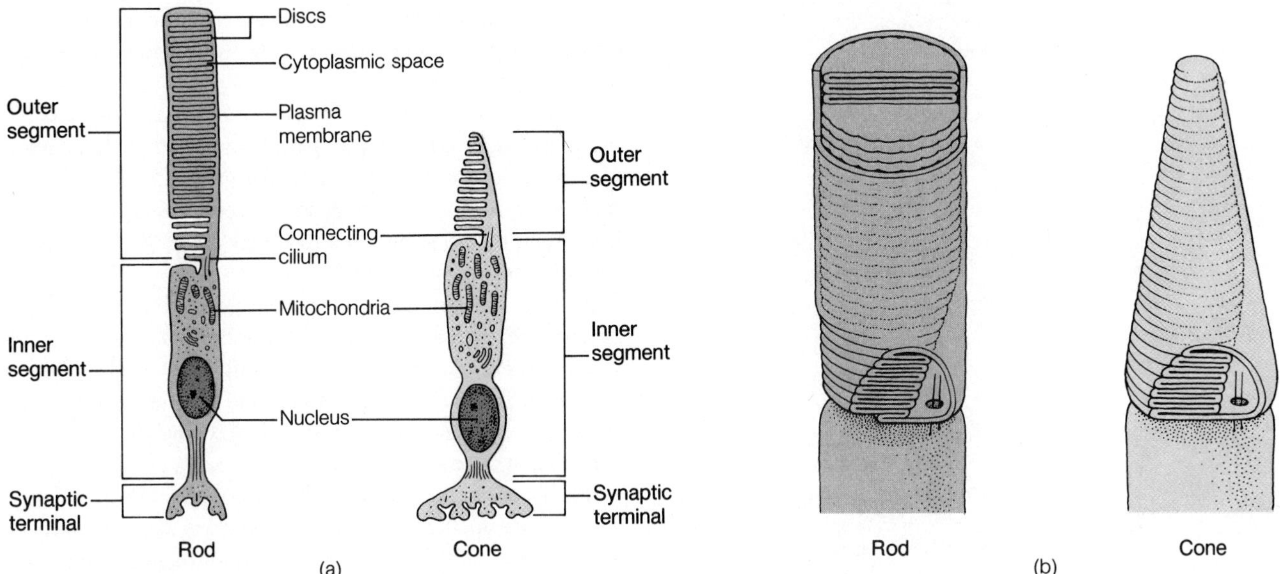

Figure 6–21 Photoreceptors *(a) Rods and cones. (b) Outer segments of photoreceptors (three-dimensional view).*

How does the retina signal the brain about light stimulation through such an inhibitory response? The photoreceptors synapse with bipolar cells. These cells in turn terminate on the ganglion cells, whose axons form the optic nerve for transmission of signals to the brain. The answer to our seeming paradox lies in the fact that the transmitter released from the photoreceptors' synaptic terminal has an inhibitory action on many of the bipolar cells. The reduction in transmitter release that accompanies light-induced receptor hyperpolarization decreases this inhibitory action on the bipolar cells. Removal of inhibition has the same effect as direct excitation of the bipolar cells. The greater the illumination on the receptor cells, the greater the removal of inhibition from the bipolar cells and the greater in effect the excitation of these next cells in the visual pathway to the brain.

Bipolar cells display graded potentials similar to the photoreceptors. Action potentials do not originate until the ganglion cells, the first neurons in the chain that must propagate the visual message over long distances to the brain.

Rods provide indistinct gray vision at night whereas cones provide sharp color vision during the day.

There are more than thirty times more rods than cones in the retina (100 million rods compared to 3 million cones per eye). Because of differential absorption of various wavelengths of light, cones provide color vision whereas rods provide vision only in shades of gray. A difference in the "wiring patterns" between these photoreceptor types and other retinal neuronal layers confers even further differences in their capabilities (Table 6–3). Cones have low sensitivity to light, being "turned on" only by bright daylight, but they have high acuity, providing sharp vision with high resolution for fine detail. Humans use cones for day vision, which is in color and distinct. Rods, on the other hand, have low acuity but have high sensitivity, so they respond to the dim light of night. We are able

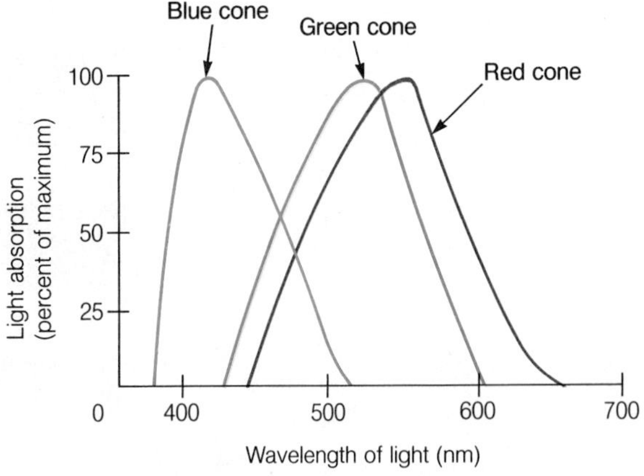

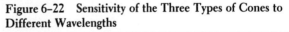

Figure 6–22 Sensitivity of the Three Types of Cones to Different Wavelengths

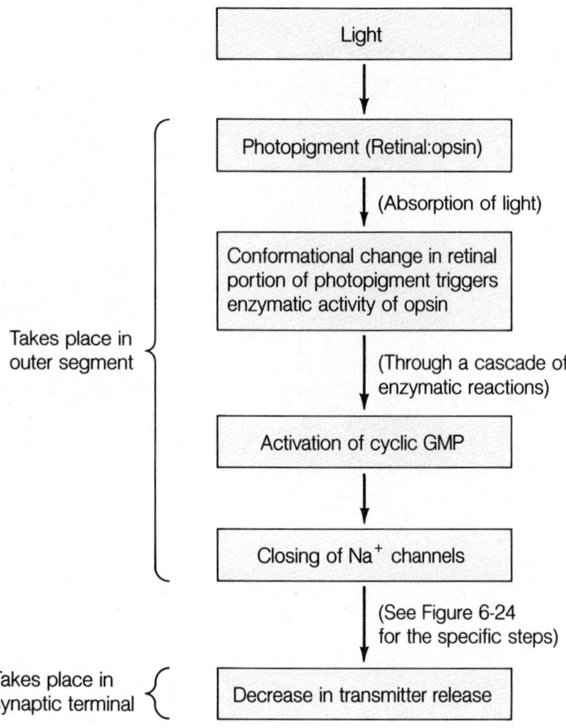
Light
Photopigment (Retinal:opsin)
(Absorption of light)
Conformational change in retinal portion of photopigment triggers enzymatic activity of opsin
Takes place in outer segment
(Through a cascade of enzymatic reactions)
Activation of cyclic GMP
Closing of Na+ channels
(See Figure 6-24 for the specific steps)
Takes place in synaptic terminal
Decrease in transmitter release

Figure 6–23 Phototransduction

to see at night with our rods but at the expense of color and distinctness. Let us see how the patterns of wiring influence sensitivity and acuity.

There is little convergence of neurons (see p. 112) in the retinal pathways for cone output. Each cone generally has a private line of connection to a particular ganglion cell. In contrast, there is much convergence in rod pathways. Output from more than one hundred rods may converge via bipolar cells on a single ganglion cell.

A ganglion cell must be brought to threshold through influence of the graded potentials in the receptors to which it is wired before it can have an action potential that will be propagated to the CNS. Because a single cone ganglion cell is often influenced by only one cone, only bright daylight is intense enough to induce a sufficient receptor potential in the cone to ultimately bring the ganglion cell to threshold. The abundant convergence in the rod visual pathways, in contrast, offers good opportunities for summation of subthreshold events in a rod ganglion cell (see p. 109). Even though a small receptor potential induced by dim light in a single cone would not be sufficient to bring its ganglion cell to threshold, similar small receptor potentials induced by the same dim light in multiple rods converging on a single ganglion cell would have an additive effect to bring the rod ganglion cell to threshold. Because rods can bring about action potentials in response to small amounts of light, they are much more sensitive than cones.

Figure 6–24 Events at Synaptic Terminal of Photoreceptor

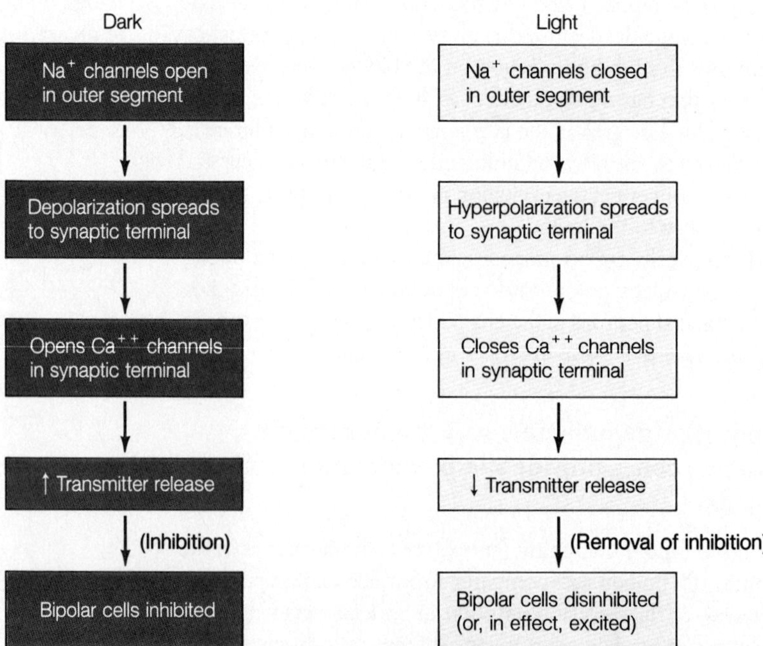
Dark
Na+ channels open in outer segment
Depolarization spreads to synaptic terminal
Opens Ca++ channels in synaptic terminal
↑ Transmitter release
(Inhibition)
Bipolar cells inhibited
Light
Na+ channels closed in outer segment
Hyperpolarization spreads to synaptic terminal
Closes Ca++ channels in synaptic terminal
↓ Transmitter release
(Removal of inhibition)
Bipolar cells disinhibited (or, in effect, excited)

Table 6–3 Properties of Rod and Cone Vision

Rods	Cones
100 million per retina	3 million per retina
Vision in shades of gray	Color vision
High sensitivity	Low sensitivity
Low acuity	High acuity
Night vision	Day vision
Much convergence in retinal pathways	Little convergence in retinal pathways
More numerous in periphery	Concentrated in fovea

However, because cones have private lines into the optic nerve, each cone transmits information about an extremely small receptive field on the retinal surface. Cones are thus able to provide finely detailed vision at the expense of sensitivity. With rod vision, acuity is sacrificed for sensitivity. Since many rods share a single ganglion cell, once an action potential is initiated it is impossible to discern which of the multiple rod inputs were activated to bring the ganglion cell to threshold. Objects appear fuzzy when rod vision is used because of this poor ability to distinguish between two nearby points.

The sensitivity of the eyes can vary markedly through dark and light adaptation.

The eyes' sensitivity to light depends on the amount of photopigment present in the rods and cones. When you first go from bright sunlight into darkened surroundings, you cannot see anything at first, but gradually you can distinguish objects via the process of **dark adaptation.** Breakdown of photopigments during exposure to sunlight tremendously decreases photoreceptor sensitivity. For example, a reduction in rhodopsin content of only 0.6% from its maximum value decreases rod sensitivity approximately three thousand times. In the dark, the photopigments broken down during light exposure are gradually regenerated. As a result, the sensitivity of your eyes gradually increases so that you can begin to see in the dark. However, only the highly sensitive, rejuvenated rods are "turned on" by the dim light.

Conversely, when you move from the dark to the light (for example, leaving a movie theater and entering the bright sunlight), your eyes are very sensitive to the dazzling light at first. With little contrast between lighter and darker parts, the entire image appears bleached. As some of the photopigments are rapidly broken down by the intense light, the sensitivity of the eyes decreases and normal contrasts can once again be detected. The rods are so sensitive to light that sufficient rhodopsin is broken down to essentially "burn out" the rods in bright light. Furthermore, a central neural adaptative mechanism switches the eye from the rod system to the cone system on exposure to bright light. Therefore, only the less-sensitive cones are used for day vision.

It is estimated that our eyes' sensitivity can change as much as one million times as they adjust to various levels of illumination via dark and light adaptation. These adaptive measures are also enhanced by pupillary reflexes that adjust the amount of available light permitted to enter the eye.

Since retinal, one of the photopigment components, is a derivative of vitamin A, adequate amounts of this nutrient must be available for the ongoing resynthesis of photopigments. **Night blindness** occurs because of dietary deficiencies of vitamin A. Although photopigment concentrations in both rods and cones are reduced in this condition, there is still sufficient cone photopigment to respond to the intense stimulation of bright light, except in the most severe cases. However, even modest reductions in rhodopsin content can decrease the sensitivity of rods so much that they are unable to respond to dim light. The person can see in the day using cones but cannot see at night because the rods are no longer functional. This is why carrots are "good for your eyes"; they are rich in vitamin A.

Color vision is dependent on the ratios of stimulation of the three cone types.

Vision is dependent on stimulation of retinal photoreceptors by light. Certain objects in the environment such as the sun, fire, and light bulbs, emit light. But how do we see objects like chairs, trees, and people, which do not emit light? The pigments in various objects selectively absorb particular wavelengths of light transmitted to them from light-emitting sources, and the unabsorbed wavelengths are reflected from the objects' surfaces. It is these reflected light rays that enable us to see the objects. An object perceived as blue absorbs the longer red and green wavelengths of light and reflects the shorter blue wavelengths. The latter can be absorbed by the photopigment in the eye's blue cones, thereby activating them.

Each cone type is most effectively activated by a particular wavelength of light in the range of color designated by its name—blue, green, or red. However, cones also respond in varying degrees to other wavelengths (Fig. 6–22). Our perception of the many colors of the world is dependent on the three cone types' various *ratios of stimulation* in response to different wavelengths. A wavelength perceived as blue does not stimulate red or green cones at all but excites blue cones maximally (percentage of maximal stimulation for red, green, and blue cones, respectively = 0:0:100). The sensation of yellow, in comparison, arises from a stimulation ratio of 83:83:0, red and green cones each being stimulated 83% of maximum while blue cones are not excited at all. The ratio for green is 31:67:36, and so on, with various combinations giving rise to the sensation of all the different colors. White is a mixture of all wavelengths of light, whereas black is the absence of light.

The extent of excitation of each of the cone types is coded and transmitted in separate parallel pathways to the brain. The brain combines and processes these inputs to generate the perception of color, taking into consideration the object in comparison with its background. The concept of color is therefore in the mind of the beholder. Most of us agree on what color we see because we have the same types of cones and use similar neural pathways for comparing their output. Occasionally, however, individuals lack a particular cone type, so their color vision is a product of the differential sensitivity of only two types of cones. This is known as **color blindness** or **color defectiveness.** Not only do color-defective individuals perceive certain colors differently, but they are also unable to distinguish as many varieties of colors. For example, people with certain color defects are unable to distinguish between red and green. At a traffic light they can tell which light is "on" by its intensity, but they must rely on the position of the bright light to know whether to stop or go.

Visual information is separated and modified within the visual pathway before it is integrated into a perceptual image of the visual field by the cortex.

The field of view that can be seen without moving the head is known as the **visual field.** The information that reaches the visual cortex in the occipital lobe is not a replica of the visual field for several reasons:

1. The image detected on the retina at the onset of visual processing is upside down and backwards because of bending of the light rays (Fig. 6–25). Once projected to the brain, the inverted image is interpreted as being in its correct orientation.
2. The information transmitted from the retina to the brain is not merely a point-to-point record of photoreceptor activation. Before ever reaching the brain, selected information is reinforced and other information is suppressed by the retinal neuronal layers beyond the rods and cones to enhance contrast. One mechanism of retinal processing is lateral inhibition, by which strongly excited cone pathways suppress activity in surrounding pathways of weakly stimulated cones. This increases the dark-bright contrast to enhance the sharpness of boundaries.

   Another mechanism of retinal processing involves differential activation of two types of ganglion cells, **on-center** and **off-center ganglion cells.** The receptive field of a cone ganglion cell is determined by the field of light detection by the cone with which it is linked. On-center and off-center cells respond in opposite ways, depending on the relative comparison of illumination between the center and periphery of their receptive fields. Think of the receptive field as a donut. An on-center ganglion cell increases its rate of firing when light is most intense at the center of its receptive field (that is, when the donut hole is lit up). In contrast, an off-center cell increases its firing rate when the periphery of its receptive field is most intensely illuminated (that is, when the donut itself is lit up). This is useful for enhancing the difference in light level between one small area at the center of a receptive field compared to the illumination immediately around it. This

Figure 6–25 Inversion of Image on Retina

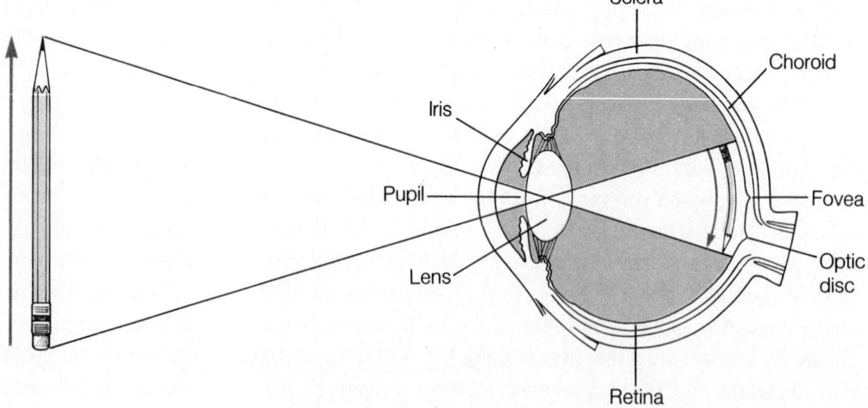

mechanism emphasizes differences in relative brightness to help define contours of images, but in so doing, information about absolute brightness is sacrificed (Fig. 6–26).

3. Various aspects of visual information such as form, color, depth, and movement are separated and projected in parallel pathways to different regions of the cortex. Only when these separate bits of processed information are integrated by higher visual regions is a reassembled picture of the visual scene perceived. This is similar to the blobs of paint on an artist's palette versus the finished portrait; the separate pigments do not represent a portrait of a face but the appropriate integration of these paints on a canvas does.

   Patients with lesions in specific visual processing regions of the brain may be unable to completely combine components of a visual impression. An example is being unable to discern movement of an object but having reasonably good vision for shape, pattern, and color. Sometimes the defect can be remarkably specific, like being unable to recognize familiar faces while retaining the ability to recognize inanimate objects.

4. Because of the pattern of wiring between the eyes and visual cortex, the left half of the cortex receives information only from the right half of the visual field as detected by both eyes, and the right half receives input only from the left half of the visual field of both eyes.

   As a result of refraction (Fig. 6–27), light rays from the left half of the visual field fall on the right half of the retina of both eyes [the medial (inner) half of the left retina and the lateral (outer) half of the right retina]. Similarly, rays from the right half of the visual field reach the left half of

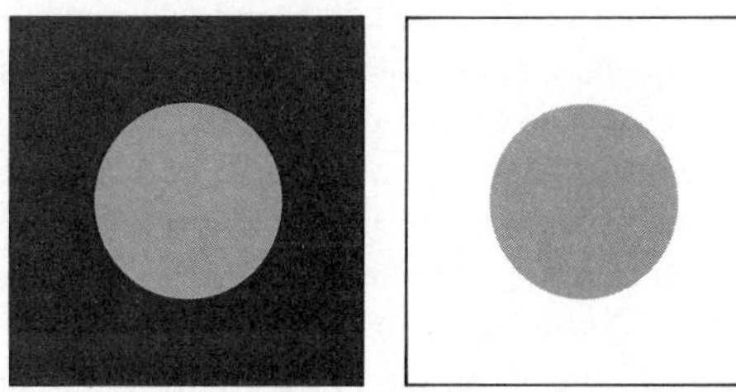

Figure 6–26 Example of Outcome of Retinal Processing by On-Center and Off-Center Ganglion Cells *The gray circles are identical (same shade and size). Emphasis of differences in relative brightness rather than absolute brightness, which helps to define contours, is accomplished largely through retinal processing by on-center and off-center ganglion cells.*

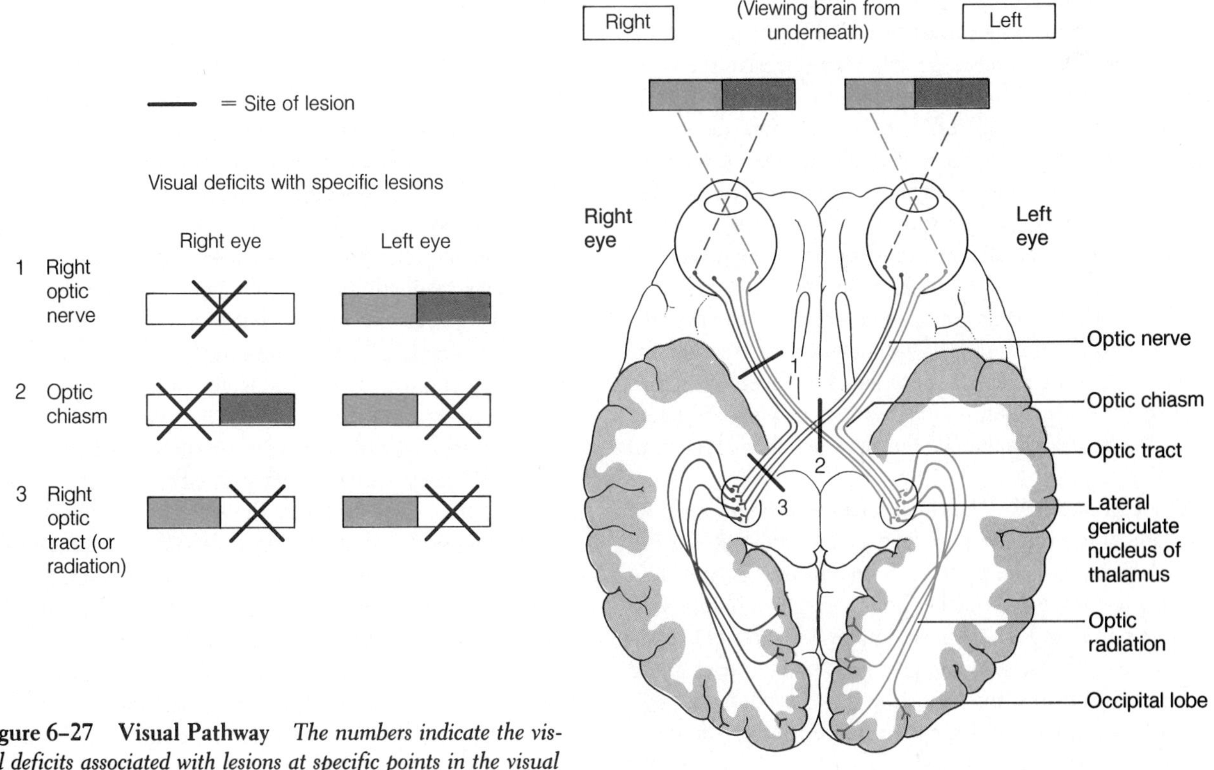

Figure 6–27 Visual Pathway *The numbers indicate the visual deficits associated with lesions at specific points in the visual pathway.*

each retina (the lateral half of the left retina and the medial half of the right retina). Each optic nerve exiting the retina carries information from both halves of the retina it serves. This information is separated as the optic nerves meet at the **optic chiasm** located underneath the hypothalamus. Within the optic chiasm, the fibers from the medial half of each retina cross to the opposite side, but those from the lateral half remain on the original side. The reorganized bundles of fibers leaving the optic chiasm are known as **optic tracts.** Each optic tract carries information from the lateral half of one retina and the medial half of the other retina. Therefore, this partial crossover brings together from the two eyes fibers that carry information from the same half of the visual field. Each optic tract, in turn, delivers to the half of the brain on its same side information about the opposite half of the visual field. A knowledge of these pathways can facilitate diagnosis of visual defects arising from interruption of the visual pathway at various points (Fig. 6–27).

The thalamus and visual cortices elaborate the visual message.

The first stop in the brain for information in the visual pathway is the **lateral geniculate nucleus** in the thalamus. It separates information received from the eyes and relays it via fiber bundles known as **optic radiations** to different zones in the cortex, each of which processes different aspects of the visual stimulus (for example, color, form, depth, movement). This sorting process is no small task, because each optic nerve contains more than 1 million fibers carrying information from the photoreceptors in one retina. This is more than all of the afferent fibers carrying somatosensory input from all the regions of the body! Yet the connections are precise. The lateral geniculate nucleus and each of the zones in the cortex that processes visual information have a topographical map representing the retina point-for-point. As with the somatosensory cortex, the neural maps of the retina are distorted. The fovea, the retinal region capable of greatest acuity, has much greater representation in the neural map than do the more peripheral regions of the retina.

Although each half of the visual cortex receives information simultaneously from the same part of the visual field as received by both eyes, the messages from both eyes are not identical. Each eye views an object from a slightly different vantage point, even though there is a tremendous area of overlap (Fig. 6–28). The overlapping area seen by both eyes at the same time is known as the **binocular** ("two-eyed") field of vision, which is important for **depth perception.** As in other areas of the cortex, the primary visual cortex is organized into

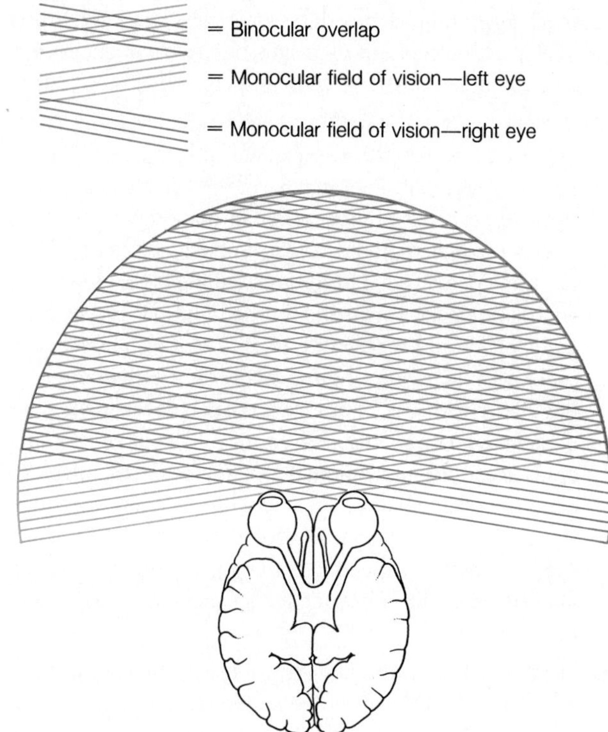

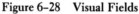

Figure 6–28 Visual Fields

functional columns, each processing information from a small region of the retina. Independent alternating columns are devoted to information about the same point in the visual field from the right and left eyes. Integration of information from the two eyes enhances depth perception. The brain uses the slight disparity in the information about this same point to estimate distance, allowing one to perceive three-dimensional objects in spatial depth. Some depth perception is possible using only one eye, based on experience and comparison with other cues. For example, if your one-eyed view includes a car and a building and the car is much larger, you correctly interpret that the car must be closer to you than the building.

Occasionally the two views are not successfully merged, because: (1) the eyes are not both focused on the same object simultaneously, since defects of the external eye muscles make fusion impossible, or (2) the binocular information is improperly integrated during visual processing. The result is double vision or **diplopia,** a condition in which the disparate views from both eyes are seen simultaneously.

Within the cortex, visual information is first processed in the primary visual cortex, then is projected to higher order visual areas for even more complex processing and abstraction. The cortex contains a hierarchy of visual cells that respond to increasingly complex stimuli. Three types of visual cortical neurons have been identified based on the complexity of stimulus requirements needed for the cell to respond; these are called **simple, complex,** and **hypercomplex cells.** Simple and complex cells are stacked on top of each other within the cortical columns of the primary visual cortex, whereas hypercomplex cells are found in the higher visual processing areas. Unlike a retinal cell, which responds to the amount of light, a cortical cell fires only when it receives a particular pattern of illumination for which it is programmed. These patterns are built up by converging connections that originate from closely aligned photoreceptor cells in the retina. For example, some simple cells fire only when a bar is viewed vertically in a specific location, others when a bar is horizontal, and others at various oblique orientations. Movement of a critical axis of orientation becomes important for response by some of the complex cells. Hypercomplex cells add a new dimension to visual processing by responding only to particular edges, corners, and curves. Each level of cortical visual neurons has increasingly greater capacity for abstraction of information built up from the increasing convergence of input from lower-order neurons. In this way, the dotlike pattern of photoreceptors stimulated to varying degrees by varying light intensities in the retinal image is transformed in the cortex into information about depth, position, orientation, movement, contour, and length. Other aspects of this information, such as color perception, are processed simultaneously through a similar hierarchical organization. How and where the entire image is finally put together is still unresolved.

Visual input goes to other areas of the brain not involved in vision perception.

Not all fibers in the visual pathway terminate in the visual cortices. Some are projected to other regions of the brain for purposes other than direct vision perception. Following are examples of non-sight activities dependent on input from the retina:

1. control of pupil size;
2. synchronization of biological clocks to cyclical variations in light intensity (for example, the sleep-wake cycle synchronized to the night-day cycle);
3. contribution to cortical alertness and attention; and
4. control of eye movements.

In the latter regard, each eye is equipped with a set of six **external eye muscles** that position and move the eye so that it can better locate, see, and track objects. Eye movements are among the fastest, most discretely controlled movements of the body.

Protective mechanisms help prevent eye injuries.

Several mechanisms help protect the eyes from injury. Except for its anterior portion, the eyeball is sheltered by the bony socket in which it is positioned. The **eyelids** act like shutters to protect the anterior portion of the eye from environmental insults. They close reflexly to cover the eye under threatening circumstances such as rapidly approaching objects, dazzling light, and instances when the cornea or eyelashes are touched. Frequent spontaneous blinking of the eyelids helps disperse the lubricating, cleansing, bactericidal ("germ-killing") tears. **Tears** are produced continuously by the **lacrimal gland** in the upper lateral corner under the eyelid. This eye-washing fluid flows across the surface of the cornea and drains into tiny canals in the corner of each eye (Fig. 6–9b), eventually emptying into the back of the nasal passageway. This drainage system cannot handle the profuse tear production during crying, so the tears overflow from the eyes. The eyes are also equipped with protective **eyelashes,** which trap fine air-borne debris such as dust before it can fall into the eye.

Ear: Hearing and Equilibrium

The ear consists of three parts: the external, the middle, and the inner ear (Fig. 6–29). The air-filled external and middle portion of the ear transmit air-borne sound waves to the fluid-filled inner ear, amplifying the sound energy in the process. The inner ear houses two different sensory systems: the *cochlea,* which contains the receptors for conversion of sound waves into nerve impulses, making hearing possible; and the *vestibular apparatus,* which is necessary for the sense of equilibrium.

Sound waves consist of alternate regions of compression and rarefaction of air molecules.

Hearing is the neural perception of sound energy. **Sound waves** are traveling vibrations of air that consist of regions of high pressure caused by compression of air molecules alternating with regions of low pressure caused by rarefaction of the molecules (Fig. 6–30a). Any device capable of producing such a disturbance pattern in air molecules is a source of sound. A simple example is a tuning fork. When a tuning fork is struck, its prongs vibrate. As a prong of the fork moves in one direction (Fig. 6–30b), air molecules ahead of it are pushed closer together, or compressed, which increases the pressure in this area. Simultaneously, the air molecules behind the prong spread out or are rarefied as the prong moves for-

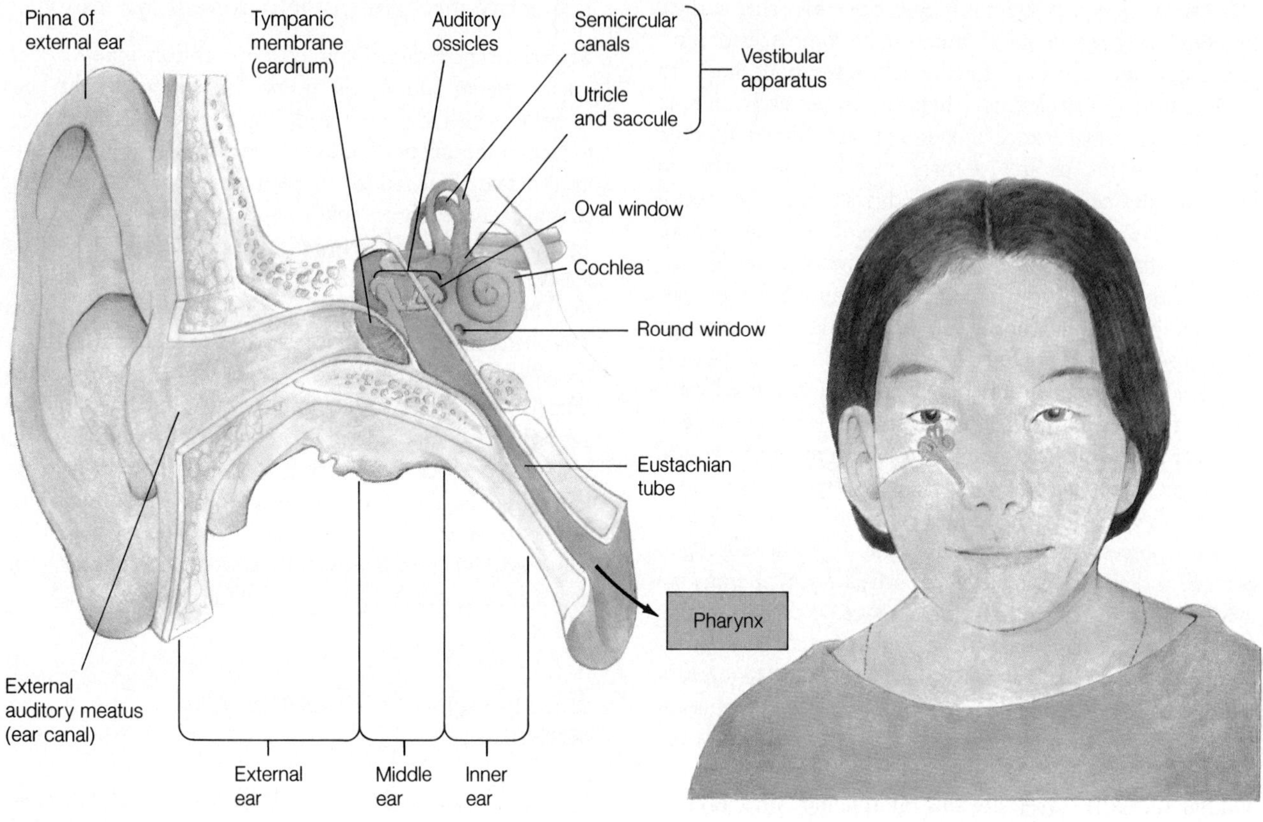

Figure 6–29 Anatomy of the Ear

ward, lowering the pressure in that region. As the prong moves in the opposite direction, an opposite wave of compression and rarefaction is created. Even though individual molecules are moved only short distances as the tuning fork vibrates, alternating waves of compression and rarefaction spread out considerable distances in a rippling fashion. Disturbed air molecules disturb other molecules in adjacent regions, which sets up new regions of compression and rarefaction, and so on (Fig. 6–30c). Sound energy is gradually dissipated as sound waves travel farther from the original sound source. The intensity of the sound decreases, until it finally dies out when the last sound wave is too weak to disturb the air molecules around it.

Sound waves can also travel through media other than air, such as water. They do so less efficiently, however; greater pressures are required to cause movements of fluid than movements of air because of the fluid's greater **inertia** (resistance to change).

Sound is characterized by its pitch (tone), intensity (loudness), and timbre (quality) (Fig. 6–31):

□ The **pitch** or **tone** of a sound (for example, whether it is a C or a G note) is determined by the *frequency* of vibrations, as expressed in cycles per second, or *Hertz* (*Hz*). The greater the frequency of vibration, the higher the pitch. Human ears can detect sound waves with frequencies from 20 to 20,000 Hz but are most sensitive to frequencies between 1,000 to 4,000 Hz.

□ The **intensity,** or **loudness,** of a sound depends on the *amplitude* of the sound waves, or the pressure difference between a high-pressure region of compression and a low-pressure region of rarefaction. Within the hearing range, the greater the amplitude, the louder the sound. Human ears can detect a wide range of sound intensities, from the slightest whisper to the painfully loud takeoff of a jet. Loudness is expressed in **decibels** (**dB**), which is a logarithmic measure of intensity compared with the faintest sound that can be heard—the **hearing threshold.** Because of the logarithmic relationship, every ten decibels indicates a tenfold increase in loudness. A few examples of common sounds illustrate the

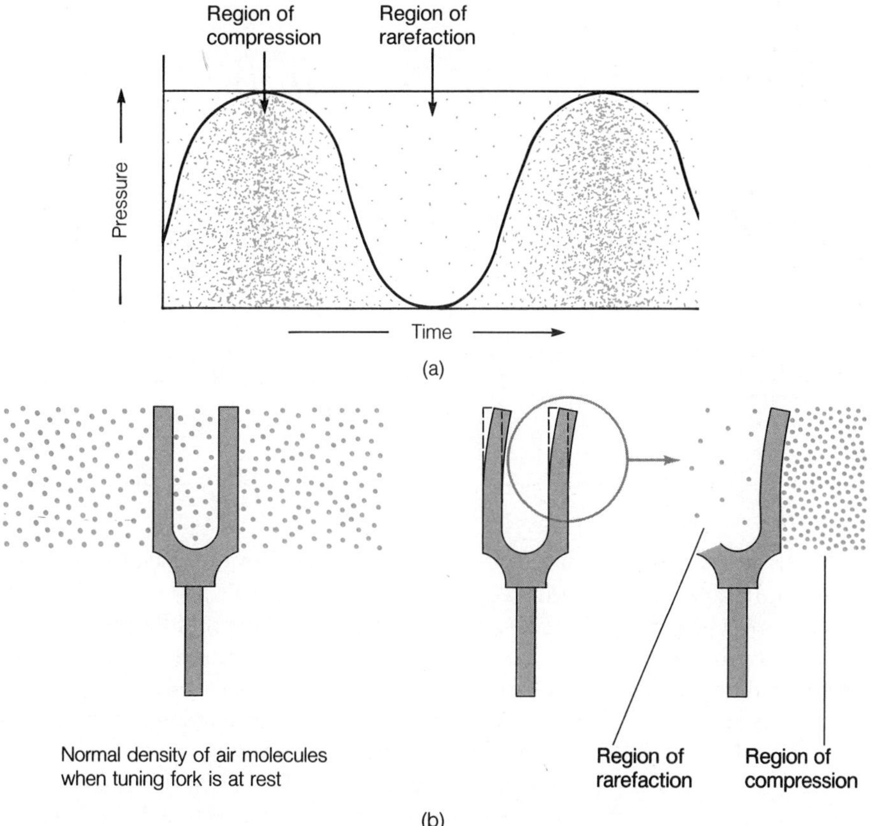

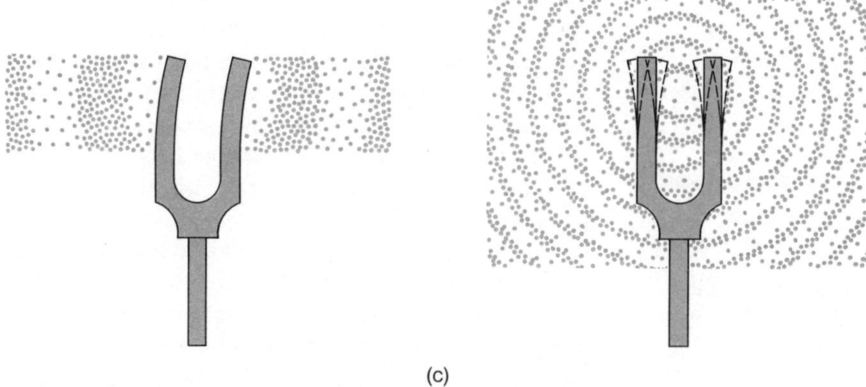

Figure 6–30 Formation of Sound Waves *(a) Sound waves are alternating regions of compression and rarefaction of air molecules. (b) A vibrating tuning fork sets up sound waves as the air molecules ahead of the advancing arm of the tuning fork are compressed while those molecules behind the arm are rarefied. (c) Disturbed air molecules bump into molecules beyond them, setting up new regions of air disturbance more distant from the original source of sound. In this way, sound waves travel progressively farther from the source, even though each individual air molecule travels only a short distance when it is disturbed. The sound wave dies out when the last region of air disturbance is too weak to disturb the region beyond it.*

significance of this (Table 6–4). Note that the rustle of leaves at 10 dB is ten times louder than hearing threshold, but the sound of a jet taking off at 150 dB is a quadrillion times, not 150 times, louder than the faintest audible sound. Sounds greater than 100 dB can permanently damage the sensitive sensory apparatus in the cochlea.

□ The **timbre,** or **quality,** of a sound depends on its *overtones*, which are additional frequencies superimposed on top of the fundamental pitch or tone. A tuning fork has a pure tone, but most sounds lack purity. For example, complex mixtures of overtones impart different sounds to different instruments playing the same note (a C note sounds different on a

Figure 6–31 Properties of Sound Waves

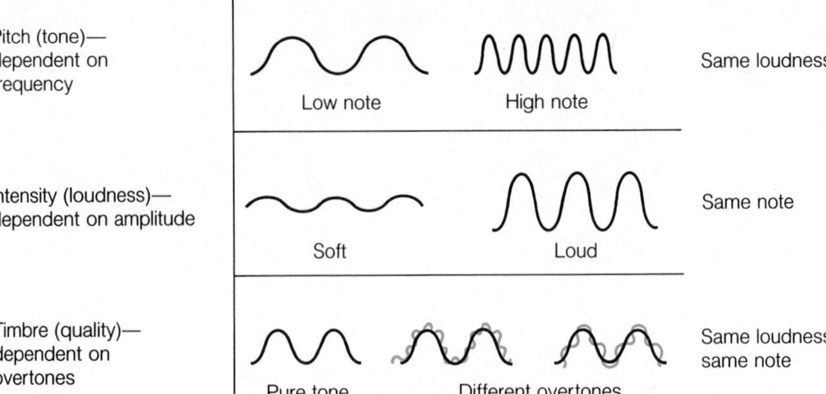

trumpet than on a piano). Overtones are likewise responsible for the characteristic differences in voices. Timbre enables the listener to distinguish the source of sound waves, because each source produces a different pattern of overtones. Thanks to timbre, you can tell whether it is your mother or girlfriend calling on the telephone before you say the wrong thing.

The external and middle ear convert air-borne sound waves into fluid vibrations in the inner ear.

The specialized receptors for sound are located in the fluid-filled **inner ear.** Air-borne sound waves must therefore be channeled toward and transferred into the inner ear, compensating in the process for the loss in sound energy that naturally occurs as sound waves pass from air into water. This is the function of the external ear and the **middle ear.**

The **external ear** (Fig. 6–29) consists of the *pinna* (*ear lobe*), *external auditory meatus* (*ear canal*), and *tympanic membrane* (*eardrum*). The **pinna,** a prominent skin-covered flap of cartilage, collects sound waves and channels them down the external ear canal. Many species (dogs, for example) can cock their ears in the direction of sound to collect more sound waves, but human ear lobes are relatively immobile. Because of its shape, the pinna partially shields sound waves that approach the ear from the rear. This helps a person distinguish whether a sound is coming from directly in front or behind. In either case the sound waves reach both ears at the same time, so the only cue is provided by the difference in approach to the ear canal afforded by the pinna's shape.

Sound localization for sounds approaching from the right or left is determined by two cues. First, the sound wave reaches the ear closer to the sound source slightly before it arrives at the farther ear. Second, the sound is less intense as it reaches the farther ear, because the head acts as a sound barrier that partially disrupts the propagation of sound waves. The auditory cortex integrates all of these cues to determine the location of the sound source. It is difficult to localize sound with only one ear.

The entrance to the **ear canal** is guarded by fine hairs. The skin lining the canal contains modified sweat glands that produce **cerumen** (earwax), a sticky secretion that traps fine foreign particles. Together the hairs and earwax help prevent air-borne particles from reaching the inner portions of the ear canal, where they could accumulate or injure the tympanic membrane and interfere with hearing.

The **tympanic membrane,** which is stretched across the entrance to the middle ear, vibrates when struck by sound

Table 6–4 Relative Magnitude of Common Sounds

Sound	Loudness in Decibels (dB)	Comparison to Faintest Audible Sound (Hearing Threshold)
Rustle of leaves	10 dB	10 times louder
Ticking of watch	20 dB	100 times louder
Normal conversation	60 dB	1 million times louder
Shouting	90 dB	1 billion times louder
Loud rock concert	120 dB	1 trillion times louder
Takeoff of jet plane	150 dB	1 quadrillion times louder

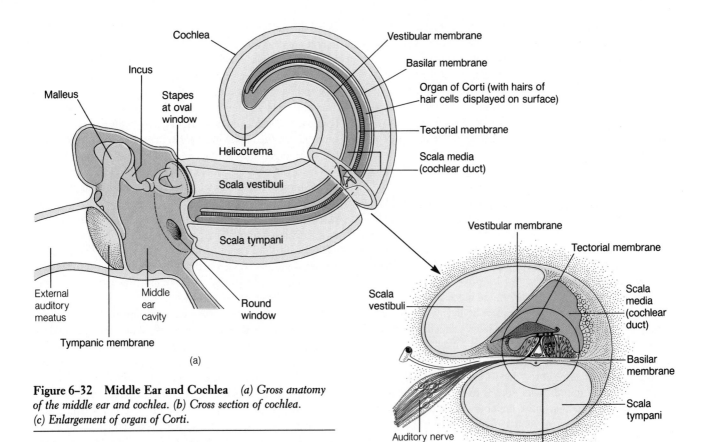

Cochlea
Incus
Malleus
Stapes at oval window
Helicotrema
Vestibular membrane
Basilar membrane
Organ of Corti (with hairs of hair cells displayed on surface)
Tectorial membrane
Scala media (cochlear duct)
Scala vestibuli
Scala tympani
External auditory meatus
Middle ear cavity
Round window
Tympanic membrane

(a)

Vestibular membrane
Tectorial membrane
Scala vestibuli
Scala media (cochlear duct)
Basilar membrane
Scala tympani
Auditory nerve

(b)

Figure 6–32 Middle Ear and Cochlea *(a) Gross anatomy of the middle ear and cochlea. (b) Cross section of cochlea. (c) Enlargement of organ of Corti.*

waves. The alternating higher- and lower-pressure regions of a sound wave cause the exquisitely sensitive eardrum to bow inward and outward in unison with the wave's frequency.

The resting air pressure on both sides of the tympanic membrane must be equal for the membrane to be free to move as sound waves strike it. The outside of the eardrum is exposed to atmospheric pressure that reaches it through the ear canal. Importantly, the inside of the eardrum facing the middle ear cavity is also exposed to atmospheric pressure via the **eustachian (auditory) tube,** which connects the middle ear to the **pharynx** (back of the throat) (Fig. 6–29). The eustachian tube is normally closed, but it can be pulled open by yawning, chewing, and swallowing. Such opening permits air pressure within the middle ear to equilibrate with atmospheric pressure so that pressures on both sides of the tympanic membrane are equal. During rapid external pressure changes (for example, during air flight), each eardrum painfully bulges as the pressure outside the ear changes while the pressure in the middle ear remains unchanged. Opening the eustachian tube by yawning allows the pressures on both sides of the tympanic membrane to equalize, relieving the pressure distortion as the eardrum "pops" back into place. Infections originating in the throat sometimes spread through the eustachian tube to the middle ear. The resultant fluid accumulation in the middle ear

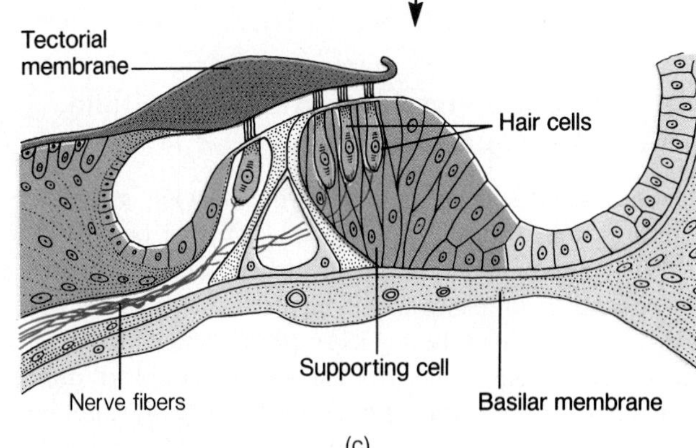

(c)

is not only painful but also interferes with conduction of sound across the middle ear.

The middle ear transfers the vibratory movements of the tympanic membrane to the fluid of the inner ear. This is facilitated by a moveable chain of three small bones or **ossicles** (the **malleus, incus,** and **stapes**) that extend across the middle ear (Fig. 6–32a). The first bone, the malleus, is attached to the tympanic membrane, and the last bone, the stapes, is attached to the **oval window,** the entrance into the fluid-filled cochlea. As the tympanic membrane vibrates in response to sound waves, the chain of bones is set into motion at the same frequency, transmitting this frequency of movement from the tympanic membrane to the oval window. The resultant pressure on the oval window with each vibration produces wave-like movements in the inner ear fluid at the same frequency as the original sound waves. However, as noted earlier, a greater pressure is required to set fluid in motion. Two mechanisms related to the ossicular system amplify the pressure of the air-borne sound waves to set up fluid vibrations in the cochlea. First, because the surface area of the tympanic membrane is much larger than that of the oval window, pressure is increased as force exerted on the tympanic membrane is conveyed to the oval window (pressure = force/unit area). Second, the lever action of the ossicles provides an additional mechanical advantage. Together, these mechanisms increase the force exerted on the oval window by twenty times compared to what it would be if the sound wave directly struck the oval window. This additional pressure is sufficient to set the cochlear fluid in motion.

Several tiny muscles in the middle ear contract reflexly in response to loud sounds (over 70 dB) to cause tightening of the tympanic membrane and to limit movement of the ossicular chain. This reduced movement of middle ear structures diminishes the transmission of loud sound waves to the inner ear to protect the delicate sensory apparatus from damage. This reflex response is relatively slow, however, happening at least 40 msec after exposure to a loud sound. It thus provides protection only from prolonged loud sounds, not from sudden loud sounds like an explosion.

Hair cells in the organ of Corti transduce fluid movements into neural signals.

The snail-shaped cochlear portion of the inner ear is a coiled tubular system lying deep within the temporal bone (Fig. 6–29). It is easier to understand the functional components of the **cochlea** by unrolling it, as in Figure 6–32a. The cochlea is divided throughout most of its length into three fluid-filled longitudinal compartments. A blind-ended **cochlear duct** (also known as the **scala media**) tunnels lengthwise through the center of the cochlea, almost but not quite reaching its end. The upper compartment, the **scala vestibuli** (Figs. 6–32a and b) follows the inner contours of the spiral, and the **scala tympani,** the lower compartment, follows the outer contours. The fluid within the cochlear duct is called **endolymph** (Fig. 6–33a). The scala vestibuli and scala tympani both contain a slightly different fluid, the **perilymph,** which is continuous beyond the tip of the cochlear duct at a region called the **helicotrema.** The scala vestibuli is sealed from the middle ear cavity by the oval window, to which the stapes is attached. Another small membrane-covered opening, the **round window,** seals the scala tympani from the middle ear. The thin **vestibular membrane** separates the cochlear duct from the scala vestibuli. The **basilar membrane** forms the floor of the cochlear duct, separating it from the scala tympani. Importantly, the basilar membrane bears the **organ of Corti,** the sense organ for hearing.

The organ of Corti, which rests on top of the basilar membrane throughout its full length, contains **hair cells** that are the receptors for sound. Hair cells generate neural signals when their actin-stiffened surface hairs (a specialized form of microvilli—see p. 39) are mechanically deformed in association with fluid movements in the inner ear. These hairs are mechanically embedded in the **tectorial membrane,** an awning-like projection overhanging the organ of Corti throughout its length (Fig. 6–32c).

The pistonlike action of the stapes against the oval window sets up pressure waves in the scala vestibuli. Because fluid is incompressible, there are two ways pressure is dissipated as the stapes causes the oval window to bulge inward: (1) displacement of the round window and (2) deflection of the basilar membrane (Fig. 6–33a). In the first pathway, the pressure wave pushes the perilymph forward in the scala vestibuli, then around the helicotrema and into the scala tympani, where it causes the round window to bulge outward into the inner ear cavity to compensate for the pressure increase. As the stapes rocks backward and pulls the oval window outward toward the middle ear, movement of the perilymph shifts in the opposite direction, displacing the round window inward. This pathway does not result in sound reception; it just dissipates pressure.

Pressure waves of frequencies associated with sound reception take a "short cut." Pressure waves in the scala vestibuli are transferred through the thin vestibular membrane into the cochlear duct and then through the basilar membrane into the scala tympani, where they cause the round window to alternately bulge outward and inward. The main difference in this pathway is that transmission of pressure waves through the basilar membrane causes this membrane to move up and down, or vibrate, in synchrony with the pressure wave. Since the organ of Corti rides on the basilar membrane, the hair cells also move up and down as the basilar membrane oscillates. Because the hairs of the receptor cells are embedded in the stiff, stationary tectorial membrane, they are bent back and forth when the oscillating basilar membrane shifts their position in relationship to the tectorial membrane (Fig. 6–34). This back-and-forth mechanical deformation of the hairs alternately opens and closes ion channels in the hair cell through unknown means, resulting in alternating depolarizing

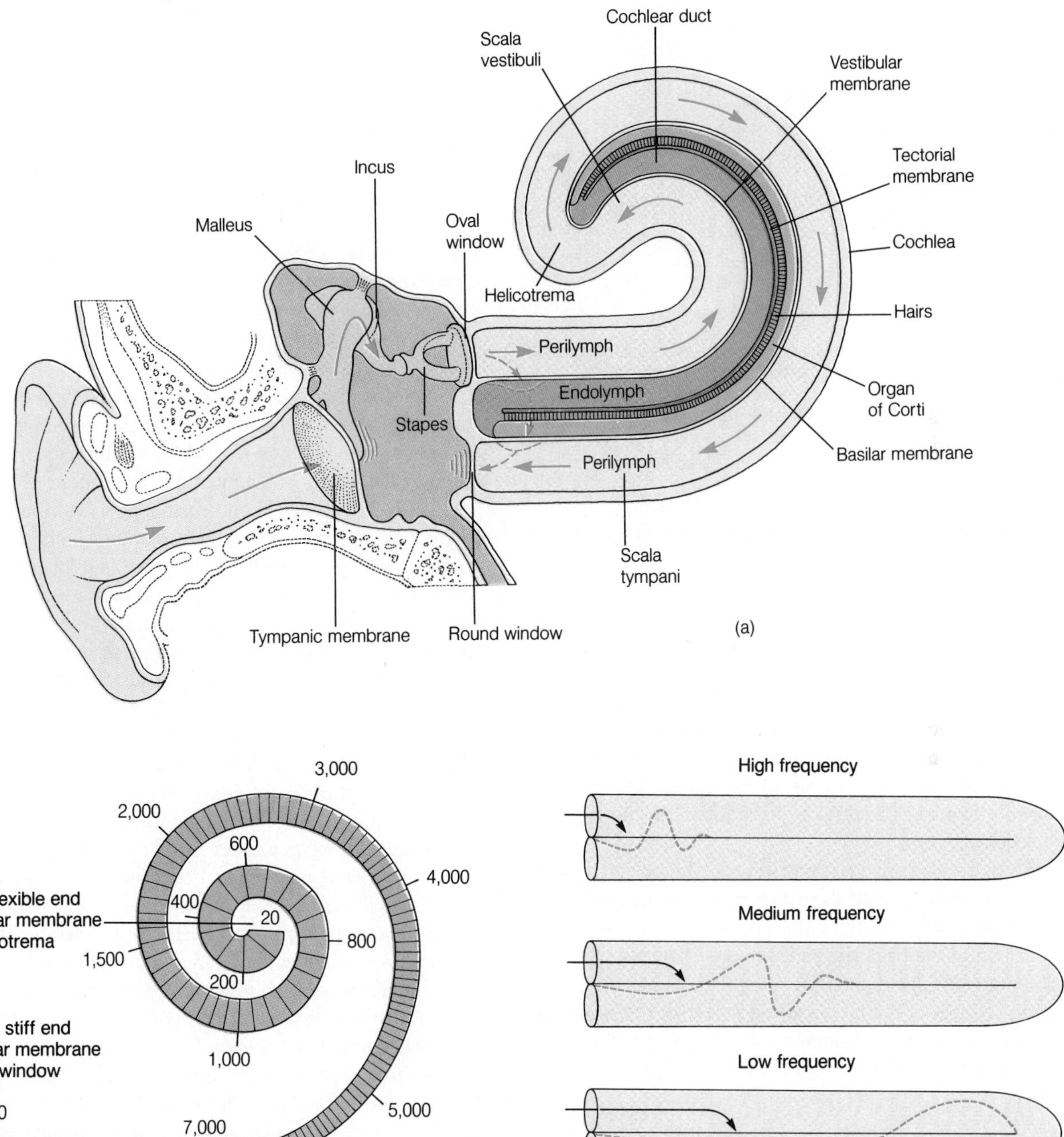

The numbers indicate the frequencies with which different regions of the basilar membrane maximally vibrate.

Figure 6–33 Transmission of Sound Waves *(a) Fluid movement within the perilymph set up by vibration of the oval window follows two pathways: (1) through the scala vestibuli, around the helicotrema, and through the scala tympani, causing the round window to vibrate; and (2) a "short-cut" from the scala vestibuli through the basilar membrane to the scala tympani. The first pathway just dissipates sound energy, but the second pathway triggers activation of the receptors for sound by bending the hairs of the hair cells as the organ of Corti on top of the vibrating basilar membrane is displaced in relation to the overlying tectorial membrane. (b) Different regions of the basilar membrane vibrate maximally at different frequencies. (c) The narrow, stiff end of the basilar membrane nearest the oval window vibrates best with high-frequency pitches. The wide, flexible end of the basilar membrane by the helicotrema vibrates best with low-frequency pitches.*

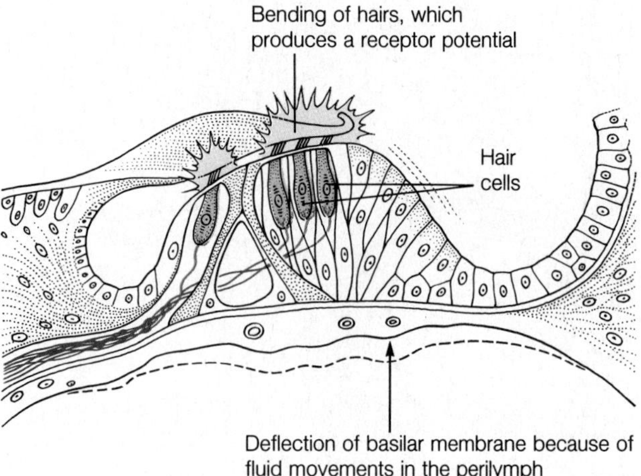

Figure 6–34 Bending of Hair Cells upon Deflection of Basilar Membrane

and hyperpolarizing potential changes—the receptor potential—at the same frequency as the original sound stimulus.

Hair cells are specialized receptor cells that communicate via a chemical synapse with the terminals of afferent nerve fibers making up the **auditory (cochlear) nerve.** Depolarization of hair cells (when the basilar membrane is deflected upward) increases their rate of transmitter release, which steps up the rate of firing in the afferent fibers. Conversely, the firing rate decreases as the hair cells release less transmitter when they are hyperpolarized upon displacement in the opposite direction.

Thus, the ear converts sound waves in the air into oscillating movements of the basilar membrane that bend the hairs of the receptor cells back and forth. This shifting mechanical deformation of the hairs alternately swings channels open and closed to bring about potential changes in the receptor, leading to changes in the rate of action potentials propagated to the brain. In this way, sound waves are translated into neural signals that can be perceived by the brain as sound sensations.

Pitch discrimination depends on the place on the basilar membrane that vibrates; loudness discrimination depends on the amplitude of the vibration.

Pitch discrimination (that is, the ability to distinguish between various frequencies of incoming sound waves) depends on the shape and properties of the basilar membrane, which is narrow and stiff at its oval window end and wide and flexible at its helicotrema end (Fig. 6–33b). Different regions of the basilar membrane naturally resonate or vibrate maximally at different frequencies; that is, each frequency displays peak vibration at different positions along the membrane. The narrow end nearest the oval window vibrates best with high-frequency pitches, whereas the wide end nearest the helicotrema vibrates maximally with low-frequency tones (Fig. 6–33c). Those pitches in between are sorted out along the length of the membrane from higher to lower frequency. As a sound wave of a particular frequency is set up in the cochlea by oscillation of the stapes, the wave travels to where the basilar membrane naturally responds maximally to that frequency. The energy of the pressure wave is dissipated with this vigorous membrane oscillation, so the wave dies out at the region of maximal displacement.

The hair cells in the region of peak vibration of the basilar membrane undergo the most mechanical deformation and, accordingly, are the most excited. This information is propagated to the CNS, which interprets the pattern of hair cell stimulation as a sound of a particular frequency. Modern techniques have determined that the basilar membrane is so fine-tuned that the peak membrane response to a single pitch extends probably no more than the width of a few hair cells. Overtones of varying frequencies cause many points along the basilar membrane to vibrate simultaneously but less intensely than the fundamental tone. This enables the CNS to distinguish the timbre of the sound.

Intensity (loudness) discrimination depends on the amplitude of vibration. As sound waves originating from louder sound sources strike the eardrum, they cause it to vibrate more vigorously (that is, bulge in and out to a greater extent) but at the same frequency as a softer sound of the same pitch. This greater tympanic membrane deflection is converted into greater amplitude of basilar membrane movement in the region of peak responsiveness. Greater basilar membrane oscillation is interpreted by the CNS as a louder sound.

The auditory system is so sensitive and it can detect sounds so faint that the distance of basilar membrane deflection is comparable to only a fraction of the diameter of a hydrogen atom, the smallest of atoms. It is no wonder that very loud sounds, which cannot be sufficiently attenuated by protective middle-ear reflexes (for example, the sounds of a typical rock concert), can set up such violent vibrations of the basilar membrane that irreplaceable hair cells are actually sheared off or permanently distorted, leading to partial hearing loss (Fig. 6–35).

The auditory cortex is mapped according to tone.

Just as various regions of the basilar membrane are associated with particular tones, the auditory cortex is also *tonotopically*

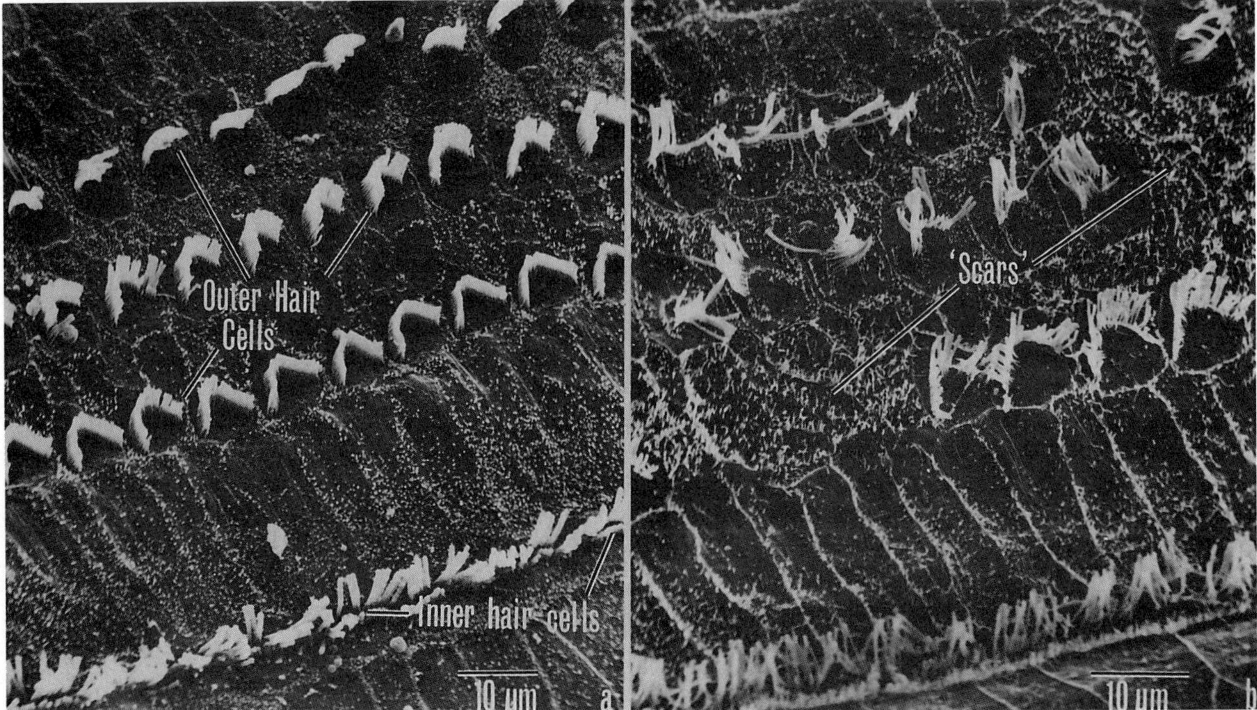

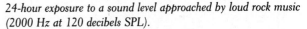

Figure 6–35 Loss of Hair Cells Because of Loud Noises

Effect of intense sound on the inner ear. (a) Organ of Corti of a normal guinea pig, showing three rows of outer hair cells and one row of inner hair cells. (b) Organ of Corti of a guinea pig after 24-hour exposure to a sound level approached by loud rock music (2000 Hz at 120 decibels SPL).

SOURCE: Scanning electron micrograph by Robert S. Preston, courtesy of Prof. J.E. Hawkins, Kresge Hearing Research Institute, University of Michigan Medical School.

organized. Each region of the basilar membrane is linked to a specific region of the auditory cortex in the temporal lobe. Accordingly, specific cortical neurons are activated only by particular tones; each region of the auditory cortex becomes excited only in response to a specific tone detected by a selected portion of the basilar membrane.

The afferent neurons that pick up the auditory signals from the hair cells exit the cochlea via the auditory nerve. The neural pathway between the organ of Corti and the auditory cortex involves several synapses en route, the most notable of which are in the brain stem and medial geniculate nucleus of the thalamus. The brain stem uses the auditory input for alertness and arousal. The thalamus sorts and relays the signals upward. Unlike the visual pathways, auditory signals from each ear are transmitted to both temporal lobes because of partial crossing over of fibers in the brain stem. For this reason, a disruption of the auditory pathway on one side beyond the brain stem does not affect hearing in either ear to any extent.

The **primary auditory cortex** is organized into columns, as elsewhere in the cortex. No hierarchy of cells within the columns that respond to increasingly complex auditory signals, similar to the visual system, has been identified. The primary auditory cortex appears to perceive discrete sounds while the surrounding higher-order auditory cortex integrates the separate sounds into a coherent, meaningful pattern. Think about the complexity of the task accomplished by your auditory system. When you are at a concert, your organ of Corti responds to the simultaneous mixture of the instruments, the applause and hushed talking of the audience, and the background noises in the theater. You are able to distinguish these separate parts of the many sound waves reaching your ears and pay attention to those of importance to you.

Deafness is caused by defects either in conduction or neural processing of sound waves.

Loss of hearing, or **deafness,** may be temporary or permanent, partial or complete. Deafness is classified into two types—conduction deafness and nerve deafness—depending on the part of the hearing mechanism that fails to function adequately. **Conduction deafness** occurs when sound waves are not adequately conducted through the external and middle portions of the ear to set the fluids in the inner ear in motion. Possible causes include physical blockage of the ear canal with earwax, rupture of the eardrum, ear infections with accompanying fluid accumulation, restriction of the ossicular movement because of adhesions between the bones, or damage to the oval window. In **nerve deafness,** the sound waves are transmitted to the inner ear, but they are not translated into nerve signals that are interpreted by the brain as sound sensations. The defect can lie in the organ of Corti, in the auditory nerves or ascending auditory pathways, or in the auditory cortex.

Hearing aids are helpful in noncurable cases of conduction deafness but are not beneficial for nerve deafness. These devices bypass a defective conduction system by amplifying and transmitting the sound waves through the bone to the inner ear to set up fluid motion. However, the receptor cell-neural pathway system must still be intact for the sound to be perceived.

The vestibular apparatus detects position and motion of the head important for equilibrium and coordination of head, eye, and body movements.

In addition to its cochlear-dependent role in hearing, the inner ear has another specialized component, the **vestibular apparatus,** which provides information essential for the sense of equilibrium and for coordinating head movements with eye and postural movements. The vestibular apparatus consists of two sets of structures lying within a tunneled-out region of the temporal bone near the cochlea—the semicircular canals and the otolith organs, namely, the utricle and saccule (Fig. 6–36a).

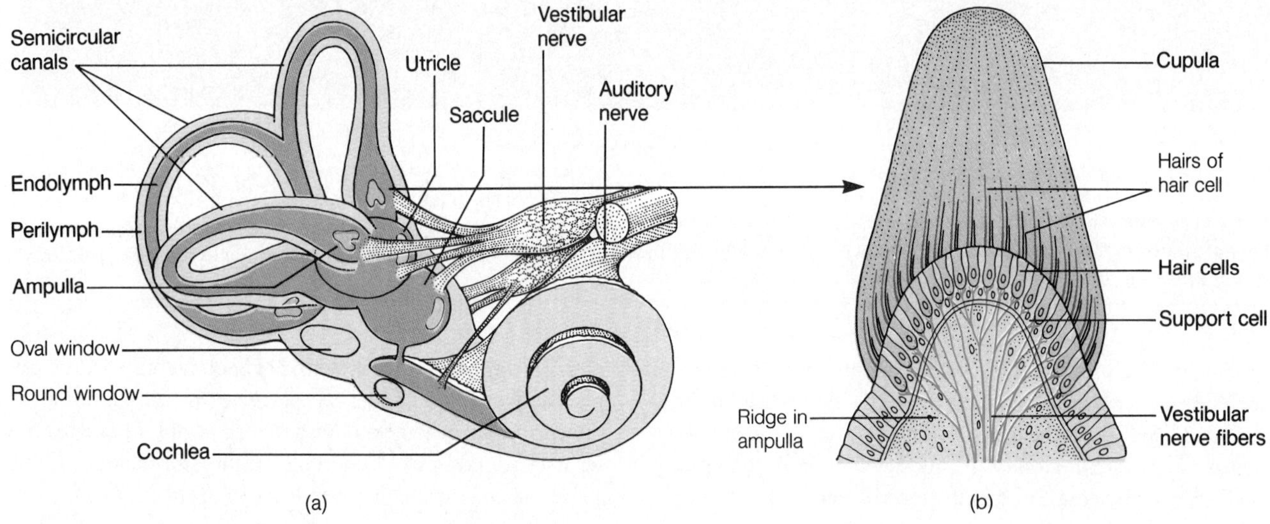

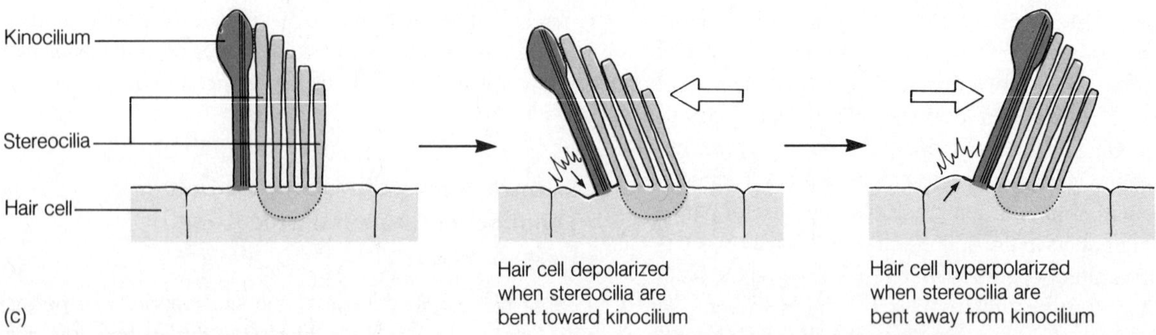

Figure 6–36 Vestibular Apparatus *(a) Gross anatomy of vestibular apparatus. (b) Receptor cell unit in ampulla of semicircular canals. (c) "Hairs" in sensory hair cells of semicircular canals.*

The vestibular apparatus detects changes in position and motion of the head. All components of the vestibular apparatus contain endolymph and are surrounded by perilymph. Similar to the organ of Corti, they each contain hair cells that respond to mechanical deformation triggered by specific movements of the endolymph. Also like the auditory hair cells, the vestibular receptors may either be depolarized or hyperpolarized, depending on the direction of the fluid movement. Unlike the auditory system, however, much of the information provided by the vestibular apparatus does not reach the level of conscious awareness.

The **semicircular canals** detect rotational or angular acceleration or deceleration of the head, such as when spinning, somersaulting, or turning the head. Each ear contains three semicircular canals arranged three-dimensionally in planes that lie at right angles to each other. The receptive hair cells of each semicircular canal are situated on top of a ridge located in the **ampulla,** a swelling at the base of the canal (Figs. 6–36a and b). The hairs are embedded in an overlying caplike layer, the **cupula,** which protrudes into the endolymph within the ampulla. The cupula sways in the direction of endolymph movement, much like seaweed leaning in the direction of the prevailing tide.

Acceleration or deceleration during rotation of the head in any direction causes endolymph movement in at least one of the semicircular canals because of their three-dimensional arrangement. As the head starts to move, the bony canal and the ridge of hair cells embedded in the cupula move with the head. However, the endolymph within the canal, not being attached to the skull, does not move intially but lags behind because of its inertia. (Because of inertia, if an object is resting, it remains at rest, whereas if it is moving, it continues to move in the same direction unless it is acted on by some external force that induces change.) When the endolymph is left behind as the head starts to rotate, this in effect shifts the endolymph that is in the same plane as the movement in the opposite direction of the head movement (similar to tilting of your body to the right as the car in which you are riding is suddenly turned to the left) (Fig. 6–37). This fluid movement causes the cupula to lean in the opposite direction of the head movement, bending the sensory hairs embedded in it. If the head movement continues at the same rate in the same direction, the endolymph catches up and moves in unison with the head so that the hairs return to their unbent position. When the head slows down and stops, the reverse situation occurs. The endolymph briefly continues to move in the direction of the rotation while the head decelerates to a stop. As a result, the cupula and its hairs are transiently bent in the direction of the preceding spin, which is opposite to the way they were bent during acceleration. When the endolymph gradually comes to a halt, the hairs straighten again. Thus the semicircular canals detect changes in the rate of rotational movement of the head. They do not respond when motionless or during circular motion at a constant speed.

The hairs of a vestibular hair cell consist of twenty to fifty **stereocilia,** which are actin-stiffened microvilli, and one cilium (see p. 35), the **kinocilium** (Fig. 6–36c). Each hair cell is oriented so that it depolarizes when its stereocilia are bent towards the kinocilium; bending in the opposite direction hyperpolarizes the cell. The hair cells form a chemically-mediated synapse with terminal endings of afferent neurons whose axons join with those of the other vestibular structures to form the **vestibular nerve.** This nerve unites with the auditory nerve from the cochlea to form the **vestibulocochlear nerve.** Depolarization of the hair cells increases the rate of firing in the afferent fibers; conversely, when the hair cells are hyperpolarized, the frequency of action potentials in the afferent fibers is reduced.

Whereas the semicircular canals provide the CNS with information about rotational changes in head movement, the **otolith organs** provide information about the position of the head relative to gravity and also detect changes in rate of linear motion (moving in a straight line regardless of direction). The **utricle** and **saccule** are saclike structures housed within a

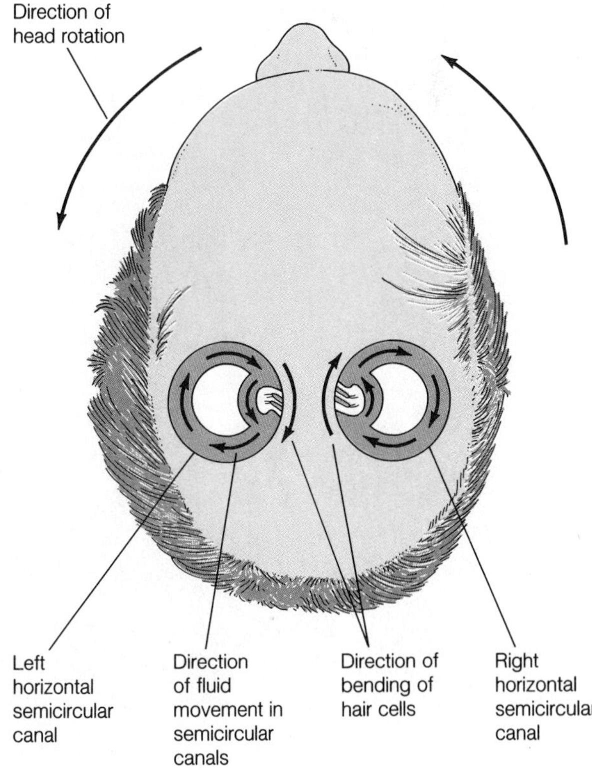

Figure 6–37 Activation of Hair Cells in Semicircular Canals

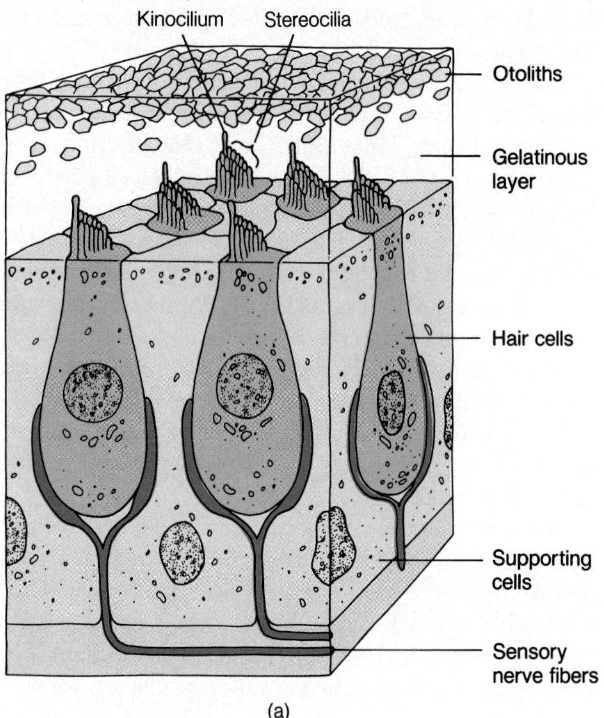

bony chamber situated between the semicircular canals and the cochlea (Fig. 6–36a). The hairs of the receptive hair cells in these sense organs also protrude into an overlying gelatinous sheet, whose movement displaces the hairs and results in changes in hair cell potential. Many tiny crystals of calcium carbonate—the ***otoliths*** ("ear stones")—are suspended within the gelatinous layer, making it heavier and giving it more inertia than the surrounding endolymph (Fig. 6–38a). When a person is in an upright position, the hairs within the utricles are oriented vertically and the saccule hairs are lined up horizontally.

Using the utricle as an example, its otolith-embedded gelatinous mass shifts positions and bends the hairs in two ways:

1. When the head is tilted in any direction other than straight up and down (Fig. 6–38b), the hairs are bent in the direction of the tilt because of the gravitational force exerted on the top-heavy gelatinous layer. Within the utricles on each side of the head, some of the hair cell bundles are oriented to depolarize and others to hyperpolarize when the head is in any position other than upright. The CNS thus receives different patterns of neural activity depending on head position with respect to gravity.

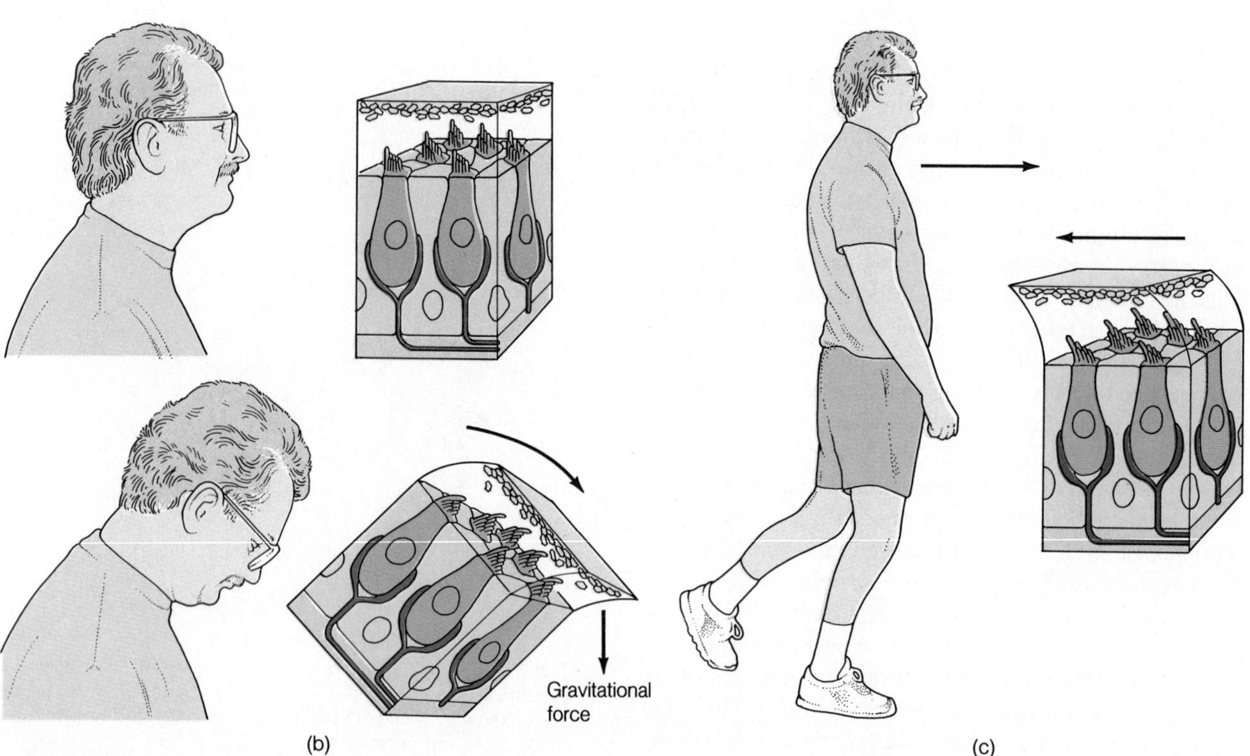

Figure 6-38 Utricle *(a) Receptor unit in utricle. (b) Activation of utricle by change of head position. (c) Activation of utricle by horizontal linear acceleration.*

2. The utricle hairs are also displaced by any change in horizontal linear motion (such as moving straight forward, backward, or to the side). As a person starts to walk forward (Fig. 6–38c), the top-heavy otolith membrane at first lags behind the endolymph and hair cells because of its greater inertia. This bends the hairs to the rear, in the opposite direction of the forward movement of the head. If the walking pace is maintained, the gelatinous layer soon catches up and moves at the same rate as the head so that the hairs are no longer bent. When the person stops walking, the otolith sheet continues to move forward briefly as the head slows and stops, bending the hairs toward the front. Thus the hair cells of the utricle detect horizontally directed linear acceleration and deceleration, but they do not provide information about movement in a straight line at constant speed.

The saccule functions similarly to the utricle, except that it responds selectively to the tilting of the head away from a horizontal position (lying down) and to vertically directed linear acceleration and deceleration (such as jumping up and down or riding in an elevator).

Signals arising from the vestibular apparatus are carried through the vestibulocochlear nerve to the **vestibular nuclei,** a cluster of neuronal cell bodies in the brain stem, and to the cerebellum. This vestibular information is integrated with cutaneous, proprioceptive, and visual input for maintaining balance and desired posture; controlling the external eye muscles so that the eyes remain fixed on the same point, despite movement of the head; and perceiving motion and orientation (Fig. 6–39).

Some individuals, for poorly understood reasons, are especially sensitive to particular motions that activate the vestibular apparatus and cause symptoms of dizziness and nausea; this sensitivity is called **motion sickness.** Occasionally, fluid imbalances within the inner ear lead to **Meniere's disease.** Not suprisingly, because both the vestibular apparatus and cochlea contain the same inner ear fluids, both vestibular and auditory symptoms occur. An afflicted individual suffers transient attacks of severe **vertigo** (dizziness) accompanied by pronounced ringing in the ears and some loss of hearing. During these episodes, the person cannot stand upright and reports feeling as though self or surrounding objects in the room are spinning around.

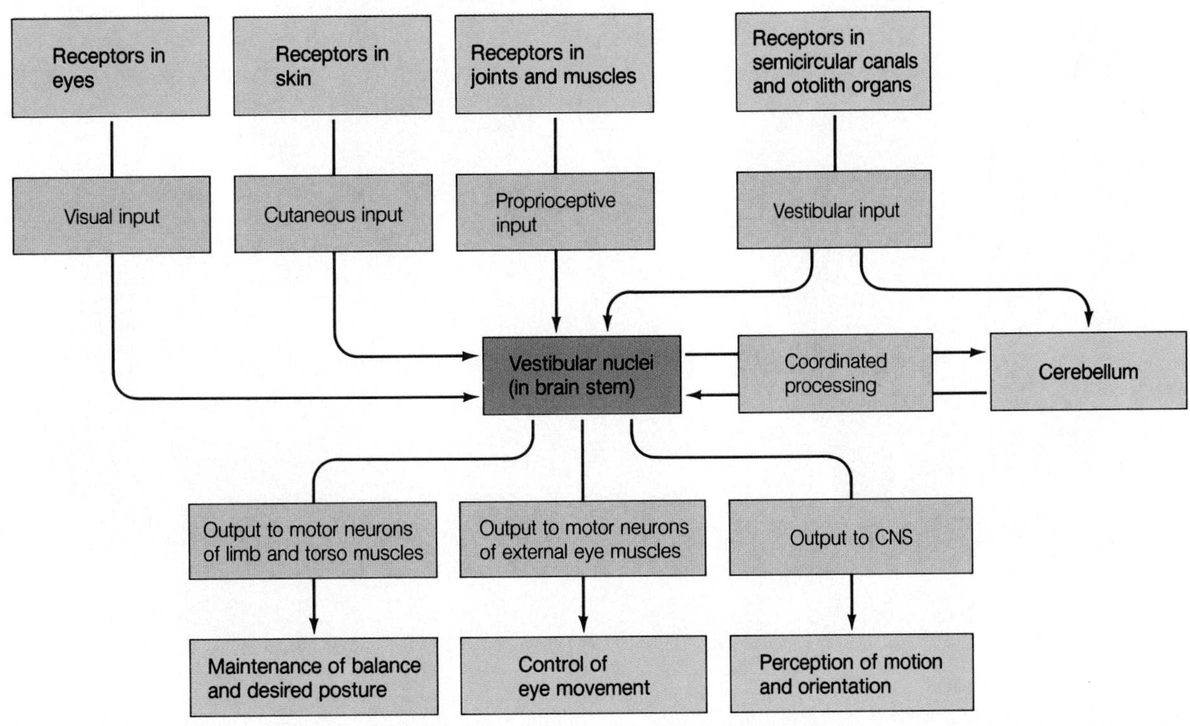

Figure 6–39 Input and Output of Vestibular Nuclei

Chemical Senses: Taste and Smell

The receptors for taste and smell are **chemoreceptors,** which generate neural signals upon binding with particular chemicals in their environment. The sensations of taste and smell in association with food intake influence the flow of digestive juices and affect appetite. Furthermore, stimulation of taste or smell receptors induces pleasurable or objectionable sensations and signals the presence of something to seek (a desirable food) or to avoid (a toxic substance). In lower animals, smell also plays a major role in finding direction, in seeking prey or avoiding predators, and in sexual attraction to a mate. The sense of smell is less sensitive in humans and much less important in influencing our behavior (although millions of dollars are spent annually on perfumes and deodorants to make us "smell" better and thereby be more socially attractive). We will first examine the mechanism of taste (**gustation**) and then turn our attention to smell (**olfaction**).

Taste sensation is coded by patterns of activity in various taste bud receptors.

The chemoreceptors for taste sensation are packaged within **taste buds** (Fig. 6–40), about ten thousand of which are present in the oral cavity and throat. The greatest percentage is found on the upper surface of the tongue. A taste bud consists of about fifty receptor cells packaged with supporting cells that are arranged together like slices of an orange. Each taste bud has a small opening, the **taste pore,** through which fluids in the mouth come into contact with the apical (top) surface of its receptor cells.

Taste receptor cells are modified epithelial cells bearing many surface folds, or microvilli, that protrude slightly through the taste pore, greatly increasing the surface area exposed to the oral contents. The plasma membrane of the microvilli contains receptor sites that bind selectively with chemical molecules in the environment. Only chemicals in solution—either ingested liquids or solids that have been dis-

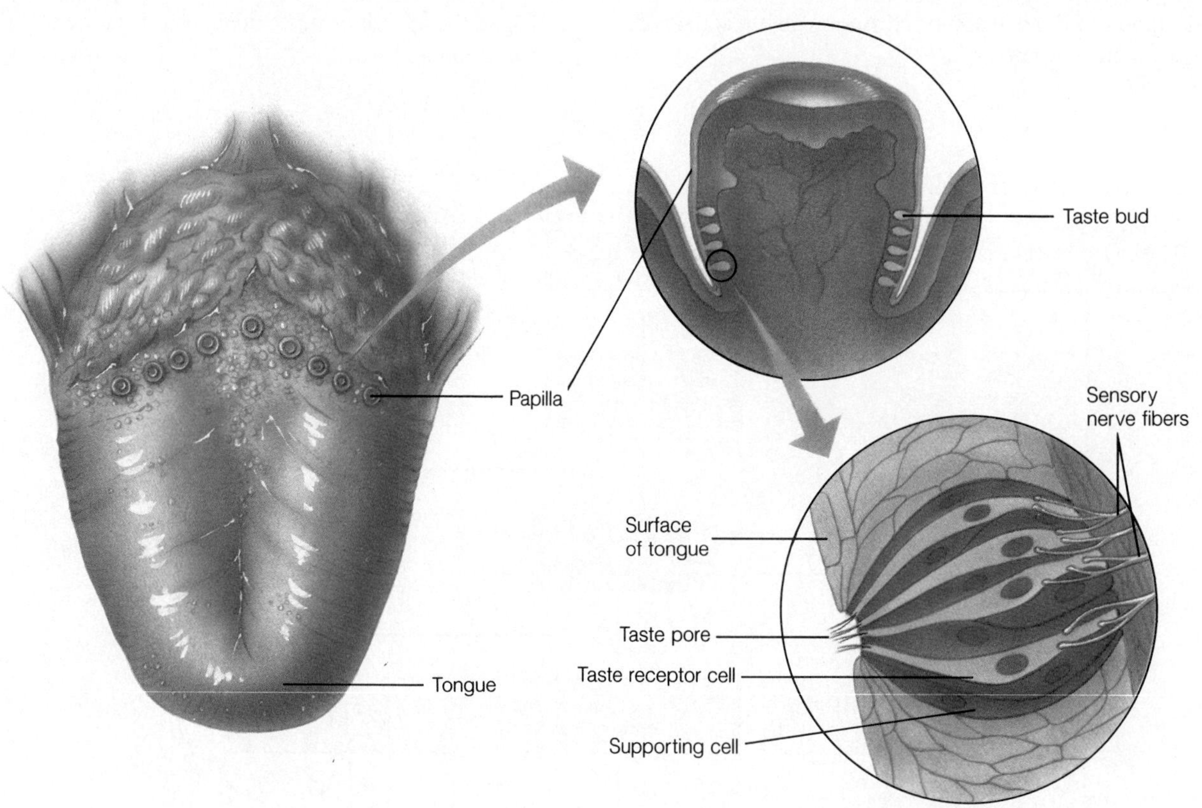

Figure 6–40 Location and Structure of Taste Buds *Taste buds are located primarily along the edges of moundlike papillae on the upper surface of the tongue. The receptor cells and supporting cells of a taste bud are arranged like slices of an orange.*

solved in saliva—can attach to receptor cells. By unknown means, binding of a taste-provoking chemical with a receptor cell alters its ionic channels to produce a depolarizing receptor potential. This receptor potential, in turn, initiates action potentials within terminal endings of afferent nerve fibers with which the receptor cell synapses.

Most receptors are carefully sheltered from direct exposure to the environment, but the taste receptor cells, by virtue of their task, frequently come into contact with potent chemicals. Unlike the eye or ear receptors, which are irreplaceable, taste receptors have a life span of about ten days. Epithelial cells surrounding the taste bud differentiate first into supporting cells and then into receptor cells to constantly renew the taste bud components.

Terminal afferent endings of several cranial nerves synapse with taste buds in various regions of the mouth. Signals in these sensory inputs are conveyed via synaptic stops in the brain stem and thalamus to the **cortical gustatory area,** a region in the parietal lobe adjacent to the "tongue" area of the somatosensory cortex. Unlike most sensory input, the gustatory pathways are primarily uncrossed. The brain stem also projects fibers to the hypothalamus and limbic system, presumably to add affective dimensions, such as whether the taste is pleasant or unpleasant, and to process behavioral aspects associated with taste and smell.

We can discriminate among thousands of different taste sensations, yet all tastes are varying combinations of four **primary tastes:** ***salty***, ***sour***, ***sweet***, and ***bitter***. Salt taste is stimulated by chemical salts, especially NaCl (table salt). Acids cause a sour taste. The citric acid content of lemons, for example, accounts for their distinctly sour taste. The sensation of sweetness is evoked by the particular configuration of glucose. Other organic molecules with similar structure can also interact with "sweet" receptor binding sites. Alkaloids (such as quinine, caffeine, nicotine, strychnine, morphine, and other toxic plant derivatives) or poisonous substances elicit a bitter taste, presumably as a protective mechanism to discourage ingestion of these potentially dangerous compounds.

Each receptor cell responds in varying degrees to all four primary tastes but is generally preferentially responsive to one of the taste modalities (Fig. 6–41). The richness of fine taste discrimination beyond the four primary tastes depends on subtle differences in the stimulation patterns of all the taste buds in response to various substances, similar to the variable stimulation of the three cone types that gives rise to the range of color sensations. Taste perception is also influenced by information derived from other receptors, especially odor. When you temporarily lose your sense of smell because of swollen nasal passageways during a cold, your sense of taste is also markedly reduced, even though your taste receptors are unaffected by the cold. Other factors affecting taste include temperature and texture of the food as well as psychological factors associated with past experiences with the food. How the cortex accomplishes the complex perceptual processing of taste sensation is beyond our scope of knowledge.

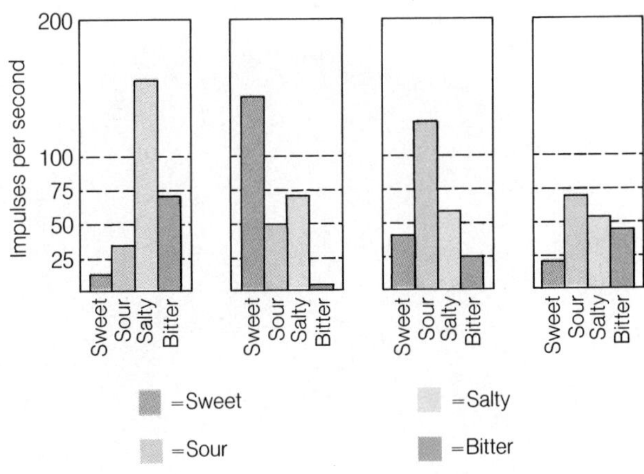

Figure 6–41 Relative Responsiveness of Different Taste Buds to Different Stimuli *Each taste bud responds in varying degrees to the four primary tastes. The pattern of stimulation for four different taste buds is depicted.*

Smell is the least understood of the special senses.

The **olfactory (smell) mucosa,** located in the ceiling of the nasal cavity, contains three cell types: olfactory receptors, supporting cells, and basal cells (Fig. 6–42). The supporting cells secrete **mucus,** which coats the nasal passages. The basal cells are precursors for new olfactory receptor cells, which are replaced about every two months. This is remarkable because, unlike the other special sense receptors, olfactory receptors are specialized endings of afferent neurons, not separate cells. The entire neuron, including its afferent axon projecting into the brain, is replaced. These are the only neurons that undergo cell division. The axons collectively form the **olfactory nerve.** The receptor portion of the olfactory neuron consists of an enlarged knob bearing several long cilia that extend to the surface of the mucosa. These cilia contain the binding sites for attachment of odoriferous molecules. Since the olfactory mucosa is above the normal path of air flow, during quiet breathing odorants typically reach the sensitive receptors only by diffusion. The act of sniffing enhances this process by drawing the air currents upward within the nasal cavity so that

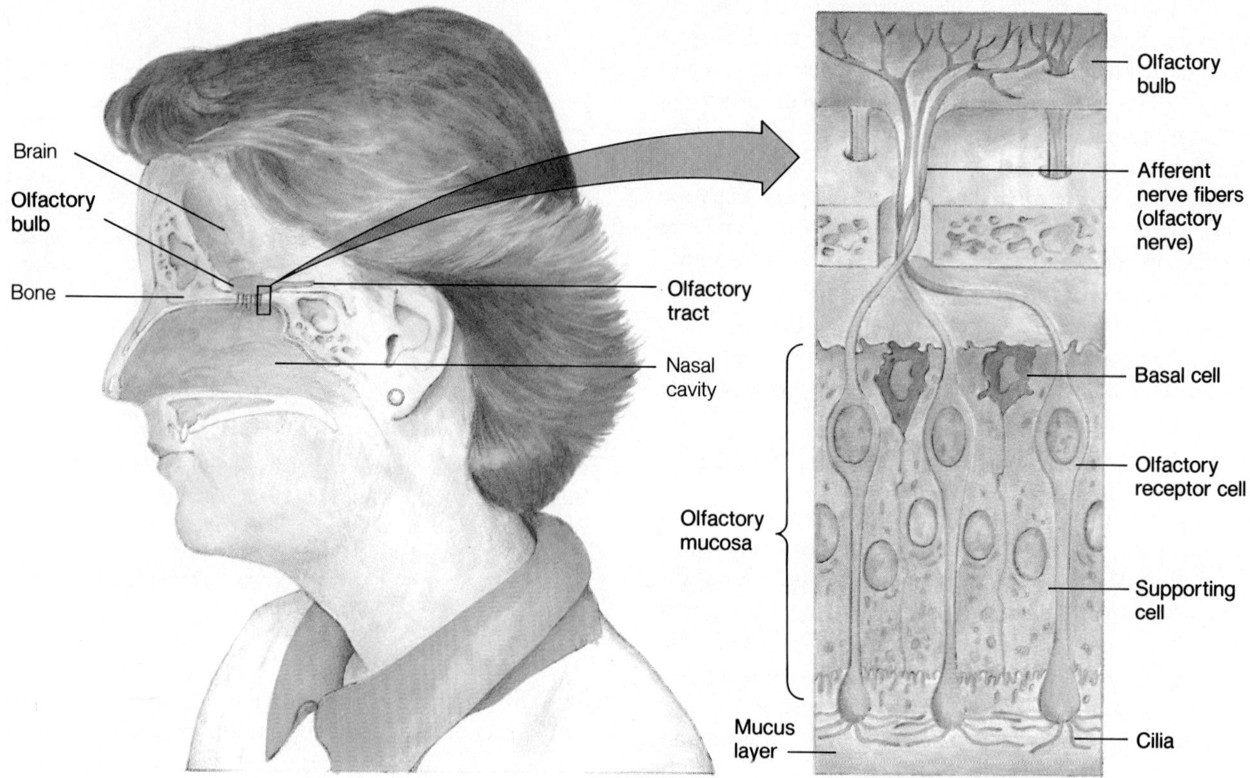

Figure 6–42 Location and Structure of the Olfactory Receptors

a greater percentage of the odoriferous molecules in the air come into contact with the olfactory mucosa.

To be smelled, a substance must be: (1) sufficiently volatile (easily vaporized) that some of its molecules can enter the nose in the inspired air, and (2) sufficiently water soluble that it can dissolve in the mucus layer coating the olfactory mucosa. As with taste receptors, molecules must be dissolved in order to be detected by olfactory receptors. Binding of an odoriferous molecule to a specialized attachment site on the cilia leads to opening of Na^+ and K^+ channels. This brings about a depolarizing receptor potential that generates action potentials in the remainder of the afferent neuron, the frequency of which depends on the concentration of the stimulating chemical molecules.

The afferent axons pass through tiny holes in the flat bone plate separating the olfactory mucosa from the overlying brain tissue (Fig. 6–42). They immediately synapse in the **olfactory bulb,** a complex neural structure containing several different layers of cells that are functionally similar to the retinal layers of the eye. Fibers leaving the olfactory bulb travel in two different routes (Fig. 6–43): (1) a *subcortical route* going primarily to regions of the limbic system, especially the lower medial sides of the temporal lobes (considered to be the **primary olfactory cortex**); and (2) a *thalamic-cortical route.* Until recently, the limbic route was thought to be the only olfactory pathway. This route, which includes hypothalamic involvement, permits close coordination between smell and behavioral reactions associated with feeding, mating, and direction orienting. The thalamic-cortical route, as with other senses, is important for conscious perception and for fine discrimination of smell.

The physiological mechanism of smell discrimination is far from being understood. Humans can distinguish tens of thousands of different odors. It is generally believed that perception of these various odors depends on combinations of **primary odors,** similar to color vision and taste discrimination. However, there is no general agreement on how many or what the primary odors are. One proposed system suggests at least seven: *floral, musky, camphoraceous, pepperminty, ethereal, pungent,* and *putrid.* According to the leading theory of odor, molecules of similar odor share a particular configuration in common, not a similar chemical composition. Accordingly,

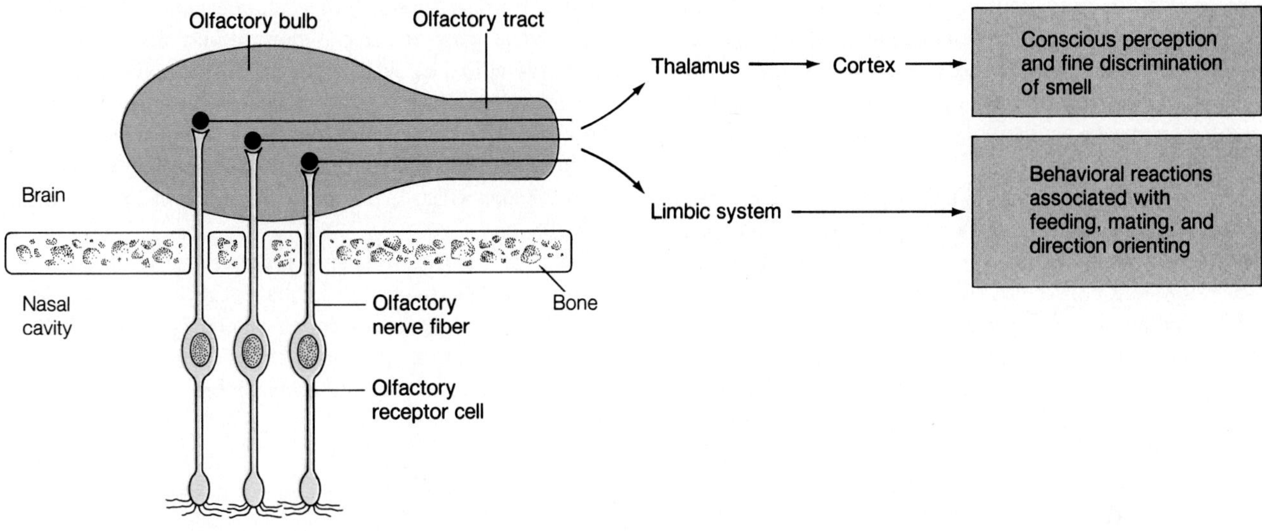

Figure 6–43 Olfactory Pathways

each type of receptor binding site is speculated to have a distinct shape and size (a lock) that matches the configuration of a particular primary odor (the key).

Whatever the mechanism for sorting out and distinguishing different odors, it is extremely effective, even in humans, who have a poor sense of smell compared to other species. A noteworthy example is our ability to detect methyl mercaptan (garlic odor) at a concentration of 1 molecule per 50,000 million molecules of air! This substance is added to odorless natural gas to enable us to detect potentially lethal gas leaks.

Even though the olfactory system is sensitive and highly discriminating, it is also quickly adaptive. Our sensitivity to a new odor rapidly diminishes after a short period of exposure to it. This reduced sensitivity does not involve receptor adaptation, as has been thought for years; actually, the olfactory receptors themselves are slowly adapting. It apparently involves some sort of adaptation process in the CNS. Adaptation is specific for a particular odor, and responsiveness to other odors remains unchanged.

Chapter in Perspective

The afferent nervous system detects, encodes, and transmits peripheral signals to the central nervous system for processing. It is the communication link by which the central nervous system is informed about the internal and external environment. Afferent information about the internal environment never reaches the level of conscious awareness, but this input to the controlling centers of the CNS is essential for the maintenance of homeostasis. Afferent input reaching the level of conscious awareness is known as sensory input and includes somesthetic and proprioceptive sensation (body sense) and special senses (vision, hearing, taste, and smell). Final perceptual processing of sensory input not only is essential for interaction with the environment for basic survival (for example, food procurement and defense from danger) but also adds immeasurably to the richness of life.

Receptors are specialized peripheral devices that respond to a specific stimulus modality and convert its energy into a graded receptor potential, which, once threshold is reached, generates action potentials in the afferent fiber. Even though all signals reaching the brain are identical (action potentials), the CNS can distinguish what is happening, where, and how much. Afferent fibers originating from various receptor types in different regions of the body travel in separate pathways to specific regions of the cortex to enable discrimination of the type and location of stimulus. Stimulus intensity is encoded by the frequency of action potentials generated. What the brain perceives from its input, however, is an abstraction, not reality. The only stimuli that can be detected are those for which receptors are present. Furthermore, as sensory signals ascend through progressively more complex processing, some of the information may be suppressed, whereas other parts of it may be enhanced.

Pain is unique among the various somatosensory modalities in that it is accompanied by motivational and emotional responses. A built-in system involving specific neural regions and endogenious opiates can suppress the sensation of pain.

Fine discrimination of special sensory input depends on the comparison of different patterns of stimulation of the receptors within the sense organ. Much of the anatomical structure of the eyes and ears is designed for collecting and transmitting stimuli to the sensitive receptor cells buried deep within these special sense organs. The light-sensitive receptors—the rods and cones—are located in the retina at the posterior aspect of the eyeball. The other structures of the eye bend incoming light rays to bring them into focus on the retina. Absorption of light causes the breakdown of photopigments, which brings about a receptor potential.

In contrast to the photoreceptors of the eye, the ear receptors located in the inner ear—the hair cells in the cochlea and vestibular apparatus—are mechanoreceptors. The external and middle parts of the ear transmit and amplify airborne sound waves to the fluid in the cochlea. This movement bends the hairs of the organ of Corti's hair cells, leading to a receptor potential. The hair cells in the vestibular apparatus are activated when their hairs are distorted by specific movements of the head, which gives rise to information important for balance.

Taste and smell are chemical senses that require attachment of specific molecules in the environment to receptor cells in order to trigger a receptor potential. Taste and olfactory receptors are continuously renewed, unlike visual and hearing receptors, which are irreplaceable.

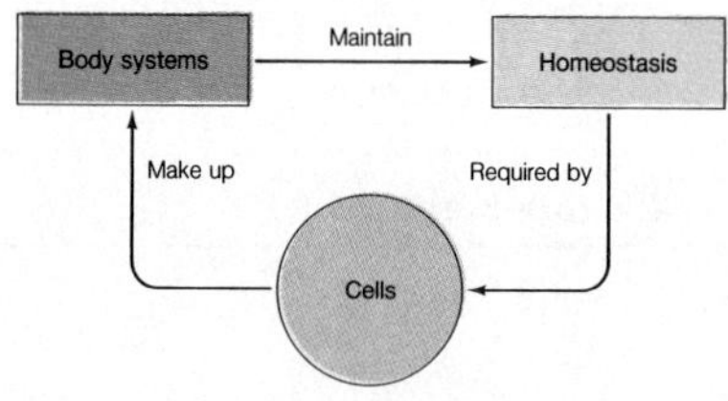

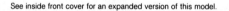

See inside front cover for an expanded version of this model.

Review Exercises

1. List and describe the receptor types according to their adequate stimulus.
2. Distinguish between a receptor potential and a generator potential.
3. Compare tonic and phasic receptors.
4. Explain how acuity is influenced by receptive field size and by lateral inhibition.
5. Compare the fast and slow pain pathways.
6. Describe the built-in analgesic system of the brain.
7. Describe the function of each of the following parts of the eye: iris, cornea, lens, and ciliary muscle.
8. Explain the defect in each of the following visual disorders: presbyopia, myopia, hyperopia, and diplopia.
9. Describe the process of phototransduction.
10. Compare the functional characteristics of rods and cones.
11. Compare the visual pathway from the eyes to the occipital lobe with the auditory pathway from the ears to the temporal lobe.
12. What are sound waves? What is responsible for the pitch, intensity, and timbre of a sound?
13. Describe the function of each of the following parts of the ear: pinna, ear canal, tympanic membrane, ossicles, oval window, and the various parts of the cochlea. Include a discussion of how sound waves are transduced into action potentials.
14. Discuss the functions of the semicircular canals, the utricle, and the saccule.
15. Describe the location, structure, and activation of the receptors for taste and smell.
16. **A point to ponder:** Patients with certain nerve disorders are unable to feel pain. Why is this disadvantageous?

CHAPTER 7

Peripheral Nervous System: Efferent Division

Chapter Contents at a Glance

Introduction

Autonomic Nervous System

Anatomy of autonomic fibers

Autonomic neurotransmitters

Dominance patterns of sympathetic and parasympathetic systems

Autonomic receptor types

CNS control of autonomic activity

Somatic Nervous System

Alpha motor neurons

Final common pathway

Neuromuscular Junction

Anatomy; sequence of events

End-plate potentials

Role of acetylcholinesterase

Influence of specific chemical agents and diseases

INTRODUCTION *Efferent neurons carry directives from the central nervous system to the effector organs. The autonomic nervous system is the efferent division that supplies cardiac muscle, smooth muscle, and exocrine glands, whereas the somatic nervous system innervates skeletal muscle. How many different neurotransmitters would you quess are released from these efferent neuronal terminals to elicit essentially all the neurally controlled effector organ responses? Only two—acetylcholine and norepinephrine! Acting independently, they bring about such wide-ranging effects as salivary secretion, bladder contraction, and voluntary motor movements. This is a prime example of multiplicity of various tissues' responses to the same chemical messenger, depending on specialization of the effector organs.*

AUTONOMIC NERVOUS SYSTEM

An autonomic nerve pathway consists of a two-neuron chain, with the terminal neurotransmitter differing between sympathetic and parasympathetic nerves.

Each autonomic nerve pathway extending from the central nervous system (CNS) to an innervated organ consists of a two-neuron chain. The cell body of the first neuron in the series is located in the CNS. Its axon, the **preganglionic fiber,** synapses with the cell body of the second neuron that lies within a ganglion outside the CNS. The axon of the second neuron, the **postganglionic fiber,** innervates the *effector organ.*

The autonomic nervous system consists of two divisions—the **sympathetic** and the **parasympathetic nervous systems** (Table 7–1). Sympathetic nerve fibers originate in the thoracic and lumbar regions of the spinal cord (Fig. 7–1). Most sympathetic preganglionic fibers are very short, synapsing with cell bodies of postganglionic neurons within ganglia that lie in a **sympathetic ganglion chain** (the **sympathetic trunk**) located along either side of the spinal cord. Long postganglionic fibers originating in the ganglion chain terminate on the effector organs. Some preganglionic fibers pass through the ganglion chain without synapsing and terminate later in sympathetic **collateral ganglia** located about halfway between the CNS and innervated organs, with postganglionic fibers traveling the remainder of the distance.

Parasympathetic preganglionic fibers arise from the cranial and sacral areas of the CNS. (Some cranial nerves contain parasympathetic fibers.) These fibers are long in comparison to sympathetic preganglionic fibers because they do not end until they reach **terminal ganglia** that lie in or near the effector organs. Very short postganglionic fibers terminate on the cells of an organ itself.

Sympathetic and parasympathetic preganglionic fibers release the same neurotransmitter, **acetylcholine (ACh)**, but these two systems differ in the neurotransmitter they use at their postganglionic endings (the neurotransmitter that influences the effector organ). Parasympathetic postganglionic fibers release acetylcholine. Accordingly, they, along with all autonomic preganglionic fibers, are called **cholinergic fibers.** Most sympathetic postganglionic fibers, in contrast, are called **adrenergic fibers** because they release **noradrenaline,** commonly known as **norepinephrine.** Both acetylcholine and norepinephrine also serve as chemical messengers elsewhere in the body (Table 7–2).

Postganglionic autonomic fibers do not end in a single terminal swelling like a synaptic knob. Instead, the terminal branches of autonomic fibers contain numerous swellings, or **varicosities,** that simultaneously release neurotransmitter over a large area of the innervated organ rather than on single cells. This, coupled with the fact that any resulting change in electrical activity is spread throughout a smooth- or cardiac-muscle mass via gap junctions (see p. 69), means that whole

Table 7–1 Distinguishing Features of the Sympathetic and Parasympathetic Nervous Systems

Feature	Sympathetic System	Parasympathetic System
Origin of preganglionic fiber	Thoracic and lumbar regions of spinal cord	Brain and sacral regions of spinal cord
Origin of postganglionic fiber (location of ganglion)	Sympathetic ganglion chain (near spinal cord) or collateral ganglia (about halfway between spinal cord and effector organs)	Terminal ganglia (in or near effector organs)
Length and type of fiber	Short cholinergic preganglionic fibers Long adrenergic postganglionic fibers (most) Long cholinergic postganglionic fibers (few)	Long cholinergic preganglionic fibers Short cholinergic postganglionic fibers
Effector organs innervated	Cardiac muscle, almost all smooth muscle, and exocrine glands	Cardiac muscle, most smooth muscle, and exocrine glands
Types of receptors for neuro-transmitters	α, β_1, β_2	Nicotinic, muscarinic
Dominance	Dominates in emergency "fight-or-flight" situations; prepares body for strenuous physical activity	Dominates in quiet, relaxed situations; promotes "general housekeeping" activities such as digestion and emptying of urinary bladder
Types of discharge	Frequently mass discharge of whole system; may involve only discrete organs	Normally involves discrete organs rather than mass discharge

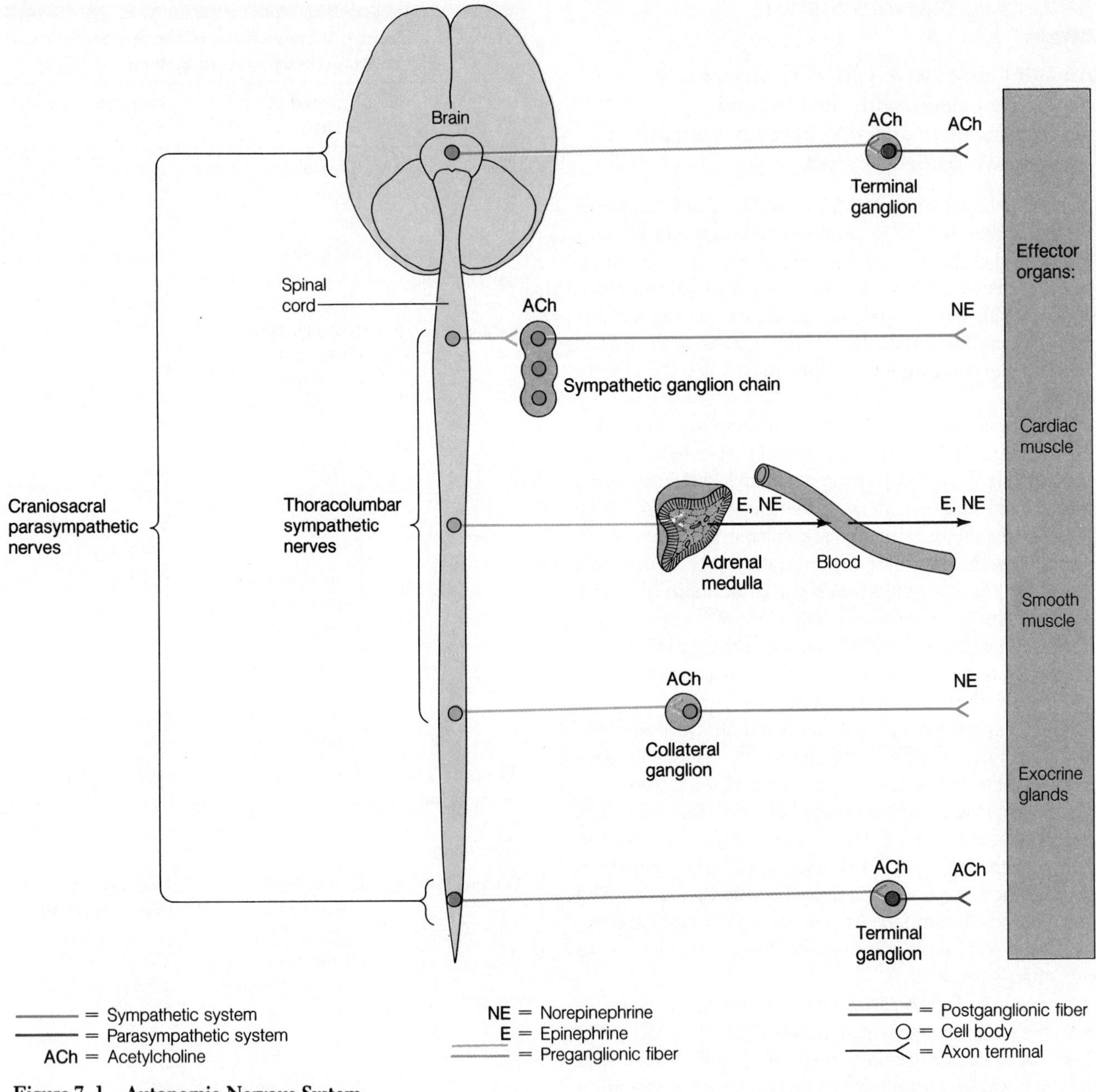

Figure 7–1 Autonomic Nervous System

organs instead of discrete cells are typically influenced by autonomic activity.

The autonomic nervous system controls involuntary visceral-organ activities.

The autonomic nervous system regulates visceral activities normally outside the realm of consciousness and voluntary control, such as circulation, digestion, sweating, and pupillary size. Accordingly, it is considered to be the *involuntary* branch of the efferent nervous system, in contrast to the somatic *voluntary* branch, which supplies skeletal muscles that can be consciously controlled. It is not entirely true, however, that an individual has no control over activities governed by the autonomic system. Visceral afferent information usually does not reach the conscious level, so individuals have no way of con-

Table 7-2 Sites of Release for Acetylcholine and Norepinephrine

Acetylcholine	Norepinephrine
All preganglionic terminals of the autonomic nervous system	Most sympathetic postganglionic terminals
All parasympathetic postganglionic terminals	Adrenal medulla
Sympathetic postganglionic terminals at sweat glands and some blood vessels in skeletal muscle	Central nervous system
Terminals of efferent neurons supplying skeletal muscle (motor neurons)	
Central nervous system	

sciously controlling the resultant efferent output. With the technique of **biofeedback,** however, people are provided with a conscious signal regarding visceral afferent information, such as a sound or a light. This enables them to voluntarily control events that are normally considered to be subconscious activities. For example, individuals have learned to consciously lower their blood pressure when they "hear" that it is elevated via special devices that convert blood-pressure levels into sound signals. Such biofeedback techniques are gaining wider acceptance and usage.

The sympathetic and parasympathetic nervous systems dually innervate most visceral organs.

Most visceral organs are innervated by both sympathetic and parasympathetic nerve fibers. Table 7-3 summarizes the major effects of these autonomic branches. Although the details of this wide array of autonomic responses are more fully described in later chapters, in which the involved organs are individually discussed, several general concepts can be derived now. As can be seen from the table, the sympathetic and parasympathetic nervous systems generally exert opposite effects in a particular organ. Sympathetic stimulation increases the heart rate, whereas parasympathetic stimulation decreases it; sympathetic stimulation slows down movement within the digestive tract, whereas parasympathetic stimulation enhances digestive motility. Note that one system is not always excitatory and the other always inhibitory. Both systems increase the activity of some organs and reduce the activity of others.

Rather than memorizing a list such as that presented in the table, it is better to logically deduce what the actions of the two systems are based on an understanding of the circumstances under which each system dominates. Usually, both systems are partially active; that is, there is normally some level of action potential activity in both the sympathetic and parasympathetic fibers supplying a particular organ. This ongoing activity is called **sympathetic** or **parasympathetic tone.** Under given circumstances, activity of one division can dominate the other. **Sympathetic dominance** to a particular organ exists when the sympathetic fibers' rate of firing to that organ increases above tonic level, coupled with a simultaneous decrease below tonic level in the parasympathetic fibers' frequency of action potentials to the same organ. The reverse situation is true for **parasympathetic dominance.** Shifts in balance between sympathetic and parasympathetic activity can be accomplished discretely for individual organs to meet specific demands, or a more generalized, widespread discharge of one system in favor of the other can be elicited to control bodywide functions. Massive widespread discharges take place more frequently in the sympathetic system. The value of this potential for massive sympathetic discharge is evident when considering the circumstances during which this system usually dominates.

The sympathetic system promotes responses that prepare the body for strenuous physical activity in the face of emergency or stressful situations, such as a physical threat from the outside environment. This is typically referred to as a **fight-or-flight response** because the sympathetic system readies the body to fight against or flee from the threat. Think about the body resources needed under such a circumstance. The heart beats more rapidly and more forcefully; blood pressure is elevated because of generalized constriction of the blood vessels; the respiratory airways open widely to permit maximal air flow; glycogen (stored sugar) and fat stores are broken down to release extra fuel into the blood; and blood vessels supplying skeletal muscles dilate (open more widely). All of these responses are aimed at providing increased flow of oxygenated, nutrient-rich blood to the skeletal muscles in anticipation of strenuous physical activity. Furthermore, the pupils dilate and the eyes are adjusted for far vision, enabling the person to quickly visually assess the entire threatening scene. Sweating is promoted in anticipation of excess heat production by the physical exertion. Because digestive and urinary activities are unessential in meeting the threat, the sympathetic system inhibits these activities.

The parasympathetic system, on the other hand, dominates in quiet, relaxed situations. Under such nonthreatening circumstances, the body can be concerned with its own "general housekeeping" activities, such as digestion and emptying of the urinary bladder. The parasympathetic system promotes these types of bodily functions while slowing down those activities that are enhanced by the sympathetic system. There is

Table 7–3 Effects of the Autonomic Nervous System on Various Organs

Organ	Type of Sympathetic Receptor	Effect of Sympathetic Stimulation	Effect of Parasympathetic Stimulation
Heart	β_1	Increased rate, increased force of contraction (of whole heart)	Decreased rate, decreased force of contraction (of atria only)
Blood vessels	α (most organs)	Constriction	Dilation of vessels supplying the penis and clitoris only
	β_2 (heart and skeletal-muscle vessels)	Dilation	
	*cholinergic (skeletal-muscle vessels)	Dilation	
Lungs	β_2 (airways)	Dilation of bronchioles (airways)	Constriction of bronchioles
	? (gland cells)	Inhibition (?) of mucus secretion	Stimulation of mucus secretion
Digestive tract	α, β_2 (organs)	Decreased motility (movement)	Increased motility
	α (sphincters)	Contraction of sphincters (to prevent forward movement of contents)	Relaxation of sphincters (to permit forward movement of contents)
	? (gland cells)	Inhibition (?) of digestive secretions	Stimulation of digestive secretions
Gall bladder	?	Relaxation	Contraction (emptying)
Urinary bladder	β_2	Relaxation	Contraction (emptying)
Eye	α (iris)	Dilation of pupil	Constriction of pupil
	β_2 (ciliary muscle)	Adjustment of eye for far vision	Adjustment of eye for near vision
Liver (glycogen stores)	β_2	Glycogenolysis (glucose released)	None
Adipose cells (fat stores)	β_2	Lipolysis (fatty acids released)	None
Exocrine glands	α (exocrine pancreas)	Inhibition of pancreatic exocrine secretion	Stimulation of pancreatic exocrine secretion (important for digestion)
	α, cholinergic (sweat glands)	Stimulation of secretion by sweat glands	None
	α (salivary glands)	Stimulation of small volume of thick saliva rich in mucus	Stimulation of large volume of watery saliva rich in enzymes
Adrenal medulla	Cholinergic	Stimulation of secretion of epinephrine and norepinephrine	None
Genitals	α	Ejaculation and orgasmic contractions (males); orgasmic contractions (females)	Erection (caused by dilation of blood vessels in penis [male] and clitoris [female])
Brain activity	?	Increased alertness	None

*Uncertain as to whether cholinergic sympathetic receptors are present in skeletal muscle vessels in humans, although they have been demonstrated in other species.

? Receptor type unknown.

no need, for example, to have the heart beating rapidly and forcefully when the person is in a tranquil setting.

What is the advantage of dual innervation of organs with nerve fibers whose actions oppose each other? It enables precise control over an organ's activity, similar to having both an accelerator and a brake to control the speed of a car. If you saw an animal suddenly start across the road as you were driving, you could eventually stop if you simply took your foot off the accelerator. However, you could come to a more rapid, controlled stop by simultaneously applying the brake as you lifted up on the accelerator. In a similar manner, a sympathetically accelerated heart rate could gradually be reduced to normal following a stressful situation by decreasing the rate of firing in the cardiac sympathetic nerve (letting up on the accelerator), but the heart rate can be reduced more rapidly by simultaneously increasing activity in the parasympathetic supply to the heart (applying the brake). Indeed, the two divisions of the autonomic nervous system are usually reciprocally controlled; increased activity in one division is accompanied by a corresponding decrease in the other.

There are several exceptions to this general rule of dual reciprocal innervation, the most notable of which are the following:

1. *Innervated blood vessels* (most arterioles and veins are innervated, arteries and capillaries are not) receive only sympathetic nerve fibers. Regulation is accomplished by increasing or decreasing the firing rate above or below the tonic level in these sympathetic fibers. The only blood vessels to receive parasympathetic fibers are those supplying the penis and clitoris. The precise vascular control this dual innervation affords these organs is important in accomplishing erection.
2. *Sweat glands* are innervated only by sympathetic nerves. The postganglionic fibers of these nerves are unusual because they secrete acetylcholine rather than norepinephrine.
3. *Salivary glands* are innervated by both autonomic divisions, but unlike elsewhere, sympathetic and parasympathetic activity is not antagonistic. Both stimulate salivary secretion, but the saliva's volume and composition differ, depending on which autonomic branch is dominant.
4. *The adrenal medulla,* the inner portion of the adrenal gland, is considered to be a modified sympathetic ganglion that does not give rise to postganglionic fibers. Instead, it secretes hormones into the blood upon stimulation by the preganglionic fiber that originates in the brain. Not surprisingly, the hormones are identical or similar to postganglionic sympathetic neurotransmitters. About 20% of adrenal medullary hormone output is norepinephrine, and the remaining 80% is the closely related substance, **epinephrine** (**adrenaline**). These hormones, in general, act to reinforce activity of the sympathetic nervous system.

There are several different types of membrane receptor proteins for each autonomic neurotransmitter.

Because each autonomic neurotransmitter and medullary hormone stimulates activity in some tissues but inhibits activity in others, the particular responses must depend on specialization of the tissue cells rather than on properties of the chemicals themselves. Responsive tissue cells possess one or more of several different types of plasma-membrane receptor proteins for these chemical messengers. Binding of a neurotransmitter to a receptor induces the tissue-specific reponse by means of a second messenger system within the cell (see p. 64).

Two types of acetylcholine (cholinergic) receptors—nicotinic and muscarinic receptors—have been identified on the basis of their response to particular drugs. **Nicotinic receptors** (activated by the tobacco plant derivative, nicotine) are found in all autonomic ganglia. They respond to acetylcholine released from both sympathetic and parasympathetic preganglionic fibers. **Muscarinic receptors** (activated by the mushroom poison, muscarine) are found on effector cell membranes (smooth muscle, cardiac muscle, and exocrine glands). They bind with acetylcholine released from parasympathetic postganglionic fibers.

There are two major classes of adrenergic receptors for norepinephrine and epinephrine based on the ability of various drugs to either initiate or prevent responses in the effector organ. These receptors are designated as **alpha (α)** and **beta (β) receptors,** with a further subclassification of the latter into **β_1 and β_2 receptors.** These various receptor types are distinctly distributed among the effector organs. Receptors of the β_2 type bind primarily with epinephrine, whereas β_1 and α receptors have about equal affinities for norepinephrine and epinephrine. Activation of α receptors usually brings about an excitatory response in the effector organ—for example, arteriolar constriction caused by increased contraction of the smooth muscle in the walls of these blood vessels. Stimulation of β_1 receptors, which are found primarily in the heart, also causes an excitatory response, namely increased rate and force of cardiac contraction. The response to β_2 receptor activation is generally inhibitory, such as arteriolar or bronchiolar (respiratory airway) dilation caused by relaxation of the smooth muscle in the walls of these tubular structures.

Because activation of various receptor types brings about different responses to the same autonomic messenger, fairly selective manipulations of these receptors by drugs are permitted. Drugs are available that selectively enhance or mimic (**ag-**

onists) or block (**antagonists**) autonomic responses at each of the receptor types. Some are only of experimental interest, but others are very important therapeutically. For example, **atropine** blocks the effect of acetylcholine at muscarinic receptors but does not affect nicotinic receptors. Since the acetylcholine released at both parasympathetic and sympathetic preganglionic fibers combines with nicotinic receptors, blockage at nicotinic synapses would knock out both of these autonomic branches. By acting selectively to interfere with acetylcholine action only at muscarinic junctions, which are the sites of parasympathetic postganglionic action, atropine effectively blocks parasympathetic effects while not influencing sympathetic activity at all. This principle is used to suppress salivary and bronchial secretions before surgery to reduce the risk of a patient inhaling these secretions into the lungs.

Likewise, drugs that act selectively at α and β adrenergic-receptor sites to either activate or block specific sympathetic effects are widely used. **Salbutamol** is an excellent example. It selectively activates β_2 adrenergic receptors at low doses, making it possible to dilate the bronchioles in the treatment of asthma without undesirably stimulating the heart (the heart has β_1 receptors). Other drugs that act selectively at α and β receptors are beneficial in manipulating blood pressure and heart rate in the treatment of hypertension and cardiac arrhythmias.

Many regions of the central nervous system are involved in the control of autonomic activities.

Messages from the CNS are delivered to cardiac muscle, smooth muscle, and exocrine glands via the autonomic nerves, but what regions of the CNS regulate autonomic output?

□ Some autonomic reflexes, such as urination, defecation, and erection, are integrated at the spinal-cord level, but all of these spinal reflexes are subject to control by higher levels of consciousness.

□ The medulla within the brain stem is the region most directly responsible for autonomic output. Centers for controlling cardiovascular, respiratory, and digestive activity via the autonomic system are located there.

□ The hypothalamus plays an important role in integrating the autonomic, somatic, and endocrine responses that automatically accompany various emotional and behavioral states. For example, the increased heart rate, blood pressure, and respiratory activity associated with anger or fear are brought about by the hypothalamus acting through the medulla.

□ Autonomic activity can also be influenced by the frontal cortex through its involvement with emotional expression characteristic of the individual's personality. An example is blushing when embarrassed, which is caused by dilation of blood vessels supplying the skin of the cheeks. Such responses are mediated through hypothalamic-medullary pathways.

Somatic Nervous System

Alpha motor neurons supply skeletal muscle.

Skeletal muscle is innervated by **alpha motor neurons,** the axons of which constitute the **somatic nervous system.** The cell bodies of these motor neurons are located within the ventral horn of the spinal cord. Unlike the two-neuron chain of autonomic nerve fibers, the axon of a motor neuron is continuous from its origin in the spinal cord to its termination on skeletal muscle. Motor-neuron axon terminals release acetylcholine, which brings about excitation and contraction of the innervated muscle fibers. Motor neurons can only stimulate skeletal muscles, in contrast to autonomic fibers, which can either stimulate or inhibit their effector organs. Inhibition of skeletal-muscle activity can be accomplished only within the CNS via activation of inhibitory synaptic input to the cell bodies and dendrites of the motor neurons supplying that particular muscle.

Alpha motor neurons are the final common pathway.

Alpha motor neurons are influenced by many converging presynaptic inputs, both excitatory and inhibitory. Some of these inputs are part of spinal-reflex pathways originating with peripheral sensory receptors. Others are part of descending pathways originating within the brain. Areas of the brain that exert control over skeletal-muscle movements include the motor regions of the cortex, the basal nuclei, the cerebellum, and the brain stem.

The somatic system is considered to be under voluntary control, but much of skeletal-muscle activity involving posture, balance, and stereotypical movements is subconsciously controlled. You may decide you want to start walking, but you do not have to consciously bring about alternate contraction and relaxation of the involved muscles because these movements are involuntarily coordinated by lower brain centers.

Alpha motor neurons are considered to be the **final common pathway** since the only way any other parts of the nervous system can influence skeletal-muscle activity is by acting on these motor neurons. The level of activity in an alpha motor neuron, and its subsequent output to the skeletal-muscle fibers it innervates, depends on the relative balance of EPSPs and IPSPs brought about by its presynaptic inputs originating from these diverse sites in the brain.

Table 7–4 Comparison of the Autonomic Nervous System and the Somatic Nervous System

Feature	Autonomic Nervous System	Somatic Nervous System
Site of origin	Brain or lateral horn of spinal cord	Ventral horn of spinal cord
Number of neurons from origin in CNS to effector organ	Two-neuron chain (preganglionic and postganglionic)	Single neuron (motor neuron)
Organs innervated	Cardiac muscle, smooth muscle, exocrine glands	Skeletal muscle
Type of innervation	Most effector organs dually innervated by the two antagonistic branches of this system (sympathetic and parasympathetic)	Effector organs innervated only by motor neurons
Neuro-transmitter at effector organs	May be acetylcholine (parasympathetic terminals) or norepinephrine (sympathetic terminals)	Only acetylcholine
Effects on effector organs	Either stimulation or inhibition (antagonistic actions of two branches)	Stimulation only (inhibition possible only centrally through IPSPs on cell body of motor neuron)
Types of control	Under involuntary control; may be voluntarily controlled with biofeedback techniques and training	Subject to voluntary control; much activity subconsciously coordinated
Higher brain centers involved in control	Spinal cord, medulla, hypothalamus, frontal cortex	Spinal cord, motor cortex, basal nuclei, cerebellum, brain stem

The cell bodies of these crucial alpha motor neurons may be selectively destroyed by polio virus. The result is paralysis of the muscles innervated by the affected neurons.

Table 7–4 summarizes the features of the two divisions of the efferent nervous system discussed in this chapter. Table 7–5 compares the distinguishing features of the three types of neurons that have been examined in the last three chapters.

Neuromuscular Junction

Acetylcholine chemically links electrical activity in motor neurons with electrical activity in skeletal-muscle cells.

An action potential in an alpha motor neuron is rapidly propagated from the CNS to the skeletal muscle along the large myelinated fiber (axon) of the neuron. As the axon approaches a muscle, it loses its myelin sheath and divides into many terminal branches. Each of these axon terminals forms a special junction, a **neuromuscular junction,** with one of the many muscle cells that compose the whole muscle. A single muscle cell, referred to as a **muscle fiber,** is long and cylindrical in shape. The axon terminal is enlarged into a knob-like structure, the **terminal button,** which fits into a shallow depression or groove in the underlying muscle fiber (Fig. 7–2). The specialized portion of the muscle-cell membrane immediately under the terminal button is known as the **motor end-plate.**

Nerve and muscle cells do not actually come into direct contact at a neuromuscular junction. The space, or cleft, between these two structures is too large to permit electrical transmission of an impulse between them (that is, the action potential cannot "jump" that far). Therefore, just as at a neuronal synapse (see p. 105), a chemical messenger is used to carry the signal between the neuron terminal and the muscle fiber. Each terminal button contains thousands of vesicles that store the chemical transmitter acetylcholine (ACh). Propagation of an action potential to the axon terminal (Step 1, Fig. 7–3 on p. 212) triggers the opening of voltage-gated Ca^{++} channels in the terminal button. This permits Ca^{++} to diffuse into the terminal button from its higher extracellular concentration (Step 2, Fig. 7–3), which in turn causes the release of ACh from several hundred of the vesicles into the cleft (Step 3, Fig. 7–3).

The released ACh diffuses across the cleft and binds with specific receptor sites, which are specialized membrane proteins unique to the motor end-plate portion of the muscle-fiber membrane (Step 4, Fig. 7–3). Binding of ACh with these receptor sites brings about the opening of chemical messenger-gated channels in the motor end-plate. These channels permit a small amount of cation traffic through them (both Na^+ and K^+) but no anions (Step 5, Fig. 7–3). Because the permeability of the end-plate membrane to Na^+ and K^+ on opening of these channels is essentially equal, the relative movement of these ions through the channels depends on their electrochemical driving forces. Recall that at resting potential, the net driving force for Na^+ is much greater than that for K^+, because the resting potential is much closer to the K^+

Table 7–5 Comparison of Types of Neurons

Feature	Afferent Neuron	Efferent Neuron		Interneuron
		Autonomic Nervous System	*Somatic Nervous System*	
Origin, structure, location	Receptor at peripheral ending; elongated peripheral axon, which travels in peripheral nerve; cell body located in dorsal root ganglion; short central axon entering spinal cord	Two-neuron chain; first neuron (preganglionic fiber) originating in CNS and terminating on a ganglion; second neuron (postganglionic fiber) originating in the ganglion	Cell body of motor neuron lying in spinal cord; long axon traveling in peripheral nerve	Various shapes; lying entirely within CNS; some cell bodies originating in brain, with long axons traveling down the spinal cord in descending pathways; some originating in spinal cord, with long axons traveling up the cord to the brain in ascending pathways; others forming short local connections
Termination	Interneurons*	Effector organs (cardiac muscle, smooth muscle, exocrine glands)	Effector organs (skeletal muscle)	Other interneurons and efferent neurons
Function	Carry information regarding the external and internal environment to the CNS	Carry instructions from CNS to effector organs	Carry instructions from CNS to effector organs	Process and integrate afferent input; initiate and coordinate efferent output; responsible for thought and other higher mental functions
Convergence of input on cell body	No (only input is through receptor)	Yes	Yes	Yes
Effect of input on neuron	Can only be excited (through receptor potential induced by stimulus; must reach threshold for action potential)	Can be excited or inhibited (through EPSPs and IPSPs at first neuron; must reach threshold for action potential)	Can be excited or inhibited (through EPSPs and IPSPs; must reach threshold for action potential)	Can be excited or inhibited (through EPSPs and IPSPs; must reach threshold for action potential)
Site of action potential initiation	First excitable portion of membrane adjacent to receptor	Axon hillock	Axon hillock	Axon hillock
Diverence of output	Yes	Yes	Yes	Yes
Effect of output on structure on which it terminates	Only excites	Either excites or inhibits	Only excites	Either excites or inhibits

*Afferent neuron terminates directly on the alpha motor neuron in the case of the monosynaptic stretch reflex.

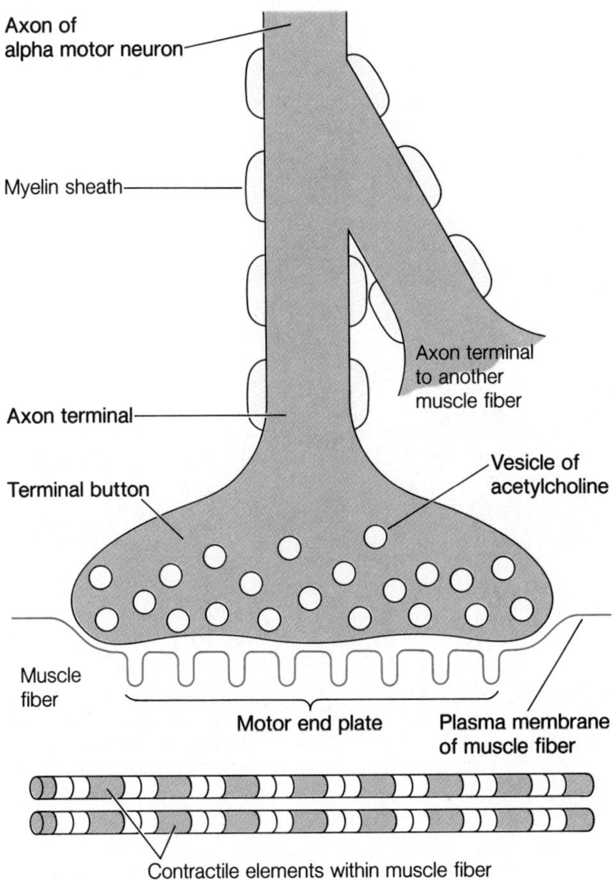

Figure 7–2 Schematic Representation of Neuromuscular Junction

equilibrium potential. Both the concentration and electrical gradients for Na^+ are inward, whereas the outward concentration gradient for K^+ is almost, but not quite, balanced by the opposing inward electrical gradient. As a result, when ACh triggers the opening of these channels, considerably more Na^+ moves inward than K^+ outward, bringing about a depolarization of the motor end-plate. This potential change is known as the **end-plate potential (EPP)**. It is similar to an EPSP (excitatory postsynaptic potential; see p. 106), except that the magnitude of an EPP is much larger because: (1) more transmitter is released from a terminal button than from a presynaptic knob in response to an action potential; (2) the motor end-plate has a larger surface area and, accordingly, more transmitter receptor sites than a subsynaptic membrane; and (3) many more ion channels are opened in response to the transmitter-receptor complex in the motor end-plate. This permits a greater net influx of positive ions and a larger depolarization. As with an EPSP, an EPP is a graded potential, the magnitude of which depends on the amount and duration of ACh at the end-plate.

The motor end-plate region itself does not have a threshold potential, so an action potential cannot be initiated at this site. However, an EPP brings about an action potential in the rest of the muscle fiber, as follows. The neuromuscular junction is usually located approximately in the middle of the long cylindrical muscle fiber. When an EPP takes place at the motor end-plate, local current flow occurs between the depolarized end plate and the adjacent resting cell membrane in both directions, reducing the potential to threshold in the adjacent areas (Step 6, Fig. 7–3). The subsequent action potential initiated at these sites is propagated throughout the muscle-fiber membrane via conduction by local current flow (see p. 98). The spread occurs in both directions away from the motor end-plate toward both ends of the fiber. This electrical activity triggers contraction of the muscle fiber. Thus, by means of ACh, an action potential in a motor neuron brings about an action potential and subsequent contraction in the muscle fiber. (See the accompanying boxed feature, A Closer Look at Exercise Physiology, on p. 213.)

Unlike synaptic transmission, the magnitude of an EPP is normally sufficient to cause an action potential in the muscle cell. Typically, therefore, one-to-one transmission of an action potential occurs at a neuromuscular junction; one action potential in a nerve cell triggers one action potential in a muscle cell that it innervates. Other comparisons of neuromuscular junctions with synapses can be found in Table 7–6.

Acetylcholinesterase terminates acetylcholine activity at the neuromuscular junction.

If ACh continued to remain in contact with the motor end-plate, the channels would remain open and the EPP would continue to exist. Because this persistent end-plate depolarization would continue to initiate action potentials in the remainder of the muscle cell membrane, the muscle fiber would remain contracted until fatigued, even in the absence of further action potentials in the motor neuron. This would not be desirable. There could be no controlled, purposeful alterations in movement. To control muscle contraction, electrical activity in the muscle fiber must be switched off promptly when there is no longer a signal from its motor neuron. The muscle's electrical response is turned off by a special enzyme present in the motor end-plate membrane. This enzyme, **acetylcholinesterase (AChE)**, inactivates ACh by splitting it into acetate and choline. The acetate diffuses away, whereas the choline is actively transported back into the nerve terminal to be reused in the synthesis of more ACh.

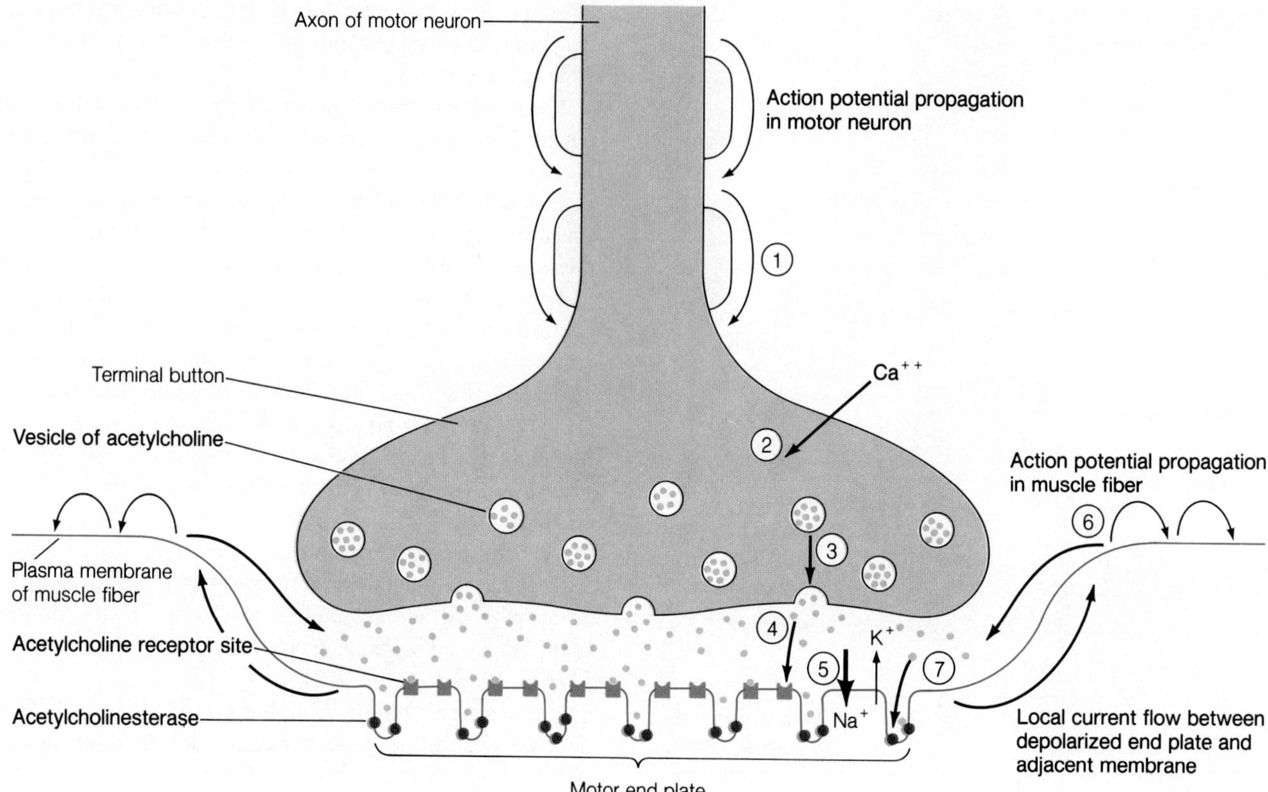

Figure 7–3 Events at a Neuromuscular Junction *See text starting on p. 209 for an explanation of the numbered items in the illustration.*

Table 7–6 Comparison of a Synapse and a Neuromuscular Junction

Similarities	Differences
Both consist of two excitable cells separated by a narrow cleft that prevents direct transmission of electrical activity between them.	A synapse is a junction between two neurons. A neuromuscular junction exists between a motor neuron and a skeletal-muscle fiber.
The axon terminals of both store chemical messengers (neurotransmitters) that are released by the Ca^{++}-induced exocytosis of storage vesicles when an action potential reaches the terminal.	There is a 1-to-1 transmission of action potentials at a neuromuscular junction, whereas one action potential in a presynaptic neuron cannot by itself bring about an action potential in a postsynaptic neuron. An action potential in a postsynaptic neuron occurs only when the summation of EPSPs brings the membrane to threshold.
In both, binding of the neurotransmitter with receptor sites in the membrane of the cell underlying the axon terminal opens specific channels in the membrane, permitting ionic movements that alter the membrane potential of the cell.	A neuromuscular junction is always excitatory (an EPP); a synapse may be either excitatory (an EPSP) or inhibitory (an IPSP). The inhibition of skeletal muscles cannot be accomplished at the neuromuscular junction; it can take place only in the CNS through IPSPs at the cell body of the motor neuron.
The resultant change in membrane potential in both cases is a graded potential.	

A CLOSER LOOK AT EXERCISE PHYSIOLOGY

Loss of Muscle Mass: A Plight of Space Flight

It's a case of "use it or lose it" with skeletal muscles. Stimulation of skeletal muscles by alpha motor neurons is essential not only for inducing contraction of the muscles but also for maintaining the muscles' size and strength. Muscles that are not routinely stimulated gradually atrophy, or diminish in size and strength.

When humans entered the weightlessness of space, it became apparent that the muscular system required the stress of work or gravity to maintain its size and strength. The effort required to move the body is remarkably less in space than on earth, and there is no need for active muscular opposition to gravity. The result is what some refer to as *functional atrophy*.

The muscles most affected are those in the lower extremities, the gluteal (buttocks) muscles, the extensor muscles of the neck and back, and the muscles of the trunk. Changes include a decrease in muscle volume and mass, decrease in strength and endurance, increased breakdown of muscle protein, and loss of muscle nitrogen (an important component of muscle protein). The exact biologic mechanisms that induce muscle atrophy are unknown, but a majority of scientists believe that the lack of customary forcefulness of contraction plays a major role.

Space programs in the United States and the Soviet Union have employed intervention techniques that emphasize both diet and exercise in an attempt to prevent muscle atrophy. Faithful performance of vigorous, carefully designed physical exercise has helped reduce the severity of functional atrophy. Studies of nitrogen and mineral balances, however, suggest that muscle atrophy continues to progress during exposure to weightlessness despite efforts to prevent it.

Acetylcholine binds very briefly (for about 1 nanosecond, 1 billionth of a second) with a receptor site, then detaches. Some of the ACh molecules quickly rebind with receptor sites, keeping the channels open, but some diffuse deeper into the folds of the motor end-plate, where AChE is located (Step 7, Fig. 7-3). Through repetition of this process, more and more ACh is inactivated until it has been virtually removed from the cleft within a few msec after its release. Removal of ACh terminates the EPP so that no more action potentials are initiated.

The neuromuscular junction is vulnerable to several chemical agents and diseases.

Several chemical agents and diseases are known to affect the neuromuscular junction by acting at different sites in the transmission process (Table 7–7). Two well-known toxins—black widow spider venom and botulinum toxin—alter the release of ACh, but in opposite directions. The venom of black widow spiders exerts its deadly effect by causing an explosive release of ACh from the storage vesicles, not only at neuromuscular junctions but at all cholinergic sites. The mechanism responsible for this release is not known. All cholinergic sites undergo prolonged depolarization, the most detrimental consequence of which is respiratory failure. Breathing is accomplished by alternate contraction and relaxation of skeletal muscles, particularly the diaphragm. Inability to relax the diaphragm in the presence of a prolonged EPP caused by excessive amounts of ACh results in respiratory paralysis.

Botulinum toxin, on the other hand, exerts its lethal blow by blocking the release of ACh from the terminal button in response to an action potential in the alpha motor neuron. This is most likely caused in some way by interference by botulinum toxin with Ca^{++}, either its entry into the terminal button or the Ca^{++}-linked release of ACh from the vesicles. *Clostridium botulinum* toxin is responsible for **botulism,** a form of food poisoning. It most frequently results from improperly canned foods contaminated with clostridial bacteria that survive and multiply within the anaerobic ("without O_2") environment of the sealed jar, producing their toxin in the process. When this toxin is consumed, it prevents muscles from responding to nerve impulses. Death is due to respiratory failure because of the inability to bring about diaphragm contraction. Botulinum toxin is one of the most lethal poisons

Table 7–7 Examples of Chemical Agents and Diseases That Affect the Neuromuscular Junction

Mechanism	Chemical Agent or Disease
Alters Release of Acetylcholine	
Causes explosive release of acetylcholine	Black widow spider venom
Blocks release of acetylcholine	*Clostridium botulinum* toxin
Blocks Acetylcholine Receptor Sites	
Reversibly binds with acetylcholine receptor sites	Curare
Very tenaciously binds with acetylcholine receptor sites	α bungarotoxin (specific snake venom)
Self-produced antibodies inactivate acetylcholine receptor sites	Myasthenia gravis
Prevents Inactivation of Acetylcholine	
Irreversibly inhibits acetylcholinesterase	Organophosphates (certain pesticides and military nerve gases)
Temporarily inhibits acetylcholinesterase	Neostigmine

known; ingestion of less than 0.0001 milligram can kill an adult.

Other chemicals interfere with neuromuscular junction activity by blocking the effect of released ACh. The best known example is **curare,** which reversibly binds to the ACh receptor sites on the motor end-plate. Unlike ACh, however, curare does not alter membrane permeability nor is it inactivated by AChE. When ACh receptor sites are occupied by curare, ACh cannot combine with these sites to open the channels that would permit the ionic movement responsible for an EPP. Consequently, because muscle action potentials cannot occur in response to nerve impulses to these muscles, paralysis ensues. When sufficient curare is present to effectively block a significant number of ACh receptor sites, the person dies from respiratory paralysis caused by an inability to contract the diaphragm. Curare was used in the past as a deadly arrowhead or blowgun dart poison. Curare and related drugs have also been used medically during surgery to help achieve more complete skeletal-muscle relaxation with less anesthetic. Under these circumstances, the amount of the agent is carefully administered, and facilities are available to maintain respiration artificially, if necessary, until the effects of the drug wear off. The venom of certain poisonous snakes (**α bungarotoxin**) similarly acts by binding tenaciously and specifically to ACh receptors.

Organophosphates are another group of chemicals that modify neuromuscular junction activity in yet another way—namely, by irreversibly inhibiting AChE. This prevents the inactivation of released ACh so excited muscles remain in a contracted state, unable to repolarize and return to resting conditions. Death from organophosphates is likewise due to respiratory paralysis. The diaphragm is unable to relax and then contract again to bring in a fresh breath of air. These toxic agents are found in some pesticides and military nerve gases.

One disease known to involve the neuromuscular junction is **myasthenia gravis** (*myasthenia* means "muscle weakness," *gravis* means "severe"), a condition characterized by extreme muscular weakness. It is an autoimmune condition (*autoimmune* means "immunity against self") in which the body erroneously produces antibodies against its own motor end-plate ACh receptors. Consequently, not all of the released ACh molecules are able to find a functioning receptor site with which to bind. As a result, much of the ACh is destroyed by AChE without ever having an opportunity to interact with a receptor site and contribute to the EPP. Treatment consists of administration of a drug that temporarily inhibits AChE (in contrast to the toxic organophosphates, which irreversibly block this enzyme). **Neostigmine** is such a short-term anti-acetylcholinesterase drug. In myasthenia gravis patients, this drug prolongs the action of ACh at the neuromuscular junction by permitting it to build up so that the resultant EPP is of sufficient magnitude to initiate action potentials and subsequent contraction in the muscle fiber, as it normally would.

Chapter in Perspective

The central nervous system controls effector organs (muscles and glands) by initiating action potentials in the cell bodies of efferent neurons whose axons terminate on these organs. Much of this efferent output is directed toward maintaining homeostasis. The efferent output to skeletal muscles, which is under voluntary control, is also directed toward nonhomeostatic activities.

There are two types of efferent output: the somatic nervous system, which is subject to voluntary control and supplies skeletal muscles, and the autonomic nervous system, which is under involuntary control and innervates cardiac and smooth muscle as well as exocrine glands. Although there are differences in the "wiring" of these two systems, they function on the same principle. They both secrete from their terminal endings neurotransmitters that combine with receptors on the effector organs to trigger the desired responses. Only two different neurotransmitters—acetylcholine and norepinephrine—are responsible for almost every different effector-organ response.

The somatic nervous system consists of alpha motor neurons, which release acetylcholine. Combination of acetylcholine with receptor sites on the motor end-plate of a skeletal-muscle fiber brings about permeability changes that lead to the initiation of a contraction-inducing action potential in the muscle fiber. The autonomic nervous system consists of two subdivisions—the parasympathetic and sympathetic nervous systems—which, in general, both innervate the same effector organs to bring about opposing responses. The parasympathetic system, which secretes acetylcholine, dominates in quiet, relaxed situations. The sympathetic system, which secretes norepinephrine, dominates in emergency fight-or-flight situations.

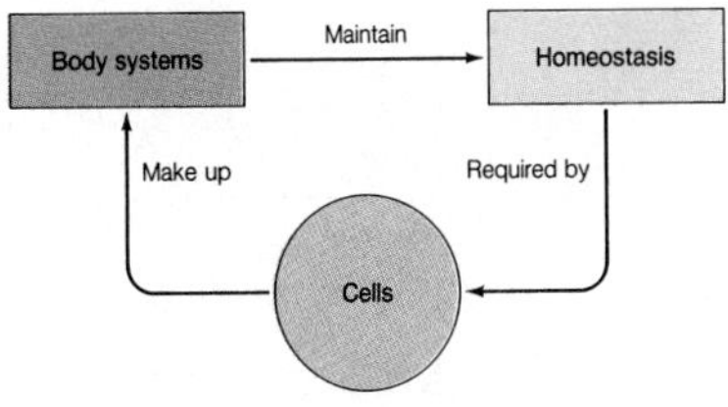

See inside front cover for an expanded version of this model.

Review Exercises

1. Distinguish between preganglionic and postganglionic fibers.
2. Compare the origin, preganglionic- and postganglionic-fiber length, and neurotransmitters of the sympathetic nervous system with those of the parasympathetic nervous system.
3. Explain the technique of biofeedback.
4. Compare the circumstances under which the sympathetic nervous system and parasympathetic nervous system dominate.
5. What is the advantage of dual innervation of many organs by both branches of the autonomic nervous system?
6. Distinguish among the following types of receptors: nicotinic receptors, muscarinic receptors, α receptors, β_1 receptors, and β_2 receptors.
7. What regions of the CNS regulate autonomic output?
8. Compare the autonomic nervous system and the somatic nervous system.
9. Why are alpha motor neurons called the final common pathway?
10. Describe the sequence of events that occur at a neuromuscular junction.
11. Discuss the effect that each of the following have at the neuromuscular junction: black-widow-spider venom, botulinum toxin, curare, α bungarotoxin, organophosphates, myasthenia gravis, and neostigmine.
12. **A point to ponder:** Explain why epinephrine, which causes arteriolar constriction in most tissues, is frequently administered in conjunction with local anesthetics.

C H A P T E R 8

Muscle Physiology

Chapter Contents at a Glance

INTRODUCTION *The Olympic athlete stands triumphantly on the winner's stand, tears of pride and joy flowing freely as the precious gold medal is slipped on. Muscles have performed skillfully and powerfully at the athlete's command. The years of training, willfully pushing the muscles to ever greater feats, have paid off. Skeletal muscles are the only organs in the body over which we have conscious, voluntary control. How does muscle excitation bring about contraction? How can contractions of varying strength in the same muscle be accomplished? How can training alter muscle performance? These are some of the questions that will be answered as you read on.*

Almost all living cells possess rudimentary intracellular machinery for producing such movement as redistributing various components of the cell during cell division. White blood cells use intracellular contractile proteins to propel themselves through their environment. The contraction specialists of the body, however, are the muscle cells. Through their highly developed ability to contract, muscle cells are capable of shortening and developing tension, which enables them to produce movement and to do work. In contrast to sensory systems, which transform other forms of energy in the environment into electrical signals, muscles in response to electrical signals convert the chemical energy of ATP into mechanical energy that can act on the environment. Controlled contraction of muscles allows: (1) purposeful movement of the whole body or parts of the body in relation to the environment (such as walking or piano playing); (2) manipulation of external objects (such as driving a car or moving a piece of furniture); (3) propulsion of contents through various hollow internal organs (such as circulation of blood or movement of materials through the digestive tract); and (4) emptying the contents of certain organs to the external environment (such as urination or giving birth).

Three types of muscle are distinguishable on the basis of their structural and functional differences: skeletal, smooth, and cardiac muscle. As implied by the name, **skeletal muscles** typically attach to the skeleton. Contraction of these muscles causes the bones to which they are attached to move in purposeful relation to one another, allowing the body to perform the great variety of coordinated motor activities in which humans engage. Some skeletal muscles do not attach to the skeleton but still produce movement; these include the muscles of the tongue, which are important in speech and eating, and the external eye muscles, which produce eye movement. A few skeletal muscles unattached to bone actually prevent movement. These are the voluntarily controlled **sphincters,** which guard the exit of urine and feces from the body. Skeletal muscles contract in response to stimulation by the alpha motor neurons of the somatic nervous system. This branch of the nervous system is under voluntary control, enabling people to use their skeletal muscles to do the things they want them to do.

Most **smooth muscle** is found as flat sheets of variable thickness in the walls of hollow organs and tubes, such as the digestive organs, urinary tract, reproductive tract, blood vessels, and respiratory airways. Contraction of such muscle sheets controls the movement of contents within these hollow structures. The contractions are not under direct conscious control and do not require nervous stimulation to initiate them.

Cardiac muscle is found only in the walls of the heart and is responsible for pumping blood into the blood vessels. It too is under involuntary control and does not require stimulation by nerves to cause its contraction.

Muscle comprises the largest group of tissues in the body, accounting for approximately half of the body's weight. Skeletal muscle alone makes up about 40% of body weight in men and 32% in women, with smooth and cardiac muscle making up another 10% of the total weight. Most of this chapter will be devoted to a detailed examination of how the most abundant and best understood muscle, skeletal muscle, works. The chapter will conclude with a discussion of the unique properties of smooth and cardiac muscle in comparison to skeletal muscle.

Structure of Skeletal Muscle

Skeletal-muscle fibers have a highly organized internal arrangement that creates a striated appearance.

Skeletal muscles are stimulated to contract via release of acetylcholine (ACh) at neuromuscular junctions between motor neuron terminals and muscle cells (fibers) (see p. 209). A basic understanding of the structural components of a skeletal-muscle fiber is essential to understanding how the muscle action potential initiated by ACh brings about contraction.

A single skeletal-muscle fiber is a relatively large, elongated, cylinder-shaped cell measuring from 10 to 100 micrometers (1 μm = 1 millionth of a meter) in diameter and up to 750,000 μm, or 2.5 feet, in length. A skeletal muscle consists of a number of muscle fibers bundled together by connective tissue (Fig. 8–1a). The fibers lie parallel to each other and extend the entire length of the muscle. During embryonic development, the huge skeletal-muscle fibers are formed by the fusion of many smaller cells; thus, one striking feature is the presence of multiple nuclei in a single muscle cell. Another feature is the abundance of mitochondria, the energy-generating organelles, as would be expected with the high energy demands of a tissue as active as skeletal muscle.

The most predominant structural feature of a skeletal-muscle fiber is the presence of numerous **myofibrils.** These specialized contractile elements, which constitute 80% of the volume of the muscle fiber, are cylinder-shaped intracellular structures 1 μm in diameter that extend the entire length of the muscle fiber (Fig. 8–1b). Each myofibril consists of a regular arrangement of highly organized cytoskeletal elements—

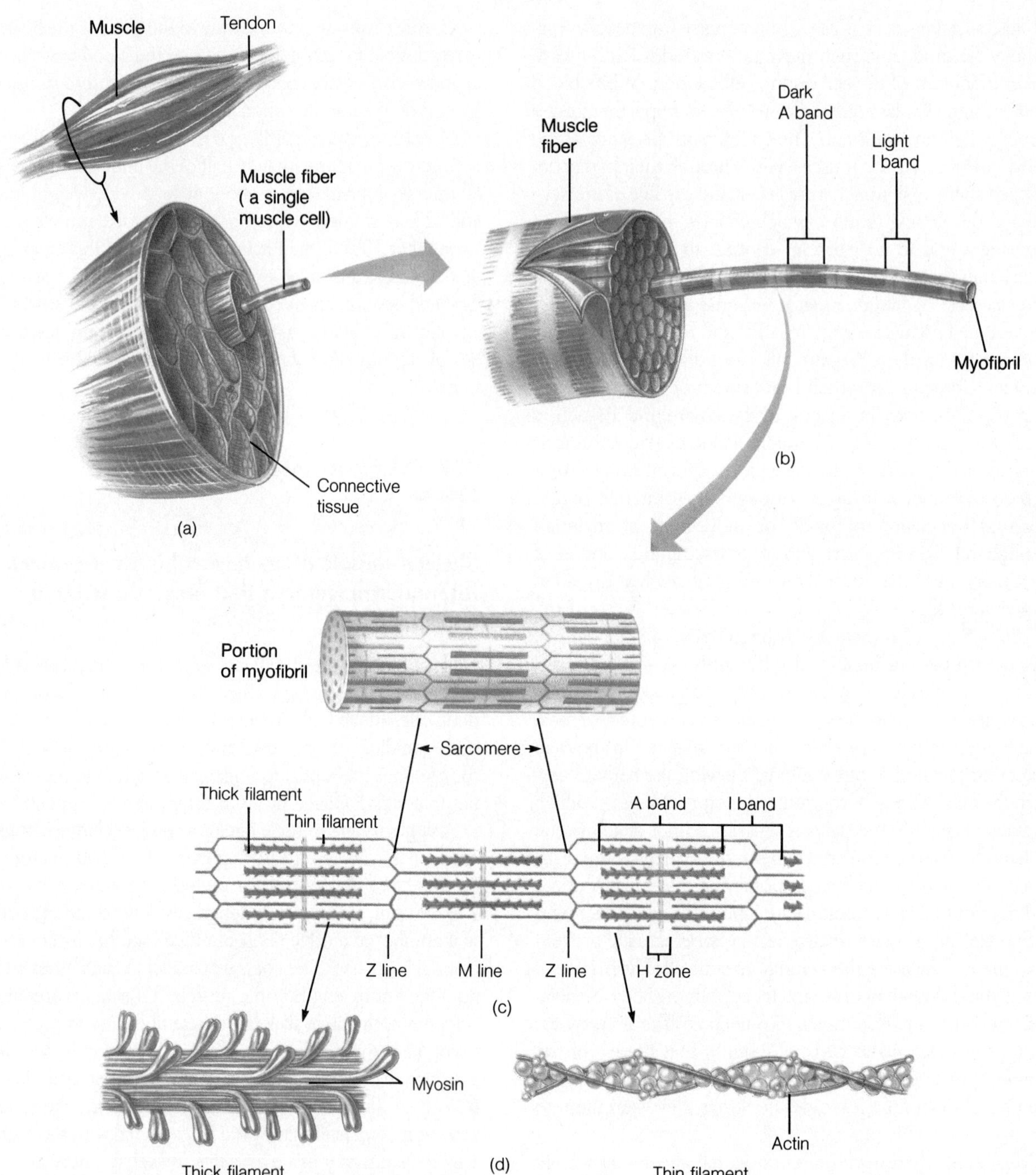

Figure 8–1 Levels of Organization in Skeletal Muscle
(a) Enlargement of a cross section of whole muscle. (b) Enlargement of a myofibril within a muscle fiber. (c) Cytoskeletal components of myofibril. (d) Protein components of thick and thin filaments.

the thick filaments, 12 to 18 nm in diameter and 1.6 μm in length, and the thin filaments, 5 to 8 nm in diameter and 1.0 μm long (Fig. 8–1c). The **thick filaments** are special assemblies of the protein myosin, whereas the **thin filaments** are made up primarily of the protein actin (Fig. 8–1d). These same proteins are found in all other cells of the body but in a less-organized fashion. Following is a review of the levels of organization in a skeletal muscle:

whole muscle →	muscle fiber →	myofibril →	thick and thin filaments →	myosin and actin
(an organ)	(a cell)	(specialized intracellular structure)	(cytoskeletal elements)	(proteins)

Viewed with a light microscope, a relaxed myofibril (Fig. 8–2a) displays alternating dark bands (the A bands) and light bands (the I bands), much like a stick of peppermint candy. The bands of all of the myofibrils lined up parallel to each other collectively lead to the *striated* appearance of a skeletal muscle fiber (Fig. 8-2b). Alternate stacked sets of thick and thin filaments that slightly overlap each other are responsible for the A and I bands (Fig. 8–1c). An **A band** consists of a stacked set of thick filaments along with the ends of the thin filaments that overlap on both sides of the thick filaments. The thick filaments are found only within the A band, extending its entire width. The lighter area within the middle of the A band, where the thin filaments do not reach, is known as the **H zone.** Only the central portions of the thick filaments are found in this region. The **I band** consists of the remaining portion of the thin filaments that do not project into the A band. Thus, the I band contains only thin filaments but not the entire length of these filaments.

Visible in the middle of each I band is a dense, vertical **Z line.** The area between two Z lines is called a **sarcomere,** which is the functional unit of skeletal muscle. A functional unit of any organ is the smallest component that can perform all of the functions of that organ. Accordingly, a sarcomere is the smallest contractile component of a muscle fiber. The Z line is actually a flattened disc-like cytoskeletal protein that connects the thin filaments of two adjoining sarcomeres. Each relaxed sarcomere is about 2.5 μm in width and consists of one whole A band and half of each of the two I bands located on either side. During growth, a muscle increases in length by the addition of new sarcomeres, not by increasing the size of each sarcomere. Just as the Z lines hold the sarcomeres together in a chain along the myofibril's length, another system of supporting proteins is believed to hold the thick filaments together within each stack. These proteins can be seen as the **M line,**

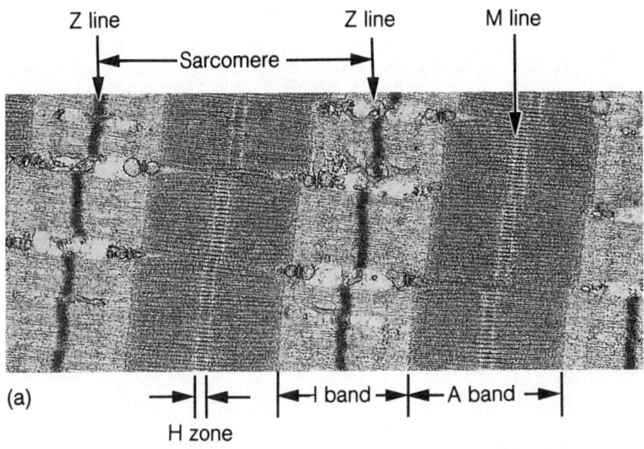

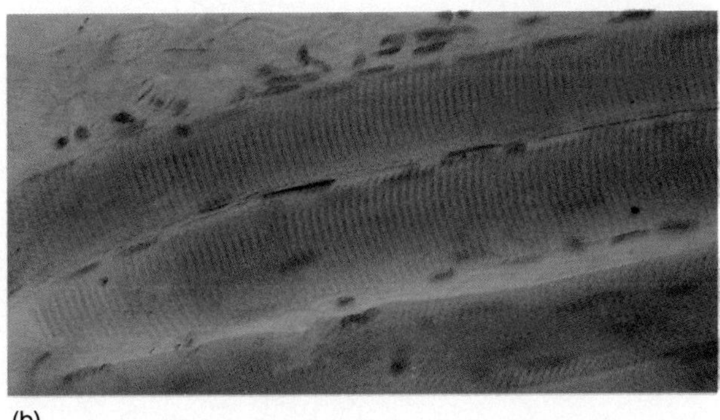

Figure 8–2 Light-Microscope View of Skeletal Muscle Components *(a) High-power light-microscope view of myofibril. (b) Low-power light-microscope view of skeletal muscle fibers. Note striated appearance.*

SOURCE: Part (a) reprinted with permission from Sydney Schochet Jr., Professor, Department of Pathology, School of Medicine, West Virginia University: *Diagnostic Pathology of Skeletal Muscle and Nerve* (East Norwalk, Connecticut: Appleton-Century-Crofts, 1986), Figure 1–13. Part (b) courtesy of Elizabeth R. Walker, Associate Professor, and Dennis O. Overman, Associate Professor, Department of Anatomy, School of Medicine, West Virginia University.

which extends vertically down the middle of the A band within the center of the H zone.

Through the use of an electron microscope, fine **cross bridges** can be seen extending from each thick filament toward the surrounding thin filaments in the regions where the thick and thin filaments overlap (Fig. 8–3a). Three-dimensionally, the thin filaments are arranged hexagonally around the thick filaments. Cross bridges project from each thick filament in all six directions toward the surrounding thin filaments. Each thin filament, in turn, is surrounded by three thick filaments (Fig. 8–3b). To give you an idea of the magni-

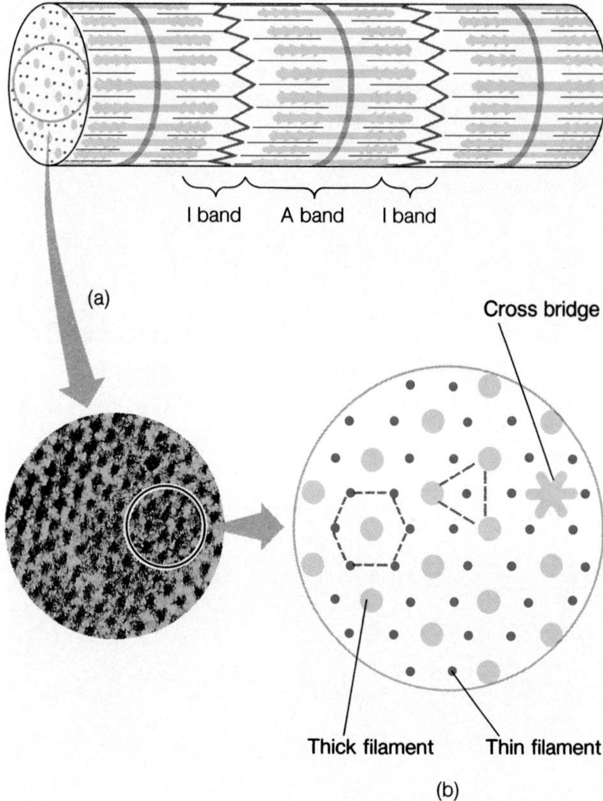

Figure 8–3 Cross-Sectional Arrangement of Thick and Thin Filaments *(a) Electron-micrograph cross section through the A band in the region of thick and thin filament overlap. Note the fine cross bridges extending from the thick filaments. (b) Schematic representation of the geometric relation among thick and thin filaments and cross bridges.*

SOURCE: Photo courtesy of Sydney Schochet Jr., Professor, Department of Pathology, School of Medicine, West Virginia University.

tude of filaments we are talking about, it is estimated that a single muscle fiber may contain 16 billion thick and 32 billion thin filaments, all arranged in this very precise pattern within the myofibrils.

Myosin forms the thick filaments whereas actin is the main structural component of the thin filaments.

Each thick filament is composed of several hundred myosin molecules packed together in a specific arrangement. A **myosin** molecule is a protein consisting of two identical subunits, each shaped somewhat like a golf club (Fig. 8–4a). The protein's tail ends are intertwined around each other, with the two globular heads projecting out at one end. The two halves of each thick filament are mirror images made up of myosin molecules lying lengthwise in a regular staggered array, with their tails oriented toward the center of the filament and their globular heads protruding outward at regularly spaced intervals (Fig. 8–4b). These heads form the cross bridges between the thick and thin filaments. Each cross bridge has two important sites crucial to the contractile process: an **actin binding site** and a **myosin ATPase site.**

Thin filaments are composed of three proteins—actin, tropomyosin, and troponin (Fig. 8–5). **Actin** molecules, the primary structural proteins of the thin filament, are spherical in shape. The backbone of a thin filament is formed by actin molecules joined into two strands twisted together in a helical structure, like two chains of pearls wrapped around each other. Each actin molecule has a special binding site for attachment with a myosin cross bridge. By a mechanism to be described shortly, binding of actin and myosin molecules at the cross bridges in the presence of energy results in contraction of the muscle fiber. Accordingly, actin and myosin are often referred to as **contractile proteins.**

In a relaxed muscle fiber, contraction does not take place; actin is not able to bind with cross bridges because of the position of the two other types of protein within the thin filament. **Tropomyosin** molecules are threadlike proteins that lie end to end alongside the groove of the actin spiral. In this position, tropomyosin covers the sites on actin for binding with the cross bridges, thus blocking the interaction that leads to muscle contraction. Tropomyosin is stabilized in this blocking position by **troponin** molecules, which fasten down the ends of each tropomyosin molecule. Troponin is a protein complex consisting of three polypeptide units: one that binds to tropomyosin, one that binds to actin, and a third that has a Ca^{++} binding site. When Ca^{++} binds to this troponin complex, the conformation of this protein is changed in such a way that tropomyosin is allowed to slide away from its blocking position (Fig. 8–6). With tropomyosin out of the way, actin and myosin can bind and interact at the cross bridges, resulting in muscle contraction. Tropomyosin and troponin are often referred to as **regulatory proteins** because of their role in covering (contraction prevented) or exposing (contraction permitted) the binding sites for cross-bridge interaction between actin and myosin.

Several important links in the contractile process remain to be discussed. How does cross-bridge interaction between actin and myosin bring about muscle contraction? How does a muscle action potential trigger this contractile process? What is the source of the Ca^{++} that physically repositions troponin and tropomyosin to permit cross-bridge binding? We will turn our attention to these topics in the next section.

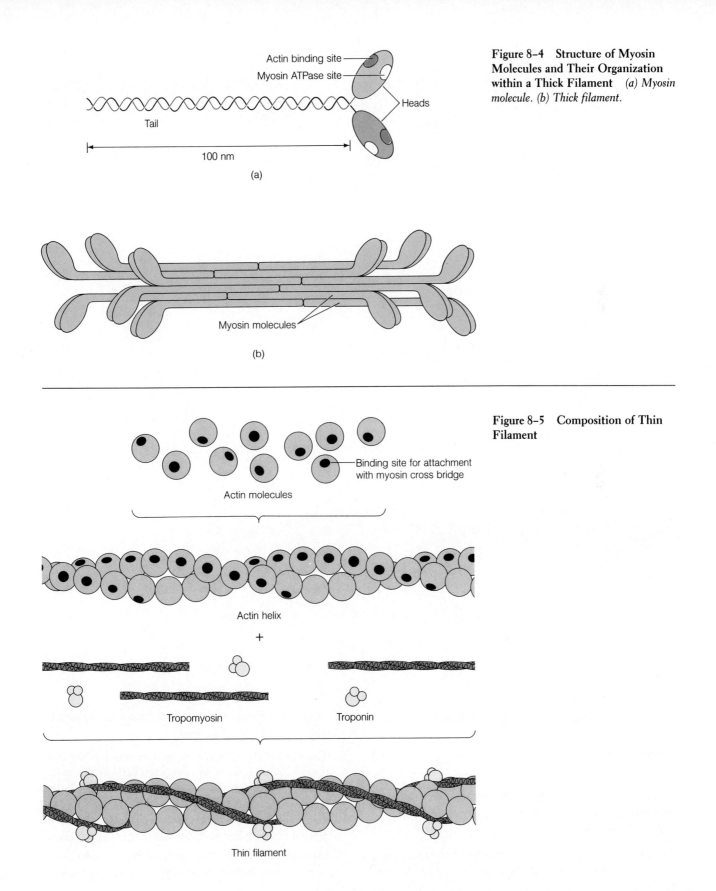

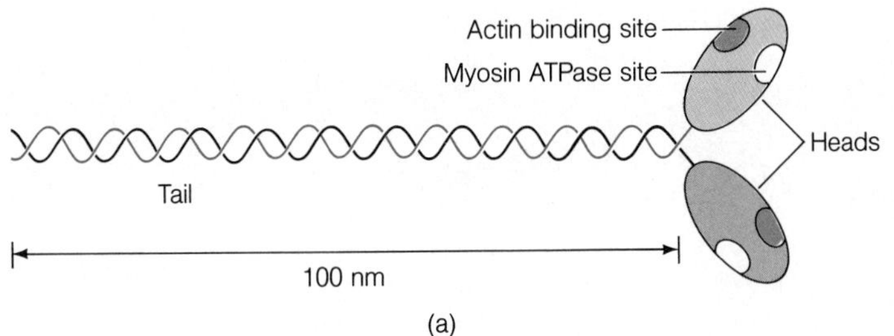

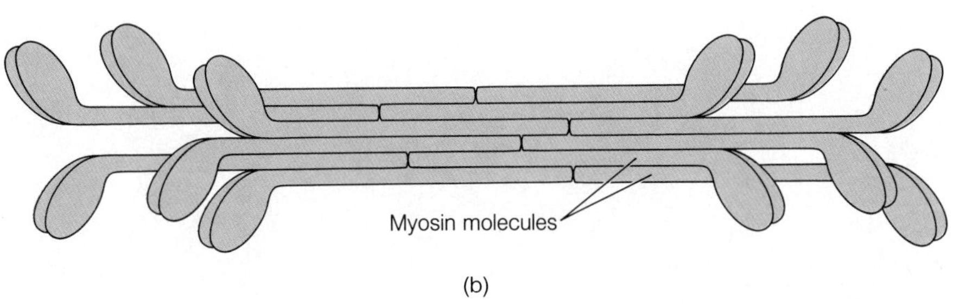

Figure 8-4 Structure of Myosin Molecules and Their Organization within a Thick Filament *(a) Myosin molecule. (b) Thick filament.*

Binding site for attachment with myosin cross bridge

Actin molecules

Actin helix

+

Tropomyosin

Troponin

Thin filament

Figure 8-5 Composition of Thin Filament

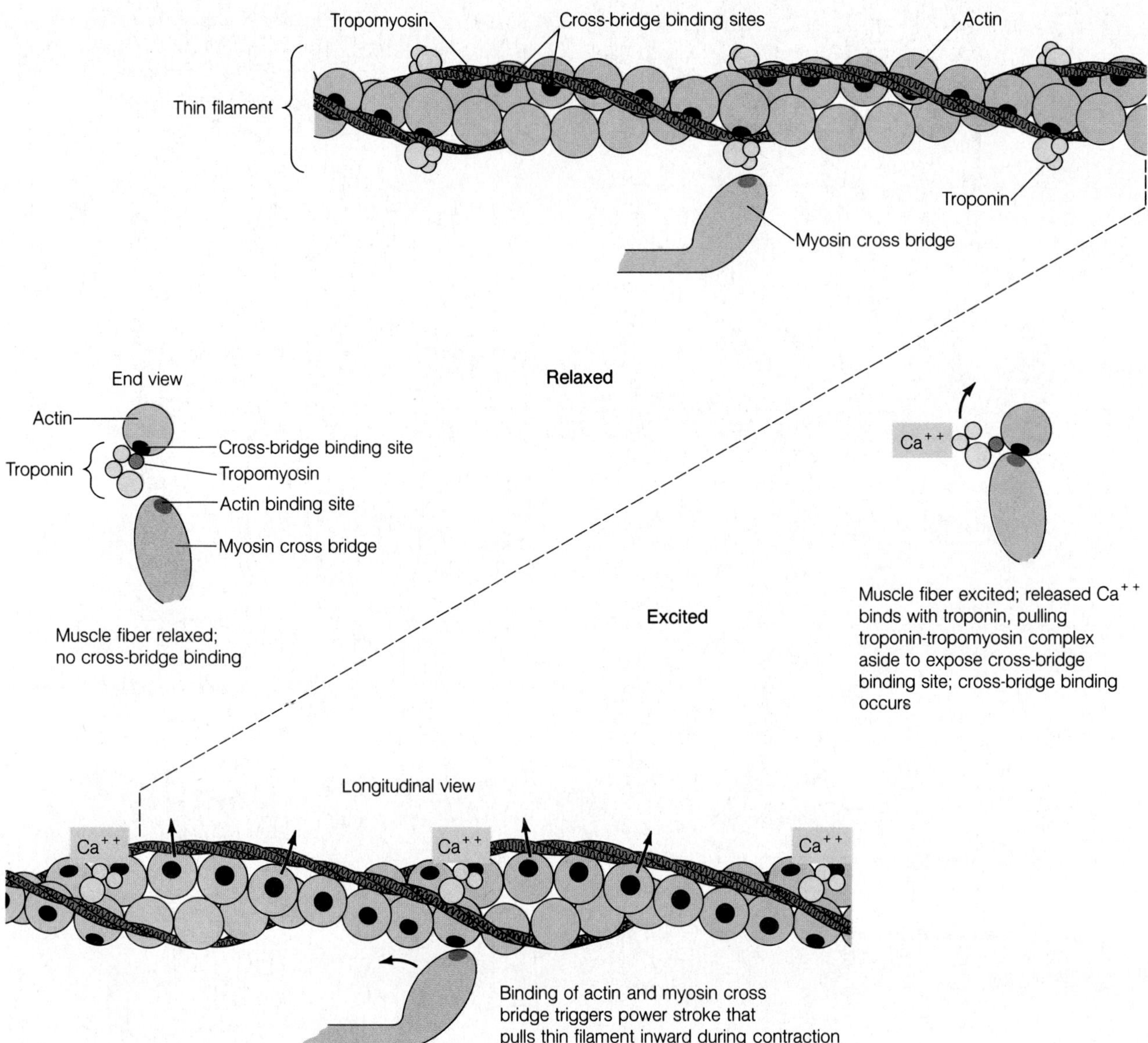

Figure 8–6 Schematic Representation of Role of Calcium in Turning on Cross Bridges

Molecular Basis of Skeletal-Muscle Contraction

Cycles of cross-bridge binding and bending pull the thin filaments closer together between the thick filaments during contraction.

Observations of alterations in the banding pattern during contraction provided the first clue about the molecular changes involved in muscle shortening. These observations and conclusions are as follows (Fig. 8–7):

1. The width of the A band remains unchanged during contraction. Because the width of this dark band is determined by the length of the thick filaments and, accordingly, by the length of the myosin molecules composing the thick filaments, the myosin molecules and thick filaments must remain at constant length during the shortening process.
2. Likewise, neither the thin filaments nor the actin molecules of which they are composed vary in size during con-

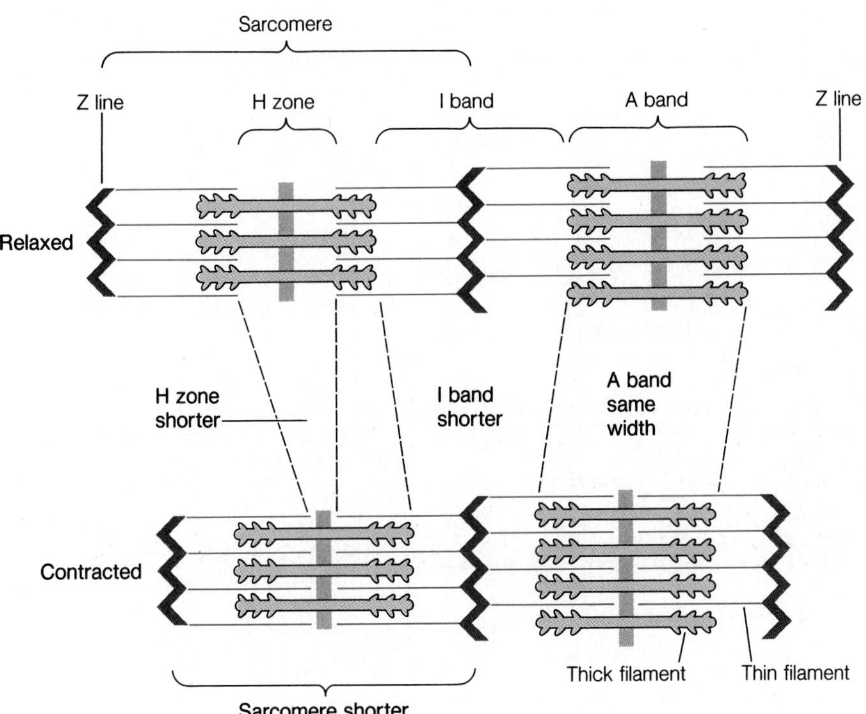

Figure 8–7 Changes in Banding Pattern during Shortening *During muscle contraction, each sarcomere shortens as the thin filaments slide closer together between the thick filaments so that the Z lines are pulled closer together.*

traction. The length of the thin filaments—the distance between a Z line and the edge of the adjacent H zone—is the same in a relaxed and contracted myofibril.

3. The alterations that do take place during contraction are: (a) the width of each sarcomere shortens (that is, the Z lines move closer together); (b) the H zones shorten; and (c) the width of each I band decreases.

These changes can be accounted for by the thin filaments on each side of a sarcomere sliding inward toward the A band's center during contraction. As they slide inward, the thin filaments pull the Z lines to which they are attached closer together, so the sarcomere shortens. As each sarcomere throughout the muscle fiber's length shortens simultaneously, the entire fiber becomes shorter. The H zone, the region in the center of the A band where the thin filaments do not reach, becomes smaller as the thin filaments approach each other when they slide more deeply inward. The H zone may even disappear if the thin filaments meet in the middle of the A band. The I band, which consists of the portions of the thin filaments that do not overlap with the thick filaments, decreases in width as the thin filaments further overlap the thick filaments during their inward slide.

For obvious reasons, this is known as the **sliding-filament mechanism** of muscle contraction. Note that neither the thick nor thin filaments decrease in length to accomplish shortening of the sarcomere. Instead, contraction is accomplished by the thin filaments sliding closer together between the thick filaments.

The thin filaments are pulled inward relative to the stationary thick filaments by cross-bridge activity. During contraction, with the tropomyosin and troponin "chaperones" pulled out of the way by Ca^{++}, the myosin cross bridges from a thick filament are able to bind with the actin molecules in the surrounding thin filaments. Concentrating on a single cross-bridge interaction (Fig. 8–8a), when myosin and actin make contact at a cross bridge, the conformation of the bridge is altered so that it bends inward as if it were on a hinge, "stroking" toward the center of the thick filament, similar to the stroking of a boat oar. This so-called **power stroke** of a cross bridge pulls the thin filament to which it is attached inward. A single power stroke pulls the thin filament inward only a small percentage of the total shortening distance. Complete shortening is accomplished by repeated cycles of cross-bridge binding and bending. At the end of one bridge cycle, the link between the myosin cross bridge and actin molecule is broken. The cross bridge returns to its original conformation and binds to the next actin molecule positioned behind its previous actin partner. The cross bridge bends once again to pull the thin filament in further, then detaches and repeats the cycle. Repeated cycles of cross-bridge binding and bending successively pull in the thin filaments, much like pulling in a rope hand

Figure 8–8 Cross-Bridge Activity *(a) During each cross-bridge cycle, the cross bridge binds with an actin molecule, bends to pull the thin filament inward during the power stroke, then detaches and returns to its resting conformation, ready to repeat the cycle. (b) The power strokes of all cross bridges extending from a thick filament are directed toward the center of the thick filament. (c) Each thick filament is surrounded by six thin filaments, all of which are pulled inward simultaneously through cross-bridge cycling during muscle contraction.*

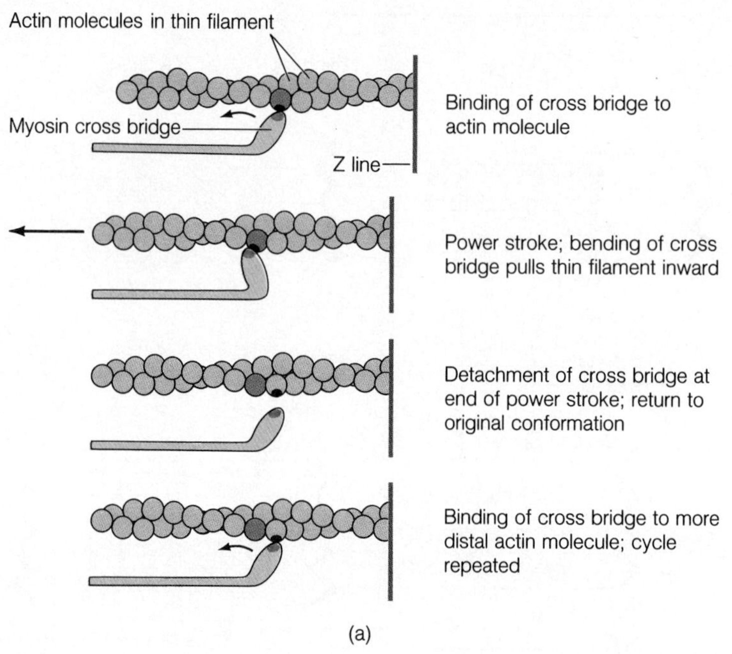

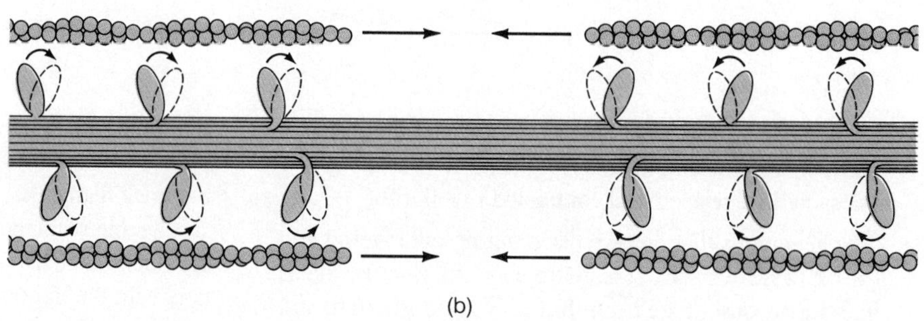

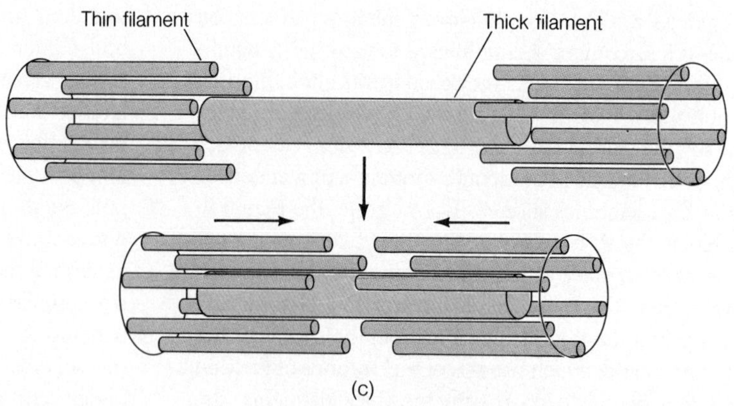

over hand. Because of the orientation of the myosin molecules within a thick filament (Fig. 8–8b), all of the cross bridges' power strokes are directed toward the center so that all six of the surrounding thin filaments are pulled inward simultaneously. (Fig. 8–8c). However, the cross bridges aligned with given thin filaments do not all stroke in unison. At any time during contraction, part of the cross bridges are attached to the thin filaments and are stroking while others are returning to their original conformation in preparation for binding with another actin molecule. Thus some cross bridges are "holding on" to the thin filaments, while others "let go" to bind with new actin. If it were not for this asynchronous cycling of the cross bridges, the thin filaments would be able to slip back toward their resting position between strokes.

Calcium is the link between excitation and contraction.

How is this cross-bridge cycling switched on by muscle excitation? **Excitation-contraction coupling** refers to the series of events linking muscle excitation (the presence of an action potential in a muscle fiber) to muscle contraction (cross-bridge activity that causes the thin filaments to slide closer together to produce sarcomere shortening).

Recall that the release of ACh from the axon terminal in response to an action potential in the motor neuron brings about permeability changes that result in the end-plate potential. Local current flow between the depolarized end plate and adjacent muscle-cell membrane initiates an action potential that is propagated over the entire surface of the membrane.

At each junction of an A band and an I band, the surface membrane dips into the muscle fiber to form a **transverse tubule** (**T tubule**), which runs perpendicularly from the surface of the muscle-cell membrane into the central portions of the muscle fiber (Fig. 8–9). Because the T tubule membrane is continuous with the surface membrane, an action potential on the surface membrane also spreads down into the T tubule, providing a means of rapidly transmitting the surface electric activity into the central portions of the fiber. The presence of a local action potential in the T tubules induces permeability changes in a separate membranous network within the muscle fiber, the sarcoplasmic reticulum.

The **sarcoplasmic reticulum** is a modified endoplasmic reticulum (see p. 22) that consists of a fine meshwork of interconnected tubules surrounding each myofibril like a sleeve (Fig. 8–9). This membranous network is not continuous down the length of the myofibril. Separate segments are wrapped around each A and I band. The ends of each segment expand to form saclike regions, the **lateral sacs,** which

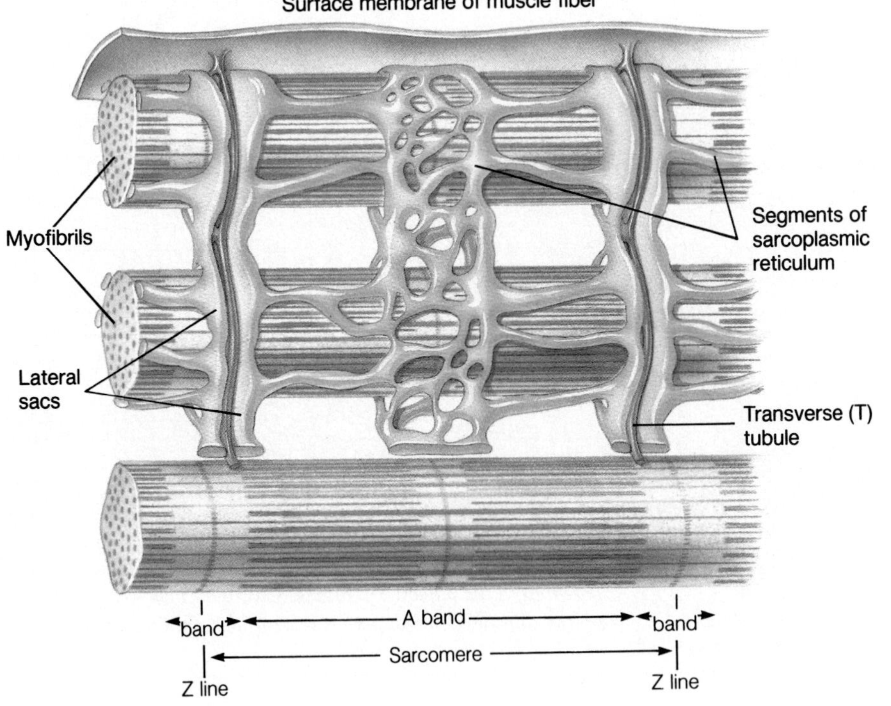

Figure 8–9 T Tubules and Sarcoplasmic Reticulum in Relationship to Myofibrils

lie in close proximity to but do not directly contact, the adjacent T tubules (Figs. 8–9 and 8–10). The sarcoplasmic reticulum's lateral sacs store the intracellular Ca^{++}. Propagation of an action potential down a T tubule triggers massive release of Ca^{++} from the sarcoplasmic reticulum into the cytosol. Recent evidence suggests that T tubule current activates a voltage-dependent enzyme that leads to the formation of a chemical messenger from a component of the T tubule membrane. In the proposed mechanism, this messenger diffuses to the adjacent lateral sacs, where it binds to membrane receptors to trigger the release of Ca^{++} from these intracellular stores. This released Ca^{++}, by slightly repositioning the troponin and tropomyosin molecules, exposes the binding sites on the actin molecules so that they can link with the myosin cross bridges at their complementary binding sites.

Recall that a myosin cross bridge has two special sites, an actin binding site and an ATPase site. The latter is an enzymatic site that can bind the universal energy carrier, ***adenosine triphosphate*** (ATP), and hydrolyze it into ***adenosine diphosphate*** (ADP) and ***inorganic phosphate*** (P_i), yielding energy in the process. In skeletal muscle, magnesium (Mg^{++}) must be attached to ATP before myosin ATPase can hydrolyze the ATP. This necessitates the presence of adequate levels of Mg^{++} for muscle contraction to proceed normally. The breakdown of ATP occurs on the myosin cross bridge before the bridge ever links with an actin molecule (Fig. 8–11, step 1). The ADP and P_i remain tightly bound to the myosin, and the generated energy is stored within the cross bridge to produce a high-energy form of myosin. To use an analogy, it is as if the cross bridge were "cocked" like a gun, ready to be fired when the trigger is pulled. When the muscle fiber is excited, Ca^{++} pulls the troponin-tropomyosin complex out of its blocking position so that the energized (cocked) myosin cross bridge can bind with an actin molecule. This contact between myosin and actin "pulls the trigger." The energy stored within the myosin cross bridge is released to cause the cross-bridge bending responsible for the power stroke that pulls the thin actin filament inward. The mechanism by which the chemical energy released from ATP is stored within the myosin cross bridge and then translated into the mechanical energy of the power stroke is not known.

Adenosine diphosphate and inorganic phosphate are also rapidly released from myosin upon contact with actin. This frees the myosin ATPase site for attachment of another ATP molecule. The actin and myosin remain linked together at the cross bridge until a fresh molecule of ATP attaches to myosin at the end of the power stroke. Attachment of the new ATP molecule permits detachment of the cross bridge, which returns to its original conformation, ready to start another cycle. The newly attached ATP is then hydrolyzed by myosin ATPase, energizing the myosin cross bridge once again. Upon binding with another actin molecule, the energized cross bridge again bends, and so on, successively pulling the thin filaments inward to accomplish contraction.

Note that fresh ATP must attach to myosin to permit breakage of the cross bridge link between myosin and actin at the end of a cycle, even though the ATP is not split during this dissociation process. This requirement for ATP to permit the

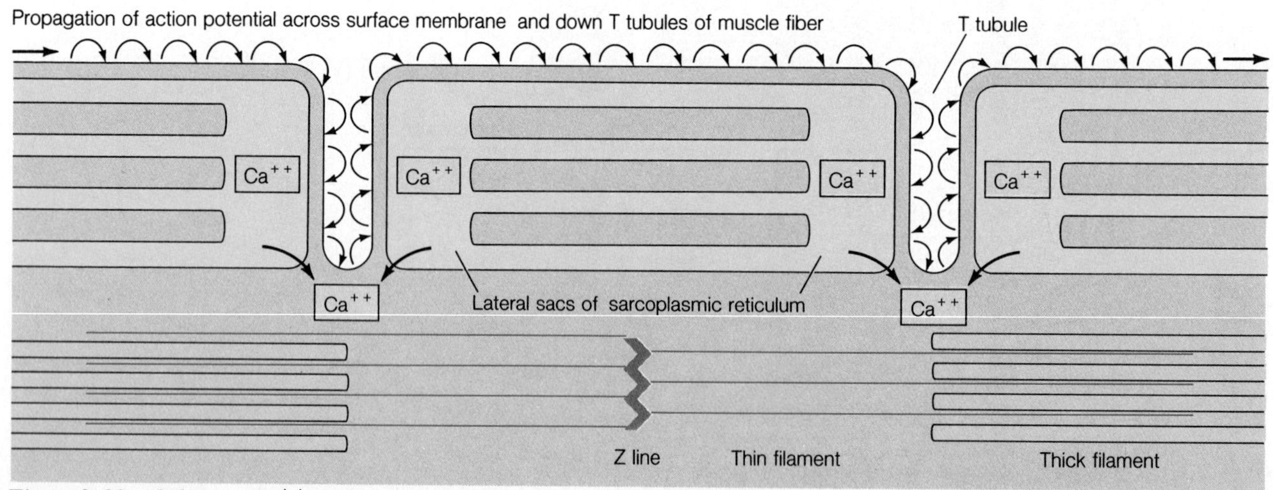

Figure 8–10 Calcium (Ca^{++}) Release in Excitation-Contraction Coupling *Calcium is released from the lateral sacs of the sarcoplasmic reticulum in response to a local action potential in the adjacent T tubule. Calcium binds with troponin on the thin filaments to permit cross-bridge binding and power stroking.*

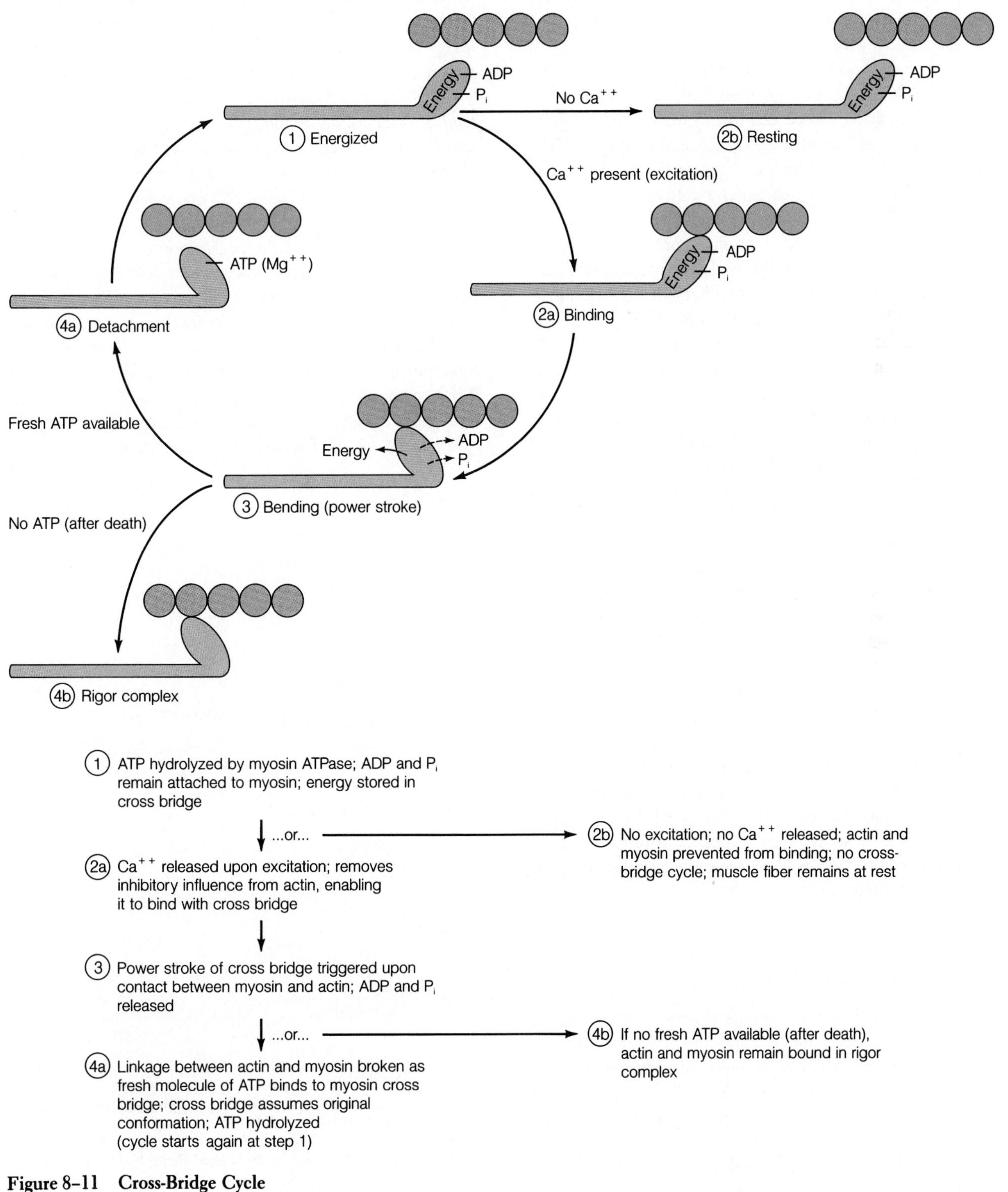
ADP
P_i
Energy
No Ca^{++}
ADP
P_i
Energy
(1) Energized
(2b) Resting
Ca^{++} present (excitation)
ATP (Mg^{++})
ADP
P_i
Energy
(4a) Detachment
(2a) Binding
Fresh ATP available
Energy
ADP
P_i
(3) Bending (power stroke)
No ATP (after death)
(4b) Rigor complex
(1) ATP hydrolyzed by myosin ATPase; ADP and P_i remain attached to myosin; energy stored in cross bridge
...or...
(2b) No excitation; no Ca^{++} released; actin and myosin prevented from binding; no cross-bridge cycle; muscle fiber remains at rest
(2a) Ca^{++} released upon excitation; removes inhibitory influence from actin, enabling it to bind with cross bridge
(3) Power stroke of cross bridge triggered upon contact between myosin and actin; ADP and P_i released
...or...
(4b) If no fresh ATP available (after death), actin and myosin remain bound in rigor complex
(4a) Linkage between actin and myosin broken as fresh molecule of ATP binds to myosin cross bridge; cross bridge assumes original conformation; ATP hydrolyzed (cycle starts again at step 1)

Figure 8–11 Cross-Bridge Cycle

separation of myosin and actin is amply demonstrated by the phenomenon of **rigor mortis.** This "stiffness of death" is a generalized contracture of the skeletal muscles that commences three to four hours after death and becomes complete in about twelve hours. Following death, the cytosolic concentration of Ca^{++} begins to rise, most likely because of the inability of the inactive muscle cell membrane to hold out extracellular Ca^{++} and perhaps also because of leaking of Ca^{++} out of the lateral sacs. This Ca^{++} moves the regulatory proteins aside, permitting actin to bind with the myosin cross bridges, which were already charged with ATP before death. Since dead cells cannot produce any more ATP, actin and myosin, once bound, are unable to detach because of the absence of fresh ATP. The thick and thin filaments thus remain linked together by the immobilized cross bridges, resulting in the stiffened condition of dead muscles (Fig. 8–11, step 4b). During the next several days, rigor mortis gradually subsides as the proteins involved in the rigor complex begin to degrade.

How is **relaxation** normally accomplished in a living muscle? Just as an action potential in a muscle fiber turns on the contractile process by triggering the release of Ca^{++} from its internal stores into the cytosol, the contractile process is turned off when Ca^{++} is returned to the lateral sacs upon cessation of local electrical activity. When acetylcholinesterase removes ACh from the end plate, the end plate channels close once again, leading to cessation of the end-plate potential and of muscle-fiber action potentials. When there is no longer a local action potential in the T tubules, the sarcoplasmic reticulum actively transports Ca^{++} back into its lateral sacs. Removal of cytosolic Ca^{++} allows the troponin-tropomyosin complex to slip back into its blocking position so that actin and myosin are no longer able to bind at the cross bridges. The thin filaments, freed from cycles of cross-bridge attachment and pulling, slide back to their resting position. Relaxation has occurred. During this relaxation process, ATP is required to operate the carrier mechanism in the sarcoplasmic reticulum membrane that drives Ca^{++} back into the lateral sacs against its concentration gradient.

Contractile activity far outlasts the electrical activity that initiated it.

A single action potential in a skeletal-muscle fiber lasts only 1 to 2 msec. The onset of the resultant contractile response lags behind the action potential because the entire excitation-contraction coupling process precedes the commencement of cross-bridge activity. In fact, the action potential is completed before the contractile apparatus even becomes operational. This time delay of a few milliseconds between stimulation and the onset of contraction is known as the **latent period** (Figure 8–12). Time is also required for the generation of tension within the muscle fiber produced by means of the sliding interactions between the thick and thin filaments through cross-bridge activity. The time from the onset of contraction until peak tension is developed—the **contraction time**—averages about 50 msec, although this time varies depending on the type of fiber, as will be described later. The contractile response does not cease until the lateral sacs have sequestered all of the Ca^{++} released in response to the action potential. This reuptake of Ca^{++} is also time-consuming. Even after Ca^{++} is removed, it takes time for the filaments to return to their resting positions. The time from peak tension until relaxation is complete, the **relaxation time,** usually lasts slightly longer than contraction time, another 50 msec or more. Consequently, the entire contractile response to a single action potential may last up to 100 msec or more; this is considerably longer than the duration of the action potential that initiated it (100 msec compared to 1 to 2 msec). This fact is important in the body's ability to produce muscle contractions of variable strength, as you will discover in the next section.

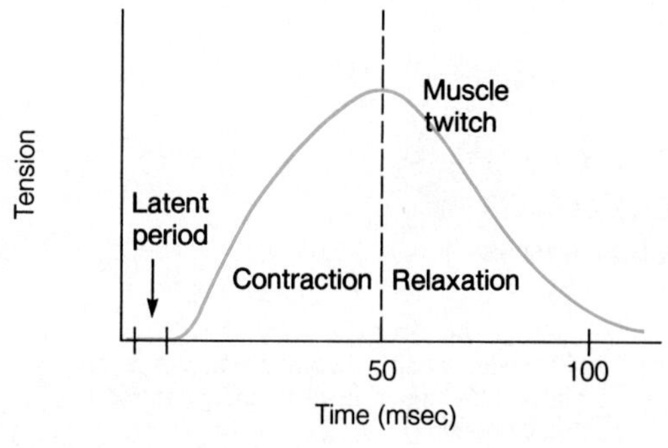

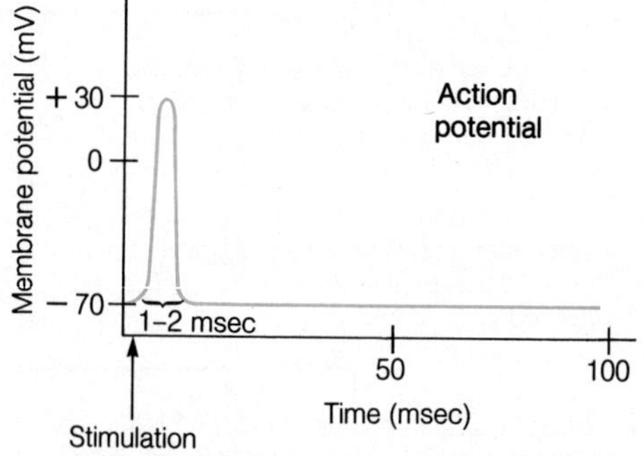

The duration of the action potential is not drawn to scale but is exaggerated.

Figure 8–12 Relationship of Action Potential to Resultant Muscle Twitch

Gradation of Muscle Contraction

Whole muscles are groups of muscle fibers bundled together by connective tissue and attached to bones by tendons.

Thus far we have described the contractile response in a single muscle fiber. In the body, however, groups of muscle fibers are organized into whole muscles. Therefore, we will now turn our attention to contraction of whole muscles. Each person has about six hundred skeletal muscles, which range in size from the delicate external eye muscles that consist of only a few hundred fibers to the large, powerful leg muscles that contain several hundred thousand fibers.

Each muscle is covered by a sheath of connective tissue that penetrates from the surface into the muscle to envelop each individual fiber and to divide the muscle into columns or bundles. The connective tissue further extends beyond the ends of the muscle to form tough, collagenous **tendons** that attach the muscle to bones. A tendon may be quite long, attaching to a bone some distance from the fleshy portion of the muscle. For example, some of the muscles involved in movement of the fingers are found in the forearm, with long tendons extending down to attach to the bones of the fingers. (You can readily observe the movement of these tendons on the top of your hand when you wiggle your fingers.) This arrangement permits greater dexterity; the fingers would be much thicker and more awkward if all the muscles involved in finger movement were actually located in the fingers.

The all-or-none law applies to individual muscle fibers but not to whole muscles.

Individual muscle fibers, like nerve fibers, obey the all-or-none law (see p. 104). Either an action potential is initiated that sweeps over the entire fiber, causing it to contract to the maximal extent possible, or the muscle membrane does not reach threshold, so no action potential occurs and none of the fiber contracts.

A single action potential in a muscle fiber produces a brief, weak contraction known as a **twitch,** which is too short and too weak to be useful and normally does not take place in the body. Muscle fibers are arranged into whole muscles where they can function cooperatively to produce graded contractions stronger than a twitch. In other words, the force exerted by the same muscle can be made to vary, depending on whether the person is picking up a piece of paper, a book, or a fifty-pound weight. Two primary factors can be adjusted to accomplish gradation of whole muscle tension: (1) the number of muscle fibers contracting within a muscle, and (2) the tension produced by each contracting fiber (Table 8–1). We will discuss each of these factors in turn.

Table 8–1 Determinants of Whole-Muscle Tension in Skeletal Muscle

Number of Fibers Contracting	Tension Developed by Each Contracting Fiber
Number of motor units recruited*	Frequency of stimulation (wave summation and tetanus; relationship between internal tension and external tension due to influence of series-elastic component)*
Number of muscle fibers per motor unit	Length of fiber at onset of contraction (length-tension relationship)
Number of muscle fibers available to contract	Extent of fatigue
Size of muscle (i.e., number of muscle fibers in muscle)	Duration of activity
Presence of disease (e.g., muscular dystrophy)	Amount of asynchronous recruitment of motor units
Extent of recovery from traumatic losses	Type of fiber (fatigue-resistant oxidative or fatigue-prone glycolytic)
	Thickness of fiber
	Type of fiber (small-diameter oxidative or large-diameter glycolytic)
	Pattern of neural activity (hypertrophy, atrophy)
	Amount of testosterone (larger fibers in males than females)

*Factors controlled to accomplish gradation of contraction.

The number of fibers contracting within a muscle depends on the extent of motor-unit recruitment.

Because the greater the number of fibers contracting, the greater the total muscle tension, larger muscles consisting of more muscle fibers are obviously capable of generating more tension than are smaller muscles with fewer fibers.

Each whole muscle is innervated by a number of different alpha motor neurons. When a motor neuron enters a muscle, it branches, with each axon terminal supplying a single muscle fiber (Fig. 8–13). One motor neuron innervates a number of muscle fibers, but each muscle fiber is supplied by only one motor neuron. When a motor neuron is activated, all of the muscle fibers it supplies are stimulated to contract simultaneously. This functional unit—one motor neuron plus all of the muscle fibers it innervates—is called a **motor unit.** The muscle fibers that compose a motor unit are dispersed throughout the whole muscle; thus their simultaneous contraction results in an evenly distributed, although weak, contraction of the whole muscle. Each muscle consists of a number of intermingled motor units. For a weak contraction of the whole muscle, only one or a few of its motor units are activated. For stronger and stronger contractions, more and more motor units are *recruited*, or stimulated to contract.

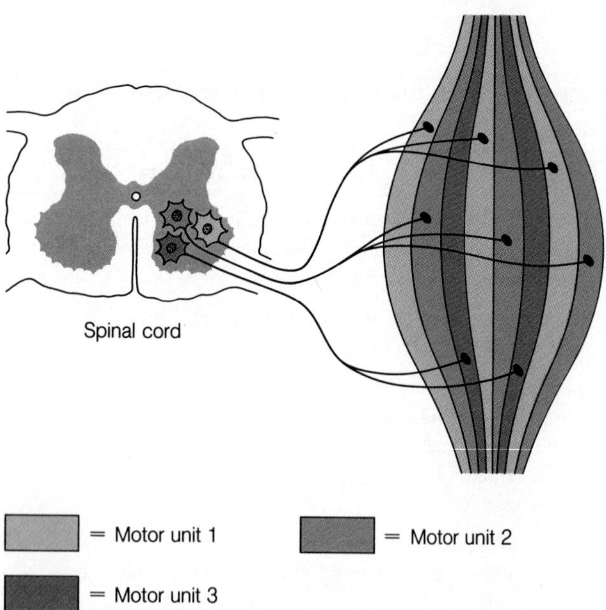

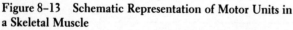

Figure 8–13 Schematic Representation of Motor Units in a Skeletal Muscle

How much stronger the contraction will be with the **recruitment** of each additional motor unit depends on the size of the motor units (that is, the number of muscle fibers controlled by a single motor neuron). The number of muscle fibers per motor unit and the number of motor units per muscle vary widely, depending on the specific function of the muscle. For muscles that produce precise, delicate movements, such as the muscles of the external eye and the hand, a single motor unit may contain as few as a dozen muscle fibers. Recruitment of each additional motor unit results in only a small additional increment in the whole muscle's strength of contraction, because so few muscle fibers are involved with each motor unit. This allows a very fine degree of control over muscle tension. However, in muscles designed for powerful, coarsely controlled movement, such as those of the legs, a single motor unit may contain 1,500 to 2,000 muscle fibers. Recruitment of motor units in these muscles results in large incremental increases in whole muscle tension. More powerful contractions occur at the expense of less precisely controlled gradations. Thus, the number of muscle fibers participating in the whole muscle's total contractile effort depends on the number of motor units recruited and the number of muscle fibers per motor unit in that muscle.

Furthermore, the type of muscle fiber that is activated varies with the extent of gradation. Most muscles consist of a mixture of fiber types that differ metabolically, some being more resistant to fatigue than others. During weak or moderate endurance-type activities (aerobic exercise), the motor units most resistant to fatigue are recruited first. The last fibers to be called into play in the face of demands for further increases in tension are those that fatigue rapidly. An individual can therefore engage in endurance activities for prolonged periods of time but can only briefly maintain bursts of all-out, powerful effort. Of course, even the muscle fibers most resistant to fatigue will eventually fatigue if required to maintain a certain level of sustained tension.

To delay or prevent fatigue during sustained submaximal contraction, as is necessary in muscles supporting the weight of the body against the force of gravity, **asynchronous recruitment of motor units** takes place. The body alternates motor-unit activity, like shifts at a factory, to give motor units that have been active an opportunity to rest while others take over. Changing of the shifts is carefully coordinated so that the sustained contraction is smooth rather than jerky. Asynchronous motor-unit recruitment is possible only for submaximal contractions, during which only a portion of the motor units are required to maintain the desired level of tension. When participation of all of the muscle fibers is essential in maximal contractions, it is impossible to alternate motor-unit activity to prevent fatigue.

The frequency of stimulation can influence the tension developed by each muscle fiber.

Whole-muscle tension depends not only on the number of muscle fibers contracting but also on the tension developed by each contracting fiber. Even though each fiber contracts to the greatest extent possible when stimulated, various factors influence the extent to which tension can be developed. These factors include the following:

1. the frequency of stimulation;
2. the length of the fiber at the onset of contraction;
3. the extent of fatigue; and
4. the thickness of the fiber.

We will examine each of these factors in turn, concentrating first on the effect of frequency of stimulation.

Even though a single action potential in a muscle fiber produces only a twitch, stronger contractions with longer duration can be achieved by repetitive stimulation of the fiber. Let us see what happens when a second action potential occurs in a muscle fiber. If the muscle fiber has completely relaxed before the next action potential takes place, a second twitch of the same magnitude as the first occurs (Fig. 8–14a). The same excitation-contraction events take place each time, resulting in identical twitch responses. If, however, the muscle fiber is stimulated a second time before it has completely relaxed from the first twitch, a second action potential occurs that causes a second contractile response, which is added "piggyback" on top of the first twitch (Fig. 8–14b). The two twitches resulting from the two action potentials add together, or sum, to produce greater tension in the fiber than that produced by a single action potential. This phenomenon is known as **wave summation.** Note that wave summation is possible only because the duration of the action potential (1 to 2 msec) is much shorter than the duration of the resultant twitch (100 msec). Remember that once an action potential has been initiated, a brief refractory period occurs during which another action potential cannot be initiated. It is therefore impossible to achieve summation of action potentials. The membrane must return to resting potential and recover from its refractory period before another action potential can occur. However, because the action potential and refractory period are over long before the resultant muscle twitch is completed, the muscle fiber may be restimulated while some contractile activity still exists to produce summation of the mechanical response. If the muscle

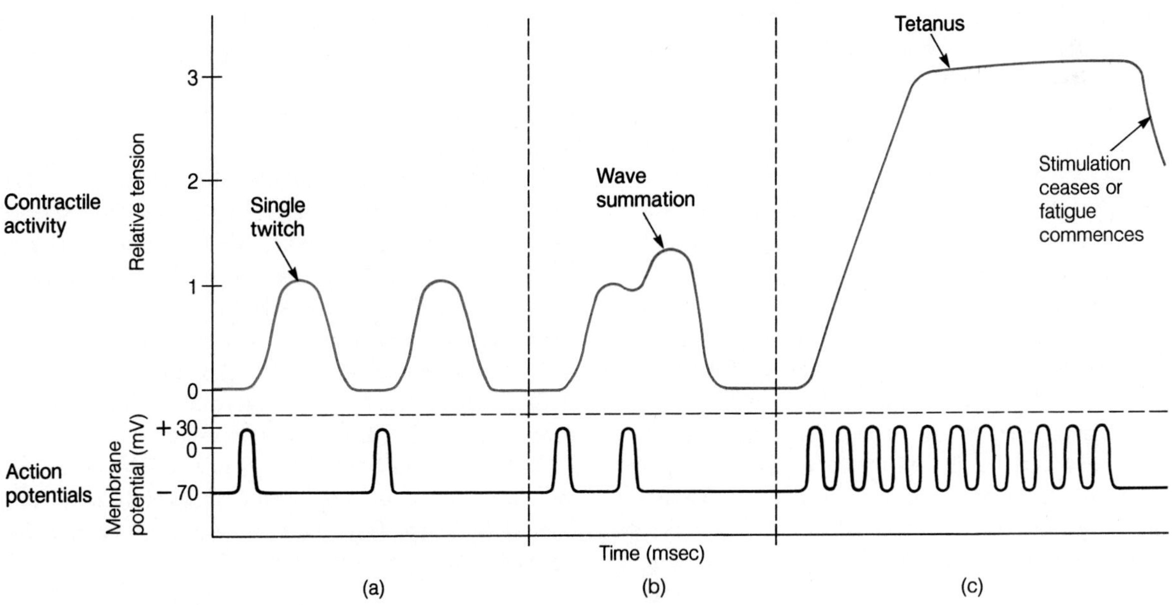

Figure 8–14 Wave Summation and Tetanus *(a) If a muscle fiber is restimulated after it has completely relaxed, the second twitch is the same magnitude as the first twitch. (b) If a muscle fiber is restimulated before it has completely relaxed, the second twitch is added on to the first twitch resulting in wave summation. (c) If a muscle fiber is stimulated so rapidly that it does not have an opportunity to relax at all between stimuli, a maximal sustained contraction known as tetanus occurs.*

fiber is stimulated so rapidly that it does not have a chance to relax at all between stimuli, a smooth, sustained contraction of maximal strength known as **tetanus** occurs (Fig. 8–14c). A tetanic contraction is usually three to four times stronger than a single twitch. (This normal physiological tetanus should not be confused with the disease tetanus see p. 111).

What is the mechanism of wave summation and tetanus at the cellular level? In response to a single action potential, sufficient Ca^{++} is released to interact with all of the troponin within the cell. As a result, all of the cross bridges are free to participate in the contractile response. If all of the cross bridges are "turned on" by a single action potential, how can repetitive action potentials bring about a greater contractile response? The difference depends on the relationship between internal tension and external tension. When cross-bridge activity shortens the sarcomeres, **internal tension** is produced within the muscle. The sarcomeres, however, are not directly attached to the bones. Instead, the internal forces generated by these contractile elements must be transmitted via the connective tissue and tendons to the bones before any **external tension** can be realized. Connective tissue, as well as other components of the muscle such as the sarcoplasmic reticulum, exhibit a certain degree of passive elasticity. These noncontractile tissues are referred to as the **series-elastic component** of the muscle, which behaves like a stretchy spring placed between the internal force-generating elements (the internal tension produced by cyclic cross-bridge activity) and the bone to be moved against an external load (Fig.8–15a). The internal tension must stretch the spring (the elastic component), which in turn develops the external tension responsible for moving the bone against a load.

An analogy of this relationship is a boat pulling a skier. As the power of the motor starts to move the boat, this force is transmitted via the ski rope to the skier and subsequently pulls the skier. The skier is not moved until the slack has been taken up in the ski rope and sufficient tension has been produced in the rope to overcome the skier's weight. The powered boat can be likened to the internal tension produced by the ATP-powered contractile elements, the tension in the ski rope to the external tension developed in the series-elastic component, and the skier to the bone to be moved against an external load. If the ski rope were stretchy, as is the case of the series-elastic component, it would take even more time to stretch the rope sufficiently to create enough tension to pull the skier at the same speed as the boat. In other words, it takes time for the internal tension to be manifested as external tension because of the intervening necessity of stretching the series-elastic component.

Now to relate this to the phenomenon of wave summation and tetanus. After Ca^{++} has been released following an action potential, cross-bridge cycling and subsequent shortening of the sarcomeres continues until the Ca^{++} is transported back into the lateral sacs. The period of time when Ca^{++} is available for triggering the development of internal tension constitutes the **active state** of the contractile elements. Because it takes time for the internal tension created during the active state to stretch the series-elastic component and produce external tension, the active state resulting from a single action potential does not last long enough for the full amount of internal tension to be transmitted to the load. Reuptake of Ca^{++} occurs before the contractile elements have fully stretched the series-elastic component, so the resultant external tension is somewhat less than the internal tension during a twitch. In wave summation, a second action potential is delivered before the Ca^{++} has all been transported back into the lateral sacs. As a result, the active state continues, cross-bridge activity persists, and the series-elastic component is stretched even further, thus creating more external tension. During tetanus, the active state is maintained long enough because of repetitive action potentials to allow the elastic component to be fully stretched, so the tension exerted on the external load equals the force generated by the internal cross-bridge activity (Fig. 8–15b). Think again about the skier. If the powerboat stops before the stretchy ski rope is fully taut (the active state ends before internal tension has been fully translated into external tension), the skier does not reach the same speed as the boat. Part of the boat's power is expended in stretching the rope. However, once the rope is fully stretched (the active state lasts long enough for full stretching of the series-elastic component), the skier moves at the same speed as the boat (external tension accurately reflects internal tension).

Because skeletal muscle requires stimulation by motor neurons to contract, the nervous system plays a key role in regulating the strength of contraction. The two main factors subject to control to accomplish gradation of contraction are the *number of motor units stimulated* and the *frequency of their stimulation*. The areas of the brain responsible for directing motor activity use a combination of tetanic contractions and precisely timed shifts of asynchronous motor-unit recruitment to execute smooth rather than jerky contractions.

Additional factors not directly under nervous control also influence the tension developed during contraction. Among these is the length of the fiber at the onset of contraction, to which we now turn our attention.

There is an optimal muscle length at which maximal tension can be developed upon a subsequent contraction.

A relationship exists between the length of the muscle before the onset of contraction and the tetanic tension that each con-

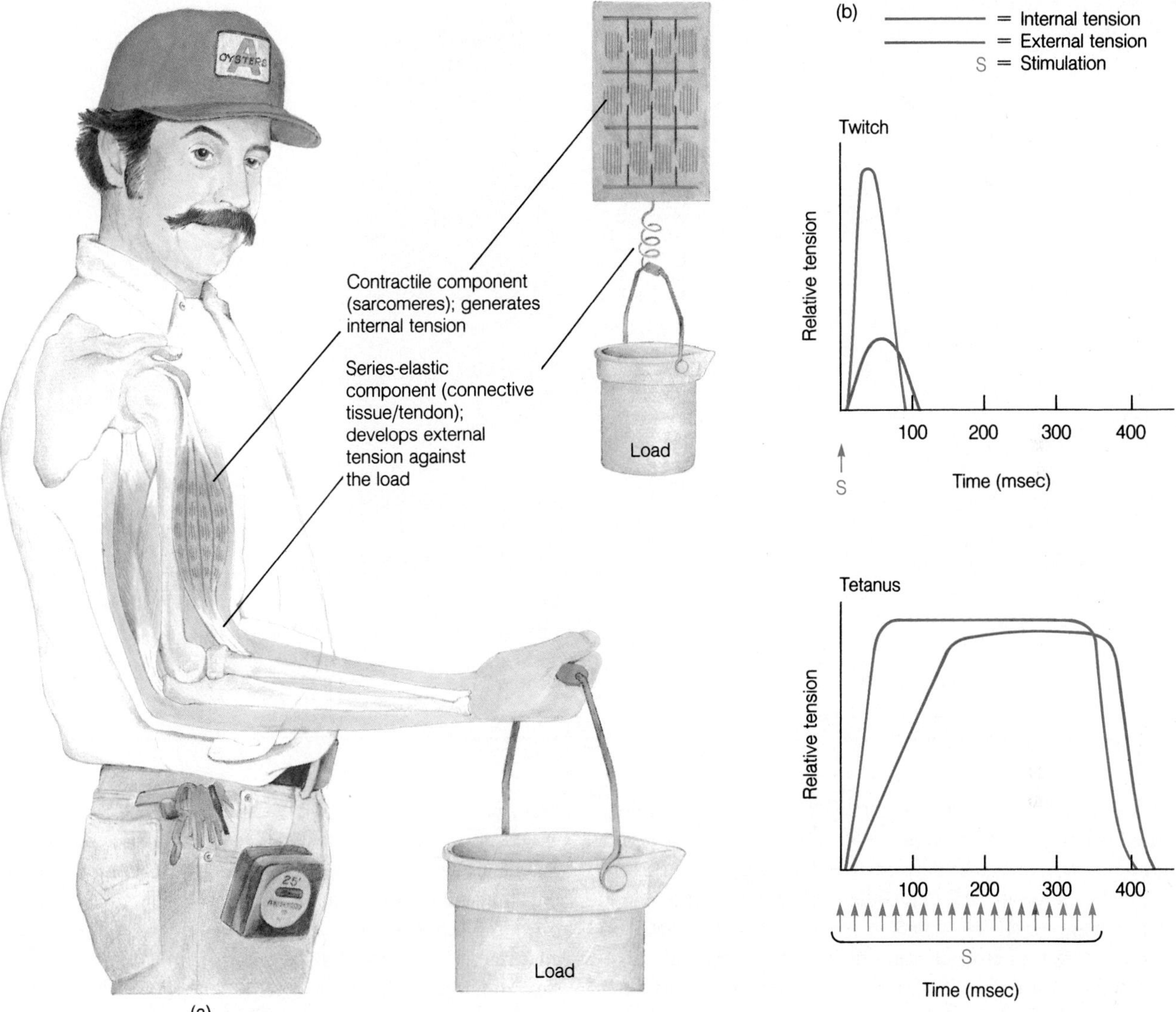

Figure 8–15 Relationship between Internal Tension and External Tension in a Twitch and Tetanus *(a) Shortening of the sarcomeres generates internal tension. External tension is developed as shortening of the sarcomeres stretches the elastic connective tissue and the tendon component of the muscle. (b) Internal tension is of the same magnitude but longer duration in tetanus compared to a twitch. Greater external tension is developed in tetanus because more time is available for internal tension to fully stretch the series-elastic component.*

tracting fiber can subsequently develop at that length. For every muscle there is an **optimal length (l_o)** at which maximal force can be achieved upon a subsequent tetanic contraction. The tension that can be achieved during tetanus at the optimal muscle length is greater than the tetanic tension that can be achieved when the contraction begins with the muscle less than or greater than its optimal length. This **length-tension relationship** can be explained by the sliding-filament mechanism of muscle contraction. At l_o when maximum tension can be developed (Fig. 8–16, point A), the thin filaments optimally overlap the regions of the thick filaments from which the cross bridges project. The central region of thick filaments is void of cross bridges; only myosin tails are found here. At l_o, because a maximal number of cross-bridge sites are accessible to the actin molecules for binding and bending, maximum tension can be developed. At greater lengths, as when a muscle is passively stretched (Fig. 8–16, point B), the thin filaments are pulled out from between the

Figure 8–16 Length-Tension Relationship

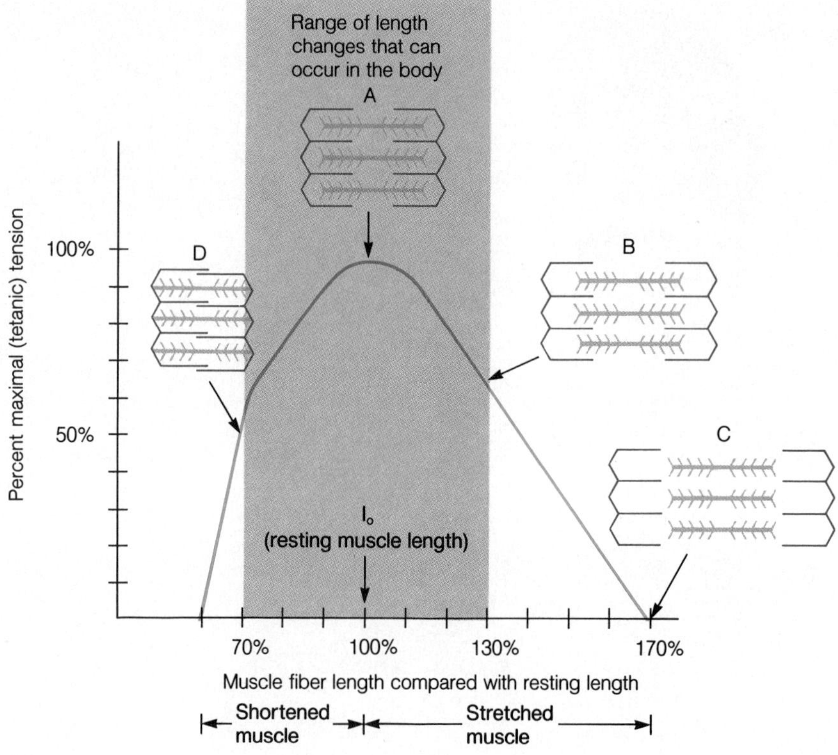

thick filaments, decreasing the number of actin sites available for cross-bridge binding. When less cross-bridge activity can occur, less tension can be developed. In fact, when the muscle is stretched to about 70% longer than its l_o (point C), the thin filaments are completely pulled out from between the thick filaments so that no cross-bridge activity and consequently no contraction can occur. If a muscle is shorter than l_o before contraction (point D), less tension can be developed for three reasons:

1. The thin filaments from the opposite sides of the sarcomere become overlapped, decreasing the number of actin sites exposed to the cross bridges.
2. The thick filaments become forced against the Z lines, so further shortening is impeded.
3. Besides these two mechanical factors, at muscle lengths less than 80% of l_o, not as much Ca^{++} is released during excitation-contraction coupling for reasons unknown. Consequently, fewer actin sites are uncovered for participation in cross-bridge activity.

The extremes in muscle length that seriously hamper the development of tension occur only under experimental conditions, when a muscle is removed and stimulated at various lengths. Fortunately, in the body the muscles are so positioned that their relaxed length is approximately their optimal length; thus they are capable of achieving near-maximal tetanic contraction most of the time. Because of limitations imposed by attachment to the skeleton, a muscle cannot be stretched or shortened more than 30% of its resting optimal length, and usually there is much less than 30% deviation from normal length. Even at the outer limits (130% and 70% of l_o), the muscles are still able to generate half their maximum tension. Furthermore, the lever action of the skeleton, by allowing a range of joint positions, provides a constantly changing mechanical advantage so that force can also vary with limb position as well as with muscle length.

The factors we have discussed thus far that influence how much tension can be developed by a contracting muscle fiber—the frequency of stimulation and the muscle length at the onset of contraction—can vary from contraction to contraction. Other determinants of muscle-fiber tension—the type of fiber (its metabolic capability relative to resistance to fatigue) and the thickness of the fiber—do not vary from contraction to contraction but can be modified over a period of time. We will consider these other factors in the next section.

METABOLISM AND TYPES OF FIBERS

Muscle fibers have alternate pathways for forming ATP.

Three different steps in the contraction-relaxation process require ATP:

1. Splitting of ATP by myosin ATPase provides the energy for the power stroke of the cross bridge.
2. Binding (but not splitting) of a fresh molecule of ATP to myosin permits detachment of the bridge from the actin filament at the end of a power stroke so that the cycle can be repeated. This ATP is subsequently split to provide energy for the next stroke of the cross bridge.
3. The active transport of Ca^{++} back into the sarcoplasmic reticulum during relaxation depends on energy derived from the breakdown of ATP.

Because ATP is the only energy source that can be directly used for these activities, ATP must constantly be supplied for contractile activity to continue. However, limited stores of ATP are immediately available in muscle tissue—only enough to fuel a maximal contraction for a few seconds. Fortunately, there are three pathways for supplying additional ATP as needed during muscle contraction: (1) transfer of a high-energy phosphate from creatine phosphate to ADP; (2) oxidative phosphorylation (the citric-acid cycle and electron-transport system); and (3) glycolysis.

Creatine phosphate is the first energy storehouse tapped at the onset of contractile activity (Fig. 8–17a). Like ATP, creatine phosphate contains a high-energy phosphate group. Just as energy is released when the terminal phosphate bond in ATP is split, similarly energy is released when the bond between phosphate and creatine is broken. The energy released from the hydrolysis of creatine phosphate, along with the phosphate, can be donated directly to ADP to form ATP. This reaction, which is catalyzed by the muscle cell enzyme **creatine kinase,** is reversible; energy and phosphate from ATP can be transferred to creatine to form creatine phosphate:

$$\text{Creatine phosphate} + \text{ADP} \overset{\text{creatine kinase}}{\rightleftharpoons} \text{creatine} + \text{ATP}$$

As the energy reserves are built up in a resting muscle, the increased concentration of ATP favors the transfer of the high-energy phosphate group to creatine phosphate, in accordance with the law of mass action (see p. 441). Thus, most energy is stored in muscle in creatine phosphate pools. A rested muscle contains about five times as much creatine phosphate as ATP. At the onset of contraction, when the meager reserves of ATP are rapidly utilized, the reaction is reversed. Additional ATP is quickly formed by the transfer of energy and phosphate from creatine phosphate to ADP. Because only one enzymatic reaction is involved in this energy transfer, ATP can be formed rapidly (within a fraction of a second) by using creatine phosphate. Thus, creatine phosphate is the immediate source for supplying additional ATP when exercise begins. Muscle ATP levels actually remain fairly constant early in contraction, but creatine phosphate stores become depleted. In fact, short bursts of high-intensity contractile effort, such as high jumps or sprints, are supported primarily by ATP derived at the expense of creatine phosphate. Other energy systems do not have a chance to become operable before the activity is over. If the energy-dependent contractile activity is to be continued, the muscle shifts to the alternate pathways of oxidative phosphorylation and glycolysis to form ATP. These multi-stepped pathways require time to pick up their rates of ATP formation to match the increased demands for energy, time that has been provided by the immediate use of energy derived from the one-step creatine phosphate system.

Oxidative phosphorylation (see p. 29) takes place within the muscle mitochondria if sufficient O_2 is present, fueled by glucose or fatty acids, depending on the intensity and duration of the activity (Fig. 8–17b). This pathway provides a rich yield of ATP molecules for each nutrient molecule processed. However, it is relatively slow because of the number of steps involved, and it necessitates a constant supply of O_2 and nutrient fuel.

The O_2 required for oxidative phosphorylation is primarily delivered by the blood. Increased O_2 is made available to muscles during exercise by several mechanisms. Deeper, more rapid breathing brings in more O_2; the heart contracts more rapidly and forcefully to pump more oxygenated blood to the tissues; more blood is diverted to the exercising muscles by dilation of the blood vessels supplying them; and the hemoglobin molecules that carry the O_2 in the blood release more O_2 in exercising muscles. (These mechanisms are discussed further in later chapters.) Furthermore, some types of muscle fibers have an abundance of **myoglobin,** which is similar to hemoglobin. Myoglobin can store small amounts of O_2, but, more importantly, it increases the rate of O_2 transfer from the blood into muscle fibers.

Glucose and fatty acids, ultimately derived from ingested food, are also delivered to the muscle cells by the blood. Excess ingested nutrients not immediately utilized are stored in the liver as glycogen (chains of glucose) and in adipose tissue as fat. Additionally, muscle cells are able to store limited quantities of glucose in the form of glycogen (Fig. 8–17c).

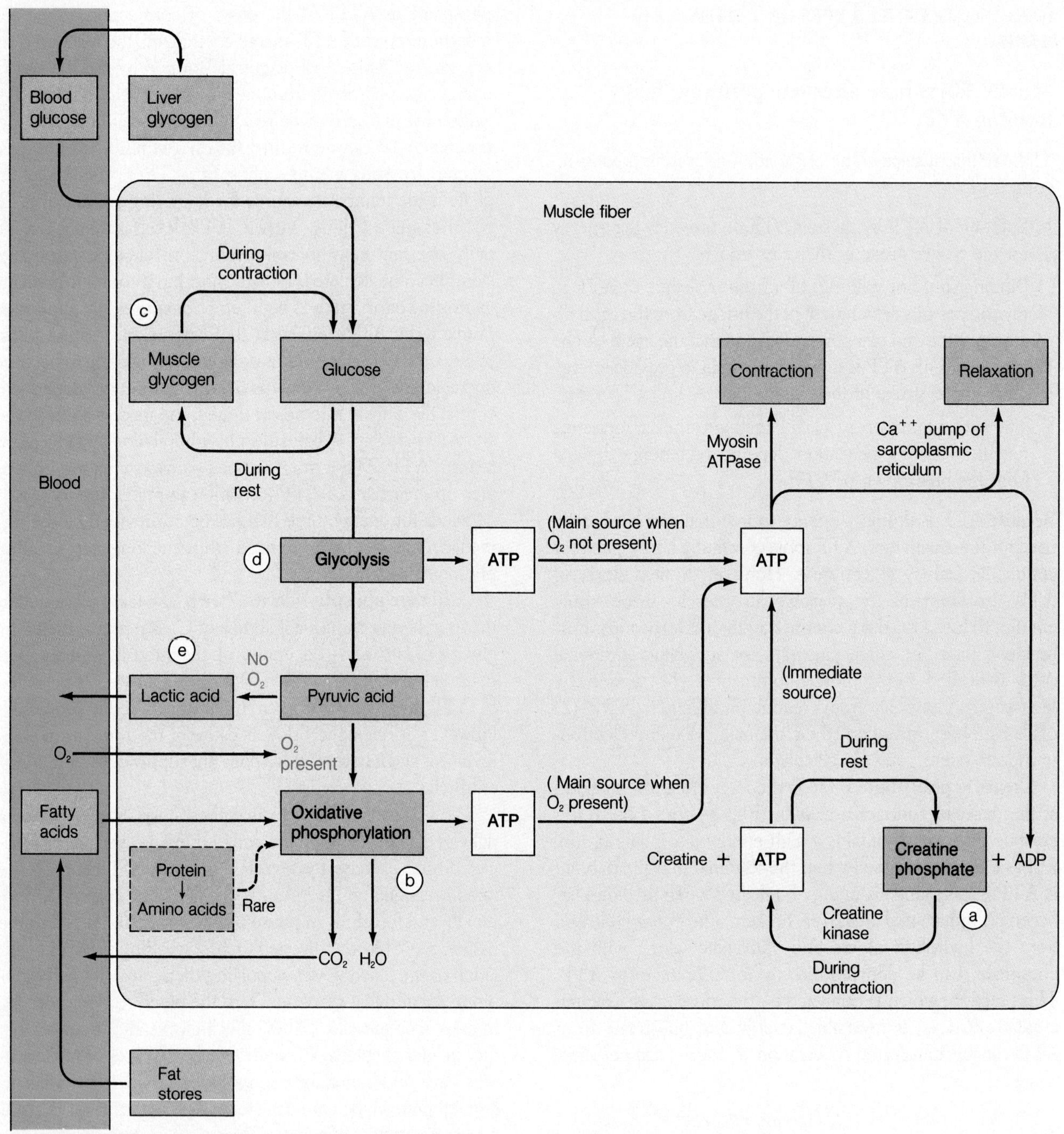

Figure 8-17 Metabolic Pathways Producing ATP Utilized during Muscle Contraction and Relaxation *See text for an explanation of the items identified by the circled letters.*

During light exercise (such as walking) to moderate exercise (such as jogging or swimming), muscle cells are able to form sufficient amounts of ATP through oxidative phosphorylation to keep pace with the modest energy demands of the contractile machinery for prolonged periods of time. They do so by drawing initially on their own limited reserves, but they then depend primarily on delivery of adequate O_2 and nutrient supplies via the circulatory system to maintain their activity. Activity that can be supported in this way is known as *endurance-type exercise* or *aerobic* ("with O_2") *exercise.*

There are cardiovascular limits to the amount of O_2 that can be delivered to a muscle. In near maximal contractions, the blood vessels that course through the muscle are almost closed by the powerful contraction, severely limiting O_2 availability to the muscle fibers. Furthermore, the relatively slow oxidative-phosphorylation system may not be able to produce ATP rapidly enough to meet the muscle's needs in intense types of activity, even when O_2 is available. When O_2 delivery or oxidative phosphorylation cannot keep pace with the demand for ATP formation as the intensity of exercise increases, the muscle fibers rely increasingly on **glycolysis** (Fig. 8–17d) to generate ATP (see p. 27). Glycolysis has two advantages. It can form ATP in the absence of O_2 (operating *anaerobically*, that is "without O_2"), and it can proceed rapidly compared to oxidative phosphorylation because of the fewer steps in glycolysis. Although glycolysis extracts much fewer ATP molecules from each nutrient molecule processed, it can proceed so much more rapidly that it can outproduce oxidative phosphorylation over a given period of time if sufficient glucose is present.

Even though anaerobic glycolysis provides a means to perform intense exercise when the O_2 delivery/oxidative-phosphorylation capacity is exceeded, there are two distinct disadvantages. First, large amounts of nutrient fuel must be processed, because glycolysis is much less efficient than oxidative phosphorylation in converting nutrient energy into the energy of ATP. Thus anaerobic glycolysis rapidly depletes the muscle's glycogen supplies. (Glycolysis yields a net of two ATP molecules for each glucose molecule degraded, forming two **pyruvic acid** molecules in the process. With adequate O_2 present, the oxidative-phosphorylation pathway can extract thirty-six molecules of ATP from each glucose molecule.) Second, the end-product of anaerobic glycolysis, pyruvic acid, is converted to **lactic acid** when it is unable to be further processed by the oxidative-phosphorylation pathway (Fig. 8–17e). Lactic-acid accumulation has been implicated in the muscle soreness that occurs during intense exercise. The delayed-onset pain and stiffness that begin the day following unaccustomed muscular exertion, however, are probably caused by reversible structural damage. Lactic acid picked up by the blood is responsible for the metabolic acidosis accompanying intense exercise. Furthermore, the fall in muscle pH caused by lactic-acid accumulation may play a role in the onset of muscle fatigue. Therefore, high-intensity exercise, being dependent on creatine-phosphate stores, anaerobic glycolysis, or both for ATP formation, can be sustained for only a short duration, in contrast to the body's prolonged ability to sustain aerobic activities.

Fatigue has multiple causes.

Contractile activity in a particular skeletal muscle cannot be maintained at a given level indefinitely. Eventually the tension in the muscle declines as **fatigue** sets in. There are apparently three different types of fatigue: muscular fatigue, neuromuscular fatigue, and central fatigue.

Muscular fatigue occurs when an exercising muscle can no longer respond to stimulation with the same degree of contractile activity. The underlying causes of muscular fatigue are unclear. The primary implicated factors include: (1) accumulation of lactic acid, which may inhibit key enzymes in the energy-producing pathways or excitation-contraction coupling process; and (2) depletion of energy reserves. The time of onset of fatigue varies with the type of muscle fiber, some fibers being more resistant to fatigue than others, and the intensity of the exercise, with more rapid onset of fatigue associated with high-intensity activities.

Evidence suggests that the limiting factor during fast, powerful activities may be at the neuromuscular junction. With **neuromuscular fatigue,** the active motor neurons are not able to synthesize acetylcholine as rapidly as is needed to sustain chemical transmission of action potentials from the motor neurons to the muscles.

Central fatigue, also known as **psychological fatigue,** occurs when the CNS no longer adequately activates the motor neurons supplying the working muscles. The person slows down or stops exercising even though the muscles are still able to perform. During strenuous exercise, central fatigue might stem from discomfort associated with the activity; it takes strong motivation (a will to win) to deliberately persevere when in pain. In less-strenuous activities, central fatigue might bring about reduced physical performance in association with boredom and monotony (such as assembly line work) or tiredness (lack of sleep). The mechanisms involved in central fatigue are poorly understood.

Increased oxygen consumption is necessary to recover from exercise.

A person continues to breathe deeply and rapidly for a period of time after having stopped exercising. The necessity for the elevated O_2 uptake during recovery from exercise is due to a variety of factors. The best known is repayment of an **oxygen debt** that was incurred during exercise, when contractile activity was being supported by ATP derived from nonoxidative sources such as creatine phosphate and anaerobic glycolysis. Oxygen is needed for recovery of the energy systems. During exercise, the creatine-phosphate stores of active muscles are reduced, lactic acid may have accumulated, and glycogen stores may have been tapped, the extent of which depends on the intensity and duration of the activity. During the recovery period, fresh supplies of ATP are formed by oxidative phosphorylation using the newly acquired O_2. Most of this ATP is used to resynthesize creatine phosphate to restore its reserves. This can be accomplished in a matter of a few minutes. Any accumulated lactic acid is converted back into pyruvic acid, part of which is used by the oxidative-phosphorylation system for ATP production. The remainder of the pyruvic acid is converted back into glucose by the liver. Most of this glucose, in turn, is used to replenish the glycogen stores drained from the muscles and liver during exercise. These biochemical transformations involving pyruvic acid require O_2 and take several hours for completion. Thus as the O_2 debt is repaid, the creatine-phosphate system is restored, lactic acid is removed, and glycogen stores are at least partially replenished. Unrelated to increased O_2 uptake, with grueling exercise in which glycogen supplies are severely depleted, such as marathon races, long-term recovery can take a day or more because the exhausted energy stores require nutrient intake for full replenishment. Thus, depending on the type and duration of activity, recovery can be complete within a few minutes or can require more than a day.

Part of the extra O_2 uptake during recovery is not directly related to the repayment of energy stores but instead is the result of a general metabolic disturbance following exercise. For example, the local increase in muscle temperature arising from heat-generating contractile activity speeds up the rate of all chemical reactions in the muscle tissue, including those dependent on O_2. Likewise, the secretion of epinephrine, a hormone that increases O_2 consumption by the body, is elevated during exercise. Until the circulating level of epinephrine returns to its pre-exercise state, O_2 uptake will be increased above normal.

There are three types of muscle fibers based on differences in ATP hydrolysis and synthesis.

Based on their biochemical capacities, there are three types of muscle fibers (Table 8–2):

1. slow-oxidative (type I) fibers;
2. fast-oxidative (type IIa) fibers; and
3. fast-glycolytic (type IIb) fibers.

As is implied by their names, the two major differences between these fiber types are their speed of contraction (slow or fast) and the type of enzymatic machinery they primarily use for ATP formation (oxidative or glycolytic). Fast fibers have higher myosin ATPase activity than slow fibers. The higher the ATPase activity, the more rapidly ATP is hydrolyzed and the faster the rate at which energy is made available for cross-bridge cycling. The result is a fast twitch, compared to the slower twitches of those fibers that split ATP more slowly. Fiber types also differ in their ATP-synthesizing ability. Those with a greater capacity to form ATP are more resistant to fatigue. Some fibers are better equipped for oxidative phosphorylation, whereas others rely primarily on anaerobic glycolysis for synthesizing ATP. Because oxidative-phosphorylation yields considerably more ATP from each nutrient molecule processed, it does not readily deplete energy stores. Furthermore, it does not result in lactic-acid accumulation. Oxidative types of muscle fibers are therefore more resistant to fatigue.

Other related characteristics distinguishing these three fiber types are summarized in Table 8–2. As would be expected, the oxidative fibers, both slow and fast, contain an abundance of mitochondria, the organelles that house the enzymes involved in the oxidative-phosphorylation pathway. Because adequate oxygenation is essential to support this pathway, these

Table 8–2 Characteristics of Skeletal Muscle Fibers

	Type of Fiber		
Characteristic	*Slow-Oxidative (Type I)*	*Fast-Oxidative (Type IIa)*	*Fast-Glycolytic (Type IIb)*
Myosin-ATPase activity	Low	High	High
Speed of contraction	Slow	Fast	Fast
Resistance to fatigue	High	Intermediate	Low
Oxidative phosphorylation capacity	High	High	Low
Enzymes for anaerobic glycolysis	Low	Intermediate	High
Mitochondria	Many	Many	Few
Capillaries	Many	Many	Few
Myoglobin content	High	High	Low
Glycogen content	Low	Intermediate	High
Fiber diameter	Small	Intermediate	Large
Intensity of contraction	Low	Intermediate	High

fibers are richly supplied with capillaries. Oxidative fibers also have a high myoglobin content, which helps support their O_2 dependency and imparts a red color to them, just as oxygenated hemoglobin is responsible for the red color of arterial blood. Accordingly, these muscle fibers often are referred to as **red fibers.**

In contrast, the fast fibers specialized for glycolysis contain few mitochondria but have a high content of glycolytic enzymes instead. Also, to supply the large amounts of glucose necessary for glycolysis, they contain an abundance of stored glycogen. Because the glycolytic fibers need relatively less O_2 to function, they receive only a meager capillary supply compared with the oxidative fibers. The glycolytic fibers contain very little myoglobin and therefore are pale in color, so they are sometimes called **white fibers.** (The most readily observable comparison between red and white fibers is the dark and white meat in poultry.) In addition to their higher myosin ATPase content, these fibers are also larger in diameter than the slow-oxidative fibers because of a greater abundance of actin and myosin filaments. With these additional power-generating filaments, the fast-glycolytic fibers are capable of rapidly producing large amounts of tension, but only for a short duration because of their dependence on glycolysis for energy production.

Fast-oxidative fibers share characteristics with each of the other two types. They have high ATPase activity like the fast-glycolytic fibers and high oxidative capacity like the slow-oxidative fibers. They contract more rapidly than the slow-oxidative fibers and can maintain the contraction for a longer period of time than the fast-glycolytic fibers. However, because their rate of ATP production by oxidative-phosphorylation cannot keep pace with the high rate of ATP splitting, they rely partially on glycolysis and are more prone to fatigue than the slow-oxidative fibers.

In humans, most of the muscles contain a mixture of all three fiber types, the percentage of each largely being determined by the type of activity for which the muscle is specialized. Accordingly, there is a high proportion of slow-oxidative fibers in muscles specialized for maintaining low-intensity contractions for long periods of time without fatigue, such as the muscles of the back and legs that support the body's weight against the force of gravity. There is a preponderance of fast-glycolytic fibers in the arm muscles, which are adapted for performing rapid, forceful movements such as lifting heavy objects.

The percentage of these various fibers differs not only between muscles within an individual but also differs considerably among individuals. Most of us have an average of about 50% each of fast and slow fibers. Those genetically endowed with a higher percentage of the fast-glycolytic fibers are good candidates for power and sprint events, whereas those with a greater proportion of slow-oxidative fibers are more likely to be successful in endurance activities such as marathon races. Of course, success in any event will depend on many factors other than genetic endowment, such as the extent and type of training and the level of dedication.

Muscle fibers adapt considerably in response to the demands placed on them.

Not only is the nerve supply to a skeletal muscle essential for initiating contraction, but also the motor neurons supplying a skeletal muscle are important in maintaining the muscles's integrity and chemical composition. Different types of exercise produce different patterns of neuronal discharge to the muscle involved. Depending on the pattern of neural activity, long-term adaptive changes occur in the muscle fibers, enabling them to most efficiently respond to the types of demands placed on the muscle. The two types of changes that can be induced in muscle fibers are changes in their ATP-synthesizing capacity and changes in their diameter. Regular endurance (aerobic) exercise (submaximal effort of long duration, such as long-distance jogging or swimming) induces changes that enable the muscles to use O_2 more efficiently. Mitochondria increase in number within the slow- and fast-oxidative fibers, which are the ones primarily recruited during aerobic exercise, and the number of capillaries supplying blood to these fibers increases. Muscles so adapted are better able to endure prolonged activity without fatiguing, but they do not change in size.

The actual size of the muscles, on the other hand, can be increased by regular bouts of short-duration, high-intensity *resistance training* (at or near maximal force of contraction against a heavy load involving tetanic contraction of almost all of the motor units of participating muscles, such as in weight training). The resulting muscle enlargement comes primarily from an increase in diameter (**hypertrophy**) of the fast-glycolytic fibers that are called into play during such powerful contractions. Most of the fiber thickening is a consequence of increased synthesis of myosin and actin filaments, which permits a greater opportunity for cross-bridge interaction and subsequently increases the muscle's contractile strength. The resultant bulging muscles are better adapted to activities that require intense strength for brief periods, but endurance has not been improved. Furthermore, **hyperplasia** (an increase in the number of muscle cells) is believed to contribute to a small extent to muscle enlargement. Muscle cells are unable to divide by mitotic cell division, but experimental evidence suggests that greatly enlarged fibers can split lengthwise down the middle, resulting in an increase in the number of fibers.

Men's muscle fibers are thicker and their muscles, accordingly, are larger and stronger than those of women, even with-

A CLOSER LOOK AT EXERCISE PHYSIOLOGY

You Be the Judge: Are Athletes Who Use Steroids to Gain Competitive Advantage Really Winners or Losers?

The testing of athletes for drugs and the exclusion from competition of those found to be using substances outlawed by sports federations has stirred considerable controversy and has been much publicized. One such group of drugs are **anabolic androgenic steroids** (*anabolic* means "build up of tissues"; *androgenic* means "male producing"; *steroids* are a class of hormone). These agents are closely related to testosterone, the natural male sex hormone, which is responsible for promoting the increased muscle mass characteristic of males.

Before their use was outlawed, these agents were taken by many athletes who specialized in power events such as weight lifting and sprinting in the hopes of increasing muscle mass and, accordingly, increasing muscle strength in an attempt to gain a competitive edge. Both male and female athletes have resorted to using these substances. The adverse effects of these drugs, however, outweigh any benefits derived. In males, who already have an abundance of their own naturally produced testosterone, sought-after increases in muscle mass, strength, and body weight are not well demonstrated. Small, variable gains occur only in conjunction with a regular muscle-building exercise program and are not experienced by all who take the agents. In fact, some suggest that the changes that do occur are largely due to the exercise alone. In females, who lack potent androgenic hormones, the effect of promoting "male type" muscle mass and strength is readily demonstrable when these agents are used. However, these drugs also "masculinize" females in other ways, such as by inducing growth of facial and overall body hair and by lowering the voice.

These are but a few of the side effects of using anabolic steroids. In both males and females, these agents adversely affect the liver and the reproductive and cardiovascular systems.

out weight training, because of the actions of testosterone, a steroid hormone secreted primarily in males. Testosterone promotes the synthesis and assembly of myosin and actin and is responsible for the naturally larger muscle mass of men. This fact has led some athletes, both males and females, to the dangerous practice of taking this or closely related steroids to increase their athletic performance. (See the accompanying boxed feature, A Closer Look at Exercise Physiology.)

Apparently, all of the muscle fibers within a single motor unit are of the same fiber type. This pattern usually is established early in life, but there is good evidence that the two types of fast-twitch fibers are interconvertible, depending on training efforts. Regular endurance activities can convert fast-glycolytic fibers into fast-oxidative fibers, whereas fast-oxidative fibers can be shifted to fast-glycolytic fibers in response to power events such as weight training. Slow and fast fibers are not interconvertible, however. Although training can induce changes in muscle fibers' metabolic support systems, whether a fiber is fast- or slow-twitch is apparently established early in development by the fiber's nerve supply. Experimental switching of motor neurons supplying slow muscle fibers with those supplying fast fibers results in a gradual reversal of the speed at which these fibers contract.

As we have seen, the demands placed on a muscle over a period of time by the pattern of neural activity to that muscle can substantially alter the muscle's capacity. The underlying cellular mechanisms responsible for these changes are unknown. Strength but not endurance improves in response to high-resistance weight training; endurance but not strength improves in response to aerobic exercise. An exercise program can be designed to improve both strength and endurance, but neither can be developed to maximum capacity under these circumstances. Therefore, an exercise program must be tailored to the desired outcome. If the established pattern of activity is stopped, the muscles will gradually return to their pre-exercise condition.

At the other extreme, if a muscle is not used, the content of actin and myosin decreases, the fibers become smaller, and the muscle accordingly decreases in mass (**atrophies**) and becomes weaker. Muscle atrophy can result in two ways. **Disuse**

Adverse Effects on the Reproductive System

In males, testosterone secretion and sperm production by the testes are normally controlled by hormones released from the anterior pituitary gland. In negative-feedback fashion, testosterone inhibits secretion of these controlling hormones so that a constant circulating level of testosterone is maintained. The anterior pituitary is similarly inhibited by androgenic steroids taken as a drug. As a result, because the testes do not receive their normal stimulatory input from the anterior pituitary, sperm production decreases, the testes atrophy, and testosterone secretion declines.

In females, inhibition of the anterior pituitary by the androgenic drugs results in repression of the hormonal output that controls ovarian function. The result is failure to ovulate, menstrual irregularities, and decreased secretion of female sex hormones. Since the latter are "feminizing hormones," their decline results in diminution in breast size and other female characteristics.

Adverse Effects on the Liver

Liver dysfunction is not uncommon with high steroid intake, perhaps because the liver, which normally inactivates steroid hormones and prepares them for urinary excretion, is overloaded by the excess steroid intake.

Adverse Effects on the Cardiovascular System

Use of anabolic steroids induces several changes in the cardiovascular system that increase the risk of developing atherosclerosis, which in turn is associated with an increased incidence of heart attacks and strokes (see p. 292). Among these adverse cardiovascular effects are: (1) a reduction in high-density lipoproteins (HDL), which are the "good" cholesterol carriers that help remove cholesterol from the body, and (2) an elevation in blood pressure. Furthermore, damage to the heart muscle itself has been demonstrated in animal studies.

Adverse Effects on Behavior

Although still controversial, anabolic steroid use appears to promote aggressive, even hostile behavior.

Thus, for health reasons, not even taking into account the ethical issues involved, athletes should not use anabolic steroids in an attempt to gain a competitive advantage. Despite these potential dangers, however, some athletes continue to illegally use these drugs and seek measures to mask their use from testing procedures.

atrophy is brought about by lack of use of a muscle for a long period of time even though the nerve supply is intact, as when a cast or brace must be worn or during prolonged bed confinement. **Denervation atrophy** occurs following the loss of the nerve supply to a muscle. If the muscle is stimulated electrically until innervation can be reestablished, such as during regeneration of a severed peripheral nerve, atrophy can be diminished but not entirely prevented. Contractile activity itself obviously plays an important role in preventing atrophy; however, poorly understood factors released from active nerve endings, perhaps packaged with the ACh vesicles, apparently contribute to the integrity and growth of muscle tissue.

Another long-term change that can occur in skeletal muscles is associated with a variety of hereditary pathological conditions grouped under the heading of **muscular dystrophy.** These conditions have in common a progressive degeneration of contractile elements, which are ultimately replaced by fibrous tissue. This gradual muscular wasting is characterized by progressive weakness over a period of years, sometimes resulting in premature death from respiratory failure if the diaphragm becomes too weak to function adequately. The underlying defect responsible for Duchenne muscular dystrophy, the most common and most devastating form of the disease, has recently been shown to be a missing protein, **dystrophin,** resulting from a defective gene. Even though this protein represents only 0.002% of the total amount of skeletal-muscle protein, its presence is crucial. Its absence is hypothesized to permit a constant Ca^{++} leakage from the lateral sacs. This Ca^{++} presumably activates an enzyme that dissolves muscle fibers, leading to the muscle wasting and ultimate fibrosis characterizing the disorder.

In a normal muscle, damaged muscle fibers are not replaced by fibrous tissue. Instead, limited repair is possible, even though muscle cells cannot divide mitotically to replace damaged cells. A small population of the same undifferentiated cells that formed the muscle during embryonic development, **myoblasts,** remain close to the muscle surface. When a muscle fiber is damaged, it can be replaced by fusion of a group of these myoblasts to form a large multinucleated cell, which immediately begins to synthesize and assemble the intracellular machinery characteristic of the muscle. With extensive injury, this limited mechanism is not adequate to completely replace all of the lost fibers, in which case the remaining fibers often hypertrophy to compensate.

Muscle Mechanics

The two primary types of contraction are isotonic and isometric.

Typically, a muscle is attached to at least two different bones across a joint by means of tendons that extend from each end of the muscle (Fig. 8–18). When the muscle shortens during contraction, the position of the joint is changed as one bone is moved in relationship to the other—for example, *flexion* of the elbow joint by contraction of the biceps muscle and *extension* of the elbow by contraction of the triceps. The end of the muscle attached to the more stationary part of the skeleton is called the **origin,** whereas the end attached to the skeletal part that moves is referred to as the **insertion.**

Not all muscle contractions result in movement. For a muscle to shorten during contraction, the external tension developed in the muscle must exceed the forces that oppose movement of the bone to which the muscle's insertion is attached. In the case of elbow flexion, the opposing force, or **load,** is the weight of an object being lifted. When you flex your elbow without lifing any external object, there is still a load, albeit minimal—the weight of your forearm being moved against the force of gravity.

There are two primary types of contraction, depending on whether or not the muscle changes length during contraction. In an **isotonic contraction,** muscle tension remains constant as the muscle changes length. In an **isometric contraction,** the muscle is prevented from shortening, so the development of tension occurs at constant muscle length. The same internal events occur in both isotonic and isometric contractions; that is, the tension-generating contractile process is turned on by muscle excitation, the cross bridges start cycling, and the internal tension created in the activated fibers is transmitted through the series-elastic component to exert external tension on the bone at the site of the muscle's insertion.

Considering your biceps as an example, assume that you are going to lift an object. When the external tension develop-

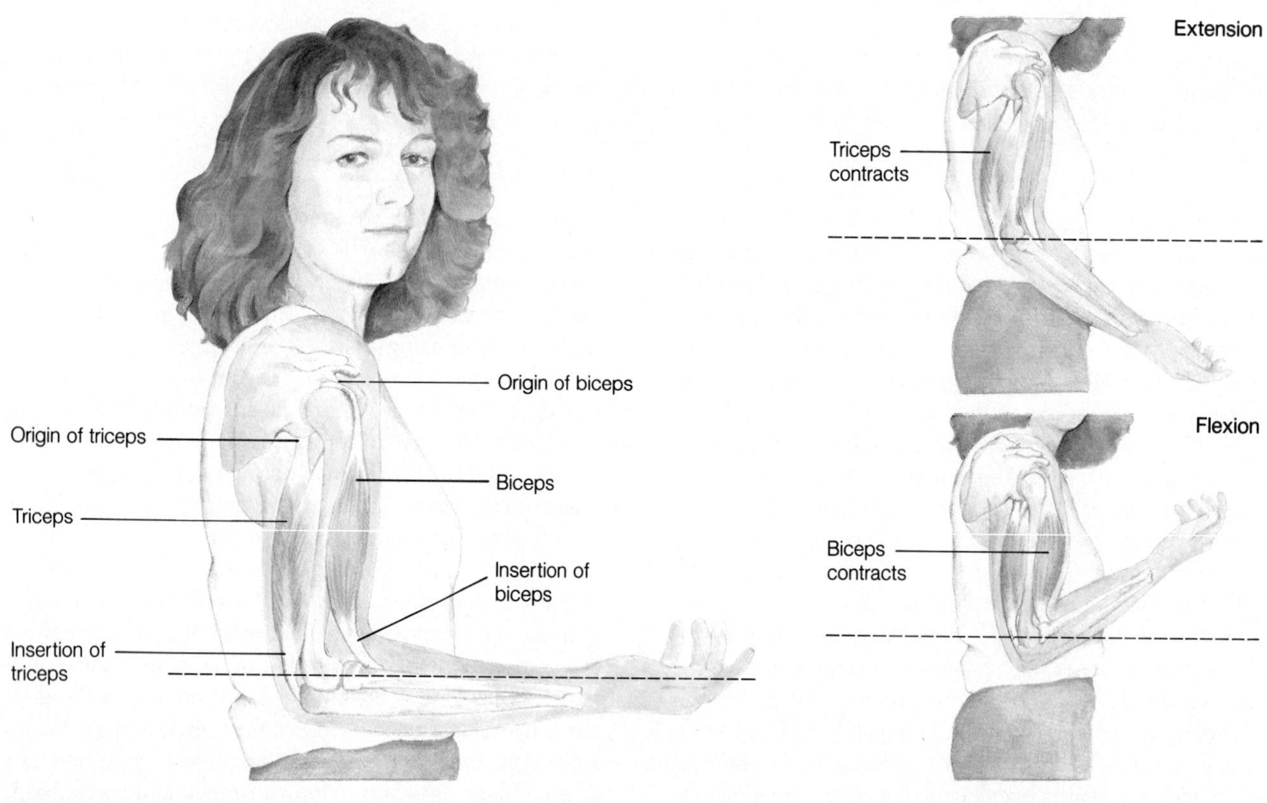

Figure 8–18 Extension and Flexion of Elbow Joint

ing in your biceps becomes great enough to overcome the weight of the object in your hand, you can lift the object, with the whole muscle shortening in the process. Because the weight of the object does not change as it is lifted, the external muscle tension, which just barely exceeds the load, remains constant throughout the period of shortening. This is an *isotonic* (literally, "constant tension") contraction. What happens if you try to lift an object too heavy for you (that is, if the external tension you are capable of developing in your arm muscles is less than that required to lift the load)? In this case, the muscle cannot shorten and lift the object but remains at constant length in spite of the development of tension, so an *isometric* ("constant length") contraction occurs. In addition to occurring when the load is too great, isometric contractions also take place when the tension developed in the muscle is deliberately less than the load to keep the muscle at fixed length although it is capable of developing more tension. Isometric contractions are important for maintaining posture (such as keeping the legs stiff while standing) and for supporting objects in a fixed position. Isotonic contractions are used for body movements and to do work by moving external objects.

There are actually two types of isotonic contraction—**concentric** and **eccentric.** In both cases the muscle changes length at constant tension. However, with concentric contractions the muscle shortens, whereas with eccentric contractions the muscle lengthens or is stretched while contracting. In the latter case, the contractile activity is resisting the stretch. An example is lowering a load to the ground. During this action, the muscle fibers in the biceps are lengthening but are still contracting in opposition to being stretched. This tension supports the weight of the object.

The body is not limited to pure isotonic and isometric contractions. Muscle length and tension frequently vary throughout a range of motion. Think about drawing a bow and arrow. The biceps' muscle tension continuously increases to overcome the progressively increasing resistance as the bow becomes further stretched. At the same time, the muscle progressively shortens as the bow is drawn farther back. Such a contraction occurs at neither constant tension nor constant length.

The velocity of shortening is related to the load.

The load is also an important determinant of the **velocity** or speed of shortening (Fig. 8–19). The greater the load, the lower the velocity at which the muscle fiber shortens during an isotonic tetanic contraction. The velocity of shortening is maximal when there is no external load, progressively decreases with an increasing load, and becomes zero (no shortening—isometric contraction) when the load equals or exceeds the maximal tetanic tension. You have frequently experienced this load-velocity relationship. You can lift light objects requiring little muscle tension quickly, whereas you can lift very heavy objects only slowly, if at all. This relationship between load and shortening velocity is a fundamental property of muscle, presumably because it takes the cross bridges longer to stroke against a greater load. Thus, there are two factors that determine the speed with which a muscle contracts: the load and the myosin ATPase activity of the contracting fibers (fast- or slow-twitch).

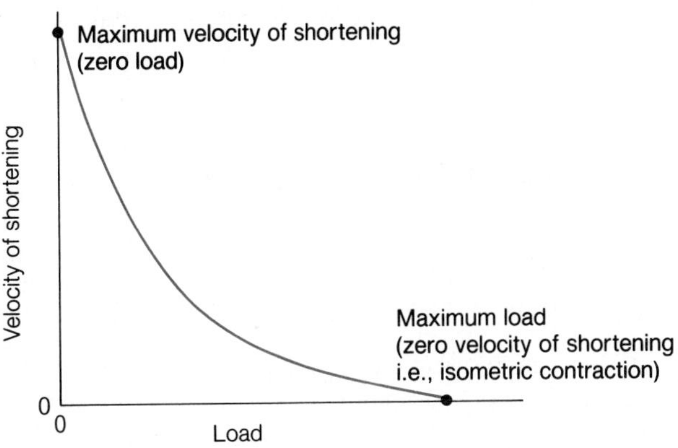

Figure 8–19 Load-Velocity Relationship

Although muscles can accomplish work, much of the energy is converted to heat.

Muscle accomplishes work in a physical sense only when an object is moved. **Work** is defined as force times distance. **Force** can be equated to the muscle tension required to overcome the load (the weight of the object). The amount of work accomplished by a contracting muscle therefore depends on how much an object weighs and how far it is moved. In an isometric contraction when no object is moved, the muscle contraction's efficiency as a producer of work is zero. All energy consumed by the muscle during the contraction is converted to heat. In an isotonic contraction, the muscle's efficiency is about 25%. Of the energy consumed by the muscle during the contraction, 25% is realized as external work whereas the remaining 75% is converted to heat. Much of this heat is not really wasted in a physiological sense because it is used in maintaining body temperature. In fact, shivering, a form of involuntarily induced skeletal muscle contraction, is a well-known mechanism to increase heat production on a cold day. Heavy exercise on a hot day, on the other hand, may overheat the body, because the normal heat-loss mechanisms may be unable to compensate for this increase in heat production (chapter 17).

Control of Motor Movement

Many inputs influence motor-unit output.

Particular patterns of motor-unit output are responsible for motor activity, ranging from maintenance of posture and balance, to stereotypical locomotor movements such as walking, to individual, highly skilled motor activity such as gymnastics. Control of any motor movement, no matter what its level of complexity, depends on converging input to the motor neurons of specific motor units. Three levels of input control motor neuron output (Fig. 8–20):

1. Input from afferent neurons, usually through intervening interneurons, at the level of the spinal cord—that is, spinal reflexes.
2. Input from the primary motor cortex. Fibers originating from cell bodies of pyramidal cells within the primary motor cortex (see p. 128) descend directly without synaptic interruption to terminate on motor neurons (or on local interneurons that terminate on motor neurons). These fibers make up the **corticospinal** (or **pyramidal**) **system.**
3. Input from the **multineuronal** (or **extrapyramidal**) **system.** The pathways composing this system include a number of synapses that involve many regions of the brain. The final link in multineuronal pathways is the brain stem, especially the reticular formation, which in turn is influenced by motor regions of the cortex, the cerebellum, and the basal nuclei. In addition, the motor cortex itself is interconnected with the thalamus plus premotor and supplementary motor areas. Only the primary motor cortex and brain stem directly influence motor neurons; the other involved brain regions indirectly regulate motor activity by adjusting motor output from the motor cortex and brain stem. There are a number of complex interactions between these various brain regions, the most important of which are represented in Figure 8–20. (See chapter 5 for further discussion of the specific roles and interactions of these brain regions.)

The corticospinal system primarily mediates performance of fine, discrete, voluntary movements of the hands and fingers, such as doing intricate needlework. Premotor and supplementary-motor areas, with input from the cerebrocerebellum, plan the voluntary motor command that is issued to the appropriate motor neurons by the primary motor cortex through this descending system. The multineuronal system, in contrast, is primarily concerned with regulation of overall body posture involving involuntary movements of large muscle groups of the trunks and limbs. Considerable complex interaction and overlapping of function exist between these two systems. For example, to voluntarily manipulate your fingers to do needlework, you subconsciously assume a particular posture of your arms to enable you to hold your work.

Some of the inputs converging on motor neurons are excitatory, whereas others are inhibitory. Coordinated movement depends on an appropriate balance of activity in these inputs. If an inhibitory system originating in the brain stem is disrupted, muscles become hyperactive (increased muscle tone; augmented limb reflexes) because of the unopposed activity in excitatory inputs to motor neurons, a condition known as **spastic paralysis.** In contrast, loss of excitatory input, such as that accompanying destruction of descending excitatory pathways exiting the primary motor cortex, brings about **flaccid paralysis** (muscles relaxed; inability to voluntarily contract muscles, although reflex activity is still present) on the opposite half of the body (**hemiplegia,** or paralysis of one side of the body). Disruption of all descending pathways, as in traumatic severence of the spinal cord, is accompanied by flaccid paralysis below the level of the damaged region—**quadriplegia** (paralysis of all four limbs) in upper spinal-cord damage and **paraplegia** (paralysis of the legs) in lower spinal-cord injury. Destruction of motor neurons—either their cell bodies or efferent fibers—causes flaccid paralysis and lack of reflex responsiveness in the affected muscles. Damage to the cerebellum or basal nuclei does not result in paralysis but instead in uncoordinated, clumsy activity and inappropriate patterns of movement. These regions normally smooth out activity initiated voluntarily. Damage to higher cortical regions involved in planning motor activity results in the inability to establish appropriate motor commmands to accomplish desired goals.

Muscle spindles and the Golgi tendon organs provide afferent information essential for controlling skeletal-muscle activity.

Coordinated, purposeful skeletal-muscle activity depends on afferent input from a variety of sources. At a simple level, afferent signals indicating your finger is touching a hot stove trigger reflex contractile activity in appropriate arm muscles to withdraw the hand from the injurious stimulus. At a more complex level, if you are going to catch a ball, the motor systems of your brain must program sequential motor commands that will move and position your body correctly for the catch based on predictions of the ball's direction and rate of move-

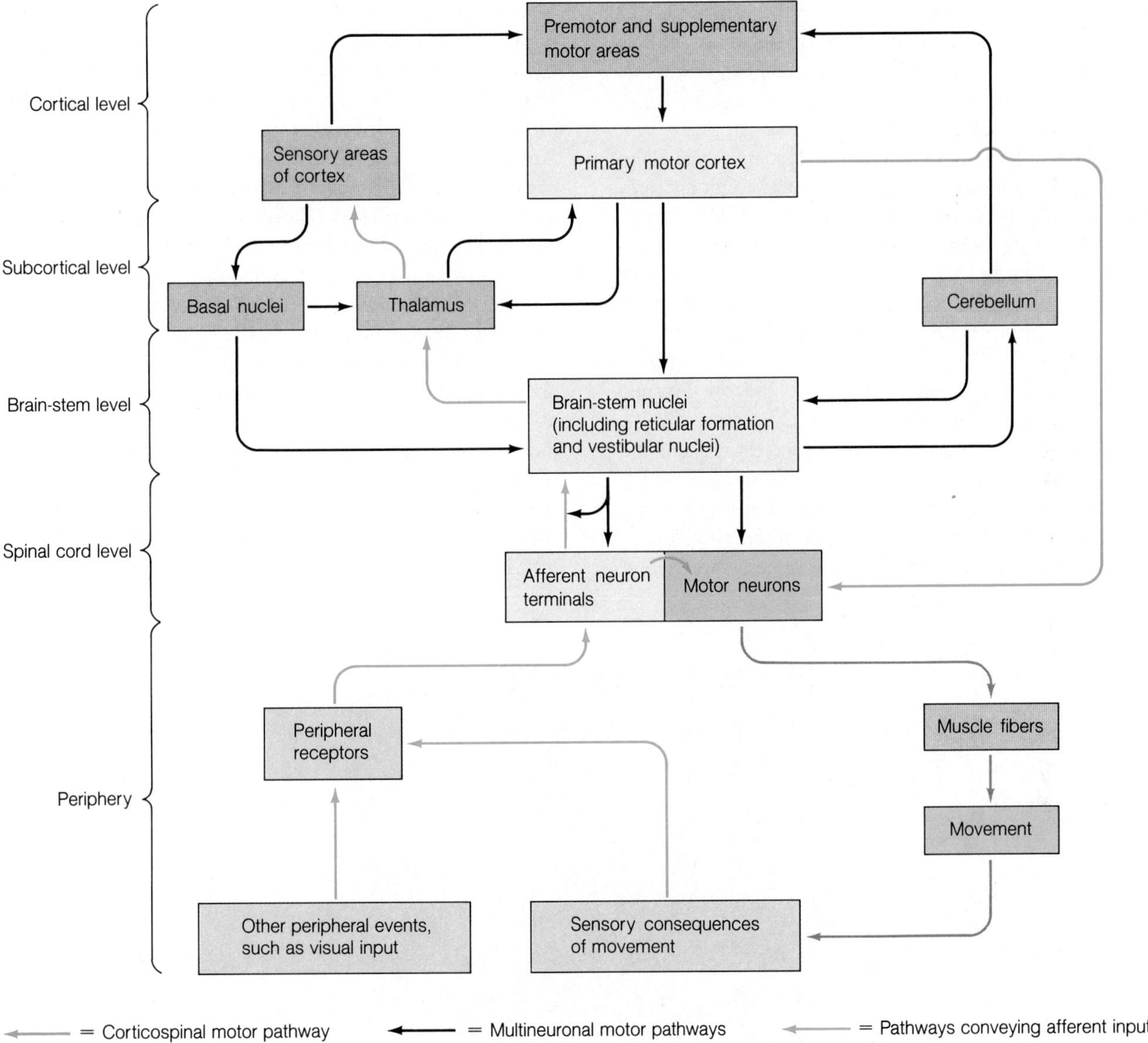

Figure 8–20 Motor Control *The CNS is constantly apprised of muscle length and tension and other peripheral events via pathways conveying afferent input, so it can program coordinated, purposeful skeletal muscle activity. Motor movement is controlled by input to the motor neurons from (1) afferent neuron terminals at the level of the spinal cord; (2) the primary motor cortex, via the corticospinal motor pathway; and (3) brain-stem nuclei, which serve as the final link in a complex system of multineuronal motor pathways involving many regions of the brain.*

ment provided by visual input. Many muscles acting simultaneously or alternately at different joints are called into play to shift your body's location and position rapidly, maintaining your balance in the process. It is critical to have ongoing input about your body position with respect to the surrounding environment, as well as the position of your various body parts in relationship to each other. This information is necessary for establishing a neuronal pattern of activity to perform the desired movement. Your CNS must know the starting position of your body to appropriately program muscle activity. Further, it must be constantly apprised of the progression of movement it has initiated so that it can make adjustments as needed. Your brain receives this information from receptors in your eyes, joints, vestibular apparatus, and skin, as well as from the muscle themselves.

Two types of muscle receptors—muscle spindles and Golgi tendon organs—monitor changes in muscle length and tension. This information is used in two ways: (1) to apprise motor areas of the brain of muscle length and tension, and (2) to control muscle length and tension in negative-feedback fashion by means of local spinal reflexes. Muscle length is monitored by muscle spindles whereas changes in muscle tension are detected by Golgi tendon organs. Both of these receptor types are activated by muscle stretch, but they are designed to convey different types of information. Let us see how.

Muscle spindles, which are distributed throughout the fleshy part of a skeletal muscle, consist of collections of specialized **intrafusal fibers** that lie within connective tissue capsules parallel to the "ordinary" **extrafusal fibers** (Fig. 8–21). Each muscle spindle has its own private afferent and efferent nerve supply. An intrafusal fiber has a noncontractile central portion and contractile elements at both ends. Two types of afferent sensory endings serve as muscle spindle receptors, both of which are activated by stretch. The **primary (annulospiral) endings** are wrapped around the central portions of the intrafusal fibers; they detect changes in the length of the fibers during stretching as well as the speed with which it occurs. The **secondary (flower-spray) endings,** which are clustered at the end segments of many of the intrafusal fibers, are sensitive only to changes in length.

Whenever the whole muscle is passively stretched, the intrafusal fibers within its muscle spindles are likewise stretched. This increases the rate of firing in the afferent nerve fibers whose terminals are wrapped around the stretched spindle fibers. The afferent neuron directly synapses on the alpha motor neuron that innervates the same muscle, resulting in contraction of that muscle. This **stretch reflex** serves as a local negative-feedback mechanism to resist any passive changes in muscle length so that optimal resting length can be maintained.

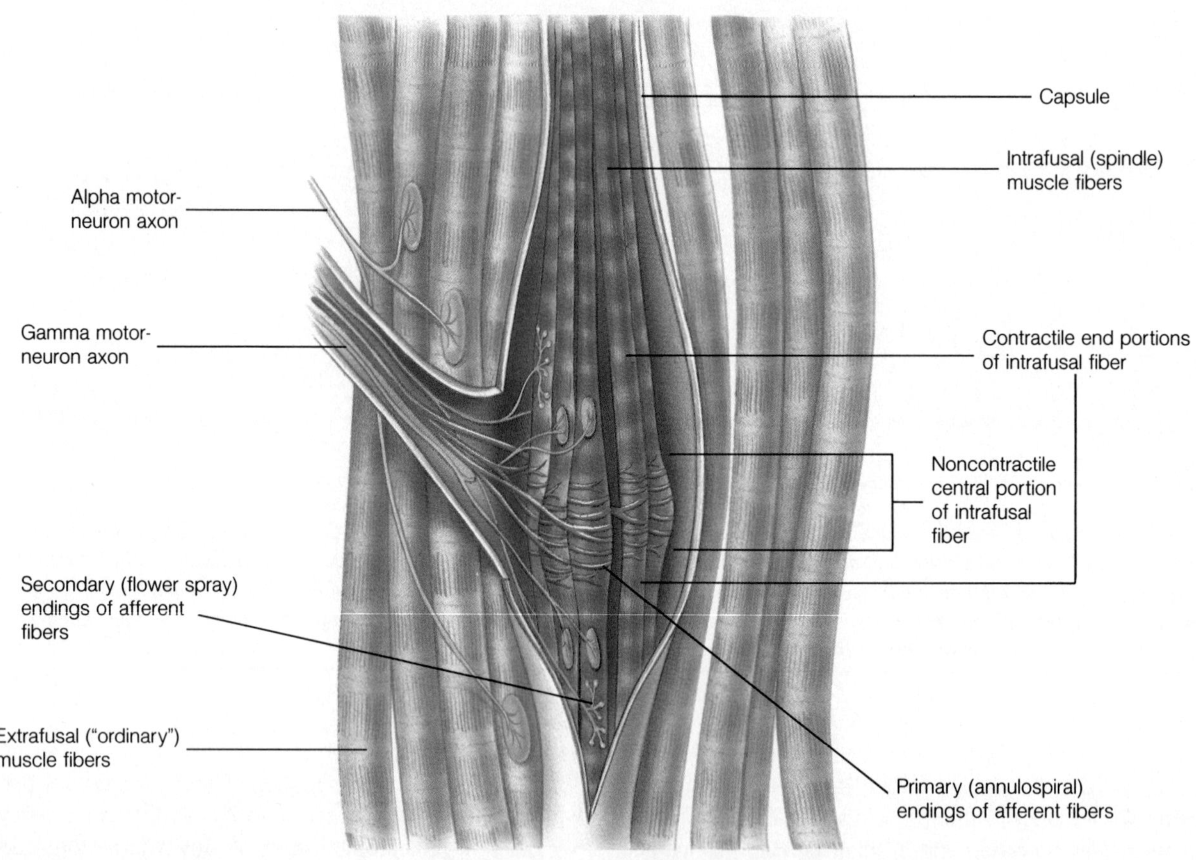

Figure 8–21 Muscle Spindle

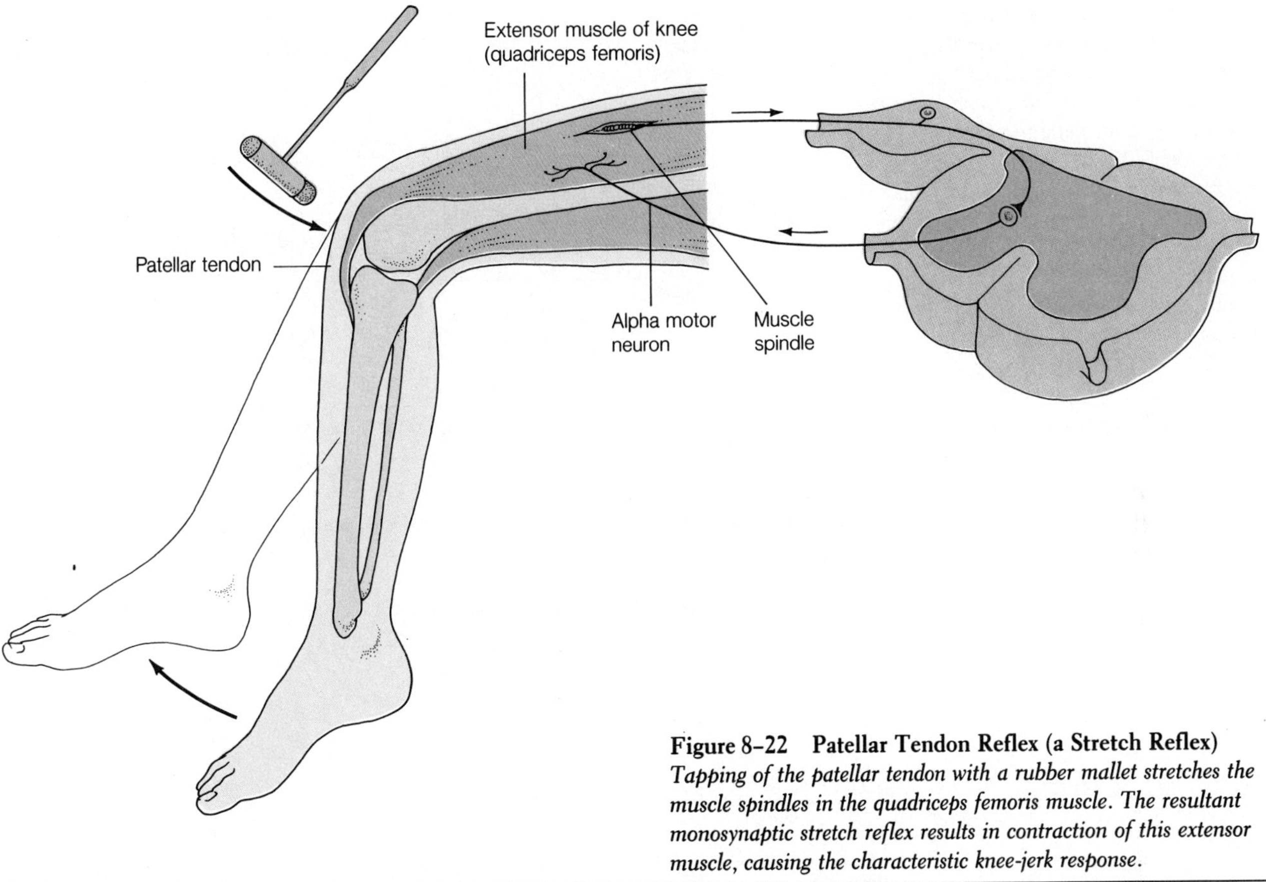

Figure 8–22 Patellar Tendon Reflex (a Stretch Reflex) *Tapping of the patellar tendon with a rubber mallet stretches the muscle spindles in the quadriceps femoris muscle. The resultant monosynaptic stretch reflex results in contraction of this extensor muscle, causing the characteristic knee-jerk response.*

The classical example of the stretch reflex is the **patellar tendon,** or **knee-jerk, reflex** (Fig. 8–22). The extensor muscle of the knee is the quadriceps femoris, which forms the anterior portion of the thigh and is attached just below the knee to the tibia (the shin bone) by the patellar tendon. Tapping of this tendon with a rubber mallet passively stretches the quadriceps muscle, activating its spindle receptors. The resultant stretch reflex brings about contraction of this extensor muscle, causing the knee to extend and raise the foreleg in the well-known knee-jerk fashion. This test is routinely performed as a preliminary assessment of nervous system function. A normal knee jerk indicates to a physician that a number of neural and muscular components—muscle spindle, afferent input, alpha motor neurons, efferent output, neuromuscular junctions, and the muscles themselves—are functioning normally. It also indicates the presence of an appropriate balance of excitatory and inhibitory input to the motor neurons from higher brain levels. Muscle jerks may be absent or depressed with loss of higher-level excitatory inputs or may be greatly exaggerated with loss of inhibitory input to the motor neurons from higher brain levels.

The primary purpose of the stretch reflex is to resist the tendency for the passive stretch of extensor muscles caused by gravitational forces when a person is standing upright. Whenever the knee joint tends to buckle because of gravity, the quadriceps muscle is stretched. The resultant enhanced contraction of this extensor muscle brought about by the stretch reflex quickly straightens out the knee, holding the limb extended so that the person remains standing.

Each muscle spindle has its own efferent nerve supply, a **gamma motor neuron,** as distinguished from the alpha motor neurons that supply the regular skeletal-muscle fibers (Fig. 8–21). Gamma motor neurons initiate contraction of the muscular end regions of intrafusal fibers. This contractile response is too weak to have any influence on whole-muscle tension, but it does have an important localized effect on the muscle spindle itself. If there were no compensating mechanisms, shortening of the whole muscle by alpha motor-neuron stimulation of extrafusal fibers would cause slack in the spindle fibers so that they would be less sensitive to stretch and therefore not as effective as muscle-length detectors (Fig. 8–23). **Coactivation** of the gamma motor-neuron system along with

Figure 8–23 Muscle-Spindle Function *(a) Pathways involved in the monosynaptic stretch reflex and coactivation of alpha and gamma motor neurons. (b) Status of muscle spindle when muscle is relaxed. (c) Status of muscle spindle when muscle is contracted upon alpha motor neuron stimulation. (d) Status of muscle spindle when both muscle and muscle spindle are contracted upon alpha and gamma motor neuron coactivation.*

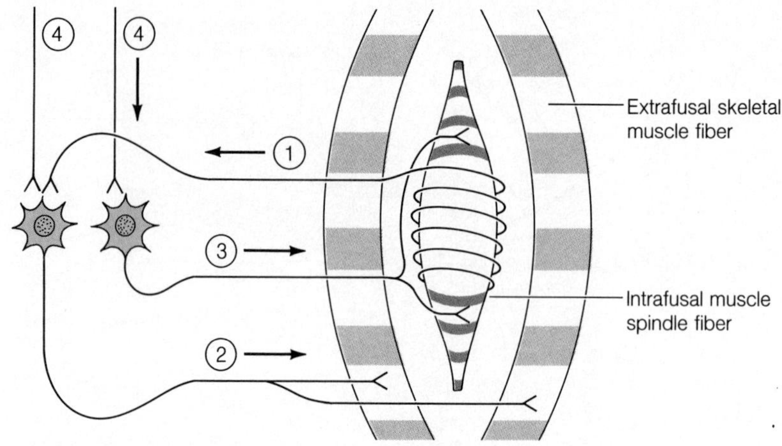

① = Afferent input from annulospiral endings of muscle spindle fiber
② = Alpha motor-neuron output to regular skeletal muscle fiber
①→② = Stretch reflex pathway
③ = Gamma motor-neuron output to contractile end portions of spindle fiber
④ = Descending pathways coactivating alpha and gamma motor neurons

(a)

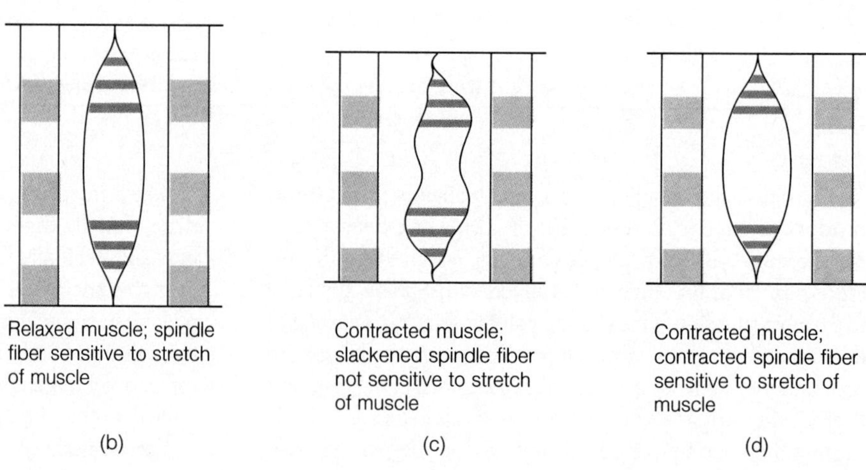

Relaxed muscle; spindle fiber sensitive to stretch of muscle

(b)

Contracted muscle; slackened spindle fiber not sensitive to stretch of muscle

(c)

Contracted muscle; contracted spindle fiber sensitive to stretch of muscle

(d)

the alpha motor-neuron system during reflex and voluntary contractions takes the slack out of the spindle fibers as the whole muscle shortens, permitting these receptor structures to maintain their high sensitivity to stretch over a wide range of muscle lengths. When gamma motor-neuron stimulation triggers simultaneous contraction of both end muscular portions of an intrafusal fiber, this pulls in opposite directions on the noncontractile central portion, tightening this region and taking out the slack. Whereas the extent of alpha motor-neuron activation depends on the intended strength of the motor response, the extent of simultaneous gamma motor-neuron activity to the same muscle depends on the anticipated distance of shortening.

In contrast to muscle spindles, which lie within the belly of the muscle, **Golgi tendon organs** are located in the tendons of the muscle, where they are able to respond to changes in the muscle's external tension rather than to changes in its length. Because a number of factors determine the tension developed in the whole muscle during contraction (for example, frequency of stimulation or length of muscle at the onset of contraction), it is essential that motor control systems be apprised of the tension actually achieved so that adjustments can be made if necessary.

The Golgi tendon organs consist of endings of afferent fibers entwined within the bundles of collagen fibers making up the tendon. When the extrafusal muscle fibers contract, the resultant pull on the tendon straightens the collagen bundles. This, in turn, increases the external tension exerted on the bone to which the tendon is attached and stretches the entwined Golgi organ afferent receptor endings in the process. The resulting distortion causes the afferent fibers to fire, the frequency of which is directly related to the tension developed.

The afferent information is sent to the brain. In addition, other branches of the afferent neuron arising from the Golgi tendon organ inhibit, by means of an interneuron, the alpha motor neurons of the same muscle. This reflex is apparently protective in nature. When the tension becomes great enough, the high level of inhibitory input from the activated Golgi tendon organs counterbalances excitatory inputs to the alpha motor neurons. This halts further contraction and brings about sudden reflex relaxation, thus helping prevent damage to muscle or tendon from excessive, tension-developing muscle contractions.

Smooth and Cardiac Muscle

Smooth and cardiac muscle share some basic properties with skeletal muscle.

The two other types of muscle—smooth and cardiac muscle—share some basic properties with skeletal muscle, but each also displays unique characteristics (Table 8–3). The three muscle types have several features in common. First, they all have a specialized contractile apparatus made up of thick myosin and thin actin filaments that interact in response to a rise in cytosolic Ca^{++} to accomplish contraction. Second, they all directly use ATP as the energy source for cross-bridge cycling. However, the structure and organization of fibers within these different types of muscles varies, as do their mechanisms of excitation and the means by which excitation and contraction are coupled. Furthermore, there are important distinctions in the contractile response itself. We will spend the remainder of this chapter highlighting unique features of smooth and cardiac muscle as compared with skeletal muscle, reserving a more detailed discussion of their function for chapters devoted to organs containing these muscle types.

Smooth-muscle cells are small and unstriated.

The majority of smooth-muscle cells are found in the walls of hollow organs and tubes. Their contraction exerts pressure on and regulates the forward movement of the contents of these structures. Smooth-muscle cells are small (2 to 10 μm in diameter and 50 to 400 μm long), elongated, and have a single nucleus, in contrast to their large, multinucleated skeletal-muscle counterparts.

Three types of filaments are found in a smooth-muscle cell, typically aligned with its long axis: (1) thick myosin filaments, which are longer than those found in skeletal muscle; (2) thin actin filaments, which lack troponin and tropomyosin; and (3), unique to smooth muscle, filaments of intermediate size, which do not appear to directly participate in the contractile process but probably serve as part of the cell's elastic component, as part of the cytoskeletal framework to support the shape of the cell, or as part of both. Smooth-muscle filaments do not appear to form myofibrils and are not arranged in the regular ordered pattern found in skeletal muscle. Thus, smooth-muscle cells do not display the banding or striation found in skeletal muscle, giving rise to the term *smooth* for this muscle type. Lacking sarcomeres, smooth muscle does not have Z lines as such, but irregularly positioned **dense bodies** containing the same protein constituent found in Z lines are present. The actin filaments are anchored either to the dense bodies or to the internal surface of the plasma membrane. Considerably more actin is present in smooth- compared to skeletal-muscle cells, with ten to fifteen thin filaments for each thick myosin filament in smooth muscle and two thin filaments for each thick filament in skeletal muscle. Exactly how these filaments are aligned in smooth muscle and how they move during contraction is unclear. According to recent evidence, a smooth-muscle cell twists in a corkscrew-like manner as it shortens, leading to speculation that the contractile units may be arranged in twisting helices.

Smooth-muscle cells are turned on by Ca^{++}-dependent phosphorylation of myosin.

Because the thin filaments of smooth muscle cells do not contain the troponin- and tropomyosin-blocking proteins, what prevents actin and myosin from binding at the cross bridges in the resting state, and how is cross-bridge activity switched on in the excited state? Smooth-muscle myosin is able to interact with actin only when the myosin is phosphorylated (that is, has a phosphate group attached to it). During excitation, the increased cytosolic Ca^{++} acts as an intracellular messenger, initiating a chain of biochemical events that results in phosphorylation of myosin (Fig. 8–24, p. 252). Compare this to the role of Ca^{++} in skeletal muscle, where it moves troponin

Table 8–3 **Comparison of Muscle Types**

Characteristic	Type of Muscle			
	Skeletal	*Multiunit Smooth*	*Single-Unit Smooth*	*Cardiac*
Location	Attached to skeleton	Large blood vessels, eye, and hair follicles	Walls of hollow organs in digestive, reproductive, and urinary tracts and in small blood vessels	Heart only
Function	Movement of body in relation to external environment	Varies with structure involved	Movement of contents within hollow organs	Pumps blood out of heart
Mechanism of contraction	Sliding-filament mechanism	Uncertain; may involve shortening of contractile unit in corkscrew fashion	Uncertain; may involve shortening of contractile unit in corkscrew fashion	Sliding-filament mechanism
Innervation	Somatic nervous system (alpha motor neurons)	Autonomic nervous system	Autonomic nervous system	Autonomic nervous system
Level of control	Under voluntary control; also subject to subconscious regulation	Under involuntary control	Under involuntary control	Under involuntary control
Initiation of contraction	Neurogenic	Neurogenic	Myogenic (pacemaker activity and slow-wave potentials)	Myogenic (pacemaker activity)
Role of nervous stimulation	Initiates contraction; accomplishes gradation	Initiates contraction; contributes to gradation	Modifies contraction; can excite or inhibit; contributes to gradation	Modifies contraction; can excite or inhibit; contributes to gradation
Modifying effect of hormones	No	Yes	Yes	Yes
Presence of thick myosin and thin actin filaments	Yes	Yes	Yes	Yes
Striated due to orderly arrangement of filaments	Yes	No	No	Yes
Presence of troponin and tropomyosin	Yes	No	No	Yes
Presence of T tubules	Yes	No	No	Yes (large)

and tropomyosin from their blocking position so that actin and myosin are free to bind with each other. Smooth-muscle Ca^{++} binds with *calmodulin*, an intracellular protein found in most cells that is structurally similar to troponin (see p. 64). This Ca^{++}-calmodulin complex binds to and activates another protein, **myosin kinase,** which in turn phosphorylates myosin. Phosphorylated myosin then binds with actin so that cross-bridge cycling can begin. When Ca^{++} is removed, myosin is dephosphorylated and can no longer interact with actin, so the muscle relaxes. Thus, smooth muscle is triggered to contract by a rise in cytosolic Ca^{++} similar to in skeletal muscle. In smooth muscle, however, Ca^{++} ultimately turns on the cross bridges by inducing a chemical change in myosin in the thick filaments, whereas in skeletal muscle it exerts its effect by invoking a physical change at the thin filaments.

The means by which excitation brings about an increase in

Table 8–3 Comparison of Muscle Types (continued)

Characteristic	Type of Muscle			
	Skeletal	*Multiunit Smooth*	*Single-Unit Smooth*	*Cardiac*
Level of development of sarcoplasmic reticulum	Well developed	Poorly developed	Poorly developed	Moderately developed
Cross bridges turned on by Ca^{++}	Yes	Yes	Yes	Yes
Source of increased cytosolic Ca^{++}	Sarcoplasmic reticulum	Extracellular fluid and sarcoplasmic reticulum	Extracellular fluid and sarcoplasmic reticulum	Extracellular fluid and sarcoplasmic reticulum
Site of Ca^{++} regulation	Troponin in thin filaments	Myosin in thick filaments	Myosin in thick filaments	Troponin in thin filaments
Mechanism of Ca^{++} action	Physically repositions troponin-tropomyosin complex to uncover actin cross-bridge binding sites	Chemically brings about phosphorylation of myosin cross bridges so they can bind with actin	Chemically brings about phosphorylation of myosin cross bridges so they can bind with actin	Physically repositions troponin-tropomyosin complex
Presence of gap junctions	No	Yes (very few)	Yes	Yes
ATP used directly by contractile apparatus	Yes	Yes	Yes	Yes
Myosin ATPase activity; speed of contraction	Fast or slow, depending on type of fiber	Very slow	Very slow	Slow
Means by which gradation accomplished	Varying number of motor units contracting (motor-unit recruitment) and frequency at which they're stimulated (wave summation)	Varying number of muscle fibers contracting and varying cytosolic Ca^{++} concentration in each fiber by autonomic and hormonal influences	Varying cytosolic Ca^{++} concentration by myogenic activity and via influences by autonomic nervous system, hormones, mechanical stretch, and local metabolites	Varying length of fiber (depending on extent of filling) and varying cytosolic Ca^{++} concentration through autonomic, hormonal, and local metabolite influence
Presence of tone in absence of external stimulation	No	No	Yes	No
Clear-cut length-tension relationship	Yes	No	No	Yes

cytosolic Ca^{++} concentration in smooth-muscle cells also differs from that for skeletal muscle. A smooth-muscle cell has no T tubules and a poorly developed sarcoplasmic reticulum. The increased cytosolic Ca^{++} that triggers the contractile response comes from two sources: some Ca^{++} is released intracellularly from the meager sarcoplasmic reticulum stores, but most enters down its concentration gradient from the ECF as Ca^{++} channels in the plasma membrane are opened. Because smooth-muscle cells are so much smaller in diameter than skeletal-muscle fibers, this Ca^{++} influx from the ECF is able to influence cross-bridge activity, even in the central portions of the cell, without the necessity of an elaborate T tubule-sarcoplasmic reticulum mechanism. Relaxation is accomplished by removal of Ca^{++} as it is actively transported back into the sarcoplasmic reticulum and out across the plasma membrane.

Figure 8–24 Calcium Activation of Myosin in Smooth Muscle

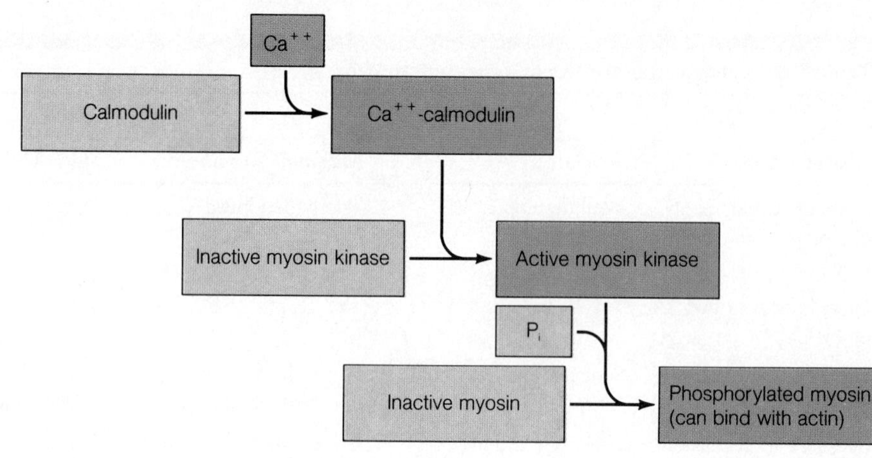

Most groups of smooth-muscle tissue are capable of self-excitation.

We still have not addressed the question of how smooth muscle becomes excited to contract; that is, what opens the Ca^{++} channels in the plasma membrane and sarcoplasmic reticulum? Smooth muscle is grouped into two categories—multiunit and single-unit smooth muscle—based on differences in how the muscle fibers become excited. **Multiunit smooth muscle** exhibits properties part way between skeletal muscle and single-unit smooth muscle. As implied by the name, a multiunit smooth muscle consists of multiple discrete units that function independently of each other and that must be separately stimulated by nerves to contract, similar to skeletal-muscle motor units. Thus, skeletal muscle and multiunit smooth muscle are both **neurogenic;** that is, they depend on their nerve supply to initiate contraction. However, in contrast to skeletal muscle, graded depolarizations rather than action potentials occur in multiunit smooth muscle in response to autonomic nervous stimulation to bring about the contractile response. The strength of contraction depends not only on the number of units stimulated and the frequency of their stimulation but also on the influence of certain circulating hormones and drugs. Multiunit smooth muscle is found: (1) in the walls of large blood vessels; (2) in large airways to the lungs; (3) in the muscle of the eye that adjusts the lens for near or far vision; (4) in the iris of the eye, which alters the size of the pupil to adjust the amount of light entering the eye; and (5) at the base of hair follicles, contraction of which causes "goose bumps."

The preponderance of smooth muscle is of the **single-unit** variety. It is alternatively called **visceral smooth muscle** because it is found in the walls of the hollow organs or viscera (for example, the digestive, reproductive, and urinary tracts and small blood vessels). The term ***single-unit smooth muscle*** is derived from the fact that the muscle fibers that make up this type of muscle become excited and contract synchronously as a single unit. The muscle fibers in single-unit smooth muscle are electrically linked together by gap junctions (see p. 69). When an action potential occurs anywhere within a sheet of single-unit smooth muscle, it is quickly propagated via these special points of electrical contact throughout the entire group of interconnected cells, which then contract as a single coordinated unit. Such a group of interconnected muscle cells that function electrically and mechanically as a unit is known as a **functional syncytium.**

Thinking about the role of the uterus during the process of labor will help you appreciate the significance of this arrangement. Muscle cells composing the uterine wall act as a functional syncytium. They repetitively become excited and contract as a unit during labor, exerting a series of coordinated "pushes" that are eventually responsible for delivering the baby. Independent, uncoordinated contractions of individual muscle cells in the uterine wall would not exert the uniformly applied pressure needed to expel the baby. A similar situation applies for single-unit smooth muscle elsewhere in the body.

Single-unit smooth muscle is **self-excitable** rather than requiring nervous stimulation for contraction. Clusters of specialized smooth-muscle cells within a functional syncytium display spontaneous electrical activity, being able to undergo action potentials without any external stimulation. In contrast to the other excitable cells we have been discussing (such as neurons, skeletal-muscle fibers, and multiunit smooth muscle), the self-excitable cells of single-unit smooth muscle do not maintain a constant resting potential. Instead, their membrane potential inherently fluctuates without any influence by

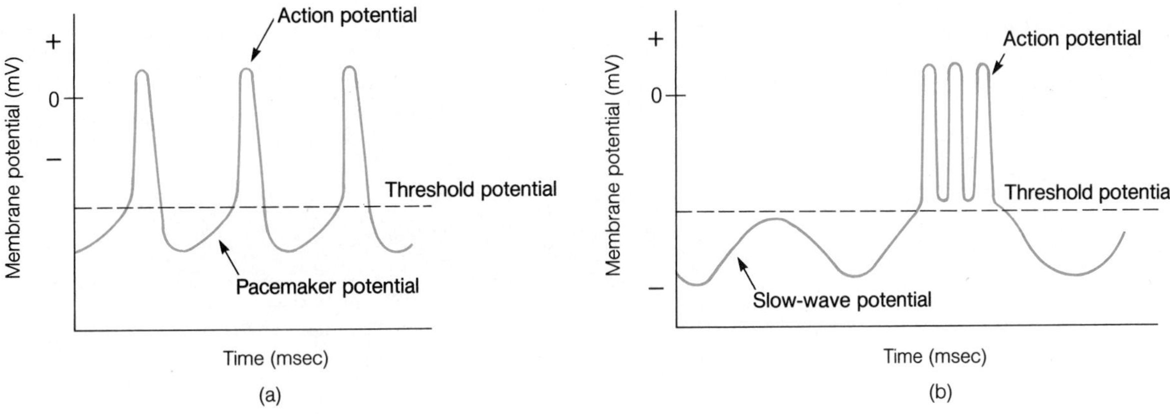

Figure 8–25 Self-Generated Electrical Activity in Smooth Muscle *(a) Pacemaker activity. (b) Slow-wave potentials.*

factors external to the cell. Two major types of spontaneous depolarizations displayed by self-excitable cells are pacemaker activity and slow-wave potentials. In **pacemaker activity** (Fig. 8–25a), the membrane potential gradually depolarizes on its own because of shifts in passive ionic fluxes accompanying automatic changes in channel permeability. When the membrane has depolarized to threshold, an action potential is initiated. After repolarizing, the membrane potential once again depolarizes to threshold, cyclically continuing in this manner to self-generate action potentials. **Slow-wave potentials** (Fig. 8–25b), on the other hand, are gradually alternating hyperpolarizing and depolarizing swings in potential caused by automatic cyclical changes in the rate at which sodium ions are actively transported across the membrane. The potential is moved farther from threshold during each hyperpolarizing swing and closer to threshold during each depolarizing swing. If threshold is reached, a burst of action potentials occurs at the peak of a depolarizing swing. Threshold is not always reached, however, so the oscillating slow-wave potentials can continue without generating action potentials. Whether or not threshold is reached depends on the starting point of the membrane potential at the onset of its depolarizing swing. The starting point, in turn, is influenced by neural and local factors.

Not all smooth-muscle cells undergo spontaneous changes in potential. However, once an action potential is initiated by a self-excitable smooth-muscle cell, it is conducted to the remaining cells of the functional syncytium via gap junctions so that the entire group of cells contracts without any nervous input (Fig. 8–26). Such nerve-independent contractile activity initiated by inherent properties of the muscle itself is called **myogenic activity,** in contrast to the neurogenic activity of skeletal muscle and multiunit smooth muscle.

The autonomic nervous system as well as other factors can modify the rate and strength of single-unit smooth-muscle contraction.

Smooth muscle is typically innervated by both branches of the autonomic nervous system. This nerve supply does not *initiate* contraction, but it can *modify* the rate and strength of contraction, either enhancing or retarding the inherent contractile activity of a given organ. Recall that the isolated motor end-plate region of a skeletal-muscle fiber interacts with ACh released from a single axon terminal of a motor neuron. In contrast, the receptor proteins that bind with autonomic transmitters are dispersed throughout the entire surface membrane of a smooth-muscle cell. Smooth-muscle cells are sensitive to varying degrees and in varying ways to autonomic transmitters, depending on their distribution of cholinergic and adrenergic receptors (see p. 207). Each terminal branch of a postganglionic autonomic fiber travels across the surface of one or more smooth-muscle cells, releasing transmitter from the vesicles within its multiple bulges (varicosities) as an action potential passes along the terminal (Fig. 8–27). The transmitter diffuses to the many receptor sites specific for it on the cells underlying the terminal. Thus, in contrast to the discrete one-to-one relationship at motor end plates, a given smooth-muscle cell can be influenced by more than one type of neurotransmitter, and each autonomic terminal can influence more than one smooth-muscle cell.

In addition to the autonomic nervous system, other factors can also influence the rate and strength of single-unit smooth muscle, including certain hormones, local metabolic intermediates, mechanical stretch, and particular drugs. Examples of such influences accompany discussions in other chapters of the various organs that contain single-unit smooth muscle.

Figure 8–26 Functional Syncytium

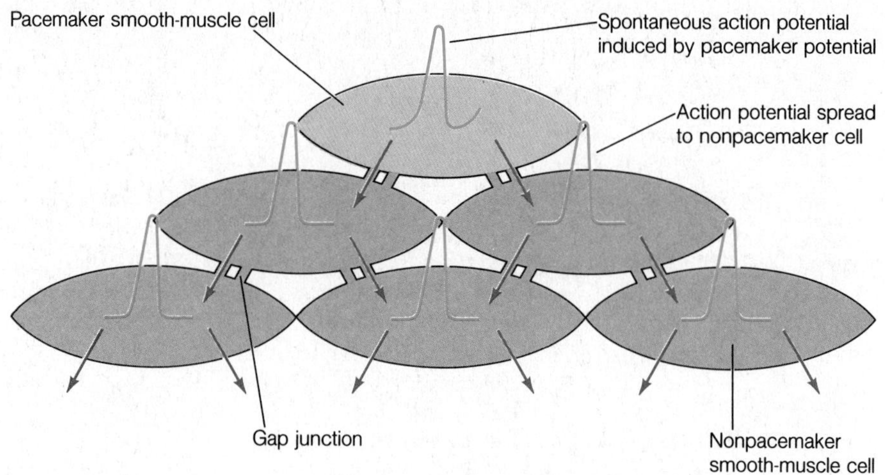

Arrows indicate spread of action potentials through gap junctions in single-unit smooth muscle.

Smooth muscle is a slow, economical contractile tissue.

A smooth-muscle contractile response proceeds at a more leisurely pace than does a skeletal-muscle twitch. A single smooth-muscle contraction may last as long as 3 seconds (3,000 msec), compared to the maximum of 100 msec required for a single contractile response in skeletal muscle. The rate of ATP splitting by myosin ATPase is much slower in smooth muscle, so cross-bridge activity and filament sliding occur more slowly. Smooth muscle also relaxes more slowly because of a slower rate of Ca^{++} removal. However, slowness should not be equated with weakness. Smooth muscle is able to generate the same contractile tension per unit of cross-sectional area as skeletal muscle, but it does so more slowly and at considerably less energy expense. Because of the low rate of cross-bridge cycling, cross bridges are maintained in the attached state for a longer period of time during each cycle compared with skeletal muscle. This so-called **latch phenomenon** enables smooth muscle to maintain tension with comparatively less ATP consumption. Smooth muscle is therefore an economical contractile tissue, making it well-suited for long-term sustained contractions with little energy consumption and without fatigue. Unlike the rapidly changing demands placed on our skeletal muscles as we maneuver through and manipulate our external environment, our smooth-muscle activities are geared for long-term duration and slower adjustments to change.

Nutrient and O_2 delivery are generally adequate to support the smooth-muscle contractile process. Smooth muscle can use a wide variety of nutrient molecules for ATP production. There are no energy storage pools comparable to creatine phosphate in smooth muscle; they are not necessary. Oxygen delivery is usually adequate to keep pace with the low rate of oxidative phosphorylation needed to provide ATP for the energy-efficient smooth muscle. If necessary, anaerobic glycolysis can sustain adequate ATP production if O_2 supplies are diminished.

Gradation of smooth-muscle contraction differs considerably from skeletal muscle.

In addition to being more leisurely and economical, visceral smooth muscle differs from skeletal muscle in the way in which gradation of contraction is accomplished. Gradation of skeletal muscle is entirely under neural control, primarily involving motor-unit recruitment and wave summation. In smooth muscle, the gap junctions assure that an entire smooth-muscle mass contracts as a single unit, making it impossible to vary the number of muscle fibers contracting. Only the tension of the fibers can be modified to achieve varying strengths of contraction of the whole organ. Changing the tension of a smooth-muscle fiber is accomplished primarily by varying the cytosolic Ca^{++} concentration. Not all of the cross bridges are switched on by a single excitation in smooth muscle, unlike skeletal muscle in which sufficient Ca^{++} is released in response to a single action potential to permit all cross bridges to cycle. In smooth muscle, the portion of cross bridges activated and the tension subsequently developed can be graded by varying the cytosolic Ca^{++} concentration. As Ca^{++} concentration increases, more cross bridges are brought into play, and greater tension is developed.

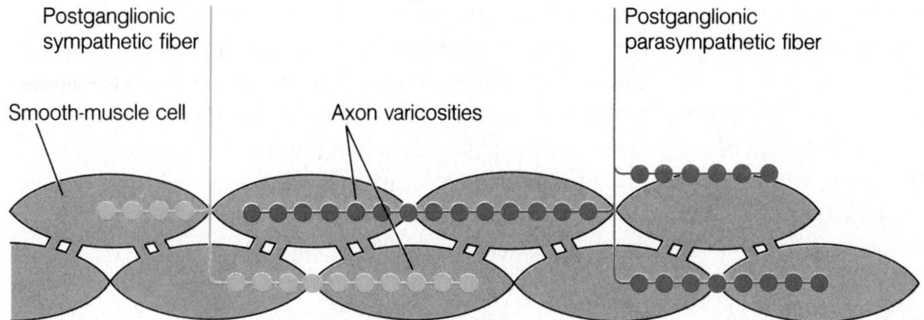

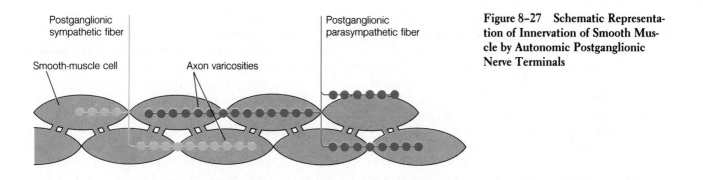

Figure 8–27 Schematic Representation of Innervation of Smooth Muscle by Autonomic Postganglionic Nerve Terminals

Many single-unit smooth-muscle cells have sufficient levels of cytosolic Ca^{++} to maintain a low level of tension, or **tone,** even in the absence of action potentials. A sudden drastic change in Ca^{++}, such as accompanies a myogenically induced action potential, brings about the characteristic slow contraction-relaxation response superimposed on the ongoing tonic tension. Besides self-induced action potentials, a number of other factors can influence contractile activity and the development of tension in smooth-muscle cells by altering their cytosolic Ca^{++} concentration. These factors, which have already been mentioned, are autonomic neurotransmitters, certain hormones, local metabolites, mechanical stretch, and specific drugs. They act by ultimately modifying the permeability of Ca^{++} channels in the plasma membrane, the sarcoplasmic reticulum, or both through a variety of mechanisms. Thus, smooth muscle is subject to more external influences than is skeletal muscle, even though smooth muscle is capable of contracting on its own, whereas skeletal muscle is not.

The relationship between the length of the muscle fibers before contraction and the tension that can be developed upon a subsequent contraction is less closely linked in smooth muscle than in skeletal muscle. The range of lengths over which a smooth-muscle fiber is able to develop near maximal tension is much greater than for skeletal muscle. Smooth muscle can still develop considerable tension even when stretched up to two and one-half times its resting length, presumably because the thin filaments still overlap with the much longer thick filaments even in the stretched-out position. In contrast, the thick and thin filaments of skeletal muscle are completely pulled apart and no longer able to interact when the muscle is stretched only three-fourths longer than its resting length.

The ability of a considerably stretched smooth-muscle fiber to still develop tension is important, because the smooth-muscle fibers within the wall of a hollow organ are progressively stretched as the volume of the organ's contents increases. Consider the urinary bladder as an example. Even though the muscle fibers in the urinary bladder are stretched as the bladder gradually fills with urine, they still maintain their tone and are even capable of developing further tension in response to inputs regulating bladder emptying. If considerable stretching prevented development of tension, as it does in skeletal muscle, a filled bladder would not be capable of contracting to empty.

In fact, when smooth muscle is suddenly stretched, it initially increases its tension, much like the tension created in a stretched rubber band. The muscle quickly adjusts to this new length, however, inherently relaxing to the tension level prior to the stretch, probably as a consequence of rearrangement of cross-bridge attachments. Smooth-muscle cross bridges detach comparatively slowly. Upon sudden stretching, it is speculated that any attached cross bridges would strain against the stretch, contributing to a passive (not actively generated) increase in tension. As these cross bridges detach, the filaments would be permitted to slide into an unstrained stretched position, restoring the tension to its original level. This inherent property of smooth muscle is called the **stress relaxation response.**

These two responses of smooth muscle to being stretched —being able to develop tension even when considerably stretched and inherently relaxing when stretched—are highly advantageous. They enable smooth muscle to exist at a variety of lengths with little change in tension. As a result, a hollow organ enclosed by smooth muscle can accommodate variable volumes of contents with little change in the pressure exerted on the contents, except when the contents are to be pushed out of the organ, at which time the tension is deliberately increased by fiber shortening. It is possible for smooth-muscle fibers to contract to one-half their normal length, enabling hollow organs to dramatically empty their contents upon increased contractile activity; thus, smooth-muscled viscera can easily accommodate large volumes but can empty to practically zero volume. This length range in which smooth muscle normally functions (anywhere from one-half to two and one-half times the normal length) is considerably greater than the limited length range within which skeletal muscle remains functional.

Because of its slowness and the less-ordered arrangement of its filaments, smooth muscle has often been mistakenly viewed as a poorly developed version of skeletal muscle, adequate only because it is not required to perform rapid, skilled movement. Actually smooth muscle is just as highly specialized for the demands placed on it; that is, being able to economically maintain tension for prolonged periods without fatigue and being able to accommodate considerable variations in the volume of contents it encloses with little change in tension. It is an extremely adaptive, efficient tissue.

Cardiac muscle blends features of both skeletal and smooth muscle.

Cardiac muscle, found only in the heart, is structurally and functionally a blend of skeletal and single-unit smooth muscle. In common with skeletal muscle, cardiac muscle is striated because its thick and thin filaments are highly organized into a regular banding pattern. Cardiac thin filaments contain troponin and tropomyosin, which constitute the site of Ca^{++} action in turning on cross-bridge activity, as in skeletal muscle. Also similar to skeletal muscle, cardiac muscle contracts according to the sliding-filament mechanism and has a clear-cut length-tension relationship. Like the oxidative skeletal-muscle fibers, cardiac fibers have an abundance of mitochondria and myoglobin. They also have T tubules (which, in fact, are larger than those in skeletal muscle) and a moderately well-developed sarcoplasmic reticulum.

In common with smooth muscle, Ca^{++} enters the cytosol from both the sarcoplasmic reticulum and the extracellular fluid during cardiac excitation. Like single-unit smooth muscle, the heart displays pacemaker (but not slow-wave) activity, initiating its own action potentials without any external influence. Cardiac cells are interconnected by gap junctions that enhance the spread of action potentials throughout the heart, just as with single-unit smooth muscle. Also similarly, the heart is innervated by the autonomic nervous system, which, along with certain hormones and local factors, can modify the rate and strength of contraction.

Unique to cardiac muscle, the muscle fibers are joined together in a branching network, and its action potentials have a much longer duration at peak reversed potential before repolarizing. Further details and the importance of cardiac muscle's features are addressed in the next chapter.

Chapter in Perspective

Muscles are the motors that produce movement for the body. Packed within all muscle cells (called muscle fibers because of their elongated shape) are contractile elements consisting of proteins that are found in most cells but that are organized into highly specialized, tension-generating units in muscle cells. These contractile elements are made up of thick myosin-containing filaments and thin actin-containing filaments. The filaments slide relative to each other during muscle contraction to generate tension and accomplish shortening. Sliding is triggered when excitation of the muscle fiber induces a rise in the level of cytosolic Ca^{++}, which turns on the ATP-driven stroking of cross bridges that extend between the thick and thin filaments. The three types of muscle—skeletal, smooth, and cardiac—vary in the ways they carry out this basic contractile scheme to accomplish their distinctive functions.

Skeletal muscle primarily produces movements of the body relative to the external environment. This type of muscle contracts only in response to stimulation by motor neurons, which are under voluntary control. A muscle action potential initiated by the motor neuron is propagated to the central portions of the muscle fiber by means of the T tubules. This action potential triggers the release of Ca^{++} into the cytosol from its intracellular stores in the sarcoplasmic reticulum. Calcium pulls aside the proteins that had been physically blocking cross-bridge interaction. Homeostatically, skeletal muscles include the respiratory muscles which are essential for breathing, and the muscles that are important in the acquisition, chewing, and swallowing of food. Skeletal muscles are also used for nonhomeostatic activities.

Smooth muscle forms the walls of hollow organs and tubes and is under involuntary control. Most smooth muscle contains self-excitable cells, which can alter their own membrane potential. Electrical activity is spread to the surrounding smooth-muscle cells in the organ by means of gap junctions, consequently inducing changes in cytosolic Ca^{++} levels and contractile activity in the entire unit without any external stimulation. However, external influences such as autonomic neurotransmitters and certain hormones can modify the self-induced contractile activity. Cytosolic Ca^{++} turns on cross-bridge activity in smooth muscle by triggering chemical changes in myosin that enable its cross bridges to bind with actin. Cross-bridge cycling and the resultant contraction occur slowly and at low energy cost in smooth muscle, making it well suited to efficiently sustain contractions for long periods of time. Smooth muscle can maintain rather constant tension at a wide range of muscle lengths, thereby accommodating substantial changes in the volume of contents of the hollow organs it encloses. Homeostatically, controlled contraction of smooth muscle is responsible for regulating movement of blood through the vasculature, food through the digestive tract, air through the respiratory airways, and urine to the exterior.

Cardiac muscle, the muscle found only in the heart, is self-excitable like most smooth muscle, but the way in which excitation brings about contraction is similar to skeletal muscle. In other ways, too, cardiac muscle blends the properties of the two other muscle types. Homeostatically, contraction of cardiac muscle drives blood forward in the circulatory system.

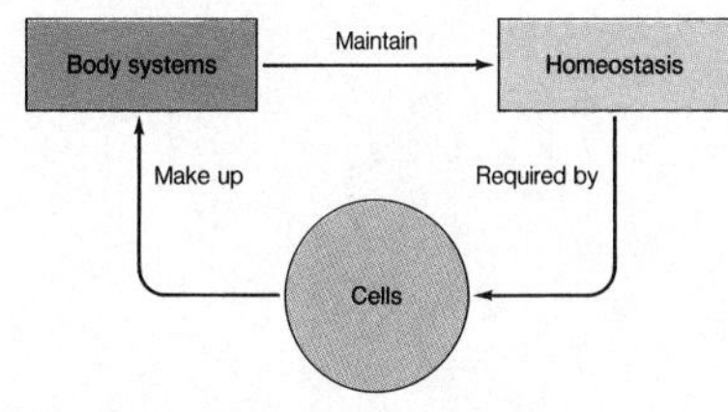

See inside front cover for an expanded version of this model.

Review Exercises

1. Compare the structural organization of skeletal, smooth, and cardiac muscle.
2. Describe the sliding filament mechanism of muscle contraction.
3. Compare the excitation-contraction coupling process in skeletal muscle with that in smooth muscle.
4. What is a motor unit? Compare the size of motor units in finely controlled muscles with those specialized for coarse, powerful contractions.
5. Distinguish between internal and external tension.
6. Discuss the means by which gradation of skeletal-muscle contraction is accomplished. How can smooth-muscle contraction be graded?
7. Describe the role of each of the following in powering skeletal-muscle contraction: ATP, creatine phosphate, oxidative phosphorylation, and glycolysis. Distinguish between aerobically and anaerobically supported exercise.
8. Compare the three types of skeletal-muscle fibers based on their biochemical properties.
9. Compare isotonic and isometric contractions.
10. What are the roles of the corticospinal system and multineuronal system in the control of motor movement?
11. Describe the structure and function of muscle spindles and Golgi tendon organs.
12. Distinguish between multiunit and single-unit smooth muscle.
13. Compare the contractile speed and relative energy expenditure of skeletal muscle with that of smooth muscle.
14. In what ways is cardiac muscle functionally similar to skeletal muscle and to smooth muscle?
15. **A point to ponder:** Why does regular aerobic exercise provide more cardiovascular benefit than weight training does? (Hint: The heart responds to the demands placed on it in a way similar to skeletal muscle.)

C H A P T E R 9

Cardiac Physiology

Chapter Contents at a Glance

INTRODUCTION *From just a matter of days following conception until death, the beat goes on. In fact, throughout an average human life span, the heart contracts about three billion times, never stopping to rest except for a fraction of a second between beats. Within about three weeks after conception, even before the mother can confirm she is pregnant, the heart of the developing embryo starts to function. It is believed to be the first organ to become functional. At this time the human embryo is only a few millimeters long, about the size of a capital letter on this page.*

Why does the heart develop so early, and why is it so crucial throughout life? It is because the circulatory system is the transport system of the body. Among its transport functions are carriage of vital nutrients, O_2, water, and electrolytes to the tissues of the body; removal of wastes for elimination from the body; and transport of hormones from their sites of production to their sites of action. A human embryo, having very little yolk available as food, depends on the prompt establishment of a circulatory system that can interact with the maternal circulation to pick up and distribute to the developing tissues the supplies so critical for survival and growth. Thus begins the story of the circulatory system, which continues throughout life as a vital pipeline for transporting materials on which the cells of the body are absolutely dependent.

The **circulatory system** consists of three basic components:

1. The **blood** serves as the transport medium within which materials being transported are dissolved or suspended. Blood, as all liquids, flows from an area of higher pressure to an area of lower pressure down a pressure gradient.
2. The **heart** serves as the pump that imparts pressure to the blood to establish the pressure gradient needed for blood to flow to the tissues.
3. The **blood vessels** serve as the passageways through which blood is directed and distributed from the heart to all parts of the body and subsequently returned to the heart.

The blood repeatedly travels through the circulatory system to and from the heart in what might be viewed as a figure-eight circuit. There are two separate vascular loops, both originating and terminating at the heart (Fig. 9–1). The **pulmonary** (lung) **circulation** consists of a closed loop of vessels carrying blood between the heart and lungs, whereas the **systemic circulation** consists of a circuit of vessels carrying blood between the heart and organ systems.

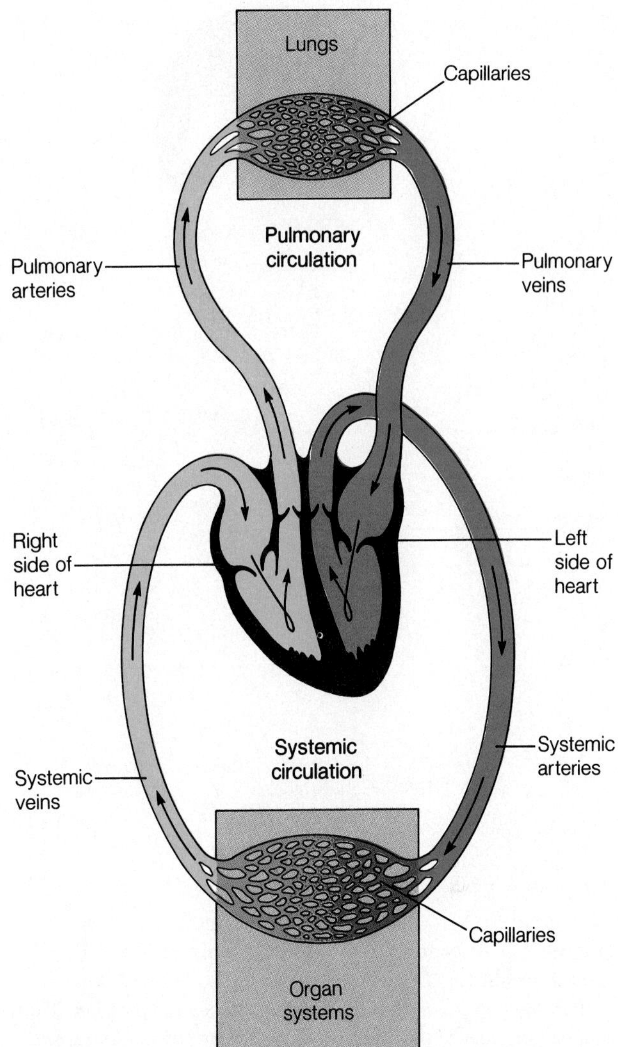

Figure 9–1 Pulmonary and Systemic Circulation in Relation to the Heart

Anatomical Considerations

The heart is located in the middle of the chest cavity.

The heart is a hollow muscular organ about the size of a clenched fist. It is located in the **thoracic** (chest) **cavity** approximately midline between the **sternum** or breastbone anteriorly and the **vertebrae** (backbone) posteriorly (Fig. 9–2a). The fact that it is positioned between these two bony structures makes it possible to manually drive blood forward from the heart when it is not pumping effectively by rhythmically depressing the sternum (Fig. 9–2b). This maneuver compresses the heart between the sternum and vertebrae so that blood is squeezed out as if the heart were beating. In many instances, this **external cardiac compression** serves as a life-saving measure until appropriate therapy can be instituted to restore the heart to normal function.

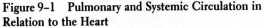

The midline location of the heart brings up a potentially confusing point. Place your hand over your heart and recite the Pledge of Allegiance. Where did you place your hand? People invariably place their hand on the left side of the chest. It is commonly believed that the heart lies on the left, even though it is actually in the middle of the chest. This common misunderstanding can be explained on the basis of the shape of the heart, the angle at which it is positioned, and the forcefulness of the apex beat. The heart has a broad base at the top and tapers to a pointed tip known as the **apex** at the bottom. It is situated at an angle under the sternum so that its base lies predominantly to the right and the apex to the left of the sternum. When the heart beats, especially when it contracts forcefully, the apex actually thumps against the inside of the chest wall on the left side. Because we become aware of the beating heart through the apex beat occurring on the left side of the chest, we have come to associate that as being the location of the entire heart.

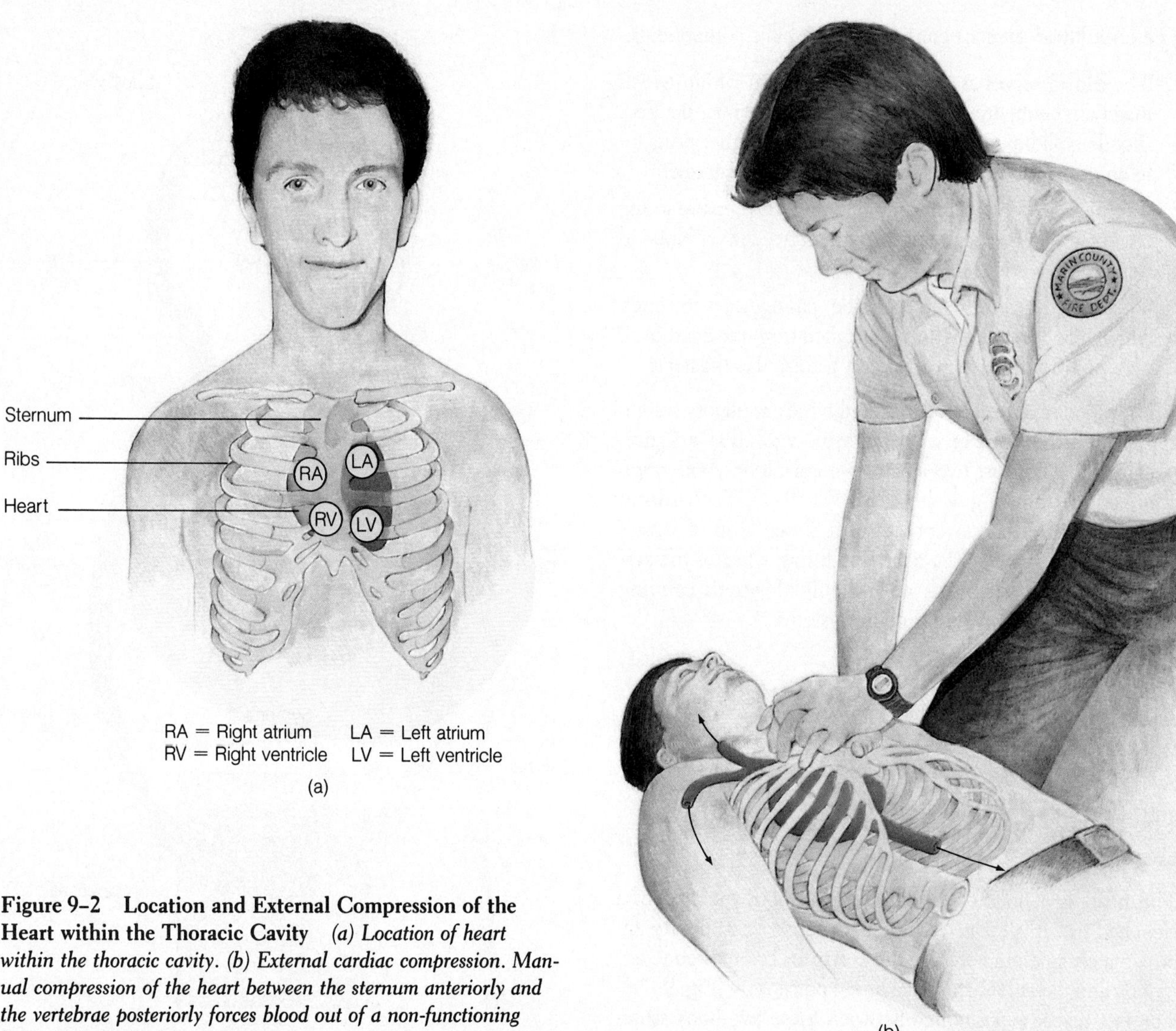

Figure 9–2 Location and External Compression of the Heart within the Thoracic Cavity *(a) Location of heart within the thoracic cavity. (b) External cardiac compression. Manual compression of the heart between the sternum anteriorly and the vertebrae posteriorly forces blood out of a non-functioning heart as if the heart were beating.*

The heart is a dual pump.

Even though anatomically the heart is a single organ, the right and left sides of the heart function as two separate pumps. The heart is divided into right and left halves and has four chambers, an upper and a lower chamber within each half (Fig. 9–3a). The upper chambers, the **atria** (**atrium,** singular) receive blood returning to the heart and transfer it to the lower chambers, the **ventricles,** which pump the blood from the heart. The vessels that return blood from the tissues to the atria are **veins,** and those that carry blood away from the heart to the tissues are **arteries.** The two halves of the heart are separated by the **septum,** a continuous muscular partition that prevents mixture of blood from the two sides of the heart. This is extremely important, because the right half of the heart is receiving and pumping low-oxygenated blood while the left side of the heart receives and pumps high-oxygenated blood.

Let us examine how the heart functions as a dual pump by tracing a drop of blood through one complete circuit (Fig. 9–3a and b). Blood returning from the systemic circulation enters the right atrium via large veins known as the **venae cavae.** The drop of blood entering the right atrium has returned from the body tissues, where O_2 has been extracted from it and CO_2 has been added to it. This partially deoxygenated blood flows from the right atrium into the right ventricle, which pumps it out through the **pulmonary artery** to the lungs. Thus, the *right side of the heart pumps blood into the pulmonary circulation*. Within the lungs, the drop of blood loses its excess CO_2 and picks up a fresh supply of O_2 before being returned to the left atrium via the **pulmonary veins.**

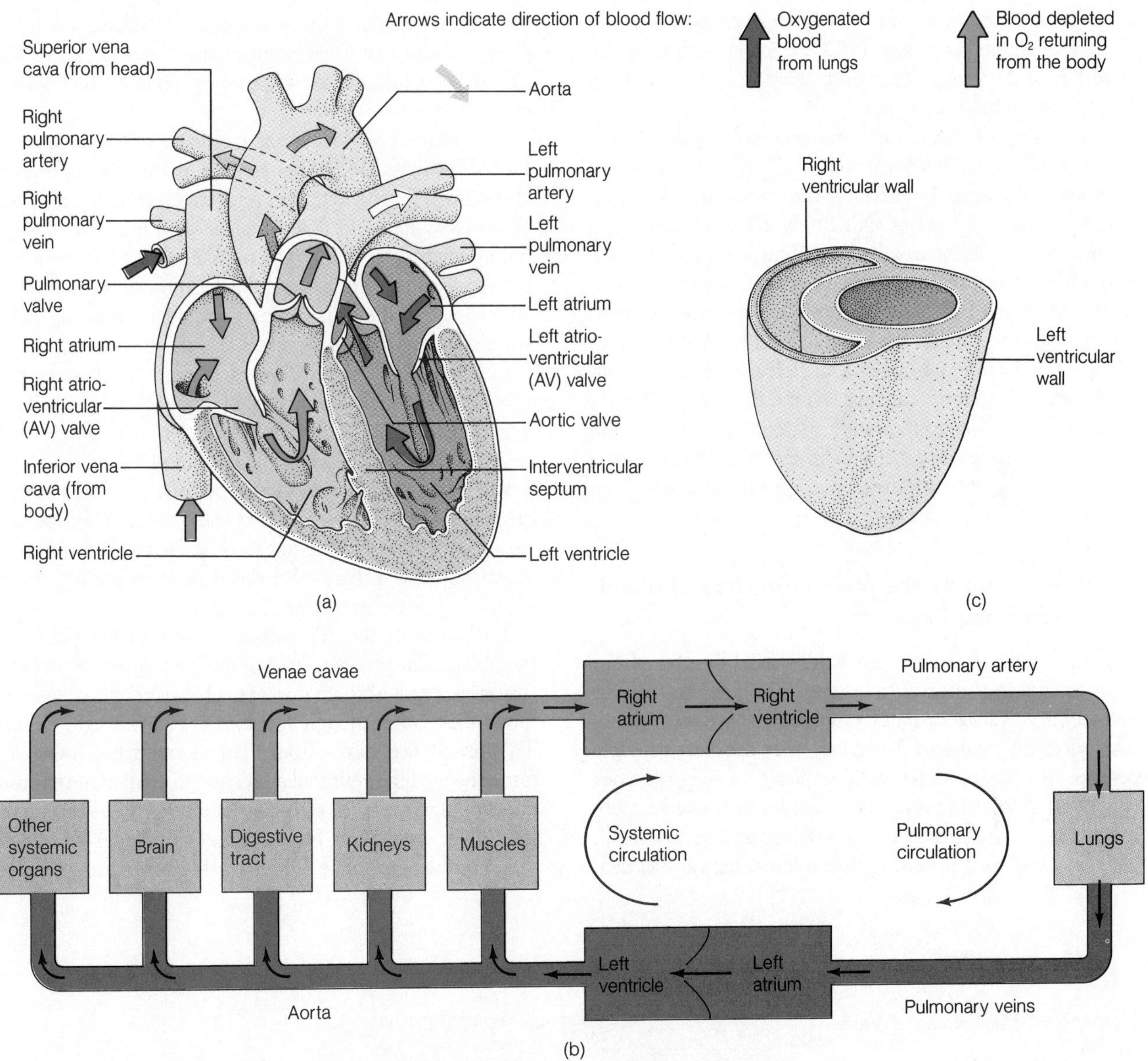

Figure 9-3 Blood Flow through and Pump Action of the Heart *(a) Blood flow through the heart. (b) Dual pump action of the heart. (c) Comparison of thickness of right and left ventricular walls.*

This richly oxygenated blood returning to the left atrium subsequently flows into the left ventricle, the pumping chamber that propels the blood to all body systems except the lungs; that is, ***the left side of the heart pumps blood into the systemic circulation***. The large artery carrying blood away from the left ventricle is the **aorta.** Major arteries branch from the aorta to supply the various tissues of the body.

In contrast to the pulmonary circulation, in which all the blood flows through the lungs, the systemic circulation may be viewed as a series of parallel pathways. Part of the blood pumped out by the left ventricle goes to the muscles, part to the kidneys, part to the brain, and so on. Thus, the output of the left ventricle is distributed so that each part of the body receives a fresh blood supply; the same arterial blood does not pass from tissue to tissue. Accordingly, the drop of blood we are tracing goes to only one of the systemic tissues. Tissue cells extract O_2 and use it to oxidize nutrients for energy production, in the process forming CO_2 as a waste product that is added to the blood. The drop of blood, now partially depleted of O_2 content and increased in CO_2 content, returns to the right side of the heart, which once again will pump it to the lungs. One circuit is complete.

Both sides of the heart simultaneously pump equal amounts of blood. The volume of O_2-poor blood being pumped to the lungs by the right side of the heart soon becomes the same volume of O_2-rich blood being delivered to the tissues by the left side of the heart. The pulmonary circulation is a low-pressure, low-resistance system, whereas the systemic circulation is a high-pressure, high-resistance system. Therefore, even though the right and left sides of the heart pump the same amount of blood, the left side performs more work, because it pumps an equal volume of blood at a higher pressure into a higher-resistance system. Accordingly, the heart wall on the left side, being a stronger pump by necessity, is much thicker than the heart wall on the right side (Fig. 9–3c).

Heart valves ensure the proper direction of blood flow through the heart.

Blood flows through the heart in one fixed direction from veins to atria to ventricles to arteries. The presence of four one-way heart valves ensures this unidirectional flow of blood. The valves are positioned so that they open and close passively because of pressure differences, similar to a one-way door (Fig. 9–4). A forward pressure gradient forces the valve open, much like you open a door by pushing on one side of it, whereas a backward pressure gradient forces the valve closed, just as you apply pressure to the opposite side of the door to close it. Note that a backward gradient can force the valve closed but cannot force it to swing open in the opposite direction.

Two of the heart valves, the **right** and **left atrioventricular (AV) valves,** are positioned between the atria and ventricle on the right and left sides, respectively (Fig. 9–5a). These valves allow blood to flow from the atria into the ventricles during ventricular filling (when atrial pressure exceeds ventricular pressure), but prevent the backflow of blood from the ventricles into the atria during ventricular emptying (when ventricular pressure greatly exceeds atrial pressure). If the AV valves were not forced to close by the rising ventricular pressure as the ventricles contracted to empty, much of the blood would inefficiently be forced back into the atria and veins instead of being pumped into the arteries. The right AV valve is also called the **tricuspid valve** because it consists of three cusps or leaflets (Fig. 9–5b). Likewise, the left AV valve, consisting of two cusps, is often called the **bicuspid valve** or, alternatively, the **mitral valve** (because of its physical resemblance to a mitre or bishop's headgear).

The cusps of the AV valves are thin-walled yet strong membranes that balloon upward like parachutes when blood surges back against them during ventricular contraction. This forces their edges snuggly together to create a tight closure. To prevent the valves from being pushed too much and forced by the high ventricular pressure to open in the opposite direction into the atria, the edges of the AV valve leaflets are fastened by tough, thin, fibrous cords of tendinous-type tissue, the **chordae tendineae.** These cords extend from the edges

Figure 9–4 Mechanism of Valve Action

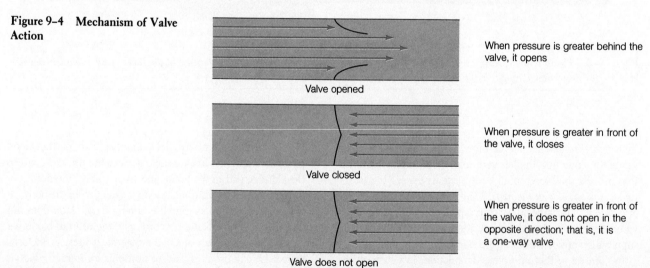

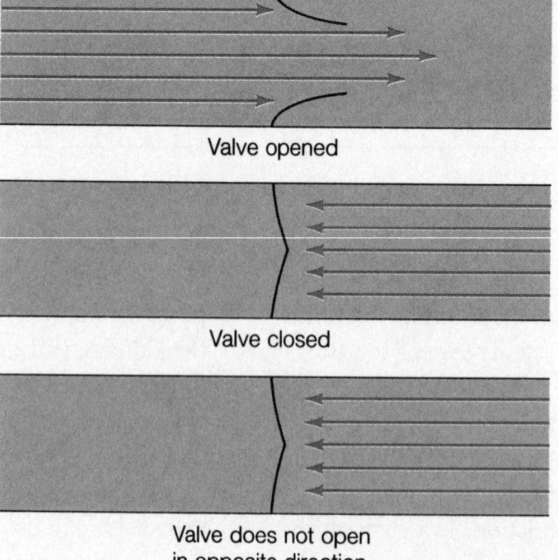
Valve opened
When pressure is greater behind the valve, it opens
Valve closed
When pressure is greater in front of the valve, it closes
Valve does not open in opposite direction
When pressure is greater in front of the valve, it does not open in the opposite direction; that is, it is a one-way valve

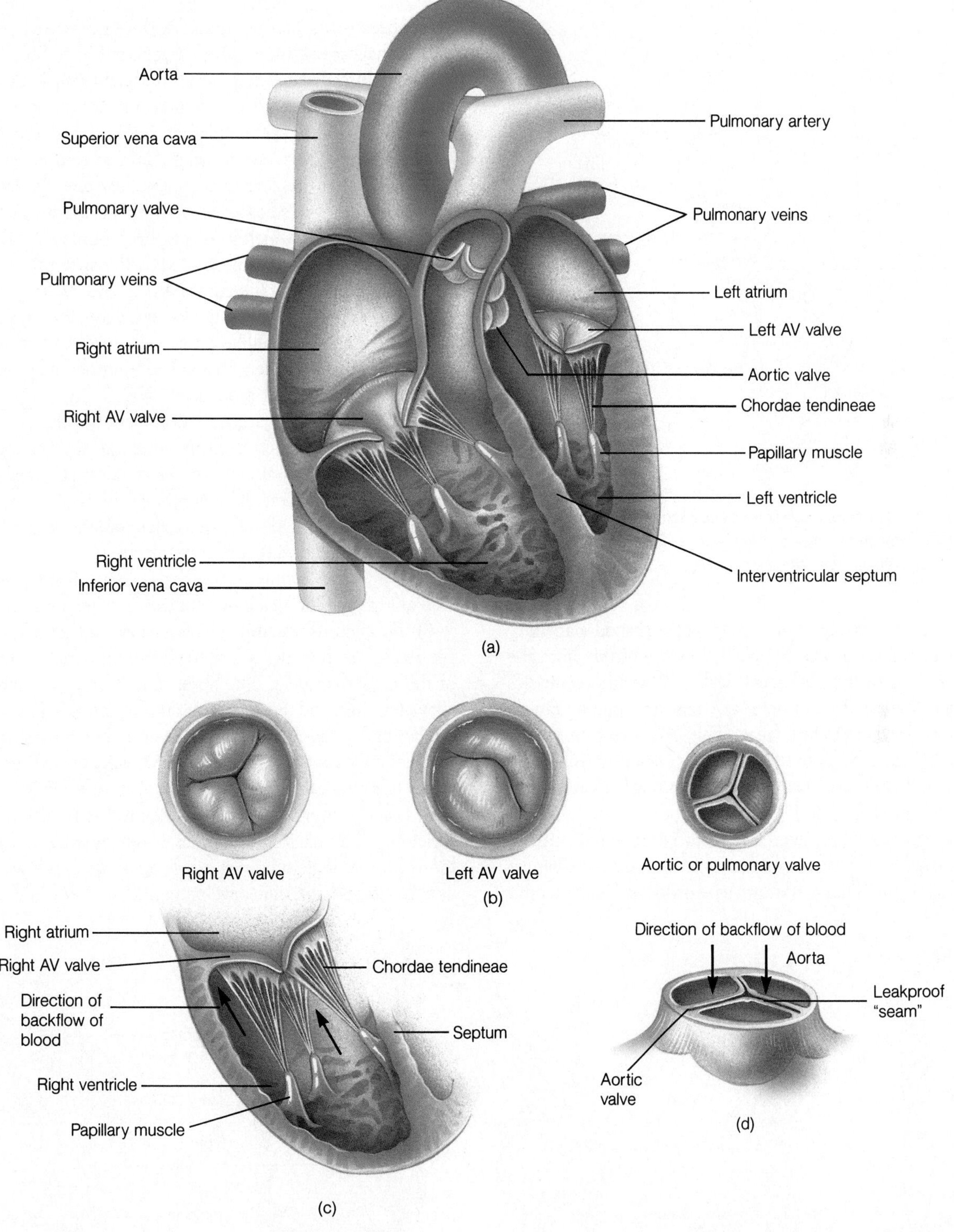

Figure 9–5 Heart Valves *(a) Longitudinal section of heart, depicting location of the four heart valves. (b) View of heart valves in closed position. (c) Prevention of eversion of AV valves. The eversion of AV valves is prevented by tension on valve leaflets exerted by chordae tendineae on contraction of papillary muscles. (d) Prevention of eversion of semilunar valves. When the semilunar valves are swept closed, their upturned edges fit together in a deep, leakproof seam that prevents valve eversion.*

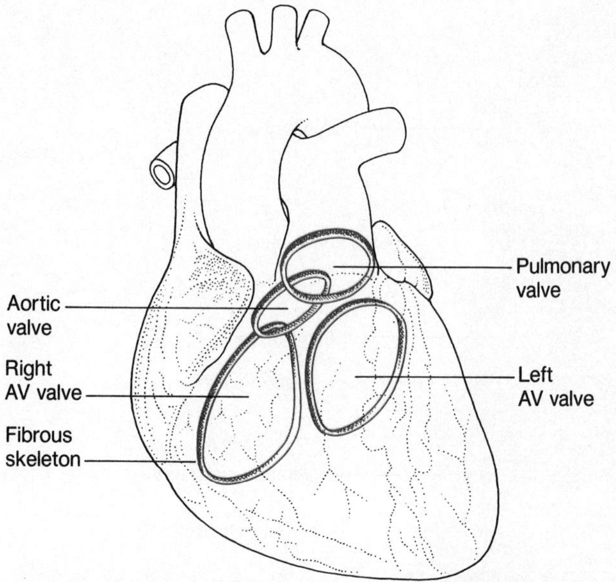

Figure 9-6 Fibrous Skeleton of the Heart *Transparent view of the heart from the front showing the location of the fibrous rings surrounding the heart valves.*

of each cusp and attach to small, nipple-shaped **papillary muscles** (*papilla* means "nipple"), which protrude from the inner surface of the ventricular walls. When the ventricles contract, the papillary muscles also contract, pulling downward on the chordae tendineae. This exerts tension on the closed AV-valve cusps to hold them in position, thus helping keep them tightly sealed in the face of a strong backward pressure gradient (Fig. 9–5c).

The two remaining heart valves, the **aortic** and **pulmonary valves,** guard the major arteries as they leave the ventricles (Fig. 9–5a). They are known as **semilunar** ("half moon") **valves** because they are composed of three cusps, each resembling a shallow half-moon–shaped pocket (Fig. 9–5b). These valves are forced open when the left and right ventricular pressures exceed the pressure in the aorta and pulmonary arteries, respectively, during ventricular contraction and emptying. Closure results when the ventricles relax and ventricular pressures fall below the aortic and pulmonary artery pressures. This prevents reverse flow of blood from the arteries back into the ventricles from which it has just been pumped. Eversion of the semilunar valves is prevented by the anatomical structure and positioning of the cusps. When a backward pressure gradient is created upon ventricular relaxation, the back surge of blood fills the pocketlike cusps and sweeps them into a closed position, with their unattached upturned edges fitting together in a deep, leak-proof seam (Fig. 9–5d).

Even though there are no valves between the atria and veins, backflow of blood from the atria into the veins is not usually a significant problem for two reasons: (1) atrial pressures usually are not much higher than venous pressures; and (2) the sites of entry of the venae cavae into the atria are partially compressed during atrial contraction.

Four interconnecting rings of dense connective tissue provide a firm base for attachment of the four heart valves (Fig. 9–6). This fibrous skeleton, which separates the atria from the ventricles, also provides a fairly rigid structure for attachment of the cardiac muscle. The atrial muscle mass is anchored above the rings and the ventricular muscle mass is attached to the bottom of the rings. It might seem rather surprising that the inlet valves to the ventricles (the AV valves) and the outlet valves from the ventricles (the semilunar valves) all lie on the same plane through the heart, as delineated by the fibrous skeleton. This relationship comes about because the heart forms from a single tube that bends upon itself and twists on its axis during embryonic development (Fig. 9–7). This turn-

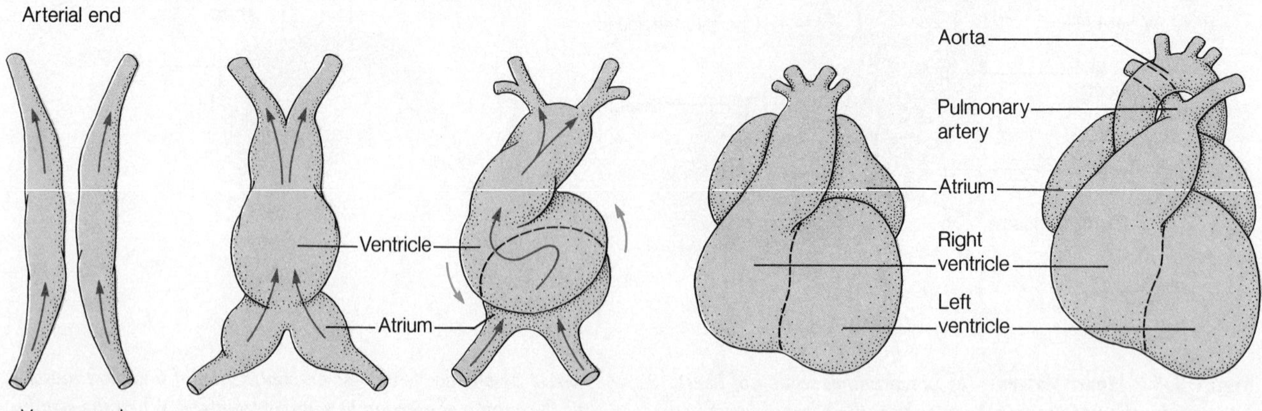

Figure 9–7 Twisting of the Embryonic Heart on Its Axis during Development

ing and twisting, although making it more difficult to study the structural relationships of the heart, has functional importance in that it helps the heart pump more efficiently. We will see how by turning our attention to the portion of the heart that actually generates the forces responsible for blood flow, the cardiac muscle within the heart walls.

The heart walls are composed primarily of spirally arranged cardiac-muscle fibers interconnected by intercalated discs.

The heart wall consists of three distinct layers:

□ The **endocardium** (*endo* means "within"; *cardia* means "heart") is a thin inner layer of **endothelium,** a unique type of epithelial tissue that lines the entire circulatory system.

□ The **myocardium** (*myo* means "muscle"), the middle layer composed of cardiac muscle, constitutes the bulk of the heart wall.

□ The **epicardium** (*epi* means "upon") is a thin external membrane covering the heart.

The myocardium consists of interlacing bundles of cardiac-muscle fibers arranged spirally around the circumference of the heart because of the heart's complex twisting during development. As a result of this arrangement, when the ventricular muscle contracts and shortens, the diameter of the ventricular chambers is reduced while the apex is simultaneously pulled upward toward the base in a rotating manner. This exerts a wringing effect, efficiently exerting pressure on the blood within the enclosed chambers and directing it upward toward the openings of the major arteries that exit at the base of the ventricles.

The individual cardiac-muscle cells are interconnected to form branching fibers, with adjacent cells joined end to end at specialized structures known as **intercalated discs.** Within an intercalated disc there are two types of membrane junctions, desmosomes and gap junctions (Fig. 9–8). A desmosome, a type of adhering junction that mechanically holds cells together, is particularly abundant in tissues, such as the heart, that are subject to considerable mechanical stress (see p. 68). At intervals along the intercalated disc, the opposing membranes approach each other very closely to form gap junctions, which are areas of low electrical resistance that allow action potentials to spread from one cardiac cell to adjacent cells (see p. 69). Cardiac muscle is capable of generating action potentials without any nervous stimulation. When one of the cardiac cells spontaneously undergoes an action potential, the electrical impulse spreads to all the other cells that are joined by gap junctions in the surrounding muscle mass so that they become excited and contract as a single functional syncytium (see p. 252). The atria and the ventricles each form a functional syncytium and contract as separate units. The synchronous contraction of the muscle cells composing the walls of each of these chambers produces the force necessary to eject the enclosed blood.

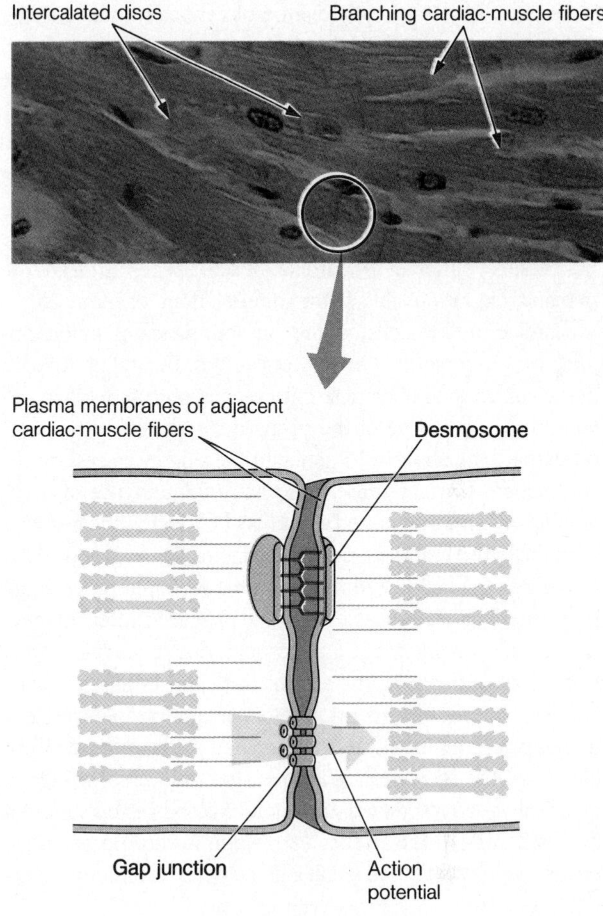

Figure 9–8 Organization of Cardiac-Muscle Fibers

SOURCE: Photo courtesy of Elizabeth R. Walker, Associate Professor, and Dennis O. Overman, Associate Professor, Department of Anatomy, School of Medicine, West Virginia University.

There are no gap junctions between the atrial and ventricular muscle masses, and, furthermore, these masses are separated by the electrically nonconductive fibrous skeleton surrounding the valves. However, an important specialized conducting system is present to facilitate and coordinate the transmission of electrical excitation from the atria to the ventricles to assure synchronization between atrial and ventricular pumping.

Because of the syncytial nature of cardiac muscle and because of the conducting system between the atria and ventricles, an impulse spontaneously generated in one part of the heart spreads throughout the entire heart. Therefore, unlike skeletal muscle, in which graded contractions can be produced by varying the number of muscle cells that are contracting within the muscle (recruitment of motor units), the heart contracts in an all-or-none fashion. A "half-hearted" contraction is not possible; either all the cardiac-muscle fibers contract to the maximal extent possible at the time, or none of them do.

Cardiac-muscle cells contain an abundance of mitochondria, the O_2-dependent energy organelles. In fact, up to 40% of the cell volume of cardiac-muscle cells is attributable to mitochondria, indicative of the strong dependence of the heart on aerobic metabolism to generate the energy necessary for contraction. Cardiac muscle also has an abundance of myoglobin, which stores limited amounts of O_2 within the heart for immediate use and furthermore facilitates the transport of O_2 arriving at the cell to the mitochondria for rapid processing.

No new cardiac-muscle cells are produced after infancy. Any cardiac enlargement that occurs is due to cardiac hypertrophy, an increase in the size of the cardiac-muscle fibers. Cardiac hypertrophy occurs in response to increased work demands placed on the heart, such as when the heart must pump blood against increased resistance (for example, through a heart valve narrowed by disease or into blood vessels stiffened by "hardening of the arteries"), or when it must pump an increased volume of blood (such as in response to a regular exercise program).

The heart is enclosed by the pericardial sac.

The heart is enclosed in the double-walled, membranous **pericardial sac** (Fig. 9–9). The outer layer of the sac is a tough, fibrous membrane that is attached to the **mediastinum,** the midline connective-tissue partition that separates the lungs. This attachment anchors the heart so that it remains properly positioned within the chest. The sac is lined by a **serous** (fluid-secreting) **membrane** whose innermost layer closely adheres to the surface of the myocardium and actually forms the heart's outermost layer, the epicardium. This membrane secretes a thin **pericardial fluid** into the sac, providing lubrication to prevent friction between the pericardial layers as these surfaces glide over each other with every beat of the heart. **Pericarditis,** an inflammation of the pericardial sac that results in a painful friction rub between the two pericardial layers, occurs occasionally because of viral or bacterial infection.

If blood hemorrhages into the pericardial sac as a result of a penetrating heart wound or rupture of the heart wall, the sac, because of its tough outer membrane, is unable to expand outward to accommodate the extra fluid volume. Instead, it distends inward, compressing the heart and restricting cardiac filling. This effect is similar to squeezing a soft rubber ball with your hand. The heart simply has less room to expand, so the amount of blood that can enter from the veins is limited. Because less blood is returned to the heart to be pumped out to supply the tissues, cardiac failure subsequently ensues. Such distension of the pericardial sac that impinges on cardiac filling is known as **cardiac tamponade.**

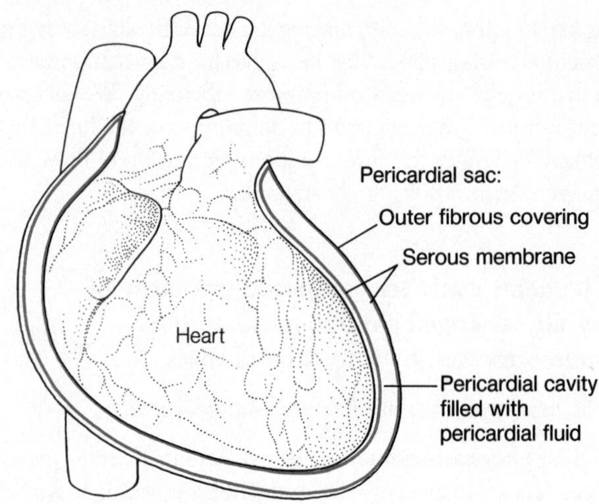

Figure 9–9 Pericardial Sac

ELECTRICAL ACTIVITY OF THE HEART

The sinoatrial node is the normal pacemaker of the heart.

Contraction of cardiac-muscle cells to bring about ejection of blood is triggered by action potentials sweeping across the muscle-cell membranes. The heart contracts or beats rhythmically as a result of action potentials that it generates by itself, a property known as **autorhythmicity.**

There are two specialized types of myocardial fibers:

1. The **contractile fibers** constitute 99% of the total cardiac musculature and do the mechanical work of pumping. These working cells normally do not initiate their own action potentials.
2. In contrast, the small but extremely important remainder of the cardiac fibers, the **autorhythmic cells,** do not contract but instead are specialized for initiating and conduct-

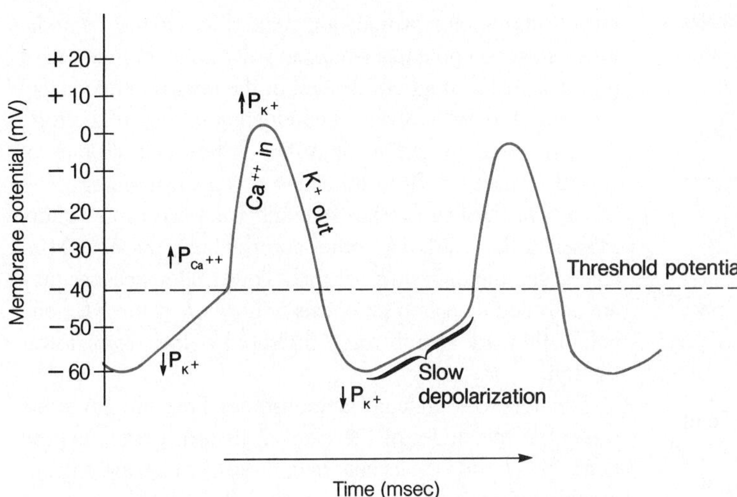

Figure 9–10 Pacemaker Potential of Autorhythmic Cells

ing the action potentials responsible for contraction of the working cells.

Let us examine the role of the specialized autorhythmic cells in the origin and spread of the heart beat. In contrast to nerve and skeletal-muscle cells, in which the membrane remains at constant resting potential unless the cell is stimulated, the cardiac autorhythmic cells do not have a resting potential. Instead they display **pacemaker activity;** that is, their membrane potential constantly depolarizes, or drifts, between action potentials until threshold is reached, at which time the membrane fires or has an action potential (Fig. 9–10). Through repeated cycles of drift and fire, these autorhythmic cells cyclically initiate action potentials, which then spread throughout the heart to trigger rhythmic beating without any nervous stimulation.

The basis for the membrane potential's slow drift to threshold is still unresolved. It is generally believed to be caused by a cyclical decrease in the passive outward flux of K^+, which we will assume in our discussion, although some recent evidence suggests that it may be due in some autorhythmic cells to a cyclical increase in the inward movement of Ca^{++}. The higher intracellular concentration of K^+ encourages the passive diffusion of K^+ out of the cells down potassium's concentration gradient. In cardiac autorhythmic cells there is no constant resting permeability to K^+ as there is in nerve and skeletal-muscle cells. Instead, membrane permeability to K^+ decreases between action potentials because of inactivation of K^+ channels, which diminishes the outflow of positive potassium ions. Since a small passive influx of cations continues, the inside gradually becomes less negative; that is, the membrane gradually depolarizes and drifts toward threshold. Once threshold is reached, the rising phase of the action potential occurs in response to activation of Ca^{++} channels, unlike skeletal muscle, in which Na^+ influx rather than Ca^{++} influx swings the potential in the positive direction. The falling phase is due, as usual, to activation of K^+ channels and a subsequent K^+ efflux. Inactivation of these channels at the end of an action potential initiates the next slow depolarization to threshold.

The cardiac cells capable of autorhythmicity are found in the following specific locations (Fig. 9–11):

1. the **sinoatrial node (SA node)**, a small specialized region in the right atrial wall near the opening of the superior vena cava;

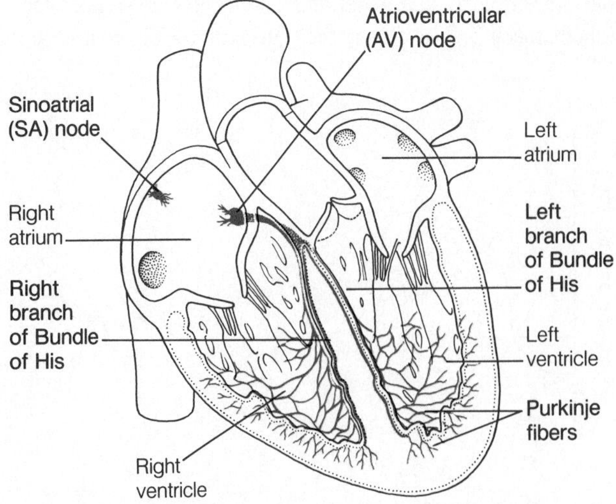

Figure 9–11 Specialized Conducting System of the Heart

Table 9–1 Inherent Rate of Action Potential Discharge in Autorhythmic Tissues of the Heart

Tissue	Action Potentials per Minute
SA node (normal pacemaker)	70–80
AV node	40–60
Bundle of His and Purkinje fibers	20–40

2. the **atrioventricular node (AV node)**, a small bundle of specialized cardiac-muscle cells located at the base of the right atrium near the septum, just above the junction of the atria and ventricles;
3. the **bundle of His (atrioventricular bundle)**, a tract of specialized cells that originates at the AV node and enters the interventricular septum, where it divides to form the right and left bundle branches that travel down the septum, curve around the tip of the ventricular chambers, and travel back toward the atria along the outer walls; and
4. **Purkinje fibers,** small terminal fibers that extend from the bundle of His and spread throughout the ventricular myocardium much like small twigs of a tree branch. This fine network of fibers is known as the **Purkinje system.**

These various autorhythmic cells differ in the inherent rate at which they are capable of generating action potentials because of differences in their rates of slow depolarization (Table 9–1). Comparing two autorhythmic cells (Fig. 9–12), cell A has a faster rate of depolarization, thus reaching threshold more quickly and generating action potentials more rapidly than cell B. The cells in the heart with the fastest inherent rate of action-potential initiation are localized in the SA node. Once an action potential occurs in any cardiac-muscle cell, it is propagated throughout the rest of the myocardium via gap junctions and the specialized conducting system. Therefore, the SA node, which exhibits the fastest natural rate of autorhythmicity at 70 to 80 action potentials/minute, drives the rest of the heart at this rate and is known as the **pacemaker** of the heart. The other autorhythmic tissues are unable to assume their own naturally slower rates, because they are activated by action potentials originating in the SA node before they are able to reach threshold at their own slower rhythm.

The following analogy demonstrates how the SA node drives the remainder of the heart at its own pace. Suppose there are a hundred cars in a train, three of which are engines capable of moving on their own; the other ninety-seven cars must be pulled in order to move (Fig. 9–13). One engine (the SA node) can travel at 70 miles/hour on its own, another engine (the AV node) at 50 miles/hour, and the last engine (the Purkinje system) at 30 miles/hour. If all of these cars are joined together, the engine capable of traveling at 70 miles/hour will pull the remainder of the cars at that speed. The engines that can travel at lesser speeds on their own will be pulled at a faster speed by the fastest engine and will therefore be unable to assume their own slower rate as long as they are being driven by a faster engine. The other ninety-seven cars (nonautorhythmic, contractile working cells), being unable to move on their own, will likewise travel at whatever speed the fastest engine pulls them.

If for some reason the fastest engine breaks down (SA node damage), the next fastest engine (AV node) takes over and the entire train will travel at 50 miles/hour; that is, if the SA node becomes nonfunctional, the AV node assumes pacemaker activity. The non-SA nodal autorhythmic tissues are **latent pacemakers** that can take over, although at a lower rate,

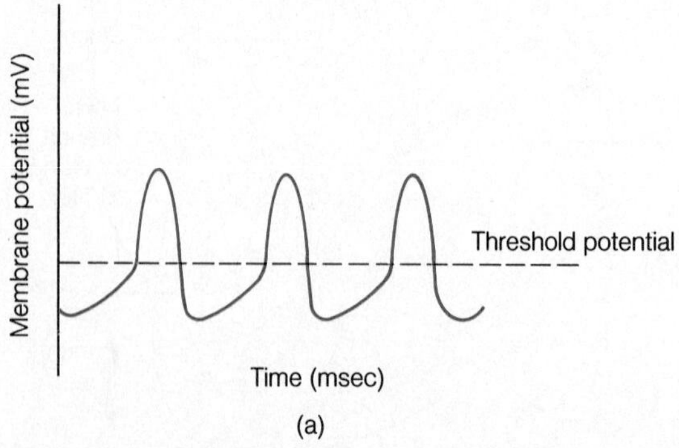

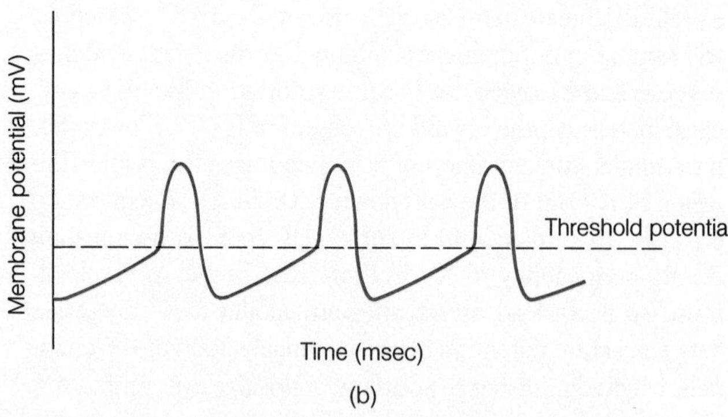

Figure 9–12 Different Autorhythmic Rates *(a) Recording from autorhythmic cell A. (b) Recording from autorhythmic cell B. See text for an explanation of the difference between these recordings.*

should the normal pacemaker fail. If conduction of the impulse becomes blocked between the atria and the ventricles, the atria continue at the normal rate of 70 beats/minute and the ventricular tissue, not being driven by the faster SA nodal rate, assumes its own much slower autorhythmic rate of about 30 beats/minute, initiated by the ventricular autorhythmic cells. This is comparable to a breakdown of the second engine (AV node) so that the lead engine (SA node) becomes discon-

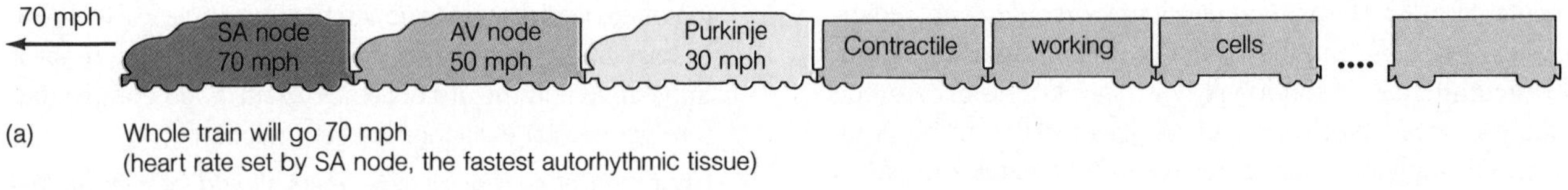

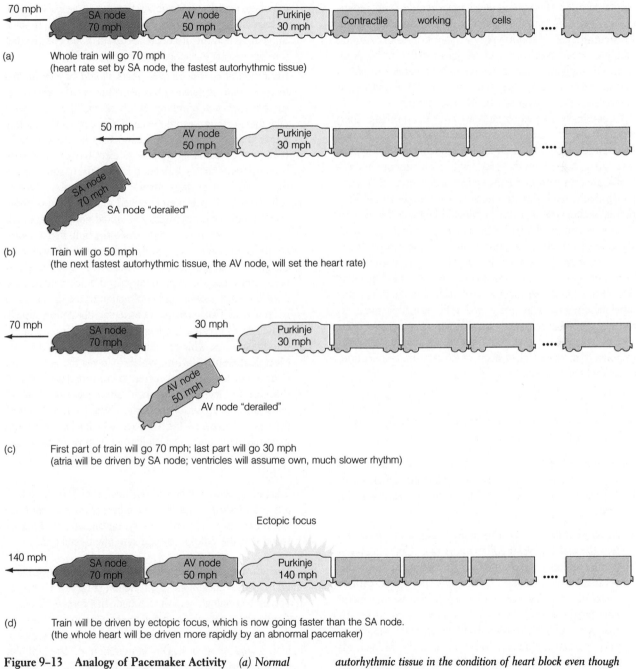

(a) Whole train will go 70 mph
(heart rate set by SA node, the fastest autorhythmic tissue)

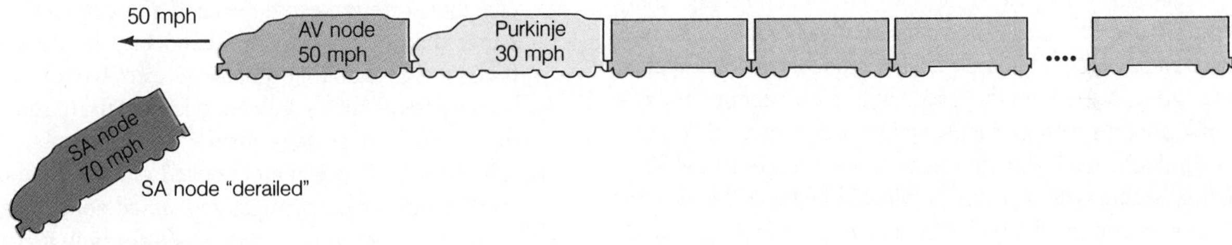

(b) Train will go 50 mph
(the next fastest autorhythmic tissue, the AV node, will set the heart rate)

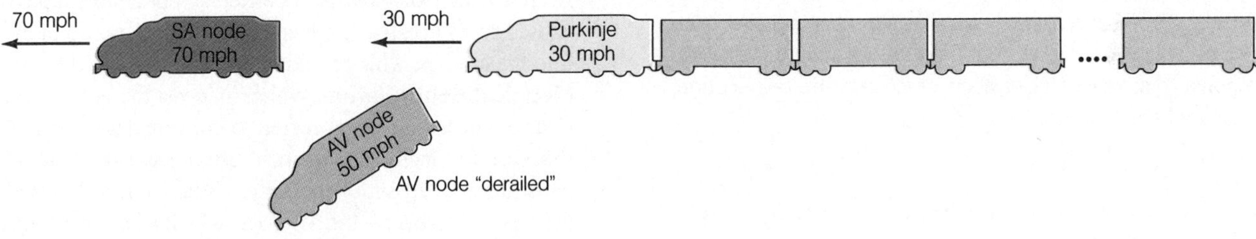

(c) First part of train will go 70 mph; last part will go 30 mph
(atria will be driven by SA node; ventricles will assume own, much slower rhythm)

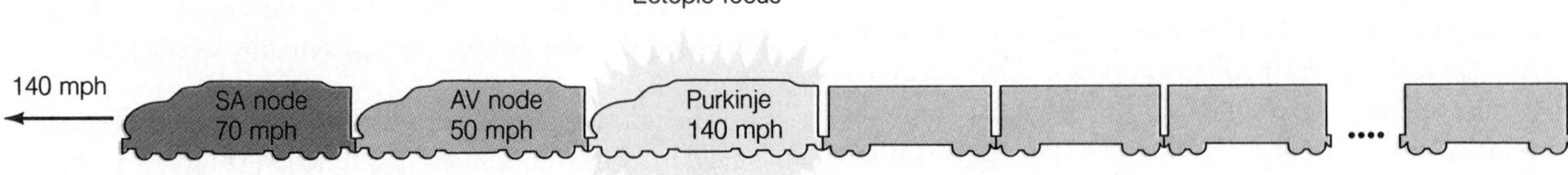

(d) Train will be driven by ectopic focus, which is now going faster than the SA node.
(the whole heart will be driven more rapidly by an abnormal pacemaker)

Figure 9-13 Analogy of Pacemaker Activity *(a) Normal pacemaker activity by the SA node. (b) Takeover of pacemaker activity by the AV node when the SA node is nonfunctional. (c) Takeover of ventricular rate by the slower ventricular autorhythmic tissue in the condition of heart block even though the SA node is still functioning. (d) Takeover of pacemaker activity by an ectopic focus.*

nected from the slow third engine (Purkinje system) and remainder of the cars. The lead engine continues at 70 miles/hour while the remainder of the train proceeds at 30 miles/hour. Such a phenomenon, known as **complete heart block,** occurs when the conducting tissue between the atria and ventricles is damaged and becomes nonfunctional. A ventricular rate of 30 beats/minute will support only a very sedentary existence; in fact, the patient usually becomes comatose. In circumstances of abnormally low heart rate, as in SA-node failure or heart block, an artificial pacemaker can be used. Such an implanted device rhythmically generates impulses that spread throughout the heart to drive both the atria and ventricles at the normal rate of 70 beats/minute.

Occasionally an area of the heart, such as a Purkinje fiber, becomes overly excitable and depolarizes at a more rapid rate than the SA node. (The slow engine suddenly has the capability of going faster than the lead engine.) This abnormally excitable area, an **ectopic focus,** initiates a premature action potential that spreads throughout the rest of the heart before a normal action potential can be initiated by the SA node. An occasional abnormal impulse from an ectopic focus produces a **premature beat,** or an **extrasystole.** If the ectopic focus continues to discharge at its more rapid rate, pacemaker activity is shifted from the SA node to the ectopic focus. The heart rate abruptly becomes greatly accelerated and continues this rapid rate for a variable time period until the ectopic focus returns to normal. Such overly irritable areas may be associated with organic heart disease, but more frequently they occur in response to anxiety, lack of sleep, or excess caffeine, nicotine, or alcohol consumption.

The spread of cardiac excitation is coordinated to assure efficient pumping.

Once initiated in the SA node, an action potential spreads throughout the rest of the heart. For efficient cardiac function, the spread of excitation should satisfy the three following criteria:

1. *Atrial excitation and contraction should be complete before the onset of ventricular contraction.* Complete ventricular filling requires that atrial contraction precedes ventricular contraction. During the period of cardiac relaxation, the AV valves are open so that venous blood entering the atria continues to flow directly into the ventricles. Almost 80% of ventricular filling occurs by this means prior to atrial contraction. When the atria do contract, additional blood is squeezed into the ventricles to complete ventricular filling. Ventricular contraction then occurs to eject blood from the heart into the arteries. If the atria and ventricles were to contract simultaneously, the AV valves would be closed immediately because ventricular pressures would greatly exceed atrial pressures. The ventricles have much thicker walls and, accordingly, can generate more pressure. Atrial contraction would be unproductive because the atria could not squeeze blood into the ventricles through closed valves. Therefore, to assure complete filling of the ventricles—to obtain the remaining 20% of ventricular filling that occurs during atrial contraction—it is imperative that the atria become excited and contract before ventricular excitation and contraction.
2. *Excitation of cardiac-muscle fibers should be coordinated to assure that each heart chamber contracts as a unit to accomplish efficient pumping.* If the muscle fibers in a heart chamber were to become excited and contract randomly rather than contracting simultaneously in a coordinated fashion, they would be unable to eject blood. A smooth, uniform ventricular contraction is essential to squeeze out the blood. As an analogy, assume you had a rubber syringe full of water. If you merely poked a finger here or there into the bulb of the syringe, you would not get out much water. However, if you compressed the bulb in a smooth, coordinated fashion, you could squeeze out the water. In a similar manner, contraction of isolated cardiac-muscle fibers is not successful in pumping blood. Such random, uncoordinated excitation and contraction of the cardiac cells is known as **fibrillation.** Ventricular fibrillation rapidly causes death because the heart is not able to pump blood into the arteries. This condition can often be corrected by **electrical defibrillation,** which involves the application of a very strong electrical current on the chest wall. When this current reaches the heart, it essentially stimulates all parts of the heart simultaneously. Usually the first part of the heart to recover is the SA node, which takes over pacemaker activity, once again initiating impulses that trigger the synchronized contraction of the remainder of the heart.
3. *The pair of atria and pair of ventricles should be functionally coordinated so that both members of the pair contract simultaneously.* This permits synchronized pumping of blood into the pulmonary and systemic circulation.

The normal spread of cardiac excitation is carefully orchestrated to assure efficient cardiac function (Fig. 9–14). An action potential originating in the SA node first spreads throughout both atria, primarily from cell to cell via gap junctions. In addition, there are several poorly delineated, specialized conduction pathways that hasten conduction of the impulse through the atria:

□ The **interatrial pathway** extends from the SA node within the right atrium to the left atrium. Because of this path-

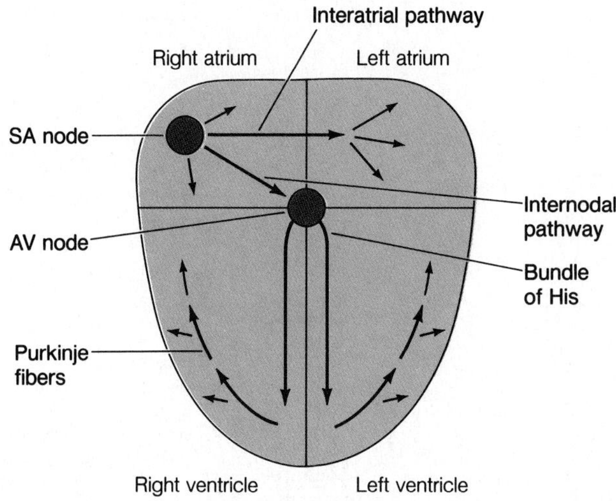

Figure 9–14 Spread of Cardiac Excitation

way, a wave of excitation can spread across the gap junctions throughout the left atrium at the same time a similar spread is being accomplished throughout the right atrium. This assures that both atria become depolarized to contract more or less simultaneously.

□ The **internodal pathway** extends from the SA node to the AV node. The AV node is the only point of electrical contact between the atria and ventricles; in other words, because the atria and ventricles are structurally connected by the electrically nonconductive fibrous skeleton, the only way an action potential in the atria can spread to the ventricles is by passing through the AV node. The internodal conduction pathway directs the spread of an action potential originating at the SA node to the AV node to assure sequential contraction of the ventricles following atrial contraction.

The action potential is conducted relatively slowly through the AV node. This is advantageous because atrial contraction must be completed before ventricular contraction for complete cardiac filling to occur. The impulse is delayed about 0.1 second (the **AV-nodal delay**), which enables the atria to become completely depolarized and to contract, emptying their contents into the ventricles, before ventricular depolarization and contraction occur.

Following this delay, the impulse rapidly travels down the bundle of His and throughout the ventricular myocardium via the Purkinje system. The network of fibers in this ventricular conduction system are specialized for rapid propagation of action potentials. Their presence hastens and coordinates the spread of ventricular excitation to assure that the ventricles contract as a unit. Although this system rapidly carries the action potential to a large number of cardiac-muscle cells, it does not terminate on every cell. The impulse quickly spreads from the excited cells to the remainder of the ventricular muscle cells by means of gap junctions.

The ventricular conduction system is more highly organized and more important than the interatrial and internodal conduction pathways. Because the ventricular mass is so much larger than the atrial mass, it is crucial that a rapid conduction system be present to expedite the spread of excitation in the ventricles. If the entire ventricular depolarization process depended on the cell-to-cell spread of the impulse via gap junctions, the ventricular tissue immediately adjacent to the AV node would become excited and contract before the impulse had even passed to the apex of the heart. This, of course, would not allow efficient pumping. The rapid conduction of the action potential down the bundle of His and its swift, diffuse distribution throughout the Purkinje network leads to almost simultaneous activation of the ventricular myocardial cells to assure a single, smooth, coordinated contraction, one that can efficiently eject blood.

The action potential of contractile cardiac-muscle cells shows a characteristic plateau.

The action potential in contractile cardiac-muscle cells, although initiated by the nodal pacemaker cells, varies considerably in ionic mechanisms and shape from the SA node potential. Unlike autorhythmic cells, the membrane of contractile cells remains essentially at rest at about -90 mV until excited by electrical activity propagated from the pacemaker. Once the membrane of a ventricular myocardial working cell is excited, the following events occur (Fig. 9–15):

1. As with neurons and skeletal-muscle cells, the rapid rising phase of the action potential is brought about by an explosive increase in membrane permeability to Na^+, with massive Na^+ influx bringing the inside of the cell to a positive potential of $+30$ mV. The Na^+ permeability then rapidly plummets to its low resting value, but, unique to these cardiac-muscle cells, the membrane potential is maintained at this positive level for several hundred milliseconds, producing a ***plateau phase*** of the action potential. This is in contrast to the short action potential of neurons and skeletal-muscle cells, which lasts less than a millisecond.
2. The sudden change in voltage occurring during the rising phase of the action potential brings about two voltage-dependent permeability changes that are responsible for maintaining this plateau: activation of "slow" Ca^{++} channels, and a marked decrease in K^+ permeability. Opening of the Ca^{++} channels results in a slow, inward diffusion of

Figure 9–15 Ventricular Myocardial Action Potential

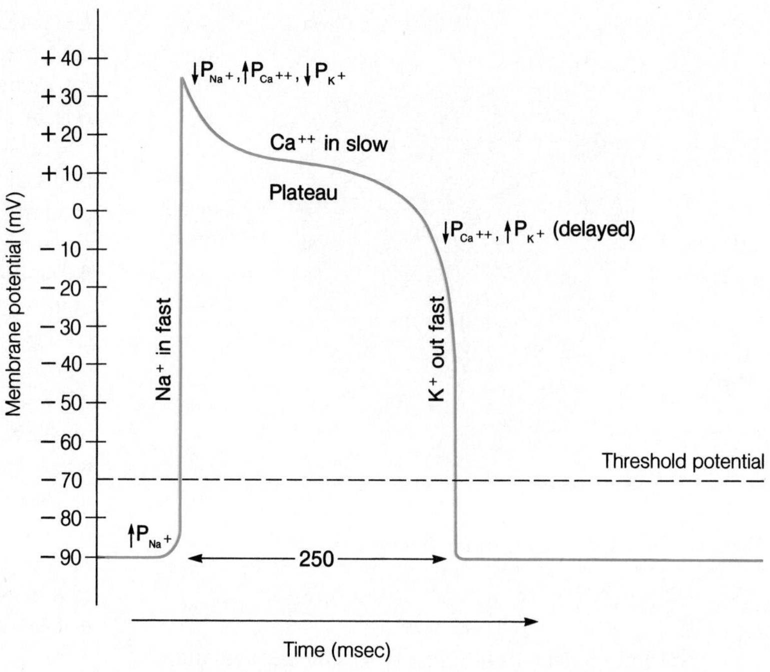

Ca^{++} because Ca^{++} is in greater concentration in the ECF. This continued influx of positively charged Ca^{++} prolongs the positivity inside the cell and is primarily responsible for the plateau portion of the action potential. This effect is enhanced by the concomitant decrease in K^+ permeability. The resultant reduction in outflux of positively charged K^+ prevents rapid repolarization of the membrane, thus contributing to prolongation of the plateau phase.

3. The rapid falling phase of the action potential results from inactivation of the Ca^{++} channels and a delayed activation of K^+ channels. The decrease in Ca^{++} permeability diminishes the slow, inward movement of positive Ca^{++}, whereas the sudden increase in K^+ permeability simultaneously promotes rapid outward diffusion of positive K^+. Thus, rapid repolarization at the end of the plateau is accomplished primarily by K^+ efflux, which once again makes the inside of the cell more negative than the outside and restores membrane potential to resting.

It should be obvious from this description that the action potential in cardiac cells is generated by a complicated interplay of permeability changes and membrane-potential changes.

The mechanism by which an action potential in a cardiac-muscle fiber brings about contraction of that fiber is quite similar to the excitation-contraction coupling process of skeletal muscle. The presence of a local action potential within the T tubules causes release of Ca^{++} into the cytosol from the intracellular stores in the sarcoplasmic reticulum. Unlike skeletal-muscle cells, during a cardiac action potential, large quantities of Ca^{++} diffuse into the cytosol across the plasma membrane from the ECF. Much of this extracellular Ca^{++} comes from within the T tubules, because they are much larger and thereby contain more ECF than their skeletal-muscle counterparts. This extra supply of Ca^{++} from the ECF is the major factor responsible for the prolongation of the cardiac action potential and subsequent lengthening of the period of cardiac contraction, which lasts about ten times longer than a single skeletal-muscle fiber contraction. This increased contractile time ensures adequate time to eject the blood.

The role of Ca^{++} within the cytosol, as with skeletal muscle, is to interfere with the inhibitory action of the troponin-tropomyosin complex. However, unlike skeletal muscle, in which sufficient Ca^{++} is always released to turn on all of the cross bridges, in cardiac muscle the extent of cross-bridge activity varies with the amount of cytosolic Ca^{++}. Removal of Ca^{++} from the cytosol by active membrane pumps restores the inhibitory action of troponin and tropomyosin so that contraction ceases and the heart muscle relaxes.

It is not surprising that changes in the ECF concentration of K^+ and Ca^{++} can have profound effects on the heart. Abnormal levels of K^+ are most important clinically, followed to a lesser extent by Ca^{++} imbalances. Changes in K^+ concentration in the ECF alter the K^+ concentration gradient between the ICF and ECF, thereby altering the rate of K^+ flux across the membrane. Normally there is substantially more K^+ inside the cells than in the ECF, but with elevated ECF K^+ levels, this gradient is reduced. Associated with this change is a reduction in "resting" potential (that is, the membrane is less negative on the inside than normal because less K^+ leaves). Among the consequences is a tendency to develop ectopic foci as well as cardiac arrhythmias. Also, because the magnitude of voltage change from the reduced "resting" state to the peak of the action potential is less than from the normal "resting" state to the peak, the resultant diminution of the action potentials' intensity causes the heart to become weak, flaccid, and dilated. At the extreme, with K^+ levels elevated two to three times the normal value, the weakened heart may actually stop pumping.

A rise in ECF Ca^{++} concentration, on the other hand, by prolonging the plateau phase of the action potential and by increasing the cytosolic concentration of Ca^{++}, augments the strength of cardiac contraction. The heart tends to contract spastically, with little time to rest between contractions. Some drugs that alter cardiac function do so by influencing Ca^{++} movement across the myocardial cell membranes. For example, Ca^{++} antagonists, such as verapamil, block Ca^{++} influx during an action potential, thereby reducing the force of cardiac contraction. Other agents, such as epinephrine, increase cardiac contractility by enhancing Ca^{++} influx.

Tetany of cardiac muscle is prevented by a long refractory period.

As with other excitable tissues, cardiac muscle has a refractory period. During the refractory period, which occurs immediately following the initiation of an action potential, an excitable membrane's responsiveness is totally abolished, making it impossible for another action potential to be generated. In skeletal muscle the refractory period is very short compared with the duration of the resultant contraction, so the fiber can be restimulated again before the first contraction is complete to produce summation of contractions. Rapidly repetitive stimulation that does not allow the muscle fiber to relax between stimulations results in a sustained, maximal contraction known as tetanus (see p. 232). In contrast, the refractory period in cardiac muscle is prolonged, lasting about 250 msec. This is almost as long as the period of contraction initiated by the action potential; a cardiac-muscle fiber contraction averages about 300 msec in duration (Fig. 9–16). Consequently, cardiac muscle cannot be restimulated until contraction is almost over, making summation of contractions and tetanus of cardiac muscle impossible. This is a valuable protective mechanism, because the pumping of blood requires alternate periods of contraction (emptying) and relaxation (filling). A prolonged tetanic contraction would prove to be fatal. The heart chambers could not be filled and emptied again.

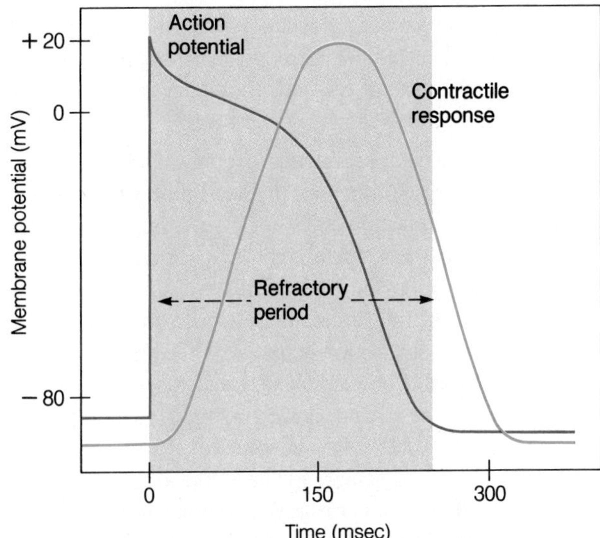

Figure 9–16 Relationship of Action Potential and Refractory Period to Duration of Contractile Response in Cardiac Muscle

The chief factor responsible for the long refractory period is inactivation during the prolonged plateau phase of the Na^+ channels that were activated during the initial Na^+ influx of the rising phase. Not until the membrane recovers from this inactivation process (when the membrane has already repolarized to resting) can the Na^+ channels be activated once again to begin another action potential.

The ECG is a record of the overall spread of electrical activity through the heart.

The electrical currents generated by cardiac muscle during depolarization and repolarization spread into the tissues surrounding the heart and are conducted through the body fluids. A small proportion of this electrical activity reaches the body surface, where it can be detected using recording electrodes. The record produced is an **electrocardiogram,** or **ECG.** (Originally the term EKG was used, because this technique was developed by a German scientist, William Einthoven, and "kardia" is the word for heart in German.) There are three important points to remember when considering what an ECG actually represents:

1. An ECG is a recording of that portion of the electrical activity induced in the body fluids by the cardiac impulse that reaches the surface of the body, not a direct recording of the actual electrical activity of the heart.
2. The ECG is a complex recording representing the overall spread of activity throughout the heart during depolarization and repolarization. It is not a recording of a single action potential at a single point in space and time. The record at any given time represents the sum of electrical activity in all of the cardiac-muscle cells, some of which may be undergoing action potentials while others may not yet be activated. For example, immediately after firing of the SA node, the atrial cells are undergoing action potentials while the ventricular cells are still at rest. At a later point, the electrical activity will have spread to the ventricular cells while the atrial cells will be repolarizing. Therefore, the overall pattern of electrical activity emanating from the heart varies with time as the impulse passes throughout the heart.
3. The recording represents comparisons between electrodes at two different points on the body surface, not the actual potential. For example, no potential is recorded at all in the ECG when the ventricular muscle is either completely depolarized or completely repolarized; both electrodes are "viewing" the same potential, so no difference in potential between the two electrodes is recorded.

The exact pattern of electrical activity recorded from the body surface depends on the orientation of the recording electrodes. Electrodes may be loosely thought of as "eyes" that "see" electrical activity and translate it into a visible recording, the ECG record. Whether an upward deflection or downward deflection is recorded is determined by the orientation of electrodes with respect to the current flow in the heart. For example, the spread of excitation across the heart is seen differently from the right arm than from the left foot, and both of these are seen differently than a recording directly over the heart. Even though the same electrical events are occurring in the heart, different wave forms representing the same electrical activity result when this activity is recorded by electrodes at different points on the body.

To provide standard comparisons, ECG records routinely consist of twelve conventional electrode systems, or leads. When an electrocardiograph machine is connected between recording electrodes at two points on the body, the specific arrangement of each pair of connections is called a **lead.** The twelve different leads each record electrical activity in the heart from different locations—six different electrical arrangements from the limbs and six chest leads at various sites around the heart. The same twelve leads are routinely used in all ECG recordings so that there is a common basis for comparison and for recognizing deviations from normal (Fig. 9–17).

Various components of the ECG record can be correlated to specific cardiac events.

Interpretation of the wave configurations recorded from each lead depends on a thorough knowledge of the sequence of the spread of cardiac excitation and the position of the heart relative to the placement of the electrodes. There are three distinct wave forms on a normal ECG: the P wave, the QRS complex, and the T wave (Fig. 9–18). (The letters do not signify anything other than the orderly sequence of the waves. Einthoven simply started in the middle of the alphabet when naming the waves.)

- ☐ The **P wave** represents atrial depolarization.
- ☐ The **QRS complex** represents ventricular depolarization.
- ☐ The **T wave** represents ventricular repolarization.

Following are further important points about the ECG record:

1. Firing of the SA node does not generate sufficient electrical activity to reach the surface of the body, so no wave is recorded for SA nodal depolarization. Therefore, the first recorded wave, the P wave, occurs when the impulse spreads across the atria.
2. In a normal ECG, there is no wave for atrial repolarization. The electrical activity associated with atrial repolarization normally occurs simultaneously with ventricular depolarization and is masked by the QRS complex.
3. The P wave is much smaller than the QRS complex because the atria consist of a much smaller muscle mass than the ventricles and consequently generate less electrical activity. Thus, less current reaches the body surface during atrial depolarization than during ventricular depolarization.
4. As a general statement, repolarization is a more gradual process than depolarization. The T wave is therefore more spread out and has less amplitude (height) than the QRS wave.
5. There are three times when no current is flowing in the heart and the ECG remains at baseline:
   a. during the AV nodal delay. This is represented by the interval of time between the end of the P wave and the

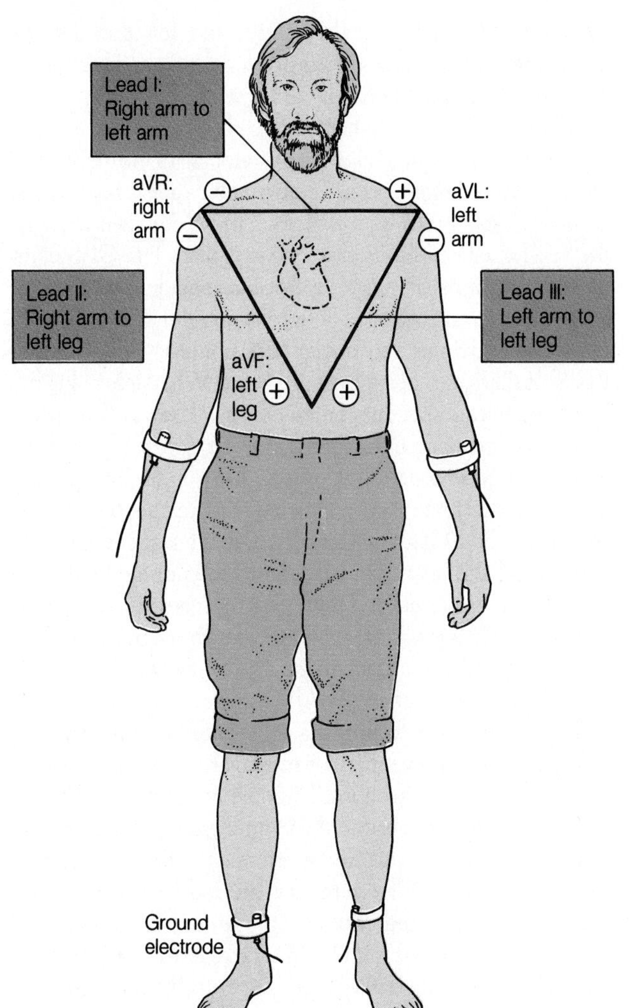

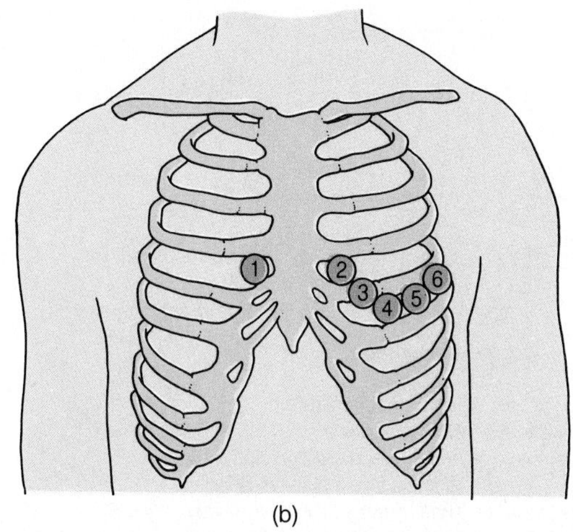

Figure 9–17 Electrocardiogram Leads *(a) Limb leads. The six limb leads include leads I, II, III, aVR, aVL, and aVF. Leads* ***I, II,*** *and* ***III*** *are* ***bipolar leads,*** *because two recording electrodes are used. The tracing records the* difference *in potential between the two electrodes. For example, lead I records the difference in potential detected at the right arm and left arm. The electrode placed on the right leg serves as a ground and is not a recording electrode. The* ***aVR, aVL,*** *and* ***aVF*** *leads are* ***unipolar leads.*** *Only the potential under one electrode, the exploring electrode, is recorded. For example, aVR records the potential reaching the right arm in comparison to the rest of the body. (b) Chest leads. The six chest leads,* ***V_1 through V_6***, *are also* ***unipolar leads.*** *The exploring electrode records mainly the electrical potential of the cardiac musculature immediately beneath the electrode in six different locations surrounding the heart.*

onset of the QRS wave, known as the **PR segment.** (The reason that this is called the PR segment rather than the PQ segment is that the Q deflection is small and sometimes absent, whereas the R deflection is the dominant wave of the complex.);

b. when the ventricles are completely depolarized before they repolarize again, represented by the **ST segment.** This segment is the interval between QRS and T, which coincides with the time during which ventricular activation is complete and the ventricles are contracting and emptying; and
c. when the heart muscle is completely at rest and ventricular filling is taking place, after the T wave and before the next P wave, called the **TP interval.**

The ECG can be useful in diagnosing abnormal heart rates, arrhythmias, and myopathies.

Evaluation of ECG patterns can provide useful information about the status of the heart, including its rate, rhythm, and the health of its musculature. Following are the principle deviations from normal that can be ascertained through electrocardiography (Table 9–2).

ABNORMALITIES IN RATE. The standard ECG recording paper is calibrated on the horizontal axis for time so that the distance between two consecutive QRS complexes represents the beat-to-beat heart rate. A rapid heart rate of more than 100 beats per minute is known as **tachycardia** (*tachy* means

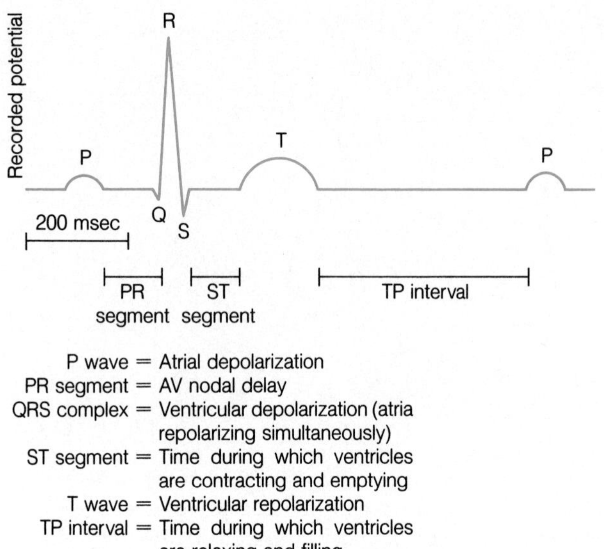

Figure 9–18 Electrocardiogram Wave Forms in Lead II

"fast"), whereas a slow heart rate of fewer than 60 beats per minute is referred to as **bradycardia** (*brady* means "slow").

ABNORMALITIES IN RHYTHM. Rhythm refers to the regularity of the ECG waves. Any variation from the normal rhythm and sequence of excitation of the heart is designated as an **arrhythmia.** It may result from alterations in SA-node pacemaker activity, the presence of ectopic foci, or interference with conduction. Oftentimes heart rate is also altered. Extrasystoles or premature beats originating from an ectopic focus are common deviations from normal rhythm. Other abnormalities in rhythm easily detected on an ECG include atrial flutter, atrial fibrillation, ventricular fibrillation, and heart block.

Atrial flutter is characterized by a rapid but regular sequence of atrial depolarizations at rates between 200 to 380 beats per minute. The ventricles rarely keep pace with the racing atria. Because the conducting tissue's refractory period is longer than that of the atrial muscle, the AV node is unable to respond to every impulse that converges on it from the atria. Maybe only one out of every two or three atrial impulses successfully passes through the AV node to the ventricles. Such a situation is referred to as a *2:1* or *3:1 rhythm*. The fact that not every atrial impulse reaches the ventricle in atrial flutter is important, because it precludes a rapid ventricular rate of more than 200 beats per minute. Such a high rate would not allow adequate time for ventricular filling between beats. In such a case, the output of the heart would be reduced to the extent that loss of consciousness or even death could result because of decreased blood flow to the brain.

Atrial fibrillation is characterized by rapid, irregular, uncoordinated depolarizations of the atria with no definite P waves. Accordingly, atrial contractions are chaotic and asynchronized. Since impulses reach the AV node erratically, the ventricular rhythm is also very irregular. The QRS complexes are normal in shape but occur sporadically. Variable lengths of time are available between ventricular beats for ventricular filling. Some ventricular beats come so close together that little filling can occur between beats. When less filling has occurred, the subsequent contraction is weaker. In fact, some of the ventricular contractions may be too weak to eject enough blood to produce a palpable wrist pulse. In this situation, if the heart rate is determined directly, either by the apex beat or via the ECG, and the pulse rate is taken concurrently at the wrist, the heart rate will exceed the pulse rate. Such a difference in heart rate and pulse rate is known as a **pulse deficit.** Normally the heart rate coincides with the pulse rate, because each cardiac contraction initiates a pulse wave as it ejects blood into the arteries.

Ventricular fibrillation is a very serious rhythmic abnormality in which the ventricular musculature exhibits uncoordinated, chaotic contractions. Multiple impulses travel erratically in all directions around the ventricles. The ECG tracing in ventricular fibrillation is very irregular with no detectable pattern or rhythm. The ventricles are ineffectual as pumps with such disorganized contractions. If circulation is not restored in less than four minutes through external cardiac compression or electrical defibrillation, irreversible brain damage occurs and death is imminent.

Another type of arrhythmia, **heart block,** arises from defects in the cardiac conducting system. The atria still beat regularly, but the ventricles occasionally fail to be stimulated and thus do not contract following atrial contraction. Impulses between the atria and ventricles can be blocked to varying degrees. In some forms of heart block, only every second or third atrial impulse is passed to the ventricles. This is known as *2:1* or *3:1 block*, which can be distinguished from the 2:1 or 3:1 rhythm associated with atrial flutter by the rates involved. In heart block, the atrial rate is normal but the ventricular rate is considerably below normal, whereas in atrial flutter, the atrial rate is very high in accompaniment with a normal or above-normal ventricular rate. Complete heart block is characterized by complete dissociation between atrial and ventricular activity, with impulses from the atria not being conducted to the ventricles at all. The atrial beat continues to be governed by the SA node, but the ventricles generate their own impulses at a rate much slower than the atria. On the ECG, the P waves

Table 9–2 Representative Heart Conditions Detectable through Electrocardiography

Condition	ECG	
Abnormalities in Rate		
Tachycardia Bradycardia	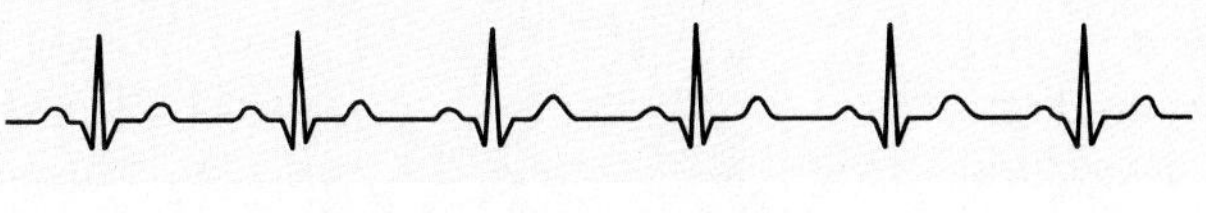	Normal rhythm
	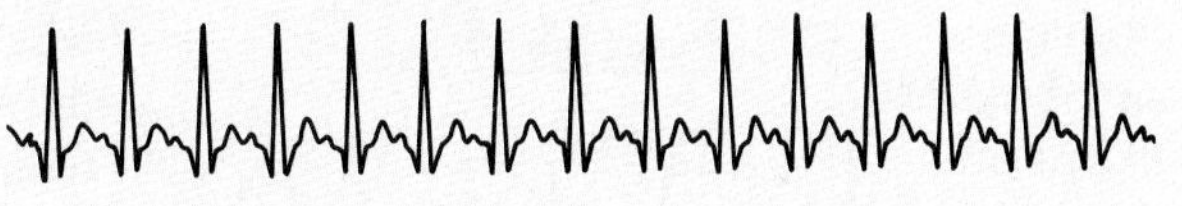	Tachycardia
Abnormalities in Rhythm		
Extrasystole (premature beat) Atrial flutter Atrial fibrillation Ventricular fibrillation Heart block	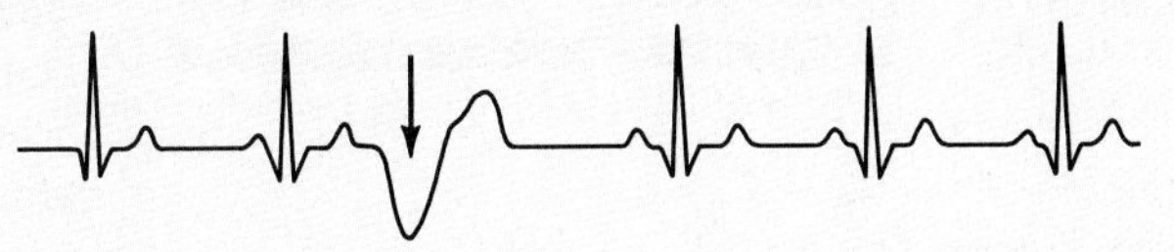	Extrasystole
	1 mV 1 sec	Ventricular fibrillation
	P P P P P P P P P QRS QRS QRS QRS	Complete heart block
Cardiac Myopathies		
Myocardial ischemia Myocardial infarction		

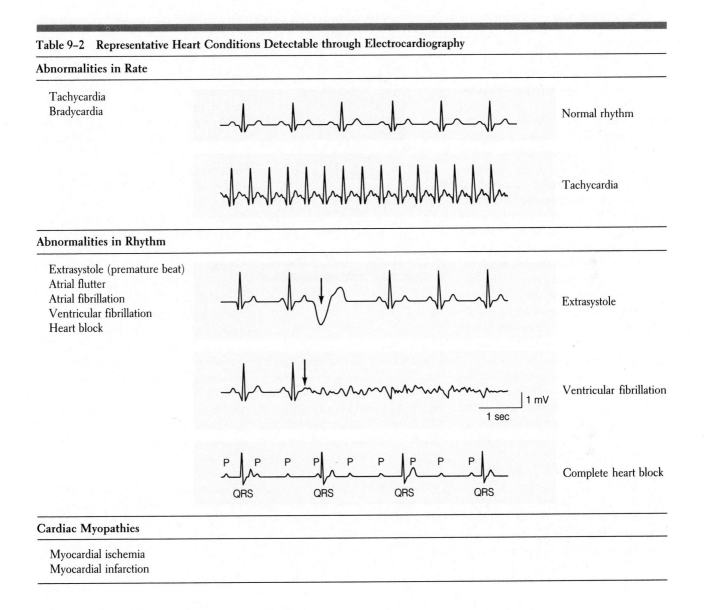

exhibit a normal rhythm. The QRS and T waves also occur regularly but at a much slower rate than the P waves and are completely independent of P wave rhythm.

CARDIAC MYOPATHIES. Abnormal ECG waves are also important in the recognition and assessment of **cardiac myopathies** (damage of the heart muscle). **Myocardial ischemia** refers to inadequate blood supply to the heart tissue. Actual death or **necrosis** of heart-muscle cells, usually caused by blockage of a blood vessel supplying that area of the heart, is termed **acute myocardial infarction,** commonly known as a **heart attack.** When a portion of the heart muscle becomes necrotic, no electrical activity exists in that area. One change in ECG pattern that may occur is an enlarged Q wave. The Q wave is due specifically to activation of the ventricular septum. Normally the Q wave is quite small; little electrical activity reaches the surface of the body from septal depolarization because the septum is electrically shielded by the surrounding ventricles, which are still at resting potential. An infarcted portion of the myocardium is no longer electrically active; that is, it exhibits no potential at all. This dead area might be visualized as a "window" within the polarized ventricular muscle mass through which the electrical activity of the septum can more easily be detected by a recording elec-

A CLOSER LOOK AT EXERCISE PHYSIOLOGY

The What, Who, and When of Stress Testing

Stress tests, or graded exercise tests, are conducted primarily to aid in diagnosing or quantifying heart or lung disease and to evaluate the functional capacity of asymptomatic individuals. The tests are usually given on motorized treadmills or bicycle ergometers (stationary, variable-resistance bicycles). Workload intensity (how hard the subject is working) is adjusted by progressively increasing the speed and incline on the treadmill or by progressively increasing the pedaling frequency and resistance on the bicycle. The test starts at a low intensity and continues until a prespecified workload is achieved, physiologic symptoms occur, or the subject is too fatigued to continue.

During diagnostic testing, the patient is monitored with an ECG, and blood pressure is taken each minute. A cardiologist, or a pulmonary specialist in the case of lung disease, is present during testing. A test is considered to be positive if ECG abnormalities occur (such as ST segment depression, inverted T waves, or dangerous arrhythmias) or if physical symptoms such as angina pectoris develop. A test that is interpreted as being positive in a person who does not have heart disease is called a false positive test. In men, this occurs only about 10% to 20% of the time, so the diagnostic stress test for men has a ***specificity*** of 80% to 90%. Women have a greater frequency of false positive tests; the specificity for them is lower at about 70%.

The ***sensitivity*** of a test means that those individuals with disease are correctly identified and there are few false negatives. The sensitivity of the stress test is reported to be 60% to 80%. This means that if one hundred individuals with heart disease were tested, sixty to eighty would be correctly identified, but twenty to forty would have a false negative test. As more is learned about what causes false positive and false negative tests, the specificity and sensitivity of stress testing should improve. Although stress testing is now an important diagnostic tool, it is just one of several tests used to determine the presence of coronary artery disease.

Graded exercise tests are also conducted on individuals not suspected of having heart or lung disease to determine their present functional capacity. These functional tests are administered in the same way as diagnostic tests, but a physician need not be present. They are conducted by exercise physiologists and are used to establish safe exercise prescriptions, to aid athletes in establishing optimal training programs, and as research tools to evaluate the effectiveness of a particular training regimen. Functional stress testing is becoming more prevalent as more people are joining hospital- or community-based wellness programs for disease prevention.

trode. The Q wave is enlarged when the recording electrode is positioned so that it can record septal activation through the infarcted area. This is one way in which the location of a myocardial infarction may be identified.

Interpretation of an ECG is a complex task requiring extensive knowledge and training. The foregoing discussion is not intended to make you an ECG expert by any means but to give you an appreciation of the ways in which the ECG can be used as a diagnostic tool, as well as to present an overview of some of the more common abnormalities of heart function. (For a further use of the ECG, see the accompanying boxed feature, A Closer Look at Exercise Physiology.)

Mechanical Events of the Cardiac Cycle

The heart alternately contracts to empty and relaxes to fill.

The cardiac cycle consists of alternate periods of **systole** (contraction and emptying) and **diastole** (relaxation and filling). Contraction occurs as a result of the spread of excitation across the heart, whereas relaxation follows the subsequent repolarization of the cardiac musculature. The following discussion correlates various events that occur concurrently during the cardiac cycle; these are ECG, pressure changes, volume changes, valve activity, and heart sounds. Reference to Figure 9–19 will facilitate this discussion. Only the events on the left side of the heart will be described, but keep in mind that identical events are occurring on the right side of the heart except that the pressures are lower. Our discussion will begin and end with ventricular diastole to complete one full cardiac cycle.

During early ventricular diastole, the atrium is still also in diastole. This corresponds to the TP interval on the ECG—the interval after ventricular repolarization and before another atrial depolarization. Because of the continuous inflow of blood from the venous system into the atrium, atrial pressure slightly exceeds ventricular pressure even though both chambers are relaxed (Fig. 9–19, point 1). Because of this pressure differential, the AV valve is open, and blood flows directly from the atrium into the ventricle throughout ventricular diastole. The ventricular volume (point 2) continues to rise even before atrial contraction takes place. Late in ventricular diastole, the SA node reaches threshold and fires. The impulse spreads throughout the atria, which is recorded on the ECG as the P wave (point 3). Atrial depolarization brings about atrial contraction, which squeezes more blood into the ventricle, causing a rise in the atrial pressure curve (point 4). The short delay between the P wave and the rise in atrial pressure is the time during which the excitation-contraction coupling process is taking place. The corresponding rise in ventricular pressure (point 5) that occurs simultaneous to the rise in atrial pressure is due to the additional volume of blood added to the ventricle by atrial contraction (point 6). Throughout atrial contraction, atrial pressure still slightly exceeds ventricular pressure, so the AV valve remains open.

Ventricular diastole ends at the onset of ventricular contraction. By this time, atrial contraction and ventricular filling are completed. The volume of blood in the ventricle at the end of diastole (point 7) is known as the **end-diastolic volume (EDV)**. No more blood will be added to the ventricle during this cycle. Therefore, the end-diastolic volume is the maximum amount of blood, which averages about 135 ml, that the ventricle will contain during this cycle.

Following atrial excitation, the impulse passes through the AV node and specialized conducting system to excite the ventricle. Simultaneously, atrial contraction is occurring. By the time ventricular activation is complete, atrial contraction is already accomplished. The QRS complex represents this ventricular excitation (point 8), which induces ventricular contraction. The ventricular pressure curve sharply increases shortly after the QRS complex, signaling the onset of ventricular systole (point 9). There is a slight delay between the QRS complex and the actual onset of ventricular systole because of the time required for the excitation-contraction coupling process to occur. As ventricular contraction begins, ventricular pressure immediately exceeds atrial pressure. This backward pressure differential forces the AV valve closed (point 9). Note the slight rise in atrial pressure (point 10) immediately following closure of the AV valve. This small increase in pressure is due to the slight bulging of the closed AV valve into the atria as ventricular pressure increases.

After ventricular pressure exceeds atrial pressure and the AV valve has closed, the ventricular pressure must continue to increase before it exceeds aortic pressure to open the aortic valve. Therefore, there is a brief period of time between closure of the AV valve and opening of the aortic valve when the ventricle remains a closed chamber (point 11). Because all valves are closed (the AV valve just closing, the aortic valve not yet open), no blood can enter or leave the ventricle during this time. This interval of time is termed the period of **isovolumetric ventricular contraction** (*isovolumetric* means "constant volume and length"). Because no blood enters or leaves the ventricle, the ventricular chamber remains at constant volume and the muscle fibers remain at constant length. This isovolumetric condition is similar to an isometric contraction in skeletal muscle. During the period of isovolumetric ventricular contraction, ventricular pressure continues to increase as the volume remains constant (point 12).

When ventricular pressure exceeds aortic pressure (point 13), the aortic valve is forced open and ejection of blood begins. The aortic pressure curve rises as blood is forced into the aorta from the ventricle faster than blood is draining off into the smaller vessels at the other end (point 14). As blood is forced into the aorta, the aortic pressure increases but still remains slightly less than ventricular pressure, so the aortic valve remains open. This allows blood to continue to pass from the ventricle into the aorta. The ventricular volume decreases substantially as blood is rapidly ejected (point 15).

The ventricle does not empty completely during ejection. Normally only about half of the blood contained within the ventricle at the end of diastole is ejected during the subsequent systole. The amount of blood remaining in the ventricle at the end of systole when ejection is complete is known as the **end-systolic volume (ESV)**, which averages about 65 ml (point 16). This is the least amount of blood that the ventricle will contain during this cycle.

The amount of blood ejected with each contraction is known as the **stroke volume (SV)**, which is equal to the end-diastolic volume minus the end-systolic volume; in other words, the difference between the volume of blood in the ventricle before contraction compared to the volume after contraction is the amount of blood ejected during the contraction. In our example, the end-diastolic volume is 135 ml, the end-systolic volume is 65 ml, and the stroke volume is 70 ml.

Atrial repolarization has already occurred simultaneous to ventricular depolarization, so the atria are in diastole through-

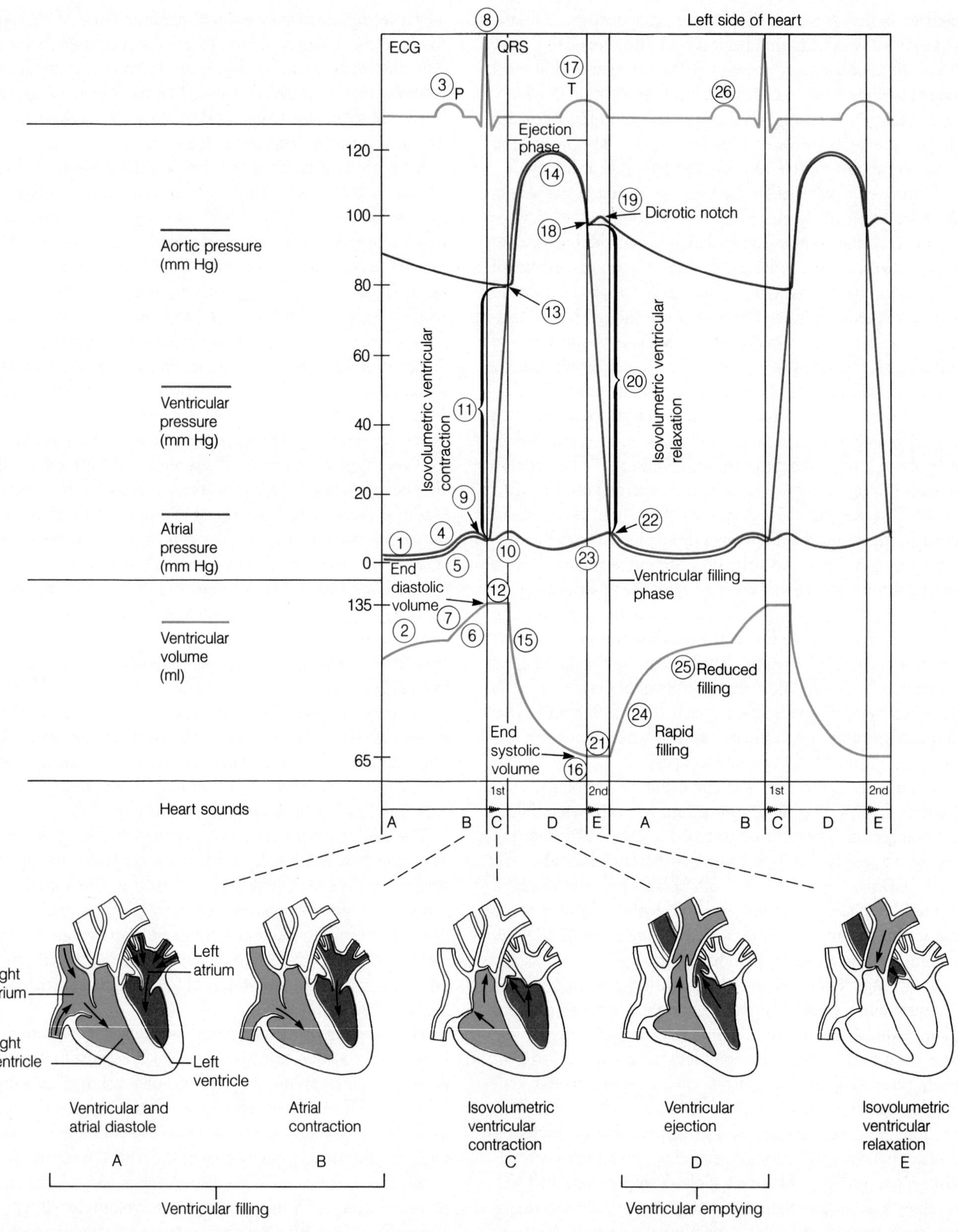

Figure 9–19 Cardiac Cycle *See text for an explanation of the circled numbers.*

out ventricular systole. The T wave signifies ventricular repolarization occurring at the end of ventricular systole (point 17). As the ventricle starts to relax upon repolarization, ventricular pressure falls below aortic pressure and the aortic valve closes (point 18). Closure of the aortic valve produces a disturbance or notch on the aortic pressure curve (point 19) that is known as the **dicrotic notch** or the **incisura.** No more blood leaves the ventricle during this cycle. The ventricular pressure has fallen below the aortic pressure, so the aortic valve has closed and no more blood can leave the ventricle. The AV valve is not yet open, however, because ventricular pressure still exceeds atrial pressure, so no blood can enter the ventricle from the atrium. Therefore, all valves are once again closed for a brief period of time known as **isovolumetric ventricular relaxation** (point 20). The muscle-fiber length and chamber volume (point 21) remain constant. No blood leaves or enters as the ventricle continues to relax and the pressure steadily falls. When the ventricular pressure falls below the atrial pressure, the AV valve opens (point 22) and ventricular filling occurs once again.

During ventricular systole, the atrium is in diastole. Blood continues to flow from the pulmonary vein into the left atrium. As this incoming blood pools in the atrium, atrial pressure continuously rises (point 23). When the AV valve opens at the end of ventricular systole, the blood that accumulated in the atrium during ventricular systole rapidly pours into the ventricle. Ventricular filling thus occurs rapidly at first (point 24) because of the increased atrial pressure resulting from the accumulation of blood in the atria. Then ventricular filling slows down (point 25) as the accumulated blood has already been delivered to the ventricle, and atrial pressure starts to fall. During this period of reduced filling, blood continues to flow from the pulmonary vein into the left atrium and through the open AV valve into the left ventricle. During late ventricular diastole, when ventricular filling is proceeding slowly, the SA node fires again (point 26), and the cardiac cycle starts over.

It is significant that much of ventricular filling occurs early in diastole during the rapid-filling phase. During times of rapid heart rates, the length of diastole is reduced to a much greater extent than is the length of systole. For example, if the heart rate increases from 75 to 180 beats per minute, the duration of diastole decreases about 75%, from 500 msec to 125 msec. This greatly reduces the time available for ventricular relaxation and filling. Fortunately, because much of ventricular filling is accomplished during early diastole, filling is not seriously impaired during periods of increased heart rate, such as during exercise (Fig. 9–20). There is a limit, however, to how rapidly the heart can beat without decreasing the period of diastole to the point that ventricular filling is severely impaired. At heart rates greater than 200 beats per minute, diastolic time is too short to allow adequate ventricular filling. With inadequate filling, the resultant cardiac output is deficient. Normally ventricular rates do not exceed 200 beats per minute because the relatively long refractory period of the AV node will not allow impulses to be conducted to the ventricles more frequently than this.

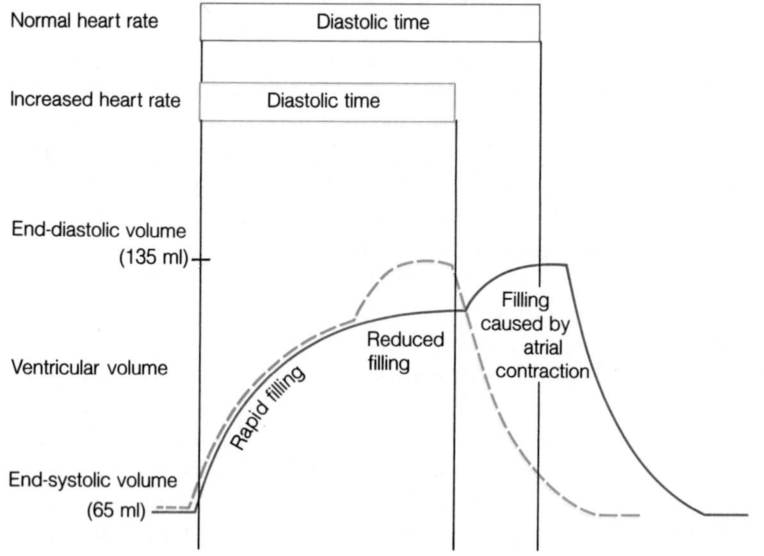

Figure 9–20 Ventricular Filling Profiles during Normal and Rapid Heart Rates

Two heart sounds associated with valve closures can be heard during the cardiac cycle.

Two major heart sounds normally can be heard during the cardiac cycle when listening with a stethoscope, a process known as **auscultation.** The **first heart sound** is low-pitched, soft, and relatively long—often said to sound like "lub." The **second heart sound** has a higher pitch, is shorter and sharper—often said to sound like "dup." Thus, one normally hears "lub, dup, lub, dup, lub, dup". . . . The first heart sound is associated with closure of the AV valves, whereas the second sound is associated with closure of the semilunar valves. Opening of valves does not produce any sound. The sounds are caused by vibrations set up within the walls of the ventricles and major arteries during valve closure, not by the valves snapping shut. Because closure of the AV valves occurs at the onset of ventricular contraction when ventricular pressure first exceeds atrial pressure, the first heart sound signals the onset of ventricular systole. Closure of the semilunar valves occurs at the onset of ventricular relaxation as the left and right ventricular pressures fall below the aortic and pulmonary artery pressures, respectively. The second heart sound, therefore, signals the onset of ventricular diastole.

Turbulent blood flow produces heart murmurs.

The heart sounds described previously are normal. Abnormal heart sounds, or **murmurs,** are usually (but not always) associated with cardiac disease. Murmurs not involving heart pathology, so-called **functional murmurs,** are more common in young people.

Blood normally flows in a *laminar* fashion, in which layers of the fluid slide smoothly over each other. Such laminar flow does not produce any sound. When blood flow becomes turbulent, however, a sound can be heard (Fig. 9–21). Such an abnormal sound is due to vibrations created in the surrounding structures by the turbulent flow.

The most common cause of turbulence is valve malfunction, either a stenotic or an insufficient valve. A **stenotic valve** is a stiff, narrowed valve that does not open completely. When valvular stenosis is present, blood cannot flow through the constricted opening in a smooth, streamlined fashion. Instead, it is forced through the constricted opening at tremendous velocity, resulting in turbulence that produces an abnormal whistling sound, similar to when you force air rapidly through narrowed lips to whistle.

An **insufficient valve,** on the other hand, is one that cannot close completely, usually because the valve edges are scarred and do not fit together properly. Less frequently, the valve leaflets themselves are normal but the papillary muscle–chordae tendineae system does not function properly, so the valve is allowed to evert into the atria to the extent that the valve edges no longer remain in contact with each other. In either case, turbulence is produced when blood flows backward through the insufficient valve and collides with blood moving in the opposite direction, creating a swishing or gurgling murmur. Such backflow of blood is known as **regurgitation.** Commonly, an insufficient heart valve is called a **leaky valve,** because it allows blood to leak back through at a time that the valve should be closed.

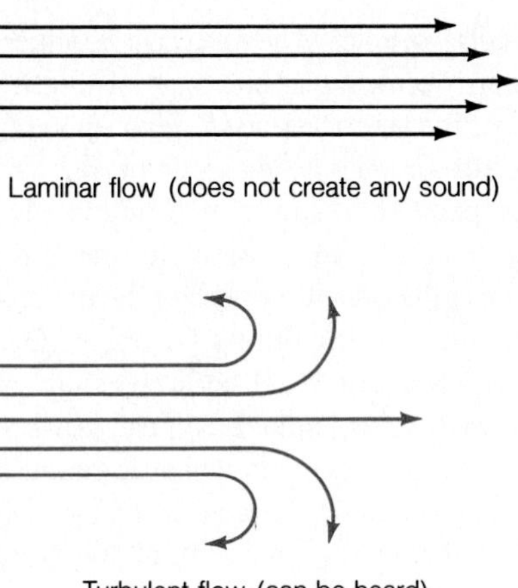

Figure 9–21 Comparison of Laminar and Turbulent Flow

Most often, both valvular stenosis and insufficiency are caused by **rheumatic fever,** an autoimmune ("immunity against self") disease triggered by a streptococcus bacterial infection. Antibodies formed against toxins produced by these bacteria interact with many of the body's own tissues, resulting in immunological damage. The heart valves are among the most susceptible tissues in this regard. Large, hemorrhagic, fibrinous lesions form along the inflamed edges of an affected heart valve, causing the valve to become thickened, stiff, and scarred. Sometimes the leaflet edges permanently adhere to each other. Depending on the extent and specific nature of the lesions, the valve may become either stenotic or insufficient or some degree of both.

The valve involved and the type of defect can usually be detected by the *location* and *timing* of the murmur. Each heart valve may be heard best at a specific location on the chest. Noting the location at which a murmur is loudest aids the diagnostician in determining which valve is involved. The timing of the murmur refers to the part of the cardiac cycle during which the murmur is heard. Recall that the first heart sound signals the onset of ventricular systole and the second heart sound signals the onset of ventricular diastole. Thus, a mur-

Table 9-3 Timing of Murmur

	Systolic Murmur	Diastolic Murmur
	First heart sound — murmur — Second heart sound	*Second heart sound — murmur — First heart sound*
Type of murmur	Whistling murmur	Whistling murmur
	↓	↓
Valve disorder	Stenotic semilunar valve	Stenotic AV valve
Type of murmur	Swishy murmur	Swishy murmur
	↓	↓
Valve disorder	Insufficient AV valve	Insufficient semilunar valve

mur occurring between the first and second heart sounds (lup-murmur-dup, lup-murmur-dup) signifies a **systolic murmur.** A **diastolic murmur,** on the other hand, occurs between the second and first heart sound (lup-dup-murmur, lup-dup-murmur). The sound of the murmur characterizes it as either a stenotic (whistling) murmur or an insufficient (swishy) murmur. Armed with these facts, it is relatively easy to determine the cause of a valvular murmur (Table 9-3). As an example, a whistling murmur (denoting a stenotic valve) occurring between the first and second heart sounds (denoting a systolic murmur) signifies the presence of stenosis in a valve that should be open during systole. This could either be the aortic or pulmonary semilunar valve through which blood is being ejected. Identifying which of these valves is stenotic is accomplished by determining the location over which the murmur is best heard. The main concern with heart murmurs, of course, is not the murmur itself but the accompanying detrimental circulatory consequences caused by the defect.

Cardiac Output and Its Control

Cardiac output is dependent on the heart rate and stroke volume.

Cardiac output (CO) is the volume of blood pumped by *each ventricle* per minute (not the total amount of blood pumped by the heart). The volume of blood flowing through the pulmonary circulation is equivalent during a period of time to the volume flowing through the systemic circulation. Therefore, the cardiac output from each ventricle normally is identical, although on a beat-to-beat basis, minor variations may occur. The two determinants of cardiac output are ***heart rate*** (beats per minute) and ***stroke volume*** (volume of blood pumped per beat or stroke).

The average heart rate is 70 beats per minute, established by SA-node rhythmicity, whereas the average stroke volume is 70 ml per beat, producing an average cardiac output of 4,900 ml/min or close to 5 liters (l)/min:

$$\text{Cardiac output} = \text{heart rate} \times \text{stroke volume}$$

$$\begin{aligned}\text{CO} &= 70 \text{ beats/min} \times 70 \text{ ml/beat} \\ &= 4{,}900 \text{ ml/min} \approx 5 \text{ l/min}\end{aligned}$$

Because the body's total blood volume averages 5 to 5.5 liters, the equivalent of the entire blood volume is pumped by each half of the heart each minute. In other words, each minute the right ventricle normally pumps 5 liters of blood through the lungs and the left ventricle pumps 5 liters of blood through the systemic circulation. At this rate, each half of the heart would pump about 2½ million liters of blood in just one year. Yet this is only the resting cardiac output! During exercise the cardiac output can increase to 20 to 25 liters per minute, and it has been recorded as high as 40 liters per minute during heavy exercise in trained athletes. The difference between the cardiac output at rest and the maximum volume of blood the heart is capable of pumping per minute is known as the **cardiac reserve.** How can cardiac output vary so tremendously, depending on the demands of the body? You can readily answer that question by thinking about how your own heart pounds rapidly (increased heart rate) and forcefully (increased stroke volume) when you engage in strenuous physical activities (need for increased cardiac output). Thus, the regulation of cardiac output is dependent on the control of both heart rate and stroke volume, topics that will be discussed next.

Heart rate is determined primarily by autonomic influences on the SA node.

The SA node is normally the pacemaker of the heart because it has the fastest spontaneous rate of depolarization to threshold. Recall that this automatic gradual reduction of membrane potential between beats is generally believed to be due to a reduction in K^+ permeability. When the SA node reaches threshold, an action potential is initiated that spreads throughout the heart, inducing the heart to contract. This happens about 70 times per minute, setting the average heart rate at about 70 beats per min.

The heart is innervated by both divisions of the autonomic nervous system, which can modify the rate (as well as strength) of contraction, even though nervous stimulation is not required to initiate contraction. The parasympathetic nerve to the heart, the **vagus nerve,** primarily supplies the atrium, especially the SA and AV nodes. Only a few vagal fibers terminate in the ventricles. The cardiac sympathetic nerves also supply the atria, including the SA and AV nodes, and richly innervate the ventricles as well.

The parasympathetic nervous system's influence on the SA node is to decrease the heart rate (Fig. 9–22 and Table 9–4). Acetylcholine released upon increased parasympathetic activity increases the permeability of the SA node to K^+. This results in a reduction in the rate at which spontaneous action potentials are initiated through a two-fold effect:

1. Enhanced K^+ permeability hyperpolarizes the SA-node membrane because more positive potassium ions leave than normal, making the inside even more negative. Because the "resting" potential starts even farther away from threshold, it takes longer to reach threshold.
2. The enhanced K^+ permeability induced by vagal stimulation also opposes the automatic reduction in K^+ permeability that is responsible for gradually depolarizing the membrane to threshold. This countering effect decreases the rate of spontaneous depolarization, prolonging the time required to drift to threshold. Therefore, the SA node reaches threshold and fires less frequently, decreasing the heart rate.

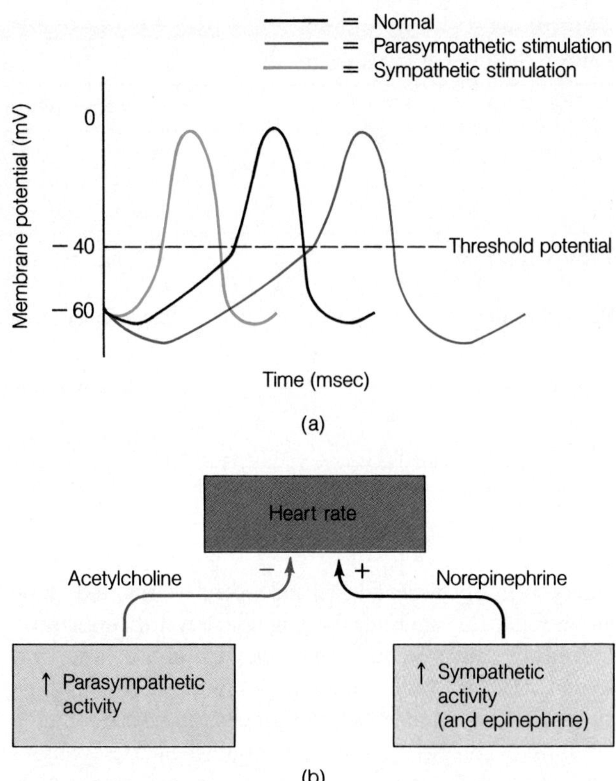

Figure 9–22 Autonomic Control of SA Node Activity and Heart Rate *(a) Autonomic influence on SA node potential. Sympathetic stimulation increases the rate of depolarization of the SA node whereas parasympathetic stimulation decreases the rate of SA-nodal depolarization. (b) Control of heart rate by the autonomic nervous system. Increased sympathetic activity increases the heart rate whereas increased parasympathetic activity decreases the heart rate.*

Parasympathetic influence on the AV node decreases the node's excitability, prolonging transmission of impulses to the ventricles even longer than the usual AV nodal delay. This effect is brought about by increasing K^+ permeability, which hyperpolarizes the membrane, thereby retarding the initiation of excitation in the AV node.

The effect of parasympathetic stimulation on the atrial contractile cells is to shorten the action potential, believed to be caused by a reduction in the slow inward current carried by Ca^{++}; that is, the plateau phase is reduced. As a result, atrial contraction is weakened.

Thus, the heart is more "leisurely" under parasympathetic influence—it beats less rapidly, the time between atrial and ventricular contraction is stretched out, and atrial contraction is weaker. This is appropriate considering that the parasympathetic system controls heart action in quiet, relaxed situations when the body is not demanding an enhanced cardiac output.

In contrast, the sympathetic nervous system, which controls heart action in emergency or exercise situations when there is a need for greater blood flow, speeds up the heart rate through its effect on the pacemaker tissue. The main effect of sympathetic stimulation on the SA node is to increase its rate of depolarization so that threshold is reached more rapidly (Fig. 9–22 and Table 9–4). Norepinephrine released from the sym-

Table 9–4 Effects of the Autonomic Nervous System on the Heart

Parasympathetic Stimulation	Sympathetic Stimulation
Decreases heart rate	Increases heart rate
Increases AV nodal delay	Increases excitability of all areas of heart, decreasing AV nodal delay and speeding up conduction of the impulse through the heart
Weakens atrial contraction	Increases contractility (strength of contraction) of atria and ventricles
	Promotes secretion of epinephrine, which augments sympathetic nervous system
	Increases venous return, which increases stroke volume via the Frank-Starling mechanism

pathetic nerve endings appears to decrease K^+ permeability by accelerating inactivation of the K^+ channels. With fewer positive potassium ions leaving, the inside of the cell becomes less negative, creating a depolarizing effect. This swifter drift to threshold under sympathetic influence permits a greater frequency of action potentials and a correspondingly more rapid heart rate.

Sympathetic stimulation of the AV node reduces the AV-nodal delay by increasing conduction velocity, presumably through enhancement of the slow, inward Ca^{++} current. A similar effect is undoubtedly responsible for acceleration of the action potential's spread throughout the specialized conducting pathway.

In the atrial and ventricular contractile cells, both of which have an abundance of sympathetic nerve endings, the sympathetic effect is one of increasing contractile strength. This effect is brought about by increasing Ca^{++} permeability, which enhances the slow Ca^{++} influx and intensifies Ca^{++} participation in the excitation-contraction coupling process.

The overall effect of sympathetic stimulation on the heart, therefore, is to improve its effectiveness as a pump by increasing the heart rate, decreasing the delay between atrial and ventricular contraction, decreasing conduction time throughout the heart, and increasing the force of contraction.

Thus, as is typical of the autonomic nervous system, parasympathetic and sympathetic effects on heart rate are antagonistic (oppose each other). At any given moment, the heart rate will be determined largely by the existing balance between the inhibitory effects of the vagus nerve and the stimulatory effects of the cardiac sympathetic nerves. Under resting conditions, parasympathetic discharge is dominant. In fact, if all autonomic nerves to the heart were blocked, the resting heart rate would increase from its normal value of 70 beats per minute to about 100 beats per minute, which is the inherent rate of the SA node's spontaneous discharge when not subjected to any nervous influence. (We used 70 beats per minute as the autonomous rate of SA-node discharge because this is the average rate under normal conditions in the body.) Alterations in heart rate beyond this resting level in either direction can be accomplished by shifting the balance of autonomic nervous stimulation. Heart rate is increased by simultaneously increasing sympathetic and decreasing parasympathetic activity; a reduction in heart rate is brought about by a concurrent rise in parasympathetic activity and decline in sympathetic activity. The relative level of activity in these two autonomic branches to the heart, in turn, is primarily coordinated by the cardiovascular control center located in the brain stem. (The role of this center will be described more thoroughly in the next chapter in conjunction with regulation of blood pressure.)

Although autonomic innervation is the primary means by which heart rate is regulated, other factors affect it as well. The most important of these is epinephrine, a hormone that is secreted into the blood from the adrenal medulla upon sympathetic stimulation and that acts in a manner similar to norepinephrine to increase the heart rate. Epinephrine therefore reinforces the direct effect that the sympathetic nervous system has on the heart.

Stroke volume is determined by the extent of venous return, sympathetic activity, and arterial afterload.

The other component determining cardiac output is stroke volume, the amount of blood pumped out by each ventricle during each beat. Because stroke volume equals end-diastolic volume minus end-systolic volume, the larger the end-diastolic volume is or the smaller the end-systolic volume is, the larger the stroke volume will be. End-diastolic volume depends on the rate of venous return, whereas end-systolic volume depends on the strength of ventricular contraction and on the pressure against which the heart is ejecting blood. Accordingly, three variables determine the amount of blood ejected: (1) the extent of filling (the **preload**); (2) the strength of contraction (**contractility**) of the heart; and (3) the average aortic or pulmonary artery pressure (the **afterload**). Let us examine each of these factors in more detail to see how they influence the stroke volume.

Increased end-diastolic volume results in increased stroke volume.

As more blood is returned to the heart, the heart ejects more blood, but the relationship is not quite as simple as it appears, because the heart does not eject all the blood it contains. The direct correlation between end-diastolic volume and stroke volume constitutes the **intrinsic control** of stroke volume, which refers to the heart's inherent ability to vary the stroke volume. This intrinsic control depends on the length-tension relationship of cardiac muscle, which is similar to that of skeletal muscle (see p. 233). For skeletal muscle, the resting-muscle length is approximately the optimal length at which maximal tension can be developed during a subsequent contraction. When the skeletal muscle is longer or shorter than this optimal length, the subsequent contraction is weaker. For cardiac muscle, the resting cardiac-muscle-fiber length is less than optimal length (Fig. 9–23). Therefore, an increase in cardiac-muscle-fiber length, by moving closer to the optimal length, increases the contractile tension of the heart on the following systole.

What causes cardiac-muscle fibers to vary in length before contraction? Skeletal-muscle length can vary before contraction by the positioning of the skeletal parts to which the muscle is attached, but cardiac muscle is not attached to any bones. The main determinant of cardiac-muscle-fiber length is the degree of diastolic filling. An analogy is a balloon filled with water—the more water you put in, the larger the balloon becomes and the more it is stretched. Likewise, the greater the extent of diastolic filling, the larger the end-diastolic volume and the more the heart is stretched. The more the heart is stretched, the longer the initial cardiac-fiber length before contraction. The increased length results in a greater force on the subsequent cardiac contraction and, consequently, a greater stroke volume. This intrinsic relationship between end-diastolic volume and stroke volume is known as the **Frank-Starling law of the heart.** Stated simply, the law says that the heart normally pumps all the blood returned to it; increased venous return results in increased stroke volume. In Figure 9–23, the purple section represents a normal Starling's curve. Assume that the end-diastolic volume increases from point A to point B. You can see that this increase in end-diastolic volume is accompanied by a corresponding increase in stroke volume from point A^1 to point B^1.

This relationship has two important advantages. First, when a larger cardiac output is needed, such as during exercise, venous return is increased through action of the sympathetic nervous system and other mechanisms to be described in the next chapter. The resultant increase in end-diastolic volume automatically increases stroke volume correspondingly. Because exercise also increases heart rate, these two factors act together to increase the cardiac output so that more

Figure 9–23 Intrinsic Control of Stroke Volume (Frank-Starling Curve)

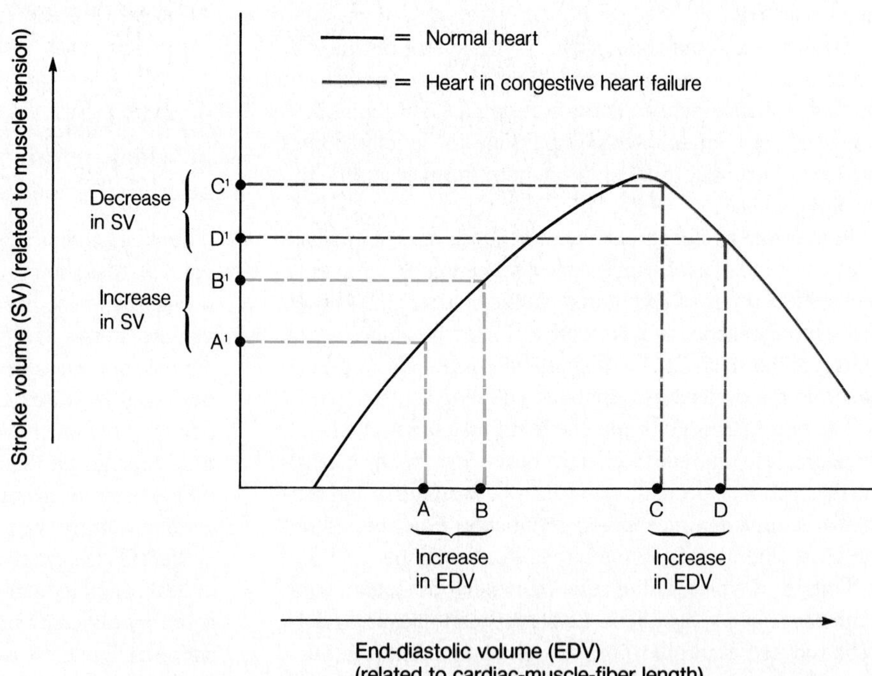

blood can be delivered to the exercising muscles. Second, one of the most important functions served by this intrinsic mechanism is equalization of output between the right and left sides of the heart. If, for example, the right side of the heart ejects a larger stroke volume, more blood enters the pulmonary circulation, so venous return to the left side of the heart is increased accordingly. The increased end-diastolic volume of the left side of the heart causes it to contract more forcefully, so it too pumps out a larger stroke volume. In this way, equality of output of the two ventricular chambers is maintained. If such equalization did not happen, excessive damming of blood would occur in the venous system preceding the ventricle with the lower output.

Such damming does occur in **congestive heart failure.** Heart failure refers to the inability of the cardiac output to keep pace with the body's demands for supplies and removal of wastes, with the veins behind the failing ventricle becoming congested with blood. Either one or both ventricles may fail. A failing ventricle becomes overstretched; when cardiac-muscle fibers are stretched beyond their optimal length, decreased force of contraction is developed during the next systole. Only up to the optimal length does a greater end-diastolic volume result in a more forceful contraction and a larger stroke volume. With any further stretching of cardiac muscle, contractile force diminishes (the green section of the curve on Fig. 9–23). Within normal physiological limits, the heart does not become overstretched to the point that contractile force is decreased. This can occur in a failing heart, however, with a vicious cycle ensuing that ends in death if there is no therapeutic intervention. As more blood is returned to a failing heart, it stretches even more, which further reduces its contractile ability (moving from point C to D on Fig. 9–23). A weaker contraction results in less blood being pumped out than was returned to the heart (moving from point C^1 to D^1), so the end-systolic volume is larger than it was during the preceding cycle. As more venous blood is returned, the heart becomes more engorged and stretches even further because it did not pump out all of the blood returned to it the previous filling. As the heart continues to be stretched, its contractile ability is even further depressed, and so on. *Backward failure* occurs as blood that cannot enter and be pumped out by the heart continues to dam up in the venous system. *Forward failure* occurs simultaneously as the heart fails to pump an adequate amount of blood forward to the tissues because the stroke volume becomes progressively smaller.

The consequences of left-sided heart failure are more serious than those of right-sided failure. Backward failure of the left side leads to pulmonary edema (excess tissue fluid in the lungs) because blood dams up in the lungs. This excess fluid accumulation in the lung tissue reduces exchange of O_2 and CO_2 between the air and blood in the lungs, leading to reduced arterial oxygenation and an elevation of acid-forming CO_2 in the blood. In addition, one of the more serious consequences of left-sided forward heart failure is an inadequate blood flow to the kidneys, which causes a two-fold problem. First, vital kidney function is depressed, and second, the kidneys, as a compensatory attempt to improve their reduced blood flow, retain extra salt and water in the body during urine formation to expand the blood volume. Excessive fluid retention further deteriorates the already existing problems of venous congestion. Treatment of congestive heart failure therefore includes measures that reduce salt and water retention and increase urinary output as well as the use of drugs that enhance the contractile ability of the weakened heart muscle—digitalis, for example. This drug increases the heart's strength by inducing an accumulation of intracellular Ca^{++}, which is so critical to the contractile process.

The contractility of the heart is increased by sympathetic stimulation.

In addition to intrinsic control, stroke volume is also subject to **extrinsic control** by factors originating outside of the heart, the most important of which are actions of the cardiac sympathetic nerves and epinephrine (Table 9–4). Sympathetic stimulation and epinephrine enhance the heart's contractility; in other words, the heart contracts more forcefully and squeezes out a greater percentage of the blood it contains, leading to more complete ejection. This increased contractility is due to the increased Ca^{++} influx triggered by norepinephrine and epinephrine. The extra cytosolic Ca^{++} allows the myocardial fibers to generate more force than they would without sympathetic influence. Normally the end-diastolic volume is 135 ml and the end-systolic volume is 65 ml for a stroke volume of 70 ml (Fig.9–24a). Under sympathetic influence, for the same end-diastolic volume of 135 ml, the end-systolic volume might be 35 ml and the stroke volume 100 ml (Fig.9–24b). In effect, sympathetic stimulation shifts the Frank-Starling curve to the left (Fig. 9–25). The curve can be shifted to varying degrees, depending on the extent of sympathetic stimulation, to a maximal increase in contractile strength of about 100% greater than normal.

Sympathetic stimulation increases stroke volume not only by strengthening cardiac contractility, but also by enhancing venous return (Fig. 9–24c). Norepinephrine constricts the veins, which squeezes more blood forward from the veins to the heart, increasing the end-diastolic volume and subsequently increasing the stroke volume even further.

The strength of cardiac-muscle contraction and, accordingly, the stroke volume, can thus be graded by: (1) varying the initial length of the muscle fibers, which in turn depends on the degree of ventricular filling before contraction (intrin-

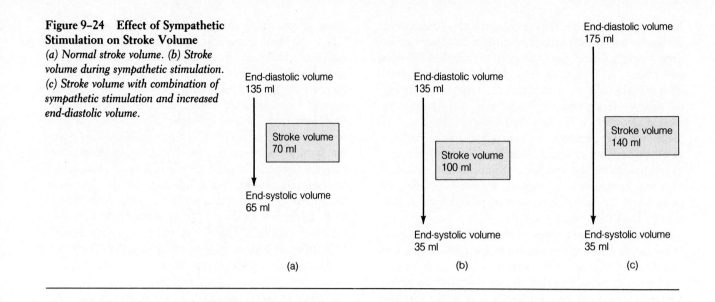

Figure 9–24 Effect of Sympathetic Stimulation on Stroke Volume *(a) Normal stroke volume. (b) Stroke volume during sympathetic stimulation. (c) Stroke volume with combination of sympathetic stimulation and increased end-diastolic volume.*

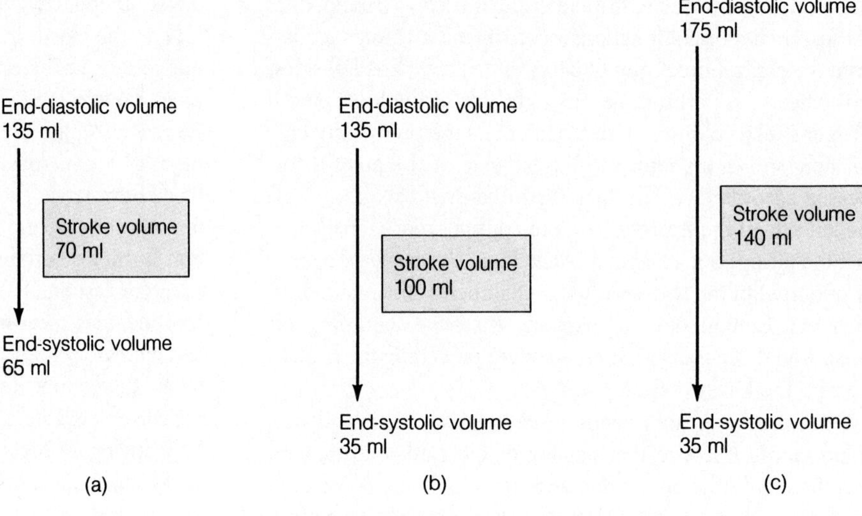

sic control); and (2) varying the extent of sympathetic stimulation (extrinsic control) (Fig. 9–26). This is in contrast to gradation of skeletal muscle. In skeletal muscle, wave summation and recruitment of motor units are employed to produce variable strength of muscle contraction, but these mechanisms are not applicable to cardiac muscle. Wave summation is impossible because of the long refractory period, and recruitment of motor units is not possible because the heart-muscle cells are arranged into functional syncytia instead of distinct motor units that can be discretely activated.

An elevation of arterial blood pressure can reduce cardiac output, especially in a diseased heart.

When the ventricles contract, they must generate sufficient pressure to exceed the blood pressure in the major arteries in order to force open the semilunar valves. The arterial blood pressure is referred to as the *afterload* because it is the workload imposed on the heart after the onset of contraction. If the arterial blood pressure is elevated or if the exit valve is stenotic,

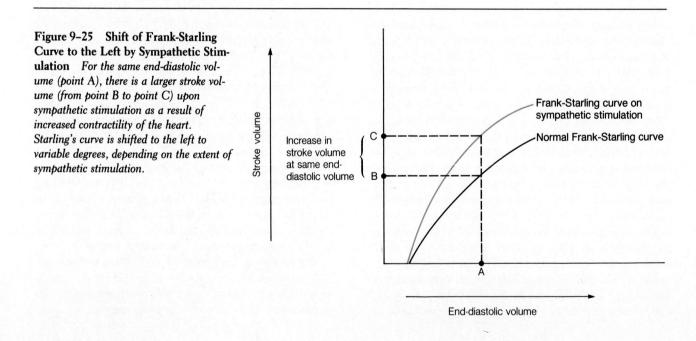

Figure 9–25 Shift of Frank-Starling Curve to the Left by Sympathetic Stimulation *For the same end-diastolic volume (point A), there is a larger stroke volume (from point B to point C) upon sympathetic stimulation as a result of increased contractility of the heart. Starling's curve is shifted to the left to variable degrees, depending on the extent of sympathetic stimulation.*

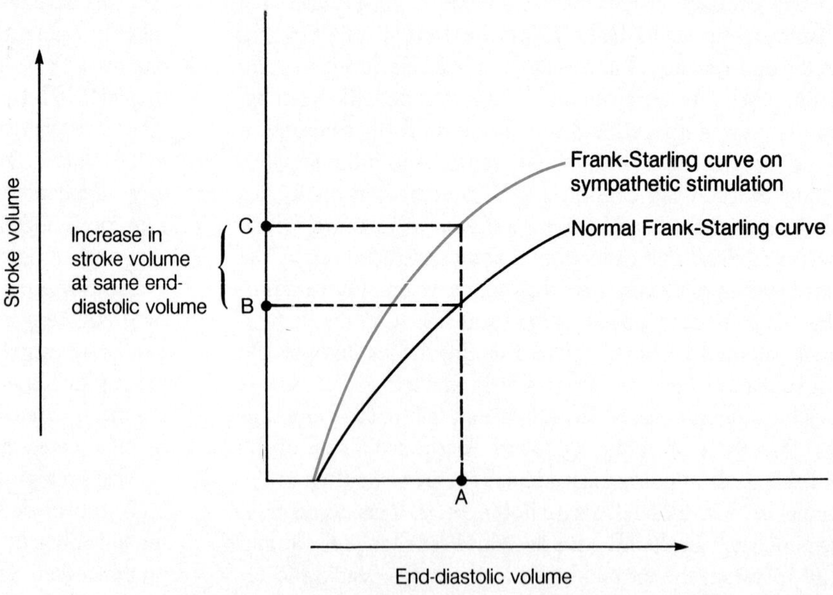

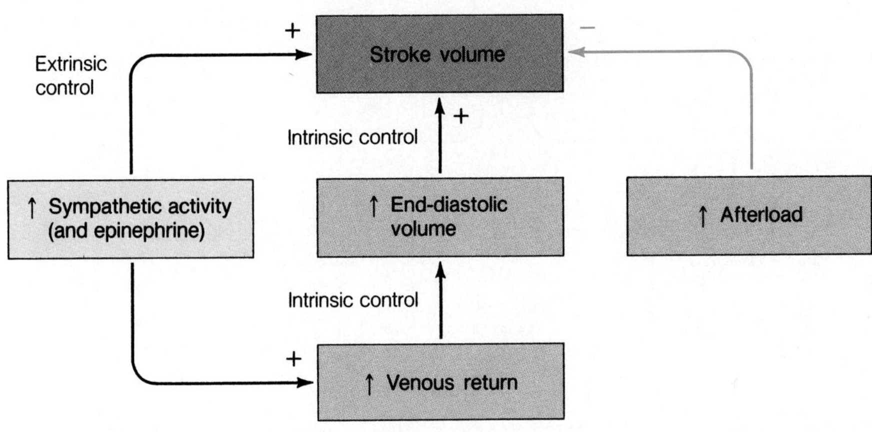

Figure 9–26 Control of Stroke Volume

the ventricle has to generate more pressure to eject blood. For example, instead of generating the normal pressure of 120 mm Hg, the ventricular pressure may need to go as high as 400 mm Hg to force blood through a narrowed aortic valve.

Cardiac muscle has a load-velocity relationship similar to skeletal muscle. Just as it takes longer to lift a heavier load, it takes longer for the ventricles to pump blood against an increased afterload. Depending on the heart rate, the ventricles only have 100 to 150 msec to eject blood. Because the heart has a limited amount of time in systole to eject blood, and because it takes longer for the ventricles to pump blood against an elevated afterload, there is a reduction in the volume that can be ejected before the end of systole when the afterload is increased. Accordingly, at any given end-diastolic volume, stroke volume is decreased as arterial blood pressure increases (Fig. 9–27). (This effect and all other factors that determine the cardiac output by influencing the heart rate or stroke volume are summarized in Figure 9–28.) The heart can compensate for the reduction in stroke volume as the afterload increases by enlarging so that it can contract more forcefully to maintain a normal stroke volume. A diseased heart may not be able to compensate, however, in which case cardiac output can be improved by reducing the blood pressure therapeutically.

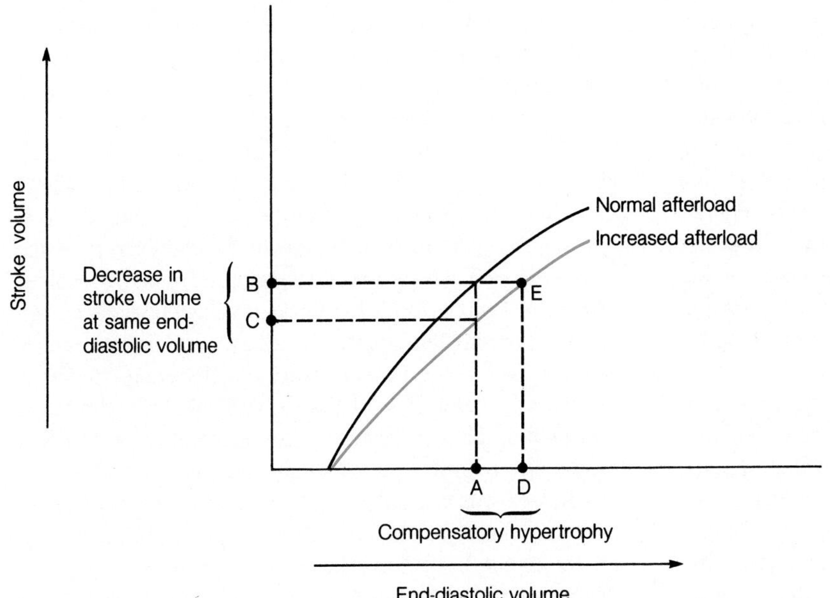

Figure 9–27 Effect of Afterload on Stroke Volume *For the same end-diastolic volume (point A), there is a smaller stroke volume (from point B to point C) with an increased afterload, caused by a reduction in the velocity of ejection. The heart can compensate by enlarging so that it can accommodate a larger end-diastolic volume (point D), thereby maintaining a normal stroke volume in spite of an increased afterload (point E).*

Figure 9–28 **Control of Cardiac Output** *Since cardiac output equals heart rate times stroke volume, this figure is a composite of Figure 9–22b (control of heart rate) and Figure 9–26 (control of stroke volume).*

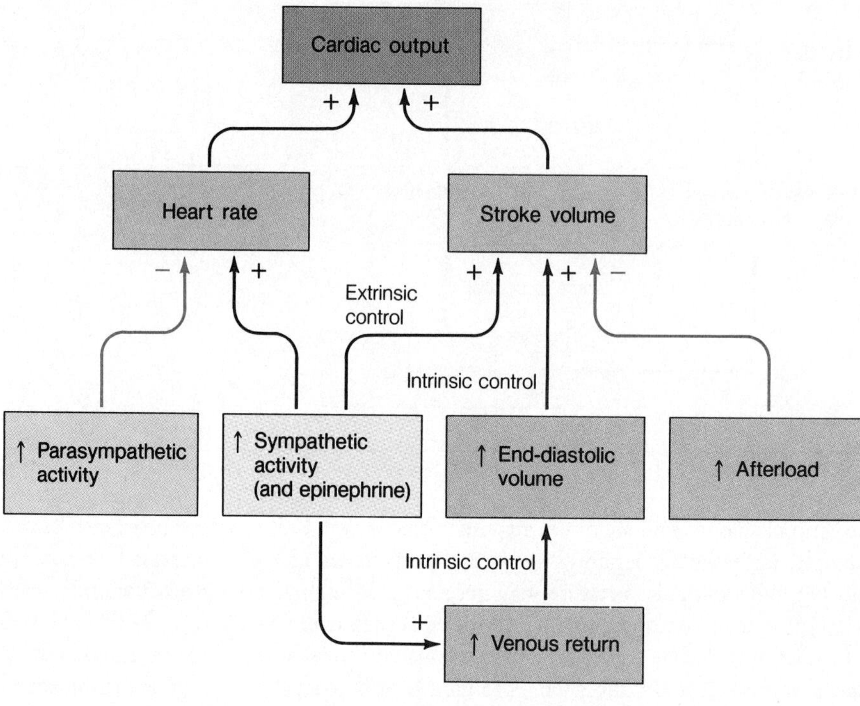

Nourishing the Heart Muscle

The heart receives most of its own blood supply through the coronary circulation during diastole.

Although all the blood passes through the heart, the heart muscle is unable to extract O_2 or nutrients from the blood within its chambers for two reasons. First, the water-tight endocardial lining does not permit passage of blood from the chamber into the myocardium. Second, the heart walls are too thick to permit diffusion of O_2 and other supplies from the blood in the chamber to the individual cardiac cells. Therefore, as with other tissues of the body, heart muscle must receive blood through blood vessels, specifically by means of the **coronary circulation.** The coronary arteries branch from the aorta just beyond the aortic valve, and the coronary veins empty into the right atrium.

Blood flow to the heart-muscle cells is substantially reduced during systole for two reasons. First, the major branches of the coronary arteries are compressed by the contracting myocardium, and second, the entrance to the coronary vessels is partially blocked by the open aortic valve. Thus, most coronary arterial flow (about 70%) occurs during diastole, driven by the aortic pressure head, with flow declining as aortic pressure drops; only about 30% of coronary arterial flow occurs during systole (Fig. 9–29). This becomes especially important during rapid heart rates, when diastolic time is substantially reduced. Just when increased demands are placed on the heart to pump more rapidly, there is less time to provide O_2 and nourishment to its own musculature to accomplish the increased work load.

Nevertheless, under normal circumstances the heart muscle does receive adequate blood flow to support its activities, even during exercise, when the rate of coronary blood flow increases up to five times its resting rate. Increased delivery of blood to the cardiac cells is accomplished primarily by vasodilation, or enlargement, of the coronary vessels, which allows more blood to flow through them, especially during diastole. The increased coronary blood flow is imperative to meet the heart's increased O_2 requirements, because, unlike most other tissues, the heart is unable to remove much additional O_2 from the blood passing through its vessels to support increased metabolic activities. Most other tissues under resting conditions extract only about 25% of the O_2 available from the blood flowing through them, leaving a considerable O_2 reserve that can be drawn on when such a tissue has increased O_2 needs; that is, the tissue can immediately increase the O_2 available to it by removing a greater percentage of O_2 from the blood passing through it. In contrast, the heart, even under resting conditions, withdraws far more O_2 from its blood supply than do

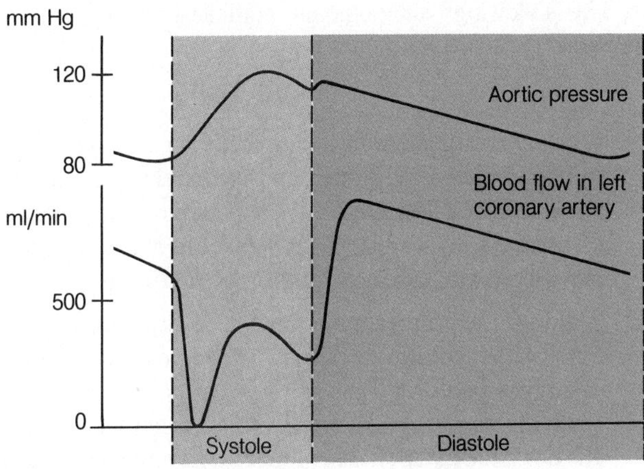

Figure 9–29 Coronary Blood Flow *Most coronary blood flow occurs during diastole because the coronary vessels are compressed almost completely closed during systole.*

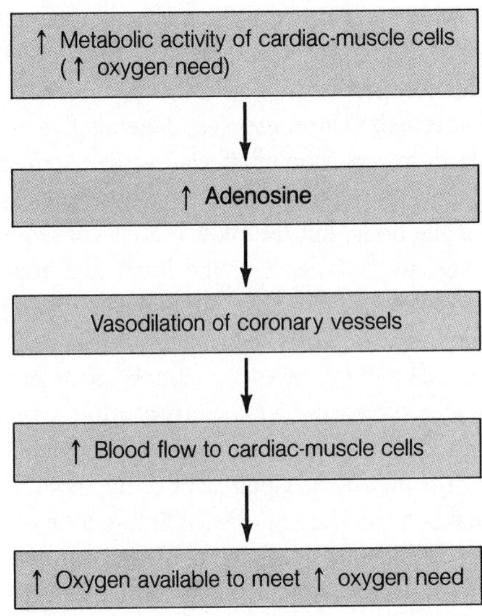

Figure 9–30 Matching of Coronary Blood Flow to Oxygen Need of Cardiac-Muscle Cells

other tissues, removing up to 65% of the O_2 available in the coronary vessels. This leaves little O_2 in reserve in the coronary blood should cardiac O_2 demands increase. Therefore, the primary means by which more O_2 can be made available to the heart muscle is by increasing coronary blood flow.

Coronary blood flow is adjusted primarily in relation to the heart's changing O_2 requirements. The link that coordinates coronary blood flow with myocardial O_2 needs is believed to be ***adenosine***, which is formed from adenosine triphosphate (ATP) during cardiac metabolic activity. Increased formation and release of adenosine from the cardiac cells occurs: (1) when there is a cardiac O_2 deficit, or (2) when cardiac activity is increased and the heart accordingly requires more O_2 and is using more ATP as an energy source. The released adenosine induces dilation of the coronary blood vessels, thereby allowing more O_2-rich blood to flow to the more active cardiac cells to meet their increased O_2 demand (Fig. 9–30). This matching of O_2 delivery with O_2 needs is critical because of the heart muscle's dependence on oxidative processes to generate energy. The heart cannot obtain sufficient ATP through anaerobic metabolism.

Although the heart has little ability to support its energy needs by means of anaerobic metabolism and must rely heavily on its O_2 supply, it is not so restricted in its nutrient supply. It primarily uses free fatty acids and, to a lesser extent, glucose and lactate as fuel sources, depending on their availability. Because of the cardiac muscle's remarkable adaptability in shifting metabolic pathways to use whatever nutrient is available, the primary danger of insufficient coronary blood flow is not in fuel shortage but in O_2 deficiency.

Atherosclerotic coronary artery disease can deprive the heart of essential oxygen.

The leading cause of O_2 deficiency to the heart is **coronary artery disease.** Complications of coronary artery disease make it the single leading cause of death in the United States. It is important to note that adequacy of coronary blood flow is relative to the heart's O_2 demands at any given moment. In the normal heart, coronary blood flow increases correspondingly as O_2 demands rise. With coronary artery disease, however, it may not be possible for coronary blood flow to keep pace with rising O_2 needs. A given rate of coronary blood flow may be adequate at rest but insufficient upon physical exertion or other stressful situations. Coronary artery disease can cause myocardial ischemia (insufficient circulation of oxygenated blood through the coronary circulation to maintain aerobic metabolism in the heart muscle) by the three following mechanisms: (1) profound vascular spasm of the coronary arteries; (2) the formation of atherosclerotic plaques; and (3) thromboembolism. We will discuss each of these in turn.

Vascular spasm is an abnormal spastic constriction that transiently narrows the coronary vessels, most often triggered by exposure to cold, physical exertion, or anxiety. Vascular spasms are associated with the early stages of coronary artery disease. The condition is reversible and usually of insufficient duration to produce damage to the cardiac muscle. Recent evidence suggests that coronary artery disease interferes with the production of **endothelium-derived relaxing factor.** This substance, which is believed to be normally produced by the endothelium (lining) of blood vessels, apparently inhibits blood vessel constriction.

Atherosclerosis is a progressive, degenerative arterial disease that leads to occlusion of affected vessels, thereby reducing blood flow through them. The disease attacks arteries throughout the body, but the most serious consequences involve damage to the vessels of the heart and brain. In the heart, atherosclerosis predisposes the individual to myocardial ischemia and its complications, whereas in the brain it is the prime cause of cerebral vascular accidents, or strokes. It is believed that atherosclerosis begins as **atheromas,** which are benign (noncancerous) tumors of smooth-muscle cells. These cells migrate from the muscular layer of the blood vessel to a position just beneath the endothelial lining, where they continue to divide and enlarge. Later, cholesterol and other lipids accumulate in the abnormal smooth-muscle cells to produce a **plaque.** The plaques bulge into the lumen of the vessel as they continue to develop (Fig. 9–31). A thickening plaque interferes with nutrient exchange for the cells of the involved arterial wall, leading to degeneration of the wall in the vicinity of the plaque. The damaged area is invaded by fibroblasts (scar tissue-forming cells), and often, in later stages of the disease, Ca^{++} precipitates in the plaque. A vessel so afflicted becomes hard and poorly distensible, a condition dubbed "hardening of the arteries."

The following are potential complications of atherosclerosis:

1. Gradual enlargement of the protruding plaque continues to narrow the vessel lumen and progressively diminishes coronary blood flow, triggering increasingly frequent bouts of transient myocardial ischemia as the ability to match blood flow with cardiac O_2 needs becomes more limited.
2. Although the heart cannot normally be "felt," pain is associated with myocardial ischemia. Such cardiac pain, known as **angina pectoris** ("pain of the chest"), can be felt beneath the sternum and is often referred to the left shoulder and down the left arm (see p. 152). The symptoms of angina pectoris appear recurrently whenever cardiac O_2 demands become too great in relation to the coronary blood flow—for example, during exertion or emotional stress. The pain is thought to result from stimulation of cardiac nerve endings by the accumulation of lactic acid when the heart shifts to its limited ability to perform anaerobic metabolism. The ischemia associated with the characteristically brief anginal attacks is usually temporary and reversible and can be relieved by rest, administration of vasodilator drugs such as nitroglycerin, or both.
3. The enlarging atherosclerotic plaque can break through the weakened endothelial lining that covers it, exposing

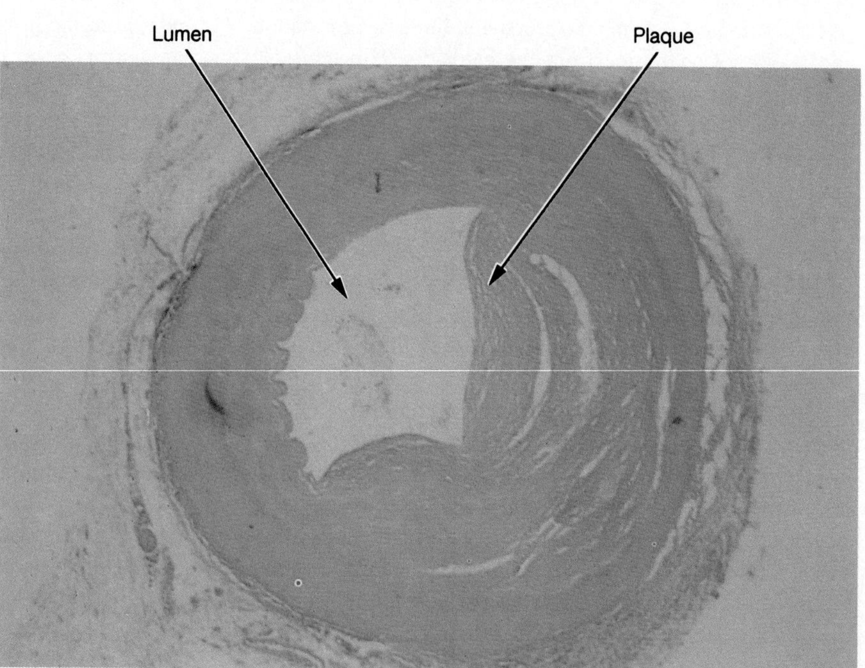

Figure 9–31 Severe Atherosclerotic Plaque in a Coronary Vessel

blood to the underlying collagen (a component of connective tissue). Blood platelets, formed elements of the blood involved in plugging vessel defects and in clot formation, do not normally adhere to smooth, healthy vessel linings. However, when platelets come into contact with collagen at the site of vessel damage, they aggregate and contribute to the formation of a blood clot. Such an abnormal clot attached to a vessel wall is known as a **thrombus.** The thrombus may enlarge gradually until it completely occludes the vessel at that site, or the continued flow of blood past the thrombus may break it loose from its attachment. Such a freely floating clot, or **embolus,** may completely plug a smaller vessel as it flows downstream (Fig. 9–32). Thus, through **thromboembolism,** atherosclerosis can result in a gradual or sudden occlusion of a coronary vessel (or any other vessel).

4. Upon complete vessel occlusion, the cardiac tissue served by the vessel undergoes sustained and extensive ischemia and will soon die from O_2 deprivation unless the area can be supplied with blood from nearby vessels. Sometimes a deprived area is fortunate enough to receive blood from more than one pathway. **Collateral circulation** exists when small terminal branches from adjacent blood vessels nourish the same area. These accessory vessels cannot develop suddenly following an acute occlusion but may be lifesaving if already developed. Such alternate vascular pathways often develop over a period of time when an atherosclerotic constriction slowly progresses, or they may be induced by sustained demands on the heart through a regular aerobic-exercise program.

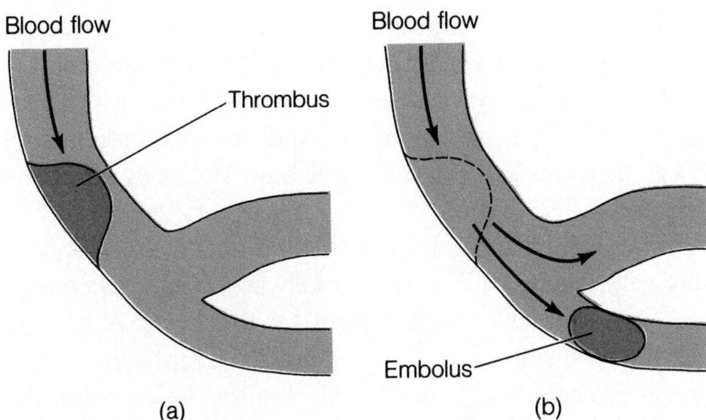

Figure 9–32 Consequences of Thromboembolism *(a) A thrombus may enlarge gradually until it completely occludes the vessel at that site. (b) A thrombus may break loose from its attachment, forming an embolus that may completely occlude a smaller vessel downstream.*

In the absence of collateral circulation, the extent of the infarcted area during a heart attack depends on the size of the occluded vessel. The larger the vessel occluded, the greater the area deprived of its blood supply. As illustrated in Figure 9–33, a blockage at point A in the coronary circulation would cause more extensive damage than would a blockage at point B. Because there are only two major coronary arteries, occlusion of either one of these main branches results in extensive myocardial damage. Left coronary artery blockage is most devastating because this vessel is responsible for supplying 85% of the cardiac tissue. There are four possible outcomes of an acute myocardial infarction: immediate death, delayed death from complications, full functional recovery, or recovery with impaired function (Table 9–5).

The amount of cholesterol carried via high-density lipoproteins versus low-density lipoproteins is linked to atherosclerosis.

The cause of atherosclerosis is still not entirely clear. Certain high-risk factors have been associated with an increased incidence of atherosclerosis and coronary heart disease. Included among them are genetic predisposition, obesity, advanced

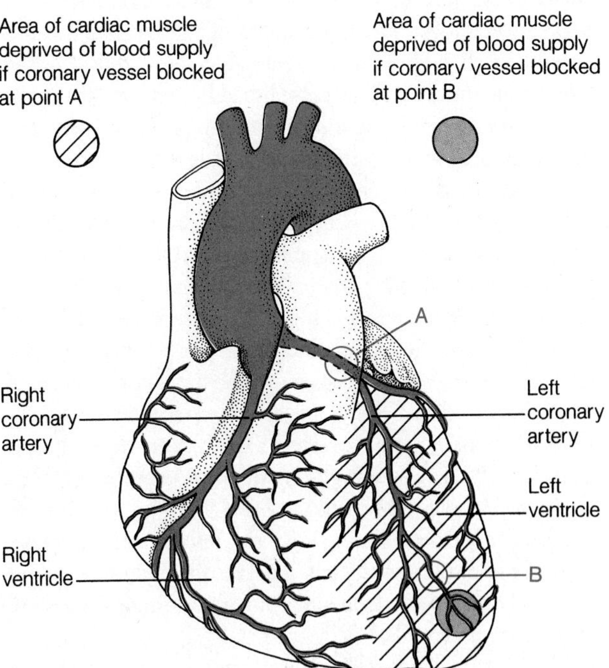

Figure 9–33 Extent of Myocardial Infarction as Function of Size of Occluded Vessel

Table 9–5 Possible Outcomes of Acute Myocardial Infarction

Immediate Death	Delayed Death from Complications	Full Functional Recovery	Recovery with Impaired Function
Acute cardiac failure occurring because the heart is too weakened to pump effectively to support the body tissues	Fatal cardiac tamponade caused by rupture of the dead, degenerating area of the heart wall	Replacement of infarcted area with a strong scar, accompanied by hypertrophy of remaining normal contractile tissue to compensate for the lost cardiac musculature	Persistence of permanent functional defects, such as bradycardia or AV-conduction blocks, caused by infarction of irreplaceable specialized autorhythmic or conductive tissues
Fatal ventricular fibrillation brought about by damage to the specialized conducting tissue	Slowly progressing congestive heart failure occurring because the weakened heart is unable to pump out all the blood returned to it		

age, smoking, hypertension, diabetes mellitus, lack of exercise, nervous tension, and, most significantly, excess cholesterol levels in the blood.

There are two sources of cholesterol for the body: (1) dietary intake of cholesterol, with animal products such as egg yolk, red meats, and butter being especially rich in this lipid (animal fats contain cholesterol, whereas vegetable fats do not); and (2) manufacture of cholesterol by many organs within the body, particularly the liver. Because of the body's ability to synthesize cholesterol, there is not a direct correlation between the amount of cholesterol ingested and the cholesterol levels in the blood, although modest reductions in blood cholesterol can be accomplished by lowering the intake of animal fats. For some individuals, drugs may be necessary to satisfactorily lower blood cholesterol levels.

Actually, it is not the total blood cholesterol level but the amount of cholesterol bound to various plasma-protein carriers that appears to be most important regarding the risk of developing atherosclerotic heart disease. Because cholesterol is a lipid, it is not very soluble in blood. Most cholesterol is carried in the blood attached to specific plasma-protein carriers in the form of lipoprotein complexes, which are soluble in blood. There are three such lipoproteins, named for their densities—**very-low-density lipoproteins (VLDL), low-density lipoproteins (LDL),** and **high-density lipoproteins (HDL).** The major cholesterol carrier in blood is LDL, which is derived from VLDL. The low-density lipoprotein complexes carry cholesterol *to* most cells, whereas the high-density lipoprotein complexes are involved in transporting cholesterol *away* from most cells.

Unlike most lipids, cholesterol is not used as a metabolic fuel by cells. Instead, it serves as an essential component of plasma membranes. In addition, a few special cell types use cholesterol as a precursor for the synthesis of secretory products, such as steroid hormones and bile salts. Although most cells are capable of synthesizing some of the cholesterol needed for their own plasma membranes, they cannot manufacture sufficient amounts and therefore must rely on supplemental cholesterol being delivered by the blood. This additional cholesterol is supplied either by the diet or by cells that specialize in cholesterol synthesis, especially liver cells.

Cells accomplish cholesterol uptake from the blood by synthesizing receptor proteins specifically capable of binding LDL and inserting these receptors into the cells' plasma membranes (Fig. 9–34). When an LDL particle binds to one of the membrane receptors, the cell engulfs the particle by endocytosis (see p. 24). Within the cell, lysosomal enzymes break down the LDL to free the cholesterol, making it available to the cell for synthesis of new cellular membrane. If too much free cholesterol accumulates in the cell, there is a shutdown of both the synthesis of LDL receptor proteins (so that less cholesterol is taken up) and the cell's own cholesterol synthesis (so that less new cholesterol is made). Faced with a cholesterol shortage, on the other hand, the cell makes more LDL receptors so that it can engulf more cholesterol from the blood.

The maintenance of a blood-borne cholesterol supply to the cells involves an interaction between dietary cholesterol and the synthesis of cholesterol by the liver. When the amount of dietary cholesterol is increased, hepatic (liver) synthesis of cholesterol is turned off because cholesterol in the blood directly inhibits a hepatic enzyme essential for cholesterol synthesis. Thus, as more cholesterol is ingested, less is produced by the liver. Conversely, when cholesterol intake from food is reduced, the liver synthesizes more of this lipid because the inhibitory effect of cholesterol on the crucial he-

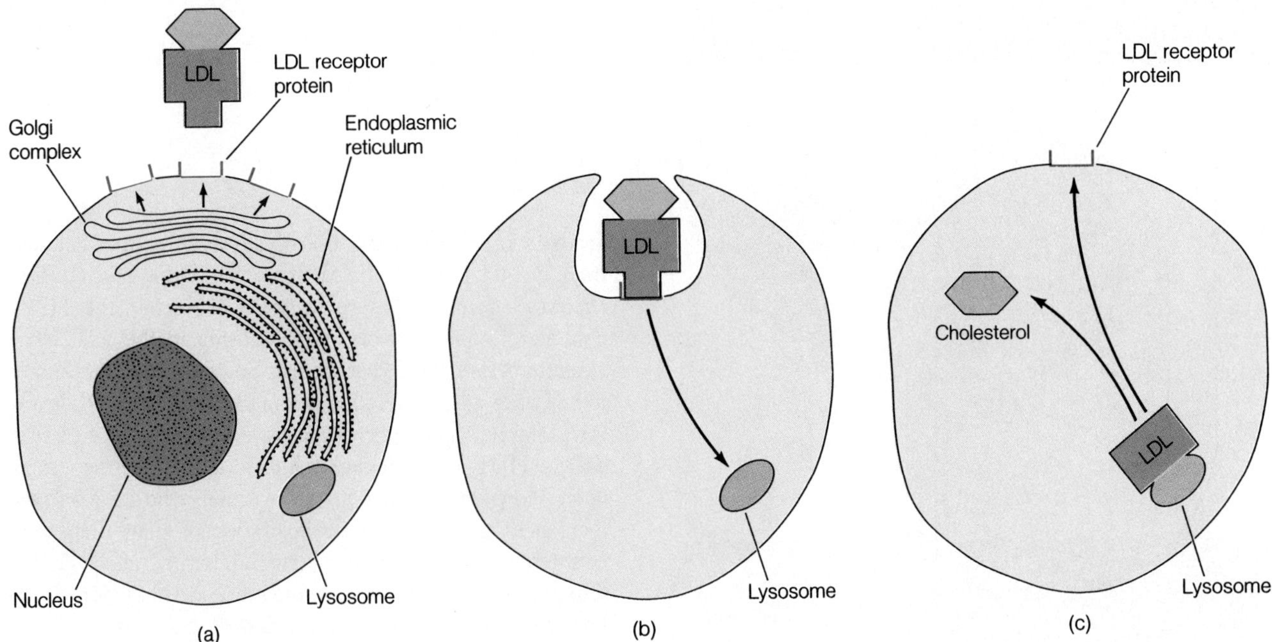

Figure 9–34 Uptake of Cholesterol by Cells *(a) A cell that needs additional cholesterol synthesizes LDL receptor proteins. (b) Receptor-bound LDL is internalized by endocytosis. (c) A lysosome frees cholesterol, and the LDL receptor is recycled back to the plasma membrane.*

patic enzyme is removed. In this way, the blood concentration of cholesterol is maintained at a fairly constant level despite changes in cholesterol intake, thereby making it difficult to significantly alter cholesterol levels in the blood by dietary manipulations.

In contrast to LDL complexes, which transport cholesterol to the cells, HDL removes cholesterol from cells and transports it to the liver. The liver, in turn, secretes cholesterol as well as cholesterol-derived bile salts into the bile. Bile enters the intestinal tract, where bile salts participate in the digestive process. Most of the secreted cholesterol and bile salts are subsequently reabsorbed from the intestinal tract into the blood to be recycled to the liver. Cholesterol molecules not reclaimed by absorption are eliminated in the feces.

Obviously, the liver has a central role in cholesterol metabolism. At any given time, the liver may be manufacturing new cholesterol; extracting old cholesterol from the blood and secreting it into the bile; or converting old cholesterol into bile salts, which likewise are secreted into the bile (Fig. 9–35). The first process is a mechanism for adding cholesterol to the blood to supplement dietary intake. The two other pathways lead to a net loss of cholesterol from the body. Thus, the liver has a primary role in determining total blood cholesterol levels, and the interplay between LDL and HDL determines the traffic flow of cholesterol between the liver and the individual cells of the body. Whenever these mechanisms are altered, blood cholesterol levels may be affected in such a way as to influence the individual's predisposition to atherosclerosis. The following are specific examples:

□ Evidence suggests that the propensity toward developing atherosclerosis substantially increases with elevated levels of LDL, the "bad" form of cholesterol. In one hereditary disease, afflicted individuals lack the genes for making LDL receptor proteins. Because their cells cannot take up LDL from the blood, the blood concentration of these cholesterol-loaded lipoproteins becomes greatly elevated. Such persons have an extraordinary tendency toward developing atherosclerosis early in life, and many die in their youth from coronary artery disease. Recent studies suggest that even in individuals with normal LDL receptors, atherosclerosis may develop if they have aberrant levels of the sole protein component of LDL,

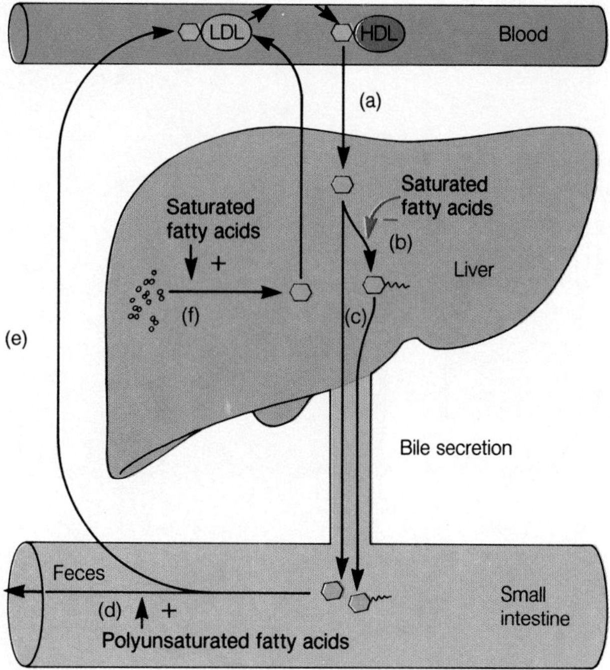

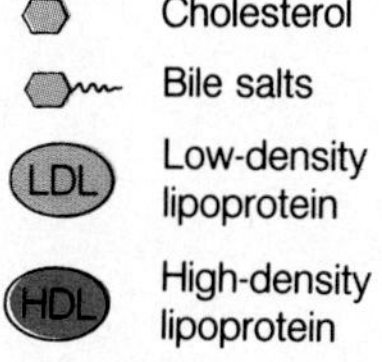

Figure 9–35 Role of the Liver in Cholesterol Metabolism *The following steps are denoted by corresponding letters in the illustration:*

(a) Liver extracts cholesterol (carried by HDL) from the blood.
(b) Liver converts most of the cholesterol into bile salts.
(c) Liver secretes bile salts and cholesterol into the bile, which empties into the intestinal lumen.
(d) Part of the cholesterol and bile salts is eliminated in the feces.
(e) Some of the bile salts are reabsorbed into the blood and recycled to the liver.
(f) Liver synthesizes new cholesterol, which is carried by LDL within the blood away from the liver to other cells.

Saturated fatty acids and polyunsaturated fatty acids are known to influence these pathways at the designated points.

apolipoprotein B-100 (**apo-B**). This is the portion of the LDL complex that interacts with the receptors.

□ The risk of atherosclerosis is inversely related to the concentration of HDL in the blood; that is, elevated levels of HDL (the "good" form of cholesterol) are associated with a low incidence of atherosclerotic heart disease. In fact, a more accurate predictor of the risk of developing atherosclerosis than the total blood cholesterol level is the ***blood HDL-cholesterol/total cholesterol ratio***. The higher the HDL-cholesterol concentration in relationship to the total blood cholesterol level, the lower the risk. Some other factors known to influence atherosclerotic risk can be related to HDL levels; for example, cigarette smoking lowers HDL in the plasma, and the HDL level is higher in individuals who exercise regularly. Moreover, premenopausal women, who have a lower incidence of atherosclerotic heart disease than their male counterparts, have a higher concentration of HDL-cholesterol, presumably because of some influence of the female sex hormone, estrogen. After production of this hormone ceases at menopause, the incidence of coronary heart disease in women parallels that in men.

□ Varying the intake of dietary fatty acids may alter total blood cholesterol levels by influencing one or more of the mechanisms involving cholesterol balance (Fig. 9–35). The blood cholesterol level tends to be raised by ingestion of saturated fatty acids found predominantly in animal fats. These fatty acids stimulate the synthesis of cholesterol and inhibit its conversion to bile salts. On the other hand, polyunsaturated fatty acids, the predominant fatty acids of plants, tend to reduce blood cholesterol levels by enhancing the elimination of both cholesterol and cholesterol-derived bile salts in the feces. Through unknown mechanisms, dietary-fiber supplements such as psyllium-seed husks have been shown to reduce the level of blood cholesterol, particularly LDL-cholesterol.

As you can see, the relationship between atherosclerosis, cholesterol, and other environmental and genetic factors is far from being clear. Much research concerning this complex disease is presently in progress, because there is such a high incidence of atherosclerosis and its consequences are potentially fatal.

Chapter in Perspective

The maintenance of homeostasis requires that essential materials are continually picked up from the external environment and delivered to the cells and that waste products are continually removed. It also necessitates the transfer of hormones, which are important regulatory chemical mediators, from their site of production to their site of action. The circulatory system, which contributes to homeostasis by serving as the body's transport system, consists of the heart, blood vessels, and blood.

The flow of blood into, through, and out of the heart is accomplished by alternate contraction and relaxation of the cardiac muscle that encloses the heart chambers. The heart serves as a dual pump to continuously circulate blood between the lungs, where O_2 is picked up, and the other body tissues, which use O_2 to support their energy-generating chemical reactions. As blood is pumped through the various tissues, other substances besides O_2 are also exchanged between the blood and tissues. For example, nutrients are picked up by the blood as it flows through the digestive organs, and other tissues remove nutrients from the blood as it flows through them.

Although all of the body tissues constantly depend on the life-supporting blood flow provided to them by the heart, the heart itself is quite an independent organ. It is able to take care of many of its own needs without any outside influence. Contraction of this magnificient muscle is self-generated through a carefully orchestrated interplay of changing ionic permeabilities. Local mechanisms within the heart assure that blood flow to the cardiac muscle normally meets the heart's need for O_2. In addition, the heart has built-in capabilities to vary its strength of contraction, depending on the amount of blood returned to it. The heart does not act entirely autonomously, however. It is innervated by the autonomic nervous system and is influenced by the hormone epinephrine, both of which can vary the rate and contractility of the heart, depending on the body's needs for blood delivery. Furthermore, as with all tissues, the cells that compose the heart depend on the other body systems for maintenance of a stable internal environment in which they can survive and function.

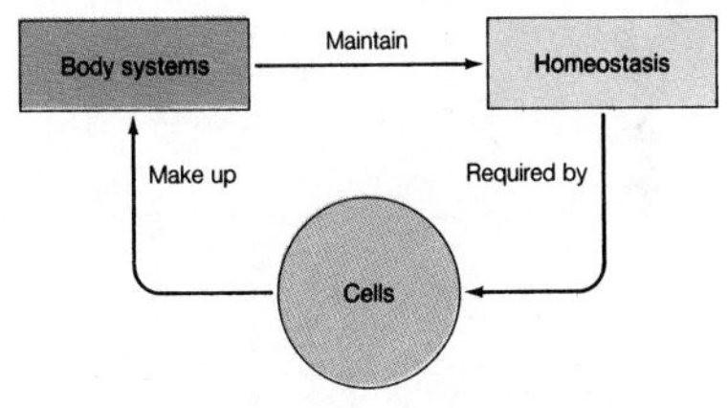

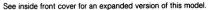

See inside front cover for an expanded version of this model.

Review Exercises

1. Trace a drop of blood through one complete circuit of the circulatory system.
2. Describe the location and function of the four heart valves.
3. Why is the SA node the pacemaker of the heart?
4. Describe the normal spread of cardiac excitation.
5. Compare the changes in ion permeability and membrane potential associated with an action potential in a nodal pacemaker cell with those in a myocardial working cell.
6. Why is tetany of cardiac muscle impossible? Why is this advantageous?
7. Draw and label the wave forms of a normal ECG. What electrical event does each component of the ECG represent?
8. Describe the mechanical events (that is, pressure changes, volume changes, valve activity, and heart sounds) that occur during the cardiac cycle. Correlate these mechanical events with the changes that take place in electrical activity.
9. Distinguish between a stenotic and an insufficient valve.
10. Define the following: end-diastolic volume, end-systolic volume, stroke volume, heart rate, and cardiac output.
11. Discuss autonomic nervous-system control of heart rate.
12. Describe the factors that determine stroke volume.
13. Why does the heart receive most of its own blood supply during diastole?
14. What are the pathological changes and consequences of coronary artery disease?
15. **A point to ponder:** Trained athletes usually have lower resting heart rates than normal (for example, 50 beats/min in an athlete compared to 70 beats/min in a sedentary individual). Considering that the resting cardiac output is 5,000 ml/min in both trained athletes and sedentary individuals, what is responsible for the bradycardia of trained athletes?

C H A P T E R 10

THE VASCULATURE AND BLOOD PRESSURE

INTRODUCTION *Jonathan Sharpel is a successful middle-aged business man. He owns a comfortable home in suburbia, is happily married to a chemical engineer, and has two children in college. He belongs to the country club, plays golf twice a week with friends, and goes to the Caribbean once a year on vacation. He feels great, and all is right with his world—or is it? Jonathan Sharpel is unknowingly being stalked by the "silent killer," alias "hypertension" or "high blood pressure." Because hypertension is typically not accompanied by any symptoms, Jonathan is not aware that he has a problem. He has been so busy and has been feeling so well that he has not thought about having a physical examination. Yet his high blood pressure is insidiously predisposing him to a heart attack or stroke.*

The causes and consequences of high blood pressure will be described in this chapter as part of our examination of the vascular components of the circulatory system. Blood vessels play a key role in determining blood pressure, the maintenance of which involves the integrated activity of the entire circulatory system.

Materials are transported within the body by the blood as it is pumped through the blood vessels.

The majority of body cells are not in direct contact with the external environment, yet these cells must make exchanges with the environment, such as picking up O_2 and nutrients and eliminating wastes. Furthermore, chemical messengers must be transported between cells to accomplish integrated activity. To accomplish these long-distance exchanges, the cells are linked with each other and with the external environment by vascular highways. Blood is transported to all parts of the body through a system of vessels that brings fresh supplies to the vicinity of all cells while removing their wastes.

All blood pumped by the right side of the heart passes through the lungs for O_2 pick-up and CO_2 removal. The blood pumped by the left side of the heart is parceled out in various proportions to the systemic organs through a parallel arrangement of vessels (Fig. 10–1). The latter arrangement assures that all organs receive blood of the same composition; that is, one organ does not receive "left-over" blood after it has passed through another organ. Because of this parallel arrangement, flow of blood through each systemic organ can be independently adjusted without directly influencing blood flow through any other organ.

Blood is constantly "reconditioned" so that its composition remains essentially constant despite an ongoing drain of supplies to support metabolic activities and the continual addition of wastes from the tissues. The organs that recondition the blood normally receive substantially more blood than is necessary to meet their basic metabolic needs so that they can perform homeostatic adjustments on the blood. Disproportionately large percentages of the cardiac output are distributed to the digestive tract (to pick up nutrient supplies), to the kidneys (to eliminate metabolic wastes and adjust water and electrolyte composition), and to the skin (to eliminate heat). Blood flow to the other organs—brain, heart, skeletal muscles, and so on—is solely for the purpose of supplying these tissues' metabolic needs and can be adjusted according to their level of activity. For example, during exercise, additional blood is delivered to the active muscles to meet their increased metabolic needs.

Because reconditioning organs receive blood flow in excess of their own needs, they can withstand temporary reductions in blood flow much better than can the other organs that do not have this extra margin of blood supply. The brain and heart suffer irreparable damage when transiently deprived of blood supply. Permanent brain damage occurs after only four minutes without O_2, and O_2 deprivation of cardiac muscle is responsible for heart attacks. Therefore, a high priority in the overall operation of the circulatory system is the constant delivery of adequate blood to the heart and brain, which can least tolerate a disruption in their blood supply. This is even more imperative because these vital organs are not enzymatically equipped to support their metabolic needs anaerobically. In contrast, the digestive organs, kidneys, and skin can tolerate significant reductions in blood flow for a considerable length of time. In fact, they frequently do so. For example, during exercise, some of the blood that normally flows through the digestive organs and kidneys is diverted instead to the skeletal muscles. Likewise, blood flow through the skin is markedly restricted during exposure to cold to conserve body heat. We will see how the distribution of cardiac output is adjusted ac-

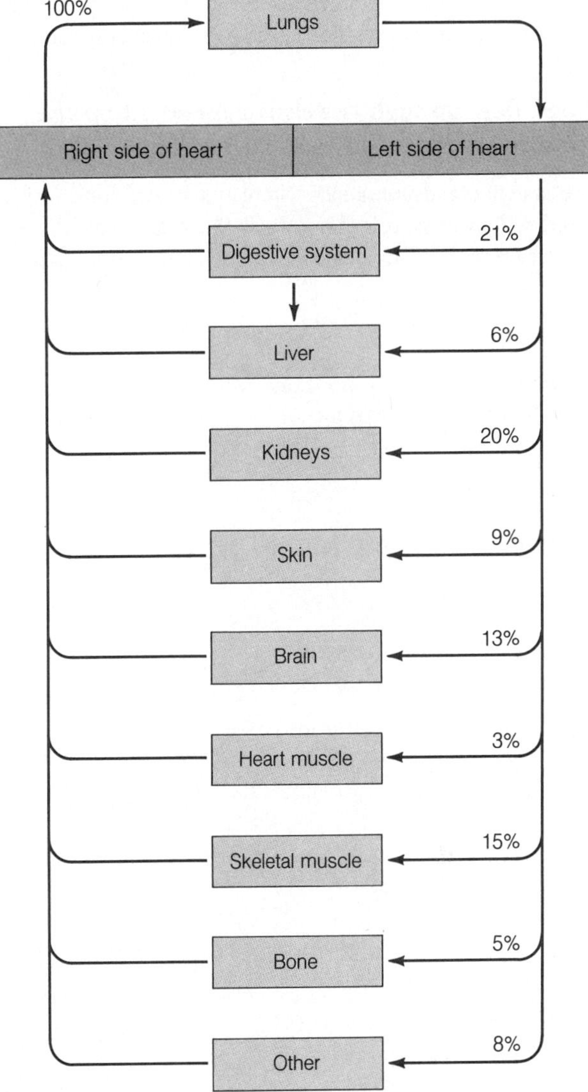

Figure 10–1 Distribution of Cardiac Output at Rest

cording to the body's momentary needs as we examine the roles of the various vessels that make up the vascular system.

Blood flow through vessels is dependent on the pressure gradient and vascular resistance.

The systemic and pulmonary circulations each consist of a closed system of vessels (Fig. 10–2). **Arteries,** which carry blood from the heart to the tissues, branch into a "tree" of progressively smaller vessels, finally giving rise to **arterioles,** which regulate the distribution of blood within the various tissues. Arterioles branch further into **capillaries,** the smallest of vessels, across which all exchanges are made with surrounding tissues. Capillary exchange is the entire purpose of the circulatory system; all other activities of the system are directed toward assuring an adequate distribution of replenished blood to capillaries for exchange with all cells. Capillaries rejoin to form small **venules,** which progressively unite to form larger **veins**

Figure 10–2 Basic Organization of the Cardiovascular System

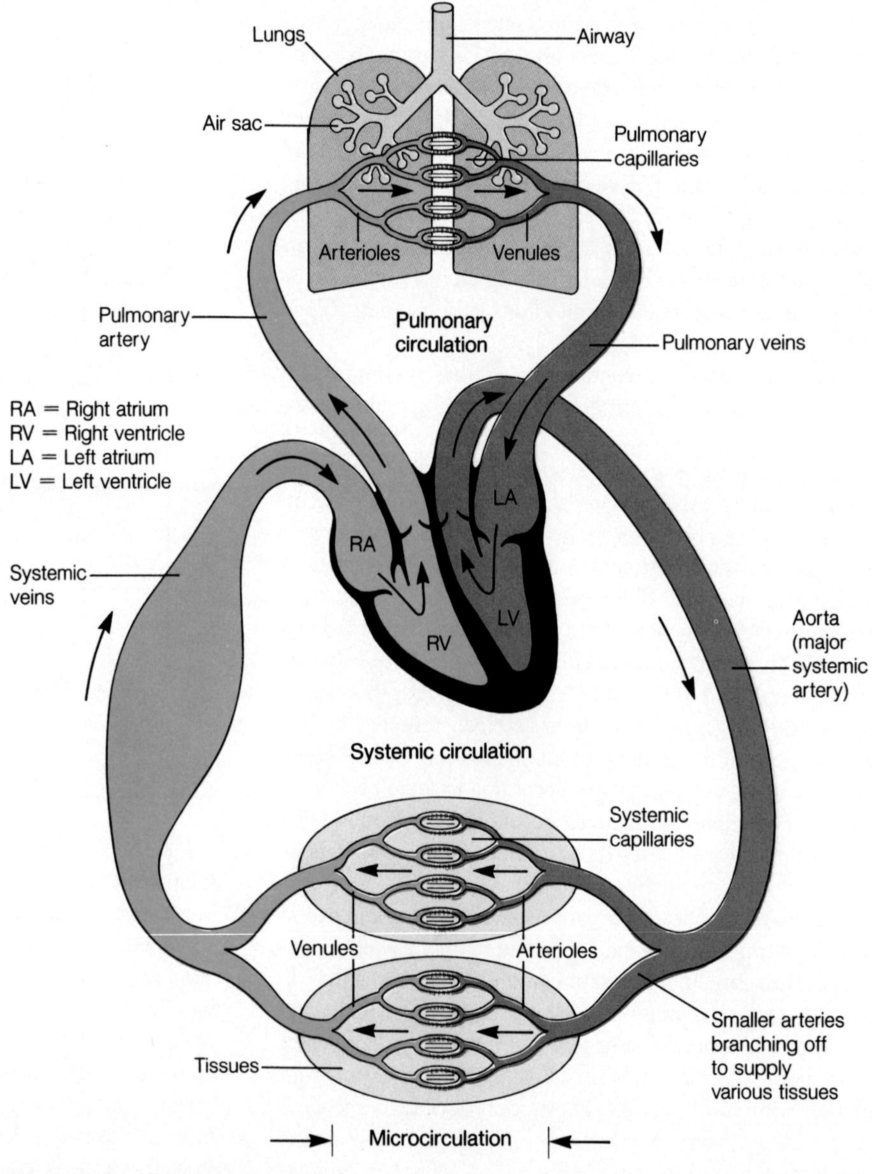

that eventually empty into the heart. The arterioles, capillaries, and venules are collectively referred to as the **microcirculation** because they are only visible microscopically.

The rate of blood flow through a vessel is directly proportional to the pressure gradient and inversely proportional to vascular resistance:

$$F = \frac{\Delta P}{R}$$

☐ F = flow rate of blood through a vessel (sometimes designated as Q)

☐ ΔP = pressure gradient

☐ R = vascular resistance

The **pressure gradient**—the difference in pressure between the beginning and end of a vessel—is the main driving force for flow through the vessel; that is, blood flows from an area of higher pressure to an area of lower pressure down a pressure gradient. Pressure is imparted to the blood by contraction of the heart, but, because of frictional losses, the pressure decreases as blood flows through a vessel. Since pressure drops throughout the vessel's length, it is higher at the beginning than at the end. This establishes a pressure gradient for forward flow of blood through the vessel. The greater the pressure gradient forcing blood through a vessel, the greater the rate of flow through that vessel. Think of a garden hose attached to a faucet. If you turn on the faucet slightly, water will flow out of the end of the hose in a small stream, because there is a slightly greater pressure at the beginning than at the end of the hose. If you open the faucet all the way, the pressure gradient will increase tremendously so that the rate of water flow through the hose will be much greater, and water will spurt forth from the end of the hose.

The other factor influencing flow rate through a vessel is the **resistance,** which is a measure of the hindrance to blood flow through a vessel caused by friction between the moving fluid and stationary vascular walls. As resistance to flow increases, it is more difficult for blood to pass through the vessel, so flow decreases (as long as the pressure gradient remains unchanged). When resistance increases, the pressure gradient must increase correspondingly to maintain the same flow rate. Accordingly, when the vessels offer more resistance to flow, the heart must work harder to maintain adequate circulation.

Resistance to blood flow depends on three factors: (1) viscosity of the blood, (2) vessel length, and (3) vessel radius, with the radius being by far the most important. **Viscosity** refers to the friction developed between the molecules of a fluid as they slide over each other during flow of the fluid. The greater the viscosity, the greater the resistance to flow. In general, the thicker a liquid, the more viscous it is. For example, molasses flows more slowly than water because molasses has greater viscosity. Viscosity of blood is determined by two factors: the concentration of plasma proteins and, more importantly, the number of circulating red blood cells. Normally these two factors are relatively constant and are not important in the control of resistance. Occasionally, however, blood viscosity and, accordingly, resistance to flow are altered because of an abnormal number of red blood cells. For example, blood flow is more sluggish than normal when excessive red blood cells are present.

Because blood "rubs" against the lining of the vessels as it flows past, the greater the vessel surface area in contact with the blood, the greater the resistance to flow. Surface area is determined by both the length (L) and radius (r) of the vessel. At a constant radius, the longer the vessel, the greater the surface area and the greater the resistance to flow. Since vessel length remains constant in the body, this is not a variable factor in the control of vascular resistance. Therefore, the major determinant of resistance to flow is the vessel's radius. Fluid passes more readily through a large tube than through a smaller tube, because a given volume of blood comes into contact with much more of the surface area of a small-radius vessel than of a larger-radius one, creating greater resistance (Fig. 10–3a). Thus, blood flows more rapidly through the large arteries and veins than through the microcirculation. Furthermore, a slight change in the radius of a vessel brings about a notable change in flow, because the resistance is inversely proportional to the fourth power of the radius (multiplying the radius by itself four times):

$$R \propto \frac{1}{r^4}$$

This means that doubling the radius (Fig. 10-3b) decreases the resistance sixteen times ($r^4 = 2 \times 2 \times 2 \times 2 = 16$; $R \propto 1/16$) and thus increases flow through the vessel sixteenfold (at the same pressure gradient). The converse is also true. Only one-sixteenth as much blood flows through a vessel at the same driving pressure when its radius is halved. It is important to note that the radius of arterioles is subject to regulation, which constitutes the most important factor in the control of resistance to blood flow throughout the vascular tree.

The factors that affect the flow rate through a vessel are integrated in an idealized way in **Poiseuille's law** as follows:

$$\text{Flow rate} = \frac{\pi \Delta P\, r^4}{8 \eta\, L}$$

where η = viscosity

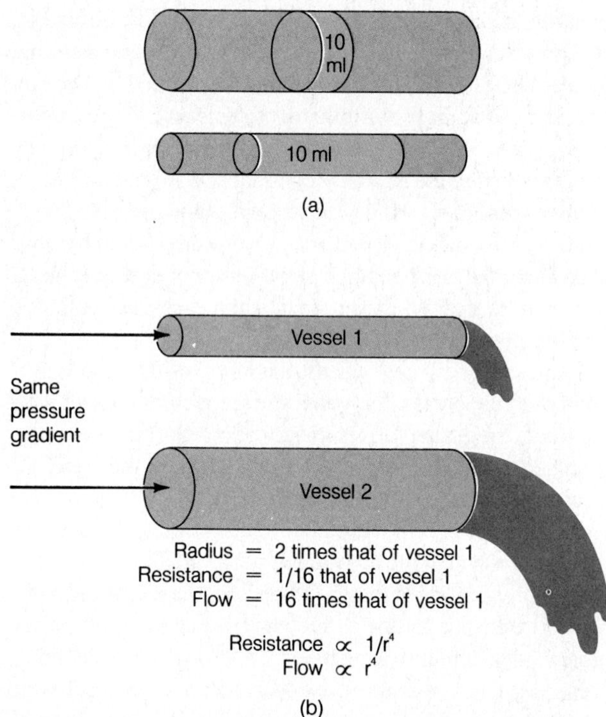

Figure 10–3 Relationship of Flow and Resistance to Vessel Radius *(a) The same volume of blood comes into contact with a greater surface area of a small-radius vessel compared to a larger-radius vessel. Accordingly, the smaller-radius vessel offers more resistance to blood flow, because the blood "rubs" against a larger surface area. (b) Doubling the radius decreases the resistance to 1/16 and increases the flow 16 times.*

The significance of the relation among flow, pressure, and resistance, as largely determined by vessel radius, will become even more apparent as we embark on a voyage through the vessels in the next section.

ARTERIES

Arteries serve as rapid-transit passageways to the tissues and as a pressure reservoir.

The consecutive segments of the vascular tree—arteries, arterioles, capillaries, and veins—are specialized to perform specific tasks (Table 10–1). Arteries are specialized to serve as rapid-transit passageways for blood to the tissues (because of their large radii, arteries offer little resistance to blood flow) and to act as a pressure reservoir to provide the driving force for blood when the heart is relaxing.

The heart alternately contracts to eject blood into the arteries and then relaxes to refill from the veins. No blood is pumped out when the heart is relaxing and refilling. However, capillary flow does not fluctuate between cardiac systole and diastole; blood flow is continuous through the capillaries supplying the tissues. The driving force for the continued flow of blood to the tissues during cardiac relaxation is provided by the elastic properties of the arterial walls. All vessels are lined with a layer of smooth, flattened **endothelial cells** that are continuous with the endocardial lining of the heart. Surrounding the arteries' endothelial lining is a thick wall containing smooth muscle and an abundance of collagen fibers, which provide tensile strength against the high driving pressure of blood ejected from the heart, and elastin fibers, which give the arterial walls elasticity so that they behave much like a balloon.

As the heart ejects blood into the arteries during ventricular systole, a greater volume of blood enters the arteries from the heart than leaves them to flow into the smaller vessels downstream, because of the smaller vessels' greater resistance to flow. The arteries' elasticity enables them to expand to temporarily hold this excess volume of ejected blood, storing some of the pressure energy imparted by cardiac contraction in their stretched walls—just as a balloon expands to accommodate the extra volume of air that you blow into it (Fig 10–4a). When the heart relaxes and no more blood is being pumped into the arteries, the stretched arterial walls passively recoil, like an inflated balloon that is released. This recoil pushes the excess blood contained within the arteries into the vessels downstream, ensuring continued blood flow to the tissues when the heart is relaxing and not pumping blood into the system (Fig. 10-4b).

Arterial pressure fluctuates in relation to ventricular systole and diastole.

Blood pressure, the force exerted by the blood against a vessel wall, depends on the volume of blood contained within the vessel and the **compliance,** or **distensibility,** of the vessel walls (how easily they can be stretched). If the volume of blood that entered the arteries were equal to the volume of blood that left the arteries during the same period, arterial blood pressure would remain constant. This is not the case, however. During ventricular systole, a stroke volume of blood enters the arteries from the ventricle while only about one-third of an equivalent volume of blood leaves the arteries to enter the arterioles. During diastole, no blood enters the arteries while blood continues to leave, driven by elastic recoil. The maximum pressure exerted in the arteries when blood is ejected into them during systole, the **systolic pressure,** averages 120 mm Hg. The minimum pressure within the arteries when blood is draining off into the remainder of the vascula-

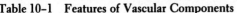

Table 10–1 Features of Vascular Components

Feature	Vessel Type: *Aorta*	*Large Arterial Branches*	*Arterioles*	*Capillaries*	*Large Veins*	*Vena Cava*
Number	One	Several hundred	Half a million	Ten billion	Several hundred	Two
Wall thickness	2 mm (2,000 μm)	1 mm (1,000 μm)	20 μm	1 μm	0.5 mm (500 μm)	1.5 mm (1,500 μm)
Internal radius	1.25 cm (12,500 μm)	0.2 cm (2,000 μm)	30 μm	3.5 μm	0.5 cm (5,000 μm)	3 cm (30,000 μm)
Total cross-sectional area	4.5 cm^2	20 cm^2	400 cm^2	6,000 cm^2	40 cm^2	18 cm^2
Special features	Thick, highly elastic walls; large radii		Highly muscular, well-innervated walls; small radii	Thin walled; large total cross-sectional area	Thin walled; highly distensible; large radii	
Functions	Passageway from heart to tissues; pressure reservoir		Primary resistance vessels; determine distribution of cardiac output	Site of exchange; determine distribution of extracellular fluid between plasma and interstitial fluid	Passageway to heart from tissues; blood reservoir	

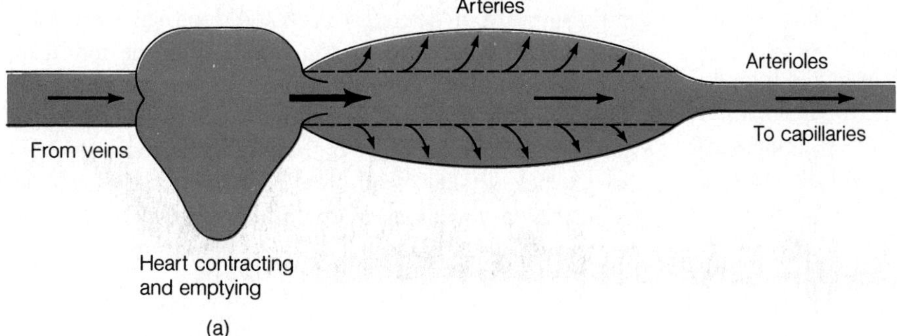

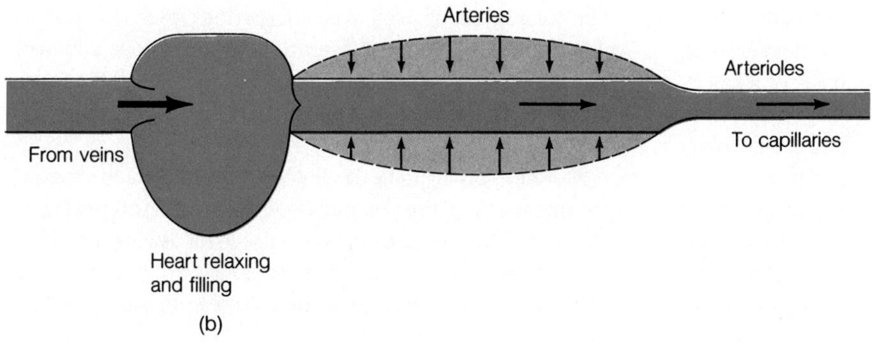

Figure 10–4 Arteries as Pressure Reservoir *Because of their elasticity, arteries act as a pressure reservoir. (a) The elastic arteries distend during cardiac systole as more blood is ejected into them than drains off into the narrow, high-resistance arterioles downstream. (b) The elastic recoil of arteries during cardiac diastole continues driving the blood forward when the heart is not pumping.*

Figure 10–5 Arterial Blood Pressure

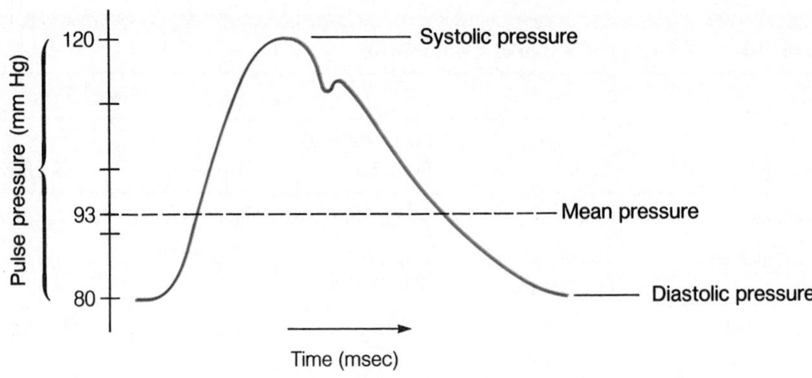

ture during diastole, the **diastolic pressure,** averages 80 mm Hg. The arterial pressure does not fall to 0 mm Hg, because the next cardiac contraction occurs and refills the arteries before all of the blood drains off (Fig. 10–5; also see Fig. 9–19).

Blood pressure can be indirectly measured through use of a sphygmomanometer.

The changes in arterial pressure throughout the cardiac cycle can be measured directly by connecting a pressure-measuring device to a needle inserted in an artery. However, it is more convenient and reasonably accurate to measure the pressure indirectly through use of a **sphygmomanometer,** an externally applied inflatable cuff attached to a pressure gauge. When the cuff is wrapped around the upper arm and then inflated with air, the pressure of the cuff is transmitted through the tissues to the underlying brachial artery, the main vessel carrying blood to the forearm (Fig. 10–6). The technique involves balancing the pressure in the cuff against the pressure in the artery. When cuff pressure is greater than the pressure in the vessel, the vessel is pinched closed so that no blood flows through it. When blood pressure is greater than cuff pressure, the vessel is open and blood flows through.

During the determination of blood pressure, a stethoscope is placed over the brachial artery just below the cuff. No sound can be detected either when blood is not flowing through the vessel or when blood is flowing in the normal, smooth laminar flow (see p. 282). Turbulent blood flow, on the other hand, creates vibrations that can be heard. At the onset of a blood-pressure determination, the cuff is inflated to a pressure greater than systolic blood pressure so that the brachial artery collapses. Because the externally applied pressure is even greater than the peak internal pressure, the artery remains completely occluded throughout the entire cardiac cycle; no sound can be heard, since no blood is passing through. As air in the cuff is slowly released, the pressure in the cuff is gradually reduced. When the cuff pressure falls to just below the peak systolic pressure, the artery transiently opens slightly when the blood pressure reaches this peak. Blood escapes through the partially occluded artery for a brief interval before the arterial pressure falls below the cuff pressure and the artery collapses once again. This spurt of blood is turbulent, so it can be heard. Thus, the highest cuff pressure at which the ***first sound*** can be heard is indicative of the ***systolic blood pressure***. As the cuff pressure continues to fall, blood intermittently spurts through the artery and produces a sound with each subsequent cardiac cycle whenever the arterial pressure exceeds the cuff pressure. When the cuff pressure finally falls below diastolic pressure, the brachial artery is no longer occluded during any part of the cardiac cycle and blood can continually flow uninterrupted through the vessel. With the return of nonturbulent blood flow, no further sounds can be heard. Therefore, the highest cuff pressure at which the ***last sound*** can be detected is indicative of the ***diastolic pressure***. In clinical practice, arterial blood pressure is expressed as systolic pressure over diastolic pressure, with the average blood pressure being 120/80 mm Hg.

Mean arterial pressure is the main driving force for blood flow.

The pulse that can be felt in an artery lying close to the surface of the skin is due to the difference between systolic and diastolic pressures. This pressure difference is known as the **pulse pressure.** When the blood pressure is 120/80, pulse pressure is 40 mm Hg (120 mm Hg − 80 mm Hg).

More important than the fluctuating systolic and diastolic pressures or the pulse pressure is the **mean arterial pressure,** which is the average pressure responsible for driving blood forward into the tissues throughout the cardiac cycle. Contrary to what you might expect, mean arterial pressure is not the half-

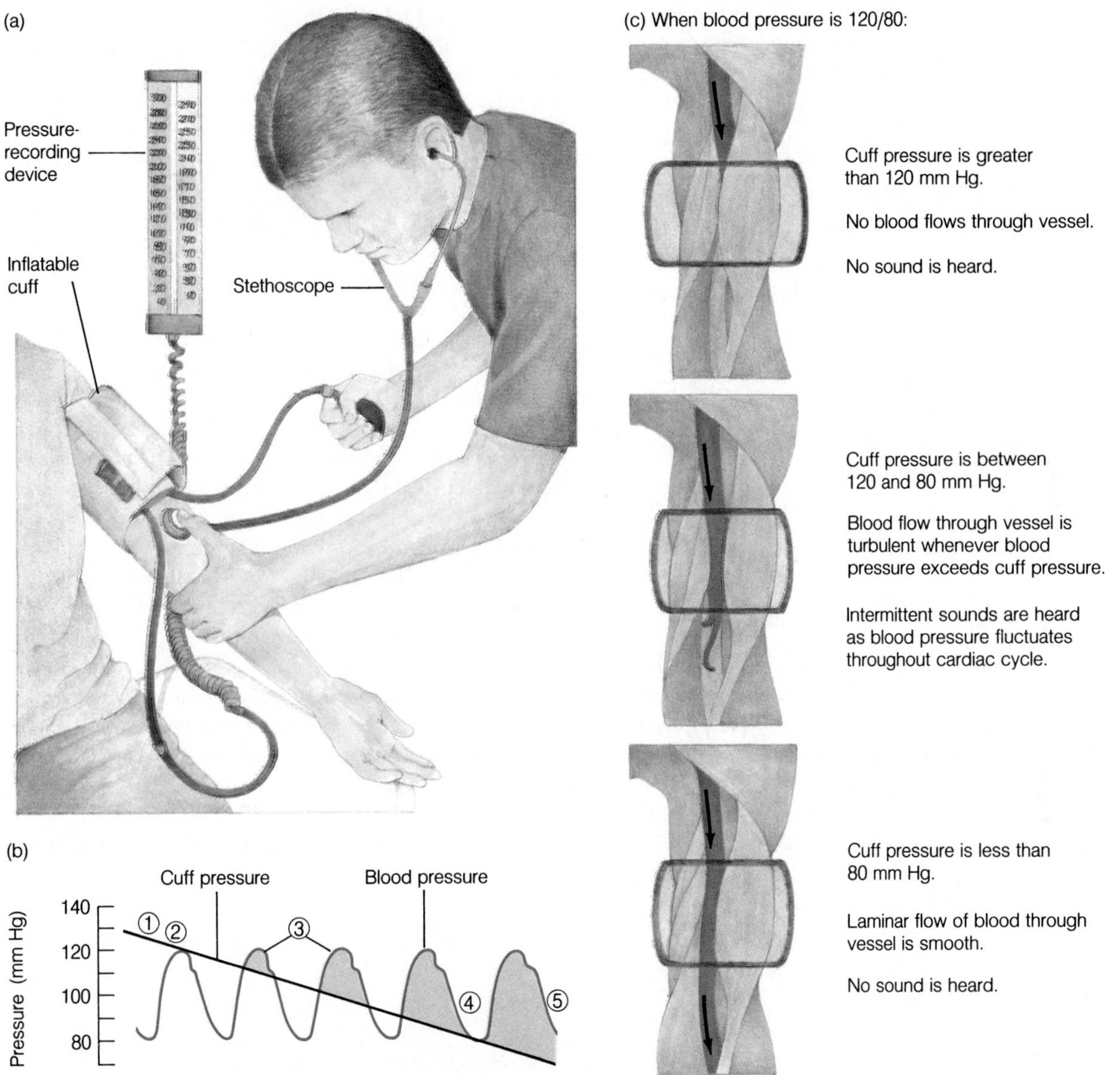
(a)
Pressure-recording device
Inflatable cuff
Stethoscope
(b)
Cuff pressure
Blood pressure
Pressure (mm Hg)
140
120
100
80
①
②
③
④
⑤
(c) When blood pressure is 120/80:
Cuff pressure is greater than 120 mm Hg.
No blood flows through vessel.
No sound is heard.
Cuff pressure is between 120 and 80 mm Hg.
Blood flow through vessel is turbulent whenever blood pressure exceeds cuff pressure.
Intermittent sounds are heard as blood pressure fluctuates throughout cardiac cycle.
Cuff pressure is less than 80 mm Hg.
Laminar flow of blood through vessel is smooth.
No sound is heard.

Figure 10-6 Sphygmomanometry *(a) Use of sphygmomanometer in determing blood pressure. The pressure in the inflatable cuff can be varied to prevent or permit blood flow in the underlying brachial artery. Turbulent blood flow can be detected through use of a stethoscope whereas smooth laminar flow and no flow are not audible. (b) Pattern of sounds in relation to cuff pressure compared with blood pressure. The numbers on the illustration refer to the following key points during a blood pressure determination: ① Cuff pressure exceeds blood pressure throughout cardiac cycle. No sound heard. ② First sound heard at peak systolic pressure. ③ Intermittent sounds heard as blood pressure cyclically exceeds cuff pressure. ④ Last sound heard at minimum diastolic pressure. ⑤ Blood pressure exceeds cuff pressure throughout cardiac cycle. No sound heard. (c) Blood flow through brachial artery in relation to cuff pressure and sounds.*

way value between systolic and diastolic pressure (for example, with a blood pressure of 120/80, mean pressure is not 100 mm Hg). The reason is that arterial pressure remains closer to diastolic than to systolic pressure for a longer portion of each cardiac cycle. At resting heart rate, about two-thirds of the cardiac cycle is spent in diastole and only one-third in systole. As an analogy, if a race car traveled 80 miles per hour (mph) for 40 minutes and 120 mph for 20 minutes, its average speed would be 93 mph, not the halfway value of 100 mph. Similarly, a good approximation of the mean arterial pressure can be determined using the following formula:

$$\text{Mean arterial pressure} = \text{Diastolic pressure} + (1/3 \times \text{Pulse pressure})$$

$$\text{At } 120/80, \text{ mean arterial pressure} = 80 \text{ mm Hg} + (1/3 \times 40 \text{ mm Hg}) = 93 \text{ mm Hg}$$

It is this mean arterial pressure, not the systolic or diastolic pressures, that is monitored and regulated by blood-pressure reflexes to be described later in the chapter.

Arterial pressure—whether it be systolic, diastolic, pulse, or mean—is essentially the same throughout the arterial tree. Because arteries offer little resistance to flow, only negligible frictional loss of pressure energy takes place in them.

Arterioles

Arterioles are the major resistance vessels.

When an artery reaches the organ it is supplying, it branches into numerous arterioles, the radii of which are small enough to offer considerable resistance to flow (Table 10–1). In fact, the arterioles are the major resistance vessels in the vascular tree. In contrast to the arteries, the high degree of arteriolar resistance causes a marked drop in mean pressure as the blood flows through these vessels. On average, the pressure falls from 93 mm Hg, the mean arterial pressure, to 37 mm Hg, the pressure at the beginning of the capillaries (Fig. 10–7). This decline in pressure helps establish the pressure differential that encourages the flow of blood from the heart to the various organs downstream. Arteriolar resistance is also responsible for converting the pulsatile systolic-to-diastolic pressure swings in the arteries into the nonfluctuating pressure present in the capillaries.

The radii (and, accordingly, the resistances) of arterioles supplying individual organs can be adjusted independently to determine the distribution of cardiac output and to regulate arterial blood pressure. We will discuss the mechanisms involved in adjusting arteriolar resistance before considering how such adjustments are important in accomplishing these two functions.

Unlike arteries, arteriolar walls contain very little elastic connective tissue. However, they do have a thick layer of smooth muscle that is richly innervated by sympathetic nerve fibers. The smooth muscle is also sensitive to many local chemical changes, to a few circulating hormones, and to mechanical stretch. The smooth-muscle layer runs circularly around the arteriole so that when it contracts, the vessel's circumference (and its radius) becomes smaller, thus increasing resistance and decreasing the flow through that vessel. **Vasoconstriction** is the term applied to such narrowing of a vessel (Fig. 10–8). **Vasodilation** refers to enlargement in the circumference and radius of a vessel as a result of relaxation of its smooth-muscle layer, which leads to decreased resistance and increased flow through that vessel.

Arteriolar smooth muscle normally displays a state of partial constriction known as **vascular tone.** This ongoing tonic ac-

Figure 10–7 Pressures throughout Systemic Circulation

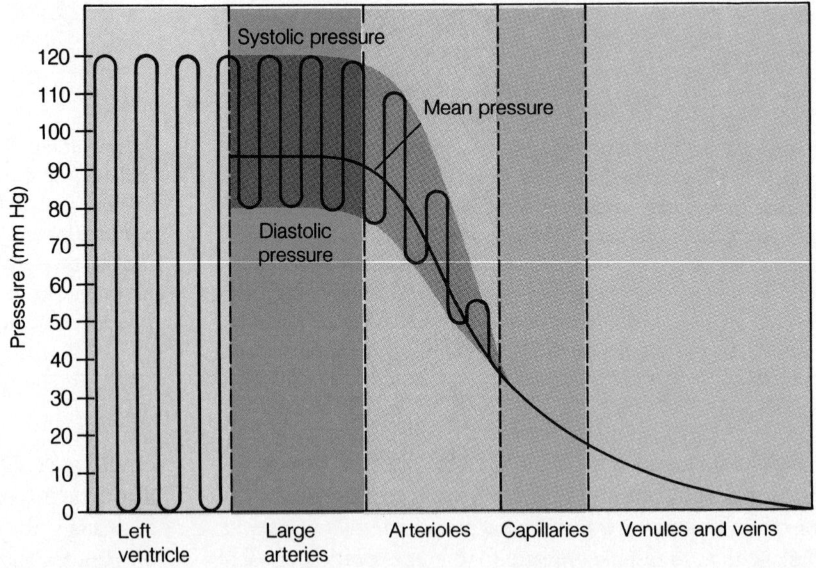

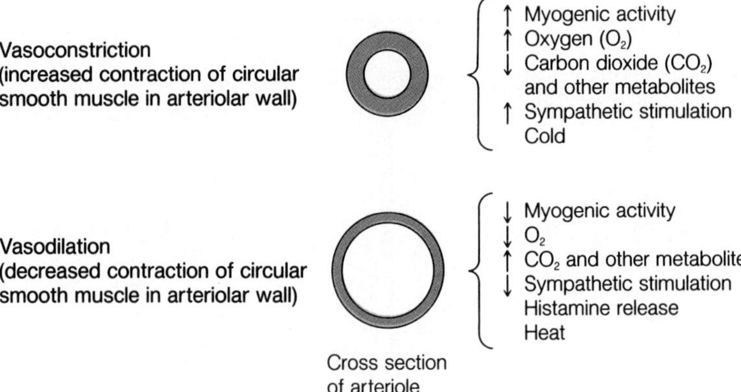

Figure 10-8 Arteriolar Vasoconstriction and Vasodilation

tivity, which establishes a baseline of arteriolar resistance, is caused by two factors. First, arteriolar smooth muscle has considerable myogenic activity; that is, its membrane potential fluctuates without any external influences, leading to self- induced contractile activity (see p. 253). Second, the sympathetic fibers supplying most arterioles continually release norepinephrine, which further enhances the vascular tone.

Because of this ongoing tonic activity, it is possible to either increase or decrease the level of contractile activity to accomplish vasoconstriction or vasodilation, respectively. Were it not for tone, it would be impossible to reduce the tension in an arteriolar wall to accomplish vasodilation; only varying degrees of vasoconstriction would be possible.

A variety of factors can influence the level of contractile activity in arteriolar smooth muscle, thereby substantially changing resistance to flow in these vessels. These factors fall into two categories: local controls important in matching blood flow to the metabolic needs of the specific tissues in which they occur; and extrinsic controls important in blood pressure regulation.

Local control of arteriolar radius is important in determining distribution of cardiac output.

The fraction of the total cardiac output delivered to each organ is not always constant; it varies, depending on the demands for blood at the time. The amount of the cardiac output received by each organ is determined by the number and caliber of the arterioles supplying that area. Recall that $F = \Delta P/R$. Because blood is delivered to all tissues at the same mean arterial pressure, the driving force for flow is identical for each organ. Therefore, differences in flow to various organs are completely determined by differences in the extent of vascularization and by differences in resistance offered by the arterioles supplying each organ. On a moment-to-moment basis, the distribution of cardiac output can be varied by differentially adjusting arteriolar resistance in the various vascular beds.

As an analogy, consider a pipe carrying water with a number of adjustable valves located throughout its length (Fig. 10–9). If one assumes a constant water pressure in the pipe, differences in the amount of water flowing into a beaker under each valve depend entirely on which valves are open and the extent to which they are open. No water enters beakers under closed valves (high resistance), and more water flows into beakers under valves opened completely (low resistance) compared with valves only partially opened (moderate resistance). Similarly, more blood flows to areas whose arterioles offer the least resistance to its passage. During exercise, for example, not only is cardiac output increased, but, because of vasodilation in skeletal muscle and the heart, a greater percentage of the pumped blood is diverted to these organs to support their increased metabolic activity. Simultaneously, blood flow to the digestive tract and kidneys is reduced as a result of arteriolar vasoconstriction in these organs (Fig. 10–10).

Only the blood supply to the brain remains remarkably constant no matter what activity the person is engaged in, be it vigorous physical activity, intense mental concentration, or sleeping. Although the *total* blood flow to the brain remains constant, new techniques demonstrate that differences in regional blood flow occur within the brain in close correlation with local neural-activity patterns.

Local controls refer to changes within a tissue that alter the radii of the vessels and hence adjust blood flow through the tissue by directly affecting the smooth muscle of the tissue's arterioles. These local influences on arteriolar radius are important in matching the blood flow through a tissue with the tissue's metabolic needs. Local controls are especially important in skeletal muscle and heart, the tissues whose metabolic

Figure 10-9 Flow Rate as Function of Resistance

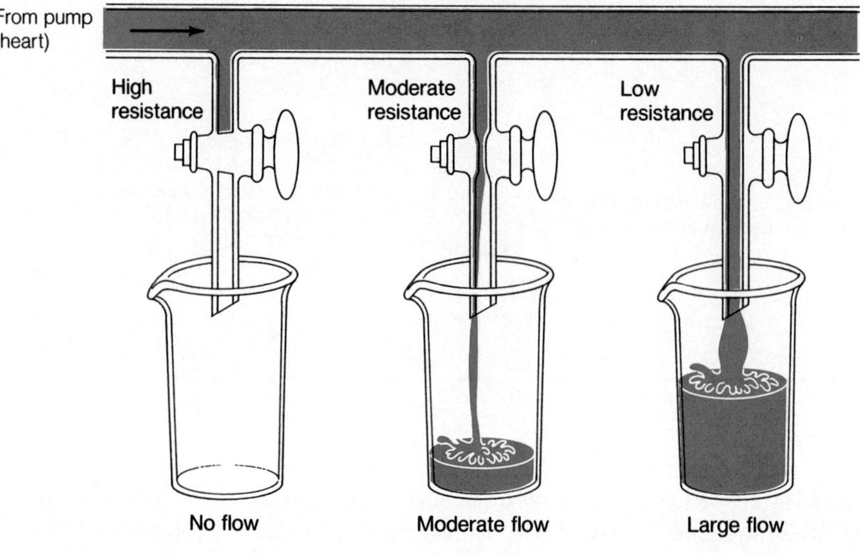

activity and need for blood supply normally vary most extensively, and the brain, whose overall metabolic activity and need for blood supply remains constant. Local controls help maintain the constancy of blood flow to the brain. Local influences may be either chemical or physical in nature.

LOCAL CHEMICAL INFLUENCES.

1. *Local metabolic changes*

   Arterioles lie within the tissue they are supplying and are exposed to the chemical composition of the interstitial fluid in the tissue. During increased metabolic activity, the local concentrations of a number of a tissue's chemicals change. Many of these chemical changes produce local arteriolar dilation, subsequently increasing blood flow to that particular area. This response is called **active hyperemia.** When cells are more active metabolically, they need more blood to bring in O_2 and nutrients and to remove metabolic wastes. The following are the local chemical factors shown experimentally to produce relaxation of arteriolar smooth muscles:

   □ *Decreased* O_2—Actively metabolizing cells use up more O_2 to support oxidative phosphorylation for ATP production.

   □ *Increased* CO_2—More CO_2 is generated as a byproduct during the stepped-up pace of oxidative phosphorylation that accompanies increased activity.

   □ *Increased acid*—More carbonic acid is generated from the increased CO_2 produced as the metabolic activity of a cell increases. Also, lactic acid accumulates if the glycolytic pathway is used for ATP production.

   □ *Increased* K^+—Repeated action potentials that outpace the ability of the Na^+-K^+ pump to restore the resting concentration gradients result in an increase in K^+ in the tissue fluid of an actively contracting muscle or a more active region of the brain.

   □ *Increased osmolarity*—Osmolarity (the concentration of osmotically active solutes) increases during elevated cellular metabolism because of increased formation of osmotically active particles.

   □ *Adenosine release*—Especially in cardiac muscle, adenosine is released in response to increased metabolic activity or O_2 deprivation (see p. 291).

   □ *Prostaglandin release*—Prostaglandins are local chemical messengers derived from fatty-acid chains within the plasma membrane. Their production and release is poorly understood.

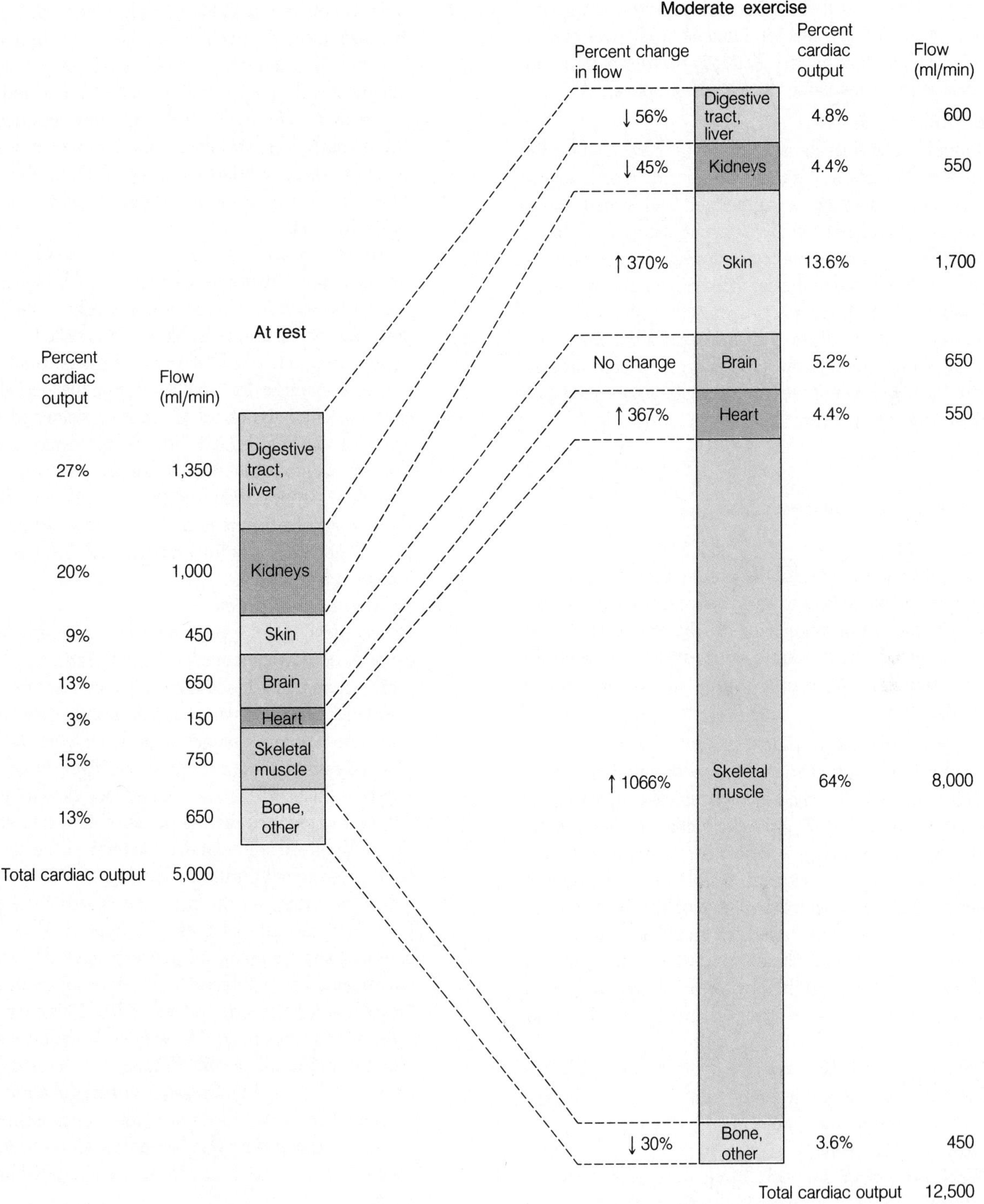

Figure 10–10 Magnitude and Distribution of Cardiac Output at Rest and during Moderate Exercise

The relative contributions of these various chemical changes (and possibly others) in the local metabolic control of blood flow in systemic arterioles is presently unclear. The important fact is that local metabolic changes can adjust local blood flow by acting directly on arteriolar smooth muscle without involving nerves or hormones.

2. *Histamine release*
   Although histamine does not normally participate in the control of blood flow, it is important in certain pathological conditions. Histamine is synthesized and stored within many tissues and in certain types of circulating white blood cells. When tissues are injured or during allergic reactions, histamine is released in the damaged region. By promoting relaxation of arteriolar smooth muscle, histamine is the major cause of vasodilation in an injured area. The resultant increase in blood flow into the area is responsible for the redness and contributes to the swelling associated with inflammatory responses.

LOCAL PHYSICAL INFLUENCES.

1. *Heat or cold*
   Heat application is a useful therapeutic agent for promoting increased blood flow to an area, because it causes localized arteriolar vasodilation. Conversely, applying ice packs to an inflamed area produces vasoconstriction, which reduces swelling by counteracting histamine-induced vasodilation.
2. *Myogenic responses to stretch*
   Evidence indicates that arteriolar smooth muscle responds to being passively stretched by myogenically increasing its tone, thereby acting to resist the initial passive stretch. Conversely, a decrease in arteriolar stretching is accompanied by a reduction in myogenic vessel tone. The extent of passive stretch varies with the volume of blood delivered to the arterioles from the arteries. An increase in mean arterial pressure drives more blood forward into the arterioles and stretches them further, whereas arterial occlusion blocks blood flow into the arterioles and reduces arteriolar stretch.

   Myogenic responses, coupled with metabolically induced responses, appear to be important in reactive hyperemia and pressure autoregulation.

   a. *Reactive hyperemia*
      When the blood supply to a region is completely occluded (blocked), arterioles in the region dilate because of: (1) myogenic relaxation, which is in response to the diminished stretch accompanying no blood flow, and (2) changes in local chemical composition. Many of the same chemical changes occur in a blood-deprived tissue that occur during metabolically induced active hyperemia. When a tissue's blood supply is blocked, O_2 levels decrease in the deprived tissue; the tissue continues to consume O_2, but no fresh supplies are being delivered. Meanwhile, the concentrations of CO_2, acid, and other metabolites increase. Even though their production is not increased as it is when a tissue is more active metabolically, these substances accumulate in the tissue when the normal amounts produced are not "washed away" by blood.

      After removal of the occlusion, blood flow to the previously deprived tissue is transiently much higher than normal because the arterioles are widely dilated. This post-occlusion increase in blood flow, called **reactive hyperemia,** can take place in any tissue. Such a response is beneficial for rapidly restoring the local chemical composition to normal. Of course, prolonged blockage of blood flow leads to irreversible tissue damage. You can demonstrate reactive hyperemia yourself by wrapping a rubber band around one of your fingers about an inch from the end for a minute. On removal, you will note that your fingertip turns red because of the increased blood flow.
   b. *Pressure autoregulation*
      When mean arterial pressure falls (for example, because of hemorrhage or a weakened heart), the driving force is reduced, so blood flow to tissues decreases. This is a milder version of what happens during vessel occlusion. The resultant changes in local metabolites and the reduced stretch in the arterioles collectively bring about arteriolar dilation to restore tissue blood flow to normal despite the reduced driving pressure. On the negative side, widespread arteriolar dilation reduces the mean arterial pressure still further, which aggravates the problem. Conversely, in the presence of sustained elevations in mean arterial pressure (hypertension), local chemical and myogenic influences triggered by the initial increased flow of blood to tissues bring about an increase in arteriolar tone and resistance. This greater degree of vasoconstriction subsequently reduces tissue blood flow toward normal despite the elevated blood pressure (Fig. 10–11). **Pressure autoregulation** is the term applied to these local arteriolar mechanisms that are aimed at keeping tissue blood flow fairly constant in spite of rather wide deviations in mean arterial driving pressure.

Active hyperemia, reactive hyperemia, and histamine release all deliberately increase blood flow to a particular

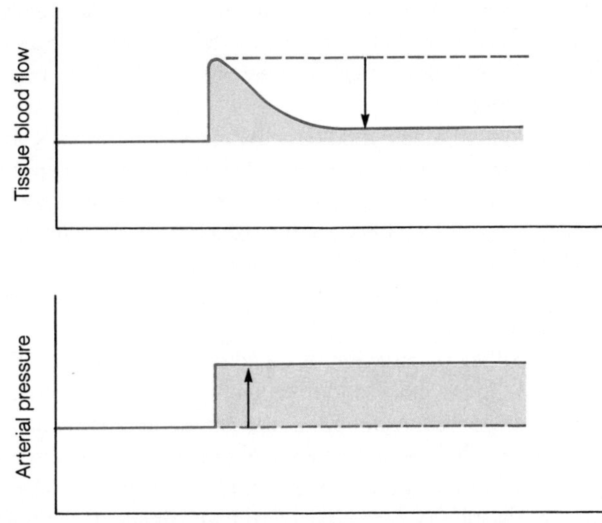

Figure 10–11 Pressure Autoregulation of Tissue Blood Flow *Even though blood flow through a tissue immediately increases in response to a rise in arterial pressure, the tissue blood flow gradually becomes reduced as a result of pressure autoregulation within the tissue, despite a sustained increase in arterial pressure.*

tissue for a specific purpose by inducing local arteriolar vasodilation. In contrast, pressure autoregulation is a means by which each tissue resists alterations in its own blood flow secondary to changes in mean arterial pressure by appropriate adjustments in arteriolar radius.

Extrinsic control of arteriolar radius is primarily important in the regulation of arterial blood pressure.

Extrinsic control of arteriolar radius includes both neural and hormonal influences, with the effects of the sympathetic nervous system being the most important. Sympathetic nerve fibers supply arteriolar smooth muscle everywhere except in the brain. A certain level of ongoing sympathetic activity contributes to vascular tone. Increased sympathetic activity produces generalized arteriolar vasoconstriction, whereas decreased sympathetic activity leads to generalized arteriolar vasodilation.

These widespread changes in arteriolar resistance bring about changes in mean arterial blood pressure. The formula $F = \Delta P/R$ applies to the entire circulation as well as to a single vessel:

- □ F: Looking at the circulatory system as a whole, flow (F) through all of the vessels is equal to the cardiac output.
- □ ΔP: The pressure gradient (ΔP) for the entire circulation is the mean arterial pressure. ΔP equals the difference in pressure between the beginning and the end of the circulatory system. Because the beginning pressure is the mean arterial pressure as the blood leaves the left ventricle at an average of 93 mm Hg, and the end pressure in the right atrium is 0 mm Hg, ΔP equals 93 mm Hg minus 0 mm Hg, or 93 mm Hg, equivalent to the mean pressure.
- □ R: By far the greatest percentage of the total resistance (R) offered by all of the peripheral vessels (**total peripheral resistance**) is due to arteriolar resistance, because arterioles are the primary resistance vessels in the vascular tree.

Therefore, for the entire system, rearranging

$$F = \Delta P/R \quad \text{to be} \quad \Delta P = F \times R,$$

makes the equation become

$$\text{Mean arterial pressure} = \text{Cardiac output} \times \text{Total peripheral resistance}$$

Thus, the extent of total peripheral resistance offered collectively by all the arterioles influences the mean arterial blood pressure immensely. An analogy to this relationship is a dam. By restricting flow of water downstream, a dam also increases the pressure upstream by elevating the water level in the reservoir behind the dam. Similarly, generalized, sympathetically induced vasoconstriction reflexly reduces blood flow downstream to the tissue cells while elevating the upstream mean arterial pressure, thereby increasing the main driving force for blood flow to all the organs.

These effects seem to be counterproductive. Why increase the driving force for flow to the organs while reducing flow to the organs? In effect, the sympathetically induced arteriolar responses help to maintain the appropriate driving pressure head to all organs. The extent to which each organ actually receives blood flow is determined by local arteriolar adjustments that override the sympathetic constrictor effect. If all arterioles were dilated, blood pressure would fall substantially, so there would not be an adequate driving force for blood flow. Thus, tonic sympathetic activity constricts most vessels (with the exception of those in the brain) to help maintain a pressure head from which organs can draw on as needed through local mechanisms that control arteriolar radius.

Cardiac and skeletal muscles have the most powerful local control mechanisms with which to override generalized sympathetic vasoconstriction. In fact, coronary vasodilation actually accompanies sympathetic stimulation, because the sympathetically induced increase in heart rate and stroke volume elevates cardiac metabolic activity, triggering a profound, locally mediated vasodilation that overpowers the weaker sympathetic vasoconstrictor effect. Likewise, increased activity in a skeletal muscle induces overriding local vasodilation in that particular muscle, despite generalized sympathetic vasoconstriction.

The norepinephrine released from sympathetic nerve endings combines with α adrenergic receptors (see p. 207) on arteriolar smooth muscle to bring about vasoconstriction. Cerebral (brain) arterioles are the only ones that do not have α receptors, so no vasoconstriction occurs in the brain. It is important that cerebral arterioles are not reflexly constricted by neural influences, since brain blood flow must remain constant despite what is going on elsewhere in the body. Cerebral vessels are almost entirely controlled by local mechanisms that maintain a constant blood flow to support a constant level of brain metabolic activity. In fact, reflex vasoconstrictor activity in the remainder of the cardiovascular system is aimed at maintaining an adequate pressure head for blood flow to the vital brain and heart.

Thus, sympathetic activity contributes in an important way to the maintenance of mean arterial pressure, assuring an adequate driving force for blood flow to the brain and heart at the expense of organs and tissues that can better withstand reduced blood flow. Other tissues that really need additional blood, such as active muscles, obtain it through local controls that override the sympathetic effect.

There is no significant level of parasympathetic innervation to arterioles, with the exception of the abundant parasympathetic vasodilator supply to the arterioles of the penis and clitoris. The rapid, profuse vasodilation induced by parasympathetic stimulation in these organs is largely responsible for accomplishing erection. Vasodilation elsewhere is produced by decreasing sympathetic vasoconstrictor activity below its tonic level.[1] When mean arterial pressure becomes elevated above normal, reflex reduction in sympathetic vasoconstrictor activity accomplishes generalized arteriolar vasodilation to help restore the driving pressure down toward normal.

The main region of the brain responsible for adjusting sympathetic output to the arterioles is the **cardiovascular control center** in the medulla of the brain stem. This is the integrating center for blood-pressure regulation. Several other brain regions also influence blood distribution, the most notable being the hypothalamus, which, as part of its temperature-regulating function, controls blood flow to the skin to adjust heat loss to the environment.

In addition to neural reflex activity, several hormones also extrinsically influence arteriolar radius. These include the adrenal medullary hormones, epinephrine and norepinephrine, which generally reinforce the sympathetic nervous system, as well as vasopressin (also known as antidiuretic hormone or ADH) and angiotensin II, which are important in the control of fluid balance.

Sympathetic stimulation of the adrenal medulla causes the release of epinephrine and norepinephrine. Adrenal medullary norepinephrine combines with the same α receptors as sympathetically released norepinephrine to produce generalized vasoconstriction. However, epinephrine, the most abundant of the adrenal medullary hormones, combines with both β_2 and α receptors and has a greater affinity for the β_2 receptors. Activation of β_2 receptors produces vasodilation, but not all tissues have β_2 receptors; they are most abundant in the heart and skeletal muscles. During sympathetic discharge, the released epinephrine combines with the β_2 receptors in the heart and skeletal muscle to reinforce local vasodilatory mechanisms in these tissues. Arterioles in digestive organs and kidneys, equipped only with α receptors, do not respond to epinephrine. Therefore, these organs, affected only by norepinephrine during generalized sympathetic discharge, undergo more profound vasoconstriction than heart and skeletal muscle do.

The two other hormones that extrinsically influence arteriolar tone are vasopressin and angiotensin II. Vasopressin is primarily involved in maintaining water balance by regulating the amount of water the kidneys retain for the body during urine formation. Angiotensin II is part of a hormonal pathway important in the regulation of the body's salt balance, which, because of the osmotic effect of salt in the ECF, has a secondary effect on water balance. Thus, both of these hormones play important roles in maintaining fluid balance in the body, which, in turn, is an important determinant of plasma volume and blood pressure. In addition, both vasopressin and angiotensin II are potent vasoconstrictors. Their role in this regard is especially important during hemorrhage. A sudden loss of blood reduces the plasma volume, which triggers increased secretion of both of these hormones to help restore plasma volume. Their vasoconstrictor effect also helps maintain blood pressure in the face of the abrupt loss of plasma volume.

This completes our discussion of the various control factors that adjust arteriolar radius. These factors are summarized in Figure 10–12.

[1]A portion of the skeletal-muscle fibers in some species are supplied by sympathetic cholinergic (ACh-releasing) fibers that bring about vasodilation in anticipation of exercise. However, the existence of such sympathetic vasodilator fibers in humans remains questionable.

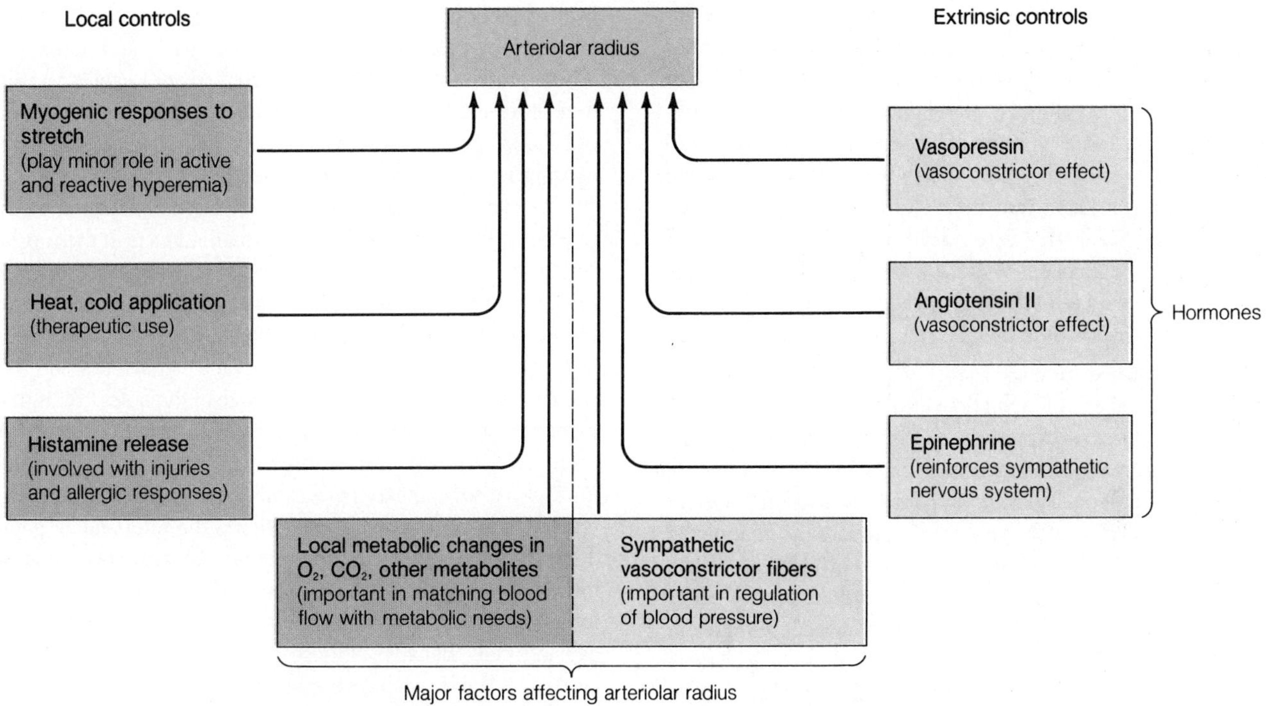

Figure 10–12 Factors Affecting Arteriolar Radius

Capillaries

Capillaries are ideally suited to serve as sites of exchange.

Capillaries, the sites for exchange[2] of materials between the blood and tissues, branch extensively to bring blood within the reach of every cell. There are no carrier-mediated transport systems across capillaries, with the exception of those in the brain that play a role in the blood-brain barrier (see p. 124). Exchange of materials across capillary walls is accomplished primarily by the process of diffusion. Capillaries are ideally suited to enhance diffusion in accordance with Fick's law of diffusion (see p. 71). They minimize diffusion distances while maximizing surface area and time available for exchange as follows:

[2]Actually, some exchange takes place across the other microcirculatory vessels, especially the postcapillary venules. The entire vasculature is a continuum and does not abruptly change from one vascular type to the other. When the term *capillary exchange* is used, it tacitly refers to all exchange at the microcirculatory level, the majority of which occurs across the capillaries.

1. Diffusing molecules have only a short distance to travel between the blood and surrounding cells because of the thin capillary wall and small capillary diameter, coupled with the close proximity of each and every cell to a capillary. This is important, because the rate of diffusion slows down as the diffusion distance increases.
   a. Capillary walls are very thin (1 μm in thickness; in comparison, the diameter of a human hair is 100 μm). Capillaries are composed of only a single layer of flattened endothelial cells—essentially, the lining of the other vessel types. No smooth muscle or connective tissue is present.
   b. Each capillary is so narrow (7 μm average diameter) that red blood cells (8 μm diameter) have to squeeze through single file. Consequently, plasma contents are either in direct contact with the inside of the capillary wall or are only a short diffusing distance from it.
   c. Because of extensive capillary branching, it is estimated that no cell is farther than 0.01 centimeter (cm; 1 cm = 4/1,000 inch) from a capillary.
2. Because capillaries are distributed in such incredible numbers (estimates range from 10 to 40 billion capillaries), a

tremendous total surface area is available for exchange (an estimated 600 square meters). In spite of this large number of capillaries, at any point in time they contain only 5% of the total blood volume (250 ml out of a total of 5,000 ml). As a result, a small volume of blood is exposed to an extensive surface area. If all the capillary surfaces were stretched out in a flat sheet and the volume of blood contained within the capillaries were spread over the top, this would be roughly equivalent to spreading a half pint of paint over the floor of a high-school gymnasium. Imagine how thin the paint layer would be!

3. The extensive capillary branching is also responsible for the slow velocity of blood flow through the capillaries. Velocity of flow is inversely proportional to the total cross-sectional area of all the vessels at any given level of the circulatory system. Even though the cross-sectional area of each capillary is extremely small compared to that of the large aorta, the total cross-sectional area of all the capillaries added together is about thirteen hundred times greater than the cross-sectional area of the aorta because there are so many capillaries. Accordingly, blood slows considerably as it passes through the capillaries (Fig. 10–13). This allows adequate time for exchange of nutrients and metabolic end products between blood and tissues, which is the sole purpose of the entire circulatory system.

Water-filled pores in the capillary wall permit passage of small, water-soluble substances that cannot cross the endothelial cells themselves.

Diffusion across capillary walls also depends on the wall's permeability to the materials being exchanged. The endothelial cells forming the capillary walls fit together in jigsaw-puzzle fashion, but the closeness of the fit varies considerably between organs. In most capillaries, narrow, water-filled clefts, or **pores,** are present at the junctions between the cells (Fig. 10–14). These pores permit passage of water-soluble substances. Lipid-soluble substances can readily pass through the endothelial cells themselves by dissolving in the lipid bilayer barrier. The size of the capillary pores varies from organ to organ. At one extreme, the endothelial cells in brain capillaries are joined by tight junctions so that pores are nonexistent. This prevents transcapillary passage of materials between the cells, which constitutes part of the protective blood-brain barrier. In most tissues, small, water-soluble substances such as ions, glucose, and amino acids can readily pass through the water-filled clefts, but large, non-lipid-soluble materials such as plasma proteins are excluded from passage. At the other extreme, liver capillaries have such large pores that even proteins pass through readily. This is adaptive, because among the liver's functions are synthesis of plasma proteins and the metabolism of protein-bound substances such as cholesterol. These proteins must all pass through the liver's capillary walls. The leakiness of various capillary beds is therefore a function of how tightly the endothelial cells are joined, which varies according to the different organs' needs.

The capillary wall has traditionally been considered a passive sieve, like a brick wall with permanent gaps in the mortar acting as pores. Recent studies, however, suggest that even under normal conditions, changes in endothelial cells are involved in the active regulation of capillary membrane permeability; that is, in response to appropriate signals, the "bricks" can readjust themselves to vary the size of the holes. Thus, the extent of leakiness does not necessarily remain constant for a given capillary bed. For example, it is speculated that histamine increases capillary permeability by triggering contractile responses in endothelial cells to widen the intercellular gaps. This is not a muscular contraction, because no smooth-muscle cells are present in capillaries. It is believed to be due to an actin-myosin contractile apparatus in the nonmuscular capillary cells. Because of these larger gaps, the affected capillary wall is leakier so that normally retained plasma proteins escape into the surrounding tissue, where they exert an osmotic effect. The resultant additional local fluid retention contributes to inflammatory swelling, along with histamine-induced vasodilation.

Frequently, endocytotic-exocytotic vesicles can be seen microscopically within capillary endothelial cells. Some researchers speculate that their presence is indicative of vesicular traffic across the capillary wall, which may play a limited role in transporting large molecules such as proteins from one side to the other. There is no general agreement on this issue, however.

Many capillaries are not open under resting conditions.

The branching and reconverging arrangement within capillary beds varies somewhat, depending on the tissue. Capillaries typically branch either directly from an arteriole or from a thoroughfare channel known as a **metarteriole,** which runs between an arteriole and a venule. Likewise, capillaries may rejoin either at a venule or at a metarteriole (Fig. 10–15). Unlike the true capillaries within a capillary bed, metarterioles are sparsely surrounded by wisps of spiraling smooth-muscle cells. An extension of this is seen as **precapillary sphincters,** each of which consists of a ring of smooth muscle around the entrance to a capillary as it arises from a metarteriole.[3] Pre-

[3]Although generally accepted, the existence of precapillary sphincters in humans has not been conclusively established.

capillary sphincters are not innervated, but they have a high degree of myogenic tone and are sensitive to local metabolic changes. They act as stopcocks to control blood flow through the particular capillary that each one guards. Arterioles perform a similar function for a small group of capillaries. Capillaries themselves have no smooth muscle, so they cannot actively participate in the regulation of their own blood flow.

Tissues that are generally more metabolically active have a greater density of capillaries. Muscles, for example, have relatively more capillaries than their tendinous attachments. Only about 10% of the precapillary sphincters in a resting muscle are open at any moment, however, so blood is flowing through only about 10% of the muscle's capillaries. As chemical concentrations start to change in a region of the muscle tis-

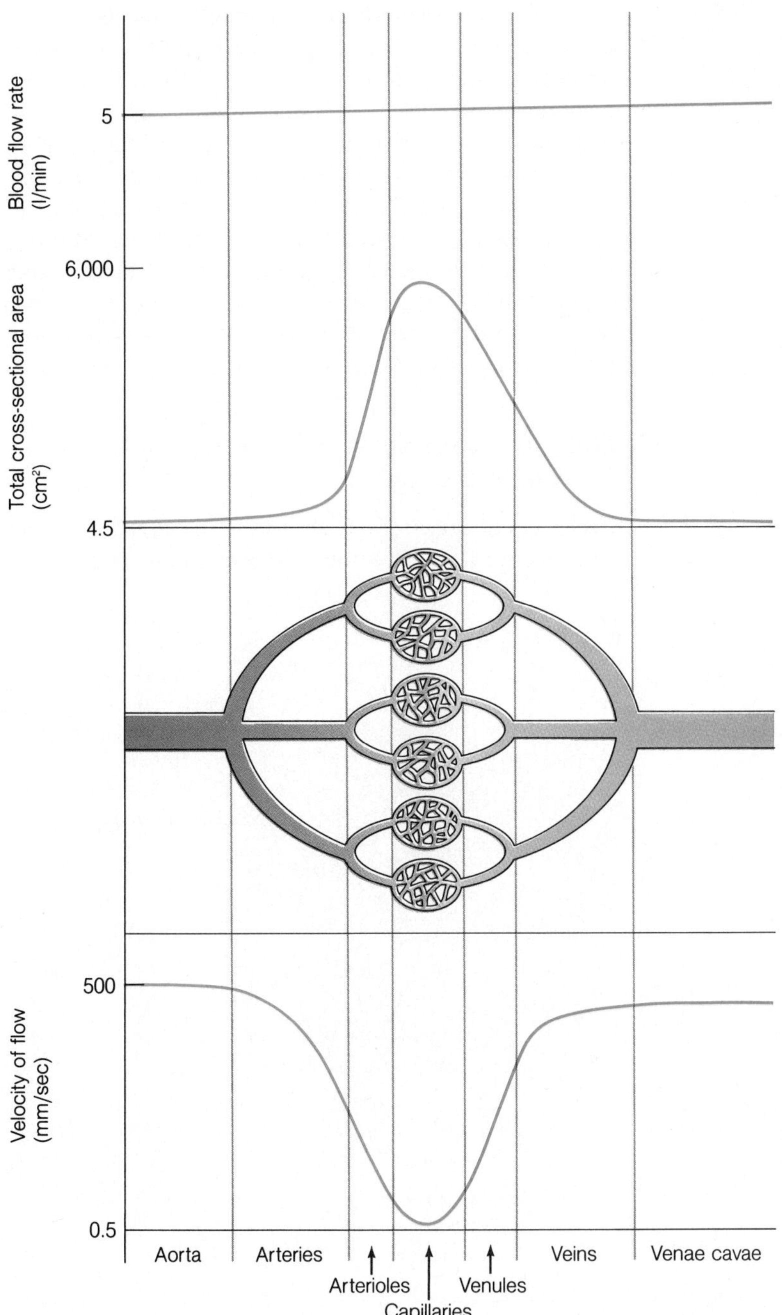

Figure 10–13 Comparison of Blood Flow and Velocity of Flow in Relation to Total Cross-sectional Area

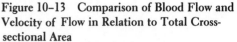

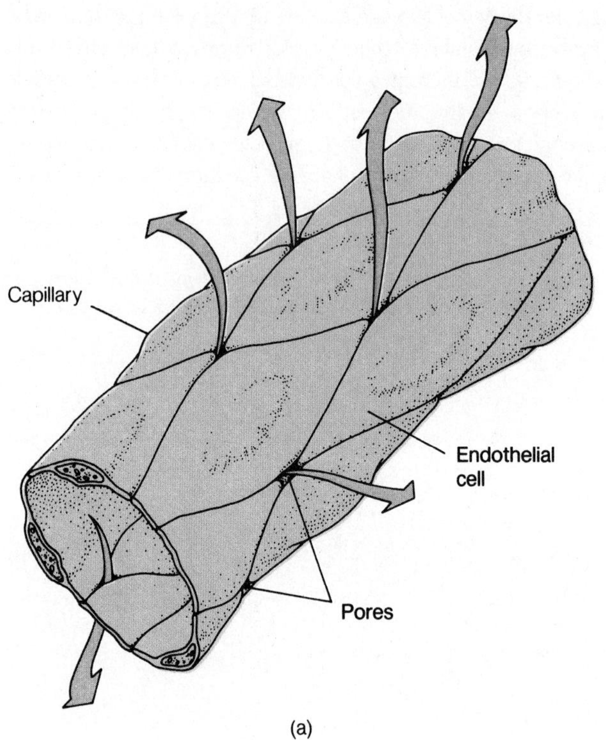

sue supplied by closed-down capillaries, the precapillary sphincters and arterioles in that region relax. Restoration of the chemical concentrations to normal as a result of increased blood flow to that region removes the impetus for vasodilation, so the precapillary sphincters close once again and the arterioles return to normal tone. In this way, blood flow through any given capillary is often intermittent as a result of arteriolar and precapillary sphincter action working in concert. When the muscle as a whole becomes more active, a greater percentage of the precapillary sphincters relax, simultaneously opening up more capillary beds, while concurrent arteriolar vasodilation increases total flow to the organ. As a result of more blood flowing through more open capillaries, the total volume and surface area available for exchange increase, and the diffusion distance between the cells and an open capillary decreases (Fig. 10–16). Thus blood flow through a particular tissue (assuming a constant blood pressure) is regulated by: (1) the degree of resistance offered by the arterioles in the organ, controlled by sympathetic activity and local factors; and (2) the number of open capillaries, controlled by action of the same local metabolic factors on precapillary sphincters.

Diffusion across the capillary wall is important in solute exchange.

Exchanges are not made directly between blood and the tissue cells. Interstitial fluid, the true internal environment in imme-

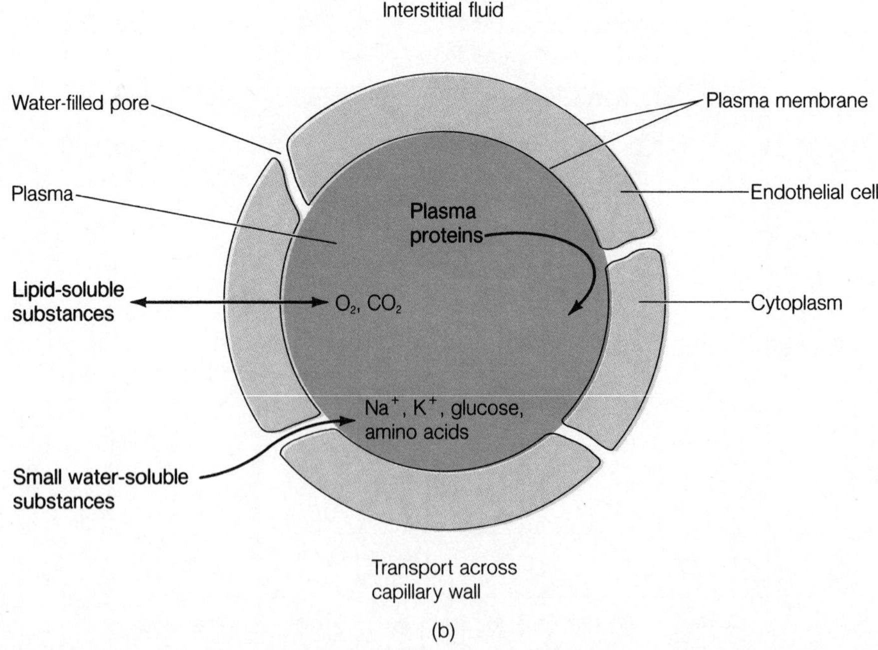

Figure 10–14 Exchanges across Capillary Wall *(a) Slitlike gaps between adjacent endothelial cells form pores within the capillary wall. (b) As depicted in this schematic representation of a cross-section of a capillary wall, small water-soluble substances are exchanged between the plasma and interstitial fluid by passing through the water-filled pores, whereas lipid-soluble substances are exchanged across the capillary wall by passing through the endothelial cells. The plasma proteins generally cannot escape from the plasma across the capillary wall.*

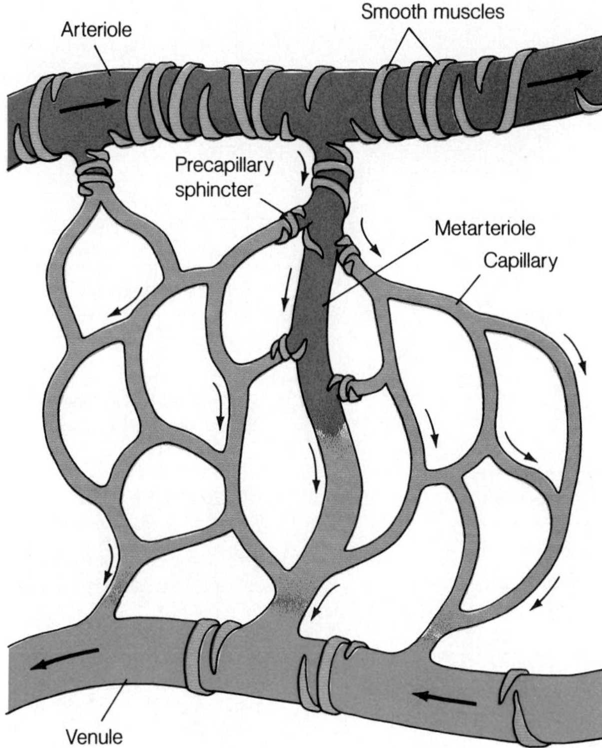

Figure 10–15 Capillary Bed

diate contact with the cells, acts as the go-between (Fig. 10–17). Only 25% of the ECF circulates as plasma. The remaining 75% consists of interstitial fluid, which bathes all of the cells in the body. Cells exchange materials directly with the interstitial fluid, the type and extent of exchange being governed by the properties of the cellular plasma membranes. Movement across the plasma membrane may involve either passive diffusion or carrier-mediated transport. In contrast, only passive exchanges occur across the capillary wall between the plasma and interstitial fluid. Because of the permeability of the capillary walls, exchange is so thorough that the interstitial fluid takes on essentially the same composition as the incoming arterial blood, with the exception of the large plasma proteins that cannot escape from the blood. Therefore, when we speak of exchanges between blood and tissue cells, we tacitly include interstitial fluid as a passive intermediary.

Exchanges between blood and surrounding tissues across the capillary walls are accomplished by: (1) passive diffusion down concentration gradients, the primary mechanism for exchange of individual solutes; and (2) bulk flow, a process that accomplishes the totally different function of determining the distribution of the ECF volume between the vascular and interstitial-fluid compartments. We will examine each of these mechanisms in more detail, starting with diffusion.

Because there are no carrier-mediated transport systems in capillary walls, solutes cross primarily by diffusion down concentration gradients. The chemical composition of arterial blood is carefully regulated to maintain the concentrations of individual solutes at levels that will promote each solute's movement in the appropriate direction across the capillary walls. The reconditioning organs primarily contribute to this homeostatic process—continuously adding nutrients and O_2 and removing CO_2 and other wastes—as blood passes through them. Meanwhile, cells are constantly using up supplies and generating metabolic wastes. Diffusion of each solute continues independently until there is no longer a concentration difference for it between the blood and surrounding cells (Fig. 10–18). This process repeats itself in an ongoing continuum. As cells use up O_2 and glucose, the blood constantly brings in fresh supplies of these vital materials, maintaining concentration gradients that favor the net diffusion of these substances from blood to cells. Simultaneously, ongoing net diffusion of CO_2 and other metabolic wastes from cells to blood is maintained by the continual production of these wastes at the cellular level and their constant removal from the tissue level by the circulating blood.

Because the capillary wall does not limit the passage of any constituent except plasma proteins, the extent of exchanges for each solute is independently determined by the magnitude of the concentration gradient that exists for it between blood and surrounding tissues. As cells increase their level of activity, they use up more O_2 and produce more CO_2, among other things. This creates larger concentration gradients for O_2 and CO_2 between cells and blood, so more O_2 diffuses out of the blood into the cells and more CO_2 proceeds in the opposite direction to help support the increased metabolic activity.

Bulk flow across the capillary wall is important in extracellular-fluid distribution.

The second means by which exchange is accomplished across capillary walls is bulk flow. A volume of protein-free plasma actually filters out of the capillary, mixes with the surrounding interstitial fluid, and is subsequently reabsorbed. This process is called **bulk flow** because the various constituents of the fluid are moving together in bulk or as a unit, in contrast to the discrete diffusion of individual solutes down concentration gradients. The capillary wall acts like a sieve because of its water-filled pores. When pressure inside the capillary exceeds pressure on the outside, fluid is pushed out through the pores,

Figure 10–16 Complementary Action of Precapillary Sphincters and Arterioles in Adjusting Blood Flow through a Tissue in Response to Changing Metabolic Needs

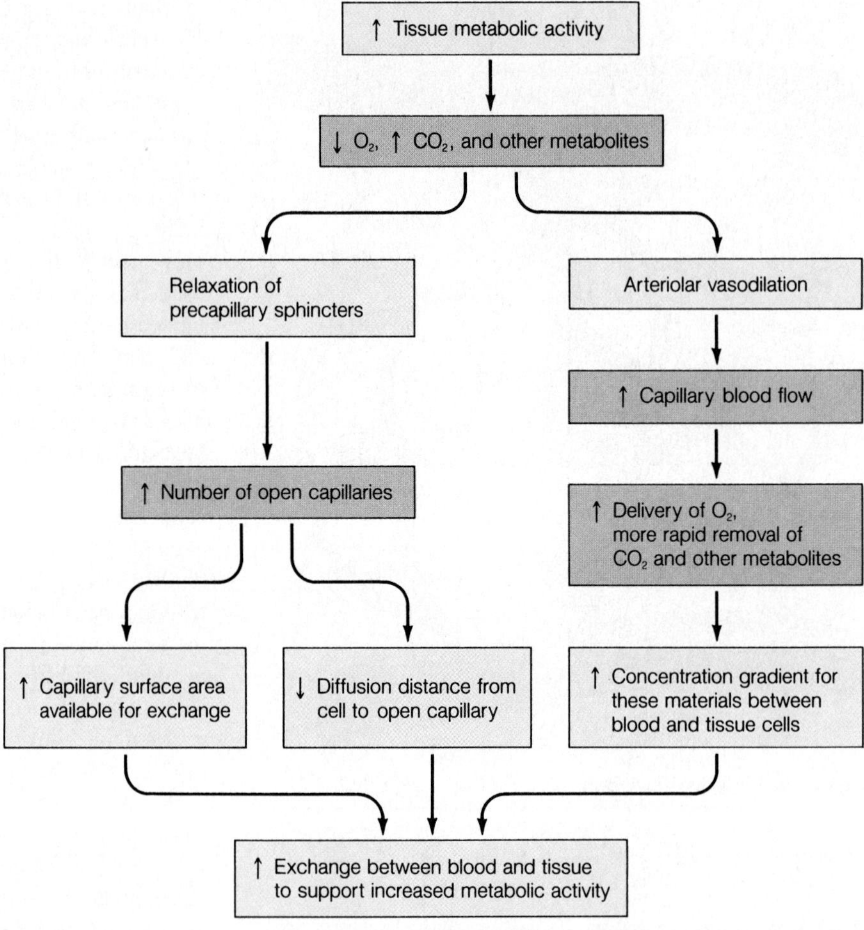

a process known as **ultrafiltration.** The majority of the plasma proteins are retained on the inside during this process because of the pores' filtering effect, although a few do escape. Since all other constituents in the plasma are dragged along as a unit with the volume of fluid leaving the capillary, the filtrate is essentially a protein-free plasma. When inward-driving pressures exceed outward pressures across the capillary wall, net inward movement of fluid from the interstitial-fluid compartment into the capillaries takes place through the pores, a process known as **reabsorption.**

Bulk flow occurs because of differences in the hydrostatic and colloid osmotic pressures between the plasma and interstitial fluid. Even though pressure differences exist between plasma and surrounding fluid elsewhere in the circulatory system, the capillaries are the only vessels that are thin enough and that have pores to allow fluids to pass through them. The four forces that influence fluid movement across the capillary wall are as follows (Fig.10–19):

1. **Capillary blood pressure (CBP)** is the fluid or hydrostatic pressure exerted on the inside of the capillary walls by the blood. This pressure tends to force fluid out of the capillaries into the interstitital fluid. Mean blood pressure has dropped substantially by the level of the capillaries because of frictional losses in pressure in the high-resistance arterioles upstream. On the average, the hydrostatic pressure is 37 mm Hg at the arteriolar end of a tissue capillary and has declined even further to 17 mm Hg at the venular end.

2. **Blood-colloid osmotic pressure (BCOP),** also known as **oncotic pressure,** is a force caused by the colloidal suspension (see p. A–7) of plasma proteins that encourages fluid movement into the capillaries. Since the plasma proteins remain in the plasma rather than entering the interstitial fluid, a protein-concentration difference exists between the plasma and interstitial fluid. Accordingly, there is also a water-concentration difference between these two re-

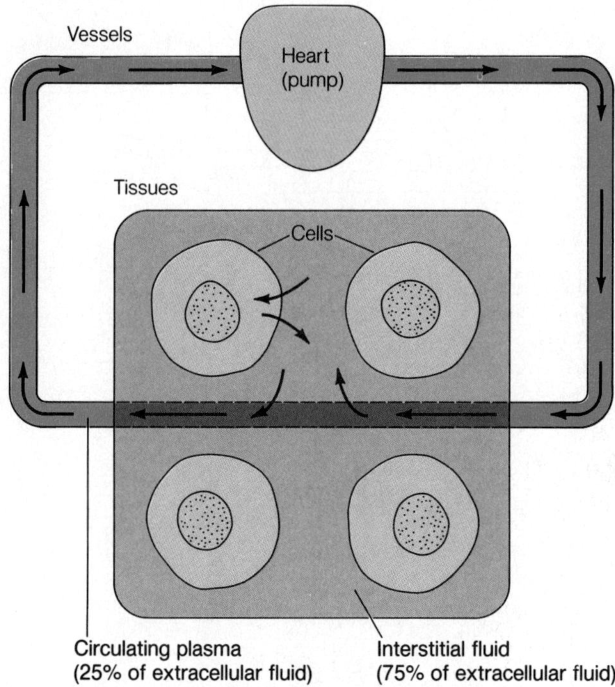

Figure 10–17 Interstitial Fluid Acting as Intermediary between Blood and Cells

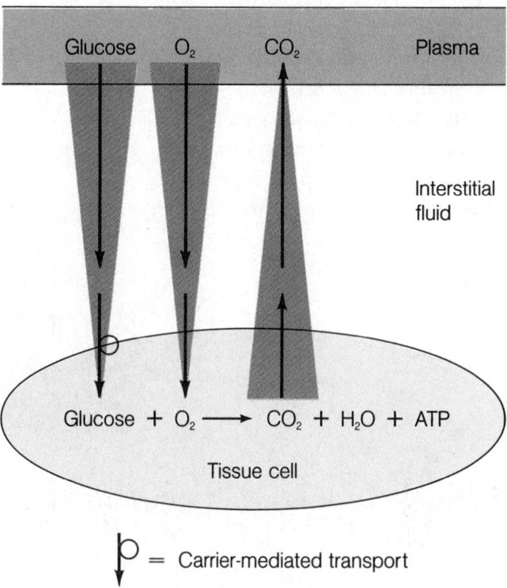

Figure 10–18 Independent Exchange of Individual Solutes Down Their Own Concentration Gradients Across the Capillary Wall

gions. The plasma has a higher protein concentration and a lower water concentration than does the interstitial fluid. This difference exerts an osmotic effect (see p. 73) that tends to move water from the area of higher water concentration in the interstitial fluid to the area of lower water concentration (or higher protein concentration) in the plasma. Thus, the plasma proteins may be thought of as "attracting" water, although this is not actually the underlying force involved. The other plasma constituents do not exert an osmotic effect because they readily pass through the capillary wall, so their concentrations are equal in the plasma and interstitial fluid. The blood-colloid osmotic pressure averages 25 mm Hg.

3. **Interstitial-fluid hydrostatic pressure** (**IFHP**) is the fluid pressure exerted on the outside of the capillary wall by the interstitial fluid. This pressure tends to force fluid into the capillaries. Because of the difficulties encountered in measuring interstitial-fluid hydrostatic pressure, the actual value of this pressure is a controversial issue. It is either at, slightly above, or slightly below atmospheric pressure. For purposes of illustration, we will say it is 1 mm Hg.

4. **Interstitial-fluid-colloid osmotic pressure** (**IFCOP**) is a force that does not normally contribute to bulk flow. The small fraction of plasma proteins that leak across the capillary walls into the interstitial spaces are normally returned to the blood by means of the lymphatic system. Therefore, the protein concentration in the interstitial fluid is extremely low, and the interstitial-fluid-colloid osmotic pressure is essentially zero. If plasma proteins pathologically leak into the interstitial fluid, however, as they do when histamine widens the intercellular clefts, the proteins exert an osmotic effect that tends to promote the movement of fluid out of the capillaries into the interstitial fluid.

Therefore, the two pressures that tend to force fluid out of the capillary are capillary blood pressure and interstitial-fluid-colloid osmotic pressure. The two opposing pressures that tend to force fluid into the capillary are blood-colloid osmotic pressure and interstitial-fluid hydrostatic pressure. We are now prepared to analyze the fluid movement that occurs across a capillary wall because of imbalances in these opposing physical forces (Fig. 10–19). At the arteriolar end of the capillary, the outward pressures total 37 mm Hg [CBP (37 mm Hg) + IFCOP (0 mm Hg)], whereas the inward pressures total 26 mm Hg [BCOP (25 mm Hg) + IFHP (1 mm Hg)], for a net outward pressure of 11 mm Hg. Ultrafiltration takes place at the beginning of the capillary as this outward pressure gradient forces a protein-free filtrate through the capillary pores.

Figure 10–19 Bulk Flow across Capillary Wall *(a) Schematic representation of ultrafiltration and reabsorption as a result of imbalances in the forces acting across the capillary wall. (b) Calculation of ultrafiltration pressure at the arteriolar end of the capillary and reabsorption pressure at venular end.*

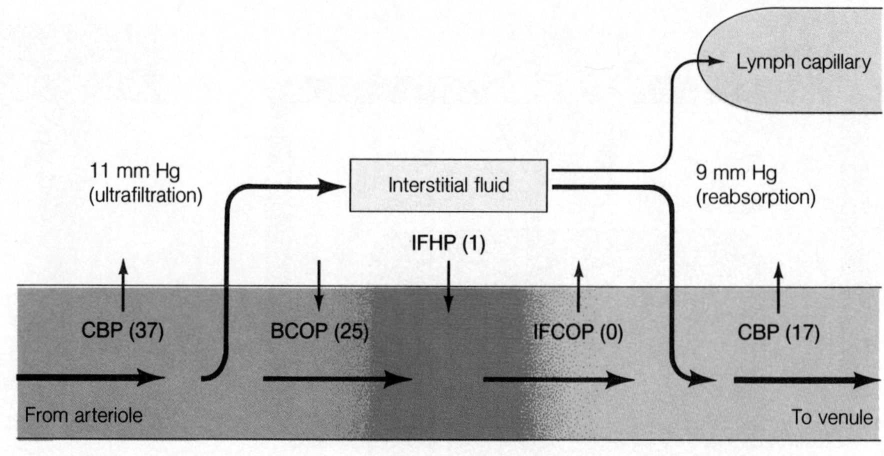

(a)

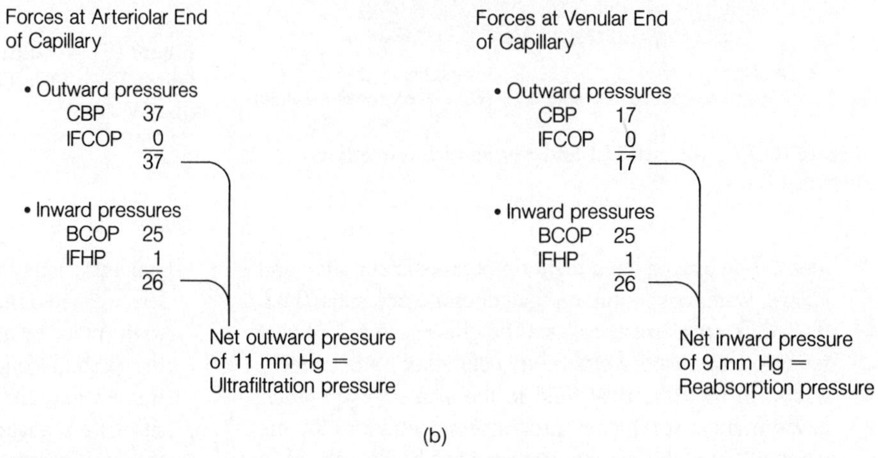

(b)

All values are given in mm Hg.

By the time the venular end of the capillary is reached, the capillary blood pressure has dropped but the other pressures have remained essentially constant. At this point the outward pressures have fallen to a total of 17 mm Hg [CBP (17 mm Hg) + IFCOP (0 mm Hg)], whereas the total inward pressure is still 26 mm Hg, for a net inward pressure of 9 mm Hg. Reabsorption of fluid takes place as this inward pressure gradient forces fluid back into the capillary at its venular end. Ultrafiltration and reabsorption, collectively known as bulk flow, are thus due to a shift in balance between the passive physical forces acting across the capillary wall. No active forces or local energy expenditures are involved in the bulk exchange of fluid between the plasma and surrounding interstitial fluid. With only minor contributions from the interstitial fluid, ultrafiltration occurs at the beginning of the capillary because capillary blood pressure exceeds blood-colloid osmotic pressure, whereas by the end of the capillary, reabsorption takes place because blood pressure has fallen below osmotic pressure.

It is important to realize that we have taken "snapshots" at two points in a hypothetical capillary—one point at the beginning and one point at the end. Actually, blood pressure gradually diminishes along the length of the capillary so that progressively diminishing quantities of fluid are filtered out in the first half of the vessel and progressively increasing quantities of fluid are reabsorbed in the last half (Fig. 10–20). Even this situation is idealized. The pressures used in this figure are average values and controversial at that. Some capillaries have such high blood pressure that filtration actually occurs

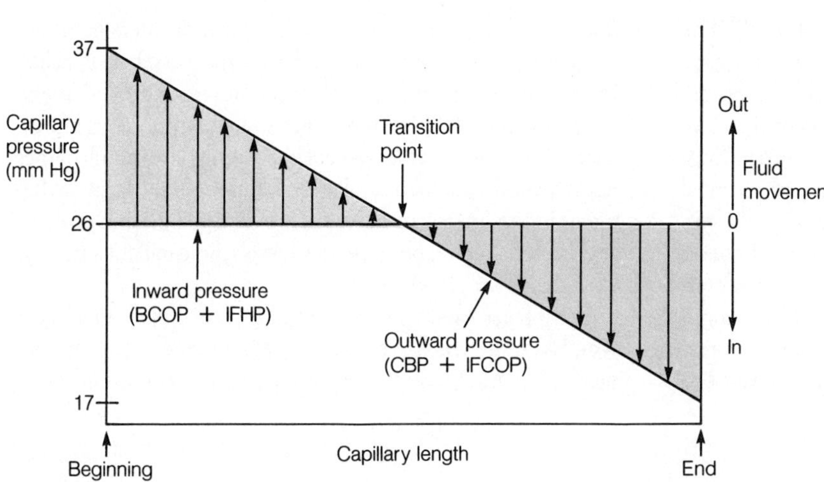

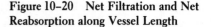

Figure 10–20 Net Filtration and Net Reabsorption along Vessel Length

throughout their entire length, whereas others have such low hydrostatic pressure that reabsorption takes place throughout their length. In fact, a recent theory that has received considerable attention is that net filtration occurs throughout the length of all ***open*** capillaries whereas net reabsorption occurs throughout the length of all ***closed*** capillaries. According to this theory, when the precapillary sphincter is relaxed, capillary blood pressure exceeds the blood osmotic pressure even at the venular end of the capillary, promoting filtration throughout the length. When the precapillary sphincter is closed, the reduction in blood flow through the capillary diminishes capillary blood pressure below the blood osmotic pressure even at the beginning of the capillary, so reabsorption takes place all along the capillary. Whichever mechanism is involved, the net effect is the same. A protein-free filtrate exits the capillaries and is ultimately reabsorbed.

Bulk flow does not play an important role in the exchange of individual solutes between blood and tissues, because the quantity of solutes moved across the capillary wall by bulk flow is extremely small compared to the much larger transfer of solutes by diffusion. If ultrafiltration and reabsorption are not important in the exchange of nutrients and wastes, then of what significance is bulk fluid exchange across the capillary wall? The answer is that it plays an extremely important role in regulating the distribution of ECF between the vascular and interstitial-fluid compartments. Maintenance of proper arterial blood pressure depends in part on an appropriate volume of circulating blood. If plasma volume is reduced (for example, by hemorrhage), blood pressure falls. The resultant lowering of capillary blood pressure alters the balance of forces across the capillary walls. Because the net outward pressure is decreased while the net inward pressure remains unchanged, extra fluid is shifted from the interstitial compartment into the plasma as a result of reduced filtration and increased reabsorption. The extra fluid soaked up from the interstitial fluid provides additional fluid for the plasma to temporarily compensate for the loss of blood. Meanwhile, reflex mechanisms acting on the heart and blood vessels (to be described later) also come into play to help maintain blood pressure until long-term mechanisms, such as thirst and reduction of urinary output, can restore the fluid volume to completely compensate for the loss.

Conversely, if the plasma volume becomes overexpanded, as with excessive fluid intake, the resultant elevation in capillary blood pressure forces extra fluid from the capillaries into the interstitial fluid. This temporarily relieves the expanded plasma volume until the excess fluid can be eliminated from the body by long-term measures.

These internal fluid shifts between the two ECF compartments, which occur automatically and immediately whenever the balance of forces acting across the capillary walls is changed, provide a temporary mechanism to help keep the plasma volume fairly constant. In the process of restoring the plasma volume to an appropriate level, the interstitial-fluid volume fluctuates, but it is much more important that the plasma volume be maintained at a constant level to ensure that the circulatory system functions effectively.

The lymphatic system is an accessory route by which interstitial fluid can be returned to the blood.

Even under normal circumstances, slightly more fluid is filtered out of the capillaries into the interstitial fluid than is reabsorbed from the interstitial fluid back into the plasma. The average net ultrafiltration pressure is 11 mm Hg, whereas the net reabsorption pressure is only 9 mm Hg (Fig. 10–19). The extra fluid filtered out as a result of this filtration-reabsorption imbalance is picked up by the **lymphatic system.** This system comprises an extensive network of one-way vessels that provide an accessory route by which fluid can be returned from the interstitial fluid to the blood. Small, blind-ended terminal lymph vessels (**lymphatic capillaries**) permeate almost every tissue of the body (Fig. 10–21a). The endothelial cells forming the walls of lymphatic capillaries are joined only near the ends of the cells, whereas their free edges overlap. This arrangement creates valvelike openings in the vessel wall. Fluid pressure on the outside of the vessel pushes the edges inward so that the valves open and permit entrance of interstitial fluid. Once interstitial fluid enters a lymphatic vessel, it is called **lymph.** Fluid pressure on the inside forces the overlapping edges together. Closing the valves so that lymph does not escape. These lymphatic valvelike openings are much larger than the pores in blood capillaries. Consequently, large particulates in the interstitial fluid, such as escaped plasma proteins and bacteria, can gain access to lymphatic capillaries but are excluded from blood capillaries.

Lymphatic capillaries converge to form larger and larger lymph vessels, which eventually empty into the venous system near the point where the blood enters the right atrium (Fig.

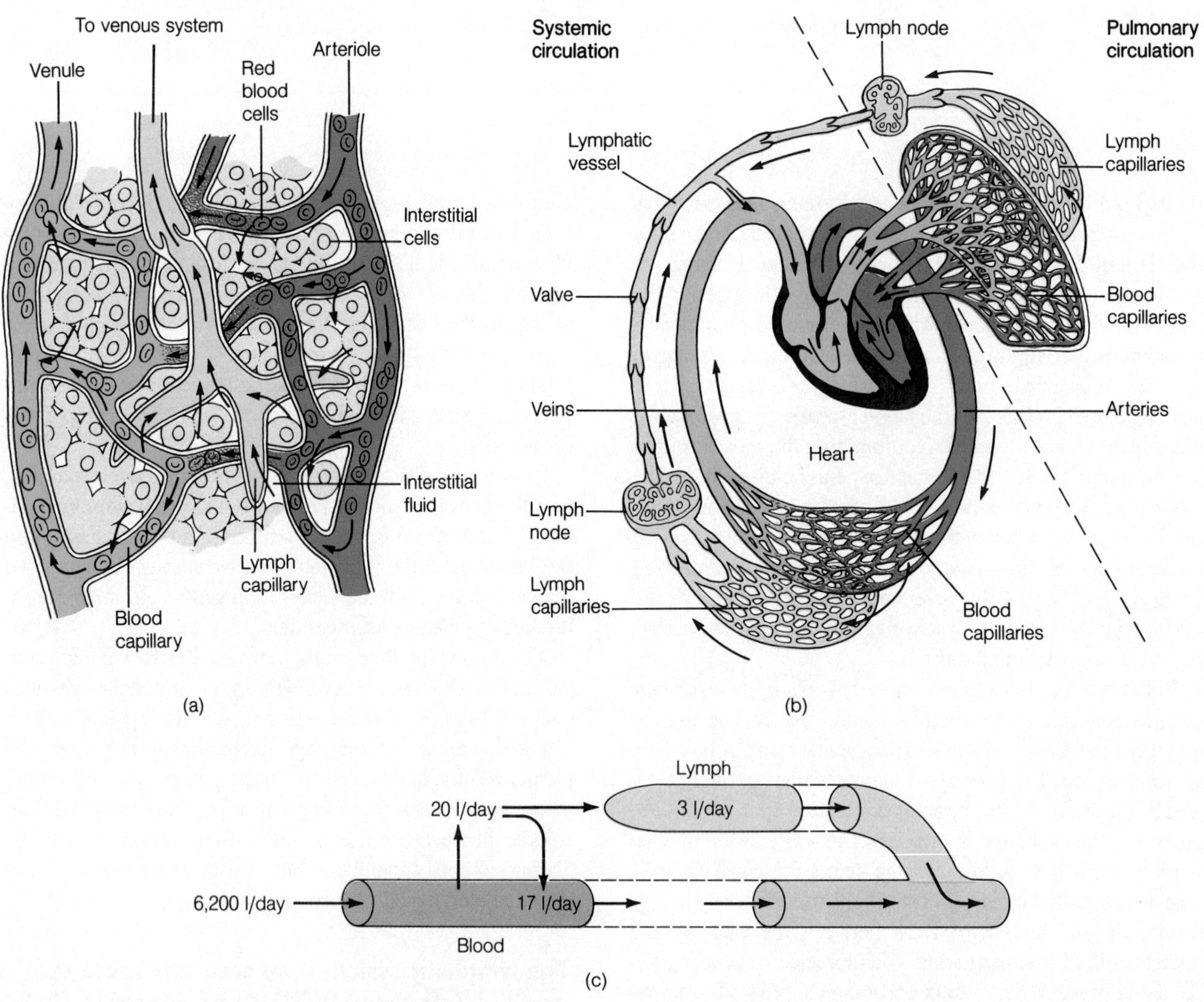

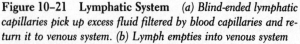

Figure 10–21 Lymphatic System *(a) Blind-ended lymphatic capillaries pick up excess fluid filtered by blood capillaries and return it to venous system. (b) Lymph empties into venous system near its entrance to right atrium. (c) Lymph flow averages 3 l/day compared with 6,200 l/day blood flow.*

10–21b). Because there is no "lymphatic heart" equivalent to the heart in the circulatory system to provide driving pressure, how is lymph directed from the tissues toward the venous system in the thoracic cavity? Lymph flow is accomplished by two mechanisms. First, since lymph vessels lie between skeletal muscles, contraction of these muscles squeezes the lymph out of the vessels. One-way valves spaced at intervals within the lymph vessels direct the flow of lymph toward its venous outlet in the chest. Second, lymph vessels beyond the lymphatic capillaries are surrounded by smooth muscle. When this muscle is stretched because the vessel is distended with lymph, the muscle myogenically contracts more forcefully, thereby propelling the lymph through the vessel.

Several important functions that are accomplished by the lymphatic system will be discussed next.

RETURN OF EXCESS FILTERED FLUID. Normally, capillary filtration exceeds reabsorption by about 3 liters per day (20 liters filtered, 17 liters reabsorbed) (Fig. 10–21c). Yet the entire blood volume is only 5 liters, and only 2.75 liters of that is plasma. (Blood cells make up the remainder of the blood volume.) With an average cardiac output, 6,200 liters of blood pass through the capillaries daily under resting conditions (more when cardiac output increases). Even though only a small fraction of the filtered fluid is not reabsorbed by the blood capillaries, the cumulative effect of this process being repeated with every heart beat results in the equivalent of more than the entire plasma volume being left behind in the interstitial fluid each day. Obviously, this fluid must be returned to the circulating plasma, which is accomplished by the lymphatic vessels. The average rate of flow through the lymph vessels is 3 liters per day, compared with 6,200 liters per day through the circulatory system.

DEFENSE AGAINST DISEASE. The lymph percolates through **lymph nodes** located en route within the lymphatic system. Passage of this fluid through the lymph nodes is an important aspect of the body's defense mechanism against disease, which functions in conjunction with the white blood cells. For example, bacteria picked up from the interstitial fluid are destroyed by special phagocytic cells located within the lymph nodes.

TRANSPORT OF ABSORBED FAT. The lymphatic system is important in the absorption of fat from the digestive tract. The end-products of the digestion of dietary fats are packaged by cells lining the digestive tract into fatty particles that are too large to gain access to the blood capillaries but that can easily enter the terminal lymphatic vessels.

RETURN OF FILTERED PROTEIN. Most capillaries permit leakage of some plasma proteins during filtration. These proteins cannot readily be reabsorbed back into the blood capillaries but they can easily gain access into the lymphatic capillaries. If the proteins were allowed to accumulate in the interstitial fluid rather than being returned to the circulation via the lymphatics, the interstitial-fluid-colloid osmotic pressure (an outward pressure) would progressively increase while the blood-colloid osmotic pressure (an inward pressure) would progressively fall. As a result, filtration forces would gradually increase and reabsorption forces would gradually decrease, resulting in progressive accumulation of fluid in the interstitial spaces at the expense of loss of plasma volume.

Edema occurs when too much interstitial fluid accumulates.

Occasionally, excessive interstitial fluid does accumulate when one of the physical forces acting across the capillary walls becomes abnormal for some reason. Swelling of the tissues because of excess interstitial fluid is known as **edema.** The causes of edema can be grouped into four general categories, as follows:

1. A *reduced concentration of plasma proteins* causes a decrease in blood-colloid osmotic pressure. Such a drop in the major inward pressure allows excess fluid to be filtered out while less-than-normal amounts of fluid are reabsorbed; hence extra fluid remains in the interstitial spaces. Edema caused by decreased concentrations of plasma proteins can arise in several different ways: excessive loss of plasma proteins in the urine caused by kidney disease; reduced synthesis of plasma proteins as a consequence of liver disease (the liver synthesizes almost all plasma proteins); a dietary deficiency of protein, which is needed to supply the amino acids for synthesis of plasma proteins; or significant loss of plasma proteins from large burned surfaces.
2. *Increased permeability of the capillary walls* allows more plasma proteins than usual to pass from the plasma into the surrounding interstitial fluid—for example, via histamine-induced widening of the capillary pores. The resultant fall in blood-colloid osmotic pressure decreases the effective inward pressure while the resultant rise in interstitial-fluid-colloid osmotic pressure caused by the abnormal presence of protein in the interstitial fluid increases the effective outward force. This imbalance contributes in part to the localized edema associated with injuries and allergic responses.

3. *Increased venous pressure*, as when blood dams up in the veins, is accompanied by an increased capillary blood pressure, since the capillaries drain into the veins. This elevation in outward pressure across the capillary walls is largely responsible for the edema seen with congestive heart failure (see p. 287). Regional edema can also occur because of localized restriction of venous return. An example is the swelling often occurring in the legs and feet during pregnancy. The enlarged uterus compresses the major veins that drain the lower extremities as these vessels enter the abdominal cavity. The resultant damming of blood in these veins causes a rise in blood pressure in the capillaries of the legs and feet, which promotes regional edema of the lower extremities.
4. *Blockage of lymph vessels* produces edema because the excess filtered fluid is retained in the interstitial fluid rather than being returned to the blood through the lymphatics. The protein accumulation in the interstitial fluid compounds the problem through its osmotic effect. Local lymph blockage can occur, for example, in the arms of women whose major lymphatic-drainage channels from the arm have been blocked as a result of lymph-node removal during surgery for breast cancer. More widespread lymph blockage occurs with filariasis, a mosquito-borne parasitic disease that is found predominantly in tropical coastal regions. In this condition, small, threadlike filaria worms infect the lymph vessels, where their presence prevents proper lymph drainage. The affected body parts, particularly the scrotum and extremities, become grossly edematous. The condition is often called elephantiasis because of the elephantlike appearance of the swollen extremities.

Whatever the cause of edema, an important consequence is a reduction in exchange of materials between the blood and cells. As excess interstitial fluid accumulates, the distance between the blood and cells across which nutrients, O_2, and wastes must diffuse is increased, so the rate of diffusion decreases. Therefore, cells within edematous tissues may not be adequately supplied.

VEINS

Veins serve as a blood reservoir as well as passageways back to the heart.

The venous system completes the circulatory circuit. Blood leaving the capillary beds enters the venous system for transport back to the heart. Veins have large radii, so they offer little resistance to flow. Furthermore, because the total cross-sectional area of the venous system gradually decreases as smaller veins converge into progressively fewer but larger vessels, the velocity of blood flow increases as the blood approaches the heart.

In addition to serving as low-resistance conduits for return of blood from the tissues to the heart, systemic veins also serve as a blood reservoir. Because of their storage capacity, veins are often referred to as ***capacitance vessels***. Veins have much thinner walls with sparser amounts of smooth muscle than do arteries. Since collagen fibers are considerably more abundant than elastin fibers in venous connective tissue, veins have very little elasticity, in contrast to arteries. Also, unlike arteriolar smooth muscle, venous smooth muscle has little inherent myogenic tone. Because of these features, veins are highly distensible, or stretchable, and have little elastic recoil. They easily distend to accommodate additional volumes of blood with only a small increase in venous pressure. Arteries stretched by an excess volume of blood recoil because of the elastic fibers in their walls, driving the blood forward. Veins containing an extra volume of blood simply stretch to accommodate the additional blood without tending to recoil. In this way veins serve as a blood reservoir; that is, when demands for blood are low, the veins can store extra blood in reserve because of their passive distensibility. Under resting conditions, the veins contain more than 60% of the total blood volume (Fig. 10–22). When the stored blood is needed, such as during exercise, extrinsic factors (soon to be described) restore the veins to their unstretched dimension, driving the extra blood from the veins

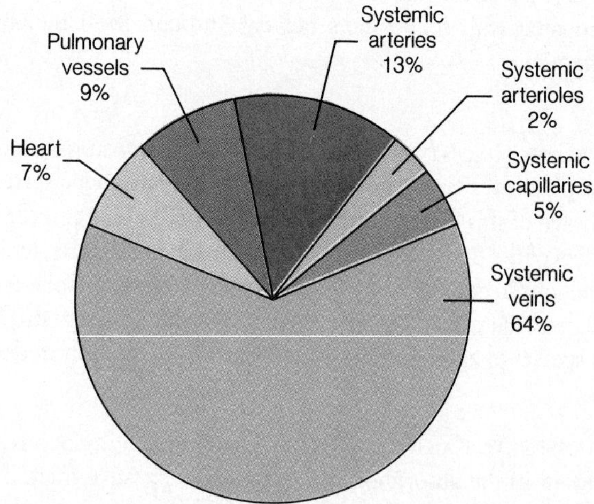

Figure 10–22 Percentage of Total Blood Volume in Different Parts of the Circulatory System

to the heart so that it can be pumped to the tissues. Increased venous return induces an increased cardiac stroke volume as in accordance with the Frank-Starling law of the heart (see p. 286). If too much blood pools in the veins instead of being returned to the heart, cardiac output is abnormally diminished. Thus, a delicate balance exists between the capacity of the veins, the extent of venous return, and the cardiac output. We will now turn our attention to the factors that affect venous capacity and contribute to venous return.

Venous return is enhanced by a number of extrinsic factors.

The **capacity** of the veins (the volume of blood that they can accommodate) depends on the distensibility of their walls (how much they can stretch to hold blood) and the influence of any externally applied pressure squeezing inwardly on the veins. At a constant blood volume, as venous capacity increases, more blood remains in the veins than is being returned to the heart. Such venous storage decreases the effective circulating volume. Conversely, when venous capacity decreases, more blood is returned to the heart and continues circulating. Thus, changes in venous capacity directly influence the magnitude of venous return, which in turn is an important (although not the only) determinant of effective circulating blood volume. The magnitude of the total blood volume is also influenced by thirst and urinary output, which are factors that control total ECF volume, and by passive shifts in bulk flow between the vascular and interstitial fluid compartments.

Venous return refers to the volume of blood entering each atrium per minute from the veins. Recall that the magnitude of flow through a vessel is directly proportional to the pressure gradient. Much of the driving pressure imparted to the blood by cardiac contraction has dissipated by the time the blood reaches the venous system because of frictional losses along the way, especially during passage through the high-resistance arterioles. By the time the blood enters the venous system, mean pressure averages only 17 mm Hg (Fig. 10–7). However, since atrial pressure is near 0 mm Hg, this provides a small but adequate driving pressure to promote the flow of blood through the large-radii, low-resistance veins. If atrial pressure becomes pathologically elevated, as in the presence of a leaky AV valve, the venous-to-atrium pressure gradient is decreased, so venous return is reduced, thus causing blood to dam up in the venous system (congestive heart failure).

In addition to the driving pressure imparted by cardiac contraction, five other factors enhance venous return: sympathetically induced venous vasoconstriction, skeletal-muscle activity, the effect of venous valves, respiratory activity, and cardiac-suction effect (Fig. 10–23). Most of these secondary factors affect venous return by influencing the pressure gradient between the veins and the heart. We will examine each in turn.

EFFECT OF SYMPATHETIC ACTIVITY ON VENOUS RETURN. Veins are not very muscular and have little inherent tone, but venous smooth muscle is abundantly supplied with sympathetic nerve fibers. Sympathetic stimulation produces venous vasoconstriction, which modestly elevates venous pressure; this, in turn, increases the pressure gradient to drive more blood from the veins into the right atrium. The veins normally have such a large diameter that the moderate vasoconstriction accompanying sympathetic stimulation has little effect on resistance to flow. Even when constricted, the veins still have a relatively large diameter and are still low-resistance vessels.

In addition to mobilizing the stored blood, venous vasoconstriction enhances venous return by decreasing venous capacity. With the filling capacity of the veins reduced, less blood draining from the capillaries remains in the veins but continues to flow instead toward the heart.

It is important to recognize the different outcomes of vasoconstriction in arterioles and veins. Arteriolar vasoconstriction *reduces* flow through these vessels because of their increased resistance, whereas venous vasoconstriction *increases* flow through these vessels because of their decreased capacity.

EFFECT OF SKELETAL-MUSCLE ACTIVITY ON VENOUS RETURN. Many of the large veins within the extremities lie between skeletal muscles so that when the muscles contract, the veins are compressed. This external venous compression decreases venous capacity and increases venous pressure, in effect squeezing blood contained within the veins forward toward the heart (Fig. 10–24). This pumping action, known as the **skeletal-muscle pump,** is one way by which extra blood stored in the veins is returned to the heart during exercise. Increased muscular activity pushes more blood out of the veins and into the heart. Increased sympathetic activity and the resultant venous vasoconstriction also accompany exercise, further enhancing venous return.

The skeletal-muscle pump also counters the effect of gravity on the venous system. The average pressures provided thus far for various regions of the vascular tree are for a person in the horizontal position. When a person is lying down, the force of gravity is uniformly applied, so it does not have to be taken into consideration. When a person stands up, however, gravitational effects are not uniform. In addition to the usual pressure that results from cardiac contraction, vessels below the level of the heart are subjected to pressure caused by the weight of the column of blood extending from the heart to the

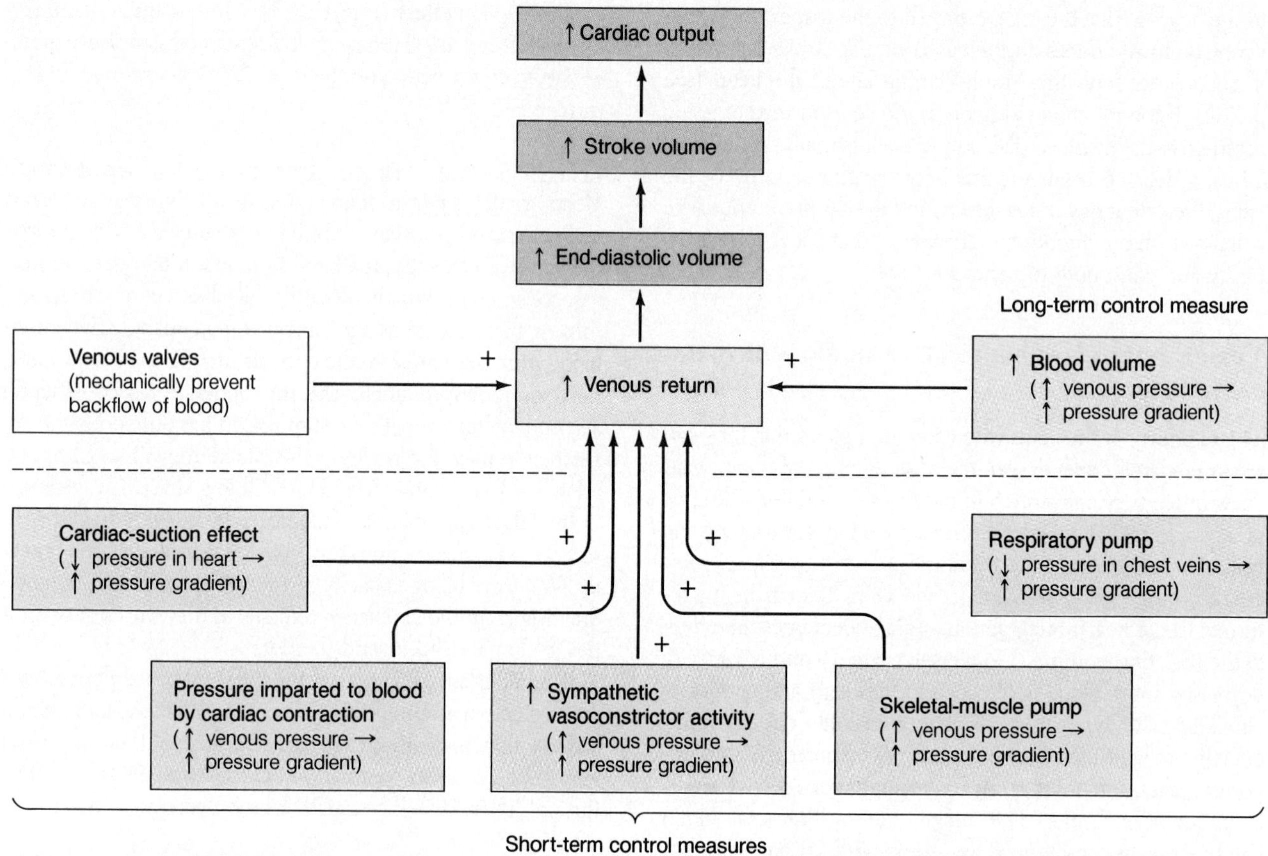

Figure 10–23 Short-Term and Long-Term Control Measures That Increase Venous Return

level of the vessel (Fig. 10–25a). There are two important consequences of this increased pressure (Fig. 10–25b). First, the marked increase in capillary blood pressure causes excessive fluid to filter out of capillary beds in the lower extremities, producing localized edema (that is, swollen feet and ankles). Second, the distensible veins yield under the increased hydrostatic pressure, further expanding so that their capacity is increased. Even though the arteries are subjected to the same gravitational effects, they do not expand like the veins because arteries are not nearly as distensible. Much of the blood entering from the capillaries tends to pool in the expanded lower-leg veins instead of returning to the heart. Because venous return is reduced, cardiac output is decreased and the effective circulating volume is reduced.

Two compensatory measures normally counteract these gravitational effects. First, the resultant fall in mean arterial pressure that occurs when a person moves from a lying-down to upright position triggers sympathetically induced venous vasoconstriction, which drives some of the pooled blood forward. Second, the skeletal-muscle pump "interrupts" the column of blood by completely emptying given vein segments intermittently so that a particular portion of a vein is not subjected to the weight of the entire venous column from the heart to its level (Figs. 10–24 and 10–26). Reflex venous vasoconstriction cannot completely compensate for gravitational effects without the assistance of skeletal-muscle activity. Therefore, when a person stands still for a long time, blood flow to the brain is reduced because of the decline in effective circulating volume, despite reflexes aimed at maintaining mean arterial pressure. Reduced cerebral flow, in turn, leads to fainting, which returns the person to a horizontal position, thereby eliminating the gravitational effects on the vascular system so that effective circulation is restored. For this reason, it is counterproductive to try to hold a fainted person upright. Fainting is the remedy to the problem, not the problem itself.

Since the skeletal-muscle pump facilitates venous return, it

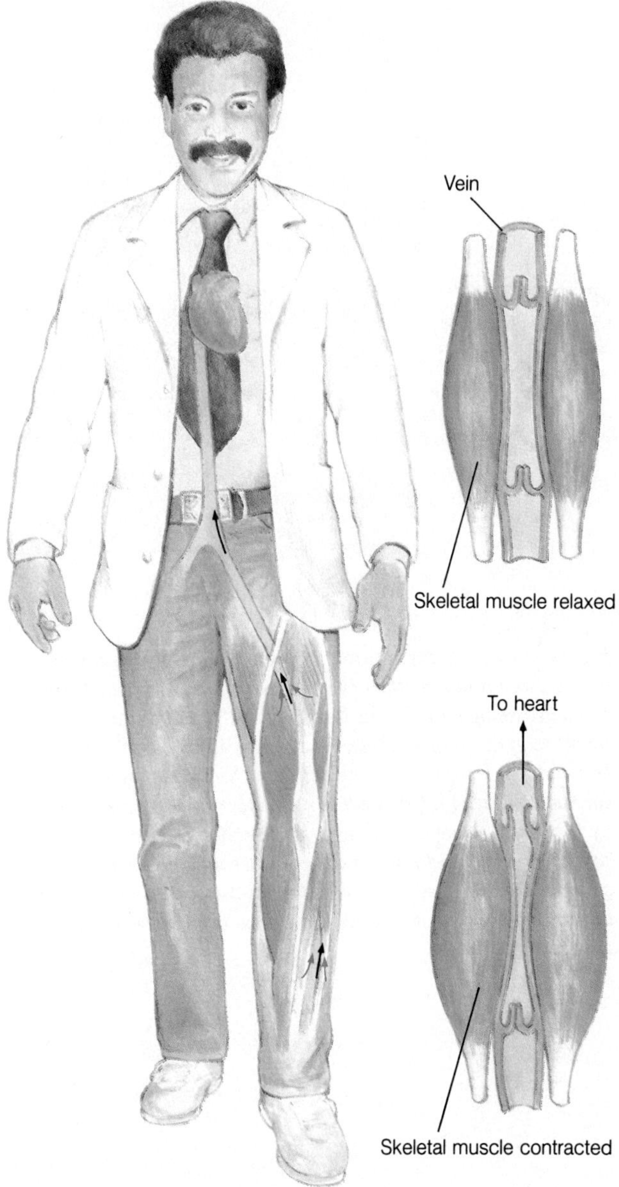

Figure 10–24 Skeletal-Muscle Pump Enhancing Venous Return

is advisable to move around when you are on your feet and to periodically get up when you are working at a desk. The mild muscular activity "gets the blood moving." It is further recommended that individuals who must be on their feet for long periods of time use elastic stockings that apply a continuous gentle external compression, similar to the effect of skeletal-muscle contraction, to further counter the effect of gravitational pooling of blood in the leg veins.

EFFECT OF VENOUS VALVES ON VENOUS RETURN. Venous vasoconstriction and external venous compression both drive blood in the direction of the heart. Yet if you squeeze a fluid-filled tube in the middle, fluid is pushed in both directions from the point of constriction (Fig. 10–27a). Why, then, isn't blood driven backward as well as forward by venous vasoconstriction and the skeletal-muscle pump? The reason is that the large veins are equipped with one-way valves spaced at 2- to 4-cm intervals that permit movement of blood forward toward the heart but prevent its movement back toward the tissues (Fig. 10–27b). These venous valves also play a role in counteracting the gravitational effects of upright posture by helping minimize the backflow of blood that tends to occur when a person stands up and by temporarily supporting portions of the column of blood when the skeletal muscles are relaxed (Fig. 10–27c).

Varicose veins occur when the venous valves become incompetent and can no longer support the column of blood above them. Individuals predisposed to this condition usually have an inherited overdistensibility and weakness of their vein walls. Aggravated by frequent, prolonged standing, the veins become so distended as blood pools in them that the edges of the valves are no longer able to meet to form a seal. Varicosed superficial leg veins become visibly overdistended and tortuous. Contrary to what might be expected, the chronic pooling of blood in the pathologically distended veins does not diminish cardiac output because of a compensatory increase in total circulating blood volume. Instead, the most serious consequence of varicosed veins is the possibility of abnormal clot formation in the sluggish pooled blood. This leads to the threat of clots breaking loose and blocking smaller vessels elsewhere, especially the pulmonary capillaries.

EFFECT OF RESPIRATORY ACTIVITY ON VENOUS RETURN. As a result of respiratory activity, the pressure within the chest cavity is *less than* atmospheric pressure (it averages 755 mm Hg). As the venous system returning blood to the heart from the lower regions of the body traverses the chest cavity, it is exposed to this subatmospheric pressure. Because the venous system in the limbs and abdomen is subjected to normal atmospheric pressure (760 mm Hg), an external pressure gradient exists between the lower veins and the chest veins that promotes increased venous return (Fig. 10–28). This mechanism of facilitating venous return is known as the **respiratory pump.** The increased respiratory activity accompanying exercise enhances venous return, along with the effects of the skeletal-muscle pump and venous vasoconstriction.

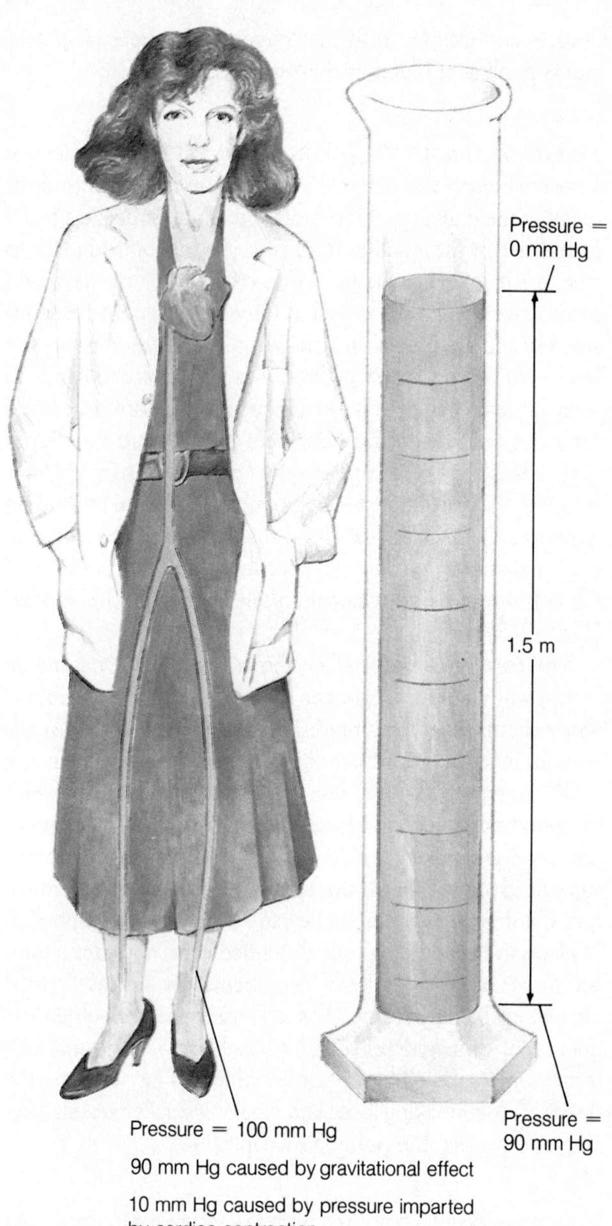

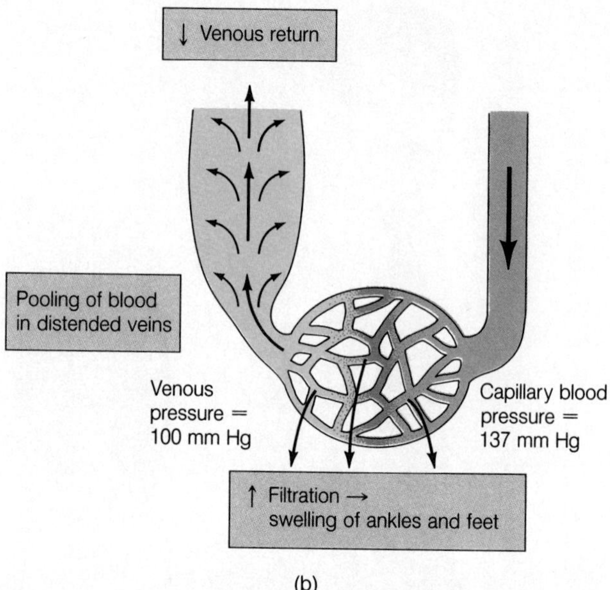

Figure 10–25 Gravitational Effects on the Vasculature *(a) As a result of the effect of gravity on the 1.5 meter column of blood extending between the heart and distal end of the lower extremities in an upright adult, the pressure in the vessels of the ankles and feet is 90 mm Hg greater than the pressure imparted by cardiac contraction. (b) Because of the increased pressure caused by the gravitational effect, increased filtration occurs across the capillary walls, resulting in swollen ankles and feet, and blood pools in the distended veins, resulting in decreased venous return unless compensatory measures are able to counteract the effect of gravity.*

EFFECT OF CARDIAC SUCTION ON VENOUS RETURN. Traditionally, the extent of cardiac filling has been considered to depend entirely on factors directly affecting the veins, with the heart being thought to play a passive role in its own filling. Recent evidence, however, suggests that the heart functions as a suction pump to facilitate venous return, thereby aiding cardiac filling. During cardiac contraction, the AV rings are drawn downward, which enlarges the atrial cavities. As a result, the atrial pressure transiently drops below 0 mm Hg, thus increasing the venous-to-atrial pressure gradient so that venous return is enhanced. In addition, the rapid expansion of the ventricular chamber during cardiac relaxation appears to create a transient negative pressure in the ventricles so that blood is "sucked in" from the atria and veins; that is, the negative ventricular pressure increases the vein-to-atria-to-ventricle pressure gradient, further enhancing venous return.

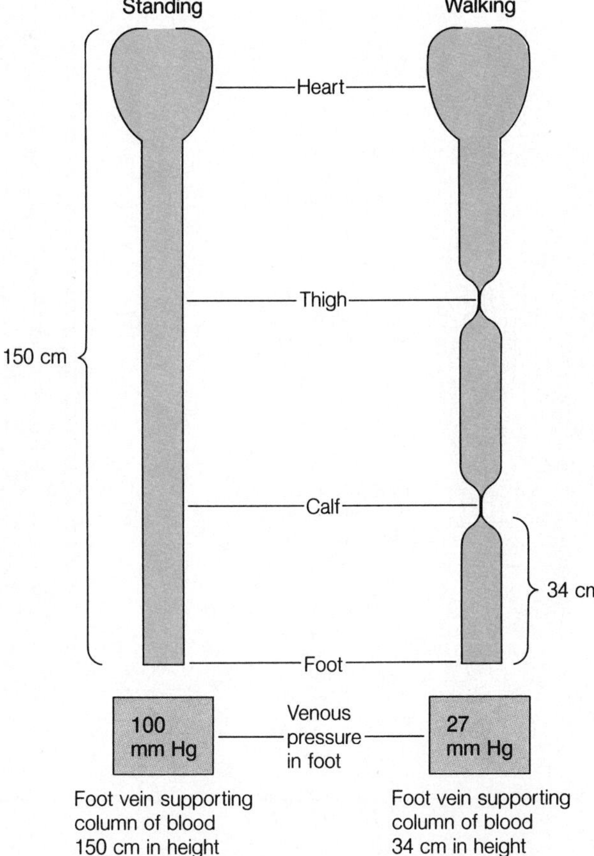

Figure 10–26 Effect of Contraction of the Skeletal Muscles of the Legs in Counteracting the Effects of Gravity *Contraction of skeletal muscles in intermittent walking completely empties given vein segments, interrupting the column of blood that must be supported by lower veins.*

SOURCE: Adapted and reproduced with permission from Little, R.C.: *Physiology of the Heart and Circulation*, 3rd edition. Copyright ©1985 by Year Book Medical Publishers, Inc., Chicago.

Blood Pressure

Regulation of mean arterial blood pressure is accomplished by controlling cardiac output, total peripheral resistance, and blood volume.

Mean arterial blood pressure is the main driving force for propelling blood to the tissues. This pressure must be closely regulated for two reasons. First, it must be high enough to assure sufficient driving pressure, without which the vital brain and heart do not receive adequate flow, no matter what local adjustments are made in the resistance of the arterioles supplying them. Second, the pressure must not be so high that it creates extra work for the heart and increases the risk of vascular damage and possible rupture of small blood vessels.

Elaborate mechanisms involving the integrated action of various components of the cardiovascular system and other body systems are vital in the regulation of this all-important mean arterial pressure (Fig. 10–29). Remember from an earlier discussion that the two determinants of mean arterial pressure are cardiac output and total peripheral resistance:

$$\text{Mean arterial pressure} = \text{Cardiac output} \times \text{Total peripheral resistance}$$

Let us analyze what this relationship means.

□ If cardiac output decreases or increases while total peripheral resistance remains constant, mean arterial pressure will decrease or increase, respectively.

□ Likewise, if total peripheral resistance decreases or increases, mean arterial pressure will do the same if cardiac output remains constant.

□ If total peripheral resistance decreases and mean arterial pressure is to remain constant, cardiac output must increase to compensate for the decreased resistance. The converse is also true. If either total peripheral resistance or cardiac output are altered, the other factor must be changed in the opposite direction to maintain a constant blood pressure.

A number of factors, in turn, determine cardiac output and total peripheral resistance. Thus one can quickly appreciate the complexity of blood-pressure regulation. Altering any of the involved factors will produce a change in blood pressure unless a compensatory change in another variable keeps the blood pressure constant. Blood flow to any given tissue depends on the degree of vasoconstriction of the tissue's arterioles and on the driving force of the mean arterial pressure. Because mean arterial pressure depends on the cardiac output and the degree of arteriolar vasoconstriction, if the arterioles in one tissue dilate, the arterioles in other tissues will have to constrict to maintain an adequate arterial blood pressure to provide a driving force to push blood not only to the vasodilated tissue but also to the vital brain and heart. Thus, the cardiovascular variables must be continuously juggled to maintain a constant blood pressure in spite of tissues' varying needs for blood.

Mean arterial pressure is constantly monitored by **baroreceptors** (pressure sensors) within the circulatory system. When deviations from normal are detected, multiple reflex responses are initiated to return the arterial pressure to its normal value. Short-term (within seconds) adjustments are accomplished by alterations in cardiac output and total periph-

Figure 10–27 Function of Venous Valves *(a) Fluid is pushed in both directions when a tube is squeezed in the middle. (b) Venous valves permit the flow of blood only toward the heart. (c) Venous valves play a role in counteracting gravitational effects by minimizing the backflow of blood and temporarily supporting portions of the column of blood.*

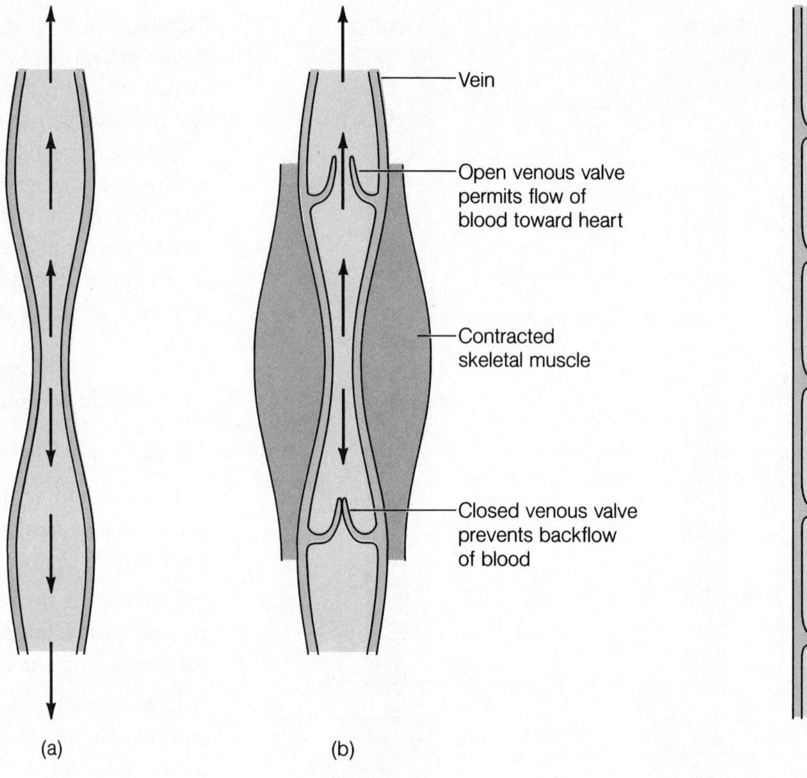

eral resistance mediated by means of autonomic-nervous-system influences on the heart, veins, and arterioles. Long-term (requiring minutes to days) control involves adjustments in total blood volume, the magnitude of which has a profound effect on cardiac output. Let us now turn our attention to the reflex mechanisms involved in the ongoing regulation of this pressure.

The baroreceptor reflex is the most important mechanism for short-term regulation of blood pressure.

Any change in mean blood pressure triggers an autonomically mediated **baroreceptor reflex** that influences the heart and blood vessels to adjust cardiac output and total peripheral re-

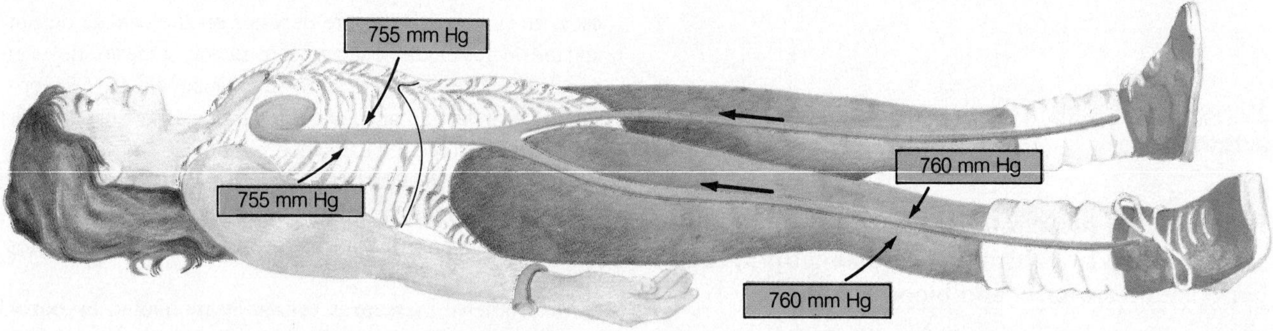

Figure 10–28 Respiratory Pump Enhancing Venous Return *As a result of respiratory activity, the pressure surrounding the chest veins is lower than the pressure surrounding the veins in the extremities and abdomen. This establishes a pressure gradient that drives blood toward the heart.*

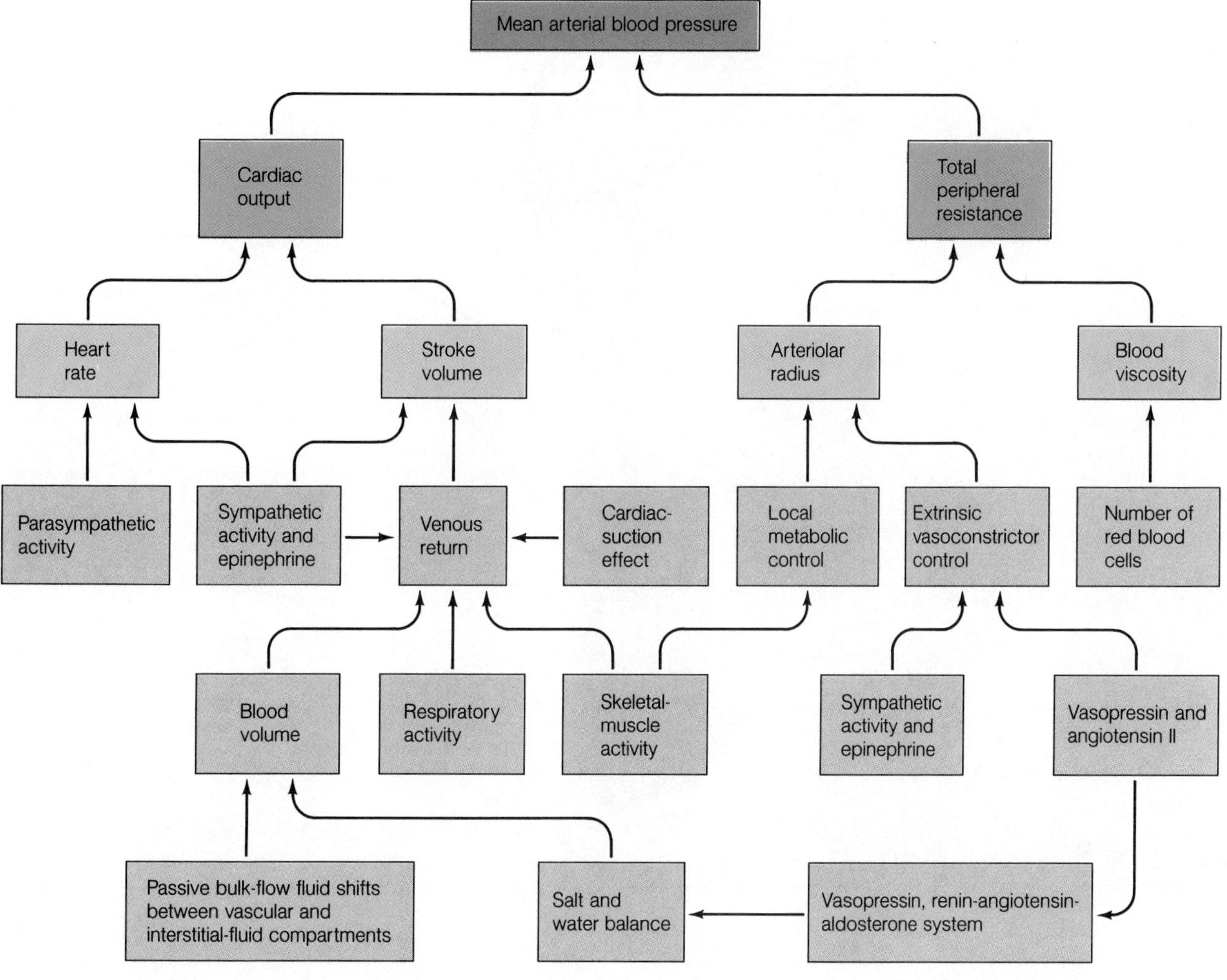

Figure 10–29 Determinants of Mean Arterial Blood Pressure

sistance in an attempt to restore blood pressure to normal. Like any reflex, the baroreceptor reflex includes a receptor, an afferent pathway, an integrating center, an efferent pathway, and effector organs.

The most important receptors involved in moment-to-moment regulation of blood pressure, the **carotid-sinus** and **aortic-arch baroreceptors,** are mechanoreceptors sensitive to changes both in mean arterial pressure and in pulse pressure. Their responsiveness to fluctuations in pulse pressure enhances their sensitivity as pressure sensors, because small changes in systolic or diastolic pressure may alter the pulse pressure without changing the mean pressure. These baroreceptors are strategically located (Fig. 10–30) to provide critical information about arterial blood pressure in the vessels leading to the brain (the carotid-sinus baroreceptor) and in the major arterial trunk before it gives off branches that supply the rest of the body (the aortic-arch baroreceptor).

The baroreceptors constantly provide information about blood pressure; in other words, they continuously generate action potentials in response to the ongoing pressure within the arteries. When arterial pressure (either mean or pulse pressure) decreases, the receptor potential of these baroreceptors decreases, thus decreasing the rate of firing in the corresponding afferent neurons. Conversely, when blood pressure increases, the rate of firing generated in the afferent neurons by the baroreceptors increases (Fig. 10–31).

The integrating center that receives the afferent impulses about the status of arterial pressure is the **cardiovascular con-**

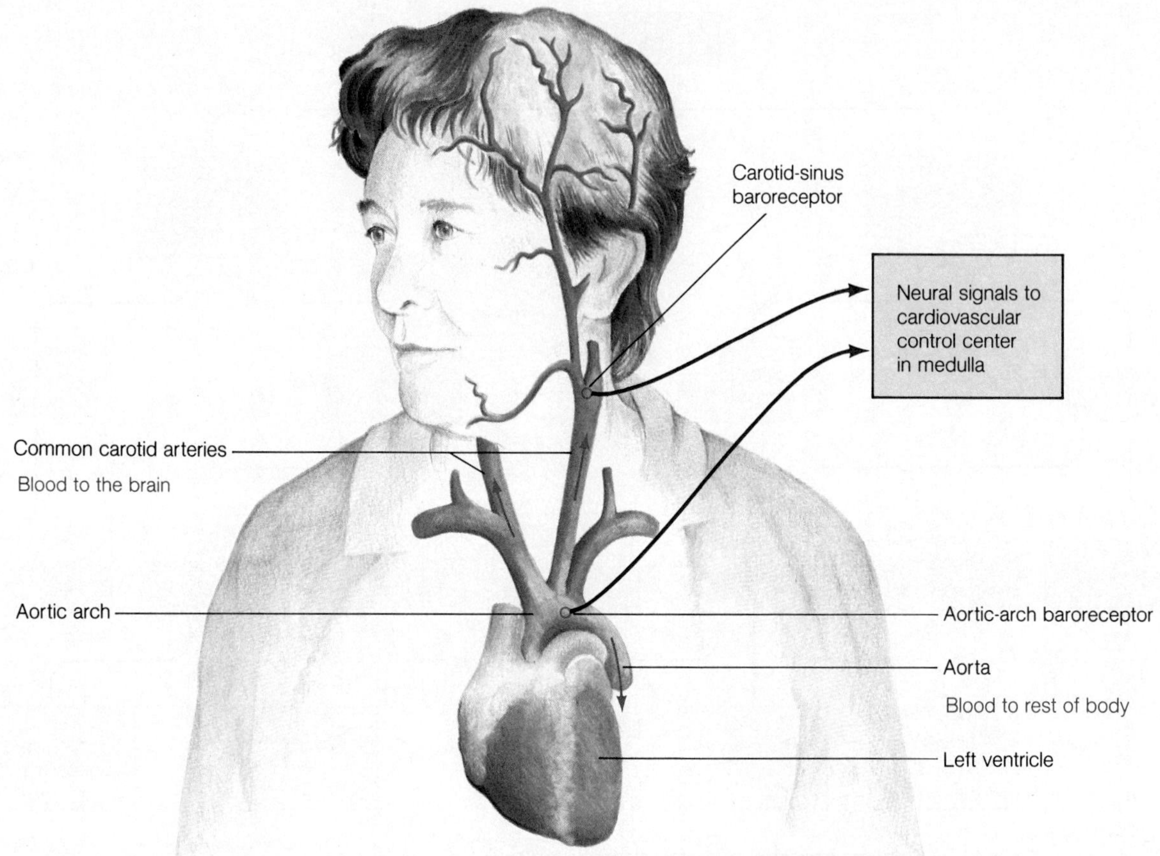

Figure 10–30 Location of Arterial Baroreceptors

trol center[4] located in the medulla within the brain stem. The efferent pathway is the autonomic nervous system. The cardiovascular control center alters the ratio between sympathetic and parasympathetic activity to the effector organs (the heart and blood vessels). To show how autonomic changes alter arterial blood pressure, Figure 10–32 provides a review of the major effects of parasympathetic and sympathetic stimulation on the heart and blood vessels.

Let us fit all the pieces of the baroreceptor reflex together now by tracing through the reflex activity occurring to compensate for an elevation or a fall in blood pressure. If for any reason arterial pressure beomes elevated above normal (Fig. 10–33a), the carotid-sinus and aortic-arch baroreceptors increase the rate of firing in their respective afferent neurons. Upon being informed by increased afferent firing that arterial pressure has become too high, the cardiovascular control center responds by increasing parasympathetic and decreasing sympathetic activity to the cardiovascular system. These efferent signals decrease heart rate, decrease stroke volume, and produce arteriolar and venous vasodilation, which in turn lead to a decrease in cardiac output and a decrease in total peripheral resistance, with a subsequent decrease in blood pressure back toward normal.

Conversely, when blood pressure falls below normal (Fig. 10–33b), baroreceptor activity decreases, inducing the cardiovascular center to decrease its parasympathetic output while increasing sympathetic cardiac and vasoconstrictor nerve activity. This efferent pattern of activity leads to an increase in heart rate and stroke volume coupled with arteriolar and venous vasoconstriction. These changes result in an increase in both cardiac output and total peripheral resistance, producing an elevation in blood pressure back toward normal.

[4]The cardiovascular control center is sometimes divided into cardiac and vasomotor centers, which are occasionally further classified into smaller subdivisions, such as cardioacceleratory and cardioinhibitory centers and vasoconstrictor and vasodilater areas. Since these regions are highly interconnected and functionally interrelated, we will refer to them collectively as the cardiovascular control center.

Other reflexes and responses influence blood pressure.

Besides the baroreceptor reflex, whose sole function is blood-pressure regulation, several other reflexes and responses have an influence on the cardiovascular system even though their primary roles are aimed at the regulation of other body functions. Some of these other influences deliberately move arterial pressure away from its normal value temporarily, overriding the baroreceptor reflex to accomplish a particular goal. These factors include the following:

1. Left atrial baroreceptors and hypothalamic osmoreceptors are primarily important in water and salt balance in the body, thereby having a bearing on blood pressure by means of controlling the plasma volume.
2. Chemoreceptors located in the carotid and aortic arteries, in close association with but distinct from the baroreceptors, are sensitive to low O_2 or high acid levels in the blood. These chemoreceptors' main function is to reflexly increase respiratory activity to bring in more O_2 or to blow off more acid-forming CO_2, but they also reflexly increase blood pressure by sending excitatory impulses to the cardiovascular center.
3. Cardiovascular responses associated with certain behaviors and emotions are mediated through the cerebral cortex-

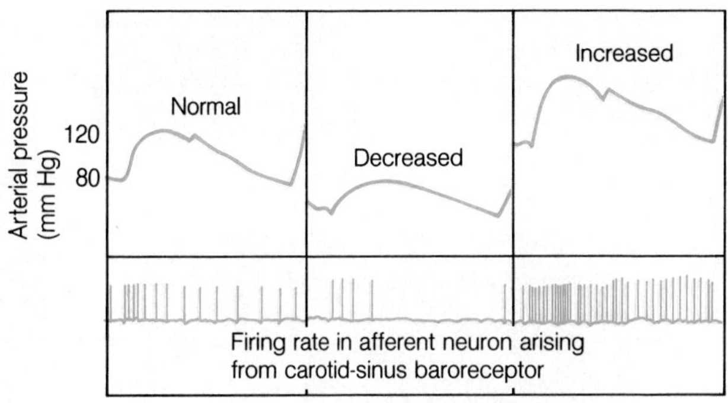

Figure 10–31 Firing Rate in Afferent Neurons from Carotid-Sinus Baroreceptor in Relation to Magnitude of Mean Arterial Pressure

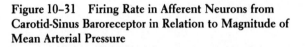

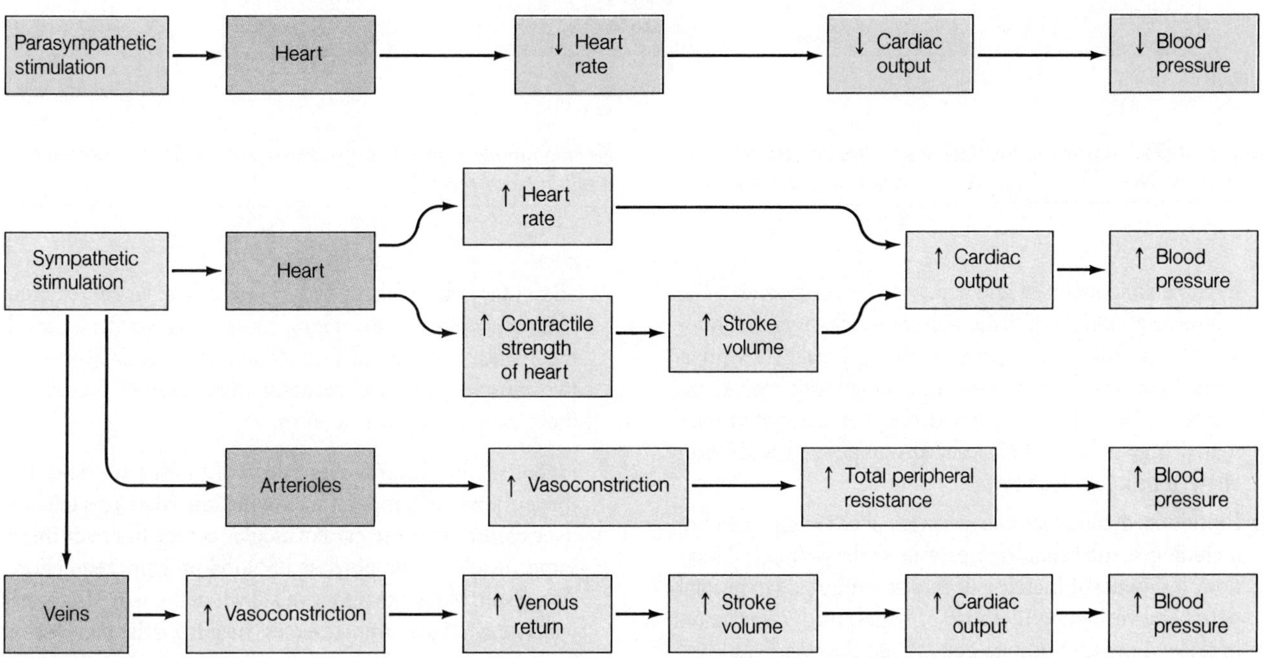

Figure 10–32 Summary of Effects of Parasympathetic and Sympathetic Nervous Systems on Factors That Influence Mean Arterial Blood Pressure

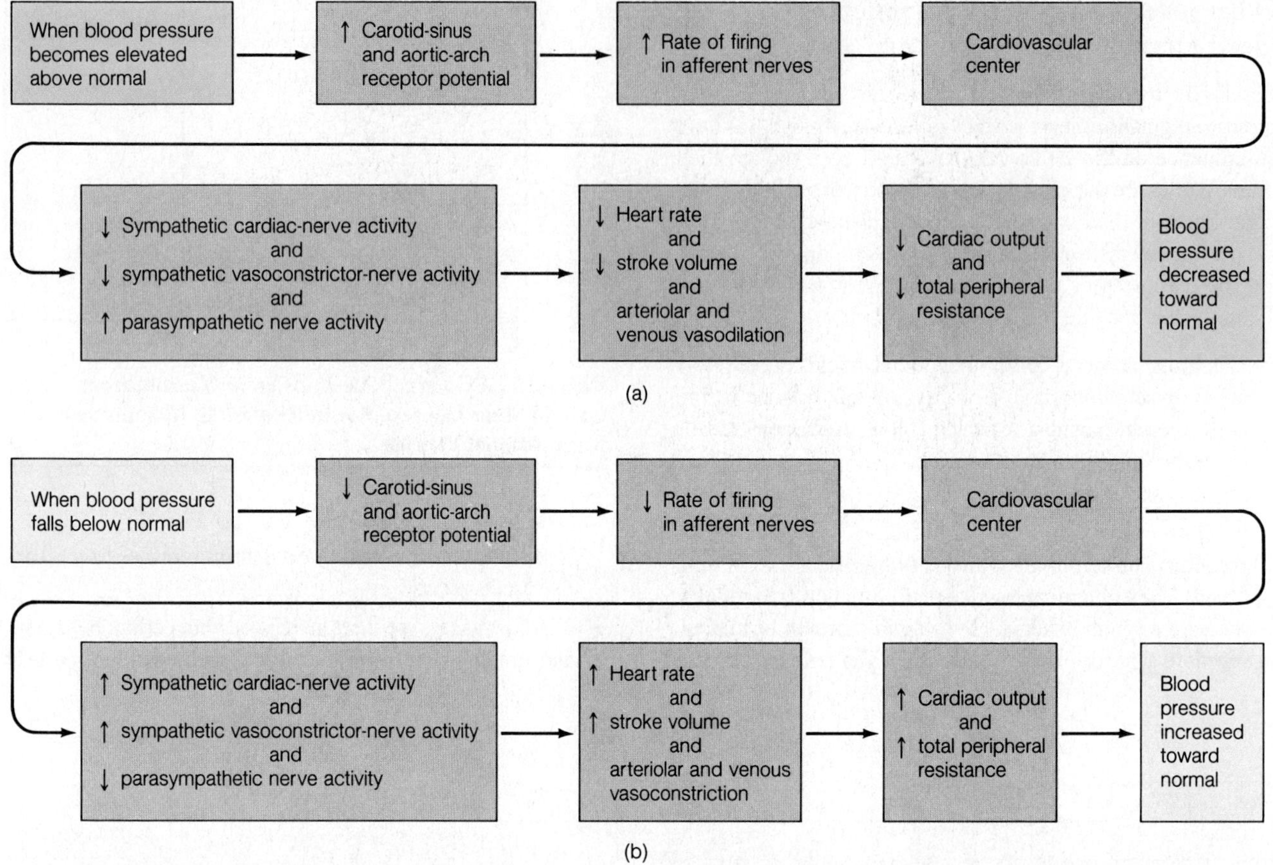

Figure 10–33 Baroreceptor Reflexes to Restore Blood Pressure to Normal *(a) Baroreceptor reflex in response to an elevation in blood pressure. (b) Baroreceptor reflex in response to a fall in blood pressure.*

hypothalamic pathway and appear to be preprogrammed. These responses include the widespread changes in cardiovascular activity accompanying the generalized sympathetic fight-or-flight response, the characteristic marked increase in heart rate and blood pressure associated with sexual orgasm, and the localized cutaneous vasodilation characteristic of blushing.

4. Pronounced cardiovascular changes accompany exercise, including a substantial increase in skeletal-muscle blood flow; a significant increase in cardiac output; a fall in total peripheral resistance (because of widespread vasodilation in skeletal muscles despite generalized arteriolar vasoconstriction in most organs); and a modest increase in mean arterial pressure. Evidence suggests that discrete exercise centers yet to be identified induce the appropriate cardiac and vascular changes at the onset of exercise or even in anticipation of exercise. These effects are then reinforced by afferent inputs to the medullary cardiovascular center from chemoreceptors in exercising muscles as well as by local mechanisms important in maintaining vasodilation in active muscles. The baroreceptor reflex further modulates these cardiovascular responses.

5. Hypothalamic control over cutaneous (skin) arterioles for the purpose of temperature regulation takes precedence over control that the cardiovascular center has over these same vessels for the purpose of blood-pressure regulation. As a result, blood pressure can fall when the skin vessels are widely dilated to eliminate excess heat from the body, even though the baroreceptor responses are calling for cutaneous vasoconstriction to help maintain adequate total peripheral resistance.

6. Recent studies suggest that numerous neurotransmitters from various regions of the brain may play a role in the

control of blood pressure. Their importance and relations to the traditional pathways for blood pressure regulation are presently unclear, as are any possible roles they may play in the poorly understood condition of hypertension.

Hypertension is a serious national public-health problem, but its causes are largely unknown.

Sometimes blood-pressure control mechanisms do not function properly or are unable to completely compensate for changes that have taken place. Blood pressure may be above the normal range (**hypertension** if above 140/90 mm Hg in individuals younger than 60 years of age or greater than 160/95 mm Hg for those older than 60) or below normal (**hypotension** if less than 100/60 mm Hg). Hypotension in its extreme form is *circulatory shock*. We will first examine hypertension before concluding this chapter with a discussion of hypotension and shock.

A definite cause for hypertension can be established in only 10% of the cases. Hypertension that occurs secondary to another primary problem is called **secondary hypertension.** The causes of secondary hypertension fall into four categories: cardiovascular hypertension, renal hypertension, endocrine hypertension, and neurogenic hypertension.

1. **Cardiovascular hypertension** is usually associated with chronically elevated total peripheral resistance caused by atherosclerosis (hardening of the arteries; see p. 292).
2. **Renal (kidney) hypertension** may occur as a result of two different kidney defects: partial occlusion of the renal arteries or diseases of the kidney tissue itself.
   a. Atherosclerotic lesions protruding into the lumen of a renal artery or external compression of the vessel by a tumor may reduce blood flow through the kidney. The kidney responds by initiating a hormonal pathway (renin-angiotensin-aldosterone pathway) that promotes salt and water retention during urine formation, thus increasing the blood volume to compensate for the reduced renal blood flow. Recall that angiotensin II is also a powerful vasoconstrictor. Although these two effects (increased blood volume and angiotensin-induced vasoconstriction) are compensatory mechanisms to improve blood flow through the narrowed renal artery, they also are responsible for elevating the arterial pressure as a whole.
   b. Renal hypertension also occurs if the kidneys are diseased and unable to eliminate the normal salt load. Salt retention induces water retention, which expands the plasma volume and leads to hypertension.
3. **Endocrine hypertension** arises from at least two different endocrine disorders: pheochromocytoma and Conn's syndrome.
   a. A pheochromocytoma is an adrenal medullary tumor that secretes excessive epinephrine and norepinephrine. Abnormally elevated levels of these hormones induce increased cardiac output and generalized peripheral vasoconstriction, both of which contribute to the hypertension characteristic of this disorder.
   b. Conn's syndrome is associated with increased production of aldosterone by the adrenal cortex. This hormone is part of the renin-angiotensin-aldosterone pathway responsible for salt and water retention by the kidneys. Once again, the excessive salt and water load in the body caused by increased aldosterone levels leads to an elevation in blood pressure.
4. **Neurogenic hypertension** occurs secondary to neural lesions.
   a. The problem may be erroneous blood-pressure control caused by a defect in the cardiovascular control center or in the baroreceptors.
   b. Neurogenic hypertension may also occur as a compensatory response to a reduction in cerebral blood flow—for example, because of compression of a major cerebral vessel by a tumor. In response to a decrease in brain blood flow, reflexes are initiated that elevate the blood pressure in an attempt to provide sufficient driving pressure to adequately supply the O_2-dependent brain tissue with blood.

The underlying cause is unknown in the remaining 90% of the cases of hypertension. Such hypertension is known as **primary (essential or idiopathic) hypertension.** Primary hypertension is undoubtedly a catch-all category for elevated blood pressure caused by a variety of unknown causes rather than a single disease entity. There is a strong genetic tendency to develop primary hypertension, which can be hastened or exacerbated by contributing factors such as obesity, stress, smoking, and excessive ingestion of salt. In fact, many researchers believe that a defect in salt management by the body may underlie many cases of primary hypertension. Disturbances in kidney function too minor to produce outward signs of renal disease could nevertheless insidiously lead to a gradual accumulation of salt and water in the body, resulting in progressive elevation in arterial pressure. Other investigators suggest that growth of smooth-muscle cells in the arterial walls leads to an elevation in total peripheral resistance by narrowing the lumen of these vessels even in the absence of vasoconstriction.

The baroreceptors do not respond to bring the blood pressure back to normal during hypertension because they adapt or are "reset" to operate at a higher level. In the presence of chronically elevated blood pressure, the baroreceptors still function to regulate blood pressure, but they maintain it at a higher mean pressure. The blood pressure of a hypertensive patient fluctuates no more than that of a normal individual.

Hypertension imposes stresses on both the heart and the blood vessels. The heart has an increased work load because it is pumping against an increased total peripheral resistance, whereas blood vessels may be damaged by the high internal pressure. Smaller vessels may burst because of the elevated pressure, particularly when the vessel wall is weakened by the degenerative process of atherosclerosis. Complications of hypertension include cerebrovascular accidents (strokes) caused by rupture of cerebral vessels or heart attacks caused by rupture of coronary vessels. Spontaneous hemorrhage that is due to bursting of small vessels elsewhere also may occur but with less serious consequences, such as rupture of blood vessels in the nose resulting in nose bleeds. Another serious complication of hypertension is renal failure caused by progressive impairment of blood flow through damaged renal blood vessels. Furthermore, retinal damage caused by changes in the blood vessels supplying the eyes may result in progressive loss of vision. When one becomes aware of these potential complications of hypertension and considers that 20% of all adults in America are estimated to be afflicted with chronic elevated blood pressure, one can appreciate the magnitude of this national health problem.

Until complications occur, hypertension is symptomless because the tissues are adequately perfused (supplied with blood). Therefore, unless blood-pressure measurements are made on a routine basis, the condition can go undetected until a precipitous complicating event results . Once hypertension is detected, therapeutic intervention can reduce the course and severity of the problem. Cornerstones of treatment include reduction in salt intake and administration of diuretics (drugs that enhance urine excretion) to reduce the salt and water load in the body, thereby decreasing plasma volume. In addition, other antihypertensive drugs reduce total peripheral resistance by manipulating some aspect of autonomic function to promote arteriolar vasodilation. No matter what the original cause, agents that reduce the plasma volume or total peripheral resistance (or both) will decrease the blood pressure toward normal. Furthermore, a regular aerobic exercise program can be employed to help reduce high blood pressure. (See the accompanying boxed feature, A Closer Look at Exercise Physiology.)

Whatever the underlying defect, once initiated, hypertension appears to be self-perpetuating. Constant exposure to elevated blood pressure predisposes vessel walls to the development of atherosclerosis, which further elevates blood pressure.

Inadequate or inappropriate autonomic activity can be responsible for fainting accompanying transient hypotension.

Hypotension, or low blood pressure, occurs either when there is a disproportion between vascular capacity and blood volume or when the heart is too weak to impart sufficient driving pressure to the blood. When blood pressure falls so low that adequate blood flow to the tissues can no longer be maintained, the condition is known as circulatory shock. We will first describe several milder, transient hypotensive conditions before turning our attention to the major causes and consequences of circulatory shock.

Two common situations in which hypotension occurs transiently are orthostatic hypotension and emotional fainting. Both are due to inappropriate autonomic nerve activity. **Orthostatic (postural) hypotension** is due to inadequate compensatory responses to the gravitational shifts in blood that occur when a person moves from a horizontal to a vertical position, especially following a prolonged bed rest. The fall in blood pressure that results from blood pooling in the leg veins upon standing up is normally detected by the baroreceptors, which initiate immediate compensatory responses to restore blood pressure to its proper level. When a long-bedridden patient first starts to rise, however, these reflex compensatory adjustments are temporarily lost or reduced because of disuse. Sympathetic control of the leg veins is inadequate, so when the patient first stands up, blood pools in the lower extremities. The resultant orthostatic hypotension and decrease in cerebral blood flow are responsible for the dizziness or actual fainting that occurs. Because postural compensatory mechanisms are depressed during prolonged bed confinement, patients sometimes are put on a tilt table so that they can be moved gradually from a horizontal to an upright position, allowing the body to slowly adjust to the gravitational shifts in blood.

Transient hypotension because of emotional stress can also cause dizziness or fainting. In this situation, higher brain centers act on the cardiovascular center to inappropriately decrease sympathetic output to the vasculature. The resultant loss of vascular tone precipitates widespread arteriolar vasodilation that leads to a decrease in total peripheral resistance. Furthermore, widespread arteriolar vasodilation causes pooling of blood in the capillaries so that venous return is decreased and cardiac output is subsequently reduced. Thus, hypotension is brought about by a decrease in both total peripheral resistance and cardiac output. As the blood pressure falls, the person becomes light-headed or faints because blood flow to the brain becomes inadequate. If the person

A CLOSER LOOK AT EXERCISE PHYSIOLOGY

The Ups and Downs of Hypertension and Exercise

When blood pressure is up, one way to bring it down is to increase the level of physical activity. An impressive epidemiologic study, which followed a group of 14,998 male graduates from Harvard for sixteen to fifty years, found that participation in collegiate sports, climbing as many as fifty stairs a day, walking five blocks, or light sports activities were not protective against the development of primary hypertension. Participation in vigorous sports such as running, swimming, handball, tennis, and cross-country skiing, on the other hand, was protective against the development of hypertension, even if other risk factors were present. Other studies have supported the finding that participation in aerobic activities is protective against the development of hypertension.

A logical question to ask is whether exercise can be used as a therapy to reduce hypertension once it has already developed. Antihypertensive medication is available to lower blood pressure in severely hypertensive patients, but sometimes undesirable side effects occur. The side effects of diuretics include electrolyte imbalances, glucose intolerance, and increased blood cholesterol levels. Side effects of drugs that manipulate total peripheral resistance, such as beta-blocker drugs, include increased blood triglyceride levels, lower HDL cholesterol levels (the "good" form of cholesterol), weight gain, sexual dysfunction, and depression.

Patients with mild hypertension, arbitrarily defined as a diastolic blood pressure between 90 and 100 mm Hg and a systolic pressure of 160 mm Hg, pose a dilemma for physicians. The risks of taking the drugs may outweigh the benefits gained from lowering the blood pressure. Because of the drug therapy's possible side effects, nondrug treatment of mild hypertension may be most beneficial. The most common nondrug therapies are weight reduction, salt restriction, and exercise. Although losing weight will almost always reduce blood pressure, research has shown that weight reduction programs usually result in the loss of only twelve pounds, and the overall long-term success in keeping the weight off is only about 20%. Salt restriction is beneficial for most hypertensives, but adherence to a low-salt diet is difficult for many people because fast foods and foods prepared in restaurants usually contain high amounts of salt. Some recent studies employing exercise as a therapeutic tool have shown that blood pressure in cases of mild to moderate hypertension has been decreased whether or not salt was restricted or weight was lowered. The preponderance of evidence in the literature suggests that moderate aerobic exercise performed three times per week for fifteen to sixty minutes is a beneficial therapy in mild to moderate hypertension. It is wise, therefore, to include a regular aerobic exercise program in conjunction with weight loss and salt reduction to optimally reduce high blood pressure without drug intervention.

faints or lies down, cardiac output is rapidly improved as the pooled blood quickly returns to the heart. The inappropriate autonomic nerve activity is only transient; when it is discontinued, vasoconstrictor tone is quickly restored. The person normally recovers quite rapidly with no after-effects. Even though fainting in reaction to a highly emotional situation is considered to be an inappropriate response, there is speculation that it may actually be adaptive, analogous to the "playing dead" ploy displayed by some animals. It temporarily relieves the person from the responsibility of coping with the emotionally charged situation.

Circulatory shock can become irreversible.

Circulatory shock, the generalized inability to deliver adequate blood for tissue use, is categorized into four main types: hypovolemic, cardiogenic, vasogenic, and neurogenic (Fig. 10-34).

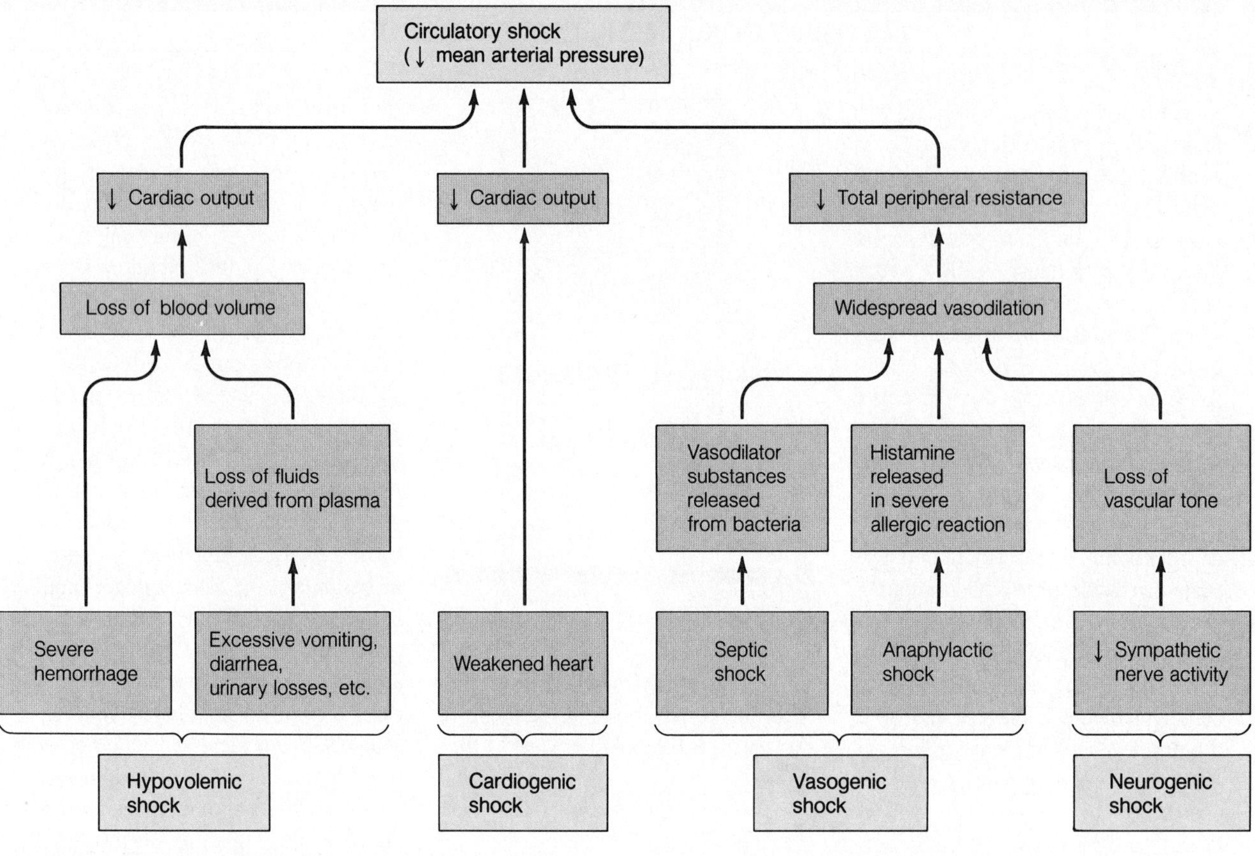

Figure 10–34 Causes of Circulatory Shock

1. **Hypovolemic shock** is induced by a fall in blood volume, which occurs either directly through severe hemorrhage or indirectly through loss of fluids derived from the plasma (for example, severe diarrhea, excessive urinary losses, or extensive sweating).
2. **Cardiogenic shock** is due to a weakened heart's failure to adequately pump blood.
3. **Vasogenic shock** is caused by widespread vasodilation triggered by the presence of vasodilator substances. There are two types of vasogenic shock: septic and anaphylactic. **Septic shock,** which may accompany massive infections, is due to vasodilator substances released from the infective agents. Similarly, extensive histamine release accompanying severe allergic reactions can cause widespread vasodilation in **anaphylactic shock.**
4. **Neurogenic shock** also involves generalized vasodilation but not by means of the release of vasodilator substances. In this case, loss of sympathetic vascular tone leads to generalized vasodilation, similar to emotional hypotension but more pronounced and prolonged. This undoubtedly is responsible for the shock accompaning crushing injuries when blood loss has not been sufficient to cause hypovolemic shock. Deep, excruciating pain apparently inhibits sympathetic vasoconstrictor activity.

We will now examine the compensations and consequences of shock, using hemorrhage as an example (Fig. 10–35). This example will also serve to pull together many of the principles discussed in this chapter. Following severe loss of blood, compensatory measures immediately attempt to maintain adequate blood flow to the brain and heart.

□ The baroreceptors reflexly increase sympathetic and decrease parasympathetic activity to the heart, increasing the heart rate to offset the reduced stroke volume brought about by the loss of blood volume. With severe fluid loss, the pulse is weak because of the reduced stroke volume but rapid because of the increased heart rate.

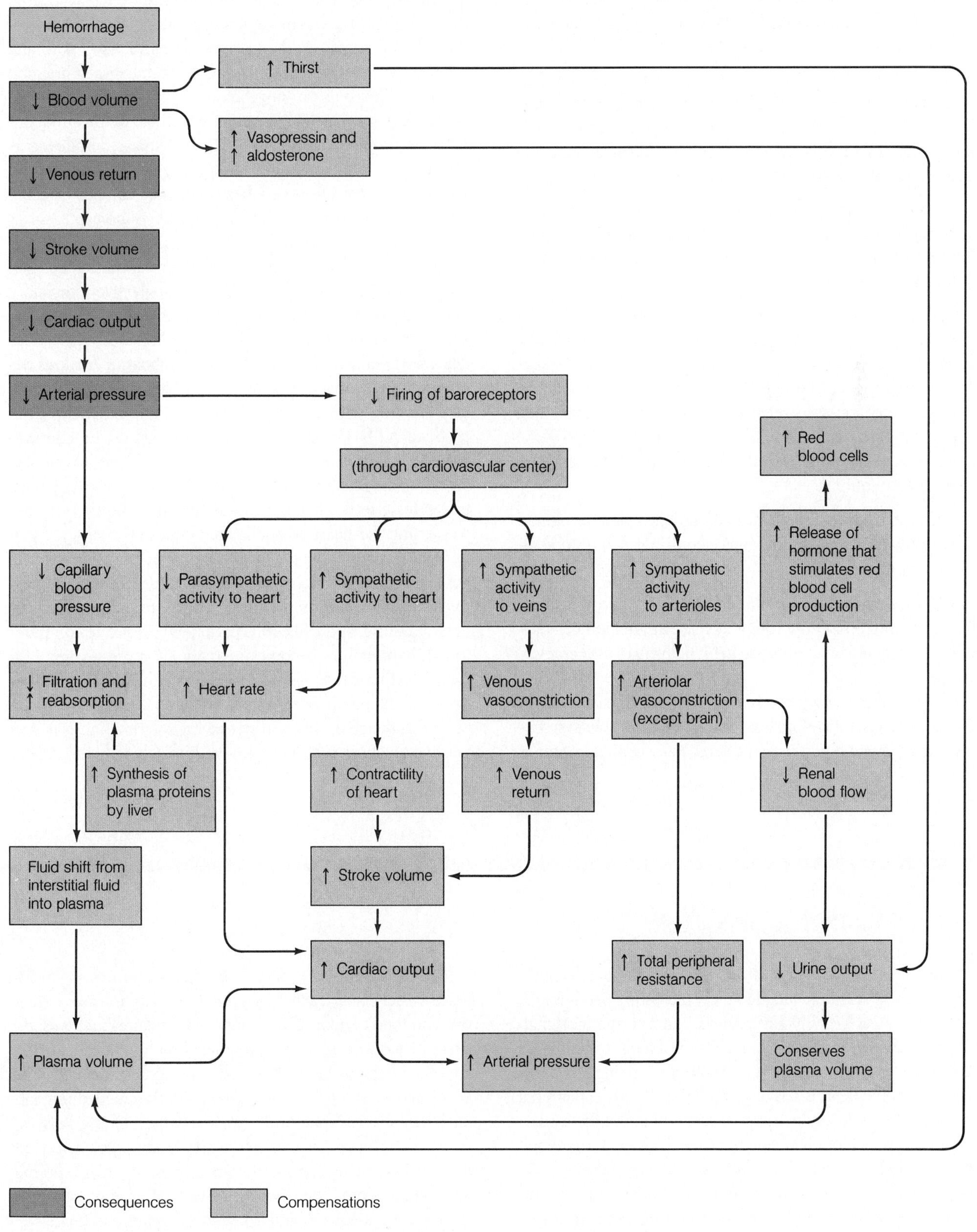

Figure 10–35 Consequences and Compensations of Hemorrhage

□ Generalized venous vasoconstriction occurs to increase venous return, and generalized arteriolar vasoconstriction occurs to increase total peripheral resistance.

□ The fall in arterial pressure is also accompanied by a fall in capillary blood pressure, which results in fluid shifts from the interstitial fluid into the capillaries to expand the plasma volume. This response is sometimes termed **autotransfusion,** because it restores the plasma volume as a transfusion does.

□ This ECF fluid shift is enhanced by plasma protein synthesis by the liver during the next few days following hemorrhage. The plasma proteins exert a colloid osmotic pressure to attract and retain extra fluid in the plasma.

□ Urinary output is reduced, thereby conserving water in the body that normally would have been lost. This additional fluid retention helps to expand the reduced plasma volume. Reduction in urinary output results from decreased renal blood flow caused by compensatory renal arteriolar vasoconstriction. The reduced plasma volume also triggers increased secretion of the hormones vasopressin and aldosterone, which further suppress urinary output.

□ Increased thirst is also stimulated by a fall in plasma volume. The resultant increased fluid intake contributes to restoration of plasma volume.

□ Over a longer course of time (a week or more), lost red blood cells are replaced through increased red-blood-cell production triggered by a reduction in O_2 delivery to the kidneys.

These compensatory mechanisms are often insufficient in the face of substantial fluid loss. Even if they are able to maintain an adequate blood-pressure level, they cannot continue indefinitely. Ultimately, fluid volume must be replaced from the outside through drinking, transfusion, or a combination of both. Blood supply to the kidneys, digestive tract, skin, and other organs can be compromised to maintain cerebral and coronary flow only so long before organ damage begins to occur. A point may be reached at which blood pressure continues to drop rapidly because of tissue damage despite vigorous therapy. This condition is frequently termed **irreversible shock,** in contrast to **reversible shock** that can be corrected by compensatory mechanisms and effective therapy.

Although the exact mechanism underlying irreversibility is not presently known, there are many logical possibilities that could contribute to the unrelenting, progressive circulatory deterioration that characterizes irreversible shock. Metabolic acidosis arises when lactic-acid production increases as blood-deprived tissues resort to anaerobic metabolism. Acidosis deranges the enzymatic systems responsible for energy production, limiting the capability of the heart and other tissues to produce ATP. Prolonged depression of kidney function results in electrolyte imbalances that may lead to cardiac arrhythmias. The blood-deprived pancreas releases a chemical that is toxic to the heart (**myocardial toxic factor**), further weakening the heart. Vasodilator substances build up within ischemic organs, inducing local vasodilation that overrides the generalized reflex vasoconstriction. As cardiac output progressively declines because of the heart's diminishing effectiveness as a pump and while total peripheral resistance continues to fall, hypotension becomes increasingly more severe. This causes further cardiovascular failure, which leads to a further decline in blood pressure. Thus, when shock progresses to the point that the cardiovascular system itself starts to fail, a vicious positive-feedback cycle ensues that ultimately results in death.

Chapter in Perspective

Homeostatically, the vasculature is designed to effectively transport and distribute blood to meet the body's needs for O_2 and nutrient delivery, waste removal, and hormonal signalling. The highly elastic arteries transport blood to the tissues and serve as a pressure reservoir to continue driving blood forward when the heart is relaxing and filling. The mean arterial pressure (the average driving pressure throughout the cardiac cycle) is closely monitored and regulated by short-term neural controls that adjust cardiac output and total peripheral resistance (the major determinants of mean arterial pressure) and by long-term renal and endocrine controls that maintain proper plasma volume by means of salt and water balance.

The highly muscular, narrow arterioles are the major resistance vessels, and, accordingly, they contribute most extensively to total peripheral resistance. Arteriolar caliber (vasodilation or vasoconstriction) is regulated both by local controls aimed at matching blood flow with a tissue's specific metabolic needs and by extrinsic sympathetic controls aimed at regulating mean arterial blood pressure. The pattern of arteriolar vasoconstriction and vasodilation is constantly juggled to shift the distribution of cardiac output to best serve the body's needs at the moment. The prime consideration is always given to providing adequate blood flow to the brain and heart.

The capillaries are the actual site of exchange between

blood and the surrounding interstitial fluid. Exchange of nutrients and metabolic end-products occurs primarily by diffusion down concentration gradients across the thin-walled, pore-lined capillary walls. In addition, bulk flow of protein-free plasma occurs across the capillary walls, primarily as a result of shifting balances between outward-directed capillary blood pressure and inward-directed blood-colloid osmotic pressure. The balance between ultrafiltration and reabsorption during this bulk flow determines the distribution of ECF between the vascular and interstitial-fluid compartments. Exchanges with cells are made with the interstitial fluid, not directly with the blood.

The highly distensible veins return the blood from the tissues to the heart and also serve as a blood reservoir. The veins' total capacity can be adjusted considerably to accommodate variations in blood volume with little change in venous pressure. Normally, 60% of the total blood volume is present in the veins. In contrast, only 5% of the blood volume is present in the capillaries, where the ultimate function of the entire system is accomplished.

Constant recycling and reconditioning of blood enables the circulatory system to use a very small volume of blood to control the chemical composition of the entire interstitial fluid upon which the cells depend for their survival.

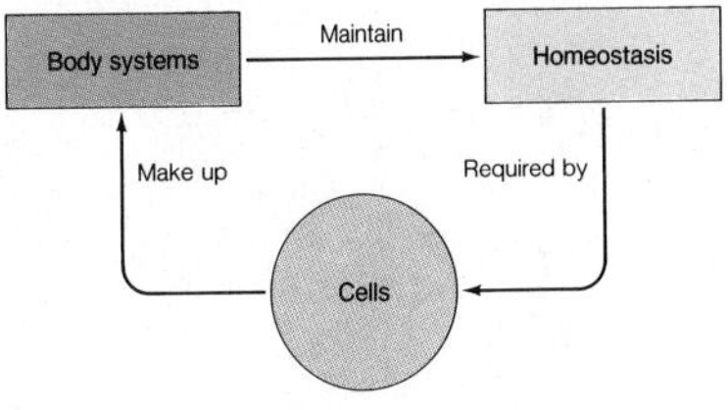

See inside front cover for an expanded version of this model.

Review Exercises

1. Compare blood flow through reconditioning organs and through organs that do not recondition the blood.
2. Discuss the relationships among flow rate, pressure gradient, and vascular resistance. What is the major determinant of resistance to flow?
3. Describe the structure and major functions of each segment of the vascular tree.
4. Describe the indirect technique of measuring arterial blood pressure by means of a sphygmomanometer.
5. Discuss the local and extrinsic controls that regulate arterial resistance.
6. What is the primary means by which individual solutes are exchanged across the capillary walls? What forces are responsible for bulk flow across the capillary walls? Of what importance is bulk flow?
7. How is lymph formed? What are the functions of the lymphatic system?
8. Compare the effect of vasoconstriction on blood flow rate in arterioles and veins.
9. Describe the factors that enhance venous return.
10. Discuss the factors that determine mean arterial pressure.
11. Review the effects on the cardiovascular system of parasympathetic and sympathetic stimulation.
12. Describe the baroreceptor reflex response that occurs when the blood pressure becomes elevated above normal.
13. List and explain the causes of secondary hypertension. What is primary hypertension? What are the potential consequences of hypertension?
14. Discuss the causes of orthostatic hypotension and of emotional fainting.
15. List and explain the causes of circulatory shock. What are the consequences and compensations of circulatory shock? What is irreversible shock?
16. **A point to ponder:** During coronary bypass surgery, a piece of vein is removed from the patient's leg and is surgically attached within the coronary circulatory system so that blood is detoured around a coronary artery segment that is extensively occluded as a result of atherosclerotic disease processes. Why must the patient for an extended period of time following surgery wear an elastic support stocking on the limb from which the vein was removed?

C H A P T E R 11

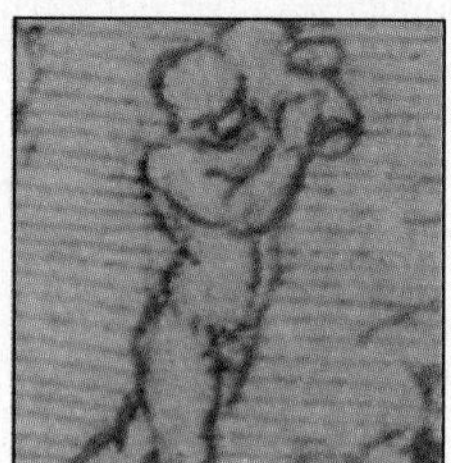

Blood

Chapter Contents at a Glance

INTRODUCTION *If a blood sample that has been treated to prevent clotting is allowed to stand, the red blood cells, being more dense than the fluid in which they are suspended, settle to the bottom of the tube, piling on top of each other like stacks of coins. However, the red blood cells, or erythrocytes, are more precious than coins because they carry life-sustaining O_2 to the tissues. About 98.5% of the total O_2 in the blood is carried within the red blood cells and is bound to the hemoglobin molecules they contain. The statistics involved are staggering. Four molecules of O_2 are able to hitch a ride with each hemoglobin molecule, and several hundred million hemoglobin molecules are packed into each erythrocyte. Each of us has a total of 25 to 30 trillion of these hemoglobin-stuffed, O_2-laden red blood cells streaming through our blood vessels at any given time (100,000 times more than the entire population of the United States)! Yet, these vital gas-transport vehicles are short-lived and must be replaced at the average rate of 2 million cells per second. During its short life span of four months, each erythrocyte travels about 700 miles as it circulates through the vasculature. The other blood cells play equally important roles and likewise must be continuously replenished, as you will see in this chapter.*

Blood represents about 8% of total body weight and has an average volume of 5 liters (in women) to 5.5 liters (in men). It consists of specialized *cellular elements* suspended in a complex liquid known as *plasma* (Table 11–1). There are three types of cellular elements: *erythrocytes* (*red blood cells* or *rbcs*); *leukocytes* (*white blood cells* or *wbcs*); and *thrombocytes* (*platelets*). More than 99% of the cellular elements are erythrocytes.

The constant movement of blood as it flows through the vascular network keeps these cellular elements rather evenly dispersed within the plasma. However, if a sample of whole blood is placed in a test tube and treated to prevent clotting, the heavier cellular elements slowly settle to the bottom and the lighter plasma rises to the top. This process can be hastened by centrifugation, which rapidly packs the cells in the bottom of the tube (Fig. 11–1). Because most of the cells are erythrocytes, the **hematocrit,** or **packed cell volume,** essentially represents the percentage of total blood volume occupied by erythrocytes. Plasma accounts for the remaining volume. The hematocrit averages 42% for women and slightly higher, at 45%, for men, with the average volume occupied by plasma being 58% for women and 55% for men. The white blood cells and platelets, which are colorless and less dense than red cells, are packed in a thin, cream-colored layer, the "buffy coat," on top of the packed red-cell column. They represent less than 1% of the total blood volume. We will first consider the properties of the largest portion of the blood, plasma, before turning our attention to the cellular elements.

PLASMA

Many of the functions of plasma are carried out by plasma proteins.

Being a liquid, **plasma** is composed of 90% water, which serves as a medium for materials being carried in the blood. Also, because water has a high capacity to hold heat, plasma is able to absorb and distribute much of the heat generated metabolically within tissues with only small changes in the temperature of the blood itself. Heat energy not needed to maintain body temperature is eliminated to the environment as the blood travels close to the surface of the skin.

A large number of organic and inorganic substances are dissolved in the plasma. The most plentiful organic constituents by weight are the plasma proteins, which compose 6 to 8% of plasma's total weight. Inorganic constituents account for approximately another 1% of plasma weight. The most abun-

Table 11–1 Blood Constituents and Their Functions

Constituent	Functions
Plasma	
Water	Transport medium; carries heat
Electrolytes	Membrane excitability; osmotic distribution of fluid between extracellular and intracellular fluid; buffering of pH changes
Nutrients, wastes, gases, hormones	No function in blood; merely being transported
Plasma proteins	In general, exert osmotic effect that is important in distribution of extracellular fluid between vascular and interstitial compartments; buffering of pH changes
Albumins	Transport of many substances; greatest contribution to colloid osmotic pressure
Globulins	
alpha and beta	Transport of many substances; clotting factors; inactive precursor molecules
gamma	Antibodies
Fibrinogen	Inactive precursor for fibrin meshwork of clot
Cellular Elements	
Erythrocytes	Oxygen and carbon dioxide transport (mainly oxygen)
Leukocytes	
Neutrophils	Phagocytes that engulf bacteria and debris
Eosinophils	Attack of parasitic worms; important in allergic reactions
Basophils	Release of histamine, which is important in allergic reactions; and heparin, which helps clear fat from blood and may function as anticoagulant
Monocytes	In transit to become tissue macrophages
Lymphocytes	
B lymphocytes	Production of antibodies
T lymphocytes	Cell-mediated immune responses
Platelets	Hemostasis

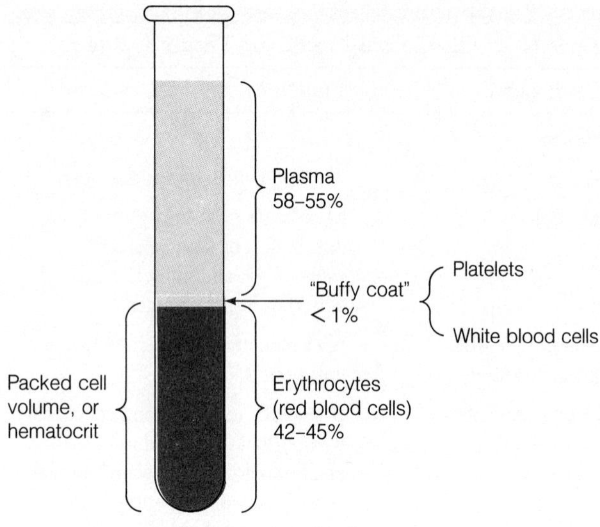

Figure 11-1 Hematocrit

dant electrolytes (ions) in the plasma are Na^+ and Cl^- (common salt). There are lesser amounts of HCO_3^-, K^+, Ca^{++}, and others. The functions of these extracellular-fluid (ECF) ions are discussed elsewhere, the most notable being their roles in membrane excitability, osmotic distribution of fluid between the ECF and cells, and buffering of pH changes. The remaining small percentage of plasma is occupied by nutrients (for example, glucose, amino acids, lipids, and vitamins); waste products (creatinine, bilirubin, and nitrogenous substances such as urea); dissolved gases (such as O_2, CO_2, and N_2); and hormones. Most of these substances are merely being transported in the plasma. For example, endocrine glands secrete hormones into the plasma, which transports these chemical messengers to their sites of action.

The **plasma proteins** are the one group of plasma constituents not present just for the ride. These important components normally remain in the plasma, where they perform many valuable functions. Because they are the largest of the plasma constituents, plasma proteins are unable to exit through the narrow pores in the capillary walls. Also, unlike other plasma constituents that are dissolved in the plasma water, the plasma proteins exist in a colloidal suspension (see p. A-7).

There are three groups of plasma proteins—albumins, globulins, and fibrinogen—which are classified according to their different physical and chemical properties. The following is an accounting of the wide range of the plasma proteins' functions, each of which is elaborated on elsewhere in the text as appropriate.

□ By virtue of their presence as a colloidal suspension in the plasma and their absence in the interstitial fluid, plasma proteins establish an osmotic gradient between blood and interstitial fluid. This colloid osmotic pressure is the primary force responsible for preventing excessive loss of plasma from the capillaries into the interstitial fluid.

□ Plasma proteins are partially responsible for the plasma's capacity to buffer changes in pH.

□ Plasma proteins contribute to blood viscosity, but erythrocytes are far more important in this regard.

□ Plasma proteins are not normally used as metabolic fuels, but in a state of starvation they can be degraded to provide energy for cells.

In addition to these general functions, each type of plasma protein performs important specific tasks:

1. **Albumins,** the most abundant of the plasma proteins, bind many substances (for example, bilirubin, bile salts, and penicillin) for transport through the plasma and contribute most extensively to the colloid osmotic pressure by virtue of their numbers.
2. There are three subclasses of **globulins: alpha** (α), **beta** (β), and **gamma** (τ).
   a. Specific alpha and beta globulins bind and transport a number of substances in the plasma, such as thyroid hormone, cholesterol, and iron.
   b. Many of the factors involved in the process of blood clotting are alpha or beta globulins.
   c. Inactive precursor protein molecules, which are activated as needed by specific regulatory inputs, belong to the alpha-globulin group (for example, angiotensinogen is activated to angiotensin, which plays an important role in the regulation of salt balance in the body).
   d. The gamma globulins are the immunoglobulins (antibodies), which are crucial to the body's defense mechanism.
3. **Fibrinogen** is a key factor in the blood-clotting process.

The plasma proteins generally are synthesized by the liver, with the exception of the gamma globulins, which are produced by lymphocytes, one of the types of white blood cells.

Erythrocytes

The structure of erythrocytes is well-suited to their primary function of oxygen transport in the blood.

The number of erythrocytes in each milliliter of blood averages about 5 billion, commonly reported clinically in a

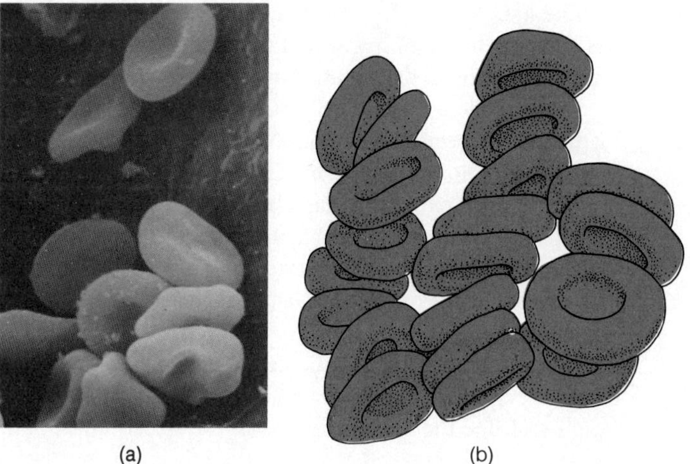

Figure 11–2 Anatomical Characteristics of Erythrocytes *(a) Scanning electron-microscopic appearance of erythrocytes. Note their biconcave shape. (b) Stacking of erythrocytes.*

red blood cell count as 5 million cells per cubic millimeter (mm^3). **Erythrocytes (red blood cells)** are flat, disc-shaped cells indented in the middle on both sides, like a doughnut with a flattened center instead of a hole (they are biconcave discs 8 μm in diameter, 2 μm thick at the outer edges, and 1 μm thick in the center) (Fig. 11–2). This unique shape contributes in two ways to the efficiency with which erythrocytes perform their main function of O_2 and CO_2 transport in the blood. First, the biconcave shape provides a larger surface area for diffusion of O_2 and CO_2 across the membrane compared to a spherical cell of the same volume. Second, the thinness of the cell enables these gases to rapidly diffuse between the exterior and innermost regions of the cell.

A second feature of erythrocytes that facilitates their transport function is their membrane flexibility, which enables them to travel through the narrow, tortuous capillaries to deliver their O_2 cargo and pick up the CO_2 waste at the tissue level without rupturing in the process. Red blood cells, whose diameter is normally 8 μm, are able to deform amazingly as they squeeze single-file through capillaries as narrow as 3 μm in diameter.

The third and most important feature of erythrocytes that enables them to transport the blood gases, especially O_2, is the **hemoglobin** they contain. A hemoglobin molecule consists of two parts (Fig. 11–3): (1) the **globin portion,** a protein made up of four highly folded polypeptide chains; and (2) four iron-containing, nonprotein, nitrogenous groups known as **heme** groups, each of which is bound to one of the polypeptides. Each of the four iron atoms can combine reversibly with one molecule of O_2; thus each hemoglobin molecule can pick up four O_2 passengers. Because O_2 is poorly soluble in the plasma, 98.5% of the O_2 is carried in the blood bound to hemoglobin. Hemoglobin is a pigment (naturally colored). Because of its iron content, it appears reddish when combined with O_2 and bluish when deoxygenated. This is why fully oxygenated arterial blood is red in color, and venous blood that has lost some of its O_2 load at the tissue level has a bluish cast.

In addition to carrying O_2, hemoglobin can also combine with the following:

1. carbon dioxide, contributing to the transport of this gas from the tissues back to the lungs;

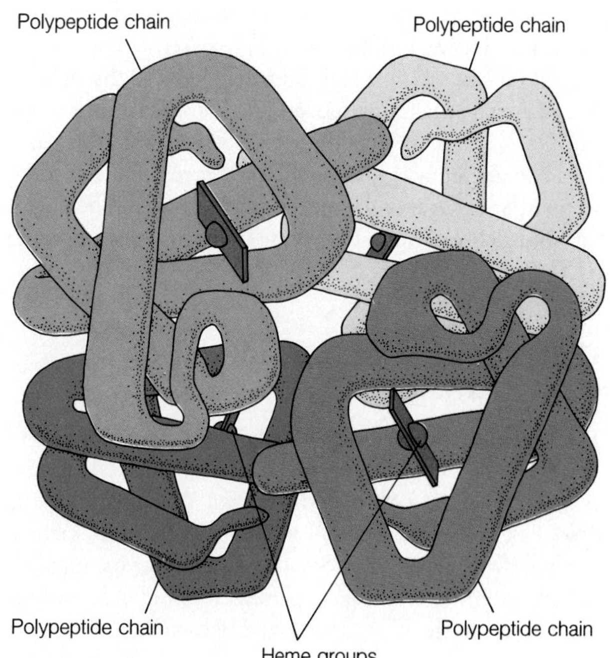

Figure 11–3 Hemoglobin Molecule

2. the acidic hydrogen-ion portion (H^+) of ionized carbonic acid, which is generated at the tissue level from CO_2, thereby buffering this acid so that it alters the pH of the blood minimally; and
3. carbon monoxide (CO), which is not normally in the blood but, if inhaled, preferentially occupies the O_2-binding sites on hemoglobin, causing carbon-monoxide poisoning.

Therefore, hemoglobin plays the key role in O_2 transport while contributing significantly to CO_2 transport and the buffering capacity of blood.

To maximize its hemoglobin content, a single erythrocyte is stuffed with several hundred million hemoglobin molecules to the exclusion of almost everything else. Erythrocytes contain no nucleus, organelles, or ribosomes. These structures are extruded during the cell's development to make room for more hemoglobin. Thus, a red blood cell is mainly a plasma membrane-enclosed sac full of hemoglobin.

Only a few crucial, nonrenewable enzymes remain within a mature erythrocyte: these are glycolytic enzymes and carbonic anhydrase. The **glycolytic enzymes** are necessary for generating the energy needed to fuel the active-transport mechanisms involved in maintaining proper ionic concentrations within the cell. Ironically, even though erythrocytes are the vehicles for transport of O_2 to all other tissues of the body, they themselves cannot use the O_2 they are carrying for energy production. Erythrocytes, lacking the mitochondria that house the enzymes for oxidative phosphorylation, must rely entirely on glycolysis for ATP formation.

The other important enzyme within red blood cells, **carbonic anhydrase,** is critical in CO_2 transport. This enzyme catalyzes a key reaction that ultimately leads to the conversion of metabolically produced CO_2 into **bicarbonate ion** (**HCO_3^-**), which is the primary form in which CO_2 is transported in the blood. Thus, erythrocytes contribute to CO_2 transport in two ways—by means of its carriage on hemoglobin and its carbonic anhydrase-induced conversion to HCO_3^-.

The bone marrow continuously replaces worn-out erythrocytes.

The price erythrocytes pay for their generous content of hemoglobin to the exclusion of the usual specialized intracellular machinery is a shortened life span. Without DNA and RNA, red blood cells cannot synthesize proteins for cellular repair, growth, and division or for renewal of enzyme supplies. Equipped only with initial supplies synthesized before extrusion of their nucleus, organelles, and ribosomes, erythrocytes are able to survive an average of only 120 days, in contrast to nerve and muscle cells, which survive during an individual's entire life. As a red blood cell ages, its nonreparable plasma membrane becomes fragile and prone to rupture as the cell squeezes through tight spots in the vascular system.

Most senile red blood cells meet their final demise in the **spleen,** an especially hazardous site for these old cells because of its narrow, tortuous capillary network. The spleen lies in the upper left part of the abdomen. In addition to removing most of the old erythrocytes from circulation, the spleen has limited ability to store healthy erythrocytes in its pulpy interior, serves as a reservoir site for platelets, and contains an abundance of lymphocytes.

Once ruptured, the fragmented portions of old red blood cells are quickly engulfed by macrophages, large phagocytic cells that line the venous cavities of the spleen, liver, bone marrow, and lymph nodes. During degradation by the macrophages, the globin portion of hemoglobin is broken down into its constituent amino acids, which are returned to the blood for use as needed anywhere in the body. The iron liberated from the heme groups is reused in the synthesis of new hemoglobin, but the noniron portion of the heme groups is not recycled. It is converted into **bilirubin,** which is eventually picked up by the liver cells and secreted into the bile for elimination from the body in the feces and urine.

Because erythrocytes cannot divide to replenish their own numbers, the old ruptured cells must be replaced by new cells produced in an erythrocyte factory—the **bone marrow**—which is the soft, highly cellular tissue that fills the internal cavities of bones. The bone marrow normally generates new red blood cells, a process known as **erythropoiesis,** at the amazing rate of 2 to 3 million per second to keep pace with the demolition of old cells.

During intrauterine development, erythrocytes are produced first by the yolk sac and then by the developing liver and spleen, until the bone marrow is formed and takes over erythrocyte production exclusively. Most of the bones in children are filled with red bone marrow that is capable of blood-cell production. As a person matures, however, red marrow is gradually replaced by fatty yellow marrow that is incapable of erythropoiesis, leaving red marrow only in the sternum (breastbone), vertebrae (backbone), ribs, base of the skull, and upper ends of the long limb bones. Red marrow not only produces red blood cells but is the ultimate source for leukocytes and platelets as well. Undifferentiated **multipotential stem cells** reside in the red marrow, where they continuously divide and differentiate to give rise to each of the types of blood cells. The different types of cells, along with the stem cells, are intermingled in the red marrow at various stages of development. The mature cells are released into the rich supply of capillaries that permeate the red marrow. Regulatory factors act on the ***hematopoietic*** ("blood producing") marrow to govern the type and number of cells generated and discharged into the blood. The mechanism regulating erythropoiesis is best understood. We will consider it now.

Erythropoiesis is controlled by erythropoietin from the kidneys.

The number of circulating erythrocytes normally remains fairly constant, indicative that erythropoiesis must be closely regulated. Because O_2 transport in the blood is the erythrocytes' primary function, you might logically suspect that the primary stimulus for increased erythrocyte production would be reduced O_2 delivery to the tissues. You would be correct, but low O_2 levels do not stimulate erythropoiesis by acting directly on the red bone marrow. Instead, reduced O_2 delivery to the kidneys stimulates them to secrete the hormone **erythropoietin** into the blood, and this hormone in turn stimulates erythropoiesis by the bone marrow (Fig. 11–4). The exact cells in the kidney responsible for erythropoietin secretion have not been identified, but this hormone's final site of action is known. Erythropoietin acts on derivatives of undifferentiated stem cells that are already committed to becoming red blood cells, stimulating their proliferation and maturation into mature erythrocytes. This increased erythropoietic activity elevates the number of circulating red blood cells, thereby increasing O_2-carrying capacity of the blood and restoring O_2 delivery to the tissues to normal. Once normal O_2 delivery to the kidneys is achieved, erythropoietin secretion is turned off until needed once again. In this way, erythrocyte production is normally balanced against destruction or loss of these cells to maintain a near-constant level of O_2-carrying capacity in the blood. (See the accompanying boxed feature, A Closer Look at Exercise Physiology.) In response to severe loss of erythrocytes, such as accompanies hemorrhage or abnormal destruction of young circulating erythrocytes, the rate of erythropoiesis can be increased by more than six times the normal level.

Several steps are involved in the preparation of an erythrocyte for its departure from the marrow, such as synthesis of hemoglobin and extrusion of the nucleus and organelles. It takes a few days for those cells closest to maturity to be "finished off" and released into the blood in response to erythropoietin, and it may take up to several weeks for less-developed and newly proliferated cells to reach maturity. Therefore, the time required for complete replacement of lost red blood cells depends on how many are needed to return the number to normal. (When you donate blood, your circulating erythrocyte supply is replenished in less than a week.) When demands for red blood cell production are high (for example,

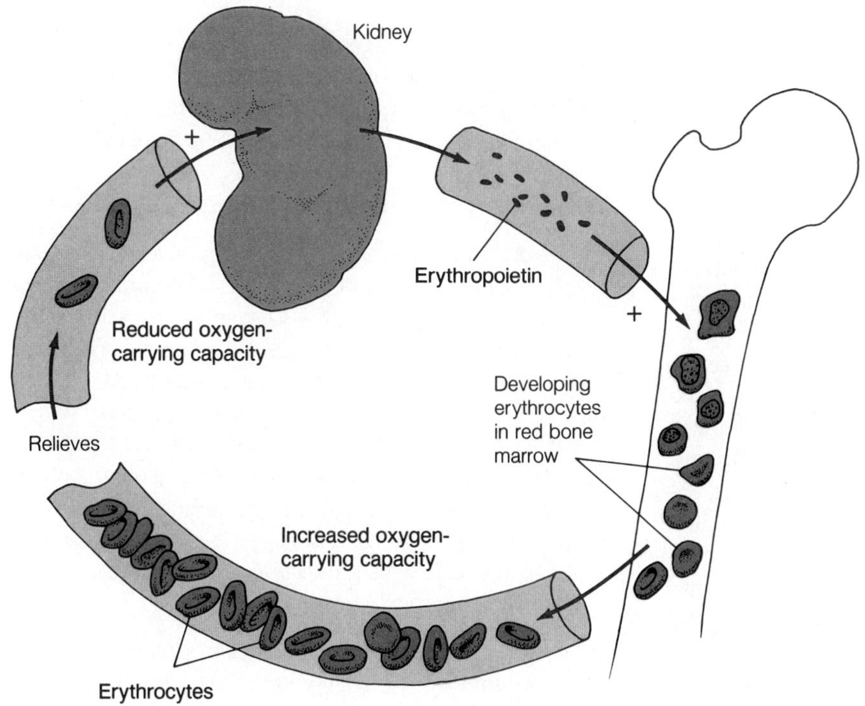

Figure 11–4 Control of Erythropoiesis

A CLOSER LOOK AT EXERCISE PHYSIOLOGY

Blood Doping: Is More of a Good Thing Better?

A continual supply of O_2 to exercising muscles is essential for generating energy to sustain endurance activities (see p. 239). Blood doping is a technique designed to temporarily increase the O_2-carrying capacity of the blood in an attempt to gain a competitive advantage. Blood doping involves removing blood from an athlete, then promptly reinfusing the plasma but freezing the red bloods cells for reinfusion one to seven days before a competitive event. One to four units of blood (one unit equals 450 ml) are usually withdrawn at three- to eight-week intervals before the competition. Increased erythropoietic activity restores the red blood cell count to a normal level in the intervening period between blood withdrawals.

Reinfusion of the stored red blood cells temporarily increases the red blood cell count and hemoglobin level above normal. Theoretically, blood doping would be beneficial to endurance athletes by improving the blood's O_2-carrying capacity. If too many red cells were infused, however, performance could suffer because the increased blood viscosity would decrease blood flow.

Early research into athletic performance after blood doping produced contradictory results. More recent research, however, indicates that, in a standard laboratory exercise test, athletes may realize a 5% to 13% increase in aerobic capacity; a reduction in heart rate during exercise as compared to that during the same exercise in the absence of blood doping; improved performance; and reduced lactic acid levels in the blood. (Lactic acid is produced when muscles resort to less efficient anaerobic glycolysis for energy production—see p. 237.) Blood doping, although probably effective, is illegal in both collegiate athletics and Olympic competition for ethical and philosophical reasons.

following hemorrhage), the bone marrow may release large numbers of immature erythrocytes, known as **reticulocytes,** into the blood to quickly meet the need. These immature cells can be recognized by staining techniques that make visible the residual ribosomes and organelle remnants that have not yet been extruded. Their presence above the normal level of a 0.5 to 1.5% of the total number of circulating erythrocytes is indicative of a high rate of erythropoietic activity. At very rapid rates, more than 30% of the circulating red cells may be at the immature reticulocyte stage.

The gene that directs erythropoietin synthesis has recently been identified and cloned so that this hormone can now be produced in abundance in a laboratory. This capability will have a profound impact on the need for blood transfusions. For example, administration of erythropoietin to surgical patients will stimulate their own red-cell production, thus reducing the need for transfused blood.

In addition to erythropoietin, which is the primary regulatory factor that adjusts the ongoing rate of red blood cell production as needed to keep erythrocyte levels constant, testosterone, the major male sex hormone, increases the basal rate of erythropoiesis. This undoubtedly is responsible, at least in part, for the normally larger hematocrit in males compared to females. The additional red blood cells in males provides extra O_2-carrying capacity to meet the needs of the larger muscle mass in men, for which testosterone is also responsible.

Anemia can be caused by a variety of disorders.

In spite of control measures, O_2-carrying capacity cannot always be maintained to meet tissue needs. **Anemia** refers to a reduction below normal in the O_2-carrying capacity of the blood, whereas **polycythemia** refers to an excess of circulating erythrocytes (Fig. 11–5). The causes and consequences of these abnormalities are discussed throughout this section.

Anemia can be caused by inadequate numbers of circulating erythrocytes, a deficiency in their hemoglobin content, or both. It can be brought about by a decreased rate of erythropoiesis or by excessive losses of erythrocytes. The various causes of anemia can be grouped into six categories.

NUTRITIONAL ANEMIA. The production of erythrocytes depends on an adequate supply of essential raw ingredients, some of which are not synthesized in the body but must be

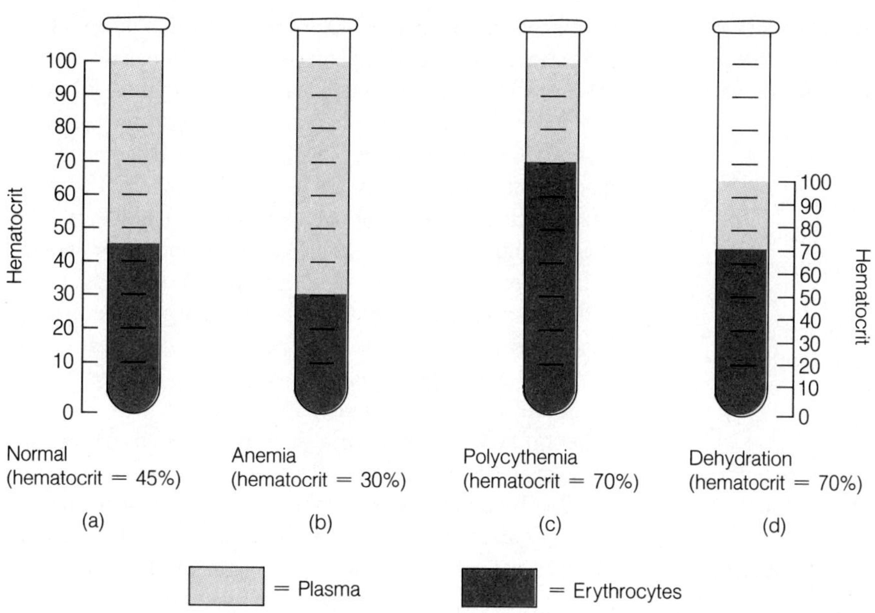

Figure 11–5 Hematocrit under Various Circumstances

provided by dietary intake. **Iron-deficiency anemia** occurs when insufficient iron is available for the synthesis of hemoglobin because of either an iron-deficient diet or poor iron absorption from the digestive tract. The usual numbers of erythrocytes are produced, but they contain less hemoglobin than normal, are subsequently smaller than usual, and are able to transport less O_2. A dietary deficiency of folic acid, a member of the B-vitamin complex, can likewise produce anemia. This vitamin is essential for the formation of DNA, which is critical in governing stem-cell division and maturation of erythrocytes. As a result of an inadequate amount of folic acid, fewer erythrocytes are formed, and those produced are larger and more fragile than normal. Despite their larger size, they contain the normal amount of hemoglobin. The resulting anemia is due to the production of fewer erythrocytes and their earlier demise because of their fragility. Many other tissues are affected by a deficiency of folic acid as well, but red blood cell production is especially sensitive because erythrocytes are among the most rapidly proliferating cells of the body.

PERNICIOUS ANEMIA. **Pernicious anemia** is caused by an inability to absorb adequate amounts of vitamin B_{12} from the digestive tract. Like folic acid, vitamin B_{12} is essential for DNA production and its associated role in the proliferation and maturation of erythrocytes. Therefore, pernicious anemia is also characterized by a reduced number of erythrocytes that are larger and more fragile than normal. Unlike the nutritional anemias, however, pernicious anemia is not due to insufficient dietary supplies of vitamin B_{12}. This nutrient is found abundantly in a variety of common foods. The problem with pernicious anemia is a deficiency of **intrinsic factor,** a special substance secreted by the lining of the stomach. Only when vitamin B_{12} is in combination with intrinsic factor can it be absorbed from the intestinal tract by special transport mechanisms. When intrinsic factor is deficient, insufficient amounts of ingested vitamin B_{12} are absorbed. The resulting impairment of red blood cell production and maturation leads to anemia. This condition is treated by administering injections of vitamin B_{12} to bypass the defective absorptive mechanism.

APLASTIC ANEMIA. **Aplastic anemia** is caused by failure of the bone marrow to produce adequate numbers of red blood cells, even though all ingredients necessary for erythropoiesis are available. Reduced erythropoietic capability can be caused by destruction of red bone marrow by toxic chemicals (such as benzene, arsenic, and certain drugs, especially chloramphenicol); by heavy exposure to radiation (fallout from a nuclear-bomb explosion, for example, or excessive exposure to x-rays); or by invasion of the marrow by cancer cells. The destructive process may selectively reduce the marrow's output of erythrocytes, or it may reduce the productive capability for leukocytes and platelets as well. The anemia's severity depends on the extent of destruction of erythropoietic tissue, with severe losses being fatal.

RENAL ANEMIA. Since erythropoietin from the kidneys is the primary stimulus for promoting erythropoiesis, inadequate erythropoietin secretion as a result of kidney disease causes **renal anemia.**

HEMORRHAGIC ANEMIA. **Hemorrhagic anemia** is caused by the loss of substantial quantities of blood. The loss can either be acute, such as that occurring because of bleeding from a wound, or chronic, such as that accompanying a history of excessive menstrual flow. Anemia is present until the lost cells are replaced by transfusion or by increased erythropoietic activity.

HEMOLYTIC ANEMIA. **Hemolysis** is the rupture of red blood cells. **Hemolytic anemia** occurs when excessive numbers of circulating erythrocytes rupture, either because the cells are defective or because otherwise normal cells are induced to rupture by external factors. Various hereditary abnormalities of red blood cells make them very fragile. The best known example is **sickle-cell anemia.** Even as young cells, the fragile, defective erythrocytes are prone to rupture as they travel through the narrow splenic capillaries. Despite an accelerated rate of erythropoiesis triggered by the constant excessive loss of red blood cells, production may not be able to keep pace with the rate of destruction and anemia may result.

Normally formed cells can also hemolyze prematurely if attacked by various external factors. An example is **malaria,** which is caused by protozoan parasites introduced into a victim's blood by the bite of a carrier mosquito. These parasites selectively invade red blood cells, where they multiply to the point that the mass of malarial organisms ruptures the cells, releasing hundreds of new active parasites that quickly invade other red blood cells. As this cycle continues and more erythrocytes are destroyed, the anemic condition progressively worsens.

Polycythemia is an excess of circulating erythrocytes.

Polycythemia, in contrast to anemia, is characterized by an excess of circulating red blood cells (Fig. 11–5c). There are two general types of polycythemia, depending on the circumstances triggering the excess red blood cell production: these are primary polycythemia, or polycythemia vera (*vera* means "true"), and secondary polycythemia, or physiological polycythemia.

Primary polycythemia is caused by a tumorlike condition of the bone marrow, in which erythropoiesis proceeds at an excessive, uncontrolled rate instead of being subject to the normal erythropoeitin regulatory mechanism. The red blood cell count may reach 11 million cells/mm^3 (normal is 5 million cells/mm^3), and the hematocrit may be as high as 70–80% (normal is 42–45%). No benefit is derived from the extra O_2-carrying capacity of the blood, because O_2 delivery is more than adequate with normal red blood cell numbers. There are adverse effects of inappropriate polycythemia, however. The excessive number of red cells increases the blood's viscosity up to five to seven times that of normal, which causes the blood to flow very sluggishly and increases the total peripheral resistance. This may elevate the blood pressure, thus increasing the work load of the heart, unless blood-pressure control mechanisms are able to compensate.

Secondary polycythemia, in contrast, is an appropriate erythropoietin-induced adaptive mechanism to improve the blood's O_2-carrying capacity in response to a prolonged reduction in O_2 delivery to the tissues. It occurs normally in persons living at high altitudes, where less O_2 is available in the atmospheric air, or in individuals in whom O_2 delivery to the tissues is impaired as a result of chronic lung disease or cardiac failure. The red cell count in secondary polycythemia is usually lower than that in primary polycythemia, typically averaging 6 to 8 million cells/mm^3. The price paid for improved O_2 delivery is an increased viscosity of the blood.

An elevated hematocrit can occur if fluid but not erythrocytes is lost from the body, as in dehydration accompanying heavy sweating or profuse diarrhea (Fig. 11–5d). This is not a true polycythemia, however, because the number of circulating red blood cells is not increased. A normal number of erythrocytes are simply concentrated in a smaller plasma volume.

Leukocytes

Leukocytes function primarily outside of the blood.

Leukocytes, or **white blood cells,** are the mobile units of the body's defense system. They defend against foreign invasion by infectious agents such as bacteria and viruses in two different ways: (1) by actually engulfing and digesting the foreigner through phagocytosis; and (2) by immune responses such as the production of antibodies, which mark invaders for destruction in more subtle ways. The leukocytes' defense tasks also include destruction of cancer cells that arise within the body. Some leukocytes also function as a "clean-up crew" that removes the body's "litter" by phagocytizing debris resulting from dead or injured cells.

To carry out their functions, the leukocytes largely employ a "seek out and attack" strategy; that is, they go to sites of invasion or tissue damage. The main reason white blood cells are present in the blood is so that they can be transported from their site of production or storage to wherever they are needed. Therefore, although the specific circulating leukocytes will be introduced here to round out our discussion of blood, we will leave a more detailed discussion of their phagocytic and immunological functions, which take place primarily in the tissues, for the next chapter, "Defense Mechanisms of the Body."

There are five different types of leukocytes.

Leukocytes, in contrast to erythrocytes, lack hemoglobin, so they are colorless (that is, are "white") unless specifically stained for microscopic visibility. Also, unlike erythrocytes—which are of uniform structure, identical function, and constant number—leukocytes vary in structure, function, and number. There are five different types of circulating leukocytes—neutrophils, eosinophils, basophils, monocytes, and lymphocytes—each with a characteristic structure and function. They are all somewhat larger than erythrocytes, ranging in size from 9 to 15 μm in diameter.

The five types of leukocytes fall into two main categories, depending on the appearance of their nuclei and the presence or absence of granules in their cytoplasm when viewed microscopically (Fig. 11–6). Neutrophils, eosinophils, and basophils are categorized as **polymorphonuclear** ("many-shaped nucleus") **granulocytes** ("granule-containing cells"). Their nuclei are segmented into several lobes of varying shapes, and their cytoplasm contains an abundance of membrane-bound granules. The three types of granulocytes are distinguished on the basis of the varying affinity of their granules for dyes: *eosinophils* have an affinity for the red dye eosin; *basophils* preferentially take up a basic blue dye; and *neutrophils* are neutral, showing no dye preference. Monocytes and lymphocytes are known as **mononuclear** ("single nucleus") **agranulocytes** ("cells lacking granules"). Both have a single, large, nonsegmented nucleus and few granules. *Monocytes* are the larger of the two and have an oval or kidney-shaped nucleus. *Lymphocytes*, the smallest of the leukocytes, characteristically have a large spherical nucleus that occupies most of the cell.

Leukocytes are produced at varying rates depending on the changing defense needs of the body.

All leukocytes ultimately originate from the same undifferentiated stem cells in the red bone marrow that also give rise to erythrocytes and platelets (Fig. 11–7). The cells destined to become leukocytes eventually differentiate into various committed cell lines and proliferate under the influence of appropriate stimulating factors. Granulocytes and monocytes are produced only in the bone marrow, which releases these mature leukocytes into the blood. Lymphocytes are originally derived from precursor cells in the bone marrow, but most new lymphocytes are actually produced by already-existing lymphocytes residing in the lymphoid (lymphocyte-containing) organs, such as the lymph nodes and tonsils.

The total number of leukocytes normally ranges from 5 to 10 million cells per milliliter of blood, with an average of 7 million cells/ml, expressed as an average **white blood cell count** of 7,000/mm^3. Leukocytes are the least numerous of the cellular elements in the blood (about 1 white blood cell for every 700 red blood cells), not because fewer are produced but because they are merely in transit while in the blood. Nor-

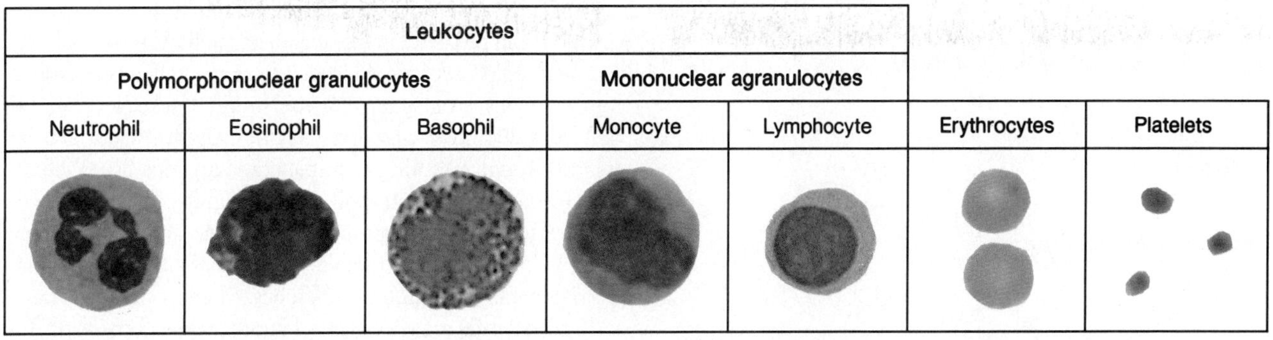

Leukocytes						
Polymorphonuclear granulocytes			Mononuclear agranulocytes			
Neutrophil	Eosinophil	Basophil	Monocyte	Lymphocyte	Erythrocytes	Platelets

Figure 11–6 Normal Blood Cellular Elements
SOURCE: Photos courtesy of Elizabeth R. Walker, Associate Professor, Dennis O. Overman, Associate Professor, and William A. Beresford, Professor, Department of Anatomy, School of Medicine, West Virginia University.

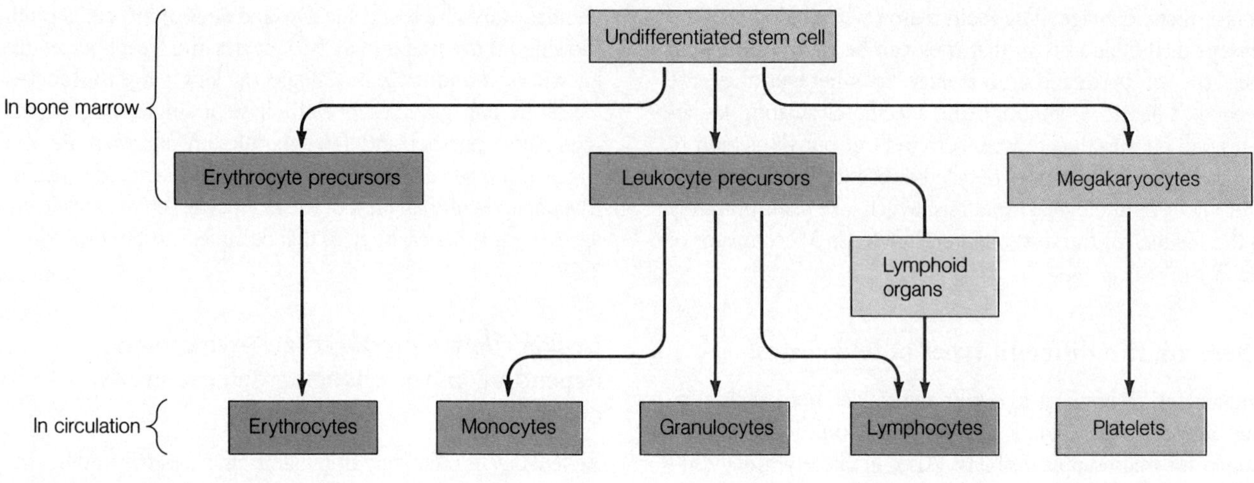

Figure 11–7 Hematopoiesis

mally, approximately two-thirds of the circulating leukocytes are granulocytes, mostly neutrophils, whereas one-third are agranulocytes, predominantly lymphocytes (Table 11–2). However, the total number of white cells and the percentage of each type may vary considerably to meet changing defense needs. Depending on the type and extent of assault the body is combating, different types of leukocytes are selectively produced at varying rates. How this is controlled is still being vigorously pursued. Chemical messengers arising from invaded or damaged tissues or from activated leukocytes themselves govern the rates of production of the various leukocytes. There is increasing evidence that a specific glycoprotein hormone analogous to erythropoietin is required to direct the differentiation and proliferation of each cell type. Some of these hormones have been identified and can be produced in the laboratory. This opens up the possibility that these hormones can be administered as a powerful new therapeutic tool to bolster a person's normal defense against infection or cancer. Examining the functions of the various types of leukocytes provides clues as to what conditions normally prompt their production.

Table 11–2 Typical Human Blood Cell Count

Total erythrocytes = 5,000,000,000 cells/ml blood	
Red blood cell count = 5,000,000/mm^3	
Total leukocytes = 7,000,000 cells/ml blood	
White blood cell count = 7,000/mm^3	
Differential white blood cell count (percent distribution of types of leukocytes)	
Polymorphonuclear granulocytes	
Neutrophils	60–70%
Eosinophils	1–4%
Basophils	0.25–0.5%
Mononuclear agranulocytes	
Lymphocytes	25–33%
Monocytes	2–6%
Total platelets = 250,000,000/ml blood	
Platelet count = 250,000/mm^3	

Neutrophils are phagocytic specialists. They invariably are the first defenders on the scene of bacterial invasion and, accordingly, are very important in inflammatory responses. Furthermore, they scavenge to clean up debris. As might be expected in view of these functions, an increase in circulating neutrophils (**neutrophilia**) typically accompanies acute bacterial infections. In fact, a differential white blood cell count (a determination of the proportion of each type of leukocyte present) can be useful in making an immediate, reasonably accurate prediction of whether an infection, such as pneumonia or meningitis, is of bacterial or viral origin. It takes several days before a definitive answer as to the causative agent can be made through culturing a sample of the infected tissue's fluid. Since an elevated neutrophil count is highly indicative of bacterial infection, it is appropriate to initiate antibiotic therapy long before the true causative agent is actually known. (Bacteria generally succumb to antibiotics, whereas viruses do not.)

Eosinophils are specialists of another type. **Eosinophilia** (an increase in circulating eosinophils) is associated with internal parasite infestations (for example, worms), allergic conditions (such as asthma and hay fever), and autoimmune diseases. Eosinophils obviously cannot engulf a much larger parasitic worm, but they do attach to the worm and secrete substances that kill it.

Basophils are the least numerous and most poorly understood of the leukocytes. They are quite similar structurally and functionally to mast cells, which never circulate in the blood but instead are dispersed in the connective tissue throughout the body. It was once believed that basophils became mast cells by migrating from the circulatory system, but it has been shown that basophils arise from the bone marrow whereas mast cells are derived from precursor cells located in the connective tissue. Both basophils and mast cells synthesize and store histamine and heparin, powerful chemical substances that can be released upon appropriate stimulation. Histamine release is important in allergic reactions, whereas heparin hastens the removal of fat particles from the blood following a fatty meal. Heparin can also prevent blood clotting (coagulation), but whether or not it plays a physiological role as an anticoagulant is still being debated.

The number of circulating granulocytes is adjusted in two ways. First, the vast majority of the mature granulocytes is stored within the bone marrow, providing a readily available pool of these cells that can be released into the blood when the need arises in response to unknown signals. Second, the actual rate of production of each type of these cells can be varied as needed. A released granulocyte usually remains in transit in the blood for less than a day before leaving the blood vessels to enter the tissues, where it survives another three to four days unless it dies sooner in the line of duty.

Monocytes, like neutrophils, are destined to become professional phagocytes. They emerge from the bone marrow while still immature and circulate for only a day or two before settling down in various tissues throughout the body. At their new residences, monocytes continue to mature and greatly enlarge, becoming the large tissue phagocytes known as **macrophages.** Macrophages' lifespan may range from months to years unless they are destroyed sooner while performing their phagocytic activity. A phagocytic cell can ingest only a limited amount of foreign material before it succumbs itself.

Lymphocytes provide immune defense against targets for which they are specifically programmed. There are two types of lymphocytes, B lymphocytes and T lymphocytes. **B lymphocytes** produce **antibodies,** which circulate in the blood. An antibody binds with and marks for destruction specific kinds of foreign matter, such as bacteria, that induced its production. **T lymphocytes** do not produce antibodies; instead they directly destroy their specific target cells, a process known as a **cell-mediated immune response.** Lymphocytes are estimated to have life spans of 100 to 300 days. During this period, the majority of them continually recycle among the lymphoid organs, lymph, and blood, spending only a few hours at a time in the blood. Therefore, only a small proportion of the total lymphocytes are in transit in the blood at any given moment.

By unknown mechanisms, the number of circulating lymphocytes is frequently elevated in association with a chronic infection. In **infectious mononucleosis,** not only is the number of lymphocytes in the blood increased but many of the lymphocytes are atypical in structure. This condition, which is caused by the Epstein-Barr virus, is characterized by pronounced fatigue, a mild sore throat, and low-grade fever. Full recovery usually requires a month or more.

Abnormal leukocyte production can occur.

Even though circulating-leukocyte levels may vary, changes in these levels are normally controlled and adjusted based on the body's needs. However, abnormalities in leukocyte production that are not subject to regulatory mechanisms can occur; that is, either too many or too few leukocytes are produced. The bone marrow can greatly slow down or even stop its production of white blood cells when it is exposed to certain toxic physical agents (such as radiation) or chemical agents (such as benzene and anticancer drugs). As you might predict, the most serious consequence is the reduction in professional phagocytes (neutrophils and macrophages), which leads to a notable diminution in the body's defense capabilities against invading microorganisms. The only defense still available when the bone marrow fails is the immune capabilities of the lymphocytes produced by the lymphoid organs.

Surprisingly, one of the major consequences of **leukemia,** a cancerous condition that involves uncontrolled proliferation of white blood cells, is inadequate defense capabilities against foreign invasion. The abnormal leukocyte proliferation can occur either in the bone marrow (*myelogenous leukemia,* which is associated with excessive granulocytes and monocytes), or it can be of lymphoid origin (*lymphogenous leukemia,* which is characterized by excessive numbers of lymphocytes). In leukemia, the white blood cell count may reach as high as 500,000/mm^3, compared to the normal of 7,000/mm^3, but because the majority of these cells are abnormal or immature, they are therefore incapable of performing their normal defense functions. Another devastating consequence of leukemia is displacement of the other hematopoietic cell lines in the bone marrow. This results in anemia because of a reduction in erythropoiesis and in internal bleeding because of a deficit of platelets. Platelets play a critical role in preventing bleeding from the myriad tiny breaks that normally occur in small blood-vessel walls. Consequently, overwhelming infections or hemorrhage are the most common causes of death in leukemic patients. The next section will examine the platelets' role in greater detail to show how they normally minimize the threat of hemorrhage.

Platelets and Hemostasis

Platelets are cell fragments derived from megakaryocytes.

In addition to erythrocytes and leukocytes, a third type of cellular element, **platelets** (also known as **thrombocytes**), are present in the blood. Platelets are not whole cells but are small, colorless cell fragments (about 2 to 4 μm in diameter) that have budded off the outer edges of extraordinarily large (up to 60 μm in diameter) bone marrow–bound cells known as **megakaryocytes.** Megakaryocytes are derived from the same undifferentiated hematopoietic precursor cells that give rise to the erythrocytic and leukocytic cell lines (Fig. 11–7). Platelets are essentially detached vesicles containing portions of megakaryocyte cytoplasm wrapped in plasma membrane. Because they are cell fragments, platelets lack nuclei. However, they are equipped with organelles and cytosolic enzyme systems for generating energy and for synthesizing secretory products, which they store in numerous granules dispersed throughout the cytosol. Furthermore, platelets contain high concentrations of actin and myosin, which enable them to contract.

An average of 250,000,000 platelets are normally present in each milliliter of blood (range of 150,000-350,000/mm^3). Platelets remain functional for an average of ten days, at which time they are removed from circulation by the tissue macrophages, especially those in the spleen and liver, and are replaced by newly released platelets from the bone marrow.

Platelets do not leave the blood as white blood cells do, but about one-third of them at any time are in storage in blood-filled spaces in the spleen. These stored platelets can be released from the spleen into the circulating blood as needed (for example, during hemorrhage) by sympathetically induced splenic contraction.

Hemostasis prevents blood loss from damaged small vessels.

Hemostasis is the arrest of bleeding from a broken blood vessel; that is, stopping of hemorrhage. (Be sure not to confuse this with the term ***homeostasis***.) For bleeding to take place from a vessel, there must be a break in the vessel wall, and the pressure inside the vessel must be greater than outside it to force the blood out through the defect. The body's inherent hemostatic mechanisms normally are adequate to seal defects and stop loss of blood through small damaged capillaries, arterioles, and venules. These microcirculatory vessels are frequently ruptured by the minor traumas of everyday life; this is the most common source of bleeding, although we usually are not even aware that any damage has taken place. Fortunately, the hemostatic mechanisms normally keep blood loss from these minor vascular traumas to a minimum.

The much rarer occurrence of bleeding from medium- to large-size vessels usually cannot be stopped by the body's hemostatic mechanisms alone. Bleeding from a severed artery is more profuse and therefore more dangerous than venous bleeding because the outward driving pressure is greater in the arteries (that is, arterial blood pressure is considerably higher than venous pressure). First-aid measures for a severed artery include application over the wound of external pressure of greater magnitude than the arterial blood pressure to temporarily halt the bleeding until the torn vessel can be surgically closed. Hemorrhage from a traumatized vein can often be stopped simply by elevating the bleeding body part to reduce gravity's effects on pressure in the vein (see p. 325). If the accompanying drop in venous pressure is not sufficient to stop the bleeding, mild external compression is usually adequate.

We will devote the remainder of the chapter to the normal ongoing hemostatic mechanisms that save us on a daily basis from crucial blood loss accompanying small-vessel damage. There are three major steps in hemostasis: (1) vascular spasm, (2) formation of a platelet plug, and (3) blood coagulation (clotting). Platelets obviously play a major part in formation of a platelet plug, but they contribute significantly to the other two steps as well.

Vascular spasm reduces blood flow through an injured vessel.

A cut or torn blood vessel immediately constricts as a result of an inherent vascular response to injury or because of sympathetically induced vasoconstriction. This constriction slows blood flow through the defect, thus minimizing blood loss. Also, as the opposing endothelial surfaces are pressed together by this initial **vascular spasm,** they become sticky and adhere to each other, further sealing off the damaged vessel. These physical measures alone cannot completely prevent further blood loss except perhaps in very small capillaries. They are important, however, in minimizing blood flow through the break in the vessel until the other hemostatic measures are able to actually plug up the defect.

Platelets aggregate to form a plug at a vessel defect.

Platelets normally do not adhere to the smooth endothelial surfaces of blood vessels, but when the endothelial surface is disrupted because of vessel injury, platelets attach to the ex-

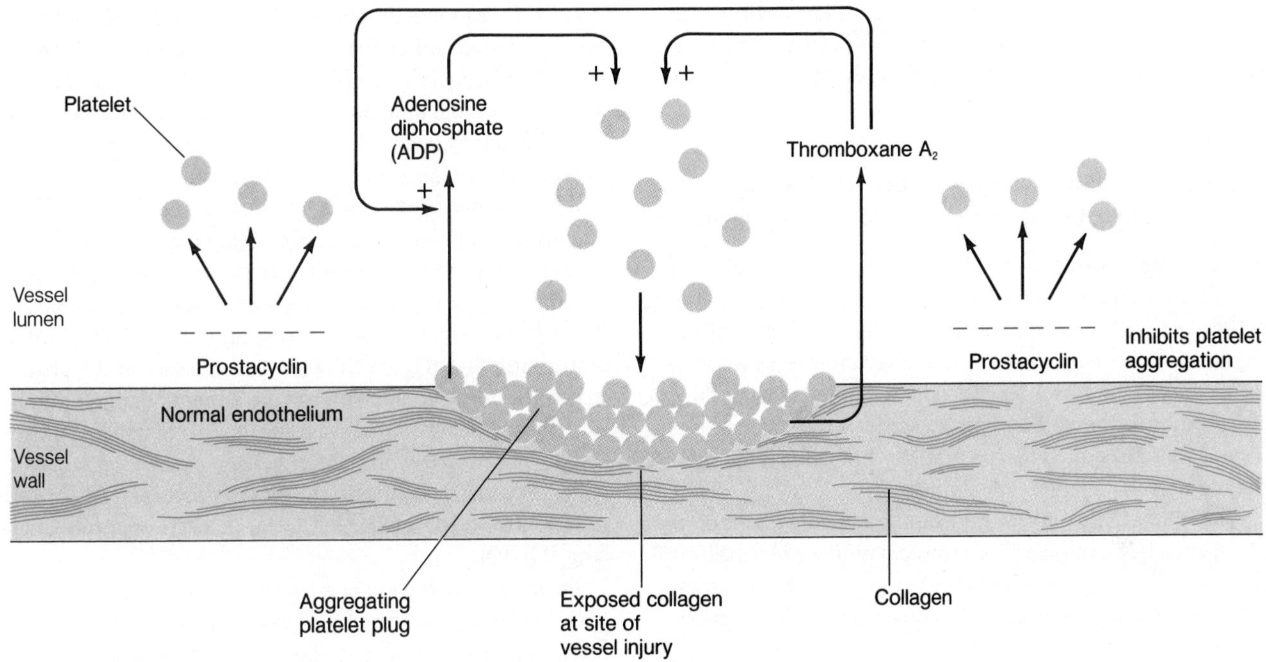

Figure 11–8 Platelet Aggregation *Platelets aggregate at vessel defect through several positive-feedback mechanisms but are prevented from aggregating at adjacent normal endothelium.*

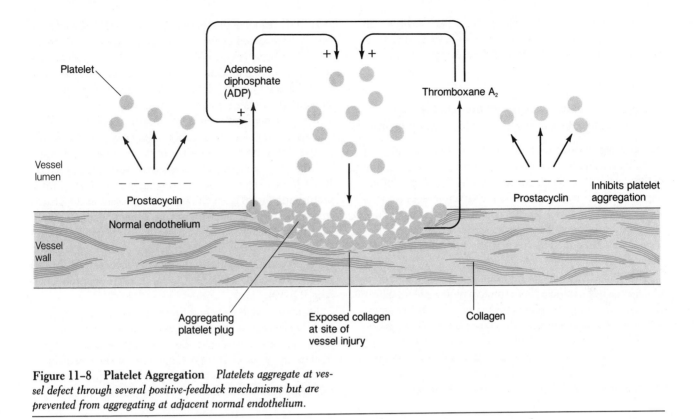

posed collagen fibers in the underlying connective tissue. Once platelets start aggregating at the site of the defect, they release several important chemicals from their storage granules (Fig. 11–8). Among these chemicals is adenosine diphosphate (ADP), which causes the surface of nearby circulating platelets to become sticky so that they adhere to the first layer of aggregated platelets. More ADP is released from these newly aggregated platelets, which causes more platelets to aggregate, and so on, rapidly building up a plug of platelets at the defect site in a positive-feedback fashion.

This aggregating process is reinforced by the formation of a chemical messenger, **thromboxane A_2**, from a component of the platelet plasma membrane upon contact with collagen. Thromboxane A_2 is closely related to the prostaglandins, a group of locally functioning chemical messengers found extensively throughout the body (see chapter 20). These local messengers are fatty-acid derivatives formed from arachidonic acid, one of the fatty acids found in membrane phospholipids. Thromboxane A_2 directly promotes platelet aggregation and further enhances it indirectly by triggering the release of even more ADP from the platelet granules.

Given all these self-perpetuating features of platelet aggregation, why is the platelet plug limited to the site of vessel injury once it is initiated? (In other words, why doesn't the platelet plug continue to develop and expand over the surface of adjacent normal endothelium?) A key reason this does not happen is that normal endothelium contains an enzyme that converts arachidonic-acid intermediates into a prostaglandin (specifically, **prostacyclin**) rather than thromboxane A_2. Whereas thromboxane A_2 (which is formed when platelets come into contact with injured vessel surfaces) induces platelet aggregation, prostacyclin (which is formed when platelets come into contact with the adjacent normal endothelium) profoundly inhibits platelet aggregation. Thus, the platelet plug is limited to the defect and does not spread to normal vascular tissue (Fig. 11–8).

The aggregated platelet plug not only physically seals the break in the vessels but also performs three other important roles. First, the actin-myosin protein complex within the aggregated platelets contracts to compact and strengthen what was originally a fairly loose plug. Second, the chemicals released from the platelet plug include several powerful vasoconstrictors (serotonin, epinephrine, and thromboxane A_2), which induce profound constriction of the affected vessel to reinforce the initial, self-induced vascular spasm. Third, the platelet plug releases other chemicals that enhance blood coagulation, the next step of hemostasis. Although the platelet-plugging mechanism alone is often sufficient to seal the myriad minute tears in capillaries and other small vessels that occur many times daily, larger holes in these vessels require the formation of a blood clot to completely stop the bleeding.

A triggered chain reaction involving clotting factors in the plasma results in blood coagulation.

Blood coagulation, or **clotting,** is the transformation of blood from a liquid into a solid gel. Formation of a clot on top of the platelet plug stengthens and supports the plug, reinforcing the seal over a break in a vessel. Furthermore, as blood in the vicinity of the vessel defect solidifies, it can no longer flow. Coagulation is the body's most powerful hemostatic mechanism, and it is required to stop bleeding from all but the most minute defects.

The ultimate step in clot formation is the conversion of fibrinogen, a large, soluble plasma protein produced by the liver and normally always present in the plasma, into **fibrin,** an insoluble, threadlike molecule. The conversion into fibrin is catalyzed by the enzyme **thrombin** at the site of vessel injury.

Fibrin molecules adhere to the damaged vessel surface, forming a loose, netlike meshwork that traps the cellular elements of the blood. The resultant mass, or **clot,** typically appears red because of the abundance of trapped red blood cells, but the foundation of the clot is formed from fibrin derived from the plasma (Fig. 11–9). Except for platelets, which play an important role in ultimately bringing about the conversion of fibrinogen to fibrin, clotting can take place in the absence of all other cellular elements in the blood.

The original fibrin web is rather weak, because the fibrin strands are only loosely interlaced. However, chemical linkages rapidly form between adjacent strands to strengthen and stabilize the clot meshwork. This cross-linkage process is catalyzed by a clotting factor known as **factor XIII (fibrin-stabilizing factor**), which normally is present in the plasma in inactive form. In addition to converting fibrinogen into fibrin, thrombin also: (1) activates factor XIII to stabilize the resulting fibrin meshwork; (2) enhances platelet aggregation, which in turn is essential to the clotting process; and (3) acts in a positive-feedback fashion to facilitate its own formation (Fig. 11–10).

Because thrombin's action converts the ever-present fibrinogen molecules in the plasma into a blood-stanching clot, thrombin must normally be absent from the plasma except in the vicinity of vessel damage. Otherwise, blood would always be coagulated, a situation incompatible with life. How can thrombin normally be absent from the plasma, yet be readily

Figure 11–9 Erythrocytes Trapped in Fibrin Meshwork of a Clot

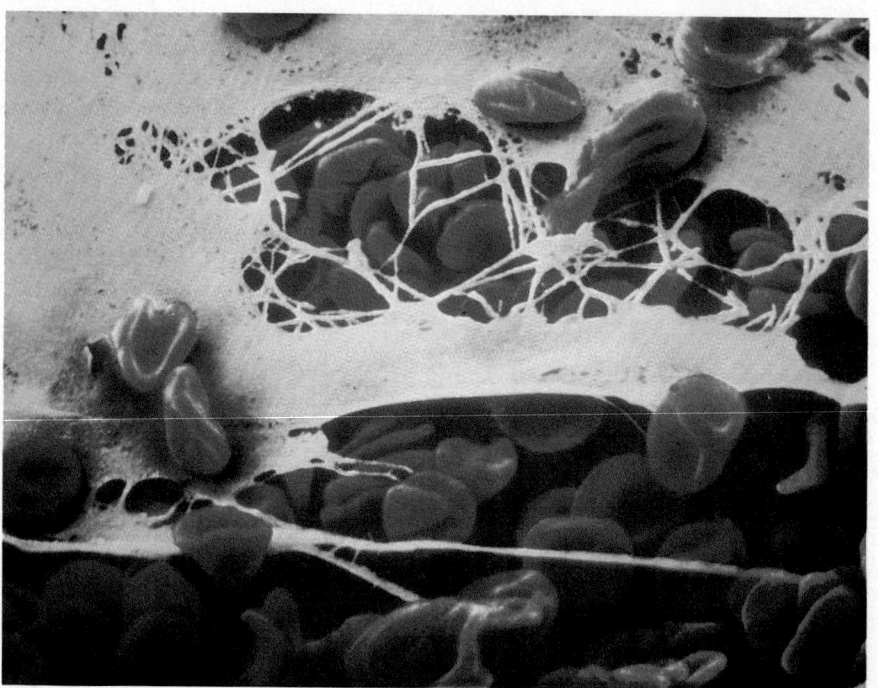

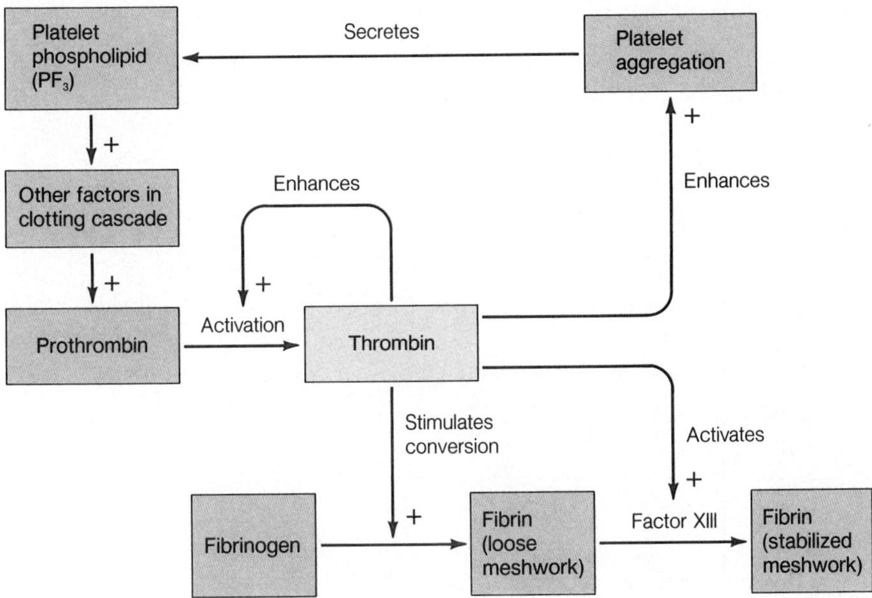

Figure 11–10 Roles of Thrombin

available to trigger fibrin formation when a vessel is injured? The solution lies in thrombin's existence in the plasma in the form of an an inactive precursor called **prothrombin.**

The next question is, what converts prothrombin into thrombin when blood clotting is desirable? Yet another activated plasma clotting factor, **factor X,** is responsible, which itself is normally present in the blood in inactive form and must be converted into its active form by still another activated factor, and so on. Altogether, twelve plasma clotting factors participate in essential steps that lead to the final conversion of fibrinogen into fibrin (Fig. 11–11). These factors are designated by Roman numerals in the order in which they were discovered, not the order in which they participate in the clotting process.[1] Most of these clotting factors are plasma proteins synthesized by the liver. Normally they are always present in the plasma in an inactive form, similar to fibrinogen and prothrombin. In contrast to fibrinogen, which is converted into insoluble fibrin strands, when prothrombin and the other precursors are converted to their active form, they act as proteolytic (protein-splitting) enzymes, which activate another specific factor in the clotting sequence. Once the first factor in the sequence is activated, it in turn activates the next factor, and so on in a series of sequential reactions known as a **cascade,** until thrombin catalyzes the final conversion of fibrinogen into fibrin. Several of these steps require the presence of plasma Ca^{++} and **PF3,** a phospholipid secreted by the aggregated platelet plug.

The clotting cascade may be triggered by the intrinsic pathway or the extrinsic pathway. All elements necessary to bring about clotting by means of the intrinsic pathway are present in the blood. This pathway precipitates intravascular clotting as well as clotting of blood samples in test tubes. The extrinsic pathway, which requires contact with tissue factors external to the blood, initiates clotting of blood that has escaped into the tissues. We will compare the steps in these two pathways as follows (Fig. 11–11):

□ The **intrinsic pathway,** which involves seven separate steps, is set off when **factor XII** (**Hageman factor**) is activated by coming into contact with either exposed collagen in an injured vessel or a foreign surface such as a glass test tube. Remember that exposed collagen also initiates platelet aggregation. Thus, formation of a platelet plug and the chain reaction leading to clot formation are simultaneously set in motion when a vessel is damaged. Furthermore, these complementary hemostatic mechanisms reinforce each other. The aggregated platelets secrete PF3, which is essential for the clotting cascade that in turn enhances further platelet aggregation (Figs. 11–10 and 11–12).

[1]The term *factor VI* is no longer used. What once was considered to be a separate factor VI has now been determined to be an activated form of factor V.

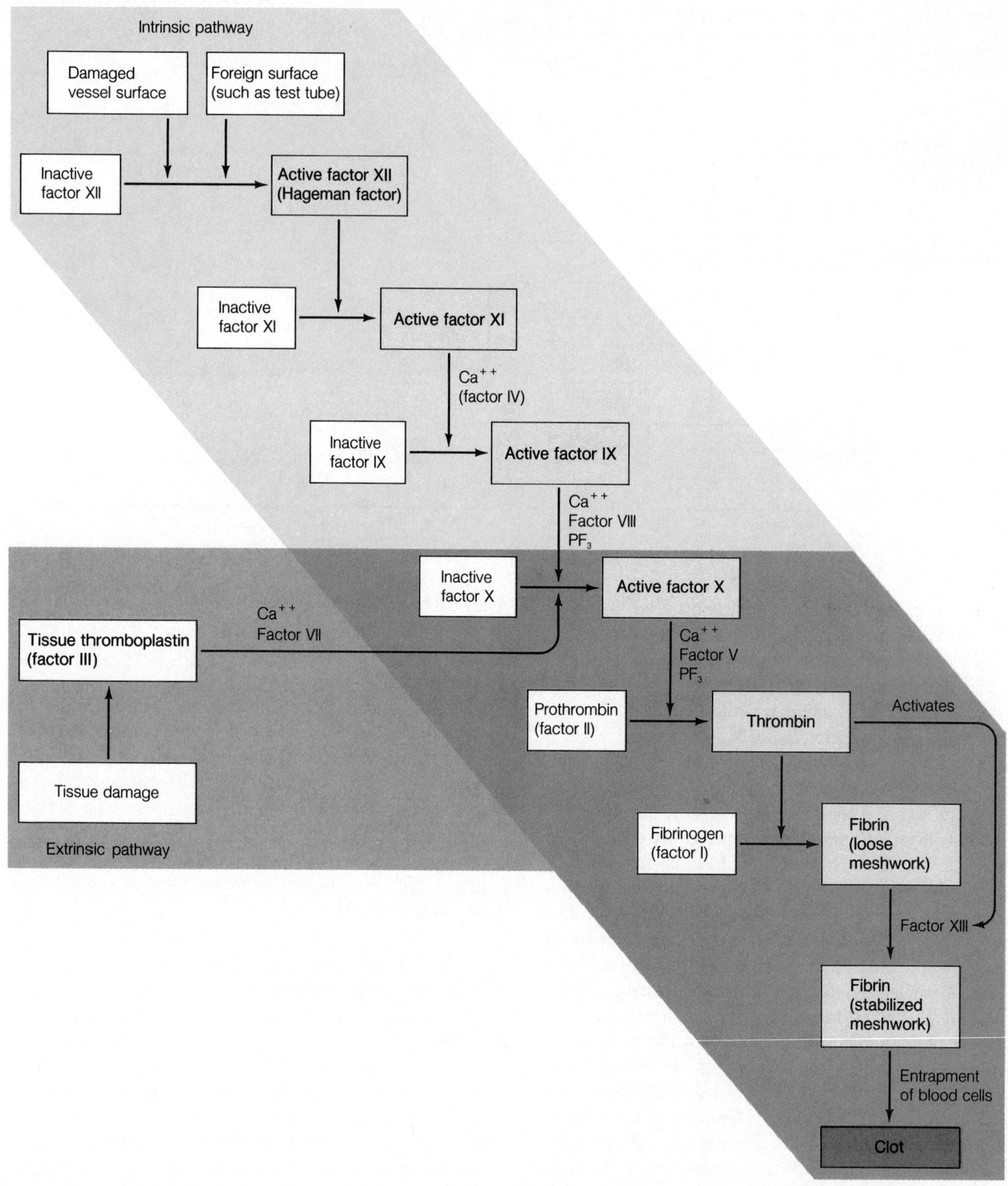

Figure 11–11 Clot Pathways

□ The **extrinsic pathway** takes a short-cut and requires only four steps. When a tissue is traumatized, it releases a protein complex known as **tissue thromboplastin.** Tissue thromboplastin, in consort with plasma **factor VII** and Ca^{++}, directly activates Factor X, thereby bypassing all preceding steps of the intrinsic pathway. From this point on, the two pathways are identical.

The intrinsic and extrinsic mechanisms usually operate simultaneously. When tissue injury involves rupture of vessels, the intrinsic mechanism stops blood in the injured vessel, whereas the extrinsic mechanism clots the blood that escaped into the tissues before the vessel was sealed off. Typically, clot formation is fully developed in three to six minutes.

Once a clot is formed, contraction of the platelets trapped within the clot shrinks the fibrin meshwork, pulling the edges of the damaged vessel closer together. During **clot retraction,** fluid is squeezed from the clot. This fluid, which is essentially plasma minus fibrinogen and other clotting precursors that have been removed during the clotting process, is called **serum.**

Although it might seem inefficient to have so many steps involved in clotting, the advantage is the amplification accomplished during many of the steps. One molecule of an activated factor can activate perhaps a hundred molecules of the next factor in the sequence, each of which can activate many more molecules of the next factor, and so on. In this way, large numbers of the final factors involved in clotting are rapidly activated as a result of the initial activation of only a few molecules in the beginning step of the sequence. How then is the clotting process, once initiated, confined to the site of vessel injury? If the activated clotting factors were allowed to circulate, they would induce inappropriate widespread clotting that would plug up vessels throughout the body. Fortunately, after participating in the local clotting process, the massive number of factors activated in the vicinity of vessel injury are rapidly inactivated by enzymes and other factors present in the plasma or tissue.

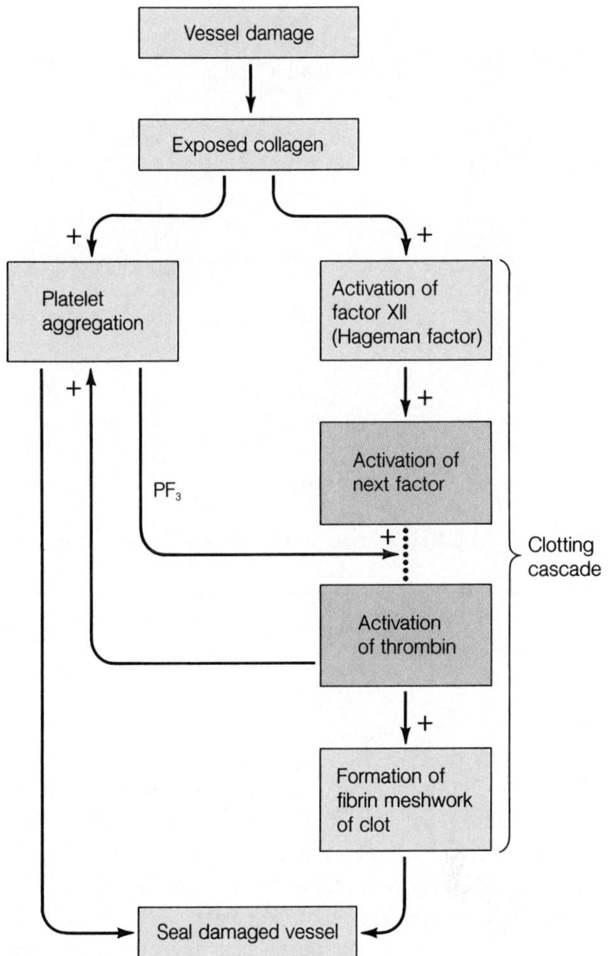

Figure 11–12 Concurrent Platelet Aggregation and Clot Formation

Fibrinolytic plasmin dissolves clots and prevents inappropriate clot formation.

A clot is not meant to be a permanent solution to vessel injury. It is a transient device to stop bleeding until the vessel can be repaired. The aggregated platelets secrete a factor that is at least partially responsible for the invasion of fibroblasts from the surrounding connective tissue into the wounded area of the vessel. Fibroblasts form a scar at the vessel defect. Simultaneous to the healing process, the clot, which is no longer needed to prevent hemorrhage, is slowly dissolved by a fibrinolytic (fibrin-splitting) enzyme called **plasmin.** If clots were not removed after they performed their hemostatic function, the vessels, especially the small ones that endure tiny ruptures on a regular basis, would eventually become occluded by clots.

Plasmin, like the clotting factors, is a plasma protein produced by the liver and present in the blood in an inactive precursor form, **plasminogen.** Plasmin is activated cascade-fashion by many factors, among them being factor XII (Hageman factor), which also triggers the chain reaction leading to clot formation (Fig. 11–13). When a clot is being formed, activated plasmin becomes trapped in the clot and

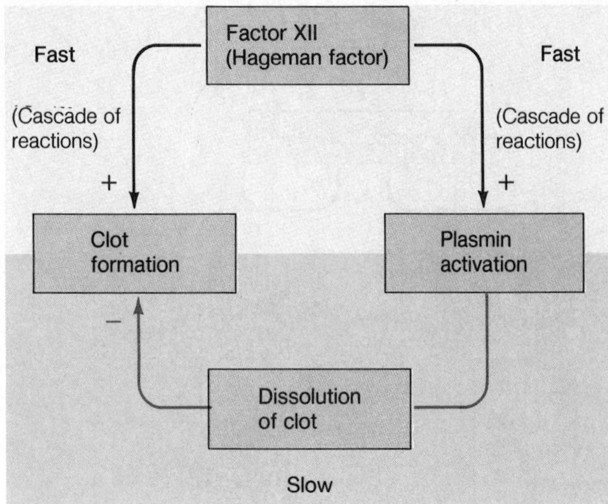

Figure 11–13 Role of Factor XII in Clot Formation and Dissolution

subsequently dissolves it by slowly breaking down the fibrin meshwork. Neutrophils and macrophages gradually remove the products of clot dissolution. You have observed the slow removal of blood that has clotted after escaping into the tissue layers of your skin following an injury. The "black and blue marks" of bruised skin result from deoxygenated clotted blood within the skin, which is eventually cleared by plasmin action, followed by the phagocytic clean-up crew.

In addition to removing clots that are no longer needed, plasmin functions continually to prevent clots from forming inappropriately. Small amounts of fibrinogen are constantly being converted into fibrin throughout the vasculature, triggered by unknown mechanisms. Clots do not develop, however, because the fibrin is quickly disposed of by plasmin that is activated by **plasminogen activators** derived from the tissues, especially the lungs. Normally, the low level of fibrin formation is counterbalanced by a low level of fibrinolytic activity, so inappropriate clotting does not occur. Only when a vessel is damaged do additional factors precipitate the explosive chain reaction that leads to more extensive fibrin formation and results in local clotting at the site of injury.

Naturally occurring anticoagulants that interfere with one or more steps in the clotting process are known to exist in the body. Whether or not they play a physiological role in clot prevention is still being debated. The best known of these is **heparin,** which is present in basophils and mast cells. It is used extensively as an anticoagulant drug as well as to prevent clotting of blood samples drawn for clinical analysis.

Inappropriate clotting is responsible for thromboembolism.

In spite of protective measures, occasionally clots form in intact vessels. Recall from chapter 9 that an abnormal intravascular clot attached to a vessel wall is known as a thrombus, and free floating clots are called emboli. Several factors, acting independently or simultaneously, can cause thromboembolism. (1) Roughened endothelial surfaces associated with atherosclerosis can lead to thrombus formation (see p. 293). (2) Imbalances in the clotting-anticlotting systems can likewise trigger clot formation. (3) Slow-moving blood is more apt to clot, probably because the small quantities of fibrin that are normally formed are allowed to accumulate in the stagnant blood. This can happen, for example, in blood pooled in varicosed leg veins. (4) Widespread clotting is triggered occasionally by the release of tissue thromboplastin into the blood from large amounts of traumatized tissue. Similar widespread clotting can occur in **septicemic shock,** in which bacteria or their toxins initiate the clotting cascade.

Hemophilia is the primary condition responsible for excessive bleeding.

In contrast to inappropriate clot formation in intact vessels, the opposite hemostatic disorder is failure of clots to form promptly in injured vessels, resulting in life-threatening hemorrhage from even relatively mild traumas. The most common cause of excessive bleeding is **hemophilia** caused by a deficiency of one of the factors in the clotting cascade. Although a deficiency of any of the clotting factors could block the clotting process, 80% of all hemophiliacs lack the genetic ability to synthesize factor VIII.

In contrast to the more profuse bleeding that accompanies defects in the clotting mechanism, individuals having a deficiency of platelets continuously develop hundreds of small, confined hemorrhagic areas throughout the body tissues as blood is permitted to leak from tiny breaks in the small blood vessels before coagulation takes place. Platelets normally are the primary sealers of these ever-occurring minute ruptures. In the skin of a platelet-deficient person, the diffuse capillary hemorrhages are visible as small, purplish blotches, giving rise to the term **thrombocytopenia purpura** ("the purple of thrombocyte deficiency") for this condition.

Vitamin K deficiency can cause a bleeding tendency. This vitamin is essential for the liver's synthesis of prothrombin and several other clotting factors. In addition to synthesizing the plasma-protein clotting factors, the liver also produces bile salts, which play an important role in the absorption of fat-soluble vitamins, including vitamin K, from the intestinal tract. Accordingly, one of the consequences of liver disease is prolonged clotting time.

Chapter in Perspective

Blood is the vehicle for long-distance mass transit of materials between the cells and the external environment or between the cells themselves. Such transport is essential for maintenance of homeostasis. Blood consists of a complex liquid plasma in which the cellular elements—erythrocytes (red blood cells), leukocytes (white blood cells), and platelets—are suspended.

Specific components of the blood perform additional activities that are unrelated to blood's transport function. Plasma proteins, which remain in the blood because they are too large to escape through the capillary pores, are important in the maintenance of plasma volume by virtue of the osmotic effect they exert. Plasma proteins also make important contributions to acid-base balance, blood clotting, and immune defense.

Even though the three types of cellular elements have very different functions, they share certain similarities in their life history. They are all derived from the same undifferentiated stem cells in the bone marrow, and they all have limited life spans and consequently must be replaced continuously by the production of new cells.

Erythrocytes are little more than plasma membrane-enclosed bags of hemoglobin that transport O_2 and to a lesser extent CO_2 and H^+. They are by far the most numerous of blood cells, not because they are more abundant but because they remain in the blood, unlike the leukocytes, which are only in transit in the blood. Erythrocytes occupy nearly half of the total blood volume, whereas leukocytes and platelets collectively occupy less than 1% of the blood volume.

Leukocytes, the body's mobile defense units, are largely in the tissues on combative or clean-up duty. Five different types of leukocytes work in coordinated ways to seek out and destroy foreign microbes that have entered the body from the external environment. Leukocytes also eliminate abnormal cells that arise from within and clean up particulate debris to pave the way for tissue repair.

Given the importance of blood, it is imperative that mechanisms exist to minimize blood loss when a vessel is injured. When a vessel is damaged, three hemostatic measures are triggered to arrest blood flow: vascular spasm, platelet plugging, and blood clotting. Platelets not only aggregate to plug the defect but also secrete chemicals that act in positive-feedback fashion to enhance all three steps of hemostasis. The blood-clotting factors are normally always present in the blood, but they exist in inactive presursor form. They are activated in cascade fashion only when a vessel is injured.

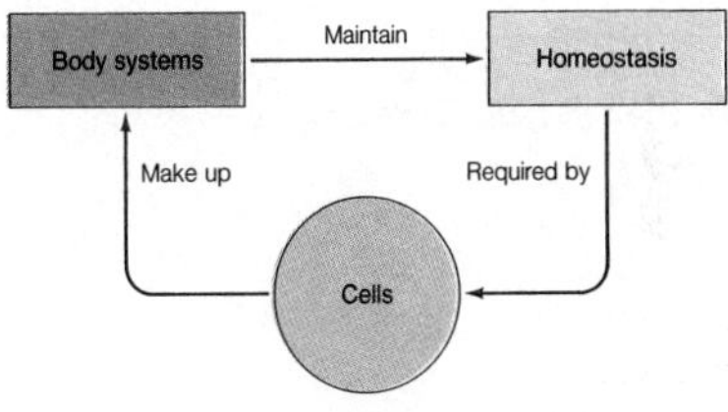

See inside front cover for an expanded version of this model.

Review Exercises

1. What is the average blood volume?
2. What is the normal percentage of blood occupied by erythrocytes and by plasma? What is the hematocrit? What is the buffy coat?
3. What is the composition of plasma?
4. List and state the functions of the three major groups of plasma proteins.
5. Describe the structure and functions of erythrocytes.
6. Why are erythrocytes able to survive for only about 120 days?
7. Describe the process and control of erythropoiesis.
8. Define and list the causes of anemia and polycythemia.
9. Describe the structure and functions of the five types of leukocytes.
10. Discuss the derivation of platelets.
11. Describe the three steps of hemostasis, including a comparison of the intrinsic and extrinsic pathways by which the clotting cascade is triggered.
12. Compare plasma and serum.
13. What factors normally prevent inappropriate coagulation in the vasculature?
14. **A point to ponder:** Why do you think it is important to closely monitor blood cell counts in cancer patients who are being treated with chemotherapeutic drugs designed to destroy rapidly multiplying cells, such as cancer cells?

CHAPTER 12

Defense Mechanisms of the Body

Chapter Contents at a Glance

INTRODUCTION *__Vaccination__ represents a modern victory over many dreaded diseases. Smallpox, diphtheria, tetanus, and poliomyelitis are examples of fatal or crippling diseases already brought under control through use of vaccines. Common childhood illnesses such as measles and mumps have likewise succumbed to this immunological tool. Modern society has come to hope and even expect that vaccines can be developed to protect us from almost any infectious agent.*

The process of vaccination is a deliberate manipulation of natural immune defense mechanisms, such as the production of antibodies that target invading __pathogens__ (disease-producing microorganisms) for destruction. Not only do some of these antibodies persist for long periods of time, but in many instances the body also "remembers" its original exposure to a particular pathogen, responding even more swiftly with new antibody production upon subsequent exposure to the same microbe. As a result, the infectious agent is destroyed before it has an opportunity to multiply and produce symptoms. The duration of protection following recovery from a disease may vary from lifelong immunity, as in the case of chickenpox, to little lasting protection, as in the case of "strep throat."

In the 1880s, Louis Pasteur, the first great experimental immunologist, demonstrated that the disease-inducing capability of organisms could be greatly reduced (attenuated) so that they could no longer produce disease but that they would still induce antibody formation when introduced into the body—the concept underlying modern-day vaccines. His first vaccine was against anthrax, a deadly disease of sheep and cows. Pasteur isolated and heated anthrax bacteria, then injected these attenuated organisms into a group of healthy sheep. A few weeks later at a gathering of fellow scientists, Pasteur injected these vaccinated sheep as well as a group of unvaccinated sheep with fully potent anthrax bacteria. The dramatic result—all vaccinated sheep survived whereas all unvaccinated sheep died. Pasteur's notorious public demonstrations, coupled with his charismatic personality, caught the attention of physicians and scientists of the time, sparking the development of modern immunology.

The immune defense system provides protection against foreign and abnormal cells and removes cellular debris.

Immunity refers to the body's ability to resist or eliminate potentially harmful foreign materials or abnormal cells. The following activities are attributable to the immune defense system, which plays a key role in recognizing and either destroying or neutralizing materials within the body that are foreign to the "normal self."

1. Defense against pathogenic microorganisms (such as viruses and bacteria).
2. Removal of "worn-out" cells (such as aged red blood cells) and tissue debris (for example, tissue damaged by trauma or disease). The latter is essential for wound healing and tissue repair.
3. Identification and destruction of abnormal or mutant cells that have originated in the body. This function, termed **immune surveillance,** is the primary internal defense mechanism against cancer.
4. Inappropriate immune responses that lead either to *allergies*, which occur when the body turns against a normally harmless environmental chemical entity, or to *autoimmune diseases*, which happens when the defense system erroneously produces antibodies against itself, leading to destruction of a particular type of the body's own cells.
5. Rejection of tissue cells of foreign origin, which constitutes the major obstacle to successful organ transplantation.

Pathogenic bacteria and viruses are the major targets of the immune defense system.

The primary foreign enemies against which the immune system defends are bacteria and viruses. **Bacteria** are nonnucleated, single-celled microorganisms self-equipped with all machinery essential for their own survival and reproduction. Pathogenic bacteria that invade the body induce tissue damage and produce disease largely by releasing enzymes or toxins that physically injure or functionally disrupt affected cells and organs. The disease-producing power of a pathogen is known as its **virulence.**

Viruses, in contrast to bacteria, are not self-sustaining cellular entities. They consist only of nucleic acids (DNA or RNA) enclosed by a protein coat. Because they lack cellular machinery for energy production and protein synthesis, viruses are unable to carry out metabolism and reproduce unless they invade a **host cell** (a body cell of the infected individual) and take over the cellular biochemical facilities for their own purposes. Not only do viruses sap the host cell's energy resources, but the viral nucleic acids also direct the host cell to synthesize proteins needed for viral replication. The effect of viral invasion and replication on the host cell varies with different types of viruses.

There are four general ways in which viruses can lead to cellular damage or death: (1) depletion of essential cellular components by the virus; (2) cellular production of substances toxic to the cell under the dictatorship of the virus; (3) transformation of normal host cells into cancer cells; and (4) incorporation of the virus into the cell so that the body's own defense mechanisms destroy the cell, because it is no longer recognized as a "normal-self" cell.

Leukocytes are the effector cells of the immune defense system.

The cells responsible for the various immune defense strategies are the leukocytes (white blood cells) and their derivatives, briefly reviewed as follows (see pp. 350–353):

1. Neutrophils are highly mobile phagocytic specialists that engulf and destroy unwanted materials.
2. Eosinophils secrete chemicals that destroy parasitic worms and are involved in allergic manifestations.
3. Basophils release histamine and heparin and also are involved in allergic manifestations.
4. Lymphocytes
   a. B lymphocytes are transformed into plasma cells, which secrete antibodies that indirectly lead to the destruction of foreign material.
   b. T lymphocytes are responsible for cell-mediated immunity involving direct destruction through nonphagocytic means of virus-invaded cells and mutant cells.

5. Monocytes are transformed into macrophages, which are large, tissue-bound phagocytic specialists.

A given leukocyte is present in the blood only transiently. Most of the leukocytes are out in the tissues on defense missions. As a result, the immune system's effector cells are widely dispersed throughout the body and are able to defend in any location.

Almost all leukocytes originate from common precursor stem cells in the bone marrow and are subsequently released into the blood. The only exception is lymphocytes, which arise in part from lymphocyte colonies in various lymphoid tissues that were originally populated by cells derived from the bone marrow. **Lymphoid tissues** refer collectively to the tissues that store, produce, or process lymphocytes. These include the lymph nodes, spleen, thymus, tonsils, adenoids, appendix; aggregates of lymphoid tissue in the lining of the gastrointestinal tract called Peyer's patches or gut-associated lymphoid tissue (GALT); and the bone marrow (Fig. 12–1). Lymphoid tissues are strategically located to intercept invading microorganisms before they have a chance to spread very far. For example, the lymphocytes populating the tonsils and adenoids are situated advantageously to respond to inhaled microbes, whereas microorganisms invading through the digestive system immediately encounter lymphocytes in the appendix and GALT. Potential pathogens that gain access to the lymph are filtered through lymph nodes, where they are exposed to lymphocytes as well as to macrophages that line the lymphatic passageways. The spleen, the largest of the lymphoid organs, performs immune functions on the blood similar to those that the lymph nodes perform on the lymph. Through actions of its lymphocyte and macrophage population, the spleen clears the blood that passes through it of microorganisms and other foreign matter and also removes worn-out red blood cells. The thymus and bone marrow play important roles in processing T and B lymphocytes, respectively, to prepare them to carry out their specific immune strategies. Table 12–1 summarizes the major functions of the various lymphoid tissues, some of which were described in chapter 11 and others of which are yet to be discussed in this chapter.

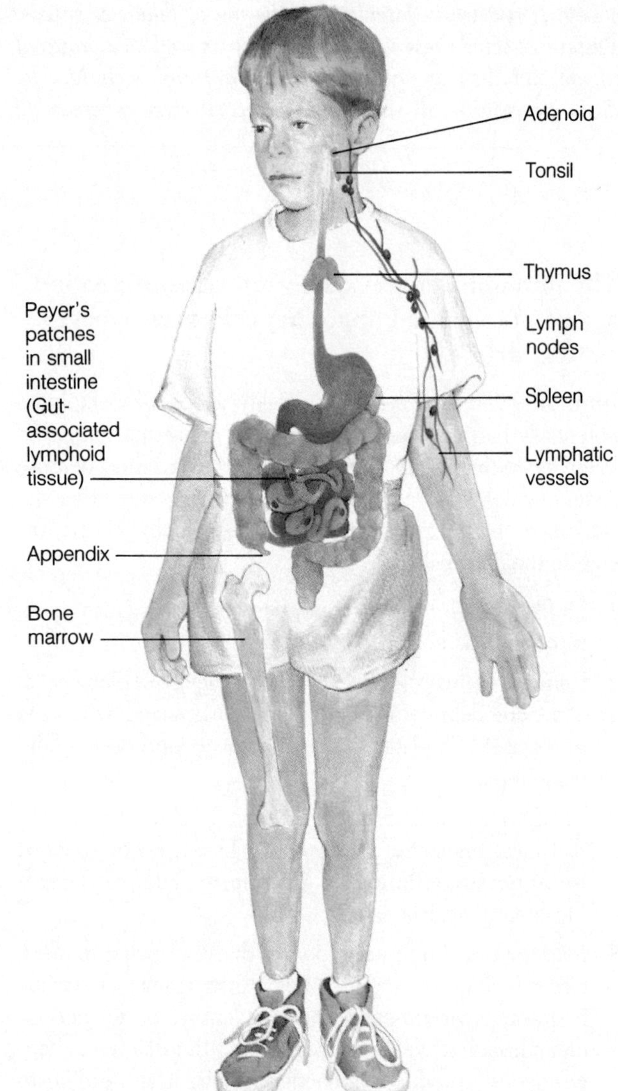

Figure 12–1 Lymphoid Tissues

Immune responses may be either specific or nonspecific.

Immune defense responses are classified as specific or nonspecific immune responses, depending on the degree of selectivity of the defense mechanism. **Specific immune responses** are selectively targeted against particular foreign material to which the body has previously been exposed. These specific responses are mediated by lymphocytes, which, upon subsequent exposure to the same offending agent, recognize and discriminately defend against it.

Nonspecific immune responses, on the other hand, are naturally inherent defense responses that nonselectively defend against foreign or abnormal material of any type, even upon initial exposure to it. Such responses provide a first line of defense against a wide range of threatening factors, including infectious agents, chemical irritants, and tissue injury accompanying mechanical trauma and burns. Nonspecific defenses that come into play whether or not there has been prior experience with the offending agent include the following:

Table 12-1 Functions of Lymphoid Tissues

Lymphoid Tissue	Function
Bone marrow	Origin of all blood cells
	Site of maturational processing for B lymphocytes
Lymph nodes, tonsils, adenoids, appendix, gut-associated lymphoid tissue	Exchange lymphocytes with lymph (remove, store, produce, and add them)
	Resident lymphocytes produce antibodies and sensitized T cells, which are released into the lymph
	Resident macrophages remove microbes and other particulate debris from lymph
Spleen	Exchanges lymphocytes with blood (removes, stores, produces, and adds them)
	Resident lymphocytes produce antibodies and sensitized T cells, which are released into the blood
	Resident macrophages remove microbes and other particulate debris, most notably worn-out red blood cells, from blood
	Stores a small percentage of red blood cells, which can be added to the blood by splenic contraction as needed
Thymus	Site of maturational processing for T lymphocytes
	Secretes the hormone thymosin

1. *inflammation*, a nonspecific response to tissue injury in which the phagocytic specialists—neutrophils and macrophages—play a major role, along with supportive input from other immune cell types;
2. *interferon*, a family of proteins that nonspecifically defend against viral infection;
3. *natural killer cells*, a special class of lymphocytelike cells that spontaneously and relatively nonspecifically lyse (rupture) and thereby destroy a variety of virus-infected host and tumor cells; and
4. *the complement system*, a group of inactive plasma proteins that, when sequentially activated, bring about destruction of foreign cells by attacking their plasma membranes. The complement system can nonspecifically be called into play by the presence of any foreign invader. It also may be activated by antibodies produced as part of the specific immune response to a particular microorganism.

The fact that the complement system is involved in both nonspecific and specific defense mechanisms illustrates an important point. There is much interaction and interdependency between the various components of the immune system, making it highly sophisticated and effective but also complex and difficult to sort out. The most significant cooperative relationships known to exist among the immune effector cells will be pointed out as the various components of the immune system are discussed separately. As further reference, Appendix B summarizes the major immune functions and interactions addressed in this chapter.

Nonspecific Immune Responses

Inflammation is a nonspecific response to foreign invasion or tissue damage.

Inflammation refers to an innate, nonspecific series of highly interrelated events that are set into motion in response to foreign invasion, tissue damage, or both. The ultimate goal of inflammation is to bring to the invaded or injured area phagocytes and plasma proteins that can: (1) isolate, destroy, or inactivate the invaders; (2) remove debris; and (3) prepare for subsequent healing and repair. The overall inflammatory response is remarkably similar no matter what the triggering event (be it bacterial invasion, chemical injury, or mechanical trauma), although some subtle differences may be evident, depending on the injurious agent or the site of damage. The following sequence of events typically occurs during the inflammatory response. As as example, we will use bacterial entry into a break in the skin.

DEFENSE BY RESIDENT TISSUE MACROPHAGES. Upon bacterial invasion through a break in the external barrier of skin, the macrophages already present in the area immediately begin phagocytizing the foreign microbes. Although the resident macrophages are usually not present in sufficient numbers to meet the challenge alone, they provide a first line of defense against infection during the first hour or so, before other mechanisms can be mobilized. Macrophages are usually rather stationary, gobbling debris and contaminants that come their way, but when necessary, they become mobile and migrate to sites of battle against invaders.

LOCALIZED VASODILATION. Almost immediately upon microbial invasion, arterioles within the area dilate, increasing blood flow to the site of injury. This localized vasodilation is induced primarily by histamine that has been released in the area of tissue damage from **mast cells,** a type of connective tissue-bound cell similar to circulating basophils. Increased local delivery of blood brings to the site more phagocytic leukocytes and plasma proteins, both of which are crucial to the immune defense system.

INCREASED CAPILLARY PERMEABILITY. Released histamine also increases the capillaries' permeability by enlarging the capillary pores (the spaces between the endothelial cells, see p. 314) so that plasma proteins that normally are prevented from leaving the blood can escape into the inflamed tissue.

LOCALIZED EDEMA. As the leaked plasma proteins accumulate in the interstitial fluid, they exert a colloid osmotic pressure. This elevation in local osmotic pressure, coupled with the increased capillary blood pressure caused by increased local blood flow, favors enhanced filtration and reduced reabsorption of fluid across the involved capillaries. The end result of this shift in fluid balance is localized edema (see p. 323). Thus, the familiar swelling that accompanies inflammation is due to histamine-induced vascular changes. Likewise, the other well-known gross manifestations of inflammation such as redness and heat are largely attributable to the enhanced flow of warm arterial blood to the damaged tissue. Pain is caused both by local distention within the swollen tissue and by the direct effect of locally produced substances on the receptor endings of afferent neurons that supply the area. Keep in mind that these observable characteristics of the inflammatory process are coincidental to the primary purpose of the vascular changes in the injured area, which is to increase the number of leukocytic phagocytes and crucial plasma proteins in the area.

WALLING-OFF OF THE INFLAMED AREA. The extravasated plasma proteins most critical to the immune response are those involved in the complement system and kinin system (to be described later) as well as clotting and anticlotting factors (see p. 359). Upon exposure to tissue thromboplastin in the injured tissue and to specific chemicals secreted by phagocytes on the scene, fibrinogen, the final factor in the clotting system, is converted into fibrin. Fibrin forms interstitial-fluid clots in the spaces around the bacterial invaders and damaged cells. This walling-off of the injured region from the surrounding tissues prevents or at least delays the spread of bacterial invaders and their toxic products. Later, the more slowly activated anticlotting factors gradually dissolve the clots after they are no longer needed. A few bacteria, such as streptococci, produce enzymes that act on plasminogen, an inactive plasma-protein precursor, to convert it into plasmin, a proteolytic enzyme that dissolves fibrin clots. This action disrupts the walling-off process and allows the streptococcal organisms to spread extensively.

EMIGRATION OF LEUKOCYTES. Within an hour after the injury, the involved area is teeming with leukocytes that have exited from the vessels. Neutrophils are the first to arrive, followed during the next eight to twelve hours by monocytes. The latter swell and mature into macrophages during another eight-to-twelve hour period.

Leukocyte emigration from the blood into the tissues involves the processes of margination, diapedesis, amoeboid movement, and chemotaxis. **Margination** refers to the sticking of blood-borne leukocytes, especially neutrophils and monocytes, to the inner endothelial lining of capillaries in the affected tissue. Soon the leukocytes start exiting by a mechanism known as **diapedesis.** Assuming amoebalike behavior (see p. 39), an adhered leukocyte pushes a long, narrow projection through a capillary pore, with the remainder of the cell flowing forward into the projection (Fig. 12–2). In this way, even though the leukocyte is much larger than the capillary pore, it is able to force its way through and move in an amoeboid fashion toward the site of tissue damage and bacterial invasion. Neutrophils arrive on the inflammatory scene earliest because they are more mobile than monocytes.

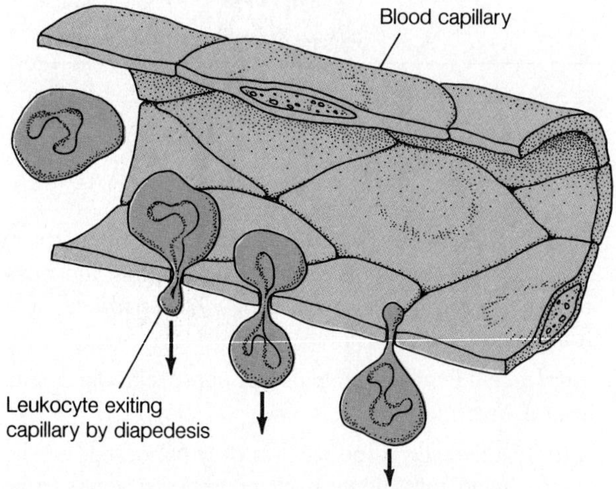

Figure 12–2 Diapedesis *Leukocytes migrate from the blood into the tissues by assuming amoebalike behavior and squeezing through the capillary pores.*

Phagocytic cells are guided in their direction of migration by attraction to certain chemical mediators, or **chemotaxins,** released at the site of damage, a process referred to as chemotaxis. Binding of chemotaxins with protein receptors on the plasma membrane of a phagocytic cell increases Ca^{++} entry into the cell. Calcium, in turn, switches on the cellular contractile apparatus that leads to amoebalike crawling. Because the concentration of chemotaxins progressively increases toward the site of injury, phagocytic cells move unerringly toward this site along a chemotaxin concentration gradient.

LEUKOCYTE PROLIFERATION. Resident tissue macrophages and leukocytes that exit from the blood and migrate to the inflammatory site are soon joined by new phagocytic recruits from the bone marrow. Within a few hours after the onset of the inflammatory response, the number of neutrophils in the blood may increase up to four to five times that of normal. This **neutrophilia** is partly caused by transfer into the blood of large numbers of preformed neutrophils stored in the bone marrow and partly caused by increased production of new neutrophils by the bone marrow. A slower-commencing but longer-lasting increase in monocyte production by the bone marrow also occurs, making larger numbers of these macrophage precursor cells available. In addition, multiplication of resident macrophages adds to the pool of these important immune cells. Proliferation of new neutrophils, monocytes, and macrophages and mobilization of stored neutrophils are stimulated by various chemical mediators released from the inflamed region.

LEUKOCYTIC DESTRUCTION OF BACTERIA. Neutrophils and macrophages clear the inflamed area of infectious and toxic agents as well as tissue debris, this being the primary function of the inflammatory response. They accomplish this by both phagocytic and nonphagocytic means.

Phagocytosis involves the engulfment and intracellular degradation (breakdown) of foreign particles and tissue debris. Macrophages can engulf a bacterium in less than 0.01 second. Phagocytic cells contain an abundance of lysosomes, which are organelles filled with hydrolytic enzymes (the same type that degrade ingested food in the digestive tract). After a phagocyte has internalized targeted material, a lysosome fuses with the membrane that encloses the engulfed matter and releases its hydrolytic enzymes within the confines of the vesicle, where the enzymes begin breaking down the entrapped material (see p. 24). Furthermore, in response to the presence of foreign material, the fused lysosome produces hydrogen peroxide and other oxygen derivatives that are highly destructive to microbes and other organic material. Many of the resultant breakdown products are low-molecular-weight organic compounds, such as amino acids and glucose, which can be used by the phagocyte itself or, in the case of macrophages, released into the extracellular fluid for use elsewhere in the body.

Nondegradable inorganic particles that have been engulfed (for example, dust, metal particles, or tattoo dyes) usually are retained indefinitely within the phagocytic cell. In fact, certain bacteria, most notably those causing tuberculosis, are likewise engulfed but not destroyed because they are resistant to lysosomal chemicals. These organisms may survive within phagocytes for many years and produce no obvious effects because they are walled-off within a fibrous capsule. They cause overt disease only if allowed to escape. Such a more-or-less permanently walled-off structure within which nondestructible offending material is imprisoned is known as a **granuloma.**

Phagocytes eventually die because of accumulation of toxic by-products from foreign particle degradation or inadvertent release of destructive lysosomal chemicals into the cytosol. Neutrophils usually succumb after phagocytizing from five to twenty-five bacteria, whereas macrophages survive much longer and can engulf up to one-hundred or more bacteria. Indeed, the longer-lived macrophages even clear the area of dead neutrophils in addition to other tissue debris. The **pus** that forms in an infected wound is a collection of these phagocytic cells, both living and dead; of necrotic (dead) tissue liquified by lysosomal enzymes released from the phagocytes; and of bacteria.

Obviously, phagocytes must be able to distinguish between normal cells and foreign or abnormal cells before accomplishing their destructive mission. Otherwise they could not selectively engulf and destroy only unwanted materials. Several different selective procedures enable phagocytes to "recognize" targets for destruction. First, dead tissue and many foreign materials have surface characteristics that differ from normal body cells. For example, the surface roughness accompanying traumatic injury to cells increases the likelihood of the cellular debris being phagocytized. Second, foreign particles are deliberately marked for phagocytic ingestion by being coated with chemical mediators generated via the immune system. Such body-produced chemicals that make bacteria more susceptible to phagocytosis are known as **opsonins.** The most important opsonins are antibodies and one of the activated proteins of the complement system.

An opsonin enhances phagocytosis by linking the foreign cell to a phagocytic cell (Fig. 12–3). One portion of an opsonin molecule binds nonspecifically to the surface of an invading bacterium, whereas another portion of the opsonin molecule binds to receptor sites specific for it on the phagocytic cell's plasma membrane. This link assures that the bacterial

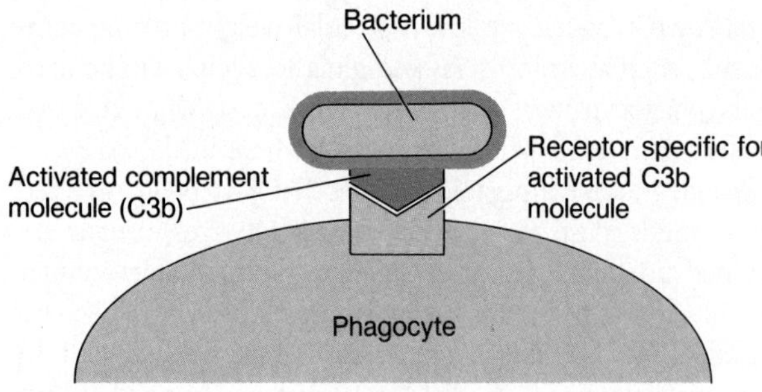

Figure 12–3 Mechanism of Opsonin Action *One of the activated complement molecules, C3b, links a foreign cell and a phagocytic cell by nonspecifically binding with the surface of the foreign cell and specifically binding with plasma membrane receptors on the surface of the phagocyte.*

victim does not have a chance to "get away" before the phagocyte can perform its lethal attack.

In addition to engulfment and intracellular destruction of foreign material, phagocytes can also kill microbes by several nonphagocytic, extracellular means. First, contact with microorganisms induces phagocytes to release some of their lysosomal enzymes and oxygen derivatives into the extracellular fluid, where these destructive chemicals directly kill invaders that have not been phagocytized. Second, as a more subtle means of destruction, neutrophils secrete **lactoferrin,** a protein that tightly binds with iron, making it unavailable for use by invading bacteria. Bacterial multiplication depends on high concentrations of available iron.

MEDIATION OF THE INFLAMMATORY RESPONSE BY PHAGOCYTE-SECRETED CHEMICALS. Microbe-stimulated phagocytes release many newly synthesized chemicals as well as preformed lysosomal chemicals, which function as mediators of the inflammatory response. These chemical mediators induce a broad range of interrelated immune activities, varying from local responses to the systemic manifestations that accompany microbe invasion. The following are among the most important functions of phagocytic secretions:

1. They directly kill microbes, as already mentioned.
2. They stimulate the release of histamine from mast cells in the vicinity. Histamine, in turn, induces the local vasodilation and increased vascular permeability that accompany inflammation.
3. They trigger both the clotting and anticlotting systems to first enhance the walling-off process and then facilitate the gradual dissolution of the fibrous clot after it is no longer needed.
4. They split **kininogens,** which are inactive precursor plasma proteins synthesized by the liver, into active **kinins.** In particular, **kallikrein,** which is released by neutrophils, can accomplish this activation. Once kinins are generated by kallikrein, they augment a variety of inflammatory events. Specifically, they: (a) stimulate several key steps in the complement system; (b) reinforce the vascular changes induced by histamine; (c) activate nearby pain receptors, being partially responsible for the soreness associated with inflammation; and d) act as powerful chemotaxins to induce phagocyte migration into the affected area. In positive-feedback fashion, the newly arriving neutrophils release kallikrein, which can generate still more kinins that chemotactically entice more neutrophils to join the battle.
5. They induce the development of fever, in particular by means of the phagocytic release of **endogenous pyrogen (EP)**. This response occurs especially when the invading organisms have spread into the bloodstream. Endogenous pyrogen is believed to cause the local release within the hypothalamus of ***prostaglandins***, which "turn up" the hypothalamic "thermostat" that regulates body temperature. The function in fighting infection served by the resultant elevation in body temperature remains unclear. The fact that fever is such a common systemic manifestation of inflammation suggests that the higher body temperature plays an important beneficial role in the overall inflammatory response. Recent evidence strongly supports this supposition. For example, higher temperatures augment the process of phagocytosis and increase the rate at which the many enzyme-dependent inflammatory activities proceed. One theory proposes that an elevated body temperature increases bacterial requirements for iron while reducing the plasma concentration of iron. The resultant iron deficiency, according to this theory, interferes with bacterial multiplication. Resolving the controversial issue of whether or not a fever can be beneficial is extremely important in view of the widespread use of drugs that suppress fever. Although a mild fever may possibly be beneficial, there is no doubt that an extremely high fever can be detrimental, particularly because of its harmful effects on the central nervous system. It is not uncommon for young children, whose temperature-regulating mechanisms are not as stable as more mature individuals, to have convulsions in association with high fevers.

6. They decrease plasma concentration of iron by altering iron metabolism within the liver, spleen, and other tissues, thus reducing the amount of iron available to support bacterial multiplication.
7. They stimulate **granulopoiesis,** the synthesis and release of neutrophils and other granulocytes by the bone marrow. This effect is especially prominent in response to bacterial infections.
8. They stimulate the release of **acute-phase proteins** from the liver. This collection of proteins, which have not yet been sorted out by scientists, exert a multitude of wide-ranging effects associated with the inflammatory process, tissue repair, and immune cell activities.

   The latter three effects (plasma iron reduction, enhanced granulopoiesis, and release of acute-phase proteins) are all attributed to **leukocyte endogenous mediator (LEM)**, a chemical mediator secreted by macrophages. Evidence suggests that LEM and EP are the same substance or at least very closely related.
9. They enhance the proliferation and differentiation of both B and T lymphocytes, which, in turn, are responsible for antibody production and cell-mediated immunity, respectively. Specifically, **interleukin 1 (IL-1)**, a secretory product released by macrophages, is responsible for this effect on lymphocytes. Amazingly, IL-1 is identical to (or closely related to) EP and LEM. Apparently, the same chemical substance is responsible for a diverse array of effects throughout the body, all of which are geared toward defending the body against infection or tissue injury. In fact, the release of EP/LEM/IL-1 can be elicited by stressful situations not associated with microbe invasion (for example, during endurance exercise). Accordingly, this mediator may be part of a generalized nonspecific protective response.

This list of events that are augmented by chemicals secreted by phagocytes is not complete, but it serves to illustrate the diversity and complexity of responses elicited by these mediators. As will soon become evident, there are other important macrophage-lymphocyte interactions that are not dependent on the release of chemicals from phagocytic cells. Thus, the effect that phagocytes, especially macrophages, ultimately have on microbial invaders far exceeds their "engulf and destroy" tactics (Fig. 12–4).

TISSUE REPAIR. The ultimate purpose of the inflammatory process is to isolate and destroy injurious agents and to clear the area for tissue repair. In some tissues (for example, skin, bone, and liver), the healthy organ-specific cells surrounding the injured area undergo cell division to replace the lost cells, often accomplishing perfect repair. In nonregenerative tissues such as nerve and muscle, however, lost cells are replaced by **scar tissue.** Fibroblasts, a type of connective-tissue cell, start to rapidly divide in the vicinity and secrete large quantities of the protein collagen (see p. 67), which fills in the region vacated by the lost cells and results in the formation of scar tissue. Even in a tissue as readily replaceable as the skin, scar formation sometimes takes place when complex underlying structures, such as hair follicles and sweat glands, are permanently destroyed by deep wounds.

Salicylates and glucocorticoid drugs suppress the inflammatory response.

Numerous drugs can suppress the inflammatory process, the most effective of which are the salicylates (aspirin-type drugs) and glucocorticoids (drugs similar to the steroid hormone cortisol that is secreted by the adrenal cortex). Salicylates interfere with the inflammatory response by decreasing histamine release, the result of which is a reduction in swelling, redness, and pain. These agents also stabilize the lysosomal membranes so that they are less likely to discharge their contents. Furthermore, salicylates reduce fever by inhibiting the production of prostaglandins, the local mediators of EP-induced fever.

Glucocorticoids, which are potent anti-inflammatory drugs, suppress almost every aspect of the inflammatory response. In addition, they destroy lymphocytes within lymphoid tissue and reduce antibody production. These therapeutic agents are useful for treating undesirable immune responses, such as allergic reactions (for example, poison ivy rash and asthma) and the inflammation associated with arthritis. Unfortunately, however, by suppressing inflammatory and other immune responses that localize and eliminate bacteria, such therapy also reduces the body's ability to resist infection. For this reason, glucocorticoids should be administered discriminately.

What about the normal role of the adrenal cortical hormone cortisol in this regard? Is cortisol secretion counterproductive to the immune defense system? Traditionally, cortisol has not been considered to display anti-inflammatory activity at normal blood concentrations. Instead, anti-inflammatory action has been attributed only to pharmacological levels of glucocorticoids (that is, at blood concentrations brought about by the administration of cortisol-like drugs that are higher than the normal physiological range). A recent hypothesis, however, suggests that cortisol, whose secretion is increased in response to any stressful situation, does exert anti-inflammatory

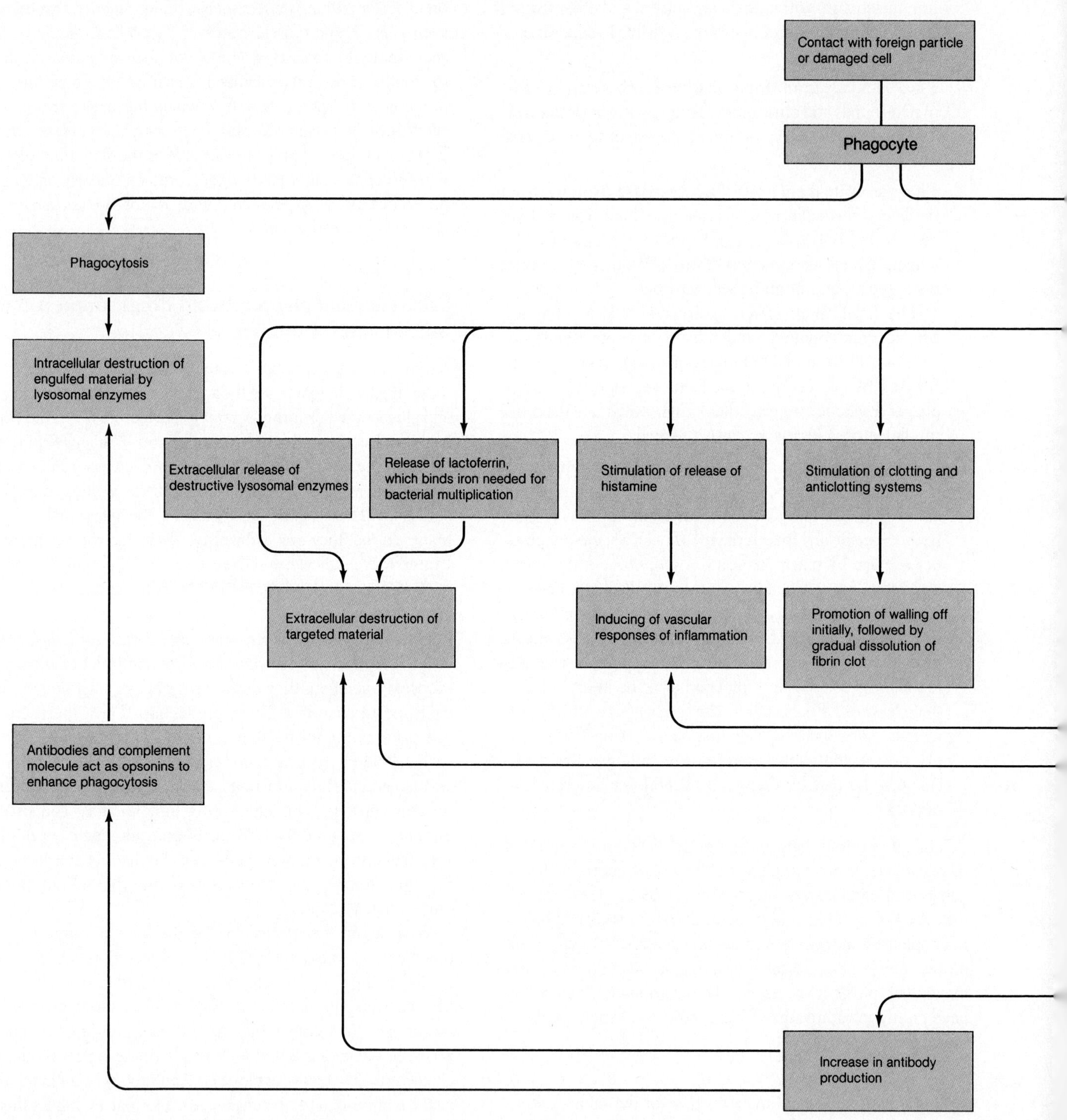

EP/lem/IL-1 = Endogenous pyrogen/leukocyte endogenous mediator/interleukin 1

Figure 12–4 Phagocyte Responsibilities

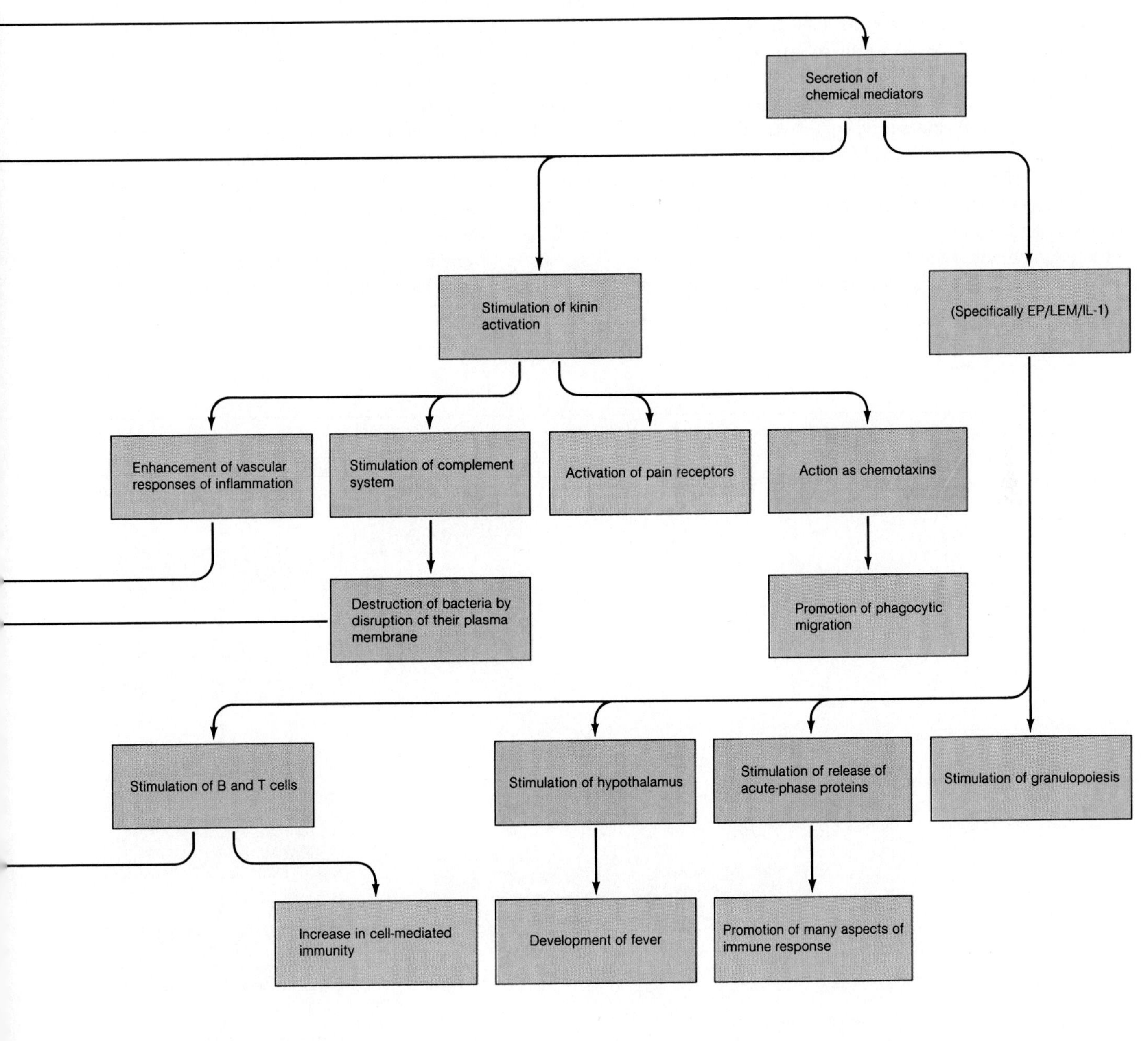
Secretion of chemical mediators
Stimulation of kinin activation
(Specifically EP/LEM/IL-1)
Enhancement of vascular responses of inflammation
Stimulation of complement system
Activation of pain receptors
Action as chemotaxins
Destruction of bacteria by disruption of their plasma membrane
Promotion of phagocytic migration
Stimulation of B and T cells
Stimulation of hypothalamus
Stimulation of release of acute-phase proteins
Stimulation of granulopoiesis
Increase in cell-mediated immunity
Development of fever
Promotion of many aspects of immune response

activity even at normal physiological levels. According to this proposal, the anti-inflammatory effect of cortisol modulates stress-activated immune responses, preventing them from overshooting and thus protecting us against damage by potentially overreactive defense mechanisms.

Interferon transiently inhibits multiplication of viruses in most cells.

Besides the inflammatory response, another nonspecific defense mechanism is the release of **interferon** from virus-infected cells. Interferon briefly provides nonspecific resistance to viral infections by transiently interfering with replication of the same or unrelated viruses in other host cells. When a cell is invaded by a virus, the presence of viral nucleic acid induces the cell's genetic machinery to synthesize interferon, which is secreted into the extracellular fluid. Macrophages, lymphocytes, fibroblasts, and many other cells each make a slightly different version of interferon, undoubtedly with slightly varying biological activities. Therefore, interferon actually encompasses a family of closely related proteins.

Once released, interferon binds with receptors on the plasma membranes of neighboring cells or even distant cells that it reaches through the bloodstream, signaling these cells to prepare for the possibility of impending viral attack. Interferon does not have a direct antiviral effect; instead, it triggers the production of viral-blocking enzymes by potential host cells. Binding with interferon induces these other cells to synthesize enzymes that can break down viral messenger RNA and inhibit protein synthesis, both of which are essential for viral replication. Although viruses are still able to invade these forewarned cells, they are unable to govern cellular protein synthesis for their own replication (Fig. 12–5).

These newly synthesized inhibitory enzymes remain inactive within the potential host cell unless it is actually invaded by virus, at which time the enzymes are activated by the presence of viral nucleic acid. This activation requirement protects the cell's own messenger RNA and protein-synthesizing machinery from unnecessary inhibition by these enzymes should viral invasion not occur. Because activation can take place only during a limited time span, this is a short-term defense mechanism.

Figure 12–5 Mechanism of Action of Interferon in Preventing Viral Replication

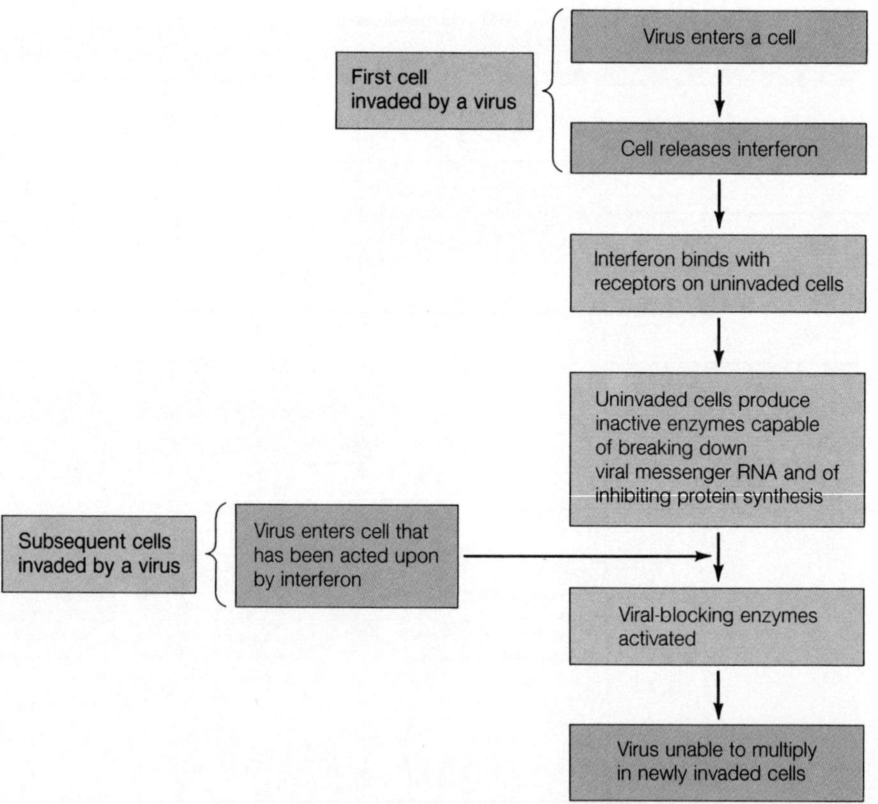

Interferon is released nonspecifically from any cell infected by any virus and, in turn, can induce temporary self-protective activity against many different viruses in any other cells that it reaches. Thus, it provides a general, rapidly responding defense strategy against viral invasion until more-specific but slower-responding immune mechanisms come into play.

In addition to facilitating inhibition of viral replication, interferon further reinforces other immune activities. It: (1) enhances macrophage phagocytic activity; (2) stimulates the production of antibodies; (3) stimulates the activity of natural killer cells; (4) stimulates the activity of special T lymphocytes (cytotoxic T cells), which directly attack and destroy virus-infected cells and cancer cells; and (5) slows cell division and suppresses tumor growth.

For the last three reasons, interferon exerts anticancer as well as antiviral effects. Fortunately, its anticancer effects are not limited to virally-induced cancers. Most types of human cancer are not caused by viruses. Interferon markedly enhances the actions of cell-killing cells—the natural killer cells and cytotoxic T cells—which attack and destroy both virus-infected cells and cancer cells.

The discovery of interferon's antiviral and anticancer effects led to great excitement and anticipation about its potential role as a miracle therapeutic weapon against enemies ranging from the common cold virus to fatal invasive cancers. For a quarter of a century after its discovery in 1957, however, it was impossible to collect human interferon in sufficient quantities to make it feasible to study its therapeutic possibilities. (Interferon is species specific; each species of higher animals produces a type of interferon capable of conferring protection only to other members of the same species.) Within the last decade, technological advancements have permitted the introduction of human interferon genes into bacterial genes so that these bacteria produce human interferon. Through this **recombinant DNA technology,** bacteria can be turned into interferon "factories" for large-scale commercial production of this valuable immune agent. Much research is currently underway to determine the potential effectiveness of interferon in treating viral diseases and cancer. The research thrust is toward treatment rather than prevention, since it would not be economically feasible to maintain a continuous preventive level of this substance in the population because of the transient nature of interferon's defense capabilities.

Natural killer cells destroy virus-infected cells and tumor cells upon first exposure to them.

Natural killer (NK) cells are naturally occurring, lymphocytelike cells that nonspecifically destroy virus-infected cells and tumor cells by directly lysing their membranes upon first exposure to them. Their mode of action and major targets are similar to cytotoxic T cells, but the latter can fatally attack only those specific types of virus-infected cells and tumor cells to which they have been previously exposed. Furthermore, following exposure, cytotoxic T cells require a maturation period before they are capable of launching their lethal assault. The natural killer cells provide an immediate, nonspecific defense against virus-invaded cells and cancer cells before the more specific and more abundant cytotoxic T cells become functional.

The complement system directly kills microorganisms on its own and in conjunction with antibodies while augmenting the inflammatory response.

The **complement system** is another defense mechanism brought into play nonspecifically in response to invading organisms. It also can be triggered by antibodies as part of the specific immune strategy. In fact, the system derives its name from the fact that it *complements* the action of antibodies, being the primary mechanism activated by antibodies to kill foreign cells.

In the same tradition as the clotting system, anticlotting system, and kinin system, the complement system consists of plasma proteins produced by the liver that circulate in the blood in inactive form. Once the first component, C1, is activated, it activates the next component, C2, and so on, in a sequential cascade of activation reactions. The five final components, C5 through C9, form a large, doughnut-shaped protein complex, the **membrane-attack complex,** which attacks the surface membrane of nearby microorganisms by imbedding itself so that a large channel is created through the microbial surface membrane (Fig. 12–6). This hole-punching technique makes the membrane extremely leaky, resulting in an osmotic flux of water into the victim cell that causes it to swell and burst. This complement-induced lysis is the major means of directly killing microbes without phagocytizing them.

Unlike the other cascade systems, in which the sole function of the various components is activation of the next precursor in the sequence, several of the activated proteins in the complement cascade perform additional important functions on their own. Besides the direct destruction of foreign cells accomplished by the membrane attack complex, various other activated complement components augment the inflammatory process by performing the following functions: (1) serving as chemotaxins, which attract and guide professional phagocytes to the site of complement activation (that is, the site of microbial invasion); (2) acting as opsonins by binding with microbes and thereby enhancing their phagocytosis; (3) promot-

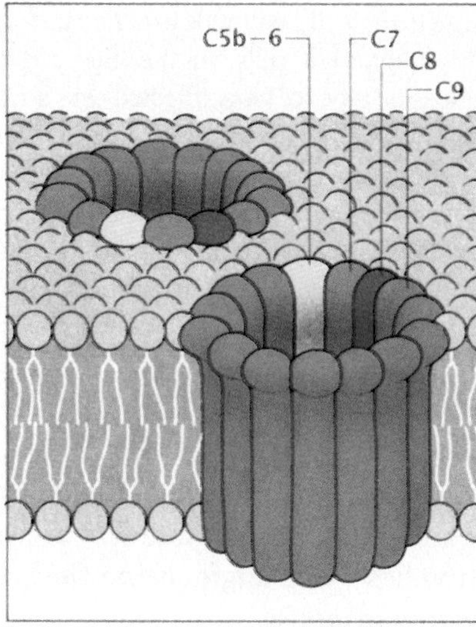

Figure 12-6 Membrane-Attack Complex (MAC) of Complement System *Activated complement proteins C5, C6, C7, C8, and a number of C9s aggregate to form a porelike channel in the plasma membrane of the target cell. The resultant leakage leads to destruction of the cell.*

SOURCE: By Dana Burns from "How Killer Cells Kill" by John Ding-E Young and Zanvil A. Kohn.

ing vasodilation and increased vascular permeability to increase blood flow to the invaded area; (4) stimulating the release of histamine from mast cells in the vicinity, which in turn enhances the local vascular changes characteristic of inflammation; and (5) activating kinins, which further reinforce inflammatory reactions.

How can this powerful complement cascade be set into motion? There are two pathways by which the complement system can be activated—the classical pathway and the alternate pathway. In the **classical pathway,** antibody specifically generated against the particular invader activates the first complement protein, initiating the chain reaction that ultimately results in the direct lysis of the invading microbes and reinforcement of other general inflammatory tactics.

Even in the absence of antibody, the **alternate pathway,** using several different protein precursors, can plug into the cascade and initiate the subsequent chain of activating reactions from C3 on. It is important to note that there are no inflammatory mediators generated within the classical pathway before the activation of C3, so essentially none of the complement system's biological activity is lost when it is activated by the alternate pathway. What triggers the alternate pathway if antibody is not the impetus? Particular carbohydrate chains present on the surfaces of microbes but not found on normal human cells can initiate the alternate pathway of complement activation. This provides a nonspecific natural resistance to infective agents that is immediately available upon first exposure to invading microorganisms even before specific antibodies are generated.

What restricts the activated complement system's destructive tactics to undesirable cells, such as invading bacteria? Several of the activated components in the cascade are very unstable, being able to activate the next component in the sequence for less than 0.1 millisecond (1/10,000 of a second) before assuming an inactive form. Because these unstable components are able to perpetuate the sequence only in the immediate vicinity in which they are activated before they decompose (that is, in the vicinity of antibody attached to a bacterium or of carbohydrate protruding from the surface of this same microorganism), the complement attack is confined to the surface membrane of the microbe whose presence initiated activation of the system. Nearby host cells are thus spared from lytic attack. To further protect against the spread of complement destruction to normal cells, specific inhibitory proteins present in the blood rapidly deactivate certain activated complement components shortly after they have participated in the chain reaction leading to microbe destruction. This deactivation process halts further complement activation, thus confining the lytic activity only to the microbial cells in the immediate vicinity.

Specific Immune Responses: General Concepts

Specific immune responses include antibody-mediated immunity accomplished by B lymphocyte derivatives and cell-mediated immunity accomplished by T lymphocytes.

A specific immune response is one aimed at the limitation or neutralization of a particular offending target for which the body has been specially prepared for selective attack following prior exposure to it. There are two classes of specific immune responses: **antibody-mediated,** or **humoral, immunity** involving the production of antibodies by B-lymphocyte derivatives known as plasma cells; and **cell-mediated immunity** involving the production of activated T lymphocytes, which directly attack unwanted cells.

B and T lymphocytes (B and T cells) have different life histories and, more importantly, different properties and functions. Both types of lymphocytes, like all blood cells, are derived from common hematopoietic precursor cells in the bone marrow. Whether or not a lymphocyte and all of its progeny are destined to be B or T cells depends on the site of final maturation and differentiation of the original cell in the lineage (Fig. 12–7). During fetal life and early childhood, some of the immature lymphocytes migrate through the blood to the thymus, where they undergo further processing to become T lymphocytes. The **thymus** is a lymphoid organ located midline within the thoracic (chest) cavity above the heart in the space between the lungs (Fig. 12–1). Lymphocytes that mature without benefit of "thymic education" become B lymphocytes. B lymphocytes were first discovered in birds, where the maturational processing takes place in a gut-related lymphoid tissue unique to birds, the bursa of Fabricius; hence the name B cells. In humans, the site of B-cell maturation and differentiation is uncertain, although it is generally assumed to be the bone marrow.

Upon being released into the blood from either the bone marrow or the thymus, mature B and T cells take up residence and establish lymphocyte colonies in the peripheral lymphoid tissues. Here, upon appropriate stimulation, they undergo mitosis to produce new generations of either B or T cells, depending on their ancestry. After early childhood, most new lymphocytes are derived from these peripheral lymphocyte colonies rather than from the bone marrow.

The role of the thymus remained obscure until recent years, because its removal from an adult animal had no obvious effect. Because most of the migration and differentiation of T cells occurs early in development, the thymus gradually atrophies and becomes less important as the individual matures. It does, however, continue to produce **thymosin,** a collection of thymic hormones important in maintaining the T-cell lineage. Thymic hormones are poorly understood, but apparently they enhance proliferation of new T cells within the peripheral lymphoid tissues and augment the immune capabilities of existing T cells. Recent evidence indicates that secretion of thymosin decreases after about thirty to forty years of age. This decline has been implicated as being a contributing factor in aging. It is further speculated that diminishing T-cell capacity with advancing age might somehow be linked to the increased susceptibility to viral infections and cancer that occur as a person ages. T cells play an especially important role in defense against viruses and viral-induced cancer.

Lymphocytes are able to specifically recognize and selectively respond to an almost limitless variety of foreign agents as well as cancer cells. The recognition and response processes are different for B and T cells. In general, B cells recognize free-existing foreign invaders such as bacteria and their toxins and a few viruses, which they combat by secreting antibodies specific for the invaders. T cells specialize in recognizing and destroying body cells gone awry, including virus-infected cells, cancer cells, and transplanted tissue cells.

Each of us has an estimated total of two trillion lymphocytes, which, if aggregated together in a mass, would be comparable to the size of the brain or liver. At any one time, the majority of these lymphocytes are concentrated in the various strategically located lymphoid tissues, but both B and T cells continually recirculate among the lymph, blood, and body tissues, where they are on constant surveillance.

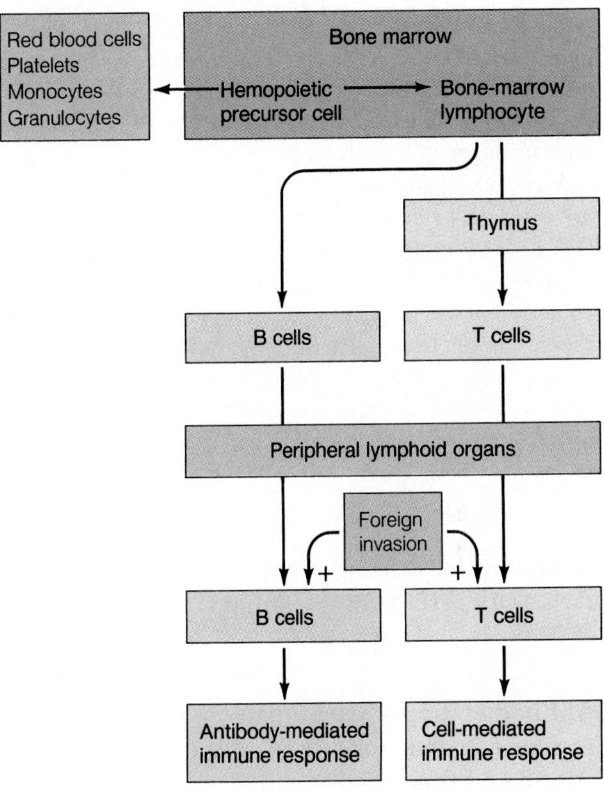

Figure 12–7 Origins of B and T Cells

An antigen induces an immune response against itself.

Both B and T cells must be able to specifically recognize unwanted cells and other material to be destroyed or neutralized as being distinct from normal-self cells. It is the presence of antigens that enables this distinction. An **antigen** is a large ($\geqslant$10,000 molecular weight), complex molecule that triggers a specific immune response against itself when it gains entry into the body. In general, the more complex a molecule, the greater its antigenicity. Antigens may exist as isolated molecules, such as bacterial toxins, or they may be an integral part of a multimolecular structure, such as being present on the surface of an invading foreign microbe. All foreign proteins are antigens. Although other macromolecules, such as large polysaccharides, can also act as antigens, proteins are more antigenic because they are structurally more complex than polysaccharides. A complex antigen may have many **antigenic determinant sites,** each capable of stimulating the production of and interacting with different antibodies. For simplicity, we will consider each antigenic determinant site to be a separate antigen.

Many low-molecular-weight organic substances that are not antigenic by themselves can become antigenic if they attach to body proteins. Such small molecules are known as **haptens.** Antibodies developed against the hapten-protein combination can react in the future against the hapten alone should it be reintroduced to the body. Examples of haptens include poison ivy toxin, various drugs (such as penicillin), and other agents that are otherwise harmless but that can elicit inappropriate immune responses known as allergies in sensitized individuals.

B Lymphocytes: Antibody-Mediated Immunity

Antibodies amplify the inflammatory response to promote destruction of the antigen that stimulated their production.

Each B and T cell has receptors on its surface for binding with one particular type of the multitude of possible antigens. In the case of B cells, binding with antigen induces the cell to differentiate into a ***plasma cell,*** which produces antibodies that are able to specifically combine with the type of antigen that stimulated the antibodies' production. During differentiation into a plasma cell, a B lymphocyte swells as the rough endoplasmic reticulum (the site for synthesis of proteins to be exported) greatly expands (Fig. 12–8). Because antibodies are proteins, plasma cells essentially become prolific protein factories, producing up to 2,000 antibody molecules per second. So great is the commitment of a plasma cell's protein- synthesizing machinery to antibody production that it is unable to maintain protein synthesis for its own viability and growth. Consequently, it dies after a brief five-to-seven day, highly productive life span.

Antibodies are secreted into the blood or lymph, depending on the location of the activated plasma cells, but all antibodies eventually gain access to the blood, where they are known as ***gamma globulins*** or **immunoglobulins.** According to differences in biological activity, immunoglobulins are grouped into the five following subclasses:

- □ **IgM** immunoglobulin serves as the B-cell surface receptor for antigen attachment and is secreted in the early stages of plasma-cell response.
- □ **IgG,** the most abundant immunoglobulin in the blood, is produced copiously upon the body's subsequent exposure to the same antigen.

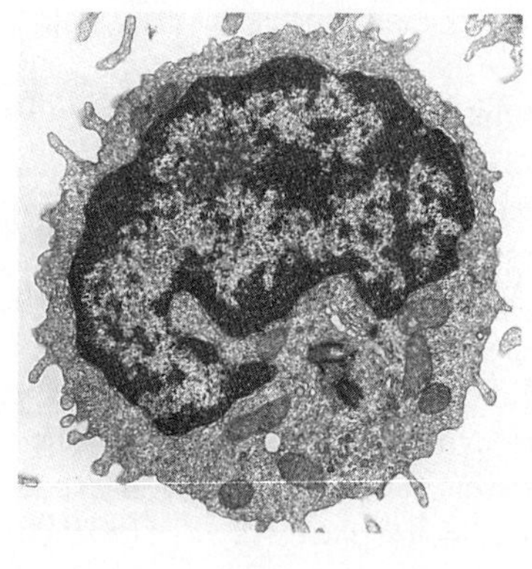
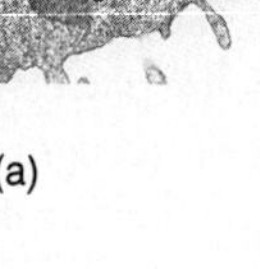
(a)

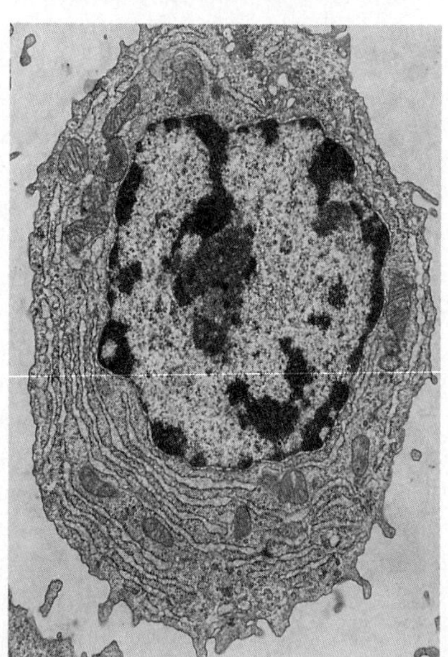
(b)

Figure 12–8 Comparison of Unactivated B Cell and Plasma Cell *Electron micrograph of (a) an unactivated B cell, or small lymphocyte, and (b) a plasma cell. A plasma cell is an activated B cell. It is filled with an abundance of rough endoplasmic reticulum distended with antibody molecules.*

SOURCE: Photos courtesy of Dr. Dorothea Zucker-Franklin, New York University School of Medicine.

Together, IgM and IgG antibodies are responsible for most specific immune responses against bacterial invaders and a few types of viruses.

□ **IgE** is the antibody mediator for common allergic responses such as hay fever, asthma, and hives.

□ **IgA** immunoglobulins are found in secretions of the digestive, respiratory, and genitourinary systems, as well as in milk and tears.

□ **IgD** is present on the surface of many B cells, but its function is uncertain.

It is important to note that this classification is based on different ways in which antibodies function. It does not imply that there are only five different antibodies. Within each functional subclass, there are millions of different antibodies, each able to bind only with a specific antigen.

Antibody proteins of all five subclasses are composed of four interlinked polypeptide chains—two long, heavy chains and two short, light chains—arranged in the shape of a "Y" (Fig. 12–9). Characteristics of the arm regions of the Y determine with what antigen the antibody can bind (that is, the *specificity* of the antibody). Properties of the tail portion of the antibody, on the other hand, determine the destiny of the antigen once it is bound (that is, the *functional properties* of the antibody). An antibody has two identical antigen-binding sites, one at the tip of each arm. These **antigen-binding fragments (Fab)** are unique for each different antibody so that each antibody can interact only with an antigen that specifically matches it, much like a lock and key. The tremendous variation between different antibodies in the amino-acid sequences of their antigen-binding fragments is responsible for the extremely large number of unique antibodies that are capable of binding specifically with millions of different antigens.

In contrast to these **variable (Fab) regions** at the arm tips, the tail portion of every antibody within each immunoglobulin subclass is identical. The antibody's so-called **constant (Fc) region** contains binding sites for particular mediators of antibody-induced activities, which vary among the different subclasses. In fact, differences in the constant region are the basis for distinguishing between the different Ig subclasses. For example, the constant tail region of IgG antibodies, when activated by antigen-binding in the variable region, binds with phagocytic cells and serves as an opsonin to enhance phagocytosis. In comparison, the constant tail region of IgE antibodies attaches to mast cells and basophils, even in the absence of antigen. When the appropriate antigen or hapten gains entry to the body and binds with the attached antibodies, this triggers the release of histamine from the affected mast cells and basophils. Histamine, in turn, induces the allergic manifestations that follow.

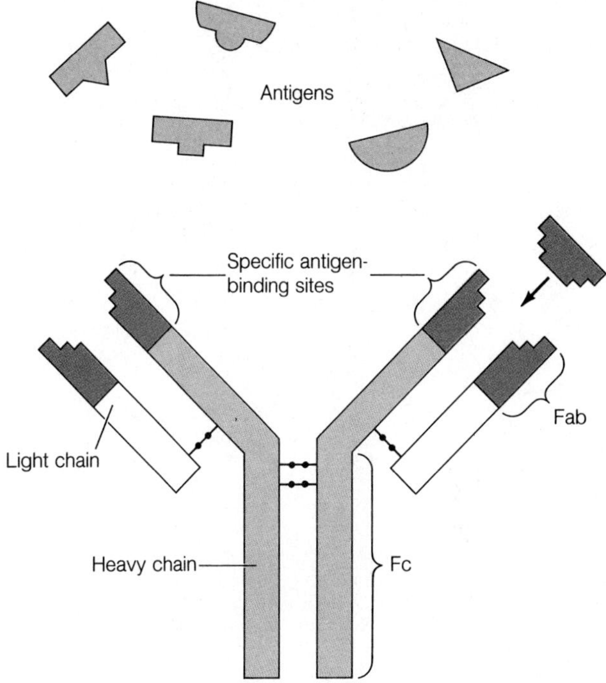

Figure 12–9 Antibody Structure

Immunoglobulins cannot directly destroy foreign organisms or other unwanted materials upon binding with antigens on their surfaces. Instead, antibodies exert their protective influence in one of two general ways: physical hindrance of antigens; and amplification of nonspecific immune responses (Fig. 12–10).

Antibodies can physically hinder some antigens from exerting their detrimental effects. For example, by combining with bacterial toxins, antibodies can prevent these harmful chemicals from interacting with susceptible cells. This process is known as **neutralization.** Similarly, antibodies are able to bind with surface antigens on some types of viruses, preventing these viruses from entering cells, where they can exert their damaging effects. Sometimes multiple antibody molecules can cross-link numerous antigen molecules into chains or lattices of antibody-antigen complexes. **Agglutination** is the term applied to foreign cells, such as bacteria or mismatched transfused red blood cells, bound together in such a clump. When linked antibody-antigen complexes involve soluble antigens, such as tetanus toxin, the lattice can become so large that it precipitates out of solution. (**Precipitation** is the process of a substance separating from a solution.) Within the body, these physical-hindrance mechanisms play only a minor protective role against invading agents. However, the tendency for certain antigens to agglutinate or precipitate upon

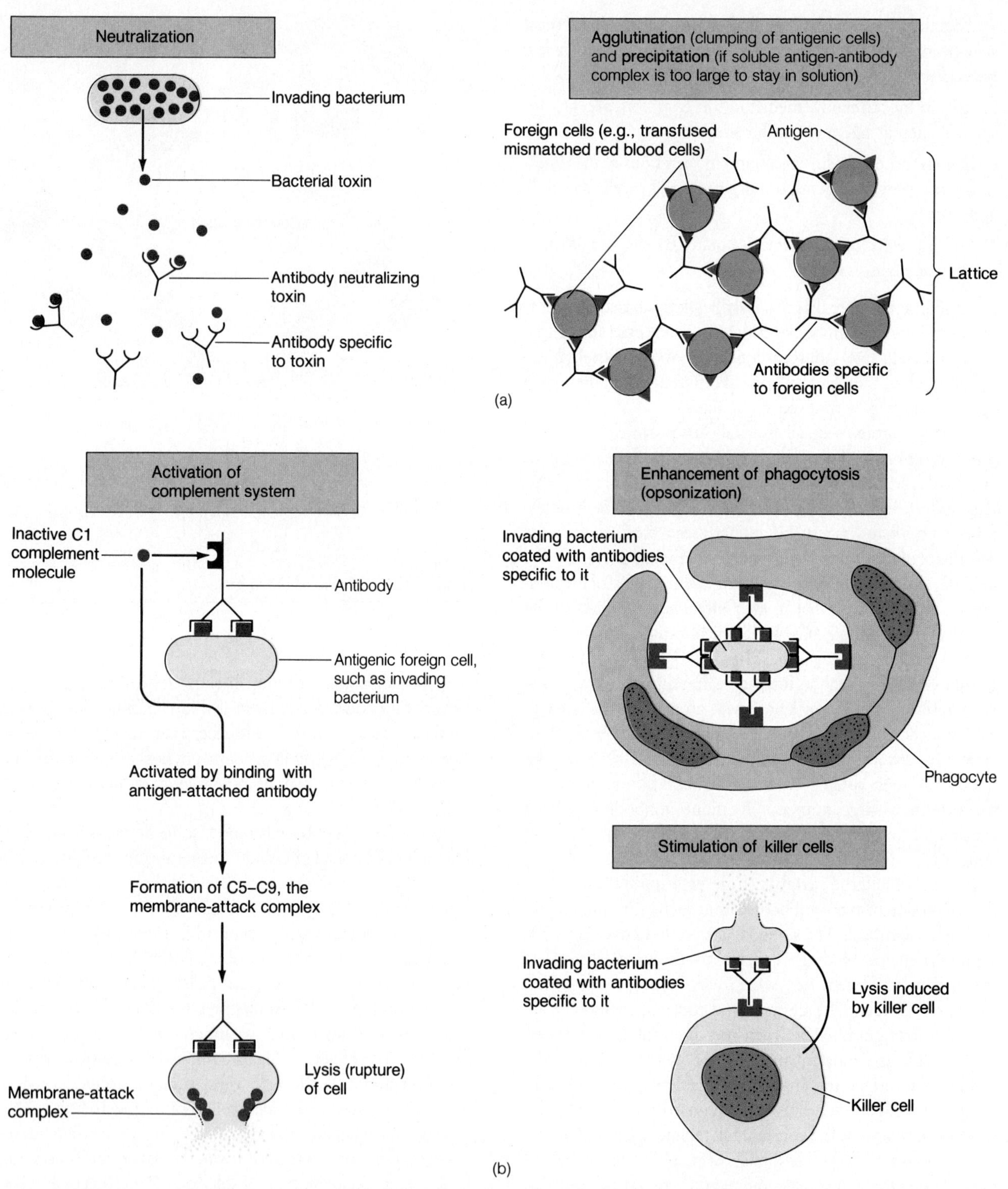

Figure 12–10 How Antibodies Help Eliminate Invading Microbes
(a) Physical hindrance of antigens. (b) Amplification of nonspecific immune responses.

forming large complexes with antibodies specific for them is useful clinically and experimentally for detecting the presence of particular antigens or antibodies. Pregnancy-diagnosis tests, for example, employ this principle to detect the presence in the urine of a hormone secreted soon after conception.

By far, antibodies' most important function is to profoundly augment the nonspecific immune responses already initiated by the invaders. Antibodies mark or identify foreign material as targets for actual destruction by the complement system, phagocytes, or killer cells while enhancing the activity of these other defense systems as follows:

1. *Activation of the complement system*
   When an appropriate antigen binds with an antibody, receptors on the tail portion of the antibody are able to bind with and activate C1, the first component of the complement system. This sets off the cascade of events leading to formation of the membrane-attack complex, which is specifically directed at the membrane of the invading cell that bears the antigen that initiated the activation process. In fact, antibody is the most powerful activator of the complement system. The biochemical attack subsequently unleashed against the invader's membrane is the most important mechanism by which antibodies exert their protective influence. Furthermore, various activated complement components enhance virtually every aspect of the inflammatory process. Note that the same complement system is activated by an antigen-antibody complex regardless of the type of antigen. Although the binding of antigen to antibody is highly specific, the outcome, which is determined by the antibody's constant tail region, is identical for all activated antibodies within a given subclass; for example, all IgG and IgM antibodies activate the same complement system.
2. *Enhancement of phagocytosis*
   As mentioned previously, antibodies, especially IgG, act as opsonins. The tail portion of antigen-bound IgG antibodies is able to bind with receptors on the surface of phagocytes and to subsequently promote the phagocytosis of the antigen-containing material attached to the antibody.
3. *Stimulation of killer (K) cells*
   The binding of antibody to antigen can also induce attack of the antigen-bearing cell by **killer (K) cells.** K cells, a specialized type of lymphocytes, are similar to natural killer (NK) cells except that K cells require the target cell to be coated with antibodies before they can destroy it by lysing its plasma membrane. K cells have receptors for the constant tail portion of antibodies.

In these ways, although antibodies are not able to directly destroy invading bacteria or other undesirable material, they bring about destruction of the antigens to which they are specifically attached by amplifying other nonspecific lethal defense mechanisms.

Occasionally, an overzealous antigen-antibody response can inadvertently cause damage to normal cells as well as to invading foreign cells. Typically, antigen-antibody complexes formed in response to foreign invasion, after having revved up nonspecific defense strategies, are removed by phagocytic cells. If large numbers of these complexes are continuously produced, however, the phagocytes are unable to effectively clear away all of the immune complexes formed. Antigen-antibody complexes that are not removed continue to activate the complement system, among other things. Excessive amounts of activated complement and other inflammatory agents may "spill over," damaging the surrounding normal cells as well as the unwanted cells. Furthermore, destruction is not necessarily restricted to the initial site of inflammation. Antigen-antibody complexes may freely circulate and become trapped in the kidneys, joints, brain, small vessels of the skin, and elsewhere, causing widespread inflammation and tissue damage. Such damage produced by immune complexes is referred to as an **immune-complex disease,** which can be a complicating outcome of bacterial, viral, or parasitic infection.

More insidiously, immune complex disease can also occur as a result of overzealous inflammatory activity prompted by the presence of immune complexes formed by "self-antigens" (proteins synthesized by the person's own body) and antibodies erroneously produced against them. Rheumatoid arthritis is believed to be brought about in this way.

Each antigen stimulates a different clone of B lymphocytes to produce antibodies.

Consider the diversity of foreign molecules that a person can potentially encounter during a lifetime. Yet each B lymphocyte is preprogrammed to respond to only one of these millions of different antigens. Different antigens cannot combine with the same B cell and induce it to secrete different antibodies. The astonishing implication is that each of us is equipped with millions of different preformed B lymphocytes, at least one for every possible antigen that we might ever encounter—including those specific for man-made substances that do not exist in nature.

According to early immunological theory, antibodies were believed to be "made to order" whenever a foreign antigen gained entry to the body. In contrast, the currently accepted **clonal selection theory** proposes that diverse B lymphocytes are produced during fetal development, each capable of synthesizing antibody against a particular antigen before ever

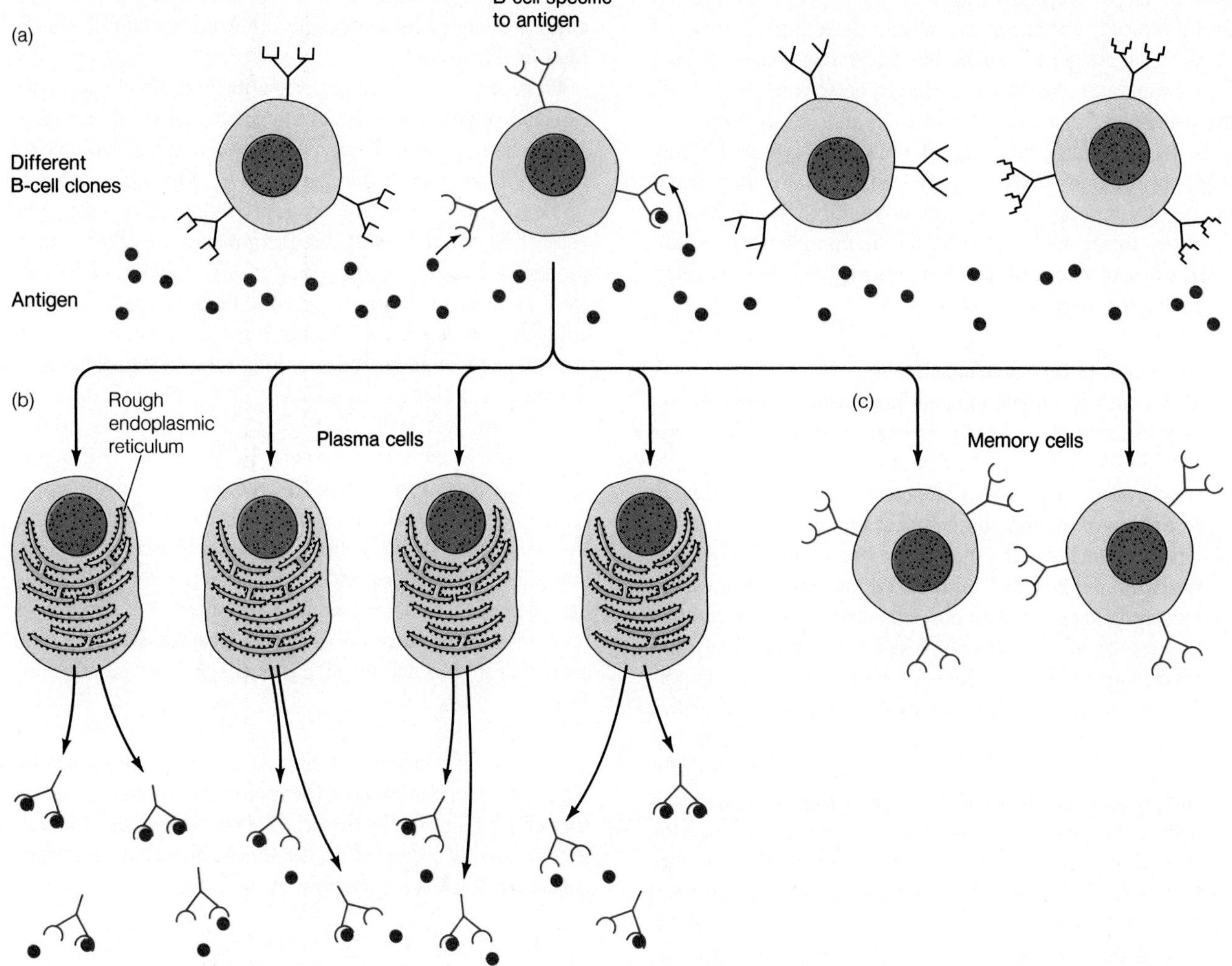

Figure 12–11 Clonal Selection Theory *(a) The B-cell clone specific to the antigen proliferates and differentiates. (b) Plasma cells secrete antibodies that bind with free antigen (antigen not attached to B cells). (c) Memory cells expand the specific clone and are primed and ready for subsequent exposure to the same antigen.*

being exposed to it. All offspring of a particular ancestral B lymphocyte form a family of identical cells, or a **clone,** which is committed to producing the same specific antibody. B cells remain dormant, not actually secreting their particular antibody product until (or unless) they come into contact with the appropriate antigen. When an antigen gains entry to the body, it activates the particular clone of B cells that bear receptors on their surface uniquely specific for that antigen (Fig. 12–11).

The first antibodies produced by a newly formed B cell (type IgM) are inserted into the cell's plasma membrane rather than being secreted. Here they serve as receptor sites for binding with a specific kind of antigen, almost like "advertisements" for the kind of antibody the cell can produce. Binding of the appropriate antigen to a B cell amounts to "placing an order" for the manufacture and secretion of large quantities of that particular antibody.

Antigen binding causes the activated B-cell clone to multiply and differentiate into two cell types—plasma cells and memory cells. Most progeny are transformed into **plasma cells,** which are prolific producers of customized antibodies that contain the same antigen-binding sites as the surface receptors. However, plasma cells switch to the production of IgG antibodies, which are secreted rather than remaining membrane-bound. In the blood, the secreted antibodies com-

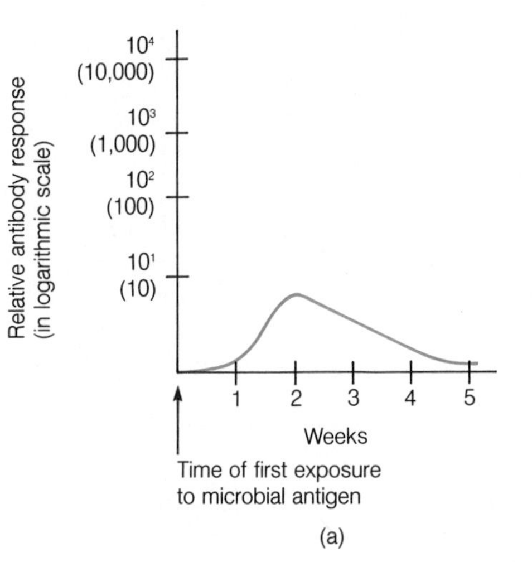

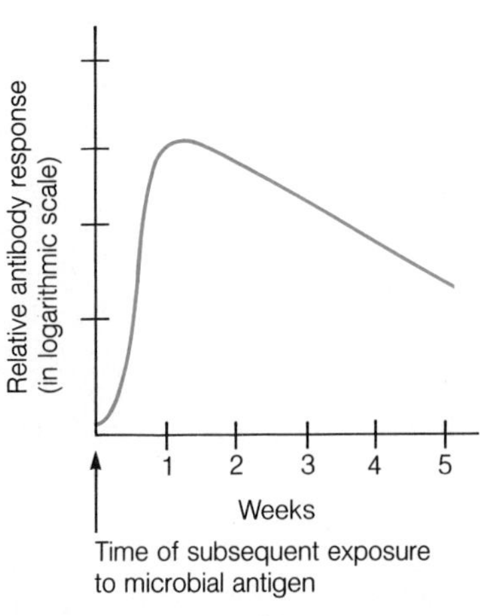

Figure 12–12 Primary and Secondary Immune Responses *(a) Primary response. (b) Secondary response. Note that the secondary response peaks in a week, whereas the primary response does not peak for several weeks. Also, the magnitude of the secondary response is 100 times that of the primary response. (The relative antibody response is designated in the logarithmic scale.)*

bine with invading free (not bound to lymphocytes) antigen, marking it for destruction by the complement system, phagocytic ingestion, or other means.

Not all of the new B lymphocytes produced by the specifically activated clone differentiate into antibody-secreting plasma cells. A small proportion of them become **memory cells,** which do not participate in the current immune attack against the antigen but instead remain dormant and expand the specific clone. Should the person ever be exposed to the same antigen again, these memory cells are primed and ready for even more immediate action than were the original lymphocytes in the clone.

During initial contact with a microbial antigen, the antibody response is delayed for several days until plasma cells are formed and does not reach its peak for several weeks (Fig. 12–12). This is known as the **primary response.** Meanwhile, symptoms characteristic of the particular microbial invasion persist either until the invader succumbs to the mounting specific immune attack against it or until the infected individual dies. The antibody levels gradually decline over a period of time, although some circulating antibody from this primary response may persist for a prolonged period. Long-term protection against the same antigen, however, is primarily attributable to the memory cells. If the same antigen ever reappears, the long-lived memory cells launch a more rapid, more potent, and longer-lasting **secondary response** than occurred during the primary response. This swifter, more powerful immune attack is frequently adequate to prevent or minimize overt infection upon subsequent exposures to the same microbes, forming the basis of long-term immunity against a specific disease. The original antigenic exposure that induced the formation of memory cells can occur either by actually having the disease or by being vaccinated (Fig. 12–13).

Even though each of us has essentially the same original pool of different B-lymphocyte clones, the pool gradually becomes appropriately biased to respond most efficiently to each individual's particular antigenic environment. Those clones specific for antigens to which an individual is never exposed remain dormant for life, whereas those specific for antigens in the individual's environment typically become expanded and enhanced through the formation of highly responsive memory cells.

Memory cells are not formed for some diseases, so no lasting immunity is conferred by an initial exposure. The course and severity of the disease are the same each time a person is reinfected with a microbe that the immune system does not "remember," regardless of the number of prior exposures.

Thinking about the millions of different antigens against which each of us has the potential to actively produce antibodies, how is it possible for an individual to have such a tremendous diversity of B lymphocytes, each capable of producing a different antibody? Antibodies are proteins that are produced in accordance with a nuclear DNA blueprint. Because all cells of the body, including the antibody-producing cells, contain the same nuclear DNA, it is hard to imagine how enough DNA could be packaged within the nuclei of every cell to code for the millions of different antibodies (a different portion of the genetic code being used by each B-cell clone), along with all of the other genetic instructions used by other cells. It is now known that only a relatively small number of gene fragments actually code for antibody synthesis, but these fragments are cut, reshuffled, and spliced in a vast number of

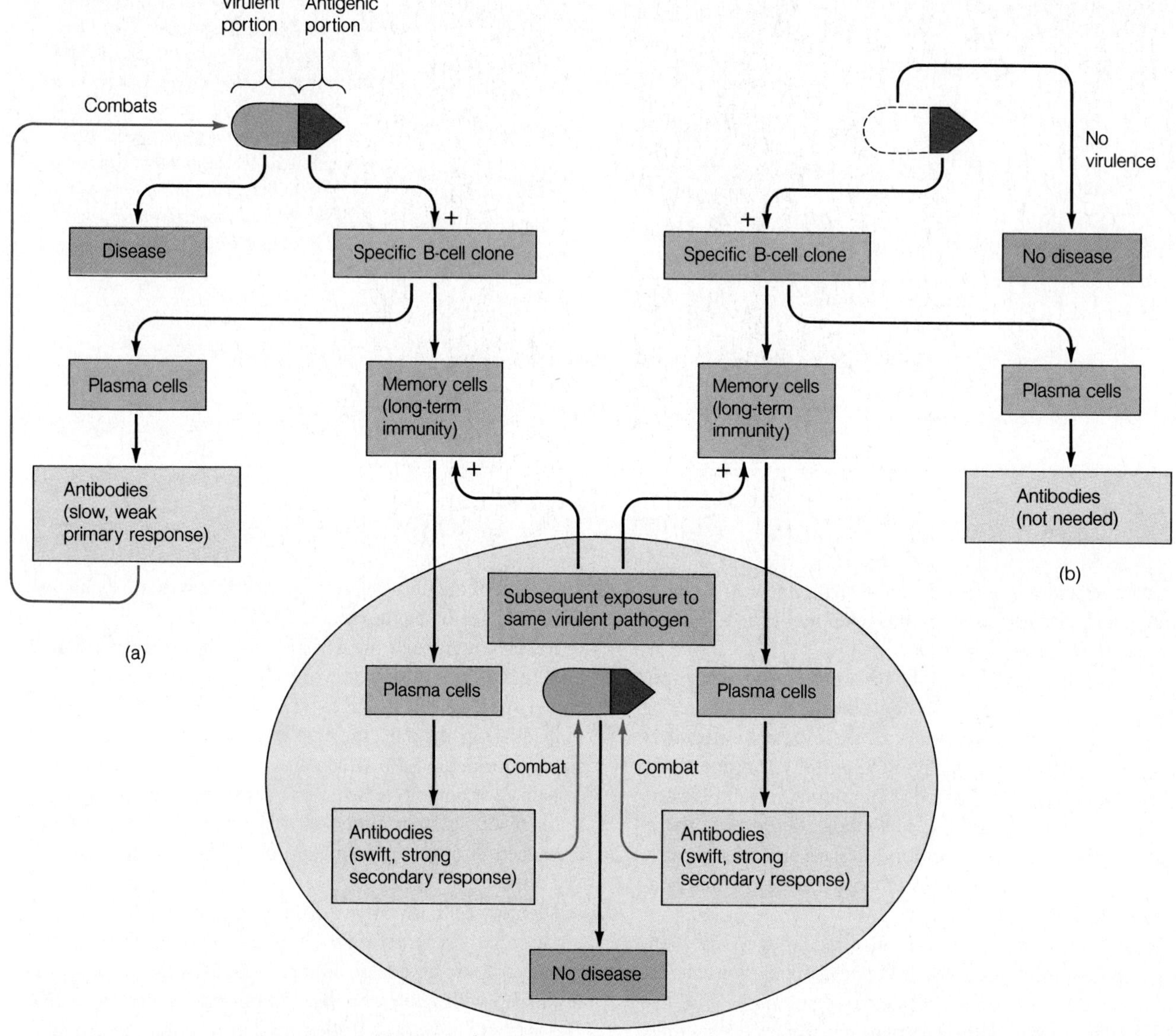

Figure 12–13 Means of Acquiring Long-Term Immunity *Long-term immunity against a pathogen can be acquired through having the disease or being vaccinated against it. (a) Exposure to virulent pathogen. (b) Vaccination with attenuated pathogen.*

different combinations during B-cell development. Each different combination gives rise to a unique B-cell clone. Even further diversification of antibody genes is subsequently accomplished by somatic mutation (see p. 56). The antibody genes of already-formed B cells are highly prone to mutations in the region that codes for the variable antigen-binding sites on the antibodies. Each different mutant cell, in turn, gives rise to a new clone. Thus, the great diversity of antibodies is made possible by the reshuffling of a small set of gene fragments during B-cell development as well as by further somatic mutation in already-formed B cells. In this way, a huge antibody repertoire is possible using only a modest share of the genetic blueprint.

Active immunity is self-generated; passive immunity is "borrowed."

The production of antibodies as a result of exposure to an antigen is referred to as **active immunity** against that antigen. A second way in which an individual can acquire antibodies is by

Table 12–2 Active versus Passive Immunity

Characteristic	Active Immunity	Passive Immunity
Exposure to antigen required for immunity to develop?	Yes, either through natural exposure during infection or through artificial exposure during vaccination	No
Source of circulating antibodies	Antibodies self-generated on exposure to antigen	Antibodies "borrowed"—that is, produced from another source and transferred to the individual, either naturally across the placenta or through the milk from mother to offspring, or artificially through injection of antibodies harvested from vaccinated animal
Injection required?	Injection of attenuated antigen required for artificial active immunity	Injection of borrowed antibodies required for artificial passive immunity
Time to develop resistance to antigen exposure	Several weeks (primary response to antigen) to several days (secondary response to antigen) required for antibody production	Immediate on acquisition of borrowed antibodies
Duration of resistance	Long—may be lifelong	Short—less than a month
Indication for artificial acquisition of this type of immunity	Well before potential exposure to virulent pathogen to allow enough time for appropriate B-cell clone to respond; boosters as needed to maintain active immunity	Immediately after exposure to extremely virulent virus or potentially lethal toxin when protection is needed immediately (rather than waiting for the more slowly responding active immunity to develop)

the direct transfer of antibodies actively formed by another person (or animal). The immediate "borrowed" immunity conferred upon receipt of preformed antibodies is known as **passive immunity** (Table 12–2). Such transfer of antibodies of the IgG class normally occurs from the mother to the fetus across the placenta during intrauterine development. In addition, a mother's colostrum (first milk) contains IgA antibodies that provide further protection for breast-fed babies. Passively transferred antibodies are usually broken down in less than a month, but meanwhile the newborn is provided important immune protection (essentially the same as its mother's) until it can begin actively mounting its own immune responses. Antibody-synthesizing ability does not develop for about a month after birth.

Passive immunity is sometimes employed clinically to provide immediate protection or to bolster resistance against an extremely virulent infectious agent or potentially lethal toxin to which a person has been exposed (for example, rabies virus, tetanus toxin in nonimmunized individuals, and poisonous snake venom). Typically the administered preformed antibodies have been harvested from another source (often nonhuman) that has been exposed to an attenuated form of the antigen. Frequently horses or sheep are used in the deliberate production of antibodies to be collected for passive immunizations. Although injection of serum containing these antibodies (**antiserum** or **antitoxin**) is beneficial in providing immediate protection against the specific disease or toxin, the recipient may mount an immune response against the injected antibodies themselves because they are foreign proteins. The result may be a severe allergic reaction to the treatment, a condition known as **serum sickness.**

Natural immunity is actually a special case of actively acquired immunity.

Besides active and passive immunity, certain antibodies were once thought to occur naturally in the blood. The classic example of "natural antibodies" are those associated with blood types. The surface membranes of human erythrocytes contain inherited antigens that vary depending on blood type. With the major blood-group system, the **ABO system,** the erythrocytes of individuals with type-A blood contain A antigens; those with type-B blood contain B antigens; those with type-AB blood have both A and B antigens; and those with type-O blood do not have any A or B red blood cell surface antigens. Antibodies against erythrocyte antigens not present on the body's own erythrocytes begin to appear in human plasma after a person is about six months of age. Accordingly, the plasma of type-A blood contains anti-B antibodies; type-B blood contains anti-A antibodies; no antibodies related to the ABO system are present in type AB-blood; and both anti-A and anti-B antibodies are present in type-O blood. Typically one would expect antibody production against A or B antigen to be induced only if blood containing the alien antigen were injected into the body. However, high levels of these antibodies are found in the plasma of individuals who have never been exposed to a different type blood. Consequently, these were considered to be naturally occurring antibodies—that is, produced without any known exposure to the antigen. It is now known that individuals are unknowingly exposed at an early age to small amounts of A- and B-like antigens associated with common intestinal bacteria. Antibodies produced against these foreign antigens coincidentally also interact with a nearly identical foreign-blood-group antigen, even upon first exposure to it.

If a person is administered blood of an incompatible type, two different antigen-antibody interactions take place. By far the most serious consequences arise from the effect of the antibodies in the recipient's plasma on the incoming donor erythrocytes. The effect of the donor's antibodies on the recipient's erythrocyte-bound antigens is of less importance unless a large amount of blood is transfused, because the donor's antibodies are so diluted by the recipient's plasma that little red blood cell damage takes place in the recipient.

Antibody interaction with erythrocyte-bound antigen may result in agglutination (clumping) or hemolysis (rupture) of the attacked red blood cells. Agglutination and hemolysis of donor red blood cells by antibodies in the recipient's plasma can lead to a sometimes fatal **transfusion reaction.** Agglutinated clumps of incoming donor cells can plug small blood vessels. In addition, one of the most lethal consequences of mismatched transfusions is acute kidney failure caused by the release of large amounts of hemoglobin from damaged donor erythrocytes. If the free hemoglobin in the plasma rises above a critical level, it will precipitate within the kidneys and block the urine-forming structures, leading to acute kidney shutdown.

Because type-O individuals do not have any A or B antigens, their erythrocytes will not be attacked by either anti-A or anti-B antibodies, so they are considered to be **universal donors.** Their blood can be transfused into persons of any blood type. However, type-O individuals can receive only type-O blood because the anti-A and anti-B antibodies present in their plasma will attack either A or B antigens in incoming blood. In contrast, type-AB individuals are called **universal recipients.** Lacking both anti-A and anti-B antibodies, they can accept donor blood of any type, although they can donate blood only to other AB persons. Because their erythrocytes possess both A and B antigens, their cells would be attacked if transfused into individuals with antibodies against either of these antigens.

The terms *universal donor* and *universal recipient* are somewhat misleading, however. In addition to the ABO system, there are numerous other erythrocyte antigens and plasma antibodies that can cause transfusion reactions, the most important of which is the Rh factor. Individuals who possess the **Rh factor** (an erythrocyte antigen first observed in rhesus monkeys, hence the term *Rh*) are said to have **Rh-positive** blood, whereas those lacking the Rh factor are considered to be **Rh-negative.** In contrast to the ABO system, no naturally occurring antibodies develop against the Rh factor. Anti-Rh antibodies are produced only by Rh-negative individuals when (and if) they are first exposed to the foreign Rh antigen present in Rh-positive blood. The subsequent transfusion of Rh-positive blood could produce a transfusion reaction in such a sensitized Rh-negative person. Rh-positive individuals, in contrast, never produce antibodies against the Rh factor that they themselves possess. Therefore, Rh-negative individuals should be administered only Rh-negative blood, whereas Rh-positive persons can safely receive either Rh-negative or Rh-positive blood. The Rh factor is of particular medical importance when an Rh-negative mother develops antibodies against the erythrocytes of an Rh-positive fetus she is carrying, a condition known as **erythroblastosis fetalis,** or **hemolytic disease of the newborn.**

Except in extreme emergencies, it is safest to individually cross-match blood before a transfusion is undertaken even though the ABO and Rh typing is already known, because there are approximately twelve other minor human erythrocyte antigen systems. Compatibility is determined by mixing the red blood cells from the potential donor with plasma from the recipient. If no clumping occurs, the blood is considered to be adequately matched for transfusion.

In addition to being an important consideration in transfusions, the various blood-group systems are also of legal importance in disputed paternity cases, because the erythrocyte antigens are inherited.

Lymphocytes respond only to antigens that have been processed and presented to them by macrophages.

B cells cannot typically perform their task of antibody production without assistance from macrophages and, in most cases, T cells as well (Fig. 12–14). Relevant B-cell clones are not able to recognize and produce antibodies in response to "raw" foreign antigens entering the body; a B-cell clone must be formally "introduced" to the antigen before it will react to it. Invading organisms or other antigens are first engulfed by macrophages, which cluster around the appropriate B-cell clone

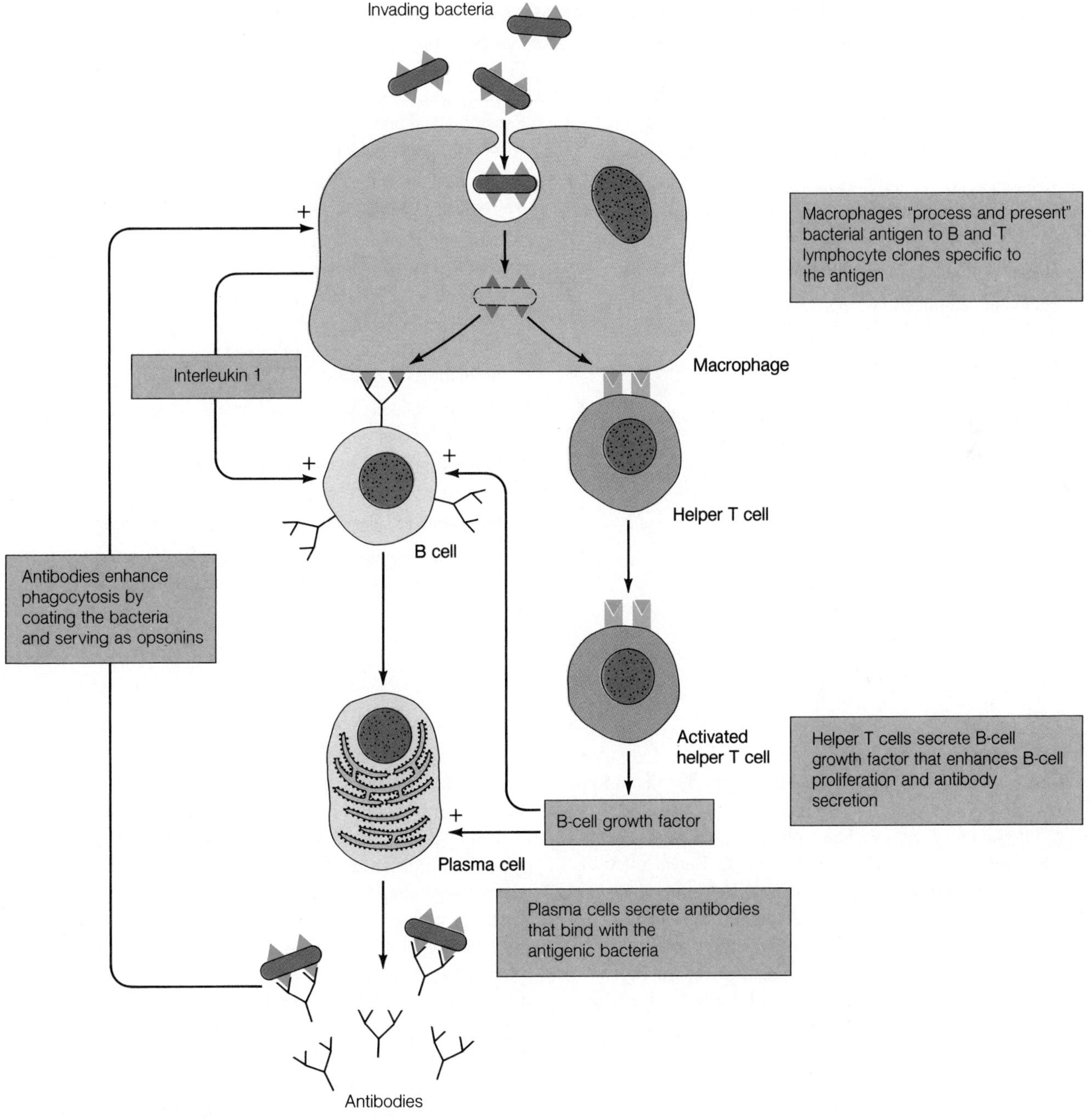

Invading bacteria
Macrophages "process and present" bacterial antigen to B and T lymphocyte clones specific to the antigen
+
Interleukin 1
Macrophage
+
+
Helper T cell
B cell
Antibodies enhance phagocytosis by coating the bacteria and serving as opsonins
Activated helper T cell
Helper T cells secrete B-cell growth factor that enhances B-cell proliferation and antibody secretion
+
B-cell growth factor
Plasma cell
Plasma cells secrete antibodies that bind with the antigenic bacteria
Antibodies

Figure 12-14 Synergistic Interactions among Macrophages, B Cells, and Helper T Cells

and handle the formal introduction. During phagocytosis, the macrophages somehow "process and present" the antigen in such a way that the adjacent B cells can then recognize and be activated by it. In addition, these antigen-presenting macrophages secrete interleukin 1, a multi-purpose chemical mediator that enhances the differentiation and proliferation of the now-activated B-cell clone. Interleukin 1 (alias endogenous pyrogen/leukocyte endogenous mediator) is also largely responsible for the fever and malaise accompanying many infections. In collaborative fashion, activated lymphocytes secrete antibodies that, among other things, enhance further phagocytic activity.

Table 12–3 Nonspecific and Specific Immune Responses to Bacterial Invasion

Nonspecific Immune Mechanisms	Specific Immune Mechanisms
Inflammation	Processing and presenting of bacterial antigen by macrophages to B cells specific to the antigen
Engulfment of invading bacteria by resident tissue macrophages	Proliferation and differentiation of activated B-cell clone into plasma cells and memory cells
Histamine-induced vascular responses to enhance delivery of increased blood flow to area, bringing in additional immune-effector cells and plasma proteins	Secretion by plasma cells of customized antibodies, which specifically bind to invading bacteria
Walling off of invaded area by fibrin clot	Enhancement by interleukin 1 secreted by macrophages
Emigration of neutrophils and monocytes/macrophages to the area to engulf and destroy foreign invaders and to remove cellular debris	Enhancement by helper T cells, which have been activated by the same bacterial antigen processed and presented to them by macrophages
Secretion of chemical mediators by phagocytic cells, which enhance both nonspecific and specific immune responses and induce local and systemic symptoms associated with infection	Binding of antibodies to invading bacteria and enhancement of nonspecific mechanisms that lead to their destruction
Nonspecific Activation of the Complement System	Action as opsonins to enhance phagocytic activity
Formation of hole-punching membrane-attack complex that lyses bacterial cells	Activation of lethal complement system
Enhancement of many steps of inflammation	Stimulation of killer cells, which directly lyse bacteria
Stimulation of phagocytosis	Persistence of memory cells that are capable of responding more rapidly and more forcefully should the same bacteria be encountered again
Action as chemotaxins to attract new phagocytes	
Stimulation of histamine secretion	

Many antigens are similarly presented to T cells. One specialized class of T lymphocytes, called helper T cells, upon being activated by macrophage-presented antigen, help B cells. The helper T cells secrete a chemical mediator, **B-cell growth factor,** which further contributes to B-cell function in concert with the interleukin 1 secreted by macrophages. Therefore, mutually supportive interactions among macrophages, B cells, and helper T cells synergistically reinforce the phagocytic-antibody immune attack against the foreign intruder. Table 12–3 summarizes the nonspecific and specific immune strategies that defend against bacterial invasion.

T Lymphocytes: Cell-Mediated Immunity

The three types of T cells are specialized to kill virus-infected host cells and to help or suppress other immune cells.

As important as B lymphocytes and their antibody products are in specific defense against invading bacteria and other foreign material, they represent only half of the body's specific immune-defense armamentarium. The T lymphocytes are equally important in defense against most viral and fungal infections and also play an important regulatory role in immune mechanisms (Table 12–4).

Unlike B cells, which secrete antibodies that can attack antigen at long distances, T cells do not secrete antibodies. Instead, they themselves must be in direct contact with their targets, a process known as ***cell-mediated immunity***. Like B cells, T cells are clonal and exquisitely antigen-specific. On its plasma membrane, each T cell bears unique receptor proteins that are produced under the direction of gene segments distinct from those involved in antibody synthesis. Importantly, unlike B cells, T cells are activated by foreign antigen only when it is present on the surface of a cell that also carries a marker of the individual's own identity; that is, both foreign antigens and ***self-antigens*** must be present on a cell's surface before a T cell can bind with it (with one important exception being whole transplanted foreign cells). It is during thymic education that T cells learn to recognize foreign antigens only in combination with the individual's own tissue antigens, a lesson that is passed on to all T cells' future progeny. The importance of this dual antigen requirement and the nature of the self-antigens will be described shortly.

There is generally a delay of a few days following exposure to the appropriate antigen before **sensitized,** or **activated, T**

Table 12-4 B versus T Lymphocytes

Characteristic	B Lymphocytes	T Lymphocytes
Ancestral origin	Bone marrow	Bone marrow
Site of maturational processing	Bone marrow	Thymus
Receptors for antigens	Antibodies inserted in plasma membrane, which serve as surface receptors; highly specific	Surface receptors present but differing from antibodies; highly specific
Bind with	Extracellular antigens such as bacteria, free viruses, and other circulating foreign material	Foreign antigen in association with self-antigen, such as virus-infected cells
Antigen must be processed and presented by macrophages	Yes	Yes
Types of active cells	Plasma cells	Cytotoxic T cells, helper T cells, suppressor T cells
Formation of memory cells	Yes	Yes
Type of immunity	Antibody-mediated immunity	Cell-mediated immunity
Secretory product	Antibodies	Lymphokines
Function	Help eliminate free foreign invaders by enhancing nonspecific immune responses against them; provide immunity against most bacteria and a few viruses	Lyse virus-infected cells and cancer cells; provide immunity against most viruses and fungi and a few bacteria; aid B cells in antibody production
Life-span	Short	Long

cells are prepared to launch a cell-mediated immune attack. When exposed to a specific antigen combination, cells of the complementary T-cell clone proliferate and differentiate for several days, yielding large numbers of activated T cells that carry out various cell-mediated responses. There are three subpopulations of T cells, depending on their roles when activated by antigen:

1. **Cytotoxic T cells,** which destroy host cells bearing foreign antigen, such as body cells invaded by viruses, cancer cells, and transplanted cells.
2. **Helper T cells,** which enhance the development of antigen-stimulated B cells into antibody-secreting cells, enhance activity of the appropriate cytotoxic and suppressor T cells, and activate macrophages.
3. **Suppressor T cells,** which suppress both B-cell antibody production and cytotoxic and helper T-cell activity.

The vast majority of the billions of T lymphocytes are believed to be of the helper or suppressor varieties, which do not directly participate in the immune destruction of invading pathogens. Collectively, these subpopulations are referred to as **regulatory T cells,** because they modulate the activities of B cells and cytotoxic T cells as well as their own activities and those of macrophages.

Like B cells, not all activated T-cell progeny become effector T cells. A small proportion of them remain dormant, serving as a pool of memory T cells that are primed and ready to respond even more swiftly and vigorously should the same foreign antigen ever reappear within a body cell. T cells, even activated ones, generally have long-life spans, in contrast to B cells, which rapidly work themselves to death producing antibodies once they have been converted into plasma cells upon antigen stimulation. Thus active immunity for cell-mediated responses is similar to that for antibody responses, but it is generally of longer duration. Passive cell-mediated immunity against a particular pathogen can also be conferred through administration of activated T lymphocytes that have been harvested from another individual or animal who possesses active immunity against the particular microbe.

Exposure to antigen frequently activates both the B- and T-cell mechanisms simultaneously. Just as the regulatory T cells can facilitate or suppress the secretion of antibody by B cells, antibodies may either enhance or block the ability of cytotoxic T cells to destroy a victim cell, depending on the circumstances. Most of the effects exerted by lymphocytes on

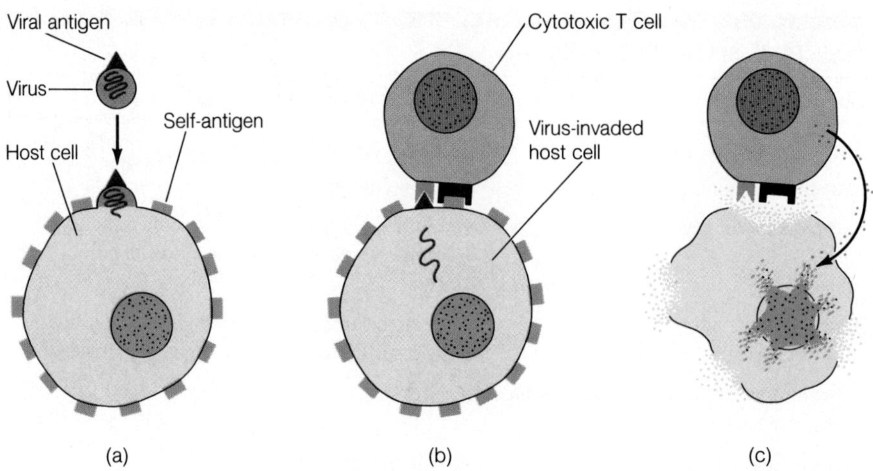

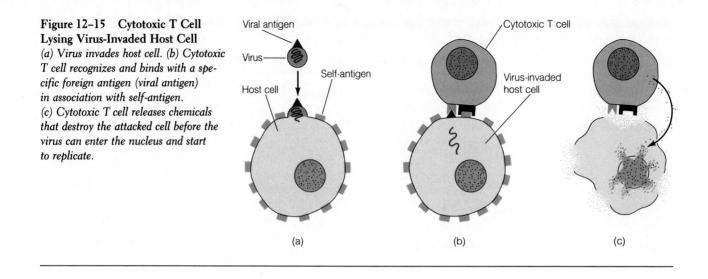

Figure 12–15 Cytotoxic T Cell Lysing Virus-Invaded Host Cell *(a) Virus invades host cell. (b) Cytotoxic T cell recognizes and binds with a specific foreign antigen (viral antigen) in association with self-antigen. (c) Cytotoxic T cell releases chemicals that destroy the attacked cell before the virus can enter the nucleus and start to replicate.*

other immune cells (such as other lymphocytes and macrophages) are mediated by means of the secretion of chemical messengers. All chemicals other than antibodies that are secreted by lymphocytes are collectively called **lymphokines,** the majority of which are produced by T cells. Unlike antibodies, lymphokines do not directly interact with the antigen responsible for inducing their production. The roles of specific lymphokines will be elucidated in the following discussion of each of the T-cell subpopulations.

CYTOTOXIC T CELLS. The targets of cytotoxic T cells most frequently are host cells infected with viruses. When a virus invades a body cell, as it must to survive, the envelope of antigenic proteins surrounding the virus are incorporated into the host cell's surface membrane (Fig. 12–15). To attack the intracellular virus, cytotoxic T cells must destroy the infected host cell in the process. Cytotoxic T cells of the clone specific for this particular virus recognize and bind to the viral antigens and self-antigens on the surface of the infected cell. Thus sensitized by viral antigen, a cytotoxic T cell destroys the victim cell by releasing chemicals that lyse the attacked cell before viral replication can begin.

Recent studies demonstrate that one means by which cytotoxic T cells and natural killer cells destroy a targeted cell is by releasing **perforin** molecules, which penetrate into the target cell's surface membrane and join together to form porelike channels (Fig. 12–16). This technique of killing a cell by punching holes in its membrane is similar to the method employed by the membrane-attack complex of the complement cascade. The virus released upon destruction of the host cell is then directly destroyed in the extracellular fluid by phagocytic cells, neutralizing antibodies, and the complement system. Meanwhile, the cytotoxic T cell can move on to kill other infected host cells, not being harmed itself in the process. Usually, not many of the host cells have to be destroyed to halt a viral infection, but if the virus has had a chance to multiply, with replicated virus leaving the original cell and spreading to other host cells, so many of the host cells may be sacrificed by the cytotoxic T cell defense mechanism that serious malfunction may ensue.

Recall that other nonspecific defense mechanisms also come into play to combat viral infections, most notably natural killer cells and interferon. As usual, an intricate web of interplay exists among the immune defenses that are launched against viral invaders (Table 12–5).

HELPER T CELLS. Even more than cytotoxic T cells, helper T cells modulate many aspects of the immune response, primarily by secretion of lymphokines. The following are among the best-known of these T-cell chemical messengers:

1. As noted earlier, helper T cells secrete B-cell growth factor, which enhances the antibody-secreting ability of the activated B-cell clone. Antibody secretion is greatly diminished in the absence of helper T cells, even though T cells themselves do not produce antibodies.
2. Helper T cells similarly secrete **T-cell growth factor,** also known as **interleukin 2 (IL-2)**, which augments the activity of cytotoxic T cells, suppressor T cells, and even other helper T cells responsive to the invading antigen. In typical interplay fashion, interleukin 1 secreted by macrophages not only enhances the activity of both the appropriate B- and T-cell clones but also stimulates the secretion of interleukin 2 by the activated helper T cells.

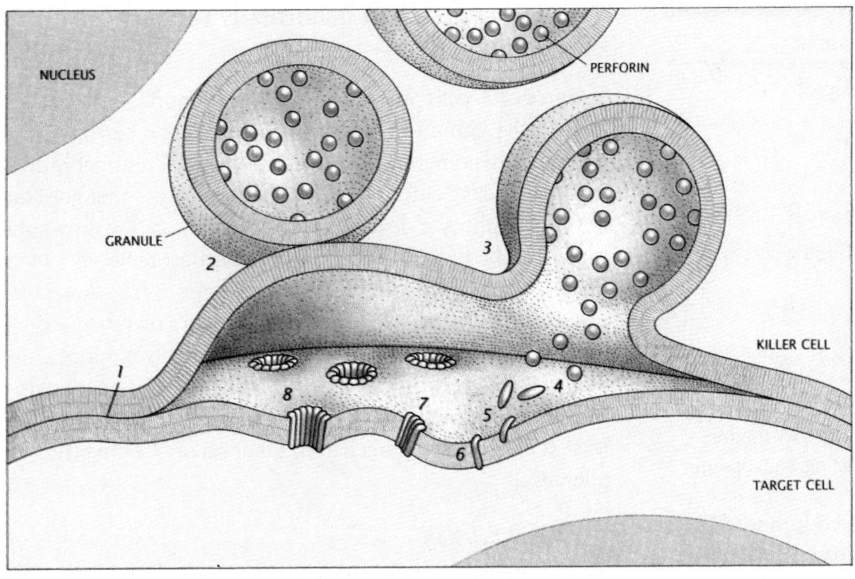

(a)

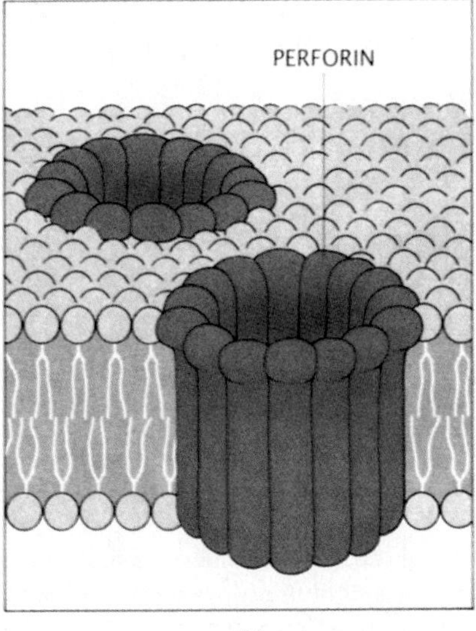

(b)

Figure 12–16 Mechanism of Killing by Killer Cells
(a) Details of the killing process. A rise in the lymphocyte's calcium-ion level, apparently triggered by the receptor-mediated binding of the killer cell to its target (1), brings about exocytosis, in which the granules fuse with the cell membrane (2) and disgorge their perforin (3) into the small intercellular space abutting the target. Calcium there changes the conformation of the individual perforin molecules, or monomers (4), which then bind to the target-cell membrane (5) and insert into it (6). The monomers polymerize like staves of a barrel (7) to form pores (8) that admit water and salts and kill the cell. (b) Enlargement of perforin-formed pores in a target cell. Note the similarity to the membrane attack complex formed by complement molecules (see Figure 12–6).

SOURCE: By Dana Burns from "How Killer Cells Kill" by John Ding-E Young and Zanvil A. Kohn.

3. Some chemicals secreted by T cells act as chemotaxins to lure more neutrophils and macrophages-to-be to the invaded area.
4. Once macrophages are attracted to the area, another important lymphokine released from helper T cells, called **macrophage-migration inhibition factor,** keeps these large phagocytic cells in the region by inhibiting their outward migration. As a result, a great number of chemotactically attracted macrophages accumulate in the infected area. This factor also confers greater phagocytic power on the gathered macrophages. These so-called **angry macrophages** have more powerful destructive ability and are especially important in defending against the bacteria that cause tuberculosis, because such microbes are able to survive simple phagocytosis by nonactivated macrophages.

Helper T cells are by far the most numerous of the T cells, making up 60–80% of circulating T cells. Because of the important role these cells play in "turning on" the full power of all of the other activated lymphocytes and macrophages, helper T cells may constitute the immune system's "master switch." It is for this reason that **acquired immune defiency syndrome (AIDS)** is so devastating to the immune defense system. The AIDS virus selectively invades helper T cells, destroying or incapacitating those cells that normally orchestrate much of the immune response (Fig. 12–17).

SUPPRESSOR T CELLS. Much less is known about suppressor T cells than the other T-cell subpopulations. They apparently serve to limit immune reactions in a "check and balance" relationship with the other lymphocytes. Whereas B cells, cytotoxic T cells, and especially helper T cells reinforce each other's immune activities, suppressor T cells limit the responses of all the other immune cells. In negative-feedback fashion, helper T cells prod suppressor T cells into action; the suppressor T cells, in turn, inhibit the helper T cells and the other cells revved up for duty by helper T-cell influence. Because of this feedback loop, the immune response tends to be self-limiting. Such a dampening effect by the suppressor T

Table 12–5 Defenses against Viral Invasion

When the virus is free in the extracellular fluid,

Macrophages:

Destroy the free virus by phagocytosis

Process and present the viral antigen to both B and T cells

Secrete interleukin 1, which activates B- and T-cell clones specific to the viral antigen

Plasma cells derived from B cells specific to the viral antigen secrete antibodies that:

Neutralize the virus to prevent its entry into a host cell

Activate the complement cascade that directly destroys the free virus and enhances phagocytosis of the virus by acting as an opsonin

When the virus has entered a host cell (which it must do to survive and multiply, with the replicated viruses leaving the original host cell to enter the extracellular fluid in search of other host cells),

Interferon:

Is secreted by virus-infected cells, macrophages, natural killer (NK) cells, and cytotoxic T cells

Binds with and prevents viral replication in other host cells

Enhances killing power of macrophages, cytotoxic T cells, and NK cells

Natural killer cells:

Nonspecifically lyse virus-infected host cells

Cytotoxic T cells:

Specifically sensitized by the viral antigen lyse the infected host cells before the virus has a chance to replicate

Helper T cells:

Secrete lymphokines, which enhance cytotoxic T-cell activity and B-cell antibody production

When a virus-infected cell is destroyed, the free virus is released into the extracellular fluids, where it is attacked directly by macrophages, antibodies, and the activated complement components

cells helps prevent excessive immune reactions that might be detrimental to the body. Suppressor T cells generally increase in number more slowly in response to a viral infection than the cytotoxic and helper T cells do, so the suppressor cells help shut down the immune response after it has already served its purpose.

The immune system is normally tolerant of self-antigens.

Suppressor T cells probably also play an important role in preventing the immune system from attacking the person's own tissues, a phenomenon known as **tolerance.** Presumably during the genetic "cut, shuffle, and paste process" that goes on during lymphocyte development, some B and T cells would by chance be formed that could react against the body's own tissue antigens. If these lymphocyte clones were allowed to function, they would destroy the individual's own body. Fortunately, the immune system does not normally produce antibodies or activated T cells against its own body antigens but instead directs its destructive tactics only at foreign antigens. At least three different mechanisms appear to be involved in tolerance:

1. *Clonal deletion*
   In response to continuous exposure to body antigens early in development, lymphocyte clones specifically capable of attacking these self-antigens in some cases are permanently destroyed by unknown means.
2. *Inhibition by suppressor T cells*
   Lymphocyte clones specific for the body's own tissues that are not eliminated during early development are presumably inhibited throughout life by suppressor T cells.
3. *Antigen sequestering*
   Some self molecules are normally hidden from the immune system because they never come into direct contact with the extracellular fluid in which the immune cells and their products circulate. Examples of such segregated antigens are the proteins of the cornea (the clear anterior portion of the eye) and thyroglobulin, a complex protein sequestered within the hormone-secreting structures of the thyroid gland.

Occasionally, the immune system fails to make the distinction between self-antigens and foreign antigens, unleashing its deadly powers against one or more of the body's own tissues. A condition in which the immune system fails to recognize and tolerate self-antigens associated with particular tissues is known as an **autoimmune disease,** of which myasthenia gravis is an example. People with this condition erroneously produce antibodies against the acetylcholine receptors on their own skeletal-muscle fibers (see p. 214).

Autoimmune diseases may arise from a number of different causes:

1. A reduction in suppressor T-cell activity or an imbalance in the ratio of suppressor to helper T cells specific for self-antigens may be responsible for the development of some autoimmune conditions.

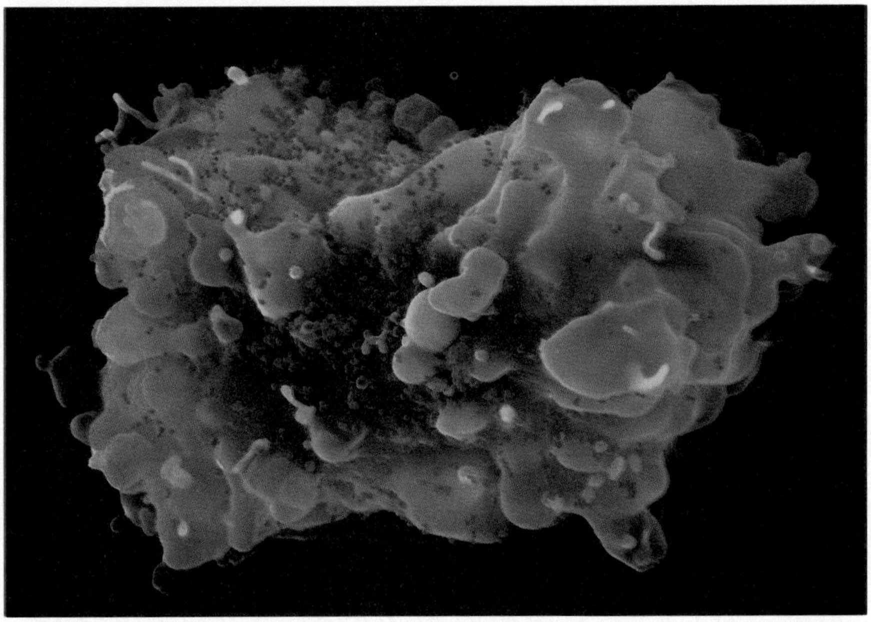

Figure 12–17 AIDS Virus
AIDS viruses (in purple) on a helper T lymphocyte.

2. Normal self-antigens may be modified by factors such as drugs, environmental chemicals, viruses, or genetic mutations so that they are no longer recognized and tolerated by the immune system.
3. Exposure of normally inaccessible self-antigens sometimes induces an immune attack against these antigens. Since the immune system is usually never exposed to hidden self-antigens, it does not "learn" to tolerate them. Inadvertent exposure of these normally inaccessible antigens to the immune system because of tissue disruption caused by injury or disease can lead to a rapid immune attack against the affected tissue, just as if these self-proteins were foreign invaders. **Hashimoto's disease,** which involves the production of antibodies against thyroglobulin and destruction of the thyroid gland's hormone-secreting capacity, is one such example.
4. Exposure of the immune system to a foreign antigen almost structurally identical to a self-antigen may induce the production of antibodies or activated T lymphocytes that not only interact with the foreign antigen but also cross-react with the closely similar body antigen. An example is the streptococcal bacteria responsible for "strep throat." The bacteria possess antigens that are structurally very similar to self-antigens in the tissue covering the heart valves of some individuals, in which case the antibodies produced against the streptococcal organisms may also bind with this heart tissue. The resultant inflammatory response is responsible for the heart-valve lesions associated with rheumatic fever.

The major histocompatibility complex codes for surface membrane-bound human leukocyte-associated antigens unique for each individual.

What is the nature of the **self-antigens** that the immune system learns to recognize as markers of a person's own cells? These self-antigens are plasma membrane-bound glycoproteins; they are known as **human leukocyte-associated antigens** or **HLA antigens** because they were first discovered in leukocytes, but actually they are present in all of the body's nucleated cells (that is, in all cells except erythrocytes, which lack nuclear DNA to direct HLA antigen synthesis). Each person has a group of genes, the **major histocompatibility complex** or **MHC,** which codes for self-antigens. Although more than one-hundred different HLA antigens have been identified in human tissue, each individual can only code for four of these possible antigens. Because of the tremendous number of different combinations possible, the exact pattern of HLA antigens varies from one individual to another, except in identical twins, who share in common the same MHC-encoded HLA antigens.

T cells typically bind with HLA self-antigens only when they are in association with a foreign antigen, such as a viral protein, also displayed on the cell surface. In the case of cytotoxic T cells, the outcome of this binding is destruction of the infected body cell. Since T cells do not bind to HLA self-antigens in the absence of foreign antigen, normal body cells are protected from lethal immune attack. T cells do bind, however, with HLA antigens present on the surface of transplanted cells in the absence of foreign antigen. The ensuing destruction of the transplanted cells is responsible for rejection of transplanted or grafted tissues. Presumably, some of the T cells "mistake" the foreign HLA antigens of the donor cells because they resemble the combination of a conventional foreign antigen complexed with self HLA antigens.

To minimize the rejection phenomenon, the tissues of donor and recipient are matched according to HLA antigens as closely as possible. Therapeutic procedures to suppress the immune system then follow. In the past, the primary immunosuppressive tools included radiation therapy and drugs aimed at destroying the actively multiplying lymphocyte populations, plus anti-inflammatory drugs that suppressed the growth of all lymphoid tissue. However, these measures not only suppressed the T cells that were primarily responsible for rejecting transplanted tissue but also depleted the antibody-secreting B cells. Unfortunately, the treated individual was left with little specific immune protection against bacterial and viral infections. In recent years a new therapeutic agent, cyclosporin, has become extremely useful in selectively depressing T cell–mediated immune activity while leaving B-cell humoral immunity essentially intact. Specifically, cyclosporin blocks interleukin 2, the lymphokine secreted by helper T cells that is required for expansion of the selected cytotoxic T-cell clone.

The major histocompatibility complex was so named because these genes and the self-antigens they encode were first discerned in relation to tissue typing (similar to blood typing), which is done to obtain the most compatible matches for tissue grafting and transplantation. It is important to realize, however, that transfer of tissue from one individual to another does not normally occur in nature. The natural function of HLA antigens lies in their ability to direct the responses of T cells, not in their artificial role in rejecting transplanted tissue.

Each individual has two main classes of MHC-encoded glycoproteins (HLA antigens) that are differentially recognized by cytotoxic T and helper T cells (Fig. 12–18). Cytotoxic T cells are able to respond to foreign antigen only in association with class I MHC glycoproteins, which are found on the surface of virtually all nucleated body cells. Class II MHC glycoproteins, which are recognized by helper T cells, are restricted to the surface of a few special types of immune cells, such as B cells, cytotoxic T cells, and macrophages. The class I and II markers serve as signposts to guide cytotoxic and helper T cells to the precise cellular locations where their immune capabilities can be most effective. The binding of suppressor T cells is less well understood.

Because cytotoxic T cells cannot dispose of free viruses, bacteria, or other antigens circulating within the host's body fluids, it would be inefficient for this subpopulation of T cells to recognize and bind with such extracellular antigens. To carry out their role of dealing with pathogens that have invaded host cells, it is appropriate that cytotoxic T cells bind only with cells of the organism's own body that have been infected by viruses. Since any nucleated body cell can be in-

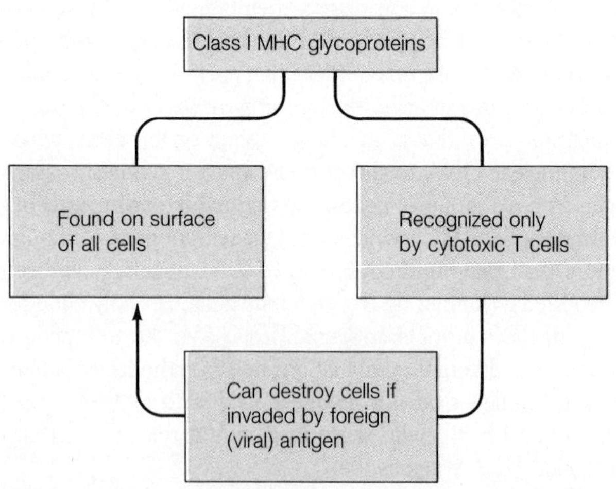

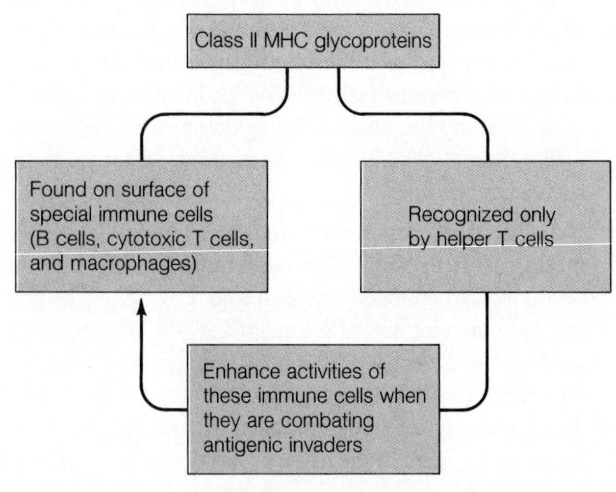

Figure 12–18 Distinctions between Class I and Class II Major Histocompatibility Complex (MHC) Glycoproteins

vaded by viruses, essentially all cells display class I MHC glycoproteins, enabling cytotoxic T cells to attack any invaded host cell. In contrast, helper T cells can bind with foreign antigen only when it is found on the surfaces of immune cells with which it interacts—namely, those bearing class II glycoproteins. These include the macrophages, which "present" antigen to helper T cells; and B and T cells, whose activity is enhanced by helper T cells. The capabilities of helper T cells would be squandered if these cells were able to bind with body cells other than these special immune cells. Thus, specific binding requirements for the various T cells help ensure the appropriate T-cell responses.

Immune surveillance against cancer cells involves an interplay among cytotoxic T cells, NK cells, macrophages, and interferon.

Besides destruction of virus-infected host cells, another important function generally attributed to the T-cell system is its role in recognizing and destroying newly arisen, potentially cancerous tumor cells before they have a chance to multiply and spread. Any normal cell may be transformed into a cancer cell if mutations occur within its genes responsible for controlling cell division and growth. Such mutations may occur by chance alone, or, more frequently, by exposure to **carcinogenic** (cancer-causing) factors such as ionizing radiation, certain environmental chemicals, certain viruses, or physical irritants.

Cellular multiplication and growth are normally under strict control, but the regulatory mechanisms are largely unknown. Cell multiplication in an adult is generally restricted to the replacement of lost cells. Furthermore, cells normally respect their own place and space in the body's society of cells. If a cell that has been transformed into a tumor cell manages to escape destruction, however, it defies the normal controls on its proliferation and position. Unrestricted multiplication of a single tumor cell results in a **tumor** that consists of a clone of cells identical to the original mutated cell.

If the mass is slow-growing, stays put in its original location, and does not infiltrate into the surrounding tissue, it is considered to be a **benign** tumor. In contrast, the transformed cell may rapidly multiply and form an invasive mass that lacks the altruistic behavior characteristic of normal cells. Such invasive tumors are known as **malignant** tumors, commonly referred to as **cancer.** Malignant tumor cells usually do not adhere well to the neighboring normal cells, often resulting in some of the cancer cells breaking away from the parent tumor. These "emigrant" cancer cells are transported through the blood to alien territories, where they continue to proliferate, forming multiple malignant tumors. **Metastasis** is the term applied to this spread of cancer to other parts of the body.

If a malignant tumor is detected early, before it has metastasized, it can be removed surgically. Once cancer cells have dispersed and seeded multiple cancerous sites, surgical elimination of the malignancy is impossible. In this case, agents that interfere with rapidly dividing and growing cells, such as certain chemotherapeutic drugs and ionizing radiation, are employed in an attempt to destroy the malignant cells. Unfortunately, these agents are also detrimental to normal body cells, especially rapidly proliferating cells such as blood cells and the cells lining the digestive tract.

Untreated cancer is eventually fatal in most cases for several interrelated reasons. The uncontrollably growing malignant mass crowds out normal cells by vigorously competing with them for space and nutrients, yet the cancer cells are unable to take over the functions of the cells they are destroying. Cancer cells typically remain immature and do not become specialized, often instead resembling embryonic cells (Fig. 12–19). Such dedifferentiated malignant cells lack the ability to perform the specialized functions of the normal cell-type from which they mutated. Affected organs gradually become disrupted to the point that they are no longer able to perform their life-sustaining functions, and death results.

Even though many body cells undergo mutations throughout a person's lifetime, most of these mutations do not result in malignancy for three reasons:

1. Only a fraction of the mutations involve loss of control over the cell's growth and multiplication. More frequently, other facets of cellular function are altered.
2. Evidence suggests that a cell usually becomes cancerous only after an accumulation of multiple independent mutations. A single mutation generally is not sufficient. This requirement could contribute at least in part to the much higher incidence of cancer in older individuals, in whom there has been more time for mutations to accumulate in a single cell lineage. Alternatively, a few cancers have been shown to be caused by tumor viruses, which permanently alter particular DNA sequences of the cells they invade.
3. Potentially cancerous cells that do arise are usually destroyed by the immune system early in their development. Presumably, cancer cells are recognized by the immune system because they bear new and different surface antigens alongside the cell's normal HLA antigens, either because of genetic mutation or invasion by a tumor virus. Immune surveillance against cancer depends on an interplay among three types of immune cells—cytotoxic T cells, natural killer cells, and macrophages—as well as interferon. Not only are these immune cells able to directly at-

Figure 12–19 Comparison of Normal and Cancerous Cells in the Large Respiratory Airways *The normal cells (top) display numerous specialized cilia, which constantly contract in whiplike motion to sweep debris and microorganisms from the respiratory airways so they do not gain entrance to the deeper portions of the lungs. The cancerous cells (bottom) are not ciliated, so they are unable to perform this specialized defense task.*

tack and destroy cancer cells, but all three of them also secrete interferon. Interferon, in turn, inhibits multiplication of cancer cells and increases the killing ability of the immune cells (Fig. 12–20).

Because natural killer cells, unlike the cytotoxic T cells, do not require prior exposure and sensitization to a cancer cell before being able to launch a lethal attack, they are probably the single most important weapon the body has in its war against cancer. Cytotoxic T cells are believed to be especially important in defending against the few kinds of virus-induced cancer. Upon contacting a cancer cell, both of these killer cells release perforin and other toxic chemicals that destroy the targeted mutant cell (Fig. 12–21). Macrophages, in addition to clearing away the remains of the dead victim cell, are themselves able to engulf and destroy cancer cells intracellularly.

The fact that cancer does sometimes occur means that cancer cells must occasionally be able to escape these immune mechanisms. Why or how immune surveillance fails to nip newly formed cancer cells in the bud remains unclear. Other facets of the immune system may even be involved. For example, **blocking antibodies** that interfere with T-cell function have been identified. Although B cells and antibodies are not believed to play a direct role in cancer defense, B cells, upon viewing a mutant cancer cell as an alien to normal self, may produce antibodies against it. These antibodies, for unknown reasons, do not activate the complement system, which could destroy the cancer cells. Instead, the antibodies are able to bind with the antigenic sites on the cancer cell, "hiding" these sites from recognition by cytotoxic T cells. The coating of a tumor cell by blocking antibodies thus protects the harmful cell from attack by deadly T cells. Other mechanisms undoubtedly are involved in cancer development to account for the fact that NK cells, macrophages, and interferon, which are not dependent on binding with specific antigenic sites, also occasionally fail to stop the relentless growth and proliferation of mutant cancer cells.

It appears that there is a regulatory loop between the immune system and the nervous and endocrine systems.

From the preceding discussion, it is obvious that complex controlling factors operate within the immune system itself. Until recently, the immune system was believed to function independently of other control systems in the body. Investigations along a number of lines indicate, however, that there are important links between the immune system and the body's two

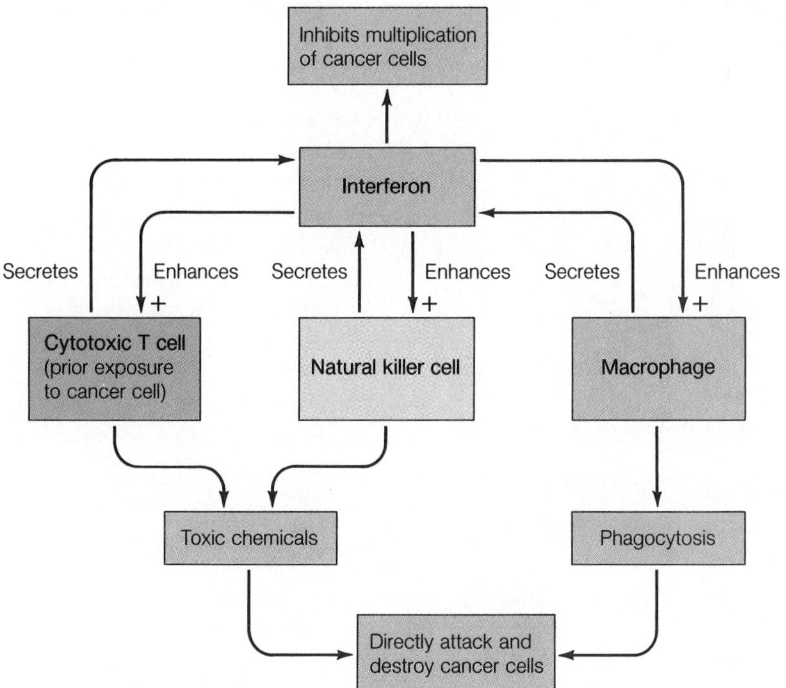

Figure 12–20 Immune Surveillance against Cancer *Anticancer interactions of cytotoxic T cells, natural killer cells, macrophages, and interferon.*

major control systems, the nervous and endocrine systems. Apparently the immune system both influences and is influenced by the nervous and endocrine systems. For example, interleukin 1 can turn on the stress response by activating a sequence of nervous and endocrine events that result in the secretion of cortisol, one of the major hormones released during stress. In the reverse direction, lymphocytes are responsive to blood-borne signals from the nervous system and from certain endocrine glands. These important immune cells possess receptors for a wide variety of neurotransmitters, hormones, and other chemical mediators. Among the mediators that influence lymphocyte activity are endorphins and enkephalins, a group of related compounds that serve as neurotransmitters in the CNS and are secreted into the blood as part of the neuroendocrine pathway involved in the control of cortisol release. (For a discussion of the possible effects of exercise on immune defense, see the accompanying boxed feature, A Closer Look at Exercise Physiology.)

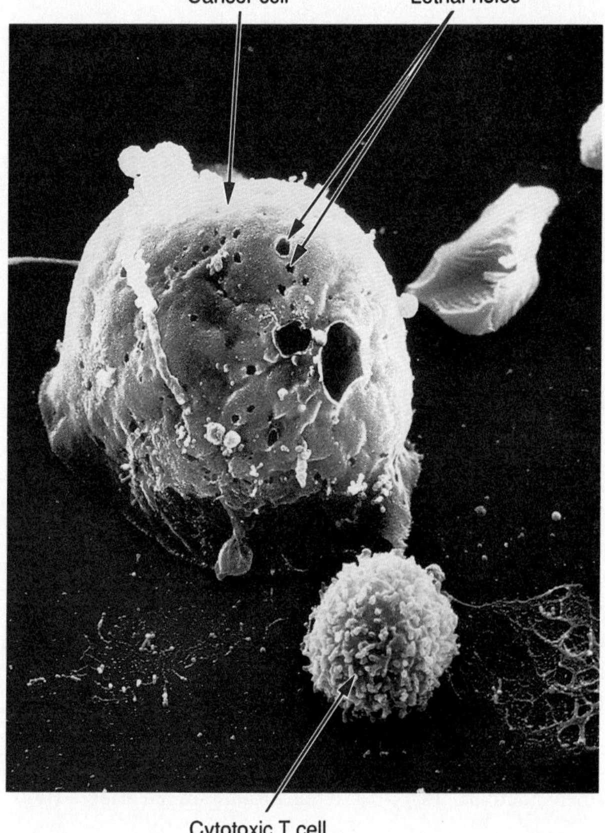

Figure 12–21 Cytotoxic T Cell Destroying Cancer Cell *Upon contacting a cancer cell with which it can specifically bind, a cytotoxic T cell releases toxic chemicals such as perforin, which destroy the cancer cell.*

A CLOSER LOOK AT EXERCISE PHYSIOLOGY

Exercise: A Help or Hindrance to Immune Defense?

Although many people who exercise claim that they have fewer infections when they are in good aerobic condition, few scientific studies have been conducted that can substantiate or refute this claim. On the basis of existing evidence, it cannot be stated conclusively that exercise influences resistance to disease. However, the studies that have been done suggest that exercise does have an impact on immune defense but that the intensity of exercise is an important factor in the effects on the immune system. The bulk of evidence indicates that moderate exercise may have beneficial effects on the immune system, whereas exhaustive exercise may be accompanied by reduced immune-system activity.

Animal studies have shown that high-intensity exercise after experimentally induced infection results in a more severe infection. Moderate exercise training done prior to infection or to tumor implantation, on the other hand, results in less severe infection and slower tumor growth in experimental animals.

In humans, moderate exercise training results in an increase in the circulating numbers of granulocytes and lymphocytes and an increase in cytotoxic and natural-killer-cell activity. One study showed, on the other hand, that maximal exercise suppresses natural-killer-cell activity for one to two hours after exercise. It is possible that intense physical activity induces the stress response, which in turn suppresses immune functions. Athletes with long, difficult training schedules and intense competition have increased incidence of respiratory infections. Studies have shown that these athletes have lower resting salivary IgA levels compared with control subjects and that their mucosal immunoglobulins are decreased after prolonged exhaustive exercise. These levels return to normal after twenty-four hours. It is possible that the athlete who is training with frequent bouts of exhaustive exercise may have a sustained reduction in mucosal immunity and therefore lower resistance to respiratory infection.

Immune Diseases

Immune deficiency diseases reduce resistance to foreign invaders.

There are two general ways in which abnormal functioning of the immune system can lead to immune diseases: deficiency diseases and inappropriate immune attacks. Deficiency diseases occur when the immune system fails to respond adequately to foreign invasion. The condition may be congenital (present at birth) or acquired, and it may specifically involve impairment of either antibody-mediated immunity, cell-mediated immunity, or both. In a rare hereditary condition known as **severe combined immunodeficiency,** both B and T cells are lacking. Its victims have extremely limited defenses against pathogenic organisms and usually die in infancy unless maintained in a germ-free environment. Acquired (nonhereditary) immune deficiency states can arise from inadvertent destruction of lymphoid tissue during prolonged therapy with anti-inflammatory agents, such as cortisol derivatives, or from cancer therapy aimed at destroying rapidly dividing cells (which unfortunately includes lymphocytes as well as cancer cells). The most recent and tragically the most common acquired immune deficiency disease is AIDS, which, as described earlier, is caused by a virus that knocks out the critical helper T cells.

Inappropriate immune attacks against harmless environmental substances are responsible for allergies.

The other category of immune diseases involves inappropriate specific immune attacks that cause reactions harmful to the body. These include: (1) autoimmune responses, in which the immune system turns against one of the body's own tissues;

(2) immune-complex diseases, which involve overexuberant antibody responses that "spill over" and damage normal tissue; and (3) allergies. The first two conditions have been described earlier in this chapter, so we will now concentrate on allergies.

An **allergy** is the acquisition of an inappropriate specific immune reactivity, or **hypersensitivity,** to a normally harmless environmental substance, such as dust, pollens, food constituents, drugs, or other chemicals. The offending agent, known as an **allergen,** may itself be an antigen, or it may be a hapten that becomes antigenic only after it combines with a body protein. Subsequent reexposure of a sensitized individual to the same allergen elicits an immune attack, which may vary from a mild, annoying reaction to a severe, body-damaging reaction that may even be fatal.

Allergic responses can be classified into two different categories: immediate hypersensitivity and delayed hypersensitivity (Table 12–6). In **immediate hypersensitivity,** the allergic response appears within about twenty minutes after exposure of a sensitized individual to an allergen, whereas in **delayed hypersensitivity** the reaction is not generally manifested until a day or so following exposure. The difference in timing is due to the different mediators involved. A particular allergen may activate either B-cell or T-cell responses. Immediate allergic reactions involve B cells and are elicited by antibody interactions with an allergen; delayed reactions involve T cells and the more-slowly responding process of cell-mediated immunity against the allergen. Let us examine the causes and consequences of each of these reactions in more detail.

In **immediate hypersensitivity,** the antibodies involved and the events that ensue upon exposure to an allergen differ from the typical antibody-mediated response to bacteria. For unclear reasons, allergens bind to and elicit the synthesis of IgE antibodies rather than the IgM and IgG antibodies associated with bacterial antigens. When an individual with an allergic tendency is first exposed to a particular allergen, compatible B cells synthesize IgE antibodies specific for it. More importantly, memory cells are also formed that are primed for a more powerful response on subsequent reexposure to the same allergen.

In contrast to the antibody-mediated response elicited by bacterial antigens, IgE antibodies do not freely circulate. Instead, their tail portions attach to mast cells and basophils. Binding of an appropriate allergen with the attached IgE antibodies triggers the release of several chemical mediators from the involved mast cells and basophils. A single mast cell (or basophil) may be coated with a number of different IgE antibodies, each able to bind with a different allergen. Thus, the mast cell can be triggered to release its chemical products by any one of a number of different allergens (Fig. 12–22). These released chemicals are responsible for the reactions that characterize immediate hypersensitivity. Following are among the most important chemicals released during immediate allergic reactions:

Table 12–6 Immediate versus Delayed Hypersensitivity Reactions

Characteristic	Immediate Hypersensitivity Reaction	Delayed Hypersensitivity Reaction
Time of onset of symptoms after exposure to allergen	Within 20 min	Within 1 to 3 days
Type of immune response involved	Antibody-mediated immunity against allergen	Cell-mediated immunity against allergen
Immune effectors involved	B cells, IgE antibodies, mast cells, basophils, histamine, slow-reactive substance of anaphylaxis, eosinophil chemotactic factor	T cells
Allergies commonly involved	Hayfever, asthma, hives, and other allergic responses to inhaled or ingested allergens; anaphylactic shock in extreme cases	Contact allergies such as allergies to poison ivy, cosmetics, and household cleaning agents
Treatment	Antihistamines (partially effective); adrenergic drugs to counteract effects of histamine and slow-reactive substance of anaphylaxis; anti-inflammatory agents such as cortisol derivatives	Anti-inflammatory agents such as cortisol derivatives

1. **histamine** and other vasoactive chemicals, which bring about vasodilation and increased capillary permeability.
2. **slow-reactive substance of anaphylaxis (SRS-A)**, which induces prolonged and profound contraction of smooth muscle, especially of the small respiratory airways, and also increases capillary permeability.
3. **eosinophil chemotactic factor,** which specifically attracts eosinophils. Interestingly, eosinophils release enzymes that inactivate SRS-A and that may also inhibit histamine, perhaps serving as an "off switch" to limit the allergic response.

Symptoms vary depending on the site, allergen, and mediators involved. Most frequently, the reaction is localized to the body site in which the IgE-bearing cells first come into contact with the allergen. If the reaction is limited to the upper respiratory passages after a person inhales an allergen such as ragweed pollen, the released chemicals bring about the symptoms characteristic of **hayfever**—for example, nasal congestion caused by histamine-induced localized edema and sneezing and runny nose caused by increased mucus secretion in response to local irritation. If the reaction is concentrated primarily within the bronchioles (the small respiratory airways that lead to the tiny air sacs within the lungs), **asthma** results.

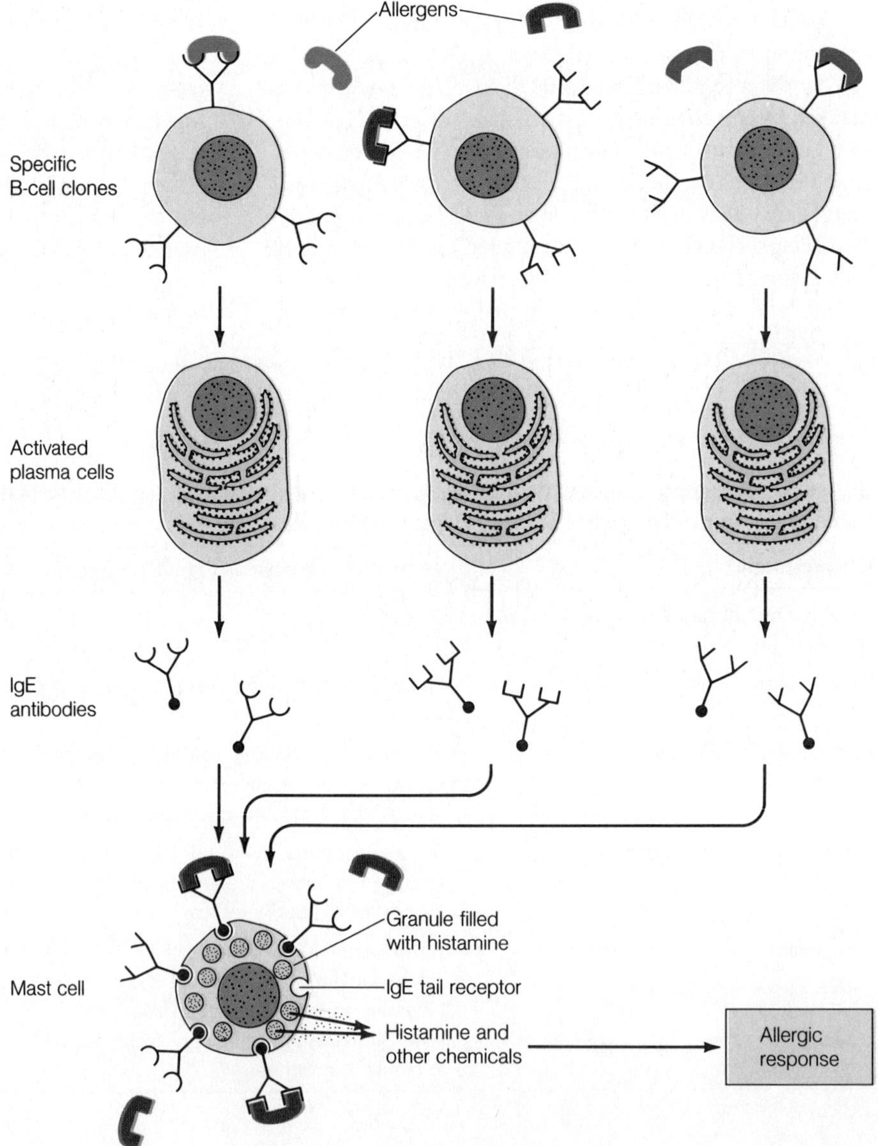

Figure 12–22 Role of IgE Antibodies and Mast Cells in Immediate Hypersensitivity *B-cell clones are converted into plasma cells, which secrete IgE antibodies on contact with the allergen for which they are specific. The Fc tail portion of all IgE antibodies, regardless of the specificity of their Fab arm regions, binds to receptor proteins specific for IgE tails on mast cells and basophils. Unlike B cells, each mast cell bears a variety of antibody surface receptors for binding different allergens. When an allergen combines with an IgE receptor specific for it on the surface of a mast cell, the mast cell releases histamine and other chemicals by exocytosis. These chemicals elicit the allergic response.*

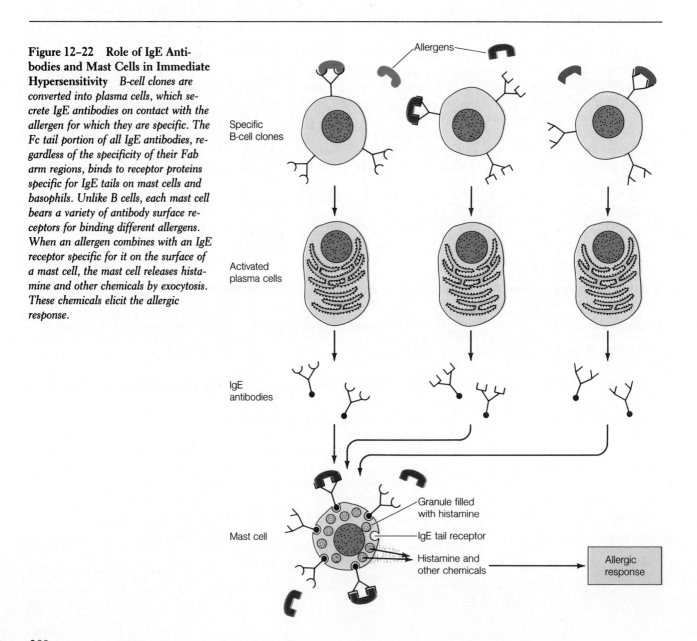

Contraction of the smooth muscle in the walls of the bronchioles narrows or constricts these passageways, making breathing difficult. Localized swelling in the skin because of allergy-induced histamine release causes **hives.**

Treatment of localized immediate allergic reactions with antihistamines often offers only partial relief of the symptoms, because some of the manifestations are invoked by other chemical mediators not blocked by these drugs. For example, antihistamines are not particularly effective in treating asthma, the most serious symptoms of which are invoked by SRS-A. Adrenergic drugs (which mimic the sympathetic nervous system) are helpful through their vasoconstrictor-bronchodilator actions in counteracting the effects of both histamine and SRS-A. Anti-inflammatory drugs such as cortisol derivatives may be necessary to block the inflammatory response.

A life-threatening systemic reaction can occur if the allergen becomes blood-borne or if very large amounts of chemicals are released from the localized site into the circulation. When large amounts of these chemical mediators gain access to the blood, the extremely serious systemic (involving the entire body) reaction known as **anaphylactic shock** occurs. Widespread vasodilation and a massive shift of plasma fluid into the interstitial spaces as a result of a generalized increase in capillary permeability cause severe hypotension (see p. 338) that can lead to circulatory failure. Concurrently, pronounced bronchiolar constriction can lead to respiratory failure. The victim may suffocate because of an inability to move air through the tightened airways. Unless countermeasures, such as injection of an adrenergic vasoconstrictor-bronchodilator drug, are undertaken immediately, anaphylactic shock is frequently fatal. This is why even a single bee sting or a single dose of penicillin can be so dangerous in individuals sensitized to these allergens.

The following are among the most common allergens that provoke immediate hypersensitivities, some of which have already been mentioned: pollen grains, bee stings, certain drugs such as penicillin, various foods, molds, dust, feathers, and animal fur. Actually, people allergic to cats are not allergic to the fur itself. The true allergen is in the cat's saliva deposited on the fur during licking. Likewise, people are not allergic to dust or feathers per se, but to tiny mites that inhabit the dust or feathers and eat the scales constantly being shed from the skin.

Still other allergens invoke a ***delayed hypersensitivity***, T cell-mediated immune response rather than an immediate, B cell-IgE antibody response. Among these are poison ivy toxin and certain chemicals to which the skin is frequently exposed, such as cosmetics and household cleaning agents. Most commonly, the response is characterized by a delayed skin eruption that reaches its peak intensity one to three days following contact with an allergen to which the T system has previously been sensitized. To illustrate, poison ivy toxin is a hapten that may bind with skin proteins with which it comes into contact. The toxin itself does not harm the skin, but it does activate T cells specific for the toxin, including formation of a memory component. Upon subsequent exposure to the toxin, activated T cells diffuse into the skin within a day or two, combining with the poison ivy toxin that is present. The resultant interaction gives rise to the tissue damage and discomfort typically associated with the condition. The best relief is obtained from application of anti-inflammatory preparations, such as those containing cortisol derivatives.

More long-term relief from both immediate and delayed hypersensitivities can sometimes be obtained by **desensitization injections** (**allergy shots**). This therapeutic regimen involves regular injections of small but increasing quantities of the offending allergen. Through this process, the individual (to varying degrees of success) gradually becomes less and less sensitive to natural exposure to the allergen. It seems ironic that deliberately exposing a person to a known allergen can make that person less sensitive to the allergen. The mechanism by which desensitization is accomplished is presently uncertain, but the leading theory proposes that it occurs because of the production of blocking IgG antibodies specific for the allergen injected (Fig. 12–23). When the individual is naturally exposed to the allergen again, complementary blocking IgG antibodies bind with the allergen so that it is not available to bind with the IgE antibodies attached to mast cells and basophils. Thus, these histamine-laden cells are not stimulated to release their symptom-causing chemicals. An alternative proposal suggests that desensitization is brought about by activation of suppressor T cells, which thwart the synthesis of IgE antibodies specifically directed against the allergen.

External Defenses

The body's defenses against foreign microbes are not limited to the intricate, interrelated immune mechanisms that destroy the microorganisms that have actually invaded the body. In addition to the internal immune-defense system, the body is equipped with external defense mechanisms designed to prevent microbial penetration wherever body tissues are exposed to the external environment. The most obvious external defense is the **skin,** or **integument,** which covers the outside of the body.

The skin consists of an outer protective epidermis and an inner connective-tissue dermis.

The skin, which is the largest organ of the body, not only serves as a mechanical barrier between the external environment and the underlying tissues but is dynamically involved in defense mechanisms and other important functions as well.

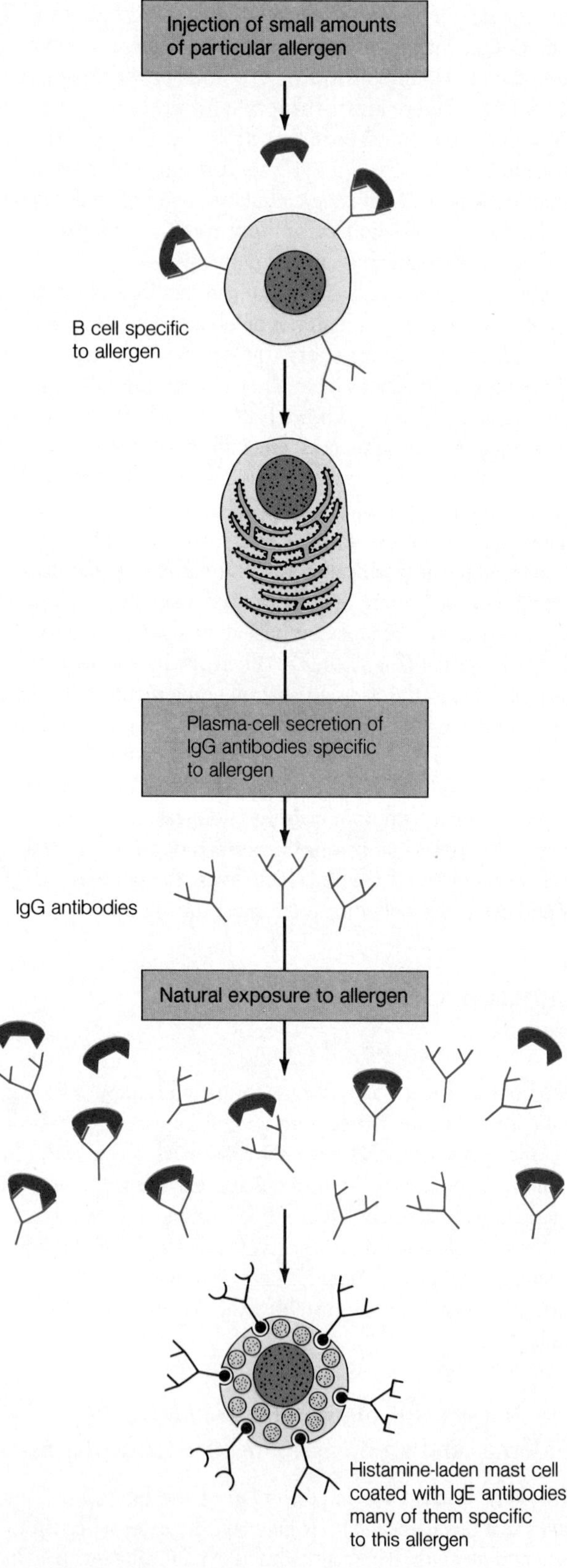

Figure 12–23 Proposed Mechanism of Desensitization *On natural exposure to an allergen to which a person has been desensitized, IgG antibodies bind with newly entering allergens, thereby blocking their attachment to mast-cell IgE receptors and reducing symptoms.*

The skin consists of two layers, an outer epidermis and an inner dermis (Fig. 12–24).

The **epidermis,** which varies from 0.5 to 4 mm in thickness, is further organized into numerous layers of epithelial cells. The inner epidermal layers are composed of cells that are living and rapidly dividing. In contrast, the cells in the outer layers are dead and flattened. The epidermis has no direct blood supply. Its cells are nourished only by diffusion of nutrients from a rich vascular network in the underlying dermis. The newly forming cells in the inner layers constantly push the older cells more superficially, farther and farther from their nutrient supply. This, coupled with the fact that the outer layers are continuously subjected to pressure and "wear and tear," causes these older cells to die and become flattened. Epidermal cells are tightly bound together by spot desmosomes (see p. 68), which interconnect with intracellular keratin filaments (see p. 40) to form a strong, cohesive covering. During maturation of a keratin-producing cell, keratin filaments progressively accumulate and cross-link with each other within the cytoplasm. As the outer cells die, only this fibrous keratin core remains, forming flattened, hardened scales that provide a tough, protective **keratinized,** or **cornified, layer.** As the scales of the outermost keratinized layer slough or flake off through abrasion, they are continuously replaced by means of mitotic activity in the deeper epidermal layers. The rate of cell division, and consequently the thickness of the keratinized layer, varies in different regions of the body. It is thickest in the areas where the skin is subjected to the most pressure.

This keratinized layer is airtight, fairly waterproof, and impervious to most substances. It serves to resist passage in both directions between the body and the external environment. For example, it minimizes loss of water and other vital constituents from the body. This protective layer's value in holding in body fluids becomes obvious in severe burns. Bacterial infections can ensue in the unprotected underlying tissue, but even more serious are the systemic consequences of loss of body water and plasma proteins, which escape from the exposed, burned surface. The resultant circulatory disturbances can be life-threatening.

Likewise, the skin barrier impedes passage into the body of most materials that come into contact with the body surface, including bacteria and toxic chemicals. In many instances, the skin modifies compounds that come into contact with it. For

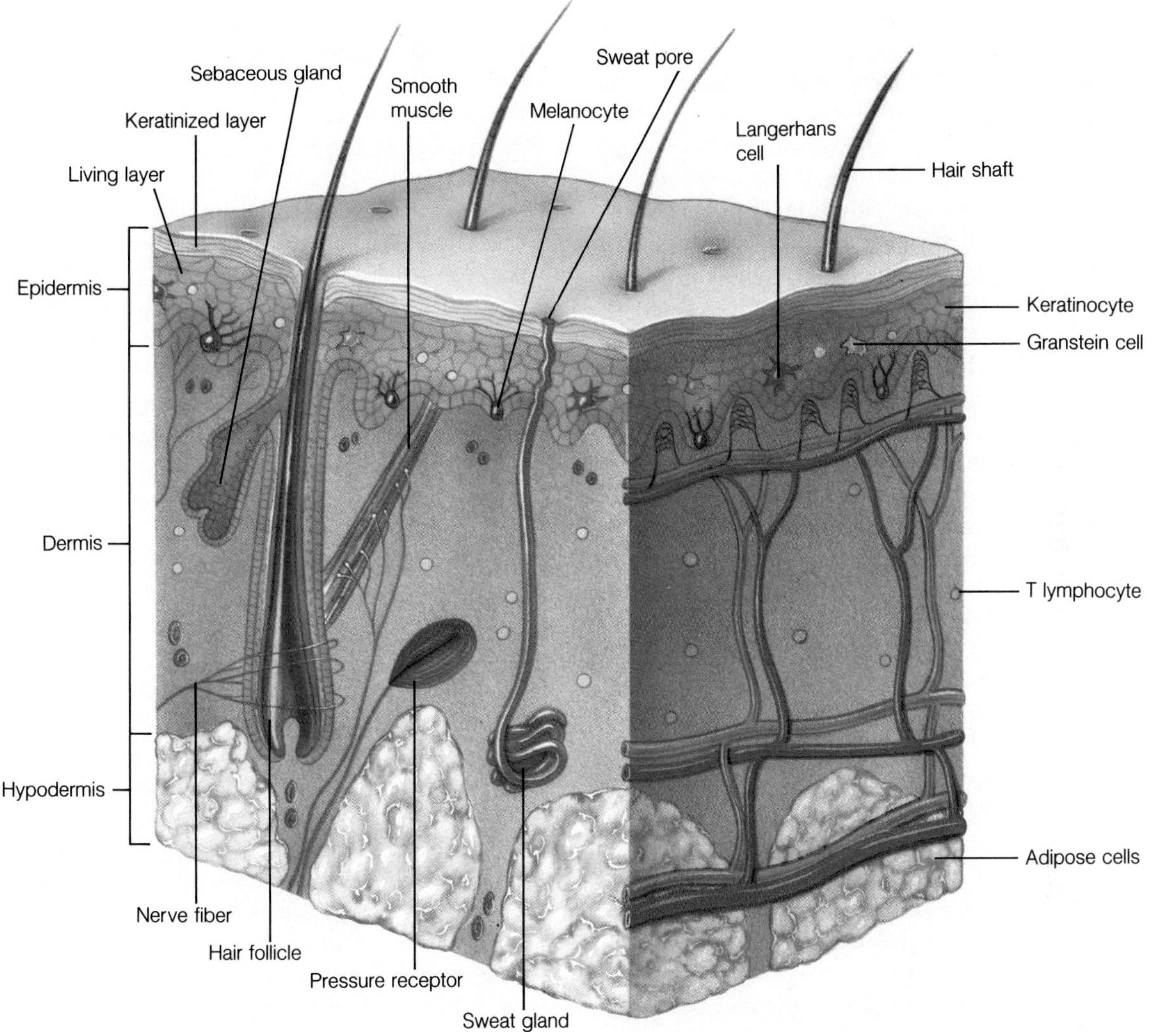

Figure 12–24 Anatomy of the Skin *The skin consists of two layers, a keratinized outer epidermis and a richly vascularized inner connective-tissue dermis. Special infoldings of the epidermis form the sweat glands, sebaceous glands, and hair follicles. The epidermis contains four types of cells: keratinocytes, melanocytes, Langerhans cells, and Granstein cells. The skin is anchored to underlying muscle or bone by the hypodermis, a loose, fat-containing layer of connective tissue.*

example, epidermal enzymes are able to convert many potential carcinogens into harmless compounds. Some substances, however, are able penetrate intact skin, especially lipid-soluble substances. Drugs that can be absorbed by the skin are sometimes administered in the form of a cutaneous "patch" impregnated with the drug.

The **dermis** is a connective-tissue matrix that contains many elastin fibers (for stretch) and collagen fibers (for strength), as well as an abundance of blood vessels and specialized nerve endings. The dermal blood vessels not only supply both the dermis and epidermis but also play a major role in temperature regulation. The caliber of these vessels, and hence the volume of blood flowing through them, is subject to control to vary the amount of heat exchange between these skin-surface vessels and the external environment. Efferent nerve endings in the dermis control blood-vessel caliber, hair erection, and exocrine-gland secretion.

Special infoldings of the epithelium into the underlying dermis form the skin's exocrine glands—the sweat glands and sebaceous glands—as well as the hair follicles. **Sweat glands,** which are located over the majority of the body, release a dilute salt solution through small openings, the sweat pores, onto the surface of the body. Evaporation of this sweat cools the skin and is important in temperature regulation. The amount of sweat produced is subject to regulation and depends on the environmental temperature, the amount of heat-generating muscular activity, and various emotional factors (for example, a person often sweats when nervous). A special type of sweat gland located in the axilla (armpit) and pubic region produces a protein-rich sweat that supports surface bacterial growth, which gives rise to a characteristic odor. In contrast, most sweat as well as the secretions from the sebaceous glands contain chemicals that are generally highly toxic to bacteria.

The cells of the **sebaceous glands** produce an oily secretion known as **sebum** that is released into adjacent hair follicles. From there the oily sebum flows to the surface of the skin, oiling both the hairs and the outer keratinized layers of the skin to help waterproof them and prevent them from drying and cracking. Insufficient protection by sebum is evidenced by chapped hands or lips. The sebaceous glands are particularly active during adolescence, causing the oily skin prominent in teenagers.

Each **hair follicle** is lined by special keratin-producing cells that secrete proteins, which form the hair shaft. Hairs serve the purpose of increasing the sensitivity of the skin's surface to tactile (touch) stimuli. In some lower species, this function is more exquisitely fine-tuned. For example, the whiskers on a cat are extremely sensitive in this regard. An even more important role of hair in lower species is heat conservation, but this function is not of importance in us relatively hairless humans. Similar to hair, the nails are another special keratinized product derived from living epidermal structures, the nail beds.

The skin is anchored to the underlying tissue (muscle or bone) by the **hypodermis** or **subcutaneous tissue,** a loose layer of connective tissue. Most fat cells in the body are housed within the hypodermis. These subcutaneous fat deposits throughout the body are collectively referred to as **adipose tissue.**

Specialized cells in the epidermis produce keratin and melanin and participate in immune defense.

The epidermis contains four distinct resident-cell types—melanocytes, keratinocytes, Langerhans cells, and Granstein cells—plus transient T lymphocytes that are scattered throughout the epidermis and dermis. Each of these resident-cell types performs specialized functions.

Melanocytes produce the brown pigment **melanin,** the amount of which is responsible for the different shades of brown color in the skin of various races. In addition to hereditary determination of melanin content, the amount of this pigment can be increased transiently in response to exposure to ultraviolet light rays from the sun. This additional melanin, the outward appearance of which constitutes a "tan," performs the protective function of absorbing harmful ultraviolet light rays.

The most abundant epidermal cells are the **keratinocytes,** which, as the name implies, are specialists in keratin production. As they die, they form the outer protective keratinized layer. They are also responsible for generating hair and nails. A surprising function recently discovered is that keratinocytes are also important immunologically. They secrete interleukin 1 (a product also secreted by macrophages), which influences the maturation of T cells that tend to localize in the skin. Interestingly, the epithelial cells of the thymus have been shown to bear anatomical, molecular, and functional similarities to those of the skin. Apparently, some post-thymic steps in T-cell maturation take place in the skin under keratinocyte guidance.

The two other epidermal cell types also play a role in immunity. Both **Langerhans cells,** which migrate to the skin from the bone marrow, and **Granstein cells,** the most recently discovered and least understood type of epidermal cells, serve as antigen-presenting cells. Langerhans cells present antigen to helper T cells, thereby facilitating their responsiveness to skin-associated antigens. In contrast, it appears that Granstein cells interact with suppressor T cells, probably serving as a "brake" on skin-activated immune responses. It is significant that Langerhans cells are less resistant to damage by ultraviolet radiation than are Granstein cells. Loss of Langerhans cells as a result of exposure to ultraviolet radiation can detrimentally lead to a predominate suppressor signal rather than the normally dominant helper signal.

It has been suggested that the various epidermal components of the immune system be collectively termed **skin-associated lymphoid tissue** or **SALT** (Fig. 12–25). Recent research suggests that the skin probably plays an even more elaborate role in specific immune defense than described here. This is appropriate, because the skin serves as the major interface with the external environment.

Protective measures within body cavities that communicate with the external environment discourage pathogen invasion into the body.

The human body's defense system must guard against entry of potential pathogens not only through the outer surface of the body but also through the internal cavities that communicate directly with the external environment—namely, the digestive system, the genitourinary system, and the respiratory sys-

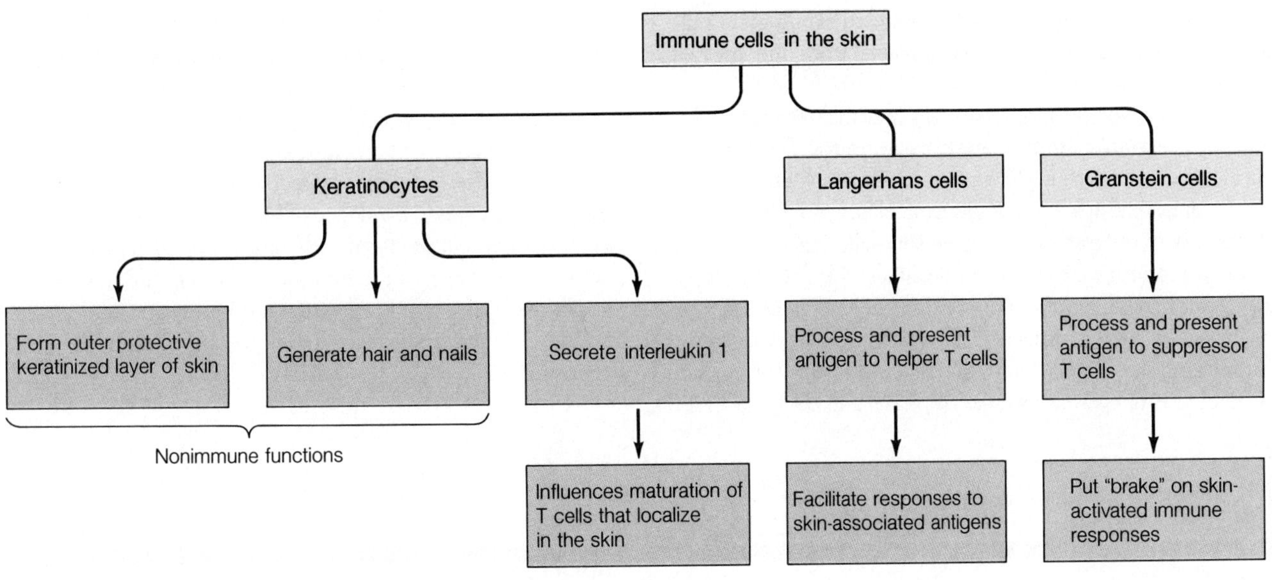

Figure 12–25 Skin-Associated Lymphoid Tissue (SALT)

tem. These systems employ various strategies to destroy microorganisms entering through these routes.

Saliva secreted into the mouth at the entrance of the digestive system contains an enzyme that lyses certain bacteria. Many of the surviving bacteria that are swallowed are killed by the strongly acidic gastric juice that they encounter in the stomach. Farther down the tract, the intestinal lining is endowed with gut-associated lymphoid tissue. These defensive mechanisms are not 100% effective, however. Some bacteria do manage to survive and reach the large intestine (the last portion of the digestive tract), where they continue to flourish. Surprisingly, this normal microbial population provides a natural barrier against infection within the lower intestine. These harmless resident flora competitively suppress the growth of potential pathogens that have managed to escape the antimicrobial measures of other parts of the digestive tract. Occasionally, orally administered antibiotic therapy against one infection within the body may actually induce another infection in the intestinal tract. By knocking out some of the normal intestinal flora, an antibiotic may permit an antibiotic-resistant pathogenic species to overgrow.

Within the genitourinary system, would-be invaders encounter hostile conditions in the acidic urine and acidic vaginal secretions. The genitourinary organs also produce a sticky mucus, which, like flypaper, entraps small invading particles. Subsequently, the particles are either engulfed by phagocytes or are swept out as the organ empties (for example, they are flushed out with urine flow).

The respiratory system is likewise equipped with several important defense mechanisms against inhaled particulate matter. The respiratory system is the largest surface of the body that comes into direct contact with the increasingly polluted external environment. The surface area of the respiratory system exposed to the air is thirty times that of the skin. Larger air-borne particles are filtered out of the inspired air by hairs at the entrance of the nasal passages. The tonsils and adenoids provide immunological protection against inspired pathogens near the beginning of the respiratory system. Farther down in the respiratory airways, millions of tiny hairlike projections known as **cilia** (see p. 36) constantly beat in an outward direction. The respiratory airways are coated with a layer of thick, sticky mucus secreted by epithelial cells within the airway lining. This mucus sheet, laden with any inspired particulate debris (such as dust) that adheres to it, is constantly moved upward to the throat by ciliary action. This moving "staircase" of mucus is known as the **mucus escalator.** The dirty mucus is either expectorated (spit out) or swallowed, with any undigestible foreign particulate matter being eliminated in the feces. Besides keeping the lungs clean, this mechanism is an important defense against bacterial infection, because many bacteria enter the body on dust particles. Also contributing to defense against respiratory infections are IgA antibodies secreted in the mucus. In addition, an abundance of phagocytic specialists called the **alveolar macrophages** scavenge within the air sacs (alveoli) of the lungs. Further respiratory defenses include coughs and sneezes. These commonly experienced reflex mechanisms involve forceful outward expulsion of material in an attempt to remove irritants from the trachea (coughs) or nose (sneezes).

Cigarette smoking suppresses these normal respiratory defenses. The smoke from a single cigarette can paralyze the cilia for several hours, with repeated exposure eventually leading to ciliary destruction. Failure of ciliary activity to constantly sweep out a stream of particulate-laden mucus enables inspired carcinogens to remain in contact with the respiratory airways for prolonged periods. Furthermore, cigarette smoke incapacitates alveolar macrophages. Particulates in cigarette smoke not only overwhelm the macrophages, but certain components of cigarette smoke also have a direct toxic effect on the macrophages, reducing their ability to engulf foreign material. In addition, noxious agents in tobacco smoke irritate the mucous linings of the respiratory tract, resulting in excess mucus production, which may partially obstruct the airways. "Smoker's cough" is an attempt to dislodge this excess stationary mucus. These and other direct toxic effects on lung tissue lead to the increased incidence of lung cancer and chronic respiratory diseases associated with cigarette smoking. Air pollutants include some of the same substances found in cigarette smoke and can similarly affect the respiratory system. We will examine the respiratory system in greater detail in the next chapter.

Chapter in Perspective

Humans are constantly coming into contact with external agents that could be harmful if they gained entry into the body, the most serious of which are pathogenic (disease-causing) microorganisms. The body surfaces that are exposed to the external environment, such as the skin and the linings of the body cavities that communicate with the outside, serve as a first line of defense to resist penetration by foreign microbes. They act as mechanical barriers, produce substances toxic to pathogens, or produce sticky mucus, which entraps would-be invaders.

In the event that a foreign microbe succeeds in gaining entry, the body is equipped with a complex, multifaceted internal defense system—the immune system—which provides continual protection against invasion by foreign agents. The immune system also guards against abnormal cancer-causing cells that arise within the body. Furthermore, it cleans up debris and paves the way for tissue repair following injury. The effectors of the immune system are the leukocytes (white blood cells) and several collections of plasma proteins.

When a foreign microbe gains entry, two different categories of responses—nonspecific and specific immune defenses—are triggered. Nonspecific immune defenses nonselectively attack foreign material of any type, whether or not the body has ever been exposed to it. Simultaneously, specific immune responses are initiated. These specific responses are targeted against a particular foreign invader after they have been specially prepared for selective attack following exposure to the invader. Since there is a time delay before specific immune responses can gear up for action upon first exposure to a particular pathogen, only the nonspecific measures are available early in the course of an initial infection. If these nonspecific mechanisms succeed in rapidly eliminating the invading microorganism, specific responses may never play an active role in this initial bout of a particular infection. If the nonspecific mechanisms are not able to resist the invader completely, the more slowly developing but more powerful specific immune mechanisms are prepared within a few days to contribute to the battle. Because the specific immune system "remembers" each infection, a subsequent invasion by the same organism is dealt with more swiftly and even more forcefully by specific defense mechanisms. Often this immediate specific attack is sufficient to squelch the invader before it does any harm, thereby providing the person with long-lasting protection or active immunity against that particular organism.

Nonspecific immune responses include: (1) inflammation, which brings an abundance of immune effector cells and plasma proteins to an invaded site; (2) interferon, which is released from any virally infected cell and which prevents viral replication in other cells and enhances the activity of several types of killer cells that can destroy the infected body cells; (3) natural killer cells, which nonspecifically lyse virally infected cells or mutant cells; and (4) the complement system, a group of inactive plasma proteins that, once activated, similarly lyse invading bacteria through formation of a hole-punching membrane-attack complex. The complement system can be activated either by the presence of bacteria themselves or by the presence of antibodies specifically produced in response to these bacteria.

There are two classes of specific immune responses: antibody-mediated immunity accomplished by B lymphocytes and cell-mediated immunity accomplished by T lymphocytes. Each B and T cell is preprogrammed by genetic recombination mechanisms to recognize and respond to the presence of only one specific, complex, nonself molecule known as an antigen from among an almost limitless variety of foreign materials. B and T cells react differently when they detect their complementary antigen. B cells, which respond primarily to extracellular foreign material such as bacteria, secrete antibodies that specifically attach to the invaders. Antibodies do not directly kill the bacteria; instead they step up the ability of nonspecific lethal mechanisms to destroy material that has been specifically marked by the antibodies. Cell-mediated immunity involves the direct destruction of virally invaded host cells by cytotoxic T cells. Two other classes of T cells—helper T cells and suppressor T cells—regulate both B and T cell activity.

The various immune effector cells support each other in a variety of ways by means of the release of numerous chemical mediators. Thus, even though each component of the body's immune system employs a limited and specialized defense strategy, the effectors of immunity collectively and cooperatively save us on a daily basis from certain death by infection.

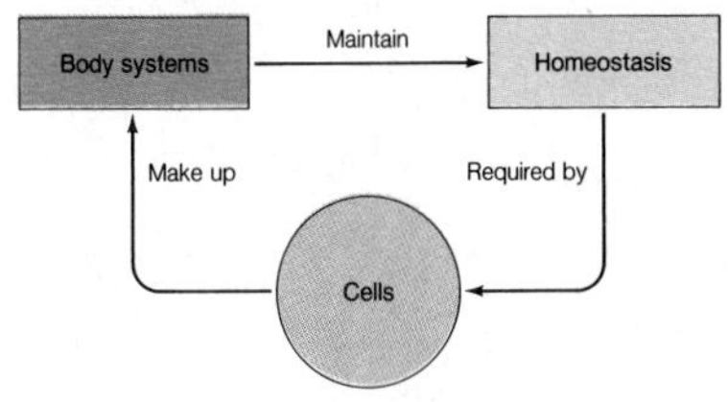

See inside front cover for an expanded version of this model.

Review Exercises

1. Distinguish between bacteria and viruses.
2. Summarize the functions of each of the lymphoid tissues.
3. Distinguish between specific and nonspecific immune responses.
4. List and describe each of the nonspecific immune responses.
5. Compare the life history of B cells and T cells.
6. What is an antigen?
7. Describe the structure of an antibody. List and describe the five subclasses of immunoglobulins.
8. In what ways do antibodies exert their effect?
9. Describe the clonal selection theory.
10. Compare the functions of B cells and T cells. What are the roles of the three types of T cells?
11. Summarize the functions of macrophages in immune defense.
12. What mechanisms are believed to be involved in tolerance?
13. What is the importance of class I and class II MHC glycoproteins?
14. Describe the factors that contribute to immunological surveillance against cancer cells.
15. Distinguish among immune deficiency disease, autoimmune disease, immune complex disease, immediate hypersensitivity, and delayed hypersensitivity.
16. What are the immune functions of the skin?
17. **A point to ponder:** Why does the fact that HIV, the virus that causes AIDS, frequently mutates make vaccine development against this virus difficult?

C H A P T E R 13

Respiratory System

Chapter Contents at a Glance

Introduction

It is hard to imagine, but:

□ *It takes 250 times more pressure to move air through a smoker's pipe than through the respiratory airways at the same flow rate.*

□ *One hundred times more distending pressure is required to inflate a child's toy balloon than to inflate the lungs.*

□ *The lungs cyclically inflate during* **inspiration** *(breathing in) and deflate during* **expiration** *(breathing out) because of alternate contraction and relaxation of respiratory muscles, yet no muscles act directly on the lungs.*

□ *The O_2 and CO_2 concentration within the lungs normally remains essentially constant, rather than fluctuating with inspiration and expiration.*

□ *Just as much blood flows through the lungs each minute as through all of the rest of the body.*

□ *Because of the structural organization of the lung, about 7,500 times more surface area is available for exchange of O_2 and CO_2 between the blood and air than there would be if the lungs were a hollow sac of the same volume.*

□ *Oxygen is poorly soluble in blood, yet the blood can carry about seventy times more O_2 than can be dissolved in the blood.*

□ *The O_2 concentration in the lungs can be reduced by 40% without noticeably altering delivery of O_2 to the tissues.*

□ *The primary regulator of respiratory activity is not the amount of O_2 in the blood but the amount of acid in the extracellular fluid of the brain.*

You will learn the explanations for these surprising facts in this chapter on respiration.

The respiratory system does not participate in all steps of respiration.

The primary function of **respiration** is to obtain O_2 for use by the body's cells and to eliminate the CO_2 the cells produce. All body cells ultimately need an adequate supply of O_2 to support their energy-generating metabolic reactions. Brain cells, which are especially dependent on a continual supply of O_2, die if deprived of O_2 for more than four minutes. Even those cells that can resort to anaerobic metabolism for energy production, such as strenuously exercising muscles, can do so only transiently by incurring an O_2 debt that ultimately must be repaid (see p. 237). As a result of oxidation of foodstuffs to generate energy, large quantities of CO_2 are produced that must continually be eliminated from the body:

$$\left.\begin{matrix}\text{Carbohydrate}\\ \text{Fat}\\ \text{Protein}\end{matrix}\right. + O_2 \rightarrow CO_2 + H_2O + \text{Energy}$$

Most people think of respiration as the process of breathing in and breathing out. In physiology, however, respiration has a much broader meaning than this. **Internal** or **cellular respiration** refers to the intracellular metabolic processes carried out within the mitochondria, which use O_2 and produce CO_2 during the derivation of energy from nutrient molecules (see p. 29). The **respiratory quotient** (**R.Q.**), the ratio of CO_2 produced to O_2 consumed, varies depending on the foodstuff consumed. When carbohydrate is being used, the R.Q. is 1; that is, for every molecule of O_2 consumed, one molecule of CO_2 is produced: $C_6H_{12}O_6 + 6O_2 \rightarrow 6CO_2 + 6H_2O +$ ATP. For fat utilization, the R.Q. is 0.7; for protein, it is 0.8. On a typical American diet consisting of a mixture of these three nutrients, resting O_2 consumption averages about 250 ml/min and CO_2 production averages about 200 ml/min, for an average R.Q. of 0.8:

$$\text{R.Q.} = \frac{CO_2 \text{ produced}}{O_2 \text{ consumed}} = \frac{200 \text{ ml/min}}{250 \text{ ml/min}} = 0.8$$

External respiration refers to the entire sequence of events involved in the exchange of O_2 and CO_2 between the cells of the body and the external environment (Fig. 13–1). External respiration, the topic of this chapter, encompasses four steps:

1. Air is alternately moved in and out of the lungs so that exchange of air can occur between the atmosphere (external environment) and the air sacs (**alveoli**) of the lungs. This is accomplished by the mechanical act of **breathing,** or **ventilation.** The rate of ventilation is regulated so that the flow of air between the atmosphere and the alveoli is adjusted according to the body's metabolic needs for O_2 uptake and CO_2 removal.
2. Oxygen and CO_2 are exchanged between air in the alveoli and blood within the pulmonary capillaries by the process of diffusion. To assure efficient exchange between air and blood, a reasonable matching of air-flow (ventilation) to blood flow (perfusion) must exist. Thus, diffusion gradients as well as ventilation-perfusion relationships are important in determining the total amount of gas that is exchanged between the lungs and blood.
3. Oxygen and CO_2 are transported by the blood between the lungs and tissues.
4. Exchange of O_2 and CO_2 takes place between the tissues and the blood by the process of diffusion across the systemic (tissue) capillaries.

The respiratory system does not accomplish all the steps of respiration; it is involved only with ventilation and the exchange of O_2 and CO_2 between the lungs and blood. The circulatory system carries out the remainder of the respiratory process.

The respiratory system additionally performs several nonrespiratory functions as follows:

□ It provides a route for water and heat elimination. Inspired atmospheric air is humidified and warmed by the respiratory airways before it is expired.

□ It enhances venous return (see the "respiratory pump," p. 327).

□ It contributes to the maintenance of normal acid-base balance by altering the amount of H^+-generating CO_2 exhaled.

□ It enables speech, singing, and other vocalization.

□ It removes, modifies, activates, or inactivates various materials passing through the pulmonary circulation. All blood returning to the heart from the tissues must pass through the lungs before being returned to the systemic circulation. The lungs, therefore, are uniquely situated to partially or completely remove specific materials that have been added to the blood at the tissue level before they have a chance to reach other parts of the body by means of the arterial system. For example, prostaglandins, a collection of chemical messengers released in numerous tissues to mediate particular local responses, may spill into the blood, but they are inactivated during passage through the lungs so that they cannot exert

Figure 13–1 External and Internal Respiration

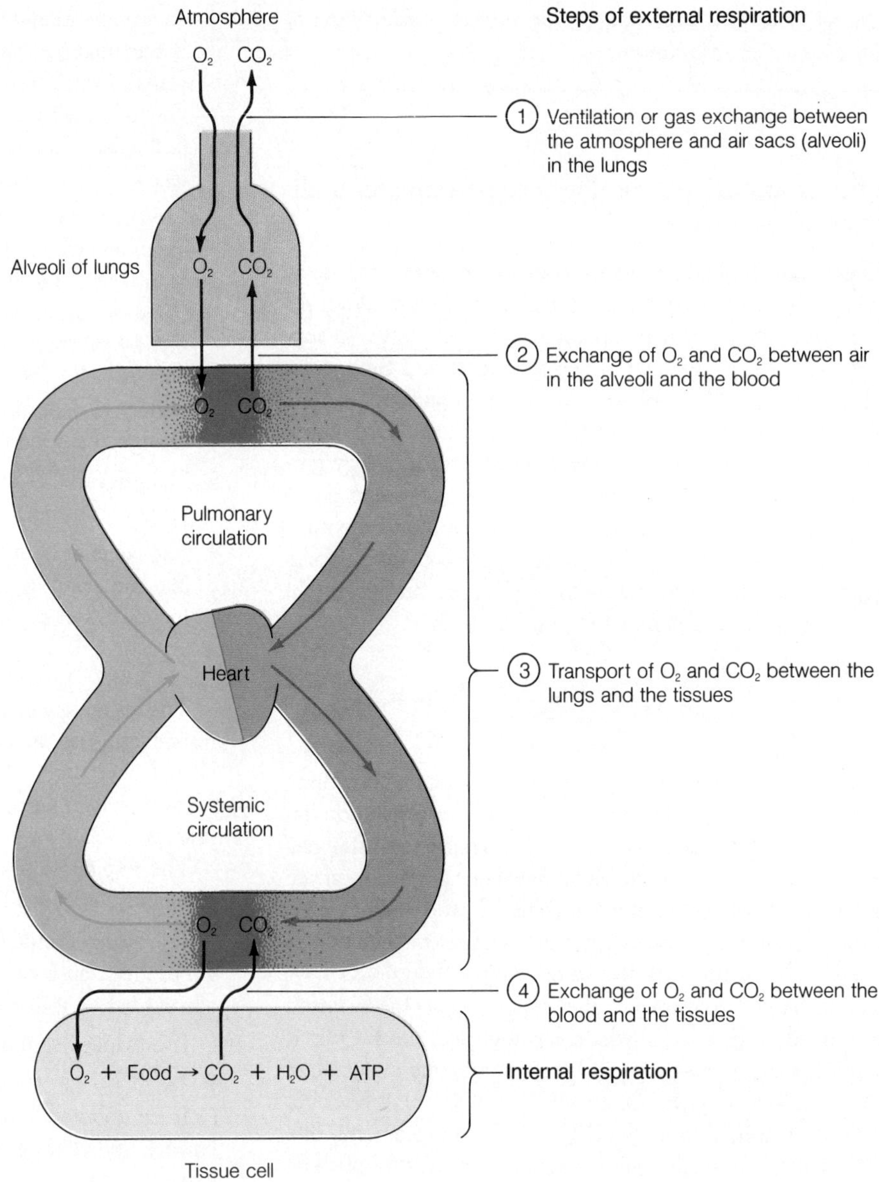

Atmosphere
O_2 CO_2
Steps of external respiration
1 Ventilation or gas exchange between the atmosphere and air sacs (alveoli) in the lungs
Alveoli of lungs
O_2 CO_2
2 Exchange of O_2 and CO_2 between air in the alveoli and the blood
O_2 CO_2
Pulmonary circulation
Heart
3 Transport of O_2 and CO_2 between the lungs and the tissues
Systemic circulation
O_2 CO_2
4 Exchange of O_2 and CO_2 between the blood and the tissues
O_2 + Food → CO_2 + H_2O + ATP
Internal respiration
Tissue cell

systemic effects. On the other hand, the lungs activate angiotensin II, a hormone that plays an important role in regulating the concentration of Na^+ in the extracellular fluid.

The respiratory airways conduct air between the atmosphere and alveoli.

The **respiratory system** includes the respiratory airways leading into the lungs, the lungs themselves, and the structures of the thorax (chest) involved in producing movement of air through the airways into and out of the lungs. The **respiratory airways** are ducts that carry air between the atmosphere and the alveoli, the latter being the only site where exchange of gases can take place between air and blood. The airways (Fig. 13–2a) begin with the **nasal passages** (**nose**). The respiratory functions of the nose and the remainder of the airways include: (1) defense against inhaled foreign matter (by means of the filtering action of nasal hairs and the cilia-activated mucus escalator—see p. 403); (2) warming of inspired air to body temperature; and (3) humidification of incoming

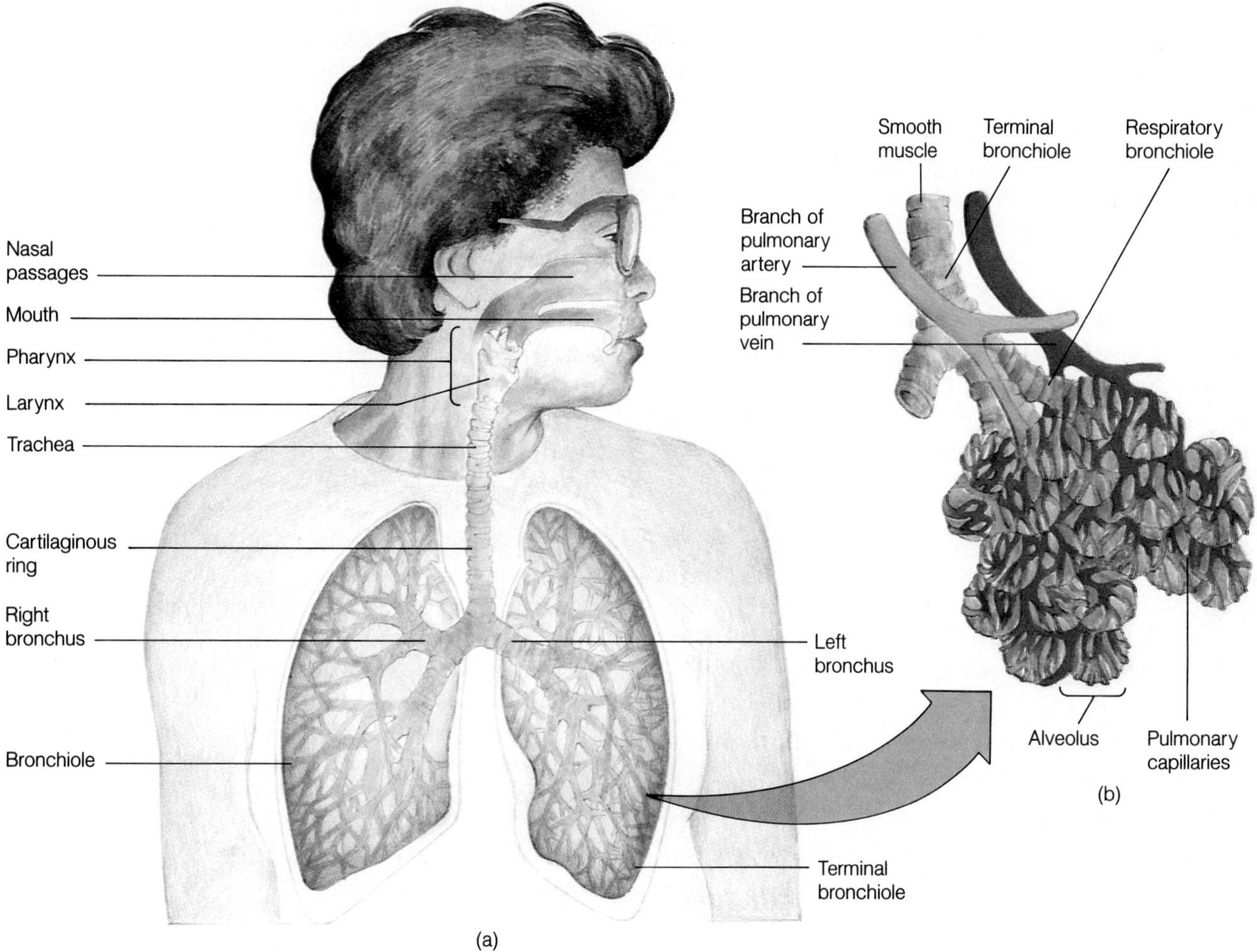

Figure 13–2 Anatomy of the Respiratory System *(a) The respiratory airways. (b) Enlargement of the alveoli (air sacs) at the terminal end of the airways.*

air. Moistening of inspired air is essential to prevent the alveolar linings from drying out. Diffusion of O_2 and CO_2 cannot occur through dry membranes. The nose also has the nonrespiratory function of serving as the organ of olfaction (smell).

The nasal passages open into the **pharynx** (throat), which serves as a common passageway for both the respiratory and digestive systems. Two tubes lead from the pharynx—the **trachea** (**windpipe**), through which air is conducted to the lungs, and the **esophagus,** the tube through which food passes to the stomach. Air normally enters the pharynx through the nose, but it can enter by the mouth as well when the nasal passages are congested. Because the pharynx serves as a common passageway for food and air, reflex mechanisms exist to close off the trachea during swallowing so that food enters the esophagus and not the airways. The esophagus remains closed except during swallowing to prevent air from entering the stomach during breathing.

Located at the entrance of the trachea is the **larynx,** or **voice box,** the anterior protrusion of which forms the "Adam's apple." The **vocal cords,** two strong bands of elastic tissue that lie across the opening of the larynx, can be stretched and positioned in different shapes by laryngeal muscles. As air is moved past the taut vocal cords, they vibrate to initiate the many different sounds of speech. Articulation of the sounds into recognizable sound patterns is accomplished by the lips, tongue, and soft palate. During swallowing, the vocal cords assume a function not related to speech; they are brought into tight apposition to each other to close off the entrance to the trachea.

Beyond the larynx, the trachea divides into two main branches, the right and left **bronchi,** which enter the right and left lungs, respectively. Within each lung, the bronchus continues to branch into progressively narrower, shorter, and more numerous airways, much like the branching of a tree. The smaller branches are known as **bronchioles.** Clustered at the ends of the terminal bronchioles are the alveoli, the tiny air sacs where gas exchange between air and blood takes place (Fig. 13– 2b).

To permit airflow in and out of the gas-exchanging portions of the lungs, the continuum of conducting airways from the entrance through the terminal bronchioles to the alveoli must remain open. The trachea and larger bronchi are fairly rigid, nonmuscular tubes encircled by a series of cartilaginous rings that prevent the tubes' compression. The smaller bronchioles have no cartilage to hold them open. Their walls contain smooth muscle that is innervated by the autonomic nervous system and is sensitive to certain hormones and local chemicals. These factors, by varying the degree of contraction of bronchiolar smooth muscle and hence the caliber of these small terminal airways, are able to regulate the amount of air passing between the atmosphere and each cluster of alveoli.

The gas-exchanging alveoli are small, thin-walled, inflatable air sacs encircled by a jacket of pulmonary capillaries.

The alveoli may be visualized as clusters of thin-walled, inflatable, grapelike sacs at the terminal branches of the conducting airways (Fig. 13–2b). A few individual alveoli bud off laterally along the last portion of the airways. Because gas exchange can take place across them, these smallest of airways are termed **respiratory bronchioles.**

The alveolar walls consist of a single layer of flattened **Type I alveolar cells** (Fig. 13–3a). The dense network of pulmonary capillaries encircling each alveolus is also only one cell-layer thick. The interstitial space between an alveolus and the surrounding capillary network forms an extremely thin barrier, with only 0.2 μm separating the air within the alveoli from the blood within the pulmonary capillaries. (A sheet of tracing paper is about fifty times thicker than this air-to-blood barrier.)

The alveolar air-blood interface presents a tremendous surface area for exchange. There are about 300 million alveoli in the lungs, each about 300 μm (⅓ mm) in diameter. So dense are the pulmonary capillary networks that each alveolus is encircled by an almost continuous sheet of blood. The total surface area thus exposed between alveolar air and pulmonary capillary blood is about 75 square meters (about the size of a tennis court). In contrast, if the lungs consisted of a single hollow chamber of the same dimensions instead of being divided into myriad alveolar units, the total surface area would be only about 1/100 square meter. Recall that according to Fick's law of diffusion (see p. 71), the rate of transfer of a diffusing molecule through a sheet of tissue is inversely proportional to the thickness of the tissue and directly proportional to its surface area. The lungs, therefore, are ideally structured for their function of gas exchange.

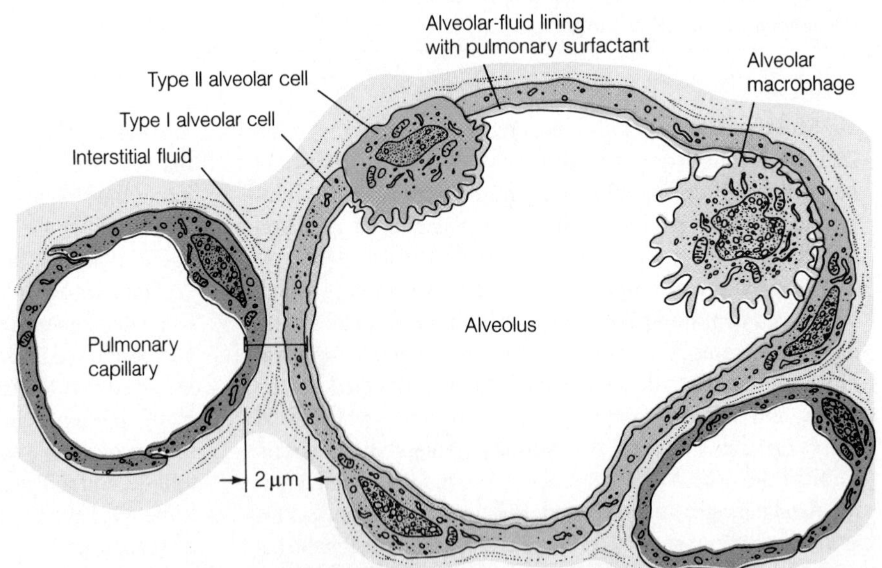

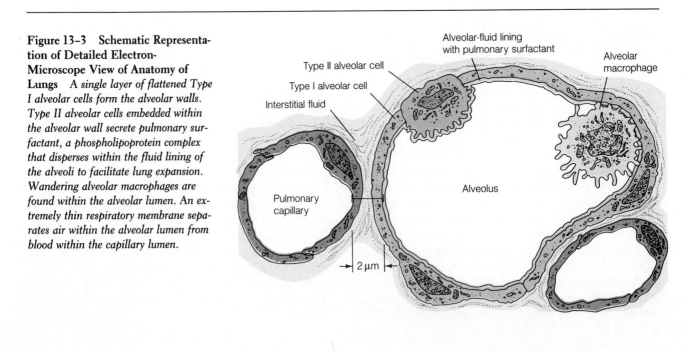

Figure 13–3 Schematic Representation of Detailed Electron-Microscope View of Anatomy of Lungs *A single layer of flattened Type I alveolar cells form the alveolar walls. Type II alveolar cells embedded within the alveolar wall secrete pulmonary surfactant, a phospholipoprotein complex that disperses within the fluid lining of the alveoli to facilitate lung expansion. Wandering alveolar macrophages are found within the alveolar lumen. An extremely thin respiratory membrane separates air within the alveolar lumen from blood within the capillary lumen.*

In addition to the thin, wall-forming Type I cells, the alveolar epithelium also contains **Type II alveolar cells** (Fig. 13–3), which secrete **pulmonary surfactant,** a phospholipoprotein complex that facilitates lung expansion (to be described later). Also present within the lumen of the air sacs are the defensive macrophages (see p. 353).

Minute **pores of Kohn** present in the alveolar walls permit air flow between adjacent alveoli, a process known as **collateral ventilation.** These passageways are especially important in allowing fresh air to enter an alveolus whose terminal conducting airway is blocked because of disease.

The lungs occupy much of the thoracic cavity.

There are two **lungs,** each divided into several lobes and each supplied by one of the bronchi. The lung tissue itself consists of the series of highly branching airways, the alveoli, the pulmonary vasculature, and large quantities of elastic connective tissue. The only muscle within the lungs is the smooth muscle in the walls of the arterioles and bronchioles, both of which are subject to control. There is no muscle within the alveolar walls to cause them to inflate and deflate during the breathing process. Rather, it is through changes in the dimensions of the thorax that corresponding changes in lung volume are produced.

The lungs occupy most of the volume of the **thoracic (chest) cavity,** the only other structures in the chest being the heart and associated vessels, the esophagus, the thymus, and some nerves. The outer chest wall is formed by twelve pairs of curved **ribs,** which join the **sternum** (breastbone) anteriorly and the **thoracic vertebrae** (backbone) posteriorly (Fig. 13–4). The rib cage provides bony protection for the lungs and heart. Through action of the **intercostal muscles,** which lie between the ribs, the dimensions of the thoracic cavity can be altered. Even more profound changes in thoracic volume can be accomplished by contraction of the diaphragm. The **diaphragm,** which forms the floor of the thorax, is a large, dome-shaped sheet of skeletal muscle that completely separates the thoracic cavity from the abdominal cavity. It is penetrated only by the esophagus and blood vessels traversing between the thoracic and abdominal cavities. The thoracic cavity is enclosed at the neck by muscles and connective tissue. The only communication between the thorax and the atmosphere is through the respiratory airways into the alveoli. As with the

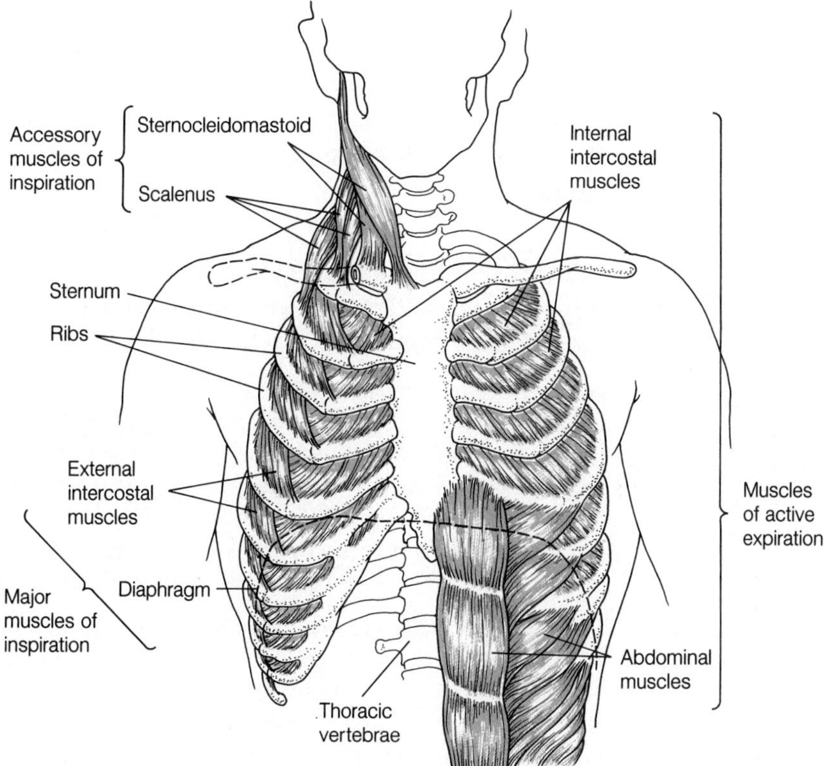

Figure 13–4 The Thorax

Figure 13–5 **Pleural Sac** *(a) Pushing a lollipop into a water-filled balloon produces a relationship analogous to that of the double-walled closed pleural sac surrounding each lung and separating it from the thoracic wall. (b) Schematic representation of the relationship of the pleural sac to the lungs and thorax. The relative size of the pleural cavity is grossly exaggerated for the purpose of visualization.*

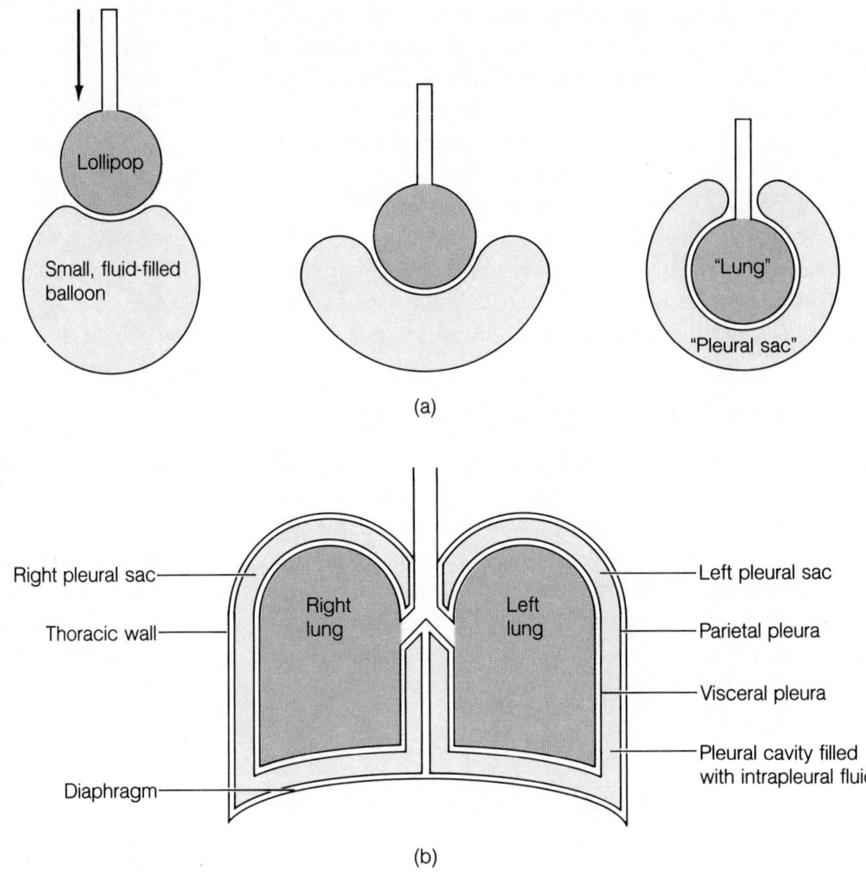

lungs, the chest wall contains considerable amounts of elastic connective tissue.

A tightly adherent pleural sac separates each lung from the thoracic wall.

Separating each lung from the thoracic wall and other surrounding structures is a double-walled, closed sac called the **pleural sac** (Fig. 13–5). One layer of the pleura, the **visceral pleura,** closely adheres to the surface of the lung, then reflects back on itself to form another layer, the **parietal pleura,** which lines the interior surface of the thoracic wall. The dimensions of the **pleural cavity** between these two layers are greatly exaggerated in the illustration to aid visualization; in reality the layers of the pleural sac are in close contact with one another. The surfaces of the pleura secrete a thin **intrapleural fluid,** which lubricates the pleural surfaces as they slide past each other during respiratory movements. **Pleurisy,** an inflammation of the pleural sac, is accompanied by painful breathing because of a "friction rub" with each inflation and deflation of the lungs.

Respiratory Mechanics

Interrelationships among atmospheric, intra-alveolar, and intrapleural pressures are important in respiratory mechanics.

Air tends to move from a region of higher pressure to a region of lower pressure, the difference in pressure being a **pressure gradient.** Air flows in and out of the lungs during the act of breathing by moving down alternately reversing pressure gradients established between the alveoli and the atmosphere by cyclical respiratory-muscle activity. Three different pressure considerations are important in ventilation (Fig. 13–6):

1. **Atmospheric (barometric) pressure** is the pressure exerted by the weight of the air in the atmosphere on objects on the earth's surface. At sea level it equals 760 mm Hg (Fig. 13–7). Atmospheric pressure diminishes with increasing altitude above sea level as the column of air above the earth's surface correspondingly decreases. Minor fluc-

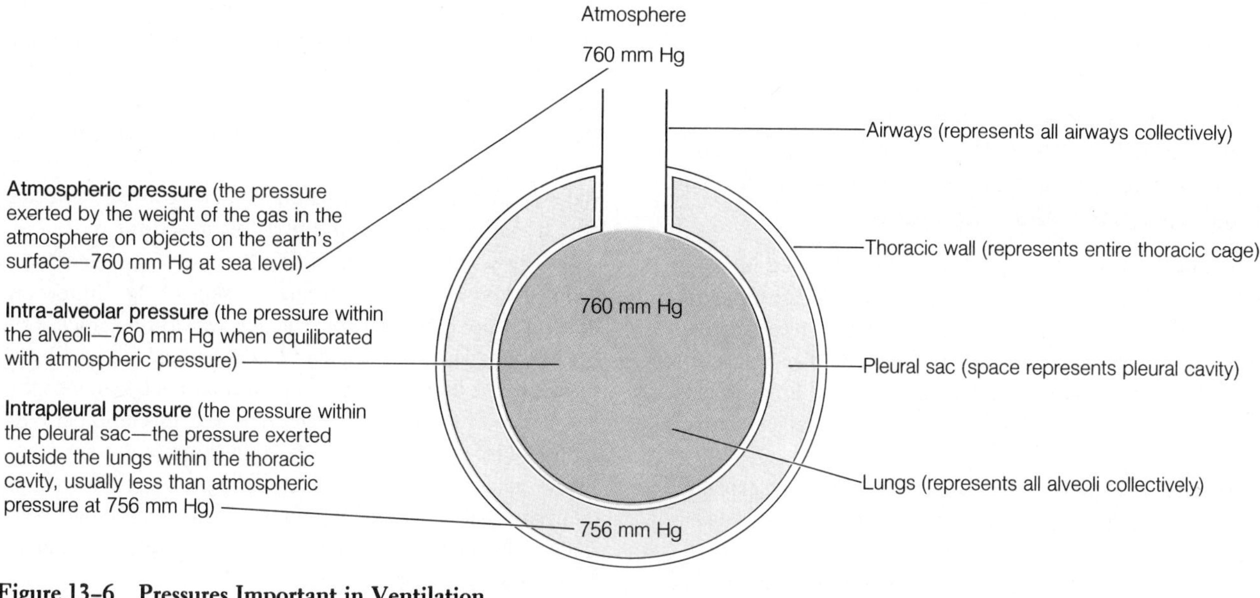
Atmosphere
760 mm Hg
Airways (represents all airways collectively)
Atmospheric pressure (the pressure exerted by the weight of the gas in the atmosphere on objects on the earth's surface—760 mm Hg at sea level)
Thoracic wall (represents entire thoracic cage)
760 mm Hg
Intra-alveolar pressure (the pressure within the alveoli—760 mm Hg when equilibrated with atmospheric pressure)
Pleural sac (space represents pleural cavity)
Intrapleural pressure (the pressure within the pleural sac—the pressure exerted outside the lungs within the thoracic cavity, usually less than atmospheric pressure at 756 mm Hg)
Lungs (represents all alveoli collectively)
756 mm Hg

Figure 13–6 Pressures Important in Ventilation

tuations in barometric pressure occur at any height because of changing weather conditions.

2. **Intra-alveolar pressure,** also known as **intrapulmonary pressure,** is the pressure within the alveoli. Because the alveoli communicate with the atmosphere through the conducting airways, air quickly flows down its pressure gradient any time intra-alveolar pressure changes compared to atmospheric pressure until the two pressures have equilibrated (become equal).
3. **Intrapleural pressure** is the pressure within the pleural sac. Also known as **intrathoracic pressure,** it is the pressure exerted outside the lungs within the thoracic cavity. The intrapleural pressure is usually less than atmospheric pressure, being 756 mm Hg at rest. Using atmospheric pressure as a reference point, 756 mm Hg is sometimes referred to as a pressure of -4 mm Hg, although there is really no such thing as an ***absolute*** negative pressure. A pressure of -4 mm Hg is just negative when compared to the normal atmospheric pressure of 760 mm Hg. To avoid confusion, we will use absolute positive values throughout our discussion. Intrapleural pressure does not equilibrate with atmospheric or intra-alveolar pressure, because there is no direct communication between the pleural cavity and either the atmosphere or the lungs. Since the pleural sac is a closed sac with no openings, air cannot enter or leave de-

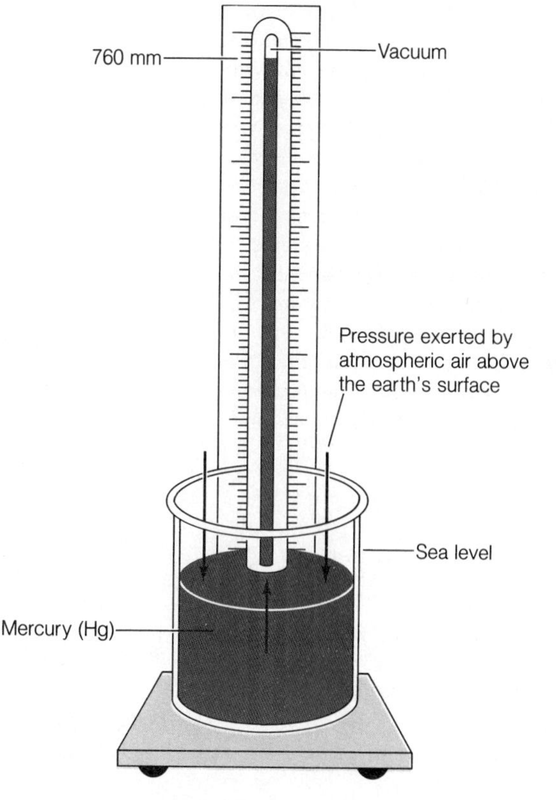
760 mm
Vacuum
Pressure exerted by atmospheric air above the earth's surface
Sea level
Mercury (Hg)

Figure 13–7 Atmospheric Pressure *The pressure exerted on objects by the atmospheric air above the earth's surface at sea level can push a column of mercury to a height of 760 mm. Therefore, atmospheric pressure at sea level is considered to be 760 mm Hg.*

spite any pressure gradients that might exist between it and surrounding regions.

The intrapleural-fluid surface tension and a transmural pressure gradient hold the lungs and thoracic wall in tight apposition, even though the lungs are smaller than the thorax.

The thoracic cavity is larger than the unstretched lungs because the thoracic cage grows more rapidly than the lungs during development. However, two important forces—the surface tension of the intrapleural fluid and the transmural pressure gradient—hold the thoracic wall and lungs in close apposition, stretching the lungs to fill the larger thoracic cavity.

Because of surface tension, a concept that will be explained more fully later, the molecules at the surface of a liquid resist being pulled apart. The surface tension developed by the intrapleural fluid tends to hold the pleural surfaces together. The intrapleural fluid can be considered very loosely as a "stickiness" or "glue" between the lining of the thoracic wall and the lung. Have you ever tried to pull apart two smooth surfaces held together by a thin layer of liquid, such as two wet glass slides? If so, you know that the two surfaces act as if they are stuck together by the thin layer of water. Even though you can easily slip the slides back and forth relative to each other (just as the intrapleural fluid facilitates movement of the lungs against the interior surface of the chest wall), you can pull the slides apart only with great difficulty, because the molecules within the intervening liquid resist being separated.

Similarly, the molecules within the intrapleural fluid resist being pulled apart. The surface tension of the intrapleural fluid helps hold the lungs and thoracic wall in close apposition, because one layer of the pleural sac is closely adhered to the surface of the lung and the other is firmly attached to the internal surface of the thoracic wall. This relationship is partly responsible for the fact that changes in thoracic dimension are always accompanied by corresponding changes in lung dimension; that is, when the thorax expands, the lungs, being stuck to the thoracic wall by virtue of the intrapleural-fluid surface tension, do likewise.

The other reason that the lungs follow the movements of the chest wall is the **transmural pressure gradient,** which exists across the wall of the lung (Fig. 13–8). The intra-

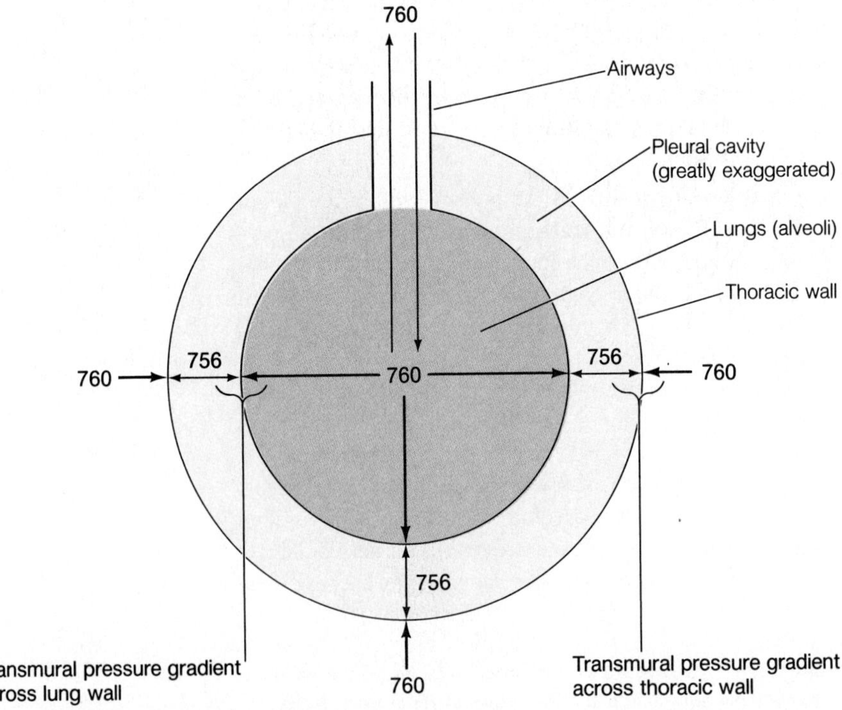

Figure 13–8 Transmural Pressure Gradient *Across the lung wall, the intra-alveolar pressure of 760 mm Hg pushes outward whereas the intrapleural pressure of 756 mm Hg pushes inward. This 4 mm Hg difference in pressure constitutes a transmural pressure gradient that pushes out on the lungs, stretching them to fill the larger thoracic cavity. Across the thoracic wall, the atmospheric pressure of 760 mm Hg pushes inward whereas the intrapleural pressure of 756 mm Hg pushes outward. This 4 mm Hg difference in pressure constitutes a transmural pressure gradient that pushes inward and compresses the thoracic wall.*

alveolar pressure, equilibrated with atmospheric pressure at 760 mm Hg, is greater than the intrapleural pressure of 756 mm Hg, so there is a greater pressure pushing outward than pushing inward across the lung wall. This net outward pressure differential, the transmural pressure gradient, pushes out on the lungs, stretching or distending them. Because of this pressure gradient, the lungs are always forced to expand to fill the thoracic cavity.

A similar transmural pressure gradient exists across the thoracic wall. The atmospheric pressure pushing inward on the thoracic wall is greater than the intrapleural pressure pushing outward on this same wall, so the chest wall tends to be "squeezed in" or compressed compared to what it would be in an unrestricted state. The effect of the transmural pressure gradient across the lung wall is much more pronounced, however, because the highly distensible lungs are influenced by this modest pressure differential to a much greater extent than is the more rigid thoracic wall.

Because neither the lungs nor the thoracic wall are in their natural position when they are held in apposition to each other, they constantly try to assume their own inherent dimensions. The stretched lungs have a tendency to pull inward away from the thoracic wall, whereas the compressed thoracic wall tends to move outward away from the lungs. These structures are prevented from pulling away from each other except to the slightest degree, however, because of the forces of the transmural pressure gradient and intrapleural fluid surface. Even so, the resultant ever-so-slight expansion of the pleural cavity is sufficient to drop the pressure in this cavity by 4 mm Hg, bringing the intrapleural pressure to the subatmospheric level of 756 mm Hg.

To understand how the slight increase in intrapleural volume causes a drop in intrapleural pressure, you must be aware of **Boyle's law.** Boyle's law states that at any constant temperature, the pressure exerted by a gas varies inversely with the volume of the gas (Fig. 13–9); that is, as the volume of a gas increases, the pressure exerted by the gas decreases proportionately, and conversely, the pressure increases proportionately as the volume decreases. Because of this inverse volume-pressure relationship, when the volume of the pleural cavity is slightly increased by the pull in opposite directions of the lungs and thoracic wall, the intrapleural pressure is slightly decreased below atmospheric pressure.

It is important to recognize the interrelationship between the transmural pressure gradient and the subatmospheric intrapleural pressure. The lungs are stretched and the thorax is compressed because a transmural pressure gradient exists across their walls due to the presence of a subatmospheric intrapleural pressure. The intrapleural pressure, in turn, is subatmospheric because the stretched lungs and compressed

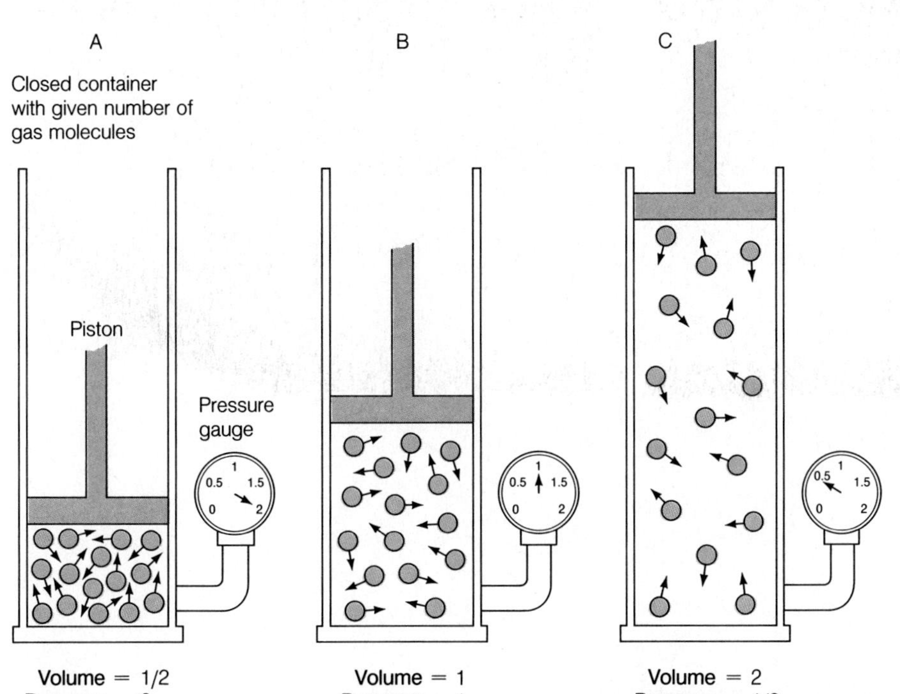

Figure 13–9 Boyle's Law *Each of the containers has the same number of gas molecules. Given the random motion of gas molecules, the likelihood of a gas molecule striking the interior wall of the container and exerting pressure varies inversely with the volume of the container at any constant temperature. The gas in container B exerts more pressure than the same gas in larger container C but less pressure than the same gas in smaller container A. This is stated as Boyle's law:* $P_1V_1 = P_2V_2$. *As the volume of a gas increases, the pressure of the gas decreases proportionately; conversely, the pressure increases proportionately as the volume decreases.*

thorax tend to pull away from each other, slightly expanding the pleural cavity and dropping the intrapleural pressure below atmospheric pressure.

If the intrapleural pressure were ever to equilibrate with atmospheric pressure, the transmural pressure gradient would be abolished. As a result, the lungs and thorax would separate and assume their own inherent dimensions. This is exactly what happens if air is permitted to enter the pleural cavity, a condition known as **pneumothorax** ("air in the chest").

Normally air is not allowed to enter the pleural cavity because there is no communication between it and either the atmosphere or alveoli. However, if the chest wall is punctured (for example, by a stab wound or a broken rib), air rushes into the pleural space from the higher atmospheric pressure down the air's pressure gradient (Fig. 13–10a). Intrapleural and intra-alveolar pressure are now both equilibrated with atmospheric pressure, so a transmural pressure gradient no longer exists across either the wall of the lung or the chest wall. With no force present to stretch the lung, it collapses to its unstretched size, a condition known as **atelectasis** (Fig. 13–10b). The thoracic wall likewise springs outward to its unrestricted dimensions, but this is of much less consequence than the collapse of the lung. Pneumothorax and atelectasis can similarly ensue if air enters the pleural cavity through a hole in the lung, produced, for example, by a disease process (Fig. 13–10c).

Bulk flow of air into and out of the lungs occurs because of cyclical intra-alveolar-pressure changes brought about indirectly by respiratory-muscle activity.

Because air flows down a pressure gradient, the intra-alveolar pressure must be less than atmospheric pressure for air to flow into the lungs during inspiration. Similarly, the intra-alveolar pressure must be greater than atmospheric pressure for air to flow out of the lungs during expiration. Intra-alveolar pressure can be changed by altering the volume of the lungs, in accordance with Boyle's law. The respiratory muscles that accomplish breathing do not act directly on the lungs to change their volume. Instead, these muscles change the volume of the thoracic cavity, which indirectly brings about a corresponding

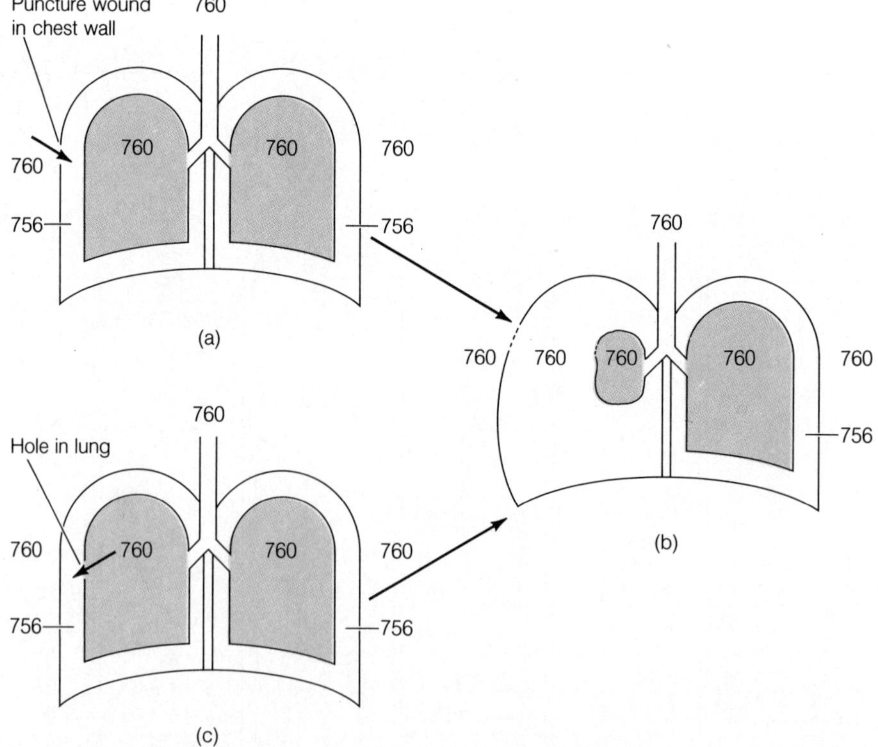

Figure 13–10 Pneumothorax *(a) Traumatic pneumothorax. A puncture in the chest wall permits air to enter the pleural cavity from the atmosphere down the air's pressure gradient, abolishing the transmural pressure gradient. (b) Atelectasis. When the transmural pressure gradient is abolished, the lung collapses (atelectasis) and the chest wall springs outward. (c) Spontaneous pneumothorax. A hole in the lung wall permits air to enter the pleural cavity from the lungs down the air's pressure gradient, abolishing the transmural pressure gradient.*

Table 13–1 Muscles of Respiration

Muscles	Result of Muscle Contraction	Timing of Stimulation to Contract
Inspiratory Muscles		
Diaphragm	Descends downward, increasing vertical dimension of thoracic cavity	Every inspiration; primary muscle of inspiration
External intercostal muscles	Elevate ribs upward and outward, enlarging thorax in both anteroposterior and lateral dimensions	Every inspiration; play secondary complementary role to primary action of diaphragm
Scalenus, sternocleidomastoid muscles	Raise sternum and elevate first two ribs, enlarging upper portion of thoracic cavity	Only during forceful inspiration; accessory inspiratory muscles
Expiratory Muscles		
Abdominal muscles	Increase intra-abdominal pressure, which exerts upward force on diaphragm to decrease vertical dimension of thoracic cavity	Only during active (forced) expiration
Internal intercostal muscles	Flatten thorax by pulling ribs downward and inward, decreasing transverse dimension of thoracic cavity	Only during active (forced) expiration

change in lung volume because the thoracic cage and lungs are linked together by the intrapleural-fluid surface tension and transmural pressure gradient.

Let us follow the changes that occur during one respiratory cycle—that is, one breath in (inspiration) and out (expiration). Before the beginning of inspiration, the respiratory muscles are relaxed, no air is flowing, and intra-alveolar pressure is equal to atmospheric pressure. At the onset of inspiration, the **inspiratory muscles**—the diaphragm and external intercostal muscles (Table 13–1; see also Fig. 13–4)—are stimulated to contract, which results in enlargement of the thoracic cavity. The major inspiratory muscle is the diaphragm, a sheet of skeletal muscle innervated by the **phrenic nerve.** The relaxed diaphragm assumes a dome shape that protrudes upward into the thoracic cavity. When the diaphragm contracts upon stimulation by the phrenic nerve, it descends downward, enlarging the thoracic cavity's volume by increasing its vertical dimension (Fig. 13–11). The abdominal wall, if relaxed, can be seen to bulge outward during inspiration as descent of the diaphragm pushes the abdominal contents downward and forward. Upon contraction of the **external intercostal muscles,** whose fibers run downward and forward between adjacent ribs, the ribs are elevated upward and outward. This enlarges the thorax in both the anteroposterior (front-to-back) and lateral (side-to-side) dimensions. **Intercostal nerves** activate the external intercostal muscles simultaneous to phrenic-nerve stimulation of the diaphragm.

As the thoracic cavity enlarges, the lungs are also forced to expand. As the lungs enlarge, the intra-alveolar pressure drops because the same number of air molecules now occupy a larger lung volume. In a typical inspiratory excursion, the intra-alveolar pressure drops 1 mm Hg to 759 mm Hg (Fig. 13–12a). Since the intra-alveolar pressure is now less than atmospheric pressure, air flows into the lungs down the pressure gradient from higher to lower pressure. Air continues to enter the lungs until no further gradient exists—that is, until intra-alveolar pressure equals atmospheric pressure. Thus lung expansion is not caused by movement of air into the lungs, but instead air flows into the lungs because of the fall in intra-alveolar pressure brought about by lung expansion.

During inspiration, the intrapleural pressure falls even lower, because the lungs are stretched even further and therefore tend to pull away from the thoracic wall even more. Typically, intrapleural pressure falls to 754 mm Hg during inspiration. This even greater transmural pressure gradient during inspiration assures that the lungs are stretched to fill the expanded thoracic cage.

At the end of inspiration, the inspiratory muscles relax (Fig. 13–11b). The diaphragm assumes its original dome-shaped position when it relaxes; the elevated rib cage falls because of gravity when the external intercostals relax; and the chest wall and stretched lungs recoil to their preinspiratory size because of their elastic properties, much like a stretched balloon would upon release. As the lungs recoil and become smaller in volume, the intra-alveolar pressure rises, because the greater number of air molecules contained within the larger lung volume at the end of inspiration are now compressed into a smaller volume. In a resting expiration, the intra-alveolar pres-

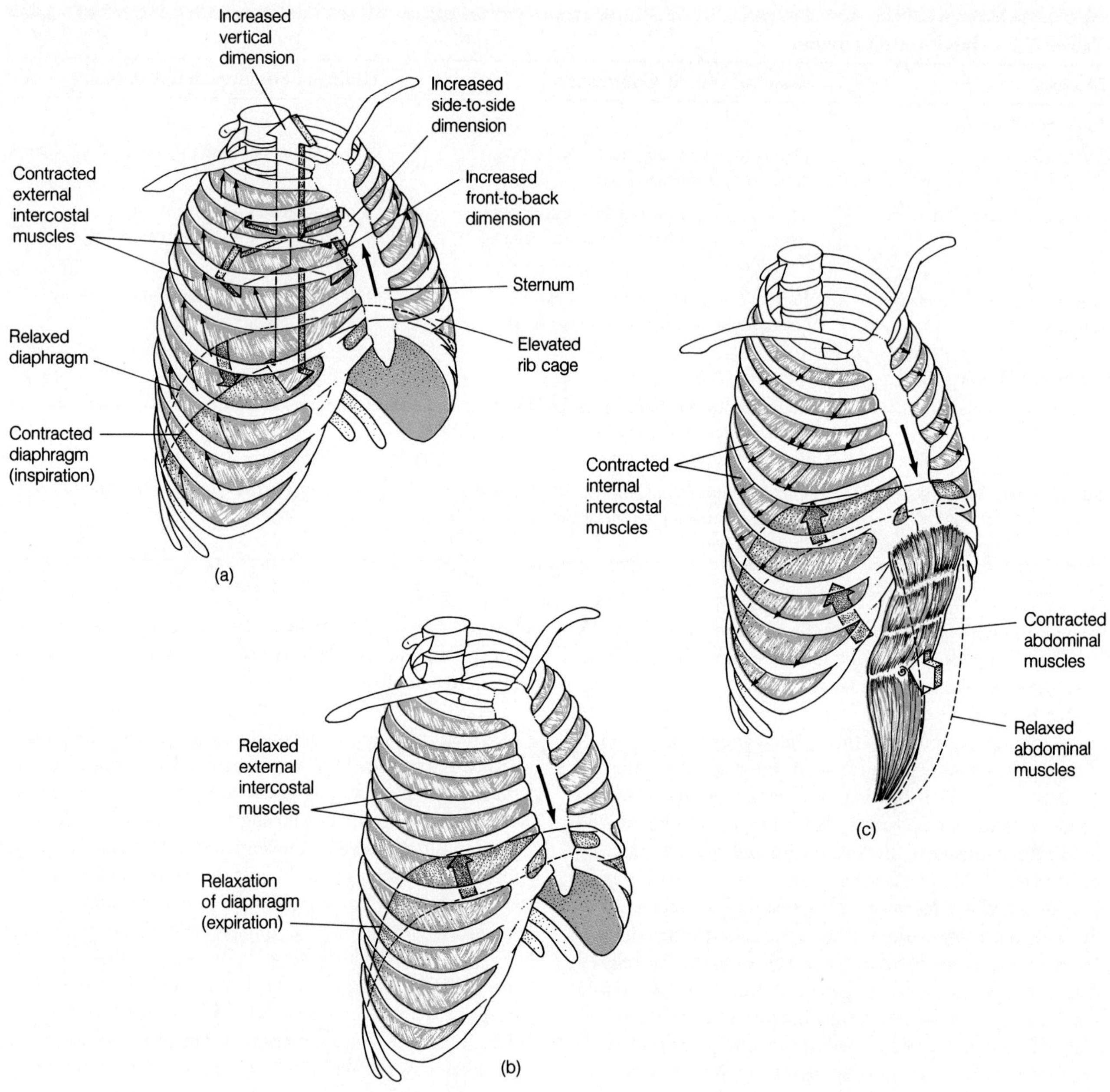

Figure 13–11 Respiratory Muscle Activity during Inspiration and Expiration *(a) Inspiration, during which the diaphragm descends on contraction, increasing the vertical dimension of the thoracic cavity. Contraction of the external intercostal muscles elevates the ribs to enlarge the thoracic cavity from front to back and side to side. (b) Quiet passive expiration, during which the diaphragm relaxes, reducing the volume of the thoracic cavity from its peak inspiratory size. As the external intercostal muscles relax, the elevated rib cage falls because of the force of gravity. This also reduces the volume of the thoracic cavity. (c) Active expiration, during which contraction of the abdominal muscles increases the intra-abdominal pressure, exerting an upward force on the diaphragm. This further reduces the vertical dimension of the thoracic cavity in comparison to the reduction with quiet passive expiration. Contraction of the internal intercostal muscles decreases the front-to-back and side-to-side dimensions by flattening the ribs.*

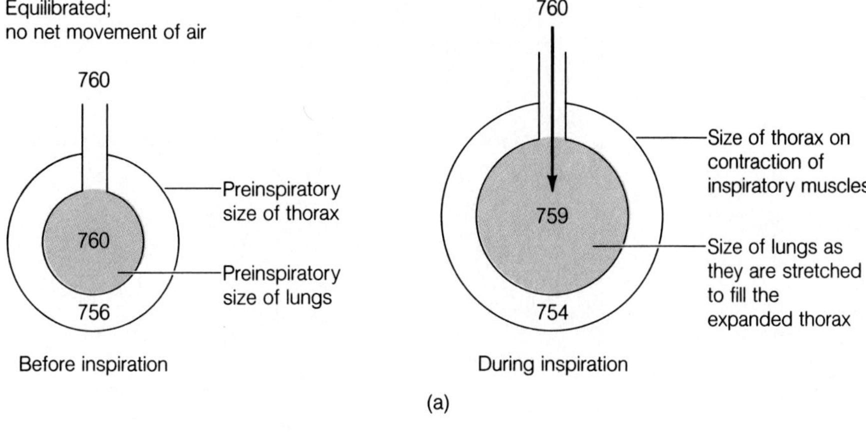

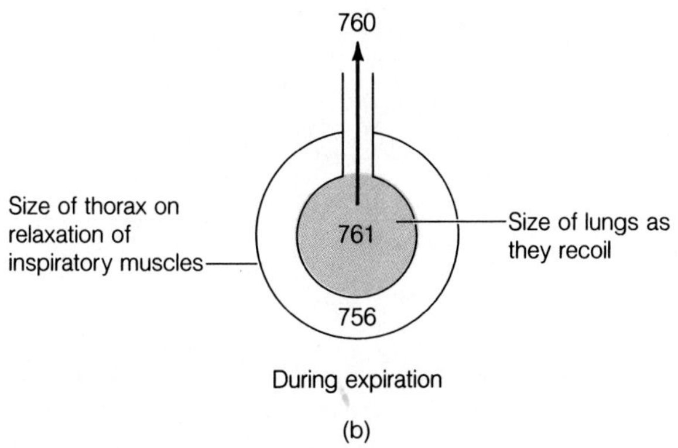

Figure 13–12 Changes in Lung Volume and Intra-alveolar Pressure during Inspiration and Expiration *(a) Inspiration. As the lungs increase in volume during inspiration, the intra-alveolar pressure decreases, establishing a pressure gradient that favors the flow of air into the alveoli from the atmosphere; that is, an inspiration occurs. (b) Expiration. As the lungs recoil to their preinspiratory size upon relaxation of the inspiratory muscles, the intra-alveolar pressure increases, establishing a pressure gradient that favors the flow of air out of the alveoli into the atmosphere; that is, an expiration occurs.*

sure increases about 1 mm Hg above atmospheric level to 761 mm Hg (Fig. 13–12b). Air now leaves the lungs down its pressure gradient from high intra-alveolar pressure to lower atmospheric pressure. Outward flow of air ceases when intra-alveolar pressure becomes equal to atmospheric pressure, so a pressure gradient no longer exists. Figure 13–13 summarizes the intra-alveolar and intrapleural-pressure changes that take place during one respiratory cycle.

Expiration is normally a ***passive*** process during quiet breathing since it is accomplished by relaxation of the inspiratory muscles, with no muscular exertion or energy expenditure required. In contrast, inspiration is *always active*, because it is brought about only by contraction of inspiratory muscles at the expense of energy utilization. For more forceful breathing excursions, such as is required by exercise, expiration does become active. In order for more air to be expired, the intra-alveolar pressure must be increased even further above atmospheric pressure than can be accomplished by simple relaxation of the inspiratory muscles. To produce such a **forced** or **active expiration** requires the contraction of **expiratory muscles** to further reduce the volume of the thoracic cavity and lungs (Table 13–1). The most important expiratory muscles are (though it may seem unbelievable at first thought) the muscles of the abdominal wall. As the abdominal muscles contract, the resultant increase in intra-abdominal pressure exerts an upward force on the diaphragm, pushing it further up into the thorax than its relaxed position (Fig. 13–11c). Elevation of the diaphragm decreases the vertical dimension of the thoracic cavity even more. The other expiratory muscles are the **internal intercostal muscles,** whose contraction pulls the ribs downward and inward, flattening the chest wall and decreasing the transverse dimension of the thoracic cavity, in an action that is just the opposite that of the external intercostal muscles.

As the thoracic dimensions become reduced because of active contraction of the expiratory muscles, the lungs also be-

Figure 13–13 Intra-alveolar and Intrapleural Pressure Changes throughout Respiratory Cycle

- *During inspiration, intra-alveolar pressure is less than atmospheric pressure.*
- *During expiration, intra-alveolar pressure is greater than atmospheric pressure.*
- *At the end of inspiration and at the end of expiration, intra-alveolar pressure is equal to atmospheric pressure, because the alveoli are in direct communication with the atmosphere and air continues to flow down its pressure gradient until the two pressures equilibrate.*
- *Throughout the respiratory cycle, the intrapleural pressure is lower than the intra-alveolar pressure. Thus there is always a transmural pressure gradient and the lung is always stretched to some degree, even during expiration.*

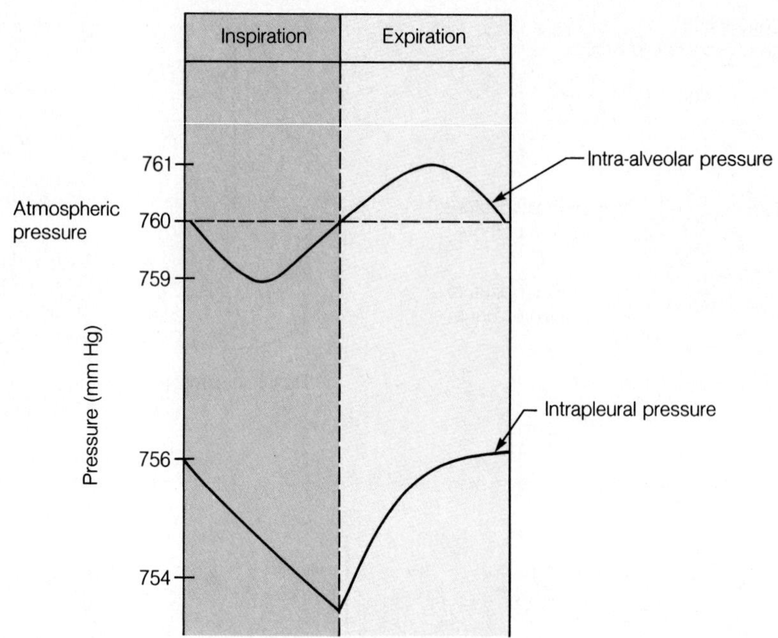

come reduced in volume because they do not have to be stretched as much to fill the smaller thoracic cavity. The intra-alveolar pressure increases as the air in the lungs is confined within an even smaller volume. The differential between intra-alveolar and atmospheric pressure is even greater now than during passive expiration, so more air leaves down the pressure gradient before equilibration is achieved. In this way, more air is expired during forceful, active expiration than during quiet, passive expiration.

During forceful expiration, the intrapleural pressure exceeds atmospheric pressure but the lungs do not collapse. Because the intra-alveolar pressure is also increased correspondingly, a transmural pressure gradient still exists across the walls of the lungs, keeping them stretched to fill the thoracic cavity. For example, if the pressure within the thorax increases 10 mm Hg, the intrapleural pressure becomes 766 mm Hg and the intra-alveolar pressure becomes 770 mm Hg—still a 4 mm Hg pressure difference.

Even though inspiration is normally active even in quiet breathing, deeper breaths can be accomplished by contracting the diaphragm and external intercostal muscles more forcefully and by bringing the **accessory inspiratory muscles** into play to further enlarge the thoracic cavity. Contraction of these accessory muscles, which are located in the neck (Table 13–1), raises the sternum and elevates the first two ribs, enlarging the upper portion of the thoracic cavity. As the thoracic cavity increases in volume, the lungs likewise expand more, dropping the intra-alveolar pressure even further. Consequently, a larger inward flow of air occurs before equilibration with atmospheric pressure is achieved; that is, a deeper breath occurs.

Because the diaphragm is the major inspiratory muscle and its relaxation also causes expiration, paralysis of the intercostal muscles alone does not seriously influence quiet breathing. Disruption of diaphragm activity caused by nerve or muscle disorders, however, leads to respiratory paralysis. Fortunately, the phrenic nerve arises from the spinal cord in the neck region (cervical segments 3, 4, and 5) and then descends to the diaphragm at the floor of the thorax, instead of arising from the thoracic region of the cord as might be expected. For this reason, individuals completely paralyzed below the neck as a result of traumatic severance of the spinal cord are still able to breathe, even though they have lost use of all other skeletal muscles in their trunk and limbs.

Airway resistance becomes an especially important determinant of airflow rates in chronic obstructive pulmonary diseases.

Thus far we have discussed airflow in and out of the lungs as being a function of varying the magnitude of the pressure gradient between the alveoli and the atmosphere, which is changed through alterations in the dimensions of the thorax and the lungs. However, just as flow (F) of blood through the vasculature depends not only on the pressure gradient (ΔP) but also on the resistance (R) to flow offered by the vessels, so it is too with airflow:

$$F = \Delta P/R$$

F = Airflow

ΔP = Difference between atmospheric and intra-alveolar pressure

R = Resistance of airways, determined by their radii

The primary determinant of resistance to airflow is the radius of the conducting airways. We ignored airway resistance in our preceding discussion of pressure gradient–induced airflow rates because, in a healthy respiratory system, the radius of the conducting system is sufficiently large that resistance remains extremely low. Therefore, the pressure difference between the alveoli and the atmosphere is normally the primary factor determining the airflow rate. Indeed, because of the low resistance offered by the airways, very small pressure gradients of only 1–2 mm Hg need normally be created to achieve adequate rates of airflow in and out of the lungs.

Resistance becomes an extremely important impediment to airflow, however, when airway lumens become narrowed as a result of pathological or physiological factors (Table 13–2). Most significantly, **chronic obstructive pulmonary disease (COPD)** is characterized by increased airway resistance. When airway resistance increases, a larger pressure gradient must be established to maintain even a normal airflow rate. For example, if resistance is doubled by narrowing of airway lumens, ΔP must be doubled through increased respiratory-muscle exertion to induce the same flow of air in and out of the lungs as a normal individual accomplishes during quiet breathing. Accordingly, patients with COPD have to work harder to breathe.

Chronic obstructive pulmonary disease encompasses three chronic (long-term) diseases: asthma, chronic bronchitis, and emphysema. In **asthma,** airway obstruction is due to: (1) profound constriction of the smaller airways caused by allergy-induced spasm of the smooth muscle in the walls of these airways (see p. 398); (2) plugging of the airways by excess secretion of a very thick mucus; and 3) histamine-induced edema of the walls of the airways. Occasional acute (short-term) asthmatic attacks generally have no permanent effect, but chronic obstruction associated with frequent, prolonged bouts of asthma can gradually lead to hypertrophy (thicken-

Table 13–2 Factors Affecting Airway Resistance

Status of Airways	Effect on Resistance	Factors Producing the Effect
Bronchoconstriction	↓ radius, ↑ resistance to airflow	*Pathological factors:*
		Allergy-induced spasm of airways caused by:
		Slow-reactive substance of anaphylaxis
		Histamine
		Physical blockage of airways caused by:
		Excess mucus
		Edema of walls
		Airway collapse
		Physiological control factors:
		Neural control: parasympathetic stimulation
		Local chemical control: ↓ CO_2 concentration
Bronchodilation	↑ radius, ↓ resistance to airflow	*Pathological factors: none*
		Physiological control factors:
		Neural control: sympathetic stimulation
		Hormonal control: epinephrine
		Local chemical control: ↑ CO_2 concentration

ing) of airway smooth muscle and permanent narrowing of the air passageways.

Chronic bronchitis is a long-term inflammatory condition of the lower respiratory airways, generally triggered by frequent exposure to irritating cigarette smoke, polluted air, or allergens. In response to the chronic irritation, the airways become narrowed because of prolonged edematous thickening of the airway linings, coupled with overproduction of a thick mucus by hypertrophied mucous glands. Despite frequent coughing associated with the chronic irritation, the plugged mucus often cannot be satisfactorily removed, especially since the ciliary mucus escalator is immobilized by the irritants. Pulmonary bacterial infections frequently occur, because the accumulated mucus serves as an excellent medium for bacterial growth.

Emphysema is characterized by collapse of the smaller airways and a breakdown of alveolar walls. This irreversible condition can arise in several different ways. Most commonly, it begins with chemically-induced chronic inflammation and obstruction of the respiratory airways (for example, chronic bronchitis or recurrent episodes of obstructive asthma). For reasons to be described shortly, expiration is more difficult to accomplish than inspiration when airway resistance is increased. Inspired air that fails to be expired becomes entrapped in the alveoli, causing them to become overdistended. Chronic overstretching, coupled with the high incidence of lung infections, gradually causes destruction of the alveolar walls.

Less frequently, emphysema arises from a genetic deficiency of a pulmonary enzyme that normally prevents destruction of lung tissues by chemicals released from alveolar macrophages. The unprotected alveolar walls gradually disintegrate under the influence of these proteolytic (protein-digesting) chemicals.

Why is expiration impeded more than inspiration with obstructive lung diseases? The smaller airways, lacking the cartilaginous rings that hold the larger airways open, are held open by the same transmural pressure gradient that distends the alveoli. Expansion of the thorax during inspiration indirectly dilates the airways even further than their expiratory dimensions, similar to alveolar expansion, so airway resistance is lower during inspiration than during expiration. In a normal individual, the airway resistance is always so low that the slight variation occurring between inspiration and expiration is not noticeable. When airway resistance has substantially increased, however, such as during an asthmatic attack, the difference between inspiration and expiration is quite noticeable. Thus, an asthmatic has more difficulty expiring than inspiring, giving rise to the characteristic "wheeze" as air is forced out through the narrowed airways.

The autonomic nervous system, epinephrine, and local CO_2 levels normally regulate airway resistance.

In a normal situation, airway size is adjusted depending on the body's needs. Controlled adjustment of airway caliber is accomplished by autonomic nervous-system regulation and by local controls mediated via CO_2 levels (Table 13–2).

Parasympathetic stimulation, which occurs in quiet, relaxed situations when there is not a high demand for airflow, promotes bronchiolar smooth-muscle contraction. This increases airway resistance by producing **bronchoconstriction.** In contrast, sympathetic stimulation and probably even more importantly, its associated hormone, epinephrine, bring about **bronchodilation** and decreased airway resistance by promoting bronchiolar smooth-muscle relaxation. Thus, during periods of sympathetic domination, when increased demands for O_2 uptake are actually or potentially placed on the body, bronchodilation occurs to assure that the pressure gradients established by respiratory-muscle activity are able to achieve maximum air flow rates with minimum resistance. Because of this bronchodilator action, epinephrine or similar drugs are useful therapeutic tools to counteract airway constriction in patients with bronchial spasms.

We have been talking about overall airway resistance in the lungs, but just as with arteriolar smooth muscle, bronchiolar smooth muscle is sensitive to local changes within its immediate environment, particularly to local CO_2 levels. If an alveolus is receiving too little airflow in comparison to its blood supply, CO_2 levels will increase in the alveolus and surrounding tissue as more CO_2 is dropped off by the blood than is being exhaled into the atmosphere. This local increase in CO_2 produces relaxation of the airway supplying the underaerated alveolus by acting directly on the involved bronchiolar smooth muscle. The resultant decrease in airway resistance leads to an increased airflow (for the same ΔP) to the involved alveolus, so its airflow now matches its blood supply (Fig. 13–14). The converse is also true. A localized decrease in CO_2 associated with an alveolus that is receiving too much air for its blood supply causes constriction of the airway supplying this overaerated alveolus by directly increasing contractile activity of the involved airway smooth muscle. The result is a reduction in airflow to the overaerated alveolus.

A similar locally induced effect on pulmonary vascular smooth muscle also takes place simultaneously to maximally match blood flow to airflow. Just as in the systemic circulation, distribution of the cardiac output to different alveolar capillary networks can be controlled by adjusting the resistance to blood flow through specific pulmonary arterioles. If the blood flow is too large for a given underaerated alveolus, the O_2 level in the

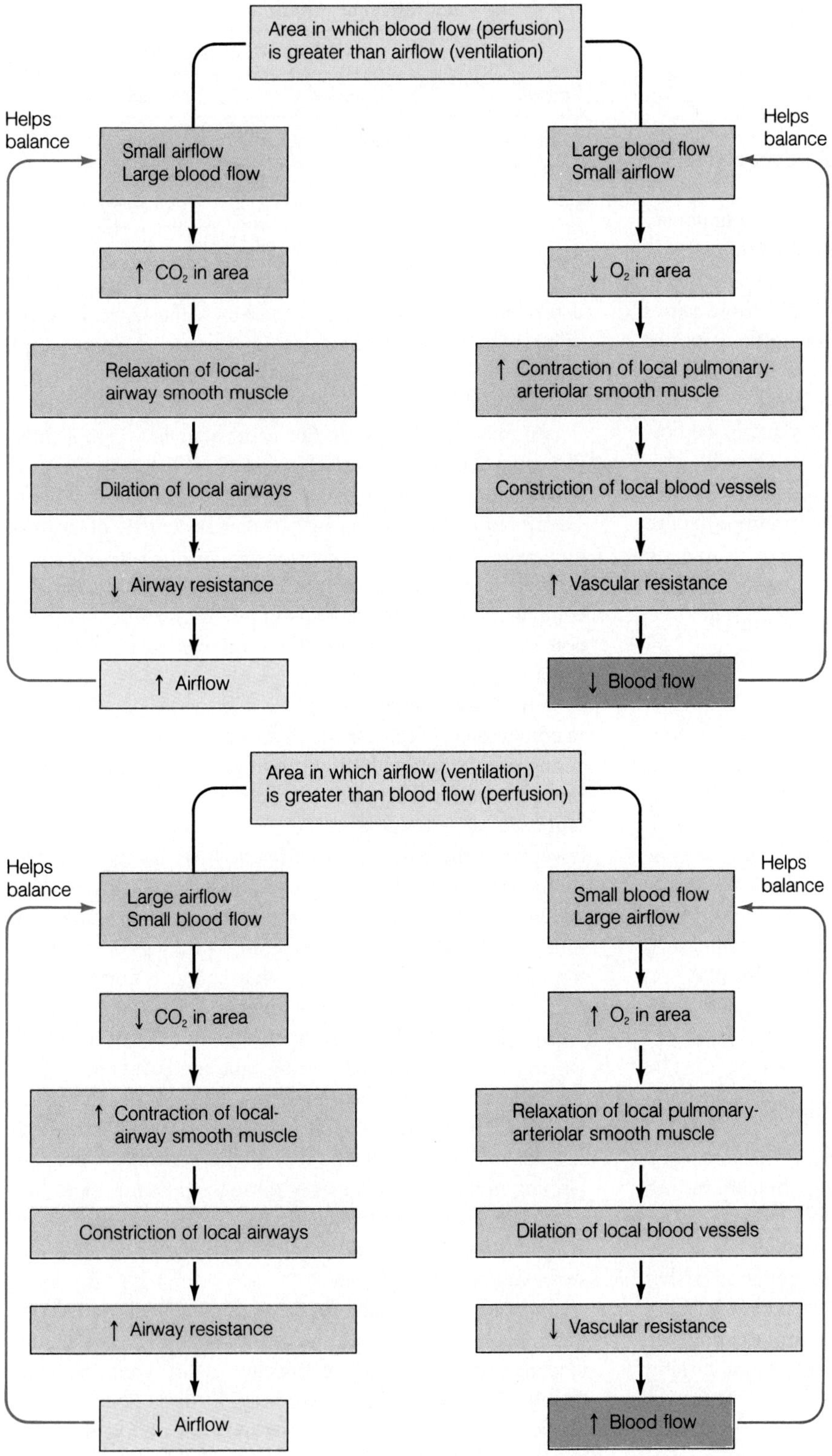

Figure 13-14 Local Controls to Match Airflow and Blood Flow to an Area

alveolus and surrounding tissues will fall below normal as more O_2 than usual is extracted from the alveolus by the overabundance of blood. The local decrease in O_2 concentration causes vasoconstriction of the pulmonary arteriole supplying this particular capillary bed, the result of which is a reduction in blood flow to match the reduced airflow. Conversely, an increase in alveolar O_2 concentration caused by a mismatched large airflow and small blood flow brings about pulmonary vasodilation, which increases blood flow to match the larger airflow. Note that the local effect of O_2 on pulmonary arteriolar smooth muscle is appropriately just the opposite of its effect on systemic arteriolar smooth muscle (Table 13–3). In the systemic circulation, a decrease in O_2 in a tissue causes localized vasodilation to increase blood flow to the deprived area, and vice versa, which is important in matching blood supply to local metabolic needs (see p. 308).

The two mechanisms for matching airflow and blood flow function concurrently, so normally very little air or blood is wasted in the lung. Airflow and blood flow at a particular alveolar interface are usually appropriately matched to accomplish efficient exchange of O_2 and CO_2.

Table 13–3 Effects of Local Changes in O_2 on Pulmonary and Systemic Arterioles

	Effect of Local Change in O_2	
Vessels	*Decreased O_2*	*Increased O_2*
Pulmonary arterioles	Vasoconstriction	Vasodilation
Systemic arterioles	Vasodilation	Vasoconstriction

Elastic behavior of the lungs is due to elastic connective-tissue fibers and alveolar surface tension.

You learned that during the respiratory cycle, the lungs alternately expand during inspiration and recoil during expiration. What properties of the lungs enable them to behave like balloons, being able to be stretched and then snapping back to their resting position when the stretching forces are removed? Two interrelated concepts are involved in pulmonary elasticity: elastic recoil and compliance.

Elastic recoil refers to how readily the lungs rebound after having been stretched. It is responsible for the lungs snapping back to their preinspiratory volume when the inspiratory muscles relax at the end of inspiration.

Compliance refers to how easily the lungs can be stretched or distended. Specifically, it is a measure of the magnitude of change in lung volume accomplished by a given change in the transmural pressure gradient, the force that stretches the lungs. A highly compliant lung stretches further for a given increase in the pressure differential than does a less compliant lung. Stated another way, the lower the compliance of the lungs, the larger the transmural pressure differential that must be created during inspiration to produce normal lung expansion. In turn, a greater-than-normal transmural pressure gradient during inspiration can be achieved only by making the intrapleural pressure more subatmospheric than usual. This is accomplished by greater expansion of the thorax through more vigorous contraction of the inspiratory muscles. Therefore, the less compliant the lungs, the more work required to produce a given degree of inflation. A poorly compliant lung is referred to as a "stiff" lung, because it lacks normal stretchability.

A number of factors can decrease respiratory compliance, including alterations in the elastic tissue of the lungs; pulmonary congestion and edema occurring, for example, with left-sided congestive heart failure; and replacement of normal lung tissue with fibrous connective tissue. The latter is caused by breathing in irritants such as asbestos fibers. Pulmonary compliance is also reduced when there is inadequate production of pulmonary surfactant for reasons to be described shortly.

Pulmonary compliance is actually increased above normal in emphysema, because some of the tissues responsible for resisting expansion are destroyed by the disease process. Even though less effort is required to inflate such lungs, serious consequences arise because elastic recoil is correspondingly reduced. For the same reason that emphysematous lungs are more stretchable, they also have less elastic recoil; that is, they do not resist being stretched by bouncing back toward their unstretched size. Because more compliant lungs do not tend to pull away from the thoracic wall as much as normal lungs do, the transmural pressure gradient is lower than normal. This pressure differential is responsible for holding the small airways open, so these airways become markedly reduced in caliber or may even collapse, which in turn disturbs the distribution of inspired gas. Alveoli served by collapsed airways are not able to directly exchange air with the atmosphere, accomplishing meager exchange only through the pores of Kohn. Furthermore, the considerable increase in airway resistance seriously hampers expiratory efforts, as is characteristic of obstructive pulmonary disease.

Pulmonary elastic behavior depends mainly on two factors: highly elastic connective tissue in the lungs and alveolar surface tension. Pulmonary connective tissue contains large quantities of elastin fibers. Not only do these fibers exhibit elastic properties themselves, but they are arranged into a meshwork that further increases their elastic behavior, much like the threads in a piece of stretch-knit fabric. The entire piece of fabric (or lung) is stretchier and tends to bounce back to its original shape more than the individual threads (elastin fibers) of which the fabric is woven.

An even more important factor influencing elastic behavior of the lungs is the **alveolar surface tension** displayed by the thin liquid film that lines each alveolus. At an air-water interface, the H_2O molecules at the surface are more strongly attracted to other surrounding H_2O molecules than to the air above the surface. This unequal attraction produces a force at the surface of the liquid known as surface tension, which is responsible for a two-fold effect. First, the liquid layer resists any force that increases its surface area; that is, it opposes expansion of the alveolus because the surface H_2O molecules oppose being pulled apart. Accordingly, the greater the surface tension, the less compliant the lungs. Second, the liquid surface area tends to become as small as possible because the surface H_2O molecules, being preferentially attracted to each other, try to get as close together as possible. Thus the surface tension of the liquid lining an alveolus tends to reduce the size of the alveolus, squeezing in on the air within it (Fig. 13–15). This, along with the rebound of the stretched elastin fibers, is responsible for the lungs' elastic recoil back to their pre-inspiratory size when inspiration is over.

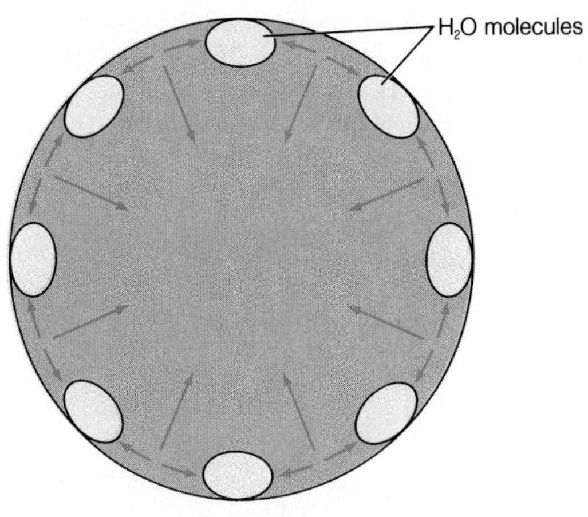

Figure 13–15 Alveolar Surface Tension *The attractive forces between the H_2O molecules in the liquid film that lines the alveolus are responsible for surface tension. Because of its surface tension, an alveolus: (1) resists being stretched, (2) tends to be reduced in surface area or size, and (3) tends to recoil after being stretched.*

Pulmonary surfactant decreases surface tension and contributes to lung stability.

Because of the strong cohesive forces between H_2O molecules, if the alveoli were lined with H_2O alone, the surface tension would be so great that the lungs would collapse, because the elastic-recoil force would exceed the opposing stretching force of the transmural pressure gradient. Furthermore, the lungs would be very poorly compliant, so exhausting muscular efforts would be required to accomplish stretching and inflation of the alveoli.

The tremendous surface tension of pure H_2O is normally counteracted by secretion of pulmonary surfactant by the Type II alveolar cells (Fig. 13–3). Pulmonary surfactant intersperses between the H_2O molecules in the fluid lining the alveoli and lowers the alveolar surface tension by reducing the cohesive forces between the H_2O molecules. By lowering the alveolar surface tension, pulmonary surfactant provides two important benefits: (1) it reduces elastic recoil so that the lungs do not collapse as readily; and (2) it increases pulmonary compliance, thus reducing the work of inflating the lungs.

Related to its ability to reduce elastic recoil of alveoli, pulmonary surfactant is also important in helping maintain lung stability by discouraging alveolar collapse. Although the fact that the lung is divided into myriad tiny air sacs provides the advantage of a tremendously increased surface area for exchange of O_2 and CO_2, it presents the problem of maintaining the stability of all of these alveoli. Recall that the pressure generated by alveolar surface tension is directed inwards, squeezing in on the air in the alveoli. If the alveoli are visualized as spherical bubbles, according to **LaPlace's law,** the magnitude of the inward-directed collapsing pressure is directly proportional to the surface tension and inversely proportional to the radius of the bubble:

$$\text{Pressure} = \frac{4 \times \text{Surface tension}}{\text{Radius}}$$

Because the collapsing pressure is inversely proportional to the radius, the smaller the alveolus, the smaller its radius and the greater its tendency to collapse at a given surface tension. Accordingly, if two alveoli of unequal size but the same surface tension are connected by the same terminal airway, the smaller alveolus—because it generates a larger collapsing pressure—has a tendency to collapse and empty its air into the larger alveolus (Fig. 13–16). Small alveoli normally do not collapse and blow up larger alveoli, however, because of pulmonary surfactant, which decreases surface tension substantially. Because the collapsing pressure is directly proportional to surface tension, when surface tension is low, the collapsing pressure remains low no matter what the radius. Pulmonary surfactant therefore helps stabilize the sizes of the alveoli and helps keep them open and available to participate in gas exchange.

Figure 13–16 Tendency for Small Alveoli to Collapse into Larger Alveoli *According to the law of LaPlace, if two alveoli of unequal size but the same surface tension are connected by the same terminal airway, the smaller alveolus—because it generates a larger inward-directed collapsing pressure—has a tendency (without pulmonary surfactant) to collapse and empty its air into the larger alveolus.*

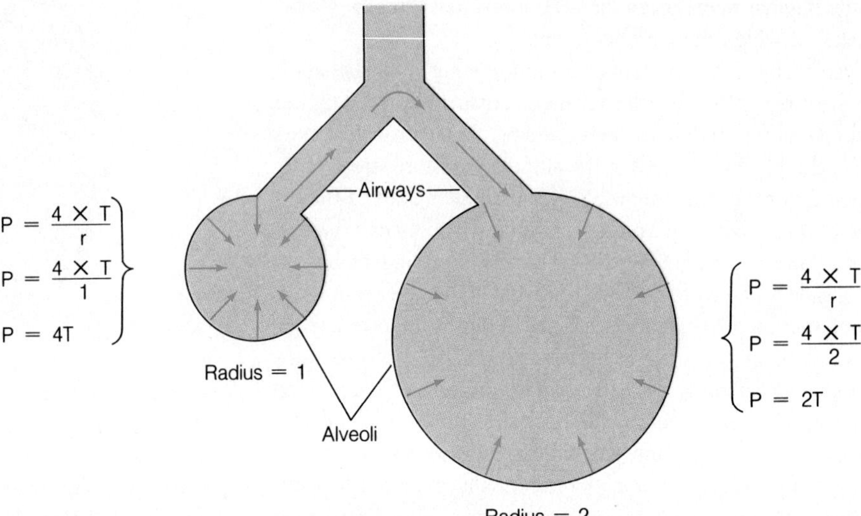

A second factor that contributes to alveolar stability is the **interdependence** of neighboring alveoli. Each alveolus is surrounded by other alveoli with which it is interconnected by connective tissue. If an alveolus starts to collapse, the surrounding alveoli are stretched as their walls are pulled in the direction of the caving-in alveolus (Fig. 13–17a). In turn, these neighboring alveoli, by recoiling in resistance to being stretched, exert expanding forces on the collapsing alveolus

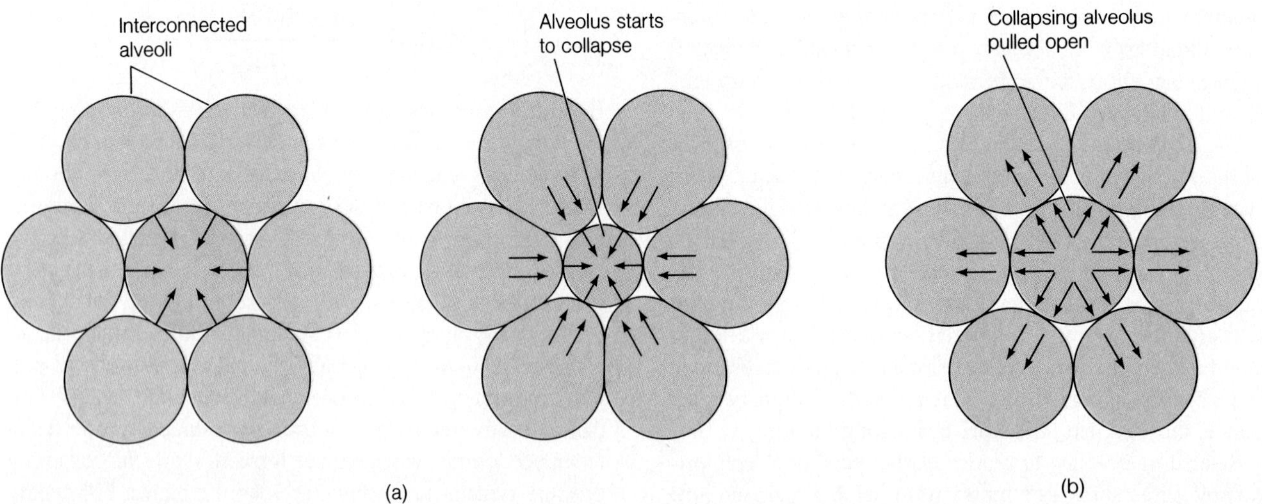

Figure 13–17 Alveolar Interdependence *(a) Surrounding alveoli are stretched by collapsing alveolus. (b) As neighboring alveoli recoil in resistance to being stretched, they pull outward on the collapsing alveolus. This expanding force pulls open the collapsing alveolus.*

Table 13–4 Opposing Forces Acting on the Lung

Forces Keeping the Alveoli Open	Forces Promoting Alveolar Collapse
Transmural pressure gradient	Elasticity of stretched pulmonary connective-tissue fibers
Pulmonary surfactant (which opposes alveolar surface tension)	Alveolar surface tension
Alveolar interdependence	

and thereby help keep it open (Fig. 13–17b). This phenomenon, which can be likened to a stalemated "tug of war" between adjacent alveoli, is termed *interdependence*.

The opposing forces acting on the lung (that is, the forces keeping the alveoli open and the countering forces that promote alveolar collapse) are summarized in Table 13–4.

A deficiency of pulmonary surfactant is responsible for newborn respiratory distress syndrome.

The ability of the developing fetal lungs to synthesize pulmonary surfactant does not normally occur until late in pregnancy. Especially in an infant born prematurely, there may be insufficient pulmonary surfactant to reduce the alveolar surface tension to manageable levels. The resultant collection of symptoms that develop are referred to as **newborn respiratory distress syndrome.** Very strenuous inspiratory efforts are required to overcome the high surface tension in an attempt to inflate the poorly compliant lungs. Adding to the dilemma, the work of breathing is further increased because the alveoli , in the absence of surfactant, tend to collapse almost completely during each expiration. It is more difficult (requires a greater transmural pressure differential) to expand a collapsed alveolus by a given volume than to increase an already partially expanded alveolus by the same volume. A similar situation exists when blowing up a new balloon. It takes more effort to blow in that first breath of air when starting to blow up a new balloon than to blow additional breaths into the already partially expanded balloon. With newborn respiratory distress syndrome, it is as though the infant must start blowing up a new balloon with every breath. Lung expansion may require transmural pressure gradients of 20 to 30 mm Hg (compared to the normal of 4 to 6 mm Hg) to overcome the collapse tendency of surfactant-deprived alveoli.

The problem is compounded by the fact that the newborn's muscles are still weak. The respiratory distress associated with surfactant deficiency may soon lead to death as breathing efforts become exhausting or inadequate to support sufficient gas exchange.

This life-threatening condition affects 30,000 to 50,000 newborns, primarily premature infants, each year in the United States. Until the surfactant-secreting cells mature sufficiently, therapy often includes forcing air into the baby's lungs at greater-than-atmospheric, or "positive," pressure. By artificially increasing the atmospheric pressure, a sufficient pressure gradient can be established to drive air into the lungs. Recent clinical studies have also demonstrated success in treating the condition by surfactant replacement.

The work of breathing normally requires only about 3% of total energy expenditure.

During normal quiet breathing, inspiration requires the respiratory muscles to work to expand the lungs against their elastic forces and to overcome airway resistance, whereas expiration is a passive process. Normally, the lungs are highly compliant and airway resistance is low, so only about 3% of the total energy expended by the body is used to accomplish quiet breathing. The work of breathing may be increased in four different situations:

1. *When pulmonary compliance is decreased*, more work is required to expand the lungs.
2. *When airway resistance is increased*, more work is required to achieve the greater pressure gradients necessary to overcome the resistance so that adequate airflow can occur.
3. *When elastic recoil is decreased*, passive expiration may be inadequate to expel the volume of air normally exhaled during quiet breathing. Thus the abdominal muscles must work to aid in emptying the lungs, even when the person is at rest.
4. *When there is a need for increased ventilation*, such as during exercise, more work is required to accomplish both a greater depth of breathing (a larger volume of air moving in and out with each breath) and a faster rate of breathing (more breaths per minute).

During strenuous exercise, the amount of energy required to power pulmonary ventilation may increase up to twenty-five-fold. However, because total energy expenditure by the body is increased up to fifteen-to twenty-fold during heavy exercise, the energy used to accomplish the increased ventilation still represents only about 5% of total energy expended. In contrast, in patients with poorly compliant lungs or obstructive lung disease, the energy required for breathing even at rest may increase to 30% of total energy expenditure. In such cases the individual's exercise ability is severely limited, breathing itself being an exhausting exercise.

Normally, the lungs contain about 2 to 2½ liters of air during the respiratory cycle but can be filled to over 5½ liters or emptied to about 1 liter.

On the average in healthy young adults, the maximum amount of air that the lungs can hold is about 5.7 liters in males and 4.2 liters in females. Anatomical build, age, the distensibility of the lungs, and the presence or absence of respiratory disease affect this total lung capacity. Normally, during quiet breathing, the lungs are not anywhere near maximally inflated nor are they deflated to their minimum volume. Thus the lungs normally remain moderately inflated throughout the respiratory cycle. At the end of a normal quiet expiration, the lungs still contain about 2,200 ml of air. During each normal breath, about 500 ml of air are inspired and the same quantity is expired, so during quiet breathing the lung volume varies between 2,200 ml at the end of expiration to 2,700 ml at the end of inspiration (Fig. 13–18). During maximal expiration, lung volume can be decreased to 1,200 ml in males and 1,000 ml in females, but the lungs can never be completely deflated.

The advantage in not being able to completely empty the lungs is that, even during maximal expiratory efforts, gas exchange can still continue between blood flowing through the lungs and the remaining alveolar air. Instead of the wide fluctuations that would occur in O_2 uptake and CO_2 removal by the blood if the lungs were to completely fill and empty with each breath, the gas content of the blood leaving the lungs for delivery to the tissues normally remains remarkably constant throughout the respiratory cycle.

The lungs cannot be completely emptied during maximal expiratory efforts for two reasons. First, when a person expires very forcefully, the pressures within the chest increase substantially—for example, an additional 30 mm Hg, bringing the intra-alveolar pressure to 791 mm Hg and the intrapleural pressure to 786 mm Hg (Fig. 13–19b). A transmural pressure gradient that exceeds the collapsing forces of the lung still exists, however, despite the fact that the intrapleural pressure greatly exceeds atmospheric pressure. Second, the small airways are compressed closed by the elevated intrapleural pressure, trapping air in the alveoli. As air flows through the airways to the outside down the air's pressure gradient, the pressure steadily diminishes downstream because of frictional losses. At the point at which intrapleural pressure exceeds the pressure within a small, nonrigid bronchiole, the airway is compressed shut. When all the small bronchioles are collapsed, no further air can be expired, even with additional expiratory effort.

Figure 13–18 Normal Range and Extremes of Lung Volume in Adult Male

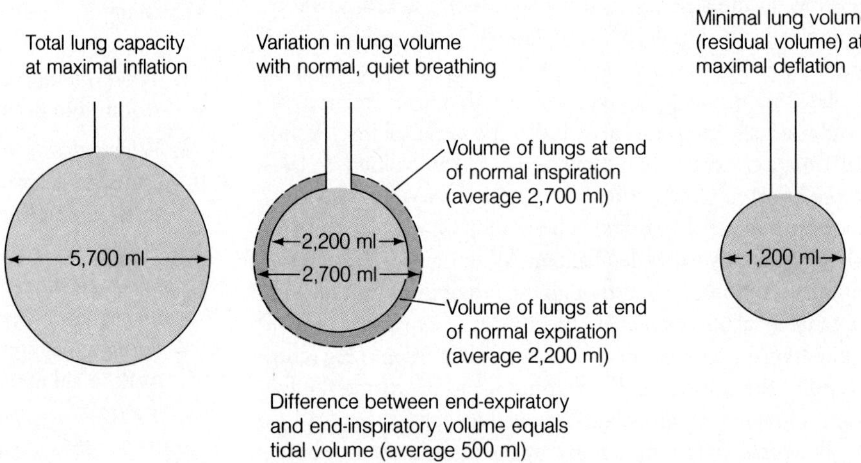

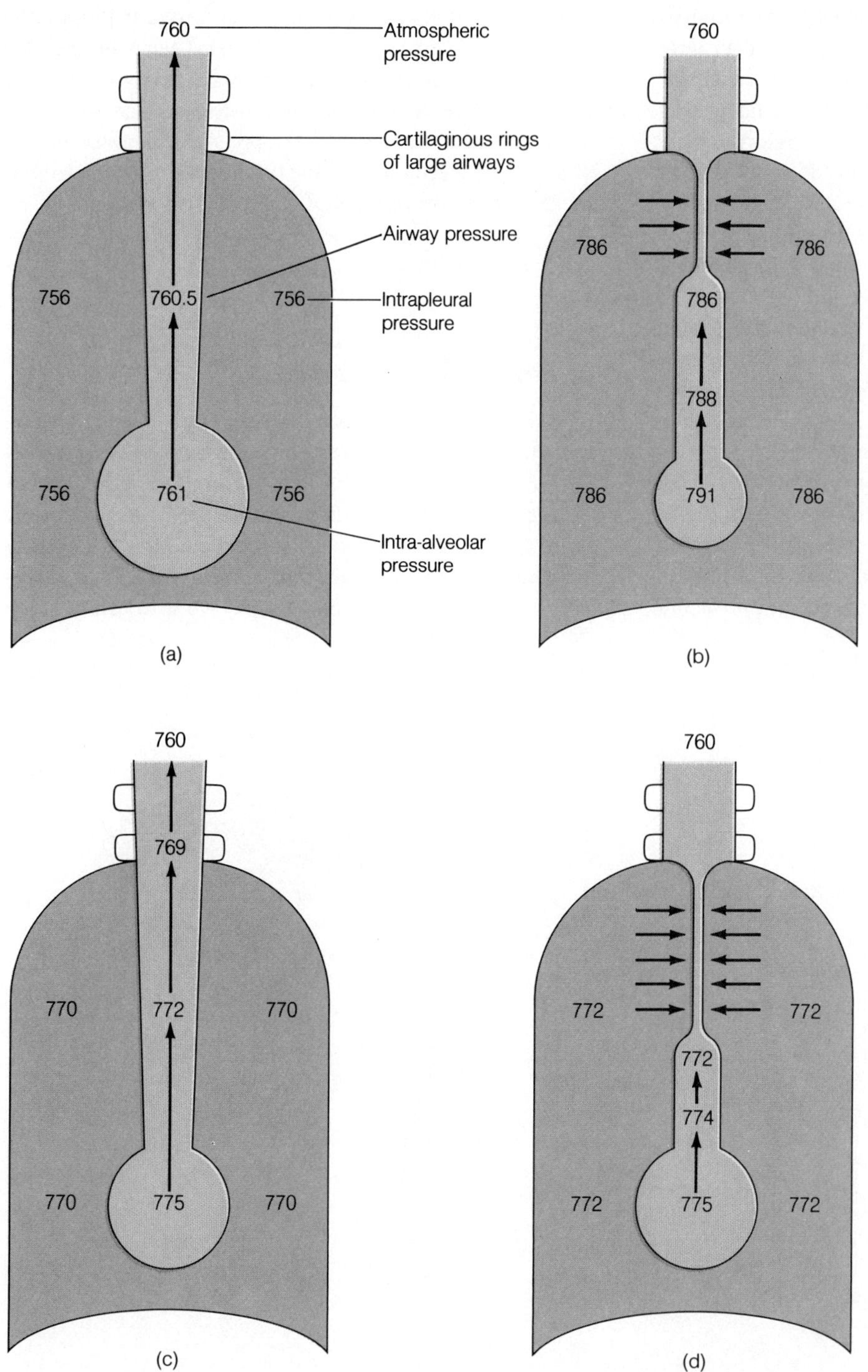

Figure 13–19 Airway Collapse during Forced Expiration *(a) Normal quiet breathing, during which airway resistance is low so there is little frictional loss of pressure within the airways. Intrapleural pressure remains less than airway pressure throughout the length of the airways, so the airways remain open. (b) Maximal forced expiration, during which both intra-alveolar and intrapleural pressures are markedly increased. When the airway pressure falls because of frictional losses to below the surrounding elevated intrapleural pressure, the small nonrigid airways are compressed closed, blocking further expiration of air through the airway. This occurs only at very low lung volumes in normal individuals. (c) Routine exercise. Even though intrapleural pressure is elevated during the active expiration accompanying routine vigorous activity, the airway pressure does not drop to below the intrapleural pressure until the level at which the airways are held open by cartilaginous rings, so airway collapse does not occur. (d) Obstructive lung disease. Premature airway collapse occurs for two reasons: (1) the pressure drop along the airways is magnified, and (2) the loss of lung elasticity, as in emphysema, causes the intrapleural pressure to be higher than normal.*

During normal respiration, the intrapleural pressure remains subatmospheric, even during expiration. Because the pressure within the bronchioles remains greater than intrapleural pressure throughout the entire length of the airway, airway collapse does not occur during quiet breathing (Fig. 13–19a). Even during routine vigorous respiratory activity, such as during exercise, airway pressure does not drop to below intrapleural pressure until the level at which the airways are held open by cartilaginous rings (Fig. 13–19c). Therefore, in a healthy lung, airway collapse occurs only at low lung volumes during maximal expiratory efforts.

This outflow-limiting mechanism can be exaggerated, however, by several factors that lead to premature airway compression and trapping of excessive amounts of air in the alveoli. First, the pressure drop along the airways is magnified by the increased resistance associated with obstructive lung diseases. Second, when the lung loses some of its elastic recoil (as occurs with emphysema), the difference between the intrapleural pressure and intra-alveolar pressure is reduced. Because the more compliant lungs do not pull away from the thoracic wall as vigorously as healthy lungs do, the intrapleural pressure is higher than normal in comparison to the intra-alveolar pressure. Consequently, as air leaves the lungs, the pressure in the airways is more likely to fall below the elevated intrapleural pressure within the nonrigid airway level of the respiratory tree, so these small airways are more apt to be pinched closed (Fig. 13–19d). Trapping of excessive amounts of air in the alveoli behind the compressed bronchiolar segments reduces the amount of gas exchanged between the alveoli and atmosphere. Therefore, less alveolar air is "freshened" with each breath.

Various lung volumes and capacities can be determined by spirometry.

The changes in lung volume that occur with different respiratory efforts can be measured using a **spirometer.** Basically, a spirometer consists of an air-filled drum floating in a water-filled chamber. As the person breathes air in and out of the drum through a tube connecting the mouth to the air chamber, the drum rises and falls in the water chamber (Fig. 13–20). This rise and fall can be recorded as a **spirogram,** which

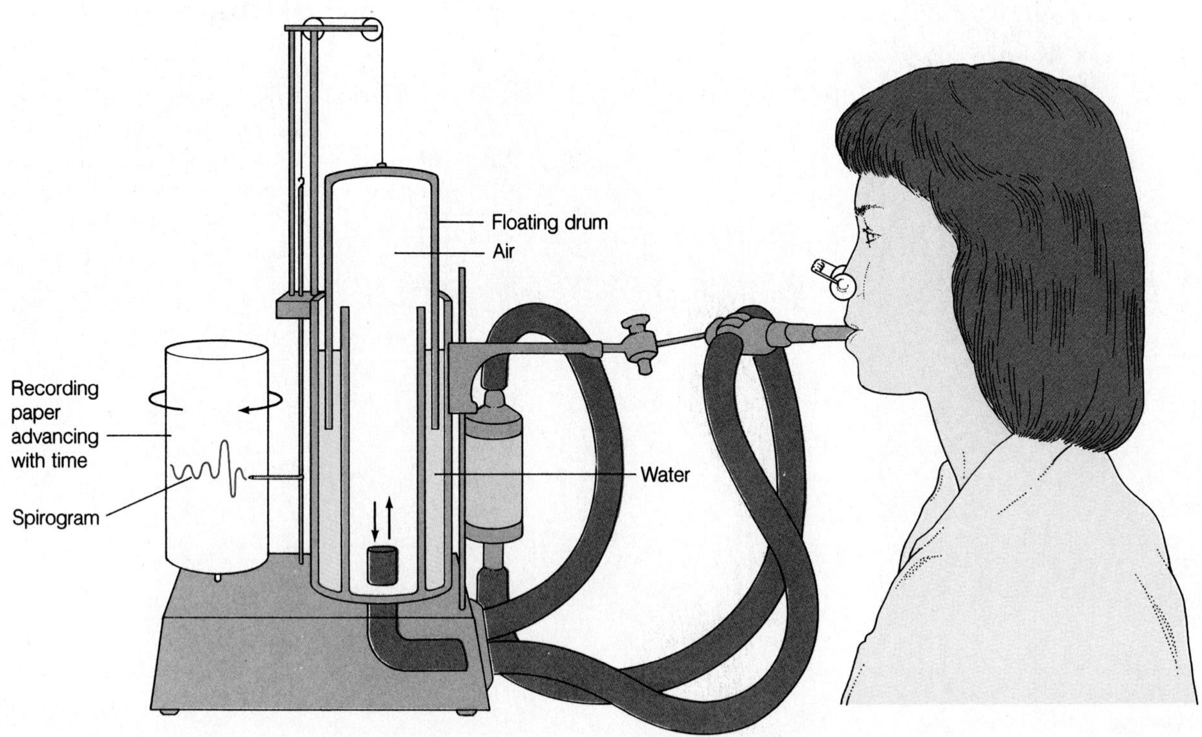

Figure 13–20 A Spirometer *See text for an explanation of this device.*

is calibrated to volume changes. The pen records inspiration as an upward deflection, whereas the pen moves downward when the subject expires.

Figure 13–21 is a hypothetical example of a spirogram in a healthy young adult male. Generally, the values are lower for females. The following lung volumes and lung capacities (a lung capacity is a sum of two or more lung volumes) can be determined:

□ **Tidal volume (TV)**—the volume of air entering or leaving the lungs during a single breath. Average value under resting conditions = 500 ml.

□ **Inspiratory reserve volume (IRV)**—the extra volume of air that can be maximally inspired over and above the tidal volume, accomplished by maximal contraction of the diaphragm, external intercostal muscles, and accessory inspiratory muscles. Average value = 3,000 ml.

□ **Inspiratory capacity (IC)**—the maximum volume of air that can be inspired at the end of a normal expiration (IC = IRV + TV). Average value = 3,500 ml.

□ **Expiratory reserve volume (ERV)**—the extra volume of air that can be actively expired by maximal contraction of the expiratory muscles beyond that normally passively expired at the end of a tidal volume. Average value = 1,000 ml.

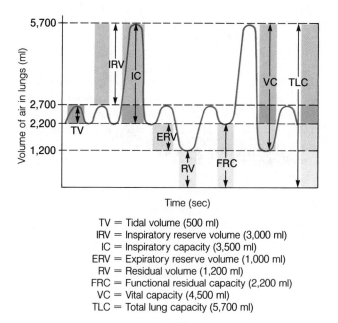

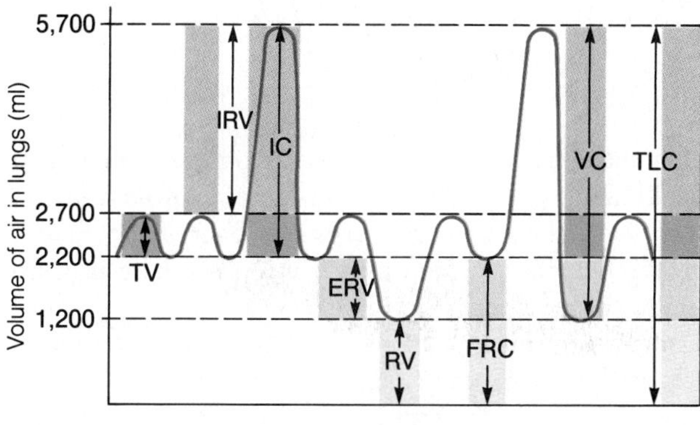

Figure 13–21 Normal Spirogram of Healthy Young Adult Male

□ **Residual volume (RV)**—the minimum volume of air remaining in the lungs even after a maximal expiration. Average value = 1,200 ml. The residual volume cannot be measured directly with a spirometer because this volume of air does not move in and out of the lungs. It can be determined indirectly, however, through gas-dilution techniques involving inspiration of a known quantity of a harmless tracer gas such as helium.

□ **Functional residual capacity (FRC)**—the volume of air in the lungs at the end of a normal passive expiration (FRC = ERV + RV). Average value = 2,200 ml.

□ **Vital capacity (VC)**—the maximum volume of air that can be moved in and out during a single breath. The subject first inspires maximally, then expires maximally (VC = IRV + TV + ERV). The vital capacity represents the maximum volume change possible within the lungs. It is rarely used because the maximal muscle contractions involved become exhausting, but it is a useful measure in ascertaining the functional capacity of the lungs. Average value = 4,500 ml.

□ **Total lung capacity (TLC)**—the maximum volume of air that the lungs can hold (TLC = VC + RV). Average value = 5,700 ml.

□ **Forced expiratory volume in one second (FEV_1)**—the volume of air that can be expired during the first second of expiration in a vital-capacity determination. Usually FEV_1 is about 80% of vital capacity; that is, normally 80% of the air that can be forcibly expired from maximally inflated lungs can be expired within 1 second. This measurement gives an indication of the possible airflow rate from the lungs.

Measurement of the lungs' various volumes and capacities is of more than pure academic interest, because such determinations provide a useful tool to the diagnostician in various respiratory-disease states. Two general categories of respiratory dysfunction yield abnormal results during spirometry—*obstructive* and *restrictive* lung disease (Fig. 13–22). However, you should not have the impression that these are the only two categories of respiratory dysfunction nor that spirometry is the only pulmonary-function test. Other conditions affecting respiratory function include: (1) diseases impairing diffusion of O_2 and CO_2 across the pulmonary membranes; (2) reduced ventilation because of mechanical failure, as with neuromuscular disorders affecting the respiratory muscles; (3) failure of adequate pulmonary blood flow; or (4) ventilation/perfusion abnormalities involving a poor matching of air and blood so that efficient gas exchange cannot occur. Some lung diseases are actually a complex mixture of different types of functional disturbances. The diagnostician relies on a variety of respiratory-function tests in addition to spirometry to evaluate what abnormalities are present, including x-ray examina-

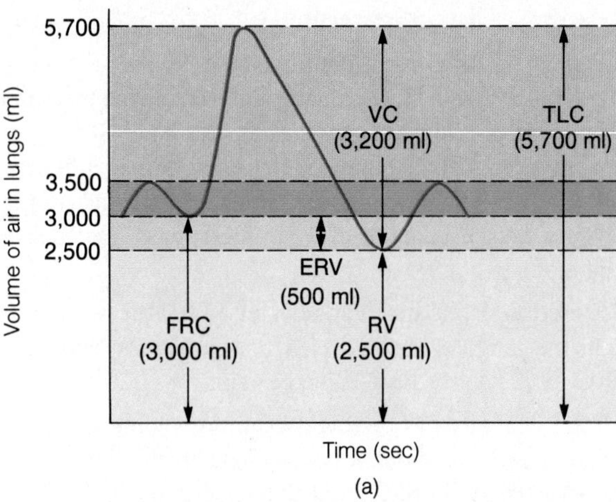

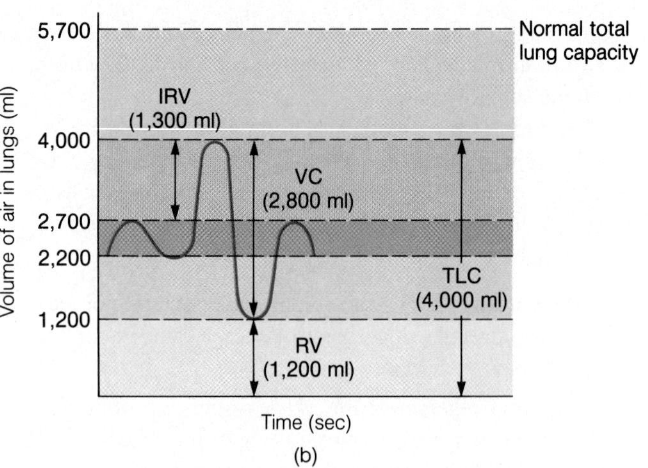

Figure 13–22 Categories of Respiratory Dysfunction That Yield Abnormal Spirograms *(a) Spirogram in obstructive lung disease. Since a patient with obstructive lung disease experiences more difficulty in emptying the lungs than in filling them, the total lung capacity is essentially normal but the functional residual capacity and the residual volume are elevated as a result of the additional air trapped in the lungs during expiration. Because the residual volume is increased, the vital capacity is reduced. With more air trapped in the lungs, less of the total lung capacity is available to be used in exchanging air with the atmosphere. Another common finding is a markedly reduced* FEV_1*, since the airflow rate is reduced by the airway obstruction. Even though both the vital capacity and the* FEV_1 *are reduced, the* FEV_1 *is reduced more markedly than is the vital capacity. As a result, the* FEV_1*-to-VC ratio is much lower than the normal 80%; that is, much less than 80% of the reduced vital capacity can be blown out during the first second. (b) Spirogram in restrictive lung disease. In this disease the lungs are less compliant than normal. Total lung capacity, inspiratory capacity, and vital capacity are reduced, since the lungs cannot be expanded as normal. Because the vital capacity is reduced, less air is available to be forced out in the first second—that is, there is simply less air to expire—so the* FEV_1 *may also be reduced. However, the percentage of the vital capacity that can be exhaled within 1 sec is the normal 80% or an even higher percentage, because air can flow freely in the airways. Therefore, the* $FEV_1/VC\%$ *is particularly useful in distinguishing between obstructive and restrictive lung disease. Also, unlike in obstructive lung disease, the residual volume is usually normal in restrictive lung disease.*

tion, blood-gas determinations, and tests to measure the diffusion capacity of the alveolar capillary membrane.

Alveolar ventilation is less than pulmonary ventilation because of the presence of dead space.

Various changes in volume represent only one factor in the determination of **pulmonary ventilation,** which is the amount of air breathed in and out in one minute. The other important factor is **respiratory rate,** which averages twelve breaths per minute.

$$\underset{\text{(ml/min)}}{\text{Pulmonary ventilation}} = \underset{\text{(ml/breath)}}{\text{Tidal volume}} \times \underset{\text{(breaths/min)}}{\text{Respiratory rate}}$$

At an average tidal volume of 500 ml/breath and a respiratory rate of 12 breaths per min, pulmonary ventilation is 6,000 ml or 6 liters of air breathed in and out in one minute under resting conditions. For a brief period of time, a healthy young adult male can voluntarily increase his total pulmonary ventilation twenty-five-fold, to 150 liters/min. To increase pulmonary ventilation, both tidal volume and respiratory rate increase, but depth of breathing is increased more than frequency of breathing.

When increasing pulmonary ventilation, it is more advantageous to have a greater increase in tidal volume than in respiratory rate because of the presence of anatomic dead space. Not all of the inspired air gets down to the site of gas exchange in the alveoli. Part of it remains in the conducting airways, where it is not available for gas exchange. The volume of the conducting passages in an adult averages about 150 ml. This volume is considered to be **anatomic dead space** because air within these conducting airways is useless for exchange purposes. Anatomic dead space has a pronounced effect on the efficiency of pulmonary ventilation. In effect, even though 500 ml are moved in and out with each breath, only 350 ml of air are actually exchanged between the atmosphere and al-

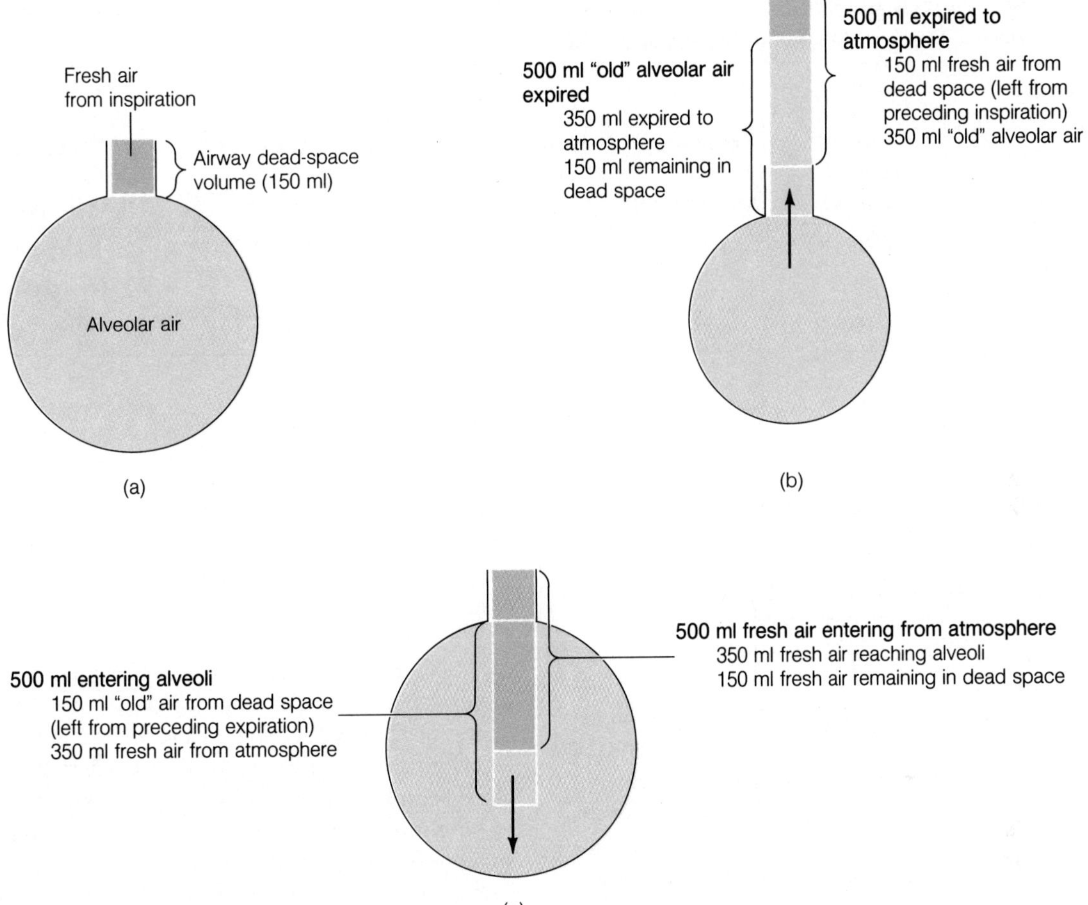

Figure 13–23 Effect of Dead-Space Volume on Exchange of Tidal Volume between Atmosphere and Alveoli *Even though 500 ml of air move in and out between the atmosphere and respiratory system and 500 ml move in and out of the alveoli with each breath, only 350 ml are actually exchanged between the atmosphere and the alveoli because of the presence of anatomic dead space (the volume of air in the respiratory airways). (a) After inspiration, before expiration. At the end of inspiration, the respiratory airways are filled with 150 ml of fresh atmospheric air from the inspiration. (b) During expiration. During the subsequent expiration, 500 ml of air are expired to the atmosphere. The first 150 ml expired are of the fresh air that was retained in the airways and never used. The remaining 350 ml are of "old" alveolar air that has participated in gas exchange with the blood. During the same expiration, 500 ml of gas also leave the alveoli. The first 350 ml are expired to the atmosphere; the other 150 ml of old alveolar air never reach the outside but remain in the conducting airways. (c) During inspiration. On the next inspiration, 500 ml of gas enter the alveoli. The first 150 ml to enter the alveoli are of the old alveolar air that remained in the dead space during the preceding expiration. The other 350 ml entering the alveoli are of fresh air inspired from the atmosphere. Simultaneously, 500 ml of air enter from the atmosphere. The first 350 ml of atmospheric air reach the alveoli; the other 150 ml remain in the conducting airways to be expired without benefit of being exchanged with the blood, as the cycle repeats itself.*

veoli for every tidal volume because of the 150 ml volume occupied by the anatomic dead space (Fig. 13–23).

Since the amount of atmospheric air actually available for exchange with the blood is of more importance than the total amount breathed in and out, **alveolar ventilation**—the volume of air exchanged between the atmosphere and alveoli per minute—is more important than pulmonary ventilation. In determining alveolar ventilation, the amount of wasted air moved in and out through the anatomic dead space must be taken into account, as follows:

Table 13–5 Effect of Different Breathing Patterns on Alveolar Ventilation

Breathing Pattern	Tidal Volume (ml/min)	Respiratory Rate (breaths/min)	Dead-Space Volume (ml)	Pulmonary Ventilation (ml/min) Equals Tidal Volume Times Respiratory Rate	Alveolar Ventilation (ml/min) Equals (Tidal Volume Minus Dead-Space Volume) Times Respiratory Rate
Normal, quiet breathing	500	12	150	6,000	4,200
Deep, slow breathing	1,200	5	150	6,000	5,250
Shallow, rapid breathing	150	40	150	6,000	0

$$\text{Alveolar ventilation} = (\text{Tidal volume} - \text{Dead-space volume}) \times \text{Respiratory rate}$$

With quiet breathing, alveolar ventilation is 4,200 ml/min [(500 ml/breath − 150 ml dead-space volume) × 12 breaths/min = 4,200 ml/min], whereas pulmonary ventilation is 6,000 ml/min.

To emphasize how important dead-space volume is in determining the magnitude of alveolar ventilation, examine the effect of various breathing patterns on alveolar ventilation in Table 13–5. If a person deliberately breathes deeply (for example, a tidal volume of 1,200 ml) and slowly (for example, a respiratory rate of 5 breaths/min), pulmonary ventilation is 6,000 ml/min, the same as when breathing normally, but alveolar ventilation is increased to 5,250 ml/min compared to the normal of 4,200 ml/min. In contrast, if a person were to deliberately breath shallowly (for example, a tidal volume of 150 ml) and rapidly (a frequency of 40 breaths/min), pulmonary ventilation would still be 6,000 ml/min; however, alveolar ventilation would be 0 ml/min. In effect the person would only be drawing air in and out of the anatomic dead space without any atmospheric air being exchanged with the alveoli, where it could be useful. Such a breathing pattern could be voluntarily maintained for only a few minutes before the person lost consciousness, at which time normal breathing would resume.

When pulmonary ventilation is increased during exercise, it should now be apparent why it is valuable that a correspondingly larger increase in depth of breathing than in rate of breathing is reflexly brought about. It is the most efficient means of elevating alveolar ventilation. When tidal volume is increased, the entire increase goes toward elevating alveolar ventilation, yet an increase in respiratory rate does not go entirely toward increasing alveolar ventilation. When respiratory rate is increased, the frequency with which air is wasted in the dead space is also increased, because a portion of *each* breath must move in and out of the dead space. As needs vary, ventilation is normally adjusted to a tidal volume and respiratory rate that meet those needs most efficiently in terms of energy cost.

We have assumed that air entering the alveoli exchanges O_2 and CO_2 with pulmonary blood. However, the match between air and blood is not always perfect, because not all alveoli are equally ventilated with air and perfused with blood. Any ventilated alveoli that do not participate in gas exchange with blood because they are inadequately perfused are considered to be **alveolar dead space. Total** or **physiological dead space** refers to the anatomic dead space plus alveolar dead space, both of which represent ventilated space that is functionally wasted. In normal persons, alveolar dead space is quite small and of no importance, but it can be increased to even lethal levels in several types of pulmonary disease. When alveolar dead space is increased, the tidal volume must be correspondingly increased, if possible, so that the functional portion of the alveoli (that which is non dead space) can still receive a normal tidal volume's worth of air.

Gas Exchange

Gases move down partial pressure gradients.

The ultimate purpose of ventilation is to provide a continual supply of fresh O_2 for pick-up by the blood and to constantly remove CO_2 unloaded from the blood. The blood acts as a transport system for O_2 and CO_2 between the lungs and tissues, with the tissue cells extracting O_2 from the blood and eliminating CO_2 into it. Gaseous exchange at both the pulmonary-capillary and tissue-capillary level involves simple passive diffusion of O_2 and CO_2 down partial pressure gradients. There are no active transport mechanisms for these gases.

Atmospheric air is a mixture of gases that contains about 80% nitrogen (N_2) and 20% O_2, with CO_2, H_2O vapor, other gases, and pollutants normally being almost negligible percentage-wise in dry air. Altogether, these gases exert a total atmospheric pressure of 760 mm Hg at sea level. According to **Dalton's law,** this total pressure is equal to the sum of the pressures that each gas in the mixture partially contributes. The pressure exerted by a particular gas is directly proportional to the percentage of that gas in the total air mixture. Every gas molecule, no matter what its size, exerts the same amount of pressure; for example, a N_2 molecule exerts the same pressure as an O_2 molecule. Since 80% of the air consists of N_2 molecules, 80% of the 760 mm Hg atmospheric pressure, or 608 mm Hg, is exerted by the N_2 molecules. Similarly, since O_2 represents 20% of the atmosphere, 20% of the 760 mm Hg atmospheric pressure, or 152 mm Hg, is exerted by O_2 (Fig. 13–24). The individual pressure exerted independently by a particular gas within a mixture is known as its **partial pressure,** designated by P_{gas}. Thus the partial pressure of O_2 in atmospheric air, $\mathbf{P_{O_2}}$, is normally 152 mm Hg. The atmospheric partial pressure of CO_2, $\mathbf{P_{CO_2}}$, is negligible at 0.3 mm Hg.

Gases dissolved in a liquid such as blood or another body fluid are also considered to exert a partial pressure. The amount of a gas that will dissolve in the blood depends on the solubility of the gas in blood and on the partial pressure of the gas in the alveolar air to which the blood is exposed. Because the solubility of O_2 and CO_2 in blood remains constant, the amount of O_2 and CO_2 dissolved in the pulmonary-capillary blood is directly proportional to the alveolar P_{O_2} and P_{CO_2}. The alveolar partial pressure of a particular gas can be thought of as "holding" that gas in solution in the blood.

If, as is the case with O_2, the alveolar partial pressure of a gas is ***higher*** than the partial pressure of that gas in the blood entering the pulmonary capillaries, the higher alveolar partial pressure drives more O_2 into the blood. Oxygen diffuses from the alveoli and dissolves in the blood until blood P_{O_2} becomes equal to alveolar P_{O_2}. Conversely, if the alveolar partial pressure of a gas is ***lower*** than its partial pressure in the entering blood—the situation that exists for CO_2—the lower alveolar partial pressure permits some of the CO_2 to escape from solution (that is, to no longer be dissolved) in the blood. As CO_2 comes out of solution, it diffuses into the alveoli until blood P_{CO_2} equilibrates with alveolar P_{CO_2}. Such a difference in partial pressure between pulmonary blood and alveolar air is known as a **partial pressure gradient.** A gas always diffuses down its partial pressure gradient from the area of higher to the area of lower partial pressure, similar to diffusion down a concentration gradient.

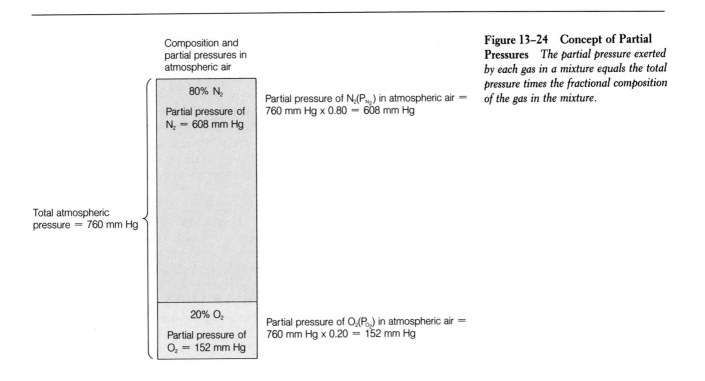

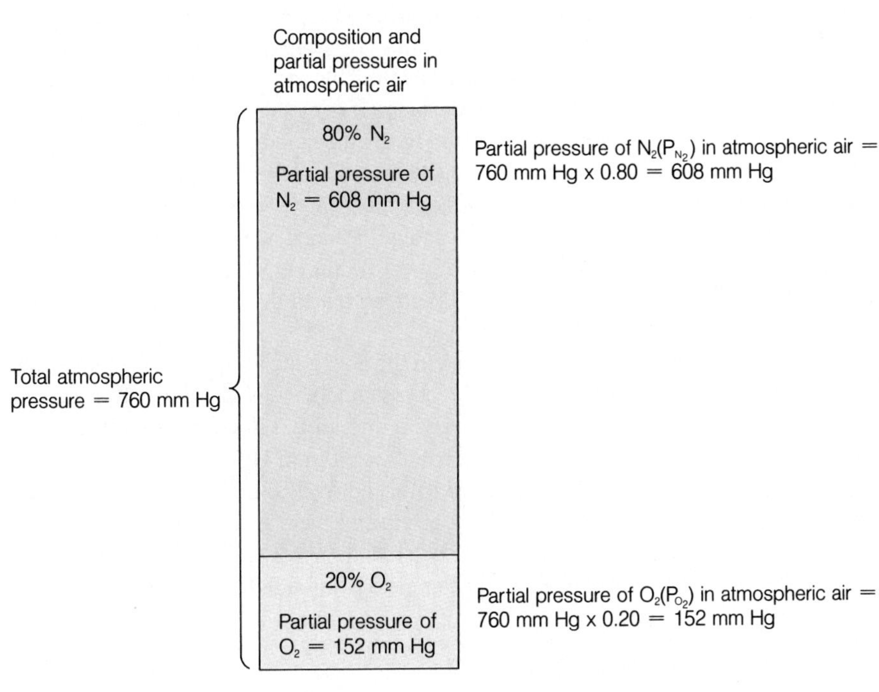

Figure 13–24 Concept of Partial Pressures *The partial pressure exerted by each gas in a mixture equals the total pressure times the fractional composition of the gas in the mixture.*

Oxygen enters and carbon dioxide leaves the blood in the lungs passively down partial pressure gradients.

Alveolar air is not of the same composition as inspired atmospheric air for two reasons. First, as soon as atmospheric air enters the respiratory passages, it becomes saturated with H_2O by exposure to the moist airways. Water vapor exerts a partial pressure just the same as any other gas. At body temperature, the partial pressure of H_2O vapor is 47 mm Hg. Humidification of inspired air in effect "dilutes" the partial pressure of the inspired gases by 47 mm Hg, because the sum of the partial pressures must total the atmospheric pressure of 760 mm Hg. In moist air, $P_{H_2O} = 47$ mm Hg, $P_{N_2} = 570$ mm Hg, and $P_{O_2} = 143$ mm Hg.

Alveolar P_{O_2} is further reduced since fresh inspired air is mixed with the old air that remained in the lungs and dead space at the end of the preceding expiration. Because of the volume of air retained in the lungs at the end of a normal expiration (the functional residual capacity), only about one-seventh of the total alveolar air is replaced by fresh atmospheric air with each normal breath. Thus at the end of inspiration, less than 15% of the air in the alveoli is fresh air. As a result of humidification and the small turnover of alveolar air, the average alveolar P_{O_2} is 100 mm Hg, compared to the atmospheric P_{O_2} of 152 mm Hg.

It is logical to think that alveolar P_{O_2} would increase during inspiration with the arrival of fresh air and would decrease during expiration. These fluctuations do not occur, however, for two reasons. First, only a small proportion of the total alveolar air is exchanged with each breath. The relatively small volume of high-P_{O_2} air that is inspired is quickly mixed with the much larger volume of retained alveolar air, which has a lower P_{O_2}. Thus the O_2 within the inspired air could only slightly elevate the total alveolar P_{O_2}. Even this potentially small elevation of P_{O_2} is diminished for another reason. Oxygen is continually moving by passive diffusion down its partial pressure gradient from the alveoli into the blood. The O_2 arriving in the alveoli in the newly inspired air simply replaces the O_2 diffusing out of the alveoli into the pulmonary capillaries. Therefore, the alveolar P_{O_2} remains relatively constant throughout the respiratory cycle. Because the pulmonary-blood P_{O_2} equilibrates with the alveolar P_{O_2}, the P_{O_2} of the blood likewise remains fairly constant. Accordingly, the amount of O_2 in the blood available to the tissues varies only slightly during the respiratory cycle.

A similar situation in reverse exists for CO_2. Carbon dioxide, which is continually produced by the body tissues as a metabolic waste product, is constantly added to the blood at the level of the systemic capillaries. In the pulmonary capillaries, CO_2 diffuses down its partial pressure gradient from the blood into the alveoli and is subsequently removed from the body during expiration. As with O_2, alveolar P_{CO_2} remains fairly constant instead of fluctuating from higher to lower levels between inspiration and expiration. This is because the tidal volume is small in relation to the total lung volume and because the rate of CO_2 entering the alveoli from the blood balances the rate at which it is removed by ventilation from the alveoli to the atmosphere. Alveolar P_{CO_2} normally remains at 40 mm Hg throughout the respiratory cycle. This is low in comparison to the alveolar P_{O_2} of 100 mm Hg. Ventilation constantly replenishes alveolar P_{O_2}, keeping it relatively high, and constantly removes CO_2, keeping alveolar P_{CO_2} relatively low. This, in turn, maintains the appropriate partial pressure gradients between the alveoli and blood to assure that O_2 enters the blood and CO_2 leaves the blood.

The blood entering the pulmonary capillaries is systemic venous blood pumped to the lungs through the pulmonary arteries. This blood, having just returned from the body tissues, is relatively low in O_2, with a P_{O_2} of 40 mm Hg, and is relatively high in CO_2, with a P_{CO_2} of 46 mm Hg. As this blood flows through the pulmonary capillaries, it is exposed to alveolar air (Fig. 13–25). Since the alveolar P_{O_2} at 100 mm Hg is higher than the P_{O_2} of 40 mm Hg in the blood entering the lungs, O_2 readily diffuses down its partial pressure gradient from the alveoli into the blood until no further gradient exists. As the blood leaves the pulmonary capillaries, it has a P_{O_2} equal to alveolar P_{O_2} at 100 mm Hg. The partial pressure gradient for CO_2 is in the opposite direction. Blood entering the pulmonary capillaries has a P_{CO_2} of 46 mm Hg, whereas alveolar P_{CO_2} is only 40 mm Hg. Carbon dioxide diffuses from the blood into the alveoli until blood P_{CO_2} equilibrates with alveolar P_{CO_2}. Thus the blood leaving the pulmonary capillaries has a P_{CO_2} of 40 mm Hg. As the blood passes through the lungs, it picks up O_2 and gives up CO_2 simply by diffusion down partial pressure gradients that exist between the blood and alveoli. After leaving the lungs, the blood, which now has a P_{O_2} of 100 mm Hg and a P_{CO_2} of 40 mm Hg, is returned to the heart to be subsequently pumped out to the body tissues as systemic arterial blood.

Note that blood returning to the lungs from the tissues still contains O_2 (P_{O_2} of systemic venous blood = 40 mm Hg) and that blood leaving the lungs still contains CO_2 (P_{CO_2} of systemic arterial blood = 40 mm Hg). The extra O_2 carried in the blood beyond that normally used represents an immediately available O_2 reserve that can be tapped by the tissue cells whenever their O_2 demands increase. The CO_2 remaining in the blood even after passage through the lungs plays an impor-

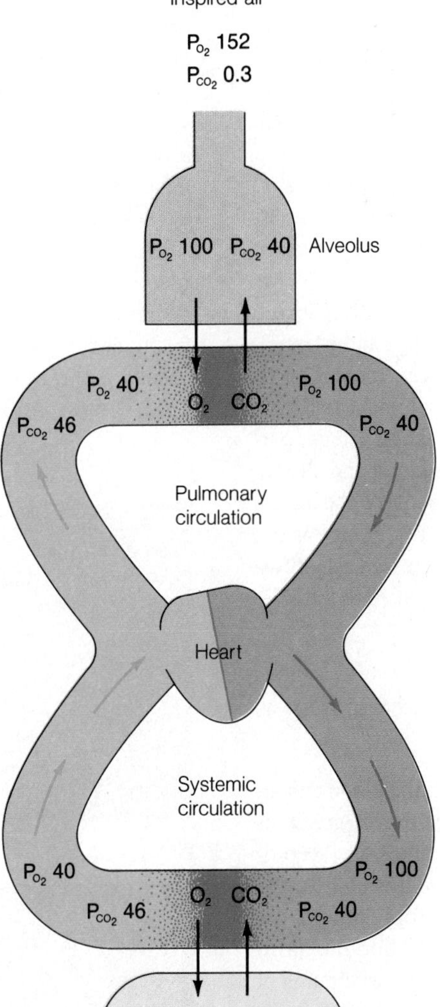

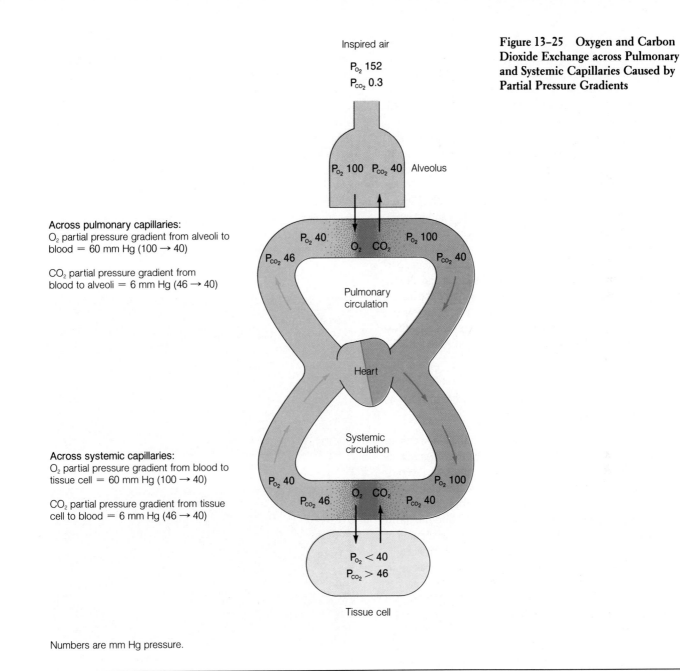

Figure 13-25 Oxygen and Carbon Dioxide Exchange across Pulmonary and Systemic Capillaries Caused by Partial Pressure Gradients

Across pulmonary capillaries:
O_2 partial pressure gradient from alveoli to blood = 60 mm Hg (100 → 40)

CO_2 partial pressure gradient from blood to alveoli = 6 mm Hg (46 → 40)

Across systemic capillaries:
O_2 partial pressure gradient from blood to tissue cell = 60 mm Hg (100 → 40)

CO_2 partial pressure gradient from tissue cell to blood = 6 mm Hg (46 → 40)

Numbers are mm Hg pressure.

---

tant role in the acid-base balance of the body, because CO_2 generates carbonic acid. Furthermore, arterial P_{CO_2} is important in driving respiration, which will be described later.

The amount of O_2 picked up in the lungs balances the amount extracted and used by the tissues. When the tissues metabolize more actively (for example, during exercise), more O_2 is extracted from the blood at the tissue level, reducing the systemic venous P_{O_2} even lower than 40 mm Hg—for example, to a P_{O_2} of 30 mm Hg. When this blood returns to the lungs, a larger-than-normal P_{O_2} gradient exists between the newly entering blood and alveolar air. The difference in P_{O_2} between the alveoli and blood is now 70 mm Hg (alveolar P_{O_2} of 100 mm Hg and blood P_{O_2} of 30 mm Hg), compared to the normal P_{O_2} gradient of 60 mm Hg (alveolar P_{O_2} of 100 mm

Hg and blood P_{O_2} of 40 mm Hg). Therefore, more O_2 diffuses from the alveoli into the blood down the larger partial pressure gradient before blood P_{O_2} equals alveolar P_{O_2}. This additional transfer of O_2 into the blood replaces the increased amount of O_2 consumed, so O_2 uptake matches O_2 use even when O_2 consumption increases. At the same time that more O_2 is diffusing from the alveoli into the blood because of the increased partial pressure gradient, ventilation is stimulated so that O_2 enters the alveoli more rapidly from the atmosphere to replace the O_2 diffusing into the blood.

Similarly, the amount of CO_2 given up to the alveoli from the blood balances the amount of CO_2 picked up at the tissues. If the tissues are metabolizing more actively, more CO_2 is picked up by the blood at the level of the systemic capillaries, increasing the systemic venous P_{CO_2} and creating a larger partial pressure gradient for CO_2 between the blood and alveoli. As a result, more CO_2 diffuses from the blood into the alveoli before blood P_{CO_2} equilibrates with alveolar P_{CO_2}. Once again, the increase in ventilation associated with increased activity assures that the increased amounts of CO_2 delivered to the alveoli are blown off to the atmosphere.

Factors other than the partial pressure gradient influence the rate of gas transfer.

We have been discussing diffusion between the blood and alveoli of O_2 and CO_2 as if these gases' partial pressure gradients were the sole determinants in their rates of diffusion. According to Fick's law of diffusion, the rate of diffusion of a gas through a sheet of tissue is not only directly proportional to the partial pressure gradient but also is proportional to the diffusion coefficient of the particular gas and to the surface area of the membrane, and it is inversely proportional to the thickness of the membrane (see p. 71). In a normal situation, changes in the rate of gas exchange are determined primarily by changes in partial pressure gradients between the blood and alveoli, because the other factors are relatively constant. The diffusion coefficient is a constant value relating to the ease with which the gas in question can dissolve in the respiratory membranes, and, except under special circumstances, the surface area and thickness of the respiratory membranes remain essentially constant.

The surface area available for exchange can be physiologically increased during exercise to enhance the rate of gas transfer. During resting conditions, some of the pulmonary capillaries are typically closed, because the normally low pressure of the pulmonary circulation is inadequate to keep all of the capillaries open. During exercise, when the pulmonary blood pressure is raised as a result of increased cardiac output, many of the previously closed pulmonary capillaries are forced open. This increases the surface area of blood available for exchange. Furthermore, the alveolar membranes are stretched further than normal during exercise because of the larger tidal volumes (deeper breathing). Such stretching increases the alveolar surface area and decreases the thickness of the alveolar membrane. Collectively, these changes expedite gas exchange during exercise.

On the other hand, several pathological conditions can markedly reduce the pulmonary surface area, which, in turn, decreases the rate of gas exchange. Most notably, surface area is reduced in emphysema because of the loss of many of the alveolar walls, which results in larger but fewer chambers (Fig. 13–26). Loss of surface area for exchange is likewise associ-

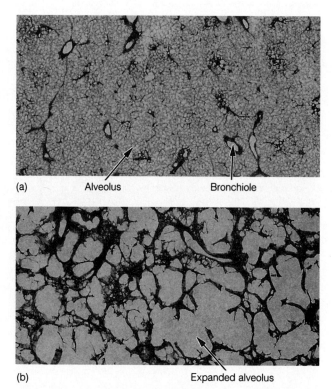

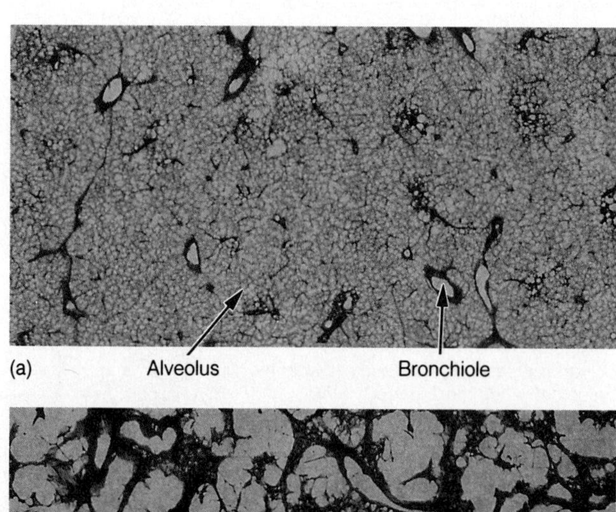

Figure 13–26 Comparison of Normal and Emphysematous Lung Tissue *(a) A thin section of lung tissue from a normal individual. (b) A thin section of lung tissue from a patient with emphysema. These are photographs of the actual tissue, unmagnified. Note the loss of alveolar walls in the emphysematous lung tissue, resulting in larger but fewer alveolar chambers.*

SOURCE: Val Vallyathan, Associate Professor, Department of Pathology, School of Medicine, West Virginia University and Research Physiologist, National Institute of Occupational Safety and Health.

ated with atelectic regions of the lung and also results when part of the lung tissue is surgically removed—for example, in the treatment of lung cancer.

Inadequate gas exchange can also occur when the thickness of the barrier separating the air and blood is pathologically increased. As the thickness increases, the rate of gas transfer decreases, because it takes longer for a gas to diffuse through the greater thickness. Thickness increases in: (1) pulmonary edema, an excess accumulation of interstitial fluid between the alveoli and pulmonary capillaries caused by pulmonary inflammation or left-sided congestive heart failure; (2) pulmonary fibrosis involving replacement of delicate lung tissue with thick fibrous tissue in response to certain chronic irritants; and (3) pneumonia, which is characterized by inflammatory fluid accumulation within or around the alveoli. Most commonly pneumonia is due to bacterial or viral infection of the lungs, but it may also arise from accidental aspiration (breathing in) of food, vomitus, or chemical agents.

The rate of gas transfer is directly proportional to the diffusion coefficient (D), a constant value related to the solubility of a particular gas in the lung tissues and to its molecular weight ($D \propto sol/\sqrt{mw}$). The diffusion coefficient for CO_2 is twenty times that of O_2, because CO_2 is much more soluble in body tissues than is O_2. The rate of CO_2 diffusion across the respiratory membranes is therefore twenty times more rapid than that of O_2 for a given partial pressure gradient. This difference in diffusion coefficients is normally offset by the difference in partial pressure gradients that exist for O_2 and CO_2 across the alveolar capillary membrane. The CO_2 partial pressure gradient is 6 mm Hg (P_{CO_2} of 46 mm Hg in the blood; P_{CO_2} of 40 mm Hg in the alveoli), compared to the O_2 gradient of 60 mm Hg (P_{O_2} of 100 mm Hg in the alveoli; P_{O_2} of 40 mm Hg in the blood).

Normally, approximately equal amounts of O_2 and CO_2 are exchanged—a respiratory quotient's worth. Even though a given volume of blood spends three-fourths of a second passing through the pulmonary capillary bed, P_{O_2} and P_{CO_2} are usually both equilibrated with alveolar partial pressures by the time the blood has traversed only one-third the length of the pulmonary capillaries. This means that the lung normally has enormous diffusion reserves, a fact that becomes extremely important during heavy exercise. The time the blood spends in transit in the pulmonary capillaries is decreased as pulmonary blood flow increases with the greater cardiac output that accompanies exercise. Even when less time is available for exchange, blood P_{O_2} and P_{CO_2} are normally able to equilibrate with alveolar levels because of the lungs' diffusion reserves.

In a diseased lung in which diffusion is impeded because the surface area is decreased or the blood-air barrier is thickened, O_2 transfer is usually more seriously impaired than is CO_2 transfer because of the larger CO_2 diffusion coefficient. By the time the blood reaches the end of the pulmonary-capillary network, it is more likely to have equilibrated with alveolar P_{CO_2} than with alveolar P_{O_2}, because CO_2 can diffuse more rapidly through the respiratory barrier. In milder conditions, diffusion of both O_2 and CO_2 might remain adequate at rest, but during exercise, when pulmonary transit time is decreased, the blood gases, especially O_2, may not have completely equilibrated with the alveolar gases before the blood leaves the lungs.

Gas exchange across the systemic capillaries also occurs down partial pressure gradients.

As at the pulmonary capillaries, O_2 and CO_2 move between the systemic capillary blood and the tissue cells by simple passive diffusion down partial pressure gradients. Refer again to Figure 13–25. The arterial blood that reaches the systemic capillaries is essentially the same blood that left the lungs by means of the pulmonary veins, because the only two places at which gas exchange can take place in the entire circulatory system are the pulmonary capillaries and the systemic capillaries. The arterial P_{O_2} is 100 mm Hg and the arterial P_{CO_2} is 40 mm Hg, the same as alveolar P_{O_2} and P_{CO_2}.

The cells constantly consume O_2 and produce CO_2 through oxidative metabolism. Cellular P_{O_2} averages about 40 mm Hg and P_{CO_2} about 46 mm Hg, although these values are highly variable, depending on the level of cellular metabolic activity. Because of the cellular plasma membrane's high permeability to O_2 and CO_2, the interstitial fluid surrounding the cells has the same partial pressures as the cells. Oxygen moves down its partial pressure gradient from the entering systemic capillary blood (P_{O_2} = 100 mm Hg) into the adjacent cells (P_{O_2} = 40 mm Hg) by diffusing through the capillary wall, the interstitial fluid, and the plasma membrane until equilibrium is reached. Therefore, the P_{O_2} of venous blood leaving the systemic capillaries is equal to the tissue P_{O_2} at 40 mm Hg. The reverse situation exists for CO_2. Carbon dioxide rapidly diffuses out of the cells (P_{CO_2} = 46 mm Hg) into the interstitial fluid and then into the entering capillary blood (P_{CO_2} = 40 mm Hg) down the partial pressure gradient created by the ongoing production of CO_2. Transfer of CO_2 continues until blood P_{CO_2} equilibrates with tissue P_{CO_2}.[1] Ac-

[1]Actually, the partial pressures of the systemic blood gases never completely equilibrate with tissue P_{O_2} and P_{CO_2}. Because the cells are constantly consuming O_2 and producing CO_2, the tissue P_{O_2} is always slightly less than the P_{O_2} of the blood leaving the systemic capillaries, and the tissue P_{CO_2} always slightly exceeds the systemic venous P_{CO_2}.

cordingly, the blood leaving the systemic capillaries has a P_{CO_2} of 46 mm Hg. This systemic venous blood that is relatively low in O_2 (P_{O_2} = 40 mm Hg) and relatively high in CO_2 (P_{CO_2} = 46 mm Hg) returns to the heart and is subsequently pumped to the lungs as the cycle repeats itself.

The more actively a tissue is metabolizing, the lower the cellular P_{O_2} falls and the higher the cellular P_{CO_2} rises. As a consequence of the larger blood-to-cell partial pressure gradients, more O_2 diffuses from the blood into the cells and more CO_2 moves in the opposite direction before blood P_{O_2} and P_{CO_2} achieve equilibrium with the surrounding cells. Thus the amount of O_2 transferred to the cells and the amount of CO_2 carried away from the cells depends on the rate of cellular metabolism.

Note that net diffusion of O_2 occurs first between the alveoli and blood and then between the blood and tissues due to the O_2 partial pressure gradients created by continuous utilization of O_2 in the cells and continuous replenishment of fresh alveolar O_2 provided by pulmonary ventilation. Net diffusion of CO_2 occurs in the reverse direction, first between the tissues and blood and then between the blood and alveoli, due to the CO_2 partial pressure gradients created by continuous production of CO_2 in the cells and the continuous removal of alveolar CO_2 through the process of pulmonary ventilation (Fig. 13–27).

Gas Transport

Most O_2 in the blood is transported bound to hemoglobin.

Oxygen picked up by the blood at the lungs must be transported to the tissues for cellular utilization. Conversely, CO_2 produced at the cellular level must be transported to the lungs for elimination. We will first examine O_2 transport in the blood.

Oxygen is present in the blood in two forms: physically dissolved and chemically bound to hemoglobin (Table 13–6). Very little O_2 is physically dissolved in the plasma water because O_2 is poorly soluble in body fluids. The amount dissolved is directly proportional to the P_{O_2} of the blood; the higher the P_{O_2}, the more O_2 dissolved. At a normal arterial P_{O_2} of 100 mm Hg, only 3 ml of O_2 can dissolve in 1 liter of blood. This means that only 15 ml of O_2/min can be dissolved in the normal pulmonary blood flow of 5 liters/min (the resting cardiac output). Even under resting conditions, the cells consume 250 ml of O_2/min, and this may increase up to twenty-five-fold during strenuous exercise. To deliver the O_2 required by the tissues even at rest, the cardiac output would have to be 83.3 liters/min if O_2 could only be transported in

Figure 13–27 Net Diffusion Gradients for O_2 and CO_2 between Lungs and Tissues

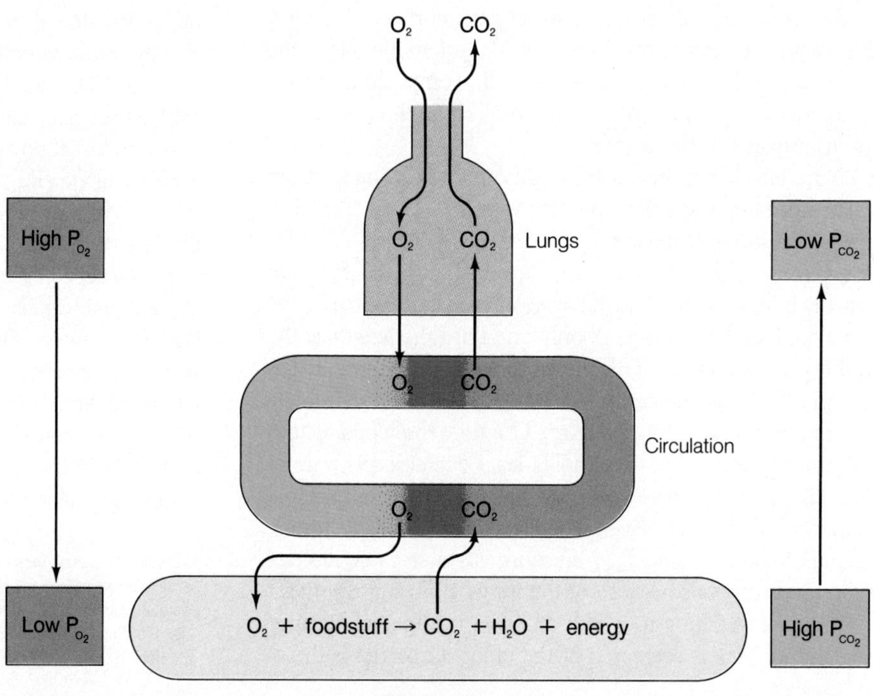

dissolved form. Obviously, there must be an additional mechanism of transporting O_2 to the tissues. This mechanism is hemoglobin (Hb). Only 1.5% of the O_2 in the blood is dissolved, the remaining 98.5% being transported in combination with hemoglobin. The O_2 bound to Hb does not contribute to the P_{O_2} of the blood. Because of this, blood P_{O_2} is not a measure of the total O_2 content of the blood but only of that portion of O_2 that is dissolved.

Hemoglobin, an iron-bearing protein molecule contained within the red blood cells, has the ability to form a loose, easily reversible combination with O_2 (see p. 345). When not combined with O_2, Hb is designated as **reduced hemoglobin;** when combined with O_2, it is called **oxyhemoglobin (HbO_2)**:

$$\underset{\text{Reduced hemoglobin}}{Hb + O_2} \rightleftharpoons \underset{\text{Oxyhemoglobin}}{HbO_2}$$

The normal Hb concentration averages 15 gm of Hb/100 ml of blood. At a normal cardiac output of 5 liters/min, the Hb contained within this blood can carry about 1,000 ml of O_2/min to the tissues, compared to only 15 ml/min as dissolved O_2.

We need to answer several important questions about the role of Hb in O_2 transport. What determines whether O_2 and Hb are combined or dissociated (separated)? Why does Hb combine with O_2 in the lungs and release O_2 at the tissues? How can a variable amount of O_2 be released at the tissue level, depending on the level of tissue activity? How can we talk about O_2 transfer between blood and surrounding tissues in terms of O_2 partial pressure gradients when 98.5% of the O_2 is bound to Hb and thus does not contribute to the P_{O_2} of the blood at all?

The P_{O_2} is the primary factor determining the percent hemoglobin saturation.

Each of the four atoms of iron within the heme portions of a hemoglobin molecule is able to combine with an O_2 molecule, so each Hb molecule can carry up to four molecules of O_2. Hemoglobin is considered to be *fully saturated* when all of the Hb present is carrying its maximum O_2 load. The **percent hemoglobin (% Hb) saturation,** a measure of the extent to which the Hb present is combined with O_2, can vary from 0% to 100%.

The most important factor determining the % Hb saturation is the P_{O_2} of the blood, which, in turn, is related to the concentration of O_2 physically dissolved in the blood. According to the **law of mass action,** if the concentration of one of the substances involved in a reversible reaction is increased, the reaction is driven toward the opposite side. Conversely, if the concentration of one of the substances is decreased, the reaction is driven toward that side. Applied to the reversible reaction involving Hb and O_2 ($Hb + O_2 \rightleftharpoons HbO_2$), when the blood P_{O_2} is increased, as it is in the pulmonary capillaries, the reaction is driven toward the right side of the equation, resulting in increased formation of HbO_2 (increased % Hb saturation). When the blood P_{O_2} is decreased, as it is in the systemic capillaries, the reaction is driven toward the left side of the equation. Oxygen is released from Hb as HbO_2 dissociates (decreased % Hb saturation). Thus, because of the difference in P_{O_2} at the lungs and other tissues, Hb automatically "loads up" on O_2 in the lungs, where fresh supplies of O_2 are continually being provided by ventilation, and "unloads" it in the tissues, which are constantly using up O_2.

Table 13–6 Methods of Gas Transport in the Blood

Gas	Method of Transport in Blood	Percent Carried in This Form
O_2	Physically dissolved	1.5
	Bound to hemoglobin	98.5
CO_2	Physically dissolved	10
	Bound to hemoglobin	30
	As bicarbonate (HCO_3^-)	60

However, the relationship between blood P_{O_2} and % Hb saturation is not directly linear, a point that is very important physiologically. Doubling the partial pressure does not double the % Hb saturation. Rather, the relationship between these variables is depicted by an S-shaped curve known as the **O_2-Hb dissociation** (or **saturation**) **curve** (Fig. 13–28). Note that at the upper end between a blood P_{O_2} of 60 and 100 mm Hg, the curve flattens off, or plateaus. Within this pressure range, a rise in P_{O_2} produces only a small increase in the extent to which Hb is bound with O_2. In contrast, in the blood P_{O_2} range of 0 to 60 mm Hg, a small change in P_{O_2} results in a large change in the extent to which Hb is combined with O_2, as depicted by the steep lower part of the curve. Both the upper plateau and lower steep portion of the curve have physiological significance.

SIGNIFICANCE OF THE PLATEAU PORTION OF THE O_2-Hb CURVE. The plateau portion of the curve is in the blood-P_{O_2} range that exists at the pulmonary capillaries where O_2 is being loaded onto Hb. The systemic arterial blood leaving the lungs, having equilibrated with alveolar P_{O_2}, normally has a P_{O_2} of 100 mm Hg. Note on the curve at a blood P_{O_2} of 100 mm Hg

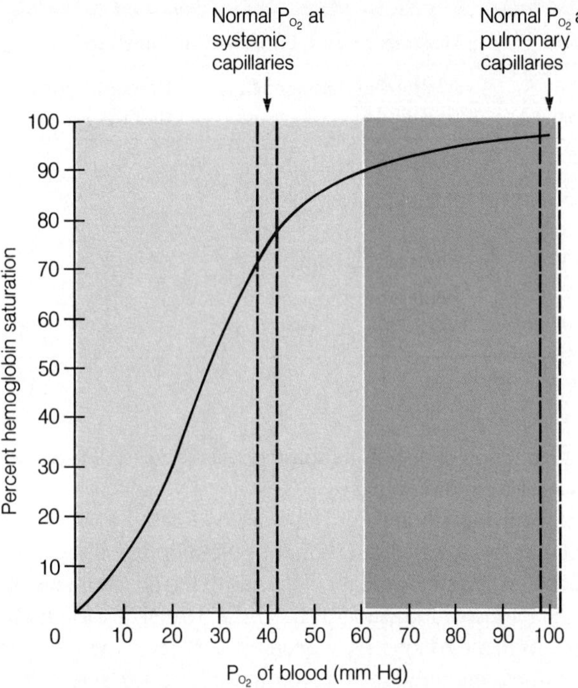

Figure 13–28 Oxygen-Hemoglobin (O_2-Hb) Dissociation (Saturation) Curve *The percent hemoglobin saturation is dependent on the P_{O_2} of the blood. The relationship between these two variables is depicted by an S-shaped curve with a plateau region between a blood P_{O_2} of 60 and 100 mm Hg and a steep portion between 0 and 60 mm Hg.*

that Hb is 97.5% saturated. Therefore, the Hb in the systemic arterial blood normally is almost fully saturated.

If the alveolar P_{O_2} and consequently the arterial P_{O_2} fall below normal, there is little reduction in the total amount of O_2 transported by the blood until the P_{O_2} falls below 60 mm Hg because of the plateau region of the curve. If the arterial P_{O_2} falls 40%, from 100 to 60 mm Hg, the concentration of dissolved O_2 as reflected by the P_{O_2} is likewise reduced 40%. At a blood P_{O_2} of 60 mm Hg, however, the % Hb saturation is still remarkably high at 90%. Accordingly, the total O_2 content of the blood is only slightly decreased despite the 40% reduction in P_{O_2} because Hb is still carrying an almost full load of O_2, and, as mentioned before, the vast majority of O_2 is transported by Hb rather than being dissolved. On the other hand, even if the blood P_{O_2} is greatly increased, say to 600 mm Hg by breathing pure O_2, very little additional O_2 is added to the blood. A small extra amount of O_2 dissolves, but the % Hb saturation can be maximally increased by only another 1½%, to 100% saturation. Therefore, in the P_{O_2} range between 60 to 600 mm Hg or even higher, there is only a 10% difference in the amount of O_2 carried by Hb. This provides a good margin of safety in O_2-carrying capacity of the blood.

Arterial P_{O_2} may be reduced because of pulmonary or circulatory diseases accompanied by ventilation/perfusion abnormalities or defective gas exchange. It may also fall in healthy individuals under two circumstances: (1) at high altitudes, where the total atmospheric pressure and hence the P_{O_2} of the inspired air are reduced; or (2) in O_2-deprived environments at sea level, such as would be encountered if someone were accidentally locked in a vault. Unless the arterial P_{O_2} becomes markedly reduced (falls below 60 mm Hg) in either pathological conditions or abnormal environmental circumstances, near-normal amounts of O_2 can still be carried to the tissues.

SIGNIFICANCE OF THE STEEP PORTION OF THE O_2-Hb CURVE. The steep portion of the curve between 0–60 mm Hg is in the blood-P_{O_2} range that exists at the systemic capillaries, where O_2 is being unloaded from Hb. In the systemic capillaries, the blood equilibrates with the surrounding tissue cells and interstitial fluid at an average P_{O_2} of 40 mm Hg. Note on Figure 13–28 that at a P_{O_2} of 40 mm Hg, the % Hb saturation is 75%. The blood arrived in the tissue capillaries at a PO_2 of 100 mm Hg with 97.5% Hb saturation. Since Hb can only be 75% saturated at the P_{O_2} of 40 mm Hg in the systemic capillaries, 22.5% of the HbO_2 must dissociate, yielding reduced Hb and O_2. This released O_2 is free to diffuse down its partial pressure gradient from the red blood cells through the plasma and interstitial fluid into the tissue cells.

The Hb in the venous blood returning to the lungs is still normally 75% saturated. If the tissue cells are metabolizing more actively, the P_{O_2} of the systemic-capillary blood falls (for example, from 40 to 20 mm Hg) because the cells are consuming O_2 more rapidly. Note on the curve that this 20% drop in P_{O_2} decreases the % Hb saturation from 75% to 30%; that is, 45% more Hb than normal gives up its O_2 for tissue use. The normal 60 mm Hg drop from a P_{O_2} of 100 to 40 mm Hg that occurs in the systemic capillaries results in less than a 25% decrease in % Hb saturation. In comparison, a further drop in P_{O_2} of only 20 mm Hg results in a further 45% reduction in % Hb saturation (and a 45% increase in the release of O_2 from HbO_2) because the O_2 partial pressures in this range are operating in the steep portion of the curve. In this range, only a small drop in systemic capillary P_{O_2} can automatically make large amounts of O_2 immediately available to meet the O_2 needs of more actively metabolizing tissues. As much as 85% of the Hb may give up its O_2 to actively metabolizing cells during strenuous exercise. In addition to this more thorough withdrawal of O_2 from the blood, even more O_2 is made available to actively metabolizing cells, such as exercising muscles, by cardiovascular and respiratory adjustments that increase the flow rate of oxygenated blood through the active tissues.

By acting as a storage depot, hemoglobin promotes the net transfer of O_2 from the alveoli to the blood.

We still have not really clarified the role of Hb in gas exchange. Because blood P_{O_2} depends entirely on the concentration of dissolved O_2, we could ignore the O_2 bound to Hb in our earlier discussion of O_2-driving P_{O_2} gradients. However, Hb does play a crucial role in permitting the transfer of large quantities of O_2 before blood P_{O_2} equilibrates with the surrounding tissues (Fig. 13–29). It does so by acting as a "storage depot" for O_2, removing the O_2 from solution as soon as it enters the blood from the alveoli. Because only dissolved O_2 contributes to the P_{O_2}, the O_2 stored in Hb cannot contribute to blood P_{O_2}. When systemic venous blood enters the pulmonary capillaries, its P_{O_2} is considerably lower than the alveolar P_{O_2}, so O_2 immediately diffuses into the blood, raising the blood P_{O_2}. As soon as the P_{O_2} of blood increases, the percentage of Hb that can bind with O_2 likewise increases, as indicated by the O_2-Hb curve. Consequently, O_2 that has diffused into the blood combines with Hb and no longer contributes to blood P_{O_2}. As O_2 is removed from solution by combining with Hb, blood P_{O_2} falls to the same level it was when the blood entered the lungs, despite the fact that the total quantity of O_2 in the blood actually has increased. Since the blood P_{O_2} is once again considerably below alveolar P_{O_2}, more O_2 diffuses from the alveoli into the blood, only to be soaked up by Hb again.

Even though we have considered this process in stepwise fashion for clarity, net diffusion of O_2 from alveoli to blood occurs continuously until Hb becomes saturated with O_2 as completely as it can be at that particular P_{O_2}. At a normal P_{O_2} of 100 mm Hg, that means Hb is 97.5% saturated. Thus, by soaking up O_2, Hb keeps blood P_{O_2} low and prolongs the existence of a partial pressure gradient so that a large net transfer

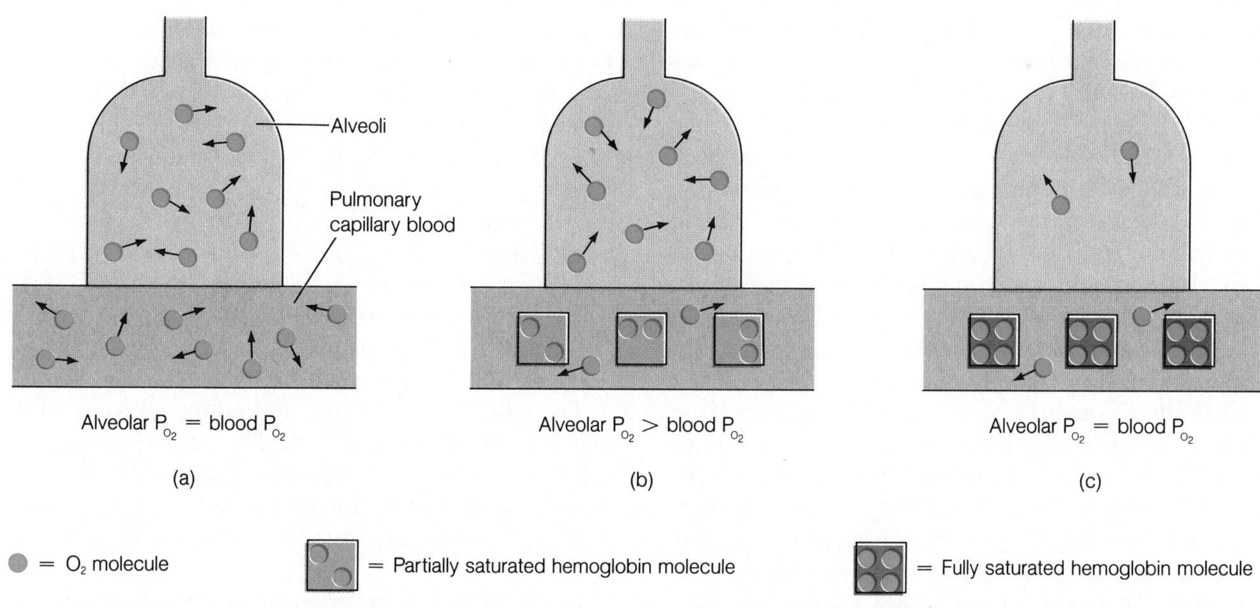

Figure 13–29 Hemoglobin Facilitating a Large Net Transfer of O_2 by Acting as a Storage Depot to Keep P_{O_2} Low *(a) In the hypothetical situation in which no hemoglobin is present in the blood, the alveolar P_{O_2} and the pulmonary capillary blood P_{O_2} are at equilibrium. (b) Hemoglobin has been added to the pulmonary capillary blood. As the hemoglobin starts to bind with O_2, it removes O_2 from solution. Since only dissolved O_2 contributes to blood P_{O_2}, the blood P_{O_2} falls below that of the alveoli, even though the same number of O_2 molecules are present in the blood as in part a. By "soaking up" some of the dissolved O_2, hemoglobin favors the net diffusion of more O_2 down its partial pressure gradient from the alveoli to the blood. (c) Hemoglobin is fully saturated with O_2, and the alveolar and blood P_{O_2} are at equilibrium again. The blood P_{O_2} resulting from dissolved O_2 is equal to the alveolar P_{O_2}, despite the fact that the total O_2 content in the blood is much greater than in part a, when blood P_{O_2} was equal to alveolar P_{O_2} in the absence of hemoglobin.*

of O_2 into the blood can take place. Not until Hb can store no more O_2 (that is, Hb is maximally saturated for that P_{O_2}) does the O_2 transferred into the blood remain dissolved and directly contribute to the P_{O_2}. Only at this time does the blood P_{O_2} rapidly equilibrate with the alveolar P_{O_2} and bring further O_2 transfer to a halt, but this point is not reached until Hb is already loaded to the maximum extent possible.

If it were not for Hb, blood P_{O_2} would be brought from a venous level of 40 mm Hg to an arterial level of 100 mm Hg by the transfer of less than 10 ml of O_2 from the alveoli into the 5-liter cardiac output of blood flowing through the pulmonary system per minute, compared with the normal resting transfer of 250 ml of O_2/min. Once the blood P_{O_2} equilibrates with the alveolar P_{O_2}, no further O_2 transfer can take place, no matter how little or how much total O_2 has already been transferred.

The reverse situation is true at the tissue level. Since the P_{O_2} of blood entering the systemic capillaries is considerably higher than the P_{O_2} of the surrounding tissue, O_2 immediately diffuses from the blood into the tissues, lowering blood P_{O_2}. When blood P_{O_2} falls, Hb is forced to unload some of its stored O_2 because the % Hb saturation is reduced. As the O_2 released from Hb dissolves in the blood, the blood P_{O_2} increases once again above the P_{O_2} of the surrounding tissues. This favors further movement of O_2 out of the blood, despite the fact that the total quantity of O_2 in the blood has already been reduced. Only when Hb is no longer able to release any more O_2 into solution (when Hb is unloaded to the greatest extent possible for the P_{O_2} existing at the systemic capillaries) can blood P_{O_2} become as low as the surrounding tissue. At this time, further transfer of O_2 ceases. Hemoglobin, because it bears a large quantity of stored O_2 that can be liberated by a slight reduction in P_{O_2} at the systemic-capillary level, permits the transfer of tremendously more O_2 from the blood into the cells than would be possible in its absence.

Thus, Hb plays an important role in the ***total quantity*** of O_2 that the blood can pick up in the lungs and drop off in the tissues. If Hb levels are reduced to one-half of normal, as in a severely anemic patient (see p. 348), the O_2-carrying capacity of the blood is reduced by 50% even though the arterial P_{O_2} is the normal 100 mm Hg with 97.5% Hb saturation. Only one-half as much Hb is available to be saturated, emphasizing once again how critical the presence of Hb is in determining the total amount of O_2 that can be picked up at the lungs and made available to the tissues.

Increased CO_2, acidity, temperature, and DPG shift the O_2-Hb dissociation curve to the right.

Even though the primary factor determining the % Hb saturation is the P_{O_2} of the blood, other factors can affect the affinity, or bond strength, between Hb and O_2 and, accordingly, can shift the O_2-Hb curve (that is, change the % Hb saturation at a given P_{O_2}). These other factors that influence the extent to which Hb and O_2 combine are: CO_2, acidity, temperature, and 2,3,-diphosphoglycerate, or DPG, which we will examine separately. The O_2-Hb dissociation curve with which you are already familiar (Fig. 13–28) is a typical curve at normal arterial CO_2 and acidity levels, normal body temperature, and normal DPG concentration.

An increase in P_{CO_2} shifts the O_2-Hb curve to the right (Fig. 13-30). The % Hb saturation still depends on the P_{O_2}, but for any given P_{O_2}, the amount of O_2 and Hb that can be combined is reduced. This is important, because the P_{CO_2} of the blood increases in the systemic capillaries as CO_2 diffuses down its gradient from the cells into the blood. The presence of this additional CO_2 in the blood in effect decreases the affinity of Hb for O_2, so Hb unloads even more O_2 at the tissue level than it would if the reduction in P_{O_2} in the systemic capillaries were the only factor affecting % Hb saturation. Furthermore, the more active a tissue, the greater are both its O_2 needs and its rate of CO_2 production. The greater increase in CO_2 shifts the O_2-Hb curve further to the right so that O_2 release from Hb is enhanced even more.

An increase in acidity also shifts the curve in the same direction as an increased P_{CO_2}. Because CO_2 generates carbonic acid (H_2CO_3), the blood becomes more acidic at the systemic-capillary level as it picks up CO_2 from the tissues. The resultant reduction in Hb affinity for O_2 in the presence of increased acidity aids in releasing even more O_2 at the tissue level for a given P_{O_2}. In actively metabolizing cells, such as exercising muscles, not only is more carbonic acid-generating CO_2 produced, but lactic acid also may be produced if the cells resort to anaerobic metabolism. The resultant local elevation of acid in the working muscles facilitates further unloading of O_2 in the very tissues that have the greatest O_2 need.

The influence of CO_2 and acid on the release of O_2 is known as the **Bohr effect.** Both CO_2 and the hydrogen-ion (H^+) component of acids are able to combine reversibly with Hb at sites other than the O_2-binding sites. The result is an alteration in the molecular structure of Hb that reduces its affinity for O_2.

In a similar manner, an elevation in temperature shifts the O_2-Hb curve to the right, resulting in more unloading of O_2 at a given P_{O_2}. An exercising muscle or other actively metabolizing cell produces heat. The resultant local elevation in temperature enhances O_2 release from Hb for use by the more active tissues.

Thus, increases in CO_2, acidity, and temperature at the tissue level, all of which are associated with increased cellular metabolism and increased O_2 consumption, enhance the effect of the drop in P_{O_2} in facilitating the release of O_2 from Hb.

These effects are largely reversed at the pulmonary level, where the extra acid-forming CO_2 is blown off and the local environment is cooler. Appropriately, therefore, Hb has a higher affinity for O_2 in the pulmonary-capillary environment, which enhances the effect of the elevation in P_{O_2} in loading O_2 onto Hb.

In addition to such changes that take place within the *environment* of the red blood cells, there is one factor *inside* the red blood cells that can affect the degree of O_2-Hb binding; this is **2,3,-diphosphoglycerate,** or **DPG.** This erythrocyte constituent, which is produced during red blood cell metabolism, can bind reversibly with Hb and reduce its affinity for O_2, similar to CO_2 and H^+. Thus, an increased level of DPG, as with the other factors, shifts the O_2-Hb curve to the right, enhancing O_2 unloading as the blood flows through the tissues. Production of DPG by red blood cells is inhibited by HbO_2. Consequently, DPG concentration gradually increases whenever Hb in the arterial blood is chronically undersaturated—that is, when arterial HbO_2 is below normal. This may occur in individuals living at high altitudes or in those suffering from certain types of circulatory or respiratory diseases or anemia. By promoting the liberation of O_2 from Hb at the tissue level, an increased amount of DPG helps maintain O_2 availability for tissue use under circumstances associated with decreased arterial O_2 supply. However, unlike the other factors—which are present only at the tissue level and thus shift the O_2-Hb curve to the right only in the systemic capillaries, where the shift is advantageous in unloading O_2—DPG is present in the red blood cells throughout the circulatory system and, accordingly, shifts the curve to the right to the same degree in both the tissues and the lungs. As a result, DPG decreases the ability to load O_2 at the pulmonary level, which is the negative side of increased DPG production.

Oxygen-binding sites on hemoglobin have a much higher affinity for carbon monoxide than for O_2.

Carbon monoxide (**CO**) and O_2 compete for the same binding sites on Hb, but the affinity of Hb for CO is 240 times that of the bond strength between Hb and O_2. The combination of CO and Hb is known as **carboxyhemoglobin** (**HbCO**). Because Hb preferentially latches onto CO, the presence of even small amounts of CO can tie up a disproportionately large share of Hb, making the latter unavailable for O_2 transport. Even though the Hb concentration and P_{O_2} are normal, the O_2 content of the blood is seriously reduced. If enough CO is present, the cells die from O_2 deprivation. Adding to the toxicity of CO, the presence of HbCO shifts the Hb-O_2 curve to the *left* (Fig. 13-30). This means that the already limited quantity of O_2-bearing Hb is unable to unload as much of its O_2 at the tissue level for a given P_{O_2}.

Fortunately, CO is not a normal constituent of inspired air. It is a poisonous gas produced during the incomplete combustion (burning) of carbon products such as automobile gaso-

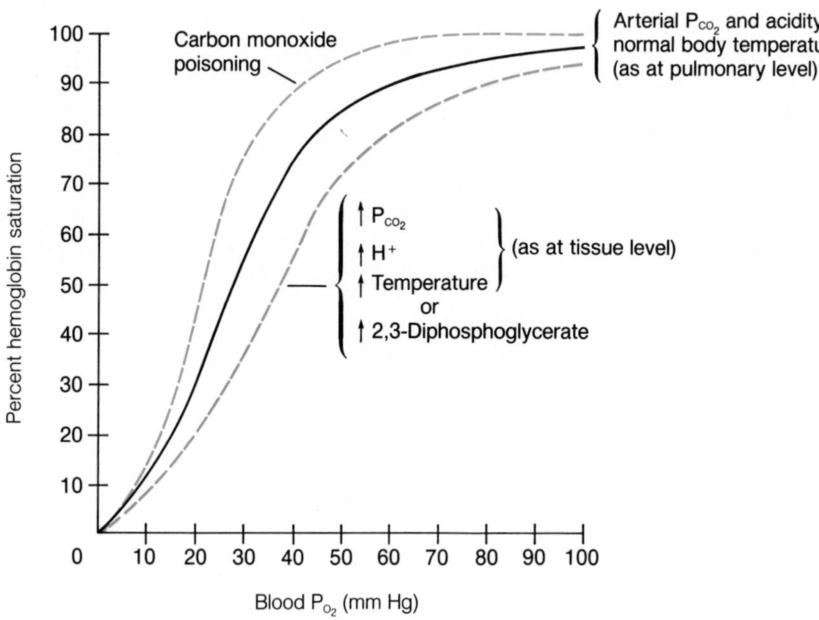

Figure 13-30 Effect of Increased P_{CO_2}, H^+, Temperature, 2,3-Diphosphoglycerate, and Carbon Monoxide on O_2-Hb Curve

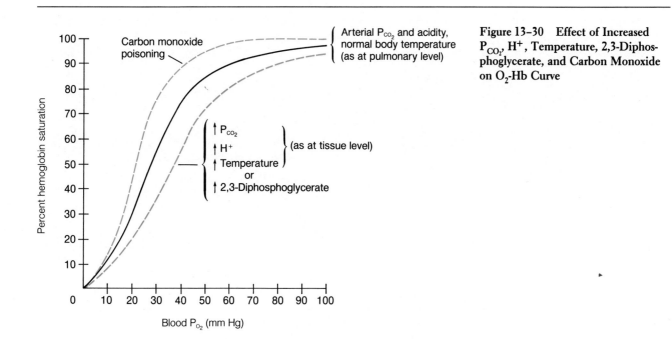

line, coal, wood, and tobacco. Carbon monoxide is especially dangerous because it is so insidious. If CO is being produced in a closed environment so that its concentration continues to increase, (for example, in a parked car with the motor running and windows closed), it can reach lethal levels without the victim ever being aware of the danger. Carbon monoxide is not detectable because it is odorless, colorless, tasteless, and nonirritating. Furthermore, for reasons to be described later, the victim has no sensation of breathlessness and makes no attempt to increase ventilation, even though the cells are O_2-starved.

The majority of CO_2 is transported in the blood as bicarbonate.

The other gas transported between the tissues and lungs is CO_2. When arterial blood flows through the tissue capillaries, CO_2 diffuses down its partial pressure gradient from the tissue cells into the blood. Carbon dioxide is transported in the blood in three ways: (1) physically dissolved, (2) bound to Hb, and (3) as bicarbonate (Fig. 13–31 and Table 13–6).

As with O_2, the amount of CO_2 ***physically dissolved*** in the blood depends on the P_{CO_2}. Because CO_2 is more soluble than O_2 in the blood, a greater proportion of the total CO_2 in the blood is physically dissolved compared to O_2. Even so, only 10% of the blood's total CO_2 content is physically dissolved at the normal systemic venous P_{CO_2} level.

Another 30% of the CO_2 combines with Hb to form **carbamino hemoglobin ($HbCO_2$)**. Carbon dioxide binds with the globin portion of Hb, in contrast to O_2, which combines with the heme portions. Reduced Hb has a greater affinity for CO_2 than does HbO_2. The unloading of O_2 from Hb in the tissue capillaries therefore facilitates the picking up of CO_2 by Hb.

By far the most important means of CO_2 transport is as **bicarbonate (HCO_3^-)**, with 60% of the CO_2 being converted into HCO_3^- by means of the following chemical reaction, which takes place within the red blood cells:

$$CO_2 + H_2O \underset{}{\overset{\text{carbonic anhydrase}}{\rightleftharpoons}} H_2CO_3 \rightleftharpoons H^+ + HCO_3^-$$

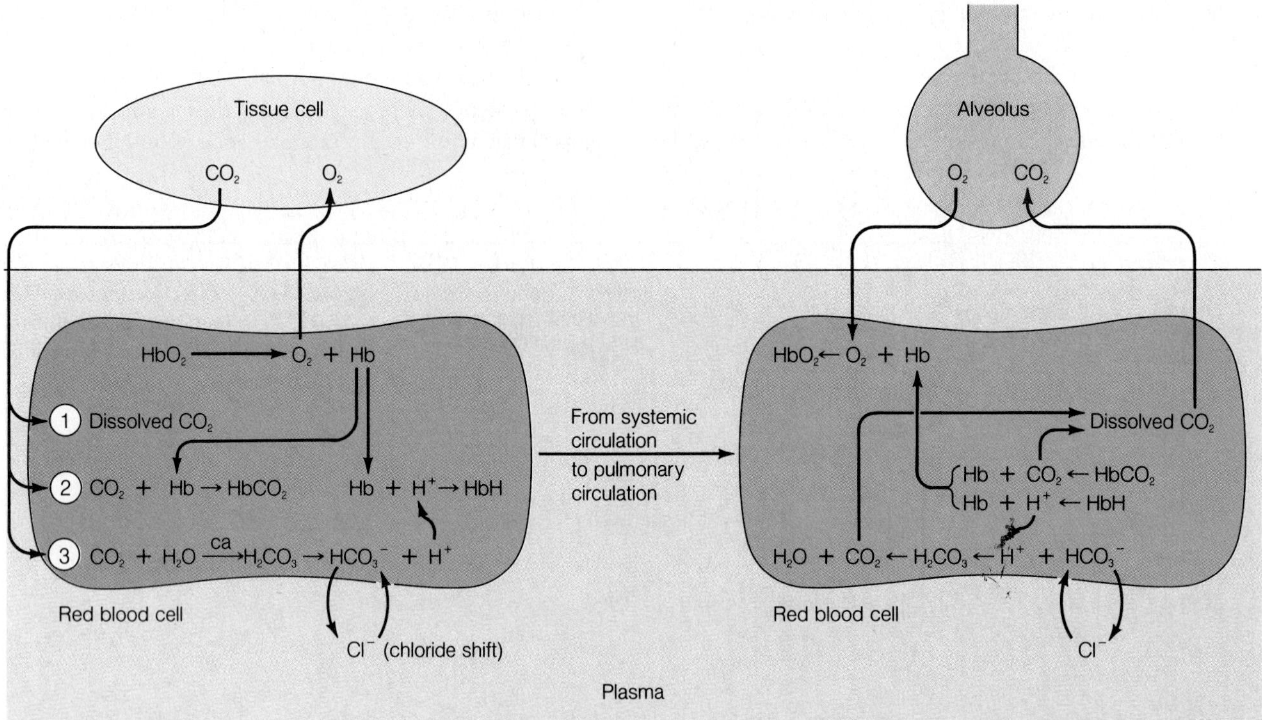

Figure 13–31 Carbon Dioxide Transport in the Blood

In the first step, CO_2 combines with H_2O to form **carbonic acid** (H_2CO_3). This reaction can occur very slowly in the plasma, but it proceeds swiftly within the red blood cells because of the presence of the erythrocyte enzyme **carbonic anhydrase,** which catalyzes (speeds up) the reaction. As is characteristic of acids, a proportion of the carbonic-acid molecules spontaneously dissociate into hydrogen ions and bicarbonate ions. The one carbon and two oxygen atoms of the original CO_2 molecule are thus present in the blood as an integral part of HCO_3^-. This is beneficial, because HCO_3^- is more soluble in the blood than CO_2.

As this reaction proceeds, HCO_3^- and H^+ start to accumulate within the red blood cells in the systemic capillaries. The red cell membrane is freely permeable to HCO_3^- but relatively impermeable to H^+. Consequently, HCO_3^- but not H^+ diffuses down its concentration gradient out of the erythrocytes into the plasma. Since HCO_3^- is a negatively charged ion, the outflux of HCO_3^- unaccompanied by a comparable outward diffusion of positively charged ions creates an electrical gradient (see p. 72). Chloride ions (Cl^-), the dominant plasma anions, diffuse into the red blood cells down this electrical gradient to restore electric neutrality. This inward shift of Cl^- in exchange for the outflux of CO_2-generated HCO_3^- is known as the **chloride (Cl^-) shift.**

Meanwhile, the vast majority of the accumulated H^+ remaining within the erythrocytes following the dissociation of H_2CO_3 becomes bound to Hb. As with CO_2, reduced Hb has a greater affinity for H^+ than does HbO_2. Therefore, the unloading of O_2 once again facilitates the pick up of CO_2-generated H^+ by Hb. Because only free dissolved H^+ contributes to the acidity of a solution, the venous blood would be considerably more acidic than the arterial blood if it were not for Hb mopping up most of the H^+ generated at the tissue level.

The fact that removal of O_2 from Hb increases the ability of Hb to pick up CO_2 and CO_2-generated H^+ is known as the **Haldane effect.** Recall that the Bohr effect refers to the reduced affinity of Hb for O_2 in the presence of increased CO_2 and H^+ at the tissue level. Thus, the Haldane effect and Bohr effect work in synchrony to facilitate O_2 liberation and the uptake of CO_2 and CO_2-generated H^+ at the tissue level. Increased CO_2 and H^+ cause increased O_2 release from Hb by means of the Bohr effect; increased O_2 release from Hb, in turn, causes increased CO_2 and H^+ uptake by Hb through the Haldane effect. This is very efficient. Reduced Hb must be carried back to the lungs to refill on O_2 anyway. Meanwhile, after O_2 is released, Hb picks up new passengers—CO_2 and H^+—that are going in the same direction to the lungs.

The reactions that occur at the tissue level as CO_2 enters the blood from the tissues are reversed once the blood reaches the lungs and CO_2 leaves the blood to enter the alveoli (Fig. 13-31). As dissolved CO_2 diffuses from the blood down the P_{CO_2} gradient, O_2 simultaneously diffuses into the blood down the P_{O_2} gradient and combines with Hb. Since oxygenated Hb has a lower affinity for CO_2 and H^+ than does reduced Hb, these latter two passengers are bumped off Hb. The freed CO_2 diffuses out of the blood into the alveoli. The freed H^+ combines with HCO_3^- to form H_2CO_3, which separates into CO_2 and H_2O, thus generating more CO_2, which diffuses out of the blood into the alveoli. This reaction is likewise catalyzed by carbonic anhydrase, with the direction of the reaction being governed by the differences in CO_2 concentration in the systemic and pulmonary capillaries as according to the law of mass action. In the systemic capillaries, the rise in blood P_{CO_2} drives the reaction to the right side of the equation, generating H^+ and HCO_3^-. On the other hand, the decline in blood P_{CO_2} in the pulmonary capillaries is responsible for the reaction proceeding to the left side of the equation to generate H_2O and CO_2. As HCO_3^- is used up within the red blood cells during the formation of CO_2, the concentration gradient thus created favors the movement of HCO_3^- back into the erythrocytes from the plasma. Chloride subsequently shifts back to the plasma to maintain electric neutrality. The returning HCO_3^- is available to combine with even more H^+ as the latter is displaced from Hb, resulting in the formation of CO_2-yielding H_2CO_3. The ultimate elimination of CO_2 by means of ventilation removes a major source of acid production from the body. For this reason, the lungs play a major role in the acid-base balance of the body.

These reactions and movements continue generating CO_2, which diffuses into the alveoli until the blood P_{CO_2} equilibrates with the alveolar P_{O_2}. The net effect is that, on the average, 200 ml/min of CO_2 are removed from the tissues and eliminated to the external environment via the alveoli. If more CO_2 is produced at the tissue level, there is a larger P_{CO_2} pressure gradient between the pulmonary-capillary blood and the alveoli, so more CO_2 can be generated from these reactions and eliminated before the blood P_{CO_2} equilibrates with alveolar P_{CO_2}.

Hypoxia can occur independently or concurrently with hypercapnia or hypocapnia.

Hyperoxia, an elevated arterial P_{O_2}, cannot occur when a person is breathing atmospheric air at sea level. However, breathing supplemental O_2 can increase alveolar and consequently arterial P_{O_2}. Because a greater percentage of the inspired air is O_2, a greater percentage of the total pressure of the inspired air is attributable to the O_2 partial pressure. Even so, as mentioned earlier, the total blood O_2 content is not significantly in-

Table 13–7 Miniglossary of Clinically Important Respiratory States

Term	Definition
Apnea	Transient cessation of breathing
Asphyxia	O_2 starvation of tissues, caused by either lack of O_2 in air, respiratory impairment, or inability of tissues to utilize O_2
Dyspnea	Difficult or labored breathing
Eupnea	Normal breathing
Hypercapnia	Excess CO_2 in arterial blood
Hyperpnea	Increased pulmonary ventilation that matches increased metabolic demands, as in exercise
Hyperventilation	Increased pulmonary ventilation in excess of metabolic requirements, resulting in decreased P_{CO_2} and respiratory alkalosis
Hypocapnia	Below-normal CO_2 in arterial blood
Hypoventilation	Underventilation in relation to metabolic requirements, resulting in increased P_{CO_2} and respiratory acidosis
Hypoxia	Insufficient O_2 at the cellular level
Anemic hypoxia	Reduced O_2-carrying capacity of the blood
Circulatory hypoxia	Too little oxygenated blood delivered to the tissues; also known as stagnant hypoxia
Histotoxic hypoxia	Cells unable to utilize O_2 available to them
Hypoxic hypoxia	Low arterial blood P_{O_2} accompanied by inadequate hemoglobin saturation
Respiratory arrest	Permanent cessation of breathing (unless clinically corrected)
Suffocation	O_2 deprivation as a result of inability to breathe oxygenated air

creased because Hb is nearly fully saturated at the normal arterial P_{O_2}. In certain pulmonary diseases associated with a reduced arterial P_{O_2}, however, breathing supplemental O_2 can be beneficial in establishing a larger alveoli-to-blood driving gradient, thereby improving the arterial P_{O_2}. On the other hand, a markedly elevated arterial P_{O_2}, far from being advantageous, can be dangerous. If the arterial P_{O_2} is too high, **oxygen toxicity** can occur. Even though the total O_2 content of the blood is only slightly increased, some cells can be damaged by exposure to a high P_{O_2}. In particular, brain damage and blindness-causing damage to the retina are problems associated with O_2 toxicity. Therefore, O_2 therapy must be administered cautiously.

Hypoxia refers to insufficient O_2 at the cellular level. (Table 13–7 is a glossary of terms used to describe various states associated with respiratory abnormalities.) There are four general categories of hypoxia: (1) hypoxic hypoxia, (2) anemic hypoxia, (3) circulatory hypoxia, and (4) histotoxic hypoxia.

Hypoxic hypoxia is characterized by a low arterial blood P_{O_2} accompanied by inadequate Hb saturation caused by: (1) a respiratory malfunction involving inadequate gas exchange, typified by a normal alveolar P_{O_2} but a reduced arterial P_{O_2}; or (2) exposure to high altitude or to a suffocating environment where atmospheric P_{O_2} is reduced so that alveolar and arterial P_{O_2} are likewise reduced.

Anemic hypoxia refers to a reduced O_2-carrying capacity of the blood. This can be brought about by: (1) a decrease in circulating red blood cells; (2) an inadequate amount of Hb within the red blood cells; (3) the presence of abnormal Hb that cannot carry O_2 as effectively; or (4) CO poisoning. In all cases of anemic hypoxia, the O_2 content of the arterial blood is lower than normal because of the reduction in available Hb, although the arterial P_{O_2} is at a normal level.

Circulatory hypoxia arises when too little oxygenated blood is delivered to the tissues. Circulatory hypoxia can be restricted to a limited area as a result of a local vascular spasm or blockage. On the other hand, the body may experience circulatory hypoxia in general as a result of congestive heart failure or circulatory shock. The arterial P_{O_2} and O_2 content are typically normal, but too little oxygenated blood reaches the cells. In more severe cases, arterial P_{O_2} is reduced because of inadequate pulmonary perfusion.

In **histotoxic hypoxia,** O_2 delivery to the tissues is normal but the cells are unable to use the O_2 available to them. The classic example is cyanide poisoning. Cyanide blocks cellular enzymes essential for internal respiration.

Some, but not all, types of hypoxia are associated with **hypercapnia,** which is excess CO_2 in the arterial blood. In hypoxic hypoxia caused by **hypoventilation** (ventilation inadequate to meet the metabolic needs for O_2 delivery and CO_2 removal), CO_2 accumulation in the arterial blood occurs concomitantly with the O_2 deficit (Fig. 13–32). Both O_2 and CO_2 exchange between the lungs and atmosphere are equally affected. However, when hypoxic hypoxia is due to reduced pulmonary diffusing capacity, such as in pulmonary edema or emphysema, O_2 transfer suffers more than CO_2 transfer, because the diffusion coefficient for CO_2 is twenty times that for O_2. As a result, hypoxia occurs much more readily than hypercapnia in these circumstances. In hypoxic hypoxia due to high altitudes, hypercapnia not only does not occur, but, in fact, arterial P_{CO_2} levels actually decrease. One of the compensatory responses in acclimatization to high altitudes is reflex stimulation of ventilation as a result of the reduction in arterial P_{O_2}. This compensatory hyperventilation to obtain more O_2 blows

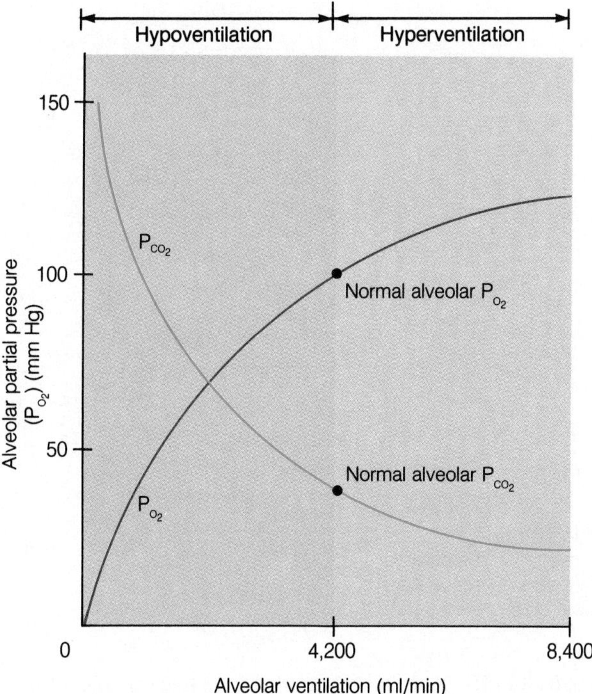

Figure 13–32 Effects of Hypoventilation and Hyperventilation on Arterial P_{O_2} and P_{CO_2}

off too much CO_2 in the process, so arterial P_{CO_2} levels decline below normal, a condition known as **hypocapnia.**

In anemic hypoxia, arterial CO_2 levels are normal. Reduced O_2-carrying capacity of the blood has no influence on blood CO_2 content. However, hypercapnia can exist in circulatory hypoxia. Just as the diminished blood flow fails to deliver adequate O_2 to the tissues, it also fails to remove sufficient CO_2. In histotoxic hypoxia, the CO_2 levels decline because oxidative metabolism is blocked by the tissue poisons, so CO_2 is not being produced. Therefore, only in hypoxia associated with inadequate ventilation or circulatory disturbances does hypercapnia also exist. In other types of hypoxia, P_{CO_2} levels are normal or even reduced.

Blood CO_2 levels can likewise be altered without blood O_2 content being affected during **hyperventilation,** the reverse of hypoventilation. Hyperventilation occurs when the rate of ventilation is in excess of the body's metabolic needs. It is invoked by anxiety states, fever, and aspirin poisoning. Alveolar P_{O_2} increases as more fresh O_2 is delivered to the alveoli from the atmosphere than is extracted by the blood for tissue consumption. Simultaneously, alveolar P_{CO_2} decreases as CO_2 is blown off to the atmosphere more rapidly than it is produced in the tissues and delivered to the alveoli (Fig. 13–32). The hyperventilatory rise in alveolar P_{O_2} is accompanied by a corresponding rise in arterial P_{O_2}. However, since Hb is almost fully saturated at the normal arterial P_{O_2}, very little additional O_2 is added to the blood. Except for the small extra amount of dissolved O_2, blood O_2 content remains essentially unchanged during hyperventilation. In contrast, the fall in alveolar P_{CO_2} during hyperventilation is accompanied by a corresponding decline in arterial P_{CO_2} as well as total CO_2 content.

Increased ventilation is not synonymous with hyperventilation. Increased ventilation that matches an increased metabolic demand, such as the increased need for O_2 delivery and CO_2 elimination in conjunction with exercise, is termed **hyperpnea.** During exercise, alveolar P_{O_2} and P_{CO_2} remain constant, the increased atmospheric exchange just keeping pace with the increased O_2 consumption and CO_2 production.

The consequences of reduced O_2 availability to the tissues during hypoxia is apparent. The cells need adequate O_2 supplies to sustain their energy-generating metabolic activities. The consequences of abnormal blood CO_2 levels are less obvious. The primary effect of changes in blood CO_2 concentration is the resultant impact on acid-base balance. Hypercapnia results in an elevated production of carbonic acid. The subsequent generation of excess H^+ produces an acidic condition termed respiratory acidosis. Conversely, less-than-normal amounts of H^+ are generated via carbonic-acid formation in conjunction with hypocapnia. The resultant alkalotic (less-acidic-than-normal) condition is called respiratory alkalosis (chapter 15).

Control of Respiration

Respiratory centers in the brain stem establish a rhythmic breathing pattern.

Breathing, like the heart beat, must occur in a continuous, cyclical pattern to sustain life processes. Cardiac muscle must rhythmically contract and relax to alternately empty and fill the heart with blood. Similarly, inspiratory muscles must rhythmically contract and relax to alternately fill and empty the lungs with air. Both these activities are accomplished automatically without conscious effort. However, the underlying mechanisms and control of these two systems are remarkably different. Whereas the heart is able to generate its own rhythm by means of its intrinsic pacemaker activity, the respiratory muscles, being skeletal muscles, require nervous stimulation to bring about their contraction. The rhythmic pattern of breathing is established by cyclical neural activity to the respiratory muscles. In other words, the pacemaker activity that establishes the rhythmicity of breathing resides in the respiratory control centers in the brain, not in the lungs or respiratory muscles themselves. The nerve supply to the heart, not being necessary to initiate the heart beat, only serves to modify the rate and strength of cardiac contraction. Actually, the heart continues to perform satisfactorily without any nerve supply, as is attested to by transplanted hearts that serve their recipients well without any neural influence. In contrast, the nerve supply to the respiratory system is absolutely essential in maintaining breathing and in reflexly adjusting the level of ventilation to match changing needs for O_2 uptake and CO_2 removal. Furthermore, unlike cardiac activity, which is not subject to voluntary control, respiratory activity can be voluntarily modified to accomplish speaking, singing, whistling, playing a wind instrument, or holding one's breath while swimming.

There are three distinct components of neural control of respiration: (1) the factors responsible for generating the alternating inspiration/expiration rhythm; (2) the factors that regulate the magnitude of ventilation (that is, the rate and depth of breathing) to match body needs; and (3) the factors that modify respiratory activity to serve other purposes. The latter may either be voluntary modification, as in the breath control required for speech, or involuntary, as in the respiratory maneuvers involved in a cough or sneeze.

Respiratory control centers housed in the brain stem are responsible for generating the rhythmic pattern of breathing. The primary respiratory control center consists of several aggregations of neuronal cell bodies within the medulla known as the dorsal respiratory group and ventral respiratory group. This **medullary respiratory center** provides output to the respiratory muscles. In addition, there are two other respiratory centers higher in the brain stem in the pons—the **apneustic center** and **pneumotaxic center.** These pontine centers influence the output from the medullary respiratory center (Fig. 13–33). Exactly how these various regions interact to establish respiratory rhythmicity is unclear, but the following factors are believed to contribute.

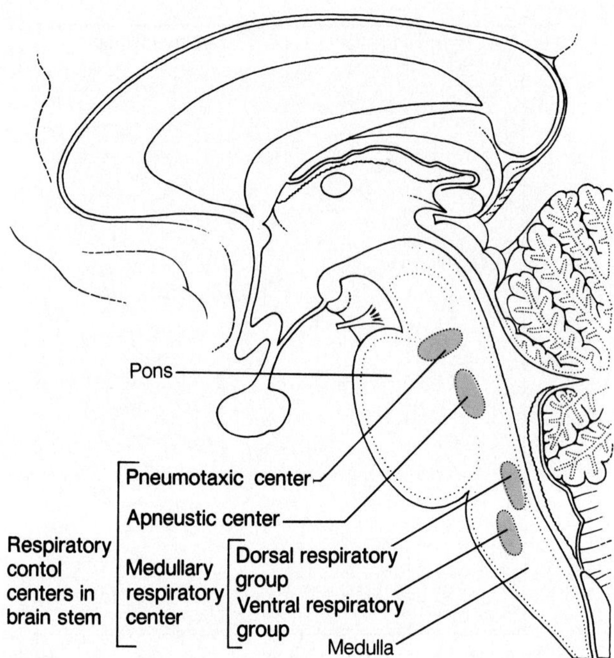

Figure 13–33 Respiratory Control Centers in Brain Stem

INSPIRATORY NEURON PACEMAKER ACTIVITY. We rhythmically breathe in and out during quiet breathing because of alternate contraction and relaxation of the inspiratory muscles, namely, the diaphragm and external intercostal muscles, supplied respectively by the phrenic nerve and intercostal nerves. The cell bodies for the neuronal fibers composing these nerves are located in the spinal cord. Impulses originating in the medullary center terminate on these motor-neuron cell bodies (Fig. 13–34). When these motor neurons are activated, they in turn stimulate the inspiratory muscles, leading to inspiration; when these neurons are not firing, the inspiratory muscles relax and expiration takes place.

The **dorsal respiratory group (DRG)** consists mostly of inspiratory neurons whose descending fibers terminate on the motor neurons that supply the inspiratory muscles. These inspiratory neurons are believed to display pacemaker activity, repetitively undergoing self-induced action potentials similar to the SA node of the heart. Some evidence alternatively suggests that, rather than the inspiratory neurons being pacemakers themselves, they may be driven by other pacemaker neurons that synapse on the inspiratory neurons. Whichever is the case, when the DRG inspiratory neurons fire, inspiration takes place; when they cease firing, expiration occurs. Expiration is brought to an end as the inspiratory neurons once again reach threshold and fire. Accordingly, the DRG is generally regarded as being responsible for the basic rhythm of ventilation. The rate at which the inspiratory neurons rhythmically fire, however, is influenced by synaptic input from other areas of the brain and from elsewhere in the body. Thus the on-off nature of the respiratory cycle is complex because of the DRG's interactions with these other regions.

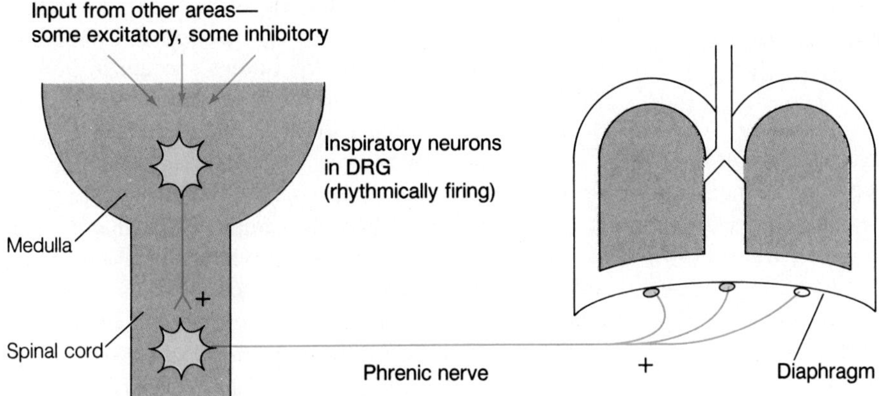

Figure 13-34 Medullary Dorsal Respiratory Group (DRG) Control of Inspiration *Inspiration takes place when the inspiratory neurons are firing and activating the motor neurons that supply the inspiratory muscles. Expiration takes place when the inspiratory neurons cease firing, so that the motor neurons supplying the inspiratory muscles are no longer activated.*

INTERACTION BETWEEN INSPIRATORY AND EXPIRATORY NEURONS. The DRG has important interconnections with the **ventral respiratory group** (**VRG**). The VRG is composed of inspiratory neurons and expiratory neurons, both of which remain inactive during normal quiet breathing. This region is called into play by the DRG as an "overdrive" mechanism during periods when demands for ventilation are increased. It is especially important in active expiration. No impulses are generated in the descending pathways from the expiratory neurons during quiet breathing. Only during active expiration do the expiratory neurons stimulate the motor neurons supplying the expiratory muscles (those being the abdominal and internal intercostal muscles). Furthermore, the VRG inspiratory neurons, when stimulated by the DRG, augment inspiratory activity.

In addition to the positive reinforcement between the DRG and VRG during periods of more vigorous breathing, a negative-feedback loop between DRG inspiratory neurons and VRG expiratory neurons contributes to ongoing breathing rhythmicity. When the inspiratory neurons fire, they stimulate, through separate branches, both the medullary expiratory neurons and the spinally located motor neurons of the inspiratory muscles. The expiratory neurons, in turn, inhibit the inspiratory neurons. Thus, the inspiratory neurons help terminate their own firing by stimulating an inhibitory input to themselves.

INFLUENCES FROM THE PNEUMOTAXIC AND APNEUSTIC CENTERS. The pontine centers are not essential for respiration, but they do exert "fine-tuning" influences over the medullary center to help produce normal, smooth inspirations and expirations. Breathing governed by the medullary center alone would occur as abrupt gasps. The pneumotaxic center sends impulses to the DRG that help "switch off" the inspiratory neurons, thereby limiting the duration of inspiration. In contrast, the apneustic center prevents the inspiratory neurons from being switched off, thus providing an extra boost to the inspiratory drive. In this check-and-balance system, the pneumotaxic center is dominant over the apneustic center, helping to bring inspiration to a halt and allowing expiration to occur normally. Without the pneumotaxic brakes, the breathing pattern consists of prolonged inspiratory gasps abruptly interrupted by very brief expirations. This abnormal pattern of breathing is known as **apneusis;** hence, the center responsible for this type of breathing is the apneustic center. Apneusis may occur in certain types of severe brain damage.

HERING-BREUER REFLEX. When the tidal volume is large (greater than 1 liter), as during exercise, the **Hering-Breuer reflex** is triggered to prevent overinflation of the lungs. **Pulmonary stretch receptors** located within the smooth-muscle layer of the airways are activated by the stretching of the lungs at large tidal volumes. Action potentials from these stretch receptors travel through afferent nerve fibers to the medullary DRG and inhibit the inspiratory neurons. This negative feedback from the highly inflated lungs themselves helps cut inspiration short before the lungs become overly stretched. This reflex mechanism was once believed to play a major role in the normal cessation of inspiration. However, the pulmonary stretch receptors in humans do not become activated during quiet breathing because of their high threshold.

Carbon dioxide-generated hydrogen-ion concentration in the brain extracellular fluid is normally the primary regulator of the magnitude of ventilation.

No matter how much O_2 is extracted from the blood or how much CO_2 is added to it at the tissue level, the P_{O_2} and P_{CO_2} of the systemic arterial blood leaving the lungs are held remarkably constant, indicative of the fact that arterial blood-gas content is subject to precise regulation. Maintenance of arterial blood gases within the normal range is accomplished almost exclusively by varying the magnitude of ventilation to match the body's needs for O_2 uptake and CO_2 removal. If more O_2 is extracted from the alveoli and more CO_2 dropped off by the blood because the tissues are metabolizing more actively, ventilation is increased correspondingly to bring in more fresh O_2 and blow off more CO_2. The resultant constancy in alveolar partial pressures accomplished by ventilatory adjustments to match changing functional demands is, in turn, reflected by the constancy of arterial P_{O_2} and P_{CO_2}.

The medullary respiratory center receives inputs that provide information about the body's needs for gas exchange. It then responds by sending appropriate signals to the motor neurons supplying the respiratory muscles to adjust the rate and depth of ventilation to meet those needs. The two most obvious signals to increase ventilation are a decreased arterial P_{O_2} or an increased arterial P_{CO_2}. Intuitively, you would suspect that if O_2 levels in the arterial blood declined or if CO_2 accumulated, ventilation would be stimulated to obtain more O_2 or to eliminate the excess CO_2. These two factors do indeed influence the magnitude of ventilation, but not to the same degree nor through the same pathway. Also, a third chemical factor, H^+, has a notable influence on the level of respiratory activity. We will examine the role of each of these three important chemical factors in the control of ventilation (Table 13–8).

ROLE OF DECREASED ARTERIAL P_{O_2} IN REGULATING VENTILATION. Arterial P_{O_2} is monitored by **peripherial chemoreceptors** known as the **carotid bodies** and **aortic bodies,** which are located, respectively, at the bifurcation of the common carotid arteries and in the arch of the aorta (Fig. 13–35). These chemoreceptors, which respond to specific changes in the chemical content of the arterial blood that bathes them, are distinctly different from the carotid sinus and aortic arch baroreceptors located in the same vicinity. The latter, being important in the regulation of systemic arterial blood pressure, monitor pressure changes rather than chemical changes.

The peripheral chemoreceptors are not sensitive to modest reductions in arterial P_{O_2}. The arterial P_{O_2} must fall below 60 mm Hg ($>$ 40% reduction) before the peripheral chemoreceptors respond by sending afferent impulses to the DRG inspiratory neurons, thereby reflexly increasing ventilation. Because arterial P_{O_2} only falls below 60 mm Hg in the unusual circumstances of severe pulmonary disease or reduced atmospheric P_{O_2}, it does not play a role in the normal ongoing regulation of respiration. This might seem surprising at first thought, since one of the primary functions of ventilation is to provide sufficient O_2 for uptake by the blood. However, there is no need to increase ventilation until the arterial P_{O_2} falls below 60 mm Hg because of the margin of safety in % Hb saturation afforded by the plateau portion of the O_2-Hb curve. Hemoglobin is still 90% saturated at an arterial P_{O_2} of 60 mm Hg, but the % Hb saturation drops precipitously when the P_{O_2} falls below this level. Therefore, reflex stimulation of respiration by the peripheral chemoreceptors serves as an important emergency mechanism in dangerously low arterial P_{O_2} states. Indeed, this reflex mechanism is a life saver, because a low arterial P_{O_2} tends to directly depress the respiratory center, as it does all the rest of the brain (Fig. 13–36). Except for the peripheral chemoreceptors, the level of activity in all nervous tissue becomes reduced in the face of O_2 deprivation. Were it not for stimulatory intervention of the peripheral chemore-

Table 13–8 Influence of Chemical Factors on Respiration

Chemical Factor	Effect on Peripheral Chemoreceptors	Effect on Central Chemoreceptors
↓ P_{O_2} in arterial blood	Stimulates only when arterial P_{O_2} has fallen to the point of being life-threatening ($<$ 60 mm Hg); emergency mechanism	Directly depresses
↑ P_{CO_2} in arterial blood [↑ H^+ in brain extracellular fluid (ECF)]	Weakly stimulates	Strongly stimulates; is dominant control of ventilation
↑ H^+ in arterial blood	Stimulates; important in acid-base balance	Does not affect; cannot penetrate blood-brain barrier

ceptors when the arterial P_{O_2} falls threateningly low, a vicious cycle ending in cessation of breathing would ensue. Direct depression of the respiratory center by the markedly low arterial P_{O_2} would further reduce ventilation, leading to an even greater fall in arterial P_{O_2}, which would even further depress the respiratory center until ventilation ceased and death occurred.

Since the peripheral chemoreceptors respond to the P_{O_2} of the blood, *not* the total O_2 content of the blood, O_2 content in the arterial blood can fall to dangerously low or even fatal levels without the peripheral chemoreceptors ever responding to reflexly stimulate respiration. Remember that only physically dissolved O_2 contributes to blood P_{O_2}. The total O_2 content in the arterial blood can be reduced in anemic states, in which there is a reduction in O_2-carrying Hb, or in CO poisoning, when the Hb is preferentially bound to this molecule rather than to O_2. In both cases, arterial P_{O_2} is normal, so respiration is not stimulated even though O_2 delivery to the tissues may be so reduced that the person dies from cellular O_2 deprivation.

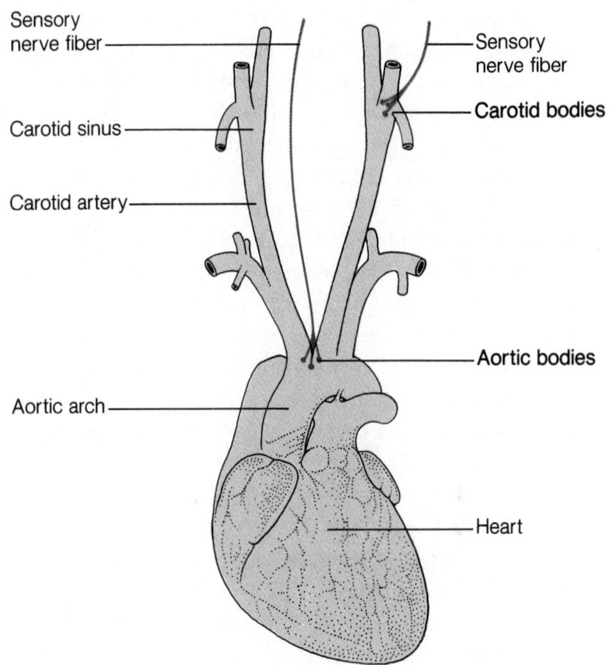

Figure 13–35 Location of Peripheral Chemoreceptors *The carotid bodies are located in the carotid sinus, and the aortic bodies are located in the aortic arch.*

ROLE OF INCREASED ARTERIAL P_{CO_2} IN REGULATING VENTILATION. In contrast to arterial P_{O_2}, which does not contribute to the minute-to-minute regulation of respiration, arterial P_{CO_2} is the most important input regulating the magnitude of ventilation under resting conditions. This is appropriate, because changes in alveolar ventilation have an immediate and pronounced effect on arterial P_{CO_2} whereas changes in ventilation have little effect on % Hb saturation and O_2 availability to the tissues until the arterial P_{O_2} falls by more than 40%. Even slight alterations in arterial P_{CO_2} from normal induce a significant reflex effect on ventilation. An increase in arterial P_{CO_2} reflexly stimulates the respiratory center, with the resultant increase in ventilation promoting elimination of the excess CO_2 to the atmosphere. Conversely, a fall in arterial P_{CO_2} reflexly reduces the respiratory drive. The subsequent decrease in ventilation allows metabolically produced CO_2 to accumulate so that P_{CO_2} can be returned to normal.

Surprisingly, given the key role of arterial P_{CO_2} in regulating respiration, there are no important receptors that monitor ar-

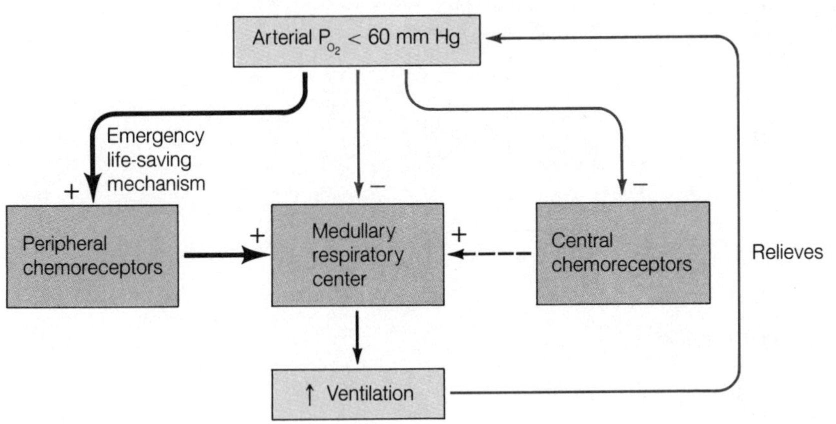

Figure 13–36 Effect of Threateningly Low Arterial P_{O_2} (< 60 mm Hg) on Ventilation

terial P_{CO_2} per se. The carotid and aortic bodies are only weakly responsive to changes in arterial P_{CO_2}, so they play only a minor role in reflexly stimulating ventilation in response to an elevation in arterial P_{CO_2}. More important in linking changes in arterial P_{CO_2} to compensatory adjustments in ventilation are the **central chemoreceptors,** located in the medulla in the vicinity of the respiratory center. These central chemoreceptors do not monitor CO_2 itself, however; they are sensitive to changes in CO_2-induced H^+ concentration in the brain extracellular fluid (ECF) that bathes them.

Unlike the free exchange of substances other than protein between the blood and surrounding tissue fluids across the capillaries elsewhere in the body, movement of materials across the brain capillaries is restricted by the blood-brain barrier (see p. 123). Because this barrier is readily permeable to CO_2, any increase in arterial P_{CO_2} causes a similar rise in brain ECF P_{CO_2} as CO_2 diffuses down its pressure gradient from the cerebral blood vessels into the brain ECF. The increased P_{CO_2} within the brain ECF causes a corresponding increase in the concentration of H^+ according to the law of mass action as it applies to this reaction: $CO_2 + H_2O \rightleftharpoons H_2CO_3 \rightleftharpoons H^+ + HCO_3^-$. An elevation in H^+ concentration in the brain ECF directly stimulates the central chemoreceptors, which in turn increase ventilation by stimulating the respiratory center through synaptic connections (Fig. 13-37). As the excess CO_2 is subsequently blown off, the arterial P_{CO_2} and the P_{CO_2} and H^+ concentration of the brain ECF are returned to normal. Conversely, a decline in arterial P_{CO_2} below normal is

Figure 13-37 Effect of Increased Arterial P_{CO_2} on Ventilation

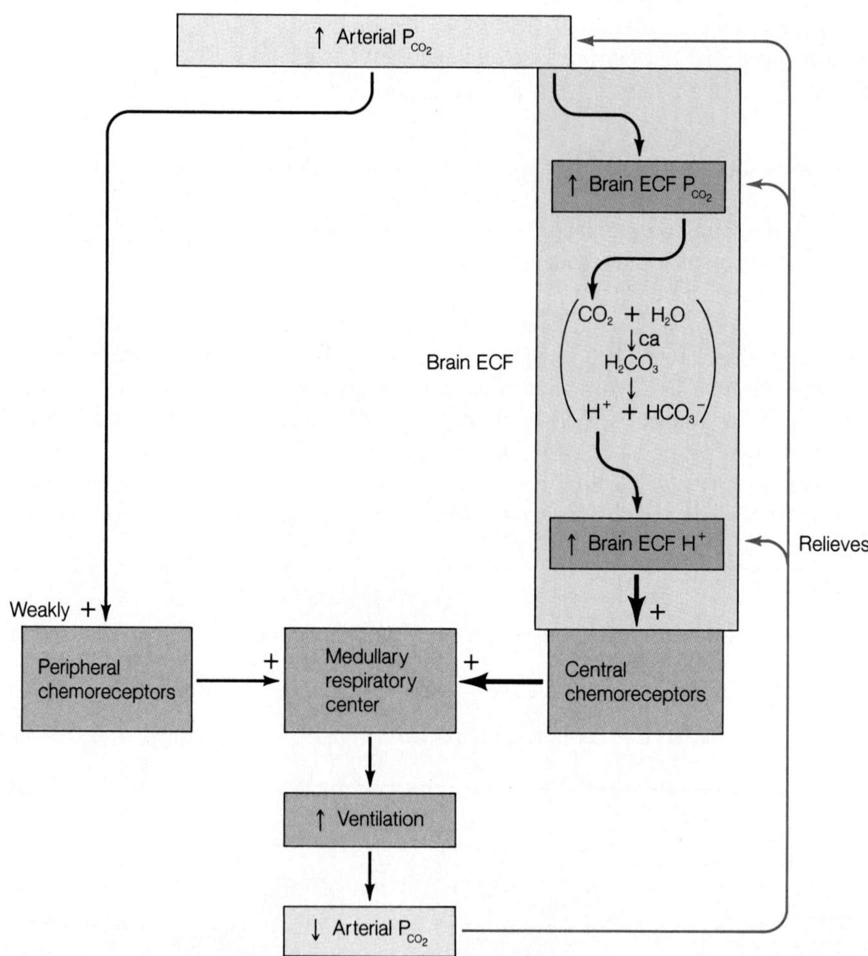

paralleled by a fall in P_{CO_2} and H^+ in the brain ECF, the result of which is a central chemoreceptor—mediated decrease in ventilation. As CO_2 produced by cellular metabolism is consequently allowed to accumulate, arterial P_{CO_2} and P_{CO_2} and H^+ of the brain ECF are restored toward normal.

Unlike CO_2, H^+ is not readily able to permeate the blood-brain barrier, so H^+ present in the plasma cannot gain access to the central chemoreceptors. Accordingly, the central chemoreceptors are responsive only to H^+ generated within the brain ECF itself as a result of CO_2 entry. Thus, the major mechanism controlling ventilation under resting conditions is specifically aimed at regulating the brain ECF H^+ concentration, which in turn is a direct reflection of the arterial P_{CO_2}. Unless there are extenuating circumstances such as reduced availability of O_2 in the inspired air, arterial P_{O_2} is coincidentally also maintained at its normal value by the brain ECF H^+ ventilatory driving mechanism.

The powerful influence of the central chemoreceptors on the respiratory center is responsible for your inability to deliberately hold your breath for more than about a minute. While you hold your breath, metabolically produced CO_2 continues to accumulate in your blood and subsequently build up the H^+ concentration in your brain ECF. Finally, the increased P_{CO_2}-H^+ stimulant to respiration becomes so powerful that central-chemoreceptor excitatory input overrides voluntary inhibitory input to respiration, so breathing resumes despite deliberate attempts to prevent it. This occurs long before arterial P_{O_2} falls to the threateningly low levels that trigger the peripheral chemoreceptors. Therefore, you cannot deliberately hold your breath long enough to create a dangerously high level of CO_2 or low level of O_2 in the arterial blood.

In contrast to the normal reflex stimulatory effect of the increased P_{CO_2}-H^+ mechanism on respiratory activity, very high levels of CO_2 directly depress the entire brain, including the respiratory center, just as very low levels of O_2 do. Up to a P_{CO_2} of 70–80 mm Hg, progressively higher P_{CO_2} levels induce correspondingly more vigorous respiratory efforts in an attempt to blow off the excess CO_2. A further increase in P_{CO_2} beyond 70–80 mm Hg, however, does not further increase ventilation but actually depresses the respiratory neurons. For this reason, CO_2 must be removed and O_2 supplied in closed environments such as closed-system anesthesia machines, submarines, or space capsules. Otherwise, CO_2 could reach lethal levels, not only because of its depressant effect on respiration but also because of the resultant severe state of respiratory acidosis.

During prolonged hypoventilation caused by certain types of chronic lung disease, an elevated P_{CO_2} occurs simultaneously with a markedly reduced P_{O_2}. In most cases, the elevated P_{CO_2} (acting via the central chemoreceptors) and the reduced P_{O_2} (acting via the peripheral chemoreceptors) are *synergistic*; that is, the combined stimulatory effect on respiration exerted by these two inputs together is greater than the sum of their independent effects. However, some patients with severe chronic lung disease lose their sensitivity to an elevated arterial P_{CO_2}. In the presence of a prolonged increase in H^+ generation in the brain ECF as a result of long-standing CO_2 retention, enough HCO_3^- may cross the blood-brain barrier to buffer, or "neutralize," the excess H^+. The additional HCO_3^- combines with the excess H^+, removing it from solution so that it no longer contributes to free H^+ concentration. By raising the brain ECF HCO_3^- concentration, the brain ECF H^+ concentration is restored to normal despite the fact that arterial P_{CO_2} and brain ECF P_{CO_2} remain high. The central chemoreceptors are no longer aware of the elevated P_{CO_2} because the brain ECF H^+ is normal. Since the central chemoreceptors no longer reflexly stimulate the respiratory center in response to the elevated P_{CO_2}, the drive to eliminate CO_2 is blunted in such patients; that is, their level of ventilation is abnormally low considering their high arterial P_{CO_2}. In these patients, the hypoxic drive to ventilation becomes their primary respiratory stimulus, in contrast to normal individuals, in whom the arterial P_{CO_2} level is the dominant factor governing the magnitude of ventilation. Ironically, administering O_2 to such patients to relieve the hypoxic condition can markedly depress their drive to breathe by elevating the arterial P_{O_2} and removing the primary driving stimulus for respiration. Because of this danger, O_2 therapy must be administered cautiously in patients with long-term pulmonary diseases.

ROLE OF INCREASED ARTERIAL H^+ CONCENTRATION IN REGULATING VENTILATION. Changes in arterial H^+ concentration cannot influence the central chemoreceptors because H^+ does not readily cross the blood-brain barrier. However, the peripheral chemoreceptors are highly responsive to fluctuations in arterial H^+ concentration, in contrast to their weak sensitivity to deviations in arterial P_{CO_2} and their unresponsiveness to arterial P_{O_2} until it falls 40% below normal.

Any change in arterial P_{CO_2} brings about a corresponding change in the H^+ concentration of the blood as well as of the brain ECF. These CO_2-induced H^+ changes in the arterial blood are detected by the peripheral chemoreceptors; the result is reflex stimulation of ventilation in response to an increase in arterial H^+ concentration and depression of ventilation in association with a decrease in arterial H^+ concentration. However, these changes in ventilation mediated by the peripheral chemoreceptors are far less important than the powerful central-chemoreceptor mechanism in adjusting ventilation in response to changes in CO_2-generated H^+ concentration.

The peripheral chemoreceptors do play a major role in adjusting ventilation in response to alterations in arterial H^+ concentration unrelated to fluctuations in P_{CO_2}. In many situations, even though P_{CO_2} is normal, arterial H^+ concentration is changed by the addition or loss of non-carbonic acid from the body. For example, arterial H^+ concentration increases during diabetes mellitus because excess H^+-generating keto acids are abnormally produced and added to the blood. A rise in arterial H^+ concentration resulting from causes other than accumulation of CO_2 reflexly stimulates ventilation via the peripheral chemoreceptors. Conversely, the peripheral chemoreceptors reflexly suppress respiratory activity in response to a fall in arterial H^+ concentration resulting from nonrespiratory causes. Changes in ventilation by this mechanism are extremely important in regulating the acid-base balance of the body. By changing the magnitude of ventilation, the amount of H^+-generating CO_2 that is eliminated can be varied. The resultant adjustment in the amount of H^+ added to the blood from CO_2 can compensate for the nonrespiratory-induced abnormality in arterial H^+ concentration that first elicited the respiratory response (Fig. 13–38).

Exercise profoundly increases ventilation but the mechanisms involved are unclear.

Alveolar ventilation may increase up to twenty-fold during heavy exercise to keep pace with the increased demand for O_2 uptake and CO_2 output. (Table 13–9 highlights the changes in O_2- and CO_2-related variables that occur during exercise.) The cause of increased ventilation during exercise is still largely speculative. It would seem logical that changes in the "big three" chemical factors—decreased P_{CO_2}, increased CO_2, and increased H^+—could account for the increase in ventilation. This does not appear to be the case, however.

□ Despite the marked increase in O_2 utilization during exercise, arterial P_{O_2} does not decrease but remains normal or may actually increase slightly. This is because the increase in alveolar ventilation keeps pace with or even slightly exceeds the stepped-up rate of O_2 consumption.

□ Likewise, despite the marked increase in CO_2 production during exercise, arterial P_{CO_2} does not increase but remains normal or decreases slightly. This is because the extra CO_2 is removed as rapidly or even more rapidly than it is produced as a result of the increase in ventilation.

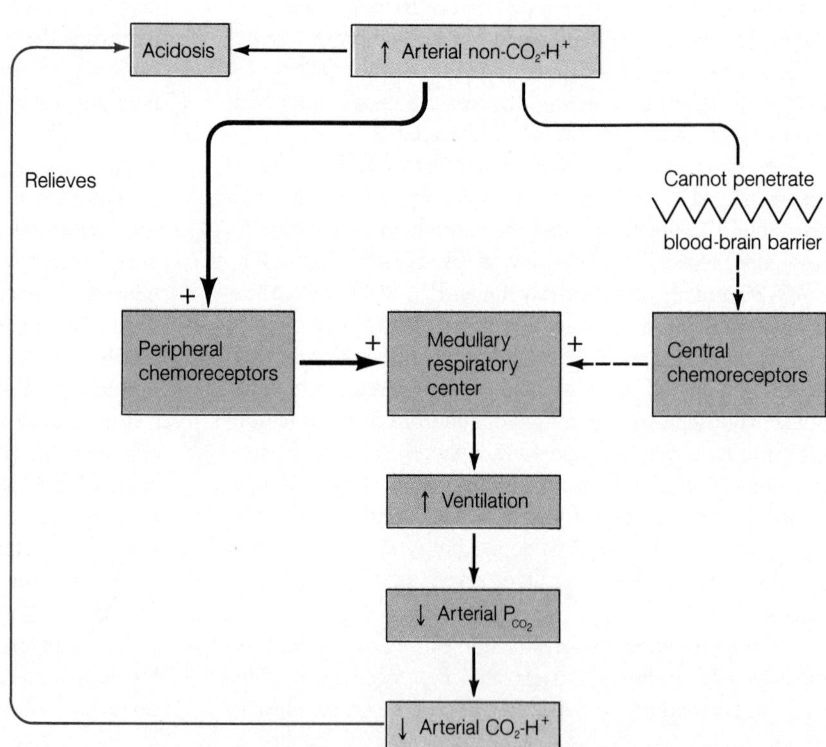

Figure 13–38 Effect of Increased Arterial Non-Carbonic Acid-Generated Hydrogen Ion (Non-CO_2-H^+) on Ventilation

Table 13-9 Oxygen- and Carbon Dioxide-Related Variables during Exercise

O_2- or CO_2-Related Variable	Change	Comment
O_2 utilization	Marked increase	Active muscles are oxidizing nutrient molecules more rapidly to meet their increased energy needs
CO_2 production	Marked increase	More actively metabolizing muscles produce more CO_2
Alveolar ventilation	Marked increase	By mechanisms not completely understood, alveolar ventilation keeps pace with or even slightly exceeds the increased metabolic demands during exercise
Arterial P_{O_2}	Normal or slight ↑	Despite marked increase in O_2 utilization and CO_2 production during exercise, alveolar ventilation keeps pace with or even slightly exceeds the stepped-up rate of O_2 consumption and CO_2 production
Arterial P_{CO_2}	Normal or slight ↓	
O_2 delivery to muscles	Marked increase	Although arterial P_{O_2} remains normal, O_2 delivery to muscles is greatly increased by the increased blood flow to exercising muscles accomplished by increased cardiac output coupled with local vasodilation of active muscles
O_2 extraction by muscles	Marked increase	Increased utilization of O_2 lowers the P_{O_2} at the tissue level, which results in more O_2 unloading from hemoglobin; this is enhanced by ↑ P_{CO_2}, ↑ H^+, and ↑ temperature
CO_2 removal from muscles	Marked increase	The increased blood flow to exercising muscles removes the excess CO_2 produced by these more actively metabolizing tissues
Arterial H^+ concentration		
Mild to moderate exercise	Normal	Since carbonic acid-generating CO_2 is held constant in arterial blood, arterial H^+ concentration does not change.
Heavy exercise	Modest increase	In heavy exercise, when muscles resort to anaerobic metabolism, lactic acid is added to the blood

□ During mild or moderate exercise, H^+ concentration does not increase because H^+-generating CO_2 is held constant. During heavy exercise, H^+ concentration does increase somewhat because of the release of H^+-generating lactic acid into the blood as a result of anaerobic metabolism in the exercising muscles. Even so, the elevation in H^+ concentration resulting from lactic-acid formation is not sufficient to account for the large increase in ventilation accompanying exercise.

Some investigators argue that the constancy of the three chemical regulatory factors during exercise is evidence that ventilatory responses to exercise are actually being controlled by these factors—particularly by P_{CO_2}, because it is normally the dominant control during resting conditions. According to this reasoning, how else could alveolar ventilation be increased in exact proportion to CO_2 production, thereby keeping the P_{CO_2} constant? This proposal, however, cannot account for the observation that during heavy exercise, alveolar ventilation may be increased relatively more than the increase in CO_2 production, thereby actually causing a slight decline in P_{CO_2}. Also, ventilation increases abruptly at the onset of exercise (within seconds), long before changes in arterial blood gases could become important influences on the respiratory center (which requires a matter of minutes).

A number of other factors have been suggested as playing a role in the ventilatory response to exercise, including the following:

1. *Reflexes originating from body movements*. Joint and muscle receptors excited during muscle contraction reflexly stimulate the respiratory center, abruptly increasing ventilation. Even passive movement of the limbs (for example, someone else alternately flexing and extending a person's knee) may increase ventilation several-fold through activation of these receptors, even though no actual exercise is occurring. Thus, the mechanical events of exercise are be-

lieved to play an important role in coordinating respiratory activity with the increased metabolic requirements of the active muscles.

2. *Increase in body temperature*. Much of the energy generated during muscle contraction is converted to heat rather than to actual mechanical work. Heat-loss mechanisms such as sweating are frequently not able to keep pace with the increased heat production that accompanies increased physical activity, so body temperature often increases slightly during exercise. Since an increase in body temperature stimulates ventilation, this exercise-related heat production undoubtedly contributes to the respiratory response to exercise. For the same reason, increased ventilation often accompanies a fever.

3. *Epinephrine release*. The adrenal medullary hormone epinephrine also stimulates ventilation. The level of circulating epinephrine is increased during exercise in response to the sympathetic nervous-system discharge that accompanies increased physical activity.

4. *Impulses from the cerebral cortex*. Especially at the onset of exercise, the motor areas of the cerebral cortex are believed to simultaneously stimulate the medullary respiratory neurons and activate the motor neurons of the exercising muscles. This is similar to the cardiovascular adjustments initiated by the motor cortex at the onset of exercise. In this way, the motor region of the brain calls forth increased ventilatory and circulatory responses to support the increased physical activity that it is about to orchestrate. These anticipatory adjustments are unusual in that regulatory steps are taken *before* any homeostatic factors have actually changed. In the usual case, regulatory adjustments take place *after* a factor has been altered in an attempt to restore homeostasis.

None of these factors or combination of factors are fully satisfactory in explaining the abrupt and profound effect exercise has on ventilation, nor can they completely account for the high degree of correlation between respiratory activity and the body's needs for gas exchange during exercise. (For a discussion of how measurement of O_2 consumption during exercise can be used to determine a person's maximum work capacity, see the accompanying boxed feature, A Closer Look at Exercise Physiology.)

Ventilation can be influenced by factors unrelated to the need to supply oxygen or remove carbon dioxide.

Respiratory rate and depth can be modified for reasons other than the need to supply O_2 or remove CO_2. Protective reflexes such as sneezing and coughing temporarily govern respiratory activity in an effort to expel irritant materials from the respiratory passages. Inhalation of particularly noxious agents frequently triggers immediate cessation of ventilation. Pain originating anywhere in the body reflexly stimulates the respiratory center (for example, one "gasps" with pain). Involuntary modification of breathing also occurs during the expression of various emotional states, such as laughing, crying, sighing, and groaning. The emotionally induced modifications are mediated through connections between the limbic system in the brain (which is responsible for emotions) and the respiratory center. In addition, the respiratory center is reflexly inhibited during swallowing, when the airways are closed to prevent food from entering the lungs.

Human beings also have considerable voluntary control over ventilation. Voluntary control of breathing is accomplished by the cerebral cortex, which does not act on the respiratory center in the brain stem but instead sends impulses directly to the motor neurons in the spinal cord that supply the respiratory muscles. We can voluntarily hyperventilate ("overbreathe") or, to the other extreme, hold our breath, but only for a brief period of time. The resulting chemical changes in the arterial blood directly and reflexly influence the respiratory center, which in turn overrides the voluntary input to the motor neurons of the respiratory muscles. Other than these extreme forms of deliberately controlling ventilation, we also control our breathing to perform such voluntary acts as speaking, singing, and whistling.

During apnea, a person subconsciously "forgets to breathe," whereas during dyspnea, a person consciously feels that ventilation is inadequate.

Apnea is the transient cessation of ventilation with the expectation that breathing will resume spontaneously. The condition is called **respiratory arrest** if breathing does not resume. Because ventilation is normally decreased and the central chemoreceptors are less sensitive to the arterial P_{CO_2} drive during sleep, especially REM sleep (see p. 149), apnea is most likely to occur during this time. Victims of **sleep apnea** may stop breathing for a few seconds up to one or two minutes as many as five-hundred times a night. Mild forms of sleep apnea are not dangerous unless the sufferer has pulmonary or circulatory disease, the consequences of which can be compounded by recurrent bouts of apnea.

In exaggerated cases of sleep apnea, the victim may be unable to recover from an apneic period, resulting in death. This is the case in **sudden-infant-death syndrome** (**SIDS**), or "crib death." With this tragic form of sleep apnea, an otherwise healthy two- to five-month-old infant is found dead in his or her crib for no apparent reason. The underlying cause of

A CLOSER LOOK AT EXERCISE PHYSIOLOGY

How to Find Out How Much Work You're Capable of Doing

The best single predictor of a person's work capacity is the determination of the maximum volume of O_2 the person is capable of using per minute to oxidize nutrient molecules for energy production. *Maximal O_2 consumption*, or *max* VO_2, is measured by having the person engage in exercise, usually on a treadmill or bicycle ergometer (a stationary bicycle with variable resistance). The work load is incrementally increased until the person becomes exhausted. Expired air samples collected during the last minutes of exercise, when O_2 consumption is at a maximum because the person is working as hard as possible, are analyzed for the percentage of O_2 and CO_2 they contain. Furthermore, the volume of air expired is measured. Equations are then employed to determine the amount of O_2 consumed, taking into account the percentage of O_2 and CO_2 in the inspired air, the total volume of air expired, and the percentage of O_2 and CO_2 in the exhaled air.

Maximal O_2 consumption depends on three systems. The respiratory system is essential for ventilation and exchange of O_2 and CO_2 between the air and blood in the lungs. The circulatory system is required for delivery of O_2 to the working muscles. Finally, the muscles must have the oxidative enzymes available to use the O_2 once it has been delivered.

Regular aerobic exercise can improve max VO_2 by making the heart and respiratory system more efficient, thereby delivering more O_2 to the working muscles. Exercised muscles themselves become better equipped to use O_2 once it is delivered. The number of functional capillaries is increased, as is the number and size of mitochondria, which contain the oxidative enzymes.

Maximal O_2 consumption is measured in liters per minute and then converted into milliliters per kilogram of body weight per minute so that large and small people can be compared. As would be expected, athletes have the highest values for maximal O_2 consumption. The max VO_2 for male cross-country skiers has been recorded to be as high as 94 ml O_2/kg/min. Distance runners maximally consume between 65–85 ml O_2/kg/min, and football players have max VO_2 values between 45–65 ml O_2/kg/min, depending on the position they play. Sedentary young men maximally consume between 25 and 45 ml O_2/kg/min. Female values for max VO_2 are 20% to 25% lower than for males when expressed as ml/kg/min of total body weight. The difference in max VO_2 between females and males is only 8% to 10% when expressed as ml/kg/min of lean body weight, however, because females generally have a higher percentage of body fat (the female sex hormone estrogen promotes fat deposition).

Available norms are used to classify people as being low, fair, average, good, or excellent in aerobic capacity for their age group. Determination of max VO_2 is used by exercise physiologists to prescribe or adjust training regimens to help people achieve their optimal level of aerobic conditioning.

SIDS is the subject of intense investigation. Most evidence suggests that the baby "forgets to breathe" as a result of the immaturity of the respiratory control mechanisms, either in the brain stem or in the chemoreceptors that monitor the body's respiratory status. For example, on autopsy, more than half the victims have been shown to have poorly developed carotid bodies, the more important of the peripheral chemoreceptors.

In contrast to sleep apnea, in which the victim unconsciously stops breathing, people who have **dyspnea** have the subjective sensation that they are not getting enough air; that is, they feel "short of breath." Dyspnea is the mental anguish associated with the unsatiated desire for more adequate ventilation. It often accompanies the labored breathing characteristic of obstructive lung disease or the pulmonary edema associated with congestive heart failure. In contrast, during exercise a person can breathe very hard without experiencing the sensation of dyspnea, because such exertion is not accompanied by a sense of anxiety over the adequacy of ventilation. Surprisingly, dyspnea is not directly related to chronic elevation of arterial P_{CO_2} or reduction of P_{O_2}. The subjective feeling of air hunger may occur even when alveolar ventilation and the blood gases are normal. Some individuals experience dyspnea when they ***perceive*** that they are short of air even though this is not actually the case, such as when they are in a crowded elevator.

Chapter in Perspective

Energy is essential for sustaining life-supporting cellular activities, such as active transport and protein synthesis. The cells of the body require a continual supply of O_2 to support their energy-generating metabolic reactions. The CO_2 produced during these metabolic reactions must be eliminated from the body at the same rate it is produced to prevent dangerous fluctuations in acid-base balance. Respiration involves the sum of the processes that accomplish ongoing passive movement of O_2 from the atmosphere to the tissue cells to support cellular metabolism, as well as the continual passive movement of metabolically produced CO_2 from the tissue cells to the atmosphere. The respiratory system contributes to homeostasis by exchanging O_2 and CO_2 between the atmosphere and blood. This exchange is accomplished across the alveoli, the lungs' tiny air sacs, each of which is surrounded by a vast network of pulmonary capillaries. Systemic venous blood is brought to one side of this blood-gas interface by the force of cardiac contraction, which drives blood through the pulmonary circulation. Atmospheric air is brought to the other side of the interface by the process of ventilation, or breathing, which is accomplished by the respiratory muscles. Inflation of the lungs causes the intra-alveolar pressure (the gas pressure within the lungs) to become less than atmospheric pressure. The consequence of this is inflow of air through the conducting respiratory airways until intra-alveolar pressure equilibrates with atmospheric pressure; that is, an inspiration occurs. Conversely, deflation of the lungs causes the intra-alveolar pressure to rise above atmospheric pressure so that air flows out of the lungs until the two pressures are equal; that is, an expiration occurs.

Another pressure gradient—the difference in the partial pressure of O_2 and CO_2 between the alveolar air and pulmonary blood—largely determines the extent of gas exchange across the blood-gas interface. The surface area of this interface is extremely large and its thickness extremely thin, making it highly conducive to gas exchange. Alveolar P_{O_2} remains relatively high and alveolar P_{CO_2} remains relatively low because a portion of the alveolar air is exchanged for fresh atmospheric air with each breath. In contrast, the systemic venous blood entering the lungs is relatively low in O_2 and high in CO_2, having given up O_2 and picked up CO_2 at the systemic capillary level. This establishes partial-pressure gradients between the alveolar air and pulmonary capillary blood that induce the passive diffusion of O_2 into the blood and CO_2 out of the blood until the blood and alveolar partial pressures become equal. The blood leaving the lungs is thus relatively high in O_2 and low in CO_2 compared to the partial pressures in the O_2-consuming, CO_2-producing tissue cells to which it is delivered. Consequently, partial-pressure gradients for gas exchange at the tissue level favor the passive movement of O_2 out of the blood into the cells to support their metabolic requirements and favor the simultaneous transfer of CO_2 into the blood. The blood then returns to the lungs to once again fill up on O_2 and dump off CO_2. The systemic arterial P_{O_2} and P_{CO_2} normally remain essentially constant, having equilibrated with the alveolar partial pressures, which remain essentially constant. In contrast, the systemic venous P_{O_2} and P_{CO_2} vary, depending on the level of metabolic activity.

Because oxygen is poorly soluble in the blood, it is transported in the blood primarily bound to hemoglobin (Hb). Carbon dioxide is transported in the blood in three ways: physically dissolved, bound to hemoglobin, and (the largest portion) as HCO_3^-. The conversion of CO_2 into HCO_3^- by the chemical reaction $CO_2 + H_2O \rightleftharpoons H_2CO_3 \rightleftharpoons H^+ + HCO_3^-$ is catalyzed by carbonic anhydrase, an erythrocytic enzyme.

The constancy of systemic arterial P_{O_2} and P_{CO_2} is attributable to respiratory control mechanisms aimed at matching the magnitude of ventilation to varying levels of body activity. Respiratory control centers in the brain stem establish the rhythmic inspiration-expiration pattern and are also responsive to inputs that monitor the need for changing the magnitude of ventilation. Determinants of the magnitude of ventilation are respiratory rate and depth of breathing, which in turn depend on the level of respiratory muscle activity. The primary input that causes the respiratory center to adjust the magnitude of ventilation by altering its output to the respiratory muscles is the arterial P_{CO_2} as reflected by the brain extracellular fluid H^+ concentration. Increased arterial P_{CO_2} stimulates ventilation to blow off the excess CO_2; decreased arterial P_{CO_2} depresses respiratory activity so that metabolically produced CO_2 can accumulate. As an emergency mechanism, a threateningly low arterial P_{O_2} reflexly stimulates ventilation. The third chemical factor to which the respiratory center is responsive is arterial H^+ concentration, the importance of which lies in respiratory compensations for nonrespiratory-induced deviations in acid-base balance.

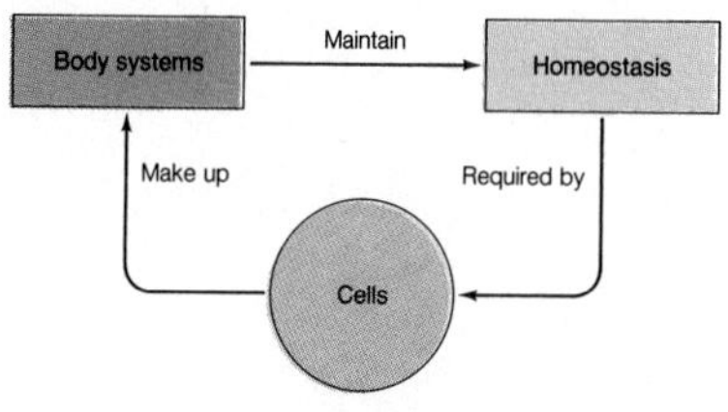

See inside front cover for an expanded version of this model.

Review Exercises

1. Distinguish between internal and external respiration. List the steps in external respiration.
2. Describe the components of the respiratory system.
3. Compare atmospheric, intra-alveolar, and intrapleural pressure.
4. Why are the lungs normally stretched, even during expiration?
5. Explain why air enters the lungs during inspiration and leaves during expiration.
6. Why is inspiration normally active and expiration normally passive?
7. Why does airway resistance become an important determinant of airflow rates in chronic obstructive pulmonary disease?
8. What factors control airway resistance?
9. Explain pulmonary elasticity in terms of elastic recoil and compliance.
10. Define the various lung volumes and capacities.
11. Compare pulmonary ventilation and alveolar ventilation. What is the consequence of anatomic and alveolar dead space?
12. What determines the partial pressures of a gas in air and in blood? What are the partial-pressure gradients for O_2 and CO_2 that are responsible for net movement of O_2 between the lungs and tissues and of CO_2 between the tissues and lungs?
13. List the methods of O_2 and CO_2 transport in the blood.
14. What is the primary factor that determines the percent hemoglobin saturation? What is the significance of the plateau and the steep portions of the O_2-Hb dissociation curve?
15. How does hemoglobin promote the net transfer of O_2 from the alveoli to the blood?
16. Explain the Bohr and Haldane effects.
17. Define the following: hypoxic hypoxia, anemic hypoxia, circulatory hypoxia, histotoxic hypoxia, hypercapnia, hypocapnia, hyperventilation, hypoventilation, hyperpnea, apnea, and dyspnea.
18. What factors contribute to rhythmicity of breathing?
19. Describe the mechanism of action and contributions of decreased arterial P_{O_2}, increased arterial P_{CO_2}, and increased arterial H^+ in control of ventilation.
20. **A point to ponder:** Why is it important that airplane interiors are pressurized?

CHAPTER 14

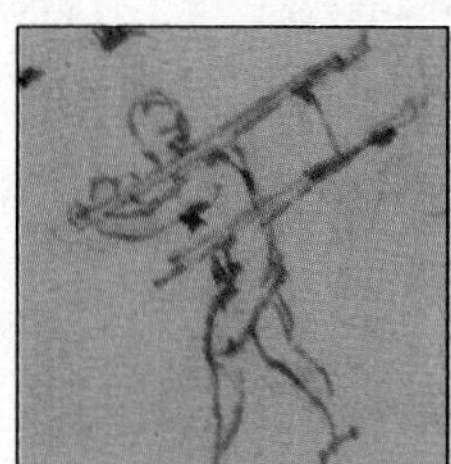

URINARY SYSTEM

CHAPTER CONTENTS AT A GLANCE

INTRODUCTION *There would be no animal life-forms on dry land today if it were not for the development of kidneys. The simplest forms of life live in an external environment of fixed composition, the sea. Likewise, the individual cells of more complex multicellular organisms are able to function and survive only in a fluid environment of essentially constant composition similar to the sea. This salty internal fluid environment is the extracellular fluid that bathes all the cells of the body.*

Simple marine organisms have no influence over the fluid environment in which they live. Exchanges between a simple life-form and its aqueous environment do not alter the external environment's composition because the volume of the organism's body is so small compared to the tremendous volume of the sea in which the organism lives. In contrast, in a complex land animal, exchanges between the cells and the extracellular fluid could notably alter the composition of this small, private internal fluid environment were it not for mechanisms to maintain its stability. To become separated from a watery environment of fixed composition and be free to move about in a dry and ever-changing external environment, land animals have internalized their own bit of sealike water and are equipped with mechanisms to maintain its constancy.

To a large extent, terrestrial animals are able to live on dry land independent of the sea because of their kidneys, the organs which, in concert with the hormonal and neural inputs that control their function, are primarily responsible for maintaining the stability of extracellular-fluid volume and electrolyte composition. By adjusting the quantity of water and vari-

ous plasma constituents that are either conserved for the body or eliminated in the urine, the kidneys are able to maintain water and electrolyte balance within the very narrow range compatible with life, despite wide variations in intake and losses of these constituents through other avenues. To illustrate, the following individuals have extracellular fluid bathing their cells that is of remarkably similar volume and salt composition: a person drinking four glasses of water a day compared to one who consumes 20 quarts of fluid while participating in a drinking contest; or a person on a low-salt diet who ingests a mere 0.5 gm of salt daily compared to a gross consumer of salt who takes in 20 gms of salt in a day.

When there is a surplus of water or a particular electrolyte such as salt in the extracellular fluid (ECF), the kidneys can eliminate the excess in the urine. If there is a deficit, the kidneys cannot actually provide additional quantities of the depleted constituent, but they can limit the urinary losses of the material in short supply and thus conserve it until more of the depleted substance can be ingested. Accordingly, the kidneys can compensate more efficiently for excesses than for deficits. This is further reflected by the fact that in some instances the kidneys cannot completely halt the loss of a particular valuable substance in the urine, even though the substance may be in short supply, a prime example being the case of a H_2O deficit. Even if a person is not consuming any H_2O, the kidneys are obligated to put out about half a liter of H_2O in the urine each day to accomplish their other role as the body's "cleaners."

In addition to the kidneys' important regulatory role in maintaining fluid and electrolyte balance, they are the primary route for elimination of potentially toxic metabolic wastes and foreign compounds from the body. These wastes cannot be eliminated in solid form; they must be excreted in solution, obligating the kidneys to produce a minimum volume of around 500 ml of waste-filled urine per day. Because the H_2O eliminated in the urine is derived from the blood plasma, a person stranded without H_2O is eventually obligated to urinate himself or herself to death by depleting the plasma volume to a fatal level as H_2O is inexorably removed to accompany the wastes. Fortunately, except under such extreme circumstances, the kidneys are able to maintain stability in the internal fluid environment despite the usual variations in intake of fluids and salts.

Not only are the kidneys able to adjust for wide variations in ingestion of H_2O, salt, and other electrolytes, but they also make adjustments in the urinary output of these ECF constituents to compensate for their abnormal losses through heavy sweating, vomiting, diarrhea, or hemorrhage. Thus urine composition varies widely as the kidneys adjust for differences in intake as well as losses of various substances in an attempt to maintain the ECF within the narrow limits compatible with life.

The kidneys perform a variety of functions aimed at maintaining homeostasis.

The following are the specific functions performed by the kidneys, most of which are directed toward preserving the internal fluid environment's constancy. The kidneys:

1. *maintain H_2O balance* in the body.
2. *regulate the quantity and concentration of most ECF ions*, including Na^+, Cl^-, K^+, HCO_3^-, Ca^{++}, Mg^{++}, $SO_4^=$, $PO_4^{\equiv}$, and H^+. Even minor fluctuations in the ECF concentrations of some of these electrolytes can have profound influences. For example, changes in the ECF concentration of K^+ can potentially lead to fatal cardiac dysfunction.
3. *maintain proper plasma volume*, thereby contributing significantly to the long-term regulation of arterial blood pressure. This is accomplished through the kidneys' regulatory role in salt (NaCl) and H_2O balance.
4. *assist in maintaining the proper acid-base balance* of the body by adjusting urinary output of H^+ and HCO_3^-.
5. *maintain the proper osmolarity* (concentration of solutes) of body fluids, primarily through regulation of H_2O balance.
6. *excrete (eliminate) the end-products (wastes) of bodily metabolism* such as urea, uric acid, and creatinine. If allowed to accumulate, these wastes are toxic, especially to the brain.
7. *excrete many foreign compounds* such as drugs, food additives, pesticides, and other exogenous non-nutritive materials that have gained entrance to the body.
8. *secrete erythropoietin*, a hormone that stimulates red blood cell production (see p. 347).
9. *secrete renin*, an enzymatic hormone that triggers a chain reaction important in the process of salt conservation by the kidneys.
10. *convert vitamin D into its active form.*

The kidneys form the urine; the remainder of the urinary system is ductwork to carry the urine to the outside.

The **urinary system** consists of the urine-forming organs—the **kidneys**—and the structures that carry the urine from the kidneys to the outside for elimination from the body (Fig. 14–1a). The kidneys are a pair of bean-shaped organs that lie in the back of the abdominal cavity, one on each side of the vertebral column slightly above the level of the waistline. Each kidney is supplied by a **renal artery** and a **renal vein,** which, respectively, enter and leave the kidney at the medial indentation that gives this organ its beanlike form. The kidney acts on the plasma flowing through it to produce urine, conserving materials to be retained in the body and eliminating unwanted materials into the urine.

After urine is formed, it drains into a central collecting cavity, the **renal pelvis,** located at the medial inner core of each kidney (Fig. 14–1b). From there urine is channeled into the **ureter,** a smooth muscle-walled duct that exits at the medial border in close proximity to the renal artery and vein. There are two ureters, one carrying urine from each kidney to the single urinary bladder.

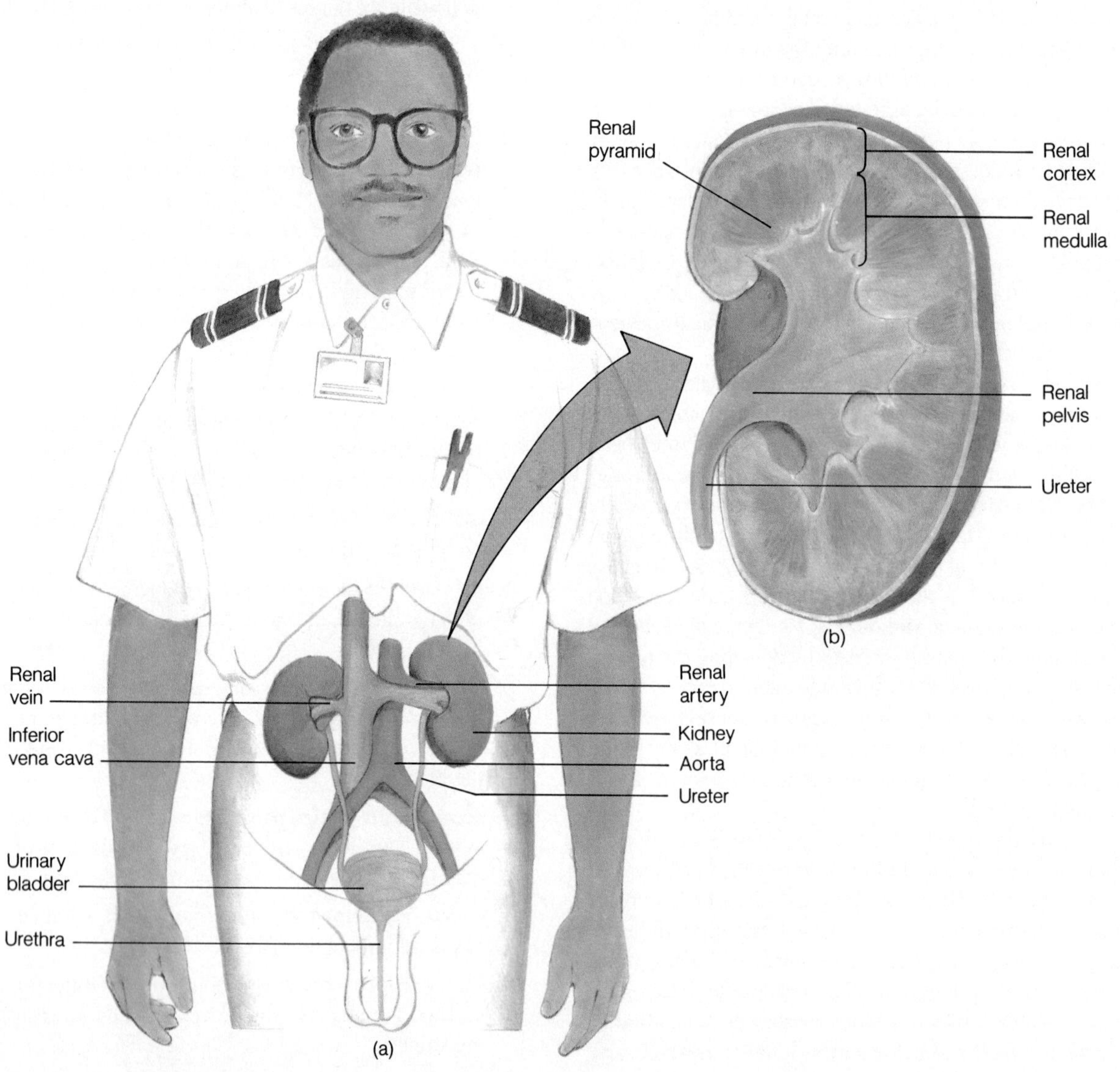

Figure 14–1 The Urinary System *(a) Location of the components of the urinary system. (b) Longitudinal section of a kidney.*

The **urinary bladder,** which temporarily stores urine, is a hollow, distensible sac whose volume can be adjusted by varying the contractile state of the smooth muscle within its walls. Periodically, urine is emptied from the bladder to the outside through another tube, the **urethra.** The urethra in females is straight and short, passing directly from the neck of the bladder to the outside (Fig. 20–2). In males the urethra is much longer and follows a curving course from the bladder to the outside, passing through both the prostate gland and penis (Fig. 14–1a; also see Fig. 20–1). The male urethra serves the dual function of providing a route for elimination of urine from the bladder and a passageway for semen from the reproductive organs. The prostate gland lies below the neck of the bladder and completely encircles the urethra. **Prostatic hypertrophy** (enlargement), which often occurs during middle to older age, can partially or completely occlude the urethra, thereby impeding the flow of urine.

The parts of the urinary system beyond the kidneys merely serve as ductwork to transport urine to the outside. Once formed by the kidneys, urine is not altered in composition or volume as it moves downstream through the remainder of the urinary system.

The nephron is the functional unit of the kidney.

Each kidney is composed of about 1 million microscopic-sized functional units known as **nephrons,** which are bound together by connective tissue. A functional unit is the smallest unit within an organ capable of performing all of that organ's functions. Because the primary function of the kidneys is to produce urine, and, in so doing, maintain constancy in the ECF composition, a nephron is the smallest unit capable of urine formation.

The arrangement of nephrons within the kidneys gives rise to two distinct regions—an outer granular-appearing region, the **renal cortex,** and an inner region of striated triangular **renal pyramids**—which collectively compose the **renal medulla** (Fig. 14–1b).

Knowledge of the structural arrangement of an individual nephron is essential for understanding the distinction between the cortical and medullary regions of the kidney and, more importantly, for understanding renal function. Each nephron consists of a *vascular component* and a *tubular component,* both of which are intimately related structurally and functionally (Fig. 14–2). The dominant portion of the vascular component is the **glomerulus,** a network or tuft of twenty to forty capillaries through which part of the water and solutes are filtered from the blood passing through. This filtered fluid, which is almost identical in composition to the plasma, then passes through the tubular component of the nephron, where it is modified by various transport processes that convert it into urine.

On entering the kidney, the renal artery systematically subdivides to ultimately form many small vessels known as **afferent arterioles,** one of which supplies each nephron. The afferent arteriole delivers blood to the glomerular capillaries, which rejoin to form another arteriole, the **efferent arteriole,** through which blood that was not filtered into the tubular component leaves the glomerulus (Fig. 14–3). The efferent arterioles are the only arterioles in the body that drain from capillaries. Typically, arterioles break up into capillaries that rejoin to form venules. At the glomerular capillaries, no O_2 or nutrients are extracted from the blood for use by the kidney tissues nor are waste products picked up from the surrounding tissue. Thus arterial blood enters the glomerular capillaries through the afferent arteriole, and arterial blood leaves the glomerulus through the efferent arteriole.

The efferent arteriole quickly subdivides into a second set of capillaries, the **peritubular capillaries,** which supply the renal tissue with blood and which are important in exchanges between the tubular system and blood during conversion of the filtered fluid into urine. These peritubular capillaries, as their name implies (*peri* means "around"), are intertwined around the tubular system. The peritubular capillaries rejoin to form venules that ultimately drain into the renal vein, by which blood leaves the kidney.

The tubular component of each nephron is a hollow, fluid-filled tube formed by a single layer of epithelial cells. Even though the tubule is continuous from its beginning in close proximity to the glomerulus to its ending at the renal pelvis, it is arbitrarily divided into various segments based on differences in structure and function that occur along its length (Figs. 14–2 and 14–4). The tubular component begins with **Bowman's capsule,** an expanded, double-walled invagination that cups around the glomerulus to collect the fluid filtered from the glomerular capillaries. Because of the intimate relationship between the glomerulus and Bowman's capsule, these structures are often referred to collectively as a **renal corpuscle.** All renal corpuscles are located in the cortex, and this is responsible for the region's granular appearance.

From Bowman's capsule, the filtered fluid passes into the **proximal tubule,** which lies entirely within the cortex and which is highly coiled or convoluted throughout much of its course. The next segment, the **loop of Henle,** forms a sharp U-shaped or hairpin loop that dips into the renal medulla. The *descending limb* of Henle's loop plunges from the cortex into the medulla; the *ascending limb* traverses back up into the cortex. The ascending limb returns to the glomerular region of its own nephron, where it passes through the fork formed by the afferent and efferent arterioles. Both the tubular and vascular

Figure 14–2 A Nephron

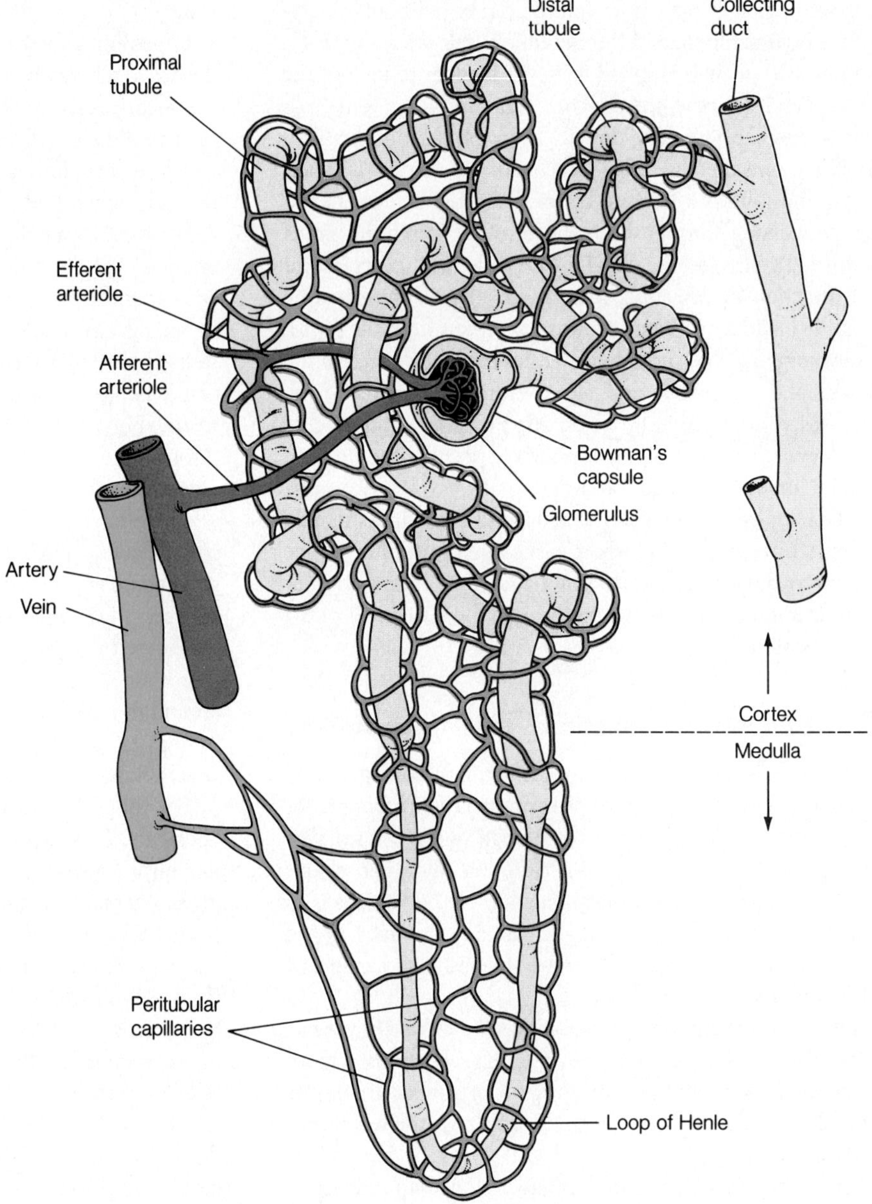

Overview of Functions of Parts of a Nephron

Vascular component

- Afferent arteriole—carries blood to glomerulus
- Glomerulus—tuft of capillaries that filters a protein-free plasma into the tubular component
- Efferent arteriole—carries blood from glomerulus
- Peritubular capillaries—supply renal tissue; involved in exchanges with fluid in tubular lumen

Tubular component

- Bowman's capsule—collects glomerular filtrate
- Proximal tubule—uncontrolled reabsorption and secretion occur here
- Loop of Henle—establishes osmotic gradient in renal medulla that is important in kidney's ability to produce urine of varying concentration
- Distal tubule—controlled reabsorption and secretion occur here
- Collecting tubule—variable H_2O reabsorption occurs here; fluid leaving the collecting tubule is urine, which enters the renal pelvis

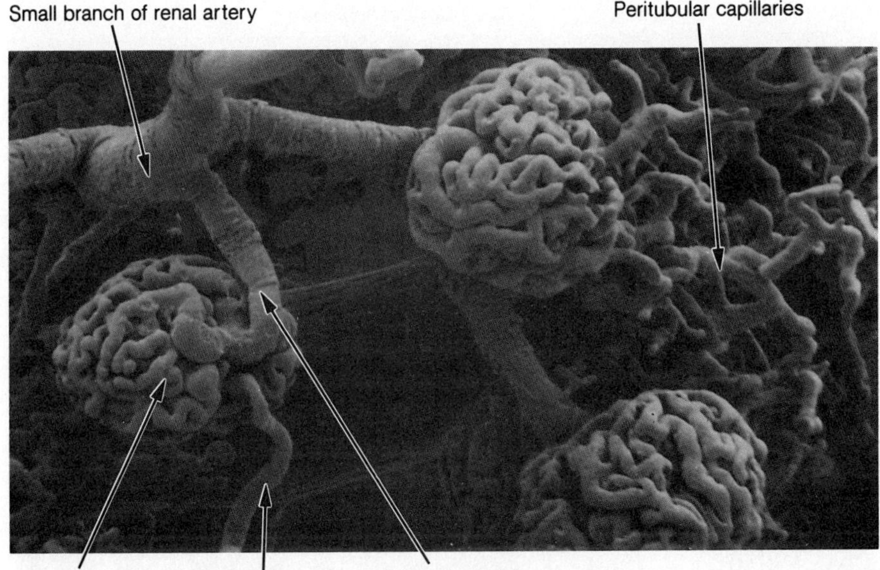

Figure 14–3 Glomerulus and Associated Arterioles

SOURCE: From *Tissues and Organs: A Text-Atlas of Scanning Electron Microscopy* by Richard G. Kessel and Randy H. Kardon. Copyright © 1979 W.H. Freeman and Company. Reprinted with permission.

cells at this point are specialized to form the **juxtaglomerular** (*juxta* means "next to") **apparatus,** a structure that plays several important roles in regulating kidney function. Beyond the juxtaglomerular apparatus, the tubule once again becomes highly coiled to form the **distal tubule,** which also lies entirely within the cortex. The distal tubule empties into a **collecting duct** or **tubule,** with each collecting duct draining fluid from up to eight separate nephrons. Each collecting duct plunges down through the medulla to empty its fluid contents (which have now been converted into urine) into the renal pelvis.

There are two types of nephrons—**cortical nephrons** and **juxtamedullary nephrons**—depending on the location and length of some of their structures (Fig. 14–4). (Note the distinction between *juxtamedullary* nephrons and *juxtaglomerular* apparatus.) All nephrons originate in the cortex, but the glomeruli of cortical nephrons lie in the outer layer of the cortex, whereas the glomeruli of juxtamedullary nephrons lie in the inner layer of the cortex, adjacent to the medulla. The most distinguishing feature between these two nephron types is their loops of Henle. The hairpin loop of cortical nephrons dips only slightly into the medulla. In contrast, the loop of juxtamedullary nephrons plunges through the entire depth of the medulla. Furthermore, the peritubular capillaries of juxtamedullary nephrons form hairpin vascular loops known as **vasa recta** ("straight vessels"), which run in close association with the loops of Henle. As they course through the medulla, the collecting ducts of both cortical and juxtamedullary nephrons run parallel to the ascending and descending limbs of the juxtamedullary nephrons' loops of Henle and vasa recta. This parallel arrangement of tubules and vessels creates the medullary tissue's striated appearance. More importantly, as you will see, this arrangement, coupled with the permeability and transport characteristics of the long loops of Henle and vasa recta, plays a key role in the kidneys' ability to produce urine of varying concentrations, depending on the needs of the body. About 80% of the nephrons in humans are of the cortical type. Species with greater urine-concentrating abilities than humans, such as the desert rat, have a greater proportion of juxtamedullary nephrons.

The three basic renal processes are glomerular filtration, tubular reabsorption, and tubular secretion.

There are three basic processes involved in the formation of urine: glomerular filtration, tubular reabsorption, and tubular secretion. To aid in visualizing the relationships among these renal processes, it is useful to unwind the nephron schematically, as in Figure 14–5.

As blood flows through the glomerulus, filtration of protein-free plasma occurs through the glomerular capillaries into Bowman's capsule. This process, known as **glomerular filtration,** is the first step in urine formation. On the average, 180 liters (about 47.5 gallons) of glomerular filtrate (filtered fluid) are formed each day. Considering that the average plasma volume in an adult is 2.75 liters, this means that the entire plasma volume is filtered by the kidneys about sixty-five

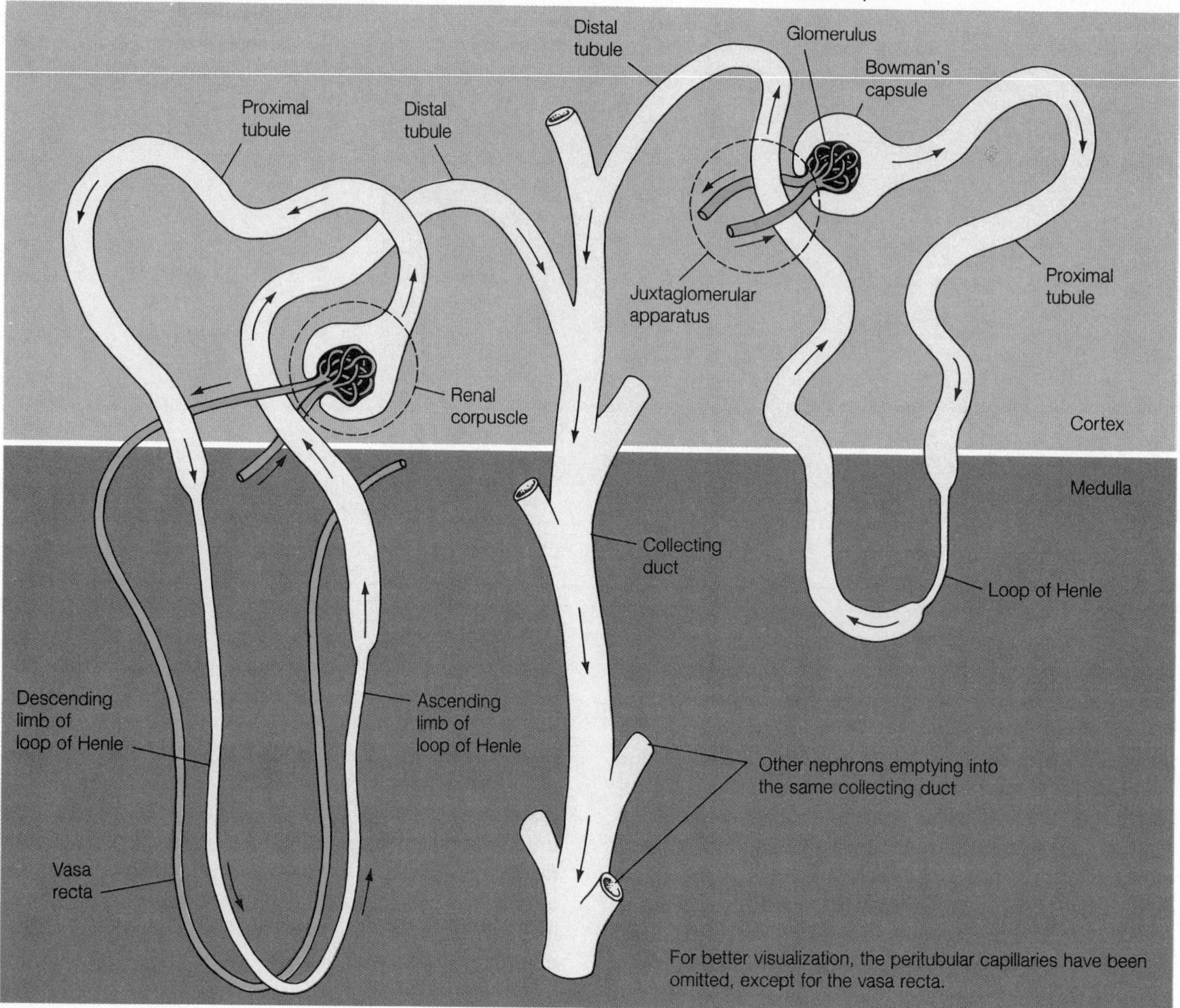

Figure 14–4 Comparison of Cortical and Juxtamedullary Nephrons

times per day. If everything filtered were to pass out in the urine, the total plasma volume would be urinated in less than half an hour! This does not happen, however, because the kidney tubules and peritubular capillaries are intimately related throughout their lengths, so transfer of materials can occur between the fluid inside the tubules and the blood within the peritubular capillaries.

As the filtrate flows through the tubules, substances of value to the body are returned to the peritubular capillary plasma. This selective movement of substances from inside the tubule (the tubular lumen) into the blood is referred to as **tubular reabsorption.** Reabsorbed substances are not lost from the body in the urine but are carried instead by the peritubular capillaries to the venous system and then to the heart to be recirculated again. Of the 180 liters of plasma filtered per day, 178.5 liters on the average are reabsorbed, with the remaining 1.5 liters passing into the renal pelvis to be eliminated as urine. In general, substances that need to be conserved by the body are selectively reabsorbed, whereas unwanted substances that need to be eliminated remain in the urine.

The third renal process, **tubular secretion,** which refers to the selective transfer of substances from the peritubular capillary blood into the tubular lumen, provides a second route for substances to enter the renal tubules from the blood. The first means by which substances move from the plasma into the tubular lumen is by glomerular filtration. However, only about 20% of the plasma flowing through the glomerular capillaries is filtered into Bowman's capsule; the remaining 80% flows on through the efferent arteriole into the peritubular capillaries. A few substances may be discriminately transferred by tubular secretion from the plasma in the peritubular capillaries into the tubular lumen. Tubular secretion provides a mechanism for more rapidly eliminating selected substances from the plasma by extracting an additional quantity of a particular substance from the 80% of unfiltered plasma in the peritubular capillaries and adding it to the quantity of the substance already present in the tubule as a result of filtration.

Urine excretion refers to the elimination of substances from the body in the urine. It is not really a separate process but is the result of the first three processes. All plasma constituents that gain access to the tubules—that is, that are filtered or secreted but not reabsorbed—remain in the tubules, passing into the renal pelvis to be excreted as urine. (Do not confuse *excretion* with *secretion*.)

Glomerular filtration is largely an indiscriminate process. With the exception of blood cells and plasma proteins, all constituents within the blood—H_2O, nutrients, electrolytes, wastes, and so on—are nonselectively filtered. The highly discriminating tubular processes then go to work on the filtrate to return to the blood a fluid of the composition and volume necessary to maintain the constancy of the internal fluid environment. The unwanted filtered material is left behind in the tubular fluid to be excreted as urine. Glomerular filtration can be thought of as pushing a portion of plasma, with all of its essential components as well as those that need to be eliminated from the body, onto a tubular "conveyor belt" that terminates at the renal pelvis, which is the collecting point within the kidney for urine. All plasma constituents that enter this conveyor belt and that are not subsequently returned to the plasma by the end of the line are spilled out of the kidney as urine. It is up to the tubular system to salvage by reabsorption the filtered materials that need to be preserved for the body while leaving behind substances that need to be excreted. In addition, some substances are not only filtered but are also secreted onto the tubular conveyor belt so that the amounts of these substances excreted in the urine are greater than the amounts that were filtered. For many substances, these renal processes are subject to physiological control. Thus the kidneys handle each constituent in the plasma in a characteristic manner by a particular combination of filtration, reabsorption, and secretion.

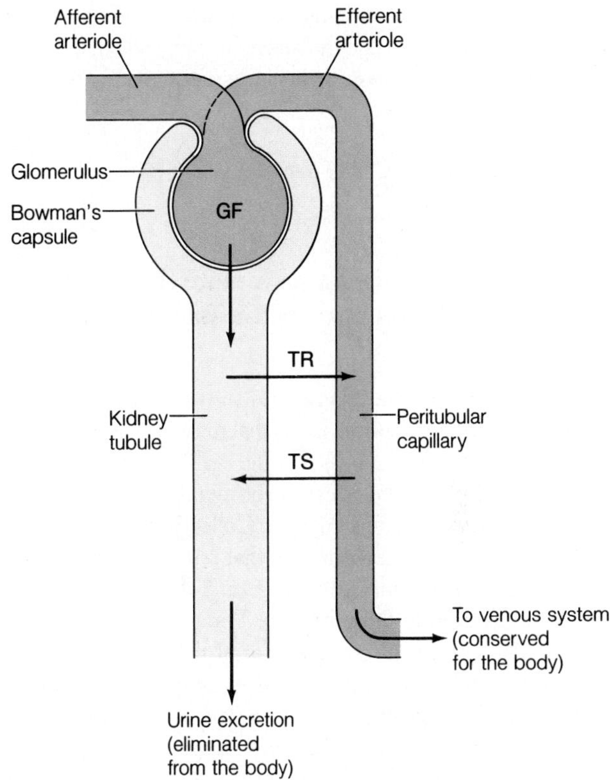

GF = Glomerular filtration, nondiscriminant filtration of a protein-free plasma from glomerulus into Bowman's capsule

TR = Tubular reabsorption, selective movement of filtered substances from tubular lumen into peritubular capillaries

TS = Tubular secretion, selective movement of nonfiltered substances from peritubular capillaries into tubular lumen

Figure 14-5 Basic Renal Processes *Anything filtered or secreted but not reabsorbed is excreted in the urine.*

The kidneys act only on the plasma, yet extracellular fluid consists of both plasma and interstitial fluid. The interstitial fluid is actually the true internal fluid environment of the body, because it is the only extracellular fluid that comes into direct contact with the cells. However, because of the free exchange between plasma and interstitial fluid across the capillary walls (with the exception of plasma proteins), interstitial-fluid composition reflects the composition of plasma. Thus, by performing their regulatory and excretory roles on the plasma, the kidneys maintain the proper interstitial-fluid environment for optimal cell function. Most of the remainder of this chapter will be devoted to considering how the basic renal processes are accomplished and the mechanisms by which they are carefully regulated to help maintain homeostasis.

Glomerular Filtration

The glomerular membrane is more than one-hundred times more permeable than capillaries elsewhere.

Fluid filtered from the glomerulus into Bowman's capsule must pass through the three layers that make up the **glomerular membrane** (Fig. 14–6): (1) the wall of the glomerular capillaries; (2) an acellular basement membrane; and (3) the inner layer of Bowman's capsule. Collectively, these layers function as a fine molecular sieve that retains the blood cells and plasma proteins but permits H_2O and solutes of small molecular dimension to filter through. The glomerular capillary wall, which consists of a single layer of flattened endothelial cells, is perforated by many large pores, or fenestrae, that render it more than one-hundred times more permeable to H_2O and solutes than capillaries elsewhere. A basement membrane composed of collagen and glycoproteins is sandwiched between the glomerulus and Bowman's capsule. The collagen provides structural strength, and the glycoproteins discourage the filtration of small plasma proteins. Even though the larger plasma proteins cannot be filtered because they cannot fit through the capillary pores, the pores are actually large enough to permit passage of albumin, the smallest of plasma proteins. The glycoproteins, however, being strongly negatively charged, repel albumin and other plasma proteins because the latter are also negatively charged. Therefore, plasma proteins are almost completely excluded from the filtrate, with less than 1% of the albumin molecules escaping into Bowman's capsule. Some renal diseases characterized by excessive albumin in the urine (albuminuria) are believed to be due to disruption of the negative charges within the glomerular membrane so that the membrane is more permeable to albumin even though the size of the pores remains constant.

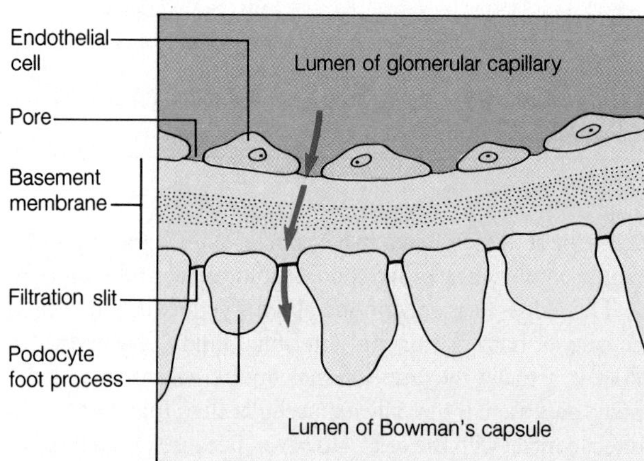

Figure 14–6 Layers of the Glomerular Membrane *To be filtered, a substance must pass through (1) the pores between the endothelial cells of the glomerular capillary, (2) an acellular basement membrane, and (3) the filtration slits between the foot processes of Bowman's capsule podocytes.*

The final layer of the glomerular membrane, which is the inner layer of Bowman's capsule, consists of **podocytes,** octopuslike cells that encircle the glomerular tuft. Each podocyte bears many elongated foot processes (*podo* means "foot") that interdigitate with foot processes of adjacent podocytes (Fig. 14–7), much as if you slid your fingers between each other as you cupped your hands around a ball. The narrow slits that exist between adjacent foot processes, known as **filtration slits,** provide a pathway through which fluid exiting the glomerular capillaries can enter the lumen of Bowman's capsule. Thus the route that filtered substances take across the glomerular membrane is completely extracellular—first through capillary pores, then through the acellular basement membrane, and finally through capsular filtration slits.

The major force responsible for inducing glomerular filtration is the glomerular-capillary blood pressure.

To accomplish glomerular filtration, a force must be present to drive a portion of the plasma in the glomerulus through the openings in the glomerular membrane. There are no active transport mechanisms or local energy expenditures involved in producing the movement of fluid from the plasma across the glomerular membrane into Bowman's capsule. Passive physical forces similar to those acting across capillaries elsewhere are responsible for glomerular filtration. Because the glomerulus is a tuft of capillaries, the same principles of fluid dynamics that are responsible for ultrafiltration across other capillaries apply (see p. 319), except for two important differences: (1) the glomerular capillaries are much more permeable than capillaries elsewhere, so more fluid is filtered for a given filtration pressure; and (2) the balance of forces across the glomerular membrane is such that filtration occurs throughout the entire length of the capillaries. This is in contrast to other capillaries, in which the balance of forces shifts so

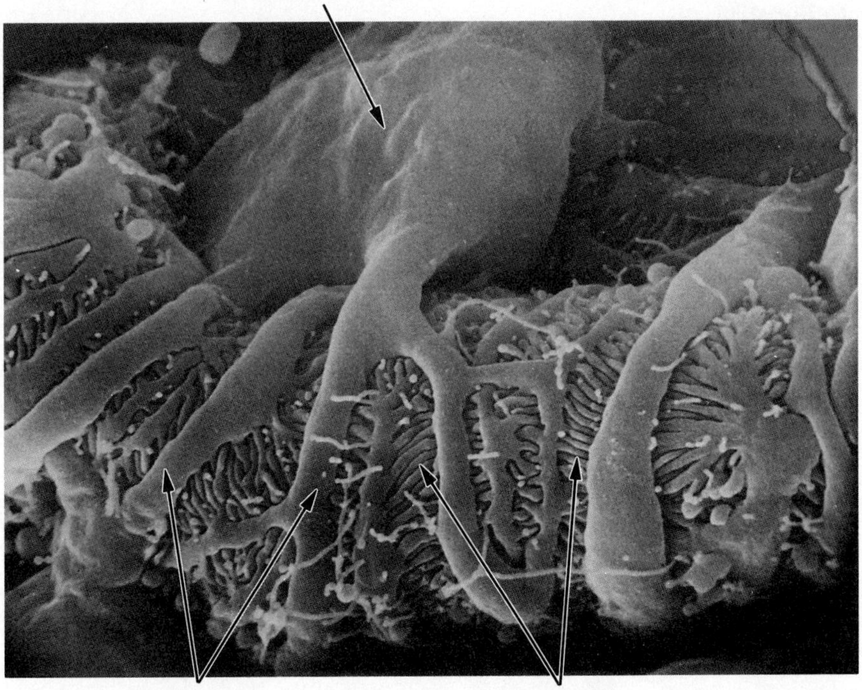

Cell body of podocyte

Foot processes

Filtration slits

Figure 14–7 Bowman's Capsule Podocytes with Foot Processes and Filtration Slits *Note the filtration slits between adjacent foot processes.*

SOURCE: From *Tissues and Organs: A Text-Atlas of Scanning Electron Microscopy* by Richard G. Kessel and Randy H. Kardon. Copyright © 1979 by W.H. Freeman and Company. Reprinted with permission.

that filtration occurs in the beginning portion of the vessel but reabsorption occurs toward the vessel's end.

Three physical forces are involved in glomerular filtration (Table 14–1): (1) glomerular-capillary blood pressure; (2) blood-colloid osmotic pressure; and (3) Bowman's capsule hydrostatic pressure. The glomerular-capillary blood pressure is the fluid pressure exerted by the blood within the glomerular capillaries. It is ultimately dependent on contraction of the heart (the source of energy that produces glomerular filtration) and the resistance to blood flow offered by the afferent and efferent arterioles. The glomerular-capillary blood pressure, at an estimated average value of 55 mm Hg, is higher than capillary blood pressure elsewhere, because the diameter of the afferent arteriole is larger than that of the efferent arteriole. Since blood can more readily enter the glomerulus through the wide afferent arteriole than it can leave through the narrower efferent arteriole, glomerular-capillary blood pressure is elevated as a result of the damming up of blood in the glomerular capillaries. Furthermore, because of the high resistance offered by the efferent arterioles, blood pressure does not have the same tendency to decrease along the length of the glomerular capillaries as it does along other capillaries. This elevated, nondecremental glomerular blood pressure tends to push fluid out of the glomerulus into Bowman's capsule along the glomerular capillaries' entire length, and it is the major force responsible for producing glomerular filtration.

Whereas glomerular-capillary blood pressure favors filtration, the two other forces acting across the glomerular membrane (blood-colloid osmotic pressure and Bowman's capsule hydrostatic pressure) oppose filtration. Blood-colloid osmotic pressure (see p. 318) is caused by the unequal distribution of plasma proteins across the glomerular membrane. Because plasma proteins cannot be filtered, they are present in the glomerular capillaries but are absent in Bowman's capsule. Accordingly, the concentration of H_2O is higher in Bowman's capsule than in the glomerular capillaries. The resultant tendency for H_2O to move by osmosis down its own concentration gradient from Bowman's capsule into the glomerulus opposes glomerular filtration. This opposing osmotic force averages 30 mm Hg, which is slightly higher than across other capillaries. It is higher because considerably more H_2O is filtered out of the glomerular blood, so the concentration of plasma proteins is higher than elsewhere.

The fluid in Bowman's capsule exerts a hydrostatic (fluid) pressure that is estimated to be about 15 mm Hg. This pressure, which tends to push fluid out of Bowman's capsule,

Table 14–1 Forces Involved in Glomerular Filtration

Force	Effect	Magnitude (mm Hg)
Glomerular-capillary blood pressure	Favors filtration	55
Blood-colloid osmotic pressure	Opposes filtration	30
Bowman's capsule hydrostatic pressure	Opposes filtration	15
Net filtration pressure (difference between force favoring filtration and forces opposing filtration)	Favors filtration	10
		55 − (30 + 15) = 10

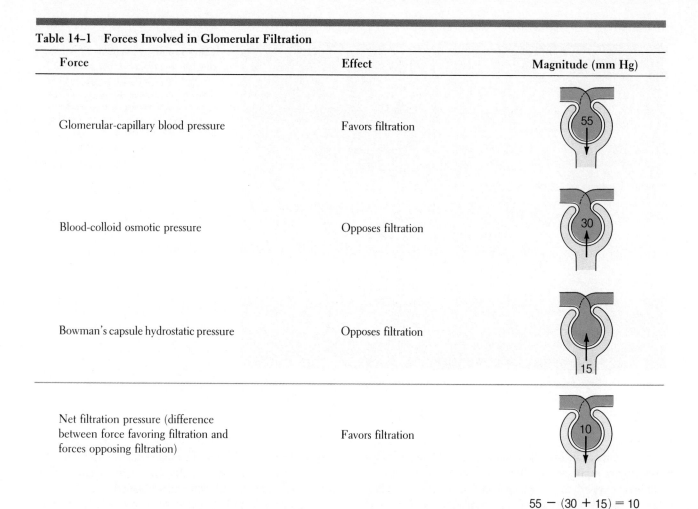

opposes the filtration of fluid from the glomerulus into Bowman's capsule.

As can be seen in Table 14–1, there is an imbalance in the forces acting across the glomerular membrane. The total force favoring filtration is attributable to the glomerular-capillary blood pressure at 55 mm Hg. The total of the two forces opposing filtration is 45 mm Hg. The net difference favoring filtration of 10 mm Hg of pressure is referred to as the **net filtration pressure.** This modest pressure is responsible for forcing large volumes of fluid from the blood through the highly permeable glomerular membrane.

The actual rate of filtration, the **glomerular filtration rate** (**GFR**), depends not only on the net filtration pressure but also on how much glomerular surface area is available for penetration and how permeable the glomerular membrane is (that is, how "holey" it is). These properties of the glomerular membrane are collectively referred to as the **filtration coefficient** (**K_f**). Accordingly:

$$\text{GFR} = K_f \times \text{Net filtration pressure}$$

Normally about 20% of the plasma that enters the glomerulus is filtered at the net filtration pressure of 10 mm Hg, producing collectively through all glomeruli 180 liters of glomerular filtrate each day and an average GFR of 125 ml/min in males and 160 liters of filtrate per day and an average GFR of 115 ml/min in females.

The most common factor resulting in a change in GFR is an alteration in the glomerular-capillary blood pressure.

Because the net filtration pressure responsible for inducing glomerular filtration is simply due to an imbalance of opposing physical forces between the glomerular-capillary plasma and Bowman's capsule fluid, alterations in any of these physical forces can affect the GFR. We will examine the effect changes in each of these physical forces have on the GFR.

Blood-colloid osmotic pressure and Bowman's capsule hydrostatic pressure are not subject to regulation and, under normal conditions, do not vary substantially. However, they can change pathologically and thus inadvertently affect the GFR. Because blood-colloid osmotic pressure opposes filtration, a decrease in plasma-protein concentration, by reducing this pressure, leads to an increase in the GFR. An uncontrollable reduction in plasma-protein concentration might occur, for example, in severely burned patients who lose a large quantity of protein-rich, plasma-derived fluid through the exposed burned surface of their skin. Conversely, in situations in which the blood-colloid osmotic pressure is elevated, such as in cases of dehydrating diarrhea, the GFR is reduced. Likewise, Bowman's capsule hydrostatic pressure can become uncontrollably elevated and filtration subsequently can decrease in the presence of a urinary tract obstruction, such as a kidney stone or prostatic hypertrophy. A damming up of fluid prior to the obstruction elevates capsular hydrostatic pressure.

Unlike these other pressures—which may be uncontrollably altered in various disease states, thereby inadvertently altering the GFR—glomerular-capillary blood pressure can be controlled to adjust the GFR to suit the body's needs. Assuming that all other factors remain constant, the magnitude of the glomerular-capillary blood pressure depends on the rate of blood flow in each of the glomeruli, which in turn is determined largely by the magnitude of the mean systemic arterial blood pressure and the resistance offered by the afferent arterioles. Control of GFR is accomplished by two mechanisms, both of which are directed at adjusting glomerular blood flow by regulating the caliber of the afferent arteriole. These are: (1) autoregulation, which is aimed at preventing spontaneous changes in GFR; and (2) extrinsic sympathetic control, which is aimed at long-term regulation of arterial blood pressure.

AUTOREGULATION OF GFR. Because the arterial blood pressure is the force that drives blood into the glomerulus, the glomerular-capillary blood pressure and, accordingly, the GFR would increase in direct proportion to an increase in arterial pressure if everything else remained constant (Fig. 14–8). Similarly, a fall in arterial blood pressure would be accompanied by a decline in GFR. Such spontaneous changes in GFR are largely prevented by intrinsic regulatory mechanisms initiated by the kidneys themselves, a process known as **autoregulation** (*auto* means "self"). The kidneys can, within limits, maintain a constant blood flow in the glomerular capillaries (and thus a constant glomerular-capillary blood pressure and a stable GFR) despite changes in the driving arterial pressure. They do so primarily by altering afferent arteriolar caliber, thereby adjusting resistance to flow through these vessels. For example, if the GFR increases as a direct result of a rise in arterial pressure, the net filtration pressure and GFR can be reduced to normal by constriction of the afferent arteriole (Fig. 14–9a), which decreases the flow of blood into the glomerulus. This local adjustment lowers the glomerular blood pressure and the GFR to normal. Conversely, when GFR falls in the presence of a decline in arterial pressure, glomerular pressure can be increased to normal by vasodilation of the afferent arteriole so that more blood enters despite the reduction in driving pressure (Fig. 14–9b). The resultant build-up of glomerular blood volume increases glomerular blood pressure, which in turn brings the GFR back up to normal.

The exact mechanisms responsible for accomplishing autoregulatory responses still elude complete understanding. Presently, two intrarenal mechanisms are thought to contribute to autoregulation: (1) a myogenic mechanism, which responds to changes in pressure within the nephron's vascular component; and (2) a tubulo-glomerular feedback mechanism, which senses changes in flow through the nephron's tubular component.

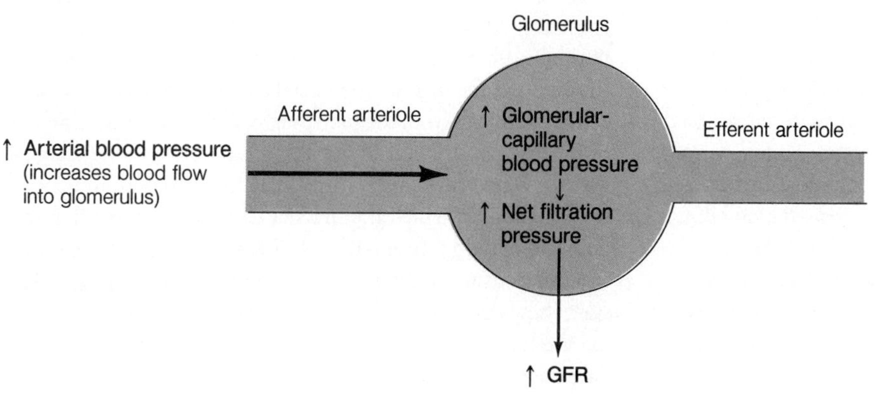

Figure 14–8 Direct Effect of Arterial Blood Pressure on Glomerular Filtration Rate (GFR)

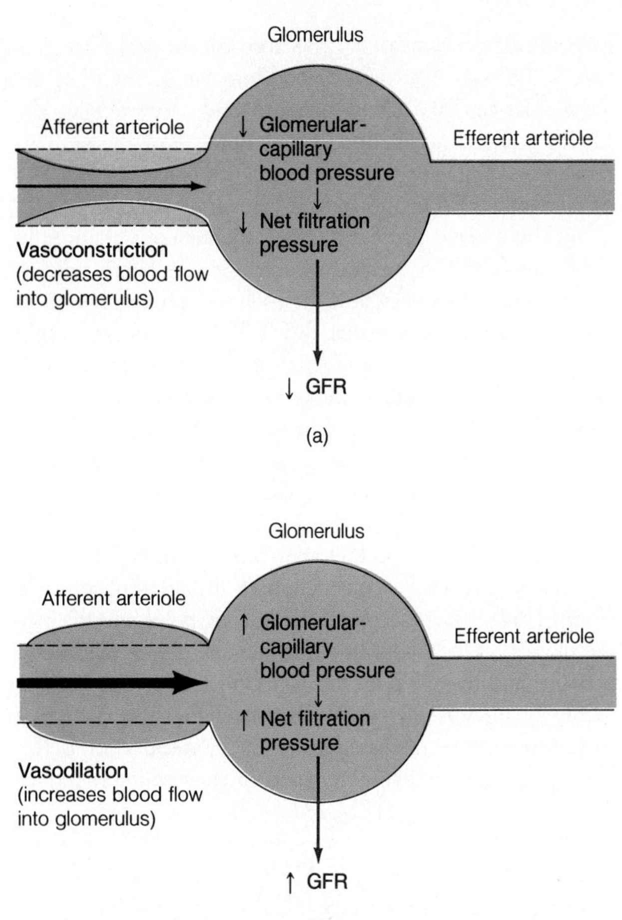

Figure 14–9 Adjustments of Afferent Arteriole to Alter GFR *(a) Arteriolar adjustment to reduce GFR. (b) Arteriolar adjustment to increase GFR.*

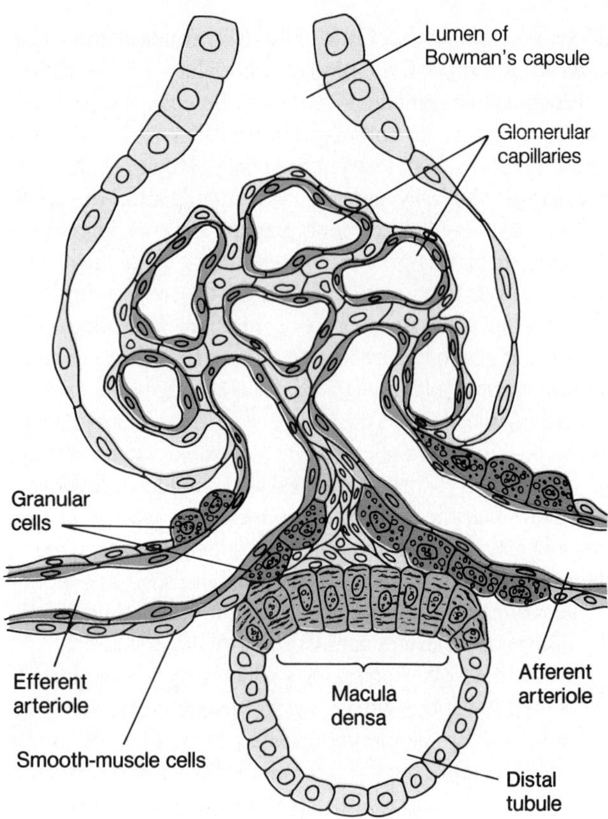

Figure 14–10 Juxtaglomerular Apparatus

The **myogenic mechanism** is a common property of vascular smooth muscle. Arteriolar vascular smooth muscle contracts inherently in response to the stretch accompanying increased pressure within the vessel. Accordingly, the afferent arteriole automatically constricts on its own when it is stretched because of an increased arterial driving pressure. This helps limit blood flow into the glomerulus to normal despite the elevated arterial pressure. Conversely, inherent relaxation of an unstretched afferent arteriole when pressure within the vessel is reduced brings about increased blood flow into the glomerulus despite the fall in arterial pressure.

The **tubulo-glomerular feedback mechanism** involves the *juxtaglomerular apparatus*, which is the specialized combination of tubular and vascular cells in the region of the nephron in which the tubule is bent back on itself, so the beginning of the distal tubule passes through the angle formed by the afferent and efferent arterioles as they join the glomerulus (Figs. 14–4 and 14–10). Within the arteriolar walls at their point of contact with the tubule, the smooth-muscle cells are specialized to form **granular cells,** so-called because they contain numerous secretory granules. Specialized tubular cells in this region are collectively known as the **macula densa.** The macula-densa cells detect changes in the rate at which fluid is flowing past them through the tubule. If the GFR is increased secondary to an elevation in arterial pressure, more fluid than normal is filtered and reaches the distal tubule (Fig. 14–11). In response, the macula-densa cells trigger the release of vasoactive chemicals from the juxtaglomerular apparatus, which in turn cause constriction of the afferent arteriole to reduce glomerular blood flow and return GFR to normal. In the opposite situation, in which the macula-densa cells detect that the flow rate of fluid through the tubules is low because of a spontaneous decline in GFR accompanying a fall in arterial pressure, these cells induce afferent arteriolar vasodilation by altering the rate of secretion of the relevant vasoactive chemicals. The resultant increase in glomerular flow rate restores the

GFR to normal. Thus, by means of the juxtaglomerular apparatus, the tubule of a nephron is able to monitor the rate of movement of fluid through it and adjust the GFR accordingly. This tubulo-glomerular feedback mechanism is initiated by the tubule to help each nephron regulate the rate of filtration through its own glomerulus.

The myogenic and tubulo-glomerular feedback mechanisms work in unison to autoregulate the GFR within the mean arterial blood pressure range of 80 to 180 mm Hg. Within this wide range, intrinsic autoregulatory adjustments of afferent arteriolar resistance can compensate for changes in arterial pressure, thus preventing inappropriate fluctuations in GFR, even though glomerular pressure tends to change in the same direction as arterial pressure. Normal mean arterial pressure is 93 mm Hg, so this range encompasses the transient changes in blood pressure that accompany daily activities unrelated to the need for the kidneys to regulate H_2O and salt excretion, such as the normal elevation in blood pressure accompanying exercise. Autoregulation is important because unintentional shifts in GFR could lead to dangerous imbalances of fluid, electrolytes, and wastes. Since at least a certain portion of the filtered fluid is always excreted, the amount of fluid excreted in the urine is automatically increased as the GFR increases. If it were not for autoregulation, the GFR would increase and H_2O and solutes would be lost needlessly as a result of the rise in arterial pressure accompanying heavy exercise. On the other hand, if the GFR were too low, the kidneys would not be able to eliminate sufficient quantities of wastes, excess electrolytes, and other materials that should be excreted. Autoregulation thus greatly blunts the direct effect that changes in arterial pressure would otherwise have on GFR and subsequently on H_2O, solute, and waste excretion.

When changes in mean arterial pressure fall outside of the autoregulatory range, these mechanisms are unable to compensate. Therefore, dramatic changes in mean arterial pressure (< 80 mm Hg or > 180 mm Hg) directly cause the glomerular-capillary pressure and, accordingly, the GFR to decrease or increase in proportion to the change in arterial pressure.

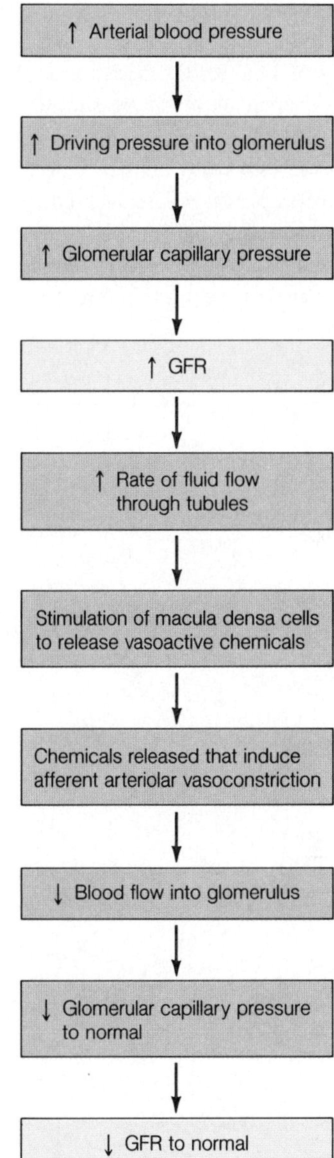

Figure 14–11 Tubulo-glomerular Feedback Mechanism of Autoregulation

EXTRINSIC SYMPATHETIC CONTROL OF THE GFR. In addition to the intrinsic autoregulatory mechanisms designed to keep the GFR constant in the face of fluctuations in arterial blood pressure, the GFR can be deliberately changed—even when the mean arterial blood pressure is within the autoregulatory range—by extrinsic-control mechanisms that override the autoregulatory responses. Extrinsic control of GFR, which is mediated by sympathetic nervous-system input to the afferent arterioles, is aimed at the regulation of arterial blood pressure.

If plasma volume is decreased—for example, because of hemorrhage—the resultant fall in arterial blood pressure is detected by the arterial carotid sinus and aortic arch baroreceptors (see p. 331), which initiate neural reflexes to increase blood pressure toward normal. These reflex responses are coordinated by the cardiovascular control center in the brain stem and are mediated primarily through increased sympathetic activity to the heart and blood vessels. Although the resultant increase in both cardiac output and total peripheral resistance helps to increase blood pressure toward normal, the plasma volume is still reduced. In the long-term, the plasma volume must be restored to normal. One of the compensations for a depleted plasma volume is a reduction in urine output so that more fluid than normal is conserved for the body. This reduction in urine output is accomplished in part by a reduction in the GFR; if less fluid is filtered, less fluid is available to be excreted.

No new mechanism is required to decrease the GFR. It is reduced as a result of the baroreceptor reflex response to a fall in blood pressure (Fig. 14–12). During this reflex, sympathetically induced vasoconstriction occurs in the majority of the arterioles throughout the body as a compensatory mechanism to increase total peripheral resistance. Among the arterioles that constrict in response to the baroreceptor reflex are the afferent arterioles carrying blood to the glomeruli. The afferent arterioles are innervated with sympathetic vasoconstrictor fibers to a far greater extent than the efferent arterioles. When the afferent arterioles constrict as a result of increased sympa-

Figure 14–12 Baroreceptor Reflex Influence on GFR in Long-Term Regulation of Blood Pressure

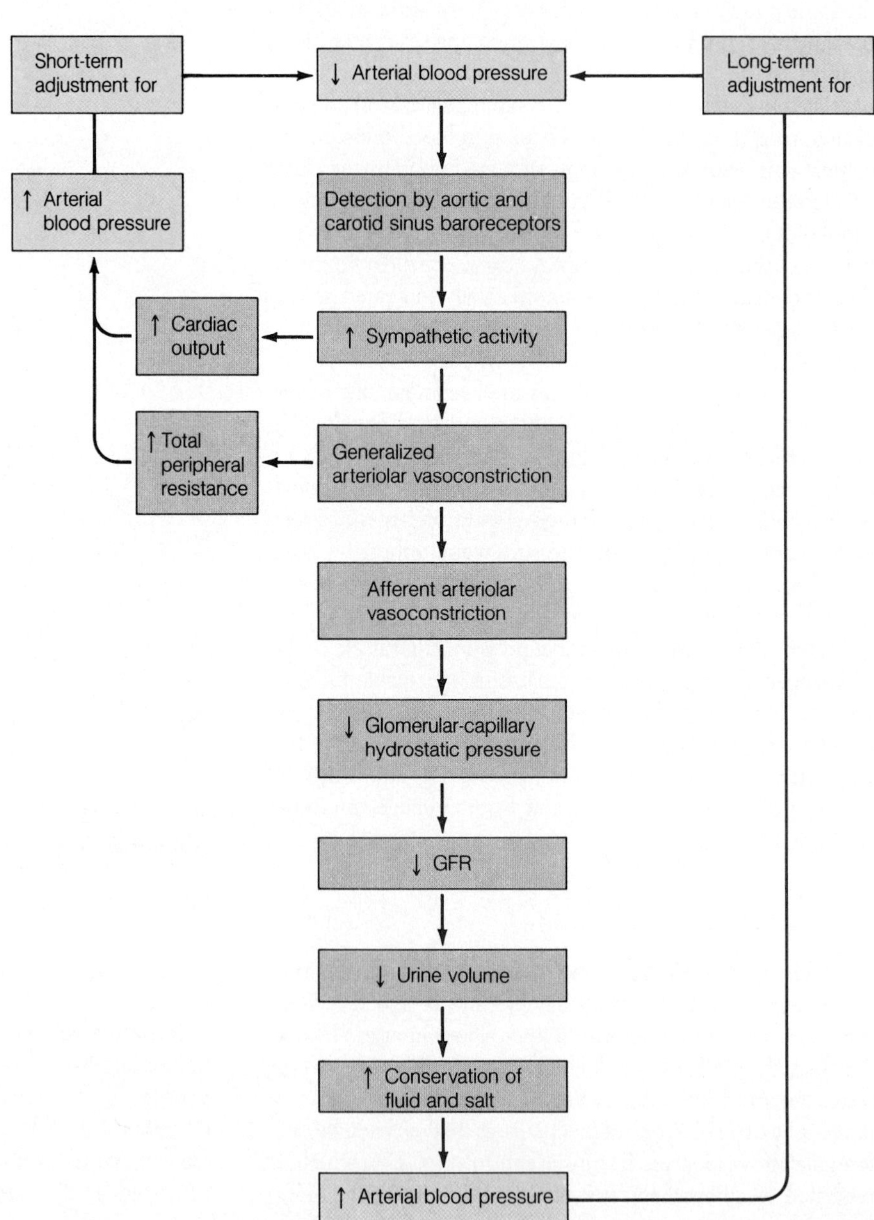

thetic activity, less blood flows into the glomeruli than normal, causing the glomerular-capillary blood pressure to fall. The resultant decrease in GFR in turn leads to a reduction in urine volume. In this way, some of the H_2O and salt that would have been lost in the urine is saved for the body. This helps in the long-term to restore the plasma volume to normal so that the short-term cardiovascular adjustments that have been made are no longer necessary. Other mechanisms described more thoroughly elsewhere, such as increased tubular reabsorption of H_2O and salt as well as increased thirst, also contribute to maintenance of blood pressure despite a loss of plasma volume.

In the converse, if blood pressure is elevated (for example, because of an expansion of plasma volume following ingestion of excessive fluid), the opposite responses occur. When a rise in blood pressure is detected by the baroreceptors, sympathetic vasoconstrictor activity to the arterioles, including the renal afferent arterioles, is reflexly reduced, allowing afferent arteriolar dilation to occur. As more blood enters the glomeruli through the dilated afferent arterioles, glomerular-capillary blood pressure rises, increasing the GFR. As more fluid is filtered, more fluid is available to be eliminated in the urine. Contributing to the increase in urine volume is a hormonally adjusted reduction in the tubular reabsorption of H_2O and salt. By these two renal mechanisms—increased glomerular filtration and decreased tubular reabsorption of H_2O and salt—urine volume is increased and the excess fluid is eliminated from the body. As with a fall in blood pressure, other compensatory measures also contribute to restoring an elevated blood pressure to normal.

Glomerular filtration rate can be influenced by changes in the filtration coefficient.

Thus far we have discussed changes in the GFR as a result of changes in the net filtration pressure. The rate of glomerular filtration, however, is dependent on the filtration coefficient (K_f) as well as on the net filtration pressure. For years K_f was considered to be a constant, except in disease situations in which the glomerular membrane becomes leakier than usual. Exciting new research to the contrary indicates that K_f is subject to change under physiological control. Both factors on which K_f depends—the surface area and the permeability of the glomerular membrane—can be modified by contractile activity within the membrane.

The surface area available for filtration within the glomerulus is represented by the inner surface of the glomerular capillaries that comes into contact with blood. Each tuft of glomerular capillaries is held together by **mesangial cells.** These cells also function as phagocytes and contain contractile elements (that is, actinlike filaments). Contraction of these mesangial cells closes off a portion of the filtering capillaries, thereby reducing the surface area available for filtration within the glomerular tuft. When the net filtration pressure remains unchanged, this reduction in K_f leads to a decrease in GFR. Evidence suggests that sympathetic stimulation causes the mesangial cells to contract, thus providing a second mechanism (besides promoting afferent arteriolar vasoconstriction) by which sympathetic activity can decrease the GFR. In addition, several hormones and local chemical mediators suspected or known to be involved in the control of other renal mechanisms, such as tubulo-glomerular feedback or tubular reabsorption, have been shown to influence mesangial-cell contractile activity.

The podocytes also possess actinlike contractile filaments, whose contraction or relaxation can, respectively, decrease or increase the number of filtration slits available in the inner membrane of Bowman's capsule by changing the shapes and proximities of the foot processes (Fig. 14–13). The number of slits is a determinant of permeability; the more slits, the greater the permeability. Contractile activity of the podocytes, which

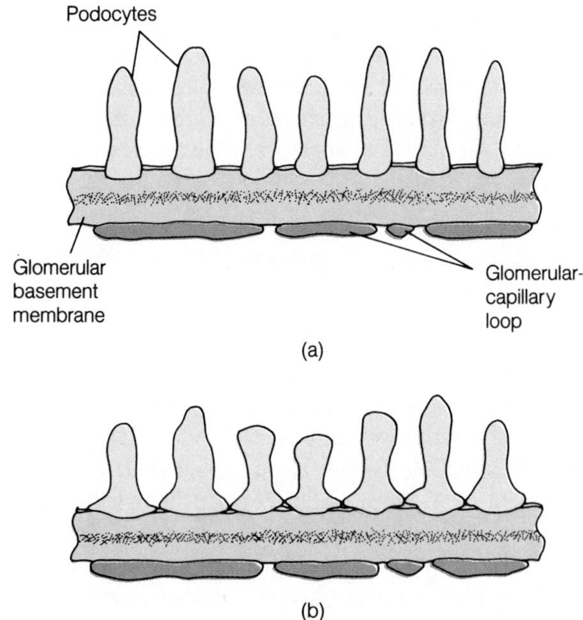

Figure 14–13 Decrease in Filtration Slits Caused by Podocyte Contraction *(a) Podocyte relaxation leads to a narrowing of bases of foot processes and an increase in the number of fully open intervening filtration slits spanning a given area. (b) Podocyte contraction leads to a flattening of foot processes and a loss of intervening filtration slits.*

SOURCE: Adapted with permission from *Federation Proceedings* 42: 3046–3052; 1983.

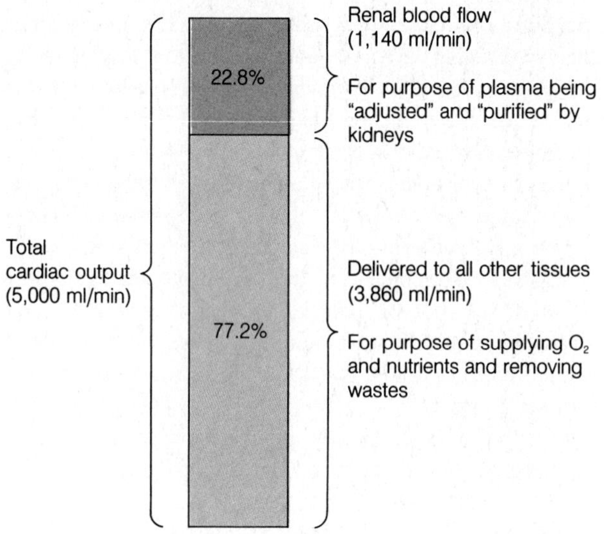

Figure 14–14 Percentage of Cardiac Output Distributed to Kidneys

in turn affects permeability and the K_f, appears to be under physiological control.

As the full extent of various regulatory mechanisms are becoming understood, it is evident that control of glomerular filtration (both through adjustments in afferent arteriolar resistance and K_f) as well as control of tubular reabsorption are complexly interrelated.

The kidneys normally receive 20% to 25% of the cardiac output.

At the average net filtration pressure and K_f, 20% of the plasma that enters the kidneys is converted into glomerular filtrate. This means that at an average GFR of 125 ml/min, the total renal plasma flow must average about 625 ml plasma/min. Because 55% of whole blood consists of plasma (that is, hematocrit = 45), the total flow of blood through the kidneys averages 1,140 ml/min. This represents about 22% of the total cardiac output of 5 liters (5,000 ml)/min for the kidneys, which compose less than 1% of total body weight.

The kidneys need to receive such a seemingly disproportionate share of the cardiac output because they must continuously perform their regulatory and excretory functions on the huge volumes of plasma delivered to them to maintain stability in the internal fluid environment. Most of the blood goes to the kidneys not to supply the renal tissue but to be adjusted and purified by the kidneys. On the average, 20% to 25% of the blood pumped out by the heart each minute "goes to the cleaners" instead of serving its normal purpose of exchanging materials with the tissues (Fig. 14–14). Only by continuously processing such a large proportion of the blood are the kidneys able to precisely regulate the volume and electrolyte composition of the internal environment and to adequately eliminate the large quantities of metabolic waste products that are constantly produced.

Tubular Reabsorption and Tubular Secretion

Tubular reabsorption is tremendous and variable.

No adjustment or purification of plasma takes place through the process of glomerular filtration. All plasma constituents except the proteins are nondiscriminantly filtered together through the glomerular capillaries. In addition to waste products and excess materials that need to be eliminated from the body, the filtered fluid also contains nutrients, electrolytes, and other substances that the body cannot afford to lose in the urine. Indeed, through the ongoing process of glomerular filtration, a greater quantity of these materials are filtered per day than are even present in the entire body. It is important that the essential materials that are filtered are returned to the blood by the process of tubular reabsorption, which refers to the discrete transfer of substances from the tubular lumen into the peritubular capillaries.

Tubular reabsorption is a highly selective process. All constituents except plasma proteins are at the same concentration in the glomerular filtrate as in the plasma. In most cases, the quantity of each material that is reabsorbed is that which is required to maintain the proper composition and volume of the internal fluid environment. In general, the tubules have a high reabsorptive capacity for substances needed by the body and a poor or no reabsorptive capacity for those substances of no value (Table 14–2). Accordingly, only a small percentage, if any, of filtered plasma constituents that are useful to the body are present in the urine, most having been reabsorbed and returned to the blood. Only excess amounts of essential materials such as electrolytes are excreted in the urine. For the essential plasma constituents regulated by the kidneys, absorptive capacity may vary depending on the body's needs. In contrast, a large percentage of filtered waste products are present in the urine. These wastes, which are useless or even potentially harmful to the body if allowed to accumulate, are not reabsorbed to any extent. Instead, they remain in the tubules to be eliminated in the urine. As H_2O and other valuable constituents are reabsorbed, the waste products remaining in the tubular fluid become highly concentrated.

When considering the magnitude of glomerular filtration, the extent of tubular reabsorption is tremendous. Just to give you an idea, the tubules typically reabsorb 99% of the filtered H_2O (47 gallons/day), 100% of the filtered sugar (2.5 lb/day), and 99.5% of the filtered salt (0.36 lb/day).

Table 14–2 Fate of Various Substances Filtered by Kidneys

Substance	Average Percentage of Filtered Substance Reabsorbed	Average Percentage of Filtered Substance Excreted
Water	99	1
Sodium	99.5	0.5
Glucose	100	0
Urea (a waste product)	50	50
Phenol (a waste product)	0	100

Tubular reabsorption involves transepithelial transport.

Throughout its entire length, the tubule is one cell-layer thick and is in close proximity with a surrounding peritubular capillary (Fig. 14–15). Adjacent tubular cells do not come into contact with each other except at their apical ends, which face the tubular lumen where they are joined by tight junctions (see p. 68). Interstitial fluid lies in the gaps between adjacent cells—the **lateral spaces**—as well as between the tubules and capillaries.

The tight junctions largely prevent movement of substances *between* the cells, so materials must pass ***through*** the cells in order to leave the tubular lumen and gain entry to the blood. To be reabsorbed, a substance must traverse five distinct barriers (Fig. 14–15):

Step 1. It must leave the tubular fluid by crossing the luminal membrane of the tubular cell (the apical membrane facing the tubular lumen);

Step 2. It must pass through the cytosol from one side of the tubular cell to the other;

Step 3. It must traverse the basolateral membrane of the tubular cell to enter the interstitial fluid;

Step 4. It must diffuse through the interstitial fluid; and

Step 5. It must penetrate the capillary wall to enter the blood plasma.

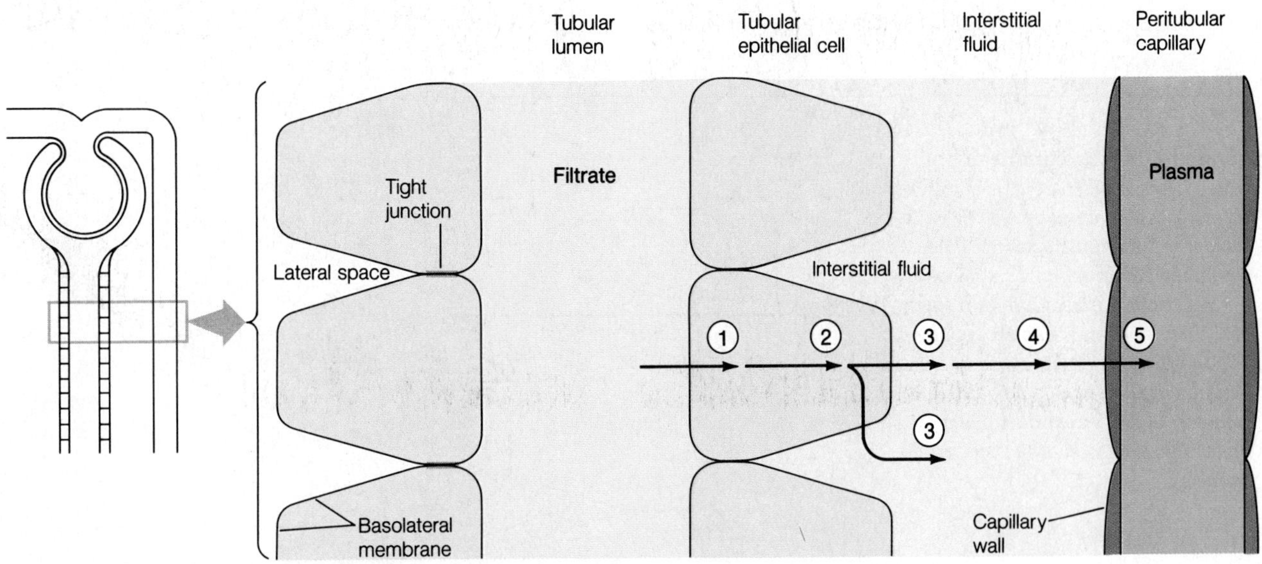

Figure 14–15 Steps of Transepithelial Transport *To be reabsorbed (move from filtrate to plasma), a substance must traverse five distinct barriers: ①—across luminal cell membrane; ②—through cytosol; ③—across basolateral cell membrane; ④—through interstitial fluid; and ⑤—across capillary wall.*

This entire sequence of steps is known as **transepithelial** ("across the epithelium") **transport.**

There are two types of tubular reabsorption—**passive reabsorption** and **active reabsorption**—depending on whether or not local energy expenditure is required for transfer of a particular substance. In passive reabsorption, *all* steps in the transepithelial transport of a substance are passive; that is, no energy is expended for the net movement of the substance from the tubular lumen to the plasma, which occurs down electrochemical or osmotic gradients. On the other hand, a substance is said to be actively reabsorbed if any one of the steps in the sequence requires energy, even if the four other steps are passive. With active reabsorption, net movement of the substance from the lumen to the plasma occurs against an electrochemical gradient. Substances that are actively reabsorbed are of particular importance to the body, such as glucose, amino acids, and other organic nutrients, as well as Na^+ and other electrolytes such as $PO_4^{\equiv}$ and Ca^{++}. Rather than specifically describing the reabsorptive process for each of the many filtered substances that are returned to the plasma, we will provide illustrative examples of the general mechanisms involved after first highlighting the unique and important case of Na^+ reabsorption.

An energy-dependent Na^+-K^+ ATPase transport mechanism in the basolateral membrane is essential for Na^+ reabsorption.

Sodium reabsorption is unique and complex. Eighty percent of the total energy requirement of the kidneys is used for Na^+ transport, indicative of the importance of this process. Unlike most filtered solutes, Na^+ is reabsorbed throughout the tubule, but to varying extents in different regions. Of the Na^+ filtered, 99.5% is normally reabsorbed, of which on the average 67% is reabsorbed in the proximal tubule, 25% in the loop of Henle, and 8% in the distal and collecting tubules. Sodium reabsorption in each of these segments plays different important roles, as will become apparent as our discussion continues.

1. Sodium reabsorption in the proximal tubule plays a pivotal role in the reabsorption of glucose, amino acids, H_2O, Cl^-, and urea.
2. Sodium reabsorption in the loop of Henle, along with Cl^- reabsorption, plays a critical role in the kidneys' ability to produce urine of varying concentrations and volumes, depending on the body's need to conserve or eliminate H_2O.
3. Sodium reabsorption in the distal portions of the nephron is variable and subject to hormonal control, being important in the regulation of ECF volume. It is also linked in part with K^+ secretion.

The active step in Na^+ reabsorption involves the energy-dependent Na^+-K^+ ATPase carrier located at the tubular cell's basolateral membrane (Fig. 14–16). This is the same carrier present in all cells that actively extrudes Na^+ from the cell (see p. 79). As this basolateral pump transports Na^+ out of the tubular cell into the lateral space, it keeps the intracellular Na^+ concentration low while it simultaneously builds up the concentration of Na^+ in the lateral space; that is, it moves

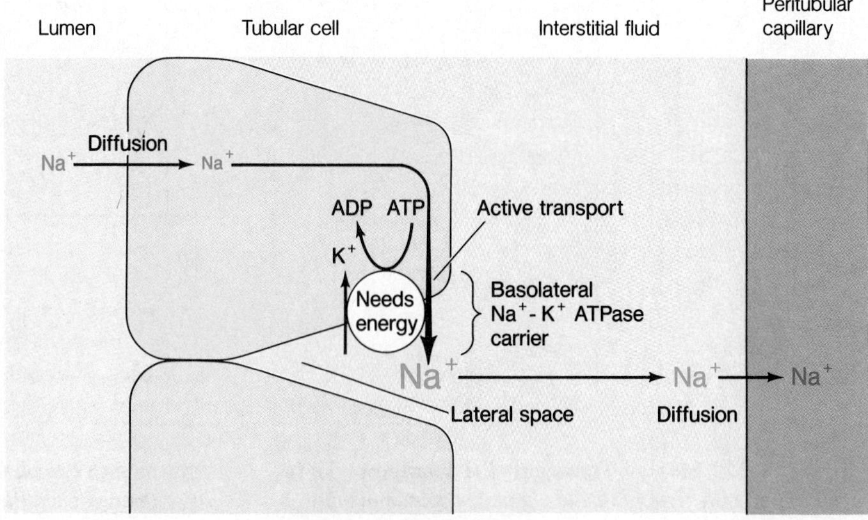

Figure 14–16 Sodium Reabsorption *The basolateral Na^+-K^+ ATPase carrier actively transports Na^+ from the tubular cell into the interstitial fluid within the lateral space. This establishes a concentration gradient for diffusion of Na^+ from the lumen into the tubular cell and from the lateral space into the peritubular capillary, accomplishing net transport of Na^+ from the tubular lumen into the blood at the expense of energy.*

Na^+ *against* a concentration gradient. Because the intracellular Na^+ concentration is kept low by basolateral-pump activity, a concentration gradient is established that favors the movement of Na^+ from its higher concentration in the tubular lumen across the luminal border into the tubular cell. Once within the cell, the Na^+ is actively extruded to the lateral space by the basolateral pump. This keeps the intracellular Na^+ concentration low, thereby favoring the net movement of Na^+ from lumen to cell to lateral space. Sodium continues to passively diffuse down a concentration gradient from its high concentration in the lateral space into the surrounding interstitial fluid and finally into the peritubular-capillary blood.

Aldosterone stimulates Na^+ reabsorption in the distal and collecting tubules; atrial natriuretic peptide inhibits it.

In the proximal tubule and loop of Henle, a constant percentage of the filtered Na^+ is reabsorbed regardless of the Na^+ load (total amount of Na^+) in the body. The reabsorption of a small percentage of the filtered Na^+ is subject to hormonal control in the distal portion of the tubule. The extent of this controlled reabsorption is inversely related to the magnitude of the Na^+ load in the body. If there is too much Na^+, little of this controlled Na^+ is reabsorbed but is lost in the urine instead, thereby removing excess Na^+ from the body. On the other hand, if Na^+ is depleted, most or all of this controlled Na^+ is reabsorbed, conserving for the body Na^+ that otherwise would be lost in the urine. The most important and best known hormonal system involved in the regulation of Na^+ is the *renin-angiotensin-aldosterone mechanism*, which stimulates Na^+ reabsorption in the distal and collecting tubules.

The Na^+ load in the body is reflected by the ECF volume. Sodium and its attendant anion Cl^- account for more than 90% of the ECF's osmotic activity. Recall that osmotic pressure can be thought of loosely as a force that attracts and holds H_2O (see p. 73). When the Na^+ load is above normal and the ECF's osmotic activity is therefore increased, the extra Na^+ "holds" extra H_2O, expanding the ECF volume. Conversely, when the Na^+ load is below normal, thereby decreasing ECF osmotic activity, less H_2O than normal can be held in the ECF, so the ECF volume is reduced. Since plasma is a component of the ECF, the most important consequence of a change in ECF volume is the corresponding change in blood pressure accompanying expansion (↑ blood pressure) or reduction (↓ blood pressure) of the plasma volume.

The granular cells of the juxtaglomerular apparatus (Fig. 14–10) secrete a hormone, **renin,** into the blood in response to a fall in NaCl/ECF volume/blood pressure. This function is in addition to the role the juxtaglomerular apparatus plays in autoregulation. These interrelated signals for increased renin secretion are all indicative of the necessity to expand the plasma volume to increase the arterial pressure to normal on a long-term basis. Increased renin secretion, through a complex series of events, brings about increased Na^+ reabsorption by the distal portion of the tubule. Chloride always passively follows Na^+ down the electrical gradient established by sodium's active movement. The ultimate benefit of this salt retention is its accompanying osmotically induced H_2O retention, which helps restore the plasma volume and blood pressure.

Let us examine the mechanism by which renin secretion ultimately leads to increased Na^+ reabsorption (Fig. 14–17). Once secreted into the blood, renin acts as an enzyme to activate **angiotensinogen** into **angiotensin I.** Angiotensinogen is a plasma protein synthesized by the liver and always present in the plasma in high concentration. On passing through the lungs via the pulmonary circulation, angiotensin I is converted into **angiotensin II** by **converting enzyme,** which is abundant in the pulmonary capillaries. Angiotensin II is the primary stimulus for secretion of the hormone **aldosterone** from the adrenal gland. The adrenal gland is an endocrine gland that produces several different hormones, each of which is secreted in response to different stimuli.

Among its actions, aldosterone increases Na^+ reabsorption by the distal and collecting tubules. It does so by stimulating the synthesis of new proteins within these tubular cells. This **aldosterone-induced protein** is believed to be involved in the formation of Na^+ channels in the luminal membranes of the distal and collecting tubular cells, which in turn allow greater passive inward movement of Na^+ from the lumen into the cells. The greater inward flux of Na^+ promotes increased active pumping of Na^+ out of the cells into the lateral spaces and hence into the plasma by the basolateral Na^+-K^+ ATPase carriers. The net result is an increase in Na^+ reabsorption.

The entire renin-angiotensin-aldosterone mechanism thus promotes salt retention and a resultant H_2O retention and elevation of arterial blood pressure. Acting in negative-feedback fashion, this mechanism alleviates the factors that triggered the initial release of renin—namely, salt depletion, plasma volume reduction, and decreased arterial blood pressure. Angiotensin II, in addition to stimulating aldosterone secretion, performs several other actions that contribute to the elevation of blood pressure. Angiotensin II is a potent constrictor of arterioles, directly increasing blood pressure by increasing total peripheral resistance. Furthermore, it stimulates thirst (increasing fluid intake) and stimulates vasopressin (a hormone that increases H_2O retention by the kidneys), both of which contribute to plasma-volume expansion and elevation of arterial pressure.

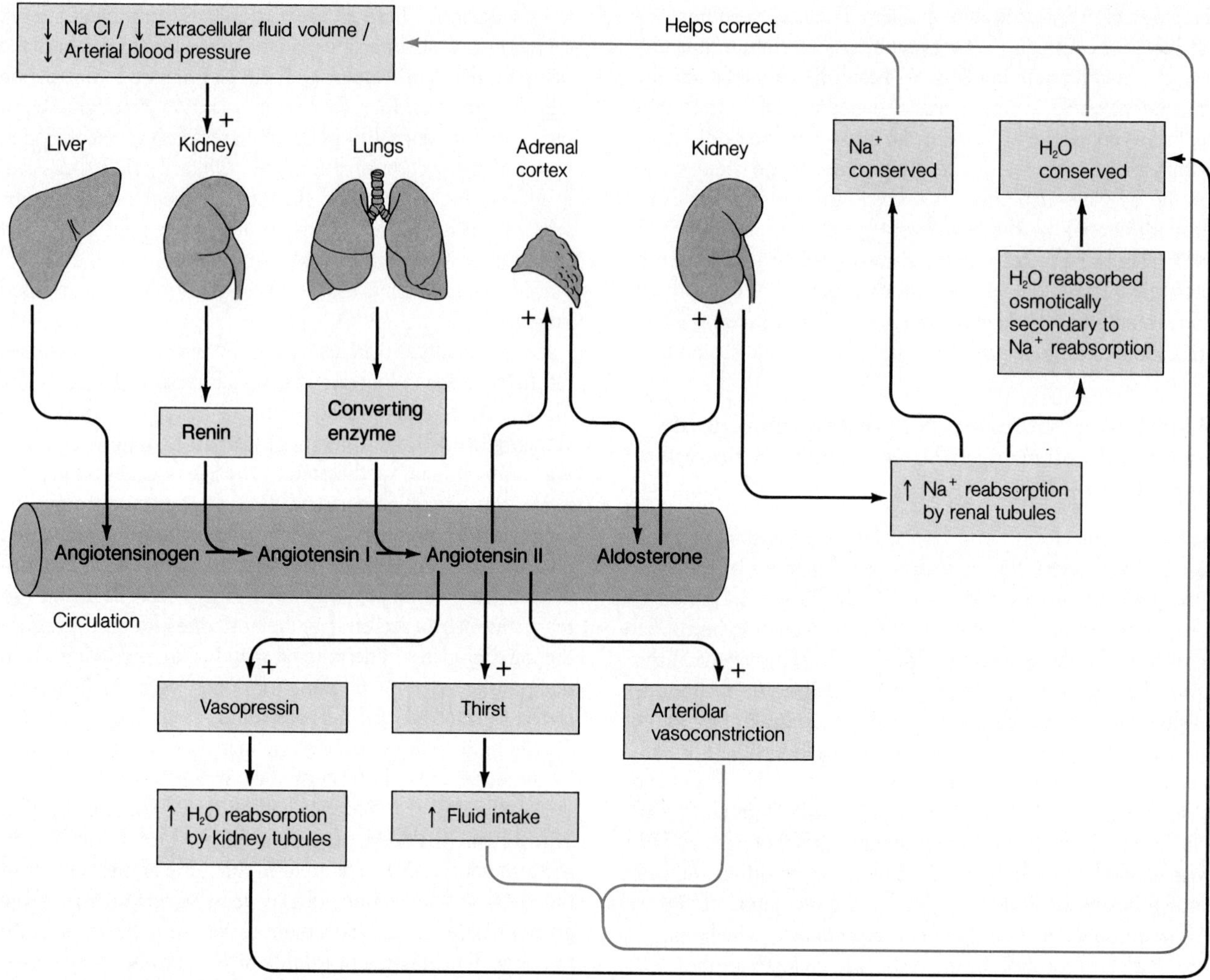

Figure 14–17 Renin-Angiotensin-Aldosterone System

The opposite situation exists when the Na^+ load, ECF and plasma volume, and arterial blood pressure are above normal. Under these circumstances, renin secretion is inhibited. Consequently, because angiotensinogen is not activated to angiotensin I and II, aldosterone secretion is not stimulated. Without aldosterone, the small aldosterone-dependent portion of Na^+ reabsorption in the distal segments of the tubule does not occur. Instead, this nonreabsorbed Na^+ is lost in the urine. In the absence of aldosterone, the ongoing loss of this small percentage of filtered Na^+ can rapidly remove excess Na^+ from the body.

Even though only about 2% of the filtered Na^+ is dependent on aldosterone for reabsorption, this small loss, multiplied manyfold as the entire plasma volume is filtered through the kidneys many times per day, can lead to a sizeable loss of Na^+. In the complete absence of aldosterone, 20 gms of salt may be excreted per day. With maximum aldosterone secretion, all of the filtered Na^+ (and, accordingly, all of the filtered Cl^-) is reabsorbed, so salt excretion in the urine is zero. The amount of aldosterone secreted, and consequently the relative amount of salt conserved versus salt excreted, usually varies between these extremes, depending on the body's needs. For example, an "average" salt consumer typically excretes about 10 gms of salt per day in the urine, a heavy salt consumer excretes more, and an individual who has lost considerable salt during heavy sweating has a lower urinary salt excretion. By varying the amount of renin and aldosterone secreted in accordance with the salt-determined fluid load in the body, the kidneys are able to finely adjust the amount of salt conserved or eliminated. In doing so, they maintain the salt load and ECF volume/arterial blood pressure at a relatively constant level in spite of wide variations in salt consumption and abnormal losses of salt-laden fluid.

It should not be surprising that abnormal increases in renin-angiotensin-aldosterone activity can contribute to the development of hypertension (high blood pressure). This system is also responsible in part for the fluid retention and edema accompanying congestive heart failure. Because of the failing heart, the cardiac output is reduced and arterial blood pressure is low in spite of a normal or even expanded plasma volume. When a fall in blood pressure is due to a failing heart rather than a reduction in the salt/fluid load in the body, the salt- and fluid-retaining reflexes triggered by the low blood pressure are inappropriate. Sodium excretion may fall to virtually zero despite continued salt ingestion and accumulation in the body. The resultant expansion of the ECF produces edema and exacerbates the congestive heart failure as the weakened heart is unable to pump the additional plasma volume.

Because their salt-retaining mechanisms are being inappropriately triggered by the reduction in blood pressure, patients with congestive heart failure are placed on low-salt diets. Often, they are treated with **diuretics,** which are therapeutic agents that increase urinary output and thus promote loss of fluid from the body. Many of these drugs function by inhibiting tubular reabsorption of Na^+. As more Na^+ is excreted, more H_2O is also lost from the body, thus helping to remove the excess ECF. The drug capoten, which blocks the action of converting enzyme, is also beneficial in the treatment of congestive heart failure as well as certain cases of hypertension. By blocking the generation of angiotensin II, capoten halts the ultimate salt- and fluid-conserving actions and arteriolar constrictor effects of the renin-angiotensin-aldosterone system.

In addition to the renin-angiotensin-aldosterone system, which is believed to exert the most powerful influence on the renal handling of Na^+, recent evidence indicates that this Na^+-retaining system is opposed by a Na^+-losing system that involves the hormone **atrial natriuretic peptide (ANP)** and perhaps other factors. Atrial natriuretic peptide is released from the cardiac atria when the ECF volume is expanded. In turn, ANP inhibits renal tubular Na^+ reabsorption by mechanisms that remain unclear. Besides its indirect effect in lowering blood pressure by reducing the Na^+ load and hence the fluid load in the body, ANP has also been shown to directly lower blood pressure by promoting arteriolar vasodilation. A similar natriuretic factor has been found in the hypothalamus. The relative contributions of these (and possibly other) salt-losing, blood pressure-lowering factors in the maintenance of salt and H_2O balance and blood pressure regulation are presently being intensely investigated. This is not purely of academic interest, because it is likely that derangements of this system will be found to contribute to hypertension or congestive heart failure. For example, a deficiency of a counterbalancing natriuretic system could theoretically cause long-term hypertension, since the powerful Na^+-conserving system would be unopposed. Even if the natriuretic factors are not found to contribute to the development of hypertension or congestive heart failure, they might prove to be useful in treating these circulatory disorders by reducing the Na^+ load in the body.

Glucose and amino acids are reabsorbed by Na^+-dependent secondary active transport.

Large quantities of nutritionally important organic molecules such as glucose and amino acids are filtered each day. Because these substances normally are completely reabsorbed back into the blood by energy- and Na^+-dependent mechanisms located in the proximal tubule, none of these materials are usually excreted in the urine. This rapid and thorough reabsorption early in the tubules protects against the loss of these important organic nutrients.

Even though glucose and amino acids are actively moved uphill against their concentration gradients from the tubular lumen into the blood until their concentration in the tubular fluid is virtually zero, no energy is directly used to operate the glucose or amino-acid carriers. Glucose and amino acids are transferred by means of secondary active transport; a specialized co-transport carrier simultaneously transfers both Na^+ and the specific organic molecule from the lumen into the cell (see p. 80). The lumen-to-cell Na^+ concentration gradient maintained by the energy-consuming basolateral Na^+-K^+ ATPase pump drives this co-transport system and pulls the organic molecule against its concentration gradient without the direct expenditure of energy. Because the overall process of glucose and amino-acid reabsorption depends on the use of energy, these organic molecules are considered to be actively reabsorbed, even though energy is not used directly to transport them across the membrane. In essence, glucose and amino acid get a "free ride" at the expense of energy already used in the reabsorption of Na^+. Secondary active transport requires the presence of Na^+ in the lumen, without which the co-transport carrier is inoperable. Once transported into the tubular cells, glucose and amino acids passively diffuse down their concentration gradients across the basolateral membrane into the plasma, facilitated by a non-energy-dependent carrier.

Actively reabsorbed substances with the exception of Na^+ exhibit a transport maximum.

All actively reabsorbed substances bind with membrane-bound carriers that transfer them across the membrane against a concentration gradient. Each carrier is limited in the types of substances it can transport; for example, the glucose carrier cannot transport amino acids, or vice versa. Since a limited number of each specific carrier type is present in the cells lining the tubules, there is an upper limit on the quantity of a particular substance that can be actively transported from the tubular fluid in a given period of time. The maximum reabsorption rate is reached when all the carriers specific for a particular substance are fully "occupied" or saturated (see p. 76) so that they cannot handle any additional passengers at that time. The **tubular maximum** or **T_m** represents this maximum amount of a substance that the tubular cells can actively transport within a given time period. All actively reabsorbed substances with the exception of Na^+ display a T_m. Any quantity of a substance filtered beyond its T_m fails to be reabsorbed but escapes instead into the urine.

The plasma concentration of some but not all substances that display carrier-limited reabsorption are regulated by the kidneys. How can the kidneys regulate some actively reabsorbed substances but not others, despite the fact that the renal tubules limit how much of each substance can be reabsorbed and returned to the plasma? We will compare a substance that has a T_m but is not regulated by the kidneys, glucose, with a T_m-limited substance that is regulated by the kidneys, phosphate.

GLUCOSE REABSORPTION. The normal plasma concentration of glucose is 100 mg of glucose/100 ml of plasma. Because glucose is freely filterable at the glomerulus, it passes into Bowman's capsule at the same concentration as it is in the plasma. Accordingly, 100 mg of glucose are present in every 100 ml of plasma filtered. With 125 ml of plasma normally being filtered each minute (average GFR = 125 ml/min), 125 mg of glucose pass into Bowman's capsule with this filtrate every minute. The quantity of any substance filtered per minute, known as its **filtered load,** can be calculated as follows:

$$\text{Filtered load of a substance} = \text{Plasma concentration of the substance} \times \text{GFR}$$

$$\text{Filtered load of glucose} = 100 \text{ mg/100ml} \times 125 \text{ ml/min} = 125 \text{ mg/min}$$

At a constant GFR, the filtered load of glucose is directly proportional to the plasma glucose concentration. Doubling the plasma glucose concentration to 200 mg/100 ml doubles the filtered load of glucose to 250 mg/min, and so on (Fig. 14–18).

The T_m for glucose averages 375 mg/min; that is, the glucose carrier mechanism is capable of actively reabsorbing up to 375 mg of glucose per minute before it reaches its maximum transport capacity. At a normal plasma glucose concentration of 100 mg/100 ml, the 125 mg of glucose filtered per minute can readily be reabsorbed by the glucose carrier mechanism, because the filtered load is well below the T_m for glucose. Ordinarily, therefore, no glucose appears in the urine because all of the filtered glucose is reabsorbed. Not until the filtered load of glucose exceeds 375 mg/min is the T_m reached. When more glucose is filtered per minute than can be reabsorbed because the T_m is exceeded, the maximum amount is reabsorbed, whereas the rest remains in the filtrate to be excreted. Accordingly, the plasma glucose concentration must be greater than 300 mg/100 ml—more than three times the normal value—before glucose starts spilling into the urine.

The plasma concentration of a particular substance at which its T_m is reached and the substance first starts appearing in the urine is known as the **renal threshold.** At the normal T_m of 375 mg/min and GFR of 125 ml/min, the renal threshold for glucose is 300 mg/100 ml. Beyond the T_m, reabsorption remains constant at its maximum rate, and any further increase in the filtered load is accompanied by a directly proportional increase in the amount of the substance excreted. For example, at a plasma glucose concentration of 400 mg/100 ml, the filtered load of glucose is 500 mg/min, 375 mg/min of which can be reabsorbed (a T_m's worth) and 125 mg/min of which is excreted in the urine. At a plasma glucose concentration of 500 mg/100 ml, the filtered load is 625 mg/min, still only 375 mg/min can be reabsorbed, and 250 mg/min spill into the urine.

The plasma glucose concentration can become extremely high in diabetes mellitus, an endocrine disorder involving deficiency of insulin, a pancreatic hormone. This hormone is important in facilitating the transport of glucose into many of the body's cells. In insulin deficiency, the glucose that cannot gain entry into the cells remains in the plasma, elevating the plasma glucose concentration. Consequently, although glucose does not normally appear in the urine, it is found in the urine of persons with diabetes when the plasma glucose concentration exceeds the renal threshold, even though there has been no change in renal function.

What happens when the plasma glucose concentration falls below normal? The renal tubules, of course, reabsorb all of the filtered glucose, because the glucose reabsorptive capacity is far from being exceeded. The kidneys cannot do anything

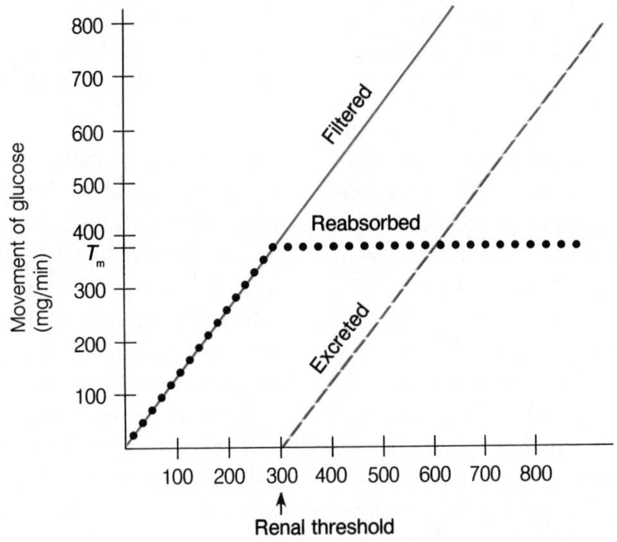

Plasma concentration of substance × GFR = Amount of substance filtered

For example, 100 mg glucose/100 ml plasma × 125 ml plasma filtered/min = 125 mg glucose filtered/min

Plasma Concentration (mg/100 ml)	GFR (ml/min)	Amount Filtered (mg/min)	T_m (mg/min)	Amount Reabsorbed (mg/min)	Amount Excreted (mg/min)
a. 80	125	100	375	100	0
b. 100	125	125	375	125	0
c. 200	125	250	375	250	0
d. 300	125	375	375	375	0
e. 400	125	500	375	375	125
f. 500	125	625	375	375	250

Figure 14–18 Renal Handling of Glucose as Function of Plasma Glucose Concentration

to raise a low plasma glucose level to normal. They simply return all of the filtered glucose to the plasma.

Thus, the kidneys do not influence the plasma glucose concentration over a wide range of values that varies from abnormally low levels to up to three times the normal level. Because the T_m for glucose is well above the normal filtered load, the kidneys usually conserve all of the glucose, thereby protecting against the loss of this important nutrient in the urine. Only when the plasma glucose concentration has more than tripled do the kidneys influence the concentration by spilling some of the excess into the urine. The kidneys do not regulate glucose because they do not maintain glucose at some specific plasma concentration; instead, this concentration is normally regulated by endocrine and liver mechanisms, with the kidneys merely maintaining whatever plasma glucose concentration is set by these other mechanisms (except when excessively high levels overwhelm the kidneys' reabsorptive capacity). The same general principle holds true for other organic plasma nutrients, such as amino acids, lactic acid, and water-soluble vitamins. For these substances, the kidneys' role is to conserve and protect against significant losses, thereby preserving whatever plasma concentration is set by the normal regulatory mechanisms.

PHOSPHATE REABSORPTION. In the case of many electrolytes, such as calcium (Ca^{++}) and phosphate ($PO_4^{=}$), the kidneys do directly contribute to their regulation, because the tubular maximums of these inorganic ions equal their normal plasma concentrations. We will use $PO_4^{=}$ as an example. Unlike the case with organic plasma solutes, the T_m for $PO_4^{=}$ is very close to the normal filtered $PO_4^{=}$ load. If the plasma $PO_4^{=}$ concentration increases even slightly above normal, which is typically the case because our diets are generally rich in $PO_4^{=}$, the excess ingested $PO_4^{=}$ is quickly spilled into the urine, restoring the plasma concentration to normal. Following a $PO_4^{=}$-rich meal, the plasma $PO_4^{=}$ concentration is transiently increased after the $PO_4^{=}$ is transferred from the digestive tract into the plasma. As soon as this plasma with its slightly elevated $PO_4^{=}$ load is filtered through the kidneys, the excess $PO_4^{=}$ is rapidly eliminated in the urine, because the tubules can reabsorb up to the normal plasma concentration's worth of $PO_4^{=}$, and no more. The more $PO_4^{=}$ ingested beyond the body's needs, the more excreted. In this way, the kidneys maintain the desired plasma $PO_4^{=}$ concentration while eliminating any excess $PO_4^{=}$ ingested. On the other hand, as with any substance, the only thing the kidneys can do if the plasma $PO_4^{=}$ concentration falls too low is to conserve all of the $PO_4^{=}$ that is filtered to minimize the extent of the depletion.

Unlike the reabsorption of organic nutrients, the reabsorption of $PO_4^{=}$ and Ca^{++} is also subject to hormonal control. Parathyroid hormone can alter the renal thresholds for $PO_4^{=}$ and Ca^{++}, thus adjusting the quantity of these electrolytes conserved depending on the body's momentary needs (chapter 19).

Active Na^+ reabsorption is responsible for the passive reabsorption of Cl^-, H_2O, and urea.

Not only is the secondary active reabsorption of glucose and amino acids linked to the basolateral Na^+-K^+ ATPase pump, but the passive reabsorption of Cl^-, H_2O, and urea also depends on this active Na^+ reabsorption mechanism.

CHLORIDE REABSORPTION. The negatively charged chloride ions are passively reabsorbed down the electrical gradient created by the active reabsorption of the positively charged sodium ions. The amount of Cl^- reabsorbed is determined by the rate of active Na^+ reabsorption instead of being directly controlled by the kidneys. This is typical of most passively reabsorbed substances, with the exception of H_2O.

WATER REABSORPTION. Water is passively reabsorbed by osmosis throughout the length of the tubule. Of the H_2O filtered, 80% is obligatorily reabsorbed in the proximal tubules and loops of Henle, osmotically following solute reabsorption. This occurs regardless of the H_2O load in the body and is not subject to regulation. Variable amounts of the remaining 20% are reabsorbed in the distal portions of the tubule, the extent of reabsorption being subject to direct hormonal control, depending on the body's state of hydration.

The driving force for H_2O reabsorption in the proximal tubule is a compartment of hypertonicity in the lateral spaces between the tubular cells that is established by the active extrusion of Na^+ by the basolateral pump (Fig. 14–19). As a result of this pump activity, the concentration of Na^+ rapidly diminishes in the tubular fluid and tubular cells while it simultaneously increases in the localized region within the lateral spaces. This osmotic gradient induces the passive net flow of H_2O from the lumen into the lateral spaces, either through the cells or intercellularly through "leaky" tight junctions. The accumulation of fluid in the lateral spaces results in a build-up of hydrostatic (fluid) pressure, which flushes H_2O out of the lateral spaces into the interstitial fluid and finally into the peritubular capillaries.

This return of filtered H_2O to the plasma is enhanced by the fact that the blood-colloid osmotic pressure is greater in the peritubular capillaries than elsewhere. The concentration of plasma proteins, which is responsible for the blood-colloid osmotic pressure, is elevated in the blood entering the peritubular capillaries because of the extensive filtration of H_2O through the glomerular capillaries upstream. The plasma proteins left behind in the glomerulus are concentrated into a smaller volume of plasma H_2O, increasing the blood-colloid osmotic pressure of the unfiltered blood that leaves the glomerulus and enters the peritubular capillaries. This force tends to "pull" H_2O into the peritubular capillaries, simultaneous to the "push" of the hydrostatic pressure in the lateral spaces that drives H_2O toward the capillaries. By these means, 65% of the filtered H_2O—117 liters per day—is passively reabsorbed by the end of the proximal tubule. No energy is directly required by the proximal tubule, or indeed by any other portion of the tubule, for this tremendous reabsorption of H_2O.

Another 15% of the filtered H_2O is obligatorily reabsorbed from the loop of Henle. Adjustable reabsorption of the remaining 20% of the filtered H_2O is accomplished in the distal

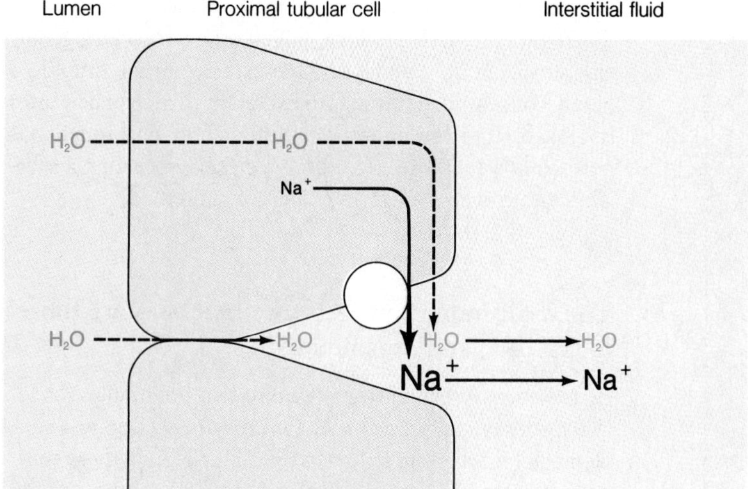

Figure 14-19 Water Reabsorption in Proximal Tubule *The force for H_2O reabsorption is the compartment of hypertonicity in the lateral spaces established by active extrusion of Na^+ by the basolateral pump. The dashed arrows show the direction of osmotic movement of H_2O.*

and collecting tubules under control of vasopressin. The mechanisms responsible for H_2O reabsorption beyond the proximal tubule will be described in later sections.

UREA REABSORPTION. In addition to Cl^- and H_2O, the passive reabsorption of urea is also indirectly linked to active Na^+ reabsorption. **Urea,** an undesirable waste product resulting from protein catabolism (breakdown), is not intentionally reabsorbed. However, the osmotically induced reabsorption of H_2O in the proximal tubule secondary to active Na^+ reabsorption produces a concentration gradient for urea that favors the inadvertent passive reabsorption of this nitrogenous waste. Because of extensive reabsorption of H_2O in the proximal tubule, the original 125 ml/min of filtrate is progressively reduced, until only 44 ml/min of fluid remains in the lumen by the end of the proximal tubule (with 65% of the H_2O in the original filtrate, or 81 ml/min, having been reabsorbed). Substances that have been filtered but not reabsorbed become progressively more concentrated in the tubular fluid as H_2O is reabsorbed but they are left behind. Urea is one such substance (Fig. 14-20). Urea's concentration as it is filtered at the glomerulus is identical to its concentration in the plasma entering the peritubular capillaries. The quantity of urea present within the 125 ml of filtered fluid at the beginning of the proximal tubule, however, is concentrated almost three-fold in the small volume of only 44 ml left at the end of the proximal tubule. As a result, the intratubular urea concentration becomes considerably greater than the plasma urea concentration in the adjacent capillaries. Therefore, a concentration gradient is created for urea to passively diffuse from the tubular lumen into the peritubular-capillary plasma. Since the walls of the proximal tubules are only moderately permeable to urea, about 50% of the filtered urea is passively reabsorbed by this means.

Even though only half of the filtered urea is eliminated from the plasma with each pass through the nephrons, this is an adequate removal rate. Only in impaired kidney function, when much less than half the urea is removed, does the urea concentration in the plasma become elevated. An elevated urea level was one of the first chemical findings to be identified in the plasma of patients with severe renal failure. Accordingly, clinical measurement of **blood urea nitrogen (BUN)** came into use as a crude assessment of kidney function. It is now known that the most serious consequences of renal failure are not attributable to the retention of urea, which itself is not especially toxic, but rather to the accumulation of other substances that are not adequately excreted because of their failure to be properly secreted—most notably H^+ and K^+. Health professionals still often refer to renal failure as **uremia** ("urea in the blood"), indicative of the presence of excess urea in the blood, even though urea retention is not this condition's major threat.

In general, unwanted waste products are not reabsorbed.

The other filtered waste products besides urea, such as phenol and creatinine, are likewise concentrated in the tubular fluid as H_2O leaves the filtrate to enter the plasma, but they are not passively reabsorbed as urea is. Urea molecules, being the

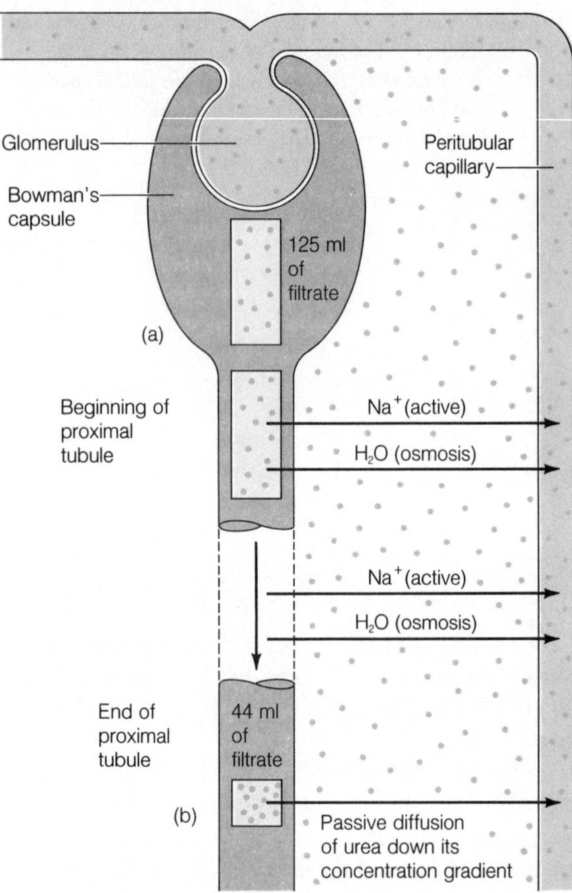

Figure 14–20 Passive Reabsorption of Urea at End of Proximal Tubule *(a) In Bowman's capsule and at the beginning of the proximal tubule, urea is at the same concentration as in the plasma and surrounding interstitial fluid. (b) By the end of the proximal tubule, 65% of the original filtrate has been reabsorbed, leading to concentration of the filtered urea in the remaining filtrate. This establishes a concentration gradient favoring the passive reabsorption of urea.*

smallest of waste products, are the only wastes able to be passively reabsorbed as a result of this concentrating effect. Even though the other wastes are also concentrated in the tubular fluid, they are unable to leave the lumen down their concentration gradients to be passively reabsorbed. They are too large and are lipid insoluble, so they are unable to permeate the tubular wall even though the concentration gradient favors their outward movement. (Recall that substances are able to passively diffuse through a membrane barrier only if they are small enough to pass through a membrane channel or are lipid soluble so they can pass through the lipid bilayer; see p. 70). Therefore, the waste products, failing to be reabsorbed, generally remain in the tubules and are excreted in the urine in a highly concentrated form. This excretion of metabolic wastes is not subject to physiological control. When renal function is normal, however, the excretory processes proceed at a satisfactory rate even though they are not controlled.

The most important secretory processes are those for H^+, K^+, and organic ions.

By providing a second route of entry into the tubules for selected substances, tubular secretion may be viewed as a supplemental mechanism to hasten the elimination of these compounds from the body. Anything that gains entry to the tubular fluid, whether by glomerular filtration or tubular secretion, and that fails to be reabsorbed is eliminated in the urine.

Tubular secretion involves transepithelial transport similar to tubular reabsorption, only the steps are reversed. As with reabsorption, tubular secretion may be active or passive. The most important substances secreted by the tubules are hydrogen ion (H^+), potassium ion (K^+), and organic anions and cations, many of which are compounds foreign to the body.

HYDROGEN ION SECRETION. Renal H^+ secretion is extremely important in the regulation of the acid-base balance in the body. Hydrogen ion can be added to the filtered fluid by being secreted by the proximal, distal, and collecting tubules. The extent of H^+ secretion depends on the acidity of the body fluids. When the body fluids are too acidic, H^+ secretion increases. Conversely, H^+ secretion is reduced when the H^+ concentration in the body fluids is too low (chapter 15).

POTASSIUM SECRETION. Potassium is an example of a substance that is selectively moved in opposite directions in different parts of the tubule. Paradoxically, K^+ is actively reabsorbed in the proximal tubule and actively secreted in the distal and collecting tubules. Potassium reabsorption early in the tubule occurs in a constant, unregulated fashion, whereas K^+ secretion later in the tubule is variable and subject to regulation. Normally, a quantity equivalent to about 10 to 15% of the filtered K^+ is excreted in the urine. However, the filtered K^+ is almost completely reabsorbed, so most of the K^+ appearing in the urine is derived from controlled K^+ secretion rather than from filtration.

During K^+ depletion, K^+ secretion in the distal portions of the nephron is reduced to a minimum, so only the small percentage of filtered K^+ that escapes reabsorption in the proxi-

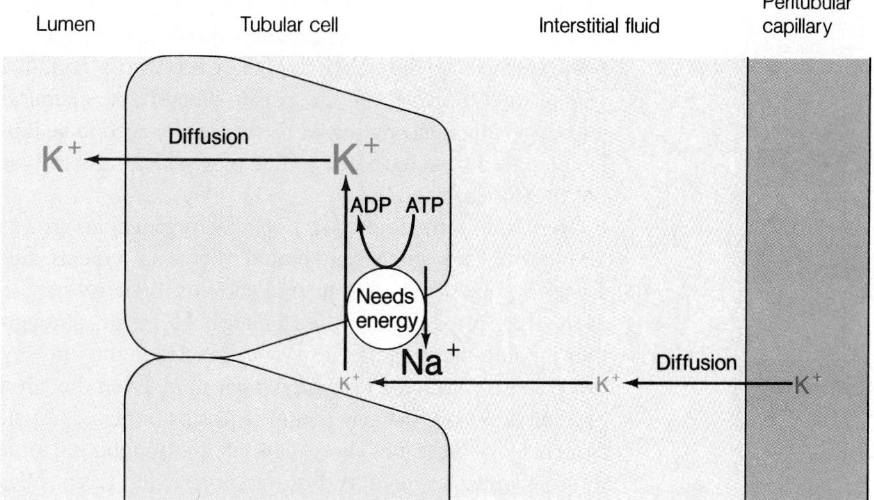

Figure 14–21 Potassium Secretion *The basolateral pump simultaneously transports Na^+ into the lateral space and K^+ into the tubular cell.*

mal tubule is excreted in the urine. This conserves for the body K^+ that normally would have been lost in the urine. On the other hand, when plasma K^+ levels are elevated, K^+ secretion is adjusted so that just enough K^+ is added to the filtrate for elimination to reduce the plasma K^+ concentration to normal. Thus, K^+ secretion, not the filtration or reabsorption of K^+, is varied in a controlled fashion to regulate the rate of K^+ excretion, and thus to maintain the desired plasma K^+ concentration.

Potassium secretion in the distal and collecting tubules is coupled in part to Na^+ absorption by means of the energy-dependent basolateral Na^+-K^+ ATPase carrier (Fig. 14–21). This pump not only moves Na^+ out into the lateral space but also transports K^+ into the tubular cells. The resultant high intracellular K^+ concentration favors the net diffusion of K^+ from the cells into the tubular lumen. Movement across the luminal membrane occurs passively through the large number of K^+ channels present in this barrier. By keeping the interstitial-fluid concentration of K^+ low as it transports K^+ into the tubular cells from the surrounding interstitial fluid, the basolateral pump also encourages the passive diffusion of K^+ out of the peritubular-capillary plasma into the interstitial fluid. Potassium exiting the plasma in this manner is subsequently pumped into the cells, from which it diffuses into the lumen. In this way, the basolateral pump actively induces the net secretion of K^+ from the peritubular-capillary plasma into the tubular lumen.

Several factors are able to alter the rate of K^+ secretion, the most important being the hormone aldosterone, which stimulates K^+ secretion by the tubular cells late in the nephron simultaneous to enhancing the cell's reabsorption of Na^+. An elevation in plasma K^+ concentration directly stimulates the adrenal cortex to increase its output of aldosterone, which in turn promotes the secretion and ultimate urinary excretion and elimination of the excess K^+. Conversely, a decline in plasma K^+ concentration causes a reduction in aldosterone secretion and a corresponding decrease in aldosterone-stimulated renal K^+ secretion.

Note that a rise in plasma K^+ concentration directly stimulates aldosterone secretion by the adrenal cortex, whereas a fall in plasma Na^+ concentration stimulates aldosterone secretion by means of the complex renin-angiotensin pathway. Thus, aldosterone secretion can be stimulated by two separate pathways (Fig. 14–22). No matter what the stimulus, however, increased aldosterone secretion always promotes simultaneous Na^+ reabsorption and K^+ secretion. For this reason, K^+ secretion can be inadvertently stimulated as a result of increased aldosterone activity brought about by Na^+ depletion, ECF volume reduction, or a fall in arterial blood pressure totally unrelated to K^+ balance. The resultant inappropriate loss of K^+ can lead to K^+ deficiency.

Except in the overriding circumstances of K^+ imbalances inadvertently induced during renal compensations for Na^+ or ECF volume deficits, the kidneys usually exert a fine degree of control over plasma K^+ concentration. This is extremely important, because even minor fluctuations in plasma K^+ concentration can have detrimental consequences. Potassium, being the most abundant cation in the intracellular fluid, plays a key role in the membrane electrical activity of excitable tissues. Both increases or decreases in the plasma (ECF) K^+

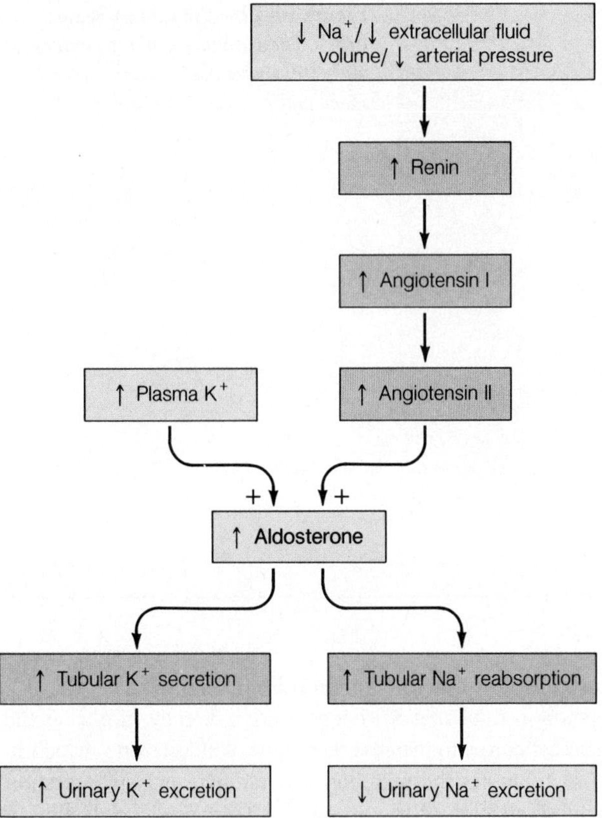

Figure 14–22 Dual Control of Aldosterone Secretion by K^+ and Na^+

concentration can alter the intracellular-to-extracellular K^+ concentration gradient, which in turn can change the resting membrane potential. A rise in ECF K^+ concentration leads to a reduction in resting potential and a subsequent increase in excitability, especially of heart muscle. This can lead to a rapid heart rate and even fatal cardiac arrhythmias. Conversely, a fall in ECF K^+ concentration results in hyperpolarization of nerve and muscle cell membranes, which reduces their excitability. The manifestations of ECF K^+ depletion are skeletal-muscle weakness, diarrhea and abdominal distention caused by smooth-muscle involvement, and abnormalities in cardiac rhythm and impulse conduction.

ORGANIC ANION AND CATION SECRETION. The proximal tubule contains two distinct types of secretory carriers, one for the secretion of organic anions and a separate system for secretion of organic cations. These systems serve several important functions. First, by adding more of a particular type of organic ion to the quantity that has already gained entry to the tubular fluid by means of glomerular filtration, these organic secretory pathways facilitate the excretion of these substances. Included among these organic ions are certain blood-borne chemical messengers that, having served their purpose, need to be rapidly removed from the blood so that their biological activity is not unduly prolonged.

Second, in some important instances, organic ions are extensively but not irreversibly bound to plasma proteins. Because they are attached to plasma proteins, these substances cannot be filtered through the glomeruli. However, although they are not filterable, they can be eliminated in the urine by being actively secreted into the tubular fluid. Even though a given organic ion is largely bound to plasma proteins, a small percentage of these ions always exists in free or unbound form in the plasma. Secretion of this free organic ion permits the "unloading" of some of the bound ion, which is then free to be secreted. This, in turn, encourages the unloading of even more organic ion, and so on.

Third, one of the biggest benefits of the organic-ion secretory systems is the elimination of many foreign compounds from the body. It is important that these secretory carriers are relatively nonspecific. The organic-anion system can secrete a large number of different organic anions, both those produced endogenously (within the body) and those foreign organic anions that have gained access to the body fluids. The same nondiscriminatory nature holds true for the organic-cation secretory system. This nonselectivity permits these organic-ion secretory systems to hasten the removal of many foreign organic chemicals, including food additives, environmental pollutants (for example, pesticides), drugs, and other nonnutritive organic substances that have gained entrance to the body.

The liver plays an important role in this regard. Many foreign organic compounds are not ionic in their original form, so they cannot be secreted by the organic-ion systems. The liver converts these foreign substances into an anionic form that facilitates their secretion by the organic-anion system and thus accelerates their elimination.

The rate of excretion of foreign organic compounds is not subject to control. Although the relatively nonselective organic-ion secretory systems enhance the removal of these substances from the body, this mechanism is not subject to physiological adjustments.

Many drugs, such as penicillin, are eliminated from the body by means of the proximal tubule organic-ion secretory system. In order to keep the plasma concentration of these drugs at effective levels, the dosage has to be repeated on a regular, frequent basis to keep pace with the rapid removal of these compounds in the urine.

Table 14–3 Summary of Transport across Various Tubular Segments

Proximal Tubule		Loop of Henle			
Reabsorption	*Secretion*	*Ascending Limb*	*Descending Limb*	**Distal Tubule**	**Collecting Duct**
All filtered glucose and amino acids reabsorbed by secondary active transport; not subject to control 67% of filtered Na^+ actively reabsorbed; not subject to control; Cl^- follows passively Variable amounts of filtered $PO_4^{\equiv}$ and other electrolytes reabsorbed; subject to control 65% of filtered H_2O osmotically reabsorbed; not subject to control 50% of filtered urea passively reabsorbed; not subject to control All filtered K^+ reabsorbed; not subject to control	Variable H^+ secretion; dependent on acid-base status of body Organic ion secretion; not subject to control	25% of filtered NaCl actively reabsorbed into medullary interstitial fluid; not subject to control; region impermeable to H_2O	15% of filtered H_2O osmotically reabsorbed into medullary interstitial fluid; not controlled Perhaps NaCl passively enters (secreted); not controlled	Variable Na^+ reabsorption; controlled by aldosterone; Cl^- follows passively Variable K^+ secretion; controlled by aldosterone Variable H_2O reabsorption; controlled by vasopressin Variable H^+ secretion; dependent on acid-base status of body	Variable H_2O reabsorption; controlled by vasopressin Variable H^+ secretion; dependent on acid-base status of body

This completes our discussion of the reabsorptive and secretory processes that occur across the various tubular segments. These processes are summarized in Table 14–3.

Urine Excretion and Plasma Clearance

On the average, one milliliter of urine per minute is excreted.

Typically, of the 125 ml of plasma filtered per minute, 124 ml/min are reabsorbed, so the final quantity of urine formed averages 1 ml/min. This equals 1.5 liters per day of urine excreted out of the 180 liters per day filtered.

Urine contains high concentrations of various waste products plus variable amounts of the substances regulated by the kidneys, with any excess quantities having spilled into the urine. Useful substances are conserved by reabsorption, so they do not appear in the urine.

A relatively small change in the quantity of filtrate reabsorbed can bring about a large change in the volume of urine formed. For example, a reduction of less than 1% in the total reabsorption rate, from 124 to 123 ml/min, increases the urinary excretion rate by 100%, from 1 to 2 ml/min.

Plasma clearance refers to the volume of plasma cleared of a particular substance per minute.

Compared to the plasma entering the kidneys through the renal arteries, the plasma leaving the kidneys through the renal veins lacks the materials that were left behind to be eliminated in the urine. By excreting substances in the urine, the kidneys clean or "clear" the plasma flowing through them of these substances. For any substance, its **plasma clearance** is defined as the volume of plasma that is completely cleared of that substance by the kidneys per minute.[1] It does not refer to

[1]Actually, plasma clearance is an artificial concept, because when a particular substance is excreted in the urine, that substance's concentration in the plasma as a whole is uniformly decreased as a result of thorough mixing in the circulatory system. However, it is useful for comparative purposes to consider clearance in effect as the volume of plasma that would have contained the total quantity of the substance that the kidneys excreted in one minute; that is, the hypothetical volume of plasma completely cleared of that substance per minute.

the *amount* of the substance removed but to the *volume of plasma* from which that amount was removed. Plasma clearance is actually a more useful measure than urine excretion; it is more important to know what effect urine excretion has on removing materials from the body fluids than to know the volume and composition of the discarded urine. Plasma clearance expresses the kidney's effectiveness in removing various substances from the internal fluid environment.

Plasma clearance can be calculated for any plasma constituent as follows:

$$\begin{array}{c}\text{Plasma clearance}\\ \text{of a substance}\\ \text{(ml/min)}\end{array} = \frac{\begin{array}{c}\text{Urine concentration}\\ \text{of the substance}\\ \text{(quantity/ml urine)}\end{array} \times \begin{array}{c}\text{urine flow rate}\\ \text{(ml/min)}\end{array}}{\begin{array}{c}\text{Plasma concentration of the substance}\\ \text{(quantity/ml plasma)}\end{array}}$$

The plasma clearance rate for different substances varies, depending on how the kidneys handle each substance.

FOR A SUBSTANCE THAT IS FILTERED BUT NOT REABSORBED OR SECRETED, ITS PLASMA CLEARANCE RATE EQUALS THE GFR. Assume that a plasma constituent, substance X, is freely filterable at the glomerulus but is not reabsorbed or secreted. As 125 ml/min of plasma is filtered and subsequently reabsorbed, the quantity of substance X originally contained within the 125 ml is left behind in the tubules to be excreted. Thus, 125 ml of plasma is cleared of substance X each minute. (Of the 125 ml/min of plasma filtered, 124 ml/min of the filtered fluid is returned through the process of reabsorption to the plasma minus substance X, thus clearing this 124 ml/min. of substance X. In addition, the 1 ml/min. of fluid lost in the urine is replaced in the long-term by an equivalent volume of ingested H_2O that is already clear of substance X. Therefore, 125 ml of plasma cleared of substance X is, in effect, returned to the plasma for every 125 ml of plasma filtered/min.)

There is no endogenous chemical with the characteristics of substance X. All substances naturally present in the plasma, even wastes, are reabsorbed or secreted to some extent. However, **inulin** (do not confuse with insulin), a harmless foreign carbohydrate produced by onions and garlic, is freely filtered and not reabsorbed or secreted—an ideal substance X. Inulin can be injected and its plasma clearance determined as a clinical means of ascertaining the GFR. Since all glomerular filtrate formed is cleared of inulin, the volume of plasma cleared of inulin per minute equals the volume of plasma filtered per minute; that is, the GFR.

Although the determination of inulin plasma clearance is accurate and straightforward, it is not very convenient because inulin must be infused continuously throughout the determination to maintain a constant plasma concentration. Therefore, the plasma clearance of an endogenous substance, **creatinine,** is often used instead to give a rough estimate of the GFR. Creatinine, an end-product of muscle metabolism, is produced at a relatively constant rate. It is freely filtered and not reabsorbed but is slightly secreted. Accordingly, creatinine clearance is not a completely accurate reflection of the GFR, but it does provide a close approximation and can be more readily determined than inulin clearance.

FOR A SUBSTANCE THAT IS FILTERED AND REABSORBED BUT NOT SECRETED, ITS PLASMA CLEARANCE RATE IS ALWAYS LESS THAN THE GFR. Some or all of a reabsorbable substance that has been filtered is returned to the plasma. Because less than the filtered volume of plasma will have been cleared of the substance, the plasma clearance rate of a reabsorbable substance is always less than the GFR. For example, the plasma clearance for glucose is normally zero. All of the filtered glucose is reabsorbed along with the rest of the returning filtrate, so none of the plasma is cleared of glucose.

For a substance that is partially reabsorbed, such as urea, only part of the filtered plasma is cleared of that substance. With about 50% of the filtered urea being passively reabsorbed, only half of the filtered plasma, or 62.5 ml, is cleared of urea each minute.

FOR A SUBSTANCE THAT IS FILTERED AND SECRETED BUT NOT REABSORBED, ITS PLASMA CLEARANCE IS ALWAYS GREATER THAN THE GFR. Tubular secretion allows the kidneys to more efficiently clear certain materials from the plasma. Only 20% of the plasma entering the kidneys is filtered. The remaining 80% passes unfiltered into the peritubular capillaries. The only means by which this unfiltered plasma can be cleared of any substance during this trip through the kidneys before being returned to the general circulation is by the process of secretion. An example is H^+. Not only will the plasma that is filtered be cleared of nonreabsorbable H^+, but the plasma from which H^+ is secreted will also be cleared of H^+. For example, if the quantity of H^+ that is secreted is equivalent to the quantity of H^+ present in 25 ml of plasma, the clearance rate for H^+ will be 150 ml/min at the normal GFR of 125 ml/min. Every minute 125 ml of plasma will lose its H^+ through the process of filtration and failure of reabsorption, and 25 more ml of plasma will lose its H^+ through the process of secretion. The plasma clearance for a secreted substance is always greater than the GFR.

Just as inulin can be used clinically to determine the GFR, the plasma clearance of another foreign compound, the organic anion **para-aminohippuric acid (PAH)**, can be used to measure renal plasma flow. Like inulin, PAH is freely filterable and nonreabsorbable. However, it differs in that the PAH in all of the plasma that escapes filtration is secreted from the peritubular capillaries. Thus PAH is removed from *all* of the plasma that flows through the kidneys—both from the plasma that is filtered and subsequently reabsorbed without its PAH, and from the unfiltered plasma that continues on in the peritubular capillaries and loses its PAH by means of active secretion into the tubules. Because all of the plasma that flows through the kidneys is cleared of PAH, the plasma clearance for PAH is a reasonable estimate of the rate of plasma flow through the kidneys. Typically, renal plasma flow averages 625 ml/min.

The total renal blood flow can be calculated using the renal plasma-flow determination and taking into account the hematocrit. At a normal hematocrit of 45, 55% of the blood is plasma. A renal plasma flow of 625 ml/min thus represents a renal blood flow of 1,140 ml/min (55% of 1,140 ml/min = 625 ml/min).

Knowing PAH clearance (renal plasma flow) and inulin clearance (GFR), you can easily determine the **filtration fraction,** or that fraction of the plasma flowing through the glomeruli that is filtered into the tubules:

$$\text{Filtration fraction} = \frac{\text{GFR (Plasma inulin clearance)}}{\begin{array}{c}\text{Renal plasma flow}\\ \text{(Plasma PAH clearance)}\end{array}} = \frac{125 \text{ ml/min}}{625 \text{ ml/min}} = 20\%$$

Thus 20% of the plasma that enters the glomeruli is filtered.

The ability to excrete urine of varying concentrations depends on the medullary countercurrent system and vasopressin.

Having considered how the kidneys deal with a variety of solutes in the plasma, we will now concentrate on renal handling of plasma H_2O. The ECF osmolarity (solute concentration) depends on the relative amount of H_2O compared to solute. At normal fluid balance and solute concentration, the body fluids are said to be **isotonic** at an osmolarity of 300 milliosmoles/liter (mosm/l). If there is too much H_2O relative to the solute load, the body fluids are **hypotonic,** which means that they are too dilute at an osmolarity less than 300 mosm/l. On the other hand, if a H_2O deficit exists relative to the solute load, the body fluids are too concentrated or are **hypertonic,** having an osmolarity greater than 300 mosm/l.

Generally speaking, the osmolarity of the ECF is uniform throughout the body. Knowing that the driving force for H_2O reabsorption throughout the entire length of the tubules is an osmotic gradient between the tubular lumen and surrounding interstitial fluid, you would expect, according to osmotic considerations, that the kidneys could not excrete urine more or less concentrated than the body fluids. Indeed, this would be the case if the interstitial fluid surrounding the tubules in the kidneys were identical in osmolarity to the remaining body fluids. Water reabsorption could proceed only until the tubular fluid equilibrated osmotically with the interstitial fluid. There would be no way to eliminate excess H_2O when the body fluids were hypotonic or to conserve H_2O in the presence of hypertonicity.

Fortunately, a large vertical osmotic gradient is uniquely maintained in the interstitial fluid of the medulla of each kidney. The concentration of the interstitial fluid progressively increases from the cortical boundary down through the depth of the renal medulla until it reaches a maximum in humans of 1,200 mosm/l at the junction with the renal pelvis (Fig. 14–23). This vertical osmotic gradient remains constant regardless of the fluid balance of the body.

The presence of this gradient enables the kidneys to produce urine that ranges in concentration from 100 to 1,200 mosm/l, depending on the body's state of hydration. When the body is in ideal fluid balance, 1 ml/min of isotonic urine is formed. When the body is overhydrated (too much H_2O), the kidneys are able to produce a large volume of dilute urine (up to 25 ml/min and hypotonic at 100 mosm/l), thus eliminating the excess H_2O in the urine. Conversely, the kidneys are able

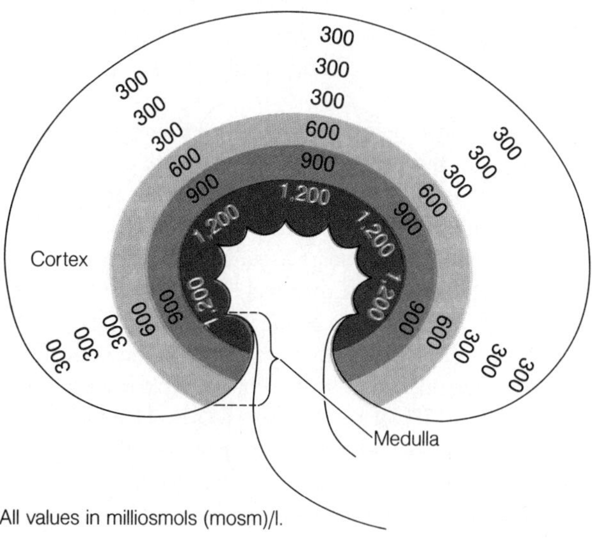

Figure 14–23 Vertical Osmotic Gradient in Renal Medulla

to put out a small volume of concentrated urine (down to 0.3 ml/min and hypertonic at 1,200 mosm/l) when the body is dehydrated (too little H_2O), thus conserving H_2O for the body.

Unique anatomical arrangements and complex functional interactions between the various nephron components present in the renal medulla are responsible for the establishment and utilization of the vertical osmotic gradient. As a brief preview, the long loops of Henle of the juxtamedullary nephrons establish the vertical osmotic gradient; the vasa recta of these same nephrons prevent the dissolution of this gradient while providing blood to the renal medulla; and the collecting tubules of *all* nephrons use the gradient, in conjunction with vasopressin, to produce urine of varying concentrations. Collectively, this entire functional organization is known as the **medullary countercurrent system.** We will examine each of its facets in greater detail.

THE LOOPS OF HENLE OF JUXTAMEDULLARY NEPHRONS ESTABLISH THE MEDULLARY VERTICAL OSMOTIC GRADIENT BY MEANS OF COUNTERCURRENT MULTIPLICATION. Reviewing the anatomical arrangement of the renal medulla (Fig. 14–4), the loops of Henle of juxtamedullary nephrons are long, hair-pin loops that dip deeply into the medulla. The peritubular capillaries of the same nephrons form vascular hair-pin loops, the vasa recta, that course along in close proximity to the loops of Henle. Flow in both the loops of Henle and vasa recta is considered to be *countercurrent* because of the flow in opposite directions in the two closely adjacent limbs of the loop. Also running through the medulla in the descending direction only on their way to the renal pelvis are the collecting tubules of both cortical and juxtamedullary nephrons.

We will follow the filtrate through a juxtamedullary nephron to see how this structure establishes a vertical osmotic gradient in the medulla. Immediately after the filtrate is formed, uncontrolled osmotic reabsorption of filtered H_2O occurs in the proximal tubule secondary to active Na^+ reabsorption. As a result, by the end of the proximal tubule about 65% of the filtrate has been reabsorbed, but the 35% remaining in the tubular lumen still has the same osmolarity as the body fluids. Therefore, the fluid entering the loop of Henle is still isotonic. An additional 15% of the filtered H_2O is obligatorily reabsorbed from the loop of Henle during the establishment and maintenance of the vertical osmotic gradient, with the osmolarity of the tubular fluid being altered in the process.

The following functional distinctions between the descending limb of Henle's loop (which carries fluid from the proximal tubule down into the depths of the medulla) and the ascending limb (which carries fluid up and out of the medulla into the distal tubule) are critical to the establishment of the incremental osmotic gradient in the medullary interstitial fluid.

The descending limb:

1. is highly permeable to H_2O.
2. does not actively extrude Na^+ (being the only segment of the tubule that does not do so); and
3. may be passively permeable to salt. (This is still controversial.)

The ascending limb:

1. actively transports NaCl out of the tubular lumen into the surrounding interstitial fluid (whether it is Na^+ or Cl^- that is actively transported, with the other ion following passively down the resulting electrical gradient, is still being debated); and
2. is always impermeable to H_2O, so salt leaves the tubular fluid without concomitant H_2O reabsorption.

The close proximity and countercurrent flow of the two limbs allow important interactions to occur between them. Even though the flow of fluid is continuous through the loop of Henle, we will visualize what happens step by step, much like an animated movie film run so slowly that each individual frame can be viewed.

INITIAL SCENE (FIG. 14–24A). Before the vertical osmotic gradient is established, the medullary interstitial-fluid concentration is uniformly 300 mosm/l, as is the remainder of the body fluids.

STEP 1 (FIG. 14–24B). The active salt pump in the ascending limb is able to transport NaCl out of the lumen until the surrounding interstitial fluid is 200 mosm/l more concentrated than the tubular fluid in this limb. When the ascending-limb pump starts actively extruding salt, the medullary interstitial fluid becomes hypertonic. Water cannot follow osmotically from the ascending limb because of the limb's impermeability to H_2O. However, net diffusion of H_2O does occur from the descending limb into the interstitial fluid. The tubular fluid entering the descending limb from the proximal tubule is isotonic. Because the descending limb is highly permeable to H_2O, net diffusion of H_2O occurs by osmosis out of the descending limb into the more concentrated interstitial fluid. (Furthermore, as a result of NaCl extrusion into the interstitial fluid by the ascending limb, the NaCl concentration in the interstitial fluid is higher than it is in the tubular fluid in the descending limb. If indeed the descending limb is permeable to NaCl, net diffusion of NaCl occurs from the interstitial fluid into the descending limb.) The passive movement of H_2O out of (and perhaps the passive diffusion of NaCl into) the descending limb continues until the osmolarities of the fluid in the descending limb and interstitial fluid become equilibrated. Thus the tubular fluid entering the loop of Henle immediately starts to become more concentrated as it loses H_2O (and perhaps gains salt). At equilibrium, the osmolarity of the ascending-limb fluid is 200 mosm/l, and the interstitial fluid and descending-limb fluid are equal at 400 mosm/l. The 200 mosm/l concentration difference between the ascending limb and surrounding interstitial fluid is maintained by active extrusion of salt from the ascending limb. The identical 200 mosm/l gradient between the ascending limb and its adjacent descending limb derives from the passive fluxes across the descending limb so that the fluid concentration in the descending limb is always the same as that of the surrounding interstitial fluid.

STEP 2 (FIG. 14-24C). If we now advance the entire column of fluid in the loop of Henle several "frames," a mass of 200 mosm/l fluid exits from the top of the ascending limb into the distal tubule, and a new mass of isotonic fluid at 300 mosm/l enters the top of the descending limb from the proximal tubule. At the bottom of the loop, a comparable mass of 400 mosm/l fluid from the descending limb moves forward around the tip into the ascending limb, placing it opposite a 400 mosm/l region in the descending limb. Note that the 200 mosm/l concentration difference has been lost both at the top and bottom of the loop.

STEP 3 (FIG. 14-24D). The ascending-limb pump again transports NaCl out while passive fluxes occur across the descending limb until a 200 mosm/l difference is reestablished between the ascending limb and both the interstitial fluid and descending limb at each horizontal level. Note, however, that the concentration of tubular fluid in the descending limb is progressively increasing whereas that in the ascending limb is progressively decreasing.

STEP 4 (FIG. 14-24E). As the tubular fluid is advanced still farther forward, the 200 mosm/l concentration gradient is disrupted once again at all horizontal levels.

STEP 5 (FIG. 14-24F). Again, active extrusion of NaCl from the ascending limb, coupled with the net diffusion of H_2O out of (and perhaps NaCl into) the descending limb, reestablishes the 200 mosm/l gradient at each horizontal level.

STEPS 6 AND ON (FIG. 14-24G). As the fluid flows slightly forward again and as this stepwise process continues, the fluid in the descending limb becomes progressively more hypertonic until it reaches a maximum concentration of 1,200 mosm/l at the bottom of the loop, four times the normal concentration of body fluids. Because the interstitial fluid always achieves equilibrium with the descending limb, an incremental vertical concentration gradient ranging from 300 to 1,200 mosm/l is likewise established in the medullary interstitium. In contrast, the concentration of the tubular fluid progressively decreases in the ascending limb as salt is pumped out but H_2O is unable to follow. In fact, the tubular fluid even becomes hypotonic as it leaves the ascending limb to enter the distal tubule at a concentration of 100 mosm/l, one-third the normal concentration of body fluids.

Note that, although a gradient of only 200 mosm/l exists between the ascending limb and surrounding fluids at each medullary horizontal level, a much larger vertical gradient exists from the top to the bottom of the medulla. Even though the ascending-limb pump can generate a gradient of only 200 mosm/l, this effect is multiplied into a large vertical gradient because of the countercurrent flow within the loop. This concentrating mechanism accomplished by the loop of Henle is known as **countercurrent multiplication.**

We have artificially described countercurrent multiplication in a "stop" and "flow," stepwise fashion to facilitate understanding. It is important to realize that, once the incremental medullary gradient is established, it remains constant because of the continuous flow of fluid coupled with the ongoing ascending-limb active-transport activity and accompanying descending-limb passive fluxes.

If you consider only what happens to the tubular fluid as it flows through the loop of Henle, the whole process seems to be an exercise in futility. The isotonic fluid that enters the loop becomes progressively more concentrated as it flows down the descending limb, achieving a maximum concentration of 1,200 mosm/l, only to become progressively more dilute as it flows up the ascending limb, finally leaving the loop at a minimum concentration of 100 mosm/l. What is the point of concentrating the fluid four-fold and then turning around and diluting it until it leaves at one-third the concentration at which it entered? Such a mechanism has a two-fold benefit. First, this mechanism establishes a vertical osmotic gradient in the medullary interstitial fluid. This gradient, in turn, is used by the collecting duct to concentrate the tubular fluid so that a urine *more concentrated* than normal body fluids can be excreted. Second, the fact that the fluid is hypotonic as it enters the distal portions of the tubule enables the kidneys to excrete a urine *more dilute* than normal body fluids.

COUNTERCURRENT EXCHANGE WITHIN THE VASA RECTA ENABLES THE MEDULLA TO BE SUPPLIED WITH BLOOD WHILE CONSERVING THE MEDULLARY VERTICAL OSMOTIC GRADIENT. Obviously, the renal medulla must be supplied with blood to nourish the tissues in this area as well as to transport water that

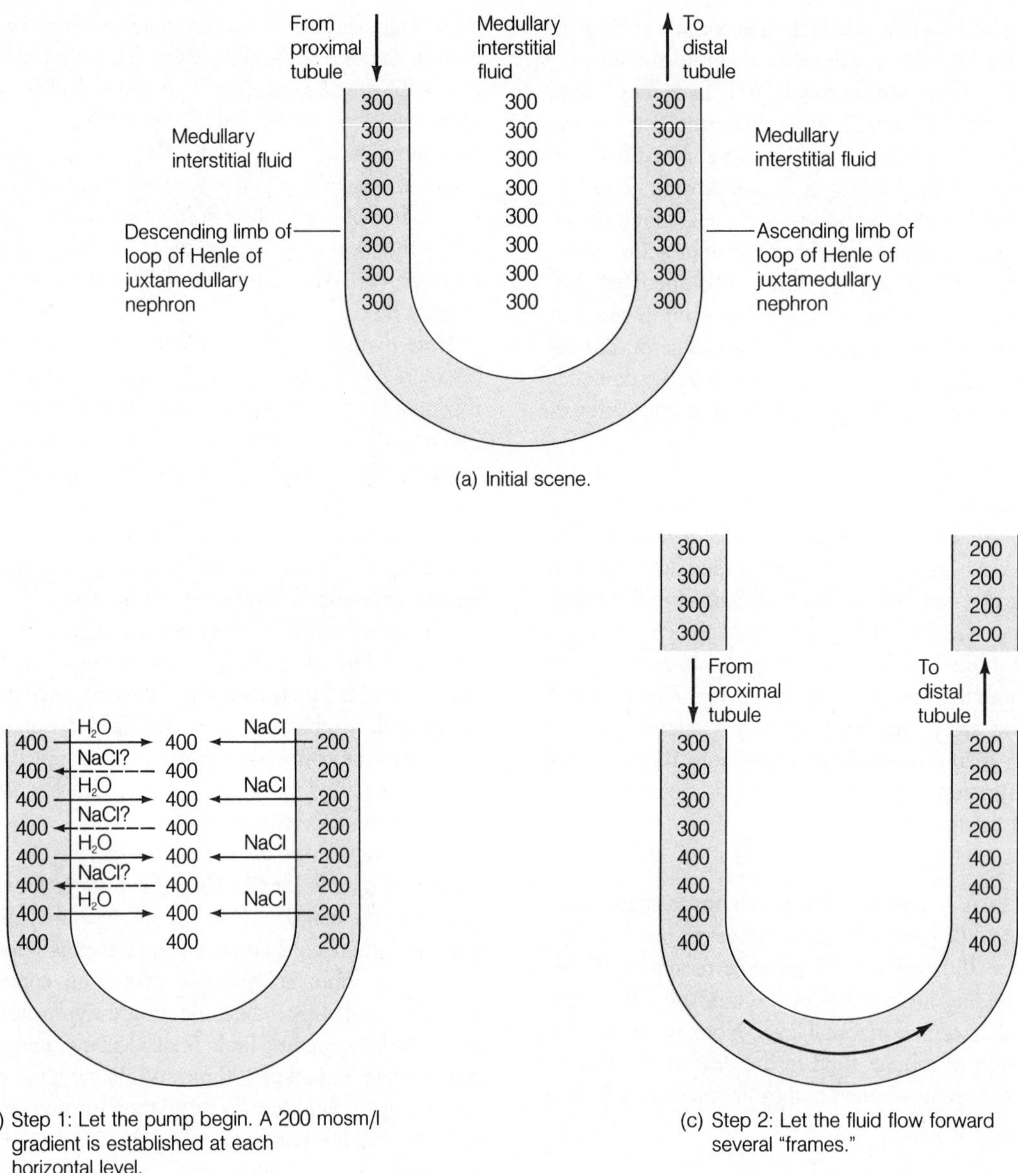

Figure 14–24 Countercurrent Multiplication

is reabsorbed by the loops of Henle and collecting tubules back to the general circulation. In doing so, however, it is important that circulation of blood through the medulla does not disturb the vertical gradient of hypertonicity established by the loops of Henle. Consider the situation if blood were to flow straight through from the cortex to the inner medulla and then directly into the renal vein (Fig. 14–25a). Since capillaries are freely permeable to NaCl and H_2O, the blood would progressively pick up salt and lose H_2O through passive fluxes down concentration and osmotic gradients as it flowed through the depths of the medulla. Isotonic blood entering the medulla, upon equilibrating with each medullary level, would leave the medulla very hypertonic at 1,200 mosm/l. It would be impossible to establish and maintain the medullary hypertonic gradient, because the NaCl pumped into the medullary interstitial fluid would continuously be carried away by the circulation.

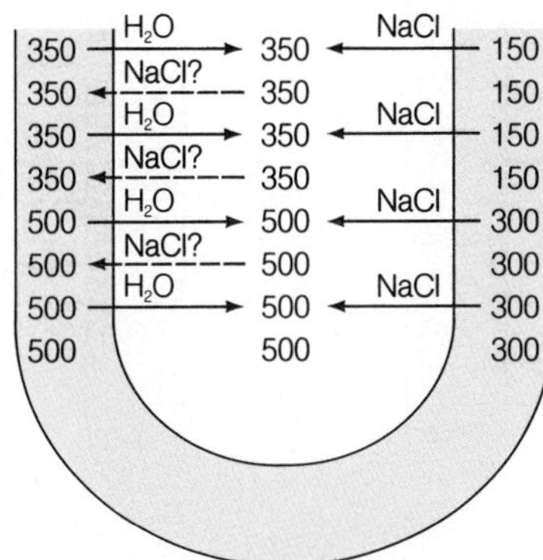

(d) Step 3: The ascending limb pump and descending limb passive fluxes reestablish the 200 mosm/l gradient at each horizontal level.

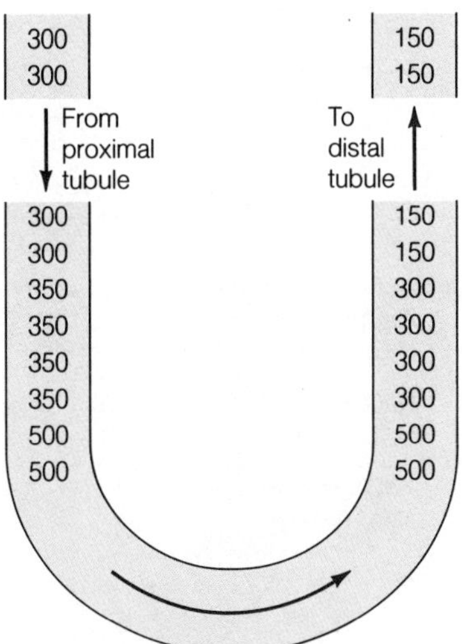

(e) Step 4: Let the fluid flow forward several "frames" once again.

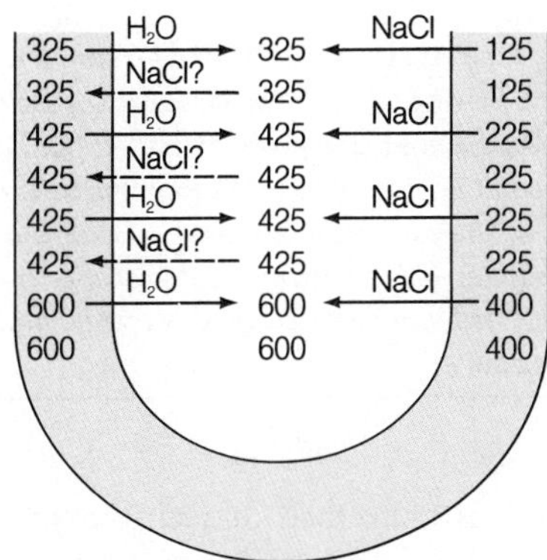

(f) Step 5: The 200 mosm/l gradient at each horizontal level is established once again.

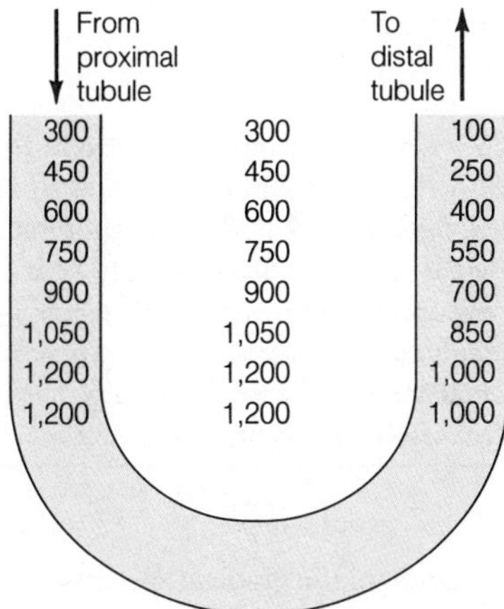

(g) Steps 6 and on: The final vertical osmotic gradient is established and maintained by the ongoing countercurrent multiplication of the long loops of Henle.

All values in mosm/l.

This dilemma is avoided by the hairpin construction of the vasa recta, which, by looping back through the concentration gradient in reverse, allows the blood to leave the medulla and enter the renal vein essentially isotonic to incoming arterial blood (Fig. 14–25b). As blood passes down the descending limb of the vasa recta, equilibrating with the progressively increasing concentration of the surrounding interstitial fluid, it picks up salt and loses H_2O until it is very hypertonic by the bottom of the loop. Then, as blood flows up the ascending limb, salt diffuses back out into the interstitium and H_2O reenters the vasa recta as progressively decreasing concentrations are encountered in the surrounding interstitial fluid. This passive exchange of solutes and H_2O between the two limbs of the vasa recta and the interstitial fluid is known as **countercurrent exchange.** It does not ***establish*** the concentration gradient, as does countercurrent multiplication. Rather, it ***pre-***

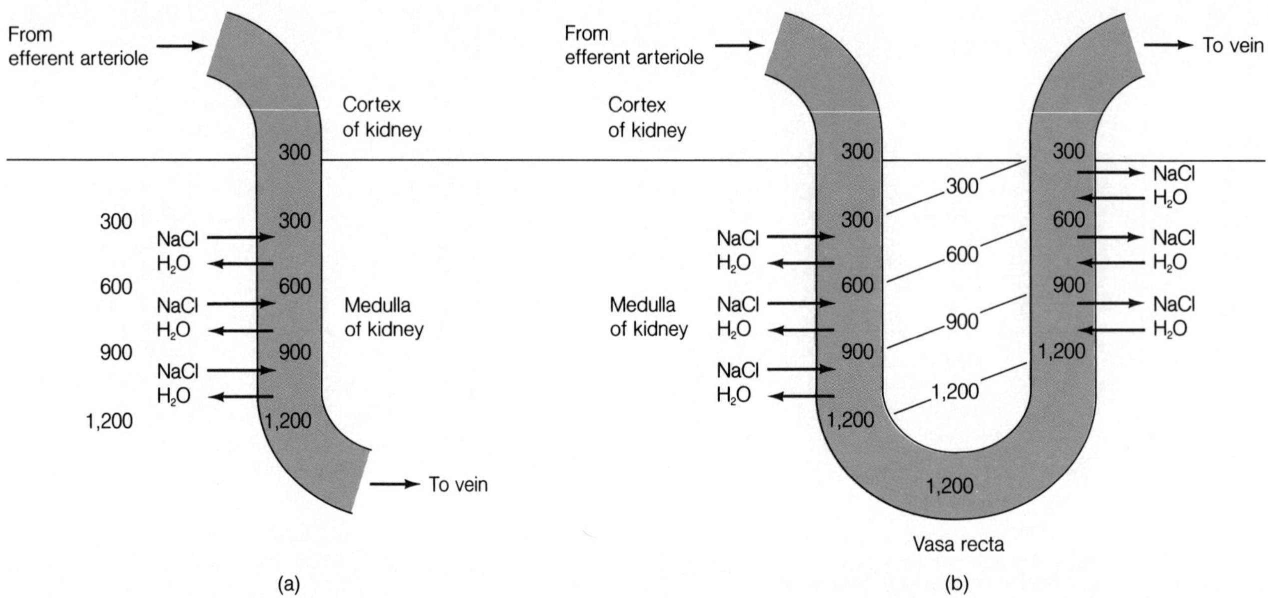

Figure 14–25 Countercurrent Exchange in the Renal Medulla *(a) Hypothetical pattern of blood flow. If the blood supply to the renal medulla were to flow straight through from the cortex to the inner medulla, the blood would be isotonic on entering but very hypertonic on leaving, having picked up salt and lost H_2O as it equilibrated with the surrounding interstitial fluid at each incremental horizontal level. It would be impossible to maintain the vertical osmotic gradient, because the salt pumped out by the ascending limb of Henle's loop would continuously be flushed away by blood flowing through the medulla. (b) Actual pattern of blood flow. Blood equilibrates with interstitial fluid at each incremental horizontal level in both the descending limb and the ascending limb of the vasa recta, so blood is isotonic as it enters and leaves the medulla. This countercurrent exchange prevents the dissolution of the medullary osmotic gradient while providing blood to the renal medulla.*

vents the dissolution of the gradient. Because blood enters and leaves the medulla at the same osmolarity as a result of countercurrent exchange, the medullary tissue is nourished with blood, yet the incremental gradient of hypertonicity in the medulla is preserved.

THE MEDULLARY VERTICAL OSMOTIC GRADIENT PERMITS EXCRETION OF URINE OF DIFFERING CONCENTRATIONS BY MEANS OF VASOPRESSIN-CONTROLLED, VARIABLE H_2O REABSORPTION FROM THE FINAL TUBULAR SEGMENTS. Following obligatory H_2O reabsorption from the proximal tubule (65% of the filtered H_2O) and loop of Henle (15% of the filtered H_2O), 20% of the filtered H_2O remains in the lumen to enter the distal and collecting tubules for variable reabsorption that is under hormonal control. This is still a large volume of filtered H_2O subject to regulated reabsorption; it constitutes 20% × GFR(180 liters/day) = 36 liters per day that is reabsorbed to varying extents, depending on the body's state of hydration, which is more than thirteen times the amount of plasma H_2O in the entire circulatory system.

The fluid leaving the loop of Henle enters the distal tubule at 100 mosm/l, so it is hypotonic to the surrounding isotonic (300 mosm/l) interstitial fluid of the renal cortex through which the distal tubule passes. The distal tubule then empties into the collecting tubule, which is bathed by progressively increasing concentrations (300 to 1,200 mosm/l) of surrounding interstitial fluid as it descends through the medulla.

In order for H_2O reabsorption to occur across a segment of the tubule, two critera must be met: (1) an osmotic gradient must exist across the tubule; and (2) the tubular segment must be permeable to H_2O. The distal and collecting tubules are *impermeable* to H_2O except in the presence of **vasopressin,** also known as **antidiuretic hormone** (*anti* means "against"; *diuretic* means "increased urine output"), which increases their permeability to H_2O. Vasopressin secretion increases when the body is dehydrated and H_2O must be conserved.

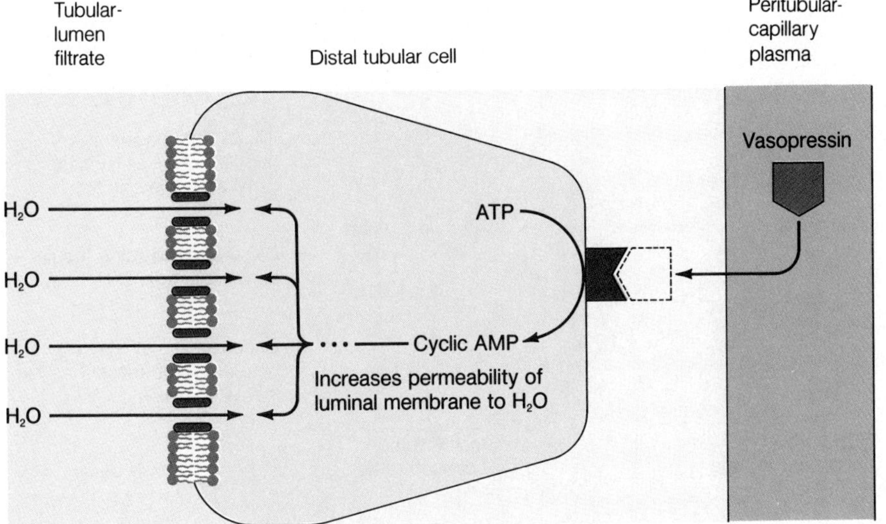

Figure 14–26 Mechanism of Action of Vasopressin

This hormone reaches the basolateral membrane of the tubular cells lining the distal and collecting tubules through the circulatory system, whereupon it binds with receptors specific for it (Fig. 14–26). This binding activates the cyclic AMP second-messenger system within the tubular cells (see p. 64), which ultimately brings about changes in the opposite luminal membrane to increase its permeability to H_2O. The tubular response to vasopressin is graded; the more vasopressin present, the greater the permeability of the distal and collecting tubules to H_2O. It has been proposed that, in response to vasopressin activation of the cyclic AMP system, proteins in the luminal membrane are rearranged to increase the number of H_2O-filled channels. This would permit more H_2O to permeate from the lumen, thus increasing H_2O reabsorption. The structural rearrangement of luminal-membrane proteins is not permanent, however. It is reversed when vasopressin secretion decreases and cyclic AMP activity is similarly decreased. Accordingly, H_2O permeability is reduced when vasopressin secretion decreases.

In the presence of vasopressin, the hypotonic tubular fluid entering the distal tubule is able to lose progressively more H_2O by osmosis into the interstitial fluid as it flows first through the isotonic cortex and then is exposed to the ever-increasing osmolarity of the medullary interstitial fluid as it plunges toward the renal pelvis (Fig. 14–27a). Under the influence of maximum levels of vasopressin, it is possible to concentrate the tubular fluid up to 1,200 mosm/l by the end of the collecting tubule. No further modification of the tubular fluid occurs beyond the collecting tubule, so what remains in the tubules at this point is urine. As a result of this extensive vasopressin-promoted reabsorption of H_2O in the late segments of the tubule, a small volume of urine concentrated up to 1,200 mosm/l can be excreted. As little as 0.3 ml of urine may be formed each minute, less than one-third the normal urine flow rate of 1 ml/min. The reabsorbed H_2O entering the medullary interstitial fluid is subsequently picked up by the vasa recta and returned to the general circulation, thus being conserved for the body.

During times of H_2O deprivation, the body is still generating metabolic wastes and has excess electrolytes such as H^+ that must be eliminated. Under such circumstances, it is advantageous for the body to excrete the materials that need to be discarded in as small a volume of urine as possible to conserve H_2O. Collectively, the waste products and other constituents eliminated in the urine average 600 mosm each day. Because the maximum urine concentration is 1,200 mosm/l, the minimum volume of urine that is required to excrete these wastes is 500 ml/day (600 mosm of wastes/day ÷ 1,200 mosm/l of urine = 0.5 l, or 500 ml/day, or 0.3 ml/min). Thus, under maximal vasopressin influence, 99.8% of the 180 liters of plasma H_2O filtered per day is returned to the blood, with an obligatory H_2O loss of half a liter. It is important to realize that, although vasopressin promotes H_2O conservation by the body, it cannot completely halt urine production because of this minimum volume of H_2O that must be excreted with the solute wastes. A person not taking in any H_2O

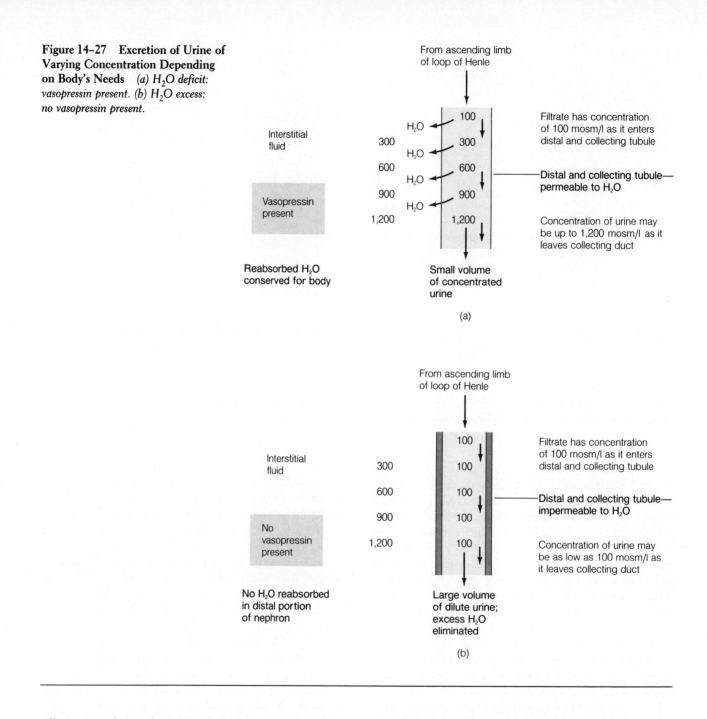

Figure 14-27 Excretion of Urine of Varying Concentration Depending on Body's Needs *(a) H_2O deficit: vasopressin present. (b) H_2O excess: no vasopressin present.*

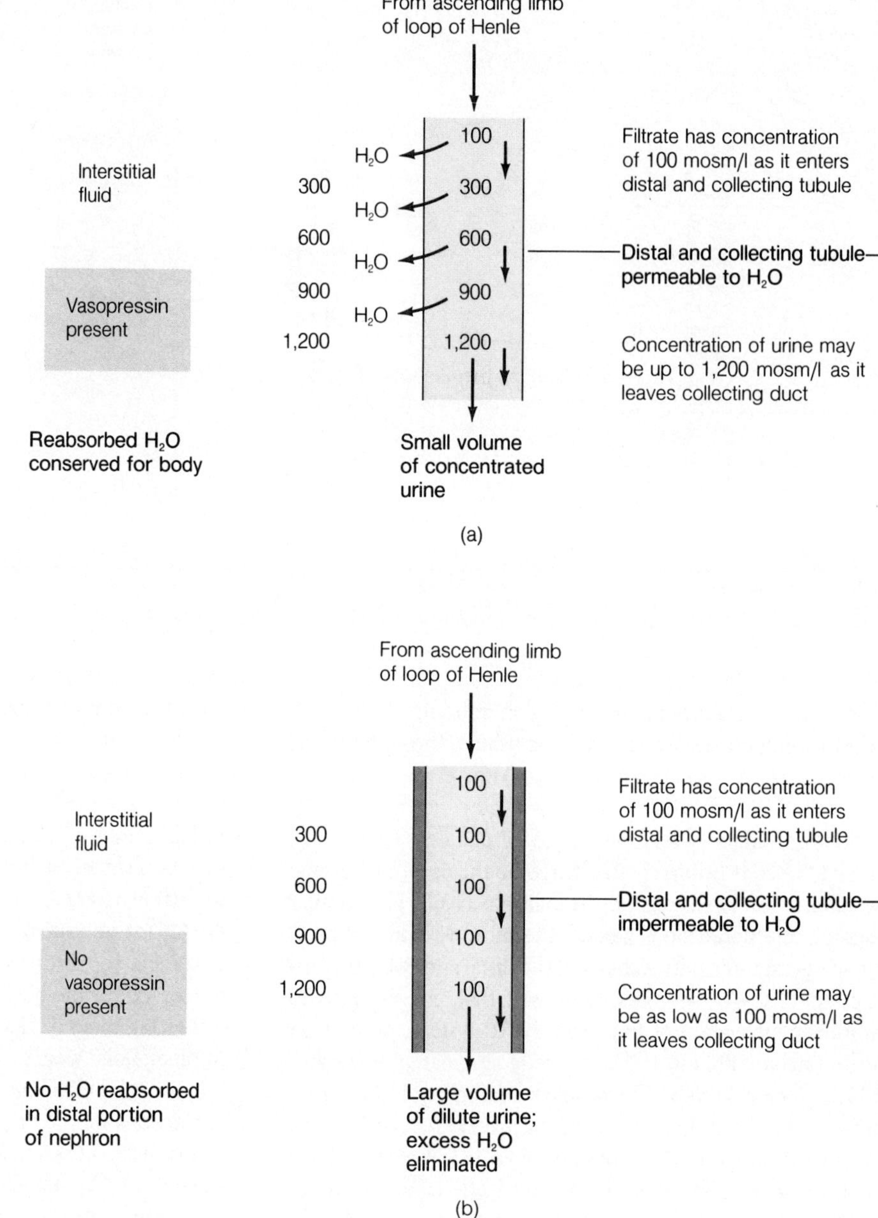
From ascending limb of loop of Henle
Interstitial fluid
300
600
900
1,200
H₂O
100
300
600
900
1,200
Vasopressin present
Filtrate has concentration of 100 mosm/l as it enters distal and collecting tubule
Distal and collecting tubule—permeable to H₂O
Concentration of urine may be up to 1,200 mosm/l as it leaves collecting duct
Reabsorbed H₂O conserved for body
Small volume of concentrated urine
(a)
From ascending limb of loop of Henle
Interstitial fluid
300
600
900
1,200
100
No vasopressin present
Filtrate has concentration of 100 mosm/l as it enters distal and collecting tubule
Distal and collecting tubule—impermeable to H₂O
Concentration of urine may be as low as 100 mosm/l as it leaves collecting duct
No H₂O reabsorbed in distal portion of nephron
Large volume of dilute urine; excess H₂O eliminated
(b)

still progressively loses H_2O from the body through this obligatory H_2O loss in the urine, as well as from evaporative losses from the body surface.

The kidneys' ability to tremendously concentrate urine to minimize H_2O loss when necessary is possible only because of the presence of the osmotic gradient in the medulla. Were it not for this gradient, the kidneys could produce a urine no more concentrated than the body fluids no matter how much vasopressin was secreted, because the only driving force for H_2O reabsorption is a concentration differential between the tubular fluid and interstitial fluid.

Conversely, when a person consumes large quantities of H_2O, the excess H_2O must be removed from the body without simultaneously losing solutes that are critical to the maintenance of homeostasis. Under these circumstances, no vasopressin is secreted, so the distal and collecting tubules remain impermeable to H_2O. The tubular fluid entering the distal tubule is hypotonic (100 mosm/l), having lost salt without a concomitant loss of H_2O in the ascending limb of Henle's loop. As this hypotonic fluid passes through the distal and collecting tubules (Fig. 14–27b), the medullary osmotic gradient is unable to exert any influence because of the late tubular segments' impermeability to H_2O. In other words, none of the H_2O remaining in the tubules can leave the lumen to be reabsorbed even though the tubular fluid is less concentrated than the surrounding interstitial fluid. Thus, in the absence of vasopressin, the 20% of the filtered fluid that reaches the distal tubule fails to be reabsorbed. Meanwhile, excretion of wastes and other urinary solutes remains constant. The net result is a large volume of dilute urine, which helps rid the body of excess H_2O. Urine osmolarity may be as low as 100 mosm/l, the same as when the fluid entered the distal tubule. Urine flow may be increased up to 25 ml/min in the absence of vasopressin, compared to the normal urine production of 1 ml/min.

Note that it would be impossible to produce urine less concentrated than the body fluids were it not for the fact that the tubular fluid is hypotonic as it enters the distal portion of the nephron. This dilution was accomplished in the ascending limb as NaCl was actively extruded but H_2O could not follow. Therefore, the loop of Henle, by simultaneously establishing the medullary osmotic gradient and diluting the tubular fluid before it enters the distal segments, plays a key role in allowing the kidneys to excrete urine that ranges in concentration from 100 to 1,200 mosm/l.

Changes in urine concentration are indicative of how much of the variably reabsorbed H_2O has been conserved for the body. Production of a large volume of dilute urine means little or none of the 20% of filtered H_2O subject to control has been returned to the plasma. In contrast, excretion of a small volume of concentrated urine implies extensive reabsorption of the controllable portion of the filtered H_2O. The extent of reabsorption varies directly with the amount of vasopressin secreted, which in turn is dependent on the body's state of hydration. Varying the amount of vasopressin secreted in proportion to the body's need for H_2O conservation enables the fine adjustments in H_2O reabsorption and excretion that are necessary to maintain proper fluid balance. Vasopressin influences H_2O permeability only in the distal and collecting tubules. It has no influence over the 80% of the filtered H_2O that is obligatorily reabsorbed without control in the proximal tubule and loop of Henle.

Vasopressin secretion is stimulated by both hypertonicity and hypotension.

Vasopressin is produced by several specific neuronal cell bodies in the hypothalamus, a portion of the brain. In negative-feedback fashion, vasopressin secretion is stimulated by a H_2O deficit in the body and is inhibited by a H_2O excess (Fig. 14–28). Two major inputs to the hypothalamus govern vasopressin secretion. First, **hypothalamic osmoreceptors** located near the vasopressin-secreting cells monitor the osmolarity of the fluid surrounding them, which in turn reflects the concentration of the entire internal fluid environment. As the osmolarity increases (too little H_2O) and the need for H_2O conservation increases accordingly, vasopressin secretion increases. Vasopressin, in turn, promotes H_2O reabsorption by the distal portions of the nephron, thus helping to alleviate the hypertonicity. Conversely, vasopressin secretion is inhibited and distal H_2O reabsorption suppressed when the body fluids are hypotonic as a result of excess H_2O gains. Second, **left atrial baroreceptors** monitor the arterial blood pressure, which is a reflection of the plasma volume. When the plasma volume and arterial blood pressure are too low, signifying a need for H_2O conservation, vasopressin secretion is stimulated. As a result, reabsorption of H_2O in the distal and collecting tubules is increased. Furthermore, vasopressin, in addition to having this effect on the kidney tubules, exerts a potent vasoconstrictor effect on arterioles (thus giving rise to its name). Both by helping expand the plasma volume and by increasing the total peripheral resistance, vasopressin helps relieve the hypotension that elicited its secretion. Conversely, vasopressin secretion is inhibited when the plasma volume and arterial blood pressure are elevated. The absence of vasopressin leads to the elimination of the excess plasma volume, helping to restore the blood pressure to normal.

Recall that low plasma volume and low arterial blood pressure also reflexly increase aldosterone secretion. The resultant increase in Na^+ reabsorption also secondarily increases H_2O reabsorption by osmosis. In fact, aldosterone-controlled Na^+ reabsorption is the most important factor in regulating ECF volume, with the vasopressin mechanism playing only a supportive role. Baroreceptor input to vasopressin secretion becomes important only in response to fairly serious changes in plasma volume and arterial pressure. Osmoreceptor input is normally the dominant factor controlling vasopressin secretion. Accordingly, vasopressin-controlled H_2O reabsorption is primarily important in regulating ECF osmolarity.

Water reabsorption versus excretion is only partially coupled with solute reabsorption versus excretion.

It is important to distinguish between H_2O reabsorption that mandatorily follows solute reabsorption and reabsorption of "pure" H_2O not linked to solute reabsorption.

1. *In the tubular segments that are permeable to H_2O, solute reabsorption is* ***always*** *accompanied by comparable H_2O*

Figure 14–28 Control of Increased Vasopressin Secretion during H_2O Deficit

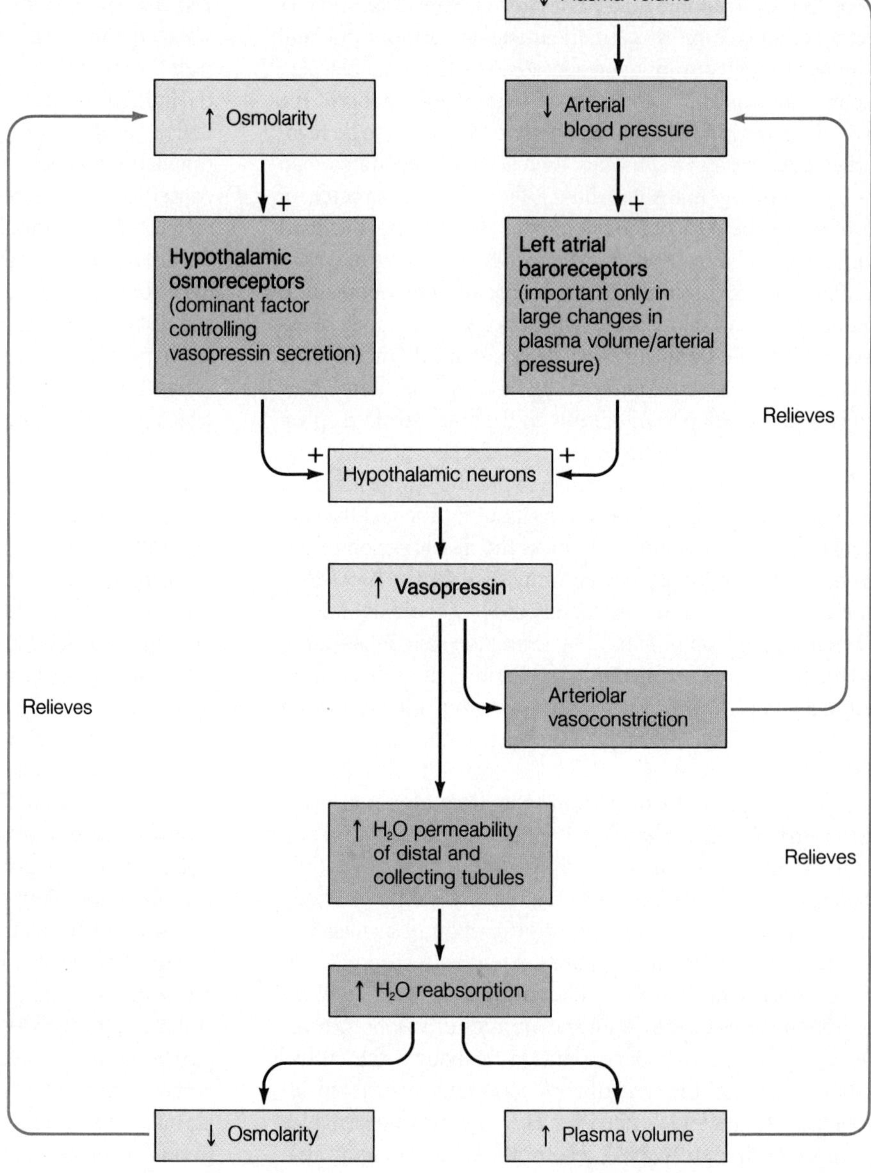

reabsorption because of osmotic considerations. Therefore, the total volume of H_2O reabsorbed is determined in large part by the total mass of solute reabsorbed; this is especially true of NaCl, because it is the most abundant solute in the ECF. The greater the salt retention within the body, the greater the H_2O retention and the larger the ECF volume. For this reason, the regulation of Na^+ reabsorption by aldosterone (followed passively by Cl^-) is the predominant factor in determining the total ECF volume.

2. By the same token, *solute excretion is **always** accompanied by comparable H_2O excretion because of osmotic considerations*. This is responsible for the obligatory excretion of at least a minimal volume of H_2O, even when a person is severely dehydrated.

For the same reason, when excess unreabsorbed solute is present in the tubular fluid, its presence exerts an osmotic effect to hold excessive H_2O in the lumen. This phenomenon is known as osmotic diuresis. **Diuresis** refers to increased urinary excretion, of which there are two types:

water diuresis and osmotic diuresis. **Water diuresis** is increased urinary output of H_2O with little or no increase in excretion of solutes. **Osmotic diuresis,** on the other hand, involves increased excretion of both H_2O and solute caused by excess unreabsorbed solute in the tubular fluid, such as occurs in diabetes mellitus. The large quantity of unreabsorbed glucose that remains in the tubular fluid in persons with diabetes osmotically drags H_2O with it into the urine. Some **diuretic drugs** (drugs that promote diuresis) act by blocking specific solute reabsorption so that extra H_2O spills into the urine along with the unreabsorbed solute.

3. A *pure loss or gain of H_2O that is not accompanied by comparable solute deficit or excess in the body leads to changes in ECF osmolarity*. Such an imbalance is corrected by partially dissociating H_2O reabsorption from solute reabsorption in the distal portions of the nephron through the combined effects of osmoreceptor-controlled vasopressin secretion and the medullary osmotic gradient. Through this mechanism, "free" H_2O can be reabsorbed without comparable solute reabsorption to ameliorate the body fluids' hypertonicity. Conversely, a large quantity of free H_2O can be excreted unaccompanied by comparable solute excretion, (that is, water diuresis) to rid the body of excess pure H_2O, thus correcting for hypotonicity of the body fluids. Water diuresis is normally a compensatory measure in response to ingestion of too much H_2O. However, excessive water diuresis accompanies alcohol ingestion. Because alcohol inhibits vasopressin secretion, the kidneys inappropriately lose too much H_2O. Typically, more fluid is lost in the urine than consumed in the alcoholic beverage, so the body becomes dehydrated in spite of substantial fluid ingestion.

Renal failure has wide-ranging consequences.

Urine excretion and the resultant clearance from the plasma of wastes and excess electrolytes is critical to the maintenance of homeostasis. When the functions of both kidneys are disrupted to the point that these organs are unable to perform their regulatory and excretory functions sufficiently to maintain homeostasis, **renal failure** is said to exist. Renal failure has a variety of causes, some of which begin elsewhere in the body and affect renal function secondarily. Among the causes of renal disease or renal failure are the following: (1) *infectious organisms*, either blood-borne or gaining entrance to the urinary tract through the urethra; (2) *toxic agents*, such as lead, arsenic, pesticides, or even long-term exposure to high doses of aspirin; (3) *inappropriate immune responses*, such as **glomerulonephritis,** which occasionally follows streptococcal throat infections as antigen-antibody complexes leading to localized inflammatory damage are deposited in the glomeruli (see p. 379); (4) *obstruction of urine flow* because of the presence of kidney stones, tumors, or enlargement of the prostate gland, with the back-pressure reducing glomerular filtration as well as damaging renal tissue; and (5) an *insufficient renal blood supply* that leads to inadequate filtration pressure. The latter can occur secondary to circulatory disorders such as heart failure, hemorrhage, shock, or narrowing and hardening of the renal arteries as a result of atherosclerosis.

Although the origin of these conditions may be different, almost all of them can cause some degree of nephron damage. The glomeruli and tubules may be independently affected or both may be dysfunctional. Regardless of the cause, renal failure can manifest itself either as *acute renal failure*, characterized by a sudden onset with a rapid reduction in urine formation until less than the essential minimum of around 500 ml of urine is being produced per day; or *chronic renal failure*, characterized by slow, progressive, insidious loss of renal function. A person may die from acute renal failure, or the condition may be reversible and lead to full recovery. Chronic renal failure, in contrast, is not reversible. Gradual, permanent destruction of renal tissue eventually proves to be fatal. Chronic renal failure is insidious because up to 75% of the kidney tissue can be destroyed before the loss of kidney function is even noticeable. Because of the abundant reserve of kidney function, only 25% of the kidney tissue can adequately maintain all the essential renal excretory and regulatory functions. With less than 25% of functional kidney tissue remaining, however, renal insufficiency becomes apparent. *End-stage renal failure* ensues when 90% of kidney function has been lost.

We will not sort out the different stages and symptoms associated with various renal disorders, but Table 14–4, which summarizes the potential consequences of renal failure, will give you an idea of the broad effects that kidney impairment can have. This should not be surprising, considering the central role the kidneys play in maintaining homeostasis. When the kidneys are unable to maintain a normal internal environment, widespread disruption of cellular activities can bring about abnormal function in other organ systems as well. By the time end-stage renal failure occurs, literally every body system has become impaired to some extent. (One of the symptoms of renal disease occurs during strenuous exercise, but it is transient and harmless. See the accompanying boxed feature, A Closer Look at Exercise Physiology.)

Since chronic renal failure is irreversible and eventually fatal, treatment is aimed at maintaining renal function by alternative methods, such as dialysis and kidney transplantation. The process of **dialysis** bypasses the kidneys to artificially maintain normal fluid and electrolyte balance and to remove

Table 14-4 Potential Ramifications of Renal Failure

***Uremic toxicity** caused by retention of waste products*

Nausea, vomiting, diarrhea and ulcers caused by toxic effect on digestive system

Bleeding tendency arising from toxic effect on platelet function

Mental changes—such as reduced alertness, insomnia, and shortened attention span, progressing to convulsions and coma—caused by toxic effect on central nervous system

Abnormal sensory and motor activity caused by toxic effect on peripheral nerves

Metabolic acidosis *caused by inability of kidneys to adequately secrete H^+ that is continually being added to body fluids as a result of metabolic activity*

Altered enzyme activity

Depression of central nervous system

Potassium retention *resulting from inadequate tubular secretion of K^+*

Altered cardiac excitability

Altered neural excitability

***Sodium imbalances** caused by the inability of the kidneys to adjust Na^+ excretion to balance changes in Na^+ consumption*

Elevated blood pressure, generalized edema, and congestive heart failure as a result of Na^+ surplus

Hypotension and, if severe enough, circulatory shock as a result of Na^+ deficiency

***Phosphate and calcium imbalances** arising from impaired reabsorption of these electrolytes*

Disturbances in skeletal structures caused by abnormalities in deposition of calcium phosphate crystals, which harden bone

***Loss of plasma proteins** as a result of increased "leakiness" of the glomerular membrane*

Edema caused by reduction in plasma colloid osmotic pressure

***Inability to vary urine concentration** as a result of impairment of countercurrent system*

Hypotonicity of body fluids if too much H_2O is ingested

Hypertonicity of body fluids if too little H_2O is ingested

***Hypertension** arising from combined effects of salt and fluid retention and vasoconstrictor action of excess angiotensin II*

***Anemia** caused by inadequate erythropoietin production*

***Depression of immune system** most likely caused by toxic levels of wastes and acids*

Increased susceptibility to infections

*Among the most life-threatening consequences of renal failure.

wastes. In one method of dialysis, a patient's blood is pumped through cellophane tubing that is surrounded by a large volume of fluid similar in composition to that of normal plasma. Following dialysis, the blood is returned to the patient's circulatory system. Like capillaries, cellophane is highly permeable to most plasma constituents but is impermeable to plasma proteins. As blood flows through the tubing, solutes move across the cellophane down their individual concentration gradients, except for plasma proteins, which remain in the blood. Urea and other wastes, which are absent in the dialysis fluid, diffuse out of the plasma into the surrounding fluid, thus cleansing the blood of these wastes. Plasma constituents that are not regulated by the kidneys and are at normal concentration, such as glucose, do not move across the cellophane into the dialysis

A CLOSER LOOK AT EXERCISE PHYSIOLOGY

When Protein in the Urine Does Not Mean Kidney Disease

Usually, urinary loss of proteins signifies kidney disease (nephritis). However, a urinary protein loss similar to that of nephritis often occurs following exercise, but the condition is harmless, transient, and reversible. The term *athletic pseudonephritis* is used to describe this postexercise (after exercise) proteinuria (protein in the urine). Studies indicate that 70% to 80% of athletes have proteinuria after very strenuous exercise. This occurs in participants in both noncontact and contact sports, so the condition does not arise from physical trauma to the kidneys. In one study, subjects who engaged in maximal short-term running excreted more protein than when they were bicycling, rowing, or swimming at the same work intensity. The reason for this difference is unknown.

Usually only a very small fraction of the plasma proteins that enter the glomerulus are filtered; those that are filtered are reabsorbed in the tubules, so no plasma proteins normally appear in the urine. Two basic mechanisms can cause proteinuria: (1) increased glomerular permeability with no change in tubular reabsorption or (2) impairment of tubular reabsorption. Research has shown that during mild to moderate exercise, the proteinuria that occurs results from changes in glomerular permeability, whereas during short-term exhaustive exercise, the proteinuria seems to be caused by both increased glomerular permeability and tubular dysfunction.

This reversible kidney dysfunction is believed to result from circulatory and hormonal changes that occur with exercise. Several studies have shown that renal blood flow is reduced during exercise as the renal vessels are constricted and blood is diverted to the exercising muscles. This reduction is positively correlated with the intensity of the exercise. With intense exercise, the renal blood flow may be reduced to 20% of normal. The GFR is also reduced, but not to the same extent as renal blood flow, presumably because of autoregulatory mechanisms. Some investigators propose that decreased glomerular blood flow enhances diffusion of proteins into the tubular lumen because, as the more slowly flowing blood spends more time in the glomerulus, a greater proportion of the plasma proteins have time to escape through the glomerular membrane. Hormonal changes that occur with exercise may also affect glomerular permeability. For example, renin injection is a well recognized way to experimentally induce proteinuria. Plasma renin activity is increased during strenuous exercise and may contribute to postexercise proteinuria. It is also hypothesized that maximal tubular reabsorption is reached during severe exercise, which could result in impaired protein reabsorption.

fluid because there is no driving force to produce their movement. (The dialysis fluid's glucose concentration is the same as normal plasma glucose concentration.) Electrolytes, such as K^+ and $PO_4^{\equiv}$, which are above their normal plasma concentrations because of the inability of the diseased kidneys to eliminate excess quantities of these substances, move out of the plasma until equilibrium is achieved between the plasma and dialysis fluid. Because the dialysis fluid's solute concentrations are maintained at normal plasma values, the solute concentration of the blood returned to the patient following dialysis is essentially normal.

In a second method of dialysis, the dialysis fluid is injected into the patient's abdominal cavity. After a sufficient time period for the plasma to have equilibrated with the dialysis fluid, the fluid, which now contains the wastes and excess electrolytes that have diffused into it from the plasma, is removed from the abdominal cavity. Dialysis is repeated as often as necessary to maintain the plasma composition within an acceptable range.

Transplantation of a healthy kidney from a donor is another option for treating chronic renal failure. A kidney is one of the few transplants that can be provided by a living donor. Because 25% of the total kidney tissue can maintain the body, both the donor and recipient have ample renal function with only one kidney each. The biggest problem with transplantation is the possibility of the organ's rejection by the patient's immune system (see p. 392). This can be minimized by matching the tissue types of the donor and recipient as closely as possible (the best donor choice is usually a close relative), coupled with the use of immunosuppressive drugs.

Urine is temporarily stored in the bladder, from which it is emptied by the process of micturition.

Once urine has been formed by the kidneys, it is transmitted through the ureters to the urinary bladder. Urine does not flow through the ureters by gravitational pull alone. Peristaltic contractions of the smooth muscle within the ureteral wall propel the urine forward from the kidneys to the bladder. The ureters penetrate the wall of the bladder obliquely, coursing through the wall several centimeters before they open into the bladder cavity. This anatomical arrangement prevents backflow of urine from the bladder to the kidneys when pressure builds up in the bladder. As the bladder fills, the ureteral ends within its wall are compressed closed. Urine can still enter, however, because ureteral contractions generate sufficient pressure to overcome the resistance and to push urine through the occluded ends.

The bladder wall is composed of smooth muscle lined by a special type of epithelium. It was once assumed that the bladder was an inert sac. However, both the epithelium and smooth muscle actively participate in the bladder's ability to accommodate large fluctuations in urine volume. Recently it has been learned that the epithelial lining is able to increase and decrease in surface area by the orderly process of vesicular traffic as the bladder alternately fills and empties. Membrane-bound cytoplasmic vesicles are inserted by means of exocytosis into the surface membrane of the epithelial cells to expand their surface area during bladder filling, then are withdrawn by endocytosis to shrink the surface area following emptying (see p. 83). As is characteristic of smooth muscle, bladder muscle is able to stretch tremendously without a build-up in bladder-wall tension. In addition, the highly folded bladder wall flattens out during filling to increase bladder storage capacity. Since urine is continuously being formed by the kidneys, it is important that the bladder has storage capacity to preclude the necessity for continual evacuation of the urine.

The bladder smooth muscle is richly supplied by parasympathetic fibers, stimulation of which causes bladder contraction. If the passageway through the urethra to the outside is open, bladder contraction brings about emptying of urine from the bladder. The exit from the bladder, however, is guarded by two sphincters, the internal urethral sphincter and the external urethral sphincter. A **sphincter** is a ring of muscle that, when contracted, closes off passage through an opening. The **internal urethral sphincter**—which is composed of smooth muscle and, accordingly, is under involuntary control—is not really a separate muscle but instead consists of the last portion of the bladder. Although it is not a true sphincter, it performs the same function as a sphincter. When the bladder is relaxed, the anatomical arrangement of the internal urethral sphincter region closes the outlet of the bladder.

Farther down the passageway, the urethra is encircled by a layer of skeletal muscle, the **external urethral sphincter.** This sphincter is reinforced by the entire **pelvic diaphragm,** a skeletal-muscle sheet that forms the floor of the pelvis and helps support the pelvic organs. The motor neurons that supply the external sphincter and pelvic diaphragm are continuously firing at a moderate rate unless they are inhibited, thus keeping these muscles tonically contracted so that they prevent urine from escaping through the urethra. Normally, when the bladder is relaxed and filling, closure of both the internal and external urethral sphincters prevents urine from dribbling out. Furthermore, because they are skeletal muscles, the external sphincter and pelvic diaphragm are under voluntary control. They can be deliberately tightened to prevent urination from occurring even when the bladder is contracting and the internal sphincter is open.

Micturition, or **urination,** the process of bladder emptying, is governed by two mechanisms: the micturition reflex and voluntary control. The **micturition reflex** is initiated when stretch receptors within the bladder wall are stimulated (Fig. 14–29). The bladder in an adult can accommodate up to 250 to 400 ml of urine before the tension within its walls begins to rise sufficiently to activate the stretch receptors. The greater the distention, the greater the extent of receptor activation. Afferent fibers from the stretch receptors carry impulses into the spinal cord and eventually, by means of interneurons, stimulate the parasympathetic supply to the bladder and inhibit the motor-neuron supply to the external sphincter. Parasympathetic stimulation of the bladder causes it to contract. No special mechanism is required to open the internal sphincter; changes in the shape of the bladder during contraction mechanically pull the internal sphincter open . Simultaneously, the external sphincter relaxes as its motor-neuron supply is inhibited. Now both sphincters are open and urine is expelled through the urethra by the force of bladder contraction. This micturition reflex, which is entirely a spinal reflex, governs bladder emptying in infants. As soon as the bladder fills sufficiently to trigger the reflex, the baby automatically wets.

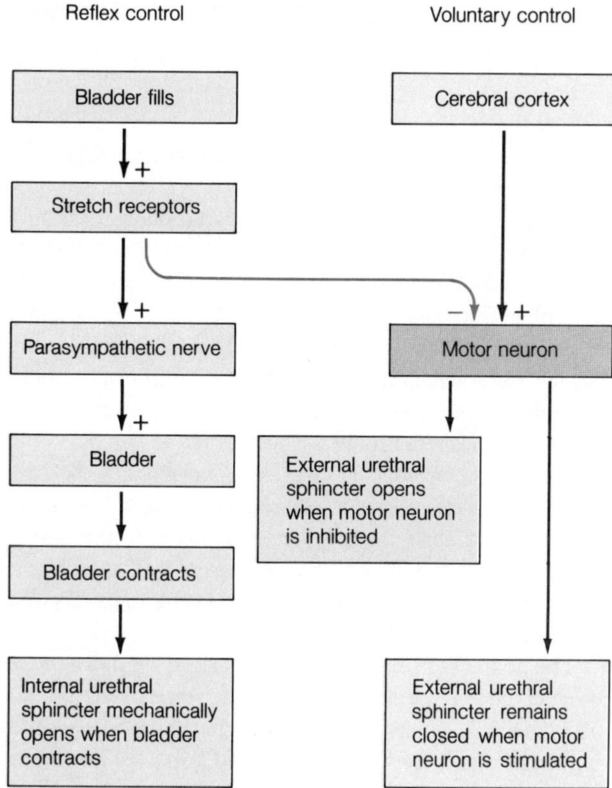

Figure 14-29 Reflex and Voluntary Control of Micturition

Bladder filling, in addition to triggering the micturition reflex, also gives rise to the conscious urge to urinate. The perception of bladder fullness appears before the external sphincter reflexly relaxes, thus providing a "warning" that micturition is imminent. As a result, voluntary control of micturition, learned during the process of toilet training in early childhood, can override the micturition reflex so that bladder emptying can take place at the person's convenience rather than at the time bladder filling first reaches the point of activating the stretch receptors. If the time is inopportune for urination when the micturition reflex is initiated, bladder emptying can be voluntarily prevented by deliberate tightening of the external sphincter and pelvic diaphragm. Voluntary excitatory impulses originating from the cerebral cortex override the reflex inhibitory input from the stretch receptors to the involved motor neurons (the relative balance of EPSPs and IPSPs), thus keeping these muscles contracted so that no urine is expelled.

Urinary incontinence, or the inability to prevent the discharge of urine, occurs as a result of disruption of the descending pathways in the spinal cord that mediate voluntary control of the external sphincter and pelvic diaphragm. In this case, because the components of the micturition reflex arc are still intact in the lower spinal cord, bladder emptying becomes governed by an uncontrollable spinal reflex, as it is in infants. A lesser degree of incontinence characterized by urine escaping when the bladder pressure suddenly increases transiently, such as during coughing or sneezing, can result from impairment of sphincter function. This is not uncommon in women who have borne children or in men whose sphincters have been injured during prostate surgery.

Even under normal circumstances, urination cannot be delayed indefinitely. As the bladder continues to fill, reflex input from the stretch receptors increases with time. Finally, reflex inhibitory input to the external-sphincter motor neuron becomes so powerful that it can no longer be overridden by voluntary excitatory input, so the sphincter relaxes and the bladder uncontrollably empties.

Micturition can also be deliberately initiated, even though the bladder is not distended, by voluntary relaxation of the external sphincter and pelvic diaphragm. Lowering of the pelvic floor allows the bladder to drop downward, which simultaneously pulls open the internal urethral sphincter and stretches the bladder wall. The subsequent activation of the stretch receptors brings about bladder contraction by means of the micturition reflex. Voluntary bladder emptying may be further assisted by contraction of the abdominal wall and respiratory diaphragm. The resultant increase in intra-abdominal pressure "squeezes down" on the bladder to facilitate its emptying.

Chapter in Perspective

The cells of the body depend not only on receiving a continual supply of nutrients and O_2 but also on continual removal of metabolic wastes and the maintenance of stable conditions in the internal fluid environment that bathes them. The kidneys play a major role in maintaining a stable internal environment by eliminating all of the waste products of bodily metabolism, with the exception of respiration-removed CO_2, and by regulating the concentration of many of the plasma constituents, especially the electrolytes.

The kidneys consist of several hundred million nephrons, the microscopic functional units that collectively clear or rid the plasma of unwanted materials by excreting them in the urine. For the plasma constituents subject to renal regulation, the kidneys are able to constantly adjust the output of these substances in the urine to reflect the momentary needs of the body. Each nephron consists of a vascular component and a tubular component. The vascular component consists basically of two capillary networks in series: the glomerular capillaries, through which a protein-free plasma is nondiscriminately filtered into the tubular system; and the peritubular capillaries, which intertwine around the tubular system. Differential transport and permeability properties of various portions of the tubular component permit highly selective transfer of particular substances between the tubular fluid and peritubular capillary plasma.

Three basic renal processes are involved in the production of urine: glomerular filtration, tubular reabsorption, and tubular secretion. Between 20% to 25% of the total cardiac output (all of the blood pumped out by the heart each minute) is delivered to the kidneys for the purpose of being purified and adjusted. Of this blood flowing through the kidneys, a substantial fraction (20%) of the plasma is filtered intact through the porous glomerular membranes, with the exception of plasma proteins, which are generally too large to penetrate the membrane pores. This glomerular filtration, which averages 125 ml/min of protein-free plasma entering the tubular system, is accomplished passively in response to a net outward imbalance between the hydrostatic and osmotic forces acting across the glomerular membrane.

By means of the process of tubular reabsorption, the tubular cells selectively return to the plasma the volume and composition of filtered fluid required to maintain the desired volume and composition of the ECF, which is the body's internal fluid environment. Various tubular transport mechanisms—some passive, some active, and some subject to hormonal and other control—are responsible for the differential reabsorption of an assortment of filtered solutes as well as H_2O.

For a few selected substances, most notably H^+, K^+, and organic ions such as foreign compounds, some of the tubular cells have the ability to add these substances to the tubular fluid from the peritubular capillary plasma by the process of tubular secretion. This provides a means to more rapidly clear these materials from the plasma.

The nonreabsorbed filtered solutes, which are mostly waste products and excess electrolytes, and the secreted constituents become highly concentrated in the urine. This is because they remain behind in a small volume of H_2O, most of the filtered H_2O having been reabsorbed. On the average, of the 125 ml/min filtered, 124 ml/min of the H_2O and valuable solutes are reabsorbed, leaving only 1 ml/min of highly concentrated, waste-filled urine to be excreted. Once formed, the urine is transported to the bladder, where it is temporarily stored until it is eliminated to the exterior by the micturition reflex, a spinal reflex that is subject to voluntary control.

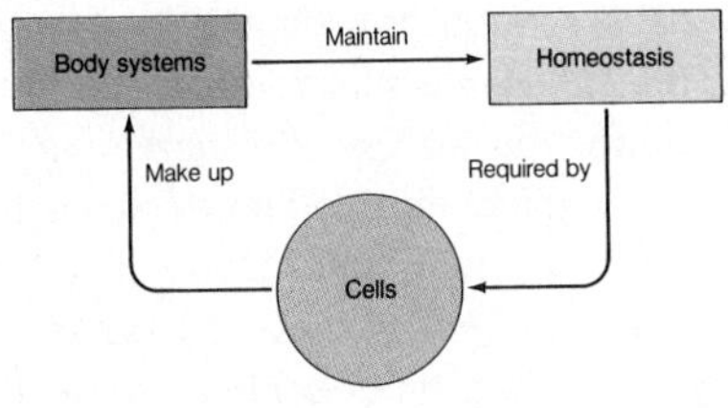

See inside front cover for an expanded version of this model.

Review Exercises

1. List the functions of the kidneys.
2. Describe the anatomy of the urinary system. Compare the structure of cortical and juxtamedullary nephrons.
3. Describe the three basic renal processes; indicate how they relate to urine excretion.
4. Discuss the forces involved in glomerular filtration. What is the average GFR? How is GFR regulated?
5. Why do the kidneys receive a seemingly disproportionate share of the cardiac output? What percentage of renal blood flow is normally filtered?
6. Distinguish between active and passive reabsorption.
7. Describe all of the tubular transport processes that are linked to the basolateral Na^+-K^+ ATPase carrier.
8. Describe the renin-angiotensin-aldosterone system.
9. To what do the terms ***tubular maximum*** (T_m) and ***renal threshold*** refer? Compare two substances that display a T_m, one that is and the other that is not regulated by the kidneys.
10. What is the importance of tubular secretion?
11. What is the average rate of urine formation?
12. Define plasma clearance.
13. What is responsible for the presence of a vertical osmotic gradient in the medullary interstitial fluid? Of what importance is this gradient?
14. Describe the function and control of vasopressin.
15. Describe the transfer of urine to, the storage of urine in, and the emptying of urine from the bladder.
16. **A point to ponder:** The juxtamedullary nephrons of animals adapted to survive with minimum water consumption, such as desert rats, have relatively much longer loops of Henle than humans have. Of what benefit would these longer loops be?

CHAPTER 15

Fluid Balance and Acid-Base Balance

Chapter Contents at a Glance

INTRODUCTION *The amount of fat stored in your body is the primary determinant of how "dry" you are. Water is by far the most abundant component of the human body, constituting an average of 60% of body weight but ranging from 40% to 80%. The H_2O content of an individual remains fairly constant over a period of time, largely because of the kidneys' efficiency in regulating H_2O balance. The reason for the wide range in body H_2O is the variability between individuals, depending on the amount of adipose tissue (fat) they have (see the accompanying boxed feature, A Closer Look at Exercise Physiology, p. 512). Adipose tissue has a low H_2O content compared to other tissues. Plasma, as you might suspect, is more than 90% H_2O. Even the soft tissues such as skin, muscles, and internal organs consist of 70% to 80% H_2O. The relatively drier skeleton is only 22% H_2O. Fat, however, is the driest tissue of all, having only 10% H_2O content. Accordingly, a high body H_2O content is associated with leanness and a low body H_2O content with obesity, since a larger proportion of the body is composed of relatively dry fat in overweight individuals. The percentage of body H_2O is also influenced by the sex and age of the individual. Women have a lower body H_2O content than men, primarily because the female sex hormone, estrogen, promotes fat deposition in the breasts, buttocks, and elsewhere. This not only gives rise to the typical female figure, but also endows women with a higher proportion of adipose tissue and, therefore, a lower body H_2O content. The percentage of body H_2O also decreases progressively with age.*

Input must equal output if balance is to be maintained.

The cells of complex multicellular organisms are able to survive and function only within a very narrow range of composition of the extracellular fluid (ECF), the internal fluid environment that bathes them. The quantity of any particular substance in the ECF is considered to be a readily available internal **pool.** More of the substance may be added to the pool either by being transferred in from the external environment (most commonly by ingestion) or by being metabolically produced within the body (Fig. 15–1). Substances may be removed from the body by excretion to the outside or by consumption in a metabolic reaction. If the quantity of a substance is to remain stable within the body, its input by means of ingestion or metabolic production must be balanced by an equal output by means of excretion or consumption. This relationship, known as the **balance concept,** is extremely important in the maintenance of homeostasis.

The ECF pool can further be altered by transferring a particular ECF constituent into storage within the cells or bones. If the body as a whole has a surplus or deficit of a particular stored substance, the storage site can be expanded or partially depleted to maintain the ECF concentration of the substance within homeostatically prescribed limits. For example, following absorption of a meal, when more glucose is entering the plasma than is being consumed by the cells, the surfeit of glucose can be temporarily stored in muscle and liver cells in the form of glycogen. This storage depot can then be tapped between meals as necessary to maintain the plasma glucose level when no new nutrients are being added to the blood by eating. It is important to recognize, however, that internal storage capacity is limited. Although an internal exchange between the ECF and storage depot can temporarily restore the plasma concentration of a particular substance to normal, in the long run any excess or deficit of that constituent must be compensated for by appropriate adjustments in total body input or output.

Another possible internal exchange between the pool and the remainder of the body is the reversible incorporation of certain plasma constituents into more complex molecular structures. For example, iron is incorporated into hemoglobin within the red blood cells during their synthesis but is released intact back into the body fluids when the red cells degenerate (see p. 346). This process differs from metabolic consumption of a substance, in which the substance is irretrievably converted into another form—for example, glucose converted into CO_2 plus H_2O plus energy. It also differs from storage in

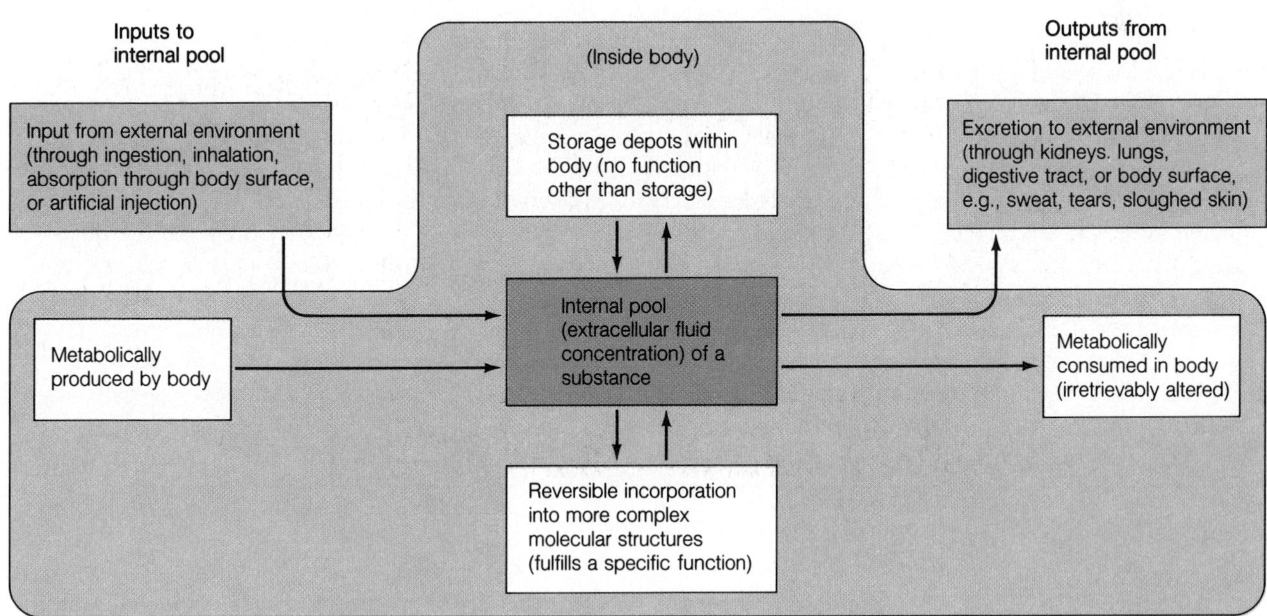

Figure 15–1 Inputs to and Outputs from the Internal Pool of a Body Constituent

What the Scales Don't Tell You

Body composition refers to the percentage of body weight that is composed of lean tissue and of adipose tissue. The assessment of body composition is an important component in evaluating the health status of individuals. The age-height-weight tables used by insurance companies can be misleading for determining healthy body weight. Many athletes, for example, would be considered overweight by these charts. A football player may be 6′5″ tall, weigh 300 lbs, but have only 12% body fat. This player's extra weight is muscle, not fat, and therefore is not a detriment to his health. A sedentary person, on the other hand, may be normal on the height-weight charts but have 30% fat. This person should maintain body weight while increasing muscle mass and decreasing fat. Ideally, men should have 15% fat or less and women should have 20% fat or less.

The most accurate method for assessing body composition is underwater weighing. This technique is based on the fact that lean tissue is denser than water and fat tissue is less dense than water. You can readily demonstrate this for yourself by dropping a piece of lean meat and a piece of fat in a glass of water; the lean meat will sink and the fat will float. The most common method of underwater weighing requires the person to expel all the air from his or her lungs and then completely submerge in a tank of water while sitting in a swing that is attached to a scale. Because a person learns to expel more air from the lungs with a little practice, this procedure is repeated ten to twelve times. When the results from trial to trial become consistent, the last few trials are averaged and used to determine body density by means of equations that take into consideration the density of water, the difference of the person's weight in air and underwater, and the residual volume of air remaining in the lungs. Because of the difference in density between lean and fat tissue, people who have more fat have a lower density and weigh relatively less underwater than in air compared to their lean counterparts. Percentage fat is determined by a simplistic equation developed by Dr. William Siri:

$$\text{Percentage fat} = 495/\text{Body density} - 450.$$

Another common way to assess body composition if laboratory facilities for underwater weighing are not available is skinfold thickness. Because approximately half of the body's total fat content is located just beneath the skin, total body fat can be estimated from measurement of skinfold thickness taken at various sites on the body. Skinfold thickness is determined by pinching up a fold of skin at one of the designated sites and measuring its thickness by means of a caliper, a hinged instrument that fits over the fold and is calibrated to measure thickness. Mathematical equations specific for the person's age and sex can be used to predict the percentage of fat from the skinfold-thickness scores. If the investigator is skilled and a valid equation is used to predict percentage of fat, skinfold determinations are a valuable tool for assessing the body composition of large numbers of people. A major criticism of skinfold assessments is that accuracy depends on the investigator's skill.

Exercise physiologists often assess body composition as an aid in prescribing and evaluating exercise programs. Exercise generally reduces the percentage of body fat and, by increasing muscle mass, increases the percentage of lean tissue.

that the latter serves no purpose other than storage, whereas reversible incorporation into a more complex structure serves a specific purpose.

When total body input of a particular substance equals its total body output, a **stable balance** exists. When the gains via input for a substance exceed its losses via output, a **positive balance** exists. The result is an increase in the total amount of the substance in the body. In contrast, when the losses for a substance exceed its gains, a **negative balance** exists and the total amount of the substance in the body decreases.

Not all input and output pathways are applicable for every body fluid constituent.

1. Organic nutrients' only avenue of input is ingestion, because they cannot be synthesized anew (although interconversions between some nutrient molecules are possible). Organic nutrients' output is usually by consumption, generally for the purpose of producing energy for the cells to expend. Nutrient molecules are not normally excreted in the urine or feces. If balance and, accordingly, body weight are to be maintained, input in the form of eating must match output in the form of energy expenditure. (The body's handling of organic nutrients and the related concept of energy balance are covered in more detail in later chapters.)
2. In the case of an electrolyte such as salt, stability of plasma NaCl concentration depends entirely on a balance between salt ingestion versus salt excretion. Salt is not synthesized or consumed by the body, and it lacks storage sites outside the ECF.
3. For H_2O and H^+, all input and output pathways to and from the ECF pool are possible. The amount of either substance in the ECF can be altered both by exchange between the ECF and external environment and by exchange between the ECF and cells.

Changing the magnitude of any of the input or output pathways for a given substance can alter its plasma concentration. In order to maintain homeostasis, any change in input must be balanced by a corresponding change in output (for example, increased salt intake must be matched by a corresponding increase in salt output in the urine), and, conversely, increased losses must be compensated for by increased intake. Thus maintenance of a stable balance necessitates control. However, not all input and output pathways are regulated to maintain balance. Generally, input of various plasma constituents is poorly controlled or not controlled at all. We frequently ingest H_2O and salt, for example, not because we *need* them but because we *want* them, so the intake of H_2O and salt is highly variable. Likewise, H^+ is uncontrollably generated internally and added to the body fluids. Water, salt, and H^+ can also be lost to the external environment to varying degrees through the digestive tract (vomiting), skin (sweating), and elsewhere (H_2O vapor loss from the respiratory passages) without regard for H_2O, salt, or H^+ balance in the body. Compensatory adjustments in the urinary excretion of these substances are responsible for maintaining the body fluids' volume and salt and acid composition within the extremely narrow homeostatic range compatible with life despite wide variations in input and unregulated losses of these plasma constituents. This chapter will be devoted to the regulation of fluid balance (maintenance of H_2O and salt balance) and acid-base balance (maintenance of H^+ balance).

Fluid Balance

Body water is distributed between the intracellular- and extracellular-fluid compartments.

Water is the most abundant molecular component of the body, accounting on the average for 60% of total body weight. In an adult this averages about 42 liters of H_2O. Water is an ideal solvent for most of the solutes in the body, and, in general, it permits their rapid diffusion.

Body H_2O is distributed between two major fluid compartments: the fluid within the cells, ***intracellular fluid*** (*ICF*), and the fluid surrounding the cells, ***extracellular fluid*** (*ECF*) (Table 15–1). (The terms "H_2O" and "fluid" are commonly used interchangeably. Although this usage is not entirely accurate because it ignores the solutes in the body fluids, it is acceptable when one considers the total volume of the fluids, since the major proportion of these fluids consists of H_2O.) The intracellular-fluid compartment comprises about two-thirds of the total body H_2O. Even though each cell contains its own unique mixture of constituents, there are sufficient similarities between these trillions of minute fluid compartments to consider them collectively as one large fluid compartment. The remaining one-third of the body H_2O found in the extracellular-fluid compartment is further subdivided into plasma and interstitial fluid. The ***plasma***, which makes up about one-fifth of the ECF volume, is the fluid portion of the blood. The ***interstitial fluid***, which represents the other four-fifths of the extracellular-fluid compartment, is the fluid that lies in the spaces between the cells. Interstitial fluid, sometimes also known as ***tissue fluid***, constitutes the true internal environment in that it is the fluid that bathes the tissue cells. Actually, only a minute portion of interstitial fluid is freely flowing. More than 99% of it is held in gel form in the extracellular matrix (see p. 67).

Two other minor categories are included in the ECF compartment: lymph and transcellular fluid. ***Lymph*** is fluid being returned from the interstitial fluid to the plasma by means of the lymphatic system, meanwhile being filtered through lymph nodes for immune-defense purposes (see p. 323). **Transcellular fluid** consists of a number of small specialized fluid volumes, all of which are secreted by specific cells into a particular body cavity to perform some specialized function. Transcellular fluid includes: ***cerebrospinal fluid*** (surrounding, cushioning, and nourishing the brain and spinal cord); ***intraocular fluid*** (maintaining the shape of and nourishing the eye); ***synovial fluid*** (lubricating and serving as a shock absorber for the joints); ***pericardial***, ***pleural***, and ***peritoneal fluids*** (lubricating movements of the heart, lungs, and intestines,

Table 15–1 Classification of Body Fluid

Compartment	Volume of Fluid (in liters)	Percentage of Body Fluid	Percentage of Body Weight
Total body fluid	42	100%	60%
Intracellular fluid (ICF)	28	67	40
Extracellular fluid (ECF)	14	33	20
Plasma	2.8	6.6 (20% of ECF)	4
Interstitial fluid	11.2	26.4 (80% of ECF)	16
Lymph	negligible	negligible	negligible
Transcellular fluid (cerebrospinal fluid, intraocular fluid, synovial fluid, pericardial fluid, pleural fluid, peritoneal fluid, digestive juices)	negligible	negligible	negligible

respectively); and the *digestive juices* (digesting ingested foods). Although these fluids are extremely important functionally, they represent an insignificant fraction of the total body H_2O. Furthermore, the transcellular compartment as a whole usually does not reflect changes in the body's fluid balance. For example, the cerebrospinal fluid does not decrease in volume when the body as a whole is experiencing a negative H_2O balance. This is not to say that these fluid volumes never change. Localized changes in a particular transcellular-fluid compartment can occur pathologically (such as too much intraocular fluid accumulating in the eyes of persons with glaucoma), but such a localized fluid disturbance does not affect the fluid balance of the body. Therefore, the transcellular compartment can usually be ignored when one is dealing with problems of fluid balance. The main exception to this generalization is when digestive juices are abnormally lost from the body during heavy vomiting or diarrhea, which can bring about a fluid imbalance.

The plasma and interstitial fluid are separated by the blood vessel walls, whereas the ECF and ICF are separated by cellular plasma membranes.

Several barriers separate the body-fluid compartments, limiting the movement of H_2O and solutes between the various compartments to differing degrees. The two components of the ECF—plasma and interstitial fluid—are separated by the walls of the blood vessels. However, H_2O and all plasma constituents with the exception of plasma proteins are continuously and freely exchanged between the plasma and interstitial fluid by passive means across the thin, pore-lined capillary walls. Accordingly, plasma and interstitial fluid are nearly identical in composition, except that interstitial fluid lacks plasma proteins. Any change in one of these ECF compartments is quickly reflected in the other compartment because they are constantly mixing.

In contrast to the very similar composition of the vascular and interstitial-fluid compartments, the composition of the ECF differs considerably from that of the ICF (Fig. 15–2). Each cell is surrounded by a highly selective plasma membrane that permits passage of certain materials while excluding others. Movement through the membrane barrier occurs by both passive and active means and may be highly discriminating. Among the major differences between the ECF and ICF are: (1) the presence of cellular proteins that are unable to permeate the enveloping membranes to leave the cells, and (2) the unequal distribution of Na^+ and K^+ and their attendant anions as a result of the action of the membrane-bound Na^+-K^+ ATPase pump that is present in all cells. This pump actively transports Na^+ out of and K^+ into cells; for this reason, Na^+ is the primary ECF cation, and K^+ is primarily found in the ICF. This unequal distribution of Na^+ and K^+, coupled with differences in membrane permeability to these ions, is responsible for the electrical properties of cells. This is particularly important in the initiation and propagation of action potentials in excitable tissues (see chapters 3 and 4).

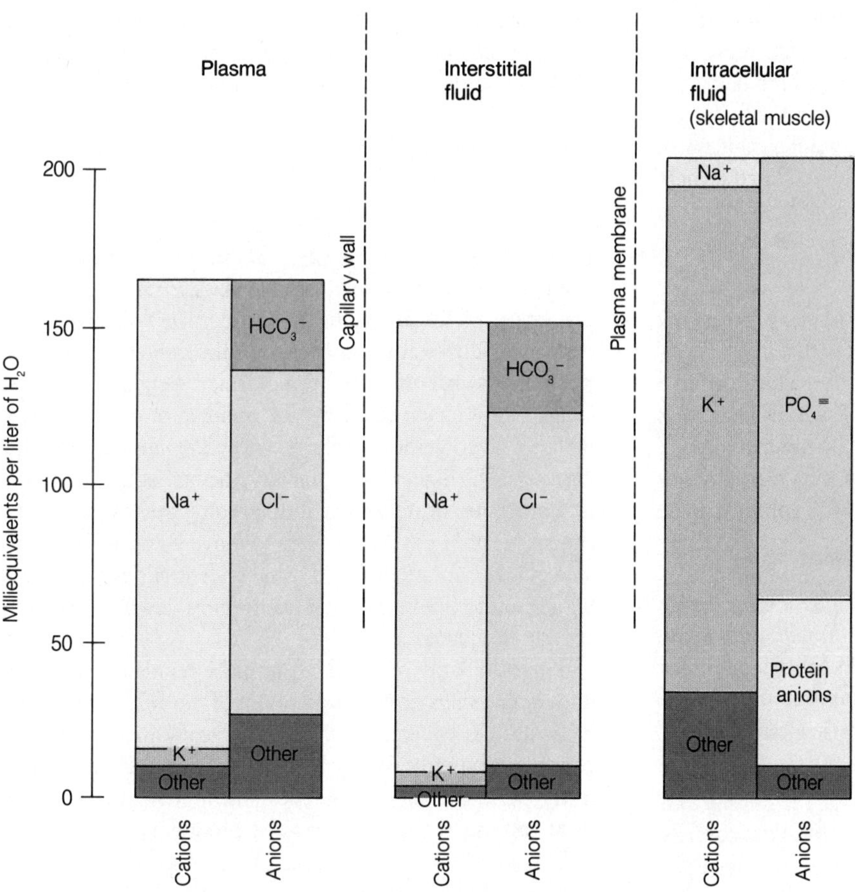

Figure 15-2 Ionic Composition of Major Body-Fluid Compartments

Except for the extremely small, electrically imbalanced portion of the total intracellular and extracellular ions that are involved in membrane potential, the majority of the ECF and ICF ions are electrically balanced. In the ECF, Na^+ is accompanied primarily by the anion Cl^- (chloride) and to a lesser extent by HCO_3^- (bicarbonate). The major intracellular anions are $PO_4^{\equiv}$ (phosphate) and the negatively charged proteins trapped within the cell.

Although all cells' plasma membranes display selective permeability, all cells are freely permeable to H_2O. The movement of H_2O between the plasma and interstitial fluid across capillary walls is governed by relative imbalances between capillary blood pressure (a fluid or hydrostatic pressure) and colloid osmotic pressure (see p. 319). In contrast, the net transfer of H_2O between the interstitial fluid and intracellular fluid across the cellular plasma membranes occurs as a result of osmotic effects alone. The hydrostatic pressures of the interstitial fluid and intracellular fluid are both extremely low and fairly constant.

Fluid balance is maintained by regulating ECF volume and ECF osmolarity.

Extracellular fluid serves as an intermediary between the cells and the external environment. All exchanges of H_2O and other constituents between the ICF and the external world must occur through the ECF. Water added to the body fluids always enters the ECF compartment first, and fluid always leaves the body by way of the ECF.

Plasma is the only fluid that can be directly acted on to control its volume and composition. However, because of the free exchange across the capillary walls, if the volume and composition of the plasma are regulated, the volume and composition of the interstitial fluid bathing the cells are likewise regulated. Thus, any control mechanism that operates on the plasma in effect regulates the entire ECF. The ICF, in turn, is influenced by changes in the ECF to the extent permitted by the permeability of the membrane barriers surrounding the cells.

The factors regulated to maintain fluid balance in the body are ECF volume and ECF osmolarity. Although regulation of these two parameters is closely interrelated, both being dependent on the relative NaCl and H_2O load in the body, it is important that they are closely controlled for different reasons. Extracellular-fluid volume must be closely regulated to help maintain blood pressure, whereas ECF osmolarity must be regulated to prevent the cells from swelling or shrinking.

CONTROL OF ECF VOLUME IS IMPORTANT IN THE LONG-TERM REGULATION OF BLOOD PRESSURE. A reduction in ECF volume, by decreasing the plasma volume, causes a fall in arterial blood pressure. Conversely, a rise in ECF volume increases the arterial blood pressure by expanding the plasma volume. Two compensatory measures that come into play to transiently adjust the blood pressure until the ECF volume can be restored to normal are as follows:

1. *baroreceptor-reflex mechanisms that alter both cardiac output and total peripheral resistance through autonomic nervous-system effects on the heart and blood vessels* (see p. 332). These immediate cardiovascular responses are designed to minimize the effect a deviation in circulating volume has on blood pressure.
2. *temporary, automatic fluid shifts between the plasma and interstitial fluid.* A reduction in plasma volume is partially compensated for by a shift in fluid out of the interstitial compartment into the vasculature to expand the circulating plasma volume at the expense of the interstitial compartment. Conversely, when the plasma volume is too large, much of the excess fluid is shifted into the interstitial compartment. These shifts occur immediately and automatically as a result of changes in the balance of hydrostatic and osmotic forces acting across the capillary walls that arise when plasma volume deviates from normal (see p. 321). These fluid shifts provide a temporary mechanism to help keep the plasma volume fairly constant.

These short-term compensatory measures are limited in their ability to minimize a change in blood pressure. If the plasma volume is too inadequate, the blood pressure remains too low no matter how vigorous the pump action of the heart, how constricted the resistance vessels, or what proportion of interstitial fluid shifts into the vasculature. Conversely, if the plasma volume is greatly overexpanded, blood pressure can not be restored down to normal even with maximum dilation of the resistance vessels and other short-term measures.

It is important, therefore, that other compensatory measures come into play in the long run to restore the ECF volume to normal. This responsibility for long-term regulation of blood pressure rests with the kidneys and the thirst mechanism, which control urinary output and fluid intake, respectively. In so doing, they accomplish needed fluid exchanges between the ECF and the external environment to regulate the body's total fluid volume. Accordingly, they have an important long-term influence on arterial blood pressure.

CONTROL OF ECF OSMOLARITY PREVENTS CHANGES IN ICF VOLUME. The **osmolarity** of a fluid is a measure of the concentration of the individual solute particles dissolved in it. Each individual solute particle exerts the same osmotic effect, no matter what its size or molecular weight. Therefore, osmolarity is determined by ***the number*** of solute particles within a given volume of fluid, not by ***the mass*** of the solute particles. The higher the osmolarity, the higher the concentration of solutes or, to look at it differently, the lower the concentration of H_2O. Water tends to move by osmosis from an area of lower solute (higher H_2O) concentration to an area of higher solute (lower H_2O) concentration down its own concentration gradient.

The osmolarity of the ECF must be regulated to prevent undesirable shifts of H_2O into or out of the cells. As far as the ECF itself is concerned, it does not really matter what the concentration of its solutes is. However, since H_2O is able to freely permeate the membranes surrounding the cells and since osmosis is the primary force producing movement of H_2O across these membranes, it is crucial that ECF osmolarity be maintained within very narrow limits to prevent the cells from swelling (by osmotically gaining H_2O from the ECF) or shrinking (by osmotically losing H_2O to the ECF).

Osmosis occurs only when there is a difference in concentration of nonpenetrating solutes (and, therefore, H_2O) between two areas separated by a barrier. Penetrating solutes quickly become equally distributed between the two areas and thus do not contribute to osmotic differences. The osmotic activity across the capillary wall is due to the unequal distribution of plasma proteins. Plasma proteins are present only in the plasma (as their name implies), because they are too large to penetrate the capillary walls to enter the interstitial fluid. All other solutes are in essentially the same concentration in the plasma and interstitial fluid, so they do not contribute to any unequal distribution of H_2O that would induce osmosis across the capillary wall.

In contrast, osmotic activity across the cellular plasma membranes is directly related to any differences in ionic concentration between the ECF and ICF. Plasma proteins play no role in the osmosis of H_2O across the cell membranes because plasma proteins are not present in either the interstitial fluid or intracellular fluid that are separated by the cell membranes.

Sodium and its attendant anions, being by far the most abundant solutes in the ECF in terms of numbers of particles, account for the vast majority of the ECF's osmotic activity. In contrast, K^+ and its accompanying intracellular anions are responsible for the ICF's osmotic activity. Even though Na^+ and K^+ are able to penetrate the plasma membrane, they behave as if they are nonpenetrating because of Na^+-K^+ pump activity. Any Na^+ that passively diffuses into the cell down its electrochemical gradient is promptly pumped back outside, with the result being the same as if Na^+ were barred from the cells. The same in reverse holds true for K^+; it in effect remains trapped within the cells. The resulting unequal distribution of Na^+ and K^+ and their accompanying anions between the ECF and the ICF is responsible for the osmotic activity of these two fluid compartments.

The osmolarities of the ECF and ICF usually are the same, because the total concentration of K^+ and other effectively non-penetrating solutes inside the cells is equal to the total concentration of Na^+ and other effectively nonpenetrating solutes in the fluid surrounding the cells. Even though the nonpenetrating solutes differ in the ECF and ICF, their concentrations are normally identical, and it is the number, not the nature, of the unequally distributed particles per volume that determines the fluid's osmolarity. Because the osmolarity of the ECF and ICF are normally equal, no net flux of H_2O usually occurs into or out of the cells. However, a variety of circumstances can induce a change in the osmolarity of the ECF. When the ECF osmolarity changes with respect to the ICF osmolarity, osmosis does take place, with H_2O either leaving or entering the cells, depending on whether the ECF is more or less concentrated than the ICF. We will examine the fluid shifts that occur between the ECF and ICF when the ECF osmolarity becomes abnormally high (that is, is hypertonic) or abnormally low (that is, is hypotonic) relative to ICF. We will also compare these situations with what happens when the ECF compartment gains or loses fluid that has an osmolarity equal to the normal osmolarity of the body fluids (that is, is isotonic).

ECF HYPERTONICITY. Hypertonicity of the ECF, or the excessive concentration of ECF solutes, is usually associated with **dehydration,** or a negative H_2O balance. Dehydration with accompanying hypertonicity can be brought about in three major ways: (1) insufficient H_2O intake, such as might occur during desert travel or might accompany difficulty in swallowing; (2) excessive H_2O loss, such as might occur during heavy sweating, vomiting, or diarrhea (even though both H_2O and solutes are lost during these conditions, relatively more H_2O is lost, so the remaining solutes become more concentrated); and (3) **diabetes insipidus,** a disease characterized by a deficiency of vasopressin (antidiuretic hormone). In this condition, the kidneys cannot concentrate urine because they are unable to reabsorb H_2O from the distal portions of the nephron tubules to take advantage of the renal medullary interstitial fluid's concentrated environment (see p. 499). Such patients typically produce up to 20 liters of very dilute urine per day, compared to the normal average of 1.5 liters per day. Unless H_2O intake keeps pace with this tremendous loss of H_2O in the urine, the person quickly dehydrates. Patients complain about the extraordinary amount of time day and night that they spend going to the bathroom and getting drinks. Fortunately, they can be treated with replacement vasopressin administered by nasal spray.

On rare occasions, the ECF becomes hypertonic in the absence of dehydration because of the abnormal accumulation of osmotically active solutes that do not normally contribute significantly to the ECF osmotic activity. This occurs, for example, with the high blood urea levels in uremia (kidney failure).

Whenever the ECF compartment becomes hypertonic, H_2O moves out of the cells by osmosis into the more concentrated interstitial fluid until the ICF osmolarity equilibrates with the ECF. As H_2O leaves the cells, the ICF osmolarity increases because the solutes remaining behind in the cell become more concentrated. Movement of H_2O by osmosis through the plasma membranes occurs so rapidly that osmotic equilibrium is usually established within a minute or less after the ECF osmolarity first deviates from normal. A relative H_2O deficit in the ECF is almost immediately distributed evenly between the ECF and ICF compartments. At equilibrium, both compartments are reduced in volume. The cells shrink as H_2O leaves them. Of particular concern is the fact that considerable shrinking of brain cells causes disturbances in brain function, which can be manifested as mental confusion and irrationality in moderate cases and which can bring about possible delirium, convulsions, or coma in more severe hypertonic conditions. Rivaling the neural symptoms in seriousness are the circulatory disturbances that arise from a reduction in plasma volume when hypertonicity is accompanied by dehydration. Circulatory problems may range from a slight reduction in blood pressure to circulatory shock and death.

Other more common symptoms become apparent, even in mild cases of dehydration. Dry skin and sunken eyeballs are indications of loss of H_2O from the underlying soft tissues. The tongue is dry and parched because of suppressed salivary secretion. Urinary output is decreased to conserve H_2O, while thirst is reflexly increased to encourage fluid intake.

ECF HYPOTONICITY. Hypotonicity of the ECF is usually associated with **overhydration;** that is, excess free H_2O (H_2O

not accompanied by comparable solutes) is present. When a positive H_2O balance exists, the ECF is less concentrated (more dilute) than normal. Usually, any surplus free H_2O is promptly excreted in the urine, so hypotonicity generally does not occur. However, hypotonicity can arise in three ways. (1) Patients with renal failure who are unable to excrete a dilute urine become hypotonic when they consume relatively more H_2O than solutes. (2) Hypotonicity can occur transiently in healthy people if H_2O is rapidly ingested to such an excess that the kidneys are unable to respond quickly enough to eliminate the extra H_2O. (3) Hypotonicity can occur when excess H_2O without solute is retained in the body as a result of inappropriate secretion of vasopressin. Vasopressin is normally secreted in response to a H_2O deficit, which is relieved by increasing H_2O reabsorption in the distal portion of the nephrons. However, vasopressin secretion, and therefore hormonally controlled tubular H_2O reabsorption, can be increased in response to pain, fear, acute infections, trauma, and other stressful situations, even when there is no H_2O deficit in the body. The increase in vasopressin secretion and resultant H_2O retention elicited in response to stress are appropriate in anticipation of potential blood loss in the stressful situation; the extra retained H_2O could minimize the effect a loss of blood volume would have on blood pressure. However, because modern-day stressful situations generally are not accompanied by blood loss, the increased vasopressin secretion is inappropriate as far as the body's fluid balance is concerned. The reabsorption and retention of too much H_2O leads to dilution of the body's solutes.

Whichever way it is brought about, excess free H_2O retention first dilutes the ECF compartment, making it hypotonic. The resultant difference in osmotic activity between the ECF and ICF induces the movement of H_2O by osmosis from the more dilute interstitial fluid into the cells. This H_2O flux into the cells dilutes the intracellular solutes, so both ECF and ICF compartments are hypotonic at osmotic equilibrium. Furthermore, the cells swell as H_2O moves into them osmotically. As with shrinking of cerebral neurons, pronounced swelling of brain cells also leads to brain dysfunction. Symptoms include confusion, irritability, lethargy, headache, dizziness, vomiting, drowsiness, and, in severe cases, even convulsions, coma, and death. Non-neural symptoms of overhydration include weakness caused by swelling of muscle cells and circulatory disturbances, including hypertension and edema, caused by expansion of the plasma volume.

The condition of overhydration, hypotonicity, and cellular swelling resulting from excess free H_2O retention is known as **water intoxication.** It should not be confused with the fluid retention that occurs with excess salt retention. In the latter case, the ECF is still isotonic because the increase in salt is matched by a corresponding increase in H_2O. Because the interstitial fluid is still isotonic, there is no osmotic gradient to drive the extra H_2O into the cells. The excess salt and H_2O burden is therefore confined to the ECF compartment, with circulatory consequences being the most important concern. In the case of water intoxication, the body's H_2O content is increased relative to the salt content, and the excess H_2O becomes evenly distributed between the ECF and ICF compartments. In addition to any circulatory disturbances, symptoms caused by cellular swelling become a problem. Because cells, especially the brain neurons, do not function properly when they are either shrunk or swollen, it is important that ECF osmolarity is closely regulated to prevent osmotic fluid shifts between the ECF and ICF.

ISOTONIC FLUID GAIN OR LOSS. Let us contrast the situations of hypertonicity and hypotonicity with what happens as a result of isotonic fluid gain or loss. An example of an isotonic fluid gain is therapeutic intravenous administration of an isotonic solution, such as isotonic saline. When the isotonic fluid is injected into the ECF compartment, the ECF volume increases but the concentration of ECF solutes remains unchanged; in other words, the ECF is still isotonic. Because there is no change in the ECF's osmolarity, the ECF and ICF are still in osmotic equilibrium, so there is no net fluid shift between the two compartments. The ECF compartment has increased in volume without causing a shift of H_2O into the cells. Thus, intravenous fluid therapy should be isotonic to prevent fluctuations in intracellular volume and possible neural symptoms.

Similarly, in the case of an isotonic fluid loss such as occurs in hemorrhage, the loss is confined to the ECF with no corresponding loss of fluid from the ICF. A shift of fluid out of the cells does not occur because the ECF remaining within the body is still isotonic, so there is no osmotic gradient to draw H_2O out of the cells. Of course, many other mechanisms come into play to counteract the loss of blood, but the ICF compartment is not directly affected by the loss.

Thus, when the ECF and ICF are in osmotic equilibrium, no net movement of H_2O into or out of the cells occurs regardless of whether the ECF volume increases or decreases. Shifts in H_2O between the ECF and ICF take place only when the ECF becomes more or less concentrated than the cells, and this usually arises from a loss or gain, respectively, of free H_2O.

Control of salt balance is primarily important in regulating ECF volume.

Maintenance of fluid balance in the body depends on the regulation of both ECF volume and osmolarity, which in turn is accomplished by controlling salt balance and H_2O balance. Although the mechanisms involved in regulating ECF volume and osmolarity are intricately interrelated, they can be separated as follows for discussion purposes: *maintenance of salt balance is of primary importance in the long-term regulation of ECF volume, whereas maintenance of H_2O balance is of primary importance in the regulation of ECF osmolarity.*

Sodium and its attendant anions account for more than 90% of the ECF's osmotic activity. Because osmotic activity can be equated with "water-holding power," the total Na^+ load in the ECF determines the total amount of H_2O that will be osmotically retained in the ECF. The total mass of Na^+ salts in the ECF therefore determines the ECF's volume, and, appropriately, regulation of ECF volume depends primarily on controlling salt balance.

To maintain salt balance at a prescribed set level, salt input must equal salt output, thus preventing salt accumulation or deficit in the body. The only avenue for salt input is ingestion, which typically is well in excess of the body's need for replacement of obligatory salt losses. A half gram of salt per day is adequate to replace the small amounts of salt usually lost in the feces and sweat. In our example of a typical daily salt balance (Table 15–2), salt intake is 10.5 grams per day. (The average American salt intake is 10 to 15 grams per day, although many people are consciously reducing their salt intake.)

Since we typically consume salt in excess of our needs, it is obvious that salt intake in humans is not well-controlled. Carnivores (meat eaters) and omnivores (eaters of meat and plants, like humans), which naturally get sufficient salt in fresh meat (meat contains an abundance of salt-rich ECF), do not normally manifest a physiological appetite to seek additional salt. In contrast, herbivores (plant eaters), which lack salt naturally in their diets, develop a salt hunger and will travel miles to a salt lick. Humans generally have a hedonistic rather than a regulatory appetite for salt; we consume salt because we like it rather than because we have a physiological need, except in the unusual circumstances of severe salt depletion caused by a deficiency of aldosterone, the salt-conserving hormone.

The excess ingested salt must be excreted in the urine to maintain salt balance. The three avenues for salt output are obligatory loss of salt in sweat and feces and controlled excretion of salt in the urine (Table 15–2). Salt loss in sweat is only partially regulated. The total amount of sweat produced is unrelated to salt balance, being determined instead by factors that control body temperature. Aldosterone can reduce the sweat's salt content, however, thus helping to conserve salt in a hot environment. The small salt loss in the feces is not subject to control. Except when sweating heavily or during diarrhea, the body normally uncontrollably loses only about 0.5 gm of salt per day. This is actually the only salt that normally needs to be replaced by salt intake. However, because salt consumption is typically far in excess of the meager amount needed to compensate for uncontrolled losses, the kidneys precisely excrete the excess salt in the urine in order to maintain salt balance. In our example, 10 gms of salt is eliminated in the urine per day so that total salt output exactly equals salt input. By regulating the rate of urinary salt excretion (that is, by regulating the rate of Na^+ excretion, with Cl^- following along), the kidneys normally keep the total Na^+ mass in the ECF constant despite any notable changes in dietary intake of salt or unusual losses through sweating, diarrhea, or other means. As a reflection of keeping the total Na^+ mass in the ECF constant, the ECF volume in turn is maintained within the narrowly prescribed limits essential for normal circulatory function.

Table 15–2 Daily Salt Balance

Salt Input		Salt Output	
Avenue	*Amount (gm/day)*	*Avenue*	*Amount (gm/day)*
Ingestion	10.5	Obligatory loss in sweat and feces	0.5
		Controlled excretion in urine	10.0
Total input	10.5	Total output	10.5

Deviations in ECF volume accompanying changes in salt load are responsible for triggering renal compensatory responses that quickly bring the Na^+ load and ECF volume back into line. The kidneys adjust the amount of salt excreted by controlling two processes: (1) the glomerular filtration rate (GFR) and (2) the tubular reabsorption of Na^+. Sodium is freely filtered at the glomerulus and actively reabsorbed but not secreted by the tubules, so the amount of Na^+ excreted represents the amount of Na^+ filtered that is not subsequently reabsorbed:

$$\text{Amount of } Na^+ \text{ excreted} = \text{Amount of } Na^+ \text{ filtered} - \text{Amount of } Na^+ \text{ reabsorbed}$$

Both the amount of Na^+ filtered and the amount reabsorbed are partially subject to control in order to maintain Na^+ balance, with regulation of Na^+ reabsorption being the most important regulatory factor.

CONTROL OF THE AMOUNT OF NA^+ FILTERED THROUGH REGULATION OF THE GFR. Because Na^+ is freely filterable, the amount of Na^+ filtered is equal to the plasma Na^+ concentration times the GFR. At any given plasma Na^+ concentration, any alteration in the GFR will correspondingly alter the amount of Na^+ filtered. Thus control of the GFR can adjust the amount of Na^+ filtered each minute.

The GFR is deliberately changed to alter the amount of salt and fluid filtered as part of the general baroreceptor reflex response to a change in blood pressure (see p. 476). The afferent arterioles that supply the renal glomeruli are constricted as part of the generalized vasoconstriction aimed at elevating a reduction in blood pressure (Fig. 15–3). As a result of reduced blood flow into the glomeruli, GFR decreases and, accordingly, the amount of Na^+ and accompanying fluid that are filtered decreases. Consequently, excretion of salt and fluid is diminished. The conserved salt and fluid that otherwise would have been filtered and excreted help minimize the reduction in fluid volume and contribute to long-term restitution of blood pressure. Conversely, an increase in Na^+ intake and the subsequent elevation in ECF volume and arterial blood pressure are reflexly countered by a baroreceptor reflex response that leads to an increase in GFR, which in turn results in enhanced salt and fluid excretion. The extra elimination of salt and fluid that otherwise would have been conserved helps relieve the expanded plasma volume.

Note that adjustments in the amount of salt filtered are accomplished as part of the general blood pressure-regulating reflexes. Changes in Na^+ load in the body are not sensed as such; instead they are monitored indirectly through the effect that Na^+ ultimately has on blood pressure via its role in determining the ECF volume. Fittingly, baroreceptors that monitor fluctuations in blood pressure are responsible for bringing about adjustments in the amount of Na^+ filtered and eventually excreted.

CONTROL OF THE AMOUNT OF NA^+ REABSORBED THROUGH THE RENIN-ANGIOTENSIN-ALDOSTERONE SYSTEM. The total amount of Na^+ reabsorbed also depends on regulatory systems that play an important role in controlling blood pressure. Although Na^+ is reabsorbed throughout the length of the tubule, only its reabsorption in the distal portions of the tubule is subject to control. The dominant factor controlling the extent of Na^+ reabsorption in the distal tubule is the powerful renin-angiotensin-aldosterone system, which promotes Na^+ reabsorption and thereby Na^+ retention (see p. 481). Renin release is coupled with the arterial blood pressure, which in turn is strongly linked to the total Na^+ mass in the ECF (Fig. 15–4). Renin, by means of angiotensin activation, ultimately brings about secretion of the hormone aldosterone from the adrenal cortex. Aldosterone, in turn, stimulates the reabsorption of Na^+ from the distal portion of the tubule. As more Na^+ is reabsorbed, less Na^+ is excreted. The conserved Na^+ and the H_2O that osmotically accompanies it help restore the fallen blood pressure to normal. The converse is also true. Increased blood pressure induced by excessive salt intake is quickly countered by a decrease in renin secretion, leading to a decline in aldosterone secretion and a subsequent reduction in Na^+ reabsorption. The resultant loss of extra salt and its accompanying H_2O help alleviate the salt excess, expanded plasma volume, and elevated blood pressure.

It should be apparent that control of GFR and Na^+ reabsorption are highly interrelated and that both are intimately tied in with long-term blood pressure-regulating reflexes. Specifically, a fall in arterial blood pressure brings about a twofold effect in the renal handling of Na^+ (Fig. 15–5): (1) a reflex reduction in the GFR to decrease the amount of Na^+ filtered, and (2) a hormonally adjusted increase in the amount of Na^+ absorbed. Together these effects reduce the amount of Na^+ excreted, thereby conserving for the body the Na^+ and accompanying H_2O necessary to compensate for the fall in arterial pressure. Conversely, an increase in arterial blood pressure brings about: (1) an increase in salt (and fluid) excretion by increasing the amount of Na^+ filtered via an elevation in the GFR, and (2) a decrease in salt (and fluid) reabsorption by a reduction in renin-aldosterone activity.

Control of water balance by means of vasopressin and thirst is of primary importance in regulating ECF osmolarity.

Just as control of salt balance is important in regulating ECF volume, control of water balance is crucial for regulating ECF osmolarity. Free H_2O gains (H_2O retention without accompanying solute) cause the ECF to become too dilute, or hypotonic, whereas deficits of free H_2O (loss of H_2O without corresponding loss of solute) cause the ECF to become too concentrated, or hypertonic. The osmolarity of the extracellular fluid must be immediately corrected by restoring stable balance of free H_2O to avoid osmotically induced fluxes of H_2O into or out of the cells.

The sources of input and output are more varied for H_2O than for salt. The skin and other epithelial surfaces—such as the lining of the digestive tract, respiratory tract, and urinary tract—act as barriers between the body fluids and external environment. They also serve as avenues for external exchanges of H_2O and other constituents between the body and its surroundings, some of which are subject to control. Most H_2O enters the body by being ingested and then absorbed from the digestive system into the blood plasma. Movement of fluid also occurs between the digestive tract and plasma in the opposite direction. About 7 liters of digestive juices are secreted from the plasma into the digestive lumen each day. Most of

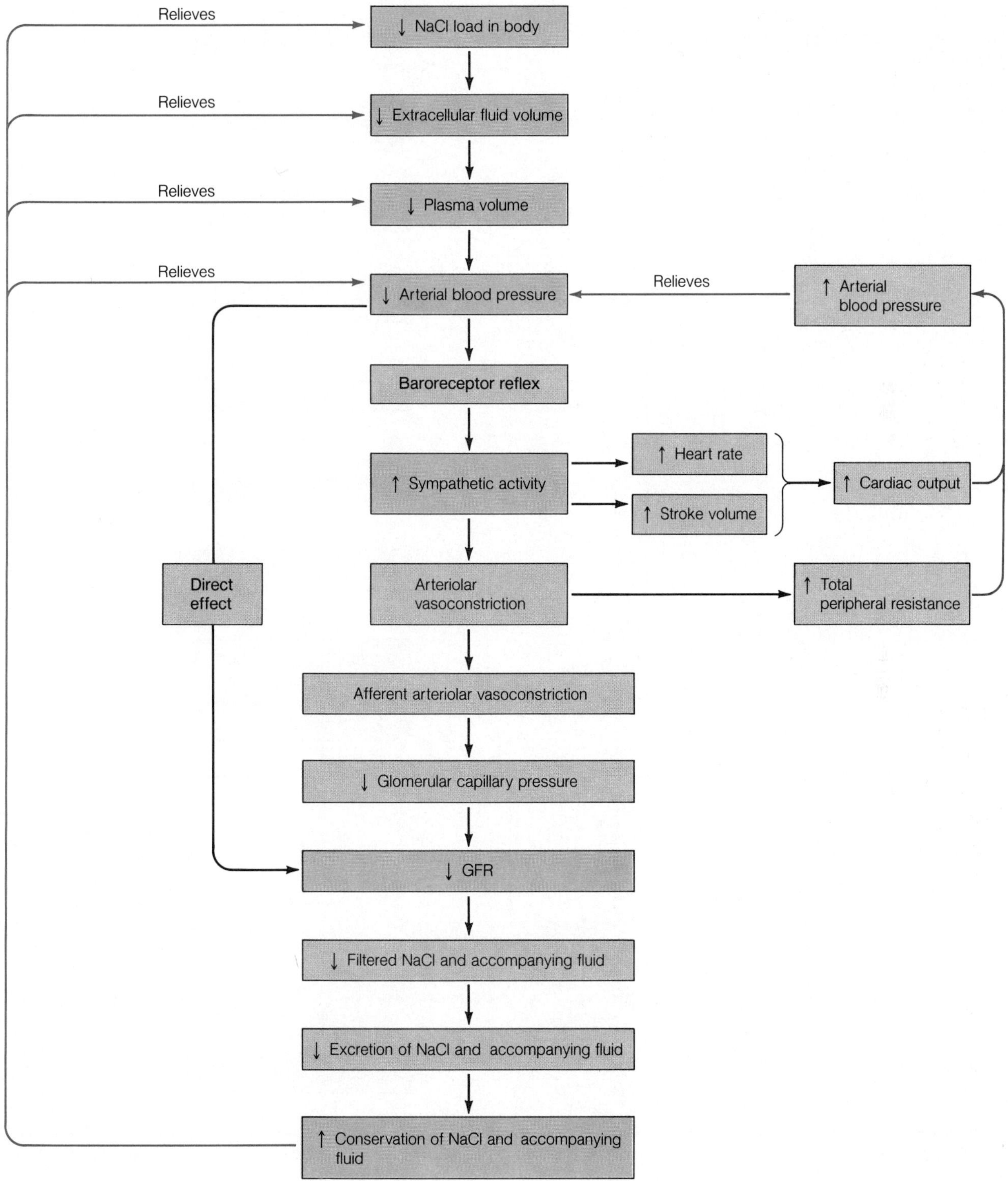

Figure 15-3 Control of Amount of Na^+ Filtered through Regulation of Glomerular Filtration Rate (GFR) in Response to Fall in Arterial Blood Pressure

Figure 15-4 Control of Amount of Na^+ Reabsorbed through Renin-Angiotensin-Aldosterone System in Response to Fall in Arterial Blood Pressure

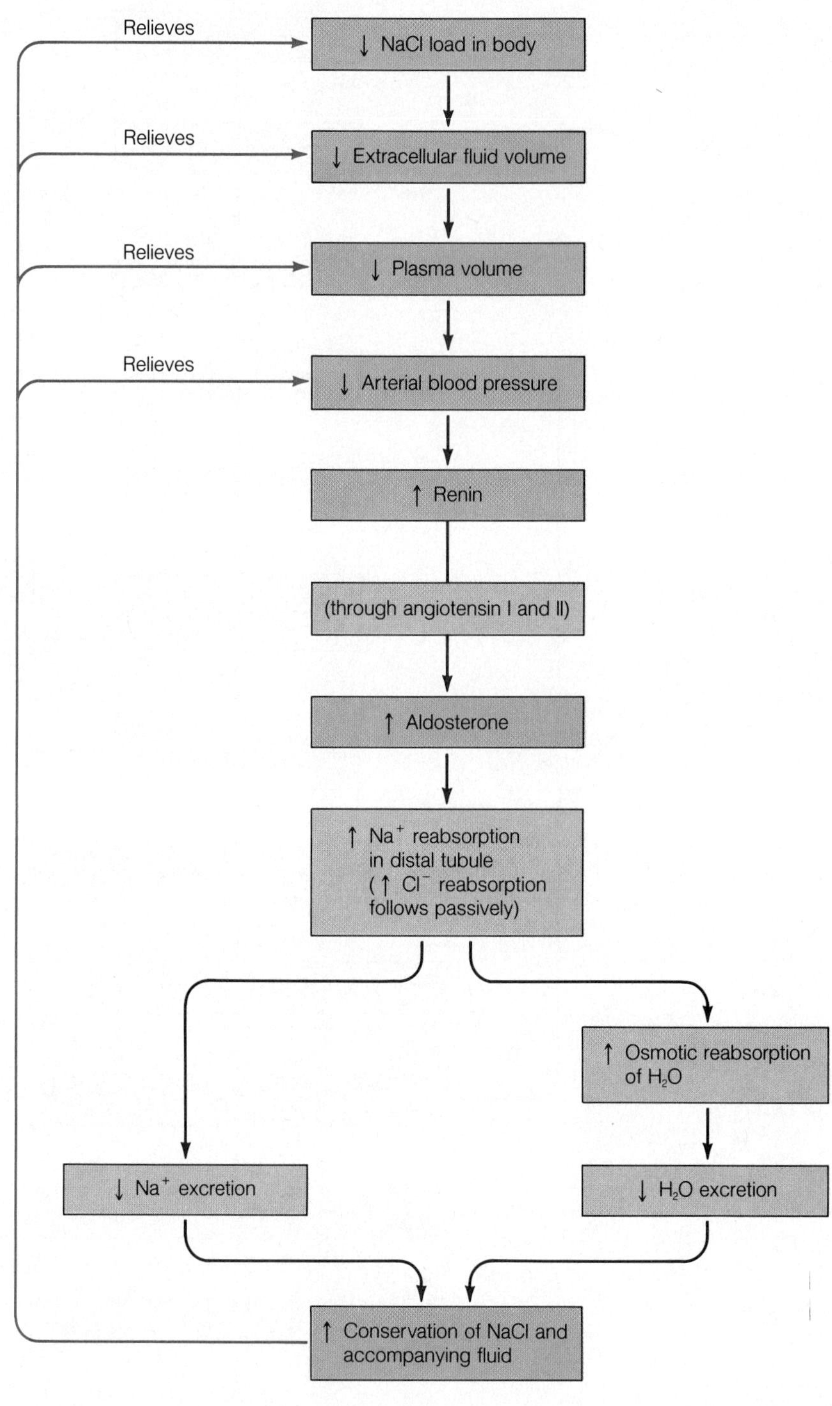

these secretions are subsequently reabsorbed after they have performed their digestive functions, so fluid normally is not lost from the body through the digestive system. However, these secreted fluids as well as the ingested fluids can be lost from the body during vomiting and diarrhea, leading the body rapidly into a negative fluid balance.

In our example of a person's typical daily H_2O balance (Table 15-3), a little more than a liter of H_2O is added to the

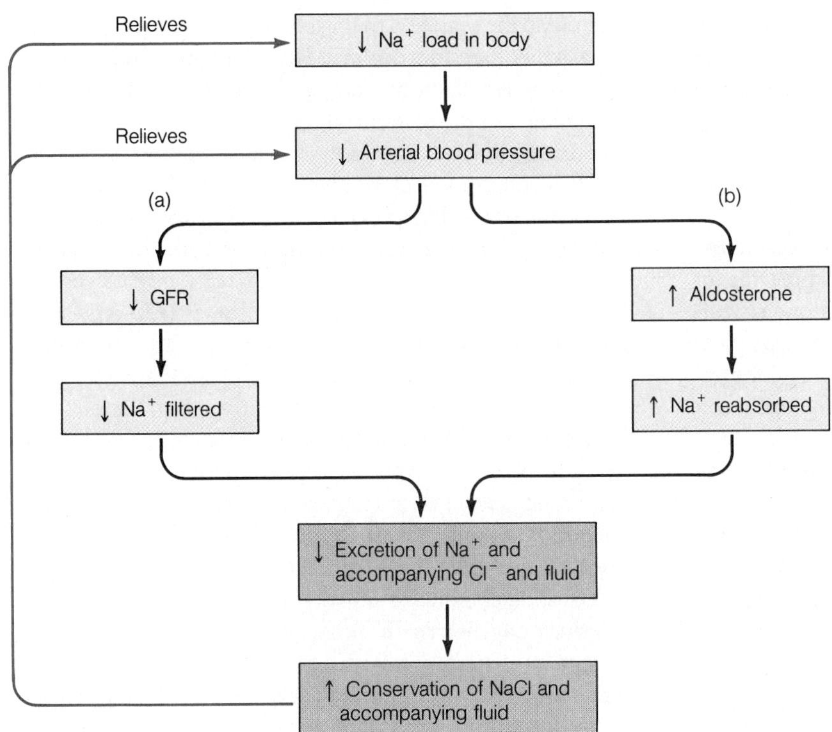

Figure 15-5 Dual Effect of Fall in Arterial Blood Pressure on Renal Handling of Na^+

body by drinking liquids. Surprisingly, an amount almost equal to that is obtained from eating solid food. Recalling that muscles consist of about 75% H_2O, meat is 75% H_2O because it is animal muscle. Likewise, fruits and vegetables consist of 60% to 90% H_2O. Therefore, people normally obtain almost as much H_2O from solid foods as from the liquids they drink. The third source of H_2O input is metabolically produced H_2O. Chemical reactions within the cells convert food and O_2 into energy, producing CO_2 and H_2O in the process. This **metabolic H_2O** produced during cellular metabolism and released into the ECF averages about 350 ml/day. The average H_2O intake from these three sources totals 2,600 ml/day. Another source of H_2O often employed therapeutically is intravenous infusion of fluid.

On the output side of the H_2O balance tally, close to a liter of H_2O is lost from the body daily without the person's awareness. This so-called **insensible loss** occurs from the lungs and nonsweating skin. During the process of respiration, inspired air becomes saturated with H_2O within the airways. This H_2O is lost when the moistened air is subsequently expired. Normally we are not aware of this H_2O loss, but it can be recognized on cold days when the H_2O vapor condenses so that we can "see our breath." The other insensible loss is the continual loss of H_2O from the skin even in the absence of sweating. Water molecules are able to diffuse through the cells of the

Table 15-3 Daily Water Balance

Water Input		Water Output	
Avenue	*Quantity (ml/day)*	*Avenue*	*Quantity (ml/day)*
Fluid intake	1,250	Insensible loss (from lungs and nonsweating skin)	900
H_2O in food intake	1,000	Sweat	100
Metabolically produced H_2O	350	Feces	100
		Urine	1,500
Total input	2,600	Total output	2,600

skin and evaporate without being noticed. Fortunately, the skin is fairly waterproof because of its keratinized exterior layer, which protects against a much greater loss of H_2O (see p. 400). When this protective surface layer is lost, such as when a person has extensive burns, fluid loss from the burned surface can increase to the point of causing serious fluid-balance problems.

Sensible loss of H_2O from the skin by means of sweat represents another avenue of H_2O output. At an air temperature of 68°F, an average of 100 ml of H_2O is lost daily through sweating. This figure can vary substantially, of course, depending on the environmental temperature and humidity and the degree of physical activity; it may range from zero up to as much as 2 liters per hour in very hot weather.

Another passageway for H_2O loss from the body is through the feces. Normally only about 100 ml of H_2O are lost via this route each day. During the process of fecal formation in the large intestine, most of the H_2O is absorbed out of the digestive-tract lumen into the blood, thereby conserving fluid and solidifying the digestive tract's contents for elimination. Additional losses of H_2O can occur from the digestive tract through vomiting or diarrhea.

By far the most important output mechanism is urine excretion, with 1,500 ml of urine being produced daily on the average. The total H_2O output is 2,600 ml/day, the same as the volume of H_2O input in our example. This is not by chance. Normally, H_2O output matches H_2O intake so that the H_2O in the body remains in balance.

Of the many sources of H_2O input and output, only two can be regulated to maintain H_2O balance. On the intake side, thirst influences the amount of fluid ingested, and on the output side, the kidneys can adjust the magnitude of urine formation. Some of the other factors are regulated but not for the purpose of maintaining H_2O balance. Food intake is subject to regulation to maintain energy balance, whereas control of sweating is important in the maintenance of body temperature. Metabolic H_2O production and insensible losses are completely unregulated.

Thirst is the subjective feeling that drives one to ingest H_2O. A **thirst center** is located in the hypothalamus in close proximity to the vasopressin-secreting cells. Vasopressin promotes H_2O reabsorption and hence H_2O conservation for the body by increasing the permeability of the distal portions of the renal tubules to H_2O. The hypothalamic osmoreceptors that monitor the ECF's osmolarity provide the predominant excitatory input to both the thirst center and vasopressin-secreting cells. When a negative H_2O balance exists, as manifested by an increase in ECF osmolarity, thirst and vasopressin secretion are both stimulated. This is adaptive, because the same circumstances that call for a reduction in urinary output (via increased vasopressin activity) to conserve body H_2O also give rise to the sensation of thirst to replenish body H_2O. Both act together to restore depleted H_2O stores, thus relieving the hypertonic condition by diluting the solutes to normal concentration. On the other hand, H_2O excess, as manifested by a reduction in ECF osmolarity, elicits increased urinary output (via suppression of vasopressin secretion) simultaneous with suppression of thirst to reduce the H_2O load in the body.

Even though the major stimulus for both thirst and increased vasopressin secretion is an increase in ECF osmolarity, other factors also influence these H_2O-regulating mechanisms (Fig. 15-6). The thirst center and vasopressin-secreting cells are both influenced to a moderate extent by changes in ECF volume mediated by input from the left atrial baroreceptors, which monitor venous pressure as a reflection of ECF volume. As expected, a volume deficit increases both thirst and vasopressin (thus decreasing urinary output), whereas a volume excess suppresses both thirst and vasopressin secretion.

Yet another stimulus for increasing both thirst and vasopressin is angiotensin II. When the renin-angiotensin-aldosterone mechanism is activated to conserve Na^+, urinary output is reduced as H_2O is osmotically conserved. Angiotensin II, in addition to stimulating aldosterone secretion, acts directly on the brain to give rise to the urge to drink while it stimulates vasopressin concurrently to enhance renal H_2O reabsorption. The resultant increased H_2O intake and decreased urinary output help to correct the reduction in ECF volume that triggered the renin-angiotensin-aldosterone system.

Several factors affect vasopressin secretion but not thirst. As described earlier, vasopressin is stimulated by stress-related inputs such as pain, fear, and trauma that have nothing directly to do with maintaining H_2O balance. In fact, H_2O retention as a result of the inappropriate secretion of vasopressin can bring about a hypotonic H_2O imbalance. On the other hand, alcohol inhibits vasopressin secretion and can lead to ECF hypertonicity by promoting excessive free H_2O excretion.

One stimulus that promotes thirst but not vasopressin secretion is a direct effect of dryness of the mouth. Nerve endings in the mouth are directly stimulated by dryness, which causes an intense sensation of thirst that can often be relieved merely by moistening the mouth even though no H_2O is actually ingested. A dry mouth can exist when salivation is suppressed by factors unrelated to the body's H_2O content, such as nervousness, excessive smoking, or certain drugs.

Factors that affect vasopressin secretion or thirst but that have nothing directly to do with the body's need for H_2O are usually short-lived. It is important to recognize that the dominant, long-standing control of vasopressin and thirst *is* directly correlated with the body's state of H_2O—namely, by the status of the ECF osmolarity and, to a lesser extent, by the ECF volume.

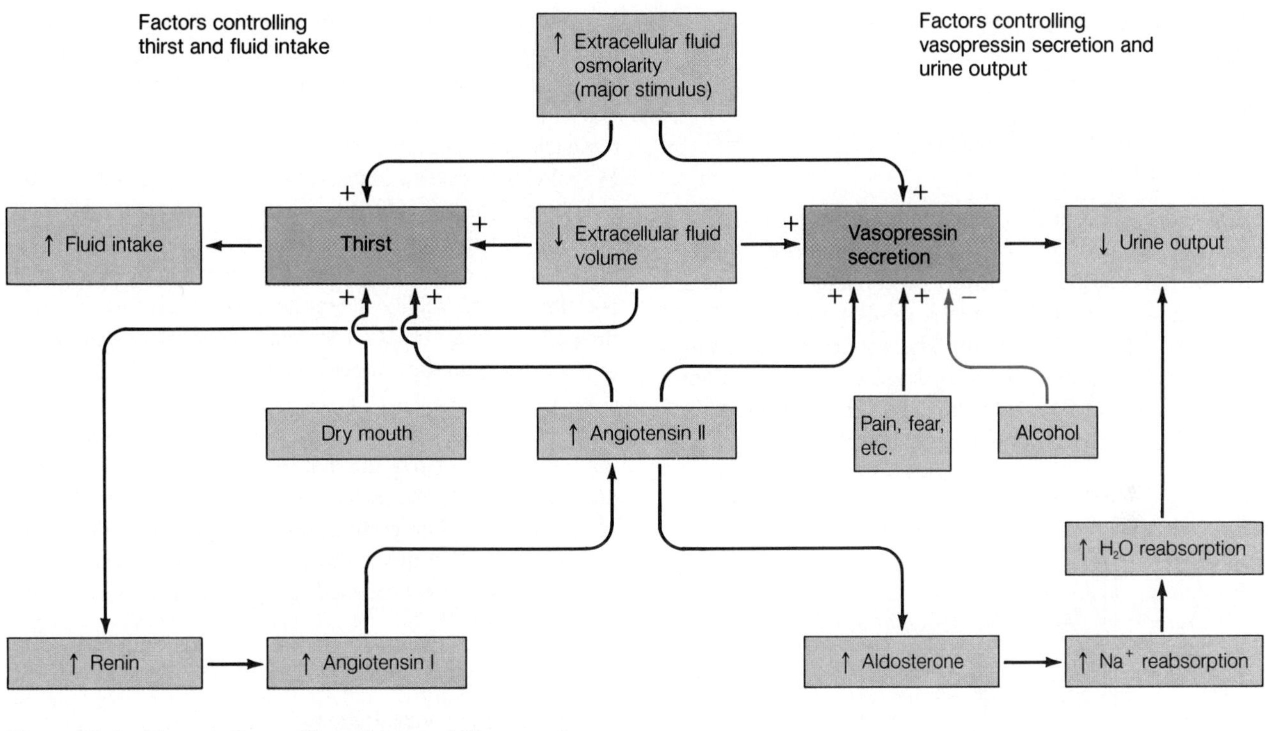

Figure 15–6 Factors Controlling Thirst and Vasopressin Secretion

In addition, there is evidence of some kind of "oral H_2O metering," at least in animals. A thirsty animal will rapidly drink only enough H_2O to satisfy its H_2O deficit. It stops drinking before the ingested H_2O has had time to be absorbed from the digestive tract to actually return the ECF compartment to normal. Exactly what factors are involved in signaling that enough H_2O has been consumed is still uncertain. It might be a learned anticipatory response based on past experience. This mechanism seems to be less effective in humans, because we frequently drink more than is necessary to meet the needs of our bodies or, conversely, may not drink enough to make up a deficit.

In fact, fluid intake by humans is often influenced more by habit and sociological factors than by the need to regulate H_2O balance. Thus, even though H_2O intake is critical in maintaining fluid balance, it is not precisely controlled in humans, who err especially on the side of excess H_2O consumption. We usually drink when we are thirsty, but we often drink even when we are not thirsty because, for example, we are on a coffee break. With H_2O intake being inadequately regulated and indeed even contributing to H_2O imbalances in the body, the primary factor involved in maintaining H_2O balance is urinary output regulated by the kidneys.

Fluctuations in ECF osmolarity caused by imbalances between H_2O input and output are quickly compensated for by adjusting the urinary excretion of H_2O without changing the usual excretion of salt; that is, H_2O reabsorption and excretion are partially dissociated from solute reabsorption and excretion, so the amount of free H_2O retained or eliminated can be varied to quickly restore ECF osmolarity to normal. Adjustments in free H_2O reabsorption and excretion are accomplished through changes in vasopressin secretion. Throughout most of the nephron, H_2O reabsorption is important in regulating ECF volume because salt reabsorption is accompanied by comparable H_2O reabsorption. In the distal portions of the tubule, however, variable free H_2O reabsorption can take place without comparable salt reabsorption because of the presence of a vertical osmotic gradient in the renal medulla to which the collecting tubule is exposed (see p. 499). Vasopressin increases this late portion of the tubule's permeability to H_2O. Depending on the amount of vasopressin present, the amount of free H_2O reabsorbed can be adjusted as necessary to restore ECF osmolarity to normal. As a further supportive measure to compensate for deviations in ECF osmolarity, the intake of free H_2O can be varied by means of the thirst mechanism to the extent it is heeded.

ACID-BASE BALANCE

Acids liberate free hydrogen ions whereas bases accept them.

Acid-base balance actually refers to the precise regulation of **free hydrogen-ion (H^+)** concentration in the body fluids. **Acids** are a special group of hydrogen-containing substances that *dissociate*, or separate, when in solution to liberate free H^+ and anions (see p. A–8). Many other substances (for example, carbohydrates) also contain hydrogen, but they are not classified as acids because the hydrogen is tightly bound within their molecular structure and is never liberated as free H^+.

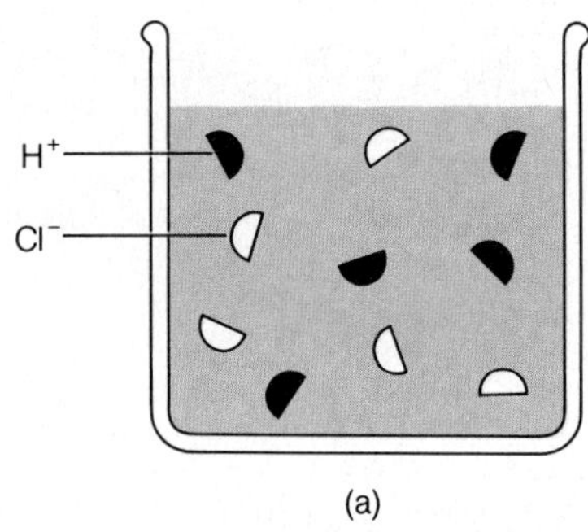

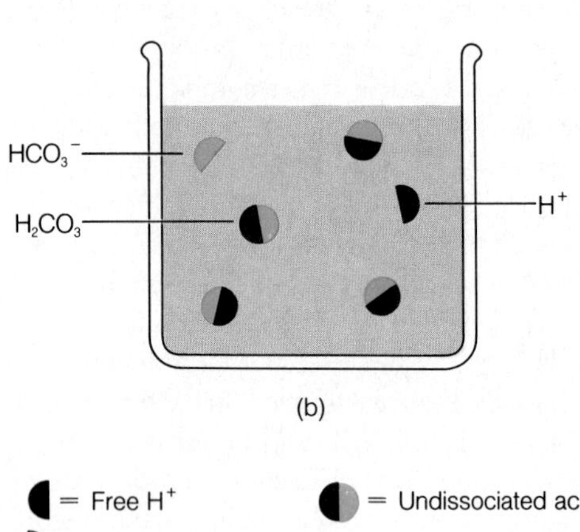

Figure 15–7 Comparison of a Strong and a Weak Acid *(a) Five molecules of a strong acid. A strong acid such as HCl (hydrochloric acid) completely dissociates into free H^+ and anion in solution. (b) Five molecules of a weak acid. A weak acid such as H_2CO_3 (carbonic acid) only partially dissociates into free H^+ and anion in solution.*

A strong acid has a greater tendency to dissociate in solution than a weak acid does; that is, a greater percentage of a strong acid's molecules separate into free H^+ and anions. Hydrochloric acid (HCl) is an example of a strong acid; every HCl molecule dissociates into free H^+ and Cl^- when dissolved in H_2O. With a weaker acid such as carbonic acid (H_2CO_3), only a portion of the molecules dissociate in solution into H^+ and HCO_3^- (bicarbonate anions). The remaining H_2CO_3 molecules remain intact. Since it is only the free hydrogen ions that contribute to the acidity of a solution, H_2CO_3 is a weaker acid than HCl because H_2CO_3 does not yield as many free hydrogen ions per number of acid molecules present in solution (Fig. 15–7).

The extent of dissociation for a given acid is always constant; that is, when in solution, the same proportion of a particular kind of acid's molecules always separate to liberate free H^+, with the other portion always remaining intact. The constant degree of dissociation for a particular acid (in this example H_2CO_3) is expressed by its **dissociation constant (K)** as follows:

$$[H^+]\,[HCO_3^-]/[H_2CO_3] = K$$

- □ [ ] designates concentration
- □ $[H^+]\,[HCO_3^-]$ represents the concentration of ions resulting from H_2CO_3 dissociation
- □ $[H_2CO_3]$ represents the concentration of intact (undissociated) H_2CO_3

The dissociation constant varies for different acids.

A **base** is a substance that can combine with a free H^+ and thus remove it from solution. A strong base is able to bind H^+ more readily than a weak acid can.

The pH designation is used to express hydrogen-ion concentration.

The $[H^+]$ in the ECF is normally 4×10^{-8} or 0.00000004 equivalents per liter. The concept of pH has been developed to more conveniently express $[H^+]$. Specifically, **pH** equals the logarithm (log) to the base 10 of the reciprocal of the hydrogen-ion concentration:

$$pH = \log 1/[H^+]$$

Two important points should be noted about this formula:

1. Because $[H^+]$ is in the denominator, *a high $[H^+]$ corresponds to a low pH, and a low $[H^+]$ corresponds to a high pH.*
2. *Every unit change in pH actually represents a tenfold change in $[H^+]$* because of the logarithmic relationship.

A log to the base 10 indicates how many times 10 must be multiplied by itself to produce a given number. For example, the log of 10 = 1, whereas the log of 100 = 2. Ten must be multiplied by itself twice to yield 100 (10 x 10 = 100). For numbers less than 10, their log is less than 1. For numbers between 10 and 100, their logs are between 1 and 2, and so on. Accordingly, each unit of change in pH is indicative of a tenfold change in $[H^+]$. For example, a solution with a pH of 7 has a $[H^+]$ ten times less than that of a solution with a pH of 6 (1 pH-unit difference) and one hundred times less than that of a solution with a pH of 5 (2 pH-unit difference).

The pH of pure H_2O is 7.0, which is considered to be chemically neutral. An extremely small proportion of H_2O molecules dissociate into hydrogen ions and hydroxyl (OH^-) ions. Because OH^- has the ability to bind with H^+ to once again form a H_2O molecule, it is considered to be basic. Since an equal number of acidic hydrogen ions and basic hydroxyl ions are formed, H_2O is neutral, being neither acidic nor basic.

Solutions having a pH that is less than 7.0 contain a higher $[H^+]$ than pure H_2O and are considered to be **acidic.** Conversely, solutions having a pH value that is greater than 7.0 have a lower $[H^+]$ and are considered to be **basic** or **alkaline** (Fig. 15–8a).

The pH of arterial blood is normally 7.45 and the pH of venous blood is 7.35, for an average blood pH of 7.4. The pH of venous blood is slightly lower than that of arterial blood because of H^+ generated by the formation of H_2CO_3 from CO_2 picked up at the tissue capillaries. **Acidosis** exists whenever the blood pH falls below 7.35, whereas **alkalosis** occurs when the blood pH is above 7.45 (Fig. 15–8b). Note that the reference point for determining the body's acid-base status is not the chemically neutral pH of 7.0 but the normal plasma pH of 7.4. Thus a plasma pH of 7.2 is considered acidotic even though in chemistry a pH of 7.2 is considered basic.

Death occurs if arterial pH falls outside of the range of 6.8 to 8.0 for more than a few seconds, because an arterial pH of less than 6.8 or greater than 8.0 is not compatible with life.

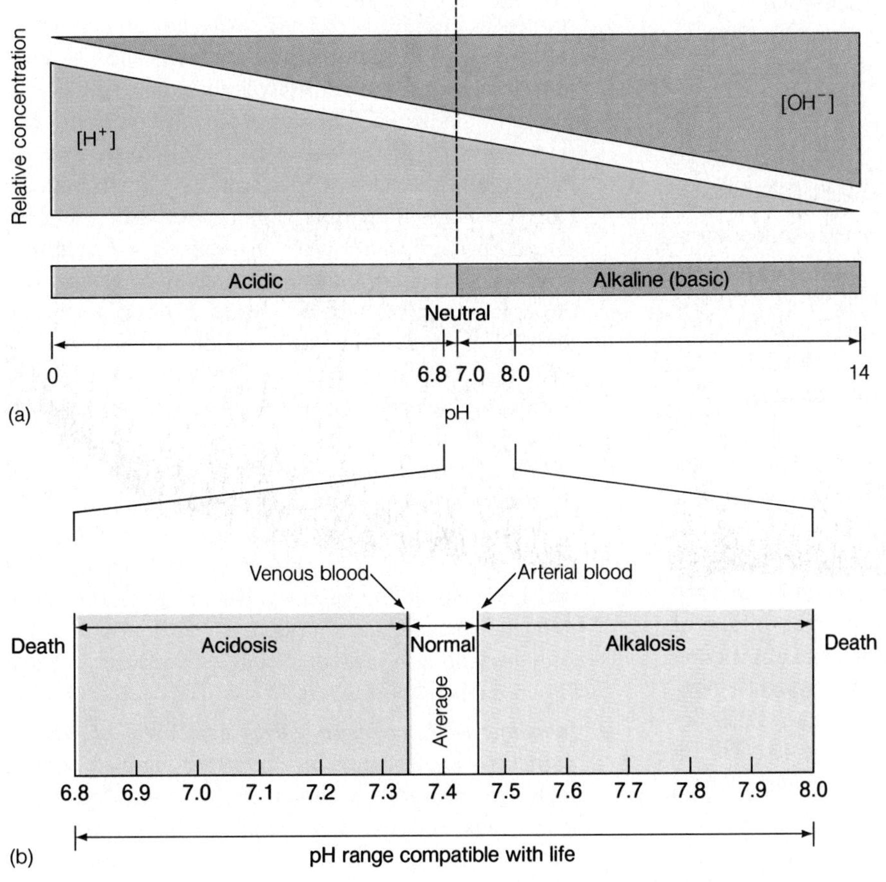

Figure 15–8 pH Considerations in Chemistry and Physiology *(a) Relationship of pH to relative concentrations of H^+ and base under neutral, acidic, and alkaline conditions in chemistry. (b) Plasma pH range under normal, acidosis, and alkalosis conditions.*

Obviously, therefore, $[H^+]$ in the body fluids must be carefully regulated.

Fluctuations in hydrogen-ion concentration have profound effects on body chemistry.

Only a narrow pH range is compatible with life because even small changes in $[H^+]$ have dramatic effects on normal cell function. The prominent consequences of fluctuations in $[H^+]$ include the following:

1. Changes in excitability of nerve and muscle cells are among the major clinical manifestations of pH abnormalities. The major clinical effect of increased $[H^+]$ (acidosis) is depression of the central nervous system. Acidotic patients become disoriented and, in more severe cases, eventually die in a state of coma. In contrast, the major clinical effect of decreased $[H^+]$ (alkalosis) is overexcitability of the nervous system, first the peripheral nervous system and later the central nervous system. Peripheral nerves become so excitable that they fire even in the absence of normal stimuli. Such overexcitability of the afferent (sensory) nerves gives rise to abnormal "pins and needles" tingling sensations. On the other hand, overexcitability of efferent (motor) nerves brings about muscle twitches and, in more pronounced cases, severe muscle spasms. Death may occur in extreme alkalosis because of spasm of the respiratory muscles. Alternatively, severely alkalotic patients may die of convulsions resulting from overexcitability of the central nervous system. In less serious situations, CNS overexcitability is manifested as extreme nervousness.
2. Hydrogen ion concentration exerts a marked influence on enzyme activity. Even slight deviations in $[H^+]$ alter the shape and activity of protein molecules. Since enzymes are proteins, a shift in the body's acid-base balance disturbs the normal pattern of metabolic activity catalyzed by these enzymes. Some cellular chemical reactions are accelerated; others are depressed.
3. Changes in $[H^+]$ influence K^+ levels in the body. When reabsorbing Na^+ from the filtrate, the renal tubular cells secrete either K^+ or H^+ in exchange. Normally they secrete a preponderance of K^+ compared to H^+. Because of the intimate relationship between secretion of H^+ and K^+ by the kidneys, an increased rate of secretion of one of these ions is accompanied by a decreased rate of secretion of the other. For example, if more H^+ than normal is eliminated by the kidneys, as occurs when the body fluids become acidotic, less K^+ than usual can be excreted. The resultant K^+ retention can affect cardiac function, among other detrimental consequences.

Hydrogen ions are continually being added to the body fluids as a result of metabolic activities.

As with any other constituent, in order to maintain a constant $[H^+]$ in the body fluids, input of hydrogen ions must be balanced by an equal output. On the input side, only a small amount of acid capable of dissociating to release H^+ is taken in with food, such as the weak citric acid found in oranges and other citrus fruits. Most H^+ in the body fluids is generated internally from metabolic activities. Hydrogen ion is normally continually being added to the body fluids from the three following sources:

1. *Carbonic acid formation*—The major source of H^+ is through H_2CO_3 formation from metabolically produced CO_2. Cellular oxidation of nutrients yields energy, with CO_2 and H_2O as end products. Catalyzed by the enzyme carbonic anhydrase, CO_2 and H_2O form H_2CO_3, which then partially dissociates to liberate free H^+ and HCO_3^-:

$$CO_2 + H_2O \overset{ca}{\rightleftharpoons} H_2CO_3 \rightleftharpoons H^+ + HCO_3^-$$

   These reactions are reversible because they can proceed in either direction, depending on the concentrations of the substances involved as dictated by the law of mass action (see p. 441). Within the systemic capillaries, the CO_2 level in the blood increases as metabolically produced CO_2 enters from the tissues. This drives the reaction to the acid side, generating H^+ as well as HCO_3^- in the process. In the lungs, the reactions are reversed: CO_2 diffuses from the blood flowing through the pulmonary capillaries into the alveoli (air sacs), from which it is expired to the atmosphere. The resultant reduction in CO_2 in the blood drives the reactions toward the CO_2 side. Hydrogen ion and HCO_3^- form H_2CO_3, which rapidly decomposes into CO_2 and H_2O once again. The CO_2 is exhaled while the hydrogen ions generated at the tissue level are incorporated into H_2O molecules.

   When the respiratory system is able to keep pace with the rate of metabolism, there is no net gain or loss of H^+ in the body fluids from metabolically produced CO_2. When the rate of CO_2 removal by the lungs does not match the rate of CO_2 production at the tissue level, however, the resultant accumulation or deficit of CO_2 in the body leads to an excess or shortage, respectively, of free H^+ in the body fluids.
2. *Inorganic acids produced during catabolism of organic nutrients*—Dietary proteins and other ingested organic molecules that are found abundantly in meat contain a large quantity of sulfur and phosphorus. When these mol-

ecules are catabolized, sulfuric acid and phosphoric acid are produced as by-products. Being moderately strong acids, these two inorganic acids dissociate to a large extent, liberating free H^+ into the body fluids. In contrast, the breakdown of fruits and vegetables produces bases that neutralize the acids derived from protein metabolism to some extent. Generally, however, more acids than bases are produced during catabolism of ingested food, which leads to an excess of these inorganic acids.

3. *Organic acids from intermediary metabolism*—Numerous organic acids are produced during normal intermediary metabolism; for example, fatty acids are liberated during fat metabolism and lactic acid is produced by muscles during heavy exercise. These organic acids partially dissociate to yield free H^+.

Hydrogen-ion generation therefore normally goes on continuously as a result of ongoing metabolic activities. Furthermore, in certain disease states, additional acids may be produced that further contribute to the total body pool of H^+. For example, in diabetes mellitus, large quantities of keto acids may be produced as a result of abnormal fat metabolism. Some types of acid-producing medications may also add to the total load of H^+ that must be handled by the body. Thus input of H^+ is unceasing, highly variable, and essentially unregulated.

The crux of H^+ balance is maintaining the normal alkalinity of the ECF (pH 7.4) despite this constant onslaught of acid. The generated free H^+ must be largely removed from solution while in the body and ultimately must be eliminated from the body so that the pH of the body fluids can remain within the narrow range compatible with life. Mechanisms also must exist to rapidly compensate for the occasional situation in which the ECF becomes too alkaline.

Three lines of defense against changes in $[H^+]$ operate to maintain the $[H^+]$ of body fluids at a nearly constant level despite unregulated input. These are: (1) the chemical buffer systems, (2) the respiratory mechanism of pH control, and (3) the renal mechanism of pH control. We will look at each of these methods separately.

Chemical buffer systems immediately minimize changes in hydrogen-ion concentration, thereby acting as the first line of defense.

A **chemical buffer system** is a mixture of two (or perhaps more) chemical compounds present in a solution that minimizes changes in the pH when either an acid or a base is added to or removed from the solution. A buffer system consists of a pair of substances involved in a reversible reaction—one substance that can yield free H^+ as the $[H^+]$ starts to fall and another that can bind with free H^+ (thus removing it from solution) when $[H^+]$ starts to rise. An important example of such a system is the carbonic acid: bicarbonate (H_2CO_3:HCO_3^-) buffer pair, which is involved in the following reversible reaction:

$$H^+ + HCO_3^- \rightleftharpoons H_2CO_3$$

When a strong acid such as HCl is added to an unbuffered solution, all the dissociated H^+ remains free in the solution (Fig. 15–9a). In contrast, when HCl is added to a solution containing the H_2CO_3:HCO_3^- buffer pair (Fig. 15–9b), the HCO_3^- immediately binds with the free H^+. The resultant weak H_2CO_3 dissociates only slightly compared to the marked reduction in pH that occurred when the buffer system was not present and the additional H^+ remained unbound. On the other hand, when the pH of the solution starts to rise because of the addition of base or loss of acid, the H^+-yielding member of the buffer pair, H_2CO_3, releases H^+ to minimize the rise in pH.

There are four major buffer systems present in the body: (1) H_2CO_3:HCO_3^- buffer system, (2) the protein buffer system, (3) the hemoglobin buffer system, and (4) the phosphate buffer system. Each serves a different important role (Table 15–4).

THE H_2CO_3: HCO_3^- BUFFER PAIR IS THE PRIMARY ECF BUFFER FOR NON-CARBONIC ACIDS. The H_2CO_3: HCO_3^- buffer pair is the most important buffer system in the ECF for buffering pH changes induced by causes other than fluctuations in CO_2. It is a very effective ECF buffer system for two reasons. First, H_2CO_3 and HCO_3^- are abundant in the ECF, so this system is readily available to resist changes in pH. Second and more importantly, each component of this buffer pair is closely regulated. The kidneys regulate HCO_3^-, whereas the respiratory system regulates CO_2, which in turn generates H_2CO_3. Thus, in the body the H_2CO_3: HCO_3^- buffer system includes involvement of CO_2 via the following reaction with which you are already familiar:

$$H^+ + HCO_3^- \rightleftharpoons H_2CO_3 \rightleftharpoons CO_2 + H_2O$$

When new H^+ is added to the plasma from any source other than CO_2 (for example, through lactic acid released into the ECF from exercising muscles), the preceding reaction is driven toward the right side of the equation. As the extra H^+ binds with HCO_3^-, it no longer contributes to the acidity of the body fluids, so the rise in $[H^+]$ is abated. In the converse situation, when the plasma $[H^+]$ occasionally falls below normal for some reason other than a change in CO_2 (such as the loss of plasma-derived HCl in the gastric juices during vomit-

Figure 15-9 Action of Chemical Buffers *(a) Addition of HCl to an unbuffered solution. (b) Addition of HCl to a buffered solution.*

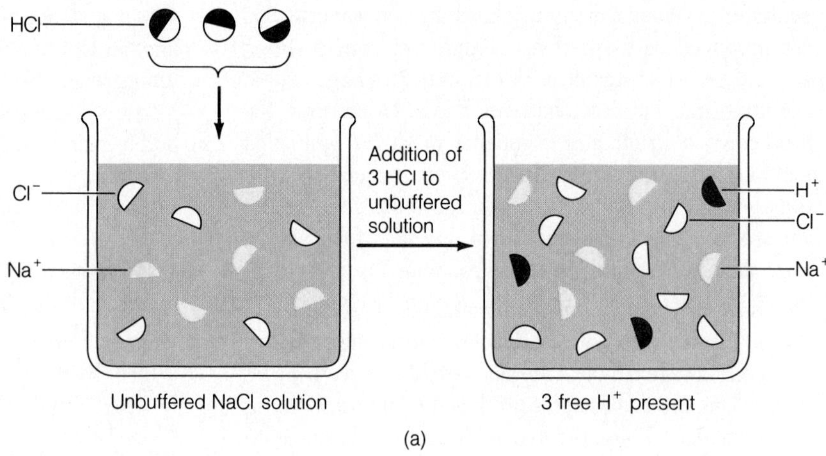

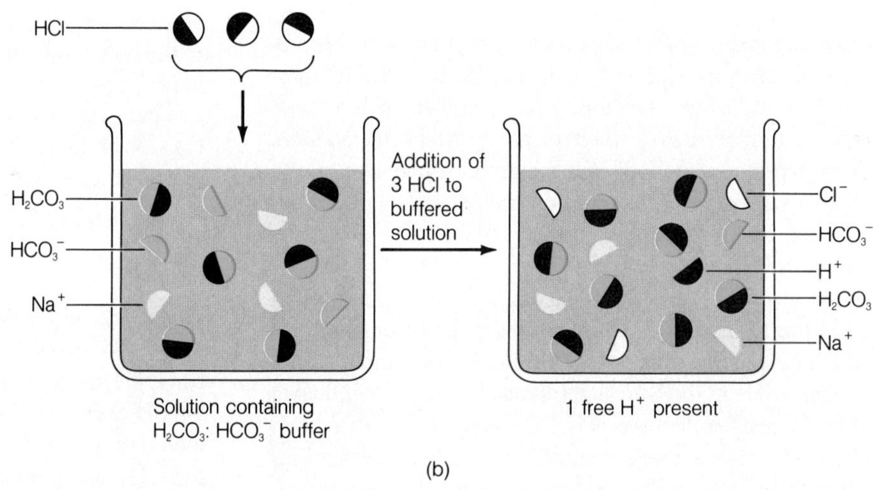

Table 15-4 Chemical Buffers and Their Primary Roles

Buffer System	Major Functions
Carbonic acid: bicarbonate buffer system	Primary buffer against non-carbonic acid changes
Protein buffer system	Primary intracellular fluid buffer; also buffers extracellular fluid
Hemoglobin buffer system	Primary buffer against carbonic acid changes
Phosphate buffer system	Important urinary buffer; also buffers intracellular fluid

ing), the reaction is driven toward the left side of the equation. Dissolved CO_2 and H_2O in the plasma form H_2CO_3, which generates additional H^+ to make up for the H^+ deficit. In so doing, the H_2CO_3: HCO_3^- buffer system resists the fall in $[H^+]$.

This buffer system is extremely useful for buffering changes in pH in the ECF resulting from any cause other than a change in CO_2/H_2CO_3, but it cannot buffer changes in pH induced by fluctuations in CO_2. A buffer system cannot buffer itself. Consider, for example, the situation in which the plasma $[H^+]$ is elevated because of CO_2 retention associated with a respiratory problem. The rise in CO_2 drives the reaction to the left according to the law of mass action, resulting in an elevation in $[H^+]$. The increase in $[H^+]$ occurs as a result of the reaction being driven to the *left* because of an increase in CO_2, so the elevated $[H^+]$ cannot drive the reaction to the *right* to buffer the increase in $[H^+]$. Only if the increase in $[H^+]$ is brought about by some mechanism other than CO_2 accumulation can this buffer system be shifted to the CO_2 side of the equation and be effective in reducing the $[H^+]$. Likewise, in the opposite situation, the H_2CO_3: HCO_3^- buffer system cannot compensate for a reduction in $[H^+]$ caused by a deficit of CO_2 by generating more H^+-yielding H_2CO_3 when the problem in the first place is a shortage of H_2CO_3-forming CO_2. Other mechanisms, to be described shortly, are available for resisting fluctuations in pH caused by changes in CO_2 levels.

The relationship between $[H^+]$ and the members of a buffer pair can be expressed according to the **Henderson-Hasselbalch equation**, which, for the H_2CO_3: HCO_3^- buffer system, is as follows:

$$pH = pK + \log [HCO_3^-]/[H_2CO_3]$$

Although you do not need to know the mathematical manipulations involved, it will be helpful for you to understand how this formula is derived. Recalling that the dissociation constant K for H_2CO_3 acid is

$$[H^+]\,[HCO_3^-]/[H_2CO_3] = K,$$

and knowing that the relationship between pH and $[H^+]$ is

$$pH = \log 1/[H^+],$$

by solving the dissociation constant formula for $[H^+]$ (that is, $[H^+] = K \times [H_2CO_3^-]/[HCO_3])$ and replacing this value for $[H^+]$ in the pH formula, one comes up with the Henderson-Hasselbalch equation.

Practically speaking, $[H_2CO_3]$ is a direct reflection of the concentration of dissolved CO_2, henceforth referred to as $[CO_2]$, because most of the CO_2 in the plasma is converted into H_2CO_3. (The dissolved CO_2 concentration is equivalent to P_{CO_2}, as described in the chapter on respiration.) Therefore, the equation becomes

$$pH = pK + \log [HCO_3^-]/[CO_2]$$

The pK is the logarithm of 1/K and, like K, always remains a constant for any given acid. For H_2CO_3, the pK is 6.1. Since the pK is always a constant, changes in pH are associated with changes in the ratio between $[HCO_3^-]$ and $[CO_2]$. Normally the ratio between $[HCO_3^-]$ and $[CO_2]$ in the ECF is 20 to 1; that is, there is twenty times more HCO_3^- than CO_2. Plugging this ratio into our formula:

$$\begin{aligned} pH &= pK + \log [HCO_3^-]/[CO_2] \\ &= 6.1 + \log 20/1 \end{aligned}$$

The log of 20 is 1.3. Therefore, pH = 6.1 + 1.3 = 7.4, which is the normal pH of plasma. When the ratio of $[HCO_3^-]$ to $[CO_2]$ increases above 20/1, pH increases. Accordingly, either a rise in $[HCO_3^-]$ or a fall in $[CO_2]$, both of which increase the $[HCO_3^-]/[CO_2]$ ratio if the other component remains constant, shifts the acid-base balance toward the alkaline side. In contrast, when the $[HCO_3^-]/[CO_2]$ ratio decreases below 20/1, pH decreases toward the acid side. This can occur either if the $[HCO_3^-]$ decreases or the $[CO_2]$ increases while the other component remains constant. Because $[HCO_3^-]$ is regulated by the kidneys and $[CO_2]$ by the lungs, the pH of the plasma can be shifted up and down by kidney and lung influences. Renal and respiratory regulation of pH operates to correct changes in $[H^+]$ largely through the kidney's and lung's control of plasma $[HCO_3^-]$ and $[CO_2]$, respectively, to restore the ratio to normal. Accordingly,

$$pH \propto \frac{[HCO_3^-] \text{ controlled by kidney function}}{[CO_2] \text{ controlled by respiratory function}}$$

Because of this relationship, not only do kidneys and lungs both normally participate in pH control, but renal or respiratory dysfunction can also induce acid-base imbalances by altering the $[HCO_3^-]/[CO_2]$ ratio.

THE PROTEIN BUFFER SYSTEM IS PRIMARILY IMPORTANT INTRACELLULARLY. The most plentiful buffers of the body fluids are the proteins, including the intracellular proteins and the plasma proteins. Proteins contain both acidic and basic groups that can give up or take up H^+, thus serving as excellent buffers. Quantitatively, the protein system is most important in buffering changes in $[H^+]$ in the intracellular fluid because of the sheer abundance of the intracellular proteins. The more limited number of plasma proteins reinforces the H_2CO_3: HCO_3^- system in extracellular buffering.

THE HEMOGLOBIN BUFFER SYSTEM BUFFERS CARBONIC ACID-GENERATED HYDROGEN ION. One intracellular protein—hemoglobin (Hb)—is of such importance in terms of buffering that it deserves special attention. Hemoglobin buffers the H^+ generated from metabolically produced CO_2 in transit between the tissues and lungs. At the systemic capillary level, CO_2 continuously diffuses into the blood from the tissue cells where it is being produced. The greatest percentage of this CO_2 forms H_2CO_3, which in turn partially dissociates into H^+ and HCO_3^-. Simultaneously, some of the oxyhemoglobin (HbO_2) releases O_2, which diffuses into the tissue. Reduced (unoxygenated) Hb has a greater affinity for H^+ than HbO_2 does. Therefore, most of the H^+ generated from CO_2 at the tissue level becomes bound to reduced Hb and no longer contributes to the acidity of the body fluids:

$$H^+ + Hb \rightleftharpoons HHb$$

At the lungs the reactions are reversed. As Hb picks up O_2 diffusing from the alveoli into the red blood cells, the affinity of Hb for H^+ is decreased, so H^+ is released. This liberated H^+ combines with HCO_3^- to yield H_2CO_3, which in turn produces CO_2, which is exhaled. Meanwhile, the hydrogen has been reincorporated into neutral H_2O molecules. Were it not for Hb, the blood would become much too acidic after picking up CO_2 at the tissue level. With the tremendous buffering capacity of the Hb system, venous blood is only slightly more acidic than arterial blood despite the large volume of H^+-generating CO_2 carried in the venous blood.

THE PHOSPHATE BUFFER SYSTEM IS AN IMPORTANT URINARY BUFFER. The remaining important buffer system, the phosphate buffer system, is composed of an acid phosphate salt (NaH_2PO_4) that can donate a free H^+ when the $[H^+]$ falls and a basic phosphate salt (Na_2HPO_4) that can accept a free H^+ when the $[H^+]$ rises. Basically, this buffer pair can alternately switch a H^+ for a Na^+ as demanded by the $[H^+]$.

Even though the phosphate pair is a good buffer, its concentration in the ECF is rather low, so quantitatively it is not very important as an ECF buffer. Because phosphates are more abundant within the cells, this system contributes significantly to intracellular buffering, being rivaled only by the more plentiful intracellular proteins. Even more importantly, the phosphate system serves as an excellent urinary buffer. Humans normally consume more phosphate ($PO_4^{\equiv}$) than needed. The excess $PO_4^{\equiv}$ filtered through the kidneys is not reabsorbed but remains in the tubular fluid to be excreted. This excreted $PO_4^{\equiv}$ buffers the urine as it is being formed by removing from solution the H^+ secreted into the tubular fluid. None of the other body-fluid buffer systems are present in the tubular fluid to play a role in buffering urine during its formation. Most or all of the filtered HCO_3^- and CO_2 (alias H_2CO_3) are reabsorbed, whereas Hb and plasma proteins are not even filtered.

All chemical buffer systems act immediately, within fractions of a second, to minimize changes in pH. When $[H^+]$ is altered, the involved buffer systems' reversible chemical reactions are shifted at once in favor of compensating for the change in $[H^+]$. Accordingly, the buffer systems are the ***first line of defense*** against changes in $[H^+]$ since they are the first mechanism to respond.

Through the mechanism of buffering, most hydrogen ions seem to "disappear" from the body fluids between the time of their generation and elimination. It must be emphasized, however, that none of the chemical buffer systems actually ***eliminates*** H^+ from the body. These ions are merely removed from solution by being incorporated within one of the members of the buffer pair, thus preventing the hydrogen ions from contributing to the body fluids' acidity. Because each buffer system has a limited capacity to "soak up" H^+, the H^+ that is unceasingly produced must be ultimately removed from the body. If H^+ were not eventually eliminated, soon all the body fluid buffers would already be bound with H^+ and there would be no further buffering ability.

The respiratory and renal mechanisms of pH control actually eliminate acid from the body instead of merely suppressing it, but they respond more slowly than the chemical buffer systems. We will now turn our attention to these other defenses against changes in acid-base balance.

The respiratory system regulates hydrogen-ion concentration by controlling the rate of carbon-dioxide removal from the plasma through adjustments in pulmonary ventilation.

The respiratory system plays an important role in acid-base balance through its ability to alter pulmonary ventilation and, consequently, to alter the excretion rate of H^+-generating CO_2. Because of this ability, it should not be surprising that the level of respiratory activity is governed at least in part by the arterial $[H^+]$ (Fig. 15–10). When arterial $[H^+]$ increases, the respiratory center in the brain stem is reflexly stimulated (see p. 456) to increase pulmonary ventilation (the rate at which gas is exchanged between the lungs and atmosphere). As the rate and depth of breathing increase, more CO_2 than usual is blown off, so less H_2CO_3 than normal is added to the body fluids. Since CO_2 forms acid, the removal of CO_2 in essence removes acid from the body. Conversely, when arterial $[H^+]$ falls, pulmonary ventilation is reduced. As a result of slower, shallower breathing, metabolically produced CO_2 diffuses from the cells into the blood faster than it is removed from the blood by the lungs, so higher-than-usual amounts of acid-forming CO_2 accumulate in the blood, thus restoring $[H^+]$ toward normal.

The lungs are extremely important in maintaining the $[H^+]$ of the plasma. Every day they remove from the body fluids what amounts to one-hundred times more H^+ derived from carbonic acid than the kidneys remove from non-carbonic acid sources. Furthermore, the respiratory system, through its ability to regulate arterial $[CO_2]$, can adjust the amount of H^+ added to the body fluids from this source as needed to restore pH toward normal when fluctuations in $[H^+]$ from non-carbonic acid sources occur.

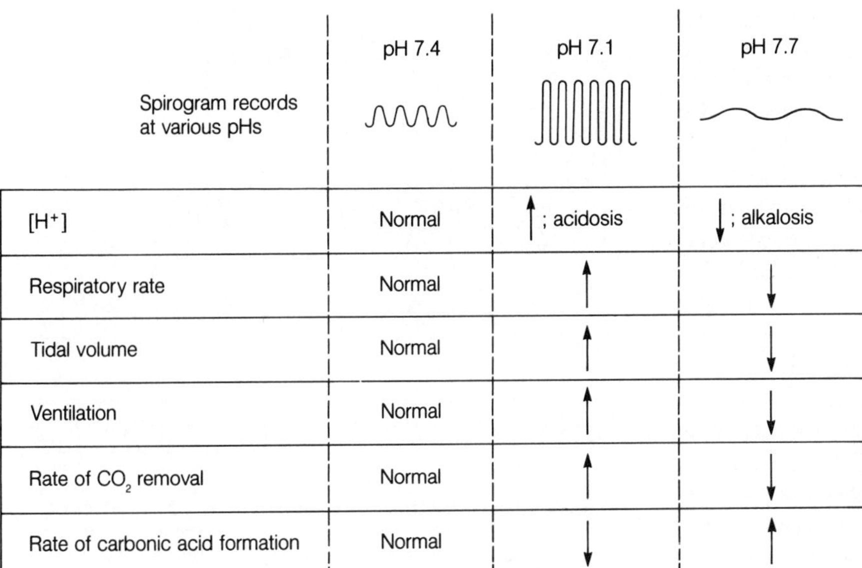

	pH 7.4	pH 7.1	pH 7.7
Spirogram records at various pHs			
$[H^+]$	Normal	↑; acidosis	↓; alkalosis
Respiratory rate	Normal	↑	↓
Tidal volume	Normal	↑	↓
Ventilation	Normal	↑	↓
Rate of CO_2 removal	Normal	↑	↓
Rate of carbonic acid formation	Normal	↓	↑

Figure 15–10 Respiratory Adjustments to Acidosis and Alkalosis Induced by Nonrespiratory Causes

Respiratory regulation acts at a moderate speed, coming into play only when the chemical buffer systems alone are unable to minimize $[H^+]$ changes. When deviations in $[H^+]$ occur, the buffer systems respond immediately, whereas adjustments in ventilation require a few minutes to be initiated. If a deviation in $[H^+]$ is not swiftly and completely corrected by the buffer systems, the respiratory system comes into action a few minutes later, thus serving as the *second line of defense* against changes in $[H^+]$.

Of course, when changes in $[H^+]$ occur because of fluctuations in $[CO_2]$ that arise from respiratory abnormalities, the respiratory mechanism cannot contribute at all to the control of pH; for example, if acidosis exists because of CO_2 accumulation caused by lung disease, the impaired lungs cannot possibly compensate for the acidosis by increasing the rate of CO_2 removal. The buffer systems (other than the H_2CO_3: HCO_3^- pair) plus renal regulation are the only mechanisms available for defending against respiratory-induced acid-base abnormalities.

The kidneys contribute powerfully to control of acid-base balance by excreting hydrogen ion while adding new bicarbonate to the blood.

The kidneys are the *third line of defense* against changes in $[H^+]$ in the body fluids; they require hours to days to compensate for changes in body-fluid pH, compared to the immediate responses of the buffer systems and the few-minute delay before the respiratory system responds. However, the kidneys are the most potent acid-base regulatory mechanism; not only can they vary H^+ removal, but they also can variably conserve or eliminate HCO_3^-, depending on the acid-base status of the body. For example, during renal compensation for acidosis, for each H^+ excreted in the urine, a new HCO_3^- is added to the plasma to buffer by means of the H_2CO_3: HCO_3^- system yet another H^+ that still remains in the body fluids. By simultaneously removing acid (H^+) from and adding base (HCO_3^-) to the body fluids, the kidneys are able to restore the pH toward normal more effectively than the lungs, which can adjust only the amount of H^+-forming CO_2 in the body.

Also contributing to the kidneys' acid-base regulatory potency is their ability to return the pH almost exactly to normal. In contrast, when some nonrespiratory abnormality has altered the $[H^+]$, the respiratory system alone can return the pH to only between 50% to 75% of the way toward normal; this is because the driving force governing the compensatory ventilatory response is diminished as the pH moves toward normal. As an example, ventilation is increased in response to a rise in arterial $[H^+]$, but as the $[H^+]$ is gradually reduced because of the stepped-up removal of H^+-forming CO_2, the ventilatory response is also gradually reduced. In comparison, the kidneys continue to respond to a change in pH until compensation is essentially complete.

The kidneys control the pH of the body fluids by adjusting three interrelated factors: (1) H^+ excretion, (2) HCO_3^- excretion, and (3) ammonia (NH_3) secretion.

HYDROGEN-ION EXCRETION. Carbonic, sulfuric, phosphoric, lactic, and fatty acids, among others, are continuously being added to the body fluids as a result of metabolic activities, yet the generated H^+ must not be allowed to accumulate. Although the body's buffer systems can resist changes in pH by removing H^+ from solution, the persistent production of acidic metabolic products would eventually overwhelm the limits of this buffering capacity. The constantly generated H^+ therefore must ultimately be eliminated from the body. The lungs are only able to remove carbonic acid from the body through the elimination of CO_2. The task of eliminating H^+ derived from sulfuric, phosphoric, and organic acids rests with the kidneys. Since the kidneys normally excrete H^+, urine is usually acidic, having an average pH of 6.0. Not only do the kidneys continuously eliminate the normal amount of H^+ that is constantly being produced from non-carbonic acid sources, but they also alter their rate of H^+ secretion to compensate for changes in $[H^+]$ arising from abnormalities in the concentration of carbonic acid.

Almost all of the excreted H^+ enters the urine by means of secretion. Recall that the filtration rate of H^+ equals plasma $[H^+]$ times GFR. Since plasma $[H^+]$ is extremely low (less than in pure H_2O except during extreme acidosis, when the pH falls below 7.0), the filtration rate of H^+ is likewise extremely low. This minute amount of filtered H^+ is excreted in the urine. However, the majority of excreted H^+ gains entry into the tubular fluid by being secreted.

The magnitude of H^+ secretion depends on a direct effect of the plasma's acid-base status on the kidneys' tubular cells (Fig. 15–11). No neural or hormonal control is involved. When the $[H^+]$ of the plasma passing through the peritubular capillaries is elevated above normal, the tubular cells respond by secreting greater-than-usual amounts of H^+ from the plasma into the tubular fluid to be excreted in the urine. Thus the kidneys reduce the plasma $[H^+]$ toward normal by excreting a more acidic urine than usual. Conversely, when plasma $[H^+]$ is lower than normal, the kidneys conserve H^+ by reducing its secretion and subsequent excretion in the urine. As a result of lowered H^+ excretion, the conserved H^+ helps restore plasma $[H^+]$ toward normal. Unlike the case with Na^+ and other mineral electrolytes, the kidneys are unable to raise plasma $[H^+]$ by reabsorbing more of the filtered H^+ because there are no reabsorptive mechanisms for H^+. The only way the kidneys can reduce H^+ excretion is by secreting less H^+.

The H^+ secretory process begins in the tubular cells with CO_2 that has come from three sources: CO_2 that has diffused into the tubular cells from either the plasma or tubular fluid or CO_2 that has been metabolically produced within the tubular cells. Under the influence of carbonic anhydrase, CO_2 and H_2O form H_2CO_3, which dissociates into H^+ and HCO_3^-.

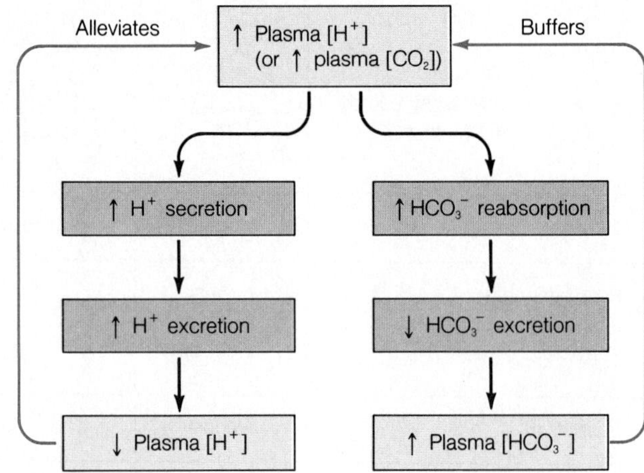

Figure 15–11 Control of Rate of Tubular H^+ Secretion

An energy-dependent carrier in the luminal membrane then transports H^+ out of the cell into the tubular lumen. In part of the nephron, this carrier transports Na^+ derived from the glomerular filtrate in the opposite direction, so H^+ secretion and Na^+ reabsorption are partially linked.

Because the chemical reactions for H^+ secretion begin with CO_2, the rate at which they proceed is influenced by $[CO_2]$. When plasma $[CO_2]$ increases, these reactions proceed more rapidly and the rate of H^+ secretion speeds up (Fig. 15–11). Conversely, the rate of H^+ secretion is reduced when plasma $[CO_2]$ falls below normal. This is especially important in renal compensations for acid-base abnormalities involving a change in H_2CO_3 caused by respiratory dysfunction. The kidneys can therefore adjust H^+ excretion to compensate for changes in both carbonic and non-carbonic acids.

BICARBONATE EXCRETION. Before being eliminated by the kidneys, the H^+ generated from non-carbonic acids is buffered to a large extent by plasma HCO_3^-. Appropriately, therefore, renal handling of acid-base balance also involves the adjustment of HCO_3^- excretion, depending on the H^+ load in the plasma.

The kidneys regulate plasma $[HCO_3^-]$ by two interrelated mechanisms: (1) variable reabsorption of the filtered HCO_3^- back into the plasma, and (2) variable addition of new HCO_3^- to the plasma. Both of these mechanisms are inextricably linked with H^+ secretion by the kidney tubules. Every time a H^+ is secreted into the tubular fluid, a HCO_3^- is simultaneously transferred into the peritubular-capillary plasma. Whether a filtered HCO_3^- is reabsorbed or a new HCO_3^- is added to the plasma in accompaniment with H^+ secretion de-

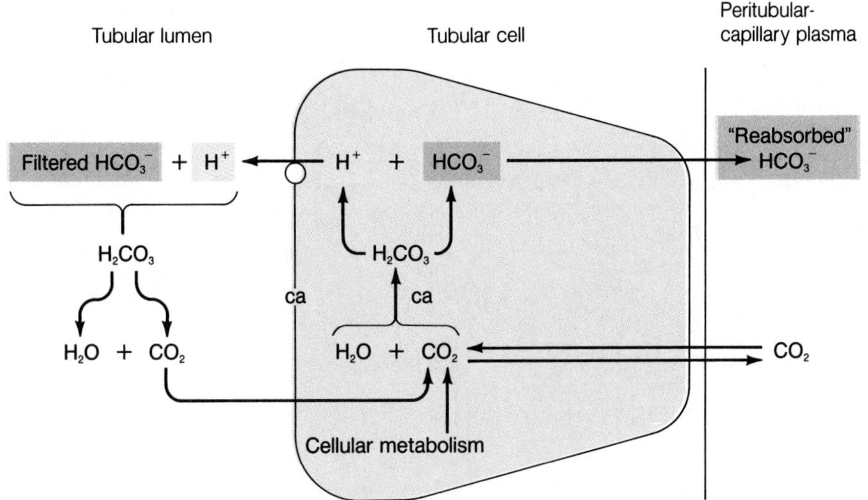

Figure 15–12 Bicarbonate Reabsorption Coupled with H^+ Secretion *Since the disappearance of a HCO_3^- from the tubular fluid is coupled with the appearance of another HCO_3^- in the plasma, HCO_3^- is considered to have been "reabsorbed."*

pends on whether or not filtered bicarbonate ions are present in the tubular fluid to react with secreted hydrogen ions.

Bicarbonate is freely filtered, but because the luminal membranes of the tubular cells are impermeable to the filtered HCO_3^-, it cannot diffuse into these cells. Therefore, reabsorption of HCO_3^- must occur indirectly (Fig. 15–12). Hydrogen ion secreted into the tubular fluid combines with the filtered HCO_3^- to form H_2CO_3. Under the influence of carbonic anhydrase, which is present on the surface of the luminal membrane, H_2CO_3 decomposes into CO_2 and H_2O within the filtrate. Unlike HCO_3^-, CO_2 can easily penetrate the tubular cell membranes. Within the cells, CO_2 and H_2O, under the influence of intracellular carbonic anhydrase, form H_2CO_3, which dissociates into H^+ and HCO_3^-. Because HCO_3^- can permeate the tubular cells' basolateral membrane, it passively diffuses out of the cells and into the peritubular-capillary plasma. Meanwhile, the generated H^+ is actively secreted. Since the disappearance of a HCO_3^- from the tubular fluid is coupled with the appearance of another HCO_3^- in the plasma, a HCO_3^- has, in effect, been "reabsorbed." Even though the HCO_3^- entering the plasma is not the same HCO_3^- that was filtered, the net result is the same as if HCO_3^- were directly reabsorbed.

Normally, slightly more hydrogen ions are secreted into the tubular fluid than bicarbonate ions are filtered. Accordingly, all of the filtered HCO_3^- is usually reabsorbed because secreted H^+ is available in the tubular fluid to combine with it to form highly reabsorbable CO_2 (Fig. 15–13). The vast majority of the secreted H^+ combines with HCO_3^- and is not excreted because it is "used up" in the process of HCO_3^- reabsorption. However, the slight excess of secreted H^+ that is not matched by filtered HCO_3^- is excreted in the urine. This normal H^+ excretion rate keeps pace with the normal rate of noncarbonic acid H^+ production.

Secretion of H^+ that is *excreted* is coupled with the ***addition of new HCO_3^-*** to the plasma, in contrast to the secreted H^+ that is coupled with ***HCO_3^- reabsorption*** and is ***not excreted***, instead being incorporated into H_2O molecules. When all of the filtered HCO_3^- has been reabsorbed, additional secreted H^+ is generated by the dissociation of H_2CO_3, which has been formed in the tubular cells from H_2O plus CO_2 derived from the plasma or metabolically produced within the cell (Fig. 15–14). The HCO_3^- produced by this reaction diffuses into the plasma as a "new" HCO_3^-, because its appearance in the plasma is not associated with reabsorption of filtered HCO_3^-. Meanwhile, the secreted H^+ combines with urinary buffers, especially basic phosophate ($HPO_4^{=}$), and is excreted.

During alkalosis, the rate of H^+ secretion diminishes while the rate of HCO_3^- filtration increases compared to normal (Fig. 15–15). When the plasma $[H^+]$ is below normal, a smaller proportion of the HCO_3^- pool is tied up buffering H^+, so the plasma $[HCO_3^-]$ is elevated above normal. As a result, the rate of HCO_3^- filtration is correspondingly increased. Not all of the filtered HCO_3^- is reabsorbed, because bicarbonate ions are in excess of secreted hydrogen ions in the tubular fluid and HCO_3^- cannot be reabsorbed without first reacting with H^+. The excess HCO_3^- is left in the tubular fluid to be excreted in the urine, thus reducing the plasma

Figure 15-13 Fate of Secreted H^+ and Filtered HCO_3^- at Normal pH *Most secreted H^+ combines with filtered HCO_3^- and is consumed in the process of HCO_3^- "reabsorption." (See Figure 15-12 for the reactions involved.) Only H^+ secreted in excess of filtered HCO_3^- is excreted in the urine. Secretion of H^+ is accompanied by the addition of new HCO_3^- to the plasma. (See Figure 15-14 for the reactions involved.)*

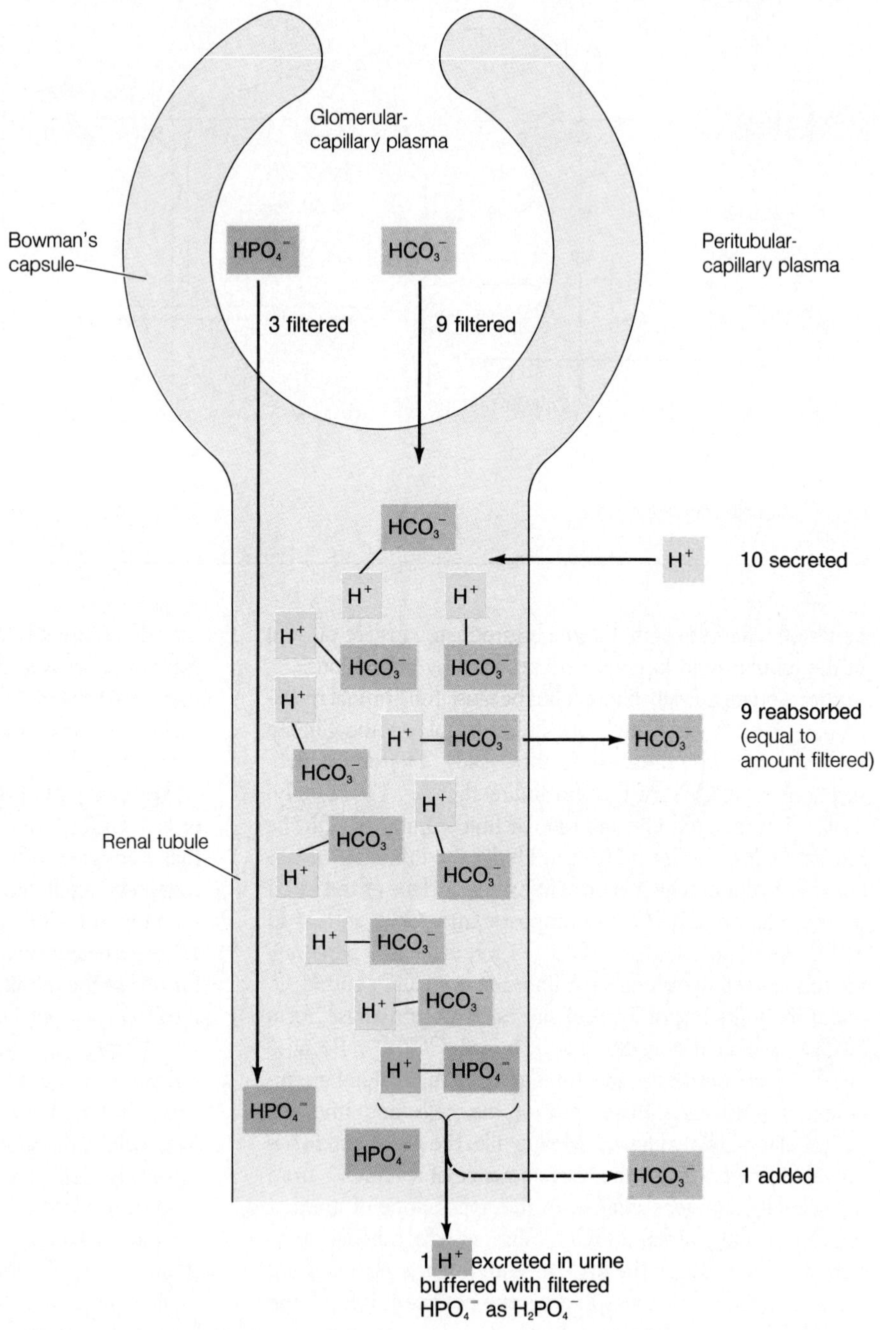

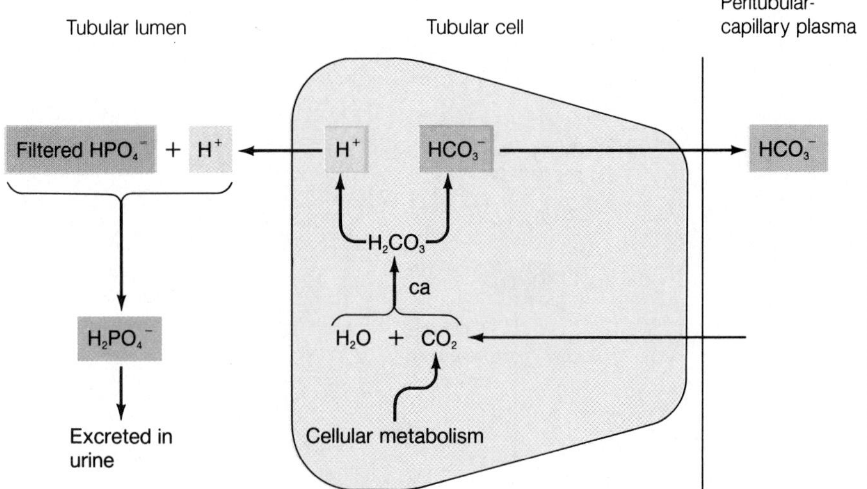

Figure 15–14 Hydrogen Ion Secretion and Excretion Coupled with Addition of New HCO_3^- to Plasma *Secreted H^+ does not combine with filtered $HPO_4^=$ and is not subsequently excreted until all of the filtered HCO_3^- has been "reabsorbed," as depicted in Figure 15–12. Once all of the filtered HCO_3^- has combined with secreted H^+, further secreted H^+ is excreted in the urine, primarily in association with urinary buffers such as basic phosphate. Excretion of H^+ is coupled with the net gain of new HCO_3^- to the plasma, instead of being merely a replacement for filtered HCO_3^-.*

$[HCO_3^-]$ while causing the urine to become alkaline. This reduction in the plasma concentration of the alkaline member of the H_2CO_3: HCO_3^- buffer pair is important in permitting the continuation of the buffering response in the internal environment. When plasma $[H^+]$ decreases, the reaction $H^+ + HCO_3^- \rightleftharpoons H_2CO_3 \rightleftharpoons CO_2 + H_2O$ is driven toward the left side of the equation to generate more H^+. However, since additional HCO_3^- is generated at the same time, the plasma HCO_3^- pool is expanded. The resultant increase in plasma $[HCO_3^-]$ drives the reaction toward the right side of the equation in opposition to the driving force established by the decrease in $[H^+]$. This counterproductive effect is corrected by the kidneys' removal of HCO_3^-. In fact, plasma $[HCO_3^-]$ is even reduced below normal. The decreased $[HCO_3^-]$ drives the reaction to the left, in the same direction as the decreased $[H^+]$, to facilitate the generation of H^+ from dissolved CO_2. Meanwhile, respiratory activity is suppressed by the alkalotic condition, so metabolically produced CO_2 is allowed to accumulate and be available for H^+ generation by the H_2CO_3: HCO_3^- buffer system. If $[CO_2]$ were to become reduced, further H^+ generation would cease. Collectively, these mechanisms raise the plasma $[H^+]$ to normal.

The opposite situation occurs during acidosis. When the plasma $[H^+]$ is elevated, more H^+ is secreted than normal. At the same time, less HCO_3^- is filtered than normal because more of the plasma HCO_3^- is used up in buffering the excess H^+ as HCO_3^- plus H^+ is converted to H_2CO_3 (Fig. 15–16). There are two consequences of this greater-than-usual inequity between filtered HCO_3^- and secreted H^+. First, more of the secreted H^+ is excreted in the urine because more hydrogen ions are entering the tubular fluid at a time when fewer of them are needed to reabsorb the reduced quantities of filtered HCO_3^-. In this way, extra H^+ is eliminated from the body, making the urine more acidic than normal. Second, because the excretion of H^+ is linked with the addition of new HCO_3^- to the plasma, more HCO_3^- than usual enters the plasma that is passing through the kidneys. This additional HCO_3^- is available to buffer the excess H^+ present in the body, helping to shift the $H^+ + HCO_3^- \rightleftharpoons CO_2 + H_2O$ reaction (ignoring the intervening H_2CO_3 step) to the CO_2 side so that H^+ is safely incorporated within H_2O molecules. The generated CO_2 is then removed from the body by means of increased ventilation. If CO_2 were allowed to accumulate, it would oppose the drive of the buffer reaction away from the acid side. Mutually, these responses decrease plasma $[H^+]$ to normal.

In short, when plasma $[H^+]$ is reduced below normal during ***alkalosis***, renal responses include the following (Table 15–5):

1. decreased secretion and subsequent reduced excretion of H^+ in the urine, resulting in conservation of H^+ and increased plasma $[H^+]$; and
2. incomplete reabsorption of filtered HCO_3^- and subsequent increased excretion of HCO_3^-, resulting in reduction of plasma $[HCO_3^-]$.

Figure 15–15 Fate of Secreted H^+ and Filtered HCO_3^- during Alkalosis *Since less H^+ is secreted than HCO_3^- filtered, all of the secreted H^+ is consumed in the process of HCO_3^- "reabsorption." Filtered HCO_3^- that fails to be reabsorbed because it does not have a secreted H^+ with which to react is excreted in the urine. Compared to normal (see Figure 15–13), urinary excretion of H^+ decreases and excretion of HCO_3^- increases during alkalosis.*

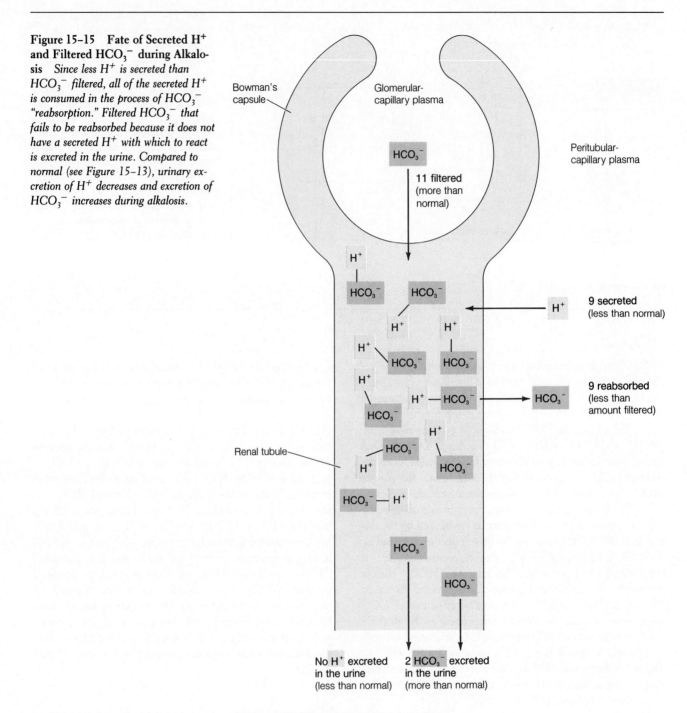

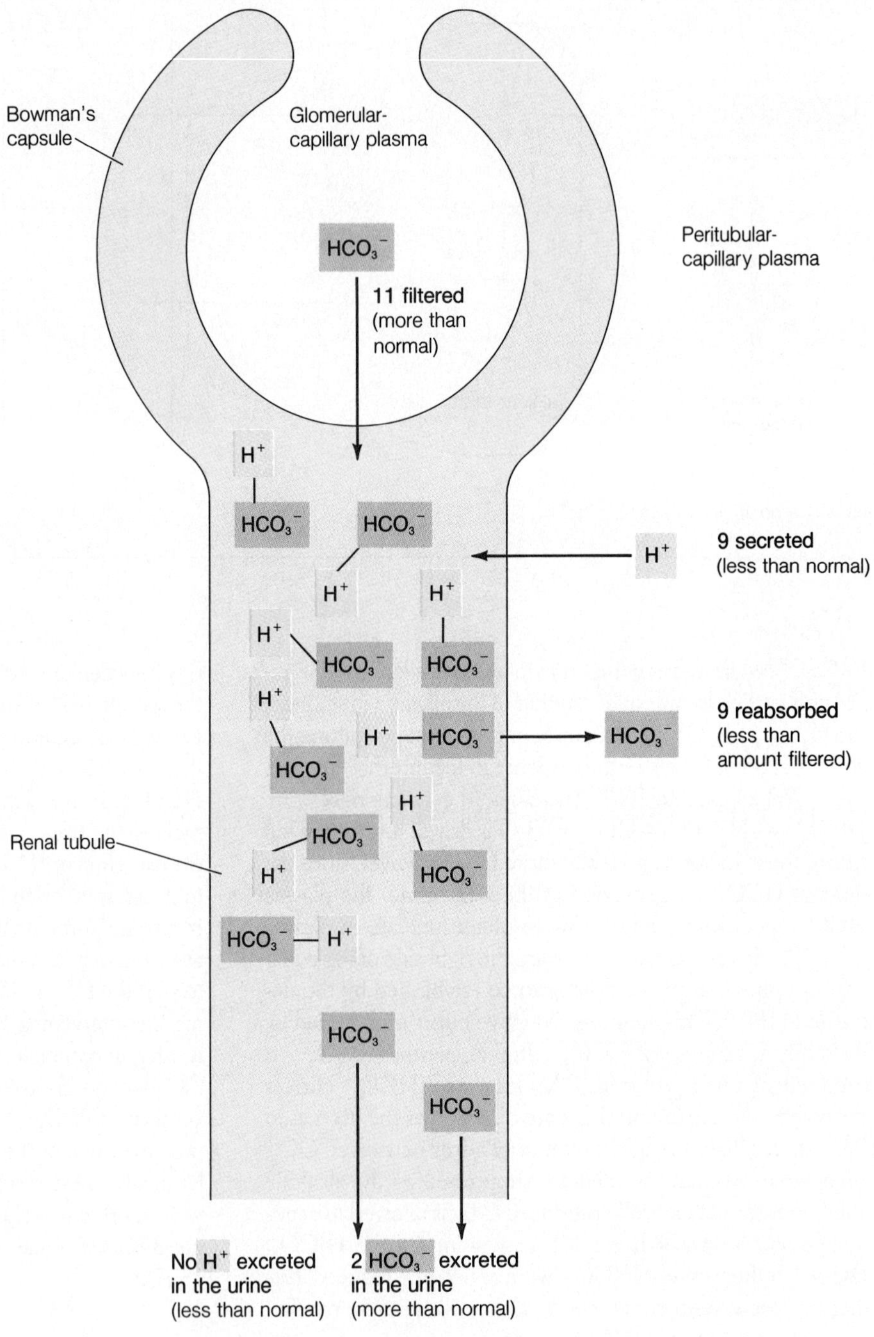

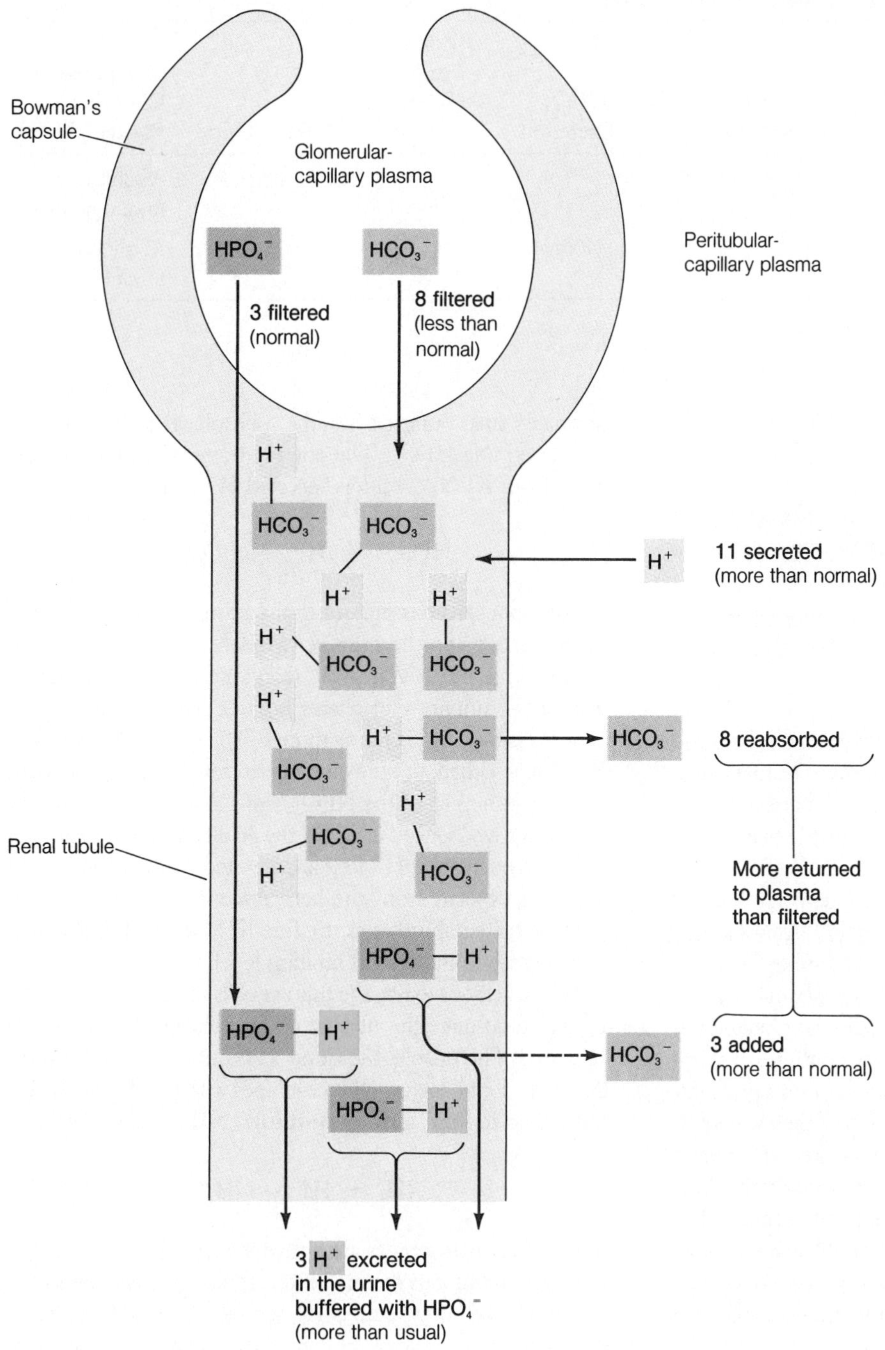

Figure 15–16 Fate of Secreted H^+ and Filtered HCO_3^- during Acidosis *Since less HCO_3^- is filtered than normal, less secreted H^+ is required to "reabsorb" filtered HCO_3^-, just at a time when extra H^+ is being secreted. Therefore, additional secreted H^+ is excreted. Excretion of extra secreted H^+ is coupled with addition of extra HCO_3^- to the plasma. Compared to normal (see Figure 15–13), urinary excretion of H^+ increases and addition of HCO_3^- to the plasma increases during acidosis.*

Table 15-5 Summary of Renal Responses to Alkalosis and Acidosis

Acid-Base Abnormality	H^+ Secretion	H^+ Excretion	HCO_3^- Reabsorption	HCO_3^- Excretion	Addition of New HCO_3^- to Plasma	pH of Urine	Compensatory Change in Plasma pH
Alkalosis	↓	↓	↓	↑	↓	Alkaline	Acidification toward normal
Acidosis	↑	↑	↑	Normal	↑	Acidic	Alkalinization toward normal

When plasma $[H^+]$ is increased above normal during ***acidosis***, renal compensation includes the following:

1. increased secretion and subsequent increased excretion of H^+ in the urine, thereby eliminating the excess H^+ and decreasing plasma $[H^+]$; and
2. reabsorption of all of the filtered HCO_3^-, plus addition of new HCO_3^- to the plasma, resulting in increased plasma $[HCO_3^-]$.

Thus, to compensate for alkalosis, the kidneys make the urine alkaline in the process of acidifying the plasma to restore a normal pH. In the opposite case of acidosis, the kidneys acidify the urine and alkalinize the plasma to bring the pH to normal.

AMMONIA SECRETION. The energy-dependent H^+ carriers in the tubular cells are capable of secreting H^+ against a concentration gradient until the tubular fluid (urine) becomes eight hundred times more acidic than the plasma. At this point, further H^+ secretion ceases, because the gradient becomes too great for the secretory process to continue. It is impossible for the kidneys to acidify the urine beyond a gradient-limiting urinary pH of 4.5. Only about 1% of the excess H^+ typically excreted daily, if left unbuffered as free H^+, would produce a urinary pH of this magnitude at normal urine flow rates. Elimination of the other 99% of the usually secreted H^+ load would be prevented, a situation that would be intolerable. For H^+ secretion to proceed, the majority of secreted H^+ must be buffered in the tubular fluid so that it does not exist as free H^+ and, accordingly, does not contribute to tubular acidity.

Bicarbonate cannot buffer urinary H^+ as it does in the ECF, because HCO_3^- is not excreted in the urine simultaneously with H^+. Whichever one of these substances is in excess in the tubular fluid is excreted in the urine. There are, however, two important urinary buffers: (1) filtered phosphate buffers and (2) secreted ammonia (NH_3). Secreted H^+ is normally first buffered by the phosphate buffer system, NaH_2PO_4: Na_2HPO_4. The anionic portion of ionized basic phosphate, $HPO_4^=$, buffers secreted H^+ as follows:

$$HPO_4^= + H^+ \rightarrow H_2PO_4^-$$

Basic phosphate is present in the tubular fluid by virtue of dietary excess, not because of any deliberate mechanism to buffer secreted H^+. When H^+ secretion is high, the buffering capacity of urinary phosphates is exceeded, but the kidneys cannot respond by excreting more $PO_4^{\equiv}$. Only the quantity of $PO_4^{\equiv}$ reabsorbed, not the quantity excreted, is subject to control. As soon as all of the $HPO_4^=$ ions that are coincidentally excreted have "soaked up" H^+, the acidity of the tubular fluid quickly rises as more H^+ ions are secreted. Without additional buffering capacity from another source, H^+ secretion would soon be halted abruptly as the free $[H^+]$ in the tubular fluid quickly rose to the critical limiting level.

When acidosis exists, the tubular cells secrete NH_3 into the tubular fluid once the normal urinary phosphate buffers are saturated. This enables the kidneys to continue secreting additional H^+ ions because NH_3 combines with free H^+ in the tubular fluid to form **ammonium ion (NH_4^+)** as follows:

$$NH_3 + H^+ \rightarrow NH_4^+$$

The tubular membranes are not very permeable to NH_4^+, so the ammonium ions remain in the tubular fluid and are lost in the urine, each one taking a H^+ with it. Thus NH_3 secretion serves to buffer excess H^+ in the tubular fluid, so large amounts of H^+ can be secreted into the urine before the pH falls to the limiting value of 4.5. Were it not for NH_3 secretion, the extent of H^+ secretion would be limited to whatever phosphate buffering capacity coincidentally happened to be present as a result of dietary excess.

In contrast to the phosphate buffers, which are in the tubular fluid because they have been filtered but not reabsorbed, NH_3

is deliberately synthesized from the amino acid glutamine within the tubular cells, from which it readily diffuses down its concentration gradient into the tubular fluid. The rate of NH_3 secretion is controlled by a direct effect on the tubular cells of the amount of excess H^+ to be transported in the urine. When an individual has been acidotic for more than two or three days, the production rate of NH_3 increases substantially. This extra NH_3 provides additional buffering capacity to allow H^+ secretion to continue after the normal phosphate buffering capacity is overwhelmed during renal compensation for acidosis.

Acid-base imbalances can arise from either respiratory dysfunction or metabolic disturbances.

Deviations from normal acid-base status are divided into four general categories, depending on the source and direction of the abnormal change in $[H^+]$. These categories are respiratory acidosis, respiratory alkalosis, metabolic acidosis, and metabolic alkalosis.

Because of the relationship between $[H^+]$ and the concentrations of the members of a buffer pair, changes in $[H^+]$ are reflected by changes in the ratio of $[HCO_3^-]$ to $[CO_2]$. Determinations of $[HCO_3^-]$ and $[CO_2]$ provide more meaningful information about the underlying factors responsible for a particular acid-base status than do direct measurements of $[H^+]$ alone. The following rules of thumb apply when examining acid-base imbalances *before any compensations take place:*

1. A change in pH that has a respiratory cause will be associated with an abnormal $[CO_2]$, giving rise to a change in carbonic acid-generated H^+. In contrast, a deviation in pH of metabolic origin will be associated with an abnormal $[HCO_3^-]$ as a result of the participation of HCO_3^- in buffering abnormal amounts of H^+ generated from non-carbonic acids.
2. Anytime the $[HCO_3^-]/[CO_2]$ ratio falls below 20/1, an acidosis exists; anytime the ratio exceeds 20/1, an alkalosis exists.

Putting these two points together:

□ *Respiratory acidosis* displays a ratio of less than 20/1 arising from an increase in $[CO_2]$.

□ *Respiratory alkalosis* has a ratio of greater than 20/1 because of a decrease in $[CO_2]$.

□ *Metabolic acidosis* presents a ratio of less than 20/1 associated with a fall in $[HCO_3^-]$.

□ *Metabolic alkalosis* has a ratio of greater than 20/1 arising from an elevation in $[HCO_3^-]$.

We will examine each of these categories separately in more detail, paying particular attention to possible causes and the compensations that occur. The "balance beam" concept, as presented in Figure 15–17 in conjunction with the Henderson-Hasselbalch equation, will help you to better visualize the contributions of the lungs and kidneys in terms of the causes and compensations of various acid-base disorders. The normal situation is represented in Figure 15–17a.

RESPIRATORY ACIDOSIS. **Respiratory acidosis** is the result of abnormal CO_2 retention arising from hypoventilation (see p. 448). As less-than-normal amounts of CO_2 are lost through the lungs, the resultant increase in H_2CO_3 formation and dissociation leads to an elevated $[H^+]$. Possible causes include lung disease, depression of the respiratory centers by drugs or disease, nerve or muscle disorders that reduce respiratory muscle ability, or (transiently) even the simple act of holding one's breath.

In uncompensated respiratory acidosis (Fig. 15–17b), $[CO_2]$ is elevated (in our example, it is doubled) while $[HCO_3^-]$ is normal, so the ratio is 20/2 (10/1) and pH is reduced. Compensatory measures act to restore pH to normal. The chemical buffers immediately take up additional H^+, but the respiratory mechanism is usually not able to respond with compensatory increased ventilation because an impairment in respiratory activity is the problem in the first place. Thus the kidneys are most important in compensating for respiratory acidosis. They conserve all of the filtered HCO_3^- and add new HCO_3^- to the plasma while they simultaneously secrete and, accordingly, excrete more H^+.

As a result, HCO_3^- stores in the body become elevated. In our example (Fig. 15–17c), the plasma $[HCO_3^-]$ is doubled, so the $[HCO_3]/[CO_2]$ ratio is 40/2 rather than 20/2 as it was in the uncompensated state. A ratio of 40/2 is equivalent to a normal 20/1 ratio, so the pH is once again the normal 7.4. Enhanced renal conservation of HCO_3^- has fully compensated for CO_2 accumulation, thus restoring the pH to normal, although both the $[CO_2]$ and $[HCO_3^-]$ are now distorted. Note that maintenance of normal pH depends on preserving a normal ratio between $[HCO_3^-]$ and $[CO_2]$, no matter what the absolute values of each of these buffer components are. (Compensation is never fully complete because the pH can be restored close to but not all the way to normal. However, in our examples, we assume full compensation for ease in mathematical calculations. Also bear in mind that the values used are only representative. There is actually a range of pH deviations and variations in the degree to which compensation can be accomplished.)

RESPIRATORY ALKALOSIS. The primary defect in **respiratory alkalosis** is excessive loss of CO_2 from the body as a result of hyperventilation. When pulmonary ventilation increases out of proportion to the rate of CO_2 production, too much CO_2 is blown off. Consequently, less H_2CO_3 is formed and $[H^+]$ decreases.

Possible causes of respiratory alkalosis include fever, anxiety, and aspirin poisoning, all of which excessively stimulate ventilation without regard to the status of O_2, CO_2, or H^+ in the body fluids. Respiratory alkalosis also occurs as a result of physiological mechanisms at high altitude. When the low concentration of dissolved O_2 in the arterial blood reflexly stimulates ventilation in an attempt to obtain more O_2 for the body, too much CO_2 is blown off in the process, inadvertently leading to an alkalotic state.

Looking at the biochemical abnormalities in uncompensated respiratory alkalosis (Fig. 15–17d), the increase in pH reflects a reduction in $[CO_2]$ (half the normal value in our example) while the $[HCO_3^-]$ remains normal. This yields an alkalotic ratio of 20/0.5, which is comparable to 40/1.

Compensatory measures act to shift the pH back toward normal. The chemical buffer systems liberate H^+ to diminish the severity of the alkalosis. As plasma $[CO_2]$ and $[H^+]$ fall below normal because of excessive ventilation, two of the normally potent stimuli for driving ventilation are removed. This effect tends to "put brakes" on the extent to which some nonrespiratory related factor such as fever or anxiety can overdrive ventilation. Therefore, hyperventilation does not continue completely unabated. If the situation continues for a few days, the kidneys compensate by conserving H^+ and excreting more HCO_3^-. If, as in our example (Fig. 15–17e), the HCO_3^- stores are reduced in half by loss of HCO_3^- in the urine, the $[HCO_3^-]/[CO_2]$ ratio becomes 10/0.5, equivalent to the normal 20/1. Therefore, the pH is restored to normal by reducing the HCO_3^- load to compensate for the CO_2 loss.

METABOLIC ACIDOSIS. **Metabolic acidosis** (also known as **nonrespiratory acidosis**) encompasses all types of acidosis besides that caused by excess CO_2 in the body fluids. In the uncompensated state (Fig. 15–17f), metabolic acidosis is always characterized by a reduction in plasma $[HCO_3^-]$ (in our example it is halved), while $[CO_2]$ remains normal, producing an acidotic ratio of 10/1. The problem may arise from excessive loss of bicarbonate-rich fluids from the body or from an accumulation of non-carbonic acids. In the latter case, plasma HCO_3^- is used up in the process of buffering the additional H^+.

This type of acid-base disorder is the one more frequently encountered. The following are the most common causes.

1. *Severe diarrhea*—During the process of digestion, large amounts of $NaHCO_3$ normally are secreted into the digestive tract and are subsequently reabsorbed back into the plasma when digestion is completed. During diarrhea, this $NaHCO_3$ is lost from the body rather than being reabsorbed. The reduction in plasma $[HCO_3^-]$ without a corresponding reduction in $[CO_2]$ lowers the pH. Looking at the situation differently, loss of HCO_3^- shifts the $H^+ + HCO_3^- \rightleftharpoons CO_2 + H_2O$ reaction to the left to compensate for the HCO_3^- deficit, increasing $[H^+]$ above normal in the process.
2. *Diabetes mellitus*—Abnormal fat metabolism resulting from the inability of cells to preferentially use glucose in the absence of insulin results in the formation of excess keto acids, whose dissociation causes an increase in plasma $[H^+]$.
3. *Strenuous exercise*—When muscles resort to anaerobic glycolysis during strenuous exercise (see p. 237), excess lactic acid is produced, leading to a rise in plasma $[H^+]$.
4. *Uremic acidosis*—In severe renal failure (uremia), the kidneys are unable to rid the body of even the normal amounts of H^+ generated from the non-carbonic acids formed by the body's ongoing metabolic processes, so H^+ starts to accumulate in the body fluids. Also, the kidneys are unable to conserve an adequate amount of HCO_3^- to be used for buffering the normal acid load. The resultant fall in $[HCO_3^-]$ without a concomitant reduction in $[CO_2]$ is correlated with a decline in pH to an acidotic level.

Except in uremic acidosis, metabolic acidosis is compensated for by both respiratory and renal mechanisms as well as by chemical buffers. The buffers take up extra H^+, the lungs blow off additional H^+-generating CO_2, and the kidneys excrete more H^+ and conserve more HCO_3^-. In our example (Fig. 15–17g), these compensatory measures restore the ratio to normal by reducing $[CO_2]$ to 75% of normal and by improving the reduction in $[HCO_3^-]$ halfway back toward normal (from 50% up to 75% of the normal value). This brings the ratio to 15/0.75 (equivalent to 20/1).

Note that in compensating for metabolic acidosis, the lungs deliberately displace $[CO_2]$ from normal in an attempt to restore $[H^+]$ toward normal. Whereas in respiratory-induced

Figure 15–17 Schematic Representation of Relationship of HCO_3^- and CO_2 Concentrations to pH in Various Acid-Base Statuses *(a) Normal acid-base balance. (b) Uncompensated respiratory acidosis. (c) Compensated respiratory acidosis. (d) Uncompensated respiratory alkalosis. (e) Compensated respiratory alkalosis. (f) Uncompensated metabolic acidosis. (g) Compensated metabolic acidosis. (h) Uncompensated metabolic alkalosis. (i) Compensated metabolic alkalosis.*

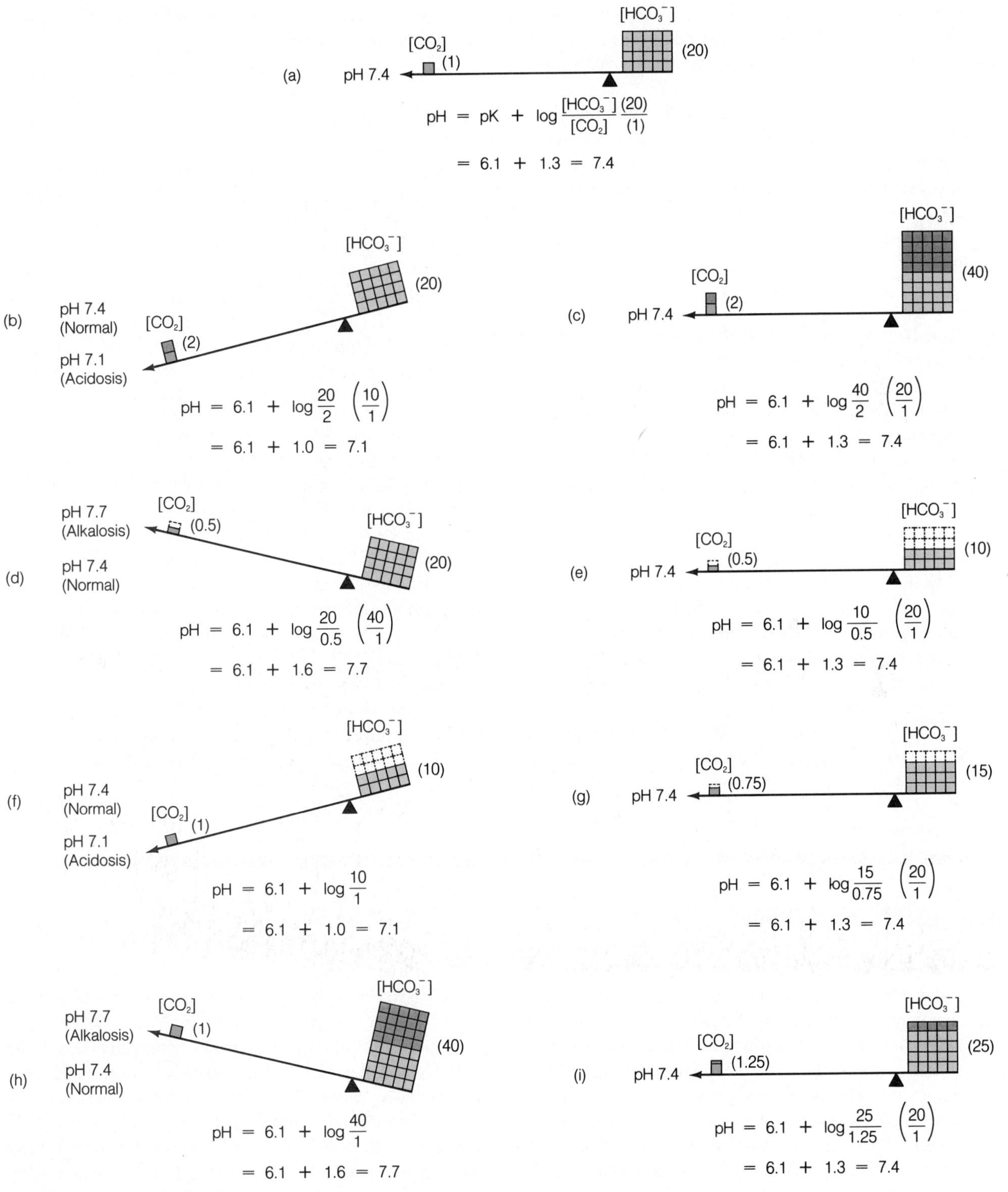

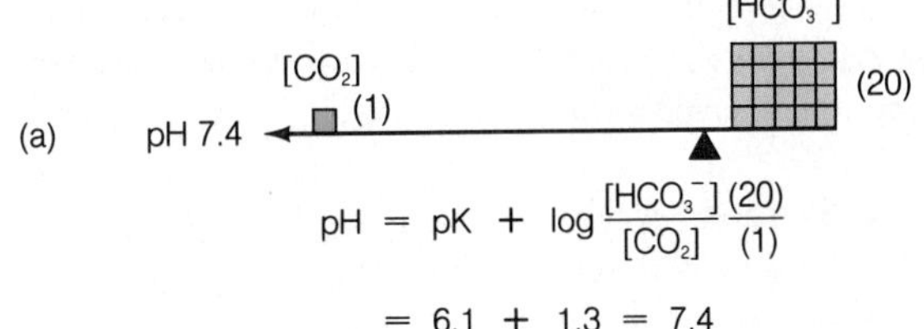

(a)
pH 7.4
[CO2]
(1)
[HCO3−]
(20)
pH = pK + log [HCO3−]/[CO2] (20)/(1)
= 6.1 + 1.3 = 7.4
(b)
pH 7.4
(Normal)
pH 7.1
(Acidosis)
[CO2]
(2)
[HCO3−]
(20)
pH = 6.1 + log 20/2 (10/1)
= 6.1 + 1.0 = 7.1
(c)
pH 7.4
[CO2]
(2)
[HCO3−]
(40)
pH = 6.1 + log 40/2 (20/1)
= 6.1 + 1.3 = 7.4
(d)
pH 7.7
(Alkalosis)
pH 7.4
(Normal)
[CO2]
(0.5)
[HCO3−]
(20)
pH = 6.1 + log 20/0.5 (40/1)
= 6.1 + 1.6 = 7.7
(e)
pH 7.4
[CO2]
(0.5)
[HCO3−]
(10)
pH = 6.1 + log 10/0.5 (20/1)
= 6.1 + 1.3 = 7.4
(f)
pH 7.4
(Normal)
pH 7.1
(Acidosis)
[CO2]
(1)
[HCO3−]
(10)
pH = 6.1 + log 10/1
= 6.1 + 1.0 = 7.1
(g)
pH 7.4
[CO2]
(0.75)
[HCO3−]
(15)
pH = 6.1 + log 15/0.75 (20/1)
= 6.1 + 1.3 = 7.4
(h)
pH 7.7
(Alkalosis)
pH 7.4
(Normal)
[CO2]
(1)
[HCO3−]
(40)
pH = 6.1 + log 40/1
= 6.1 + 1.6 = 7.7
(i)
pH 7.4
[CO2]
(1.25)
[HCO3−]
(25)
pH = 6.1 + log 25/1.25 (20/1)
= 6.1 + 1.3 = 7.4

The lengths of the arms of the balance beams are not to scale.

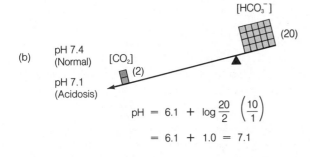

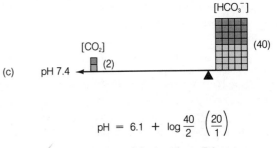

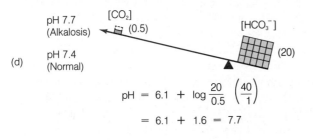

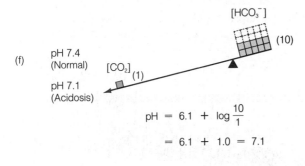

acid-base disorders, an abnormal [CO_2] is the *cause* of the [H^+] imbalance, in metabolic acid-base disorders, [CO_2] is intentionally shifted from normal as an important *compensation* for the [H^+] imbalance.

When kidney disease is the cause of metabolic acidosis, complete compensation is not possible because the renal mechanism is not available for pH regulation. Recall that the respiratory system is capable of compensating only up to 75% of the way toward normal. Uremic acidosis is very serious, because the kidneys cannot help to restore the pH all the way to normal.

METABOLIC ALKALOSIS. **Metabolic alkalosis (nonrespiratory alkalosis)** is a reduction in plasma [H^+] caused by a relative deficiency of non-carbonic acids. This acid-base disturbance is associated with an increase in [HCO_3^-], which, in the uncompensated state, is not accompanied by a change in [CO_2]. In our example (Fig. 15–17h), [HCO_3^-] is doubled, producing an alkalotic ratio of 40/1. This condition arises most commonly from the following:

1. Ingestion of alkaline drugs, such as when baking soda ($NaHCO_3$) is used as a self-administered remedy for the treatment of gastric hyperacidity. By neutralizing excess acid in the stomach, HCO_3^- relieves the symptoms of stomach irritation and heartburn, but when more HCO_3^- is ingested than needed, the extra HCO_3^- is absorbed from the digestive tract and increases plasma [HCO_3^-]. An alkalosis results because the rise in [HCO_3^-] drives the $H^+ + HCO_3^- \rightleftharpoons CO_2 + H_2O$ reaction to the right, using up some of the H^+ present in the plasma from non-carbonic acid sources.
2. Abnormal loss of H^+ from the body as a result of vomiting of acidic gastric juices. The process of HCl secretion in the stomach involves the addition of HCO_3^- to the plasma. This HCO_3^- is neutralized by H^+ as the gastric secretions are eventually reabsorbed back into the plasma, so normally there is no net addition of HCO_3^- to the plasma from this source. However, when this acid is lost from the body during vomiting, not only is plasma [H^+] decreased, but also reabsorbed H^+ is no longer available to neutralize the extra HCO_3^- added to the plasma during gastric HCl secretion. Thus loss of HCl in effect increases plasma [HCO_3^-]. (On the other hand, with "deeper" vomiting, $NaHCO_3$ from the upper intestine may be lost in the vomitus, resulting in an acidosis instead of an alkalosis.)

In metabolic acidosis, the chemical buffer systems immediately liberate H^+, and ventilation is reduced so that extra H^+-generating CO_2 is retained in the body fluids. If the condition persists for several days, the kidneys conserve H^+ and excrete the excess HCO_3^- in the urine. The resultant compensatory increase in [CO_2] (up 25% in our example—Fig. 15–17i) and the partial reduction in [HCO_3^-] (75% of the way back down toward normal in our example) together restore the [HCO_3^-]/[CO_2] ratio back to the equivalent of 20/1 at 1.25/25.

It should be obvious that assessing an individual's acid-base status cannot be made on the basis of pH alone. Even though the pH is essentially normal, determinations of [HCO_3^-] and [CO_2] can reveal compensated acid-base disorders.

Chapter in Perspective

Homeostasis depends on maintaining a balance between the input and output of all constituents present in the internal fluid environment. Regulation of fluid balance involves two separate components: control of salt balance and control of H_2O balance. Control of salt balance is primarily important in the long-term regulation of blood pressure because of the effect of the body's salt load on the osmotic determination of the ECF volume, of which plasma volume is a part. Salt output in the urine is constantly adjusted to match unregulated, variable salt intake, primarily by controlling the amount of salt reabsorbed from the renal tubules. Tubular reabsorption of salt is subject to control by the adrenal-cortex hormone aldosterone.

Control of H_2O balance is important in preventing changes in ECF osmolarity, which would induce detrimental shifts of H_2O between the cells and the ECF. As with salt balance, H_2O balance is largely maintained by controlling the volume of H_2O lost in the urine to compensate for uncontrolled losses of variable volumes of H_2O from other avenues, such as through sweating or diarrhea, and for poorly regulated H_2O intake. Even though a thirst mechanism exists to control H_2O intake based on need, the amount drunk is often influenced by social custom and habit instead of thirst alone. The volume of H_2O lost in the urine is regulated by the hypothalamic hormone vasopressin, which determines the extent of H_2O reabsorption from the renal tubules.

Similar to the case of salt and H_2O balance, control of H^+ output by the kidneys is the main regulatory factor in achieving a balance between input and output of H^+ to maintain acid-base balance within the narrow limits compatible with life. Hydrogen ion is uncontrollably, continually being added to the body fluids as a result of ongoing metabolic activities, yet the plasma pH must be kept constant at a slightly alkaline level of 7.4 for optimal body function. Deviations in the internal fluid environment's pH lead to altered neuromuscular excitability, to changes in enzymatically controlled metabolic activity, and to K^+ imbalances. These effects are fatal if the pH falls outside of the range of 6.8 to 8.0.

By means of changes in the rate of tubular H^+ secretion and elimination, which is tightly coupled with the reabsorption and conservation of the important buffer component HCO_3^-, the kidneys can almost completely compensate for a wide range of acid-base abnormalities. Assisting the kidneys in eliminating H^+ are the lungs, which can adjust their rate of excretion of H^+-generating CO_2. The concept that the same input-output balance that applies to salt and H_2O is essential for maintaining H^+ homeostasis is obscured by the fact that H^+ is buffered in the body. Such buffering prevents H^+ ions from producing harmful fluctuations in pH while they are present in the body. Furthermore, the buffer mechanism can also generate "new" H^+ to temporarily compensate for a H^+ deficit in the body fluids. Such a buffering system, which can can take up or liberate H^+ to transiently keep its concentration constant within the body until its output can be brought into line with its input, is not available for salt or H_2O balance.

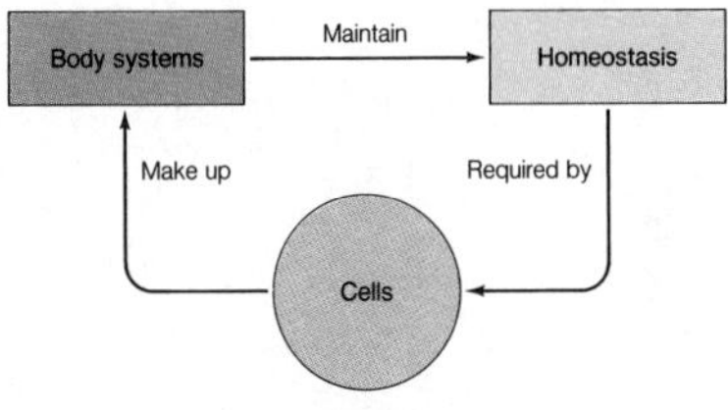

See inside front cover for an expanded version of this model.

Review Exercises

1. Explain the balance concept. What factors can alter the internal pool of a particular substance?
2. Outline the distribution of body H_2O. What factors are responsible for accomplishing exchange across the barriers that separate the body-fluid compartments?
3. What factors are regulated to maintain the body's fluid balance?
4. Why is regulation of ECF volume important? How is it regulated?
5. Why is regulation of ECF osmolarity important? How is it regulated?
6. What are the causes and consequences of ECF hypertonicity and of ECF hypotonicity? Why is there no fluid shift between the ECF and ICF with an isotonic fluid gain or loss?
7. Outline the sources of input and output in a daily H_2O balance and a daily salt balance. Which are subject to control to maintain the body's fluid balance?
8. What is the relationship between $[H^+]$ and pH? How are $[HCO_3^-]$ and $[CO_2]$ related to pH?
9. What is the normal pH of body fluids? How does this compare to the pH of H_2O? Define acidosis and alkalosis.
10. What are the consequences of fluctuations in $[H^+]$?
11. What are the body's sources of H^+?
12. Describe the three lines of defense against changes in $[H^+]$ in terms of mechanisms and speed of action.
13. What are the causes, consequences, and compensations of the four categories of acid-base imbalances?
14. **A point to ponder:** If a person loses 1,500 ml of salt-rich sweat and drinks 1,000 ml during the same time period, why will the vasopressin-secreting cells receive conflicting inputs from the hypothalamic osmoreceptors and left atrial baroreceptors? Why is it, therefore, important to replace both the water and salt?

C H A P T E R 16

Digestive System

Chapter Contents at a Glance

INTRODUCTION *A good example of the philosophy, "When life hands you lemons, make lemonade," took place in 1822 when a hunter, Alexis St. Martin, was accidentally shot in the stomach. The wound did not heal properly, forming a flap that prevented the stomach contents from falling out but leaving an open passageway under the flap from the interior of the stomach to the outside of the abdominal wall. Such an abnormal passage leading to the surface of the body is known as a* ***fistula.*** *Rather than bemoaning his fate, St. Martin agreed to join forces with his physician, American surgeon William Beaumont, to take advantage of the situation. Beaumont was easily able to insert various foods into St. Martin's stomach and to extract them after variable periods of* *time to see what digestion had taken place. He was also able to collect gastric (stomach) juice through the fistulous tract under various controlled conditions. This living laboratory enabled Beaumont to study the composition, functions, and regulation of gastric secretions. These experimental studies on the stomach were the beginnings of our knowledge of digestive physiology. In fact, they led to Beaumont's recognition as America's "father of physiology." They demonstrated a phenomenon that has been much more thoroughly investigated since then: digestive activities are normally regulated carefully to provide optimal conditions for the digestion and absorption of ingested food, no matter what kind or how much is eaten.*

The primary function of the **digestive system** (**alimentary system**) is to transfer nutrients (after modifying them), water, and electrolytes from the food we eat into the body's internal environment. Ingested food is essential as an energy source from which the cells can produce ATP to carry out their particular energy-dependent activities, such as active transport, contraction, synthesis, and secretion. Food is also a source of building supplies for the renewal and addition of body tissues.

Plants can capture the sun's energy and manufacture the organic molecules they need from inorganic compounds such as CO_2 and H_2O through the process of **photosynthesis:**

$$\underset{\text{(from sun)}}{\text{Energy}} + CO_2 + H_2O \xrightarrow{\text{plant photosynthesis}} \underset{\text{molecules}}{\text{Organic}} + O_2$$

Humans cannot harvest energy from sunlight directly, so they have to survive on second-hand energy by eating either plants or other animals that have eaten plants. In turn, humans use the organic molecules (in food) and O_2 to produce energy, with CO_2 and H_2O being formed as by-products in the process:

$$\underset{\text{molecules (in food)}}{\text{Organic}} + O_2 \xrightarrow{\text{human cell metabolism}} \underset{\text{(for use by human cells)}}{\text{Energy}} + CO_2 + H_2O$$

Note that these fundamental reactions in plants and humans (as well as other animals) are the reverse of each other, thus providing a natural balance between these forms of life. (Both plants and animals perform additional chemical reactions that involve these organic molecules as well.)

The act of eating does not automatically make the preformed organic molecules available to the body cells as a source of fuel or as building blocks. The food first must be digested, or broken down into small, simple molecules that can be absorbed from the digestive tract into the circulatory system for distribution to the cells. Normally, about 95% of the ingested food is made available for the body's use.

Unlike most body systems, the digestive system performs only limited regulatory functions aimed at maintaining homeostasis. It does not vary nutrient, water, or electrolyte uptake based on body needs (with few exceptions), but, rather, it optimizes conditions for digesting and absorbing what is ingested. Truly, what you eat is what you get (with the exception of the cellulose in the walls of plant cells, which constitutes dietary fiber, or "bulk," since it cannot be digested by the human digestive system.) The digestive system is subject to many regulatory processes, but these are not influenced by the nutritional state of the body. Instead, these control mechanisms are governed by the composition and volume of the digestive-tract contents so that the rate of motility and secretion of digestive juices are appropriate for digestion and absorption of the ingested food.

Because the preformed organic food molecules acquired from plants and animals are similar to the organic molecules that compose the human body, the secretions that digest food are also capable of digesting the person's own tissues. You will learn in this chapter about the clever devices by which this self-digestion is normally prevented. We will first provide an overview of the digestive system, examining the features shared in common by the various components of the system, before we commence on a detailed tour of the tract from beginning to end.

The digestive system performs four basic digestive processes.

There are four basic digestive processes: motility, secretion, digestion, and absorption.

MOTILITY. **Motility** refers to the muscular contractions that mix and move forward the contents of the digestive tract. As with vascular smooth muscle, the smooth muscle in the walls of the digestive tract maintains a constant low level of contraction known as **tone.** Tone is important in maintaining a steady pressure on the digestive-tract contents as well as in preventing the walls of the digestive tract from remaining permanently stretched following distention.

Superimposed on this ongoing tonic base are two basic types of digestive motility: propulsive movements and mixing movements. *Propulsive movements* propel or push the contents foward through the digestive tract at varying speeds, with the rate of propulsion depending on the functions accomplished by the different regions; that is, food is moved forward in a given segment at an appropriate velocity to allow that segment to "do its job." For example, transit of food through the esophagus is rapid, which is appropriate because this structure merely serves as a passageway from the mouth to the stomach. In comparison, in the small intestine, the major site of digestion and absorption, the contents are moved forward slowly, allowing sufficient time for the breakdown and absorption of food to be accomplished.

Mixing movements serve a two-fold function. First, by mixing food with the digestive juices, these movements promote digestion of the food. Second, they facilitate absorption by exposing all portions of the intestinal contents to the absorbing surfaces of the digestive tract.

Movement of material through most of the digestive tract is accomplished by contraction of the smooth muscle within the walls of the digestive organs, with the exception that motility at both ends of the tract—the mouth through the early portion of the esophagus at the beginning and the external anal sphincter at the end—involves skeletal-muscle rather than smooth-muscle activity. Accordingly, the acts of chewing, swallowing, and defecation have voluntary components since skeletal muscle is under voluntary control, whereas motility accomplished by smooth muscle throughout the remainder of the tract is controlled by complex involuntary mechanisms.

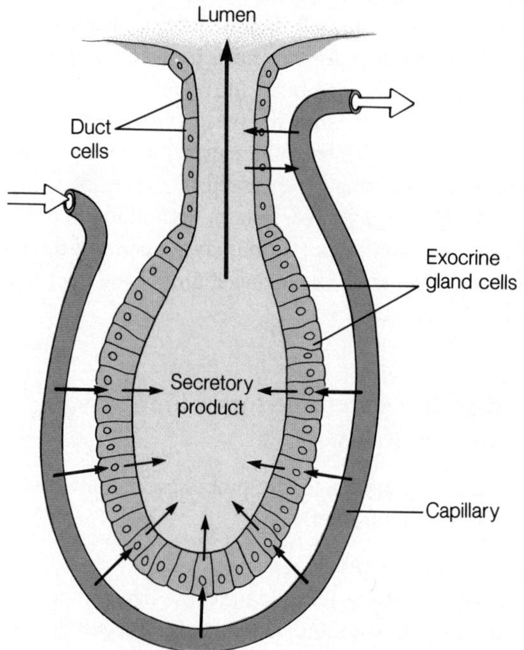

Figure 16–1 General Mode of Exocrine Gland Secretion *Exocrine gland cells extract from the plasma by both active and passive means the raw materials that they need to produce their secretory product. This product is emptied into ducts, which lead, in the case of the digestive system, to the lumen of the digestive tract. Frequently the secretion is modified as it moves through the duct by active and passive transport mechanisms within the membranes of the cells lining the duct.*

SECRETION. A number of digestive juices are secreted into the digestive-tract lumen by exocrine glands (see p. 5) located along the route, each with its own specific secretory product or products. Each **digestive secretion** consists of water, electrolytes, and specific organic constituents that are important in the digestive process, such as enzymes, bile salts, or mucus. The secretory cells extract from the plasma large volumes of water and those raw materials necessary to produce their particular secretion (Fig. 16–1). These exocrine cells are richly endowed with mitochondria to support the extensive energy requirement necessary for secretion. Secretion of all digestive juices requires energy, both for active transport of some of the raw materials into the cell (others diffuse in passively) and for synthesis of secretory products by the endoplasmic reticulum. The secretions are released into the digestive-tract lumen upon appropriate neural or hormonal stimulation. Normally the digestive secretions are reabsorbed in one form or another back into the blood after their participation in digestion. Failure to do so (because of vomiting or diarrhea, for example) results in loss of this fluid that has been "borrowed" from the plasma.

DIGESTION. **Digestion** refers to the breaking-down process whereby the structurally complex foodstuffs of the diet are converted into smaller absorbable units by the enzymes produced within the digestive system. Humans consume three different biochemical categories of energy-rich foodstuffs: carbohydrates, proteins, and fats. These large molecules are unable to cross plasma membranes intact to be absorbed from the lumen of the digestive tract into the blood or lymph. The process of digestion serves to degrade these large food molecules into smaller nutrient molecules that can be absorbed (Table 16–1).

The simplest form of **carbohydrates** are simple sugars, or **monosaccharides** ("one sugar" molecules), such as **glucose, fructose,** and **galactose,** very few of which are normally found in the diet. Most ingested carbohydrate is in the form of **polysaccharides** ("many sugar" molecules), which consist of chains of interconnected glucose molecules. The most common polysaccharide consumed is **starch** derived from plant sources. Additionally, meat contains **glycogen,** the polysaccharide storage form of glucose in muscle. **Cellulose,** another

Table 16-1 Process of Digestion

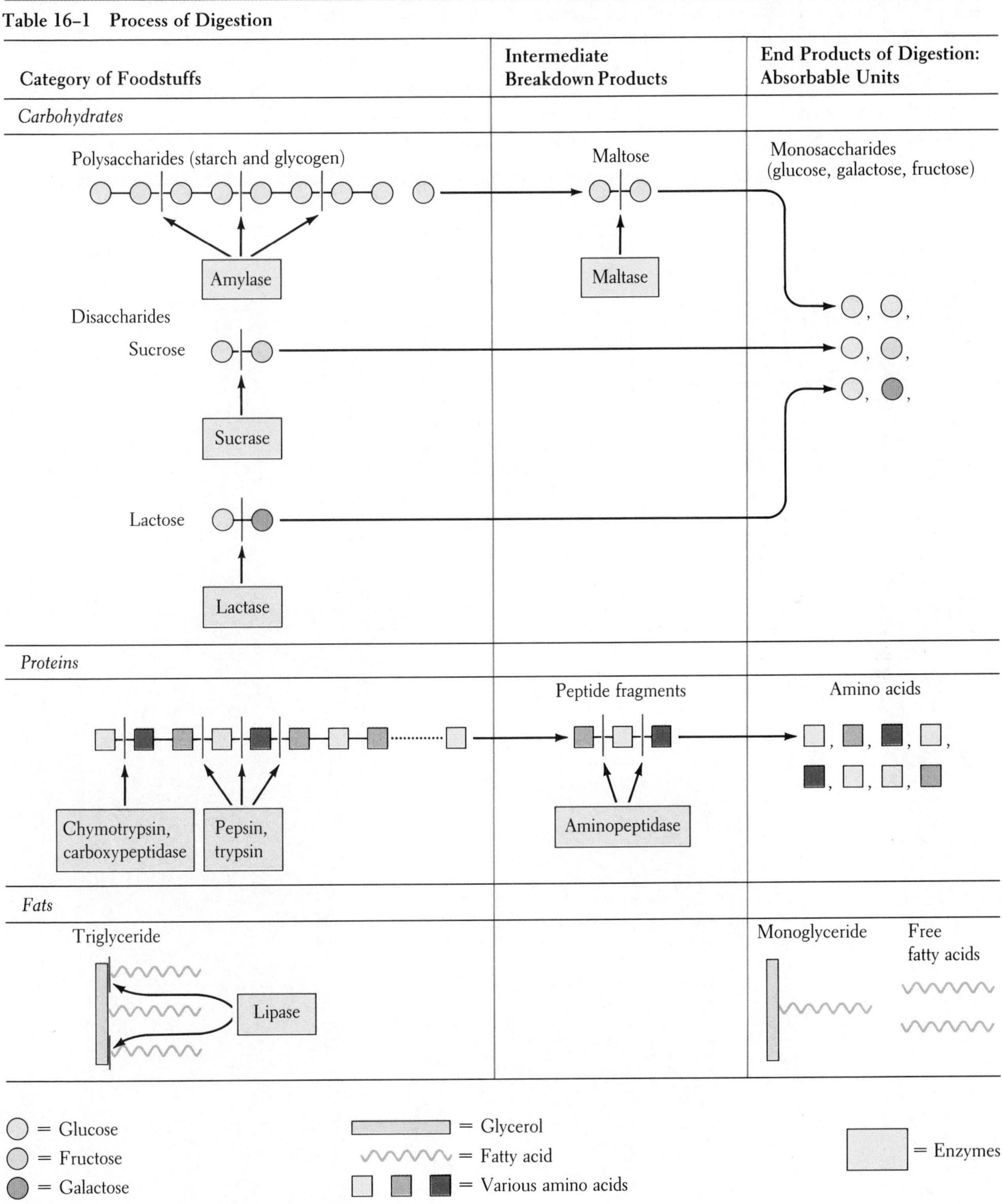

dietary polysaccharide that is found in plant walls, cannot be digested into its constituent monosaccharides by the digestive juices secreted in humans; thus it represents the undigested fiber in our diets, as mentioned earlier. Besides polysaccharides, a lesser source of dietary carbohydrate is in the form of **disaccharides** ("two sugar" molecules), including **sucrose** (table sugar, which consists of one glucose and one fructose molecule) and **lactose** (milk sugar made up of one glucose and one galactose molecule).

Starch, glycogen, and dissacharides are converted through the process of digestion into their constituent monosaccharides, principally glucose with small amounts of fructose and galactose. These monosaccharides are the absorbable units for carbohydrates.

Table 16–2 Anatomy and Functions of Components of the Digestive System

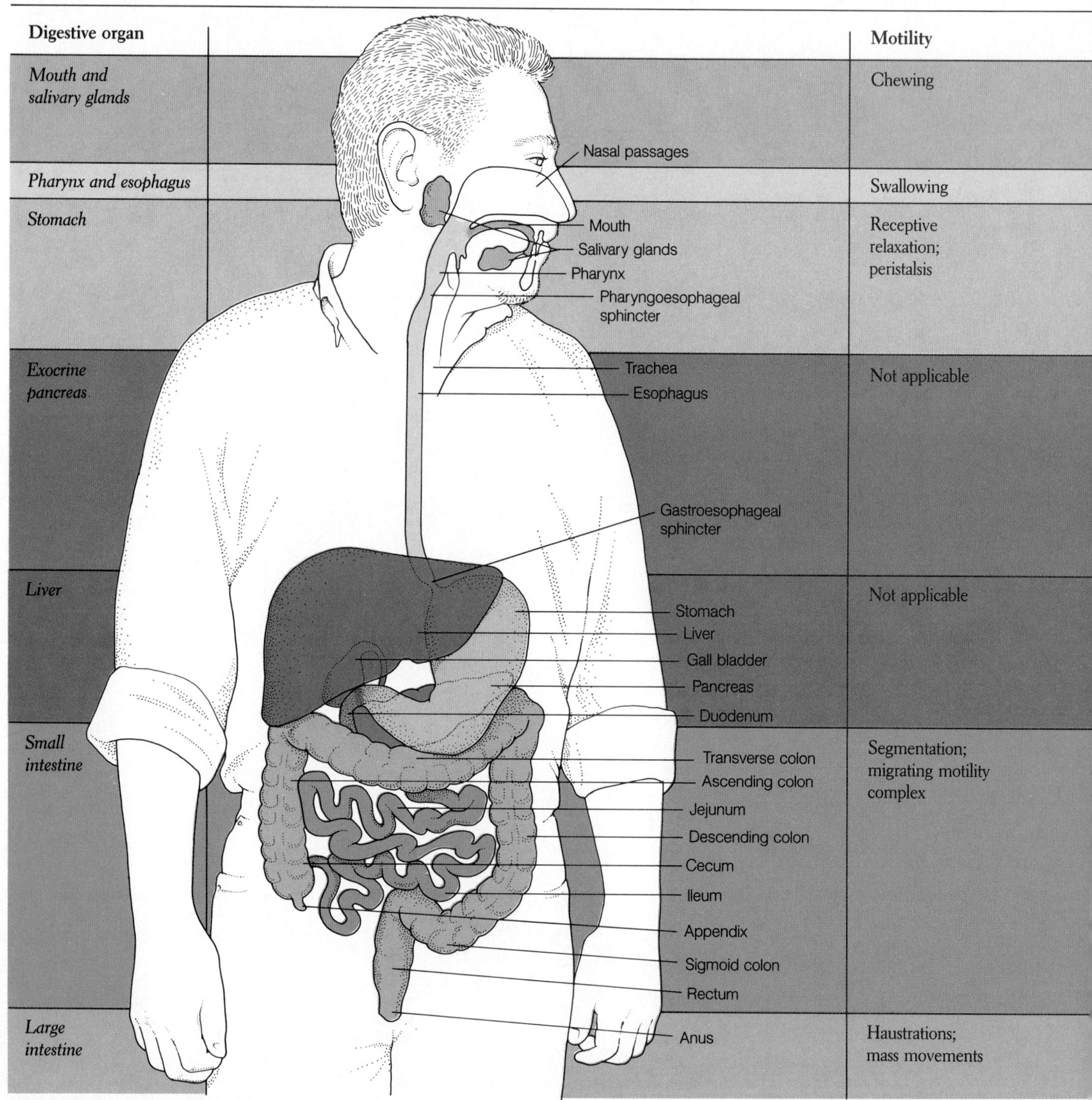

Digestive organ		Motility
Mouth and salivary glands		Chewing
Pharynx and esophagus		Swallowing
Stomach		Receptive relaxation; peristalsis
Exocrine pancreas		Not applicable
Liver		Not applicable
Small intestine		Segmentation; migrating motility complex
Large intestine		Haustrations; mass movements

The second category of foodstuffs is **proteins,** which consist of various combinations of **amino acids** held together by peptide bonds (see p. A–12). Through the process of digestion, proteins are degraded primarily into their constituent **amino acids** as well as a few **small polypeptides** (several amino acids linked together by peptide bonds), both of which are the absorbable units for protein.

Secretion	Digestion	Absorption
Saliva - Amylase - Mucus - Lysozyme	Carbohydrate digestion begins	No foodstuffs; a few medications—for example, nitroglycerine
Mucus	None	None
Gastric juice - HCl - Pepsin - Mucus - Intrinsic factor	Carbohydrate digestion continues in body of stomach; protein digestion begins in antrum of stomach	No foodstuffs; a few lipid-soluble substances, such as alcohol and aspirin
Pancreatic digestive enzymes - Trypsin, chymotrypsin, carboxypeptidase - Amylase - Lipase Pancreatic aqueous $NaHCO_3$ secretion	These pancreatic enzymes accomplish digestion in duodenal lumen	Not applicable
Bile - Bile salts - Alkaline secretion - Bilirubin	Bile does not digest anything, but bile salts facilitate fat digestion and absorption in duodenal lumen	Not applicable
Succus entericus - Mucus - Salt (Small intestine enzymes are not secreted but function intracellularly in the brush border—disaccharidases and aminopeptidases)	In lumen, under influence of pancreatic enzymes and bile, carbohydrate and protein digestion continue and fat digestion is completely accomplished; in brush border, carbohydrate and protein digestion completed	All nutrients, most electrolytes, and water
Mucus	None	Salt and water, converting contents to feces

Fats represent the third category of foodstuffs. Most dietary fat is in the form of **triglycerides** (**triacylglycerols**), which are neutral fats, each consisting of a combination of **glycerol** with three (*tri* means "three") **fatty-acid** molecules attached. During digestion, two of the fatty-acid molecules are split off, leaving a **monoglyceride** (**monoacylglycerol**), a glycerol molecule with one (*mono* means "one") fatty-acid molecule attached. Thus, the end products of fat digestion are monoglycerides and free fatty acids, which are the absorbable units of fat.

Digestion is accomplished by enzymatic **hydrolysis** (see p. A–13). By adding H_2O at the bond site, enzymes in the digestive secretions break down the bonds that hold the small molecular subunits within the nutrient molecules together, thus setting the small molecules free. These small subunits are joined to form nutrient molecules in the first place by the removal of H_2O at the bond sites. Hydrolysis replaces the H_2O and frees the small absorbable units. Digestive enzymes are specific in the bonds they can hydrolyze. As food moves through the digestive tract, it is subjected to various enzymes, each of which breaks down the food molecules even further. In this way, the conversion of large food molecules to simple absorbable units occurs in a progressive, step-wise fashion as the digestive tract contents are propelled forward.

ABSORPTION. Digestion is completed and most absorption occurs in the small intestine. Through the process of **absorption,** the small absorbable units that result from digestion, along with water, vitamins, and electrolytes, are transferred from the digestive-tract lumen into the blood or lymph.

As we examine the digestive tract from beginning to end, we will discuss the four processes of motility, secretion, digestion, and absorption as they take place within each digestive organ (Table 16–2).

The digestive tract and accessory digestive organs make up the digestive system.

The digestive system consists of the digestive (gastrointestinal) tract plus the accessory digestive organs. The **accessory digestive organs** include the *salivary glands*, the *exocrine pancreas*, and the *biliary system*, the latter being composed of the *liver* and *gall bladder*. These exocrine organs are located outside of the wall of the digestive tract and empty their secretions through ducts into the digestive-tract lumen. They develop from outpouchings of the embryonic digestive tube and maintain their connection with the digestive tract through the ducts that are formed.

The **digestive tract** is essentially a tube about 30 feet (9 m) in length that runs through the middle of the body from the mouth to the anus. (This is the length in a cadaver; it is about half that length in a living person because of ongoing contractions of its muscular walls.) The digestive tract includes the following organs: *mouth*; *pharynx* (throat); *esophagus*; *stomach*; *small intestine* (consisting of the *duodenum*, *jejunum*, and *ileum*); *large intestine* (composed of the the *cecum*, the *appendix*, the *colon* and the *rectum*); and the *anus*. It should be noted that these organs are continuous with each other and are discussed as separate entities only because of their regional modifications, which allow them to specialize in particular digestive activities.

Since the digestive tract is continuous from the mouth to the anus, the lumen of this tube, like the lumen of a straw, is continuous with the external environment. As a result, any contents within the lumen of the digestive tract are technically outside the body, just as the soda that you suck through a straw is not part of the straw. Only after a substance has been absorbed from the lumen across the intestinal wall is it considered to have become a part of the body. This fact is important because conditions essential to the digestive process can be tolerated in the digestive-tract lumen that could not be tolerated in the body proper. Consider the following examples:

1. The pH of the stomach contents falls as low as 2 as a result of the gastric secretion of hydrochloric acid (HCl), yet the range of pH in the body fluids compatible with life is 6.8 to 8.0.
2. The harsh digestive enzymes that hydrolyze food could also destroy the body's own tissues that produce them. Therefore, once they are synthesized in inactive form, these enzymes are not activated until they reach the lumen, where they actually attack the food outside of the body, thereby protecting the body tissues against self-digestion.
3. The lower portion of the intestinal tract is inhabited by millions of living microorganisms that are normally harmless and even beneficial, yet if these same microorganisms enter the body proper (as may happen with a ruptured appendix), they may be extremely harmful or even lethal.

The wall of the digestive tract has the same general structure throughout most of its length from the esophagus to the anus, with some local variations characteristic for each region. A cross section of the digestive tube (Fig. 16–2) reveals four major tissue layers. From the innermost layer of the tract outwards they are the mucosa, the submucosa, the muscularis externa, and the serosa.

The **mucosa** lines the luminal surface of the digestive tract. It is divided into three layers:

□ an inner epithelial layer, or **mucous membrane,** which serves as a protective surface as well as being modified in particular areas for secretion of digestive juices, secretion of gastrointestinal hormones, and absorption of luminal contents.

□ the **lamina propria,** a thin middle layer of connective tissue on which the epithelium rests; through which small blood vessels, lymph vessels, and nerve fibers pass; and in which is housed the gut-associated lymphoid tissue (GALT) that is important in the defense against intestinal bacteria (see p. 364).

□ the **muscularis mucosa,** a sparse outer layer of smooth muscle that lies adjacent to the submucosa.

The mucosal surface is generally not flat and smooth but is highly folded with many ridges and valleys that greatly increase the surface area available for absorption. The degree of folding varies in different areas of the digestive tract, being most extensive in the small intestine, where maximum absorption occurs, and least extensive in the esophagus, which merely serves as a transit tube. The pattern of surface folding can be modified by contraction of the muscularis mucosa. This is important primarily in exposing different areas of the absorptive surface to the luminal contents.

The **submucosa** ("under the muscosa") is a thick layer of connective tissue that provides the digestive tract with its distensibility and elasticity. It contains the larger blood and lymph vessels, both of which send branches inward to the mucosal layer and outward to the surrounding thick muscle layer. Also lying within the submucosa is a nerve network known as the *submucous plexus*, which helps control local activities of each gut region.

Surrounding the submucosa is the *muscularis externa*, the major smooth-muscle coat of the digestive tube. In most parts of the tract, it consists of two layers: an inner circular layer and an outer longitudinal layer. The fibers of the inner smooth-muscle layer (adjacent to the submucosa) run circularly around the circumference of the tube. Contraction of these circular fibers constricts or decreases the diameter of the lumen at the point of contraction. Contraction of the fibers in the outer layer, which run longitudinally along the length of the tube, accomplishes shortening of the tube. Together, contractile activity of these smooth-muscle layers produces the propulsive and mixing movements. Lying between the two muscle layers is another nerve network, the *myenteric plexus*, which, along with the submucous plexus, helps to regulate local gut activity.

The outer connective-tissue covering of the digestive tract is the **serosa,** which secretes a watery serous fluid that lubricates and prevents friction between the digestive organs and surrounding viscera. Throughout much of the tract, the serosa is continuous with the **mesentery,** which suspends the digestive organs from the inner wall of the abdominal cavity like a sling (Fig. 16–2). This attachment provides relative fixation, so that the digestive organs are supported in proper position, while still allowing them freedom for mixing and propulsive movements. One cause of *hernias*, or protrusions of an organ through the muscular wall of the cavity that contains them, is mesenteric tearing. Such tearing allows a portion of the diges-

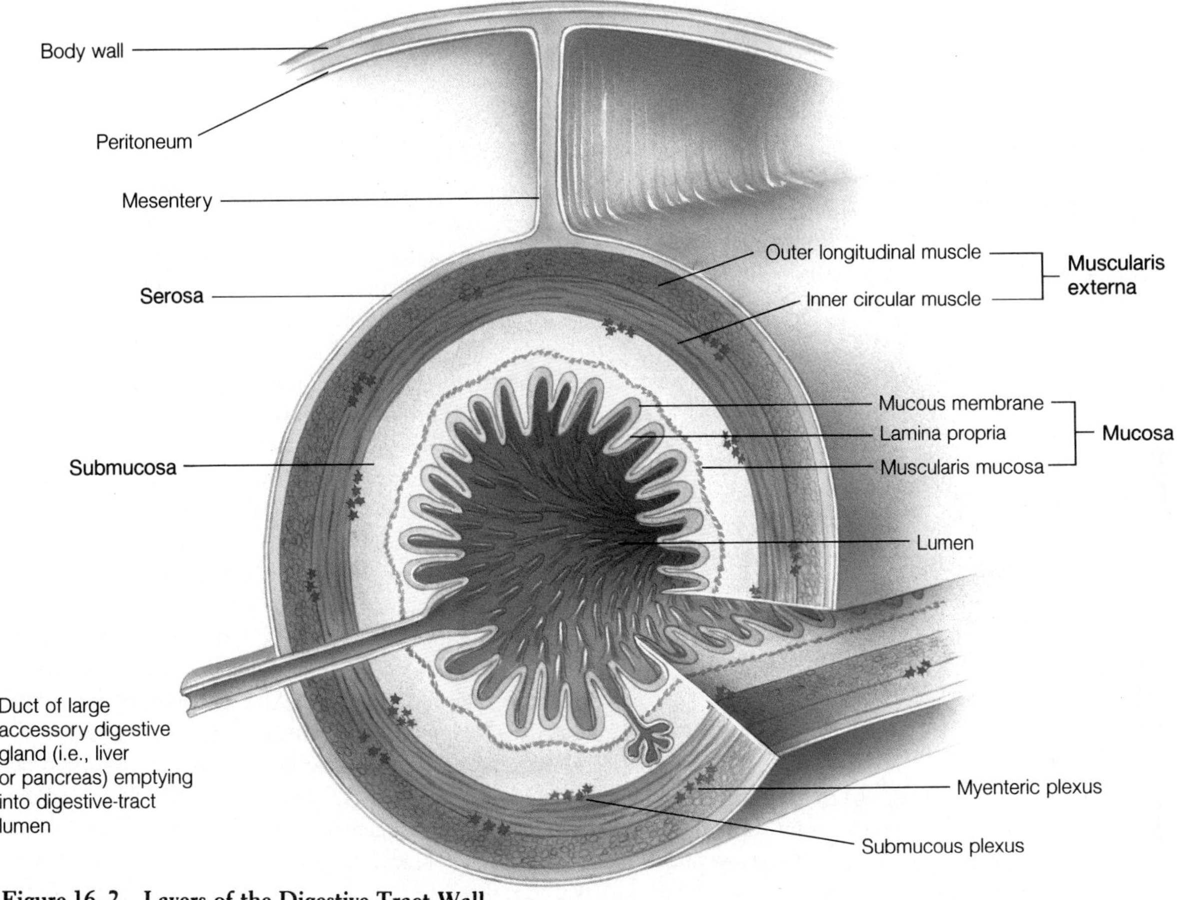

Figure 16–2 Layers of the Digestive Tract Wall

tive tube to fall free from its attachment and bulge through the abdominal wall.

Regulation of digestive function is complex and synergistic.

Digestive motility and secretion are carefully regulated to maximize digestion and absorption of the ingested food. There are four factors involved in the regulation of gastrointestinal function: (1) autonomous smooth-muscle function; (2) internal or intrinsic nerve plexuses; (3) external or extrinsic nerves; and (4) gastrointestinal hormones.

AUTONOMOUS SMOOTH-MUSCLE FUNCTION. Like self-excitable cardiac-muscle cells, some smooth-muscle cells do not have a constant resting potential but rather display rhythmic, spontaneous variations in membrane potential. The prominent train of self-induced electrical activity in digestive smooth muscle is **slow-wave potentials** (see p. 253), alternatively referred to as **basic electrical rhythm** (**BER**) or **pacesetter potential.** Slow waves are not action potentials and do not directly induce muscle contraction; they are rhythmic, wavelike fluctuations in membrane potential that cyclically bring the membrane closer to or farther from threshold. These slow-wave oscillations are believed to be due to cyclical variations in the rate at which the Na^+ pump transports Na^+ out of the pacesetter cell. Should these waves reach threshold at the peaks of depolarization, a volley of action potentials is triggered at each peak, which results in repeating, rhythmical cycles of muscle contraction.

As with cardiac muscle, sheets of smooth-muscle cells are connected by gap junctions (see p. 69), which serve as points of low electrical resistance so that slow-wave electrical activity initiated in a digestive-tract pacesetter cell can spread to adjacent smooth-muscle cells. If threshold is reached and action potentials are triggered, the whole muscle sheet behaves as a functional syncytium, becoming excited and contracting as a unit. If threshold is not achieved, the oscillating electrical activity continues to sweep across the muscle without being accompanied by contractile activity.

Whether or not threshold is reached depends on the effect of various mechanical, nervous-system, and hormonal factors that influence the "resting" potential, or the starting point around which the slow-wave rhythm oscillates. If the starting point is nearer the threshold level, the depolarizing slow-wave peak is more likely to reach threshold, so action-potential frequency and its accompanying contractile activity increases. Conversely, if the starting point is farther from threshold, there is less likelihood of reaching threshold, so action-potential frequency is lowered and contractile activity is reduced.

The pace or rate of rhythmic digestive contractile activities, such as peristalsis in the stomach, depends on the inherent rate established by the involved pacesetter cells. The intensity of the contractions depends on the number of action potentials that occur when the slow-wave potential reaches threshold—the greater the number of action potentials, the higher the cytosolic Ca^{++} concentration, the greater the cross-bridge activity, and the stronger the contraction. Thus, the level of contractility can range from low-level tone to vigorous mixing and propulsive movements by varying the cytosolic Ca^{++} concentration.

INTRINSIC NERVE PLEXUSES. The second factor involved in the regulation of gastrointestinal function is the internal or **intrinsic nerve plexuses.** A nerve plexus is an interconnecting network of nerve cells. Two major networks of nerve fibers form the plexuses of the digestive tract: the **myenteric (Auerbach's) plexus,** which is located between the longitudinal and circular smooth-muscle layers; and the **submucous (Meissner's) plexus,** which is located in the submucosa. These two plexuses are known as intrinsic plexuses because they are located entirely within the digestive-tract wall. They run the entire length from the esophagus to the anus. Thus, unlike any other organ system, the digestive tract has its own intramural ("within wall") nervous system, which endows the tract with a considerable degree of self-regulation. These nerve plexuses, which can function independently of any nerve fibers from the central nervous system, are evolutionary descendants of the primitive nervous system in the wall of such tubular animals as the hydra.

The intrinsic plexuses influence all facets of gastrointestinal activity. Through innervation of the smooth-muscle cells and exocrine and endocrine cells of the digestive tract, the intrinsic plexuses directly affect digestive-tract motility, secretion of digestive juices, and secretion of gastrointestinal hormones. These intrinsic nerve networks are primarily responsible for coordinating local activity within the digestive tract. For example, if a large piece of food gets stuck in the esophagus, local contractile responses coordinated by the intrinsic plexuses are initiated to push the food forward. Intrinsic nerve activity can, in turn, be influenced by the external or extrinsic nerves.

EXTRINSIC NERVES. **Extrinsic innervation** refers to the nerves originating outside of the digestive tract that innervate the various digestive organs—namely, nerve fibers from both branches of the autonomic nervous system. The autonomic nerves influence gut motility and secretion either by modifying on-going activity in the intrinsic plexuses, altering the level of gastrointestinal hormone secretion, or, in some instances, acting directly on the smooth muscle and glands.

Recall that, in general, the sympathetic and parasympathetic nerves supplying any given tissue exert opposing actions on that tissue. The sympathetic system, which dominates in fight-or-flight situations, tends to inhibit or slow down digestive tract contraction and secretion. This is appropriate considering that digestive processes are not of highest priority when the body is faced with an emergency or threat from the external environment. The sympathetic system decreases motility by driving the "resting" potential of the digestive smooth muscle farther from threshold so that the depolarizing slow-wave peak is less likely to reach threshold, thus reducing contractile activity. The parasympathetic nervous system, on the other hand, dominates in quiet, relaxed situations when general maintenance types of activities such as digestion can proceed optimally. Accordingly, the parasympathetic nerve fibers supplying the digestive tract, which arrive primarily by way of the vagus nerve, tend to increase smooth-muscle motility and promote secretion of digestive enzymes and hormones. The effect on smooth muscle is accomplished by bringing the starting point nearer to threshold, thereby increasing the frequency with which the oscillating slow-wave potential reaches threshold and brings about a contractile response.

The extrinsic nerves can modify ongoing digestive-tract activity but do not absolutely control it. In addition to being called into play during generalized sympathetic or parasympathetic discharge, the autonomic nerves supplying the digestive system can be discretely activated to modify only digestive activity. One of the major purposes of specific activation of extrinsic innervation is the correlation of activity between different regions of the digestive system; for example, the act of chewing food reflexly increases not only salivary secretion but also stomach, pancreatic, and liver secretion via vagal reflexes in anticipation of the arrival of food. Another purpose of specific activation is the provision of a pathway by which factors outside of the digestive system can influence digestion, as, for example, with the vagally mediated increase in digestive juices that occurs in anticipation of a meal when a person sees or smells food.

GASTROINTESTINAL HORMONES. The fourth factor influencing gut activity is hormonal control. Tucked within the mucosa of certain regions of the digestive tract are endocrine gland cells that release hormones into the blood upon appropriate stimulation. These **gastrointestinal hormones** are carried through the blood to other areas of the digestive tract, where they exert either excitatory or inhibitory influences on smooth muscle, exocrine gland cells, or even other endocrine gland cells. Gastrointestinal hormones are released primarily in response to specific local changes in the intraluminal contents (such as the presence of protein, fat, or acid), acting either directly on the endocrine gland cells or indirectly through the intrinsic plexuses or external autonomic nerves. Interestingly, many of these same hormones have been found in the brain, where they act as neurotransmitters and neuromodulators. The endocrine cells of the digestive system migrate from developing neural tissue during embryonic development.

Receptor activation alters digestive activity through short and long reflexes and hormonal pathways.

The wall of the digestive tract contains three different types of sensory receptors that respond to local chemical or mechanical changes in the digestive tract: (1) ***chemoreceptors*** sensitive to chemical components within the lumen; (2) ***mechanoreceptors*** (pressure receptors) sensitive to stretch or tension within the wall; and (3) ***osmoreceptors*** sensitive to the osmolarity of the luminal contents. Stimulation of these receptors elicits neural reflexes or secretion of hormones, both of which alter the level of activity in the digestive system's effector cells. These effector cells include smooth-muscle cells (for modifying motility); exocrine gland cells (for controlling secretion of digestive juices); and endocrine gland cells (for varying secretion of gastrointestinal hormones (Fig. 16–3).

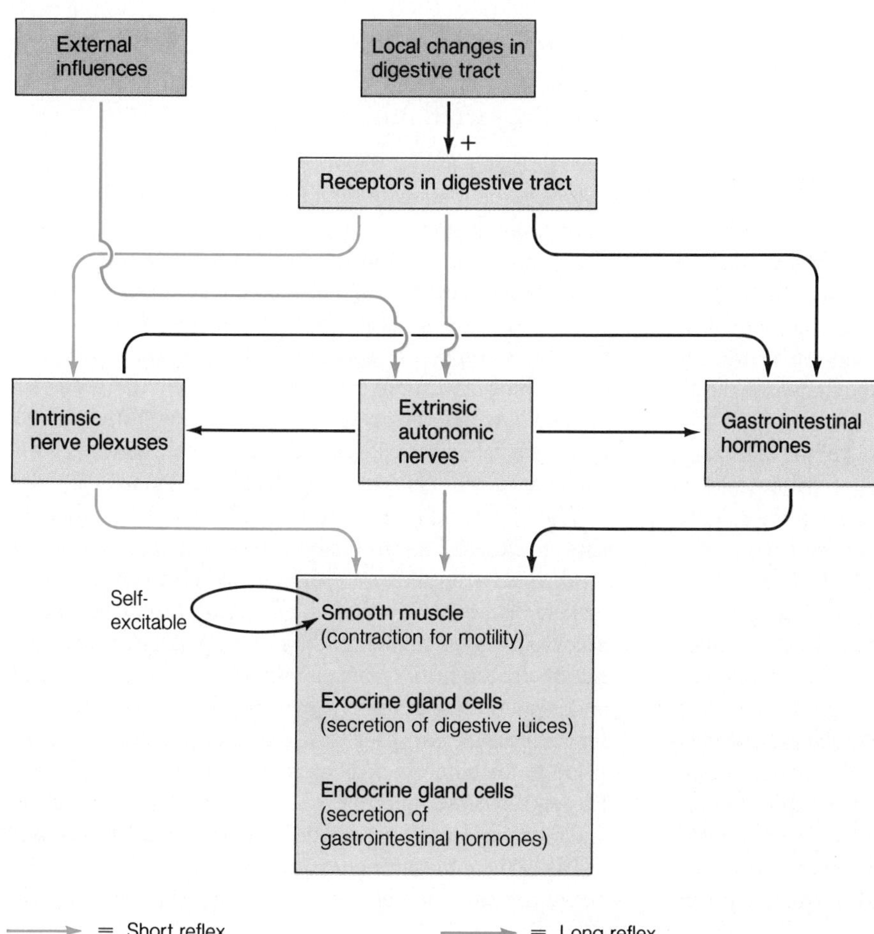

Figure 16–3 Summary of Pathways Controlling Digestive-System Activities

Two types of neural reflexes—short reflexes and long reflexes—may be brought forth by receptor activation. When the intrinsic nerve networks influence local motility or secretion in response to specific local stimulation, such a reflex, in which all elements of the reflex arc are located within the wall of the gut itself, is known as a **short reflex.** Extrinsic autonomic nervous activity can be superimposed on the local controls to modify smooth-muscle and glandular responses, either to correlate activity between different regions of the digestive system or to modify gut activity in response to external influences. Because the autonomic reflexes involve long pathways between the central nervous system and digestive system, they are known as **long reflexes.** In addition to these neural reflexes, gut activity is also coordinated by the secretion of gastrointestinal hormones, which are triggered directly by local gut changes or by short or long reflexes.

From this overview, it should be apparent that regulation of gastrointestinal function is very complex, being modulated by many synergistic, interrelated pathways designed to assure that the appropriate responses occur to digest and absorb the ingested food. Nowhere else in the body is such a level of overlapping control exercised.

MOUTH

The oral cavity is the entrance to the digestive tract.

Entry to the digestive tract is through the **mouth** or **oral cavity.** The opening is formed by the muscular **lips,** which help procure, guide, and contain the food in the mouth. The lips also serve nondigestive functions; they are important in speech (articulation of many sounds depends on a particular lip formation) and as a sensory receptor in interpersonal relationships (for example, as in kissing).

The **palate,** which forms the arched roof of the oral cavity, separates the mouth from the nasal passages. Its presence allows breathing and chewing or sucking to take place simultaneously. Embryologically, the palate is derived from projections that grow inward from the jaws on both sides and that fuse in the midline of the oral cavity. Failure of these projections to properly unite results in a cleft palate, which can interfere with sucking, eating, and speech if not surgically corrected. Toward the front of the mouth, the palate is made of bone, forming what is known as the **hard palate.** No bone is deposited in the palate toward the rear of the mouth; this region is called the **soft palate.** Hanging down from the soft palate in the rear of the throat is a dangling conical projection, the **uvula,** which plays an important role in sealing off the nasal passages during swallowing. (This is the structure you elevate when you say "Ahhh" so that the physician can better see your throat.)

The **tongue,** which forms the floor of the oral cavity, is composed of voluntarily controlled skeletal muscle. Movements of the tongue are not only important in guiding food within the mouth during chewing and swallowing, but they also play an important role in speech. Embedded within the tongue are **taste buds** (see p. 196), which are dispersed on the soft palate, throat, and linings of the cheeks as well.

The **pharynx** is the cavity at the rear of the throat that acts as a common passageway for both the digestive system and respiratory system. It serves as the link between the mouth and esophagus for food and provides access between the nasal passages and trachea for air. Although this arrangement allows you to breathe through your mouth to get air into your lungs when your nasal passages are blocked because of a cold, it also necessitates mechanisms (to be described shortly) to guide food and air into the proper passageways beyond the pharynx. Housed within the side walls of the pharynx are the **tonsils,** which are lymphoid organs that are part of the body's defense team.

The teeth accomplish chewing, which breaks up food, mixes it with saliva, and stimulates digestive secretions.

The first step in the digestive process is **mastication,** or **chewing,** the motility of the mouth that involves the slicing, tearing, grinding, and mixing of ingested food by the **teeth.** The teeth are firmly embedded in and protrude from the jaw bones. The exposed portion of a tooth is covered by **enamel,** the hardest structure of the body. Enamel is formed prior to the tooth's eruption by special cells that are lost as the tooth erupts. Since it cannot be regenerated after the tooth has erupted, any defects ("cavities") that develop in the enamel must be patched by artificial "fillings" or else the surface will continue to erode into the underlying living pulp.

The upper and lower teeth normally fit together when the jaws are closed. This **occlusion** allows food to be ground and crushed between the tooth surfaces. When the teeth do not make proper contact with each other, they cannot adequately accomplish their normal cutting and grinding action. Such **malocclusion** results from abnormal positioning of the teeth and is often caused either by overcrowding of teeth too large for the available jaw space or by one jaw being displaced in relation to the other. In addition to inefficient chewing, malocclusion can cause abnormal wearing of affected tooth surfaces and dysfunction and pain of the **temporomandibular joint (TMJ),** where the jaw bones articulate with each other. Malocclusions can often be corrected by applying braces, which exert prolonged gentle pressure against the teeth to gradually move them to the desired position.

The purposes of chewing are to grind and break food up into smaller pieces to facilitate swallowing, to mix food with saliva, and to stimulate the taste buds. This not only gives rise to the pleasurable subjective sensation of taste but also reflexly increases salivary, gastric, pancreatic, and bile secretion to prepare for the arrival of food.

The act of chewing can be voluntary, but most chewing during a meal is a rhythmic reflex brought about by activation of the skeletal muscles of the jaws, lips, cheeks, and tongue in response to the pressure of food against the oral tissues.

The teeth can exert forces much greater than those necessary to eat ordinary food. For example, the molars in an adult man can exert a crushing force of up to 200 pounds, which is sufficient to crack a hard nut, but these powerful forces are not ordinarily used. In fact, the degree of occlusion is more important than the force of the bite in determining the efficiency of chewing.

Saliva begins carbohydrate digestion but plays more important roles in oral hygiene and in facilitating speech.

Saliva is the secretion associated with the mouth. There are three major pairs of salivary glands—the **sublingual,** the **submandibular,** and the **parotid glands**—each of which secretes into a duct that conducts the saliva into the oral cavity. In addition, minor salivary glands, the **buccal glands,** are located in the mucosa lining the cheeks.

Saliva is composed of about 99.5% H_2O and 0.5% protein and electrolytes. The most important salivary proteins—amylase, mucus, and lysozyme—contribute to the functions of saliva as follows:

1. Saliva begins digestion of carbohydrate in the mouth through action of **salivary amylase,** an enzyme that catalyzes the hydrolysis of polysaccharides into disaccharides.
2. Saliva facilitates swallowing by moistening food particles, thereby holding them together, and by providing lubrication through the presence of **mucus,** which is thick and slippery.
3. Saliva exerts some antibacterial action by means of a twofold effect—first by **lysozyme,** an enzyme that lyses or destroys certain bacteria, and second, by rinsing away material that may serve as a food source for bacteria.
4. Saliva serves as a solvent for molecules that stimulate the taste buds. Only molecules in solution can react with taste-bud receptors. You can demonstrate this for yourself. Dry your tongue and then drop some sugar on it. You cannot taste the sugar until it is moistened. Glucose is never present in saliva and the salivary salt concentration is considerably below that in the plasma. This fact is probably important in the perception of sweet and salty tastes.
5. Saliva aids speech by facilitating movements of the lips and tongue. It is difficult to talk when the mouth feels dry.
6. Saliva plays an important role in oral hygiene by helping to keep the mouth and teeth clean. The constant flow of saliva helps to flush away food residues, shed epithelial cells, and foreign particles. Saliva's contribution in this regard is apparent to anyone who has experienced a foul taste in the mouth when salivation is suppressed for a while, such as during a fever or states of prolonged anxiety.
7. Buffers in the saliva in the form of bicarbonate neutralize acids in food as well as acids produced by bacteria in the mouth, thereby helping to prevent dental caries.

In spite of these many functions, saliva is not essential for the digestion and absorption of foods, because enzymes produced by the pancreas and small intestine can complete the digestion of food even in the absence of salivary and gastric secretion. The main problems associated with diminished or arrested salivary secretion, a condition known as **xerostomia,** are difficulty in chewing and swallowing, inarticulate speech unless frequent sips of water are taken when talking, and a rampant increase in dental caries.

The continuous low level of salivary secretion can be increased by simple and conditioned reflexes.

On the average, about 1 to 2 liters of saliva are secreted per day, ranging from a continuous spontaneous basal rate of 0.5 ml/min to a maximum flow rate of about 5 ml/min in response to a potent stimulus such as sucking on a lemon. The continuous spontaneous secretion of saliva, even in the absence of apparent stimuli, is due to constant low-level stimulation by the parasympathetic nerve endings that terminate in the salivary glands. This basal secretion is important in keeping the mouth and throat moist at all times.

In addition to this continuous, low-level secretion, salivary secretion may be enhanced by two different types of salivary reflexes (Fig. 16–4): (1) the simple, or unconditioned, salivary reflex and (2) the acquired, or conditioned, salivary reflex. The **simple,** or **unconditioned, salivary reflex** occurs when chemoreceptors and pressurereceptors within the oral cavity respond to the presence of food. On activation, these receptors initiate impulses in afferent nerve fibers that carry the information to the **salivary center** located in the medulla of the brain stem. The salivary center, in turn, sends impulses over the extrinsic autonomic nerves to the salivary glands to pro-

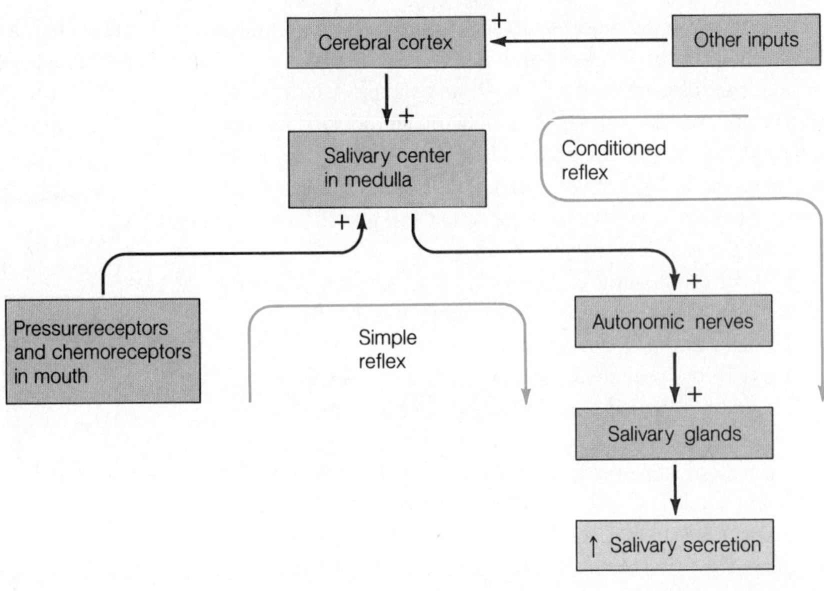

Figure 16-4 Control of Salivary Secretion

mote increased salivation. Dental procedures promote salivary secretion in the absence of food in the mouth because these manipulations activate pressurereceptors in the mouth.

With the **acquired,** or **conditioned, salivary reflex,** salivation occurs without oral stimulation. Just thinking about, seeing, smelling, or hearing the preparation of pleasant food initiates salivation through this reflex. All of us have experienced such "mouth watering" in anticipation of something delicious to eat. This reflex is a learned response based on previous experience. Inputs arising outside of the mouth that are mentally associated with the pleasure of eating act through the cerebral cortex to stimulate the medullary salivary center.

The salivary center controls the degree of salivary output by means of the autonomic nerves that supply the salivary glands. Unlike the autonomic nervous system elsewhere in the body, sympathetic and parasympathetic responses in the salivary glands are not antagonistic. Both sympathetic and parasympathetic stimulation increase salivary secretion, but the quantity, characteristics, and mechanisms are different. Parasympathetic stimulation, which exerts the dominant role in salivary secretion, produces a prompt and abundant flow of watery saliva that is rich in enzymes. Sympathetic stimulation, on the other hand, produces a much smaller volume of thick saliva that is rich in mucus. Because of the smaller volume of saliva elicited by sympathetic stimulation, the mouth feels drier than usual during circumstances when the sympathetic system is dominant, such as during stress situations. This accounts for the dry feeling in the mouth that people experience when they are nervous about giving a speech.

It is important to note that salivary secretion is entirely under neural control. All other digestive secretions are regulated by both nervous-system reflexes and hormones.

Digestion in the mouth is minimal, and no absorption of nutrients occurs.

Digestion in the mouth involves the hydrolysis of polysaccharides into disaccharides by amylase. However, most digestion by this enzyme is accomplished in the body of the stomach after the food mass and saliva have been swallowed. Acid inactivates amylase, but in the center of the food mass, where stomach acid has not yet reached, this salivary enzyme continues to function for several more hours.

No absorption of foodstuff occurs from the mouth. Importantly, some therapeutic agents can be absorbed by the oral mucosa, a prime example being a vasodilator drug, nitroglycerin, which is used by certain cardiac patients to relieve anginal attacks associated with myocardial ischemia (see p. 277).

Pharynx and Esophagus

Swallowing is a sequentially programmed all-or-none reflex.

The motility associated with the pharynx and esophagus is **swallowing,** or **deglutition.** Most of us think of swallowing

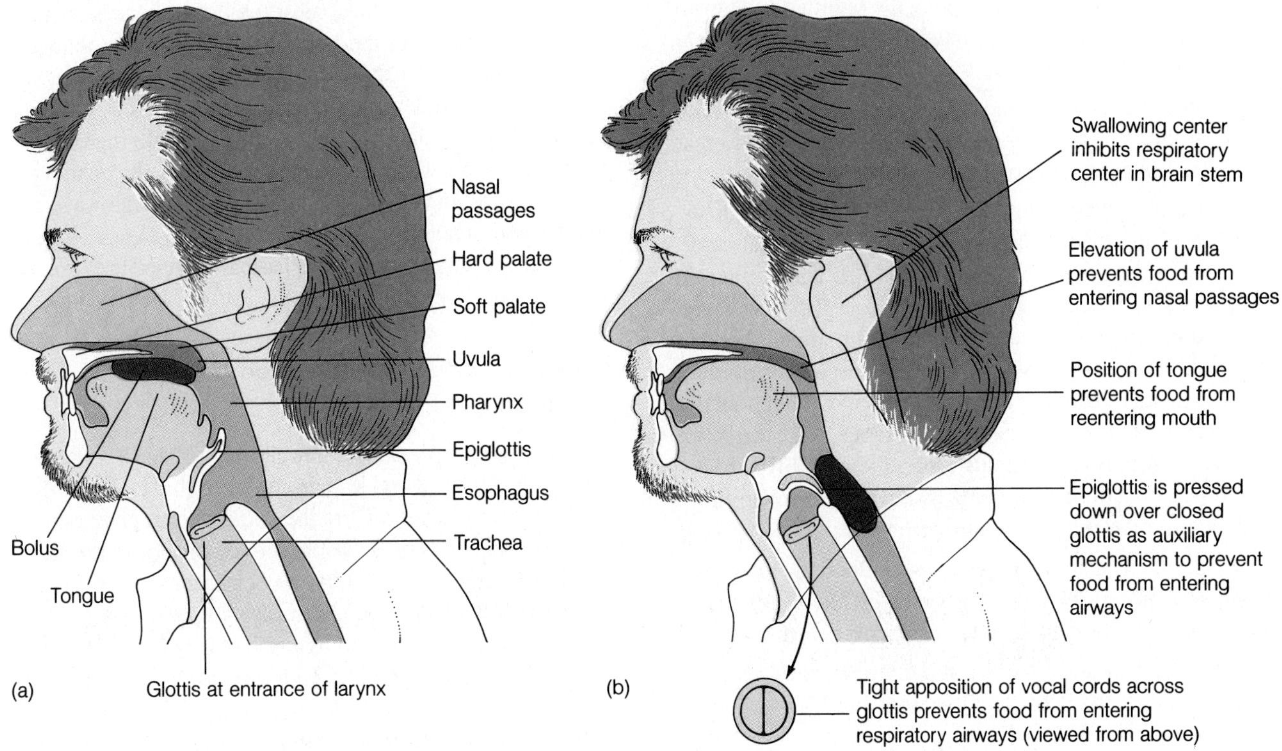
Nasal passages
Hard palate
Soft palate
Uvula
Pharynx
Epiglottis
Esophagus
Trachea
Bolus
Tongue
(a)
Glottis at entrance of larynx
Swallowing center inhibits respiratory center in brain stem
Elevation of uvula prevents food from entering nasal passages
Position of tongue prevents food from reentering mouth
Epiglottis is pressed down over closed glottis as auxiliary mechanism to prevent food from entering airways
(b)
Tight apposition of vocal cords across glottis prevents food from entering respiratory airways (viewed from above)

Figure 16–5 Oropharyngeal Stage of Swallowing *(a) Position of oropharyngeal structures at rest. (b) Changes that occur during the oropharyngeal stage of swallowing to prevent the bolus of food from entering the wrong passageways.*

as the limited act of moving food out of the mouth into the esophagus. However, swallowing actually refers to the entire process of moving food from the mouth through the esophagus into the stomach. Swallowing is initiated when a **bolus,** or ball of food, is voluntarily forced to the rear of the mouth by the tongue. The tongue gathers the food from around the mouth, forms it into a ball, and then presses upward and backward against the hard palate, squeezing or rolling the food mass back into the pharynx. This act is voluntary, even though the person is not really thinking about all of these tongue manipulations.

Once the bolus is pushed to the rear of the mouth by the tongue, the pressure of the bolus stimulates pressurereceptors in the pharynx, which send afferent impulses to the **swallowing center** located in the medulla. The swallowing center then reflexly activates in the appropriate sequence the muscles that are involved in swallowing. Swallowing is an example of a sequentially programmed all-or-none reflex in which multiple responses are triggered in a specific temporal sequence; that is, a number of highly coordinated activities are initiated in a regular pattern over a period of time to accomplish the act of swallowing. Swallowing is initiated voluntarily, but once it is initiated it cannot be stopped. Perhaps you have experienced this when a large piece of hard candy inadvertently slipped to the rear of your throat, triggering an unintentional swallow.

During the oropharyngeal stage of swallowing, food is directed into the esophagus and is prevented from entering the wrong passageways.

Swallowing is arbitrarily divided into two stages: the oropharyngeal stage and the esophageal stage. The **oropharyngeal stage** lasts about one second and consists of moving the bolus from the mouth through the pharynx and into the esophagus. When the bolus enters the pharynx during swallowing, it must be directed into the esophagus and prevented from entering the other openings that communicate with the pharynx. In other words, food must be prevented from reentering the mouth, from entering the nasal passages, and from entering the trachea. All of this is accomplished by the following coordinated activities that occur during the oropharyngeal stage (Fig. 16–5):

□ Food is prevented from reentering the mouth during swallowing by the position of the tongue against the hard palate and by contraction of skeletal muscles in the pharynx.

□ The uvula is elevated and lodges against the back of the throat, sealing off the nasal passage from the pharynx so that food does not enter the nose.

□ Food is prevented from entering the trachea primarily by elevation of the larynx and tight closure of the vocal cords across the laryngeal opening, or **glottis.** The first portion of the trachea is the *larynx* or *voice box*, across which are stretched the *vocal cords* (see p. 409). During swallowing, the vocal cords serve a purpose unrelated to speech. Contraction of laryngeal muscles aligns the vocal cords in tight apposition to each other, thus sealing the glottis entrance. Also, the bolus tilts a small flap of cartilaginous tissue, the **epiglottis,** backward down over the closed glottis as further protection from food entering the respiratory airways.

□ Because the respiratory passages are temporarily sealed off during swallowing, respiration is briefly inhibited so that the individual does not attempt futile respiratory efforts.

□ With the larynx and trachea sealed off, pharyngeal muscles contract to force the bolus into the esophagus.

The esophagus is guarded by sphincters at both ends.

The **esophagus** is a fairly straight muscular tube that extends between the pharynx and stomach (Table 16–2). Lying for the most part in the thoracic cavity, it penetrates the diaphragm at an opening called the **esophageal hiatus** and joins the stomach in the abdominal cavity a few centimeters below the diaphragm. Occasionally, a portion of the stomach herniates through the esophageal hiatus and protrudes into the thoracic cavity, a condition known as a **hiatal hernia.**

The esophagus is guarded at both ends by sphincters. A sphincter is a ringlike muscular structure that, when closed, prevents passage through the tube it guards. The upper esophageal sphincter is the **pharyngoesophageal sphincter,** and the lower sphincter is the **gastroesophageal sphincter.** Since the esophagus is exposed to subatmospheric intrapleural pressure (see p. 413), a pressure gradient exists between the atmosphere and the esophagus. Except during a swallow, the pharyngoesophageal sphincter keeps the entrance to the esophagus closed to prevent large volumes of air from entering the esophagus and stomach during breathing. Instead, the air is directed only into the respiratory airways. Were it not for the pharyngoesophageal sphincter, the digestive tract would be subjected to large volumes of gas, which would lead to excessive **eructation** (burping). Air present in the stomach is largely swallowed air. Unlike most sphincters, because of passive elastic tensions in the walls of the pharyngoesophageal sphincter, the esophagus is closed when this sphincter is relaxed. During swallowing, this sphincter contracts, opening the sphincter and allowing the bolus to pass into the esophagus. Once the bolus has entered the esophagus, the pharyngoesophageal sphincter closes, the respiratory airways are opened, and breathing resumes. The oropharyngeal phase is complete, and about one second has passed since the swallow was first voluntarily initiated.

Peristaltic waves push the food through the esophagus.

The **esophageal phase** of the swallow now begins. The swallowing center initiates a **primary peristaltic wave** that sweeps from the beginning to the end of the esophagus, forcing the bolus ahead of it through the esophagus to the stomach. **Peristalsis** refers to ringlike contractions of the circular smooth muscle that move progressively forward with a stripping motion, pushing the bolus ahead of the contraction (Fig. 16–6). Thus, propulsion of food through the esophagus is an active process that does not rely on gravity. Food can be pushed to the stomach even during a head stand. The peristaltic wave moves relatively slowly down the esophagus, taking about five to nine seconds to reach the lower end of the esoph-

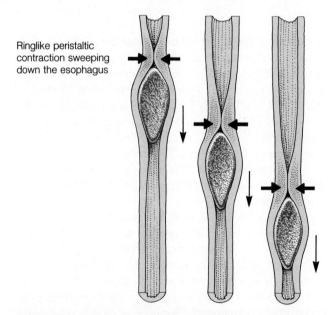

Figure 16–6 Peristalsis in the Esophagus *As the wave of peristaltic contraction sweeps down the esophagus, it pushes the bolus ahead of it toward the stomach.*

agus. Progression of the wave is controlled by the swallowing center, with innervation being by means of the vagus. The esophagus is activated by vagal fibers in a sequential manner from beginning to end so that peristalsis proceeds smoothly.

If a large or sticky swallowed bolus, such as a bite of peanut butter sandwich, fails to be carried along to the stomach by the primary wave of peristalsis, the lodged bolus distends the esophagus, stimulating pressurereceptors within its walls. This initiates a second, more forceful peristaltic wave that is mediated by the intrinsic nerve plexuses at the level of the distention. These **secondary peristaltic waves** do not involve the swallowing center, nor is the person aware of their occurrence. Distention of the esophagus also reflexly increases salivary secretion. The trapped bolus is eventually dislodged and moved forward through the combination of lubrication by the extra swallowed saliva and the forceful secondary peristaltic waves.

Liquids, not being held up by friction against the esophageal wall, fall quickly to the lower esophageal sphincter by gravity and then must wait about five seconds until the primary peristaltic wave finally arrives before they can pass through the gastroesophageal sphincter.

The gastroesophageal sphincter prevents reflux of gastric contents.

Except during swallowing, the gastroesophageal sphincter remains tonically contracted to maintain a barrier between the stomach and esophagus, thus reducing the possibility of reflux of acidic gastric contents into the esophagus. If gastric contents do flow back into the esophagus in spite of the sphincter, the acidity of these contents irritates the esophagus, causing the esophageal discomfort known as **heartburn.** (The heart itself is not involved at all.)

The gastroesophageal sphincter relaxes reflexly as the peristaltic wave sweeps down the esophagus so that the bolus can pass into the stomach. After the bolus has entered the stomach, the gastroesophageal sphincter again contracts.

In a condition known as **achalasia,** the lower esophageal sphincter fails to relax during swallowing but instead contracts more vigorously. Food accumulates in the esophagus, which causes the esophagus to become enormously distended as the food's passage into the stomach is greatly delayed. People with this condition are prone to aspiration pneumonia because of the increased likelihood that some of the detained meal in the esophagus may spill into the pharynx and accidentally be aspirated (sucked) into the lungs. The underlying defect is apparently the result of damage to the myenteric nerve plexus in the region of the gastroesophageal sphincter.

Esophageal secretion is entirely protective.

Esophageal secretion is entirely mucus. In fact, mucus is secreted throughout the length of the digestive tract. By providing lubrication for passage of food, esophageal mucus lessens the likelihood that the esophagus will be damaged by any sharp edges in the newly entering food. Furthermore, it protects the esophageal wall from acid and enzymes in gastric juice if gastric reflux should occur.

The entire transit time in the pharynx and esophagus averages a mere 6 to 10 seconds, too short a time for any digestion or absorption to occur in this region.

STOMACH

The stomach stores food and begins protein digestion.

The **stomach** is a J-shaped saclike chamber lying between the esophagus and small intestine. It is arbitrarily divided into three sections based on anatomical, histological, and functional distinctions (Fig. 16–7). The **fundus** is the portion of the stomach that lies above the esophageal opening. The middle or main part of the stomach is the **body.** The smooth-muscle layers in the fundus and body are relatively thin, but the lower portion of the stomach, the **antrum,** has much heavier musculature. There are also glandular differences in the mucosa between these regions, as will be described later. The terminal portion of the stomach consists of the **pyloric**

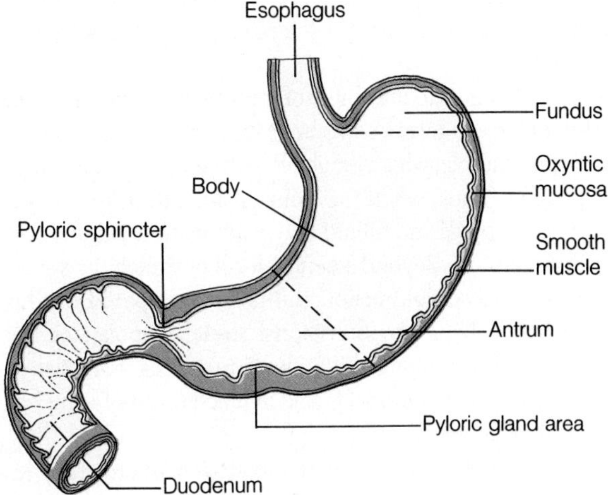

Figure 16–7 Anatomy of the Stomach

sphincter, which acts as a barrier between the stomach and the upper part of the small intestine, the duodenum.

The stomach performs several functions. The most important function is to store ingested food until it can be emptied into the small intestine at a rate appropriate for optimal digestion and absorption. It takes hours to digest and absorb a meal that was consumed in only a matter of minutes. Because the small intestine is the primary site for this digestion and absorption to take place, it is important that the stomach stores the food and meters it into the duodenum at a rate that does not exceed the small intestine's capacities. A second function of the stomach is to secrete hydrochloric acid (HCl) and enzymes that begin protein digestion. Finally, through the stomach's mixing movements, the boluses of ingested food are broken down and mixed with gastric secretions to produce a thick liquid mixture known as **chyme.**

$$\text{Food} + \text{Gastric juice} \xrightarrow{\text{Gastric mixing}} \text{Chyme}$$

Stomach motility is complex and subject to multiple regulatory inputs.

There are four aspects of gastric motility: (1) gastric filling, (2) gastric storage, (3) gastric mixing, and (4) gastric emptying.

GASTRIC FILLING. When empty, the stomach has a volume of about 50 ml, but it can fill to a capacity of about 1 liter during a meal. Accommodation of such a twenty-fold change in volume would create tension in the walls of the stomach and greatly increase intragastric pressure were it not for two factors: (1) the basic property of plasticity of the stomach smooth musculature, and (2) the phenomenon of receptive relaxation.

Plasticity refers to the ability of smooth muscle to maintain a constant tension over a wide range of lengths, unlike skeletal and cardiac muscle, which exhibit length-tension relationships (see p. 233). Thus, when the stomach smooth-muscle fibers are stretched by gastric filling, they yield without an increase in muscle tension. Beyond a certain level of stretch, however, a stretch-activated contraction can be superimposed on this passive plastic behavior. Substantial stretch depolarizes the pacesetter cells, bringing them closer to resting potential so that the slow-wave potential is able to reach threshold and initiate contractile activity.

This basic behavior of smooth muscle is augmented by **receptive relaxation.** The interior surface of the stomach is thrown into deep folds known as **rugae.** During a meal, the folds get smaller and flatten out as the stomach relaxes slightly with each mouthful, much like the gradual expansion of a collapsed ice bag as it is being filled. This enhances the stomach's ability to accommodate the extra volume of food with little rise in stomach pressure. Of course, if more than 1 liter of food is consumed, the stomach becomes overdistended, and the person experiences discomfort. Receptive relaxation is mediated by the vagus nerve. The receptor for the reflex is associated with eating, and may possibly involve stimulation of the taste buds.

GASTRIC STORAGE. Recall that some smooth-muscle cells are capable of rhythmic, autonomous, partial depolarization. One such group of these pacesetter cells is located high on the greater curvature of the stomach. These cells generate slow-wave potentials that sweep down the length of the stomach toward the pyloric sphincter at a rate of three per minute. This rhythmic train of spontaneous depolarizations, the basic electrical rhythm or BER of the stomach, occurs continuously and may or may not be accompanied by contraction of the stomach's circular smooth-muscle layer. Depending on the level of excitability in the smooth muscle, it may be brought to threshold by this flow of current and undergo action potentials, which in turn initiate muscle contractions recognized as peristaltic waves that sweep over the stomach in pace with the BER at a rate of three per minute.

Once initiated, the peristaltic wave spreads over the fundus and body to the antrum and pyloric sphincter. Because the muscle layers are thin in the fundus and body, the peristaltic contractions in this region are weak. When the waves reach the antrum, they become much stronger and more vigorous because the muscle there is much thicker.

Since only feeble mixing movements occur in the body and fundus, food emptied into the stomach from the esophagus is stored in the relatively quiet body without being mixed. The fundic area usually does not store food but contains only a pocket of gas. Food is gradually fed from the body into the antrum, where mixing does take place.

GASTRIC MIXING. The strong antral peristaltic contractions are responsible for the mixing of food with gastric secretions to produce chyme. Each antral peristaltic wave propels chyme forward toward the pyloric sphincter. Tonic contraction of the pyloric sphincter normally keeps it almost, but not completely, closed. The opening is large enough for water and other fluids to pass through with ease but too small for the thicker chyme to pass through except when a strong peristaltic antral contraction pushes it through. Even then, of the 30 ml of chyme that the antrum can hold, usually only a few ml of antral contents are forced into the duodenum with each peristaltic wave. Before more chyme can be squeezed out, the peristaltic wave reaches the pyloric sphincter and causes it to contract more forcefully, sealing off the exit and blocking further passage into the duodenum. The bulk of the antral chyme that was being propelled forward but failed to be pushed into the duodenum is abruptly halted at the closed sphincter and is tumbled back into the antrum, only to be propelled forward and tumbled back again as the next peristaltic wave advances. This process, called **retropulsion,** accomplishes thorough mixing of the chyme in the antrum.

GASTRIC EMPTYING. The antral peristaltic contractions, in addition to being responsible for gastric mixing, provide the driving force for gastric emptying. The amount of chyme that escapes into the duodenum with each peristaltic wave before the pyloric sphincter closes tightly depends largely on the strength of peristalsis. The intensity of antral peristalsis can vary markedly under the influence of different signals from both the stomach and the duodenum; thus gastric emptying is regulated by both gastric and duodenal factors. By slightly depolarizing or hyperpolarizing the gastric smooth muscle, these factors influence the muscle's excitability, which in turn is a determinant of the degree of antral peristaltic activity. The greater the excitability, the more frequently the BER will generate action potentials, the greater the degree of peristaltic activity in the antrum, and the faster the rate of gastric emptying (Table 16–3).

Factors in the stomach that influence the rate of gastric emptying. The main gastric factor that influences the strength of contraction is the amount of chyme in the stomach. Other things being equal, the stomach empties at a rate proportional to the volume of chyme in it at any given time. Distention of the stomach triggers increased gastric motility through a direct effect of stretch on the smooth muscle as well as through involvement of the intrinsic plexuses, the vagus nerve, and the stomach hormone gastrin.

Furthermore, the degree of fluidity of the chyme in the stomach influences gastric emptying. The stomach contents must be converted into a finely divided, thick liquid form before emptying. The sooner the appropriate degree of fluidity can be achieved, the more rapidly the contents are ready to be evacuated.

Factors in the duodenum that influence the rate of gastric emptying. In spite of these gastric influences, factors in the

Table 16–3 Factors Regulating Gastric Motility and Emptying

Factors	Mode of Regulation	Effects on Gastric Motility and Emptying
Within Stomach		
Volume of chyme	Distention has direct effect on gastric smooth-muscle excitability, as well as acting through intrinsic plexuses, vagus nerve, and gastrin	Increased volume stimulates motility and emptying
Degree of fluidity	Direct effect; contents must be in fluid form to be evacuated	Increased fluidity allows more rapid emptying
Within Duodenum		
Presence of fat, acid, hypertonicity, or distention	Initiates enterogastric reflex or triggers release of enterogastrones (cholecystokinin, secretin, gastric inhibitory peptide)	These factors in duodenum inhibit further gastric motility and emptying until duodenum has coped with factors already present
Outside Digestive System		
Emotion	Alters autonomic balance	Stimulates or inhibits motility and emptying
Intense pain	Increases sympathetic activity	Inhibits motility and emptying
Decreased glucose utilization in hypothalamus	Increases vagal activity	Stimulates motility; accompanied by hunger pangs

duodenum are of primary importance in controlling the rate of gastric emptying. The duodenum must be ready to receive the chyme and can act to delay gastric emptying by reducing peristaltic activity in the stomach until the duodenum is ready to accommodate more chyme. Even if the stomach is distended and its contents are in a liquid form, it cannot empty until the duodenum is ready to deal with the chyme.

The four most important factors in the duodenum that influence gastric emptying are fat, acid, hypertonicity, and distention. The presence of one or more of these stimuli in the duodenum activates appropriate duodenal receptors, thereby triggering either a neural or hormonal response that puts brakes on gastric motility and subsequently delays gastric emptying by reducing the excitabilitiy of the gastric smooth muscle. The *neural response* is mediated through both the intrinsic nerve plexuses (short reflex) and the autonomic nerves (long reflex). Collectively, these reflexes are called the **enterogastric reflex.** The *hormonal response* involves the release from the duodenal mucosa of several hormones collectively known as **enterogastrones.** These hormones are transported by the blood to the stomach, where they inhibit antral contractions to reduce gastric emptying. Three of these enterogastrones have been clearly identified: **secretin, cholecystokinin,** and **gastric inhibitory peptide.** *Secretin* was the first hormone discovered (in 1902). Because it was a secretory product that entered the blood, it was termed *secretin.* The name *cholecystokinin* derives from the fact that this same hormone is also responsible for contraction of the bile-containing gall bladder (*chole* means "bile"; *cysto* means "bladder"; and *kinin* means "contraction"). The name *gastric inhibitory peptide* is self-explanatory; it is a peptide hormone that inhibits the stomach.

Let us examine why it is important that each of these stimuli in the duodenum (fat, acid, hypertonicity, and distention) delays gastric emptying (acting through the enterogastric reflex or one of the enterogastrones).

□ *Fat.* Fat is digested and absorbed more slowly than the other nutrients. Furthermore, fat digestion and absorption take place only within the small-intestine lumen. Therefore, when fat is already present in the duodenum, further gastric emptying of more fatty stomach contents into the duodenum is prevented until the small intestine has processed the fat already there. In fact, fat is the most potent stimulus for inhibition of gastric motility. This is evident when one compares the rate of emptying of a high-fat meal (after six hours some of a bacon and eggs meal may still be found in the stomach) with that of a protein and carbohydrate meal (a lean meat and potatoes meal may empty in three hours).

□ *Acid.* Since the stomach secretes HCl, highly acidic chyme is emptied into the duodenum, where it is neutralized by sodium bicarbonate ($NaHCO_3$) secreted into the duodenal lumen from the pancreas. Unneutralized acid in the duodenum inhibits further emptying of acidic gastric contents until complete neutralization can be accomplished.

□ *Hypertonicity.* As molecules of protein and starch are digested in the duodenal lumen, large numbers of amino acid and glucose molecules are released. If absorption of these amino acid and glucose molecules does not keep pace with the rate at which protein and carbohydrate digestion proceeds, these large numbers of molecules remain in the chyme and increase the osmolarity of the duodenal contents. Because osmolarity is dependent on the number of molecules present, not on their size, one protein molecule may be hydrolyzed into several hundred amino acid molecules, each of which has the same osmotic activity as the original protein molecule. The same holds true for one large starch molecule, which yields many smaller but equally osmotically active glucose molecules. Since water is freely diffusible across the duodenal wall, water enters the duodenal lumen from the plasma as the duodenal osmolarity rises. Large volumes of water entering the intestine from the plasma lead to intestinal distention and, more importantly, circulatory disturbances ensue because of the reduction in plasma volume. To prevent this from occurring, when the osmolarity of the duodenal contents starts to rise, gastric emptying is reflexly inhibited. This reduces the amount of food entering the duodenum for further digestion into a multitude of additional osmotically active particles until absorption processes have had an opportunity to catch up.

□ *Distention.* Too much chyme in the duodenum inhibits the emptying of even more gastric contents. This allows the distended duodenum time to cope with the excess volume of chyme it already contains before it receives an additional quantity.

Peristaltic contractions occur in the empty stomach before the next meal.

After a meal is finally emptied from the stomach, there are no more gastric factors to enhance gastric excitability, so peristaltic contractions eventually die out and the stomach remains quiet for a while. In conjunction with the sensation of hunger before the next meal, however, peristaltic contractions commence again, sweeping over the nearly empty antrum. This arousal of stomach motility appears to be mediated by increased parasympathetic activity, perhaps brought about by the hypothalamus in response to a fall in hypothalamic glucose utilization as the time for the next meal approaches. One of the leading theories of why we get hungry also involves decreased use of glucose by the hypothalamus (chapter 17). An individual may experience the sensation of hunger pangs

A CLOSER LOOK AT EXERCISE PHYSIOLOGY

Pregame Meal: What's In and What's Out?

Many studies have been done to determine the effect of the pregame meal on athletic performance. Although substances such as caffeine have been shown to improve endurance in laboratory studies, no food substance has been identified that will greatly enhance performance. The athlete's prior training is the most important determinant of performance.

The greatest benefit of the pregame meal is to prevent hunger during competition. Because it can take from one to four hours for the stomach to empty, an athlete should eat at least three to four hours before competition begins. Excessive quantities of food should not be consumed before competition. Food that remains in the stomach during competition may cause nausea and possibly vomiting. This condition can be aggravated by nervousness, which slows digestion and delays gastric emptying by means of the sympathetic nervous system.

Foods that are slowly digested, such as high-fat fried foods, should be limited or avoided in the pregame meal. High-carbohydrate foods are recommended because they are removed from the stomach more easily than fat or protein are. Carbohydrates do not inhibit gastric emptying by means of cholecystokinin release, whereas fat and protein do.

Beverages or foods high in sugar should be avoided before competition because they trigger insulin release. Insulin is the hormone that enhances glucose entry into most body cells. Once the person begins exercising, insulin sensitivity increases (see p. 79), which results in a decrease in the plasma glucose level. A lowered plasma glucose level induces feelings of fatigue and an increased use of muscle glycogen stores, which can limit performance in endurance events such as the marathon. It is best for the athlete to only drink plain water within an hour of competition.

when these peristaltic contractions are occurring, but the contractions themselves are not responsible for the sensation. Rather, both the sensation of hunger and the increased peristaltic activity are triggered simultaneously by the reduced amount of glucose being metabolized by the brain.

Emotions can influence gastric motility.

Other factors unrelated to digestion can also alter gastric motility by acting through the autonomic nerves to influence the degree of gastric smooth-muscle excitability. Even though the effect of emotions on gastric motility varies from one individual to another and is not always predictable, sadness and fear generally tend to decrease motility whereas anger and aggression tend to increase it. In addition to emotional influences, intense pain from any part of the body tends to inhibit motility, not just in the stomach but throughout the digestive tract. This response is brought about by increased sympathetic activity and a corresponding decrease in parasympathetic activity. (For a discussion of the pregame meal before participation in an athletic event, see the accompanying boxed feature, A Closer Look at Exercise Physiology.)

The body of the stomach does not actively participate in the act of vomiting.

Vomiting, or **emesis,** the forceful expulsion of gastric contents out through the mouth, is generally perceived as being caused by abnormal gastric motility. However, vomiting is not accomplished by reverse peristalsis, as might be predicted. Actually, the stomach itself does not actively participate in the act of vomiting. The stomach, the esophagus, the gastroesophageal sphincter, and the pyloric sphincter are all relaxed during vomiting. The major force for expulsion comes, surprisingly, from contraction of the respiratory muscles—namely, the diaphragm (the major inspiratory muscle) and the abdominal muscles (the muscles of active expiration).

Vomiting begins with a deep inspiration and closure of the glottis. The contracting diaphragm descends downward on the stomach while simultaneous contraction of the abdominal muscles compresses the abdominal cavity, increasing the intra-abdominal pressure and forcing the abdominal viscera upward. As the flaccid stomach is squeezed between the diaphragm from above and the compressed abdominal cavity from below, the gastric contents are forced out through the re-

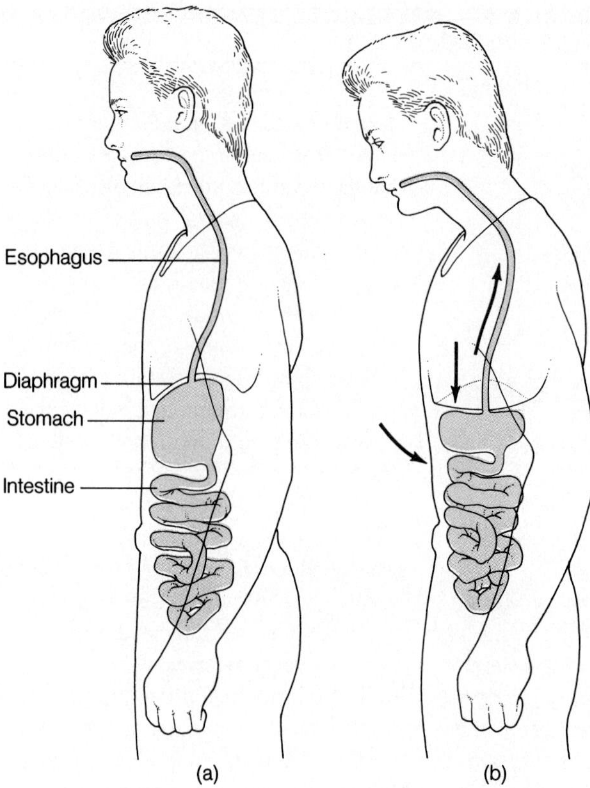

Figure 16–8 Muscular Contractions Responsible for Vomiting *(a) Position of abdominal organs at rest. (b) Position of abdominal organs during vomiting. The contents of the relaxed stomach are squeezed out as the contracting diaphragm descends downward on the stomach and the intestines are pushed upward on the stomach upon contraction of the abdominal muscles.*

laxed gastroesophageal sphincter into the esophagus (Fig. 16–8). The glottis is closed, so vomited material does not enter the respiratory airways. Also, the uvula is elevated to close off the nasal cavity.

Often when vomitus first enters the esophagus, the pharyngoesophageal sphincter remains closed so that no gastric contents enter the mouth. Distention of the esophagus by the vomitus induces secondary peristaltic waves that force the gastric contents back into the stomach. This cycle repeats itself as the contents are immediately squeezed up into the esophagus again. This is the act of *retching* or *heaves*. After a series of heaves, when the pressure becomes great enough, the person thrusts out the jaw, pulling open the pharyngoesophageal sphincter. The gastric contents are then forced through the esophagus, through the pharyngoesophageal sphincter, and out through the mouth. During this time the duodenum contracts strongly, which may force some of the intestinal contents back into the stomach and out with the vomitus. The vomited material may therefore be bile-stained, the yellowish bile having entered the small intestine from the liver and gall bladder. The vomiting cycle may be repeated several times until the stomach is emptied. Vomiting is usually preceded by profuse salivation, sweating, rapid heart rate, and the sensation of nausea, all of which are characteristic of a generalized discharge of the autonomic nervous system.

This complex act of vomiting is coordinated by a **vomiting center** in the medulla. Nausea, retching, and vomiting can be initiated by afferent input to the vomiting center from a number of receptors throughout the body. The causes of vomiting include the following:

1. Tactile (touch) stimulation of the back of the throat, which is one of the most potent stimuli. For example, sticking a finger in the back of the throat or even the presence of a tongue depressor or dental instrument in the back of the mouth is sufficient stimulation to cause gagging and even vomiting in some people.
2. Irritation or distention of the stomach and duodenum.
3. Elevated intracranial pressure, such as that caused by cerebral hemorrhage. This is why vomiting following a head injury is considered to be a bad sign; it suggests swelling or bleeding within the cranial cavity.
4. Rotation or acceleration of the head producing dizziness, such as that occurring in motion sickness.
5. Intense pain arising from a variety of organs, such as that accompanying passage of a kidney stone.
6. Chemical agents, including drugs or noxious substances that initiate vomiting (that is, **emetics**) either by acting in the upper portions of the gastrointestinal tract or by stimulating chemoreceptors in a specialized **chemoreceptor trigger zone** in the brain. Activation of this zone triggers the vomiting reflex.
7. Psychic vomiting induced by emotional factors, such as those accompanying nauseating sights and odors and even including vomiting before taking an examination or in other stressful situations.

Excessive vomiting leads to large losses from the body of secreted fluids and acids that normally would be reabsorbed. The resultant reduction in plasma volume can lead to severe dehydration and circulatory problems while the loss of acid from the stomach simultaneously can lead to metabolic alkalosis (see p. 544).

Vomiting is not always detrimental, however. Limited vomiting brought about by irritation of the digestive tract can provide a useful service in removing noxious material from the stomach rather than allowing it to be retained and absorbed. In fact, emetics are frequently given in the case of accidental ingestion of a poison to quickly remove the offending substance from the body.

Gastric pits are the source of gastric digestive secretions.

Each day the stomach secretes about 2 liters of gastric juice. The cells responsible for gastric secretion are located in the lining of the stomach, the gastric mucosa, which is divided into two distinct areas: (1) the **oxyntic glandular mucosa,** which covers the body and fundus, and (2) the **pyloric gland area** (**PGA**), which covers the antrum. The muscosal gland cells are found in **gastric pits** (Fig. 16–9), which are invaginations or deep pockets in the luminal surface of the stomach.

Three types of secretory cells are found in the walls of the pits in the oxyntic mucosa. The neck or entrance of the gastric pit is lined by **mucous cells,** which secrete a thin, watery mucus. The deeper portions of the pit are lined by **chief cells,** which secrete the enzyme precursor, ***pepsinogen***. On the outer wall of the gastric pits are found the **parietal** or **oxyntic cells,** which secrete HCl and intrinsic factor. (*Parietal* means "wall," in reference to the location of these cells. *Oxyntic* means "sharp," in reference to these cells' potent HCl secretory product.) Although the parietal cells are separated from the lumen of the gastric pit by the chief cells, they transmit their HCl secretion into the lumen through fine channels, or **canaliculi,** that pass between the chief cells. Between the gastric pits, the gastric mucosa is covered by surface epithelial cells, which secrete a thick, viscous, alkaline mucus that forms a visible layer several millimeters thick over the surface of the mucosa.

The mucous neck cells rapidly divide and serve as the parent cells of all new cells of the gastric mucosa. The daughter cells that result from cell division either migrate out of the pit to become surface epithelial cells or migrate down into the deeper parts of the pit to differentiate into chief or parietal cells. Through this activity, the entire stomach mucosa is replaced about every three days.

The gastric pits of the pyloric gland area, in contrast to the oxyntic mucosa, primarily secrete mucus and a small amount of pepsinogen, with no acid being secreted in this area. More importantly, endocrine cells in the PGA secrete the hormone **gastrin** into the blood. Thus the most important gastric digestive secretions produced within the body and fundus are HCl, pepsinogen, mucus, and intrinsic factor, whereas the most important product of the pyloric gland area is the hormone gastrin, each of which we will examine in more detail.

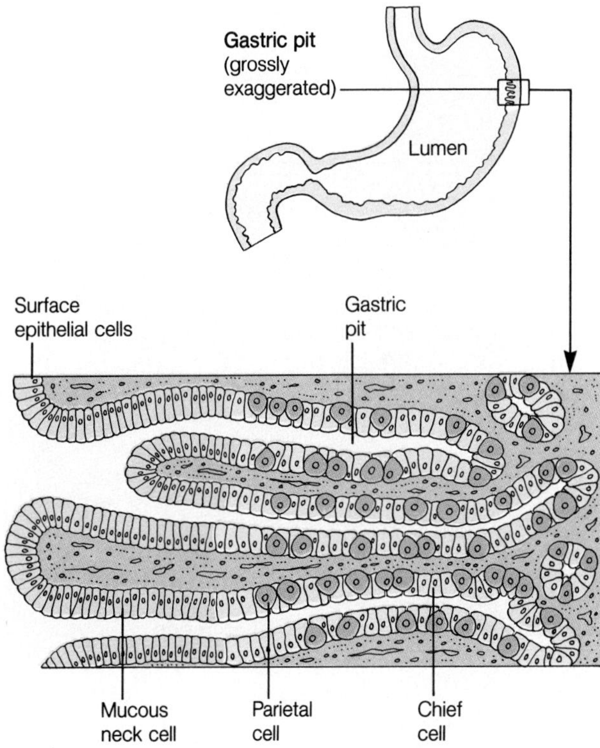

Figure 16–9 Gastric Pits

HYDROCHLORIC ACID SECRETION. The parietal cells actively secrete HCl into the lumen of the gastric pits, which in turn empty into the lumen of the stomach. Hydrogen ion (H^+) and chloride ion (Cl^-) appear to be actively transported by separate pumps in the parietal cell's plasma membrane. Hydrogen ion is actively transported against a tremendous concentration gradient, with the H^+ concentration being as much as 3 to 4 million times greater in the lumen than in the blood. Because of the high energy expenditure needed to move H^+ against such a large gradient, parietal cells have an abundance of mitochondria, the energy organelles. Chloride is also actively secreted but against a much smaller concentration gradient of only one and a half times.

The secreted H^+ is not transported from the plasma but is derived instead from metabolic processes within the parietal cell (Fig. 16–10). Whenever a H^+ is secreted, neutrality of the interior of the cell is maintained by generation of a new H^+ from carbonic acid (H_2CO_3) to replace the secreted H^+. The parietal cells contain an abundance of the enzyme carbonic anhydrase (ca). In the presence of carbonic anhydrase, H_2O readily combines with CO_2, which either has been produced within the parietal cell by metabolic processes or has diffused in from the blood. The combination of H_2O and CO_2 results in the formation of H_2CO_3, which partially dissociates to yield H^+ and HCO_3^-:

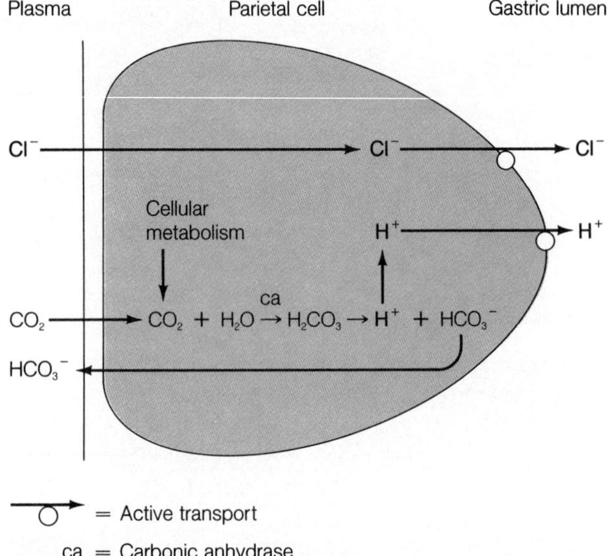

Figure 16–10 Mechanism of HCl Secretion

$$CO_2 + H_2O \xrightarrow{ca} H_2CO_3 \rightarrow H^+ + HCO_3^-$$

The generated H^+ replaces the one secreted, and the HCO_3^- passes into the plasma to maintain electrical neutrality by replacing the Cl^- that has been actively secreted from the plasma into the gastric lumen.

Although HCl does not actually digest anything and is not absolutely essential to gastrointestinal function, it does perform several functions that assist digestion. Hydrochloric acid: (1) activates the enzyme precursor pepsinogen to an active enzyme, pepsin, and provides an acid medium that is optimal for pepsin activity; (2) aids in the breakdown of connective tissue and muscle fibers, thereby reducing large food particles into smaller particles; and (3) along with salivary lysozyme, kills most of the microorganisms ingested with the food, although some do escape and continue to grow and multiply in the large intestine.

PEPSINOGEN SECRETION. The major digestive constituent of gastric secretion is **pepsinogen,** an inactive enzymatic molecule synthesized and packaged by the endoplasmic reticulum and Golgi complex of the chief cells. Pepsinogen is stored in the cytoplasm within secretory vesicles known as **zymogen granules,** from which it is released by exocytosis (see p. 24) upon appropriate stimulation. When pepsinogen is secreted into the gastric lumen (Fig. 16–11), HCl cleaves off a small fragment of the molecule, converting it to the active form of the enzyme, **pepsin.** Once formed, pepsin acts on other pepsinogen molecules to produce more pepsin. A mechanism such as this whereby an active form of an enzyme activates other molecules of the same enzyme is referred to as an **autocatalytic** ("self-activating") **process.**

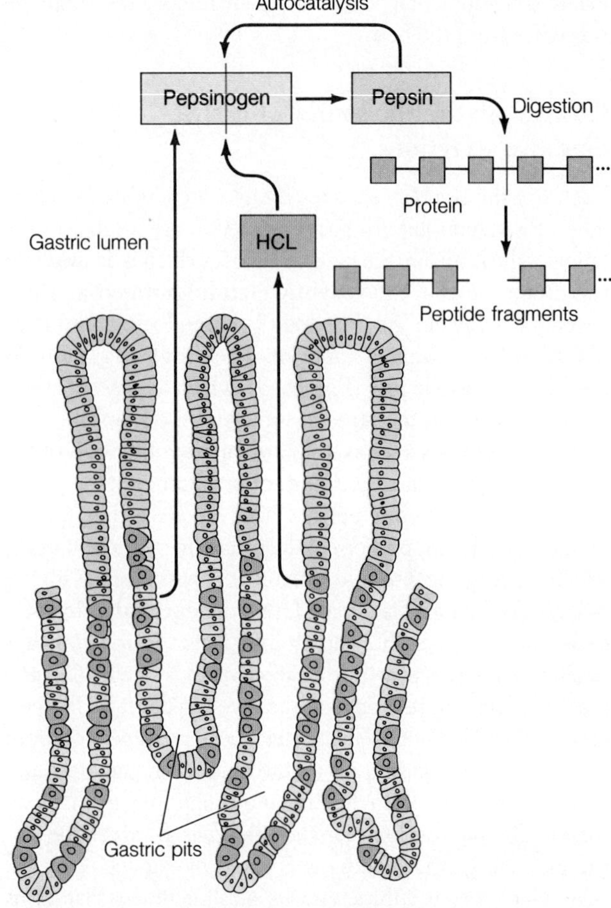

Figure 16–11 Pepsinogen Activation in Stomach Lumen
Hydrochloric acid activates pepsinogen to its active form, pepsin, within the lumen by cleaving off a small fragment. Once activated, pepsin can autocatalytically activate more pepsinogen and begins protein digestion. Secretion of pepsinogen in the inactive form prevents it from digesting the protein structures of the cells within which it is produced. Its activation process does not commence until it reaches the lumen and comes into contact with HCl secreted by a separate cell in the gastric pit.

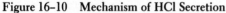

Pepsin initiates protein digestion by splitting certain amino acid linkages in proteins to yield peptide fragments (small amino acid chains), which it does most effectively in the acid environment provided by HCl. Because pepsin can digest protein, it must be stored and secreted in an inactive form so that it does not digest the cells in which it is formed. (The primary structural component of cells is protein.) Therefore, pepsin is maintained in the inactive form of pepsinogen until it reaches the gastric lumen, where it is activated by HCl.

MUCUS SECRETION. Mucus, which is derived from the surface epithelial cells and mucous neck cells, serves as a protective barrier against several forms of potential injury to the gastric mucosa.

□ By virtue of its lubricating properties, mucus protects the gastric mucosa against mechanical injury.

□ It helps protect the stomach wall from self-digestion because pepsin is inhibited when it comes in contact with the mucus layer coating the stomach. (However, mucus does not affect pepsin activity in the lumen, where digestion of dietary protein proceeds without interference.)

□ Being alkaline, mucus helps protect against acid injury by neutralizing HCl in the vicinity of the gastric lining.

INTRINSIC FACTOR. **Intrinsic factor,** another secretory product of the parietal cells besides HCl, is important in the absorption of vitamin B_{12}. Only when in combination with intrinsic factor can this vitamin be absorbed by a special transport mechanism in the terminal portions of the ileum . Vitamin B_{12} is essential for the normal formation of red blood cells. In the absence of intrinsic factor, vitamin B_{12} fails to be absorbed, so erythrocyte production is defective, resulting in pernicious anemia (see p. 349).

Occasionally, the oxyntic glandular mucosa atrophies or degenerates. With the loss of chief and parietal cells, the stomach is unable to secrete pepsinogen, HCl, and intrinsic factor. Although pepsin and acid normally begin protein digestion in the stomach, they are not absolutely essential to protein digestion. If necessary, the pancreatic and small intestinal enzymes can completely digest proteins. The most detrimental consequence of **gastric mucosal atrophy** is the loss of intrinsic factor and the subsequent development of pernicious anemia unless therapeutic injection of vitamin B_{12} is administered. The exact cause of gastric mucosal atrophy is uncertain, although it is suspected to be an autoimmune response (see p. 390), because many individuals with this condition have antibodies to oxyntic cells in their blood.

GASTRIN SECRETION. Special endocrine cells, the **G cells,** located in the pyloric gland area of the stomach secrete gastrin into the blood upon appropriate stimulation. After being carried by the blood back to the oxyntic mucosa, gastrin stimulates the parietal and chief cells, thereby promoting further secretion of a highly acidic gastric juice. Gastrin is also ***trophic*** (growth-promoting) to the mucosa of the stomach and small intestine, thereby maintaining their secretory capabilities.

Control of gastric secretion involves three phases.

The rate of gastric secretion can be influenced by: (1) factors arising before food ever reaches the stomach; (2) factors resulting from the presence of food in the stomach; and (3) factors in the duodenum after food has left the stomach. Accordingly, gastric secretion is divided into three phases—the cephalic, the gastric, and the intestinal phases (Table 16–4).

Table 16–4 Stimulation of Gastric Secretion

Phase	Stimuli	Excitatory Mechanism for Enhancing Gastric Secretion
Cephalic phase	Stimuli in head—seeing, smelling, tasting, chewing, swallowing food	→ (+) Vagus → (+) Intrinsic nerves → (+) Parietal and chief cells; Vagus → (+) Pyloric gland area → ↑ Gastrin → (+) Parietal and chief cells
Gastric phase	Stimuli in stomach—protein, (peptide fragments), distention, caffeine, alcohol	→ (+) Vagus → (+) Intrinsic nerves → (+) Parietal and chief cells; Vagus → (+) Pyloric gland area → ↑ Gastrin → (+) Parietal and chief cells; stimuli → (+) Intrinsic nerves; stimuli → (+) Pyloric gland area
Excitatory intestinal phase	Stimuli in duodenum—digested protein products	→ ↑ Intestinal gastrin → (+) Parietal and chief cells

CEPHALIC PHASE. The cephalic phase of gastric secretion refers to the increased secretion of HCl and pepsinogen that occurs in response to stimuli acting in the head (***cephalic*** refers to "head") even before food reaches the stomach. Thinking about, tasting, smelling, chewing, and swallowing food increase gastric secretion by means of vagal nerve activity in two ways. First, vagus stimulation acts through the intrinsic plexuses to promote increased secretion of HCl and pepsinogen by the secretory cells. Second, vagal stimulation of the PGA causes the release of gastrin, which in turn further enhances secretion of HCl and pepsinogen.

GASTRIC PHASE. The gastric phase of gastric secretion occurs when food actually reaches the stomach. Stimuli acting in the stomach—namely, protein, especially peptide fragments; distention; caffeine; or alcohol—increase gastric secretion by means of overlapping efferent pathways. For example, protein in the stomach, the most potent stimulus, initiates short local reflexes in the intrinsic nerve plexuses to stimulate the secretory cells. Furthermore, protein initiates long reflexes so that extrinsic vagal fibers to the stomach are activated. Vagal activity further enhances intrinsic nerve stimulation of the secretory cells and triggers the release of gastrin. Protein also directly stimulates the release of gastrin. Gastrin, in turn, is a powerful stimulus for further acid and pepsinogen secretion. Through these synergistic and overlapping pathways, protein induces the secretion of a highly acidic, pepsin-rich gastric juice, which continues the digestion of the protein that first initiated the process.

When the stomach is distended with protein-rich food that needs to be digested, these secretory responses are appropriate. Caffeine and, to a lesser extent, alcohol also stimulate the secretion of a highly acidic gastric juice, even when no food is present. This unnecessary acid can irritate the linings of the stomach and duodenum. For this reason, caffeinated and alcoholic beverages should be avoided by persons with ulcers or gastric hyperacidity.

INTESTINAL PHASE. The third phase of gastric secretion, the intestinal phase, encompasses those factors originating in the small intestine that influence gastric secretion. There are two components of the intestinal phase—the excitatory and the inhibitory components (Fig. 16–12).

To a limited extent, the presence of the products of protein digestion in the duodenum stimulates further gastric secretion, acting through release of an **intestinal gastrin** that is carried by the blood to the stomach. This is the *excitatory component* of the intestinal phase of gastric secretion. It is as if the small intestine, on noting the arrival of protein fragments coming down from the stomach, offers the stomach a "helping hand" in digesting the protein by enhancing gastric secretion.

Figure 16–12 Excitatory and Inhibitory Components of Intestinal Phase of Gastric Secretion

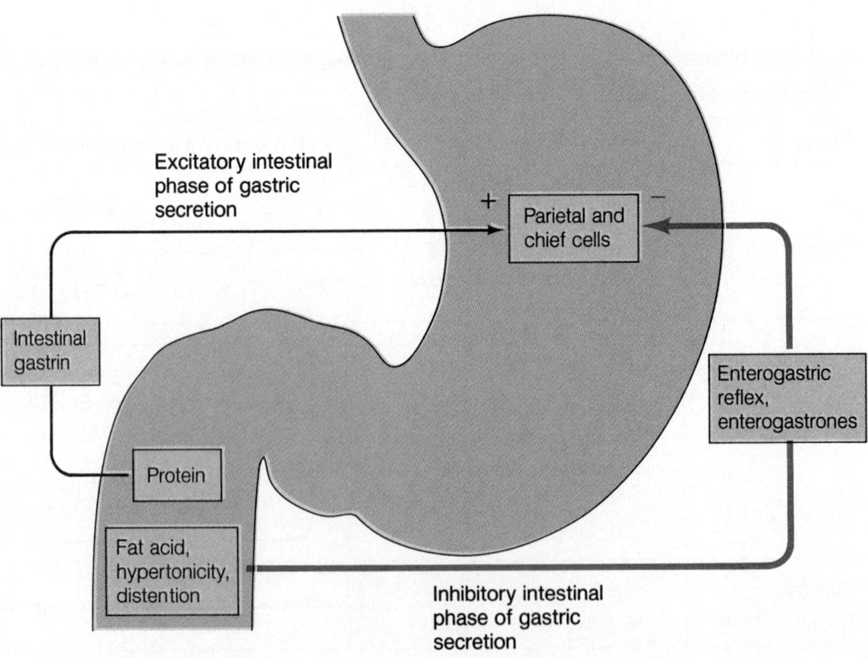

Table 16–5 Inhibition of Gastric Secretion

Region	Stimuli	Inhibitory Effect on Gastric Secretion
Body and antrum	Removal of protein and distention as stomach empties	→ (−) Intrinsic nerves; → (−) Vagus; → (−) Pyloric gland area ↓ Gastrin } → ↓ Gastric secretion
Duodenum (inhibitory intestinal phase of gastric secretion)	Fat Acid Hypertonicity Distention	→ (+) Enterogastric reflex; → ↑ Enterogastrones (cholecystokinin, secretin, gastric inhibitory peptide) } → ↓ Gastric secretion
Antrum	Accumulation of acid	→ (−) Pyloric gland area → ↓ Gastrin → ↓ Gastric secretion

However, the ***inhibitory component*** is the dominant component of the intestinal phase; that is, inhibitory influences on gastric secretion arising from the small intestine dominate over excitatory influences. This is important in helping to shut off the flow of gastric juices as chyme begins to be emptied into the small intestine. Gastric secretion is gradually reduced by three different means as the stomach empties (Table 16–5):

□ The same stimuli in the duodenum that inhibit gastric motility (fat, acid, hypertonicity, or distention) inhibit gastric secretion as well; the enterogastric reflex and the enterogastrones suppress the gastric secretory cells while they simultaneously reduce the excitability of the gastric smooth-muscle cells.

□ As the meal is gradually emptied into the duodenum, the major stimulus for enhanced gastric secretion—the presence of protein in the stomach—is withdrawn.

□ After foods leave the stomach and gastric juices accumulate to such an extent that gastric pH falls very low, gastric secretion is inhibited because a high concentration of H^+ directly inhibits the PGA from releasing gastrin. As gastrin secretion declines, the most potent stimulant of gastric secretion is withdrawn.

The stomach lining is protected from gastric secretions by the gastric mucosal barrier.

How can the stomach contain strong acid contents and proteolytic enzymes without destroying itself? We already learned that mucus provides a protective coating. In addition, other barriers to mucosal acid damage are provided by the mucosal lining itself (Fig. 16–13). First, the luminal membranes of the gastric mucosal cells are almost impermeable to H^+, so acid cannot penetrate *into* the cells and cause cellular damage. Furthermore, the lateral edges of these cells are joined together near their luminal borders by tight junctions (see p. 68), so acid cannot diffuse *between* the cells from the lumen into the underlying submucosa. The properties of the gastric mucosa that enable the stomach to contain acid without injuring itself are referred to as the **gastric mucosal barrier.** These protective mechanisms are further enhanced by the fact that the entire stomach lining is replaced every three days. Because of rapid mucosal turnover, cells are usually replaced before they are exposed to the wear and tear of harsh gastric conditions long enough to produce cellular damage.

In spite of the protection provided by mucus, by the gastric mucosal barrier, and by the frequent turnover of cells, the barrier occasionally is broken so that the gastric wall is injured by its acidic and enzymatic contents. When this occurs, an erosion, or **peptic ulcer,** of the stomach wall results. Ulceration caused by acid and pepsin can also occur in the duodenum and the esophagus; these areas do not have the same mucosal barriers as the stomach, although they are protected to some extent. Both the esophagus and duodenum secrete mucus, and the duodenum is further protected by pancreatic secretions that empty into the duodenal lumen and neutralize the acidic contents that arrive from the stomach. However, excessive gastric reflux into the esophagus and dumping of exces-

Figure 16–13 Gastric Mucosal Barrier

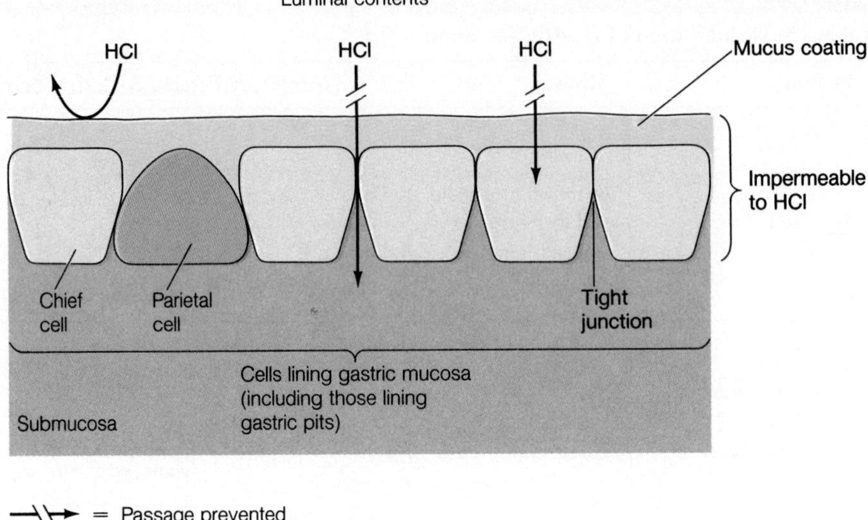

sive acidic gastric contents into the duodenum can lead to peptic ulcers in these locations. In fact, duodenal ulcers are considerably more common than gastric ulcers; esophageal ulcers are less common.

As might be expected, patients with duodenal ulcers have a high duodenal acid content that overwhelms the protective mechanisms, but, surprisingly, individuals with gastric ulcers usually have a low acid content in their stomachs. This is presumably due either to a weak mucosal barrier that is vulnerable to even normal levels of acid or to an originally high level of acid that has overwhelmed the normal defensive mechanisms so that acid diffuses out of the lumen into the damaged stomach wall, leaving behind a deceptively low level of acid in the lumen.

The exact cause of ulcers is uncertain. Some substances can break the gastric mucosal barrier, the most important of which are ethyl alcohol, aspirin, and bile salts, the latter being regurgitated from the duodenum into the stomach. The barrier frequently breaks in patients with preexisting debilitating conditions, such as severe injuries or infections. It appears that excessive acid and pepsin secretion sometimes overwhelm otherwise normal protective mechanisms. For example, the persistence of stressful situations is frequently associated with ulcer formation, presumably because of excessive stimulation of gastric secretion brought about by the emotional response related to the stress, although this potential relationship is still being debated.

When the barrier is broken (either because it is weak or damaged or because it is overwhelmed by excessive secretion), acid and pepsin diffuse into the mucosa with serious pathophysiologic consequences (Fig. 16–14). Acid triggers the release of histamine, a potent acid stimulant that is produced and stored in large amounts in the mucosa. (It is uncertain whether histamine plays a role under normal circumstances.) Released histamine stimulates further acid secretion, which can diffuse back into the mucosa to stimulate further histamine release, which triggers more acid release, and so on, thus establishing a vicious cycle. The ulcer continues to enlarge under the influence of increasing levels of acid and pepsin. Two of the most serious consequences of ulcers are: (1) hemorrhage resulting from damage of submucosal capillaries, and (2) perforation of the stomach wall due to complete erosion through the wall caused by HCl and pepsin action, resulting in the escape of potent gastric contents into the abdominal cavity.

One of the leading treatments for ulcers is an antihistamine (cimetidine) that specifically blocks H-2 receptors, the type of receptors that bind histamine released from the stomach. These receptors differ from H-1 receptors that bind the histamine involved in allergic respiratory disorders. Accordingly, traditional antihistamines used for respiratory allergies (such as hay fever and asthma) are not effective against ulcers, nor is cimetidine useful for respiratory problems. Before the discovery of the key role of these unique histamine receptors in ulcer formation, the treatment of ulcers was aimed at either neutralizing stomach acidity through the use of antacids or removing factors known to enhance gastric secretion. This included: (1) a bland diet void of caffeine and alcohol, (2) cutting the vagus nerve supply to the stomach, and (3) removal of the stomach antrum to eliminate the source of gastrin.

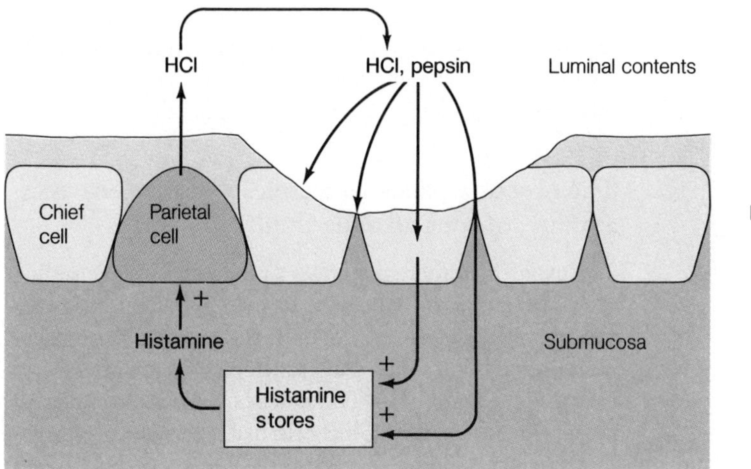

Figure 16–14 Ulcer Formation

Carbohydrate digestion continues in the body of the stomach, whereas protein digestion commences in the antrum.

Two separate digestive processes take place within the stomach. Food in the body of the stomach remains as a semisolid mass, because peristaltic contractions in this region are too weak for mixing to occur. Because food is not mixed with gastric secretions in the body of the stomach, very little protein digestion occurs here. Acid and pepsin are only able to attack the surface of the food mass. Carbohydrate digestion, however, continues in the interior of the mass under the influence of salivary amylase. Even though acid inactivates salivary amylase, the unmixed interior of the food mass is free of acid.

Digestion by the gastric juice itself is accomplished primarily in the antrum of the stomach, where the food is thoroughly mixed with HCl and pepsin, thereby commencing protein digestion in this region.

The stomach absorbs alcohol and aspirin but no food.

No food is absorbed into the blood from the stomach mucosa. Carbohydrate and protein digestion have not been completed in the stomach. These large, partially digested substances are not lipid soluble, so they cannot penetrate the cell membranes. Furthermore, there are no special transport mechanisms present in the stomach wall to facilitate the absorption of these nutrients. Fat digestion has not even begun in the stomach. Although it is theoretically possible for fat molecules, which are lipid soluble, to pass through the lipid portions of the cell membranes that line the stomach, this does not occur to any appreciable extent because dietary fat in the gastric lumen separates into large fat droplets that float in the chyme. This reduces the possibility that fat molecules will come into contact with the gastric mucosa. Ingested electrolytes such as Na^+ and Ca^{++} exist as nonlipid-soluble ions that can be absorbed only through use of special transport systems, which exist in the small intestine but not in the stomach. Likewise, H_2O exchange does not take place across the gastric mucosa because the stomach is impermeable to H_2O.

Even though none of the ingested food is absorbed from the stomach, ethyl alcohol and aspirin are absorbed directly by the stomach. Alcohol is water soluble and is also lipid soluble to a degree, so it can diffuse through the lipid membranes of the epithelial cells that line the stomach and enter the blood through the submucosal capillaries. Yet although alcohol can be absorbed by the gastric mucosa, it can be absorbed even more rapidly by the small intestinal mucosa, because the surface area available for absorption in the small intestine is much greater than in the stomach. Thus, alcohol absorption occurs more slowly if gastric emptying is delayed so that the alcohol remains in the stomach longer. Since fat is the most potent duodenal stimulus for inhibiting gastric motility, consumption of fat-rich foods (for example, milk or pizza) before or during alcohol ingestion delays gastric emptying and prevents the alcohol from producing its effects as rapidly.

Another category of substances absorbed by the gastric mucosa includes weak acids, most notably acetylsalicylic acid (aspirin). In the highly acidic environment of the stomach lumen, weak acids are almost totally unionized; that is, the H^+ and associated anion of the acid are bound together. In an unionized form, these weak acids are lipid soluble, so they can be absorbed quickly by crossing the plasma membranes of the epithelial cells that line the stomach.

Pancreatic and Biliary Secretions

When gastric contents are emptied into the small intestine, they are mixed not only with juice secreted by the small intestine mucosa but also with the secretions of the exocrine pancreas and liver that are emptied into the duodenal lumen. We will discuss the roles of each of these accessory digestive organs before we examine the contributions of the small intestine itself.

The pancreas is a mixture of exocrine and endocrine tissue.

The **pancreas** is an elongated gland that lies behind and below the stomach, above the first loop of the duodenum (Fig. 16–15). It is a mixed gland that contains both exocrine and endocrine tissue. The exocrine portion consists of grapelike clusters of secretory cells that form sacs known as **acini,** which connect to ducts that eventually empty into the duodenum. The endocrine portion consists of isolated islands of endocrine tissue, the **islets of Langerhans,** which are dispersed throughout the pancreas. The most important hormones secreted by the islet cells are insulin and glucagon (chapter 19). The exocrine and endocrine pancreas have nothing in common except for sharing the same location.

The exocrine pancreas secretes digestive enzymes and an aqueous alkaline fluid.

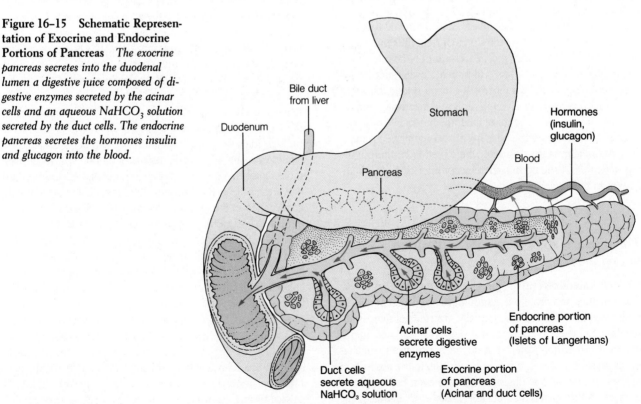

The **exocrine pancreas** secretes a pancreatic juice consisting of two components—a potent enzymatic secretion and an aqueous (watery) alkaline secretion that is rich in sodium bicarbonate ($NaHCO_3$). The pancreatic enzymes are actively secreted by the acinar cells. The aqueous $NaHCO_3$ component is actively secreted by the duct cells that line the early portions of the pancreatic ducts, and it then is modified as it passes down the ducts.

As with pepsinogen, pancreatic enzymes are synthesized by the endoplasmic reticulum and Golgi complex of the acinar cells, then are stored within zymogen granules and released by exocytosis as needed. The acinar cells secrete three different types of enzymes that are capable of digesting all three categories of foodstuffs. These pancreatic enzymes are important be-

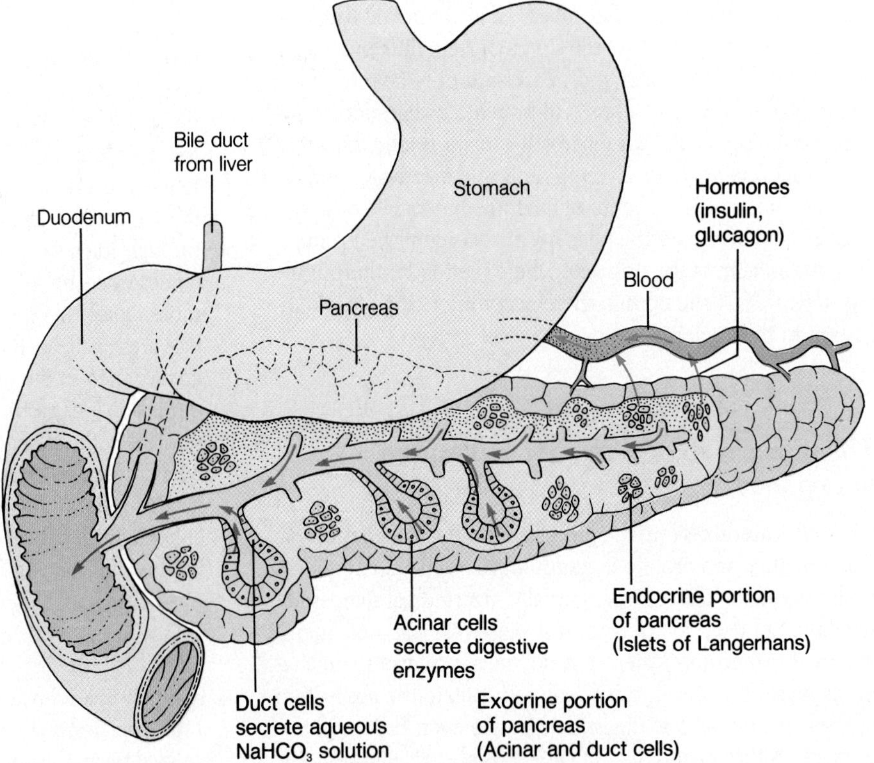

Figure 16–15 Schematic Representation of Exocrine and Endocrine Portions of Pancreas *The exocrine pancreas secretes into the duodenal lumen a digestive juice composed of digestive enzymes secreted by the acinar cells and an aqueous $NaHCO_3$ solution secreted by the duct cells. The endocrine pancreas secretes the hormones insulin and glucagon into the blood.*

cause they are capable of almost completely digesting food in the absence of all other digestive secretions. The three types of pancreatic enzymes are: (1) **proteolytic enzymes,** which are involved in protein digestion; (2) **pancreatic amylase,** which contributes to carbohydrate digestion in a way similar to salivary amylase; and (3) **pancreatic lipase,** the only enzyme important in fat digestion.

PANCREATIC PROTEOLYTIC ENZYMES. The three major proteolytic enzymes secreted by the pancreas are trypsinogen, chymotrypsinogen, and procarboxypeptidase, each of which is secreted in an inactive form. When **trypsinogen** is secreted into the duodenal lumen, it is activated to its active enzyme form, **trypsin,** by **enterokinase,** an enzyme embedded in the luminal border of the cells that line the duodenal mucosa. Trypsin then autocatalytically activates more trypsinogen. As with pepsinogen, trypsinogen must remain inactive within the pancreas to prevent this proteolytic enzyme from digesting the cells in which it is formed. Trypsinogen remains inactive, therefore, until it reaches the duodenal lumen, where enterokinase triggers the activation process, which then proceeds autocatalytically. As further protection, the pancreatic tissue also produces a chemical known as **trypsin inhibitor,** which blocks trypsin's actions should spontaneous activation of trypsinogen inadvertently occur within the pancreas.

Chymotrypsinogen and **procarboxypeptidase,** the other pancreatic proteolytic enzymes, are converted by trypsin to their active forms, **chymotrypsin** and **carboxypeptidase,** respectively, within the duodenal lumen. Thus, once enterokinase has activated some of the trypsin, trypsin is then responsible for the remainder of the activation process.

Each of these proteolytic enzymes attacks different peptide linkages. The end products that result from this action are a mixture of amino acids and small peptide chains. Mucus secreted by the intestinal cells provides protection against digestion of the small intestine wall by the activated proteolytic enzymes.

PANCREATIC AMYLASE. Like salivary amylase, pancreatic amylase plays an important role in carbohydrate digestion by converting polysaccharides into disaccharides and monosaccharides. Amylase is secreted in the pancreatic juice in an active form because there is no danger to the secretory cells from being in contact with active amylase.

PANCREATIC LIPASE. Pancreatic lipase is extremely important because it is the only enzyme secreted throughout the entire digestive system that can accomplish digestion of fat. Pancreatic lipase hydrolyzes dietary triglycerides into monoglycerides and free fatty acids, which are the absorbable units of fat. As with amylase, lipase is secreted in its active form because there is no risk of pancreatic self-digestion by lipase.

When pancreatic enzymes are deficient, digestion of food is incomplete. Because the pancreas is the only significant source of lipase, pancreatic enzyme deficiency results in serious maldigestion of fats. The principal clinical manifestation of pancreatic exocrine insufficiency is **steatorrhea,** or excessive undigested fat in the feces. Up to 60% to 70% of the ingested fat may be excreted in the feces. Digestion of protein and carbohydrates is impaired to a lesser degree because salivary, gastric, and small intestinal enzymes contribute to the digestion of these two foodstuffs.

Pancreatic enzymes function best in a neutral or slightly alkaline environment, yet the highly acidic gastric contents are emptied into the duodenal lumen right in the vicinity of pancreatic enzyme entry into the duodenum. It is imperative that the acidic chyme be quickly neutralized in the duodenal lumen, not only to allow optimal functioning of the pancreatic enzymes but also to prevent acid damage to the duodenal mucosa. Therefore, the alkaline ($NaHCO_3$-rich) fluid secreted by the pancreas into the duodenal lumen serves the important function of neutralizing the acidic chyme as the latter is emptied into the duodenum from the stomach. This aqueous $NaHCO_3$ secretion is by far the largest component of pancreatic secretion. The volume of pancreatic secretion ranges between 1 to 2 liters per day, depending on the types and degree of stimulation.

Pancreatic secretion is hormonally regulated to maintain neutrality of the duodenal contents and to optimize digestion.

Pancreatic secretion is regulated primarily by hormonal mechanisms. A small amount of parasympathetically induced pancreatic secretion occurs during the cephalic phase of digestion, with a further token increase occurring during the gastric phase in response to gastrin. However, the predominant stimulation of pancreatic secretion occurs during the intestinal phase. The release of the two major enterogastrones, secretin and cholecystokinin (CCK), in response to chyme in the duodenum plays the central role in the control of pancreatic secretion.

Of the factors that stimulate enterogastrone release, the primary stimulus specifically for secretin release is acid in the duodenum. Secretin, in turn, is carried by the blood to the pancreas where it stimulates the duct cells to markedly increase their secretion of a HCO_3^--rich aqueous fluid into the duodenum. Even though other stimuli may cause the release of secretin, it is appropriate that the most potent stimulus is acid because secretin promotes the alkaline pancreatic secretion that neutralizes the acid. This provides a control system for maintaining neutrality of the chyme in the intestine. The amount of secretin released is proportional to the amount of acid that enters the duodenum, so the amount of $NaHCO_3$ secreted parallels the duodenal acidity.

Cholecystokinin, on the other hand, is important in the regulation of pancreatic enzyme secretion. The main stimulus for release of CCK from the duodenal mucosa is the presence of the products of fat and protein—namely, fatty acids and amino acids. The circulatory system transports CCK to the pancreas, where it stimulates the pancreatic acinar cells to increase enzyme secretion. Among these enzymes are lipase and the proteolytic enzymes, which appropriately bring about further digestion of the fat and protein that initiated the response and also help digest carbohydrate. In contrast to fat and protein, carbohydrate apparently does not have any direct influence on pancreatic secretion.

Just as gastrin is trophic to the stomach and small intestine, CCK and secretin exert trophic effects on the exocrine pancreas to maintain its integrity.

The liver performs various important functions including bile production.

Besides pancreatic juice, the other secretory product that is emptied into the duodenal lumen is **bile.** The **biliary system** includes the liver, the gall bladder, and associated ducts.

The **liver** is the largest and most important metabolic organ in the body. Its importance to the digestive system is its secretion of *bile salts*, but the liver performs a wide variety of other functions, including the following:

1. metabolic processing of the major categories of nutrients (carbohydrates, proteins, and lipids) after their absorption from the digestive tract.
2. detoxification or degradation of body wastes and hormones as well as drugs and other foreign compounds;
3. synthesis of plasma proteins, including those necessary for the clotting of blood and those that transport steroid and thyroid hormones and cholesterol in the blood;
4. storage of glycogen, fats, iron, copper, and many vitamins;
5. activation of vitamin D, which the liver accomplishes in conjunction with the kidneys;
6. removal of bacteria and worn-out red blood cells, thanks to its resident macrophages (see p. 353); and
7. excretion of cholesterol and bilirubin, the latter being a breakdown product derived from the destruction of worn-out red blood cells.

Given this wide range of complex functions, there is amazingly little specialization of cells within the liver. Each liver cell, or **hepatocyte,** appears to be able to perform the same wide variety of metabolic and secretory tasks, with the exception of the phagocytic activities carried out by the resident macrophages. The specialization comes from the highly developed organelles within each hepatocyte.

To carry out these wide-ranging tasks, the anatomical organization of the liver permits each hepatocyte to be in direct contact with blood from two sources: venous blood coming directly from the digestive tract and arterial blood coming from the aorta. Venous blood enters the liver by means of a unique and complex vascular connection between the digestive tract and the liver that is known as the **hepatic portal system** (Fig. 16–16). The veins draining the digestive tract do not directly join the inferior vena cava, the large vein that returns blood to the heart. Instead, the veins from the stomach and intestine enter the hepatic portal vein, which carries the products absorbed from the digestive tract directly to the liver for processing, storage, or detoxification before they gain access to the general circulation. Within the liver, the portal vein once again breaks up into a capillary network (the liver sinusoids) to permit exchange between the blood and hepatocytes before draining into the hepatic vein, which joins the inferior vena cava. The hepatocytes also are provided with fresh arterial blood, which provides their oxygen supply and delivers blood-borne metabolites for hepatic processing.

The liver lobules are delineated by vascular and bile channels.

The liver is organized into functional units known as **lobules,** which are hexagonal arrangements of tissue surrounding a central vein, like a six-sided angel food cake with the hole representing the central vein (Fig. 16–17a). At the outer edge of each "slice" of the lobule, there are three vessels: a branch of the hepatic artery, a branch of the portal vein, and a bile duct. Blood from the branches of both the hepatic artery and portal vein flows from the periphery of the lobule into large, expanded capillary spaces called **sinusoids,** which run between rows of liver cells to the central vein like spokes on a bike tire (Fig. 16–17b). Lining the sinusoids are **Kupffer cells,** which are the large resident macrophages that engulf and destroy old red blood cells, microorganisms, and other foreign particulate matter that passes through in the blood. The hepatocytes are arranged in plates two cell layers thick between the sinusoids

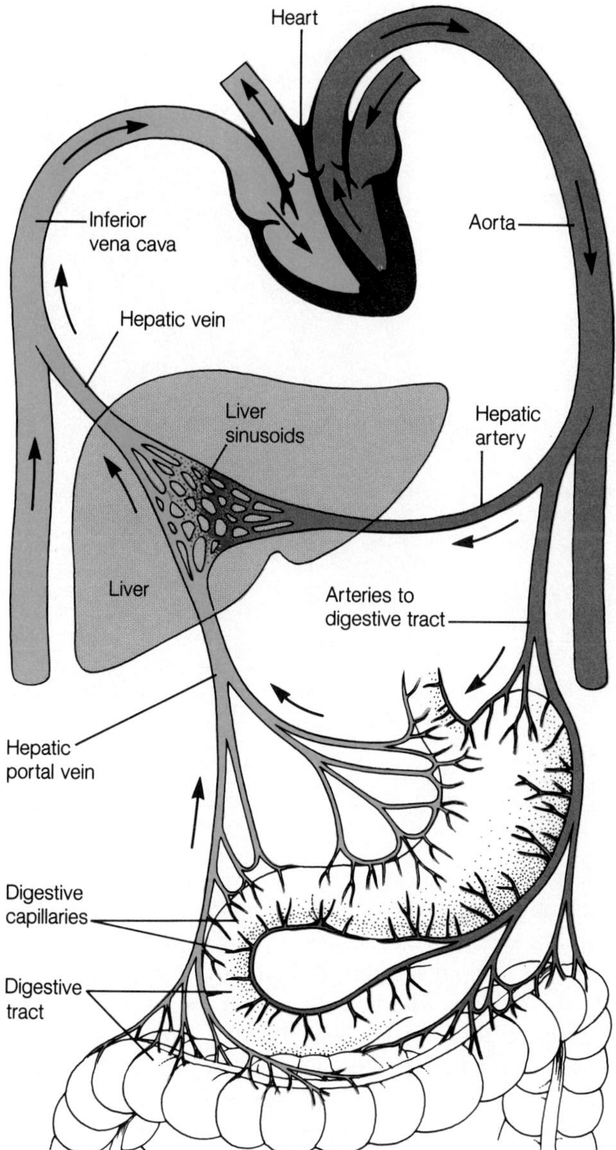

Figure 16–16 Schematic Representation of Liver Blood Flow

so that each lateral edge faces a sinusoidal pool of blood. The central veins of all of the liver lobules converge to form the hepatic vein, which carries the blood away from the liver. A thin bile-carrying channel, a **bile canaliculus,** runs between the cells within each hepatic plate. Hepatocytes continuously secrete bile into these thin channels, which carry the bile to a bile duct at the periphery of the lobule. The bile ducts from the various lobules converge to eventually form the hepatic duct, which transports the bile from the liver to the duodenum. Each hepatocyte is in contact with a sinusoid on one side and a bile canaliculus on the other side.

Bile does not flow directly from the liver to the gall bladder.

The opening of the bile duct into the duodenum is guarded by the **sphincter of Oddi,** which prevents the entry of bile into the duodenum except during digestion of meals (Fig. 16–18). When this sphincter is closed, most of the bile secreted by the liver is diverted back up into the **gall bladder,** a small sac-like structure tucked beneath but not directly connected to the liver. The bile is subsequently stored and concentrated in the gall bladder between meals. After a meal, bile enters the duodenum as a result of the combined effects of gall bladder emptying and increased bile secretion by the liver. The amount of bile secreted per day ranges from 250 ml to 1 liter, depending on the degree of stimulation.

Bile salts are recycled through the enterohepatic circulation.

Bile consists of an aqueous alkaline fluid similar to the pancreatic $NaHCO_3$ secretion as well as several organic constituents, including bile salts, cholesterol, lecithin, and bilirubin. The organic constituents are derived from hepatocyte activity, whereas the water, $NaHCO_3$, and other inorganic salts are added by the duct cells. Even though bile does not contain any digestive enzymes, it is important for the digestion and absorption of fats, primarily through the activity of the bile salts.

Bile salts are derivatives of cholesterol. They are actively secreted into the bile and eventually enter the duodenum along with the other biliary constituents. Following their participation in fat digestion and absorption, most bile salts are reabsorbed back into the blood by special active-transport mechanisms located in the **terminal ileum,** the last portion of the small intestine. From here the bile salts are returned by means of the hepatic portal system to the liver, which resecretes them again into the bile. This recycling of bile salts (and some of the other biliary constituents) between the small intestine and liver is referred to as the **enterohepatic circulation** (*entero* means "intestine"; *hepatic* means "liver") (Fig. 16–18).

The total amount of bile salts in the body averages about 3–4 gm, yet 3–15 gm of bile salts may be emptied into the duodenum in a single meal. Obviously, the bile salts must be recycled many times per day. Usually only about 5% of the secreted bile salts escapes into the feces daily. These lost bile salts are replaced by new bile salts synthesized by the liver; thus the size of the pool of bile salts is kept constant.

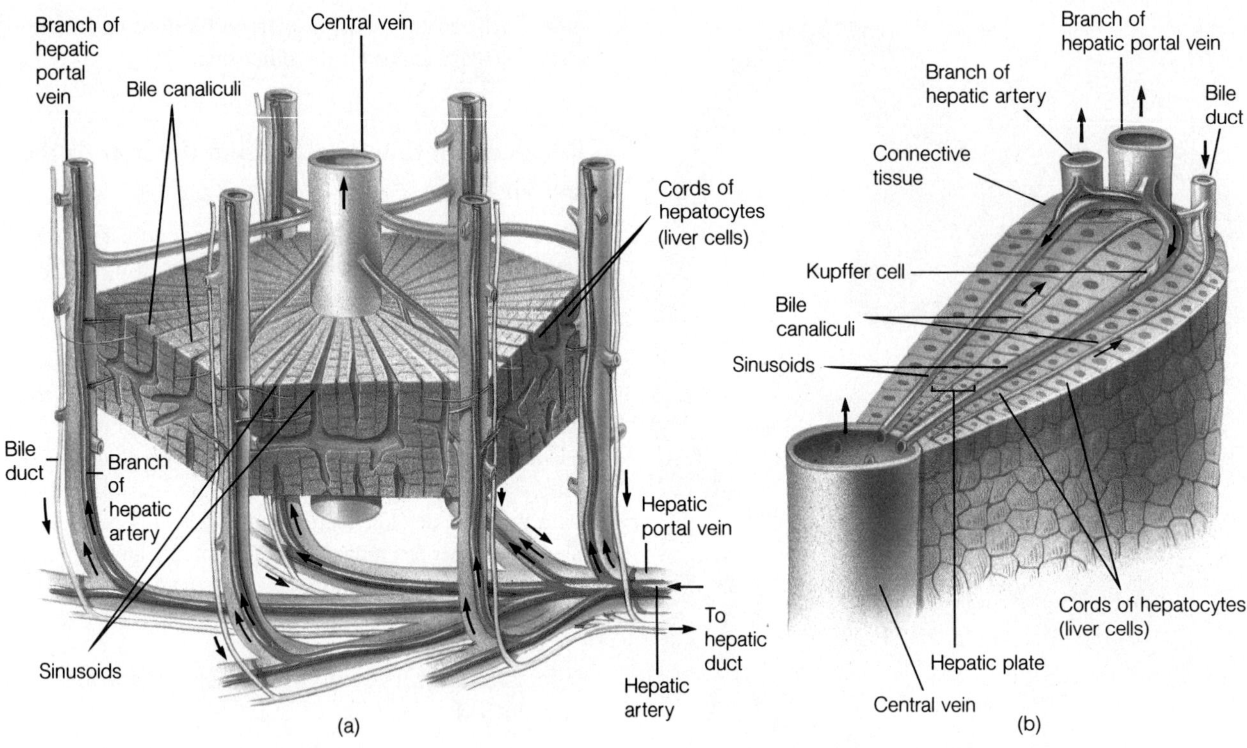

Figure 16–17 Anatomy of the Liver *(a) Hepatic lobule. (b) Wedge of hepatic lobule.*

Bile salts aid fat digestion and absorption through their detergent action and micellar formation, respectively.

Bile salts aid fat digestion through their detergent action (emulsification) and facilitate fat absorption through their participation in the formation of micelles. Both functions are related to the structure of bile salts.

DETERGENT ACTION OF BILE SALTS. **Detergent action** refers to bile salts' ability to convert large fat globules into a **lipid emulsion** that consists of many small fat droplets suspended in the aqueous chyme, which increases the surface area available for attack by pancreatic lipase. In order for lipase to digest fat, it must come into direct contact with the triglyceride molecules. Because fat molecules are not soluble in water, they tend to aggregate into large droplets in the aqueous environment of the small-intestine lumen. If bile salts did not emulsify these large droplets, lipase could act on the lipids only at the surface of the droplet, greatly prolonging triglyceride digestion.

Bile salts exert a detergent action similar to that of the detergent you use to break up grease when you wash dishes. A bile-salt molecule contains a lipid-soluble portion (a steroid derived from cholesterol) plus a negatively charged water-soluble portion. The lipid-soluble portion dissolves in a fat droplet, leaving the charged water-soluble portion projecting from the surface of the droplet (Fig. 16–19a). Intestinal mixing movements break up large fat droplets into smaller ones. These small droplets would quickly recoalesce were it not for bile salts adsorbing on their surface and creating a "shell" of water-soluble negative charges on the surface of each little droplet. Because like charges repel, these negatively charged groups on the droplet surfaces cause the fat droplets to repel each other (Fig. 16–19b). This electric repulsion prevents the small droplets from recoalescing into large fat droplets, thus producing a lipid emulsion that increases the surface area available for lipase action. This is extremely important in accomplishing rapid completion of fat digestion; without bile salts, digestion of fat would be very slow.

MICELLAR FORMATION. Bile salts—along with cholesterol and lecithin, which are also constituents of the bile—play an important role in facilitating fat absorption through micellar formation. Lecithin has both a lipid-soluble and water-soluble portion similar to bile salts, whereas cholesterol is almost totally insoluble in water. In a **micelle,** the bile salts and lecithin aggregate in small clusters with their fat-soluble portions huddled together in the middle to form a hydrophobic ("water-fearing") core while their water soluble portions form an outer hydrophilic ("water-loving") shell (Fig. 16–20). A micellar aggregate is about one-millionth the size of an emulsified lipid droplet. Micelles, themselves being water-soluble by virtue of their hydrophilic shells, can dissolve water-insoluble (and hence lipid-soluble) substances in their lipid-soluble cores. This provides a handy vehicle for carrying water-insoluble substances through the watery luminal contents. The most important lipid-soluble substances thus carried are the products of fat digestion (monoglycerides and free fatty acids) as well as fat-soluble vitamins, which are all transported to their sites of absorption by means of the micelles. Without hitching a ride in the water-soluble micelles, these nutrients would float on the surface of the aqueous chyme (just like oil floats on the top of water), never reaching the absorptive surfaces of the small intestine.

In addition, cholesterol, a highly water-insoluble substance, becomes dissolved in the micelle's hydrophobic core, a mechanism that is important in cholesterol homeostasis. The amount of cholesterol that can be carried in micellar formation depends on the relative amount of bile salts and lecithin in comparison to cholesterol. When cholesterol secretion by the liver is out of proportion to bile-salt and lecithin secretion (either too much cholesterol or too little bile salts and lecithin), the excess cholesterol in the bile precipitates into microcrystals that can aggregate into **gall stones.** One treatment of cholesterol-containing gall stones is ingestion of bile salts to increase the bile-salt pool in an attempt to dissolve the cholesterol stones. Only about 75% of the cases of gall stones are derived from cholesterol, however. The other 25% are made up of abnormal precipitates of another bile constituent, bilirubin.

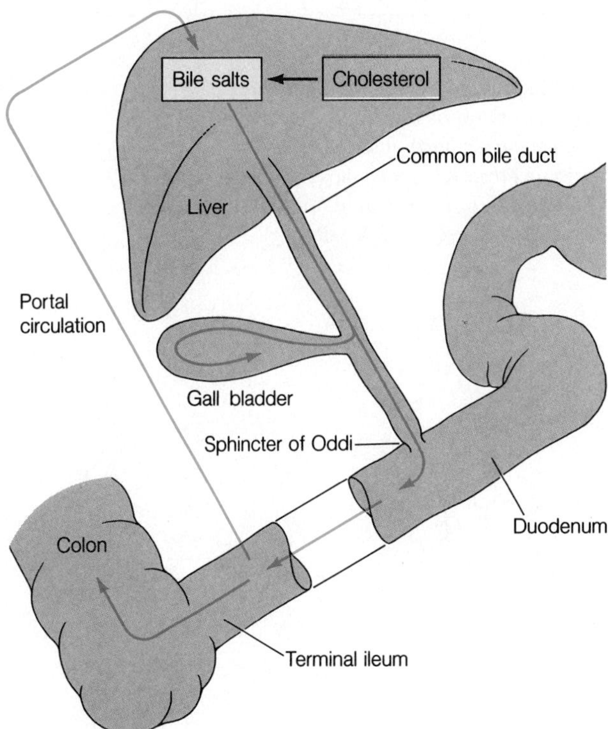

Figure 16–18 Enterohepatic Circulation of Bile Salts

Bilirubin is a waste product excreted in the bile.

Bilirubin, the other major constituent of bile, does not play a role in digestion at all but instead is one of the few waste products excreted in the bile. Bilirubin is the primary bile pigment derived from the breakdown of worn-out red blood cells. The typical life span of a red blood cell in the circulatory system is 120 days. Worn-out red blood cells are removed from the blood by the macrophages that line the liver sinusoids and reside in other areas in the body. Bilirubin is the end product resulting from the degradation of the heme (iron-containing) portion of the hemoglobin contained within these old red blood cells. This bilirubin is extracted from the blood by the hepatocytes and is actively excreted into the bile.

Bilirubin is a yellow pigment that gives bile its yellow color. Within the intestinal tract, this pigment is modified by bacterial enzymes, giving rise to the characteristic brown color of the feces. When bile secretion does not occur, such as if the bile duct is completely obstructed by a gall stone, the feces are grayish-white. A small amount of bilirubin is normally reabsorbed by the intestine back into the blood, and when it is eventually excreted in the urine, it is largely responsible for the urine's yellow color. The kidneys are unable to excrete bilirubin until after it has been modified during its passage through the liver and intestine.

If bilirubin is formed more rapidly than it can be excreted, this pigment accumulates in the body and causes **jaundice.** Patients with this condition appear yellowish, with this color being seen most easily in the whites of their eyes. Jaundice can be brought about in three different ways:

1. ***Pre-hepatic*** or ***hemolytic jaundice*** is due to excessive breakdown (hemolysis) of red blood cells so that the liver is presented with more bilirubin than it is capable of excreting.
2. *Hepatic jaundice* occurs when the liver is diseased and is unable to deal with even the normal load of bilirubin.

Figure 16–19 Schematic Structure and Function of Bile Salts *(a) Schematic representation of structure of bile salts and their adsorption on surface of fat droplet. (b) Formation of lipid emulsion through action of bile salts.*

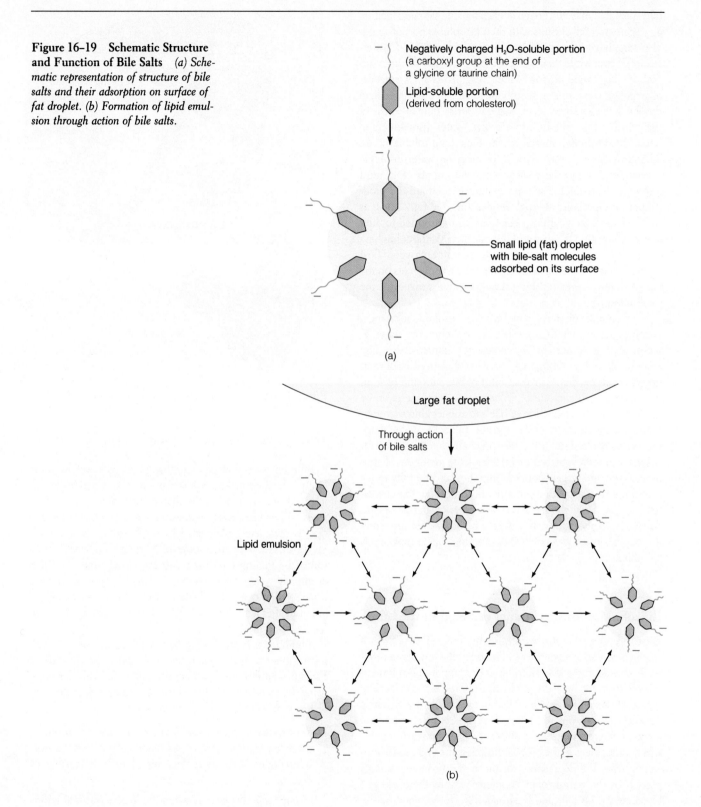

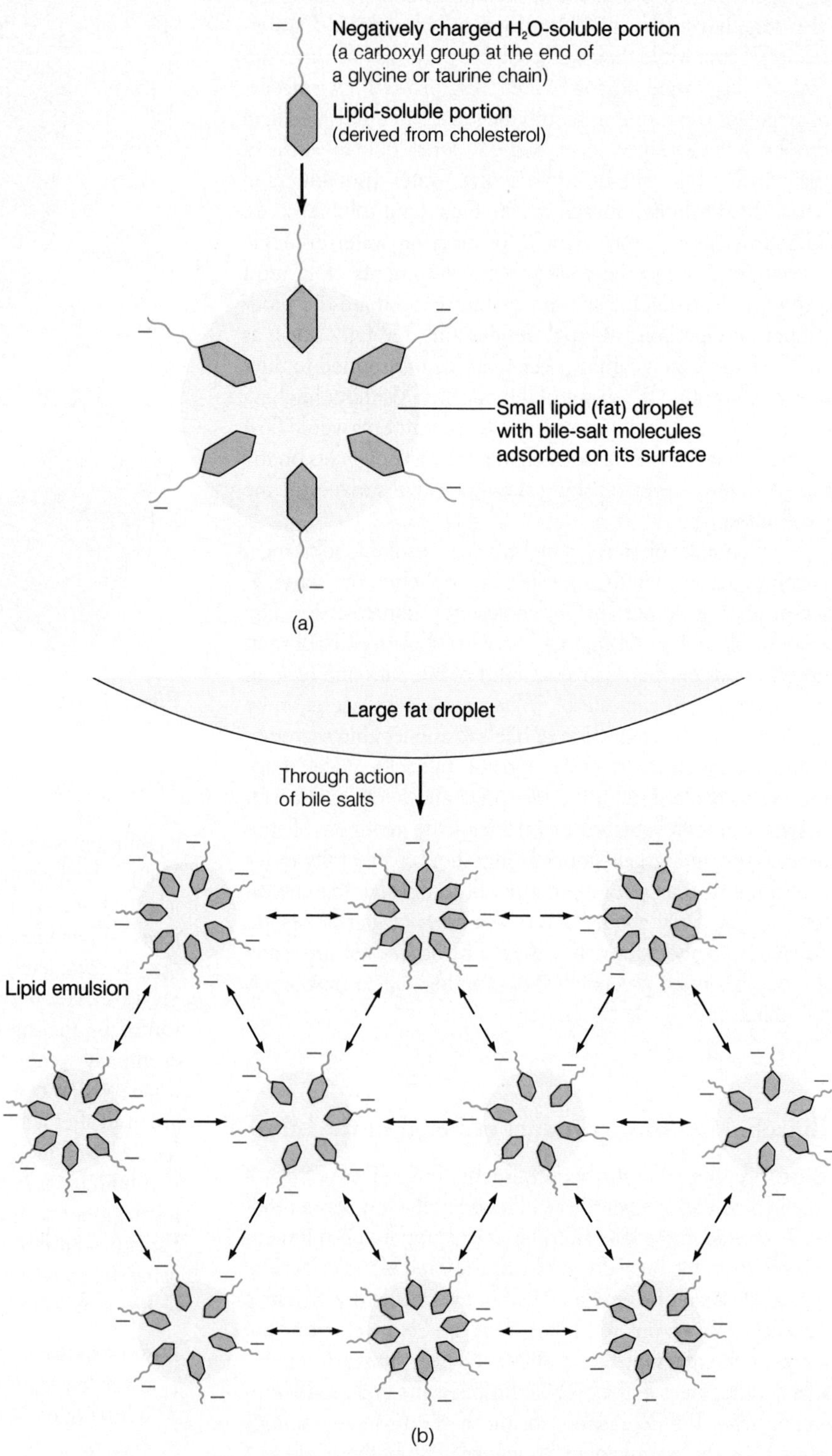

3. *Post-hepatic* or *obstructive jaundice* occurs when the bile duct is obstructed, such as by a gall stone, so that bilirubin cannot be eliminated in the feces.

Bile salts are the most potent stimulus for increased bile secretion.

Bile secretion may be increased by chemical, hormonal, and neural mechanisms.

□ *Chemical mechanism* (*bile salts*)—Any substance that increases bile secretion by the liver is called a **choleretic.** The most potent choleretic is bile salts themselves. Between meals bile is stored in the gall bladder, but during a meal bile is emptied into the duodenum as the gall bladder contracts. After bile salts participate in fat digestion and absorption, they are reabsorbed and returned by the enterohepatic circulation to the liver, where they act as potent choleretics to increase further bile secretion. Therefore, during a meal, when bile salts are needed and being used, bile secretion by the liver is enhanced.

□ *Hormonal mechanism* (*secretin*)—Besides increasing the aqueous $NaHCO_3$ secretion by the pancreas, secretin also stimulates an aqueous alkaline bile secretion by the liver ducts without any corresponding increase in bile salts.

□ *Neural mechanism* (*vagus nerve*)—Vagal stimulation of the liver plays a minor role in bile secretion during the cephalic phase of digestion, promoting an increase in liver bile flow before food ever reaches the stomach or intestine.

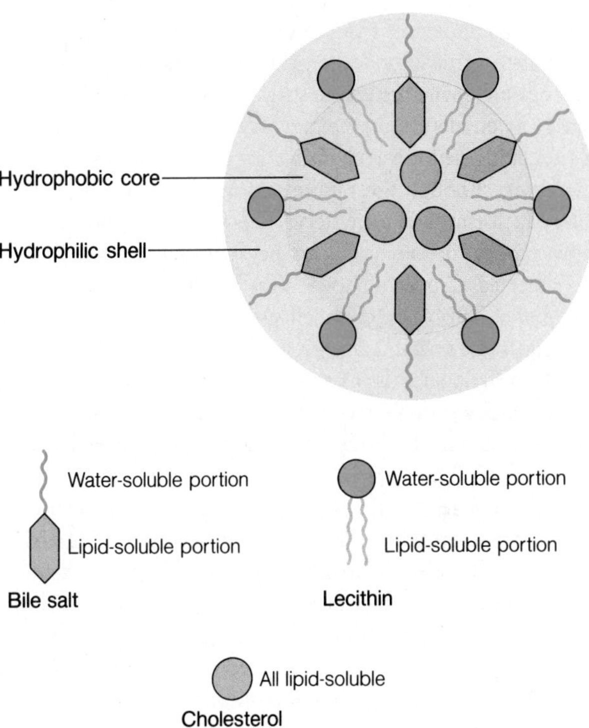

Figure 16–20 Schematic Representation of a Micelle

The gall bladder stores and concentrates bile between meals and empties during meals.

Even though the factors just described increase bile secretion by the liver during and after a meal, bile secretion by the liver occurs continuously. Between meals the secreted bile is shunted into the gall bladder, where it is stored and concentrated. Concentration is accomplished by active transport of salt out of the gall bladder, with water following osmotically, resulting in a five to ten times concentration of the organic constituents. Since the gall bladder stores this concentrated bile, it is the primary site for precipitation of concentrated bile constituents into gall stones. Fortunately, the gall bladder does not play an essential digestive role, so its removal in the presence of gall stones or other gall-bladder disease presents no particular problem. The bile secreted between meals is stored instead in the common bile duct, which becomes dilated.

During a meal, when chyme reaches the small intestine, the presence of food, especially fat products in the duodenal lumen, triggers the release of CCK. This hormone stimulates contraction of the gall bladder and relaxation of the sphincter of Oddi, so bile is discharged into the duodenum, where it appropriately aids in the digestion and absorption of the fat that initiated the release of CCK.

Hepatitis and cirrhosis are the most common liver disorders.

Hepatitis is an inflammatory disease of the liver that results from a variety of causes, including viral infection or exposure to toxic agents. With viral hepatitis, the viruses replicate in the nuclei of the hepatocytes, bringing about injury, inflammation, and even death of the infected cells. Toxic hepatitis occurs as a result of injury and destruction of liver cells from abusive exposure to toxic chemicals or drugs, including alcohol, carbon tetrachloride, and certain tranquilizers. Hepatitis ranges in severity from mild, reversible symptoms to acute massive liver damage, with possible imminent death resulting from acute hepatic failure.

Repeated or prolonged hepatic inflammation, usually in association with chronic alcoholism, can lead to **cirrhosis,** a condition in which damaged hepatocytes are permanently replaced by connective tissue. Liver tissue has the ability to regenerate, normally undergoing a gradual turnover of cells. If part of the hepatic tissue is destroyed, the lost tissue can be replaced by an increase in the rate of cell division. It appears that a circulating factor is responsible for regulating liver-cell proliferation, although the nature and mechanism of this regulatory factor remain a mystery. There is a limit, however, to how rapidly hepatocytes can be replaced. In addition to hepatocytes, dispersed between the hepatic plates are a small number of fibroblasts (connective-tissue cells) that form a supporting framework for the liver. If the liver is repeatedly exposed to toxic substances such as alcohol so often that new hepatocytes are unable to be generated rapidly enough to replace the damaged cells, the sturdier fibroblasts take advantage of the situation and overproduce. This extra connective tissue leaves little space for the hepatocytes' regrowth. Because nodules of hepatocytes that do manage to regenerate are often less functional and are frequently separated from an adequate blood supply by the thick bands of connective tissue, even the new hepatocyte growth may die. Thus, as cirrhosis develops slowly over time, there is a gradual reduction of active liver tissue, which eventually leads to chronic liver failure. Symptoms of hepatic malfunction may not appear until 70 to 80% of the liver tissue is destroyed, demonstrating that, just as with the kidneys, our bodies are endowed with a generous reserve of hepatic tissue.

Small Intestine

The **small intestine** is the site at which most digestion and absorption take place. No further digestion is accomplished after the luminal contents pass beyond the small intestine, and no further absorption of nutrients occurs, although the large intestine does absorb small amounts of salt and water. The small intestine is a tube about 21 ft (6.3 m) long with a small diameter of 1 in (2.5 cm). It lies coiled within the abdominal cavity extending between the stomach and the large intestine. It is arbitrarily divided into three segments—the **duodenum** [the first 8 in (20 cm)], the **jejunum** [8 ft (2.5 m)], and the **ileum** [12 ft (3.6 m)].

Segmentation contractions mix and slowly propel the chyme.

Segmentation, the small intestine's primary method of motility, accomplishes both mixing and slow propulsion of the chyme. Segmentation consists of oscillating, ringlike contractions of the circular smooth muscle along the length of the small intestine, with the relaxed areas betweeen the contracted segments containing a small bolus of chyme. The contractile rings occur every few centimeters apart, dividing the small intestine into segments like a chain of sausages. These contractile rings do not sweep along the length of the intestine like peristaltic waves do. Rather, after a brief period of time, the contracted segments relax, and ringlike contractions appear in the previously relaxed areas (Fig. 16–21). The chyme that was contained within a previously relaxed segment is forced by the new contraction in both directions into the now relaxed adjacent segments. A newly relaxed segment therefore receives chyme from both the contracting segment immediately ahead of it and the one immediately behind it. Shortly thereafter the areas of contraction and relaxation alternate again. In this way, the chyme is chopped, churned, and thoroughly mixed. These contractions can be compared to squeezing a pastry tube with your hands to mix the contents. This mixing serves the dual functions of mixing the chyme with the digestive juices secreted into the small-intestine lumen and of exposing all of the chyme to the absorptive surfaces of the small-intestinal mucosa.

Segmentation contractions are initiated by small-intestinal pacesetter cells, which produce a basic electrical rhythm (BER) similar to the gastric BER responsible for peristalsis in the stomach. If the small-intestine BER brings the circular smooth-muscle layer to threshold, segmentation contractions are induced, with the frequency of segmentation following the frequency of the BER. The circular smooth muscle's degree

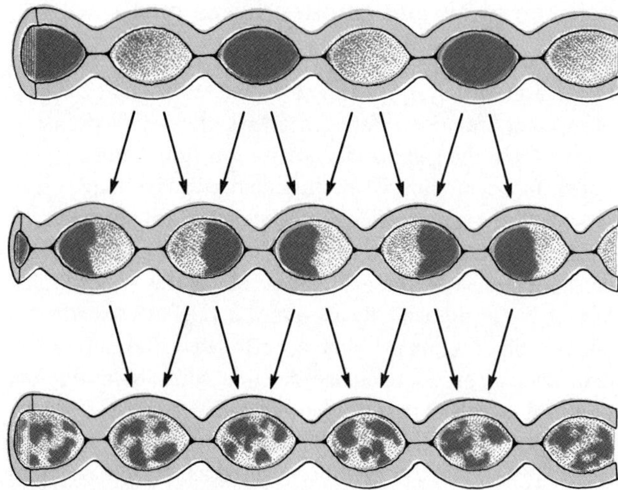

Figure 16–21 Segmentation *Segmentation contractions thoroughly mix the chyme within the small-intestine lumen.*

of responsiveness and thus the intensity of the segmentation contractions can be influenced by distention of the intestine, by extrinsic-nerve activity, and by the hormone gastrin. Stretch of the circular smooth-muscle cells because of the presence of chyme tends to enhance contraction by bringing the "resting" potential closer to threshold. Thus, segmentation can proceed in the absence of extrinsic-nerve stimulation. Extrinsic nerves can modify the strength of these contractions, however. Parasympathetic stimulation enhances segmentation, whereas sympathetic stimulation depresses segmental activity.

Segmentation is slight or absent between meals but becomes very vigorous immediately after a meal. Both the duodenum and ileum start to segment simultaneously when the meal first enters the small intestine. The duodenum starts to segment primarily in response to local distention caused by the presence of chyme. Segmentation of the empty ileum, on the other hand, appears to be brought about by gastrin, which is secreted in response to the presence of chyme in the stomach, a mechanism known as the **gastroileal reflex.**

Segmentation not only accomplishes mixing but also is the primary factor responsible for slowly moving chyme through the small intestine. How can this be, if with each segmental contraction chyme is propelled both forward and backward? Slow propulsion is due to a decline in the frequency of segmentation down the length of the small intestine. The pacesetter cells in the duodenum spontaneously depolarize at a faster rate than those found farther down the tract, with the pacesetter cells in the terminal ileum exhibiting the slowest rate of spontaneous depolarization. The rate of segmentation in the duodenum is twelve contractions per minute, compared to a rate of only nine per minute in the terminal ileum. Because segmentation occurs with greater frequency in the upper part of the small intestine than in the lower part, more chyme tends to be pushed forward than pushed backward. As a result, chyme is moved very slowly from the upper to the lower part of the small intestine, being shuffled back and forth to accomplish thorough mixing and absorption in the process. This slow propulsive mechanism is advantageous to allow ample time for the digestive and absorptive processes to take place. It usually takes three to five hours for the contents to move through the small intestine.

The migrating motility complex sweeps the intestine clean between meals.

When absorption of the meal has been mostly accomplished, segmentation contractions cease and are replaced between meals by the **migrating motility complex,** or **"intestinal housekeeper."** This between-meal motility consists of weak, repetitive peristaltic waves that move a short distance down the intestine before dying out. The waves start at the stomach and, for unknown reasons, migrate down the intestine; that is, each new peristaltic wave is initiated at a site a little farther down the small intestine. It takes about 100–150 min for these short peristaltic waves to gradually migrate from the stomach to the end of the small intestine, with each contraction "sweeping" any remnants of the preceding meal plus mucosal debris and bacteria forward toward the colon, just like a good "intestinal housekeeper." After the end of the small intestine is reached, the cycle begins again and continues to repeat itself until the next meal. It is speculated that the unconfirmed hormone **motilin** might play a role in regulating the migrating motility complex. When the next meal arrives, segmental activity is triggered again, and the migrating motility complex ceases.

The ileocecal juncture prevents contamination of the small intestine by colonic bacteria.

At the juncture between the small and large intestines, the last portion of the ileum empties into the cecum (Fig. 16–22). Two factors contribute to this region's ability to act as a barrier between the small and large intestine. First, the anatomical arrangement is such that valve-like folds of tissue protrude from the ileum into the lumen of the cecum. When the ileal contents are pushed forward, this **ileocecal valve** is easily pushed open, but the folds of tissue are forcibly closed when the cecal contents attempt to move backward. Second, the smooth muscle within the last several centimeters of the ileal wall just preceding the ileocecal valve is thickened, forming a sphincter that is under neural and hormonal control. Most of the time this **ileocecal sphincter** remains at least mildly constricted. Pressure on the cecal side of the sphincter causes it to contract more forcibly; distention of the ileal side causes the sphincter to relax, a reaction that is presumably mediated by short reflexes within the myenteric plexus. In this way, the ileocecal sphincter prevents the bacteria-laden contents of the large intestine from contaminating the small intestine and at the same time allows the ileal contents to pass into the colon. If the colonic bacteria were to gain access to the nutrient-rich small intestine, they would rapidly multiply. Relaxation of the sphincter is enhanced through the release of gastrin at the onset of a meal, when increased gastric activity is taking place. This relaxation allows the undigested fibers and unabsorbed solutes from the preceding meal to be moved forward as the new meal enters the tract.

Figure 16–22 Control of Ileocecal Valve/Sphincter

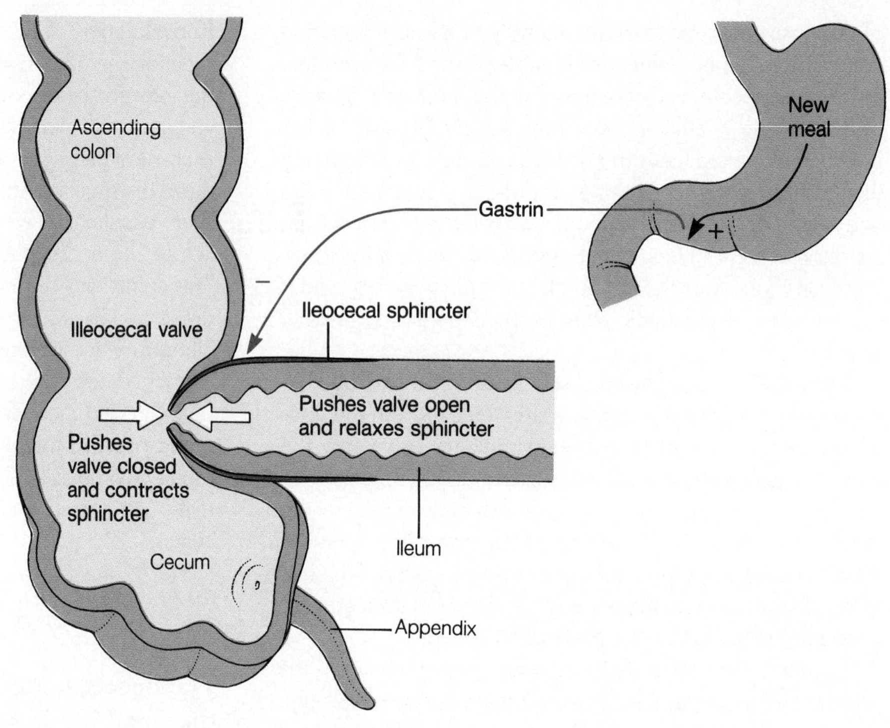

Small-intestine secretions do not contain any digestive enzymes.

Each day the exocrine glands located in the mucosa of the small intestine secrete into the lumen about 1.5 liters of an aqueous salt and mucus solution known as the **succus entericus** (*succus* means "juice"; *entericus* means "of the intestine"). No digestive enzymes are secreted into this intestinal juice. The small intestine does synthesize digestive enzymes, but they act intracellularly within the borders of the epithelial cells that line the lumen instead of being secreted directly into the lumen.

Are any functions served by this small-intestine secretion, since no digestive enzymes are involved? The mucus in the secretion provides protection and lubrication. Furthermore, this aqueous secretion provides an abundance of H_2O to participate in the enzymatic digestion of food. Recall that digestion involves hydrolysis—bond breakage by reaction with H_2O—which proceeds most efficiently when all the reactants are in solution. The succus entericus provides H_2O in the area where most of these digestive chemical reactions occurs.

The regulation of small-intestine secretion is not clearly understood. Secretion of succus entericus does increase after a meal. The most effective stimulus for secretion appears to be local stimulation of the small-intestinal mucosa by the presence of chyme. Vagal control and hormones may play a minor role.

Digestion in the small-intestine lumen is accomplished by pancreatic enzymes, whereas the small-intestine enzymes act intracellularly.

Digestion within the small-intestine lumen is accomplished by the pancreatic enzymes, with fat digestion being enhanced by bile secretion. As a result of pancreatic enzymatic activity, fats are completely reduced to their absorbable units of monoglycerides and free fatty acids, proteins are broken down into small peptide fragments and some amino acids, and carbohydrates are reduced to disaccharides and some monosaccharides. Thus fat digestion is completed within the small-intestine lumen, but carbohydrate and protein digestion have not been brought to completion.

Special actin-stiffened, hairlike projections on the luminal surface of the small-intestinal epithelial cells form the **brush border** (see p. 39), which contains three different categories of enzymes: (1) **enterokinase,** which activates the pancreatic enzyme trypsinogen; (2) the **disaccharidases** (**sucrase, maltase,** and **lactase**), which complete carbohydrate digestion by hydrolyzing the remaining disaccharides (sucrose, maltose, and lactose, respectively) into their constituent monosaccharides; and (3) the **aminopeptidases,** which hydrolyze the small peptide fragments into their amino acid components, thereby completing protein digestion. Thus carbohydrate and protein digestion are completed intracellularly within the confines of the brush border.

A fairly common disorder, **lactose intolerance,** involves a deficiency of lactase, the disaccharidase specific for the digestion of lactose, or milk sugar. Most children under four years of age have adequate lactase activity, but this may be gradually lost so that in many adults, lactase activity is diminished or absent. When milk or dairy products (except those in which lactose has already been digested by bacterial action during processing, such as some kinds of cheese and yogurt) are consumed by an individual with lactase deficiency, the undigested lactose remains in the lumen. This brings about several related consequences. First, accumulation of undigested lactose creates an osmotic gradient that draws H_2O into the intestinal lumen. Second, bacteria residing in the large intestine possess lactose-splitting ability, so they eagerly attack the lactose as an energy source, producing large quantities of CO_2 and methane gas in the process. Distention of the intestine by both fluid and gas produces pain (cramping) and diarrhea. Depending on the extent of the lactase deficiency and the quantity of lactose ingested, symptoms can vary from mild abdominal discomfort to severe dehydrating diarrhea. Occasionally, infants have a genetic defect that prevents them from synthesizing any lactase. Such infants may die from excessive fluid loss and malnutrition unless they are provided with milk that has been especially treated before consumption to prehydrolyze lactose, or they are placed on a milk-substitute formula such as soy formula.

The small intestine is remarkably well adapted for its primary role in absorption.

All products of carbohydrate, protein, and fat digestion, as well as most of the ingested electrolytes, vitamins, and water, are normally absorbed by the small intestine indiscriminately. Usually only the absorption of calcium and iron is adjusted to the body's needs. Thus, the more food that is consumed, the more that will be digested and absorbed, as those who are trying to control their weight are all too painfully aware.

Most absorption occurs in the duodenum and jejunum; very little occurs in the ileum, not because the ileum does not have absorptive capacity but because most absorption has already been accomplished before the intestinal contents reach the ileum. The small intestine has an abundant reserve absorptive capacity. In fact, about 50% of the small intestine can be removed with little interference to absorption—with one exception. If the terminal portion of the ileum is removed, vitamin B_{12} and bile salts are not properly absorbed because the specialized transport mechanisms for these two substances are located only in this region. All other substances can be absorbed throughout the length of the small intestine.

The mucous lining of the small intestine is remarkably well adapted for its special absorptive function for two reasons: (1) it has a very large surface area, and (2) the epithelial cells in this lining possess a variety of specialized transport mechanisms. The following special modifications of the small-intestine mucosa greatly increase the surface area available for absorption (Fig. 16–23).

- ☐ The inner surface of the small intestine is thrown into circular folds that increase the surface area threefold and that are visible to the naked eye. The folds are not permanently positioned but change patterns as a result of contractions of the muscularis mucosa, the thin ribbon of smooth muscle tucked within the mucosa.
- ☐ Projecting from this folded surface are microscopic fingerlike projections known as **villi,** which give the lining a velvety appearance and which increase the surface area by another ten times. The surface of each villus is covered by epithelial cells interspersed occasionally with mucous cells.
- ☐ Even smaller hairlike projections known as **microvilli** (or the brush border) arise from the luminal surface of these epithelial cells, increasing the surface area another twentyfold. Each epithelial cell has as many as 3,000 to 6,000 of these microvilli, which are visible only with an electron microscope. It is within the membrane of this brush border that the small-intestine enzymes perform their functions.

Altogether, the folds, villi, and microvilli provide the small intestine with a luminal surface area six-hundred times greater than it would have if it were a tube of the same length and diameter but lined by a flat surface. In fact, if the surface area of the small intestine were spread out flat, it would cover an entire tennis court.

Malabsorption (impairment of absorption) may be caused by damage to or reduction of the surface area of the small intestine. One of the most common causes is **gluten enteropathy.** In this condition, the individual's small intestine is abnormally sensitive to gluten, a protein constituent of wheat and some other grains. Exposure to gluten appears to damage the intestinal villi so that the normally luxuriant array of villi is reduced, the mucosa becomes flattened, and the brush border becomes short and stubby (Fig. 16–24). Because this loss of villi decreases the surface area available for absorption, absorption of all nutrients is impaired. The mechanism whereby gluten induces this damage of the small- intestinal mucosa is not

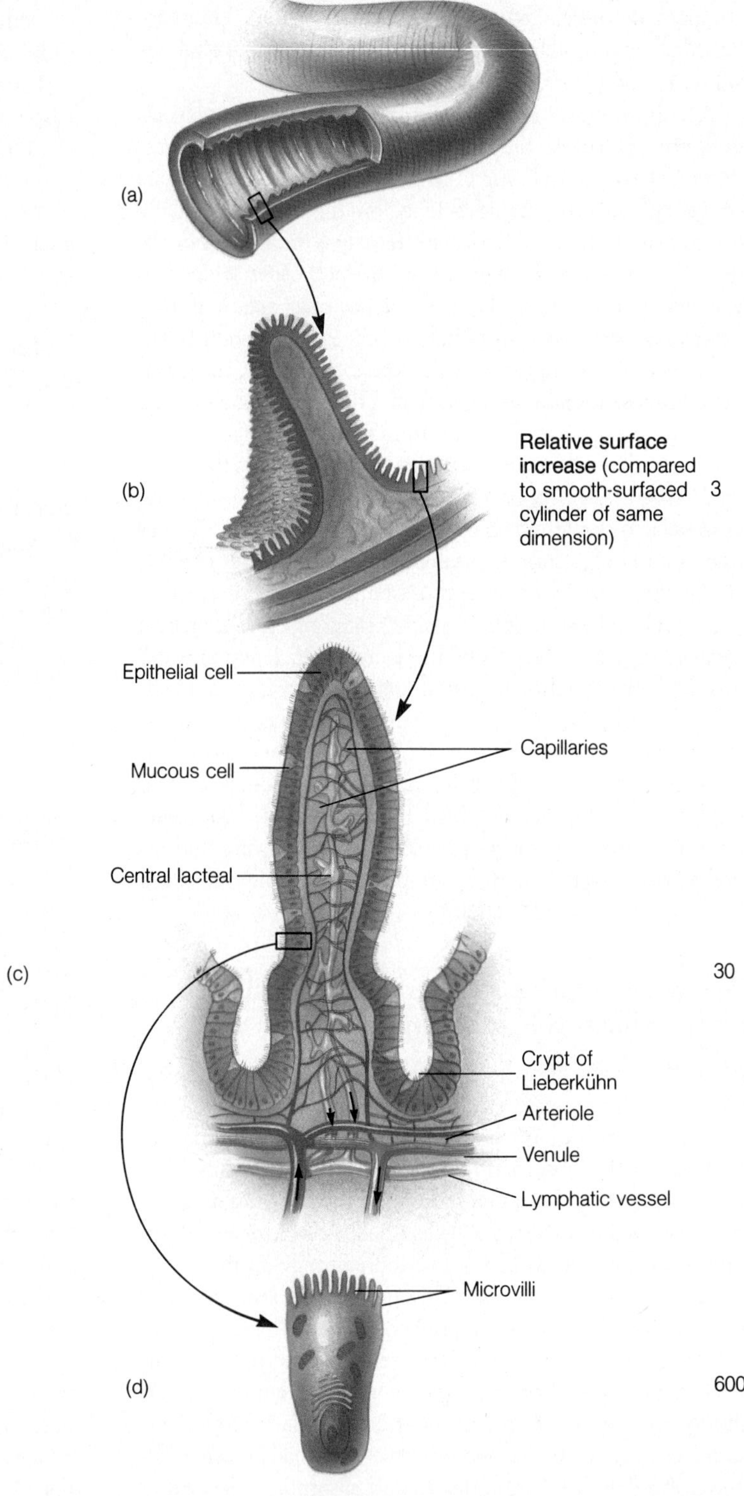

Figure 16–23 Small-Intestine Absorptive Surface *(a) Gross structure of the small intestine. (b) Circular folds of small-intestine mucosa, which increase the absorptive surface area threefold. (c) Microscopic fingerlike projection known as a villus. Collectively, the villi increase the surface area another tenfold. (d) Electron-microscope view of villus epithelial cell depicting the presence of microvilli on its luminal border; the microvilli increase the surface area another twentyfold.*

clearly understood. The condition is treated by eliminating gluten from the diet.

Absorption across the digestive-tract wall involves transepithelial transport similar to movement of material across the kidney tubules (see p. 480). The epithelial cells that cover the surface of each villus are joined at their lateral borders by tight junctions, which limit passage of luminal contents between the cells, although the tight junctions in the small intestine are "leakier" than those in the stomach. The connective-tissue core of the villus is formed by the lamina propria. Each villus is supplied by an arteriole that breaks up into a capillary network within the villus (Fig. 16–23c). The capillaries rejoin to form a venule that drains away from the villus. Each villus is also provided with a terminal lymphatic vessel known as a **central lacteal.** It is into this capillary network or central lacteal that digested substances enter during the process of absorption. To be absorbed, a substance must pass completely through the epithelial cell, diffuse through the interstitial fluid within the connective-tissue core, and then cross the wall of a capillary or lymph vessel. As with renal transport, intestinal absorption may be an active or passive process, with active absorption involving energy expenditure for at least one of the steps involved in the transepithelial transport process.

There is rapid turnover of the mucosal lining.

Dipping down into the mucosal surface between the villi are shallow invaginations known as the **crypts of Lieberkühn** (Fig. 16–23c). Unlike the gastric pits, these intestinal crypts do not secrete digestive enzymes but function as "nurseries." The epithelial cells lining the small intestine slough off and are replaced at a rapid rate as a result of high mitotic activity in the crypts. New cells that are continually being produced at the bottom of the crypts migrate up the villi and, in the process, push off the older cells at the tips of the villi into the lumen. In this manner, more than 100 million intestinal cells are shed per minute. The entire trip from crypt to tip averages about three days, so the epithelial lining of the small intestine is replaced approximately every three days. Because of this high rate of cell division, the cells within the crypts are very sensitive to damage by radiation and anti-cancer drugs, both of which may inhibit cell division.

The cells undergo several changes as they migrate up the villus. The concentration of brush-border enzymes increases and the capacity for absorption improves, so the cells at the tip of the villus have the greatest digestive and absorptive capability. Just at their peak, these cells are pushed off by the newly migrating cells. Thus the luminal contents are constantly exposed to cells that are optimally equipped to complete the digestive and absorptive functions efficiently. Furthermore, the rapid turnover of cells in the small intestine, just as in the stomach, is essential because of the harsh luminal conditions. Cells exposed to the abrasive and corrosive gut contents are easily damaged and cannot live for long, so they must be continually replaced by a fresh supply of newborn cells.

The old cells that are sloughed off into the lumen are not entirely lost to the body. These cells are digested, with the cell constituents being absorbed into the blood and reclaimed for synthesis of new cells, among other things.

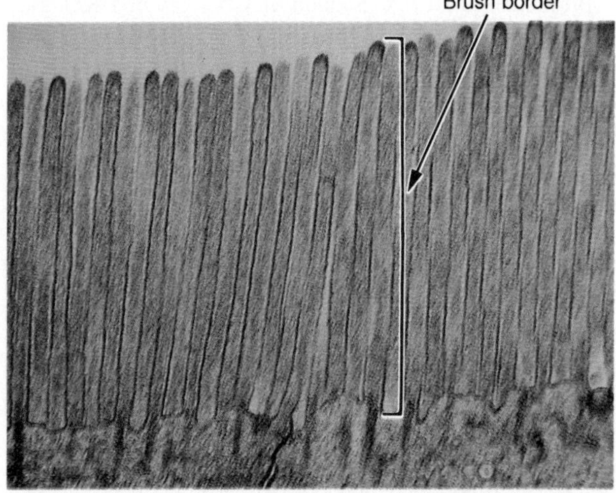

(a)

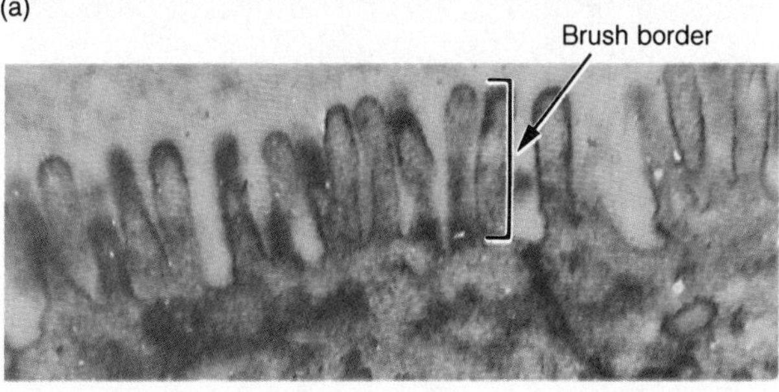

(b)

Figure 16–24 Reduction in Brush Border with Gluten Enteropathy *(a) Electron micrograph of the brush border of a small intestine epithelial cell in a normal individual. (b) Electron micrograph of a short, stubby brush border of a small intestine epithelial cell in a patient with gluten enteropathy.*

SOURCE: Thomas W. Sheehy, M.D.; Robert L. Slaughter, M.D., "The Malabsorption Syndrome," by Medcom, Inc. Reprinted by permission of Medcom, Inc.

Special mechanisms facilitate absorption of most nutrients.

We will now turn our attention to the mechanisms by which absorption of the specific dietary constituents is normally accomplished.

SODIUM ABSORPTION. Sodium may be absorbed both passively and actively. When the electrochemical gradient favors movement of Na^+ from the lumen to the blood, passive diffusion of Na^+ can occur ***between*** the intestinal epithelial cells through the "leaky" tight junctions. Movement of Na^+ ***through*** the cells is energy-dependent and involves at least two different carriers, similar to the process of Na^+ reabsorption across the kidney tubules (see p. 483). Sodium may passively enter the epithelial cells at their luminal surface by itself, or it may be co-transported with glucose or amino acid. It is then actively pumped out of the cell at the basolateral border into the lateral spaces between the cells where they are not joined by tight junctions. From here Na^+ eventually diffuses into the capillaries. As with the renal tubules in the early portion of the nephron, the absorption of H_2O, glucose, and amino acids is linked to this energy-dependent Na^+ absorption.

WATER ABSORPTION. For years physiologists were puzzled about the mechanism of H_2O absorption. More H_2O was absorbed than could be accounted for by the osmotic gradient created by the quantity of absorbed digestive products and electrolytes. Active absorption was ruled out because H_2O transport across membranes is not known to require energy anywhere in the body. The answer to this puzzle is another example of the ingenious mechanisms with which the body is endowed. The same mechanism is now known to also facilitate H_2O reabsorption across the kidney tubules (see p. 486).

Most H_2O absorption in the digestive tract depends on the active carrier that pumps Na^+ into the lateral spaces, resulting in a concentrated area of high osmotic pressure in that localized region between the cells. This localized high osmotic pressure induces movement of water from the lumen through the cell (and possibly from the lumen through the leaky tight junction) into the lateral space. Water entering the space reduces the osmotic pressure but raises the hydrostatic (fluid) pressure. As a result, H_2O is flushed out of the lateral space into the interior of the villus, where it is picked up by the capillary network. Meanwhile, more Na^+ is pumped into the lateral space to encourage more water absorption.

CARBOHYDRATE ABSORPTION. Dietary carbohydrate is presented to the small intestine for absorption mainly in the forms of the disaccharides maltose (the product of polysaccharide digestion), sucrose, and lactose (Table 16–1 and Fig. 16–25). The disaccharidases located in the brush borders of the small-intestinal cells further reduce these disaccharides into the absorbable monosaccharide units of glucose, fructose, and galactose.

Glucose and galactose are both absorbed by Na^+- and energy-dependent secondary active transport, with both sugars competitively sharing the same co-transport carrier system. The co-transport carrier at the luminal border has two binding sites, one for the monosaccharide and one for Na^+. This non-energy-dependent carrier is driven to transport glucose (or galactose) and Na^+ from the lumen into the interior of the cell by the Na^+ concentration gradient established by the active basolateral Na^+ pump (see p. 80). As the co-transport carrier transfers glucose and Na^+ into the cell, the glucose concentration within the cell becomes progressively higher than in the lumen, but the intracellular Na^+ concentration remains low because the basolateral pump actively transports Na^+ out. In effect, glucose absorption is achieved against a concentration gradient, but energy is not directly expended to accomplish this glucose transport. The energy required in the overall process of glucose (or galactose) absorption is that necessary to run the basolateral Na^+ pump, which maintains the Na^+ concentration gradient that drives the co-transport carrier. Thus the same energy that accomplishes Na^+ absorption and most H_2O absorption also efficiently drives most carbohydrate absorption.

The only form of carbohydrate absorption not Na^+- or energy-dependent is fructose, which is absorbed solely by facilitated diffusion (passive carrier-mediated transport).

PROTEIN ABSORPTION. Not only are ingested proteins absorbed, but **endogenous proteins** that have entered the digestive-tract lumen from the three following sources are absorbed as well:

1. digestive enzymes, all of which are proteins, that have been secreted into the lumen;
2. proteins within the cells that are pushed off from the villi into the lumen during the process of mucosal turnover; and
3. small amounts of plasma proteins that normally leak from the capillaries into the digestive-tract lumen.

About 20–40 gm of endogenous protein enter the lumen each day from these three sources. This can amount to more than half of the protein presented to the small intestine for digestion and absorption. All endogenous proteins must be digested and absorbed along with the dietary proteins to prevent depletion of the body's protein stores. The amino acids absorbed from both food and the endogenous protein are used primarily to synthesize new protein in the body.

The protein presented to the small intestine for absorption is primarily in the form of amino acids and a few small peptide fragments (Fig. 16–26). Amino acids are absorbed across the intestinal cells by secondary active transport, similar to glucose and galactose absorption. Thus amino acids also get a "free ride" in on the energy expended for Na^+ transport. Small pep-

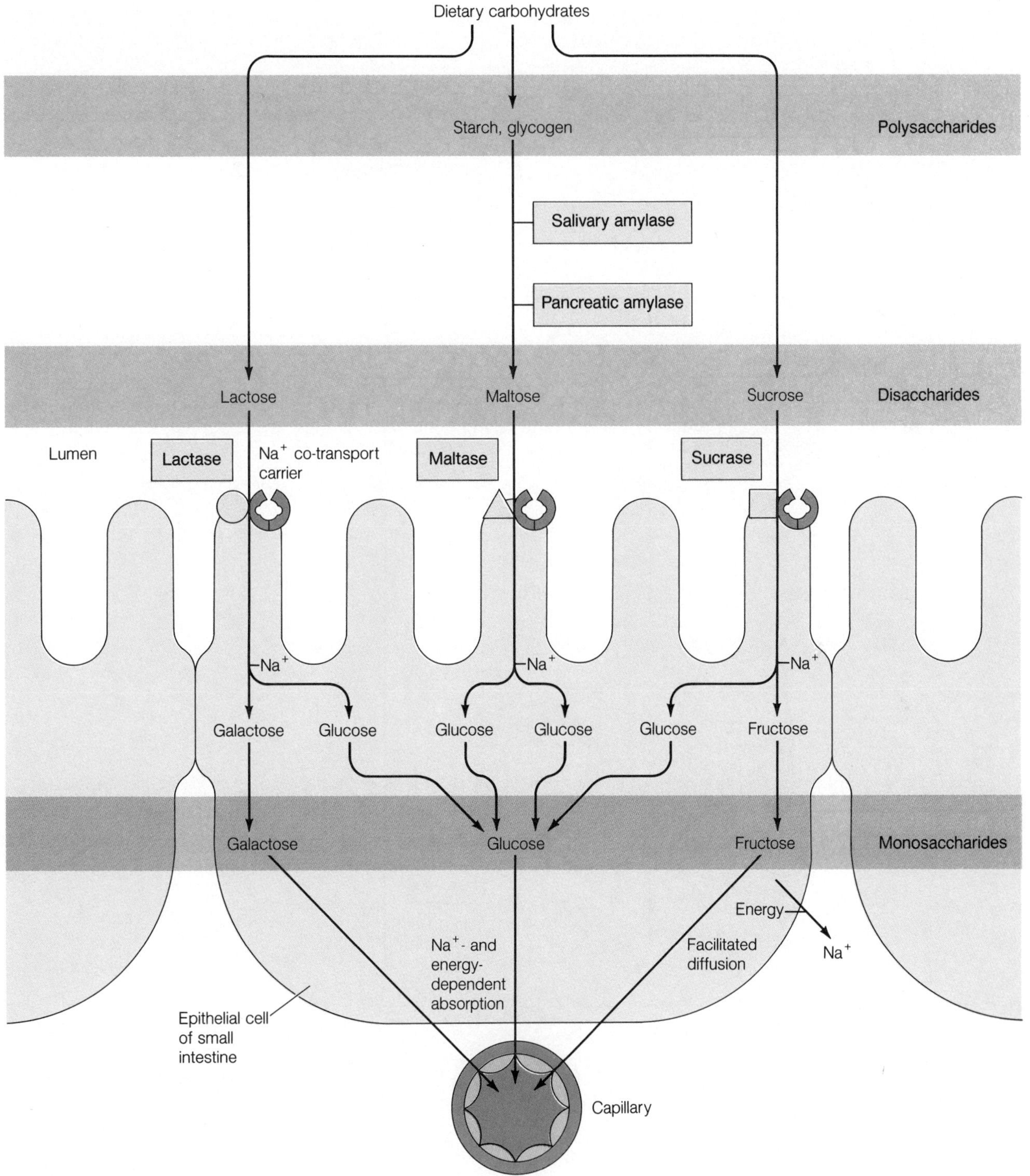

Figure 16–25 Carbohydrate Digestion and Absorption

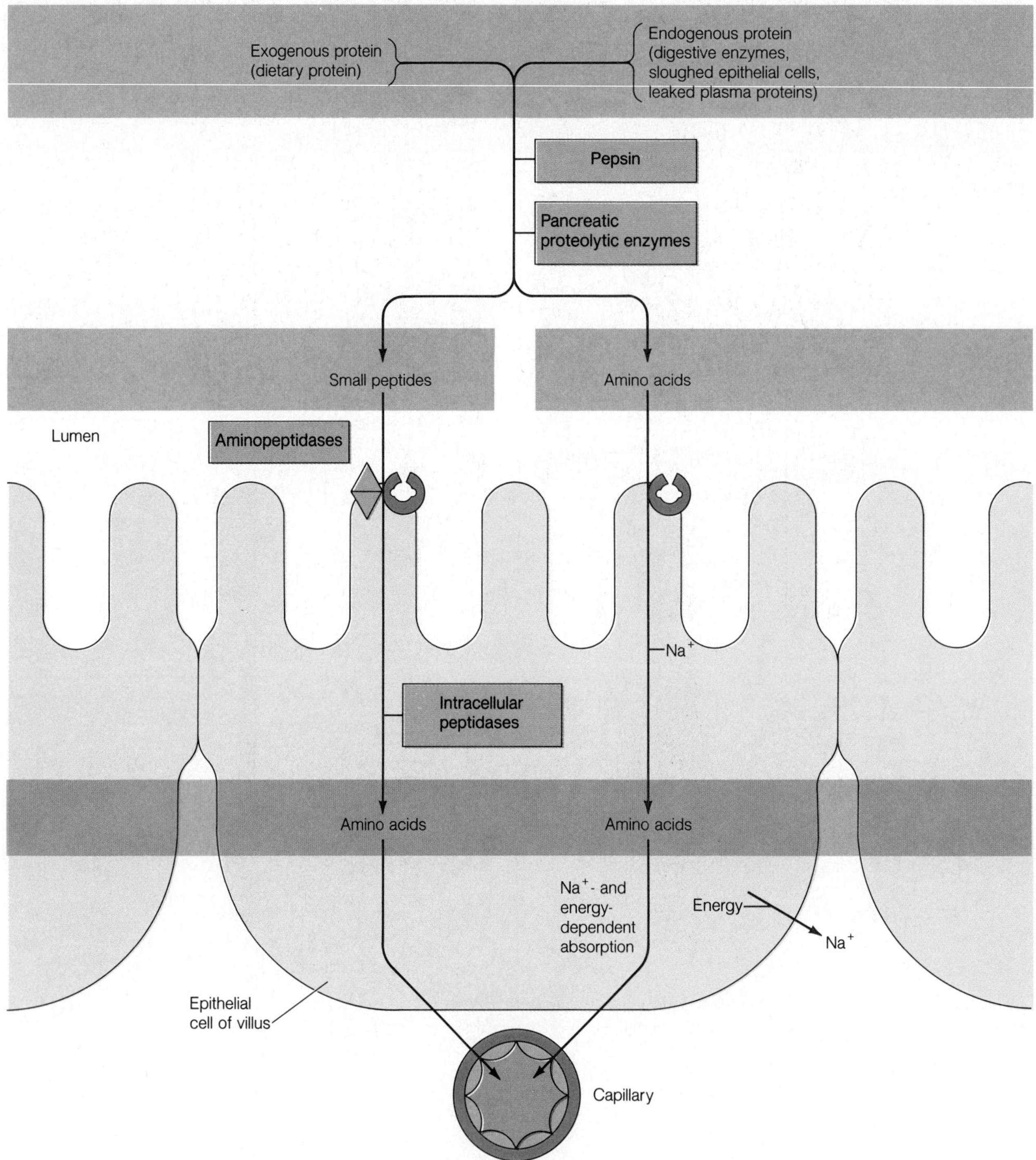

Figure 16–26 Protein Digestion and Absorption

tides appear to gain entry by means of a different carrier. They are broken down into their constituent amino acids by the aminopeptidases in the brush borders or by intracellular peptidases. As with monosaccharides, amino acids enter the capillary network within the villus.

Thus, absorption of the digestion end products of both carbohydrates and proteins involves special carrier-mediated transport systems that require energy expenditure and Na^+ co-transport, and both categories of end products are absorbed into the blood.

FAT ABSORPTION. Fat absorption is quite different than carbohydrate and protein absorption because the H_2O-insolubility of fat presents a special problem. Fat must be transferred from the watery chyme through the watery body fluids despite the fact that it is not water soluble. Therefore, fat must undergo a series of transformations to circumvent this problem during its digestion and absorption (Fig. 16–27).

When the stomach contents are emptied into the duodenum, the ingested fat is aggregated into large, oily triglyceride droplets that float in the chyme. Recall that through bile salts' detergent action within the small intestine, the large droplets are dispersed into a lipid emulsification of small droplets, thereby exposing a much greater surface area of fat available for digestion by pancreatic lipase. The products of lipase digestion (monoglycerides and free fatty acids) are also not very water soluble, so very little of these end products of fat digestion can diffuse through the aqueous chyme to reach the absorptive lining. However, biliary components facilitate absorption of these fatty end products through formation of micelles. Remember that micelles are water-soluble particles that can carry the end products of fat digestion within their lipid-soluble interiors. Once these micelles reach the luminal membranes of the epithelial cells, the monoglycerides and free fatty acids passively diffuse from the micelles through the lipid component of the epithelial cell membranes to enter the interior of these cells. As these fat products leave the micelles and are absorbed across the epithelial cell membranes, the micelles are able to pick up more monoglycerides and free fatty acids, which have been released from digestion of further triglyceride molecules in the fat emulsion.

Bile salts continuously repeat their fat solubilizing function down the length of the small intestine until all the fat is absorbed. Then the bile salts themselves are reabsorbed in the terminal portion of the ileum by special active transport. This is an efficient process, because relatively small amounts of bile salts can facilitate digestion and absorption of large amounts of fat, with each bile salt repeatedly performing its ferrying function before it is reabsorbed.

Once within the interior of the epithelial cells, the monoglycerides and free fatty acids are resynthesized into triglycerides. These triglycerides conglomerate into droplets and are coated with a layer of lipoprotein (synthesized by the endoplasmic reticulum of the epithelial cell), which renders the fat droplets water soluble. These large, coated fat droplets, known as **chylomicrons,** are extruded by exocytosis from the epithelial cells into the interstitial fluid within the villus. The chylomicrons subsequently enter the central lacteals rather than the capillaries because of the structural differences between these two vessels. Capillaries have a **basement membrane** (an outer layer of polysaccharides) that prevents chylomicrons from entering, but the lymph vessels do not have this barrier. Thus fat can be absorbed into the lymphatics but not directly into the blood. (Fatty acids with short- or medium-length carbon chains can enter the blood, but very few of these are eaten in the normal diet.)

The actual transfer of monoglycerides and free fatty acids from the chyme across the cell membranes of the intestinal epithelial cells is a passive process, because the lipid-soluble fatty end products merely dissolve in and pass through the lipid portions of the membrane. Fat absorption is therefore said to be a passive process. However, the overall sequence of events necessary for fat absorption does require energy. For example, bile salts are actively secreted by the liver, and the resynthesis of triglycerides and formation of chylomicrons within the epithelial cells are active processes.

VITAMIN ABSORPTION. Water-soluble vitamins are primarily absorbed passively with water, whereas fat-soluble vitamins are carried in the micelles and absorbed passively with the end products of fat digestion. Absorption of some of the vitamins can also be accomplished by carriers, if necessary. Vitamin B_{12} is unique in its requirement to be in combination with gastric intrinsic factor for absorption by special transport in the terminal portion of the ileum, presumably by the process of endocytosis.

IRON AND CALCIUM ABSORPTION. In contrast to the almost complete and unregulated absorption of other ingested electrolytes, dietary iron and calcium may not be completely absorbed, being subject to regulation that depends on the body's needs for these electrolytes.

Iron is essential for hemoglobin production. The normal iron intake is typically 15–20 mg/day, yet a man usually absorbs only about 0.5–1 mg/day into the blood, and a woman takes up slightly more at 1.0–1.5 mg/day (women's need for iron is greater because of the periodic loss of iron in the menstrual blood flow).

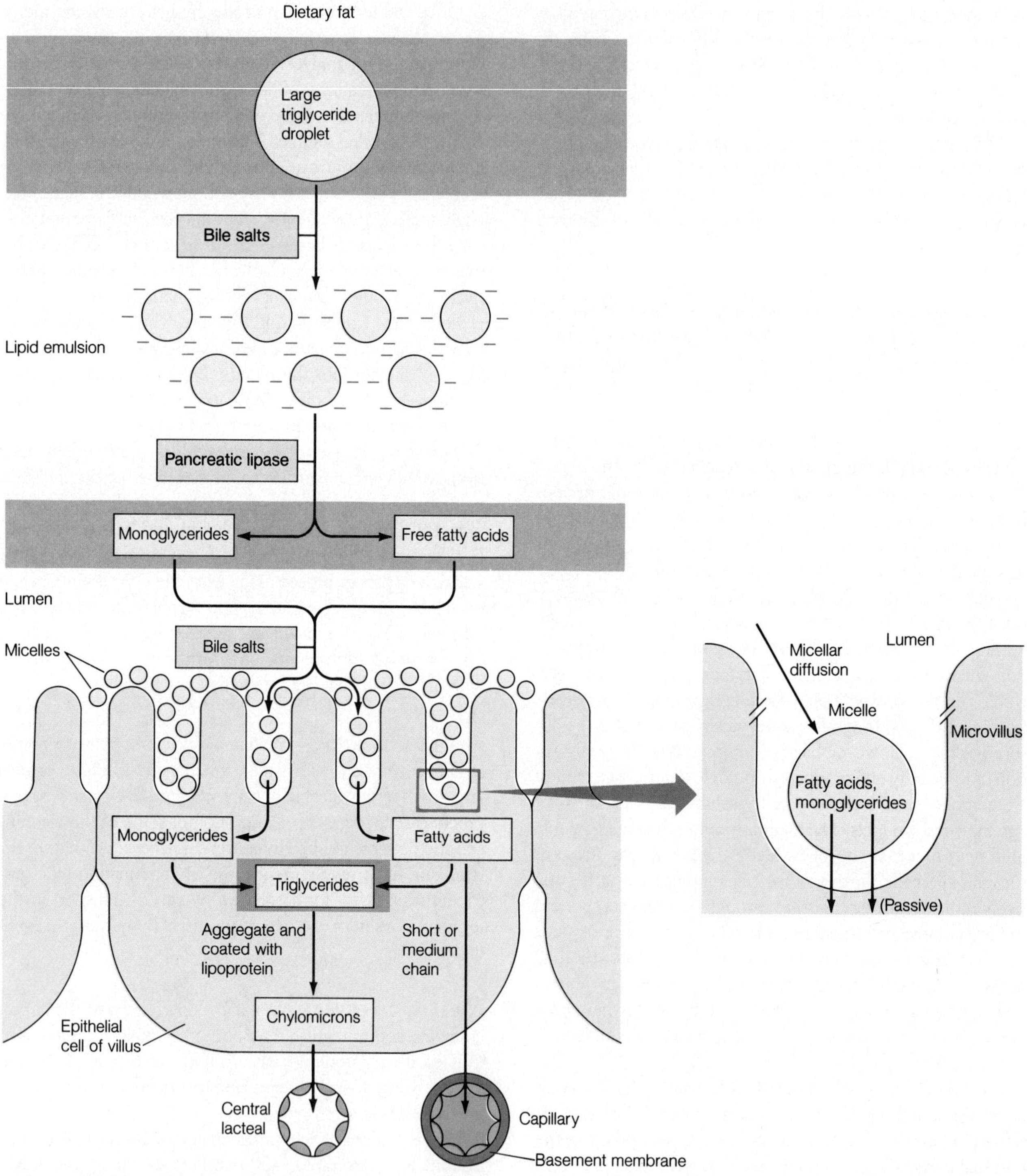

Figure 16–27 Fat Digestion and Absorption

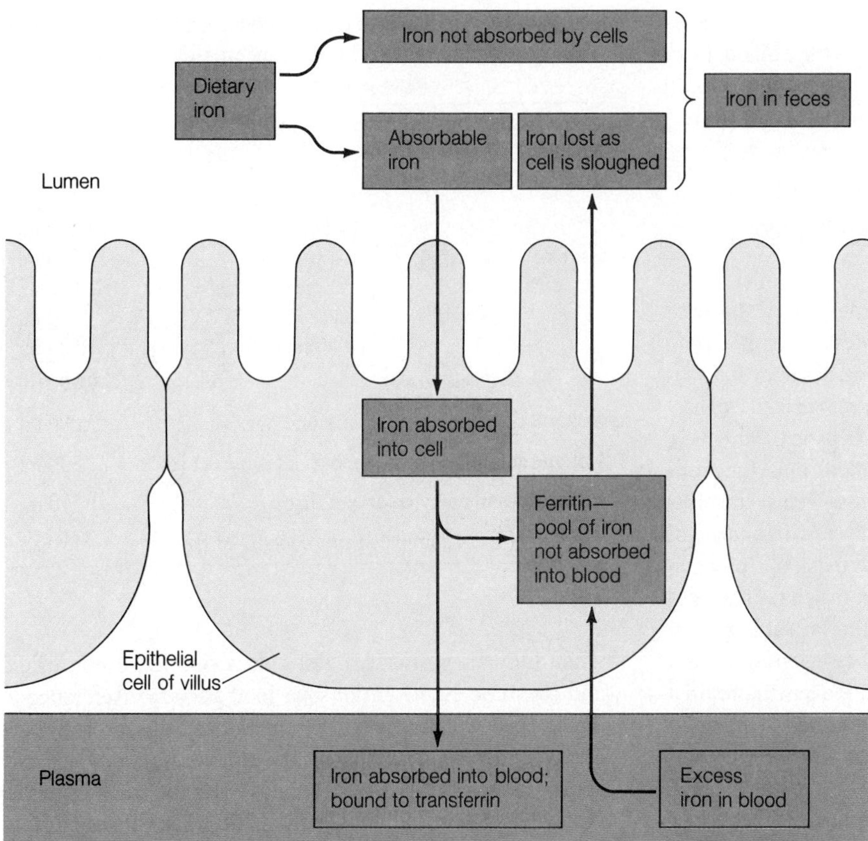

Figure 16–28 Iron Absorption

Two main steps are involved in the absorption of iron into the blood: (1) absorption of iron from the lumen into the intestinal epithelial cells, and (2) absorption of iron from the epithelial cells into the blood (Fig. 16–28).

Iron is actively transported from the lumen into the epithelial cells, with women having about four times more active transport sites for iron than men. The extent to which ingested iron is taken up by the epithelial cells depends on the type of iron consumed (ferrous iron, Fe^{++}, is absorbed more easily than ferric iron, Fe^{+++}), and the presence of other substances in the lumen that can either promote or reduce iron absorption. For example, ascorbic acid (vitamin C) increases iron absorption, primarily by reducing ferric to ferrous iron. Phosphate and oxalate, on the other hand, combine with ingested iron to form insoluble iron salts that cannot be absorbed.

After active absorption into the small-intestine epithelial cells, iron has two possible fates:

1. If iron is immediately needed for the production of red blood cells, it can be absorbed into the blood for delivery to the bone marrow, the site of red blood cell production. Iron is transported in the blood by a plasma protein carrier known as **transferrin.** The hormone responsible for stimulating red blood cell production, erythropoietin (see p. 347), is believed to also enhance iron absorption from the intestinal cells into the blood. The absorbed iron is then used in the synthesis of hemoglobin for the newly produced red blood cells.
2. If there is no immediate need for iron, it remains stored within the epithelial cells in a nonabsorbable granular form called **ferritin.** If the blood level of iron is too high, excess iron may be dumped from the blood into this unabsorbable pool of ferritin in the intestinal epithelial cells. Iron stored as ferritin is lost in the feces within three days as the epithelial cells containing these granules are sloughed off during mucosal regeneration. Large amounts of iron in the feces give them a dark, almost black color.

The amount of calcium (Ca^{++}) absorbed is also regulated. Absorption of Ca^{++} is accomplished partly by passive diffusion but mostly by active transport. Vitamin D greatly stimulates this active transport, a process that is enhanced by parathyroid hormone. Secretion of parathyroid hormone increases in response to a fall in Ca^{++} concentration in the blood. Normally, of the average 1,000 mg of Ca^{++} taken in daily, only about two-thirds is absorbed in the small intestine, with the rest passing out in the feces.

Most absorbed nutrients immediately pass through the liver for processing.

The venules that leave the small-intestinal villi, along with those from the remainder of the gastrointestinal tract, empty into the portal vein, which carries the blood to the liver. Consequently, anything absorbed into the gastrointestinal capillaries first must pass through the impressive hepatic biochemical factory before entering the general circulation. Thus the products of carbohydrate and protein digestion as well as the electrolytes and water are channeled into the liver, where many of these products are subjected to immediate metabolic processing. Furthermore, harmful substances that may have been absorbed are detoxified by the liver before they are able to gain access to the general circulation. After passing through the portal circulation, the venous blood from the gastrointestinal system is emptied into the vena cava and returned to the heart to be distributed throughout the body, carrying glucose and amino acids for use by the tissues.

Fat, which cannot penetrate the intestinal capillaries, is picked up by the central lacteal and enters the lymphatic system instead, thereby bypassing the hepatic portal system. Contractions of the villi, presumably accomplished by the muscularis mucosa, periodically compress the central lacteal and "milk" the lymph out of this vessel. Increased contractions of the villi are known to occur following a meal, and are perhaps mediated by an unconfirmed hormone from the duodenal mucosa, **villikinin.** The lymph vessels eventually converge to form the thoracic duct, a large lymph vessel that empties into the venous system within the chest. By this means, fat ultimately gains access to the circulatory system. The absorbed fat is carried by the systemic circulation to the liver and to other tissues of the body. Therefore, the liver does have a chance to act on the digested fat, but not until the fat has been diluted by the blood in the general circulatory system. This dilution of fat presumably protects the liver from being inundated with more fat than it is capable of handling at one time.

Extensive absorption by the small intestine keeps pace with secretion.

The small intestine normally absorbs about 9 liters of fluid per day in the form of H_2O and solutes, including the absorbable units of nutrients, vitamins, and electrolytes. How can that be when humans normally ingest only about 1,250 ml of fluid and consume 1,250 gm of solid food per day? Reference to Table 16–6 will illustrate the tremendous daily absorptive accomplishments performed by the small intestine. Each day about 9,500 ml of H_2O and solutes enter the small intestine. Note that of this 9,500 ml, only 2,500 ml are ingested from the external environment. The remaining 7,000 ml (7 liters) of fluid consist of digestive juices that are essentially derived from plasma. Recall that plasma is the ultimate source of digestive secretions because the secretory cells extract the necessary raw materials for their secretory product from the plasma. Considering that the entire plasma volume is only about 2.75 liters, it is obvious that absorption must closely parallel secretion to prevent the plasma volume from falling sharply. Of the 9,500 ml of fluid entering the small-intestine lumen per day, about 95%, or 9,000 ml of fluid, is normally absorbed by the small intestine back into the plasma, with only 500 ml of the small-intestinal contents passing on into the colon. Thus, the digestive juices are not lost from the body. After the constituents of the juices are secreted into the digestive-tract lumen and perform their function, they are returned to the plasma. The only secretory product that escapes from the body is bilirubin, a waste product that must be eliminated.

Table 16–6 Volumes Absorbed by Small and Large Intestine per Day

Volume entering small intestine per day			
Sources	Ingested	Food eaten	1,250 gm
		Fluid drunk	1,250 ml
	Secreted from plasma	Saliva	1,500 ml
		Gastric juice	2,000 ml
		Pancreatic juice	1,500 ml
		Bile	500 ml
		Intestinal juice	1,500 ml
			9,500 ml
Volume absorbed by small intestine per day			9,000 ml
Volume entering colon from small intestine per day			500 ml
Volume absorbed by colon per day			350 ml
Volume of feces eliminated from colon per day			150 gm

Biochemical balance among the stomach, pancreas, and small intestine is normally maintained.

Since the secreted juices are normally absorbed back into the plasma, the acid-base balance of the body is not altered by digestive processes. When secretion and absorption do not parallel each other, however, acid-base abnormalities can result. Figure 16–29 is a summary of the biochemical balance that normally exists among the stomach, pancreas, and small intestine. The arterial blood entering the stomach contains Cl^-, CO_2, H_2O, and Na^+, among other things. During HCl secretion, the gastric parietal cells extract Cl^-, CO_2, and H_2O from the plasma (the CO_2 and H_2O being essential for H^+ secretion) and add HCO_3^- to it (the HCO_3^- being formed in the process of generating H^+). The HCO_3^- diffuses into the plasma to replace the secreted Cl^- and to electrically balance the Na^+ in the plasma. Plasma Na^+ levels are not altered by gastric secretory processes. Because HCO_3^- is an alkaline ion, the venous blood leaving the stomach is more alkaline than the arterial blood delivered to it. The overall acid-base balance of the body is not altered, however, because the pancreatic duct cells extract a comparable amount of HCO_3^- (along with Na^+) from the plasma to neutralize the acidic gastric chyme as it is emptied into the small intestine. Within the intestinal lumen, the alkaline pancreatic $NaHCO_3$ secretion neutralizes the gastric HCl secretion, yielding NaCl and H_2CO_3. The latter molecules form Na^+ and Cl^- plus CO_2 and H_2O respectively. All four of these constituents (Na^+, Cl^-, CO_2, and H_2O) are absorbed by the intestinal epithelium into the plasma. Note that these are exactly the same constituents present in the arterial blood entering the stomach. Thus, through these interactions, there is normally no net gain or loss of acid or base from the body during digestion.

Diarrhea results in loss of fluid and electrolytes.

When vomiting or diarrhea occurs, these normal neutralization processes cannot take place. We have already described

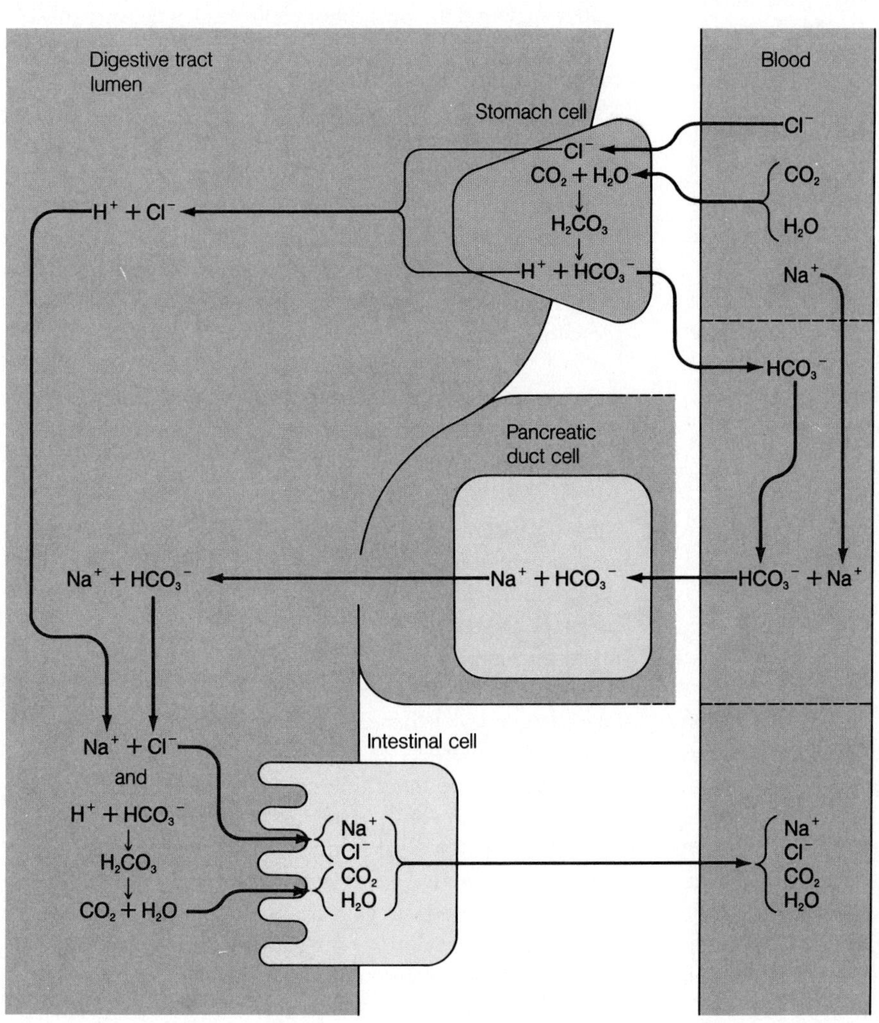

Figure 16–29 Biochemical Balance among Stomach, Pancreas, and Small Intestine *When digestion and absorption proceed normally, there is no net loss or gain of acid or base or other chemicals from the body fluids as a consequence of digestive secretions.*

vomiting in the section on gastric motility. The other common digestive-tract disturbance that can lead to a loss of fluid and an acid-base imbalance is **diarrhea.** This condition is characterized by passage of a highly fluid fecal matter, often with increased frequency of defecation. Just as with vomiting, the effects of diarrhea can be either beneficial or harmful. Diarrhea is beneficial when rapid emptying of the intestine hastens the elimination of harmful material from the body. However, excessive loss of intestinal contents causes dehydration, loss of nutrient material, and metabolic acidosis resulting from the loss of secreted HCO_3^- that normally would have been absorbed. The abnormal fluidity of the feces in diarrhea usually occurs because the intestine is unable to absorb fluid as extensively as normal. This extra unabsorbed fluid passes out in the feces.

The following are the causes of diarrhea:

1. The most common cause of diarrhea is excessive intestinal motility, which arises either from local irritation of the gut wall caused by bacterial or viral infection of the intestine or from emotional stress. Rapid transit of the intestinal contents does not allow sufficient time for adequate absorption of fluid to occur.
2. Diarrhea also occurs when excess osmotically active particles are present in the digestive-tract lumen, such as are found in lactase deficiency. These particles cause excessive fluid to enter and be retained in the lumen, thus increasing the fluidity of the feces.
3. Toxins of the bacterium *Vibrio cholerae* (the causative agent of cholera) and certain other microorganisms promote the secretion of excessive amounts of fluid by the small-intestinal mucosa, resulting in profuse diarrhea.

Large Intestine

The large intestine is primarily a drying and storage organ.

The colon normally receives about 500 ml of chyme from the small intestine each day. Since most digestion and absorption have been accomplished in the small intestine, the contents delivered to the colon consist of undigestible food residues (such as cellulose), unabsorbed biliary components, and the remaining fluid. The colon extracts more H_2O and salt from the contents. What remains to be eliminated is known as **feces.** The primary function of the large intestine is to store this fecal material before defecation. Cellulose and other indigestible substances in the diet provide bulk and help maintain regular bowel movements by contributing to the volume of the colonic contents.

The large intestine is thrown into haustral sacs.

The large intestine consists of the colon, cecum, appendix, and rectum (Fig. 16–30). The **cecum** forms a blind-ended pouch below the junction of the small and large intestines at the ileocecal valve. The small fingerlike projection at the bottom of the cecum is the **appendix,** a lymphoid organ that houses lymphocytes (see p. 364). The **colon,** which makes up most of the large intestine, is not coiled like the small intestine but consists of three relatively straight portions—the ascending colon, the transverse colon, and the descending colon. The terminal portion of the descending colon becomes S-shaped, forming the sigmoid colon (***sigmoid*** means "s-shaped"), then straightens out to form the **rectum** (*rectum* means "straight").

The outer longitudinal smooth-muscle layer does not completely surround the large intestine. Instead, it consists only of three separate, conspicuous, longitudinal bands of muscle, the **taeniae coli,** which run the length of the large intestine. These taeniae coli are shorter than the underlying circular smooth muscle and mucosal layers would be if the latter were stretched out flat. Because of this, the underlying layers are gathered into pouches or sacs called **haustra,** much like the material of a full skirt is gathered at the narrower waistband. The haustra are not merely passive permanent gathers, however; they actively change location as a result of contraction of the circular smooth muscle layer.

Haustral contractions slowly shuffle the colonic contents back and forth while mass movements propel colonic contents long distances.

Most of the time, movements of the large intestine are slow and nonpropulsive, which is appropriate for its absorptive and storage functions. The colon's primary method of motility is **haustral contractions** initiated by the autonomous rhythmicity of colonic smooth-muscle cells. These contractions are similar to small-intestinal segmentations but occur at a much lower frequency. Whereas small-intestinal segmentation occurs at rates of between nine and twelve contractions per minute, there may be thirty minutes between haustral contractions. The location of the haustral sacs gradually changes as a relaxed segment that has formed a sac slowly contracts while a previously contracted area simultaneously relaxes to form a new sac. These movements are nonpropulsive; they slowly shuffle the contents in a back-and-forth mixing movement that exposes the colonic contents to the absorptive mucosa.

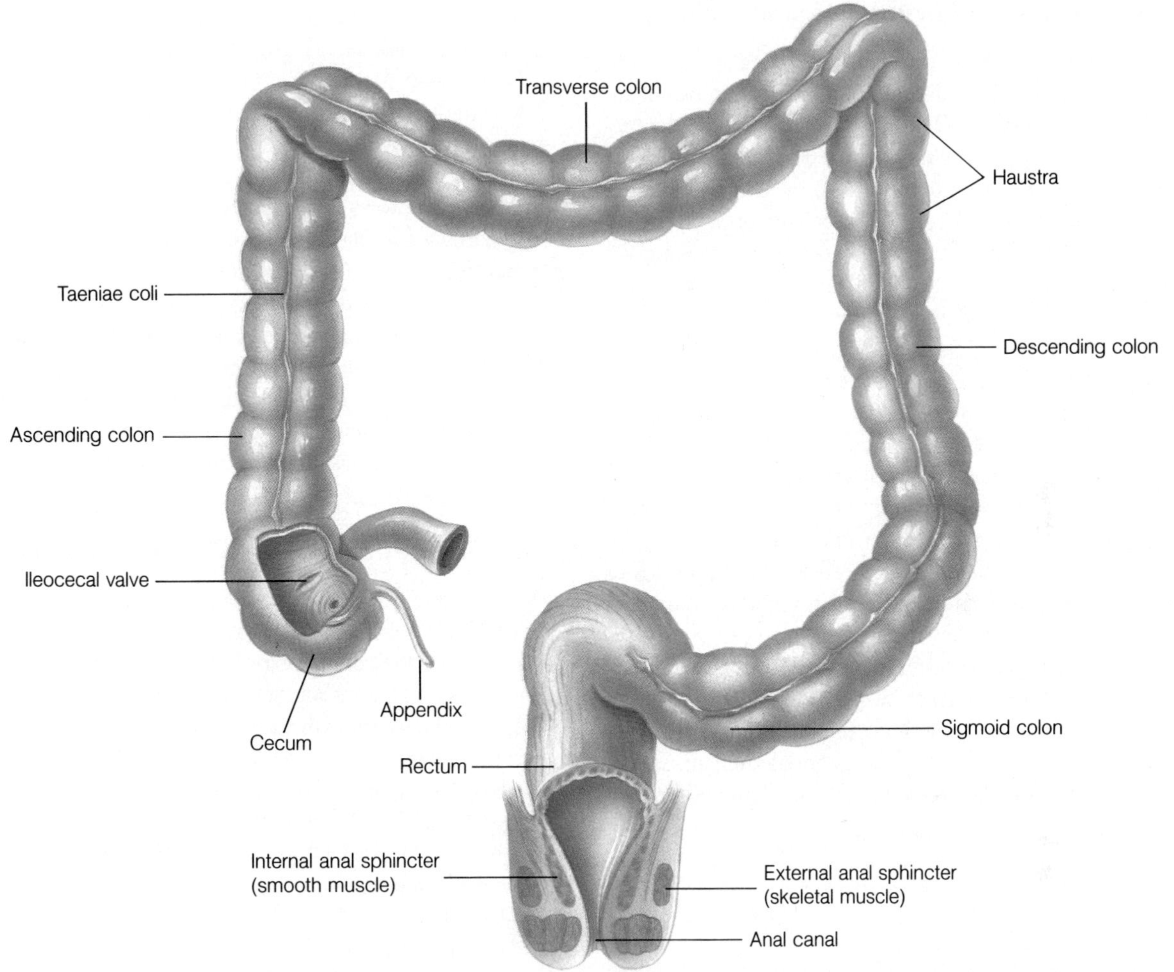

Figure 16–30 Anatomy of Large Intestine

In addition to haustral contractions, weakly propulsive peristaltic waves occur in the colon between meals to very slowly propel the contents forward. Haustral contractions and peristalsis are largely controlled by locally mediated reflexes.

Because of this slow colonic movement, bacteria have time to grow and accumulate in the large intestine; this is in contrast to the small intestine, where the contents are normally moved through too rapidly for bacterial growth to occur. Not all ingested bacteria are destroyed by salivary lysozyme and gastric HCl, so the surviving bacteria continue to thrive in the large intestine. Most colonic microorganisms are harmless in this location.

Three to four times a day, generally after meals, there is a marked increase in motility during which large segments of the ascending and transverse colon contract simultaneously, driving the feces one-third to three-fourths of the length of the colon in a few seconds. These massive contractions, appropriately called **mass movements,** drive the colonic contents into the distal portion of the large intestine, where material is stored until defecation occurs.

When food enters the stomach, mass movements occur in the colon primarily by means of the **gastrocolic reflex,** which is mediated from the stomach to the colon by gastrin and by the extrinsic autonomic nerves. In many people, this reflex is most evident after the first meal of the day and is often followed by the urge to defecate. Thus, when a new meal enters the digestive tract, reflexes are initiated to move the existing contents farther along down the tract to make way for the incoming food. The gastroileal reflex moves the remaining small-intestine contents into the large intestine, and the gastrocolic reflex pushes the colonic contents into the rectum, triggering the defecation reflex.

Feces are eliminated by the defecation reflex.

When mass movements of the colon move fecal material into the rectum, the resultant distention of the rectum stimulates stretch receptors in the rectal wall, thus initiating the **defecation reflex.** This reflex causes the **internal anal sphincter** (which is composed of smooth muscle) to relax and the rectum and sigmoid colon to contract more vigorously. If the **external anal sphincter** (which is composed of skeletal muscle) is also relaxed, defecation occurs. Being skeletal muscle, the external anal sphincter is under voluntary control. The initial distention of the rectal wall is accompanied by the conscious urge to defecate. If circumstances are unfavorable for defecation, voluntary tightening of the external anal sphincter can prevent defecation in spite of the defecation reflex. If defecation is delayed, the distended rectal wall gradually relaxes, and the urge to defecate subsides until the next mass movement propels more feces into the rectum, once again distending the rectum and eliciting the defecation reflex. During periods of nonactivity, both anal sphincters remain tonically contracted to assure fecal continence.

When defecation does occur, it is usually assisted by voluntary straining movements that involve simultaneous contraction of the abdominal muscles and a forcible expiration against a closed glottis. This maneuver brings about a large increase in intra-abdominal pressure, which assists in elimination of the feces.

Constipation occurs when the feces become too dry.

If defecation is delayed too long, **constipation** may result. When the colonic contents are retained for longer periods of time than normal, more than the usual amount of H_2O is absorbed, so the feces become hard and dry. There are normal variations in frequency of defecation among individuals, ranging from after every meal to up to once a week. When the frequency is delayed beyond that which is normal for a particular individual, constipation and its attendant symptoms may occur. These include abdominal discomfort, dull headache, loss of appetite sometimes accompanied by nausea, and mental depression. Contrary to popular belief, these symptoms are not caused by toxins absorbed from the retained fecal material. Although some potentially toxic substances are produced by bacterial metabolism in the colon, these substances normally pass through the portal system and are removed by the liver before they can reach the systemic circulation. Instead, the symptoms associated with constipation seem to be due to prolonged distention of the large intestine, particularly the rectum, because these sensations are promptly alleviated following relief from distention.

Possible causes for delayed defecation that might lead to constipation include: (1) ignoring the urge to defecate; (2) decreased colon motility accompanying aging, emotion, or a low-bulk diet; (3) obstruction of fecal movement in the large bowel caused by a local tumor or colonic spasm; and (4) impairment of the defecation reflex, such as through injury of the nerve pathways involved.

If hardened fecal material becomes lodged in the appendix, it may obstruct normal circulation and mucus secretion in this narrow, blind-ended appendage. This blockage leads to inflammation of the appendix, or **appendicitis.** The appendix often becomes swollen and filled with pus, and the inflamed tissue may die as a result of local circulatory interference. If not surgically removed, the diseased appendix may rupture, spewing its infectious contents into the abdominal cavity.

Large-intestine secretion is protective in nature.

No digestive enzymes are secreted by the large intestine. There is no need, since digestion is completed before chyme ever reaches the colon. Colonic secretion consists of an alkaline (HCO_3^-) mucus solution, the function of which is to protect the large-intestine mucosa from mechanical and chemical injury. The mucus provides lubrication to facilitate passage of the feces, whereas the HCO_3^- neutralizes irritating acids produced by local bacterial fermentation. Secretion increases in response to mechanical and chemical stimulation of the colonic mucosa mediated by short reflexes and parasympathetic innervation.

No digestion takes place within the large intestine because there are no digestive enzymes. However, the colonic bacteria do digest some of the cellulose and use it for their own metabolism.

The large intestine absorbs salt and water, converting the luminal contents into feces.

Some absorption takes place within the colon but not to the same extent as in the small intestine. Because the luminal surface of the colon is fairly smooth, it has considerably less absorptive surface area than the small intestine. Furthermore, no specialized transport mechanisms are present in the colonic mucosa for absorption of glucose or amino acids, as there are in the small intestine. When excessive small-intestinal motility delivers the contents to the colon before absorption of nutrients has been completed, the colon is unable to absorb these materials, and they are lost in diarrhea.

The colon normally absorbs some salt and H_2O. Sodium is actively reabsorbed, Cl^- follows passively down the electrical gradient, and H_2O follows osmotically. Bacteria in the colon synthesize some vitamins that the colon is capable of absorbing, but this is normally not a significant contribution, except in the case of vitamin K.

Through absorption of salt and H_2O, a firm fecal mass is formed. Of the 500 ml of material entering the colon per day from the small intestine, the colon normally absorbs about 350 ml, leaving 150 gm of feces to be eliminated from the body each day (Table 16–6). The composition of this fecal material is normally 100 gm of H_2O and 50 gm of solid, including undigested cellulose, bilirubin, bacteria, and small amounts of salt. Thus, contrary to popular thinking, the digestive tract is not a major excretory passageway for eliminating wastes from the body. The main waste product excreted in the feces is bilirubin. The other fecal constituents are unabsorbed food residues and bacteria, which were never actually a part of the body.

Intestinal gases are absorbed or expelled.

Occasionally, instead of fecal material passing from the anus, intestinal gas, or **flatus** passes out. This gas is derived primarily from two sources: (1) swallowed air, with as much as 500 ml of air being swallowed during a meal, and (2) gas produced by bacterial fermentation in the colon. The presence of gas percolating through the luminal contents gives rise to gurgling sounds known as **borborygmi.** Eructation (burping or belching) removes most of the swallowed air from the stomach, but some passes on into the intestine. Usually very little gas is present in the small intestine, because the gas is either quickly absorbed or passes on into the colon. Most gas in the colon is due to bacterial activity, with the quantity and nature of the gas being dependent on the type of food eaten and the characteristics of the colonic bacteria. Some foods, such as beans, contain types of carbohydrates for which humans lack digestive enzymes. These fermentable carbohydrates enter the colon, where they are attacked by gas-producing bacteria. Much of the gas entering or forming in the large intestine is absorbed through the intestinal mucosa. The remainder is expelled through the anus.

To accomplish selective expulsion of gas when fecal material is also present in the rectum, the abdominal muscles and external anal sphincter are voluntarily contracted simultaneously. When contraction of the abdominal muscles raises the pressure sufficiently against the contracted anal sphincter, the pressure gradient forces air out at a high velocity through a slit-like anal opening that is too narrow for solid feces to escape. This passage of air at high velocity causes the edges of the anal opening to vibrate, giving rise to the characteristic low-pitched sound accompanying passage of gas.

Overview of the Gastrointestinal Hormones

Throughout our discussion of digestion, we have repeatedly mentioned different functions of the four accepted gastrointestinal hormones: gastrin, secretin, cholecystokinin, and gastric inhibitory peptide. We will now fit all of these functions together so that you can appreciate the overall adaptive importance of these interactions.

Chyme in the stomach, especially if it contains protein, stimulates the release of gastrin, which acts on parietal and chief cells to increase secretion of HCl and pepsinogen. These two substances, in turn, are of primary importance in initiating digestion of the protein that promoted their release. Furthermore, gastrin enhances gastric motility, stimulates ileal motility, relaxes the ileocecal sphincter, and induces mass movements in the colon—functions that are all aimed at keeping the contents moving through the tract upon the arrival of a new meal. Gastrin also is trophic to not only the stomach mucosa but also to the small-intestine mucosa, helping maintain a well-developed, functionally viable digestive-tract lining. Predictably, gastrin secretion is inhibited by an accumulation of acid in the stomach and by the presence in the duodenal lumen of acid and other constituents that necessitate a delay in gastric secretion.

As the stomach empties into the duodenum, the presence of acid in the duodenum stimulates the release of secretin into the blood. Secretin performs four major interrelated functions: (1) it inhibits gastric emptying to prevent further acid from entering the duodenum until the acid that is already present is neutralized; (2) it inhibits gastric secretion to reduce the amount of acid being produced; (3) it stimulates the pancreatic duct cells to produce a large volume of aqueous $NaHCO_3$ secretion, which is emptied into the duodenum to neutralize the acid; and (4) it stimulates secretion by the liver of a $NaHCO_3$-rich bile, which likewise is emptied into the duodenum to assist in the neutralization process. Neutralization of the acidic chyme in the duodenum helps prevent damage to the duodenal walls and provides a suitable environment for the optimal functioning of the pancreatic digestive enzymes.

As chyme is emptied from the stomach, fat and the products of partial protein and carbohydrate digestion enter the duodenum. Carbohydrates do not directly serve as stimuli in the regulation of digestive activities, but fat and protein products do. Fatty acids and amino acids are particularly important in the release of cholecystokinin (CCK) from the duodenal mucosa. This hormone also performs several important interrelated functions: (1) it inhibits gastric motility and secretion, thereby allowing adequate time for the nutrients already in the duodenum to be digested and absorbed; (2) it stimulates the pancreatic acinar cells to increase secretion of pancreatic enzymes, which continue the digestion of these nutrients in the duodenum; and (3) it causes contraction of the gall bladder and relaxation of the sphincter of Oddi so that bile is emptied into the duodenum to aid fat digestion and absorption. Bile salts' detergent action is particularly important in enabling pancreatic lipase to perform its digestive task. Once again, the multiple effects of CCK are remarkably well adapted to dealing with the fat and protein products in the duodenum that triggered this hormone's release.

Furthermore, it is appropriate that both secretin and CCK, which have profound stimulatory effects on the exocrine pancreas, are trophic to this tissue.

Gastric inhibitory peptide (GIP) is the most recently recognized gastrointestinal hormone. We already mentioned its role in inhibiting gastric motility and secretion. In addition, it has been shown to stimulate insulin release by the pancreas, so it is also called **glucose-dependent insulinotrophic peptide** (once again GIP). Again, this is remarkably adaptive. As soon as the meal is absorbed, the body has to shift its metabolic gears to use and store the newly arriving nutrients. The metabolic activities of this postabsorptive phase are largely under the control of insulin (chapter 19). Stimulated by the presence of a meal in the digestive tract, GIP initiates the release of insulin in anticipation of the absorption of the meal in a "feed-forward" fashion.

This overview of the multiple, integrated, adaptive functions of the gastrointestinal hormones provides an excellent example of the remarkable efficiency of the human body. The list of hormones that regulate digestive function will undoubtedly grow, because a number of peptides found in the digestive tract are suspected of being hormones. Of these candidate hormones, motilin and villikinin have already been mentioned.

Chapter in Perspective

To maintain homeostasis, nutrient molecules used for energy production must continually be replaced by the acquisition of new, energy-rich nutrients. Similarly, water and electrolytes that are constantly lost in the urine and sweat and through other avenues must be replenished on a regular basis. The digestive system contributes to homeostasis by transferring nutrients, water, and electrolytes from the external environment to the internal environment.

The carbohydrate, protein, and fat nutrients in the ingested food must be broken down into absorbable units by the digestive system before they can be absorbed into the internal environment. The food is procured, chewed, and moistened in the mouth and then propelled by a swallow through the esophagus to the stomach, where it is stored until it can be processed by the small intestine. Token digestion of carbohydrate begins in the mouth and continues in the body of the stomach under the influence of salivary amylase. A modest percentage of protein digestion occurs in the stomach as a result of gastric pepsin secretion in concert with HCl secretion. The major site of digestion and absorption is the small intestine. Here the potent pancreatic enzymes completely digest fat, assisted by bile, and almost bring carbohydrate and protein digestion to completion. Digestion of the latter two foodstuffs is completed by enzymes embedded in the lining of the small intestine.

The small intestine absorbs all of the digested nutrients, along with much of the water and electrolytes. The undigested residues and remaining fluid pass into the large intestine, which extracts more of the water and salt. What remains is fecal material, which is periodically eliminated from the body.

Both the progression of the contents through the tract and the secretion of digestive juices are carefully regulated by complex overlapping neural and hormonal mechanisms to assure that the meal is maximally digested and absorbed.

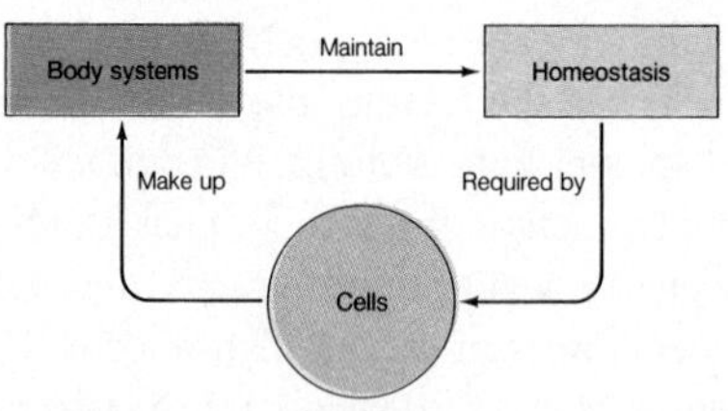

See inside front cover for an expanded version of this model.

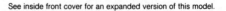

Review Exercises

1. Describe the four basic digestive processes.
2. List the three categories of energy-rich foodstuffs and the absorbable units of each.
3. List the components of the digestive system. Describe the cross-sectional anatomy of the digestive tract.
4. What four general factors are involved in the regulation of gastrointestinal function? What is the role of each?
5. Describe the factors that control motility in each component of the digestive tract.
6. Describe the factors that control each digestive secretion.
7. List the enzymes involved in the digestion of each category of foodstuff. Indicate the source and control of secretion of each of these enzymes. Why are some digestive enzymes secreted in inactive form? How are they activated?
8. What absorption processes take place within each component of the digestive tract? What special adaptations of the small intestine enhance its absorptive capacity?
9. What are the contributions of the accessory digestive organs? What are the nondigestive functions of the liver?
10. Summarize the functions of each of the gastrointestinal hormones.
11. What waste product is excreted in the feces?
12. How is vomiting accomplished? What are the causes and consequences of vomiting, diarrhea, and constipation?
13. Describe the process of mucosal turnover in the small intestine.
14. **A point to ponder:** Why do patients who have had a large portion of their stomachs removed for treatment of stomach cancer or severe peptic ulcer disease have to eat small quantities of food frequently instead of consuming three meals a day?

CHAPTER 17

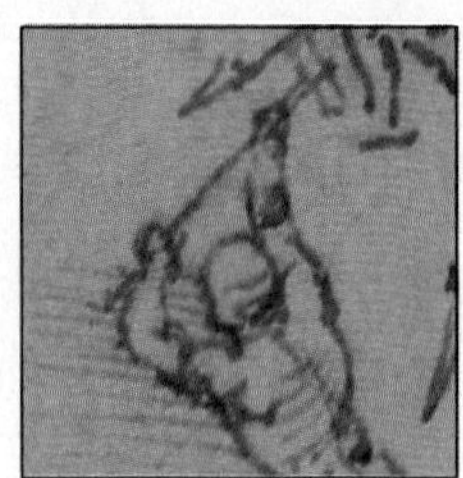

ENERGY BALANCE AND TEMPERATURE REGULATION

CHAPTER CONTENTS AT A GLANCE

INTRODUCTION *How does a bikini-clad sunbather basking on the beach compare with a snake sunning itself on a rock with regard to temperature regulation? There are important differences but also some similarities in the means by which humans and snakes adapt to the environmental temperature. Humans, other mammals, and birds are* ***thermoregulators;*** *they are able to maintain a remarkably constant, rather high internal body temperature despite the body's exposure to a wide range of environmental temperatures. To maintain thermal homeostasis, thermoregulators physiologically manipulate mechanisms within their bodies to adjust heat production, heat conservation, and heat loss. In contrast, with few exceptions reptiles, amphibians, fish, and invertebrates are* ***thermoconformers;*** *their body temperatures conform to the temperature of their surroundings. Thus their body temperatures vary capriciously with changes in the environmental temperature. Even though thermoconformers produce heat, these species cannot physiologically regulate internal heat production, nor can they control heat exchange with their environment to maintain a constant body temperature when the temperature in their surroundings rises or falls. When a snake's surroundings are cold, its body temperature falls accordingly. Because the rate of chemical reactions is temperature dependent, cooling slows all chemical activity within the body. Thermoregulators such as humans are spared this slowdown of all bodily functions during cold weather.*

Likewise, thermoregulators have an advantage over thermoconformers in hot environments. Thermoregulators have the ability, within limits, to prevent overheating of the body, for example, by using heat-loss mechanisms such as sweating.

Overheating actually presents more problems than overcooling. Irreversible protein denaturation and nerve malfunction occur with even moderate elevations in body temperature, whereas the effects of low body temperature are largely reversible except when cells are actually frozen.

The only option available to thermoconformers for adjusting their body temperatures to more optimal levels is behavioral adaptation. For example, snakes bask on sun-warmed rocks to raise their body temperatures or slither into the shade of tall grasses to cool themselves when it gets hot outside. Thermoregulators also can employ behavioral strategies to supplement their physiological thermoregulatory mechanisms. A hot sunbather can move to the shade or take a swim in the cool water to escape from the sun's heat. Adaptive behavior is the thermoregulatory mechanism shared in common by both humans and snakes.

Humans are usually in environments cooler than their bodies, so they must constantly generate heat internally to maintain their body temperatures. Heat production ultimately depends on the oxidation of metabolic fuel derived from food. In this chapter we will discuss control of food intake to maintain energy balance as well as regulation of body temperature.

Energy Balance

Most food energy is ultimately converted into heat in the body.

Each cell in the body requires energy to perform the functions essential for the cell's own survival (such as active transport and cellular repair) and to carry out its specialized contributions toward maintenance of homeostatic balance (such as gland secretion and muscle contraction). All energy used by cells is ultimately provided by food intake. Chemical energy locked in the intramolecular bonds of nutrient molecules is released when these molecules are broken down within the body. Energy harvested from biochemical processing of ingested nutrients either is used immediately to perform biological work or is stored in the body for later use as needed during periods when food is not being digested and absorbed. Thus, even though immediate energy needs can be provided by either food intake or energy stores within the body such as adipose tissue (fat), the ultimate source of all energy input to the body is food.

According to the *first law of thermodynamics*, energy can neither be created nor destroyed. Therefore, energy is subject to the same kind of input-output balance as are the chemical components of the body such as H_2O and salt (see p. 511).

The energy in ingested foodstuffs constitutes energy input to the body. Energy output or expenditure falls into two categories (Fig. 17–1): external work and internal work. **External work** refers to the energy expended by contracting skeletal muscles to move external objects or to move the body in relation to the environment. **Internal work** constitutes all other forms of biological energy expenditure that do not accomplish mechanical work outside of the body. Internal work encompasses two types of energy-dependent activities: (1) skeletal-muscle activity used for purposes other than external work, such as the contractions associated with postural maintenance and shivering, and (2) all the energy-expending activities that must go on all of the time just to sustain life. The latter include the work of pumping blood and breathing; the energy required for active transport of critical materials across plasma membranes; and the energy used during synthetic reactions essential for the maintenance, repair, and growth of cellular structures—in short, the "metabolic cost of living."

Not all energy within nutrient molecules can be harnessed to perform biological work. Energy cannot be created or destroyed, but it can be converted from one form to another. The energy in nutrient molecules that is not used to energize work is transformed into **thermal energy,** or **heat.** During biochemical processing, only about half of the energy in nutrient molecules is transferred to ATP; the other 50% of nutrient energy is immediately lost as heat. During ATP expenditure by the cells, another 25% of the energy derived from ingested food becomes heat. Since the body is not a heat engine, it cannot convert heat into work. Therefore, not more than 25% of nutrient energy is available to accomplish work, either external or internal. The remaining 75% is lost as heat during the sequential transfer of energy from nutrient molecules to ATP to cellular systems.

Furthermore, of the energy actually captured for use by the body, almost all expended energy eventually becomes heat. To exemplify, energy expended by the heart to pump blood is gradually changed by friction into heat as blood flows through the vessels. Likewise, energy used in the synthesis of cellular structural protein eventually appears as heat when that protein is degraded during the normal course of turnover of bodily constituents. Even in the performance of external work, skeletal muscles inefficiently convert chemical energy into mechanical energy, with as much as 75% of the expended energy being lost as heat. Thus all energy liberated from ingested food that is not directly used for movement of external objects or that is not stored as fat deposits (or, in the

Figure 17–1 Energy Input and Output

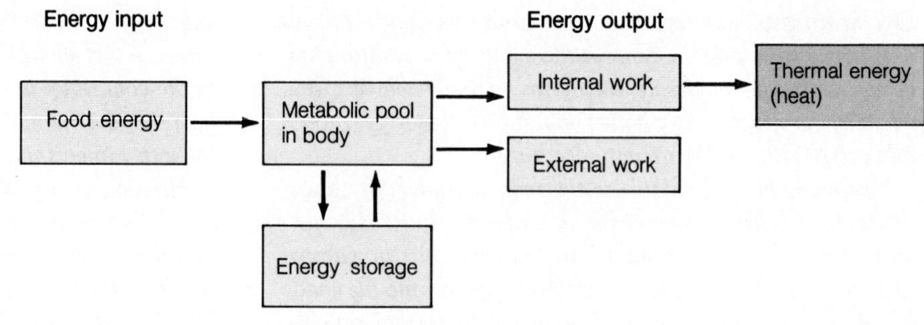

case of growth, as protein) eventually becomes body heat. This heat is not entirely wasted energy, however, because much of it is used to maintain body temperature.

The metabolic rate can be measured indirectly by means of the energy equivalent of oxygen.

The rate at which energy is expended by the body during both external and internal work is known as the **metabolic rate:**

Metabolic rate = Energy expenditure/Unit of time

Since most of the body's energy expenditure eventually appears as heat, the metabolic rate is normally expressed in terms of the rate of heat production in kilocalories per hour. The basic unit of heat energy is the **calorie,** which is the amount of heat required to raise the temperature of 1 gram of H_2O 1° C. Since this unit is too small to be convenient when discussing the human body because of the magnitude of heat involved, the **kilocalorie** or **Calorie,** which is equivalent to 1,000 calories, is used. When nutritionists speak of "calories" in quantifying the energy content of various foods, they are actually referring to kilocalories or Calories. Four kilocalories of heat energy are released when 1 gram of glucose is oxidized or "burned," whether the oxidation takes place inside or outside of the body.

The metabolic rate and consequently the amount of heat produced varies depending on the presence or absence of a variety of factors, such as exercise, food intake, shivering, and anxiety. Increased skeletal-muscle activity is the factor that can increase metabolic rate to the greatest extent. Even slight increases in muscle tone notably elevate the metabolic rate, and various levels of physical activity alter energy expenditure and heat production markedly (Table 17–1). For this reason, a person's metabolic rate is determined under standardized basal conditions established to control as many of the variables that can alter metabolic rate as possible in an attempt to evaluate the metabolic activity necessary to maintain the basic body functions at rest. This so-called **basal metabolic rate (BMR)** is a reflection of the body's "idling speed," or the minimal waking rate of internal energy expenditure. The specified conditions for measurement of the BMR are as follows:

1. The person should be at physical rest, having refrained from exercise for at least thirty minutes to eliminate any contribution of muscular exertion to heat production.
2. The individual should be at mental rest to minimize skeletal-muscle tone (people "tense up" when they are nervous) and to prevent a rise in epinephrine secretion. Stress elicits the secretion of the hormone epinephrine, which increases the metabolic rate.
3. The measurement should be performed at a comfortable room temperature so that the individual does not shiver. Shivering can markedly increase heat production because the purpose of these reflex oscillating skeletal-muscle contractions is to generate heat in response to cold exposure.
4. The subject should not have eaten any food within twelve hours before the BMR determination to avoid the **specific dynamic action of foods.** Specific dynamic action refers to the obligatory increase in metabolic rate that occurs as a consequence of food intake. This short-lived (less than twelve-hour) rise in metabolic rate is not due to digestive activities but to the increased metabolic activity associated with the processing and storage of ingested nutrients, especially by the major biochemical factory, the liver.

The rate of heat production in BMR determinations can be measured directly or indirectly. **Direct calorimetry** involves the cumbersome procedure of placing the subject in an insulated chamber with H_2O circulating through the walls. A determination of the difference in the temperature of the H_2O entering and leaving the chamber reflects the amount of heat liberated by the subject and picked up by the H_2O as it passes through the chamber. Even though this method provides a direct measurement of heat production, it is not practical because a calorimeter chamber is costly and takes up a lot of space. Therefore, a more practical alternative method of indirectly determining the rate of heat production was developed for widespread use. With **indirect calorimetry,** the only measurement made is of the subject's O_2 uptake per unit of time, which is a simple task using minimal equipment. Recall that:

$$\text{Food} + O_2 \rightarrow CO_2 + H_2O + \text{Energy (mostly transformed into heat)}$$

Accordingly, there is a direct relationship between the volume of O_2 used and the quantity of heat produced. This relationship also depends on the type of food being oxidized. Although carbohydrates, proteins, and fats require different amounts of O_2 for their oxidation and yield different amounts of kilocalories when oxidized, an average estimate can be made of the quantity of heat produced per liter of O_2 consumed on a typical mixed American diet. This approximate value, known as the **energy equivalent of O_2,** is 4.8 kilocalories of energy liberated per liter of O_2 consumed. Using this method, the metabolic rate of a person consuming 15 l/hr of O_2 can be estimated as follows:

15 l/hr	=	O_2 consumption
×4.8 kilocalories/l	=	Energy equivalent of O_2
72 kilocalories/hr	=	Estimated metabolic rate

In this way, a simple measurement of O_2 consumption can be used to reasonably approximate heat production in the determination of metabolic rate.

Once the rate of heat production is determined under the prescribed basal conditions, it must be compared with normal values for persons of the same sex, age, height, and weight because these factors all affect the basal rate of energy expenditure. For example, the actual rate of heat production is greater for a large man than for a smaller man, but when expressed in terms of total surface area (which is a reflection of height and weight), the caloric output in kilocalories per hour per square meter of surface area is normally about the same.

Thyroid hormone is the primary but not sole determinant of the rate of basal metabolism. As the level of active circulating thyroid hormone increases, the BMR increases correspondingly.

Surprisingly, the BMR is not the body's lowest metabolic rate. The rate of energy expenditure is 10% to 15% lower than the BMR during sleep, presumably because of the more complete muscle relaxation that occurs during the paradoxical stage of sleep (see p. 149).

Table 17–1 Rate of Energy Expenditure for 70-Kg Person during Different Types of Activity

Form of Activity	Energy expenditure (kcal/hour)
Sleeping	65
Awake, lying still	77
Sitting at rest	100
Standing relaxed	105
Getting dressed	118
Typewriting	140
Walking slowly on level (2.6 mi/hr)	200
Carpentry, painting a house	240
Sexual intercourse	280
Bicycling on level (5.5 mi/hr)	304
Shoveling snow, sawing wood	480
Swimming	500
Jogging (5.3 mi/hr)	570
Rowing (20 strokes/min	828
Walking up stairs	1,100

Energy input must equal energy output to maintain a neutral energy balance.

Since energy cannot be created or destroyed, energy input must equal energy output, as represented by the following equation:

$$\text{Energy input} = \text{Energy output}$$
$$\text{Energy in food consumed} = \text{External work} + \text{Internal heat production} \pm \text{Stored energy}$$

There are three possible states of energy balance.

NEUTRAL ENERGY BALANCE. If the amount of energy in food intake exactly equals the amount of energy expended by the muscles in performing external work plus the basal internal energy expenditure that eventually appears as body heat, then energy input and output are exactly in balance and body weight remains constant.

POSITIVE ENERGY BALANCE. If the amount of energy in food intake is greater than the amount of energy expended by means of external work and internal functioning, the extra energy taken in but not used is stored in the body, primarily as adipose tissue, so body weight increases.

NEGATIVE ENERGY BALANCE. Conversely, if the energy derived from food intake is less than the body's immediate energy requirements, the body must use stored energy to supply energy needs, and, accordingly, body weight decreases.

To maintain a constant body weight (with the exception of minor fluctuations caused by changes in H_2O content), energy acquired through food intake must equal energy expenditure by the body. Because the average adult maintains a fairly constant weight over long periods of time, this implies that precise homeostatic mechanisms exist to maintain a long-term balance between energy intake and energy expenditure. Theoretically, total body energy content could be maintained at a constant level by regulating the magnitude of food intake, physical activity, or internal work and heat production. Control of food intake to match changing metabolic expenditures is the major means of maintaining a neutral energy balance. The level of physical activity is principally under voluntary control. For example, there are no regulatory factors that automatically impel an obese person to exercise; in fact, in most cases overweight people are even less inclined to engage in physical activities than are their slimmer counterparts. There are mechanisms that can alter the degree of internal work and heat production, but these control measures are aimed primarily at regulating body temperature rather than total energy balance. Some studies suggest however, that after several weeks of eating less or more than desired, small counteracting changes in metabolism may occur. For example, a compensatory increase in the body's efficiency of energy use in response to underfeeding may partially explain why some dieters become stuck at a plateau after having lost the first ten or so pounds of weight fairly easily. Similarly, a compensatory reduction in the efficiency of energy use in response to overfeeding may account in part for the difficulty experienced by very thin people who are deliberately trying to gain weight. Despite the possibility of compensatory changes in metabolism, there is no question that regulation of food intake is the most important factor in the long-term maintenance of energy balance and body weight.

Food intake is controlled primarily by the hypothalamus, but none of the proposals for the mechanism(s) involved is fully satisfactory.

Control of food intake is primarily a function of the hypothalamus. Classically, the hypothalamus is considered to house a pair of **feeding** or **appetite centers** located in the lateral regions of the hypothalamus, one on each side, and another pair of **satiety centers** located in the ventromedial area of the hypothalamus along the midline. The functions of these areas have been elucidated by a series of experiments that involve either destruction or stimulation of these specific regions. Stimulation of the clusters of nerve cells designated as feeding or appetite centers makes the animal hungry, driving it to eat voraciously, whereas selective destruction of these areas suppresses eating and food-intake behavior to the point that the animal starves itself to death. In contrast, stimulation of the satiety centers signals satiety, or the feeling of having had enough to eat. Consequently, the stimulated animal refuses to eat, even if previously deprived food. As expected, destruction of this area produces the opposite effect—profound overeating and obesity because the animal never achieves a feeling of being full (Fig. 17–2). Thus the feeding centers tell us to eat, whereas the satiety centers tell us when we have had enough. Although it is convenient to consider these specific areas as exciting and inhibiting feeding behavior respectively, this approach is probably too simplistic. Other areas of the brain are believed to play important roles in controlling food intake, but their contributions and interrelationships with the hypothalamus remain obscure.

Whatever the site of final integration, a major interest is determining what input factors to these integrating centers govern feeding onset and termination. Exactly what switches feeding behavior on and off is still unclear. Even though food intake is adjusted to balance changing energy expenditures over a period of time, there are no calorie receptors per se to monitor energy input, energy output, or total body energy content. A number of proposals have been set forth, each supported by some experimental findings, but none of them alone is able to account for all observed feeding behavior. Undoubtedly, control of food intake does not depend on changes in a single signal but is determined by the integration of many inputs that provide information about the body's energy status. Some information is apparently used for short-term regulation of food intake, helping to control meal size and frequency. Even so, over a twenty-four-hour period the energy in ingested food rarely matches energy expenditure for that day. The correlation between total caloric intake and total energy output is excellent, however, over long periods of time. As a result, the total energy content of the body—and, consequently, body weight—remain relatively constant on a long-term basis. The following sections discuss some of the proposals for regulation of food intake (Fig. 17–3).

ROLE OF GASTROINTESTINAL DISTENTION. Initial proposals inspired by common sense were based on the idea that cues of

Figure 17–2 Comparison of Normal Rat with Rat Whose Satiety Center Has Been Destroyed *Several months after destruction of the satiety center in the ventromedial area of the hypothalamus, the rat on the right had gained considerable weight as a result of overeating compared to its normal litter mate on the left. Rats sustaining lesions in this area also display less grooming behavior, accounting for the soiled appearance of the fat rat.*

SOURCE: Photo courtesy of Wilbert E. Gladfelter, Associate Professor, Department of Physiology, School of Medicine, West Virginia University.

emptiness or fullness of the digestive tract could signal hunger or satiety, respectively. However, several bits of information suggest that the degree of stomach distention is not very important. If an experimental animal is provided food mixed with calorie-free material, it will increase its food intake in direct relation to the caloric dilution; thus, quality of food, not quantity, is apparently important. Along the same line, we all are aware that calorie-rich, heavy foods are much more "filling" than low calorie, light foods, even though the same quantity of each may be consumed. The best demonstration of the relative unimportance of the volume of food in the gastrointestinal tract in determining feeding comes from experiments in which rats hold their body weight constant by feeding themselves appropriate nutrient intake through intravenous tubes by pressing a bar, even though the food never enters their digestive tracts. Similarly, complete denervation of the stomach and small intestine does not interfere with normal maintenance of food intake. Thus physical cues from the digestive tract cannot constitute a major input for regulating food intake, although they probably play minor modifying roles.

GLUCOSTATIC AND LIPOSTATIC THEORIES. If one eliminates signals from the digestive tract as major inputs controlling food intake, it becomes evident that some internal signal reflecting the depletion or availability of energy-producing substances must be involved in controlling the initiation and cessation of eating. Two major theories—the glucostatic and lipostatic theories—have been proposed based on the premise that decreases or increases in a particular nutrient molecule in the body would induce hunger or satiety, respectively. The **glucostatic theory** proposes that satiety is signaled by increased glucose utilization, such as would occur during a meal when glucose is being absorbed from the digestive tract and the level of blood glucose is increasing. Conversely, when no new glucose is entering the blood after absorption of a meal is complete, the resultant reduction in the cells' glucose use would arouse the sensation of hunger. According to this proposal, the degree of activation of specific glucose receptors in the brain would govern eating behavior. The glucostatic theory is generally considered to be the leading proposal for short-term regulation of the timing and amount eaten in a meal. A major flaw of the glucostatic theory is that it cannot account for the fact that the feeling of satiation occurs when most of the meal is still in the lumen of the digestive tract instead of being used by the cells. Thus, when feeding ceases, the metabolic deficit has not yet been corrected, although the food in the digestive tract will eventually reverse the deficit as the nutrients are absorbed and used.

The **lipostatic theory** is generally considered to be responsible for the long-term matching of food intake to energy expenditure so that total body energy content remains balanced and body weight remains constant. According to the lipostatic theory, increased fat storage in adipose tissue signals satiety. It is proposed that glycerol serves as a blood-borne signal between fat stores and the areas of the brain controlling food intake. The amount of glycerol in the blood appears to be an excellent indicator of the total amount of triglyceride fat stored in the adipose tissue. One problem with the lipostatic theory is that overweight people who have an abundance of adipose tissue still get hungry. Perhaps the percentage of filling of each adipose cell is important as well. Accordingly, persons with a greater number of adipose cells may be overweight in comparison to normal standards, yet they may still be hungry because their adipose cells are not "full."

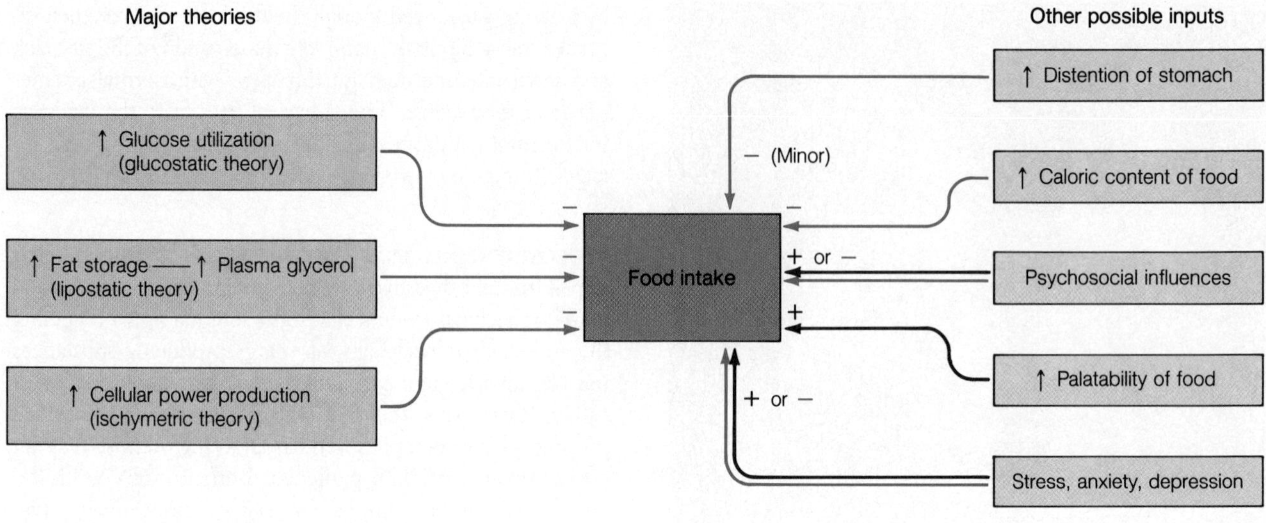

Figure 17–3 Possible Inputs Controlling Food Intake

ISCHYMETRIC THEORY. The most recent proposal, the **ischymetric theory** (*ischys* means "power"; *metric* means "measure"), suggests that the signal for short-term control of food intake is not a deficit or surfeit of any particular major nutrient such as glucose but is linked instead to the magnitude of cellular power (ATP) production. Changes in the availability of one or all of the nutrients to a cell may result in decreases or increases in the rate of ATP/ADP turnover, which in turn could be transduced into some sort of blood-borne or neural signal of low power (hunger) or high power (satiety).

PSYCHOSOCIAL INFLUENCES. Thus far we have described possible involuntary signals that might automatically occur to control food intake. However, as with H_2O intake, people's eating habits are also shaped by psychological and social factors. We often do not eat merely because we are hungry or stop because we are full. Frequently we eat out of habit (eating three meals a day on schedule no matter what our status on the hunger-satiety continuum) or because of social custom (food often plays a prime role in entertainment, leisure, and business activities).

Furthermore, the amount of pleasure derived from eating can reinforce feeding behavior. Eating foods with an enjoyable taste, smell, and texture can increase appetite and food intake. This has been demonstrated in an experiment in which rats were offered their choice of highly palatable human foods. They overate by as much as 70% to 80% and became obese. When the rats returned to eating their regular monotonous but nutritionally balanced rat chow, their obesity was rapidly reversed as their food intake was controlled once again by physiological drives rather than by hedonistic urges for the tastier offerings.

Stress, anxiety, depression, and boredom have also been shown to alter eating behavior in ways that are unrelated to energy needs in both experimental animals and humans. Thus, any comprehensive explanation of control of food intake must also take into account these voluntary consummatory acts that can reinforce or override the internal signals governing eating behavior.

Obesity occurs when more calories are consumed than are burned up.

Obesity is defined as excessive fat content in the adipose tissue stores. The arbitrary boundary for obesity is generally considered to be greater than 20% overweight compared to normal standards. Obesity occurs when, during a period of time, more kilocalories are ingested in food than are used to support the body's energy needs, with the excessive energy being stored as triglycerides in adipose tissue. The causes of obesity are many and obscure. Some factors that may be involved include: (1) emotional disturbances in which overeating replaces other gratifications; (2) development of an excessive number of fat cells as a result of overfeeding; (3) certain endocrine disorders such as hypothyroidism; (4) disturbances of the satiety-appetite regulatory centers in the hypothalamus; (5) hereditary tendencies; (6) the palatability of available food; and (7) lack of physical exercise.

Numerous studies have shown that, on the average, fat people do not eat any more than thin people. One possible explanation is that overweight persons do not overeat but "underexercise." It has been shown that very low levels of physical activity are not accompanied by comparable reductions in food intake. Another explanation is that excess energy input occurs only during the time that obesity is actually developing. Some investigators suggest that the hypothalamic centers determining long-term satiety are "set at a higher level" in obese persons. Thus overweight people do tend to maintain their weight, but at a higher set point than normal. Once obesity has developed, all that is required to maintain the condition is that energy input equals energy output.

Other studies further complicate the weight-control issue by suggesting that even though the total caloric content of food remains the same, the frequency of eating might be important in the distribution of the nutrients. In rat and monkey studies, nibblers had relatively more body protein and less body fat than animals given the same food in several large meals each day. Further studies suggest that, even though total caloric input for the day remains unchanged, foods consumed before bedtime are more likely to be put into storage. In this regard, bedtime snacks are more fattening than they would be if they were eaten earlier in the day.

Thus, our knowledge about the causes and control of obesity are still rather limited, as evidenced by the number of people who are constantly trying to "battle the bulge" to stabilize their weight at a more desirable level. This is important from more than an aesthetic viewpoint. It is known that obesity, especially of the android type, can predispose an individual to illness and premature death from a multitude of diseases (see the accompanying boxed feature, A Closer Look at Exercise Physiology, p. 610).

Persons suffering from anorexia nervosa have a pathological fear of gaining weight.

The converse of obesity is generalized nutritional deficiency. The obvious causes for reduction of food intake below energy needs are lack of availability of food, interference with the swallowing or digestive mechanism, and impairment of appetite. Chronic diseases such as renal failure, cancer, and tuberculosis are commonly accompanied by a lack of appetite, which contributes to the subsequent weight loss characteristic of these "wasting" diseases. The mechanisms for this appetite suppression are obscure.

Another poorly understood disorder in which lack of appetite is a prominent feature is **anorexia nervosa.** Patients with this disorder, most commonly adolescent girls and young women, have a morbid fear of becoming fat. As a result of having a distorted body image, they tend to visualize themselves as being much heavier than they actually are. Because they have an aversion to food, they eat very little and consequently lose considerable weight, perhaps even starving themselves to death. Other characteristics of the condition include altered secretion of many hormones, absence of menstrual periods, and low body temperature. It is unclear whether these symptoms occur secondarily as a result of general malnutrition or arise independently of the eating disturbance as part of a primary hypothalamic malfunction. Many investigators think the underlying problem may be psychological rather than biological.

Temperature Regulation

There is no one "normal" body temperature.

Because cellular function is sensitive to fluctuations in internal temperature, humans homeostatically maintain body temperature at a level that is optimal for cellular metabolism to proceed in a stable fashion. Even moderate elevations of body temperature begin to cause nerve malfunction and irreversible protein denaturation. Most people suffer convulsions when the internal body temperature reaches about 106° F (41° C); 110° F (43.3° C) is considered to be the upper limit compatible with life. On the other hand, most of the body's tissues can transiently withstand substantial cooling. Knowledge of this fact is used during cardiac surgery, when the heart must be stopped. The cooled tissues need less nourishment than when they are at at normal body temperature because of their pronounced reduction in metabolic activity. The lower O_2 need of cooled tissues also accounts for the occasional survival of drowning victims who have been submerged in icy water considerably longer than one can normally survive without O_2.

Normal body temperature is traditionally considered to be 98.6° F (37° C). Yet there is no one "normal" body temperature because the temperature varies from organ to organ, from person to person, and from time to time. From a thermoregulatory viewpoint, the body may conveniently be viewed as a central core surrounded by an outer shell. The temperature within the inner core, which consists of the abdominal and thoracic organs, the central nervous system, and the skeletal muscles, generally remains fairly constant. It is this **core temperature** that is considered to be the body temperature and that is subject to precise regulation to maintain its homeostatic constancy. The core tissues function best at a relatively constant temperature of around 100° F. The skin and subcutaneous fat constitute the outer shell. In contrast to the constant high temperature in the core, the temperature within the shell

A CLOSER LOOK AT EXERCISE PHYSIOLOGY

All Fat Is Not Created Equal

There are different ways to be fat, and one way is more dangerous than the other. Obese patients can be classified into two categories—*android*, a male-type of adipose tissue distribution, and *gynoid*, a female-type distribution—based on the anatomical distribution of adipose tissue measured as the ratio of waist circumference to hip circumference. Android obesity is characterized by abdominal fat distribution or upper body obesity, whereas gynoid obesity is characterized by fat distribution in the hips and thighs. Females can have android obesity as well as males, and males can exhibit gynoid obesity.

Android obesity is associated with a number of disorders, including insulin resistance, excessive insulin secretion, Type II (adult-onset) diabetes mellitus, excess blood lipid levels, hypertension, coronary heart disease, and stroke. Gynoid obesity is not associated with high risk for these diseases.

Because android obesity is associated with increased risk for disease, it is most important for overweight individuals with this type of adipose distribution to reduce their fat stores. Research on the success of weight-reduction programs indicates that it is very difficult for people to lose weight, but when weight loss occurs, it is from the area of increased stores. Recent interesting research has shed some light on the problems of obesity. Studies indicate that the resting metabolic rate might be an inherited trait. In one study, at three months of age, babies of obese parents showed 20% less energy expenditure than babies of lean parents. Also, diet-induced thermogenesis (DIT), which is the increase in metabolism that follows consumption of carbohydrate or protein, has been shown to be lower in those who have been obese since childhood. In other words, people who have been obese since childhood are very efficient at storing the excess calories they ingest. Lean people may metabolize more of the calories they ingest, with the energy being given off as heat.

Even after obese people reduce their weight to normal levels, their DIT remains lower than someone of the same weight who has always been lean. This would be an admirable physiologic trait in times of food deprivation, but in times of food abundance, low DIT and a low metabolic rate can predispose an individual to obesity. A person with these traits has to eat less than his or her lean counterpart to maintain normal weight.

Because very-low-calorie diets are difficult to maintain, an alternative to severely cutting caloric intake to lose weight is to increase energy expenditure through physical exercise. An aerobic-exercise program further helps reduce the risk of the disorders associated with android obesity and helps to reduce fat stores.

---

is generally cooler and may vary substantially. For example, skin temperature may fluctuate between 68° and 104° F without damage. In fact, as you will see, the temperature of the skin is deliberately varied as a control measure to help maintain the core's thermal constancy.

We are accustomed to thinking in terms of oral or rectal temperature because these are easy sites for monitoring body temperature. The average resting oral temperature is 98.6° F, with a normal range of 97 to 99° F. Rectal temperatures average about 1° F higher at 99.7° F (37.6° C), ranging from 97 to 100° F. Neither of these measurements is an absolute indication of the internal-core temperature, which averages about 100° F. Even though the core temperature is held relatively constant, several factors cause it to vary slightly:

1. Most people's core temperature normally varies about 1.8° F (1° C) during the day, with the lowest level occurring in early morning before arising (6 to 7 A.M.) and the highest point occurring in late afternoon (5 to 7 P.M.). This variation is due to an innate biological rhythm or "biological clock."

2. Women also experience a monthly rhythm in core temperature in connection with their menstrual cycle. The core temperature averages 0.9° F (0.5° C) higher during the last half of the cycle from the time of ovulation to menstruation. This mild sustained elevation in temperature was once thought to be caused by the increased secretion of progesterone, one of the ovarian hormones, during this

time period, but now this is not believed to be the case. The actual cause is still undetermined.

3. Core temperature increases during exercise because of the tremendous increase in heat production by the contracting muscles. During hard exercise, core temperature may increase to as much as 104° F (40° C). This temperature would be considered to be a fever in a resting person, but it is normal during strenuous exercise.
4. Because the temperature-regulating mechanisms are not 100% effective, core temperature may vary slightly with exposure to extremes of temperature. For example, core temperature may fall several degrees in cold weather or rise a degree or so in hot weather.

Thus core temperature can vary at the extremes between about 96° to 104° F but usually deviates less than a few degrees. This relative constancy of core temperature is made possible by multiple thermoregulatory mechanisms coordinated by the hypothalamus.

Heat gain must balance heat loss to maintain a stable core temperature.

The core temperature is a reflection of the body's total heat content. To maintain a constant total heat content and thus stability of the core temperature, heat input to the body must balance heat output (Fig. 17–4). *Heat input* occurs by way of heat gain from the external environment and internal heat production, the latter being the most important source of heat for the body. Recall that most of the body's energy expenditure ultimately appears as heat. This heat is important in the maintenance of core temperature. In fact, most of the time more heat is generated than is required to maintain the body temperature at a normal level, so the excess heat must be eliminated from the body. *Heat output* occurs by way of heat loss from exposed body surfaces to the external environment.

Balance between heat input and output is frequently disturbed by: (1) changes in heat production for purposes unrelated to regulation of body temperature, most notably by exercise, which markedly increases heat production, and (2) changes in the temperature of the external environment that influence the degree of heat gain or heat loss that occurs between the body and its surroundings. To maintain body temperature within narrow limits in spite of changes in metabolic heat production and changes in environmental temperature, compensatory adjustments must take place in heat-loss and heat-gain mechanisms.

If the core temperature starts to fall, heat production is increased; at the same time, heat loss is minimized so that normal temperature can be restored in negative-feedback fashion.

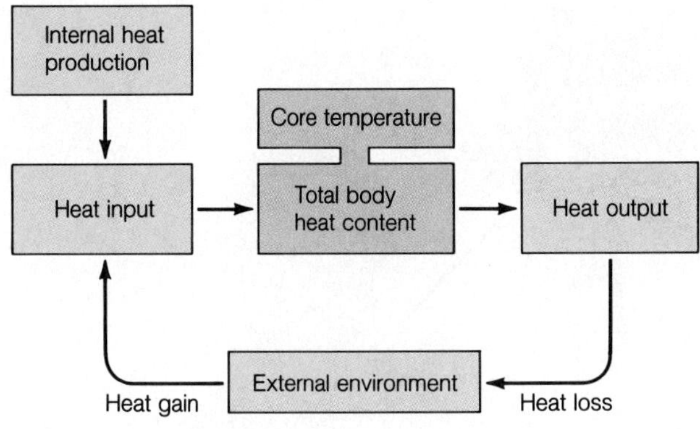

Figure 17–4 Heat Input and Output

Conversely, if the temperature starts to rise above normal, it can be corrected by increasing heat loss while simultaneously reducing heat production.

We will now elaborate on the means by which heat gains and losses can be adjusted to maintain body temperature, commencing with a discussion of the methods of heat exchange between the body and its surroundings.

Heat exchange between the body and environment takes place by radiation, conduction, convection, and evaporation.

All heat loss or heat gain between the body and external environment must take place between the body surface and its surroundings. The same physical laws of nature that govern heat transfer between inanimate objects also control the transfer of heat between the body surface and environment. The temperature of an object may be thought of as a measure of the concentration of heat within the object. Accordingly, heat always moves down its concentration gradient; that is, down a **thermal gradient** from a warmer to a cooler region.

Four mechanisms of heat transfer are used by the body: radiation, conduction, convection, and evaporation. **Radiation** is the emission of heat energy from the surface of a warm body in the form of **electromagnetic waves** or heat waves, which travel through space (Fig. 17–5a). When radiant energy strikes an object and is absorbed, the energy of the wave motion is transformed into heat within the object. The human body both emits (source of heat loss) and absorbs (source of heat gain) radiant energy. Whether or not the body loses or gains heat by radiation depends on the difference in temperature between the skin surface and the surfaces of various other objects in the body's environment. Because net transfer of heat by radiation is always from warmer objects to cooler ones,

Figure 17–5 Mechanisms of Heat Transfer *(a) Radiation—the transfer of heat energy from a warmer object to a cooler object in the form of electromagnetic waves ("heat waves"), which travel through space. (b) Conduction—the transfer of heat from a warmer object to a cooler object that is in direct contact with the warmer one, through the movement of thermal energy from molecule to adjacent molecule. (c) Convection—the transfer of heat energy by air currents. Cool air warmed by the body via conduction rises and is replaced by more cool air. This process is enhanced by the forced movement of air across the body surface. (d) Evaporation, during which the heat of vaporization required to convert a liquid such as sweat into a gaseous vapor is absorbed from the skin.*

the body gains heat by radiation from objects warmer than the skin surface, such as the sun, a radiator, or burning logs. On the other hand, the body loses heat by radiation to objects in its environment whose surfaces are cooler than the surface of the skin, such as building walls, furniture, or trees. Whether or not there is *net* gain or loss of heat by radiation between the body and its surroundings depends on whether the average temperature of all surrounding surfaces is warmer or cooler than the body surface. On the average, humans lose close to half of their heat energy through radiation.

Conduction is the transfer of heat between objects of differing temperatures that are in *direct contact* with each other (Fig. 17–5b). In this case, heat moves down its thermal gradient from the warmer to the cooler object by being transferred from molecule to molecule. All molecules are constantly in vibratory motion, with warmer molecules moving faster than cooler ones. When molecules of differing heat content touch each other, the faster moving warmer molecule agitates the cooler molecule into more rapid motion, thereby warming up the cooler molecule. During this process the original warmer molecule loses some of its thermal energy as it slows down and cools off a bit. Therefore, given enough time, the temperature of the two touching objects eventually equalizes.

The rate of heat transfer by conduction depends on the *temperature difference* between the touching objects and the *thermal conductivity* of the substances involved (that is, how easily heat is conducted by the molecules of the substances). Heat can be lost or gained by conduction when the bare skin is in contact with a good conductor. When you hold a snowball, for example, heat moves by conduction from your hand to the snowball so that your hand becomes cold. On the other hand, when you apply a heating pad to a body part, the part is warmed up as heat is transferred directly from the pad to the body.

Similarly, you either lose or gain heat by conduction to the layer of air in direct contact with your body, with the direction of heat transfer depending on whether the air is cooler or warmer, respectively, than your skin. Only a small percentage of total heat exchange between the skin and environment takes place by conduction alone, however, because air is not a very good conductor of heat. (This is why swimming-pool water at 80° F feels cooler than air at the same temperature; heat is conducted more rapidly from the body surface into the water, which is a good conductor, compared to air, which is a poor conductor.)

Convection refers to the transfer of heat energy by *air (or H_2O) currents*. As the body loses heat by conduction to the surrounding cooler air, the air in immediate contact with the skin is warmed. Because warm air is lighter (less dense) than cool air, the warmed air rises while the cooler air moves in next to the skin to replace the vacating warm air. The process is then repeated (Fig. 17–5c). These air movements, known as convection currents, help carry heat away from the body. If it were not for convection currents, no further heat could be dissipated from the skin by conduction once the temperature of the layer of air immediately around the body equilibrated with skin temperature.

The combined conduction-convection process of dissipating heat from the body is enhanced by forced movement of air across the body surface, either by external air movements, such as is caused by the wind or a fan, or by movement of the body through the air, such as during bicycle riding. Because forced air movement sweeps away the air warmed by conduction and replaces it with cooler air more rapidly, a greater total amount of heat can be carried away from the body over a given time period. This is why wind makes us feel cooler on hot days, and it is also why windy days in the winter are more chilling than calm days at the same cold temperature. For this reason, weather forecasters have developed the concept of *wind chill factor*.

Evaporation is the final method of heat transfer used by the body. When water evaporates from the skin surface, the heat required to transform water from a liquid to a gaseous state (the **heat of vaporation**) is absorbed from the skin, thereby cooling the body (Fig. 17–5d). Evaporative heat loss is why you feel cooler when your bathing suit is wet than when it is dry. Evaporative heat loss occurs continually from the surface of the skin and the linings of the respiratory airways. Because the skin is not completely waterproof, H_2O molecules constantly diffuse through the skin and evaporate in a way that is completely unrelated to the sweat glands. Similarly, heat is continuously lost through the H_2O vapor in the expired air as a result of the air's humidification during its passage through the respiratory system. Loss of H_2O and heat from nonsweating skin and the respiratory tract is considered to be insensible loss because the person is unaware of its occurrence. This loss involves passive processes that are not subject to physiological control. Insensible heat loss goes on even in very cold weather, when the problem is one of conserving body heat.

Sweating, on the other hand, is an active evaporative heat-loss process under sympathetic nervous control. The rate of evaporative heat loss can be deliberately adjusted by means of sweating as an important homeostatic mechanism to eliminate excess heat as needed. In fact, when the environmental temperature exceeds the skin temperature, sweating is the *only* avenue for heat loss, because the body is gaining heat by radiation and conduction under these circumstances.

Sweat is a dilute salt solution that is actively extruded to the surface of the skin by sweat glands dispersed all over the body. Sweat must be evaporated from the skin for heat loss to occur. If sweat merely drips from the surface of the skin or is wiped away, no heat loss is accomplished. The most important factor determining the extent of evaporation of sweat is the *relative humidity* of the surrounding air (the percentage of H_2O vapor actually present in the air compared to the greatest amount that the air can possibly hold at that temperature; for example, a relative humidity of 70% means that the air contains 70% of the H_2O vapor it is capable of holding). When the relative humidity is high, the air is already almost fully saturated with H_2O, so it has limited ability to take up additional moisture from the skin. For this reason, little evaporative heat loss can occur on hot, humid days. The sweat glands continue to secrete, but the sweat simply remains on the skin or drips off instead of evaporating and producing a cooling effect. As a measure of the discomfort associated with combined heat and high humidity, meteorologists have devised the ***temperature-humidity index***.

The hypothalamus integrates a multitude of thermosensory inputs from both the core and surface of the body.

The hypothalamus serves as the body's thermostat. The home thermostat keeps track of the temperature in the room and triggers heating mechanisms (the furnace) or cooling mechanisms (the air conditioner) as necessary to maintain the room temperature at the indicated setting. Similarly, the hypothalamus, as the body's thermoregulatory integrating center, receives afferent information about the temperature in various regions of the body and initiates extremely complex, coordinated adjustments in heat-gain and heat-loss mechanisms as necessary to correct any deviations in core temperature from the "normal setting." The hypothalamic thermostat is far more sensitive than your home thermostat. The hypothalamus is able to respond to changes in blood temperature as small as 0.01° C. The hypothalamus's degree of response to deviations in body temperature is finely matched so that precisely enough heat is lost or generated as need be to restore the temperature to normal.

To make the appropriate adjustments in the delicate balance between the opposing heat loss mechanisms and heat producing and heat conserving mechanisms, the hypothalamus must be continuously apprised of both the skin temperature and core temperature by means of specialized temperature-sensitive receptors called ***thermoreceptors***. Peripheral warmth and cold thermoreceptors monitor skin temperature throughout the body and transmit information about changes in surface temperature to the hypothalamus. The core temperature is monitored by central thermoreceptors, which are located in the hypothalamus itself as well as elsewhere in the central nervous system and the internal organs.

Two centers for temperature regulation have been identified in the hypothalamus. The anterior region, which is activated by warmth, initiates reflexes that mediate heat loss. The posterior region is thrown into activity in response to cold and subsequently triggers reflexes that mediate heat production and heat conservation. Let us examine the means by which the hypothalamus fulfills its thermoregulatory functions.

Shivering is the primary involuntary means of increasing heat production.

Heat can be gained by the body as a result of internal heat production generated by metabolic activity or from the external environment if the latter is warmer than body temperature. Because body temperature usually is higher than environmental temperature, metabolic heat production is the primary source of body heat. In a resting person, most body heat is produced by the thoracic and abdominal organs as a result of ongoing, cost-of-living metabolic activities. Above and beyond this basal level, the rate of metabolic heat production can be variably increased primarily by changes in skeletal-muscle activity or to a lesser extent by certain hormonal actions. Thus, changes in skeletal-muscle activity constitute the major form of control of heat gain for temperature regulation.

In response to a fall in core temperature caused by exposure to cold, the hypothalamus takes advantage of the fact that increased skeletal-muscle activity generates more heat (Fig. 17–6). Acting through descending pathways that terminate on the motor neurons controlling the body's skeletal muscles, the hypothalamus first gradually increases skeletal-muscle tone. (Muscle tone refers to the constant level of tension within the muscles.) Soon, shivering begins. **Shivering** consists of rhythmic, oscillating skeletal-muscle contractions that occur at a rapid rate of ten to twenty per second. This mechanism is very effective in increasing heat production; all of the energy liberated during these muscle tremors is converted to heat because no external work is accomplished. Within a matter of seconds to minutes, internal heat production may increase two to five fold as a result of shivering.

Frequently, reflex changes in skeletal-muscle activity are augmented by increased voluntary, heat-producing actions such as bouncing up and down or hand clapping. Such behavioral responses appear to share neural systems in common with the involuntary physiological responses. The hypothalamus and the limbic system of which it is a part (see p. 138) are extensively involved with controlling motivated behavior. Therefore, one should not think of the hypothalamic homeostatic control systems as operating only at the subconscious level.

In addition to reflex and voluntary changes in muscle activity, which represent the major means of increasing the rate of heat production, **nonshivering (chemical) thermogenesis** plays a lesser role in thermoregulation. Chronic cold exposure in most experimental animals brings about an increase in metabolic heat production that is independent of muscle contraction, appearing instead to involve changes in chemical activity.

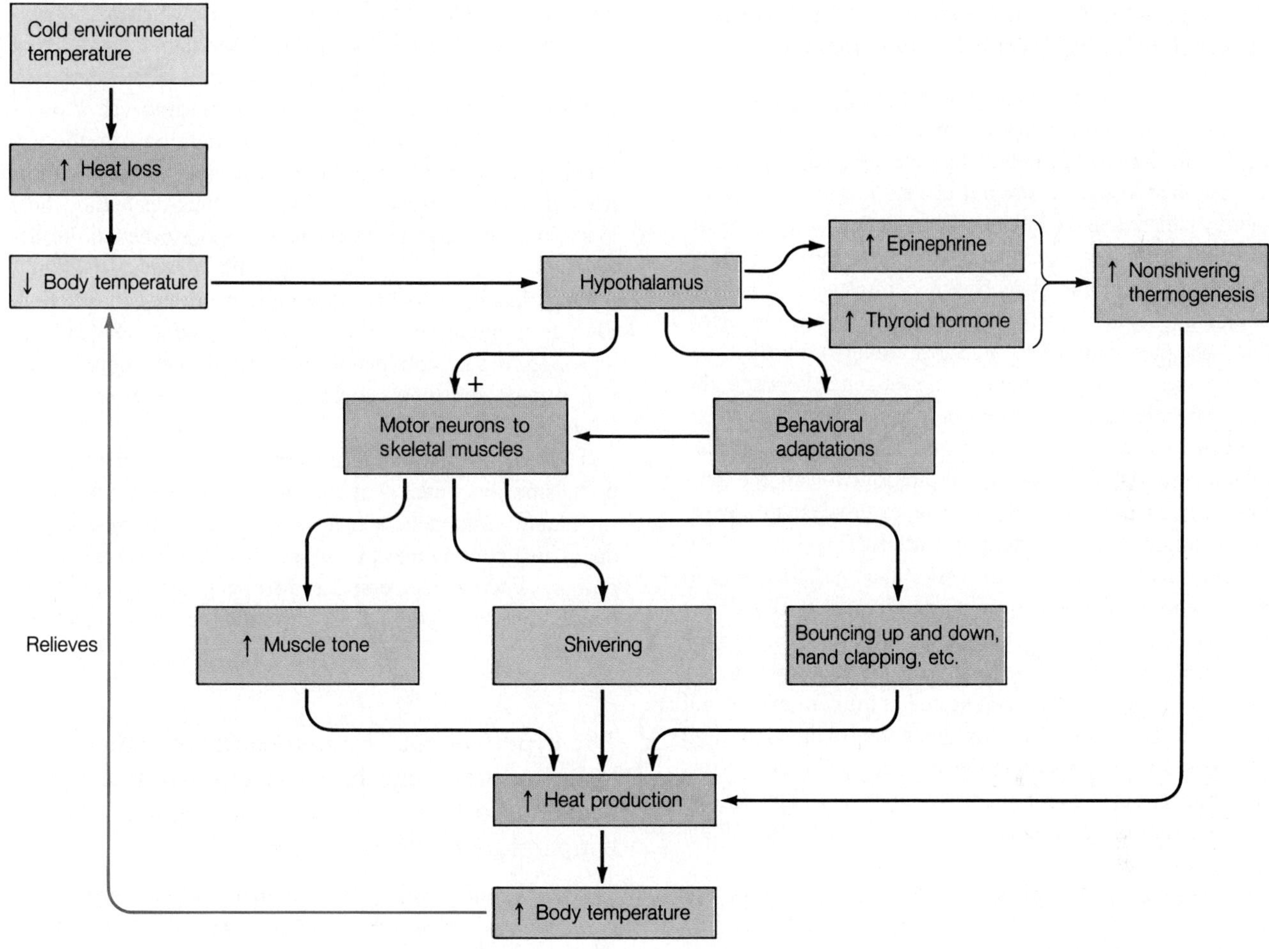

Figure 17–6 Hypothalamic Control of Heat Production in Response to Cold Exposure

In humans this nonshivering or chemical thermogenesis is most important in newborns, because they lack the ability to shiver. Nonshivering thermogenesis is mediated by the hormone epinephrine, perhaps with contributions from thyroid hormone. Both hormones increase heat production by stimulating fat metabolism. Newborns have deposits of a special type of adipose tissue known as **brown fat,** which is especially capable of converting chemical energy into heat. The role of nonshivering thermogenesis in adults remains controversial. Other roles of hormones in thermoregulation are likewise poorly defined and, at any rate, are of secondary importance to adjustments in skeletal-muscle activity in controlling heat gain.

In the situation of an elevation in core temperature caused by heat exposure, two mechanisms are employed to reduce heat-producing skeletal-muscle activity: muscle tone is reflexly reduced, and voluntary movement is curtailed. When the air becomes very warm, people often complain about it being "too hot to even move." These responses are not as effective in reducing heat production during heat exposure as are the muscular responses that increase heat production during cold exposure for two reasons. First, because muscle tone is normally quite low, the capacity to reduce it further is limited. Second, the elevated body temperature tends to increase the rate of metabolic heat production by means of the direct effect of temperature on the rate of chemical reactions.

The magnitude of heat loss can be adjusted by varying the flow of blood through the skin.

Heat-loss mechanisms are also subject to control, again largely by the hypothalamus. When we are hot, we want to increase heat loss to the environment; when we are cold, we want to decrease heat loss. The amount of heat lost to the environment by radiation and conduction-convection is largely determined by the temperature gradient between the skin and the external environment. The body's central core is a heat-generating chamber in which the temperature must be maintained at approximately 100° F. Surrounding the core is an insulating shell through which heat exchanges between the body and external environment take place. In an attempt to maintain a constant core temperature, the insulative capacity and temperature of the shell can be adjusted to vary the temperature gradient between the skin and external environment, thereby influencing the extent of heat loss.

The insulative capacity of the shell can be varied by controlling the amount of blood flowing through the skin. Blood flow to the skin serves two functions. First, it provides a nutritive blood supply to the skin. Second, as blood is pumped to the skin from the heart, it has been heated in the central core and carries this heat to the skin. Most of the flow of blood through the skin is for purposes of temperature regulation; at normal room temperature, twenty to thirty times more blood flows through the skin than is needed to meet the skin's nutritional needs.

In the process of thermoregulation, skin blood flow can vary tremendously, from 400 ml up to 2,500 ml/min. The more blood reaching the skin from the warm core, the closer the skin's temperature is to the core temperature. The cutaneous blood vessels diminish the effectiveness of the skin as an insulator by carrying heat to the surface, where it can be lost from the body by radiation and conduction-convection. Accordingly, vasodilation of the skin vessels, which permits increased flow of heated blood through the skin, increases heat loss or, if the environmental temperature is above core temperature, reduces heat gain. Conversely, vasoconstriction of the skin vessels reduces blood flow through the skin, thereby keeping the warm blood in the central core, where it is insulated from the external environment, thus decreasing heat loss. Cold, relatively bloodless skin provides excellent insulation between the core and environment. However, the skin is not a perfect insulator, even with maximum vasoconstriction. Despite minimal blood flow to the skin, some heat can still be transferred by conduction from the deeper organs to the skin surface and then can be lost from the skin to the environment.

These skin vasomotor responses are coordinated by the hypothalamus by means of sympathetic nervous-system output. Increased sympathetic activity to the cutaneous vessels produces heat-conserving vasoconstriction in response to cold exposure, whereas decreased sympathetic activity produces heat-losing vasodilation of the skin vessels in response to heat exposure.

The hypothalamus simultaneously coordinates heat-production mechanisms and heat-loss and heat-conservation mechanisms to regulate core temperature homeostatically.

Let us now pull together the coordinated adjustments in heat production as well as in heat loss and heat conservation in response to exposure to either a cold or a hot environment (Fig. 17–7 and Table 17–2). In response to cold exposure, the posterior region of the hypothalamus directs decreased heat loss and increased heat production. Because there is a limit to the body's ability to reduce skin temperature through vasocon-

Table 17–2 Coordinated Adjustments in Response to Cold or Heat Exposure

In Response to Cold Exposure (coordinated by posterior hypothalamus)		In Response to Heat Exposure (coordinated by anterior hypothalamus)	
Increased Heat Production	*Decreased Heat Loss (heat conservation)*	*Decreased Heat Production*	*Increased Heat Loss*
Increased muscle tone	Skin vasoconstriction	Decreased muscle tone	Skin vasodilation
Shivering	Postural changes to reduce exposed surface area (hunching shoulders, etc.)*	Decreased voluntary exercise*	Sweating
Increased voluntary exercise*	Warm clothing*		Cool clothing*
Nonshivering thermogenesis			

*Behavioral adaptations.

striction, even maximum vasoconstriction is not sufficient to prevent excessive heat loss when the external temperature falls too low. Accordingly, other measures must be instituted to further reduce heat loss, while increased heat production measures such as shivering must be instituted simultaneously.

In animals bearing dense fur or feathers, the hypothalamus, acting through the sympathetic nervous system, brings about contraction of the tiny muscles at the base of the hair or feather shafts to lift the hair or feathers off the skin surface. This puffing up traps a layer of poorly conductive air between the skin surface and environment, thus increasing the insulative barrier between the core and the cold air and reducing heat loss. Even though contraction of the hair-shaft muscles takes place in humans in response to cold exposure, it is an ineffective heat-retention mechanism because of the reduced density and fine texture of most human body hair. The result instead is useless goosebumps.

Thus, in humans, further heat dissipation can be prevented beyond maximum skin vasoconstriction during exposure to cold only by several behavioral adaptations. First, postural changes occur to reduce as much as possible the exposed surface area from which heat can escape. This involves maneuvers such as hunching over, clasping the arms in front of the

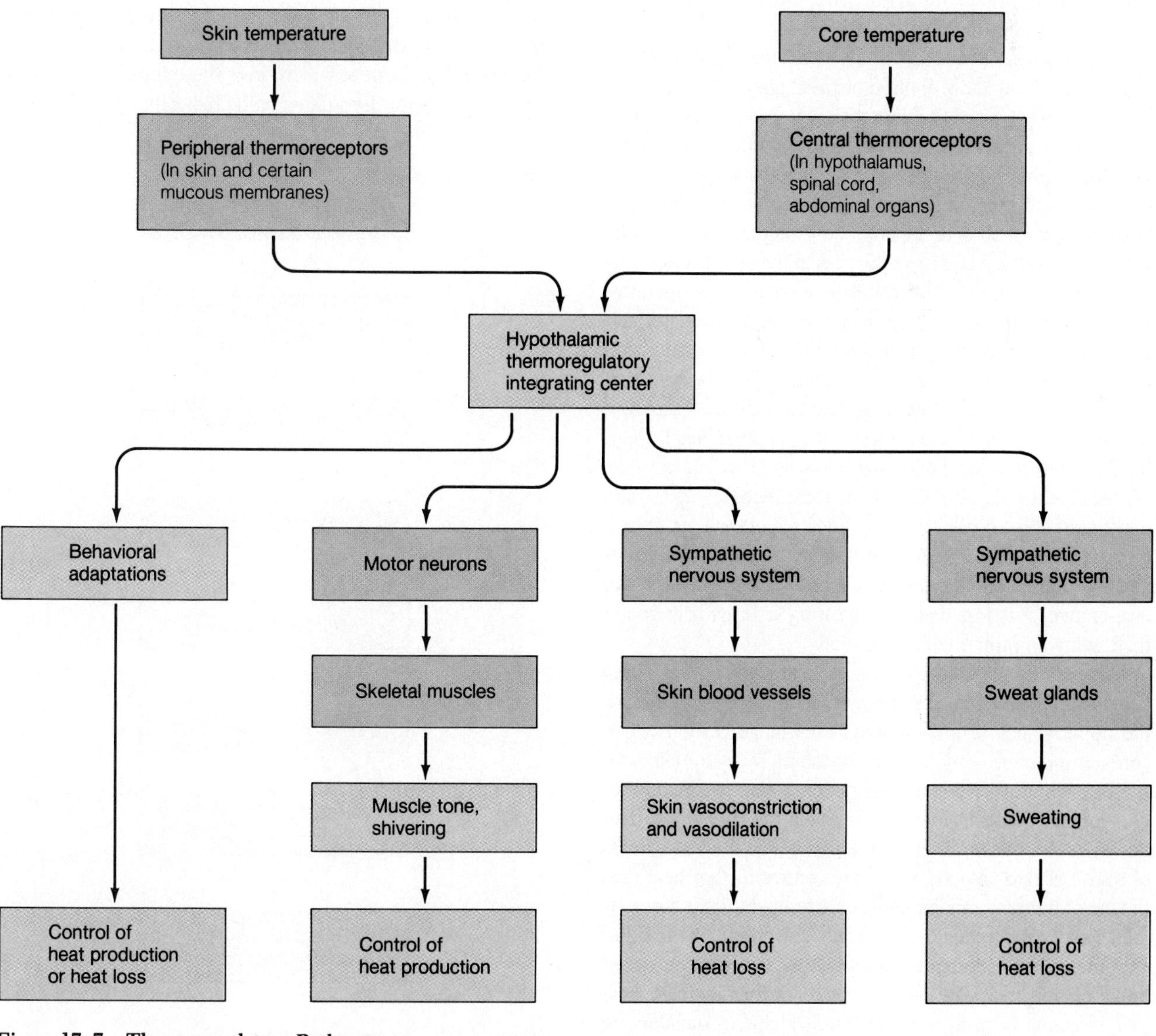

Figure 17–7 Thermoregulatory Pathways

chest, or curling up in a ball. Second, putting on warmer clothing further insulates the body from too much heat loss. As a result of clothing's ability to entrap layers of poorly conductive air between the skin surface and the environment, loss of heat by conduction from the skin to the cold external air is diminished, and the flow of convection currents is curtailed.

Under the opposite circumstance of heat exposure, the anterior part of the hypothalamus promotes increased heat loss and reduces heat production. When even maximal skin vasodilation is inadequate to rid the body of excess heat, sweating is brought into play to induce further heat loss through evaporation. In fact, if the air temperature rises above the temperature of maximally vasodilated skin (that is, the core temperature), the temperature gradient reverses itself so that heat is gained from the environment. Sweating is the only means of heat loss under these conditions.

Humans sweat all over their bodies, but many species either lack or have limited distribution of sweat glands; for example, dogs have sweat glands only on their paw pads. These species must employ alternative methods, such as panting, to enhance heat loss during heat exposure. The hypothalamus induces a shallow, rapid breathing pattern that permits movement of large volumes of air over the hot, moist tongue and respiratory airways without leading to imbalances in the blood concentration of O_2 and CO_2. The resultant increase in evaporative heat loss from the respiratory tract provides a cooling effect.

In addition to sweating, heat production is reduced and voluntary measures such as use of fans, wetting the body, drinking cold beverages , and wearing cool clothing are employed to further enhance heat loss. Contrary to popular belief, wearing light-colored, loose clothing is cooler than being nude. Naked skin absorbs almost all of the radiant energy that strikes it, whereas light-colored clothing reflects almost all of the radiant energy that falls on it. Thus, if light-colored clothing is loose and thin enough to permit convection currents and evaporative heat loss to occur, wearing it is actually cooler than going without any clothes at all.

Skin vasomotor activity is highly effective in controlling heat loss in the range of environmental temperatures between the upper sixties to mid-eighties. This range, within which core temperature can be kept constant by vasomotor responses without calling supplementary heat-production or heat-loss mechanisms into play, is called the **thermoneutral zone.** As external air temperature falls below the lower limits of maximal skin vasoconstriction to reduce further heat loss, the major burden of maintaining core temperature is borne by increased heat production, especially shivering. At the other extreme, when external air temperature exceeds the upper limits of maximal skin vasodilation to further increase heat loss, sweating becomes the dominant factor in maintaining core temperature.

During a fever, the hypothalamic thermostat is "reset" at an elevated temperature.

Fever refers to an elevation in body temperature as a result of infection or inflammation. In response to microbial invasion, certain white blood cells release a chemical known as **endogenous pyrogen,** which among its many infection-fighting effects (see p. 368) acts on the hypothalamic thermoregulatory center to raise the setting of the thermostat (Fig. 17–8). The hypothalamus now maintains the temperature at the new set level instead of maintaining normal body temperature. If, for example, endogenous pyrogen raises the set point to 102° F (as recorded orally), the hypothalamus senses that the normal prefever temperature of 98.6° F is too cold, and it initiates the cold-response mechanisms to elevate the temperature to 102° F. Shivering is initiated to rapidly increase heat production, while skin vasoconstriction is brought about to rapidly reduce heat loss, both of which drive the temperature upward. This accounts for the sudden cold chills often experienced at

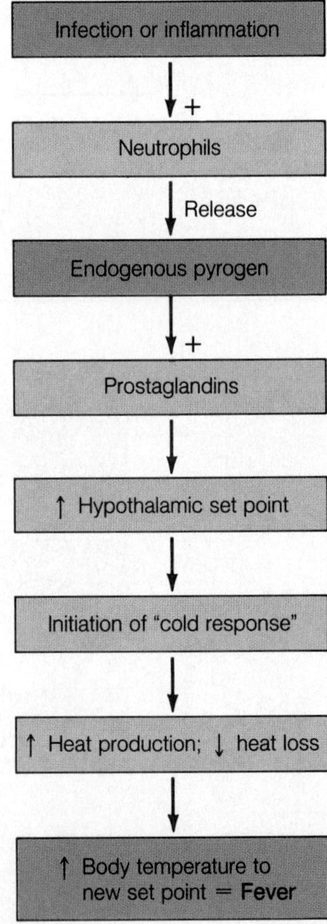

Figure 17–8 Fever Production

the onset of a fever. Because the person feels cold, he or she may put on more blankets as a voluntary mechanism that helps elevate body temperature by conserving body heat. Once the new temperature is achieved, body temperature is regulated as normal in response to cold and heat but at a higher setting. Thus fever production in response to an infection is a deliberate outcome and is not due to a breakdown of thermoregulatory mechanisms. The physiological significance of a fever is still unclear.

Also uncertain is the exact means by which endogenous pyrogen acts on the hypothalamic thermostat during fever production, although the effect is believed to be mediated through local release of prostaglandins (see p. 727). This supposition is based on the fact that aspirin, which is widely recognized as being effective in reducing a fever, is also known to inhibit the synthesis of prostaglandins. Aspirin does not lower the temperature in a nonfebrile person, probably because there are not appreciable quantities of prostaglandins present in the hypothalamus in the absence of endogenous pyrogen.

The exact molecular cause of a fever "breaking" naturally is unknown, although it presumably results from a reduction in pyrogen release or decreased prostaglandin synthesis. When the hypothalamic set point is restored to normal, the temperature at 102° F (in this example) is too high. Thus heat-response mechanisms are instituted to cool down the body. Skin vasodilation occurs and sweating commences. The person feels hot and throws off extra covers. The gearing up of these heat-loss mechanisms by the hypothalamus reduces the temperature to normal.

Hyperthermia can occur unrelated to infection.

Hyperthermia denotes any elevation in body temperature that is above the normally accepted range. The term *fever* is usually reserved specifically for an elevation in temperature caused by resetting of the hypothalamic set point by the release of endogenous pyrogen during infection or inflammation; hyperthermia refers to all other imbalances between heat gain and heat loss that increase the body temperature. There are a variety of causes for hyperthermia, some of which are normal and harmless, others of which are pathological and fatal.

The most common cause of hyperthermia is sustained exercise. As a physical consequence of the tremendous heat load generated by exercising muscles, body temperature rises during the initial stage of exercise because heat gain exceeds heat loss (Fig. 17–9). The elevation in core temperature reflexly triggers heat-loss mechanisms (skin vasodilation and sweating), which eliminate the discrepancy between heat production and heat loss. As soon as the heat-loss mechanisms are stepped up sufficiently to equalize heat production, core temperature stabilizes at a level slightly above the set point despite continued heat-producing exercise. Thus, during sustained exercise, body temperature initially rises, then is maintained at the higher level as long as the exercise continues.

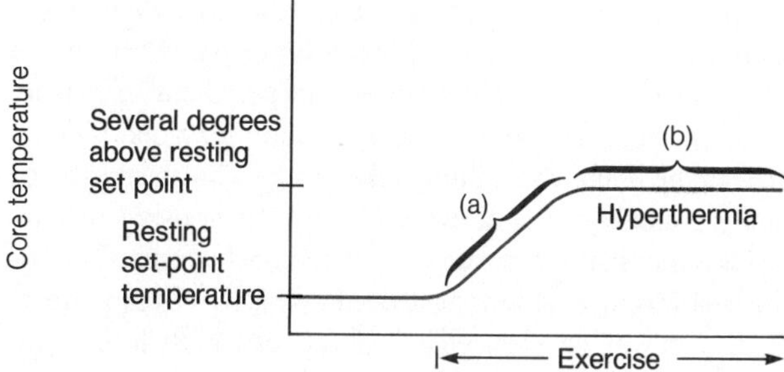

Figure 17–9 Hyperthermia in Sustained Exercise *(a) The rate of heat production initially exceeds the rate of heat loss at the onset of exercise, so that core temperature rises. (b) When heat-loss mechanisms are reflexly increased sufficiently to equalize the elevated heat production, core temperature stabilizes slightly above the resting set point for the duration of the exercise.*

A completely different way in which hyperthermia can be brought about is by excessive heat production in connection with abnormally high circulating levels of epinephrine or thyroid hormone that result from dysfunctions of the adrenal medulla or thyroid gland, respectively. Both of these hormones elevate core temperature by increasing the overall rate of metabolic activity and heat production.

Another way in which hyperthermia occurs is malfunction of the hypothalamic control centers. Certain brain lesions, for example, destroy the normal regulatory capacity of the hypothalamic thermostat. When the thermoregulatory mechanisms are not functional, lethal hyperthermia may occur very rapidly. Normal metabolism produces enough heat to kill a person in less than five hours if the heat-loss mechanisms are completely shut down. In addition to brain lesions, a breakdown in the function of hypothalamic thermoregulation can also occur upon exposure to severe, prolonged heat stress, a topic to which we now turn our attention.

The extremes of hyperthermia and hypothermia can be fatal.

Heat exhaustion refers to a state of collapse, usually manifested by fainting, that is caused by reduced blood pressure brought about as a result of overtaxing the heat-loss mechanisms. Extensive sweating reduces cardiac output by depleting the plasma volume, and pronounced skin vasodilation causes a drop in total peripheral resistance. Since blood pressure is determined by cardiac output times total peripheral resistance, blood pressure falls, an insufficient amount of blood is delivered to the brain, and fainting takes place. Thus heat exhaustion is a consequence of overactivity of the heat-loss mechanisms rather than a breakdown of these mechanisms. Because the heat-loss mechanisms have been very active, body temperature is only mildly elevated in heat exhaustion. By forcing the cessation of activity when heat-loss mechanisms are no longer able to cope with heat gain through exercise or a hot environment, heat exhaustion serves as a safety valve to help prevent the more serious consequences of heat stroke.

Heat stroke is an extremely dangerous situation that arises from the complete breakdown of the hypothalamic thermoregulatory systems. Heat exhaustion may progress into heat stroke if the heat-loss mechanisms continue to be overtaxed. Heat stroke is most likely to occur upon overexertion during a prolonged exposure to a hot, humid environment. The elderly, in whom thermoregulatory responses are generally slower and less efficient, are particularly vulnerable to heat stroke during prolonged, stifling heat waves. So too are individuals who are taking certain common tranquilizers, because these drugs interfere with the hypothalamic thermoregulatory centers' neurotransmitter activity.

The most striking feature of heat stroke is a lack of compensatory heat-loss measures such as sweating in the face of a rapidly rising body temperature. No sweating occurs despite a markedly elevated body temperature because the hypothalamic thermoregulatory control centers are not functioning properly and cannot initiate heat-loss mechanisms. During the development of heat stroke, body temperature starts to climb as the heat-loss mechanisms are eventually overwhelmed by prolonged, excessive heat gain. Once the core temperature reaches the point at which the hypothalamic temperature-control centers are damaged by the heat, the body temperature rapidly rises even higher because of the complete shutdown of heat-loss mechanisms. Furthermore, as the body temperature increases, the rate of metabolism increases correspondingly, because higher temperatures speed up the rate of all chemical reactions, resulting in even greater heat production. This positive-feedback state sends the temperature spiraling upward. Heat stroke is a very dangerous situation that is rapidly fatal if untreated. Even with treatment to halt and reverse the rampant rise in body temperature, there is still a high rate of mortality and a high rate of permanent disability in survivors because of irreversible damage caused by the high internal heat.

At the other extreme, the body can be harmed by cold exposure in two ways: frostbite and generalized hypothermia. **Frostbite** involves excessive cooling of a particular part of the body to the point of causing tissue damage in that area. If exposed tissues actually freeze, tissue damage occurs as a result of disruption of the cells by dehydration or formation of ice crystals. **Hypothermia,** a fall in body temperature, occurs when generalized cooling of the body exceeds the ability of the normal heat-producing and heat-conserving regulatory mechanisms to match the excessive heat loss. As hypothermia sets in, the rate of all metabolic processes slows down because of the declining temperature. Higher cerebral functions are the first to be affected by body cooling. This leads to loss of judgment, apathy, disorientation, and tiredness, all of which diminish the cold victim's ability to initiate voluntary mechanisms to reverse the falling body temperature. As body temperature continues to plummet, depression of the respiratory center occurs, reducing the ventilatory drive so that breathing becomes slow and weak. Activity of the cardiovascular system also is gradually reduced. The heart is slowed and cardiac output decreased. Disturbances of cardiac rhythm occur, eventually leading to ventricular fibrillation and death.

Chapter in Perspective

Because energy can be neither created nor destroyed, input must equal output in the case of both the body's total energy balance and heat balance in order for body weight and body temperature, respectively, to remain constant. If input exceeds output, the extra energy is stored in the body so that body weight or body temperature increases. Conversely, if output exceeds input, body weight decreases or body temperature falls. The hypothalamus is the major integrating center for maintenance of both a constant total energy balance (and thus a constant body weight) and a constant heat-energy balance (and thus a constant body temperature).

Chemical energy in ingested foods provides the only source of total energy input into the body. Energy is expended through external and internal work, with all energy that is not used to move objects in the external environment ultimately being converted into heat. The major factor subject to hypothalamic regulation in the maintenance of total energy balance is the control of food intake.

The primary means of heat input is internal heat production resulting from expenditure of the energy derived from foodstuffs. Skeletal-muscle activity is the most potent form of heat production. Heat output is accomplished by heat loss from the warmer body to the cooler environment by means of radiation and the combined processes of conduction and convection. Evaporation of sweat represents another means of heat loss that can occur even when environmental temperature exceeds body temperature and the body is gaining heat from the environment by radiation and conduction.

In response to cold exposure, the hypothalamus increases heat production by promoting shivering and decreases heat loss by inducing skin vasoconstriction. This vascular response reduces the flow of warm blood to the body surface, where the heat could be lost to the environment by radiation and conduction-convection. In response to heat exposure, the hypothalamus decreases heat production by reducing muscle activity and increases heat loss by inducing skin vasodilation to bring the heated blood closer to the skin surface, where it can be dissipated to the external air by radiation and conduction-convection, and by stimulating sweating to promote evaporation.

Body temperature, which is one of the homeostatically regulated factors of the internal environment, must be maintained within narrow limits because the structure and reactivity of the chemicals that make up the body are temperature sensitive. Deviations in body temperature outside a limited range result in protein denaturation and death if the temperature rises too high, or metabolic slowing and death if the temperature falls too low.

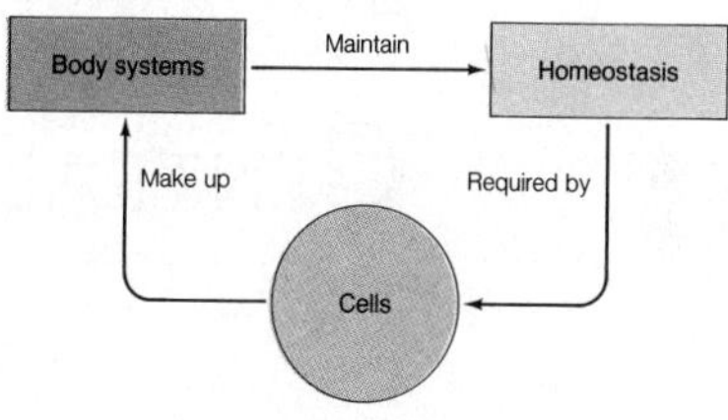

See inside front cover for an expanded version of this model.

Review Exercises

1. Differentiate between external and internal work.
2. Define metabolic rate and basal metabolic rate. Explain the process of indirect calorimetry.
3. Describe the three states of energy balance.
4. By what means is energy balance primarily maintained?
5. What factors cause small, normal fluctuations in core body temperature?
6. Describe the four mechanisms of heat transfer.
7. List the sources of heat input and output for the body.
8. What are the two hypothalamic centers for temperature regulation?
9. Discuss the compensatory measures that occur in response to a fall in core temperature as a result of cold exposure and in response to a rise in core temperature as a result of heat exposure.
10. Describe the process of fever production.
11. Describe each of the following conditions: heat exhaustion, heat stroke, frostbite, and hypothermia.
12. **A point to ponder:** Why is it dangerous to engage in heavy exercise on a hot day?

CHAPTER 18

PRINCIPLES OF ENDOCRINOLOGY; CENTRAL ENDOCRINE ORGANS

CHAPTER CONTENTS AT A GLANCE

INTRODUCTION *How many times have you heard that the key to successful relationships is* communication? *Counselors constantly remind us that keeping the lines of communication open is important for creating and maintaining mutually rewarding interpersonal relationships, such as those between husband and wife or parents and teenage children. Only by sharing information can we become aware of and responsive to the needs and concerns of others. Expanding from the familial to the global scale, survival of the planet itself is dependent on meaningful communications between nations through such vehicles as diplomatic channels, summit meetings, the United Nations, and even a hot line between the leaders of nuclear powers. Similarly, communication is critical for the survival of the society of cells that collectively compose the body. The ability of cells to communicate with each other is essential for coordination of their diverse activities to maintain homeostasis as well as to control growth and development of the body as a whole. The endocrine system, the topic introduced in this chapter, specializes in intercellular communications.*

There are three types of cell-to-cell communication (Fig. 18–1):

1. The most intimate means of intercellular communication is through gap junctions, which are minute channels that bridge the cytoplasm of neighboring cells in some types of tissue (see p. 69). Through these specialized anatomical arrangements, small molecules and ions are directly exchanged between interacting cells without ever entering the extracellular fluid (ECF). Gap junctions are especially important in permitting the spread of electrical signals from one cell to the next in cardiac and smooth muscle.
2. The presence of signaling molecules on the surface membrane of some cells permits them to directly link up and interact with other particular cells in a specialized way. This is the means by which the phagocytes of the body's defense system specifically recognize and selectively destroy only undesirable cells, such as microbial invaders, while leaving the body's own cells alone.
3. The most common means by which cells communicate with each other is through intercellular chemical messengers, of which there are four types: paracrines, neurotransmitters, hormones, and neurohormones. In each case, a specific chemical messenger is synthesized by specialized cells to serve a designated purpose. Upon being released into the ECF by appropriate stimulation, these signaling agents act on other particular target cells in a prescribed manner. The difference between these four types of chemical messengers lies in their source and the distance and means by which they get to their site of action.

□ **Paracrines** are local chemical mediators whose effect is exerted only on neighboring cells in the immediate environment of their site of secretion. Since paracrines are distributed by simple diffusion, their action is restricted to short distances, because diffusion is a slow process. They do not gain entry to the blood in any significant quantity because they are rapidly inactivated by locally existing enzymes. One example of a paracrine is histamine, which is released from mast cells during an inflammatory response within an invaded or injured tissue (see p. 366). Among other things, histamine acts on the arteriolar muscle in the vicinity to produce the local vasodilation and subsequent increased blood flow necessary to bring additional blood-borne combat supplies into the affected area.

Paracrines must be distinguished from chemicals that influence neighboring cells after being nonspecifically released during the course of cellular activity. For example, an increased local concentration of CO_2 in an exercising muscle is among the factors that promote local vasodilation by acting directly on the arteriolar smooth muscle in the vicinity (see p. 308). The resultant increased blood flow helps to meet the more active tissue's increased metabolic demands. However, CO_2 is produced by all cells and is not specifically released to accomplish this particular response, so it and similar nonspecifically released chemicals such as H^+ and K^+ are not considered to be paracrines.

□ Nerve cells (neurons) communicate directly with the cells they innervate by releasing very short-range chemical mediators known as **neurotransmitters** in response to action potentials (nerve impulses). The message that is originally encoded electrically within the neuron as an action potential is carried chemically by means of a neurotransmitter across the narrow extracellular space that separates the nerve-cell terminal from the target cell (see p. 105). Like paracrines, neurotransmitters diffuse from their site of release to act locally on only an adjoining target cell, which is either a muscle, gland, or another neuron.

□ **Hormones,** in contrast, are long-range chemical mediators that are specifically secreted into the blood by endocrine (ductless) glands (see p. 5) in response to an appropriate signal. The blood carries them to other sites in the body, where they exert their effects some distance away from their site of release.

□ **Neurohormones** are hormones released into the blood specifically by neurosecretory neurons. Similar in many ways to ordinary neurons, **neurosecretory neurons** possess axons and dendrites and can respond to and conduct action potentials. Instead of directly innervating target cells, however, a neurosecretory neuron terminates on a blood vessel, into which it directly releases its secretory product (a neurohormone) upon appropriate stimulation. The neurohormone is then distributed through the blood to a target site elsewhere in the body. Thus, like endocrine cells, neurosecretory cells communicate by releasing blood-borne chemical messengers that act over long distances. This is in contrast to the short-range neurotransmitters that are secreted into a confined space by ordinary neurons.

Neurotransmitters and neurohormones must be distinguished from neuromodulators (see p. 111). The first two chemical messengers are *secreted* by neurons. The latter messenger *acts* on neurons at nonsynaptic sites to bring about long-term biochemical changes in the nerve cell. A neuromodulator may be from any source, such as a locally acting neural secretion or a hormone acting at long distance.

The major topic of this chapter will be the endocrine system and its long-distance hormonal and neurohormonal messengers. The endocrine system is one of the two major control systems of the body, the other being the nervous system, with which you are already familiar (chapters 4–7). Throughout this and the next chapter we will point out comparisons and interactions between the endocrine and nervous systems.

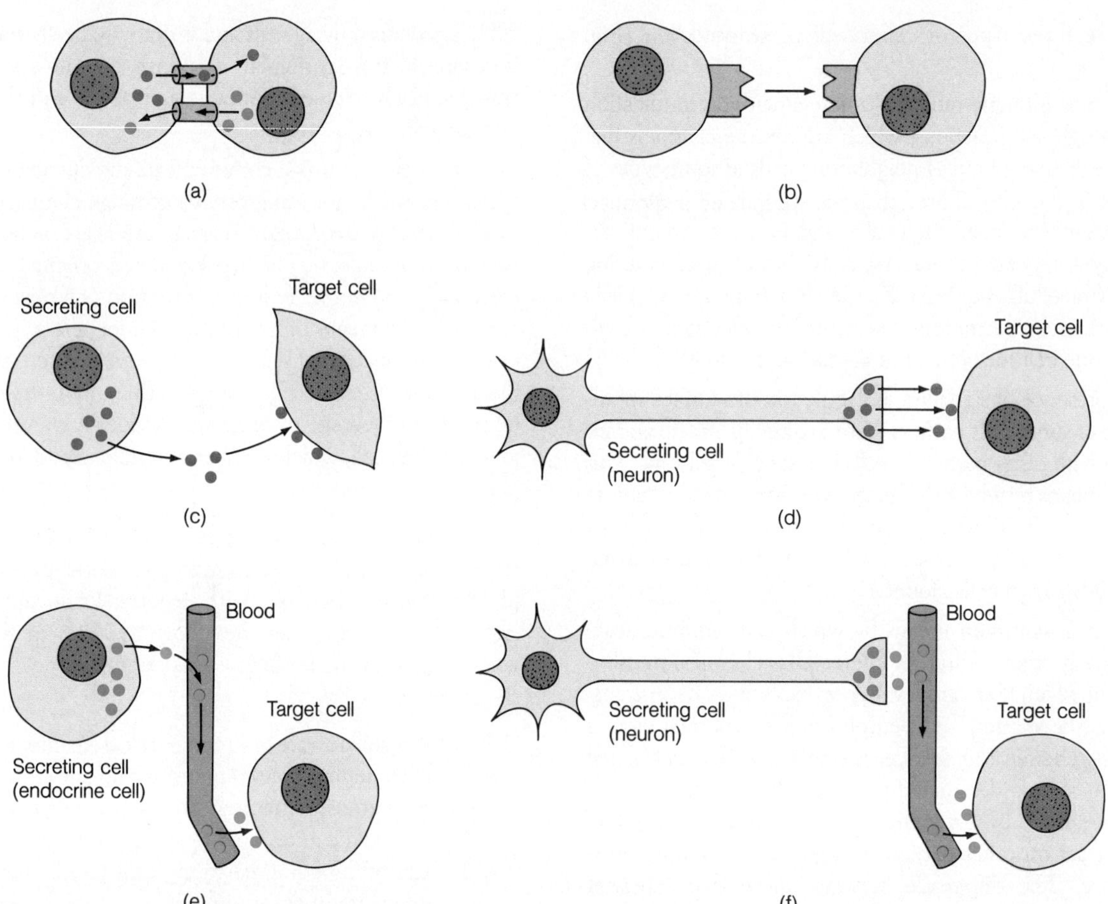

Figure 18–1 Types of Intercellular Communication *(a) Gap junctions. (b) Transient direct linkup of cells. (c) Paracrine secretion. (d) Neurotransmitter secretion. (e) Hormonal secretion. (f) Neurohormone secretion. Paracrines, neurotransmitters, hormones, and neurohormones are all intercellular chemical messengers that accomplish indirect communication between cells.*

General Principles of Endocrinology

Specificity of endocrine communication depends on specialization of target-tissue receptors.

Endocrinology is the study of the homeostatic chemical adjustments and other activities accomplished by hormones, the secretions of the body's endocrine glands. Once secreted, a hormone travels in the blood to another tissue or organ, where it regulates or directs a particular function. The specific distant site or sites on which a hormone exerts its effects are referred to as **target tissues** for that hormone. Some hormones have a single target tissue; others have many.

Because hormones travel in the blood, they are able to reach virtually all tissues. Yet despite this ubiquitous distribution, only specific target tissues are able to respond to each hormone. Specificity of hormonal action depends on specialization of target-tissue **receptors.** For a hormone to exert its effect, the essential first step is the binding of the hormone with receptors specific for it that are located only on or in the cells of the hormone's target tissues. Target-tissue receptors are highly discerning in their binding function. They will recognize and bind only a certain hormone, even though they are exposed simultaneously to many other blood-borne hormones, some of which are structurally very similar to the one that they discriminately bind. Recognition of a specific hormone by a receptor is accomplished by the fact that the conformation of a portion of the receptor molecule matches a unique portion of its binding hormone in "lock-and-key" fashion. Binding of a hormone with target-tissue receptors initiates a reaction (or series of reactions) that culminates in the hormone's final effect. The hormone cannot influence any other tissues because they lack the right binding receptors. This lock-and-key type of interaction between a hormone and its target tissue assures that the regulation of a particular tissue will be governed by the appropriate hormone.

Note that specificity of endocrine communication is built into the receiving end; that is, at the target tissue. At the transmitting end; the original hormonal messenger is widely dispersed instead of being directed specifically toward the target tissue. An analogy is the case of radio transmission and reception. Radio waves (a specific hormone) are nondiscriminately dispersed in all directions from their point of origin (an endocrine gland). Even though radio waves are present in the air surrounding all of your appliances (hormone circulating throughout the body in the blood), only your radio (the target tissue) is able to respond to them because it alone has receiving components specific for radio waves (receptors). Thus, even though your radio and coffee pot are both being equally bombarded by radio waves, only your radio can respond. Similarly, even though all tissues are equally exposed to a given hormone through the blood, only the target tissue with the appropriate receptors for that hormone is able to respond to it.

This is in strong contrast to the way in which specificity of communication is built into the nervous system. Specificity of neural communication depends on the close anatomical relationship between nerve cells and their targets so that each neuron has a very narrow range of influence. A neurotransmitter is released for restricted distribution only to a specific adjacent target tissue, then is swiftly inactivated by enzymes at the nerve–target tissue juncture or is taken back up by the nerve terminal before it is able to gain access to the blood. The target cells for a particular neuron have receptors for the neurotransmitter, but so do many other cells in other locations, and they could respond to this same mediator if it were delivered to them. For example, the entire system of nerve cells supplying all of your body's skeletal muscles (motor neurons) use the same neurotransmitter, acetylcholine (ACh), and all of your skeletal muscles bear complementary ACh receptors. Yet specific muscles are able to be discretely activated to contract because of precise structural arrangements, or wiring patterns, between motor neurons and muscle cells. You are able to specifically wiggle your big toe without influencing any of your other muscles because ACh can be discretely released from the motor neurons that are specifically wired to the muscles controlling your toe. If ACh were to be indiscriminately released into the blood, as are the hormones of the "wireless" endocrine system, all of the skeletal muscles would simultaneously respond by contracting, because they all have identical receptors for ACh. This does not happen, of course, because of the direct lines of communication between neurons and their target cells.

The endocrine system and nervous system have different modes of action and different realms of authority, yet they interact extensively.

Because neural specificity depends on the precise linking of nerve cells with their target tissues, the nervous system is highly organized structurally as well as functionally. In contrast, the endocrine system is composed of endocrine glands that are scattered throughout the body with no anatomical connection (Fig. 18–2). These glands constitute a system in a functional sense, however, because they share in common the secretion of hormones and because there are many functional interactions between various endocrine organs.

In fact, the sole function of some hormones is regulating the production and secretion of another hormone. A hormone that has as its primary function the regulation of hormone secretion by another endocrine gland is classified functionally as a **tropic hormone.** For example, the only function of thyroid-stimulating hormone (TSH) from the anterior pituitary gland is regulation of thyroid hormone secretion by the thyroid gland. A **nontropic hormone,** on the other hand, primarily exerts its effects on nonendocrine target tissues. Thyroid hormone, which increases the rate of O_2 consumption and the metabolic activity of almost every cell in the body, is an example of a nontropic hormone.

The following points add to the complexity of the endocrine system:

1. A single endocrine gland may produce multiple hormones. The anterior pituitary, for example, secretes six different hormones, each under different control mechanisms and each with different functions, some being tropic and others having nontropic effects.
2. A single hormone may be secreted by more than one endocrine gland. For example, the hypothalamus and pancreas both secrete the hormone somatostatin.
3. Frequently, a single hormone has more than one type of target tissue and therefore can induce more than one type of effect. As an example, vasopressin promotes H_2O reabsorption by the kidney tubules as well as vasoconstriction of arterioles throughout the body.
4. A single target tissue may be influenced by more than one hormone. Some cells contain an array of receptors for responding in different ways to different hormones. To illus-

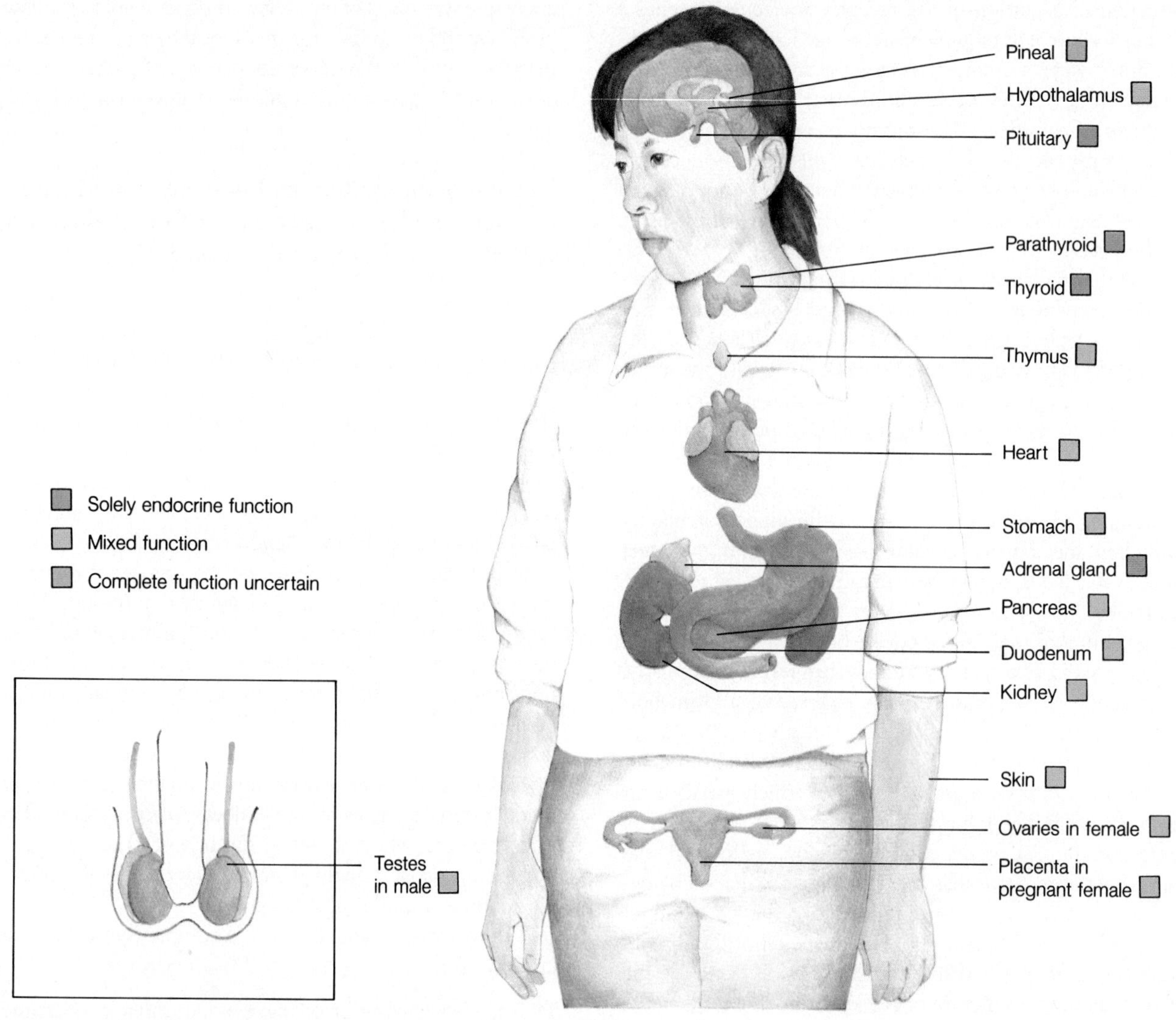

Figure 18–2 Endocrine System

trate, insulin promotes the conversion of glucose into glycogen within liver cells by stimulating one particular hepatic enzyme, whereas another hormone, glucagon, enhances the degradation of glycogen into glucose within liver cells by activating yet another hepatic enzyme.

5. The same chemical messenger may be either a hormone or a neurotransmitter, depending on its source and mode of delivery to the target tissue. Norepinephrine, which is secreted as a hormone by the adrenal medulla and released as a neurotransmitter from sympathetic postganglionic nerve fibers, is a prime example.
6. Some organs are exclusively endocrine in function (they specialize in hormonal secretion alone, the anterior pituitary and thyroid glands being examples), whereas other organs of the endocrine system perform nonendocrine functions in addition to secreting hormones. For example, the testes produce sperm and also secrete the male sex hormone testosterone. Other examples of mixed organs are the ovaries, digestive tract, pancreas, kidneys, and even the brain. In each case except the brain, mixed function comes from the fact that the organ houses nonendocrine tissue plus isolated clusters of endocrine cells that migrated to the organ during embryonic development. The endocrine function of the brain derives from the presence of neurosecretory neurons. There are no endocrine cells as such in the brain.

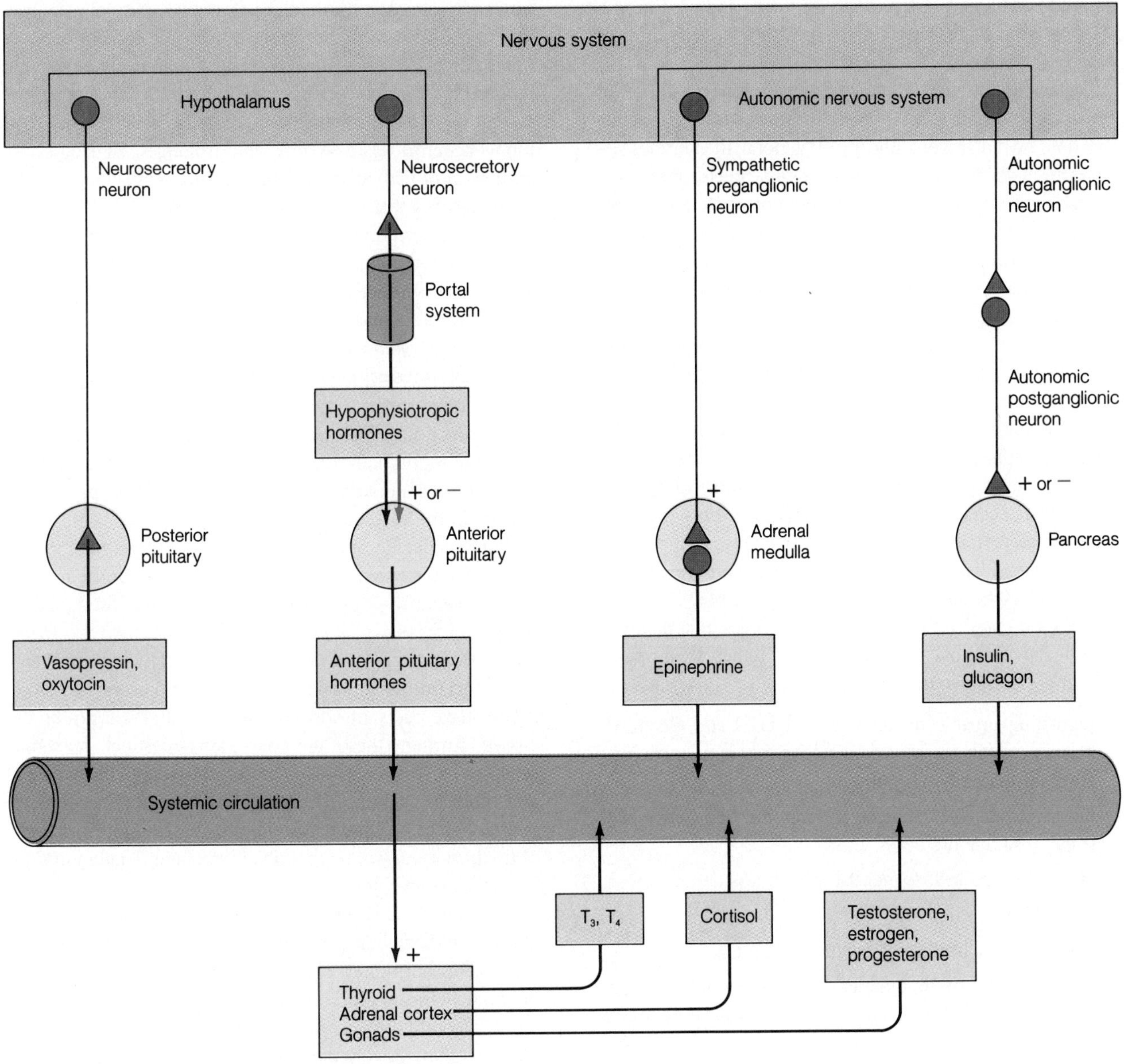

Figure 18–3 Pathways by Which Neural Input Influences Hormonal Secretion

The secretion of blood-borne neurohormones by neurosecretory neurons represents an overlap between neural and endocrine function. Many other important interfaces also exist between these two major control systems, the study of which is known as **neuroendocrinology.** The nervous system directly or indirectly controls the secretion of many hormones; certain hormones, in turn, influence neural function considerably. Therefore, the nervous and endocrine systems in part regulate each other. As an example, the brain plays a crucial role in regulating the secretion of most hormones (Fig. 18–3). In the reverse situation, hormones play an important role in modulating the excitability of the nervous system and in influencing behavior patterns. Furthermore, hormones play a critical role in the development of the brain during fetal life. In some instances, both the endocrine and nervous systems influence the same target tissue in supplementary fashion. For example, secretion of gastric digestive juices is under the control of both the vagus nerve and the stomach-derived hormone gastrin.

The endocrine and nervous systems are intimately interconnected and normally work in close concert, yet they each have their own realm of authority. In general, the nervous system is responsible for coordinating rapid, precise responses. Neural signals in the form of action potentials are rapidly propagated along nerve-cell fibers, resulting in the release at the nerve terminal of a neurotransmitter that has to diffuse only a microscopic distance to its target cell before a response is effected. Not only is a neurally mediated response rapid, but it is also brief; the action is quickly brought to a halt as the neurotransmitter is swiftly removed from the target site. This permits either termination of the response, almost immediate repetition of the response, or rapid initiation of an alternate response, depending on what is indicated by the circumstances (for example, the swift changes in commands to muscle groups needed to coordinate walking). This mode of action makes neural communication extremely rapid and precise. The target tissues of the nervous system are the muscles and glands, especially exocrine glands, of the body.

The supremacy of the endocrine system lies in its ability to control activities that require duration rather than speed, including the following:

1. regulating organic metabolism and H_2O and electrolyte balance, which are important collectively in maintaining a constant internal environment;
2. inducing adaptive changes to help the body cope with stressful situations;
3. promoting smooth, sequential growth and development;
4. controlling reproduction;
5. regulating red blood cell production; and
6. along with the autonomic nervous system, controlling and integrating both circulation and the digestion and absorption of food.

The endocrine system responds more slowly to its triggering stimuli than the nervous system does for several reasons. First, it must depend on blood flow to convey its hormonal messengers over long distances. Second, hormones' mechanism of action at their target cells is more complex than that of neurotransmitters, thus requiring more time before a response occurs. For some hormones, their ultimate effect cannot be detected until a few hours after they bind with target-tissue receptors. Also, because of the receptors' high affinity for their respective hormone, hormones often remain bound to receptors for some time, thus prolonging their biological effectiveness. Furthermore, unlike the brief, neurally induced responses that come to a halt almost immediately after the neurotransmitter is removed, endocrine effects usually last for some time after the hormone's withdrawal. Neural responses to a single burst of neurotransmitter release usually last only milliseconds to seconds, whereas the alterations in target cells induced by hormones range from minutes to days or, in the case of growth-promoting effects, even a lifetime. Thus hormonal action is relatively slow and prolonged, making endocrine control particularly suitable for the regulation of metabolic activities that require long-term stability.

Because hormones receive wide dispersal through the circulatory system, those that have multiple target tissues are able to coordinate the activities of various tissues toward a common goal. Tissues such as muscle, liver, and fat all respond to hormones in a fashion that ensures that their contributions to overall intermediary metabolism are acting in concert.

Endocrine systems also provide temporal (time) coordination of function. This is particularly apparent in the endocrine control of reproductive cycles, such as the menstrual cycle, in which normal function requires highly specific patterns of change in the secretion of various hormones.

This has been a brief overview of the general functions of the endocrine system as compared with those of the nervous system. Table 18–1 provides a summary of the distinguishing features of these two coordinating systems. Table 18–2 (pages 630 to 632) presents an overview of the most important specific functions of the major hormones. Some of these hormones have been introduced elsewhere and will not be discussed further; these are the gastrointestinal hormones (chapter 16), the renal hormones (erythropoietin in chapter 11 and renin in chapter 14), atrial natriuretic peptide from the heart (chapter 15), and thymosin (chapter 12). The remainder of the hormones will be described in greater detail in this and the next two chapters.

As extensive as Table 18–2 appears to be, it leaves out a variety of "candidate" or potential hormones that have not fully qualified as hormones, either because their characteristics do not quite fit the classic definition of a hormone or because they are so recently discovered that their hormonal status has not yet been conclusively documented. The table also excludes the hormones secreted by the effector cells of the defense system (white blood cells and macrophages) and a variety of recently revealed and poorly understood growth factors that promote growth of specific tissues, such as ***epidermal growth factor*** and ***nerve growth factor***. In addition, the table is undoubtedly incomplete in ways as yet unknown. First, it is more than likely that there are hormones yet to be discovered; the list of hormones still continues to grow as a result of the probing techniques of endocrinologists. Second, an exciting finding is that hormones with well-established functions are sometimes found to exert other, quite unrelated functions as well. As an example, antidiuretic hormone, which controls H_2O retention by the kidneys during the formation of urine, was determined to also exert vasoconstrictor effects on arteriolar

Table 18–1 Comparison of the Nervous System and the Endocrine System

Property	Nervous System	Endocrine System
Anatomical arrangement	A "wired" system; specific structural arrangement between neurons and their target cells; structural continuity in the system	A "wireless" system; endocrine organs widely dispersed and not structurally related to one another or to their target cells
Type of chemical messenger	Neurotransmitters released into synaptic cleft	Hormones released into blood
Distance of action of chemical messenger	Very short distance (diffuses across synaptic cleft)	Long distance (carried by blood)
Means of specificity of action on target cell	Dependent on close anatomical relationship between nerve cells and their target cells	Dependent on specificity of target cell binding and responsiveness to a particular hormone
Speed of response	Rapid (milliseconds)	Slow (minutes to hours)
Duration of action	Brief (milliseconds)	Long (minutes to days or longer)
Major functions	Coordinates rapid, precise responses	Controls activities that require longer duration rather than speed
Influence on other major control system?	Yes	Yes

smooth muscle, giving rise to its more commonly used alternative name of vasopressin. More recently, vasopressin has also been implicated as playing roles in fever, learning, memory, and behavior.

Hormones are chemically classified into three categories: peptides, amines, and steroids.

Hormones are not all similar chemically, but instead they fall into three distinct classes according to their biochemical structure (Table 18–3, p. 633); the classes include: (1) peptides and proteins, (2) amines, and (3) steroids. The first two categories are both amino-acid derivatives. The **peptide hormones** and **protein hormones** consist of a chain of specific amino acids of varying length, the shorter ones being peptides and the longer ones categorized as proteins. For convenience, we will refer to this entire category as peptides. The majority of hormones fall into this class, including those secreted by the hypothalamus, anterior pituitary, posterior pituitary, pancreas, parathyroid, gastrointestinal tract, kidneys, heart, and liver. The **amines** are derived from the amino acid *tyrosine* and include the hormones secreted by the thyroid gland and adrenal medulla. The adrenomedullary hormones are specifically known as *catecholamines*. The **steroids,** which include the hormones secreted by the adrenal cortex, gonads, and most placental hormones, are neutral lipids derived from cholesterol.

Minor differences in chemical structure between hormones within each category often result in profound differences in biological response. For example, note the subtle difference (Fig. 18–4, p. 634) between testosterone, the male sex hormone responsible for inducing the development of masculine characteristics, and estradiol, the predominant form of estrogen, which is the feminizing female sex hormone.

The structural classification of hormones is of more than biochemical interest. The means by which a hormone is synthesized and secreted, the way it is transported in the blood, and the mechanism by which it exerts its effects at the target cell all depend on its chemical properties, the most notable property being its solubility. The following differences in solubility of the various types of hormones are critical to their function (Table 18–3):

1. All peptides and catecholamines are polar and, accordingly, are hydrophilic (water-loving) and lipophobic (lipid-fearing); that is, they are highly H_2O soluble and have low lipid solubility.
2. All steroid and thyroid hormones are nonpolar and, accordingly, are lipophilic (lipid-loving) and hydrophobic (water-fearing); that is, they have high lipid solubility and are poorly soluble in H_2O.

The mechanisms of hormone synthesis, storage, and secretion vary according to the class of hormone.

Because of their chemical differences, the means by which the various classes of hormones are synthesized, stored, and secreted differ as follows on page 634.

Table 18–2 Summary of Major Hormones

Endocrine Gland	Hormones	Secretion Directly Controlled by	Target Cells	Major Functions of Hormones
Hypothalamus	Releasing and inhibiting hormones (TRH, CRH, GnRH, GHRH, GHIH, PRF, PIH)	Circulating levels of end-resultant hormones in negative-feedback fashion; variety of neural and chemical inputs	Anterior pituitary	Controlling release of anterior pituitary hormones
Posterior pituitary (hormones stored in)	Vasopressin (antidiuretic hormone)	Hypothalamic osmoreceptor and left atrial baroreceptor input	Kidney tubule	Increasing H_2O reabsorption
			Arterioles	Vasoconstriction
	Oxytocin	Input from birth canal and cervix	Uterus	Increasing contractility
		Input from mammary gland	Mammary glands (breasts)	Milk ejection
Anterior pituitary	Thyroid-stimulating hormone (TSH)	TRH; T_3 and T_4	Thyroid follicular cells	Stimulation of T_3 and T_4 secretion; promotion of growth of thyroid gland
	Adrenocorticotropic hormone (ACTH)	CRH; cortisol	Zona fasciculata and zona reticularis of adrenal cortex	Stimulation of cortisol secretion; promotion of growth of inner two layers of adrenal cortex
	Growth hormone	GHRH; GHIH; somatomedins	Bone; soft tissues	Essential but not sole responsibility for growth; stimulation of growth of bones and soft tissues; metabolic effects include protein anabolism, fat mobilization, and glucose conservation
			Liver	Somatomedin secretion
	Follicle-stimulating hormone	GnRH; inhibin	Females: ovarian follicles	Follicular growth and development; estrogen secretion
			Males: seminiferous tubules in testes	Sperm production
	Luteinizing hormone (LH) (interstitial cell-stimulating hormone—ICSH)	GnRH; estrogen; progesterone; testosterone	Females: ovarian follicle and corpus luteum	Ovulation; corpus luteum development; estrogen and progesterone secretion
			Males: interstitial cells of Leydig in testes	Testosterone secretion
	Prolactin	PIH; PRF	Females: mammary glands	Breast development; milk secretion
			Males: interstitial cells of Leydig in testes	Possible enhancement of action of luteinizing hormone

Table 18–2 Summary of Major Hormones (continued)

Endocrine Gland	Hormones	Secretion Directly Controlled by	Target Cells	Major Functions of Hormones
Thyroid gland follicular cells	Tetraiodothyronine (T_4 or thyroxine); triiodothyronine (T_3)	TSH	Most cells	Increasing metabolic rate; essential for normal growth and nerve development
Thyroid gland C cells	Calcitonin	Plasma calcium concentration	Bone	Decreasing plasma calcium concentration
Adrenal cortex				
Zona glomerulosa	Aldosterone (mineralocorticoid)	Renin-angiotensin system; plasma K^+ concentration	Kidney tubules	Increasing Na^+ reabsorption and K^+ secretion
Zona fasciculata and zona reticularis	Cortisol (glucocorticoid)	ACTH	Most cells	Increasing blood glucose at the expense of protein and fat stores; stress adaptation
	Androgens (dehydro-epiandrosterone)	ACTH	Females: bone and brain	Pubertal growth spurt and sex drive in females
Adrenal medulla	Epinephrine and norepinephrine	Preganglionic sympathetic nerve fibers	Sympathetic receptor sites throughout the body	Reinforcement of sympathetic nervous system; stress adaptation; blood pressure regulation; mobilization of energy supplies
Endocrine pancreas (islets of Langerhans)	Insulin (β cells)	Blood glucose concentration; other inputs related to food intake of lesser importance	Most cells	Promotion of cellular uptake, utilization, and storage of absorbed nutrients
	Glucagon (α cells)	Blood glucose concentration; other inputs of lesser importance	Most cells	Maintenance of nutrient levels in blood during postabsorptive state
	Somatostatin (D cells)	Blood nutrient concentrations; gastrointestinal hormones	Digestive system	Inhibition of digestion and absorption of nutrients
			Pancreatic islet cells	Inhibition of pancreatic hormone secretion
Parathyroid gland	Parathyroid hormone (PTH)	Plasma calcium concentration	Bone, kidneys, intestine	Increasing plasma calcium concentration; decreasing plasma phosphate concentration; stimulation of vitamin D activation
Gonads				
Female: ovaries	Estrogen (estradiol)	FSH, LH	Female sex organs; body as a whole	Follicular development; development of secondary sex characteristics; stimulation of uterine and breast growth
			Bone	Promotion of closure of epiphyseal plate
	Progesterone	LH	Uterus	Preparation for pregnancy

Table 18–2 Summary of Major Hormones (continued)

Endocrine Gland	Hormones	Secretion Directly Controlled by	Target Cells	Major Functions of Hormones
Male: testes	Testosterone	LH	Male sex organs; body as a whole	Sperm production; development of secondary sex characteristics; promotion of sex drive
			Bone	Enhancement of pubertal growth spurt; promotion of closure of epiphyseal plate
Testes and ovaries	Inhibin	Some factor associated with spermatogenesis and follicular activity	Anterior pituitary	Inhibition of secretion of follicle-stimulating hormone
Pineal gland	Melatonin	Sympathetic nerves; light-dark cycle	Reproductive organs	Believed to inhibit gonadotropins; initiation of puberty possibly caused by reduction in secretion
Placenta	Estrogen (estriol); progesterone	Unknown	Female sex organs	Maintenance of pregnancy; preparation of breasts for lactation
	Chorionic gonadotropin	Unknown	Ovarian corpus luteum	Maintenance of corpus luteum of pregnancy
Kidneys	Renin (→ angiotensin)	Extracellular fluid Na^+ load, plasma volume, arterial blood pressure	Adrenal cortex (acted on by angiotensin, which is activated by renin)	Stimulation of aldosterone secretion; promotion of growth of outer layer of adrenal cortex
	Erythropoietin	Low arterial P_{O_2}	Bone marrow	Stimulation of erythrocyte production
Stomach	Gastrin	Composition of digestive-tract contents; intrinsic nerve plexuses; parasympathetic stimulation	Digestive tract exocrine glands and smooth muscles; pancreas; liver; gall bladder	Control of motility and secretion to facilitate digestive and absorptive processes
Duodenum	Secretin; cholecystokinin; gastric inhibitory peptide			
Liver	Somatomedins	Growth hormone	Bone; soft tissues	Promotion of growth
Skin	Vitamin D	Sunlight; activation within kidneys controlled by parathyroid hormone and plasma phosphate concentration	Intestine	Increasing absorption of ingested calcium and phosphate
Thymus	Thymosin	Unknown	T lymphocytes	Enhancement of T lymphocyte proliferation and function
Heart	Atrial natriuretic peptide	Pressure in right atrium	Kidney tubules	Inhibition of Na^+ reabsorption

Table 18–3 Chemical Classification of Hormones

		Amines		
Properties	**Peptides**	*Catecholamines*	*Thyroid hormone*	**Steroids**
Structure	Chains of specific amino acids, for example: 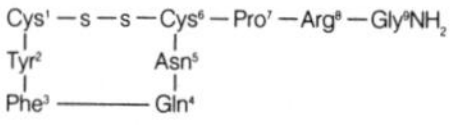 Cys[1]—S—S—Cys[6]—Pro[7]—Arg[8]—Gly[9]NH$_2$; Tyr[2], Asn[5], Phe[3]——Gln[4] (vasopressin)	Tyrosine derivative, for example: 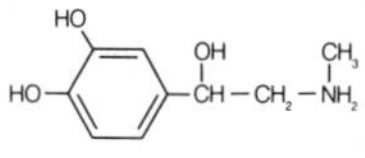 (epinephrine)	Iodinated tyrosine derivative, for example: 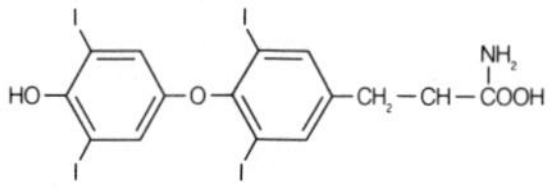 (thyroxine, T_4)	Cholesterol derivative, for example: 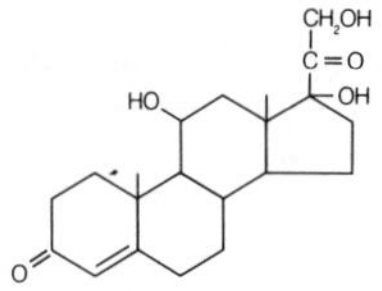 (cortisol)
Solubility	Hydrophilic (lipophobic)	Hydrophilic (lipophobic)	Lipophilic (hydrophobic)	Lipophilic (hydrophobic)
Synthesis	In rough endoplasmic reticulum; packaged in Golgi complex	In cytosol	In colloid, an inland extracellular site	Stepwise modification of cholesterol molecule in various intracellular compartments
Storage	Large amounts in secretory granules	In chromaffin granules	In colloid	Not stored; cholesterol precursor stored in lipid droplets
Secretion	Exocytosis of granules	Exocytosis of granules	Endocytosis of colloid	Simple diffusion
Transport in blood	As free hormone	Half bound to plasma proteins	Mostly bound to plasma proteins	Mostly bound to plasma proteins
Receptor site	Surface of target cell	Surface of target cell	Nucleus of target cell	Cytoplasm or possibly nucleus of target cell
Mechanism of action	Channel changes or activation of second messenger system to alter activity of preexisting proteins that produce the effect	Activation of second messenger system to alter activity of preexisting proteins that produce the effect	Activation of specific genes to produce new proteins that produce the effect	Activation of specific genes to produce new proteins that produce the effect
Hormones of this type	All hormones from the hypothalamus, anterior pituitary, posterior pituitary, pancreas, parathyroid gland, gastrointestinal tract, kidneys, liver, thyroid C cells, heart	Only hormones from the adrenal medulla	Only hormones from the thyroid follicular cells	Hormones from the adrenal cortex and gonads plus most placental hormones (vitamin D is steroidlike)

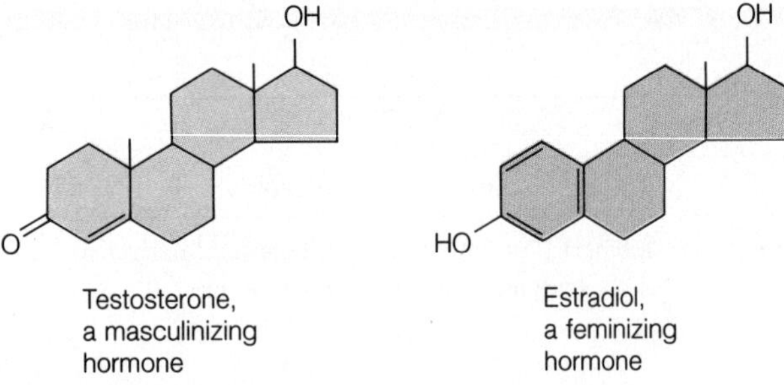

Figure 18–4 Comparison of Testosterone and Estradiol

PEPTIDE HORMONES. Peptide hormones are synthesized by the same method used for the manufacture of any protein that is to be exported (see p. 20). Since hormones are destined to be released from the endocrine cell, they must be segregated from intracellular proteins by being sequestered in a membrane-bounded compartment from the time of their synthesis until they are secreted. Briefly, this is accomplished in the following steps:

1. Peptide hormones are synthesized as larger precursor proteins, or **preprohormones,** by ribosomes on the rough endoplasmic reticulum. Preprohormones migrate to the Golgi complex in membrane-bounded vesicles that pinch off from the endoplasmic reticulum.
2. The Golgi complex concentrates the hormones, then packages them into secretory vesicles that are pinched off and stored in the cytoplasm until an appropriate signal triggers their secretion. This storage of peptide hormones in a readily releasable form allows the gland to respond rapidly to any demands for increased secretion without the necessity of first increasing hormone synthesis.
3. During their journey through the endoplasmic reticulum and Golgi complex, the large preprohormone precursor molecules are pruned first to **prohormones** and finally to **active hormones.** The peptide "scraps" that are left over as a large preprohormone molecule is cleaved to form the classic hormone are often stored and cosecreted along with the hormone. This raises the possibility that these other peptides may also exert biological effects that differ from the traditional hormonal product; that is, the cell may actually be secreting multiple hormones, but the functions of the other peptide products are unknown.
4. Upon appropriate stimulation, the secretory vesicles fuse with the plasma membrane and release their contents to the outside by the process of exocytosis (see p. 24). Such secretion usually does not go on continuously; it is triggered only by specific stimuli. The secreted hormone is subsequently picked up by the blood for distribution.

STEROID HORMONES. Similar steps are used by all *steroidogenic* (steroid-producing) cells to produce and release their hormonal product as follows:

1. Cholesterol is the common precursor for all steroid hormones. Although steroidogenic cells synthesize some cholesterol on their own, most of this raw material is derived from the low-density lipoproteins (LDL) that have been internalized into the cell and degraded by lysosomal enzymes to liberate free cholesterol (see p. 294). The uptake and degradation of LDL is regulated so that increased amounts of cholesterol are made available to steroidogenic cells at times of increased need for steroid hormones. Furthermore, unused cholesterol may be chemically modified and stored in large amounts as lipid droplets within steroidogenic cells. Conversion of this major storage form of cholesterol into free cholesterol for use in steroid-hormone production is also subject to control. Thus the provision of free cholesterol for use by steroidogenic cells can be closely coordinated with the body's overall need for the hormone product.
2. Synthesis of the various steroid hormones from cholesterol requires a series of enzymatic reactions that modify the basic cholesterol molecule by varying the type and position of side groups or the degree of saturation within the steroid ring (Fig. 18–5). Each of the conversions from cholesterol to a specific steroid hormone requires the assistance of a number of enzymes that are limited to certain steroidogenic organs. Accordingly, each steroidogenic organ is able to produce only the steroid hormone or hormones for which it has a complete set of appropriate enzymes. For example, a key enzyme necessary for the production of cortisol is found only in the adrenal cortex, so no other steroidogenic organ is able to produce this hormone. Each of the enzymes necessary for the conversion of cholesterol into a steroid hormone tends to be localized to a specific intracellular compartment, such as the mitochondria or endoplasmic reticulum. The steroid molecule is therefore shuttled back and forth by unknown means between different compartments within the steroidogenic cell for step-by-step modification until the final secretory product is formed.
3. Unlike peptide hormones, there is no mechanism for storage of steroid hormones after their formation. Once formed, steroid hormones immediately diffuse through the steroidogenic cell's lipid plasma membrane to enter the blood. Only the hormone precursor cholesterol is stored in

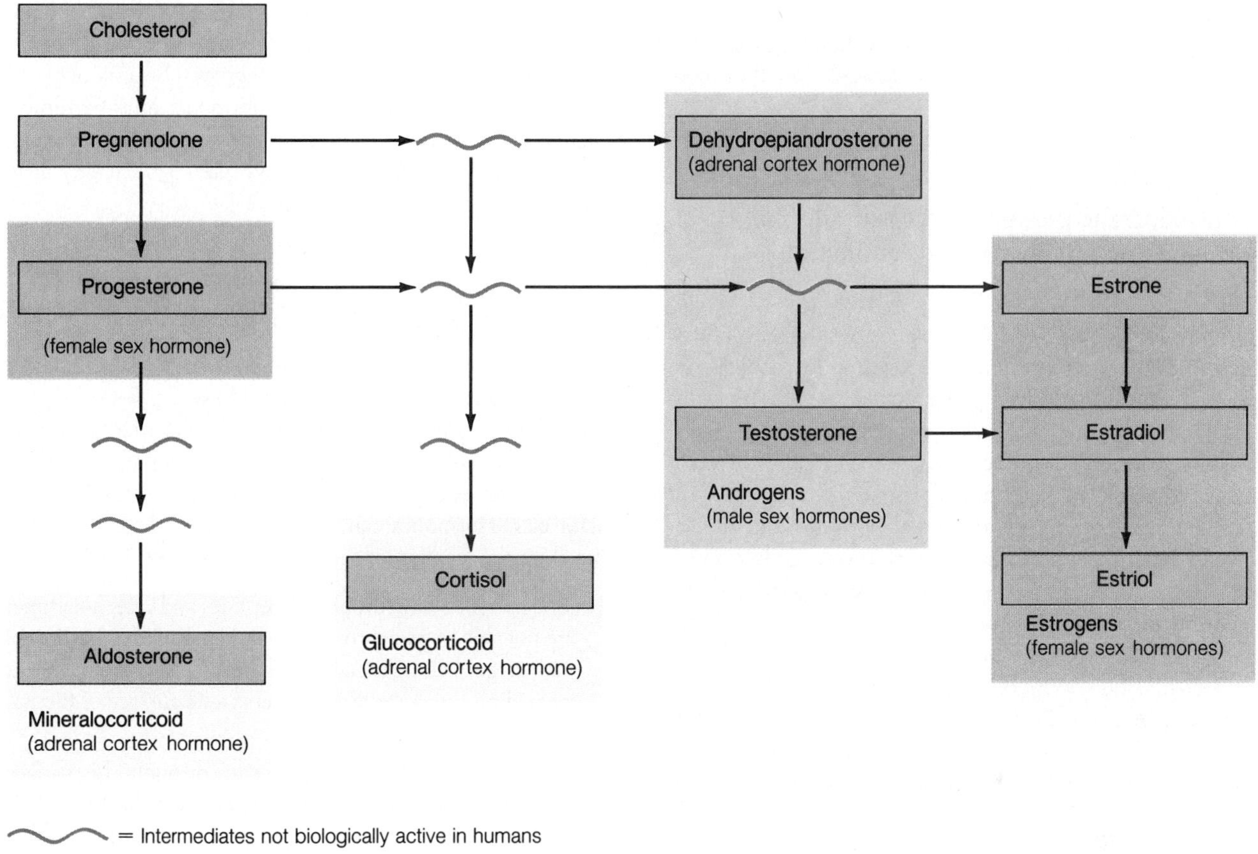

Figure 18–5 Steroidogenic Pathways for Major Steroid Hormones

significant quantities within steroidogenic cells. Accordingly, the rate of steroid-hormone secretion is controlled entirely by the rate of hormone synthesis. This is in contrast to peptide-hormone secretion, in which control is aimed primarily at the release of presynthesized, stored hormone.

4. Following their secretion into the blood, some steroid hormones undergo further interconversions within the blood or other organs, where they are converted into more potent or different hormones.

AMINES. The amines—thyroid hormone and adrenomedullary catecholamines—have unique synthetic and secretory pathways that will be thoroughly described when each of these organs is specifically addressed. However, they do share the following features in common:

1. Both are derived from the naturally occurring amino acid tyrosine.
2. None of the enzymes directly involved in the synthesis of either of these hormone types are located in organelle compartments within the secretory cells. Thyroid-hormone synthesis takes place within an extracellular site, the **colloid,** which consists of "inland pools" of protein-filled extracellular fluid completely encircled by secretory cells within the thyroid gland (see Fig. 19–1b). Catecholamine synthesis is accomplished almost entirely within the cytosol of the secretory cells, with only one step taking place within hormonal storage granules.
3. Both of these hormones are stored until they are secreted. Thyroid hormone is stored in abundance in the colloid. Finished catecholamines are actively transported into preformed vesicles for storage until they are released by

exocytosis. Secretion of thyroid hormones is complex, because it involves transfer of the finished hormone from the colloid completely across the secretory cells into the blood on the other side.

Water-soluble hormones are transported dissolved in the plasma, whereas lipid-soluble hormones are largely transported bound to plasma proteins.

All hormones are carried by the blood, but they are not all transported in the same manner. The hydrophilic peptides are transported simply dissolved in the plasma. It is physically impossible, however, for the hydrophobic steroids and thyroid hormone to dissolve in the aqueous plasma in sufficient quantities to account for their known plasma concentrations. Instead, the majority of the hydrophobic hormones circulate in the blood to their target cells reversibly bound to plasma proteins. Some are bound to specific plasma proteins that are designed to carry only one type of hormone, whereas other plasma proteins, such as albumin, nondiscriminately pick up any "hitchhiking" hormone.

In the case of some hormones, 1% or less of the total hormone remains unbound. This is a noteworthy figure, because only the small unbound, freely dissolved fraction of a hydrophobic hormone is biologically active (that is, is free to cross capillary walls and bind with target-tissue receptors to exert an effect). Once a hormone has interacted with a target cell, it is rapidly inactivated or removed so that it is no longer available to interact with another target cell. Since the carrier-bound hormone is in dynamic equilibrium with the free hormone pool, the bound form of hydrophobic hormones provides a large reserve of steroid and thyroid hormones that can be called on to replenish the active free pool. It is the magnitude of the small free effective pool rather than the total plasma concentration of a particular hormone that is monitored and adjusted to maintain normal endocrine function.

Catecholamines are unusual in that only about 50% of these hydrophilic hormones circulate as free hormone, whereas the other 50% are loosely bound to the plasma protein albumin. Because catecholamines are water soluble, the importance of this protein binding is unclear.

The chemical properties of a hormone dictate not only the means by which it is transported in the blood but also the means by which it can be artificially introduced into the blood for therapeutic purposes. Because the digestive system does not secrete enzymes that can digest steroid and thyroid hormones, these hormones can be absorbed intact from the digestive tract into the blood when taken orally as a pill. Common examples are the sex steroids contained in birth-control pills or thyroid hormone administered to individuals whose own thyroid hormone secretion is inadequate. None of the other types of hormones can be taken orally, because they would be attacked and converted into inactive fragments by proteolytic digestive enzymes. Therefore, these hormones must be administered when necessary by non-oral routes; for example, diabetes mellitus (insulin deficiency) often is treated by daily injections of insulin.

Hormones generally produce their effect by altering intracellular protein activity.

Hormones must bind with target-cell receptors specific for them in order to induce their effect. However, the location of the receptor within the target cell and the mechanism by which the target cell's response is induced by binding with its receptors varies depending on the hormone's solubility characteristics. Hormones can be grouped into three categories based on the location of their receptors (Table 18–3):

1. The lipophobic (hydrophilic) peptides and catecholamines are not able to pass through the lipid membrane barriers of their target cells. Instead they bind with specific receptors located on the *outer plasma-membrane surface* of the target cell.
2. The lipophilic steroids easily pass through the surface membrane to bind with specific receptors located inside the target cell. The steroid receptors are generally considered to be in the *cytoplasm*, although recent evidence suggests that some receptors may be located in the nucleus.
3. Lipophilic thyroid hormone also gains easy access into the target cell, binding with receptors within the *nucleus* in direct association with the genes.

Each interaction between a particular hormone and a target-cell receptor produces a highly characteristic target-cell response that differs for different hormones and between different target tissues influenced by the same hormone. For example, one of the adrenomedullary catecholamines, epinephrine, through its ubiquitous distribution and target-cell specialization, simultaneously produces such diverse effects as contraction of vascular smooth muscle, relaxation of respiratory airway smooth muscle, and breakdown of glycogen (stored glucose) in the liver.

Even though hormones elicit a wide variety of biological responses because the outcome of each hormone–target tissue interaction is highly specific, there are only three general means by which these effects are produced (Fig. 18–6 and Table 18–3):

1. A few hydrophilic hormones, upon binding with a target cell's surface receptors, bring about changes in the cell's permeability (either opening or closing channels to one

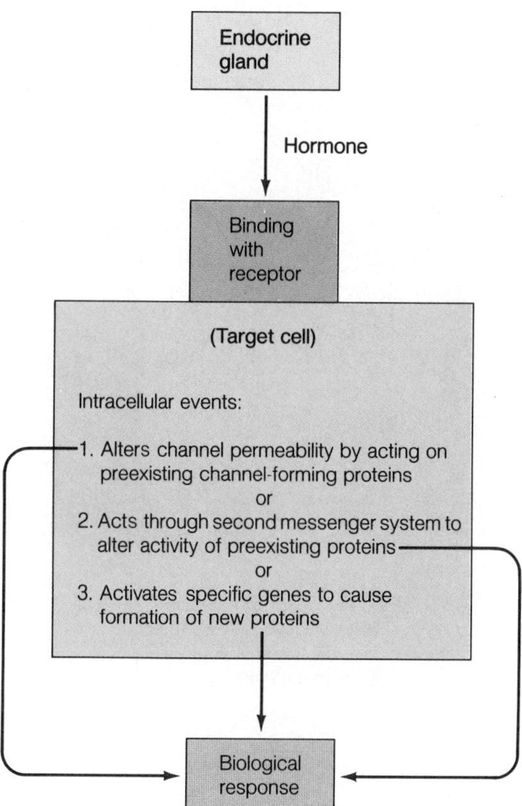
Endocrine gland
Hormone
Binding with receptor
(Target cell)
Intracellular events:
1. Alters channel permeability by acting on preexisting channel-forming proteins
or
2. Acts through second messenger system to alter activity of preexisting proteins
or
3. Activates specific genes to cause formation of new proteins
Biological response

Figure 18–6 Three General Means by Which Hormones Elicit Biological Responses

or more ions) by altering the conformation of adjacent channel-forming proteins already present in the membrane.

2. Most surface-binding hydrophilic hormones function by activating second messenger systems within the target cell. This activation directly alters the activity of preexisting intracellular proteins, usually enzymes, to produce the desired effect.
3. All lipophilic hormones, whether they initially bind with a cytoplasmic or a nuclear receptor, function by activating specific genes in the target cell to cause the formation of new intracellular proteins, which in turn produce the desired effect.

Thus hormones ultimately influence their target cells by altering intracellular protein activity. The onset of action varies depending primarily on the mode employed. Hormones that act through a second messenger system to alter a preexisting enzyme's activity elicit full action within a few minutes. In contrast, hormonal responses that require the synthesis of new protein may take up to several hours before any action is initiated. Let us examine each of the two major postreceptor events in more detail.

POSTRECEPTOR EVENTS: HYDROPHILIC HORMONES. Because of the central role played by second messenger systems in hydrophilic hormone activity, most commonly cyclic AMP (cAMP), it is worthwhile to briefly review the steps involved (Fig. 18–7):

1. The hydrophilic hormone, the first messenger, combines with its specific receptor on the outer surface of the target-cell plasma membrane.
2. Binding of the hormone to its receptor activates the membrane-bound enzyme adenylate cyclase, which is located on the cytoplasmic side of the plasma membrane.
3. Activated adenylate cyclase converts intracellular ATP to cAMP, the intracellular second messenger, within the cytoplasm.
4. Cylic AMP stimulates a second enzyme, protein kinase, which is present in inactive form in the cytoplasm.
5. Activated protein kinase phosphorylates (transfers phosphate groups to) another set of cytoplasmic proteins, usually enzymes that exist in either active or inactive form, depending on whether or not they are phosphorylated.
6. Phosphorylation causes some enzymes to become activated and others to become inactivated.
7. Altered enzyme activity brings about the target cell's ultimate physiological response to the hormone.
8. Once the hormone is removed, cAMP is converted to inactive AMP by the enzyme phosphodiesterase within the cytoplasm.

The nature of the protein kinases and the proteins whose activity they modify by phosphorylation varies in different target cells. This explains why various target cells respond very differently to the universal mechanism of hormonally induced changes in their intracellular cAMP levels. Different kinds of enzyme activity are ultimately modified in different target cells.

Most but not all hydrophilic hormones use cAMP as their second messenger. A few are known to use intracellular Ca^{++} as the second messenger; for others such as insulin, the second messenger is still unknown.

POSTRECEPTOR EVENTS: LIPOPHILIC HORMONES. The common denominator of the effects that lipophilic hormones produce in their target cells is the enhanced synthesis of new enzymatic or structural proteins. This results from stimulation of messenger RNA (see p. 46) as follows (Fig. 18–8):

Figure 18-7 Postreceptor Events for Hydrophilic Hormones *See text for an explanation of the steps involved.*

SOURCE: Adapted with permission from George A. Hedge, Howard D. Colby, and Robert L. Goodman, *Clinical Endocrine Physiology* (Philadelphia: W.B. Saunders Company), Figure 1-8, p. 18.

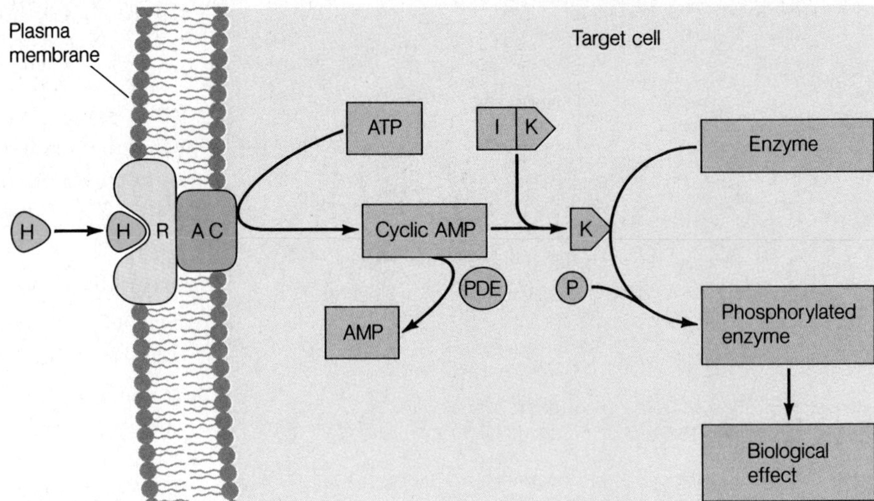

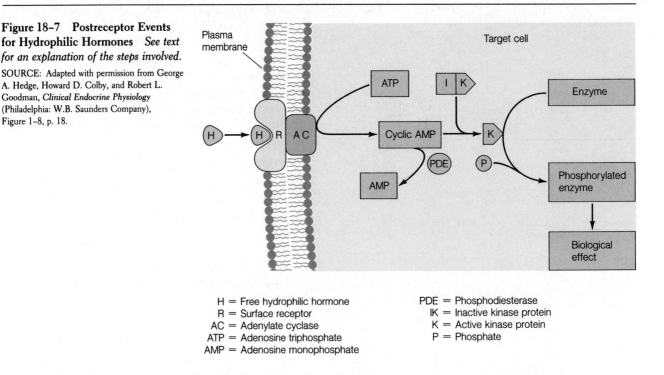

H = Free hydrophilic hormone
R = Surface receptor
AC = Adenylate cyclase
ATP = Adenosine triphosphate
AMP = Adenosine monophosphate
PDE = Phosphodiesterase
IK = Inactive kinase protein
K = Active kinase protein
P = Phosphate

Figure 18-8 Postreceptor Events for Lipophilic Hormones *See text for an explanation of the steps involved.*

SOURCE: Adapted with permission from George A. Hedge, Howard D. Colby, and Robert L. Goodman, *Clinical Endocrine Physiology* (Philadelphia: W.B. Saunders Company), Figure 1-9, p. 20.

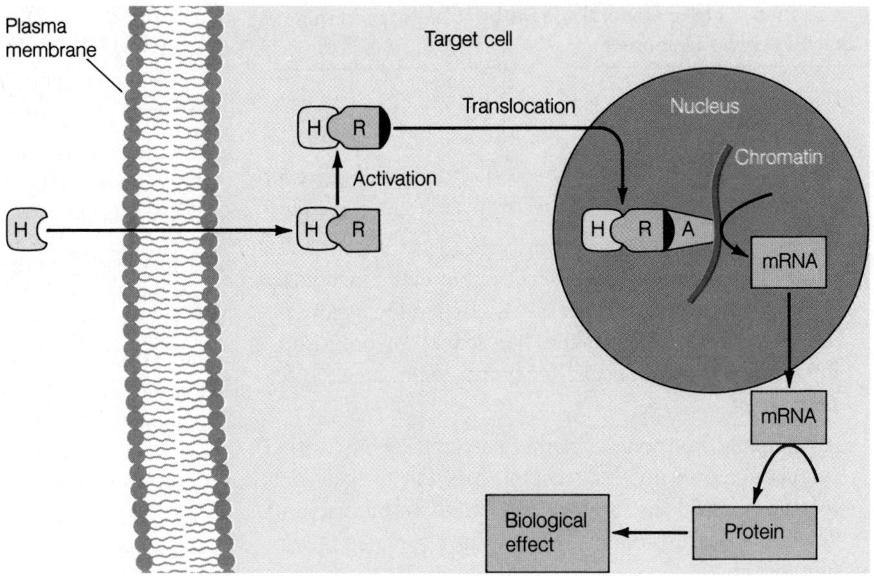

H = Free steroid hormone
R = Cytoplasmic receptor
A = Nuclear acceptor site
mRNA = Messenger RNA

1. Free steroid hormone (hormone not bound with its plasma-protein carrier) diffuses through the plasma membrane and binds with its specific receptor in the target cell's cytoplasm.
2. Hormone–receptor binding causes a conformational change in the receptor known as **activation.**
3. The affinity (attraction) of the activated steroid hormone–cytoplasmic receptor complex for chromatin (genetic material in the nucleus) is notably enhanced.
4. As a consequence, the entire complex is translocated (moved across) to the nucleus.
5. Within the nucleus, the complex binds to chromatin at a site of attachment specific for the complex known as an **acceptor site.**
6. Binding of the hormone–cytoplasmic receptor complex with nuclear acceptor sites increases transcription of specific genes (that is, "turns on" specific genes).
7. Activated genes produce complementary messenger RNA, which enters the plasma and binds to a ribosome, the cell organelle that mediates assembly of new proteins.
8. The new protein produces the target cell's final biological response to the hormone.

As a result of this mechanism, different genes are activated by different steroid hormones, which results in different biological effects.

Thyroid hormone functions similarly to steroid hormones except that, on entering the target cell, free thyroid hormone bypasses the cytoplasmic receptor step and binds directly to a nuclear receptor specific for it. Thyroid hormone activates genetic machinery for the formation of many different enzymes that collectively promote overall enhanced intracellular metabolic activity.

Despite the differences in molecular events that are induced by hydrophilic and lipophilic hormones, there are several characteristics common to both hormone types that have important implications for hormonal actions:

□ First, the actions of hormones are greatly amplified at the target tissue. Hormones are greatly diluted by the blood and thus must exert their effect at incredibly low concentrations—as low as 1 picogram (1 millionth of a millionth of a gram) per ml—as opposed to the much higher localized concentration of neurotransmitter at the target cell during neural communication. Interaction of one hormonal molecule with its receptor can result in the formation of many active protein products that ultimately carry out the biological effect. For example, one peptide hormone results in the production of numerous cAMP messengers, each in turn activating many latent enzymes (see p. 65). Similarly, one steroid-hormone activated gene induces the formation of many messenger RNA molecules, each being used to make many enzymes.

□ A second common feature of hormone action is that hormones regulate the rates of existing reactions instead of initiating new reactions. Enzymes under hormonal control usually show some activity even in the absence of the hormone.

□ Finally, as mentioned previously, hormone action is relatively slow and prolonged. In general, lipophilic hormones, which act by increasing protein synthesis, take longer to produce an effect than hydrophilic hormones, which need only to activate a preformed enzyme. Once an enzyme is activated, it no longer depends on the presence of the hormone. As a result, a hormone's effect usually lasts for some time after hormone withdrawal.

The effective plasma concentration of a hormone is normally regulated by changes in its rate of secretion.

The plasma concentration of free hormone (the free pool), and thus the hormone's availability to its receptors, may depend on several factors (Fig. 18–9): (1) its rate of secretion into the blood by the endocrine gland; (2) its rate of metabolic activation; (3) its rate of removal from the blood by metabolic

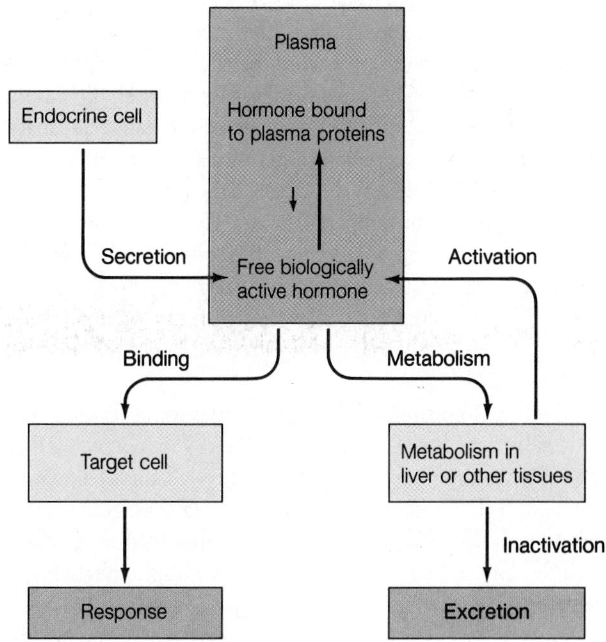

Figure 18–9 Factors Affecting the Plasma Concentration of Free Biologically Active Hormone

inactivation and excretion; and (4) the extent of its binding to plasma proteins.

The primary function of hormones is the regulation of various homeostatic activities. Because hormones' effects are proportional to their concentrations in the blood, it follows that these concentrations must be subject to control according to homeostatic need. Normally, the effective plasma concentration of a hormone is regulated by appropriate adjustments in the rate of its secretion. Endocrine glands do not secrete their hormones at a constant rate; the secretion rates of all hormones vary subject to control, often by a combination of several complex mechanisms. The regulatory system for each hormone will be considered in detail in subsequent sections. As you proceed through the chapters on the endocrine system, note the following general mechanisms of controlling secretion that are common to many different hormones.

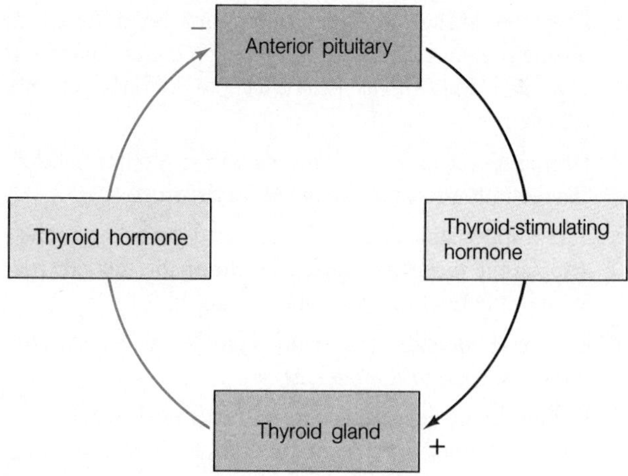

Figure 18–10 Negative-Feedback Control

NEGATIVE-FEEDBACK CONTROL. If changes in the rate of an endocrine organ's secretion are to be useful in maintaining an appropriate plasma concentration of the hormone, the endocrine organ must constantly be "aware" of the concentration of the hormone itself or of some outcome brought about by the hormone. This awareness is provided by negative-feedback loops (see p. 11), in which information about the status of the hormone product is transmitted through the blood back to its secreting organ. The endocrine organ then adjusts its rate of secretion accordingly to maintain hormone levels at a relatively constant set point.

Because negative feedback is a prominent feature of hormonal control systems, it is worth briefly reviewing this concept. Stated simply, negative feedback exists when the output of a system opposes a change in input. Negative feedback maintains the blood concentration of a hormone at a given level, similar to the way in which a home heating system maintains the room temperature at a given set point. If the temperature in a room falls below the setting on the thermostat, the thermostat activates the furnace to produce more heat. The room does not get any hotter than the temperature setting, because the system shuts itself off when the set level has been achieved.

Control of hormonal secretion provides some classic physiological examples of negative feedback. For example (Fig. 18–10), the anterior pituitary secretes thyroid-stimulating hormone (TSH), which stimulates the thyroid to secrete thyroid hormone. Thyroid hormone in turn inhibits further secretion of TSH by the anterior pituitary. Just as with the heating system, this assures that once thyroid-gland secretion has been "turned on" by TSH, it will not continue unabated but instead will be "turned off" when the appropriate level of free circulating thyroid hormone has been achieved. Thus, the effect of a particular hormone's actions can inhibit its own secretion. The feedback loops often become quite complex.

In several instances within reproductive endocrinology, positive-feedback loops are encountered. These control loops, instead of stabilizing hormone secretion, promote a build up of a particular hormonal effect until an explosive culminating point is reached, such as during the birth of a baby.

NEUROENDOCRINE REFLEXES. Many endocrine control systems involve **neuroendocrine reflexes,** which include neural as well as hormonal components. The purpose of such reflexes is to produce a sudden increase in hormone secretion (that is,"turn up the thermostat setting") in response to a specific stimulus, frequently a stimulus external to the body. Some endocrine control systems include both feedback control, which maintains a constant basal level of the hormone, and neuroendocrine reflexes, which cause sudden bursts in secretion in response to a sudden increased need for the hormone, such as during a stress response.

DIURNAL OR CIRCADIAN RHYTHMS. Although hormone secretion rates are usually regulated by some form of negative feedback, this does not imply that they are always maintained at a constant level. Instead, the secretion rates of all hormones rhythmically fluctuate up and down as a function of time. The most common endocrine rhythm is the **diurnal** ("day-night"), or **circadian** ("around a day"), **rhythm,** which is characterized by repetitive oscillations in hormone levels that are very regular and have a frequency of one cycle every twenty-four hours. This rhythmicity appears to be caused by endogenous oscillators similar to the self-paced respiratory

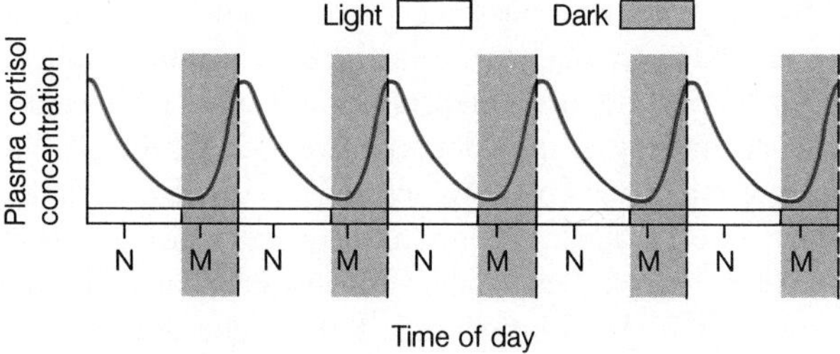

N is noon; M is midnight.

Figure 18–11 Diurnal Rhythm of Cortisol Secretion

SOURCE: Adapted with permission from George A. Hedge, Howard D. Colby, and Robert L. Goodman, *Clinical Endocrine Physiology* (Philadelphia: W.B. Saunders Company), Figure 1–13, p. 28.

neurons in the brain stem that are responsible for the rhythmic motions of breathing. However, unlike the rhythmicity of breathing, endocrine rhythms are locked on or "entrained" to external cues called **zeitgebers,** such as the light-dark cycle or the activity cycle; that is, the inherent twenty-four-hour cycles of peak and ebb of hormone secretion are set to "march in step" with cycles of light/activity and dark/inactivity. For example, cortisol secretion rises during the night, reaching its peak secretion in the morning before a person rises, then falls throughout the day to its lowest level at bedtime (Fig. 18–11). Most investigators believe that inherent hormonal rhythmicity and the entrainment of zeitgebers are not accomplished by the endocrine organs themselves, but that they are the result of the central nervous system changing the set point of these organs. Negative-feedback control mechanisms operate to maintain whatever set point is established for that time of day.

Some endocrine cycles operate on time scales other than a circadian rhythm—some much shorter than a day and some much longer. A well-known example of the latter is the monthly menstrual cycle. Periodicity is not unique to the endocrine system. Humans have similar biological clocks for other bodily functions such as temperature regulation (see p. 610). The mechanism and purpose served by these rhythmic fluctuations in biological activity are unclear, but the effect of upsetting them is well known by people who experience jet lag when their inherent rhythm is out of step with external cues.

The effective plasma concentration of a hormone can be influenced by the hormone's transport, metabolism, and excretion.

Even though the effective plasma concentration of a hormone is normally regulated by appropriate adjustments in the rate of secretion, alterations in the transport, metabolism, or excretion of a hormone can also influence the size of its effective pool, sometimes inappropriately. For example, since the liver synthesizes plasma proteins, liver disease may result in abnormal endocrine activity because of a change in the balance between free and bound pools of certain hormones.

Eventually, all hormones are metabolized by enzyme-mediated reactions that modify the hormonal structure in some way. In most cases this results in inactivation of the hormone. However, hormone metabolism is not always a mechanism for removal of used hormones. In some cases, a hormone is ***activated*** by metabolism; that is, the hormone's product has greater activity than the original hormone. For example, after the thyroid hormone thyroxine is secreted, it is converted to a more powerful hormone by enzymatic removal of one of the iodine atoms it contains. Usually the rate of such hormone activation is itself under hormonal control.

The liver is the most common site for metabolic hormonal inactivation (or activation, if that be the case), but some hormones are also metabolized in the kidneys, blood, or target tissues. Hormonal inactivation and excretion are not subject to control. The primary means of eliminating hormones and their metabolites from the blood is by urinary excretion. Because the liver and kidneys are important in the removal of hormones from the blood by means of metabolic inactivation and urinary excretion, patients with liver or kidney disease may suffer from excess activity of certain hormones solely as a result of reduced hormone elimination. On the other hand, when liver and kidney function are normal, measurement of urinary concentrations of hormones and their metabolites provides a useful noninvasive means of assessing endocrine function, because the rate of excretion of these products in the urine is a direct reflection of their rate of secretion by the endocrine glands.

The amount of time after a hormone is secreted before it is inactivated and the means by which this is accomplished differ for different classes of hormones. In general, the hydrophilic peptides and catecholamines are easy targets for blood and tissue enzymes, so they remain in the blood only briefly (a few minutes to a few hours) before being enzymatically inactivated. In the case of some peptide hormones, the target cell actually engulfs the bound hormone by endocytosis and degrades it intracellularly. In contrast, binding of hydrophobic hormones to plasma proteins renders them less vulnerable to metabolic inactivation and prevents them from escaping into the urine. Therefore, hydrophobic hormones are removed from the plasma much more slowly. They may persist in the blood for hours (steroids) or up to a week (thyroid hormone). In general, hydrophobic hormones undergo a series of reactions that reduces their biological activity and enhances their H_2O solubility so that they can be freed of their plasma-protein carriers and be eliminated in the urine.

Endocrine disorders are attributable to hormonal excess, hormonal deficiency, or decreased responsiveness of the target tissue.

From the preceding discussion, it should be apparent that abnormalities in a hormone's effective plasma concentration can arise from a variety of factors (Table 18–4). Most commonly, abnormal plasma concentrations of a hormone are due to inappropriate rates of secretion; that is, too little hormone secreted (**hyposecretion**) or too much hormone secreted (**hypersecretion**). Occasionally, endocrine dysfunction arises because of abnormal tissue responsiveness to the hormone, even though the plasma concentration of the hormone is normal.

HYPOSECRETION. If an endocrine organ is secreting too little of its hormone because of an abnormality within that organ, the condition is referred to as *primary hyposecretion*. If, on the other hand, the endocrine organ is normal but is secreting too little hormone because of a deficiency of its tropic hormone, the condition is known as *secondary hyposecretion*. The following are among the many different factors (each listed with an example) that may be responsible for hormone deficiency: (1) genetic (inborn absence of an enzyme that catalyzes synthesis of the hormone); (2) dietary (lack of iodine, which is necessary for synthesis of thyroid hormone); (3) chemical or toxic (certain insecticide residues may destroy the adrenal cortex); (4) immunologic (autoimmune antibodies may cause self-destruction of the person's own thyroid tissue); (5) other disease processes (cancer or tuberculosis may coincidentally destroy endocrine glands); (6) *iatrogenic* (physician-induced, such as surgical removal of a thyroid tumor); and (7) *idiopathic* (meaning the cause is not known).

The most common method of treating hormone hyposecretion is administration of a hormone that is the same as (or similar to, such as from another species) the one that is deficient or missing. Such replacement therapy is straightforward in theory, but there are some practical problems when it comes to the hormone's source and means of administration. The sources of hormone preparation for clinical use include: (1) endocrine tissues from domestic livestock; (2) the placental tissue and urine of pregnant women; (3) laboratory synthesis of hormones; and (4) "hormone factories," which consist of bacteria into which genes coding for the production of human hormones have been introduced. The method of choice for yielding a given hormone is determined largely by its structural complexity and degree of species specificity.

HYPERSECRETION. As with hyposecretion, hypersecretion by a particular endocrine organ is designated as primary or secondary depending on whether the defect lies in that organ or is due to excessive stimulation from the outside. Hypersecretion may be caused by: (1) tumors that ignore the normal regulatory input and continuously secrete excess hormone; and (2) immunologic factors, such as excessive stimulation of the thyroid gland by an abnormal antibody that mimics the action of TSH, the thyroid tropic hormone. Excessive levels of a particular hormone may also arise from substance abuse, such as the outlawed practice among athletes of using certain steroids that increase muscle mass by promoting protein synthesis in muscle cells (see p. 240).

There are several ways of treating hormonal hypersecretion. If a tumor is the culprit, it may be surgically removed or destroyed with radiation treatment. In some instances, hypersecretion can be limited by drugs that block hormone synthesis or inhibit hormone secretion. Sometimes the condition may be treated by giving drugs that inhibit the action of the hormone, even though nothing is done to actually reduce the excess hormone secretion.

ABNORMAL TISSUE RESPONSIVENESS. Endocrine dysfunction can also occur as a result of inadequate responsiveness of the target tissue to the hormone, even though the effective

Table 18–4 Means by Which Endocrine Disorders Can Arise

Too Little Hormone Activity	Too Much Hormone Activity
Too little hormone secreted by endocrine gland (hyposecretion)*	Too much hormone secreted by endocrine gland (hypersecretion)*
Increased removal of hormone from blood	Reduced plasma-protein binding of hormone (too much free biologically active hormone)
Abnormal tissue responsiveness to hormone	Decreased removal of hormone from blood
Lack of target-tissue receptors	Decreased inactivation
Lack of enzyme essential to target-tissue response	Decreased excretion

*Most common causes of endocrine dysfunction.

plasma concentration of a hormone is normal. This inadequate responsiveness may be caused, for example, by an inborn lack of receptors for the hormone, such as is seen in *testicular feminization syndrome*. In this condition, testosterone receptors are not produced because of a specific genetic defect. In spite of adequate testosterone availability, masculinization does not take place, just as if no testosterone were present. Abnormal responsiveness may also occur if the target tissue lacks an enzyme essential to carrying out the response.

The responsiveness of a target tissue to its hormone can be varied by regulating the number of its hormone-specific receptors.

In contrast to endocrine dysfunction being caused by *unintentional* receptor abnormalities, the target tissue receptors for a particular hormone can be *deliberately altered* subject to physiologic control. Because receptors are the primary sites of interaction between a hormone and its target tissue, they represent an important avenue for regulating the target tissue's sensitivity to the hormone. A target tissue's response to a hormone is correlated with the number of the tissue's receptors occupied by molecules of that hormone, which in turn depends on the effective plasma concentration of the hormone. Accordingly, changes in target-tissue response are directly related to changes in the hormone's secretion rate. Furthermore, the response of the target tissue to a given plasma concentration can be modulated (fine-tuned up or down) by varying the number of receptors available for hormone binding. Thus the number of receptors that a target cell has for a specific hormone does not always remain constant but can be reduced or increased.

As an illustration, when the plasma concentration of insulin is chronically elevated, the total number of target-tissue receptors for insulin is reduced as a direct consequence of the elevated level of insulin's effect on its own receptors. This phenomenon, known as **down regulation,** constitutes an important locally acting negative-feedback mechanism that prevents the target cells from overreacting to the high concentration of insulin; that is, the target cells are *desensitized* to insulin. This helps blunt the effect of insulin hypersecretion. Down regulation of insulin is accomplished by the following mechanism. The binding of insulin to its surface receptors induces endocytosis of the hormone–receptor complex, which is subsequently attacked by intracellular lysosomal enzymes. This internalization serves a two-fold purpose: it provides a pathway for degradation of the hormone, and it also plays a role in regulating the number of receptors available for binding on the target cell's surface. At high plasma insulin concentrations, the number of surface receptors for insulin is gradually reduced as a result of the accelerated rate of receptor internalization and degradation brought about by increased hormonal binding. The rate of synthesis of new receptors within the endoplasmic reticulum and their insertion in the plasma membrane does not keep pace with their rate of destruction. Over time, this self-induced loss of target-cell receptors for insulin results in a reduction in the target cell's sensitivity to the elevated hormone concentration.

Hormones frequently alter the receptors for other kinds of hormones as part of their normal physiological activity. Because of the widespread distribution of hormones through the blood, target cells may be exposed simultaneously to many different hormones, giving rise to numerous complex hormonal interactions on target cells. A hormone can influence the activity of another hormone at a given target tissue in one of three ways: permissiveness, synergism, and antagonism. In the phenomenon of **permissiveness,** one hormone must be present in adequate amounts for the full exertion of another hormone's effect, even though the first hormone does not directly elicit the response. In essence, the first hormone, by enhancing a target organ's responsiveness to another hormone, is "permitting" this other hormone to exert its full effect. For example, thyroid hormone increases the number of receptors for epinephrine in epinephrine's target tissues, thereby increasing the effectiveness of epinephrine. Epinephrine is only marginally effective in the absence of thyroid hormone.

Synergism occurs when the actions of several hormones are complementary and their combined effect is greater than the sum of their separate effects. An example is the synergistic action of several hormones (follicle-stimulating hormone and testosterone) required to maintain the normal rate of sperm production. Synergism probably results from each hormone's influence on the number or affinity of receptors for the other hormone.

Antagonism occurs when one hormone induces the loss of another hormone's receptors, which suppresses the effectiveness of the second hormone. To illustrate, progesterone (a hormone secreted during pregnancy that *reduces* uterine motility) inhibits uterine responsiveness to estrogen (another hormone secreted during pregnancy that *increases* uterine motility). Progesterone, by causing loss of estrogen receptors on uterine smooth muscle, prevents estrogen from exerting its excitatory effects during pregnancy, thus keeping the uterus a quiet (noncontracting) environment suitable for the developing fetus.

Thus, a given hormone's effects are influenced not only by the concentration of the hormone itself but also by the concentration of other hormones that interact with it. As a result, alterations in the plasma concentration of a given hormone, whether caused by disease, administration of a hormone as a drug, or surgical removal of an endocrine gland, often produce widespread and unpredictable effects. Not only are the target tissues that are directly controlled by the hormone affected, but seemingly unrelated symptoms occur because of altered permissive, synergistic, or antagonistic effects on other hormones.

Hypothalamus and Pituitary

The pituitary gland consists of well-developed anterior and posterior lobes plus a rudimentary intermediate lobe.

The **pituitary gland,** or **hypophysis,** is a small endocrine gland located in a bony cavity at the base of the brain just below the hypothalamus (Fig. 18–12). If you point one finger between your eyes and another finger toward one of your ears, the imaginary point where these lines would intersect is about the location of your pituitary. The pituitary is connected to the hypothalamus by a thin stalk, the **infundibulum,** which contains nerve fibers and small blood vessels.

The pituitary has two anatomically and functionally distinct lobes, the **posterior pituitary** and the **anterior pituitary.** The posterior pituitary, being derived embryonically from an outgrowth of the brain, is composed of nervous tissue and thus is also termed the **neurohypophysis.** The anterior pituitary, in contrast, consists of glandular epithelial tissue derived embryonically from an outpouching that buds off from the roof of the mouth. Accordingly, the anterior pituitary is also known as the **adenohypophysis** (*adeno* means "glandular"). The anterior and posterior pituitary share nothing more in common than their location. The posterior pituitary is connected to the hypothalamus by a neural pathway, whereas the anterior pituitary is connected to the hypothalamus by a vascular link.

In the adenohypophysis in some species, a third, well-defined intermediate lobe is also included, but in humans this lobe is rudimentary. In lower vertebrates, the intermediate lobe secretes several **melanocyte-stimulating hormones** or **MSHs,** which regulate skin coloration by controlling the dispersion of granules containing the pigment **melanin.** By causing variable skin darkening in certain amphibia, reptiles, and fishes, MSHs play a vital role in the camouflage of these species. In humans, the small amount of MSH secreted is generally attributed to the anterior pituitary. The function of this MSH, if any, is still unclear. It is not involved in the differences in the amount of melanin deposited in the skin of various races, nor is it associated with the process of tanning. However, excessive MSH activity does cause darkening of the skin. Some evidence implicates MSH in humans in the totally different role of influencing the excitability of the nervous system, perhaps playing a role in improving memory and learning.

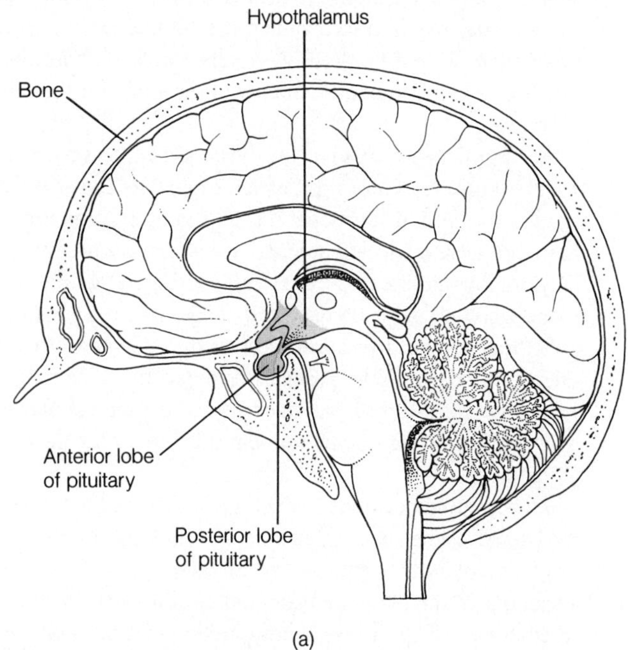

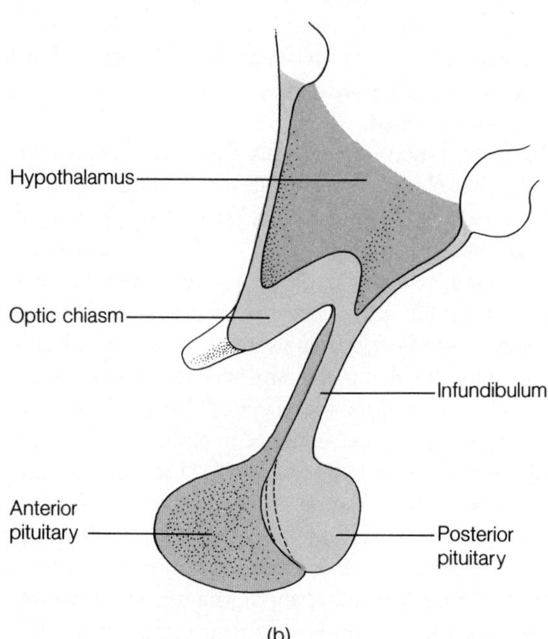

Figure 18–12 Anatomy of the Pituitary Gland *(a) Relation of the pituitary gland to the hypothalamus and to the remainder of the brain. (b) Schematic enlargement of the pituitary gland and its connection to the hypothalamus.*

The hypothalamus and posterior pituitary form a neurosecretory system that secretes vasopressin and oxytocin.

The release of hormones from both the posterior and anterior pituitary is directly controlled by the hypothalamus, but the nature of the relationship is entirely different. The hypothalamus and posterior pituitary form a neuroendocrine system that consists of a population of neurosecretory neurons whose cell bodies lie in two well-defined clusters in the hypothalamus (the **supraoptic** and **paraventricular nuclei**) and whose axons pass down through the connecting stalk to terminate on capillaries in the posterior pituitary (Fig. 18–13). The posterior pituitary consists of these neuronal terminals plus glial-like supporting cells called **pituicytes.**

Functionally as well as anatomically, the posterior pituitary is simply an extension of the hypothalamus. The posterior pituitary does not actually produce any hormones. It simply stores and, upon appropriate stimulation, releases into the blood two small peptide hormones, vasopressin (antidiuretic hormone) and oxytocin, which are synthesized by the neuronal cell bodies in the hypothalamus. Both of these hydrophilic peptides are made in both the supraoptic and the paraventricular nuclei, but a single neuron is capable of producing only one of these neurohormones. The synthesized neurohormones are packaged into secretory granules that are transported down the cytoplasm of the axon (see p. 35) to be stored in the neuronal terminals within the posterior pituitary. Each terminal stores either vasopressin or oxytocin but not both. Depending on whether vasopressin-secreting or oxytocin-secreting neurons are activated, vasopressin or oxytocin is released into the capillary blood of the posterior pituitary by Ca^{++}-induced exocytosis of the secretory granules. The hormones are released in response to action potentials that originate in the hypothalamic cell body and sweep down the axon to the neuronal terminal. As in any other neuron, action potentials are generated in neurosecretory neurons in response to synaptic input to their cell bodies.

The actions and control of vasopressin and oxytocin are briefly summarized here to make our endocrine story complete. They are described more thoroughly elsewhere—vasopressin in chapters 14 and 15 and oxytocin in chapter 20.

VASOPRESSIN. Vasopressin has two major effects that correspond to its two names: (1) it enhances the retention of H_2O by the kidneys (an antidiuretic effect) and (2) it causes contraction of arteriolar smooth muscle (a vessel pressor effect). The first effect is of greater physiological importance. Under normal conditions, vasopressin is the primary endocrine factor

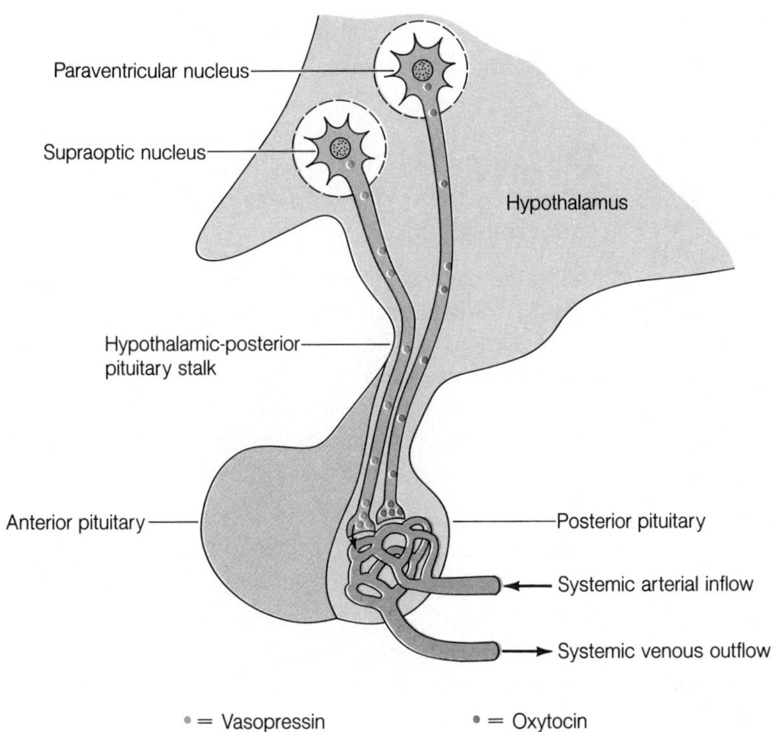

Figure 18–13 Relationship of Hypothalamus and Posterior Pituitary *The hormone, either vasopressin or oxytocin, depending on the neuron, is synthesized in the neuronal cell body in the hypothalamus and travels down the axon to be stored in the neuronal terminals within the posterior pituitary. The stored hormone is released into the systemic blood upon excitation of the neuron.*

A CLOSER LOOK AT EXERCISE PHYSIOLOGY

The Endocrine Response to the Challenge of Combined Heat and Marching Feet

When one exercises in a hot environment, maintenance of plasma volume becomes a critical homeostatic concern. Exercise in the heat results in the loss of large amounts of fluid through sweating. Simultaneously, blood is needed for shunting to the skin for cooling and for increased blood flow to nourish the working muscles. There must also be adequate venous return to maintain cardiac output. The hypothalamus–posterior pituitary neurosecretory system responds to these multiple, conflicting needs for fluid by releasing water-conserving vasopressin in an attempt to reduce urinary fluid loss to preserve plasma volume.

Studies have generally shown that exercise in heat stimulates vasopressin release, which results in decreased urinary fluid loss. In one study conducted during an eighteen-mile road march in heat, the participants' average urine output dropped to 134 ml (the normal urine output during the same time period would be about twice that much), while sweat loss averaged four liters. Overhydration prior to exercise appears to decrease the intensity of this response, suggesting that increased vasopressin release is related to plasma osmolarity. If fluid loss is not adequately replaced, plasma osmolarity increases. When the hypothalamic osmoreceptors detect this hypertonic condition, they promote increased secretion of vasopressin from the posterior pituitary. Some investigators believe, however, that increased vasopressin release results from other factors, such as changes in blood pressure or in renal blood flow. Regardless of the mechanism, vasopressin release is an important physiologic response to exercise in heat.

that regulates urinary H_2O loss and overall H_2O balance. In contrast, typical levels of vasopressin play only a minor role in regulating blood pressure by means of the hormone's pressor effect.

The major control for hypothalamic-induced release of vasopressin from the posterior pituitary is input from hypothalamic osmoreceptors, which increase vasopressin secretion in response to a rise in plasma osmolarity. A less powerful input from the left atrial baroreceptors increases vasopressin secretion in response to a fall in extracellular-fluid volume and arterial blood pressure. The osmoreceptor and baroreceptor mechanisms both function as negative-feedback systems to resist changes in extracellular fluid osmolarity and volume, respectively, by adjusting the H_2O load in the body through variable vasopressin activity (see the accompanying boxed feature, A Closer Look at Exercise Physiology).

OXYTOCIN. Oxytocin stimulates contraction of the uterine smooth muscle to aid in expulsion of the baby during childbirth, and it promotes ejection of milk from the mammary glands (breasts) during breastfeeding. Appropriately, oxytocin secretion is increased by reflexes that originate within the birth canal during childbirth and by reflexes that are triggered when the infant suckles the breast.

The anterior pituitary secretes six established hormones, many of which are tropic to other endocrine glands.

Unlike the posterior pituitary, the anterior pituitary synthesizes the hormones that it releases into the blood. Different cell populations within the anterior pituitary produce and secrete six established peptide hormones. The actions of each of these hormones will be dealt with in detail in subsequent sections. For now, a brief statement of their primary effects will be given to provide a rationale for their nomenclature (Fig. 18–14):

1. **Growth hormone (GH, somatotropin)**, the primary hormone responsible for regulating overall body growth, is also important in organic metabolism.
2. **Thyroid-stimulating hormone (TSH, thyrotropin)** stimulates secretion of thyroid hormone and growth of the thyroid gland.

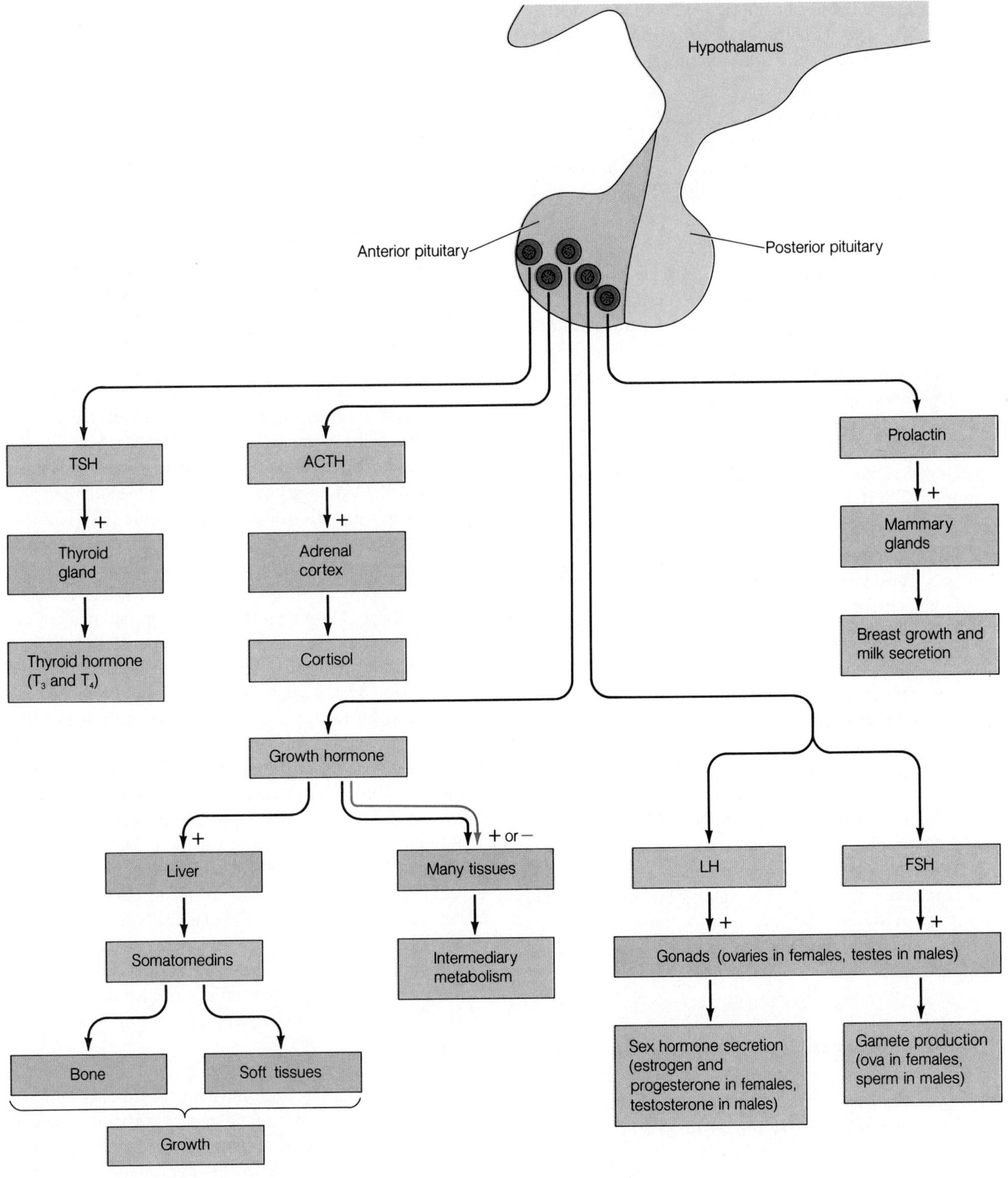

Figure 18–14 Functions of Anterior Pituitary Hormones

3. **Adrenocorticotropic hormone (ACTH, adrenocorticotropin)** stimulates hormone secretion, especially of cortisol, by the adrenal cortex and promotes growth of the adrenal cortex.
4. **Follicle-stimulating hormone (FSH)** has different functions in males and females. In females it stimulates growth and development of ovarian follicles, within which the ova, or eggs, develop. Furthermore, FSH promotes secretion of the hormone estrogen by the ovaries. In males FSH is required for sperm production.
5. **Luteinizing hormone (LH)** also functions differently in females and males. In females LH is responsible for ovulation; luteinization (that is, the formation of a postovulatory hormone-secreting corpus luteum in the ovary); and regulation of ovarian secretion of the female sex hormones, estrogen and progesterone. In males the same hormone stimulates the interstitial cells of Leydig in the testes to secrete the male sex hormone, testosterone, giving rise to its alternate name of **interstitial cell–stimulating hormone (ICSH)**.
6. **Prolactin (PRL)** enhances breast development and milk production in females. Its function in males is uncertain, although evidence indicates that it may induce the production of testicular LH receptors.

Because they each regulate the secretion of another specific endocrine gland, TSH, ACTH, FSH, and LH are all tropic hormones. More specifically, FSH and LH are collectively referred to as **gonadotropins,** because they control secretion of the sex hormones by the gonads (ovaries and testes). Since GH has recently been shown to exert its growth-promoting effects indirectly by stimulating the release of the liver hormones, the somatomedins, it too is sometimes categorized as a tropic hormone. Among the anterior pituitary hormones, PRL is the only one that does not stimulate secretion of another hormone. Of the tropic hormones, FSH, LH, and GH differ from TSH and ACTH in that the former exert nontropic functions in addition to stimulating secretion of other hormones.

Hypothalamic releasing and inhibiting hormones are delivered to the anterior pituitary by the hypothalamo-hypophyseal portal system to control anterior pituitary hormone secretion.

None of the anterior pituitary hormones are secreted at a constant rate. Even though each of these hormones has a unique control system, there are some common regulatory patterns. The two most important factors that regulate anterior pituitary hormone secretion are: (1) hypothalamic hormones, and (2) feedback by target-organ hormones.

Because the anterior pituitary secretes hormones that control the secretion of various other hormones, it has long had the undeserved title of "master gland." It is now known that the release of each of the anterior pituitary hormones is largely controlled by still other hormones produced by the hypothalamus and that secretion of these regulatory neurohormones, in turn, is controlled by a variety of neural and hormonal inputs to the hypothalamic neurosecretory cells (Fig. 18–15).

The secretion of each of the anterior pituitary hormones is stimulated or inhibited by one or more of the seven generally accepted hypothalamic **hypophysiotropic hormones,** which are listed in Table 18–5. Depending on the actions of these small peptides, they are called ***releasing hormones*** or ***inhibiting hormones***. (The term *factor* rather than *hormone* is used for those substances whose structures are as yet unknown. A substance may qualify as a full-fledged hormone only if it has been isolated and chemically identified.) In each case, the primary action of the hormone is apparent from its name. For example, **thyrotropin-releasing hormone (TRH)** stimulates the release of TSH (alias thyrotropin) from the anterior pituitary, whereas **prolactin-inhibiting hormone (PIH)** inhibits the release of prolactin from the anterior pituitary. Although it was originally speculated that there was a neat one-to-one correspondence—one hypophysiotropic hormone for each anterior pituitary hormone—it is now clear that many of the hypothalamic hormones have more than one effect, with their names indicating only the function that was initially attributed to them. Moreover, a single anterior pituitary hormone may be regulated by two or more hypophysiotropic hormones, which may even exert opposing effects. For example, **growth hormone–releasing hormone (GHRH)** stimulates growth hormone secretion, whereas **growth hormone–inhibiting hormone (GHIH)**, also known as **somatostatin,** inhibits it. The output of the anterior pituitary growth hormone–secreting cells (that is, the rate of growth hormone secretion) in response to two such opposing inputs depends on the relative concentrations of these hypothalamic hormones as well as on the intensity of other regulatory inputs. This is analogous to the output of a nerve cell (that is, the rate of action-potential propagation) being dependent on the relative magnitude of excitatory and inhibitory synaptic inputs (EPSPs and IPSPs) to it.

Chemical messengers that are identical in structure to the hypothalamic releasing and inhibiting hormones and to vasopressin are produced in many areas of the brain outside of the hypothalamus. Instead of being released into the blood, these messengers act locally as neurotransmitters and neuromodulators in these extrahypothalamic sites. They are thought to modulate a variety of functions that range from motor activity (TRH) to libido (GnRH) to learning (vasopressin). This is a further example of the multiplicity of use of the same chemical messengers.

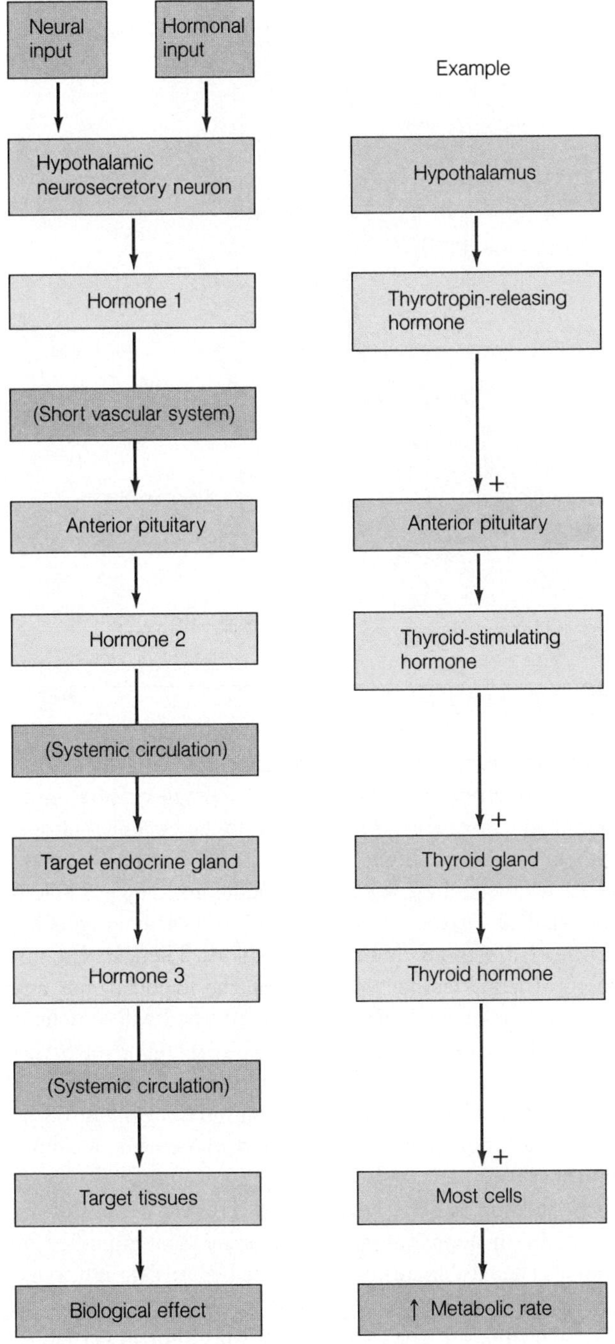

Figure 18–15 Hierarchical Chain of Command in Endocrine Control

Table 18–5 Major Hypophysiotropic Hormones

Hormone	Effect on Anterior Pituitary
Thyrotropin-releasing hormone (TRH)	Stimulates release of TSH (thyrotropin) and prolactin
Corticotropin-releasing hormone (CRH)	Stimulates release of ACTH (corticotropin)
Gonadotropin-releasing hormone (GnRH)	Stimulates release of FSH and LH (gonadotropins)
Growth hormone–releasing hormone (GHRH)	Stimulates release of growth hormone
Growth hormone–inhibiting hormone (GHIH), or somatostatin	Inhibits release of growth hormone and TSH
Prolactin-releasing factor (PRF)	Stimulates release of prolactin
Prolactin-inhibiting hormone (PIH), or dopamine	Inhibits release of prolactin

The hypothalamic regulatory hormones reach the anterior pituitary by means of a unique vascular link. In contrast to the direct neural connection between the hypothalamus and posterior pituitary, the anatomical and functional link between the hypothalamus and anterior pituitary is an unusual capillary-to-capillary connection, the **hypothalamo-hypophyseal portal system.** A portal system is a vascular arrangement in which venous blood flows directly from one capillary bed through a connecting vessel to another capillary bed without passing through the systemic circulation. The largest and best known portal system is the hepatic portal system, which drains intestinal venous blood directly into the liver for immediate processing of absorbed nutrients (see p. 576). Although much smaller, the hypothalamo-hypophyseal portal system is no less important, because it provides a critical link between the brain and much of the endocrine system. It begins in the base of the hypothalamus with a group of capillaries that recombine into small portal vessels, which pass down through the connecting stalk into the anterior pituitary. Here they branch to form most of the anterior pituitary capillaries, which in turn drain into the systemic venous system (Fig. 18–16).

Note that almost all of the blood supply to the anterior pituitary must first pass through the hypothalamus. Since materials can be exchanged between the blood and surrounding tissue only at the capillary level, the hypothalamo-hypophyseal portal system provides a route for the pick-up of releasing and inhibiting hormones at the hypothalamus and allows their immediate local delivery to the anterior pituitary at relatively high concentrations, completely bypassing the systemic circulation. This type of chemical-messenger distribution is intermediate between a classic endocrine system, in which the hormone is released into the general circulation, and a neuronal

Figure 18–16 Vascular Link between the Hypothalamus and Anterior Pituitary

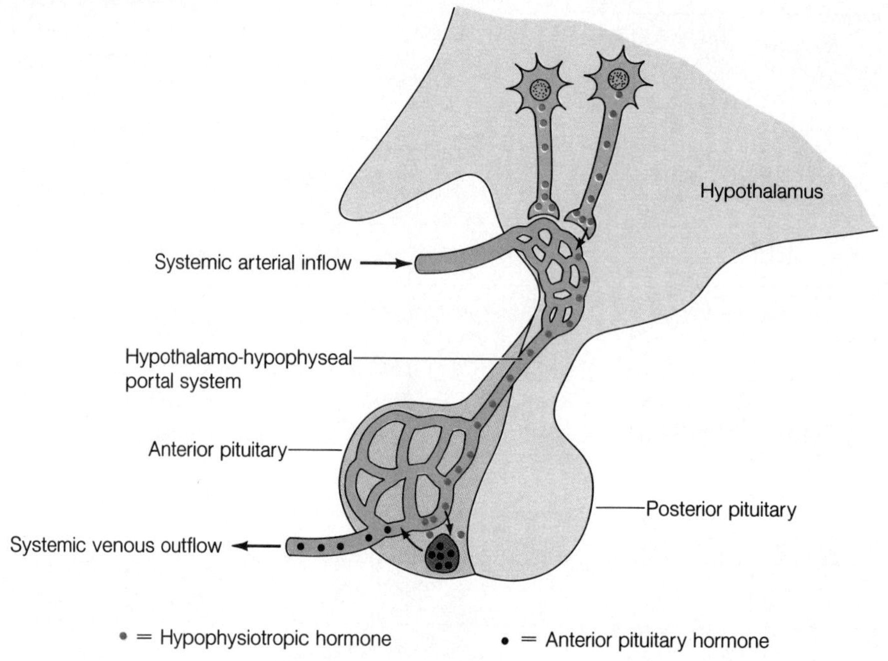

system, in which the neurotransmitter is released into a confined space (that is, a synapse). The hypophysiotropic hormones would be considerably more diluted if they were to arrive at the anterior pituitary by means of the usual systemic circulatory route.

The axons of the neurosecretory neurons that produce the hypothalamic regulatory hormones terminate on the capillaries at the origin of the portal system. These hypothalamic neurons secrete their hormones in the same way as the hypothalamic neurons that produce vasopressin and oxytocin. The hormone is synthesized in the cell body and then transported to the axon terminal. It is stored there until its release into an adjacent capillary when, upon appropriate stimulation, an action potential is generated in the neuron. The major difference is that the hypophysiotropic hormones are released into the portal vessels, which deliver them to the anterior pituitary, where they control the release of anterior pituitary hormones into the systemic circulation. In contrast, the hypothalamic hormones stored in the posterior pituitary are themselves released into the systemic circulation (Fig. 18–17).

Knowing that the secretion of anterior pituitary hormones is largely controlled by the hypothalamic releasing and inhibiting hormones, the next logical question is: What regulates the secretion of these hypophysiotropic hormones? Like other neurons, those secreting these regulatory hormones receive abundant input of information that they must integrate, both neural and hormonal and both excitatory and inhibitory. Studies are still in progress to unravel the complex neural input from many diverse areas of the brain to the hypophysiotropic secretory neurons. Some of these inputs carry information about a variety of environmental conditions. One example is the marked increase in secretion of corticotropin-releasing hormone in response to stressful situations. There are also numerous neural connections between the hypothalamus and the portions of the brain that are concerned with emotions (the limbic system—see p. 138). Thus secretion of hypophysiotropic hormones is greatly influenced by emotions. A common manifestation of this relationship is the menstrual irregularities sometimes encountered in women who are emotionally upset.

In addition to being regulated by different regions of the brain, the hypophysiotropic neurons are also controlled by various chemical inputs that reach the hypothalamus through the blood. Unlike other regions of the brain, portions of the hypothalamus are not guarded by the blood-brain barrier, so the hypothalamus can easily monitor chemical changes in the blood. In some instances, hypophysiotropic secretion is influenced by the immediate metabolic state of the individual. For example, one or more of the metabolic responses produced by growth hormone (such as an elevated blood-glucose level) negatively feeds back to the hypothalamus to control growth hormone secretion. The more common blood-borne factors

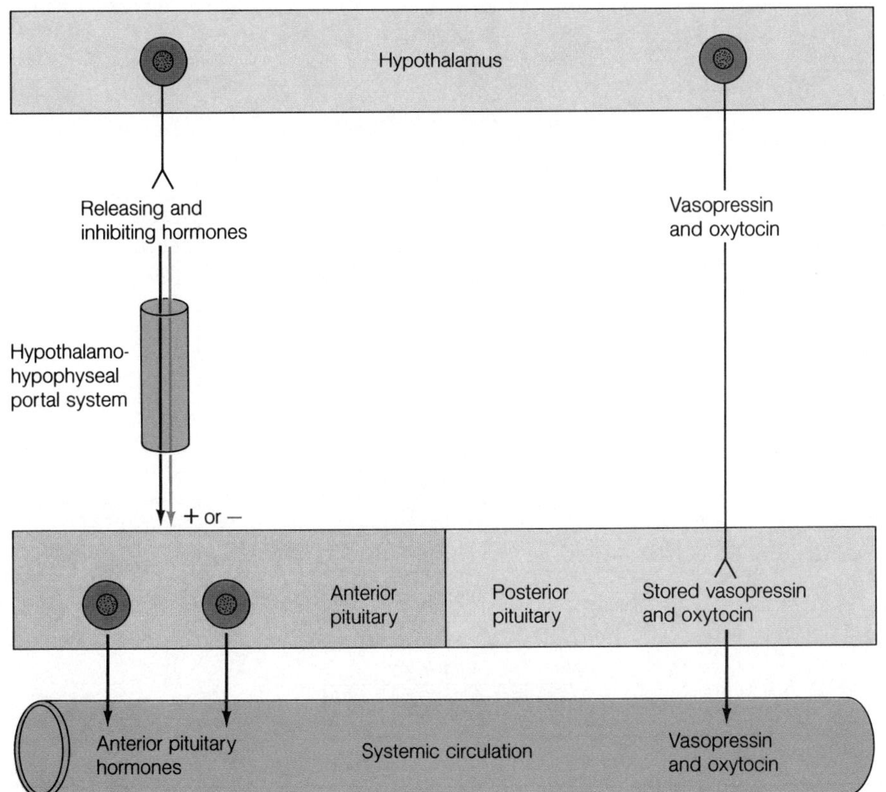

Figure 18–17 Comparison of Hypothalamic Relationship to Anterior Pituitary and Posterior Pituitary

that influence hypothalamic neurosecretion are the negative-feedback effects of either anterior pituitary or target-organ hormones, to which we now turn our attention.

In general, feedback by target-organ hormones strives to maintain relatively constant rates of anterior pituitary hormone secretion.

In most cases, hypophysiotropic hormones initiate a three hormone sequence: (1) hypothalamic releasing hormone, (2) anterior pituitary tropic hormone, and (3) peripheral target-organ hormone. In all such instances except one, **long-loop negative feedback** occurs, with the target-organ hormone acting to suppress secretion of the tropic hormone that is driving it. This is accomplished by the target-organ hormone acting either directly on the pituitary itself or on the release of hypothalamic hormones, which in turn regulate anterior pituitary function (Fig. 18–18). As an example, consider the CRH-ACTH-cortisol system. Hypothalamic CRH (corticotropin-releasing hormone) stimulates the anterior pituitary to secrete ACTH (adrenocorticotropic hormone, alias corticotropin), which in turn stimulates the adrenal cortex to secrete cortisol. The final hormone in the system, cortisol, inhibits the hypothalamus to reduce CRH secretion by suppressing the generation of action potentials in the CRH-secreting neurons, thus indirectly reducing pituitary secretion of ACTH. Cortisol also reduces the sensitivity of the ACTH-secreting cells to CRH by acting directly on the anterior pituitary. Through this double-barreled approach, cortisol exerts negative-feedback control to stabilize its own plasma concentration. If plasma cortisol levels start to rise above a prescribed set level, cortisol suppresses its own further secretion by its inhibitory actions at the hypothalamus and anterior pituitary. This assures that once a hormonal system is activated, its secretion does not continue unabated. If plasma cortisol levels fall below the desired set point, cortisol's inhibitory actions at the hypothalamus and anterior pituitary are reduced, so the driving forces for cortisol secretion (CRH-ACTH) increase accordingly.

Similarly, the other target-organ hormones act by means of long-loop negative feedback. The goal of such feedback is to maintain relatively constant levels of target-organ hormone. It

Figure 18–18 Negative Feedback in Hypothalamic-Anterior Pituitary Control Systems

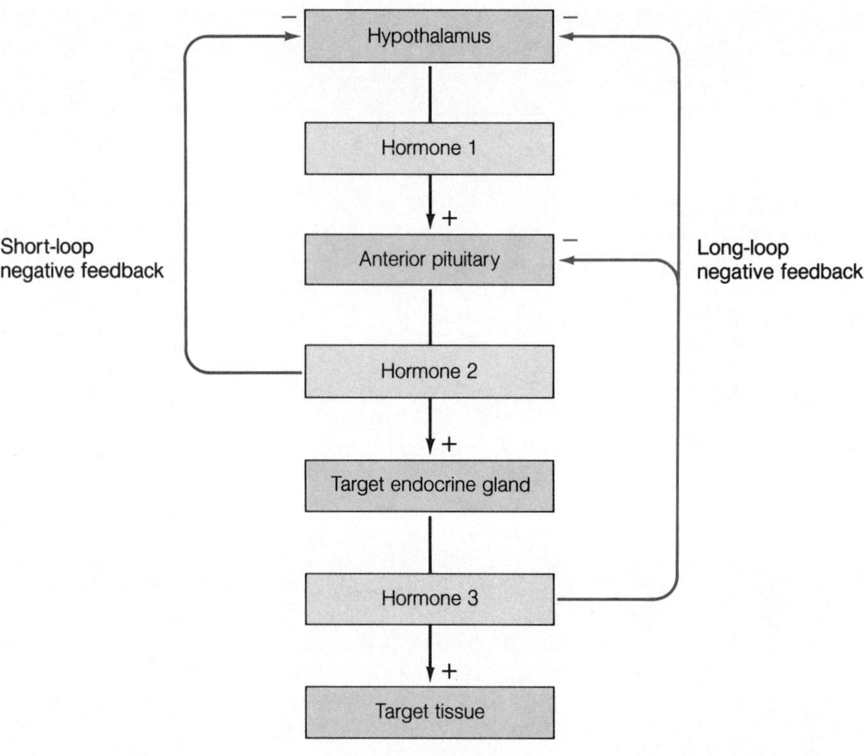

is important to remember that the diurnal rhythms are superimposed on this type of stabilizing negative-feedback regulation. Furthermore, other controlling inputs may break through the negative-feedback control to alter hormone secretion (that is, change the set level) at times of special need.

The one exception to the negative-feedback relationship just described is the long-loop positive-feedback effect of estrogen on LH secretion. This causes an extremely dramatic rise in LH secretion that is responsible for triggering ovulation.

Besides the physiologically more important long-loop feedback mechanism just described, evidence exists for **short-loop negative feedback,** which refers to a pituitary hormone's action at the hypothalamus to inhibit the release of its stimulatory neurohormone (Fig. 18–18). For example, prolactin, which is the only anterior pituitary hormone that does not participate in a three-hormone sequence, is believed to act directly on the hypothalamus to influence secretion of hypophysiotropic hormones that control prolactin secretion. It is possible that anterior pituitary tropic hormones employ similar short-loop feedback on their respective hypophysiotropic neurons. The recent discovery that a substantial portion of the blood flowing from the anterior pituitary may actually go back to the hypothalamus before entering the general circulation suggests that this may be the case. Such retrograde flow of blood makes it plausible that pituitary hormones may reach the brain in sufficiently high concentration to influence releasing-hormone secretion. The actual existence and importance of such mechanisms in humans remains uncertain.

In addition, other hormones ***outside*** of a particular sequence may also exert important influences, either stimulatory or inhibitory, on the secretion of hypothalamic or anterior pituitary hormones within a given sequence. For example, even though estrogen is not in the direct chain of command for prolactin secretion, this sex steroid notably enhances prolactin secretion by the anterior pituitary. This is but one example of the common phenomenon in the endocrine system that one seemingly unrelated hormone can have pronounced effects on the secretion or actions of another hormone.

Hormonal Control of Growth

The detailed functions and control of all of the anterior pituitary hormones with the exception of growth hormone are discussed elsewhere in conjunction with the peripheral target tissues that they influence; for example, thyroid-stimulating hormone is covered in the next chapter with the discussion of thyroid gland. Accordingly, growth hormone is the only centrally produced hormone that will be elaborated on at this time.

Growth is dependent on growth hormone but is influenced by genetic, environmental, and other hormonal factors as well.

In growing children, continuous net protein synthesis occurs under the influence of growth hormone as the body steadily gets larger. Weight gain alone is not synonymous with growth, because weight gain may occur as a result of retention of excess H_2O or fat without true structural growth of tissues. Growth requires net synthesis of proteins and includes lengthening of the long bones (the bones of the extremities) as well as increases in the size and numbers of cells in the soft tissues throughout the body.

Although, as the name implies, growth hormone, or somatotropin, is absolutely essential for growth, it alone is not wholly responsible for determining the rate and final magnitude of growth in a given individual. An individual's maximum growth capacity is ***genetically determined***. Attainment of this full growth potential further depends on the following:

1. *An* ***adequate diet***, including sufficient total protein and ample essential amino acids to accomplish the protein synthesis necessary for growth. Malnourished children never achieve their full growth potential. The growth-stunting effects of inadequate nutrition are most profound when they occur in infancy. In severe cases, the child may be locked in to irreversible stunting of body growth and brain development. About 70% of the total growth of the brain occurs in the first two years of life. On the other hand, one cannot exceed his or her genetically determined maximum by eating more than an adequate diet. The excess food intake produces obesity instead of growth.
2. *Freedom from chronic disease and stressful environmental conditions*. Stunting of growth under these circumstances is due in large part to the prolonged stress-induced secretion of cortisol from the adrenal cortex. Cortisol exerts several potent antigrowth effects. It promotes protein breakdown; inhibits growth in the long bones; inhibits DNA synthesis in a number of tissues (DNA in turn is responsible for directing protein synthesis); and blocks the secretion of growth hormone. Even though sickly or stressed children do not grow well, if the condition is rectified before adult size is achieved, they can rapidly catch up to their normal growth curve through a remarkable spurt in growth.
3. *A normal milieu of growth-influencing hormones*. In addition to the absolutely essential growth hormone, other hormones, including thyroid hormone, insulin, and the sex hormones, play secondary roles in the promotion of growth.

The rate of growth is not continuous nor are the factors responsible for promoting growth the same throughout the growth period. *Fetal growth* seems to be largely independent of hormonal control, with the size at birth being determined principally by genetic and environmental factors. Hormonal factors begin to play an important role in the regulation of growth after birth. Genetic and nutritional factors also strongly affect growth during this period.

Children display two periods of rapid growth—a ***postnatal growth spurt*** during their first two years of life and a ***pubertal growth spurt*** during adolescence (Fig. 18–19). From the age of two until puberty, there is a progressive decline in the ***rate*** of linear growth, even though the child is still growing. Before puberty there is little sexual difference in height or weight. During puberty, a marked acceleration in linear growth takes place because of the lengthening of the long bones. Puberty begins at about age eleven in girls and thirteen in boys and lasts for several years in both sexes. The mechanisms responsible for the pubertal growth spurt are not clearly understood. Apparently, both genetic and hormonal factors are involved. Some evidence indicates that growth hormone secretion is elevated during puberty and thus may contribute to the acceleration of growth during this time. Furthermore, **androgens** ("male" sex hormones), whose secretion increases dramatically at puberty, also contribute to the pubertal growth spurt by promoting protein synthesis and bone growth. The potent androgen from the male testes, testosterone, is of greatest im-

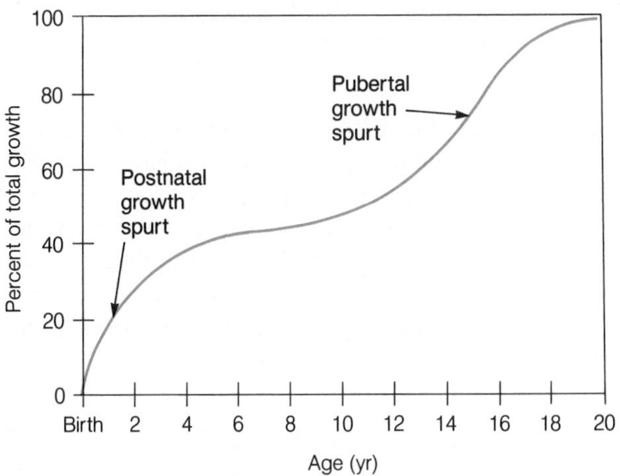

Figure 18–19 Normal Growth Curve

portance in promoting a sharp increase in height in adolescent boys, whereas the less potent adrenal androgens from the adrenal gland, which also manifest a sizable increase in secretion during adolescence, are most likely important in the female pubertal growth spurt. Although estrogen secretion by the ovaries also commences during puberty, it is unclear what role this "female" sex hormone may play in the pubertal growth spurt in girls. There is no doubt that testosterone and estrogen both ultimately act on bone to halt its further growth so that full adult height is attained by the end of adolescence.

Growth hormone is essential for growth, but it also exerts metabolic effects not related to growth.

Growth hormone is the most abundant hormone produced by the anterior pituitary, even in adults in whom growth has already ceased. The continued high secretion of growth hormone beyond the growing period implies that this hormone must have important influences other than on growth. Its growth-promoting effects are fairly well understood. Its metabolic actions not related to growth are known, but their physiological role remains somewhat nebulous.

ACTIONS UNRELATED TO GROWTH. Opposite to insulin's actions, growth hormone increases fatty acid levels in the blood by enhancing the breakdown of triglyceride fat stored in adipose tissue, and it increases blood glucose levels by decreasing glucose uptake by muscles. Muscles use the mobilized fatty acids instead of glucose as a metabolic fuel. Thus the overall metabolic effect of growth hormone is to mobilize fat stores as a major energy source while conserving glucose for glucose-dependent tissues such as the brain. Furthermore, growth hormone enhances body proteins by promoting protein synthesis. This metabolic pattern is suitable for maintaining the body during prolonged fasting, which may well be one of growth hormone's important roles.

GROWTH-PROMOTING ACTIONS. When tissues are responsive to its growth-promoting effects, growth hormone stimulates growth of both soft tissues and the skeleton. Growth of soft tissues is accomplished by promoting proliferation of connective tissue throughout the body; increasing the number of cells (**hyperplasia**) by stimulating cell division; and increasing the size of cells (**hypertrophy**) by favoring synthesis of proteins, the main structural component of cells.

ACTIONS ON PROTEIN. Growth hormone stimulates almost all aspects of protein synthesis while it simultaneously inhibits protein degradation. It promotes the uptake by cells of amino acids (the raw materials for protein synthesis), decreasing blood amino acid levels in the process. Furthermore, it stimulates the replication of nuclear DNA and the synthesis of RNA and ribosomes (the cytoplasmic factories for producing protein according to the DNA blueprint), and it increases protein synthesis by ribosomes.

ACTIONS ON BONES. Growth of the long bones resulting in increased height is the most dramatic effect of growth hormone. **Bone** is a living tissue. Being a form of connective tissue, it consists of cells and an extracellular matrix or scaffolding that is produced by the cells. The types of cells in bone that secrete the organic matrix components are known as **osteoblasts** ("bone-formers"). The matrix, which is composed of collagen fibers and mucopolysaccharide ground substance, has a rubbery consistency and is responsible for the resilience of bone when tension is applied. The unique feature of bone is that it is made hard by precipitation of calcium salts within the ground substance of the matrix. **Cartilage** is similar to bone, except that living cartilage is not calcified and has no blood vessels of its own.

A long bone basically consists of a fairly uniform cylindrical shaft, the **diaphysis,** with a flared articulating knob at either end, an **epiphysis.** In a growing bone, the diaphysis is separated at each end from the epiphysis by a layer of cartilage known as the **epiphyseal plate** (Fig. 18-20a). Growth hormone promotes lengthening of long bones by stimulating activity within the epiphyseal plates.

Growth of long bones in length is accomplished by a different mechanism than growth in circumference. Growth in circumference is achieved by the addition of new bone on top of the already existing bone on the outer surface. This occurs through activity of osteoblasts within the **periosteum,** a connective-tissue sheath that covers the outer bone surface. As new bone is being deposited by osteoblast activity on the external surface, other cells within the bone, the **osteoclasts** ("bone-breakers"), dissolve the bony tissue on the inner surface adjacent to the marrow cavity. In this way, the marrow cavity is enlarged to keep pace with the increase in the circumference of the bone shaft.

Bones grow in length as a result of proliferation of the cartilage cells in the epiphyseal plates (Fig. 18–20b). During growth, new cartilage cells (**chondrocytes**) are produced through cell division on the outer edge of the plate adjacent to the epiphysis. As new chondrocytes are being formed on the epiphyseal border, the older cartilage cells toward the diaphyseal border are enlarging. The epiphyseal plate temporarily increases in width because of this combination of proliferation of new cartilage cells and hypertrophy of maturing chondrocytes. This thickening of the intervening cartilaginous plate causes the bony epiphysis to be pushed farther away from the diaphysis. Soon the matrix surrounding the old-

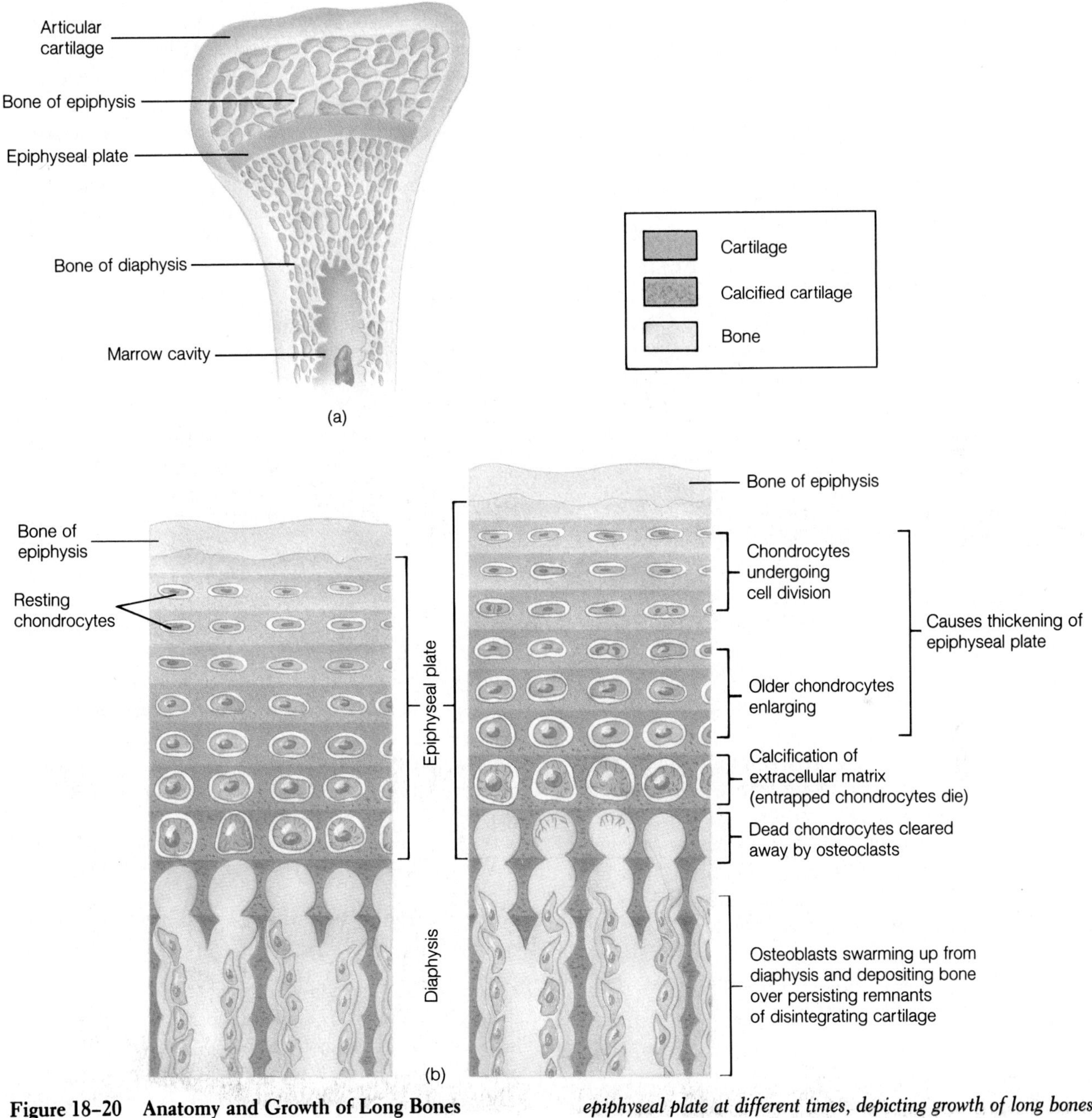

Figure 18–20 Anatomy and Growth of Long Bones *(a) Anatomy of long bones. (b) Two sections of the same epiphyseal plate at different times, depicting growth of long bones in length.*

est hypertrophied cartilage becomes calcified. Since cartilage lacks its own capillary network, the survival of cartilage cells depends on diffusion of nutrients and O_2 through the ground substance, a process that is prevented by the deposition of calcium salts. As a result, the old nutrient-deprived cartilage cells on the diaphyseal border die. As osteoclasts clear away the dead chondrocytes and the calcified matrix that imprisoned them, the area is invaded by osteoblasts, which swarm upward from the diaphysis, trailing their capillary supply with them. These new tenants lay down bone around the persisting remnants of disintegrating cartilage, until the inner region of cartilage on the diaphyseal side of the plate is entirely replaced by bone. As this **ossification** (bone-forming) process is completed, the bone on the diaphyseal side has increased in length and the epiphyseal plate has been returned to its original thickness. The cartilage that has been replaced by bone on the diaphyseal end of the plate is equivalent in thickness to the new cartilaginous growth on the epiphyseal end of the plate. Thus growth of bone is made possible by the growth and death of cartilage, which acts like a "spacer" to push the epiphysis farther out while it provides a framework for future formation of bone on the end of the diaphysis.

As the extracellular matrix produced by an osteoblast becomes calcified, the osteoblast becomes entombed by the matrix that it has deposited around itself, like its chondrocyte predecessor. Unlike chondrocytes, however, osteoblasts trapped within a calcified matrix do not die because they are supplied by nutrients transported to them through small channels, or **canaliculi.** Osteoblasts form these life-supporting channels themselves by sending out cytoplasmic processes around which the bony matrix is deposited. This leaves within the final bony product a network of permeating tunnels that radiate from each entrapped osteoblast, serving as a lifeline system for nutrient delivery and waste removal. The entrapped osteoblasts, which are now called **osteocytes,** retire from active bone-forming duty, since they are prevented from laying down any new bone because of their imprisonment. However, they are involved in the hormonally regulated exchange of calcium between bone and the blood. This exchange is under the control of parathyroid hormone (to be discussed in the next chapter), not growth hormone.

Growth hormone promotes skeletal growth by stimulating the proliferation of epiphyseal cartilage, thereby making space for more bone formation. It also stimulates osteoblast activity. Growth hormone is able to promote lengthening of long bones as long as the epiphyseal plate remains cartilaginous or is "open." At the end of adolescence, under the influence of the sex hormones, these plates completely ossify or "close" so that the bones can grow no further in length despite the presence of growth hormone. Thus, after the plates are closed, the individual does not grow any taller.

Growth hormone exerts its growth-promoting effects indirectly by stimulating somatomedins.

The growth-promoting actions of growth hormone (enhanced protein synthesis and increased cell division) are not accomplished directly by growth hormone's effect on the target tissues. These effects are directly brought about by peptide mediators known as **somatomedins,** whose synthesis in turn is induced by growth hormone. These peptides are also referred to as **insulin-like growth factors (IGF)** because they are structurally and functionally similar to insulin. Two somatomedins—IGF I and IGF II—have been identified.

The major site of somatomedin production is the liver, which releases this peptide product into the blood. However, somatomedin production has also been demonstrated in a variety of other tissues. It has been proposed that somatomedins that are produced locally in target tissues may act through paracrine means for at least some of the growth hormone–induced effects. This could account for the fact that blood levels of growth hormone are no higher, and indeed circulating somatomedin levels are lower, during the first several years of life compared to adult values, even though growth is quite rapid during the postnatal period. Local production of somatomedins in target tissues may possibly be more important than delivery of blood-borne somatomedins during this time.

To complicate the situation further, the production and activity of somatomedins are also controlled by a number of factors other than growth hormone, including age and nutritional status. As a result, changes in circulating somatomedin levels do not always coincide with changes in growth hormone secretion. For example, fasting decreases IGF I levels even though it increases growth hormone secretion. Furthermore, secretion of the two different somatomedins do not always parallel each other. A dramatic increase in IGF I levels accompanies the moderate increase in growth hormone at puberty, which may, of course, be an important factor in the pubertal growth spurt. In contrast, IGF II concentrations do not appear to increase during puberty. These few examples serve to illustrate the confusing state of knowledge regarding the role and control of these important growth-hormone mediators.

Growth hormone secretion is regulated by two hypophysiotropic hormones.

Two antagonistic regulatory hormones from the hypothalamus are involved in the control of growth hormone secretion. These are growth–hormone releasing hormone (GHRH), which is stimulatory, and growth–hormone inhibiting hormone (GHIH, somatostatin), which is inhibitory (Fig. 18–21). (Note the distinction between *somatotropin,* alias growth hormone; *somatomedin,* the liver hormone that directly mediates the effects of growth hormone; and *somatostatin,* which inhibits growth hormone secretion.) Any factor that alters growth hormone secretion could theoretically do so either by stimulating GHRH release or inhibiting GHIH release. It is not known which of these pathways is used in each specific case.

As with the other hypothalamo–anterior pituitary axes, negative-feedback loops participate in the regulation of growth hormone secretion. Both growth hormone and the

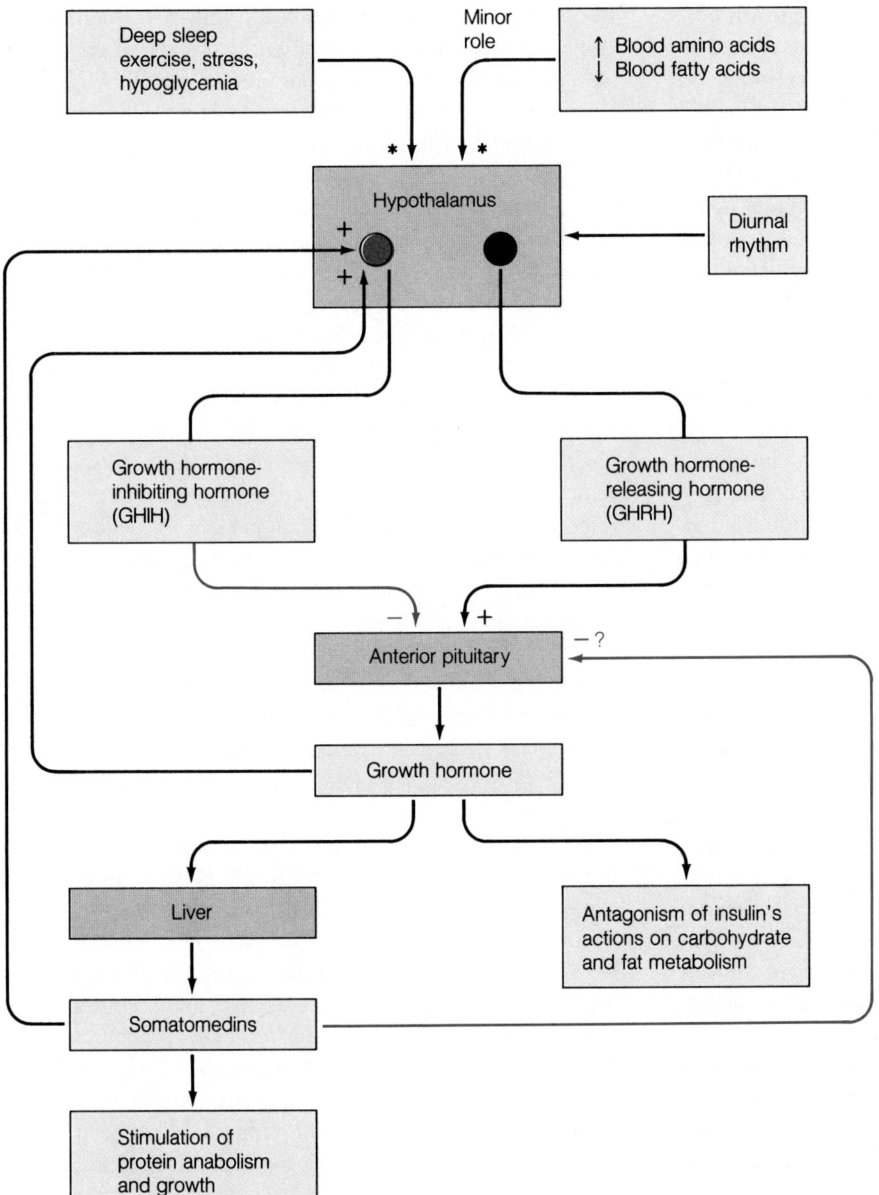

Figure 18–21 Control of Growth Hormone Secretion

somatomedins are believed to inhibit pituitary secretion of growth hormone by stimulating GHIH release from the hypothalamus. The somatomedins may also exert direct effects on the anterior pituitary to inhibit the effects of GHRH on growth hormone release.

Interacting with or overriding the basic negative-feedback control system are a number of factors that influence growth hormone secretion. Growth hormone secretion displays a well-characterized diurnal rhythm. Through most of the day, growth hormone levels tend to be low and fairly constant, although some increases may occur after meals. However, approximately one hour after the onset of deep sleep, growth hormone secretion markedly increases up to five times the daytime value, then rapidly drops over the next several hours.

Superimposed on this diurnal undulation in growth hormone secretion, further bursts in secretion can be elicited by several stimuli. Exercise, stress, and hypoglycemia (low blood glucose) all stimulate growth hormone secretion, the benefit of which is presumably the provision of fatty acids as an alternative energy source for muscle and the conservation of glucose for the brain. Also, because growth hormone uses up fat stores and promotes synthesis of body proteins, it encourages a change in body composition away from adipose deposition toward an increase in muscle protein. Accordingly, the increase in growth hormone secretion that accompanyies exercise may at least in part mediate exercise's effects in reducing the percentage of body fat while increasing the lean body mass. An abundance of blood amino acids, such as after a high protein meal, also enhances the secretion of growth hormone, which in turn promotes the use of these amino acids for protein synthesis. Growth hormone release is also stimulated by a decline in the level of blood fatty acids. Because of the fat-mobilizing actions of growth hormone, such regulation contributes to the maintenance of fairly constant blood fatty acid levels.

Note that the known regulatory inputs for growth hormone secretion are aimed at adjusting the levels of glucose, fatty acids, and amino acids in the blood. There are no known growth related signals that influence growth hormone secretion. The whole issue of what really controls growth is complicated by the fact that growth hormone levels during early childhood, a period of quite rapid linear growth, are similar to those found in normal adults. As mentioned earlier, the poorly understood control of somatomedin activity may be important in this regard. Another related question is why aren't tissues still responsive to growth hormone's growth-promoting effects in adulthood? We know that we do not grow any taller after adolescence because the epiphyseal plates have closed, but why don't the soft tissues continue to grow through hypertrophy and hyperplasia under the influence of growth hormone? One speculation is that the levels of growth hormone may only be high enough to produce its growth-promoting effects during the bursts in secretion that occur in deep sleep. It is interesting to note that time spent in deep sleep is greatest in infancy and gradually declines with age. However, even as we age we still spend some time in deep sleep, yet we do not gradually get larger. Further research will be needed to unravel some of these mysteries.

Abnormal growth hormone secretion results in aberrant growth patterns.

Diseases related to both growth hormone deficiency and excess can occur. The effects on the pattern of growth are much more pronounced than the metabolic consequences.

GROWTH HORMONE DEFICIENCY. Growth hormone deficiency may occur as an isolated entity or may be associated with deficiencies in other anterior pituitary hormones. It may be caused by a primary pituitary lesion or occur secondarily to hypothalamic dysfunction. Hyposecretion of growth hormone in a child results in **dwarfism.** The predominant feature is short stature caused by retarded skeletal growth. Less obvious characteristics include poorly developed musculature (reduction in muscle protein synthesis) and excess subcutaneous fat (less fat mobilization).

In addition, growth may be thwarted because of a failure of the tissues to respond normally to growth hormone. This is well illustrated by **Laron dwarfism.** The symptoms resemble those of severe growth hormone deficiency even though blood growth-hormone levels are actually high. In some instances, growth hormone levels are adequate and target-tissue responsiveness is normal but somatomedins are deficient. African pygmies are an interesting example. Their short stature is attributable to a genetic deficit of the most potent of the somatomedins.

The onset of growth hormone deficiency in adulthood after growth is already complete produces relatively few symptoms. Growth hormone–deficient adults tend to have reduced muscle strength (less muscle protein) as well as decreased bone density (less osteoblast activity during ongoing bone remodeling).

Growth hormone is species specific; that is, unlike most hormones, the structure and activity of GH differs among species. As a result, GH from animal sources is ineffective in the treatment of GH deficiency in humans. The only source of human GH in the past was pituitary glands from human cadavers. This source never was adequate, and then it was removed from the market because of fear of viral contamination. Recently, however, the supply of human GH has become unlimited through the technique of genetic engineering. The gene that directs the synthesis of human GH has been introduced into bacteria, converting them into GH-synthesizing factories.

GROWTH HORMONE EXCESS. Hypersecretion of growth hormone is most often caused by a functional tumor of the growth hormone–producing cells of the anterior pituitary. The symptoms that result depend on the age of the individual when the abnormal secretion commences. If overproduction of growth hormone begins in childhood before the epiphyseal plates are closed, the principal manifestation of the disorder is a rapid growth in height without distortion of body proportions. Appropriately, this condition is known as **gigantism.** If not treated by removal of the tumor or drugs that block the effect of growth hormone, the individual may reach a height of eight feet or more. All of the soft tissues grow correspondingly, so the body is still well-proportioned.

If growth hormone hypersecretion occurs after adolescence when the epiphyseal plates have already closed, further growth in height is prevented. Under the influence of excess growth hormone, however, the bones become thicker and the soft tissues, especially connective tissue and skin, proliferate. This disproportionate growth pattern produces a disfiguring condition known as **acromegaly.** Bone thickening is most obvious in the face and extremities. A marked coarsening of the features to an almost apelike appearance gradually develops as the jaws and cheekbones become more prominent because of the thickening of the facial bones and the skin (Fig. 18–22). The hands and feet enlarge, and the fingers and toes become greatly thickened. Peripheral nerve disorders often occur as a result of entrapment of nerves by overgrowth of connective tissue or bone or both.

Visual disturbances may accompany either gigantism or acromegaly. This is not due to excessive growth hormone per se but, rather, due to the fact that a growth hormone–secreting tumor causes enlargement of the anterior pituitary gland, which lies in close proximity to the optic chiasm, the point at which the nerve fibers passing from the eyes cross over on their way to the visual cortex of the brain (Fig. 18–12b).

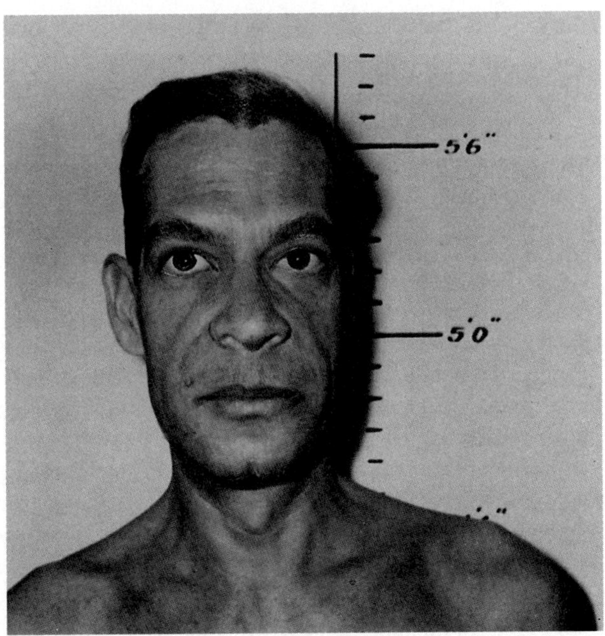

Figure 18–22 Patient with Acromegaly

Other hormones besides growth hormone are essential for normal growth.

Several other hormones in addition to growth hormone contribute in special ways to overall growth.

□ *Thyroid hormone* is essential for growth but is not itself directly responsible for producing growth-promoting effects. It plays a permissive role in promoting skeletal growth; the actions of growth hormone are fully manifested only in the presence of adequate amounts of thyroid hormone. As a result, growth is severely stunted in hypothyroid children, but hypersecretion of thyroid hormone does not cause excessive growth.

□ *Insulin* is believed to be an important growth-promoting factor, as is evidenced by a high correlation between insulin levels and growth. Growth failure often accompanies insulin deficiency, and hyperinsulinism is frequently associated with excessive growth. Since insulin promotes protein synthesis, its growth-promoting effects should not be surprising. However, insulin's growth-promoting effects may also arise from a different mechanism other than its direct effect on protein synthesis. Insulin is structurally similar to the somatomedins and may interact with the somatomedin (IGF I) receptor, which is very similar to the insulin receptor.

□ *Androgens*, which are believed to play an important role in the pubertal growth spurt, are powerful stimulants of protein synthesis in many organs. Androgens stimulate linear growth, promote weight gain, and increase muscle mass. The most potent androgen, testicular testosterone, is responsible for the development of the heavier musculature in men as compared to women. These androgenic growth-promoting effects depend on the presence of growth hormone. Androgens have virtually no effect on body growth in the absence of growth hormone, but in its presence, they synergistically enhance linear growth. Although androgens stimulate growth, they ultimately stop further growth by promoting closure of the epiphyseal plates.

□ *Estrogens*, like androgens, ultimately terminate linear growth by stimulating complete conversion of the epiphyseal plates to bone. However, the effects of estrogens on growth prior to bone maturation are not well understood. Some studies suggest that large doses of estrogen may even inhibit further body growth.

Several factors contribute to the average height differences between men and women. First, since puberty occurs about two years earlier in girls than in boys, on the average boys have two more years of prepubertal growth than girls. As a result, boys are usually several inches taller than girls at the start of their respective growth spurts. Second, the greater androgen-induced growth spurt in boys than in girls before their respective gonadal steroids seal their long bones from further growth results in greater heights in men than in women on the average. Third, recent evidence suggests that androgens "imprint" the brains of males during development, giving rise to a "masculine" secretory pattern of growth hormone characterized by higher cyclical peaks, which are speculated to contribute to the greater height of males.

In addition to these hormones that exert overall effects on body growth, a number of poorly understood peptide ***growth factors*** have been identified that stimulate mitotic activity of specific tissues (for example, epidermal growth factor).

Chapter in Perspective

The endocrine system is one of the two major control systems of the body, the other being the nervous system. The endocrine system generally regulates activities that require duration rather than speed, most of which are directed toward maintaining homeostasis. Endocrine organs release hormones, the body's "wireless" chemical messengers that are responsible for controlling growth and metabolism, H_2O and electrolyte balance, reproduction, and adaptation to stress. Each hormone is secreted by a particular endocrine gland into the blood, which disperses it throughout the body. Only the target tissues specific for each hormone are influenced by the hormone's presence in the blood, because these tissues alone have unique receptors for binding the hormone. Specificity of endocrine communication is therefore dependent on specialization of target-tissue receptors. The "lock and key" fit between a hormone and its target tissue's receptors triggers a prescribed response in the target tissue.

Hormones are grouped into three categories—peptides, amines, and steroids—based on differences in their mode of synthesis, storage, secretion, transport in the blood, and interaction with target tissues. No matter what mechanism is employed by the hormonal system, the end result is almost always the same—an alteration in protein activity, usually involving enzymes crucial to a particular metabolic pathway. Occasionally the end result is an alteration in structural proteins or a change in permeability of the target-tissue cells.

The effective plasma concentration of each hormone is normally controlled by regulated changes in the rate of hormone secretion. Secretory output of endocrine cells is influenced by one or more of three different types of direct regulatory inputs: (1) neural input, which forms one branch of neuroendocrine relationships, the other being the reverse situation of hormones influencing neural activity; (2) input from another hormone, which involves either stimulatory input from a tropic hormone or inhibitory input from a target-tissue hormone in negative-feedback fashion; and (3) changes in the plasma concentration of an organic nutrient or an electrolyte being regulated by the hormone, also acting in negative-feedback fashion. In all cases, the regulated secretory output is designed to accomplish particular adjustments needed to maintain homeostasis, promote growth, or control reproduction.

Endocrine dysfunction arises when too much or too little of any particular hormone is secreted. Symptoms are generally referable to exaggerated or insufficient target-tissue responses normally controlled by the hormone.

The hypothalamus, a portion of the brain, secretes nine peptide neurohormones, two of which (vasopressin and oxytocin) are stored in the posterior pituitary and released upon hypothalamic stimulation. Vasopressin is important for H_2O balance; oxytocin plays an important role in reproductive activities. The seven other hypothalamic hormones are carried through a special vascular link to the anterior pituitary gland, where they either stimulate or inhibit the release of particular anterior pituitary hormones.

The anterior pituitary secretes six peptide hormones. With the exception of prolactin, which stimulates milk secretion, the five other anterior pituitary hormones stimulate and maintain other endocrine tissues; that is, they are tropic. The sole function of thyroid-stimulating hormone (TSH) is to stimulate secretion of thyroid hormone, and that of adrenocorticotropic hormone (ACTH) is to stimulate secretion of cortisol by the adrenal cortex. The gonadotropic hormones—follicle-stimulating hormone (FSH) and luteinizing hormone (LH)—stimulate production of gametes (eggs and sperm) as well as secretion of sex hormones. Growth hormone stimulates growth indirectly by promoting the liver's production of somatomedins, which act directly on bone and soft tissue to cause growth. Growth hormone exerts direct metabolic effects on the liver, adipose tissue, and muscle. In general, growth hormone enhances proteins, conserves carbohydrates, and uses up fat stores.

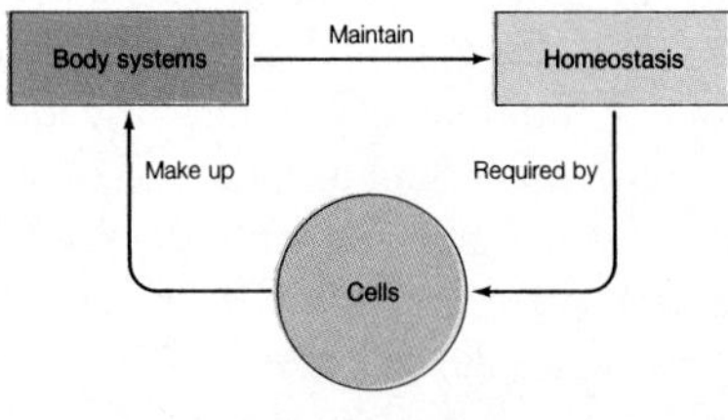

See inside front cover for an expanded version of this model.

Review Exercises

1. List and describe each type of cell-to-cell communication.
2. Compare the endocrine and nervous systems in terms of specificity of communication, anatomical organization, and mode of action.
3. List the overall functions of the endocrine system.
4. Compare the three categories of hormones in terms of chemical structure; mechanisms of synthesis, storage, and secretion; transport in the blood; and interaction with target cells.
5. How is the effective plasma concentration of a hormone normally regulated? What nonregulated factors influence a hormone's effective plasma concentration?
6. Describe the general ways in which endocrine disorders arise.
7. In what ways can the responsiveness of a target tissue to its hormones be varied?
8. Distinguish anatomically between the neurohypophysis and the adenohypophysis.
9. List and briefly state the functions of the posterior pituitary hormones.
10. List and briefly state the functions of the anterior pituitary hormones.
11. Compare the relationship between the hypothalamus and posterior pituitary with the relationship between the hypothalamus and anterior pituitary. Describe the role of the hypothalamus-hypophyseal portal system and the hypothalamic releasing and inhibiting hormones.
12. Distinguish between long-loop and short-loop negative feedback.
13. Discuss the general factors that determine the rate and final magnitude of growth in a given individual.
14. Describe the actions of growth hormone that are unrelated to growth. What are growth hormone's growth-promoting actions? What is the role of somatomedins?
15. Describe the growth of long bones in circumference and in length.
16. Discuss the control of growth hormone secretion.
17. What diseases are attributable to growth hormone dysfunction?
18. **A point to ponder:** Even though the problem of adequacy of growth hormone supply has been solved by means of genetic engineering (see p. 658), the solution has given rise to ethical problems for the medical community regarding the circumstances under which growth-hormone treatment is appropriate. Obviously, growth hormone should be administered to patients with growth hormone deficiency, but what about those who have no growth hormone deficiency yet are desirous of its growth-promoting actions for "cosmetic" reasons, such as normally growing teenagers who wish to attain even greater stature? What do you think the medical community's stance should be?

C H A P T E R 19

Peripheral Endocrine Organs

Chapter Contents at a Glance

Introduction

What do the following individuals have in common?

☐ *a person with bulging goldfish-like eyes*

☐ *a bearded woman*

☐ *a person with a "moon face" and a "buffalo hump"*

☐ *a person with "poly disease" characterized by polyuria (increased urination), polydipsia (increased thirst), and polyphagia (increased appetite)*

☐ *a person with a condition dubbed the disease of "bones, stones, and abdominal groans"*

Each of these persons is suffering from a specific endocrine disorder.

In this chapter we will examine the functions, control, and abnormalities of the peripherally located endocrine organs and will discover the underlying defect responsible for the symptoms in each of these cases.

Thyroid Gland

The major thyroid hormone secretory cells are organized into colloid-filled spheres.

The **thyroid gland** consists of two lobes of endocrine tissue joined in the middle by a narrow portion of the gland called the **isthmus,** giving it a bow tie–shaped appearance (Fig. 19–1a). The gland is even located in the appropriate place for a bow tie, lying over the trachea just below the larynx. The major thyroid secretory cells are arranged into hollow spheres, each of which forms a functional unit called a **follicle.** Consequently, these secretory cells are often referred to as *follicular cells*. On a microscopic section (Fig. 19–1b) the follicles appear as rings of follicular cells enclosing an inner lumen filled with **colloid,** a substance that serves as an "inland" extracellular storage site for thyroid hormones.

The chief constituent of the colloid is a large, complex glycoprotein known as **thyroglobulin,** within which are incorporated the thyroid hormones in their various stages of synthesis. The follicular cells produce two iodine-containing hormones derived from the amino acid tyrosine; these are **tetraiodothyronine (T_4 or thyroxine)** and **triiodothyronine (T_3)**. The prefixes "tetra" and "tri" and the subscripts "4" and "3" denote the number of iodine atoms incorporated into each of these hormones, respectively. These two hormones, collectively referred to as **thyroid hormone,** are important regulators of overall basal metabolism in most tissues.

Interspersed in the interstitial spaces between the follicles is another secretory cell type, the **C cells,** so-called because they secrete the peptide hormone **calcitonin,** which plays a role in calcium metabolism. Calcitonin is not related in any way to the two other major thyroid hormones. The follicular cells and C cells are distinctively different cell types; they have no apparent functional relationship to each other, and they secrete in response to totally different stimuli. We will reserve our discussion of thyroid hormones to the secretions of the follicular cells, deferring coverage of calcitonin until a later section dealing with endocrine regulation of calcium balance.

All of the steps of thyroid hormone synthesis occur on the large thyroglobulin molecule, which subsequently stores the hormones.

The basic ingredients for thyroid hormone synthesis are tyrosine and iodine, both of which must be sequestered from the

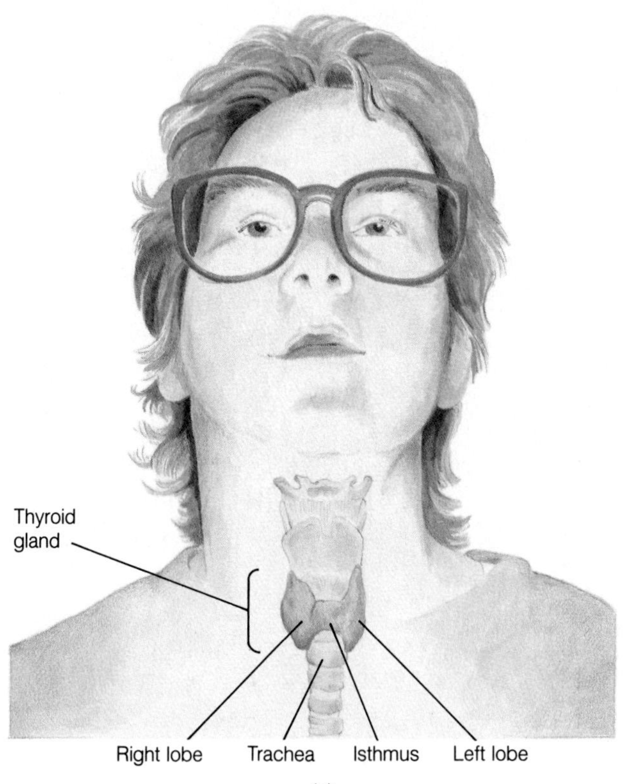

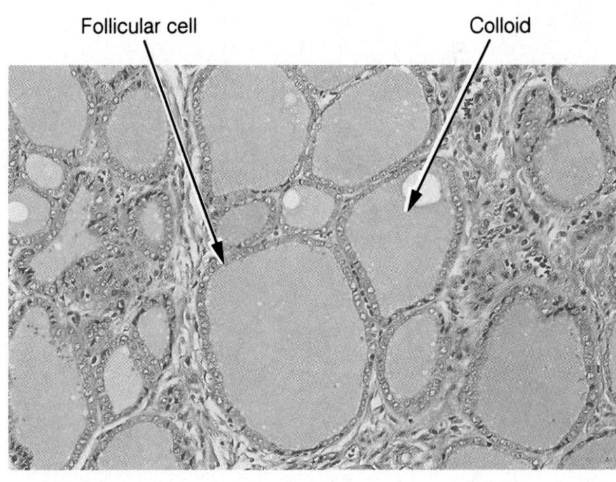

Figure 19–1 Anatomy of the Thyroid Gland *(a) Gross anatomy of thyroid gland, anterior view. The thyroid gland lies over the trachea just below the larynx and consists of two lobes connected by a thin isthmus. (b) Light-microscope appearance of thyroid gland. The thyroid gland is composed primarily of colloid-filled spheres enclosed by a single layer of follicular cells.*

SOURCE: Part (b) courtesy of Elizabeth R. Walker, Associate Professor, and Dennis O. Overman, Associate Professor, Department of Anatomy, School of Medicine, West Virginia University.

blood by the follicular cells. Tyrosine, an amino acid, is synthesized in sufficient amounts by the body, so it is not an essential dietary requirement. The tyrosine molecules used for thyroid hormone synthesis do not exist as free amino acids within the follicular cells. Instead, they become incorporated as part of the much larger thyroglobulin molecules as ***tyrosyl residues***. All of the steps of thyroid hormone synthesis take place on thyroglobulin. Thyroglobulin is produced in the rough endoplasmic reticulum of the follicular cells, then packaged into secretory vesicles in the Golgi complex before being exported into the colloid by exocytosis. Exocytosis of tyrosine-containing thyroglobulin is labeled as step 1 in Figure 19–2, which depicts the steps in the synthesis, storage, and release of thyroid hormones.

The iodine needed for thyroid hormone synthesis must be obtained from dietary intake. This mineral is usually ingested as organic iodine, which is converted to iodide (I^-), the ionized form of iodine, before it is absorbed from the intestine into the blood. Iodide is captured from the blood by the thyroid by means of a very active "iodide pump" or "iodide trapping mechanism," in reference to the powerful, energy-requiring carrier proteins located in the outer membranes of the follicular cells (step 2). Most of the I^- in the body is moved against both concentration and electrical gradients by this mechanism to become trapped in the thyroid for the purpose of thyroid hormone synthesis. Iodine serves no other purpose in the body.

The trapped I^- must be oxidized to molecular iodine (I^o) within the follicular cells before it can be incorporated into thyroid hormones. This is accomplished as I^- diffuses down its concentration gradient into the colloid by specific oxidative enzymes located at the luminal border of the follicular cells adjacent to the colloid (step 3). Once oxidized, I^o is quickly attached to a tyrosyl residue within the thyroglobulin molecule to form **monoiodotyrosine** (**MIT**) (step 4). Addition of another I^o to MIT yields **diiodotyrosine** (**DIT**) (step 5). Next, a coupling process occurs between the iodinated tyrosine molecules to form the thyroid hormones. If two DITs, each bearing two iodine atoms, are coupled, tetraiodothyronine (T_4 or thyroxine), the four-iodine-bearing form of thyroid hormone, is produced (step 6a). On the other hand, one MIT (with one iodine) and one DIT (with two iodines) yield triiodothyronine or T_3 (with three iodines) (step 6b). Coupling does not occur between two MIT molecules.

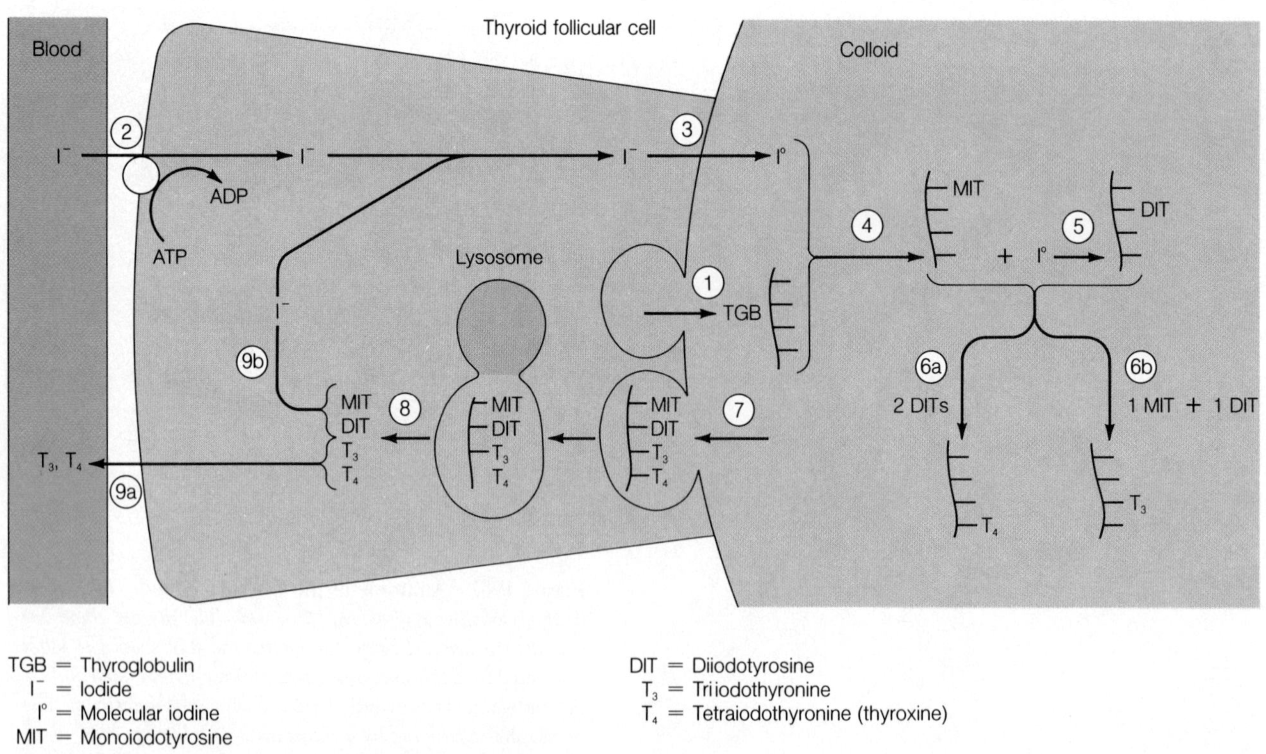

TGB = Thyroglobulin
I^- = Iodide
I^o = Molecular iodine
MIT = Monoiodotyrosine
DIT = Diiodotyrosine
T_3 = Triiodothyronine
T_4 = Tetraiodothyronine (thyroxine)

Figure 19–2 Synthesis, Storage, and Secretion of Thyroid Hormone *See text for an explanation of the numbered steps.*

Because these iodination and coupling reactions resulting in the formation of MIT/DIT and T_4/T_3, respectively, occur within the thyroglobulin molecule, all of these products remain attached as part of this large protein. Thyroid hormones remain stored in this form in the colloid until they are split off and secreted. It is estimated that sufficient thyroid hormone is normally stored in the colloid to supply the body's needs for several months.

The follicular cells phagocytize thyroglobulin-laden colloid to accomplish thyroid hormone secretion.

The release of the thyroid hormones into the systemic circulation requires a rather complex process for two reasons. First, before their release, T_4 and T_3 are still bound within the thyroglobulin molecule. Second, these hormones are stored at an inland extracellular site, the follicular lumen, from which they must be transported completely across the follicular cells before they can enter the blood vessels that course through the interstitial spaces. The process of thyroid hormone secretion essentially involves the follicular cells "biting off" a piece of colloid, breaking the thyroglobulin molecule down into its component parts, and "spitting out" the freed T_4 and T_3 into the blood. Upon appropriate stimulus for thyroid hormone secretion, the follicular cells internalize a portion of the thyroglobulin-hormone complex by phagocytizing a piece of colloid (step 7 of Fig. 19–2). Within the cells, the membrane-bounded droplets of colloid coalesce with lysosomes, whose enzymes split off the biologically active thyroid hormones, T_4 and T_3, as well as the inactive iodotyrosines, MIT and DIT (step 8). The thyroid hormones, being very lipophilic, pass freely through the outer membranes of the follicular cells and into the blood (step 9a). The MIT and DIT are of no endocrine value. If they were to escape into the blood, this would result in wastage of I^- that was originally transported into the thyroid at considerable metabolic expense. To avoid this, the follicular cells contain an enzyme that swiftly deiodinates (removes an I^- from) MIT and DIT, allowing the freed I^- to be recycled for synthesis of more hormone (step 9b). This highly specific enzyme will only deiodinate the worthless MIT and DIT, not the valuable T_4 or T_3.

Most of the secreted T_4 is converted into T_3 outside of the thyroid; T_4 and T_3 are both transported largely bound to specific plasma proteins.

About 90% of the secretory product released from the thyroid gland is in the form of T_4, yet T_3 is considerably more potent in its biological activity. However, most of the secreted T_4 is converted into T_3, or *activated*, by being deiodinated in numerous tissues outside of the thyroid, especially in the liver and kidneys. About 80% of the circulating T_3 is derived from secreted T_4 that has been peripherally stripped of one of its I atoms. Therefore, even though the thyroid gland originally synthesizes and releases all of the thyroid hormone products, extrathyroidal sites play an important role in the production of maximally functional thyroid hormone. Practically speaking, T_3 is the major biologically active form of thyroid hormone at the cellular level.

Once released into the blood, the highly lipophilic thyroid hormone molecules very quickly bind with several plasma proteins, resulting in less than 1% of the T_3 and less than 0.1% of the T_4 remaining in the unbound (free) form. This is remarkable considering that only the free portion of the total thyroid hormone pool has access to the target-tissue receptors and thus is able to exert an effect.

Three different plasma proteins are important in thyroid hormone binding: ***thyroxine-binding globulin*** selectively binds only thyroid hormones—55% of the circulating T_4 and 65% of the T_3—even though its name specifies only "thyroxine" (T_4); ***albumin*** nonselectively binds many lipophilic hormones, including 10% of the T_4 and 35% of the T_3; and ***thyroxine-binding prealbumin*** binds the remaining 35% of the T_4.

Thyroid hormones are the primary determinants of the body's overall metabolic rate and are also important for bodily growth and normal development and function of the nervous system.

Compared to other hormones, the action of thyroid hormone is "sluggish." Only after a delay of several hours is the metabolic response to thyroid hormone first detectable, and the maximal response is not evident for several days. The duration of the response is also quite long, partially because thyroid hormone is not rapidly degraded but also because the response continues to be expressed for days or even weeks after the plasma thyroid hormone concentrations have returned to normal.

Virtually every tissue in the body is affected either directly or indirectly by thyroid hormone. The effects of T_3 and T_4 (the former being about four times more potent) can be grouped into several overlapping categories, which are described in the following sections.

CALORIGENIC EFFECT. Closely related to thyroid hormone's overall metabolic effect is its **calorigenic** or, heat-producing, effect. Since metabolic activity results in heat production, a thyroid hormone–induced increase in metabolic rate brings about an increase in total heat production by the body.

EFFECT ON INTERMEDIARY METABOLISM. In addition to increasing the general metabolic rate, thyroid hormone modulates the rates of many specific reactions involved in intermediary metabolism. The effects of thyroid hormone on the metabolic fuels are multifaceted; not only can it influence both the synthesis and degradation of carbohydrate, fat, and protein, but opposite effects may be induced by small or large amounts of the hormone. For example, the conversion of glucose to glycogen, the storage form of glucose, is facilitated by small amounts of thyroid hormone, but the reverse—the breakdown of glycogen into glucose—occurs with large amounts of the hormone. Similarly, adequate amounts of thyroid hormone are essential for the protein synthesis needed for normal bodily growth, yet protein degradation effects predominate at high doses. In general, at abnormally high plasma levels of thyroid hormone, such as in thyroid hypersecretion, the overall effect is to favor consumption rather than storage of fuel, as manifested by depletion of liver glycogen stores, depletion of fat stores, and muscle wasting resulting from protein degradation. (The main structural component of cells is protein. Muscle cells are particularly rich in structural protein because they are packed full of contractile elements made of protein—that is, the actin and myosin filaments.)

SYMPATHOMIMETIC EFFECT. Many of the symptoms of thyroid hypersecretion are similar to those observed with activation of the sympathetic nervous system (a sympathomimetic effect), indicative of some kind of interaction between these two systems. A large part of this interaction is due to thyroid hormone's ability to increase tissue responsiveness to catecholamines (epinephrine and norepinephrine), the chemical messengers used by the sympathetic nervous system and its hormonal reinforcements from the adrenal medulla. This permissive action presumably is accomplished by thyroid hormone causing a proliferation of specific catecholamine target-tissue receptors.

EFFECT ON THE CARDIOVASCULAR SYSTEM. Through its effect on increasing responsiveness of the heart to circulating catecholamines, thyroid hormone increases heart rate and force of contraction, thus increasing cardiac output. In addition, peripheral vasodilation occurs in response to the heat load generated by the calorigenic effect of thyroid hormone as a means to carry the extra heat to the body surface for elimination to the environment (see p. 616).

EFFECT ON GROWTH AND THE NERVOUS SYSTEM. Thyroid hormone is essential for normal growth and CNS development in children and for normal CNS function in adults. The growth-promoting effect of thyroid hormone seems to be secondary to its effects on growth hormone. Thyroid hormone not only stimulates growth hormone secretion but also serves in a permissive capacity to promote the effects of growth hormone (or somatomedins) on the synthesis of new structural proteins and on skeletal growth. Thyroid-deficient children have stunted growth, which is reversible with thyroid replacement therapy. However, in contrast to excess growth hormone, excess thyroid hormone does not result in excessive growth.

Thyroid hormone also plays a crucial role in the normal development of the nervous system, especially the CNS, an effect impeded in children with thyroid deficiency from birth. Abnormal thyroid hormone levels are also associated with behavioral changes. Furthermore, the conduction velocity of peripheral nerves varies directly with the availability of thyroid hormones.

Thyroid hormone is regulated by the hypothalamo-pituitary-thyroid axis.

Thyroid-stimulating hormone (TSH), the thyroid tropic hormone from the anterior pituitary, is the most important physiological regulator of thyroid hormone secretion (Fig. 19-3). Almost every step of thyroid hormone synthesis and release is enhanced by TSH. In the absence of TSH, the thyroid gland secretes its hormones at a very low rate.

In addition to regulating thyroid hormone secretion, TSH is responsible for maintaining the structural integrity of the thyroid gland. In the absence of TSH, the thyroid atrophies (decreases in size). Conversely, it undergoes hypertrophy (increase in the size of each follicular cell) and hyperplasia (increase in the number of follicular cells) in response to excess TSH stimulation.

Like the other pituitary tropic hormones, TSH is subject to negative-feedback inhibition by thyroid hormone. In the hypothalamo-pituitary-thyroid axis, inhibition is exerted primarily at the level of the anterior pituitary. As with the other negative-feedback loops, this one tends to provide stability to TSH (and thus T_4 and T_3) output.

Although thyroid hormone, in negative-feedback fashion, "turns off" TSH secretion, the hypothalamic **thyrotropin-releasing hormone (TRH)**, in tropic fashion, "turns on" TSH secretion by the anterior pituitary. The hypothalamus may secrete different amounts of TRH to vary the rate of TSH synthesis and release as needed.

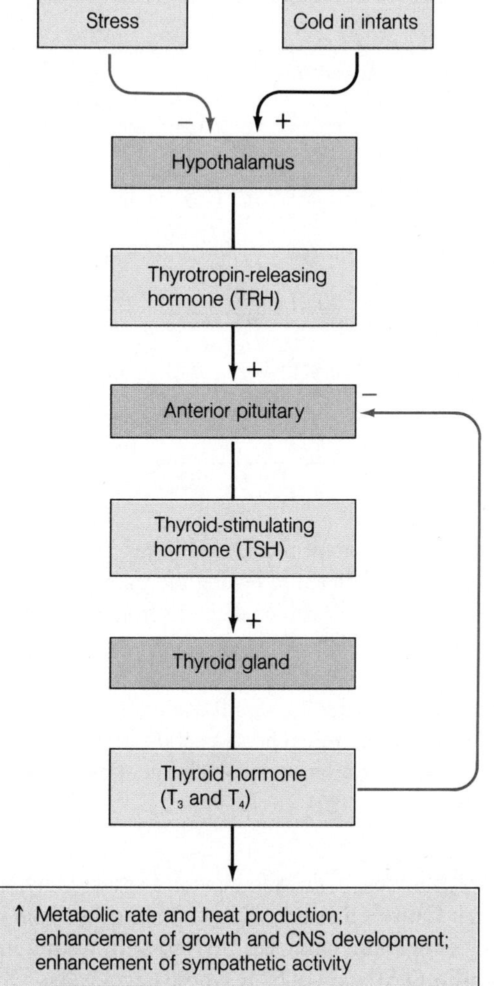

Figure 19–3 Regulation of Thyroid Hormone Secretion

It appears that the day-to-day regulation of free thyroid-hormone levels is accomplished by negative feedback between the thyroid and anterior pituitary and that long range adjustments are mediated by the hypothalamus. Unlike most of the other hormonal systems, the hormones in the thyroid axis in an adult do not normally undergo sudden, wide swings in secretion. The relatively steady rate of thyroid hormone secretion is in keeping with the sluggish, long-lasting responses that this hormone induces; there would be no adaptive value in suddenly increasing or decreasing plasma thyroid-hormone levels.

The hypothalamus is presumably responsible for mediating the few changes in TSH secretion known to be induced by certain environmental conditions. For example, TSH secretion is inhibited by various types of stress, presumably through neural influences on the hypothalamus, although the adaptive importance of this is unclear. In contrast, an example of a highly adaptive neuroendocrine mechanism is the dramatic increase in TSH and, consequently, in heat-producing thyroid hormone secretion in newborn infants in response to exposure to cold. This response is thought to contribute to the maintenance of body temperature in the face of an abrupt drop in surrounding temperature at birth from the mother's warm body to the cooler environmental air. A similar TSH response to cold exposure does not occur in adults, although it makes sense physiologically and does occur in many types of experimental animals.

Abnormalities of thyroid function include both hypothyroidism and hyperthyroidism.

Euthyroidism refers to normal circulating levels of free thyroid hormone. Abnormalities of thyroid function are among the most common of all endocrine disorders. They fall into two major categories—hypothyroidism and hyperthyroidism—reflecting deficient and excess thyroid hormone secretion, respectively. A number of specific causes can give rise to each of these conditions (Table 19–1). Whatever the cause, the consequences of too little or too much thyroid hormone secretion are largely predictable, based on a knowledge of the functions of T_3 and T_4.

HYPOTHYROIDISM. **Hypothyroidism** can result: (1) from primary failure of the thyroid gland itself; (2) secondary to a deficiency of TRH, TSH or both; or (3) from an inadequate dietary supply of iodine. The symptoms of hypothyroidism are largely referable to a reduction in overall metabolic activity. Among other things, a patient with hypothyroidism has a reduced basal metabolic rate; displays poor resistance to cold (lack of calorigenic effect); gains excessive weight (not burning fuels at normal rate); is easily fatigued (not as much energy production); has a slow, weak pulse (caused by a reduced rate and strength of cardiac contraction and a lowered cardiac output); and exhibits slow reflexes and slow mentation (because of the effect on the nervous system). The latter is characterized by diminished alertness, slow speech, and poor memory.

Another notable characteristic is an infiltration of the interstitial spaces of the skin with excess amounts of mucopolysaccharides, perhaps related to altered metabolism. By changing the osmotic relationship between the plasma and interstitial fluid, this results in an edematous condition that creates a puffy appearance, primarily of the face, hands, and feet, known as **myxedema.** In fact, the term *myxedema* is often

Table 19-1 Types of Thyroid Dysfunctions

Thyroid Dysfunction	Cause	Plasma Concentrations of Relevant Hormones	Goiter Present?
Hypothyroidism	Primary failure of thyroid gland	↓ T_3 and T_4; ↑ TSH	Yes
	Secondary to hypothalamic or anterior pituitary failure	↓ T_3 and T_4; ↓ TRH and/or ↓ TSH	No
	Lack of dietary iodine	↓ T_3 and T_4; ↑ TSH	Yes
Hyperthyroidism	Secondary to excess hypothalamic or anterior pituitary secretion	↑ T_3 and T_4; ↑ TRH and/or ↑ TSH	Yes
	Hypersecreting thyroid tumor	↑ T_3 and T_4; ↓ TSH	No
	Abnormal presence of thyroid-stimulating immunoglobulin (TSI) (Grave's disease)	↑ T_3 and T_4; ↓ TSH	Yes

used as a synonym for hypothyroidism in an adult because of the prominence of this symptom.

If an individual has hypothyroidism from birth, a condition known as **cretinism** develops. Because of the dependence of normal growth and CNS development on the presence of adequate levels of thyroid hormone, cretinism is characterized by dwarfism and mental retardation as well as other general symptoms of thyroid deficiency. The mental retardation is preventable if replacement therapy is started promptly, but it is not reversible once it has developed for a few months after birth, even with later treatment with thyroid hormone. For this reason, an important public health measure is the recent establishment of neonatal screening programs in which congenital hypothyroidism can be detected from a single drop of a newborn infant's blood.

Treatment of hypothyroidism, with one exception, consists of replacement therapy through the administration of exogenous thyroid hormone. The exception is hypothyroidism caused by iodine deficiency, in which the remedy is adequate dietary iodine.

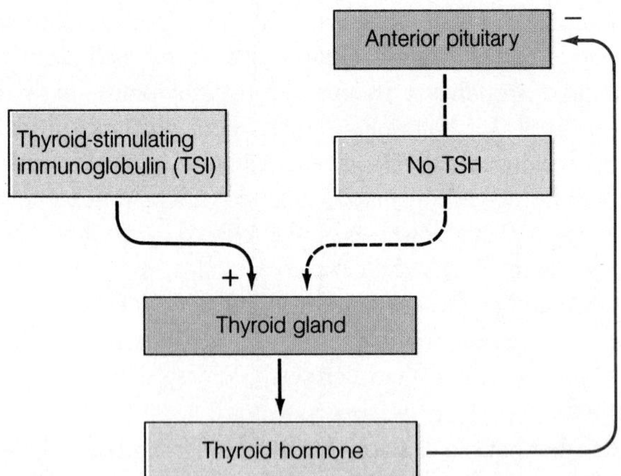

Figure 19-4 Role of Thyroid-Stimulating Immunoglobulin in Grave's Disease *Thyroid-stimulating immunoglobulin, an antibody erroneously produced in the autoimmune condition of Grave's disease, binds with the TSH receptors on the thyroid gland and continuously stimulates thyroid hormone secretion outside the normal negative-feedback control system.*

HYPERTHYROIDISM. The most common cause of **hyperthyroidism** is **Grave's disease,** an autoimmune disease in which the body erroneously produces **thyroid-stimulating immunoglobulin** (**TSI**), an antibody whose target is the TSH receptors on the thyroid cells. Thyroid-stimulating immunoglobulin stimulates both secretion and growth of the thyroid in a way similar to TSH. Unlike TSH, however, it is not subject to negative-feedback inhibition by T_3 and T_4 so thyroid secretion and growth continue unchecked (Fig. 19-4). Less frequently, hyperthyroidism occurs secondary to excess TRH or TSH or in association with a hypersecreting thyroid tumor.

As expected, the hyperthyroid patient has an elevated basal metabolic rate. The resultant increase in heat production leads to excessive perspiration and poor tolerance of heat. In spite of the increased appetite and food intake that occur in response to the increased metabolic demands, body weight still typically falls because of the net degradation of endogenous carbohydrate, protein, and fat stores. The resultant loss of skeletal-muscle protein results in weakness. Various cardiovascular abnormalities are associated with hyperthyroidism, caused both by the direct effects of thyroid hormones and by their interactions with catecholamines. Heart rate and strength of contraction may increase so much that the individual has palpitations (an unpleasant awareness of the heart's activity). In severe cases, the heart may fail to meet the body's metabolic demands in spite of increased cardiac output. Nervous-system involvement is manifested by an excessive degree of mental alertness to the point of the patient being irritable, tense, anxious, and excessively emotional.

A prominent feature of Grave's disease but not of the other types of hyperthyroidism is **exophthalmos** (bulging eyes) (Fig. 19–5). Mucopolysaccharide deposition and edema behind the eyes, which occur for reasons still being debated, push the eyeballs forward so that they protrude from their bony orbit. They may bulge so far that the lids cannot completely close, in which case the eyes become dry, irritated, and prone to corneal ulceration. Even after correction of the hyperthyroid condition, these troublesome eye symptoms may persist.

Three general methods of treatment are available for suppressing excess thyroid hormone secretion: surgical removal of a portion of the oversecreting thyroid gland; administration of radioactive iodine, which, after being concentrated in the thyroid gland by the iodide pump, selectively destroys thyroid glandular tissue; and use of anti-thyroid drugs that specifically interfere with thyroid hormone synthesis.

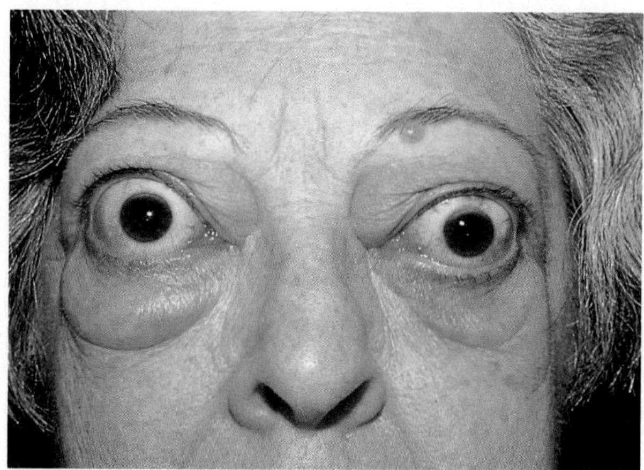

Figure 19–5 Patient Displaying Exophthalmos

A goiter may or may not accompany either hypothyroidism or hyperthyroidism.

A **goiter** refers to an enlarged thyroid gland. Because of the location of the thyroid over the trachea, a goiter is readily palpable and usually highly visible (Fig. 19–6). A goiter will occur whenever there is excessive stimulation of the thyroid gland. It may accompany hypothyroidism or hyperthyroidism, but it need not be present in either condition. Knowing the hypothalamic-pituitary-thyroid axis and feedback control, we can predict which types of hypo- and hyperthyroidism will be accompanied by a goiter (Table 19–1).

□ Hypothyroidism secondary to hypothalamic or anterior pituitary failure will not be accompanied by a goiter because the thyroid gland is not being adequately stimulated, let alone excessively stimulated.

□ In the case of hypothyroidism caused by thyroid gland failure, a goiter does develop because the circulating level of thyroid hormone is so low that there is little negative-feedback inhibition on the anterior pituitary, and TSH secretion is therefore elevated. Recall that TSH acts on the thyroid to increase the size and number of follicular cells and to increase their rate of secretion. If the thyroid cells are incapable of secreting hormone because of a lack of a critical enzyme, no amount of TSH will be able to induce these cells to secrete T_3 and T_4. However, TSH can still promote hypertrophy and hyperplasia of the thyroid with a consequent paradoxical enlargement of the gland (that is, a goiter), even though the gland is still underproducing.

□ Without adequate dietary intake of iodine, the thyroid gland is unable to synthesize and secrete normal amounts of thyroid hormone. Once again, lack of feedback inhibition results in elevated TSH secretion, which overstimulates growth of the thyroid, producing an overdeveloped but unproductive goiter.

Similarly, a goiter may or may not accompany hyperthyroidism.

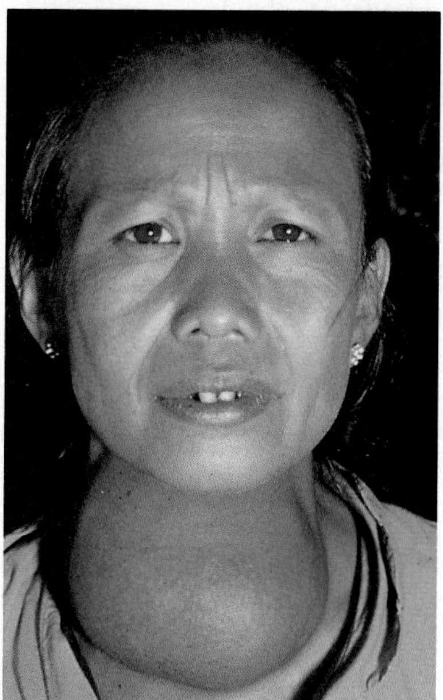

Figure 19–6 Patient with Goiter

□ Excessive TSH secretion resulting from a hypothalamic or anterior pituitary defect would obviously be accompanied by a goiter because of overstimulation of thyroid growth. Since the thyroid gland in this circumstance is also capable of responding to excess TSH with increased hormone secretion, hyperthyroidism is present with this goiter.

□ In Grave's disease, a hypersecreting goiter occurs because thyroid-stimulating immunoglobulin promotes growth of the thyroid as well as enhancing secretion of thyroid hormone. Because the high levels of circulating T_3 and T_4 inhibit the anterior pituitary, TSH secretion itself is low. In all other cases when a goiter is present, TSH levels are elevated and are directly responsible for excessive growth of the thyroid.

□ Hyperthyroidism resulting from overactivity of the thyroid in the absence of overstimulation, such as that caused by an uncontrolled thyroid tumor, is not accompanied by a goiter. The spontaneous secretion of excessive amounts of T_3 and T_4 inhibit TSH, so there is no stimulatory input to promote growth of the thyroid.

Adrenal Glands and Stress

The adrenal gland consists of an outer, steroid-secreting adrenal cortex and an inner, catecholamine-secreting adrenal medulla.

There are two **adrenal glands,** one embedded above each kidney in a capsule of fat (*adrenal* means "next to the kidney"). Each adrenal is actually composed of two endocrine organs, one surrounding the other (Fig. 19–7a). The inner portion, the **adrenal medulla,** secretes catecholamines; the outer layers composing the **adrenal cortex** secrete a variety of steroid hormones. Derived embryologically from distinct structures, the adrenal medulla and cortex secrete hormones belonging to different chemical categories whose functions, mechanisms of action, and regulation are entirely different.

The adrenal cortex secretes mineralocorticoids, glucocorticoids, and sex hormones.

About 80% of the adrenal gland is composed of the cortex, which consists of three different layers or zones: the **zona glomerulosa,** the outermost layer; the **zona fasiculata,** the middle and largest portion; and the **zona reticularis,** the innermost zone (Fig. 19–7b). The adrenal cortex produces a number of different **adrenocortical hormones,** all of which are steroids derived from the common precursor molecule, cholesterol. Slight variations in structure confer different functional capabilities on the various adrenocortical hormones. On the basis of their primary actions, the adrenal steroids can be divided into three categories: (1) **mineralocorticoids,** mainly *aldosterone,* which influence mineral (electrolyte) balance; (2) **glucocorticoids,** primarily *cortisol,* which play a major role in glucose metabolism as well as in protein and lipid metabolism; and (3) *sex hormones* identical or similar to those produced by the gonads.

The production of adrenocortical and other steroid hormones requires a step-wise series of enzymatic modifications of the cholesterol molecule. The different functional types of adrenal steroids are produced in anatomically distinct portions of the adrenal cortex. This zonation is the result of differential distribution of the enzymes that are required to catalyze the different biosynthetic pathways leading to the formation of each of these steroids. Of the two major adrenocortical hormones, aldosterone is produced exclusively in the zona glomerulosa. Cortisol synthesis is limited to the two inner layers of the cortex, with the zona fasiculata being the major source of this glucocorticoid. No other steroidogenic tissues have the capability of producing either mineralocorticoids or glucocorticoids.

In contrast, the adrenal sex hormones, also produced by the inner cortical zones, are produced in far greater abundance in the gonads. In both sexes, the zona fasiculata and zona reticularis produce both *androgens* or "male" sex hormones, and *estrogens,* or "female" sex hormones. Because the enzymes required for the production of these sex hormones are found in very low concentrations in the adrenocortical cells, androgens and estrogens are normally produced in very small quantities from this source. Their main site of production is the testes for androgens (of which testosterone is the most powerful and most abundant) and the ovaries for estrogens. Accordingly, males have a preponderance of circulating androgens, whereas in females, estrogens are predominantly found. However, there are no hormones that are unique to either males or females (except those from the placenta during pregnancy) because small amounts of the sex hormone of the opposite sex are produced by the adrenal cortex in both sexes.

Under normal circumstances, the adrenal androgens and estrogens are not sufficiently abundant or powerful to induce masculinizing or feminizing effects, respectively. The only adrenal sex hormone that has any biological importance is the androgen **dehydroepiandrosterone (DHEA).** The testes' primary androgen product is the potent testosterone, but the most abundant adrenal androgen is the much weaker DHEA. Adrenal DHEA is overpowered by testicular testosterone in males, but it is of physiological significance in females, who otherwise lack androgens. This adrenal androgen is responsible for androgen-dependent processes in the female such as growth of pubic and axillary (arm-pit) hair, enhancement of

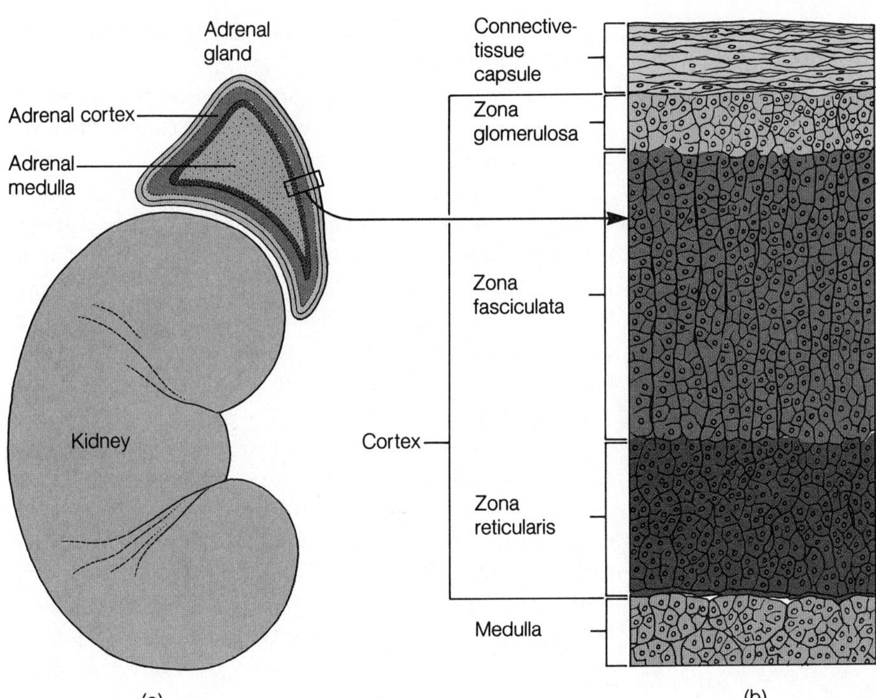

Figure 19–7 Anatomy of the Adrenal Gland *(a) Location and structure of adrenal gland. (b) Layers of adrenal cortex.*

the pubertal growth spurt, and development and maintenance of the female sex drive.

Being lipophilic, the adrenocortical hormones are all carried in the blood extensively bound to plasma proteins. About 60% of circulating aldosterone is protein bound, primarily to nonspecific albumin. Cortisol is 98% bound, mostly to a plasma protein specific for it, called **corticosteroid-binding globulin (transcortin)**. Likewise, 98% of DHEA is bound, in this case exclusively with albumin.

Mineralocorticoids' major effects are on electrolyte balance and blood-pressure homeostasis.

The actions and regulation of the primary adrenocortical mineralocorticoid, aldosterone, is described thoroughly elsewhere (chapters 14 and 15). To highlight aldosterone activity, its principal site of action is on the distal tubules of the kidney, where it promotes Na^+ retention and enhances K^+ elimination during the formation of urine. The promotion of Na^+ retention by aldosterone secondarily induces osmotic retention of H_2O, thereby causing expansion of the extracellular-fluid volume, which is important in long-term regulation of blood pressure.

Mineralocorticoids are *essential for life*. Without aldosterone, a person rapidly dies from circulatory shock because of the marked fall in plasma volume caused by excessive losses of H_2O-holding Na^+. With most other hormonal deficiencies, death is not imminent, even though a chronic hormonal deficiency may eventually lead to a premature death.

Regulation of aldosterone secretion is related to Na^+ and K^+ balance and blood-pressure regulation. The adrenal tropic hormone ACTH primarily affects the inner cortical zones and has only a weak effect in stimulating aldosterone secretion. Thus, unlike cortisol regulation, the regulation of aldosterone secretion is largely independent of anterior pituitary control. The two factors of greatest physiological importance in the regulation of aldosterone secretion are: (1) a direct effect of plasma K^+ on the zona glomerulosa, and (2) the renin-angiotensin-aldosterone system, which is activated by various factors related to a fall in plasma Na^+ concentration/ECF volume/arterial blood pressure (see p. 481). In addition to its effect on aldosterone secretion, angiotensin promotes growth of the zona glomerulosa, in a way similar to the effect of TSH on the thyroid.

Glucocorticoids influence intermediary metabolism and have an important role in adaptation to stress.

Cortisol, the primary glucocorticoid, plays an important role in intermediary metabolism, exhibits significant permissive actions for other hormonal activities, and helps people resist stress. Regarding its metabolic role, the overall effect of cortisol's actions is to increase the concentration of blood glucose at the expense of protein and fat stores. Specifically, cortisol performs the following functions:

1. It stimulates hepatic **gluconeogenesis** (*gluco* means "glucose"; *neo* means "new"; *genesis* means "production"), which refers to the conversion of noncarbohydrate sources (namely, amino acids), into carbohydrate within the liver. Between meals or during periods of fasting, when no new nutrients are being absorbed into the blood for utilization and storage, the glycogen (stored glucose) in the liver tends to become depleted as it is broken down to release glucose into the blood. Gluconeogenesis is an important factor in replenishing hepatic glycogen stores and thus in maintaining normal blood glucose levels between meals. This is essential, because the brain can use only glucose as its metabolic fuel, yet nervous tissue cannot store glycogen to any extent. Therefore, the concentration of glucose in the blood must be maintained at an appropriate level to adequately supply the brain with nutrients.
2. It inhibits glucose uptake and use by many tissues, but not the brain. This spares glucose for use by the brain, which absolutely requires it as a metabolic fuel.
3. It stimulates protein degradation in many tissues, especially muscle. By breaking down a portion of muscle proteins into their constituent amino acids, cortisol increases the concentration of blood amino acids. These mobilized amino acids are available for use in gluconeogenesis or whenever else they are needed, such as for repair of damaged tissue or synthesis of new cellular structures.
4. It facilitates lipolysis (*lysis* means "breakdown"), or the breakdown of lipid (fat) stores in adipose tissue, thus releasing free fatty acids into the blood. The mobilized fatty acids are available as an alternative metabolic fuel for the tissues that can use this energy source in lieu of glucose, thereby conserving glucose for the brain.

In addition to its direct effects on organic metabolism, cortisol is extremely important for its permissive action in allowing other hormones to fully exert their effects. For example, cortisol must be present in adequate amounts to permit the catecholamines to induce vasoconstriction. A person lacking cortisol, if untreated, may go into circulatory shock in a stressful situation that demands immediate wide-spread vasoconstriction.

Still other effects of cortisol are recognized but remain poorly understood. The most important of these is the key role cortisol plays in adaptation to *stress*. Stress of any kind is one of the major stimuli for increased cortisol secretion. Indeed, stress has come to be defined as any condition that elicits increased cortisol secretion. Cortisol is also known to influence mood and behavior.

When cortisol or synthetic cortisol-like compounds are administered to yield higher than physiological concentrations of glucocorticoids (that is, *pharmacological levels*), not only are all of the effects on organic metabolism increased in magnitude, but several important new actions not evidenced at normal physiological levels are seen. The most noteworthy of glucocorticoids' pharmacological effects are anti-inflammatory and immunosuppressive effects. Synthetic glucocorticoids have been developed that maximize the anti-inflammatory and immunosuppressive effects of these steroids while minimizing the metabolic effects.

Administration of large amounts of glucocorticoid inhibits almost every step of the inflammatory response, making these steroids effective drugs in treating conditions in which the inflammatory response itself has become a destructive process, such as rheumatoid arthritis. It is important to recognize that glucocorticoids used in this manner do not affect the underlying disease process; they merely suppress the body's responses to the disease. Because glucocorticoids also exert multiple inhibitory effects on the overall immune process, such as "knocking out of commission" the white blood cells responsible for antibody production and destruction of foreign cells, these agents have proved to be useful in the management of various allergic disorders and in the prevention of organ transplant rejections.

When these steroids are employed therapeutically, they should be used only when warranted and then only sparingly for several important reasons. First, because they suppress the normal inflammatory and immune responses that form the backbone of the body's defense system, a glucocorticoid-treated individual has limited ability to resist infections. Second, in addition to the anti-inflammatory and immunosuppressive effects readily exhibited at pharmacological levels, other less desirable effects may also be observed with prolonged exposure to supraphysiological concentrations of glucocorticoids. These include development of gastric ulcers, high blood pressure, atherosclerosis, and menstrual irregularities. Third, high levels of exogenous glucocorticoids act in negative-feedback fashion to suppress the hypothalamo-pituitary axis that drives normal glucocorticoid secretion and maintains the integrity of the adrenal cortex. Prolonged suppression of this axis can lead to permanent inability of the body to produce its own cortisol because of irreversible atrophy of the inner cortical zones of the adrenal gland.

Cortisol secretion is directly regulated by ACTH.

Cortisol secretion by the adrenal cortex is regulated by a long-loop negative-feedback system involving the hypothalamus and anterior pituitary (Fig. 19–8). The only factor known to directly stimulate the adrenal cortex to secrete cortisol is adrenocorticotropic hormone (ACTH) from the anterior pituitary. Being tropic to the zona fasiculata and zona reticularis, ACTH stimulates both the growth and secretory output of these two inner layers of the cortex. In the absence of adequate amounts of ACTH, these layers shrink considerably, and cortisol secretion is drastically reduced. Recall that angiotensin, not ACTH, is responsible for maintaining the size of the zona glomerulosa. Like the actions of TSH on the thyroid gland, ACTH enhances many steps in the synthesis of cortisol.

The ACTH-producing cells, in turn, secrete only at the command of corticotropin-releasing hormone (CRH) from the hypothalamus. The feedback control loop is completed by cortisol's inhibitory actions on CRH and ACTH secretion by the hypothalamus and anterior pituitary, respectively.

In addition to controlling cortisol secretion, ACTH (not the pituitary gonadotropic hormones) controls adrenal androgen secretion. In general, cortisol and DHEA output by the adrenal cortex parallel each other. However, adrenal androgens feed back outside of the hypothalamo-pituitary-adrenal cortex loop. Instead of inhibiting corticotropin-releasing hormone, DHEA inhibits gonadotropin-releasing hormone, just like testicular androgens. Furthermore, there are times when adrenal androgen and cortisol output diverge from each other—for example at the time of puberty when there is a marked surge in adrenal androgen secretion but no change in cortisol secretion. It is this change that initiates the development of androgen-dependent processes in females. In males the same thing is accomplished primarily by testicular androgen secretion, which is also aroused at puberty. The nature of the pubertal inputs to the adrenals and gonads is still unresolved.

The negative-feedback system for cortisol maintains a relatively constant average level of cortisol secretion that is punctuated by alternating bursts of modest secretion separated by "silent" periods of little or no secretion. The amount of cortisol secreted with each burst does not vary much from one secretory episode to the next. However, the total amount of cortisol secreted during a given period of time can be changed by varying the *frequency* of secretory bursts. Reminiscent of wave summation in skeletal-muscle contraction, one burst "adds on" to the previous burst, increasing the plasma concentration of cortisol beyond that achieved by a single burst. Thus the mean plasma concentration of cortisol can be elevated by stimulating the secretory bursts to occur more frequently (Fig. 19–9).

Superimposed on the basic negative-feedback control system are two additional factors that influence plasma cortisol concentrations by varying the frequency of secretory bursts: these are *diurnal rhythm* and *stress*. There is a characteristic

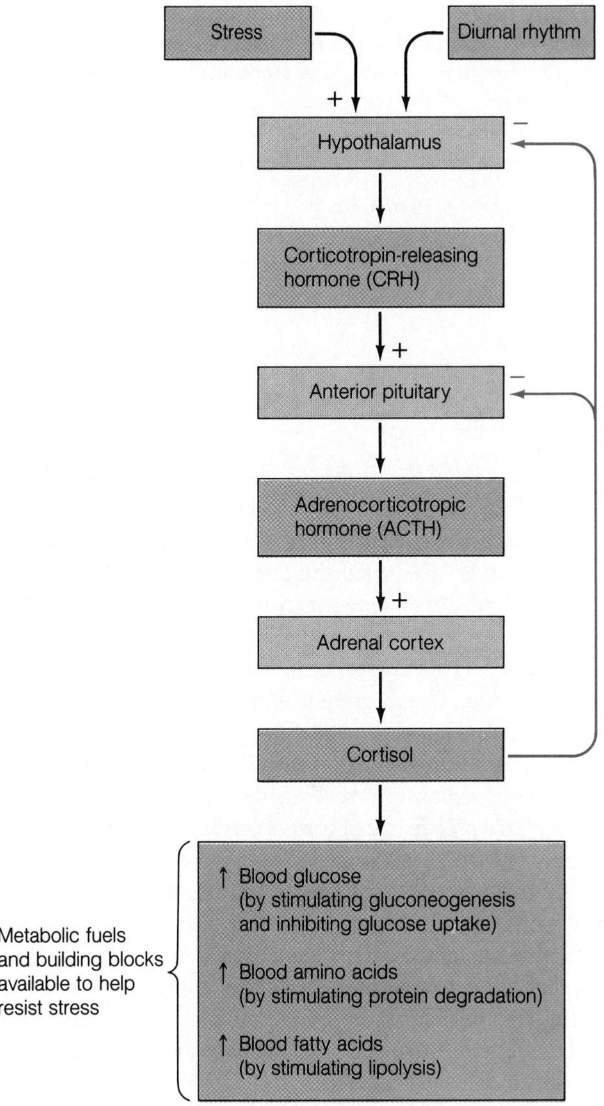

Figure 19–8 Control of Cortisol Secretion

diurnal rhythm in plasma cortisol concentration, with the highest level occurring in the morning and the lowest level being found at night (see Fig. 18–11). This diurnal rhythm, which is intrinsic to the hypothalamo-pituitary control system, is related primarily to the sleep-wake cycle. The peak and low levels are reversed in a person who works at night and sleeps during the day. Such temporal variations in secretion are of more than academic interest for several reasons. First, it is important clinically to know at what time of day a blood sample was taken when interpreting the significance of a particular value. Second, the linking of biological rhythms to day-night activity patterns raises serious questions about the common practice of swing shifts at work. Third, because cortisol helps

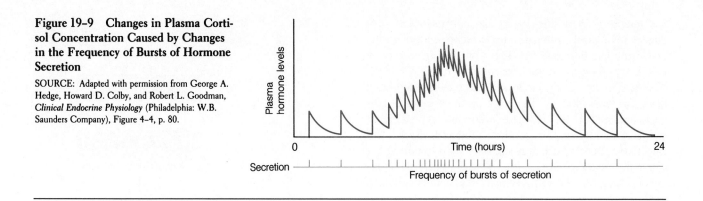

Figure 19–9 Changes in Plasma Cortisol Concentration Caused by Changes in the Frequency of Bursts of Hormone Secretion

SOURCE: Adapted with permission from George A. Hedge, Howard D. Colby, and Robert L. Goodman, *Clinical Endocrine Physiology* (Philadelphia: W.B. Saunders Company), Figure 4–4, p. 80.

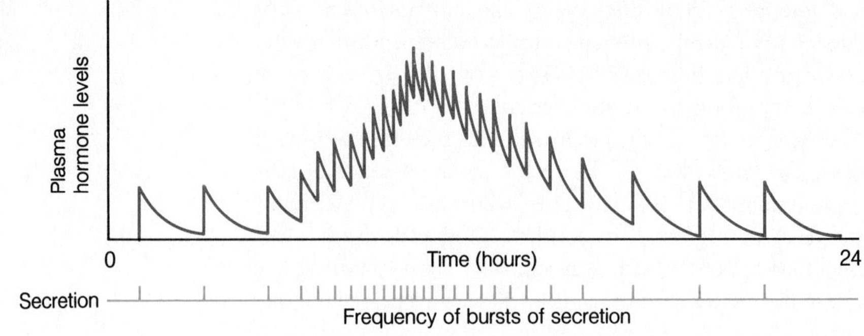

a person resist stress, increasing attention is being given to the time of day various surgical procedures are performed.

The other major factor that is independent of, and in fact can override, the stabilizing negative-feedback control is stress. Dramatic increases in cortisol secretion, mediated by the central nervous system through enhanced activity of the CRH-ACTH system, occur in response to all kinds of stressful situations. The magnitude of the increase in plasma cortisol concentration is generally proportional to the intensity of the stressful stimulation; a greater increase in cortisol levels is evoked in response to severe stress than to mild stress. Although the precise role of cortisol as a key factor in helping the body cope with stress is not known, it is presumed to be related to cortisol's effects of shifting the body from using protein and fat stores to carbohydrate stores and the sparing of carbohydrate by non-glucose dependent tissues. A logical assumption is that the increased pool of glucose, amino acids, and fatty acids are available for use as needed, such as to sustain nourishment to the brain and to provide metabolic building blocks to repair damaged tissues.

The adrenal gland may secrete too much or too little of any one of its hormones.

Although uncommon, there are a number of different disorders of adrenocortical function. Excessive secretion may occur with any of the three categories of adrenocortical hormones. Accordingly, three main patterns of symptoms resulting from hyperadrenalism can be distinguished, depending on which hormone type is in excess: these are aldosterone hypersecretion, cortisol hypersecretion, and adrenal androgen hypersecretion (Table 19–2).

ALDOSTERONE HYPERSECRETION. Excess mineralocorticoid secretion may be caused by: (1) a hypersecreting tumor made up of zona glomerulosa cells (**primary hyperaldosteronism** or **Conn's syndrome**) or (2) inappropriately high activity of the renin-angiotensin system (**secondary hyperaldosteronism**). The latter may be produced by any number of conditions that cause a chronic reduction in arterial blood flow to the kidneys, thereby excessively activating the renin-angiotensin-aldosterone system. An example is atherosclerotic narrowing of the renal arteries.

The symptoms of either primary or secondary hyperaldosteronism are related to the exaggerated effects of aldosterone—namely excessive Na^+ retention (**hypernatremia**) and K^+ depletion (**hypokalemia**). Also, high blood pressure (hypertension) is generally present, at least partially because of excessive Na^+ retention.

CORTISOL HYPERSECRETION. **Cushing's syndrome,** or excessive cortisol secretion, can be caused by: (1) overstimulation of the adrenal cortex by excessive amounts of CRH and/or ACTH; (2) adrenal tumors that uncontrollably secrete cortisol independent of ACTH; and (3) ACTH-secreting tumors located in places other than the pituitary, most commonly in the lung. Whatever the cause, the prominent findings are related to the exaggerated effects of glucocorticoid, with the main symptoms being reflections of excessive gluconeogenesis. When too many amino acids are converted into glucose, the body suffers from combined glucose excess and protein shortage. Because the resultant hyperglycemia and glucosuria mimic diabetes mellitus, the condition is sometimes referred to as adrenal diabetes. For reasons that are unclear, some of the extra glucose is deposited as body fat in locations characteristic for this disease, typically in the abdomen, face, and above the shoulder blades. This abnormal fat distribution in the latter two locations is descriptively called a "moon-face" and "buffalo hump." The appendages, in contrast, remain thin.

Besides the effects attributable to excessive glucose production, other effects arise from the widespread mobilization of

Table 19–2 Major Adrenocortical Abnormalities

Abnormality	Condition	Cause	Symptoms
Excess aldosterone	Conn's syndrome (primary hyperaldosteronism)	Hypersecreting tumor of zona glomerulosa	Hypernatremia; hypokalemia; hypertension
	Secondary hyperaldosteronism	Inappropriately high activity of renin-angiotensin system	
Excess cortisol	Cushing's syndrome	Excess corticotropin-releasing hormone (CRH) and/or adrenocorticopic hormone (ACTH) caused by hypothalamic or anterior pituitary disease; hypersecreting tumor of inner layers of adrenal cortex; ACTH-secreting tumor in lung	Glucose excess; protein shortage; abnormal fat distribution
Excess androgen	Adrenogenital syndrome	Lack of enzyme in cortisol pathway	Inappropriate masculinization in all but adult males
Deficient cortisol and aldosterone	Addison's disease (primary adrenocortical insufficiency)	Destruction or idiopathic atrophy of adrenal cortex	Related to cortisol deficiency: poor response to stress; hypoglycemia; lack of permissiveness for many metabolic activities. Related to aldosterone deficiency: hyperkalemia; hyponatremia; hypotension (if severe enough, fatal)
Deficient cortisol	Secondary adrenocortical insufficiency	Insufficient ACTH caused by hypothalamic or anterior pituitary failure	

amino acids from body proteins for use as glucose precursors. Loss of muscle protein leads to muscle weakness and fatigue. The protein-poor, thin skin of the abdomen becomes overstretched by the excessive underlying fat deposits. As a result, the subdermal tissues rupture, causing the formation of irregular, reddish-purple linear streaks. The weakness of blood vessel walls caused by depletion of structural protein results in an excessive tendency to bruise and to form ecchymoses (small patches of subcutaneous bleeding). Formation of collagen, a major structural protein of connective tissue, is depressed. This interferes with scar formation so that wounds heal poorly. Furthermore, loss of the proteinaceous collagen framework of bone causes weakness of the skeleton, so fractures may result from little or no apparent injury.

ADRENAL ANDROGEN HYPERSECRETION. Excess adrenal androgen secretion, a masculinizing condition, is more common than the extremely rare feminizing condition of excess adrenal estrogen secretion. Either condition is referred to as **adrenogenital syndrome,** emphasizing the pronounced effects excessive adrenal sex hormones have on the genitalia and associated sexual characteristics.

The symptoms that result from excess androgen secretion depend on the age and sex of the individual when the hyperactivity first begins. Because androgens exert masculinizing effects, a woman with this disease tends to develop a male pattern of body hair, a condition referred to as **hirsutism.** She usually also acquires other male secondary sexual characteristics such as deepening of the voice and more muscular arms and legs. The breasts become smaller, and menstruation may cease as a result of androgen suppression of the woman's hypothalamic-pituitary-ovarian pathway for her own female sex-hormone secretion.

Female infants born with adrenogenital syndrome manifest male-type external genitalia because excessive androgen secretion occurs early enough during fetal life to induce development of their genitalia along male lines, similar to the development of males under the influence of testicular androgen. The clitoris, which is the female homologue of the male penis, enlarges under androgen influence and takes on a penile-like appearance so that in some cases it is difficult at first to determine the child's sex. This condition is known as **female pseudohermaphroditism.** (A true hermaphrodite has the gonads of both sexes.)

Excessive adrenal androgen secretion in a young boy gives rise to different problems. Prepubertal boys develop male secondary sexual characteristics—for example, deep voice, beard, enlarged penis, sex drive—at an unusually early age. This condition is referred to as **precocious pseudopuberty** to differentiate it from true puberty, which occurs as a result of increased testicular activity. In precocious pseudopuberty, the androgen secretion from the adrenal cortex is not accompanied by sperm production or any other gonadal activity, because the testes are still in their non-functional prepubertal state.

Overactivity of adrenal androgens in adult males has no apparent effect because of the already existing male sex characteristics.

The adrenogenital syndrome is most commonly caused by an inherited enzymatic defect in the cortisol steroidogenic pathway. The pathway for synthesis of androgens branches from the normal biosynthetic pathway for cortisol (see Fig. 18–5). When an enzyme specifically essential for synthesis of cortisol is deficient, the result is decreased secretion of cortisol. The decline in cortisol secretion removes the negative-feedback effect on the hypothalmus and anterior pituitary so that levels of CRH and ACTH increase considerably (Fig. 19–10). The defective adrenal cortex is incapable of responding to this increased ACTH secretion with cortisol output, and instead shunts more of its cholesterol precursor into the androgen pathway, resulting in excess DHEA production. This excess androgen does not inhibit ACTH but, rather, inhibits the gonadotropins. As a result, individuals with adrenogenital syndrome are sterile, because gamete production is not stimulated in the absence of gonadotropins. Of course, the victims also exhibit symptoms of cortisol deficiency.

The symptoms of adrenal virilization, sterility, and cortisol deficiency are all reversed by glucocorticoid therapy. Admin-

Figure 19–10 Hormonal Interrelationships in Adrenogenital Syndrome

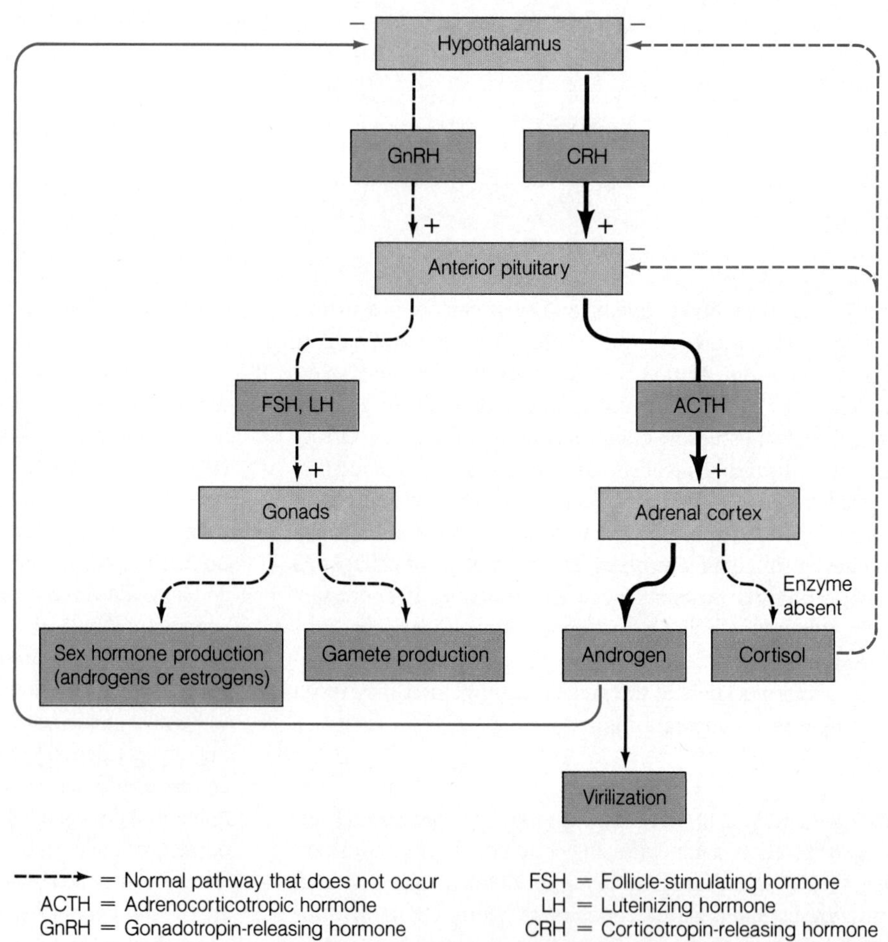

istration of exogenous glucocorticoid replaces the cortisol deficit, but, more dramatically, it also inhibits the hypothalamus and pituitary so that ACTH secretion is suppressed. Once ACTH secretion is reduced, the profound stimulation of the adrenal cortex ceases, and androgen secretion declines markedly. Removal of the large quantities of adrenal androgens from the circulation allows the masculinizing characteristics to gradually recede and normal gonadotropin secretion to resume. Without understanding the relation among these hormonal systems, it would be very difficult to comprehend how glucocorticoid administration could dramatically reverse symptoms of masculinization and sterility.

ADRENOCORTICAL INSUFFICIENCY. If one adrenal gland is nonfunctional or removed, the other healthy organ can take over the function of both through hypertrophy and hyperplasia. Therefore, both glands must be affected before adrenocortical insufficiency occurs.

In **primary adrenocortical insufficiency,** also known as **Addison's disease,** all layers of the adrenal cortex are undersecreting. This condition is most commonly due to idiopathic atrophy of the gland. Although unproven, the most probable cause is autoimmune destruction of the gland by erroneous production of adrenal cortex–attacking antibodies. **Secondary adrenocortical insufficiency** may occur because of a pituitary or hypothalamic abnormality, resulting in insufficient ACTH secretion. In Addison's disease, both cortisol and aldosterone are deficient, whereas in the secondary form of the condition, only cortisol is deficient, because aldosterone secretion does not depend on ACTH stimulation.

Symptoms associated with cortisol deficiency are as would be expected—poor response to stress, hypoglycemia caused by reduced gluconeogenic activity, and lack of permissive action for many metabolic activities. With the primary form of the disease, hyperpigmentation (darkening of the skin) resulting from excessive secretion of ACTH is also seen. Because the pituitary is normal, the decline in cortisol secretion brings about an uninhibited elevation in ACTH output. Both ACTH and melanocyte-stimulating hormone (MSH) are produced from the same large precursor molecule, **propiomelanocortin.** As a result of their close structural relationship, ACTH has slight MSH-like activity that becomes apparent when the blood levels of ACTH are very high.

The symptoms associated with aldosterone deficiency in Addison's disease are the most threatening. If severe enough, the condition is fatal because aldosterone is essential for life. However, the loss of adrenal function may develop slowly and insidiously so that aldosterone secretion may be subnormal but not totally lacking. Patients with aldosterone deficiency display K^+ retention (**hyperkalemia**) caused by reduced K^+ secretion into the urine and Na^+ depletion (**hyponatremia**) caused by excessive urinary losses of Na^+. The former results in disturbances in cardiac rhythm. The latter results in a fall in ECF volume, including a reduction in circulating blood volume, which in turn leads to low blood pressure (hypotension).

The catecholamine-secreting adrenal medulla is a modified sympathetic postganglionic neuron.

The adrenal medulla is actually a modified part of the sympathetic nervous system. Collectively, the sympathetic nervous system and adrenomedullary hormonal output mobilize the body's resources to support peak physical exertion.

A sympathetic pathway consists of two neurons in sequence—a *preganglionic neuron* originating in the central nervous system whose axonal fiber terminates on a second peripherally located *postganglionic neuron*, which in turn terminates on the effector organ (see p. 203). The neurotransmitter released by sympathetic postganglionic fibers is norepinephrine, which interacts locally with the innervated organ by binding with specific target receptors known as *adrenergic receptors* (see p. 207).

The adrenal medulla is composed of modified postganglionic sympathetic neurons. Unlike ordinary postganglionic sympathetic neurons, those in the adrenal medulla do not possess axonal fibers that terminate on effector organs. Instead, the ganglionic cell bodies within the adrenal medulla release their chemical transmitter directly into the circulation upon stimulation by the preganglionic fiber (Fig. 19–11). In this case, the transmitter qualifies as a hormone instead of a neurotransmitter. Like sympathetic fibers, the adrenal medulla does release norepinephrine, but its most abundant secretory output is a similar chemical messenger known as **epinephrine.** Both epinephrine and norepinephrine belong to the chemical class of **catecholamines,** which are derived from the naturally occurring amino acid tyrosine. Epinephrine is the same as norepinephrine except that it has a methyl group added to it.

Once produced in adrenomedullary cells, epinephrine and norepinephrine are stored in **chromaffin granules,** which are similar to the transmitter storage vesicles found in sympathetic nerve endings. Accordingly, adrenomedullary tissue is often called *chromaffin tissue*. The chromaffin granules have a very efficient, active-transport catecholamine uptake system, with the result being that the concentration of epinephrine in the chromaffin granules is at least 25,000 times greater than that in the cytosol. Segregation of catecholamines in chromaffin granules protects them from being destroyed by cytosolic enzymes during their storage. Furthermore, packaging of these hormones into granules is a prerequisite for their export from the adrenomedullary cells. Secretion of catechol-

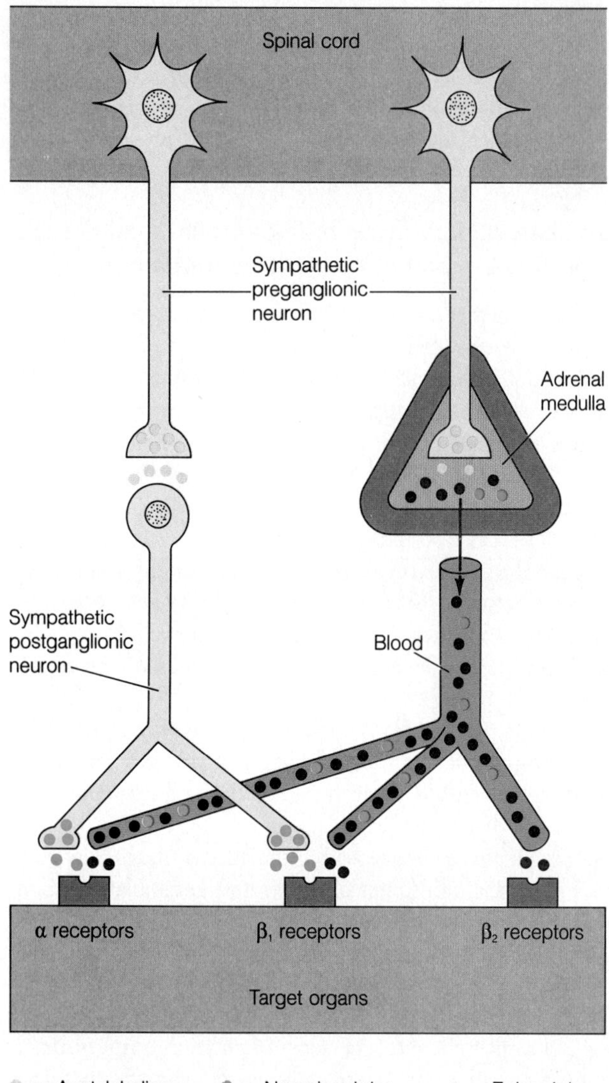

Figure 19–11 Comparison of Sympathetic Innervation and Adrenomedullary Secretion

amines into the circulation takes place by Ca^{++}-dependent exocytosis of chromaffin granules, analogous to the release mechanism for secretory vesicles that contain stored peptide hormones or the release of catecholamines at sympathetic postganglionic terminals.

Of the total adrenomedullary catecholamine output, epinephrine accounts for 80% and norepinephrine for 20%. Whereas epinephrine is produced exclusively by the adrenal medulla, the bulk of norepinephrine in the body is produced by sympathetic postganglionic fibers. Adrenomedullary norepinephrine is generally secreted in quantities too small to exert significant effects on target tissues. Therefore, for practical purposes, we can assume norepinephrine effects to be predominantly mediated directly by the sympathetic nervous system and epinephrine effects to be brought about exclusively by the adrenal medulla.

Epinephrine and norepinephrine have differing affinities for several different types of adrenergic receptors. Catecholamine actions are mediated through interactions with three distinctive receptor types: α, β_1, and β_2 adrenergic receptors (see p. 207). Some target tissues have only α receptors, some have only β_2 receptors, and some have both α and β_2 receptors; β_1 receptors are found almost exclusively in the heart. In general, the responses elicited by activation of α and β_1 receptors are excitatory in nature, whereas the responses to stimulation of β_2 receptors are inhibitory.

Neither catecholamine is totally specific for particular receptor types, but each has markedly different affinities. Norepinephrine binds predominantly with α and β_1 receptors located in the vicinity of postganglionic sympathetic-fiber terminals. Hormonal epinephrine, which is able to reach all α and β_1 receptors by means of its circulatory distribution, interacts with these receptors with approximately the same potency as neurotransmitter norepinephrine. In addition, epinephrine activates β_2 receptors, over which the sympathetic system exerts little influence. Epinephrine is at least ten times more potent than norepinephrine at β_2 receptors. In fact, many of the essentially epinephrine-exclusive β_2 receptors are located at tissues not even supplied by the sympathetic system but which epinephrine reaches through the blood. An example is skeletal muscle, where epinephrine exerts metabolic effects.

Adrenomedullary hormones are not essential for life, but virtually all organs in the body are affected by these catecholamines. They play important roles in the regulation of intermediary metabolism, in cardiovascular control, and in responses to stress. Epinephrine and norepinephrine exert similar effects in many tissues, with epinephrine generally reinforcing sympathetic activity. For example, epinephrine and norepinephrine both accelerate the heart rate by binding with cardiac β_1 receptors, and both induce generalized arteriolar vasoconstriction of the skin, digestive tract, and kidneys through their mutual α-receptor activation. Yet there are some important differences in response that can be explained on the basis of differential activation of different receptor types. As an example, epinephrine, through its exclusive β_2-receptor activation, brings about vasodilation of the blood vessels that supply skeletal muscles and the heart. This effect is in addition to its generalized vasoconstrictor effect mediated by α-receptor stimulation. Epinephrine also exerts metabolic effects, whereas norepinephrine has limited action in this regard.

It is important to realize, however, that epinephrine functions only at the bidding of the sympathetic nervous system, which is solely responsible for stimulating its secretion from the adrenal medulla. Epinephrine secretion always accompanies a generalized sympathetic discharge, so sympathetic activity indirectly exerts control over the actions performed by epinephrine. By having the more versatile circulating epinephrine at its call, the sympathetic system has a means of reinforcing its own neurotransmitter effects plus a way of exerting additional actions on tissues that it does not directly innervate.

Under conditions of fear or stress when the sympathetic system is activated, it simultaneously triggers a surge of adrenomedullary catecholamine release, flooding the circulation with up to three-hundred times the normal concentration of epinephrine. The overall effect of sympathetic stimulation accompanied by epinephrine release is to prepare the body for meeting emergency or stressful situations. The sympathetic and epinephrine actions collectively constitute a fight-or-flight response that prepares the individual to combat an enemy or flee from danger. The following sections discuss epinephrine's major effects, which it accomplishes either in collaboration with the sympathetic transmitter norepinephrine, or it performs alone to complement direct sympathetic responses in helping to prepare the body for peak physical responsiveness in the face of impending danger.

EFFECT ON ORGAN SYSTEMS. The sympathetic system and epinephrine both exert widespread effects on organ systems that are ideally suited for fight-or-flight responses (see p. 205). Under the influence of epinephrine and the sympathetic system, the rate and strength of cardiac contraction increase, resulting in increased cardiac output, and their generalized vasoconstrictor effects increase total peripheral resistance. Together these effects cause an increase in arterial blood pressure, thus assuring an appropriate driving pressure to force blood to the organs that are most vital for meeting the emergency. Meanwhile, vasodilation of coronary and skeletal-muscle blood vessels induced by epinephrine and local metabolic factors shifts blood to the heart and skeletal muscles from other vasoconstricted regions of the body. Because of their profound influence on the heart and vasculature, the adrenomedullary catecholamines and sympathetic system also play an important role in the ongoing maintenance of arterial blood pressure.

Epinephrine (but not norepinephrine) dilates the respiratory airways to reduce the resistance encountered in moving air in and out of the lungs. Epinephrine also reduces digestive activity and inhibits bladder emptying, both being activities that can be "put on hold" during a fight-or-flight situation.

METABOLIC EFFECTS. Epinephrine exerts some important metabolic effects, even at blood hormone concentrations lower than those required for eliciting the cardiovascular responses. In general, epinephrine prompts the mobilization of stored carbohydrate and fat to provide immediately available energy for use as needed to fuel muscular work. Specifically, epinephrine increases the blood glucose level by several different mechanisms. First, it stimulates both hepatic (liver) gluconeogenesis and **glycogenolysis,** the latter referring to the breakdown of stored glycogen into glucose, which is released into the blood. Epinephrine also stimulates glycogenolysis in skeletal muscles. Because of the difference in enzyme content between liver and muscle, however, muscle glycogen cannot be converted directly to glucose. Instead, lactic acid is released into the blood as a result of the breakdown of muscle glycogen. This lactic acid is removed from the blood by the liver and is converted into glucose, so epinephrine's actions on skeletal muscle indirectly contribute to an increase in blood glucose levels. Epinephrine and the sympathetic system may further add to this hyperglycemic effect by inhibiting the secretion of insulin, the pancreatic hormone primarily responsible for removal of glucose from the blood, and by stimulating glucagon, another pancreatic hormone that promotes hepatic glycogenolysis and gluconeogenesis. In addition to increasing blood glucose levels, epinephrine also increases the blood fatty acids level by promoting lipolysis.

Epinephrine's metabolic effects are appropriate for fight-or-flight situations. The elevated levels of glucose and fatty acids provide additional fuel to power the muscular movement required of the situation while they assure adequate nourishment for the brain during the crisis when no new nutrients are being consumed. Muscles can use fatty acids for energy production, but the brain cannot.

Because of its other widespread actions, epinephrine also increases the overall metabolic rate. Under the influence of epinephrine, many tissues metabolize at a faster rate. For example, the work of the heart and respiratory muscles is increased and the pace of liver metabolism is stepped up. Thus, epinephrine as well as thyroid hormone can increase the metabolic rate.

OTHER EFFECTS. Epinephrine affects the central nervous system to promote a state of arousal and increased CNS alertness. This permits "quick thinking" to help cope with the impending emergency. Many drugs that are used as stimulants or sedatives probably exert their effects by altering catecholamine levels in the CNS.

Both epinephrine and norepinephrine cause sweating, which helps the body rid itself of extra heat generated by increased muscular activity. Also, epinephrine acts on smooth muscles within the eyes to dilate the pupil and flatten the lens. These actions adjust the eyes for more encompassing vision so that the whole threatening scene can be quickly viewed.

Sympathetic stimulation of the adrenal medulla is solely responsible for epinephrine release.

Catecholamine secretion by the adrenal medulla is controlled entirely by sympathetic input to the gland. Although a number of different factors have been shown to influence adrenal catecholamine secretion, they all act by increasing preganglionic sympathetic impulses to the adrenal medulla. Among the major factors that stimulate increased adrenomedullary output are a variety of stressful conditions such as physical or psychological trauma, hemorrhage, illness, exercise, hypoxia (low arterial O_2), cold exposure, and hypoglycemia (low blood glucose). The amount of epinephrine released depends on the type and intensity of the stressful stimulus.

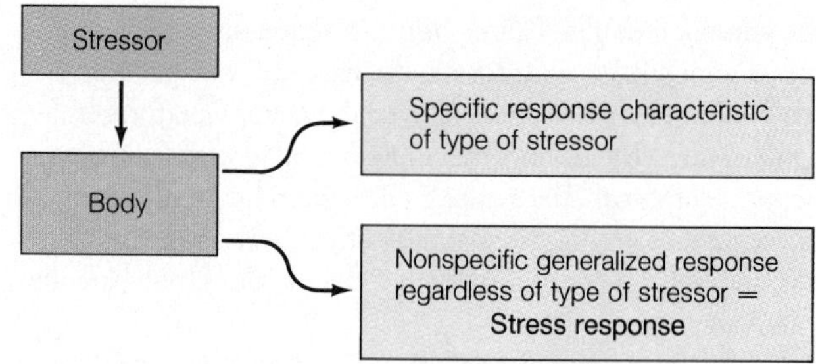

Figure 19–12 Action of Stressor on Body

Adrenomedullary dysfunction is very rare.

Adrenomedullary hyposecretion is not a recognized clinical entity. No adverse effects have been attributed to a deficiency of epinephrine, presumably not because a deficiency never occurs but because the majority of epinephrine's functions can be duplicated by activation of the sympathetic nervous system alone.

The only catecholamine disorder is a **pheochromocytoma,** a rarely occurring catecholamine-secreting tumor. These chromaffin tumors are usually, but not always, located in the adrenal medulla. Pheochromocytomas may contain up to twenty times more catecholamines per gram of tissue than normal medullary tissue, with release of these hormones not being subject to neural control. The symptoms of this condition are directly attributable to the actions of excessive amounts of catecholamines, the most common of which are hypertension, tachycardia, palpitations, excessive sweating, and hyperglycemia.

The stress response is a generalized, nonspecific pattern of neural and hormonal reactions to any situation that threatens homeostasis.

Since both components of the adrenal gland play an extensive role in responding to stress, this is an appropriate place to pull together the various major factors involved in the stress response. **Stress** refers to the generalized, nonspecific response of the body to any factor that overwhelms, or threatens to overwhelm, the body's compensatory abilities to maintain homeostasis. Contrary to popular usage, the agent inducing the response is correctly called a *stressor*, whereas *stress* refers to the state induced by the stressor.

The following types of stressors illustrate the range of noxious stimuli that can induce a stress response: *physical* (trauma, surgery, intense heat or cold); *chemical* (reduced O_2 supply, acid-base imbalance); *physiological* (heavy exercise, hemorrhagic shock, pain); *psychological* or *emotional* (anxiety, fear, sorrow); and *social* (personal conflicts, changes in life-style). Different stressors may produce some specific responses characteristic of that stressor; for example, the body's specific response to cold exposure is shivering and skin vasoconstriction, whereas the specific response to bacterial invasion includes increased phagocytic activity and antibody production. However, all stressors, in addition to their specific response, also produce a similar nonspecific, generalized response regardless of the type of stressor (Fig. 19–12).

Dr. Hans Selye was the first to recognize this commonality of responses to noxious stimuli in what he called the **general adaptation syndrome.** When a stressor is recognized, both nervous and hormonal responses are called into play to bring about defensive measures to cope with the emergency. The result is a state of intense readiness and mobilizaton of biochemical resources.

To appreciate the value of the multifaceted stress response, imagine a primitive cave dweller who has just seen a large wild beast lurking in the shadows. The major neural response to such a stressful stimulus is generalized activation of the sympathetic nervous system. The resultant increase in cardiac output and ventilation as well as the diversion of blood from vasoconstricted regions of suppressed activity, such as the digestive tract and kidneys, to the more active vasodilated skeletal muscles and heart prepare the body for a fight-or-flight response. Simultaneously, the sympathetic system calls forth hormonal reinforcements in the form of a massive outpouring of epinephrine from the adrenal medulla. Epinephrine strengthens sympathetic responses and gets into places not innervated by the sympathetic system to perform additional functions, such as mobilizing carbohydrate and fat stores.

Besides epinephrine, a number of other hormones are involved in the overall stress response (Table 19–3). The predominant hormonal response is activation of the CRH-

Table 19-3 Major Hormonal Changes during the Stress Response

Hormone	Change	Purpose Served
Epinephrine	↑	Reinforces sympathetic nervous system to prepare the body for "fight or flight"
		Mobilizes carbohydrate and fat energy stores; increases blood glucose and blood fatty acids
CRH-ACTH-cortisol	↑	Mobilizes energy stores and metabolic building blocks for use as needed; increases blood glucose, blood amino acids, and blood fatty acids
		ACTH facilitates learning and behavior
Glucagon	↑	Act in concert to increase blood glucose and blood fatty acids
Insulin	↓	
Renin-angiotensin-aldosterone	↑	Conserves salt and H_2O to expand plasma volume; helps sustain blood pressure when acute loss of plasma volume occurs
Vasopressin	↑	
		Vasopressin and angiotensin II cause arteriolar vasoconstriction to increase blood pressure
		Vasopressin facilitates learning

ACTH-cortisol system. In fact, a condition is not considered to be stressful unless it is accompanied by increased secretion of cortisol. Although cortisol's precise role in adapting to stress is not known, a speculative but plausible explanation might be as follows. A primitive man or an animal wounded or faced with a threat must forego eating. A cortisol-induced shift away from protein and fat stores in favor of expanded carbohydrate stores and increased availability of blood glucose would help protect the brain from malnutrition during the imposed fasting period. Also, the amino acids liberated by protein catabolism would provide a readily available supply of building blocks for tissue repair should physical injury occur.

Besides the effects of the end hormone, cortisol, in the hypothalamo-pituitary-adrenal cortex axis, there is evidence that the anterior pituitary secretory product ACTH may play a role in resisting stress. Since ACTH is one of several peptides that facilitate learning and behavior, it is possible that an increase in ACTH during psychosocial stress helps the body cope more readily with similar stressors in the future by facilitating the learning of appropriate behavioral responses. Furthermore, ACTH is not released alone from its anterior pituitary storage vesicles. Pruning of the large pro-opiomelanocortin precursor molecule yields not only ACTH but also several other peptides, including morphine-like β endorphin and similar compounds. These compounds are cosecreted with ACTH upon stimulation by CRH during stress. It has been hypothesized that β endorphin, as a potent endogenous opiate (see p. 166), might exert a role in mediating analgesia (reduction of pain perception) should physical injury be inflicted during stress. It is further speculated that these cosecreted peptides have possible roles in learning, mood alterations, and appetite suppression, among other things. The precise contributions of these ACTH-related compounds during stress is unclear, but their role is a subject of considerable interest.

Other hormonal responses besides those of cortisol contribute to the overall metabolic response to stress. The sympathetic nervous system as well as the epinephrine secreted at its bidding both inhibit insulin and stimulate glucagon. These hormonal changes act in concert to elevate the blood levels of glucose and fatty acids. Epinephrine and glucagon, whose blood levels are elevated during stress, promote hepatic glycogenolysis and (along with cortisol), hepatic gluconeogenesis. However, insulin, whose secretion is suppressed during stress, opposes the breakdown of liver glycogen stores. All of these effects contribute to increasing the concentration of plasma glucose. They also promote a release of fatty acids from fat stores, because lipolysis is favored by epinephrine, glucagon, and cortisol but is opposed by insulin.

The primary stimulus for insulin secretion is a rise in blood glucose; a primary effect of insulin in turn is to lower the blood glucose. If it were not for the deliberate inhibition of insulin during the stress response, the hyperglycemia caused by stress would stimulate secretion of glucose-lowering insulin. As a result, the elevation in plasma glucose could not be sustained.

In addition to the hormonal changes that mobilize energy stores during stress, other hormones are simultaneously called into play to sustain blood volume and blood pressure during the emergency. The sympathetic system and epinephrine have major responsibilities in acting directly on the heart and blood vessels to improve circulatory function. In addition, the renin-angiotensin-aldosterone system is activated as a consequence of a sympathetically induced reduction of blood supply to the kidneys (see p. 481). Vasopressin secretion is also increased during stressful situations. Collectively, these hormones expand the plasma volume by promoting retention of salt and H_2O. Presumably, the enlarged plasma volume serves as a protective measure to help sustain blood pressure should acute loss of plasma fluid occur through hemorrhage or heavy sweating during the impending period of danger. Vasopressin and angiotensin also have direct vasopressor effects, which would be of benefit in maintaining an adequate arterial pressure in the face of acute blood loss. Vasopressin is further believed to facilitate learning, which has implications for future adaptation to stress.

The multifaceted stress response is coordinated within the central nervous system.

All of the individual responses to stress just described are either directly or indirectly influenced by the hypothalamus (Fig. 19–13). The hypothalamus receives input concerning physical and emotional stressors from virtually all areas of the brain

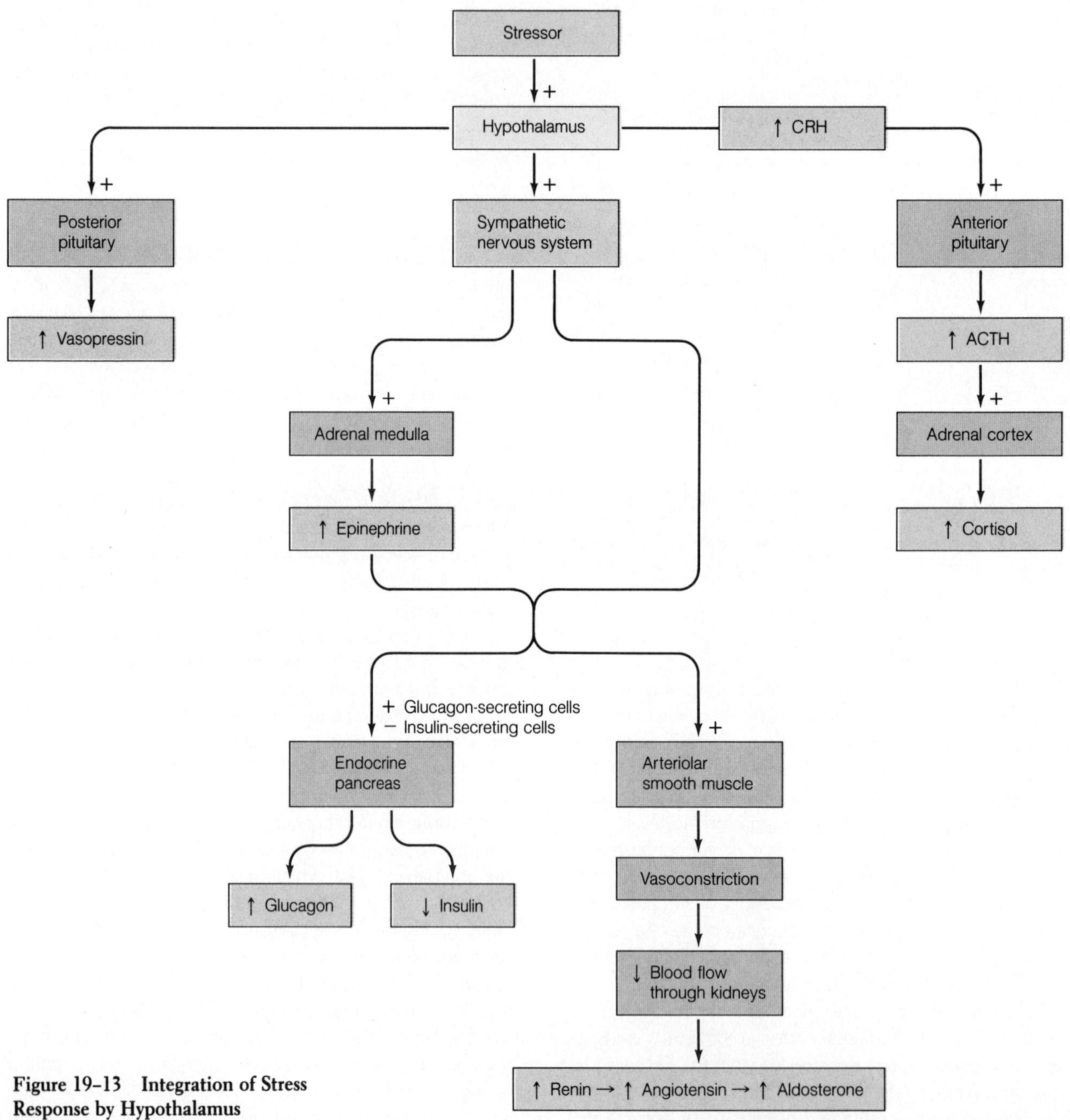

Figure 19–13 Integration of Stress Response by Hypothalamus

and from many receptors throughout the body. In response, the hypothalamus directly activates the sympathetic nervous system, secretes CRH to stimulate ACTH and cortisol release, and triggers the release of vasopressin. Sympathetic stimulation, in turn, brings about the secretion of epinephrine, with which it has a conjoined effect on the pancreatic secretion of insulin and glucagon. Furthermore, vasoconstriction of the renal afferent arterioles by the catecholamines indirectly triggers the secretion of renin by reducing the flow of oxygenated blood through the kidneys. Renin, in turn, sets in motion the renin-angiotensin-aldosterone mechanism. In this way, the hypothalamus integrates the responses of both the sympathetic nervous system and endocrine system during stress.

Activation of the stress response by chronic psychosocial stressors may be harmful.

Acceleration of cardiovascular and respiratory activity, retention of salt and H_2O, and mobilization of metabolic fuels and building blocks can be of benefit in response to a physical stressor, such as an athletic competition. However, even though the stressors in our everyday lives are mostly of a psychosocial nature, the same magnified responses are induced when the stressful situation is emotional or social instead of physical. Stressors such as anxiety about an exam, conflicts with loved ones, or impatience while sitting in a traffic jam can elicit a stress response. Although the rapid mobilization of body resources is appropriate in the face of real or threatened physical injury, it is generally inappropriate in response to nonphysical stress. If there is no extra energy demand, no tissue damage, and no blood loss, body stores are being broken down and fluid retained needlessly, probably to the detriment of the emotionally stressed individual. In fact, there is strong circumstantial evidence for a link between chronic exposure to psychosocial stressors and the development of pathological conditions such as atherosclerosis and high blood pressure, although no definitive cause-and-effect relationship has been ascertained. As a result of "unused" stress responses, could hypertension result from too much sympathetic vasoconstriction? too much salt and H_2O retention? too much vasopressin and angiotensin pressor activity? a combination of these? other factors? Recall that development of these same diseases can occur with prolonged exposure to pharmacological levels of glucocorticoids. Could long-standing lesser elevations of cortisol, such as might occur in the face of continual psychosocial stressors, do the same thing, only more slowly? Considerable work remains to be done to evaluate what contributions the stressors in our everyday life-styles have toward disease production.

Endocrine Control of Fuel Metabolism

All three classes of nutrient molecules can be used to provide cellular energy and, to a large extent, can be interconverted.

We have just discussed the metabolic changes that are elicited during the stress response. Now we will concentrate on the metabolic patterns that occur in the absence of stress, including the hormonal factors that govern this normal metabolism.

Metabolism refers to all of the chemical reactions that occur within the cells of the body. Those reactions involving the degradation and synthesis of the three classes of energy-rich organic molecules—protein, carbohydrate, and fat—are collectively known as **intermediary metabolism** or **fuel metabolism** (Table 19–4).

During the process of digestion, large nutrient molecules (**macromolecules**) are hydrolyzed into their smaller absorbable subunits as follows: proteins are converted into amino acids, complex carbohydrates into monosaccharides (mainly glucose), and triglycerides (dietary fats, also known as triacylglycerols) into monoglycerides (monoacylglycerols) and free fatty acids. These absorbable units are transferred from the digestive-tract lumen into the blood, either directly or by way of the lymph.

Constant exchange of these organic molecules occurs between the blood and the cells of the body (Fig. 19–14). On entering the cells, the small absorbed nutrient molecules have one of two fates. Either they are immediately oxidized as metabolic fuel to yield energy for the production of ATP, or they are resynthesized into larger organic macromolecules. The chemical reactions in which the organic molecules participate within the cells are categorized into two metabolic processes: anabolism and catabolism. **Anabolism** refers to the build up or synthesis of larger organic macromolecules from the small organic molecular subunits. Anabolic reactions generally require the input of energy in the form of ATP. These reactions result in either: (1) the manufacture of materials needed by the cell, such as cellular structural proteins or secretory product, or (2) storage, as in fat reservoirs, of excess ingested nutrients that are not immediately needed for energy production or needed as cellular building blocks. **Catabolism,** on the other hand, refers to the breakdown or degradation of large, energy-rich organic molecules within cells. Catabolism encompasses two levels of breakdown: (1) hydrolysis of large cellular organic macromolecules into their smaller subunits, similar to the process of digestion except that the reactions take place within the cells of the body instead of within the digestive-tract lumen (for example, release of glucose by the catabolism of stored glycogen); and (2) oxidation of the smaller subunits,

Table 19–4 **Summary of Reactions in Intermediary Metabolism of Organic Nutrients**

Metabolic Process	Reaction	Consequence
Glycogenesis	Glucose → Glycogen	↓ Blood glucose
Glycogenolysis	Glycogen → Glucose	↑ Blood glucose
Gluconeogenesis	Amino acids → Glucose	↑ Blood glucose
Protein synthesis	Amino acids → Protein	↓ Blood amino acids
Protein degradation	Protein → Amino acids	↑ Blood amino acids
Lipogenesis (triglyceride synthesis)	Fatty acids and glycerol → Triglycerides	↓ Blood fatty acids
Lipolysis (triglyceride degradation)	Triglycerides → Fatty acids and glycerol	↑ Blood fatty acids

such as glucose, to yield energy for ATP production (Fig. 19–15). Indeed, even the structural components of cells represent stored energy, albeit an "expensive" energy source, because they contain energy-rich proteins that can be cannibalized if necessary to yield energy. Alternatively to energy production, the smaller, multipotential organic subunits derived from intracellular hydrolysis may be released into the blood. These mobilized glucose, fatty acid, and amino acid molecules can then be used as needed for energy production or cellular synthesis elsewhere in the body.

In an adult, the rates of anabolism and catabolism are generally in balance, so the adult body remains in a dynamic steady state and appears unchanged even though the organic molecules that determine its structure and function are continuously being turned over. During growth, anabolism exceeds catabolism.

In addition to being able to resynthesize catabolized organic molecules back into the same type of molecules, many cells of the body, especially liver cells, have the ability to convert most types of small organic molecules into other types—as in, for example, the transformation of amino acids into glucose or fatty acids (Fig. 19–15). Because of these interconversions, adequate nourishment can be provided by a wide range of molecules present in different types of foods. There are limits, however. **Essential nutrients** such as the essential amino acids and vitamins cannot be formed in the body by conversion from another type of organic molecule.

The major fate of both ingested carbohydrates and fats is catabolism to yield energy. Amino acids are predominantly used for protein synthesis. However, amino acids can be used to supply energy after being converted to carbohydrate or fat. Thus all three categories of foodstuff can be used as fuel, and excesses of any foodstuff can be deposited as stored fuel.

At a superficial level, fuel metabolism appears relatively simple: the amount of nutrients in the diet must be sufficient to meet the body's needs for energy production and cellular synthesis. This simple relationship is complicated, however, by two important considerations. First, dietary fuel intake is intermittent, not continuous. As a result, excess energy must be absorbed during meals and stored for use during fasting periods between meals, when dietary sources of metabolic fuel are not available (Table 19–5). Excess circulating glucose is stored as ***glycogen***, a polymer of interconnected glucose molecules, in the liver and muscle. Because glycogen is a relatively small energy reservoir, less than a day's energy needs can be stored in this form. Once the liver and muscle glycogen stores are "filled up," additional glucose is transformed into fatty acids and glycerol, which are used to synthesize ***triglycerides***, primarily in adipose tissue (fat) and to a lesser extent in muscle. Excess circulating fatty acids derived from dietary intake also become incorporated into triglycerides. Excess circulating amino acids not needed for protein synthesis are not stored as extra protein but are converted to glucose and fatty acids, which ultimately end up being stored as triglycerides. Thus the major site of energy storage for excess nutrients of all three classes is adipose tissue. Normally there is sufficient stored triglyceride to provide energy for about two months, more so in an overweight individual. Consequently, during any prolonged period of fasting, the fatty acids released from triglyceride catabolism serve as the primary source of energy for most tissues. The catabolism of stored triglycerides frees glycerol as well as fatty acids, but quantitatively speaking, the fatty acids

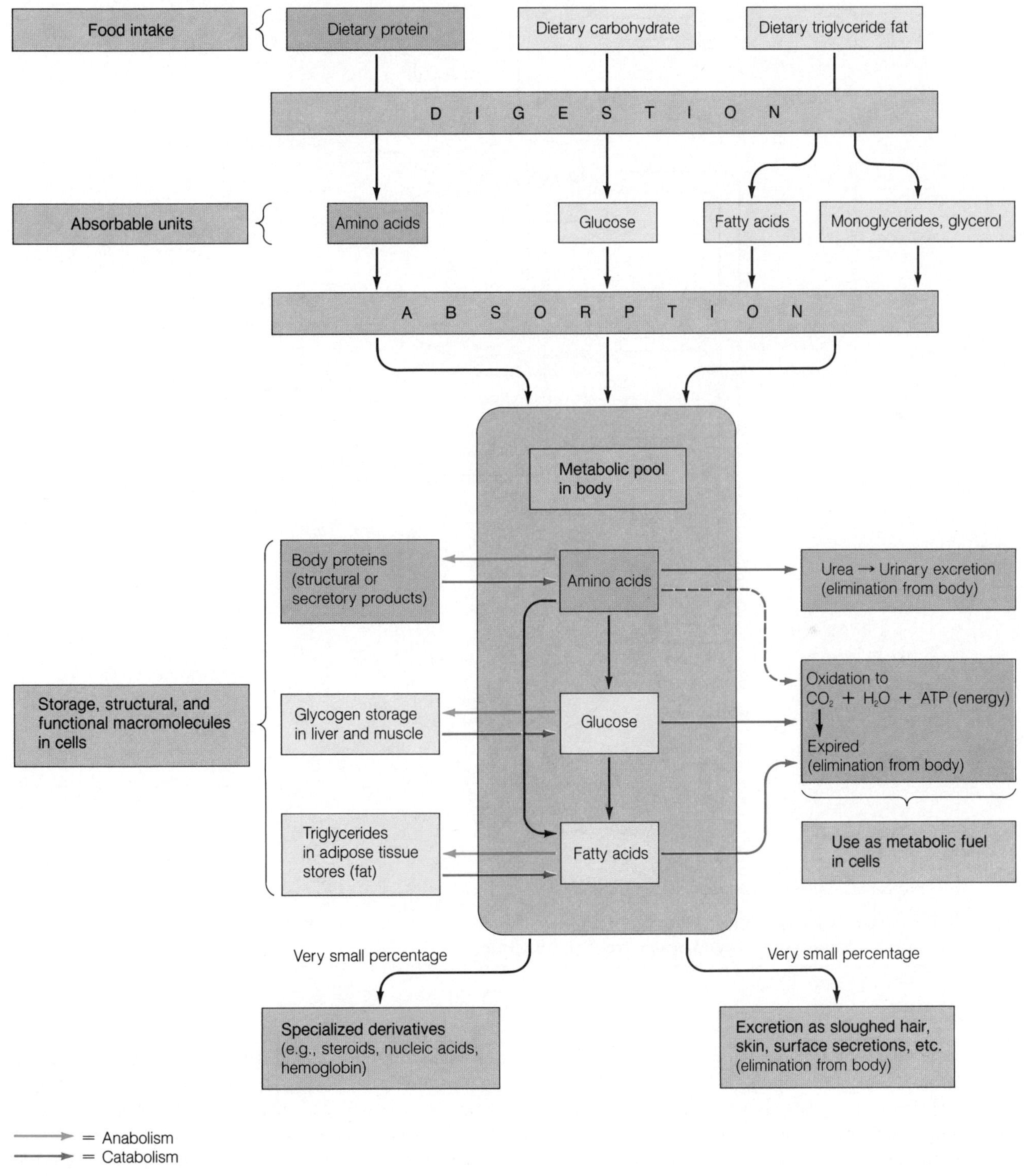

Figure 19–14 Summary of Major Pathways Involving Organic Nutrient Molecules

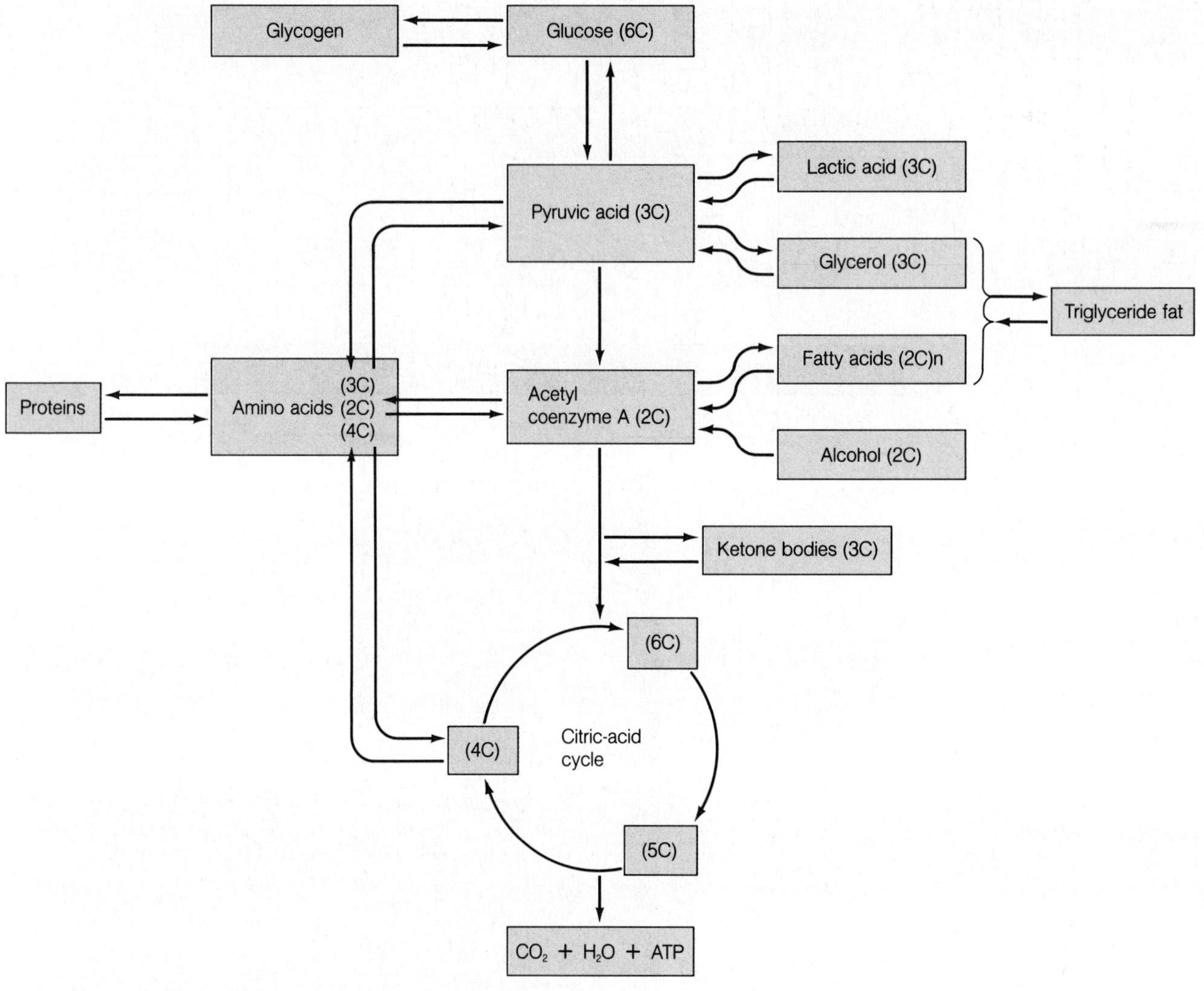

Figure 19–15 Major Pathways in Energy Production by, Storage of, and Interconversions between Organic Nutrient Molecules

are far more important. Catabolism of stored fat yields 90% fatty acids and 10% glycerol by weight. Glycerol can be converted to glucose by the liver and contributes in a small way to maintaining blood glucose during a fast.

As a third energy reservoir, a substantial amount of energy is stored as ***structural protein***, primarily in muscle, the most abundant protein mass in the body. However, protein is not a first choice to tap as an energy source because it serves other essential functions, in contrast to the glycogen and triglyceride reservoirs, which serve solely as energy depots.

The second factor complicating fuel metabolism is that the brain normally depends on the delivery of adequate blood glucose as its sole source of energy. Consequently, it is essential that the blood glucose concentration be maintained above a critical level. The blood glucose concentration is typically 100 mg glucose/100 ml plasma and is normally maintained within the narrow limits of 70–110 mg/100 ml. Liver glycogen is an important reservoir for maintaining blood glucose levels during a short fast. However, hepatic glycogen is depleted relatively rapidly, so during a longer fast, other mechanisms must

Table 19–5 Stored Metabolic Fuel in Body

Fuel	Circulating Form	Storage Form	Major Storage Site	Percentage of Total Body Energy Content (and Calories*)	Reservoir Capacity	Role
Carbohydrate	Glucose	Glycogen	Liver, muscle	1% (1,500 calories)	Less than a day's worth of energy	First energy source; essential for brain
Fat	Free fatty acids	Triglycerides	Adipose tissue	77% (143,000 calories)	About 2 months' worth of energy	Primary energy reservoir; energy source during a fast
Protein	Amino acids	Body proteins	Muscle	22% (41,000 calories)	Death results long before capacity is utilized because of structural and functional impairment	Source of glucose for brain during a fast; last resort to meet other energy needs

*Actually refers to kilocalories; see p. 604.

be used to assure that the energy requirements of the glucose-dependent brain are met. First, when new dietary glucose is not entering the blood, tissues not obligated to use glucose shift their metabolic gears to burn fatty acids instead, thus sparing glucose for the brain. Fatty acids are made available by catabolism of triglyceride stores as an alternative energy source for non-glucose-dependent tissues. Second, amino acids can be converted to glucose by gluconeogenesis (production of "new" glucose from noncarbohydrate sources), whereas fatty acids cannot. Thus, once glycogen stores are depleted despite glucose-sparing, new glucose supplies for the brain are provided by the catabolism of body proteins and conversion of the freed amino acids into glucose.

Metabolic fuels are stored during the absorptive state and are mobilized during the postabsorptive state.

From the preceding discussion, it should be obvious that the disposition of organic molecules depends on the body's metabolic state. There are two functional metabolic states related to eating and fasting cycles—the absorptive state and postabsorptive state, respectively (Table 19–6). Following a meal, ingested nutrients are being absorbed and are entering the blood during the **absorptive state.** During this time glucose is plentiful and serves as the major energy source. Very little of the absorbed fat and amino acids are used for energy during the absorptive state because most of the cells prefer to use glucose when it is available. Extra nutrients not immediately used for energy or structural repairs are channeled into storage as glycogen or triglycerides.

The average meal is completely absorbed in about four hours. Therefore, on a typical three-meal-a-day diet, no nutrients are being absorbed from the digestive tract during late morning, late afternoon, and throughout the night. These times constitute the **postabsorptive** or **fasting state.** During this state, endogenous energy stores are mobilized to provide energy, whereas gluconeogenesis and glucose-sparing are used to maintain the blood glucose at an adequate level for nourishment of the brain. The synthesis of protein and fat are curtailed. Instead, stores of these organic molecules are catabolized for glucose formation and energy production, respectively. Carbohydrate synthesis does occur through gluconeogenesis, but the utilization of glucose for energy is greatly reduced.

Note that the blood concentration of nutrients does not fluctuate markedly between the absorptive and postabsorptive states. During the absorptive state, the glut of absorbed nutrients is swiftly removed from the blood and placed into storage; during the postabsorptive state, these stores are catabolized to maintain the blood concentrations at levels necessary to sustain tissue energy demands.

Table 19-6 Comparison of Absorptive and Postabsorptive States

Metabolic Factor	Absorptive State	Postabsorptive State
Carbohydrates	Glucose providing major energy source Glycogen synthesis and storage Conversion to and storage as triglyceride fat	Glycogen degradation and depletion Glucose-sparing to conserve glucose for brain Production of new glucose through gluconeogenesis
Fats	Triglyceride synthesis and storage	Triglyceride catabolism Fatty acids providing major energy source for non-glucose-dependent tissues
Proteins	Protein synthesis Conversion to and storage as triglyceride fat	Protein catabolism Amino acids used for gluconeogenesis

As already described, during these alternating metabolic states, various tissues play different roles. The liver plays the primary role in maintenance of normal blood glucose levels. It stores glycogen when excess glucose is available, releases glucose into the blood when needed, and is the principal site for metabolic interconversions such as gluconeogenesis. Adipose tissue serves as the primary energy storage site and is important in the regulation of fatty acid levels in the blood. Muscle is the primary site of amino acid storage and is the major energy user. The brain normally can use only glucose as an energy source, yet it does not store glycogen, making it mandatory that blood glucose levels be maintained.

Several other organic intermediates play a lesser role as energy sources—namely, glycerol, lactic acid, and ketone bodies. As mentioned earlier, ***glycerol*** derived from triglyceride hydrolysis (it is the backbone to which the fatty acid chains are attached) can be converted to glucose by the liver. Similarly, ***lactic acid***, which is produced by the incomplete catabolism of glucose via glycolysis in muscle (Fig. 19–15; also see p. 237), can also be converted to glucose in the liver. ***Ketone bodies*** are produced by the liver during glucose-sparing. When the liver uses fatty acids as an energy source, unlike other tissues it oxidizes fatty acids only to acetyl CoA, which it is unable to process through the citric-acid cycle for further energy extraction. Thus the liver does not degrade fatty acids all the way to CO_2 and H_2O for maximum energy release. Instead, it partially extracts the available energy and converts the remaining energy-bearing acetyl CoA molecules into a group of compounds called **ketone bodies,** which it releases into the blood. Ketone bodies serve as an alternative energy source for tissues capable of further oxidizing them by means of the citric-acid cycle.

During long-term starvation, the brain starts using ketones instead of glucose as a major energy source. Since death resulting from starvation is usually due to protein wasting rather than hypoglycemia (low blood glucose), prolonged survival without any caloric intake requires that gluconeogenesis be kept to a minimum while the energy needs of the brain are not compromised. A sizable portion of cell protein can be catabolized without serious cellular malfunction, but a point is finally reached at which a cannabilized cell is no longer able to function adequately. To ward off the fatal point of failure as long as possible during prolonged starvation, the brain starts using ketones as a major energy source, correspondingly decreasing its use of glucose. Use by the brain of this fatty acid "table scrap" left over from the liver's "meal" limits the necessity of mobilizing body proteins for glucose production to nourish the brain. Both of the major metabolic adaptations to prolonged starvation—a decrease in protein catabolism and use of ketones by the brain—are attributable to the high levels of ketones in the blood at the time. Utilization of ketones by the brain occurs only when the blood ketone level is high. The high blood levels of ketones also directly inhibit protein degradation in muscle. Thus ketones are responsible for sparing body proteins while satisfying the brain's energy needs.

Insulin and glucagon are the most important hormonal products of the islets of Langerhans, the endocrine units of the pancreas.

How does the body "know" when to shift its metabolic gears from one of net anabolism and nutrient storage to one of net catabolism and glucose sparing? The flow of organic nutrients along metabolic pathways is influenced by a variety of hormones, including insulin, glucagon, epinephrine, cortisol, thyroid hormone, and growth hormone. Under most circumstances, the pancreatic hormones are the dominant hormonal regulators that shift the metabolic pathways back and forth from net anabolism to net catabolism and glucose sparing, depending on whether the body is in a state of feasting or fasting, respectively.

The **pancreas** is an organ composed of both exocrine and endocrine tissues. The exocrine portion of the pancreas secretes a watery alkaline solution and digestive enzymes

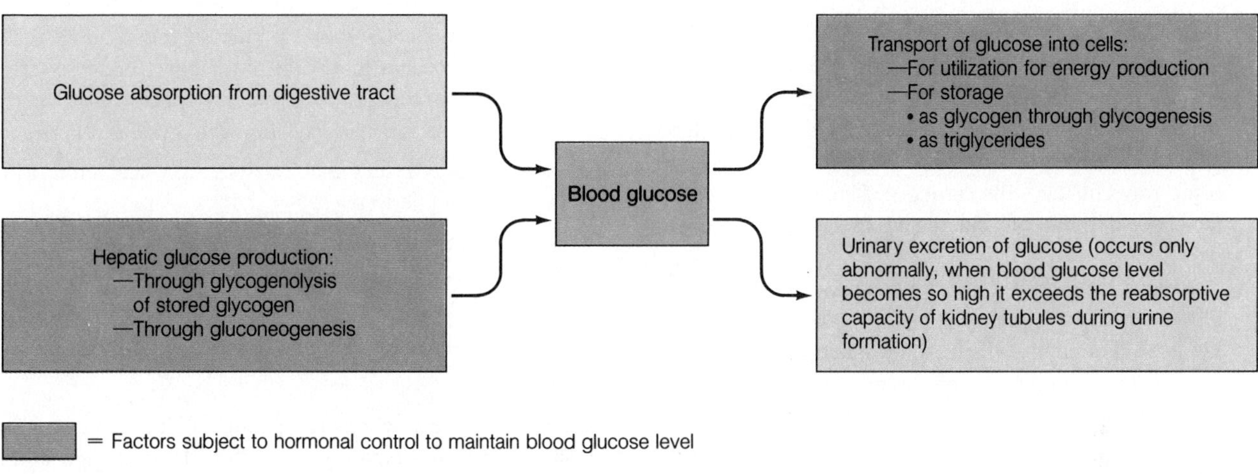

Figure 19–16 Factors Affecting Blood Glucose Concentration

through the pancreatic duct into the digestive-tract lumen. Scattered throughout the pancreas between the exocrine cells are clusters, or "islands," of endocrine cells known as the **islets of Langerhans** (see Fig. 16–15). The islets are composed of four major cell types, each producing a different hormone. The most abundant pancreatic endocrine-cell type is the β (**beta**) **cell,** the site of *insulin* synthesis and secretion. Next most important are the α (**alpha**) **cells,** which produce *glucagon*. The **D** (**delta**) **cells** are the pancreatic site of *somatostatin* synthesis, whereas the least common islet cells, the **PP cells,** secrete *pancreatic polypeptide*.

The most important pancreatic hormones in the regulation of fuel metabolism are insulin and glucagon. Accordingly, we will pay the most attention to these two pancreatic hormones, giving only a brief highlight now of the other two endocrine pancreatic secretory products. **Somatostatin** is also produced by the hypothalamus, where it serves to inhibit the secretion of growth hormone and TSH. Furthermore, somatostatin is produced by cells lining the digestive tract, where it is suspected of acting locally as a paracrine to inhibit most digestive processes. Pancreatic somatostatin also exerts a variety of inhibitory effects on the digestive system, the overall effect of which is to inhibit digestion of nutrients and to decrease nutrient absorption. Somatostatin is released from the pancreatic D cells directly in response to an increase in blood glucose and blood amino acids during absorption of a meal. By exerting its inhibitory effects, pancreatic somatostatin acts in negative-feedback fashion to put the brakes on the rate at which the meal is being digested and absorbed, thereby preventing excessive plasma levels of nutrients. In addition, pancreatic somatostatin may play an important role in the local regulation of pancreatic hormone secretion. The secretion of insulin, glucagon, pancreatic polypeptide, and somatostatin itself is decreased by the local presence of somatostatin, but the physiological importance of such paracrine function has not been determined.

Little is known about **pancreatic polypeptide.** Its effects seem to be concerned primarily with inhibiting gastrointestinal function. Like somatostatin, pancreatic polypeptide appears to have no direct effects on carbohydrate, protein, or fat metabolism.

Insulin lowers blood glucose, amino acid, and fatty acid levels and promotes anabolism of these small nutrient molecules.

Insulin has important effects on carbohydrate, fat, and protein metabolism. It lowers the blood levels of glucose, fatty acids, and amino acids and promotes their storage. As these nutrient molecules enter the blood during the absorptive state, insulin promotes their cellular uptake and conversion into glycogen, triglycerides, and protein, respectively. Insulin exerts its many effects either by altering transport into cells of specific blood-borne nutrients or by altering the activity of the enzymes involved in specific metabolic pathways.

ACTIONS ON CARBOHYDRATES. The maintenance of blood glucose homeostasis is a particularly important function of the pancreas. Circulating glucose concentrations are determined by the balance that exists among the following processes (Fig. 19–16): glucose absorption from the digestive tract; transport of glucose into cells; cellular (primarily hepatic) glucose production; and (abnormally) urinary excretion of glucose.

Insulin exerts a four-fold effect to lower blood glucose levels. Specifically, insulin does the following:

1. It facilitates glucose transport into most cells. Glucose molecules cannot readily penetrate most cell membranes in the absence of insulin. Most tissues, therefore, are highly dependent on insulin for uptake of glucose from the blood and for its subsequent use. It is believed that glucose in some way enhances the carrier-mediated mechanism for facilitated diffusion of insulin into these insulin-dependent cells.

   Several tissues are not dependent on insulin for their glucose uptake—namely, the brain, working muscles, and the liver. The brain, which requires a constant supply of glucose for its minute-to-minute energy needs, is freely permeable to glucose at all times. In comparison, skeletal-muscle cells are not freely permeable to glucose except during exercise. Resting muscle cells depend on insulin for the uptake of glucose and its conversion into stored glycogen for later use. For reasons that are unclear, glucose uptake and use by muscle cells are increased during exercise, even though insulin levels are reduced during this time. Thus working muscles are not dependent on insulin for their glucose uptake. The liver is also not dependent on insulin for glucose uptake; however, insulin does enhance the metabolism of glucose by the liver by stimulating the first step in glucose metabolism.
2. Insulin stimulates **glycogenesis,** the production of glycogen from glucose. It promotes the conversion of glucose into its storage form glycogen in both skeletal muscle and liver.
3. Insulin inhibits glycogenolysis, the breakdown of glycogen into glucose. By inhibiting the degradation of glycogen, it likewise favors carbohydrate storage and decreases glucose output by the liver.
4. Insulin inhibits gluconeogenesis, the conversion of amino acids into glucose. It further decreases hepatic glucose output by reducing the production of glucose from amino acids in the liver. It does so in two ways: by decreasing the amount of amino acids in the blood that are available to the liver for gluconeogenesis, and by inhibiting the hepatic enzymes required for the conversion of amino acids into glucose.

Thus insulin decreases the concentration of blood glucose by promoting the cells' removal of glucose from the blood for utilization and storage, while it simultaneously blocks the two mechanisms by which the liver releases glucose into the blood. Insulin is the only hormone capable of lowering the blood glucose level.

ACTIONS ON FAT. Insulin exerts multiple effects to lower blood fatty acids and promote triglyceride storage as follows.

1. It increases the transport of glucose into adipose tissue cells, as it does for most cells of the body. Glucose serves as a precursor for the formation of fatty acids and glycerol, which are the raw materials for triglyceride synthesis.
2. It activates enzymes that catalyze fatty-acid production from glucose derivatives.
3. It promotes entry of fatty acids from the blood into adipose tissue cells.
4. It inhibits lipolysis (triglyceride catabolism), thus reducing the conversion of triglycerides into fatty acids and glycerol.

Collectively, these actions favor removal of glucose and fatty acids from the blood and promote their storage as triglycerides.

ACTIONS ON PROTEIN. Insulin lowers blood amino acid levels and enhances protein synthesis through the following actions.

1. It promotes the active transport of amino acids from the blood into muscles and other tissues. This effect decreases the circulating amino acid level and provides the building blocks for protein synthesis within the cells.
2. It increases the rate of amino acid incorporation into protein by increasing the activity of the ribosomal protein-synthesizing machinery.
3. It inhibits protein degradation.

The collective result of these actions is a protein anabolic effect. For this reason, insulin is essential for normal growth.

In short, insulin stimulates biosynthetic pathways that lead to increased glucose utilization, increased carbohydrate and fat storage, and increased protein synthesis. In so doing, this hormone lowers the blood glucose, fatty acid, and amino acid levels. This metabolic pattern is characteristic of the absorptive state. Indeed, insulin secretion rises during this state and is responsible for shifting metabolic pathways to net anabolism.

When insulin secretion is low, the opposite effects occur. The rate of glucose entry into cells is reduced, and net catabolism rather than net synthesis of glycogen, triglycerides, and protein occurs. This pattern is reminiscent of the postabsorptive state; indeed, insulin secretion is reduced during the postabsorptive state. However, the other major pancreatic hormone, glucagon, also plays an important role in shifting from absorptive to postabsorptive metabolic patterns, as will be described later.

The primary stimulus for increased insulin secretion is an increase in blood glucose concentration.

The primary control of insulin secretion is a direct negative-feedback system between the pancreatic β cells and the concentration of glucose in the blood flowing to them. An elevated blood glucose level, such as occurs during absorption of a meal, directly stimulates synthesis and release of insulin by the β cells. The increased insulin, in turn, reduces the blood glucose to normal while it promotes utilization and storage of this nutrient. Conversely, a fall in blood glucose below normal, such as occurs during fasting, directly inhibits insulin secretion. Lowering the rate of insulin secretion shifts metabolism from the absorptive to the postabsorptive pattern. Thus this simple negative-feedback system is able to maintain a relatively constant supply of glucose to the tissues without requiring the participation of nerves or other hormones.

In addition to plasma glucose concentration, there are other inputs in the regulation of insulin secretion, including the following (Fig. 19–17).

□ An elevated plasma amino acid level, such as occurs following ingestion of a high-protein meal, directly stimulates the β cells to increase insulin secretion. In negative-feedback fashion, the increased insulin enhances the entry of these amino acids into the cells, lowering the blood amino acid level while promoting protein synthesis.

□ The major gastrointestinal hormones secreted by the digestive tract in response to the presence of food, especially gastric inhibitory peptide (see p. 600), stimulate pancreatic insulin secretion in addition to their direct regulatory effects on the digestive system. Through this control, insulin secretion is increased in "feed-forward" or anticipatory fashion even before nutrient absorption increases the concentration of glucose and amino acids in the blood.

□ The autonomic nervous system also directly influences insulin secretion. The islets are richly innervated by both parasympathetic (vagal) and sympathetic nerve fibers. The increase in vagal activity that occurs in response to food in the digestive tract stimulates insulin release. This, too, is a feed-forward response in anticipation of nutrient absorption. In contrast, both the activation of sympathetic input to the pancreas and the concurrent increase in epinephrine decrease insulin secretion. The reduction in insulin allows the blood glucose level to increase, which is appropriate for the circumstances under which generalized sympathetic activation occurs—namely, stress (fight or flight) and exercise. In both of these situations, extra fuel is needed for increased muscle activity.

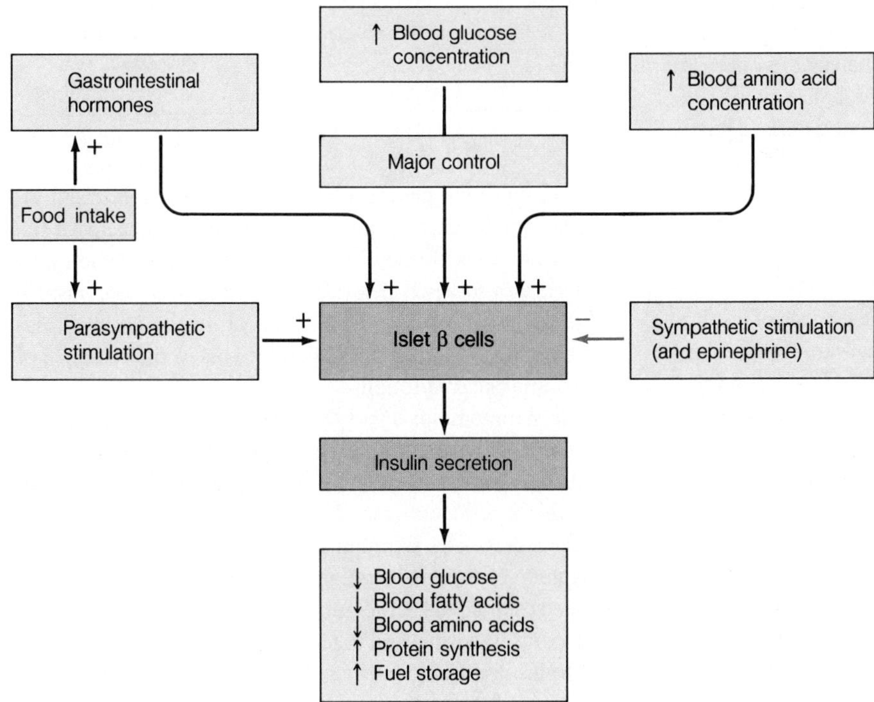

Figure 19–17 Factors Controlling Insulin Secretion

There are two types of diabetes mellitus, depending on the insulin-secreting capacity of the cells.

Diabetes mellitus is by far the most common of all endocrine disorders. The acute symptoms of diabetes mellitus are attributable to inadequate insulin action. Because insulin is the only hormone capable of lowering blood glucose levels, one of the most prominent features of diabetes mellitus is elevated blood glucose levels, or **hyperglycemia.** *Diabetes* literally means "syphon" or "running through," in reference to the large urine volume accompanying this condition. A large urine volume occurs in both diabetes mellitus due to insulin insufficiency and diabetes insipidus due to vasopressin deficiency. *Mellitus* means "sweet" ; *insipidus* means "tasteless." The urine of patients with diabetes mellitus acquires its sweetness from excess blood glucose that spills into the urine, whereas there is no sugar in the urine of patients with diabetes insipidus, so it is tasteless. (Aren't you glad you were not a health professional in the time when these two conditions were distinguished on the basis of the taste of the urine?)

There are two major variants of diabetes mellitus, differing with respect to the capacity for pancreatic insulin secretion (Table 19–7): *Type I* (*insulin-dependent* or *juvenile-onset*) *diabetes*, which accounts for 10–20% of all cases of diabetes, and *Type II* (*noninsulin-dependent* or *maturity-onset*) *diabetes*. Genetic as well as environmental factors appear to be important in the development of both types of diabetes mellitus. Although either type can first be manifested at any age, Type I has a greater prevalence in children, whereas the onset of Type II more generally occurs in adulthood, hence giving rise to the age-related designations of the two conditions. Type I diabetics suffer a total or near-total lack of insulin secretion by their pancreatic β cells; thus they require exogenous insulin for survival. Type II diabetics, on the other hand, do secrete insulin in varying amounts. In fact, insulin levels may be normal or even exceed those in nondiabetics.

The basic problem in Type II diabetes is not lack of insulin but reduced sensitivity of insulin's target cells to its presence, usually because of down-regulation (see p. 643) of insulin receptors in association with obesity. Chronic overeating by an obese person results in the secretion of increased amounts of insulin to maintain the blood glucose at normal levels by putting the excess nutrients in storage. In response to chronic hyperinsulinemia, the number of insulin receptors gradually becomes reduced over time. The resultant decrease in sensitivity to insulin in obese but otherwise normal individuals is overcome by secretion of additional insulin. In this way, the excess nutrients are stored despite the decreased availability of insulin receptors, so blood glucose homeostasis is maintained. In obese diabetes-prone individuals, however, the sustained overtasking of the pancreas by chronic excessive nutrient intake eventually exceeds the reserve secretory capacity of the genetically weak β cells. Even though insulin secretion may be normal or somewhat elevated, symptoms of insulin insufficiency develop because the amount of insulin is still inadequate to prevent significant hyperglycemia in the presence of excess nutrient absorption.

Table 19–7 Comparison of Type I and Type II Diabetes Mellitus

Characteristic	Type I Diabetes	Type II Diabetes
Level of insulin secretion	None or almost none	May be normal or exceed normal
Typical age of onset	Childhood	Adulthood
Percent of diabetics	10–20%	80–90%
Basic defect	Destruction of β cells	Reduced sensitivity of insulin's target cells
Associated with obesity?	No	Usually
Genetic and environmental factors important in precipitating overt disease?	Yes	Yes
Speed of development of symptoms	Rapid	Slow
Development of ketosis	Common if untreated	Rare
Treatment	Insulin injections; dietary management	Dietary control and weight reduction; occasionally oral hypoglycemic drugs

The symptoms in Type II diabetes are usually slower in onset and less severe than in Type I diabetes. Whereas Type I diabetics are permanently insulin-dependent, dietary control and weight reduction may be all that is necessary to completely reverse the symptoms in Type II diabetics. As the magnitude of insulin secretion decreases in connection with reduced caloric intake, the number of insulin receptors gradually returns to normal and so too does target-tissue responsiveness to insulin. Exercise is also useful in the management of both types of diabetes, because working muscles are not insulin dependent. Exercising muscles take up and use some of the excess glucose in the blood, thus reducing the overall need for insulin.

Occasionally Type II diabetics are administered drugs that accomplish removal of excess nutrients from the blood by driving the weakened β cells to secrete more insulin than they do on their own. These therapeutic agents are referred to as ***oral hypoglycemic drugs*** since they can be taken orally to bring about a reduction in blood glucose. In contrast, insulin must be administered by injection, because this peptide hormone would be digested by proteolytic enzymes if it were to be swallowed. Although oral hypoglycemics alleviate the symptoms of diabetes, their use is controversial because they overwork already weakened β cells, the end result of which may be the complete "burnout" of these cells so that they are no longer able to produce insulin at all. In such a case, the previously noninsulin-dependent patient would have to be placed on insulin therapy for life. Weight reduction and exercise are safer and more permanent solutions.

The symptoms of diabetes mellitus are characteristic of an exaggerated postabsorptive state.

The acute consequences of diabetes mellitus can be grouped according to the effects of insulin deficiency on carbohydrate, fat, and protein metabolism (Fig. 19–18). Since the postabsorptive metabolic pattern is induced by low insulin activity, the changes that occur in diabetes mellitus are an exaggeration of this state, with the exception of hyperglycemia. In the usual fasting state, the blood glucose level is slightly below normal.

The elevated blood glucose level, which is a hallmark of diabetes mellitus, arises from a combination of factors. In the face of insulin deficiency, less glucose than normal is removed from the blood for cellular use and glycogen storage. This alone leads to an increase in blood glucose. Meanwhile, hepatic output of glucose increases as the glucose-producing processes of glycogenolysis and gluconeogenesis proceed unchecked in the absence of insulin, further contributing to hyperglycemia. Because most of the body's cells are unable to use glucose without the assistance of insulin, there is an ironic extracellular glucose excess coincident with an intracellular glucose deficiency—"starvation in the midst of plenty." Even though the noninsulin-dependent brain is adequately nourished during diabetes mellitus, further consequences of the disease lead to brain dysfunction, as you will see shortly.

When the blood glucose rises to the level that the amount of glucose filtered exceeds the tubular cells' capacity for reabsorption, glucose appears in the urine (glucosuria). Glucose in the urine exerts an osmotic effect that draws H_2O with it, producing an osmotic diuresis characterized by polyuria (frequent urination). The excess fluid lost from the body leads to dehydration, which in turn can ultimately lead to peripheral circulatory failure because of the marked reduction in blood volume. Circulatory failure, if uncorrected, can lead to death because of low cerebral blood flow or because of secondary renal failure due to inadequate filtration pressure. Furthermore, cells lose water as the body becomes dehydrated as a result of an osmotic shift of water from the cells into the hypertonic extracellular fluid. Brain cells are especially sensitive to shrinking so that nervous system malfunction ensues (see p. 517). Another characteristic symptom of diabetes mellitus is polydipsia (excessive thirst), which is actually a compensatory mechanism that takes place to counteract the dehydration.

The story is still not complete. In the face of intracellular glucose deficiency, appetite is stimulated, leading to polyphagia (excessive food intake). In spite of the increased food intake, however, progressive weight loss occurs as a result of the effects of insulin deficiency on fat and protein metabolism. Triglyceride synthesis decreases while lipolysis increases, resulting in large-scale mobilization of fatty acids from triglyceride stores. The increased blood fatty acids are used to a large extent by the cells as an alternative energy source. Increased liver utilization of fatty acids results in the release of excessive ketone bodies into the blood, causing **ketosis.** Since the ketone bodies include several different acids that result from the incomplete breakdown of fat during hepatic energy production, this developing ketosis leads to progressive metabolic acidosis. Acidosis depresses the brain, and, if severe enough, can lead to diabetic coma and death.

A compensatory measure for metabolic acidosis is increased ventilation to blow off extra acid-forming CO_2. Exhalation of one of the ketone bodies, acetone, causes a "fruity" breath odor. Sometimes a patient collapsed in a diabetic coma is unfortunately mistaken by passersby as a "wino" passed out in a state of drunkenness because of this odor. (This situation is an example of the merits of medical-alert identification tags.) Persons with Type I diabetes are much more prone to develop ketosis than are Type II diabetics.

The effects of a lack of insulin on protein metabolism result in a net shift toward protein catabolism. The net breakdown of muscle proteins leads to wasting and weakness of skeletal muscles and, in child diabetics, a reduction in overall growth. Reduced amino acid uptake coupled with increased protein degradation result in **aminoacidemia** (excess amino acids in the blood). The increased circulating amino acids can be used for additional gluconeogenesis, which further aggravates the hyperglycemia.

As you can readily appreciate from this overview, diabetes mellitus is a complicated disease that can lead to disturbances in carbohydrate, fat, and protein metabolism and in fluid

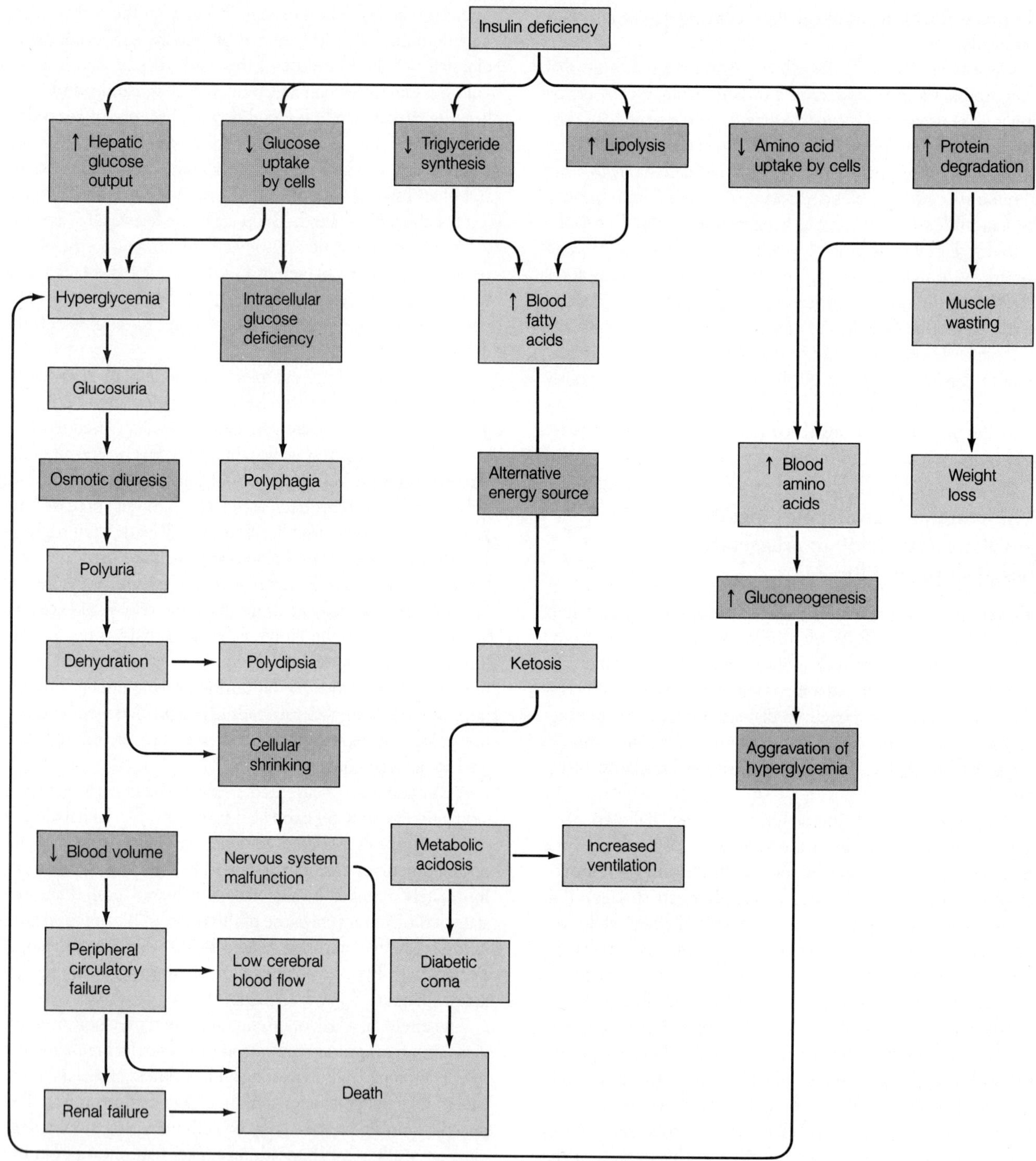

Figure 19–18 Acute Effects of Diabetes Mellitus

and acid-base balance. It can also have repercussions on the circulatory system, kidneys, respiratory system, and nervous system.

In addition to these potential consequences of untreated diabetes, which can be explained on the basis of insulin's short-term metabolic effects, numerous long-range complications of this disease frequently occur that are not as easy to explain. These chronic complications, which account for the shorter life expectancy of diabetics, primarily involve degenerative disorders of the vasculature and nervous system. Cardiovascular lesions are the most common cause of premature death in diabetics. Heart disease and strokes occur with greater incidence than in nondiabetics. Because vascular lesions often develop in the kidneys and retinas of the eyes, kidney failure is a very common long-term complication of diabetes, and diabetes is the second leading cause of blindness in the United States. Extremities may become gangrenous as a result of impaired delivery of blood to these tissues, which leads to the necessity of amputating toes or even whole limbs. In addition to the problems related to the circulatory system, degenerative lesions in nerves lead to multiple neuropathies that result in dysfunction of the brain, spinal cord, and peripheral nerves.

The mechanisms responsible for the development of these long-range vascular and neural degenerative complications are largely unknown. There is considerable disagreement about whether or not good management of the diabetic patient's condition to continuously keep the blood glucose levels within reasonable limits diminishes the incidence of the chronic abnormalities.

Current research on several fronts may dramatically change the approach to diabetic therapy. New treatments may include such things as β cell transplants and implanted, glucose-sensitive insulin-releasing devices.

Insulin excess causes brain-starving hypoglycemia.

Let us now look at the opposite of diabetes mellitus, insulin excess, which can occur in two different ways. First, insulin excess can occur in a diabetic patient when too much insulin has been injected for the person's caloric intake and exercise level. This is so-called **insulin shock.** Second, an abnormally high blood insulin level may occur in a nondiabetic individual with a β cell tumor or in whom the β cells are overresponsive to glucose. The true incidence of the latter condition, so-called **reactive hypoglycemia,** is a subject of intense controversy, because laboratory measurements to confirm the presence of low blood glucose during the time of symptoms have not been performed in the majority of people who have been diagnosed as having the condition. In mild cases, the symptoms of hypoglycemia, such as tremor, fatigue, sleepiness, and inability to concentrate, are nonspecific. Because these symptoms could also be attributable to emotional problems or other factors, a definitive diagnosis based on symptoms alone is impossible to make.

The consequences of insulin excess are primarily manifestations of the effects of hypoglycemia on the brain. Recall that the brain relies on a continuous supply of blood glucose for its nourishment and that glucose uptake by the brain does not depend on insulin. With insulin excess, more glucose than necessary is driven into the other insulin-dependent cells of the body. The result is a lowering of the blood glucose level so that not enough glucose is left in the blood to be delivered to the brain. The brain literally starves in the presence of hypoglycemia. The symptoms, therefore, are primarily referable to depressed brain function, which, if severe enough, may rapidly progress to unconsciousness and death. Persons with overresponsive β cells usually do not become sufficiently hypoglycemic to manifest these more serious consequences, but they do show milder symptoms of depressed CNS activity.

Treatment depends on the cause. In the case of a diabetic with insulin overdose, something sweet should be taken at the first indication of a hypoglycemic attack. If the patient is unconscious, glucose must be given intravenously. Prompt treatment of severe hypoglycemia is imperative if brain damage is to be prevented. Note that a diabetic can lose consciousness and die either from diabetic ketoacidotic coma caused by prolonged insulin deficiency or from acute hypoglycemia caused by insulin shock. Fortunately, the other accompanying signs and symptoms differ sufficiently between the conditions to enable medical caretakers to administer appropriate therapy, either insulin or glucose. For example, ketoacidotic coma is accompanied by deep, labored breathing and fruity breath, whereas insulin shock is not.

Ironically, in the case of reactive hypoglycemia, persons who suffer from this low-blood sugar disorder are treated with a low-carbohydrate diet. If a carbohydrate-rich meal is consumed, the overresponsive β cells "overshoot" and secrete more insulin than is necessary in response to the elevated blood glucose, thereby producing a hypoglycemic condition; that is, too much glucose is driven into the cells by the excessive insulin. With low carbohydrate intake, the blood glucose does not rise as much during the absorptive state. Since blood glucose elevation is the primary regulator of insulin secretion, the β cells are not stimulated as much with a low-carbohydrate meal as with a typical meal. Accordingly, because these cells do not have as much opportunity to overrespond with excess insulin output, reactive hypoglycemia is less likely to occur.

Giving a symptomatic individual with reactive hypoglycemia something sweet, as is done with a hypoglycemic diabetic, temporarily alleviates the symptoms, because the blood glucose level is transiently restored to normal so that the brain's energy needs are once again satisfied. As soon as the extra glucose triggers further insulin release, however, the situation is merely aggravated.

Glucagon in general opposes the actions of insulin.

Even though insulin plays a central role in controlling the metabolic adjustments between the absorptive and postabsorptive states, it is becoming increasingly apparent that the secretory product of the pancreatic islet α cells, **glucagon,** is also very important. Many physiologists now view the β and α cells as a coupled endocrine system whose combined secretory output, insulin plus glucagon, is a major factor in the regulation of intermediary metabolism.

Glucagon affects many of the same metabolic processes that are influenced by insulin, but in most cases glucagon's actions are opposite to those of insulin. The major site of action of glucagon is the liver, where it exerts a variety of effects on carbohydrate, fat, and protein metabolism.

ACTIONS ON CARBOHYDRATE. The overall effects of glucagon on carbohydrate metabolism result in an increase in hepatic glucose production and release and thus an increase in blood glucose levels. Glucagon exerts its hyperglycemic effects by decreasing glycogen synthesis, promoting glycogenolysis, and stimulating gluconeogenesis.

ACTIONS ON FAT. Glucagon also antagonizes the actions of insulin with regard to fat metabolism by promoting lipolysis and inhibiting triglyceride synthesis. Glucagon enhances hepatic ketone production (**ketogenesis**) by promoting the conversion of fatty acids to ketone bodies. Thus the blood levels of fatty acids and ketones increase under glucagon's influence.

ACTIONS ON PROTEIN. Glucagon inhibits hepatic protein synthesis and promotes degradation of hepatic protein. Stimulation of gluconeogenesis further contributes to glucagon's catabolic effect on hepatic protein metabolism. Despite these effects on liver protein metabolism, glucagon does not have any significant effect on blood amino acid levels because it does not affect muscle protein, the major protein store in the body.

Glucagon secretion is increased during the postabsorptive state.

Considering the catabolic effects of glucagon on the body's energy stores, you would be correct in assuming that glucagon secretion is increased during the postabsorptive state and decreased during the absorptive state, just the opposite of the case with insulin secretion. In fact, insulin is sometimes referred to as a "hormone of feasting" and glucagon as a "hormone of fasting." Insulin tends to put nutrients in storage when their blood levels are high, such as following a meal, whereas glucagon promotes catabolism of nutrient stores between meals to keep up the blood nutrient levels, especially blood glucose.

As with insulin secretion, the major regulator of glucagon secretion is a direct effect of the plasma glucose concentration on the pancreatic α cells. In the case of glucagon, however, an inverse relationship exists between plasma glucose levels and glucagon secretion. Glucagon secretion increases in response to a fall in blood glucose. The hyperglycemic actions of this hormone tend to restore the blood glucose level to normal. Conversely, an increase in blood glucose concentration, such as occurs after a meal, inhibits glucagon secretion, which likewise tends to restore the blood glucose level to normal. Thus there is a direct negative-feedback relationship between blood glucose concentration and the α cells' rate of secretion, but it is in the opposite direction of the effect of blood glucose on the β cells; in other words, an elevated blood glucose level inhibits glucagon secretion but stimulates insulin secretion, whereas a fall in blood glucose level leads to increased glucagon secretion and decreased insulin secretion (Fig. 19–19). Since glucagon raises blood glucose and insulin decreases blood glucose, the changes in secretion of these pancreatic hormones in response to deviations in blood glucose work together to restore normal blood glucose levels. Note that control of the secretion of both glucagon and insulin by blood glucose concentration is by simple negative feedback, even though high levels of glucose inhibit the secretion of glucagon and stimulate the release of insulin. Similarly, a fall in blood fatty acid concentration directly stimulates glucagon output and inhibits insulin output by the pancreas, both of which are negative-feedback control mechanisms to restore the blood fatty acid level to normal.

The fact that the blood concentrations of glucose and fatty acids exert opposite effects on the pancreatic α and β cells is appropriate for regulating the circulating levels of these nutrient molecules, because the actions of insulin and glucagon on carbohydrate and fat metabolism oppose one another. The effect of blood amino acid concentration on the secretion of these two hormones is a different story. A rise in blood amino acid concentration stimulates *both* glucagon and insulin secretion. Why this seeming paradox, since glucagon does not exert any effect on blood amino acid concentration? The identical effect of high blood amino acid levels on both glucagon and insulin secretion makes sense if you consider the concomitant effects these two hormones have on blood glucose

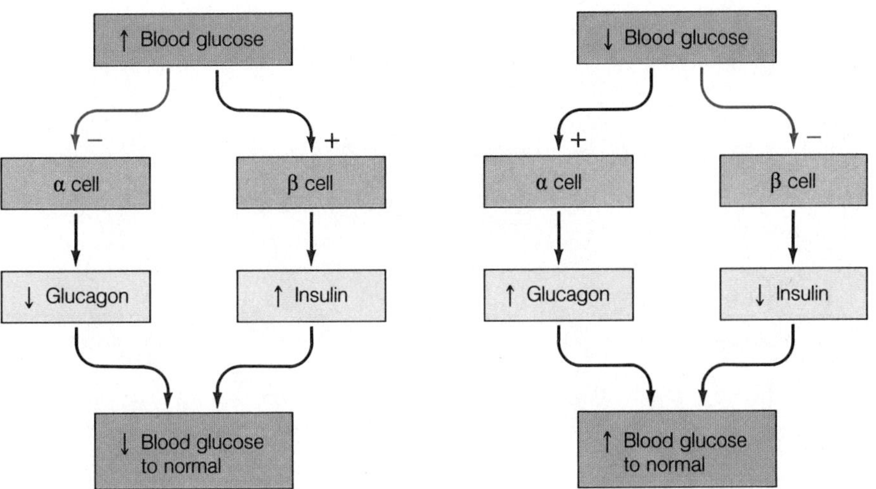

Figure 19–19 Complementary Interactions of Glucagon and Insulin

levels (Fig. 19–20). If during absorption of a protein-rich meal the rise in blood amino acids stimulated only insulin secretion, hypoglycemia might result. Because little carbohydrate is available for absorption following consumption of a high-protein meal, the amino acid–induced increase in insulin secretion would drive too much glucose into the cells, causing a sudden, inappropriate drop in the plasma glucose level. However, the simultaneous increase in glucagon secretion elicited by elevated blood amino acid levels increases hepatic glucose production. Since the hyperglycemic effects of glucagon counteract the hypoglycemic actions of insulin, the net result is maintenance of normal blood glucose levels (and prevention of hypoglycemic starvation of the brain) during absorption of a meal that is high in protein but low in carbohydrates.

Glucagon excess can aggravate the hyperglycemia of diabetes mellitus.

There are no known clinical abnormalities attributable to glucagon deficiency or excess per se. However, diabetes mellitus is frequently accompanied by excess glucagon secretion, the reason being that insulin is required for glucose to gain entry into the α cells, where it can exert control over glucagon secretion. As a result, diabetics frequently have a high rate of glucagon secretion concomitant with their insulin insufficiency because the elevated blood glucose is not able to inhibit glucagon secretion as it normally would. Since glucagon is a hormone that raises blood glucose, its excess intensifies the hyperglycemia of diabetes mellitus. For this reason, some insulin-dependent diabetics respond best to a combination of insulin and somatostatin therapy. By inhibiting glucagon secretion, somatostatin indirectly helps to achieve better reduction of the elevated blood glucose concentration than can be accomplished by insulin therapy alone.

Epinephrine, cortisol, thyroid hormone, and growth hormone also exert direct metabolic effects.

The pancreatic hormones are the most important regulators of normal fuel metabolism. However, several other hormones exert direct metabolic effects even though control of their secretion is keyed to factors other than transitions in metabolism between feasting and fasting states (Table 19–8).

The stress hormones, epinephrine and cortisol, both increase blood levels of glucose and fatty acids. In addition, cortisol mobilizes amino acids by promoting protein catabolism. Neither of these hormones play important roles in the regulation of fuel metabolism under resting conditions, but they are important for the metabolic responses to stress and exercise. Cortisol also appears to contribute to the maintenance of blood glucose concentration during long-term starvation.

Growth hormone has protein anabolic effects in muscle. In fact, this is one of its growth-promoting features. Although growth hormone can elevate the blood levels of glucose and fatty acids (see p. 654), it is normally of little importance to the overall regulation of fuel metabolism. Stress, exercise, and severe hypoglycemia stimulate growth hormone secretion, the purpose of which may be to provide fatty acids and glucose as energy sources under these circumstances. Growth hormone, like cortisol, appears to contribute to the maintenance of blood glucose concentrations during starvation.

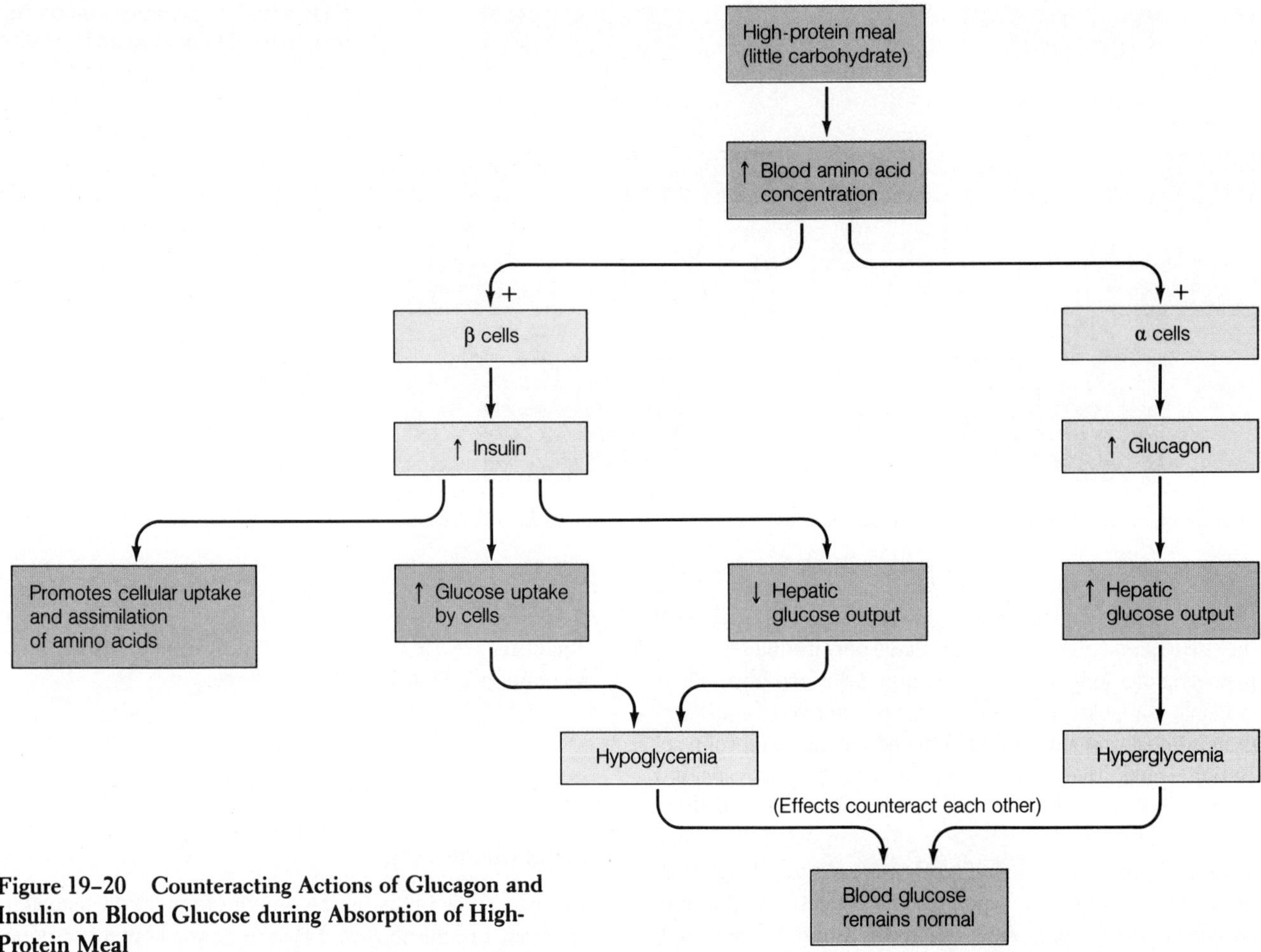

Figure 19–20 Counteracting Actions of Glucagon and Insulin on Blood Glucose during Absorption of High-Protein Meal

The effect of thyroid hormone on intermediary metabolism is a "mixed bag." It stimulates both anabolic and catabolic actions as well as the overall metabolic rate. Catabolic effects tend to dominate at higher thyroid hormone levels. However, changes in thyroid hormone secretion are usually not important for fuel homeostasis for two reasons. First, control of thyroid hormone secretion is not directed toward maintenance of nutrient levels in the blood. Second, the onset of thyroid hormone action is too slow to have any significant effect on the rapid adjustments required to maintain normal blood levels of nutrients. During starvation, less T_4 is converted to the more potent T_3. This shift in the pattern of circulating thyroid hormone is probably responsible in large part for the reduction in basal metabolic rate that usually occurs with prolonged fasting.

Note that, with the exception of the anabolic effects of growth hormone on protein metabolism, all of the metabolic actions of these other hormones are opposite to those of insulin. Insulin alone is able to reduce blood glucose and blood fatty acid levels, whereas glucagon, epinephrine, cortisol, growth hormone, and thyroid hormone all increase blood levels of these nutrients. Because of this, these other hormones are considered to be ***insulin antagonists***. This is the main reason that diabetes mellitus has such devastating metabolic consequences. There is no other control mechanism to pick up the slack to promote anabolism when insulin activity is insufficient, so the hormonally induced catabolic reactions are allowed to proceed unchecked. The only exception is protein anabolism stimulated by growth hormone.

Endocrine Control of Calcium Metabolism

Plasma calcium must be closely regulated to prevent changes in neuromuscular excitability.

Besides regulating the concentration of organic nutrient molecules in the blood by manipulation of anabolic and catabolic pathways, the endocrine system also regulates the plasma con-

Table 19-8 Summary of Hormonal Control of Metabolism

Hormone	Major Metabolic Effects				Control of Secretion	
	Effect on Blood Glucose	*Effect on Blood Fatty Acids*	*Effect on Blood Amino Acids*	*Effect on Muscle Protein*	*Major Stimuli for Secretion*	*Primary Role in Metabolism*
Insulin	↓ + Glucose uptake + Glycogenesis − Glycogenolysis − Gluconeogenesis	↓ + Triglyceride synthesis − Lipolysis	↓ + Amino acid uptake	↑ + Protein synthesis − Protein degradation	↑ Blood glucose ↑ Blood amino acids	Primary regulator of absorptive and postabsorptive cycles
Glucagon	↑ + Glycogenolysis + Gluconeogenesis − Glycogenesis	↑ + Lipolysis − Triglyceride synthesis	No effect	No effect	↓ Blood glucose; ↑ Blood amino acids	Action in concert with insulin to regulate absorptive and postabsorptive cycles; protection againt hypoglycemia
Epinephrine	↑ + Glycogenolysis + Gluconeogenesis − Insulin secretion + Glucagon secretion	↑ + Lipolysis	No effect	No effect	Sympathetic stimulation during stress and exercise	Energy for emergencies and exercise
Cortisol	↑ + Gluconeogenesis − Glucose uptake by tissues other than brain; glucose-sparing	↑ + Lipolysis	↑ + Protein degradation	↓ + Protein degradation	Stress	Mobilization of metabolic fuels and building blocks during adaptation to stress; important in starvation
Growth hormone	↑ Antagonizes insulin's actions on muscle and liver	↑ + Lipolysis	↓ + Amino acid uptake	↑ + Protein synthesis − Protein degradation + Synthesis of DNA and RNA	Stress Exercise Hypoglycemia	Promotion of growth; normally little role in metabolism; mobilization of fuels in extenuating circumstances; important in starvation

centration of a number of inorganic electrolytes. As you already know, aldosterone controls Na^+ and K^+ concentrations in the extracellular fluid (ECF). Three other hormones—parathyroid hormone, calcitonin, and vitamin D—control calcium (Ca^{++}) and phosphate ($PO_4^{\equiv}$) metabolism. These hormonal agents concern themselves with regulation of plasma Ca^{++}, and, in the process, plasma $PO_4^{\equiv}$ is also maintained. Plasma Ca^{++} concentration is one of the most tightly controlled variables in the body. The need for the precise regulation of plasma Ca^{++} stems from its critical influence on so many body activities.

About 99% of the Ca^{++} in the body is in crystalline form within the skeleton and teeth. Of the remaining 1%, about 0.9% is found intracellularly within the soft tissues; less than

0.1% is present in the ECF. Approximately half of the plasma Ca^{++} is either bound to plasma proteins and therefore is restricted to the plasma, or is complexed with $PO_4^{=}$ and is not free to participate in chemical reactions. The other half of the plasma Ca^{++} is freely diffusible and can readily pass into the interstitial fluid and interact with the cells. The free Ca^{++} in the plasma and interstitial fluid is considered to be a single pool. As with hormones, only this free Ca^{++} is biologically active and subject to regulation; it constitutes less than one-thousandth of the total Ca^{++} in the body.

This small, freely diffusible fraction of Ca^{++} in the ECF plays a vital role in a number of essential activities, including the following:

1. *Neuromuscular excitability*. Even minor variations in the concentration of free ECF Ca^{++} can have a profound and immediate impact on the sensitivity of excitable tissues. A fall in free Ca^{++} results in overexcitability of nerves and muscles, and, conversely, a rise in free Ca^{++} depresses neuromuscular excitability. This is due to the influence of Ca^{++} on membrane permeability to Na^{+}. A decrease in free Ca^{++} increases Na^{+} permeability, with the resultant influx of Na^{+} moving the resting potential closer to threshold. Consequently, in the presence of **hypocalcemia** (low blood Ca^{++}), excitable tissues may be brought to threshold by normally ineffective physiologic stimuli so that skeletal muscles discharge and contract (go into spasm) "spontaneously" (in the absence of normal stimulation). If severe enough, spastic contraction of the respiratory muscles results in death by asphyxiation. **Hypercalcemia** (elevated blood Ca^{++}), on the other hand, is also life-threatening because it causes cardiac arrhythmias accompanied by generalized depression of neuromuscular excitability.
2. *Excitation-contraction coupling in cardiac and smooth muscle*. Entry of ECF Ca^{++} into cardiac- and smooth-muscle cells, resulting from increased Ca^{++} permeability in response to an action potential, triggers the contractile mechanism. Calcium is also necessary for excitation-contraction coupling in skeletal-muscle fibers, but in this case the Ca^{++} is released from intracellular Ca^{++} stores in response to an action potential. Some of the increase in cytosolic Ca^{++} in cardiac-muscle cells is also derived from internal stores.

   Note that a *rise in cytosolic Ca^{++}* within a muscle cell causes contraction, whereas an *increase in free ECF Ca^{++}* decreases neuromuscular excitability and reduces the likelihood of contraction occurring. Unless this point is kept in mind, it is difficult to understand why low plasma Ca^{++} levels induce muscle hyperactivity when Ca^{++} is necessary to switch on the contractile apparatus. We are talking about two different pools of Ca^{++}, which exert different effects.
3. *Stimulus-secretion coupling*. The entry of Ca^{++} into secretory cells, which results from increased permeability to Ca^{++} in response to appropriate stimulation, triggers the release of the secretory product by exocytosis. This process is important for the secretion of neurotransmitters by nerve cells and for peptide and catecholamine hormone secretion by endocrine cells.
4. *Maintenance of tight junctions between cells*. Calcium forms part of the intercellular cement that holds particular cells tightly together.
5. *Clotting of blood*. Calcium serves as a co-factor in several steps of the cascade of reactions that lead to clot formation.

Because of the profound effects of deviations in free Ca^{++}, especially on neuromuscular excitability, the plasma concentration of this electrolyte is regulated with extraordinary precision. In addition, intracellular Ca^{++} (not the free ECF Ca^{++}) serves as a second messenger in many cells and is involved in cell motility and cilia action. Finally, the Ca^{++} in bone and teeth is essential for the structural and functional integrity of these tissues.

Control of calcium metabolism includes regulation of both calcium homeostasis and calcium balance.

Maintenance of the proper plasma concentration of free Ca^{++} differs from regulation of Na^{+} and K^{+} in two important regards. Sodium and potassium homeostasis is maintained primarily by regulating the urinary excretion of these electrolytes so that controlled output matches uncontrolled input. In the case of Ca^{++}, in contrast, not all of the ingested Ca^{++} is absorbed from the digestive tract, with the extent of absorption being hormonally controlled depending on the Ca^{++} status of the body. In addition, bone serves as a large Ca^{++} reservoir that can be drawn on to maintain the free plasma Ca^{++} concentration within the narrow limits compatible with life should dietary intake become too low. Exchange of Ca^{++} between the ECF and bone is also subject to control. Similar in-house stores are not available for Na^{+} and K^{+}.

Regulation of Ca^{++} metabolism depends on hormonal control of exchanges between the ECF and three other compartments: bone, kidneys, and digestive tract. Accordingly, control of Ca^{++} metabolism encompasses two aspects. First, regulation of **calcium homeostasis** involves the immediate adjustments required to maintain a constant free plasma Ca^{++} concentration on a minute-to-minute basis. This is largely accomplished by rapid exchanges between the bone and ECF and to a lesser extent by modifications in urinary excretion of Ca^{++}. Second, regulation of **calcium balance** involves the more slowly responding adjustments in intestinal Ca^{++} absorption as well as adjustments in urinary Ca^{++} excretion required to maintain a constant total amount of Ca^{++} in the body. Control of Ca^{++} balance ensures that Ca^{++} intake is equivalent to Ca^{++} excretion over the long term (weeks to months).

Parathyroid hormone (PTH), the principal regulator of Ca^{++} metabolism, acts directly or indirectly on all three of these effector sites. It is the primary hormone responsible for maintenance of Ca^{++} homeostasis and is essential for maintenance of Ca^{++} balance, although **vitamin D** also contributes in important ways to Ca^{++} balance. The third Ca^{++}-influencing hormone, **calcitonin,** is not essential for maintenance of either Ca^{++} homeostasis or balance. It serves a back-up function during the rare times of extreme hypercalcemia. We will examine the specific effects of each of these hormonal systems in more detail.

Parathyroid hormone raises free plasma calcium levels by its effects on bone, kidneys, and intestine.

Parathyroid hormone, a protein hormone, is the principal secretory product of the **parathyroid glands.** Four small parathyroid glands are located on the back surface of the thyroid gland, one in each corner. As with aldosterone, PTH *is essential for life*. The overall effect of PTH is to increase the Ca^{++} concentration of plasma (and, accordingly, of the entire ECF), thereby preventing hypocalcemia. In the complete absence of PTH, death ensues within a few days, usually because of asphyxiation caused by hypocalcemic spasm of respiratory muscles. By its actions on bone, kidney, and intestine, PTH raises the plasma Ca^{++} level when it starts to fall so that hypocalcemia and its effects are normally avoided. This hormone also acts to lower plasma $PO_4^{=}$ concentration.

ACTIONS ON BONE. Recall that 99% of the Ca^{++} in the body is found in the skeleton. (See Table 19–9 for other functions of the skeleton.) By mobilizing some of the Ca^{++} stores in the bone, PTH raises the ECF Ca^{++} concentration when it starts to fall.

Bone is a living tissue that is composed principally of an organic extracellular matrix impregnated with **hydroxyapatite crystals,** which consist primarily of precipitated $Ca_3(PO_4)_2$ (calcium phosphate) salts. Normally $Ca_3(PO_4)_2$ salts are in solution in the ECF, but the conditions within the bone are suitable for these salts to precipitate (crystallize) around the collagen fibers in the matrix. These inorganic crystals provide the bone with compressional strength, whereas collagen fibers provide the bone's tensile strength. If bones were composed entirely of inorganic salts, they would be brittle, like pieces of chalk. Bones have structural strength approaching that of reinforced concrete, yet they are not brittle and are much lighter in weight as a result of the structural blending of organic fibers and inorganic crystals.

In spite of the apparent inanimate nature of bone, bone constituents are continually being turned over. **Bone deposition** (formation) and **bone resorption** (removal) normally go on concurrently so that bone is constantly being remodeled, much as people remodel buildings by tearing down walls and replacing them. Bone remodeling serves two purposes: (1) it keeps the skeleton appropriately "engineered" for maximum effectiveness in its mechanical uses, and (2) it helps to maintain Ca^{++} homeostasis and balance.

Mechanical factors are responsible for adjusting the strength of bone in response to the demands placed on it. The greater the physical stress and compression to which a bone is subjected, the greater the rate of bone deposition. For example, the bones of athletes and heavy-duty laborers are more massive and stronger than those of sedentary individuals. On the other hand, loss of bone mass occurs in response to removal of mechanical stress, such as is seen in persons who undergo prolonged bed confinement or those in space flight. Early astronauts were noted to lose up to 20% of their bone mass during their time in orbit. Therapeutic exercises can limit or prevent such loss of bone.

The relative rates of bone resorption and deposition are also influenced by hormones. During the childhood years, growth hormone promotes deposition of bone to accomplish skeletal growth. Throughout life, parathyroid hormone uses bone as a "bank" from which it withdraws Ca^{++} as needed to maintain the plasma Ca^{++} level.

Recall that three types of bone cells are present in bone (see p. 654). The *osteoblasts* secrete the organic matrix within which the $Ca_3(PO_4)_2$ crystals precipitate. The *osteocytes* are the retired osteoblasts imprisoned within the bony wall that they have deposited around themselves. The *osteoclasts* re-

Table 19–9 Functions of the Skeleton

Support
Protection of vital internal organs
Assistance in body movement by giving attachment to muscles and providing leverage
Manufacture of blood cells (bone marrow)
Storage depot for Ca^{++} and $PO_4^{=}$, which can be exchanged with plasma to maintain plasma concentrations of these electrolytes

sorb bone in their vicinity by releasing acids that dissolve the hydroxyapatite crystals and enzymes that break down the organic matrix.

Parathyroid hormone has two major effects on bone to raise plasma Ca^{++} concentration. First, it induces a fast Ca^{++} efflux into the plasma from the small *labile pool* of Ca^{++} in the bone fluid. Second, by stimulating bone dissolution, it promotes a slow transfer into the plasma of both Ca^{++} and $PO_4^{=}$ from the *stable pool* of bone minerals in the bone itself. As a result, ongoing bone remodeling is tipped in favor of bone resorption over bone deposition.

Parathyroid hormone's earliest effect is to promote movement of Ca^{++} from the bone fluid into the plasma without the accompaniment of $PO_4^{=}$ across the osteocytic-osteoblastic bone membrane. The surface osteoblasts and entombed osteocytes are connected by an extensive network of small canals that allow substances to be exchanged between trapped osteocytes and the circulation. These small canals also contain long, filmy cytoplasmic extensions of osteocytes and osteoblasts that are connected to each other by tight junctions. The interconnecting cell network, which is called the **osteocytic-osteoblastic bone membrane,** forms a continuous membrane that separates the bone itself from the ECF within the canals. The small labile pool of Ca^{++} is present in the thin layer of **bone fluid** that separates this bone membrane from the bone (Fig.19–21). Movement of Ca^{++} out of the labile pool across the bone membrane accounts for the fast exchange between bone and ECF. Parathyroid hormone enhances the rate of this fast exchange, drawing Ca^{++} out of the "quick-cash branch" of the bone bank to rapidly increase the plasma Ca^{++} level without actually entering the bank (that is, without altering the structural integrity of the bone). Although the mechanism of Ca^{++} transport across the membrane is not fully understood, it does not involve the breakdown of bone. Under normal conditions, this exchange is much more important for the maintenance of plasma Ca^{++} concentration than is the slow exchange.

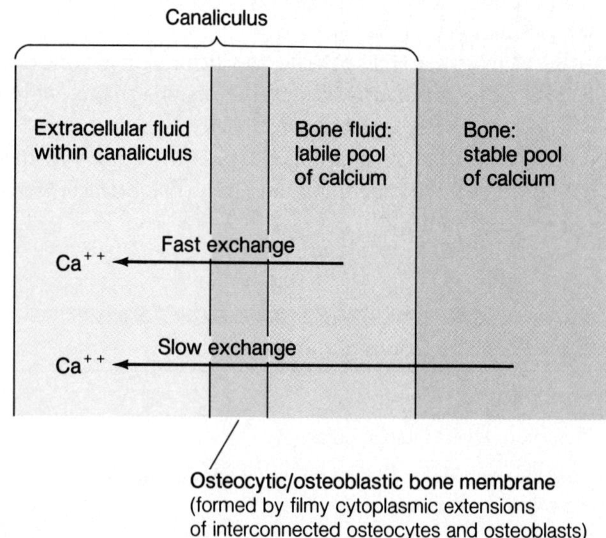

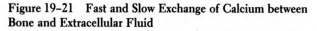

Figure 19–21 Fast and Slow Exchange of Calcium between Bone and Extracellular Fluid

Under conditions of chronic hypocalcemia, such as might occur with dietary Ca^{++} deficiency, PTH influences the slow exchange of Ca^{++} between the bone itself and the ECF by promoting actual localized dissolution of bone. It does so by stimulating osteoclasts to gobble up bone, increasing the formation of more osteoclasts, and transiently inhibiting the bone-forming activity of the osteoblasts. Bone contains such a great abundance of Ca^{++} in comparison to the ECF (more than one-thousand times as much) that even when PTH promotes increased bone resorption, there are no immediate discernible effects on the skeleton because such a minute amount of bone is affected. Yet the negligible amount of Ca^{++} "borrowed" from the bone bank can be life-saving in terms of restoring the free plasma Ca^{++} level to normal. The borrowed Ca^{++} is then redeposited in the bone at another time when Ca^{++} supplies are more abundant. Meanwhile, the free ECF Ca^{++} level has been maintained without sacrificing the integrity of the bone. However, prolonged excess PTH secretion over months or years is eventually evidenced by the formation of cavities throughout the skeleton that are filled with very large, overstuffed osteoclasts.

When PTH promotes dissolution of the $Ca_3(PO_4)_2$ crystals in bone to harvest their Ca^{++} content, both Ca^{++} and $PO_4^{=}$ are released into the plasma. An elevation in plasma $PO_4^{=}$ is undesirable, but PTH deals with this dilemma by its actions on the kidneys.

ACTIONS ON KIDNEYS. Parathyroid hormone stimulates Ca^{++} conservation and promotes $PO_4^{=}$ elimination by the kidneys during the formation of urine. Under the influence of PTH, the renal threshold for Ca^{++} is increased so that more filtered Ca^{++} is reabsorbed than escapes into the urine. This effect increases the plasma Ca^{++} level and decreases urinary Ca^{++} losses. It would be futile to dissolve bone to obtain more Ca^{++} only to lose it in the urine.

Simultaneous to stimulating renal Ca^{++} reabsorption, PTH also increases urinary $PO_4^{=}$ excretion by decreasing $PO_4^{=}$ reabsorption. As a result, PTH causes a fall in plasma $PO_4^{=}$ levels at the same time it increases Ca^{++} concentrations.

This PTH-induced removal of extra $PO_4^{=}$ from the body fluids is essential for preventing reprecipitation of the Ca^{++}

freed from the bone. Because of the solubility characteristics of calcium phosphate salt, the product of the plasma concentration of Ca^{++} times the plasma concentration of $PO_4^{=}$ must remain approximately a constant value. Therefore, an inverse relationship exists between the plasma concentrations of Ca^{++} and $PO_4^{=}$; for example, when the plasma $PO_4^{=}$ level rises, some of the plasma Ca^{++} is forced back into the bone through hydroxyapatite crystal formation, thus reducing the plasma Ca^{++} level and keeping the calcium phosphate product constant. This inverse relationship occurs because the free ions in the ECF are in equilibrium with the bone crystals.

Recall that both Ca^{++} and $PO_4^{=}$ are released from the bone when PTH promotes bone dissolution. Because PTH is secreted only when plasma Ca^{++} falls below normal, the released Ca^{++} is needed to restore plasma Ca^{++} to normal, yet the released $PO_4^{=}$ tends to increase plasma $PO_4^{=}$ levels above normal from this effect alone. If the plasma $PO_4^{=}$ level were allowed to increase above normal, this would force redeposition in the bone of some of the plasma Ca^{++} along with the $PO_4^{=}$, thereby lowering plasma Ca^{++}, to keep the calcium phosphate product constant. Therefore, PTH acts on the kidneys to decrease the reabsorption of $PO_4^{=}$ by the renal tubules. This increases the urinary excretion of $PO_4^{=}$ and lowers its plasma concentration, even though extra $PO_4^{=}$ is being released from the bone into the blood. Such action prevents the self-defeating redeposition of released Ca^{++} back into the bone.

The third important action of PTH on the kidneys (besides increasing Ca^{++} reabsorption and decreasing $PO_4^{=}$ reabsorption) is enhancement of the activation of vitamin D by the kidneys.

ACTION ON THE INTESTINE. Although PTH has no direct effect on the intestine, it indirectly increases both Ca^{++} and $PO_4^{=}$ absorption from the gut by means of its role in vitamin D activation. This vitamin, in turn, directly increases intestinal absorption of Ca^{++} and $PO_4^{=}$.

The primary regulator of PTH secretion is the plasma concentration of free calcium.

All the effects of PTH are aimed at raising the plasma Ca^{++} levels. Therefore, PTH secretion is increased in response to a fall in plasma Ca^{++} concentration and decreased by a rise in plasma Ca^{++} levels. The secretory cells of the parathyroid glands are directly and exquisitely sensitive to changes in free Ca^{++} concentration in the plasma that supplies the glands. Since PTH regulates plasma Ca^{++} concentrations, this relationship forms a simple negative-feedback loop for controlling PTH secretion without involving any nervous or other hormonal intervention.

Calcitonin lowers the plasma calcium concentration but is not important in the normal control of calcium metabolism.

Calcitonin, the hormone produced by the C cells of the thyroid gland, is another hormone that exerts an influence on plasma Ca^{++} levels. Like PTH, calcitonin has two effects on bone, but in this case, both effects decrease plasma Ca^{++} levels. First, on a short-term basis, calcitonin decreases Ca^{++} efflux across the osteocytic-osteoblastic bone membrane. Second, on a long-term basis, calcitonin decreases bone resorption by inhibiting the activity of osteoclasts. The suppression of bone resorption results in decreased plasma $PO_4^{=}$ levels as well as a reduced plasma Ca^{++} concentration. The hypocalcemic and hypophosphatemic effects of calcitonin are due entirely to this hormone's actions on bone.

As with PTH, the primary regulator of calcitonin release is the free plasma Ca^{++} concentration (Fig. 19–22), but in contrast to its effect on PTH release, an increase in plasma Ca^{++} stimulates calcitonin secretion and a fall in plasma Ca^{++} inhibits calcitonin secretion. Since calcitonin reduces plasma Ca^{++} levels, this system constitutes a second simple negative-feedback control over plasma Ca^{++} concentration, one that is opposed to the PTH system.

Most evidence suggests, however, that calcitonin plays little or no role in the normal control of Ca^{++} or $PO_4^{=}$ metabolism. Although calcitonin will protect against hypercalcemia, this condition rarely occurs under normal circumstances. Moreover, neither thyroid removal nor calcitonin-secreting tumors alter circulating levels of Ca^{++} or $PO_4^{=}$, implying that this hormone is not normally essential to the maintenance of

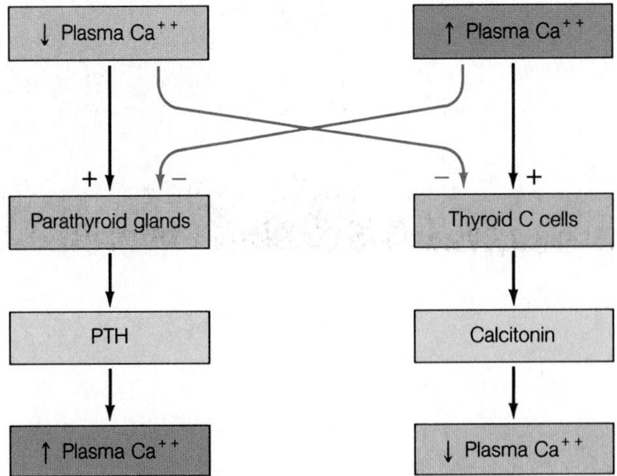

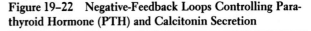

Figure 19–22 Negative-Feedback Loops Controlling Parathyroid Hormone (PTH) and Calcitonin Secretion

Ca^{++} or $PO_4^{\equiv}$ homeostasis. A possible role for calcitonin may be in protecting skeletal integrity when there is a large Ca^{++} demand, such as during pregnancy or breast-feeding.

Vitamin D is actually a hormone that increases calcium absorption in the intestine.

The final factor involved in the regulation of Ca^{++} metabolism is **cholecalciferol,** or vitamin D, a steroid-like compound that is essential for Ca^{++} absorption in the intestine. Strictly speaking, vitamin D should be considered a hormone because it can be produced in the skin from a precursor related to cholesterol (7-dehydrocholesterol) on exposure to sunlight. It is subsequently released into the blood to act at a distant target site, the intestine. The skin, therefore, is actually an endocrine gland and vitamin D a hormone. However, this mediator is traditionally considered to be a vitamin for two reasons. First, it was originally discovered and isolated from a dietary source and tagged as a vitamin. Second, even though the skin would be an adequate source of vitamin D if it were exposed to sufficient sunlight, indoor dwelling and clothing in response to cold weather and social customs preclude significant exposure of the skin to sunlight most of the time. At least part of the essential vitamin D must therefore be derived from dietary sources.

Regardless of its source, vitamin D is biologically inactive when it first enters the blood, either from the skin or the digestive tract. It must be activated by two sequential biochemical alterations that involve the addition of two hydroxyl (-OH) groups (Fig. 19–23). The first of these reactions occurs in the

Figure 19–23 Activation of Vitamin D

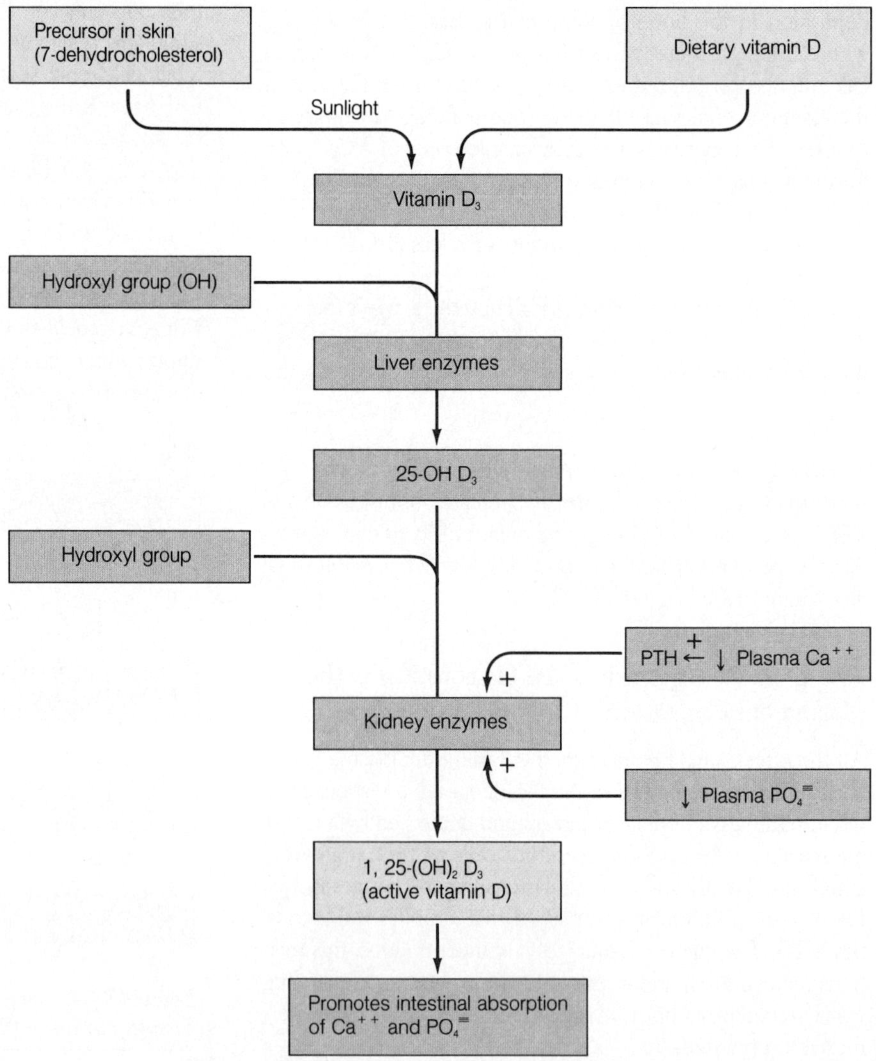

liver, the second in the kidneys. The end result is production of the active form of vitamin D, 1, 25-$(OH)_2$-vitamin D_3. The kidney enzymes that are involved in the second step of vitamin D activation are stimulated by PTH. To a lesser extent, a rise in plasma $PO_4^{\equiv}$ inhibits activation of vitamin D (whereas a fall in $PO_4^{\equiv}$ enhances the activation process) by exerting effects on these enzymes that are opposite to those exerted by PTH.

The most dramatic and biologically important effect of activated vitamin D is to increase Ca^{++} absorption in the intestine. Unlike most dietary constituents, dietary Ca^{++} is not indiscriminately absorbed by the digestive system. In fact, the majority of ingested Ca^{++} is typically not absorbed but is lost instead in the feces. When needed, more dietary Ca^{++} is absorbed into the plasma under the influence of vitamin D. Independently of its effects on Ca^{++} transport, the active form of vitamin D also increases intestinal $PO_4^{\equiv}$ absorption. Furthermore, it increases the responsiveness of bone to PTH. Thus, vitamin D and PTH have a closely interdependent relationship (Fig. 19–24).

Parathyroid hormone is principally responsible for controlling Ca^{++} homeostasis, because the actions of vitamin D are too sluggish for it to contribute substantially to the minute-to-minute regulation of plasma Ca^{++} concentration. However, both PTH and vitamin D are essential to Ca^{++} balance, the process that ensures that, over the long term, Ca^{++} input into the body is equivalent to Ca^{++} output. When there is a decrease in dietary Ca^{++} intake, the resultant transient fall in plasma Ca^{++} level stimulates PTH secretion. The increased PTH has two effects that are important for the maintenance of Ca^{++} balance: (1) it stimulates Ca^{++} reabsorption by the kidneys, thereby decreasing Ca^{++} output, and (2) it activates vitamin D, which increases the efficiency of uptake of ingested Ca^{++}. Since PTH also promotes bone resorption, a

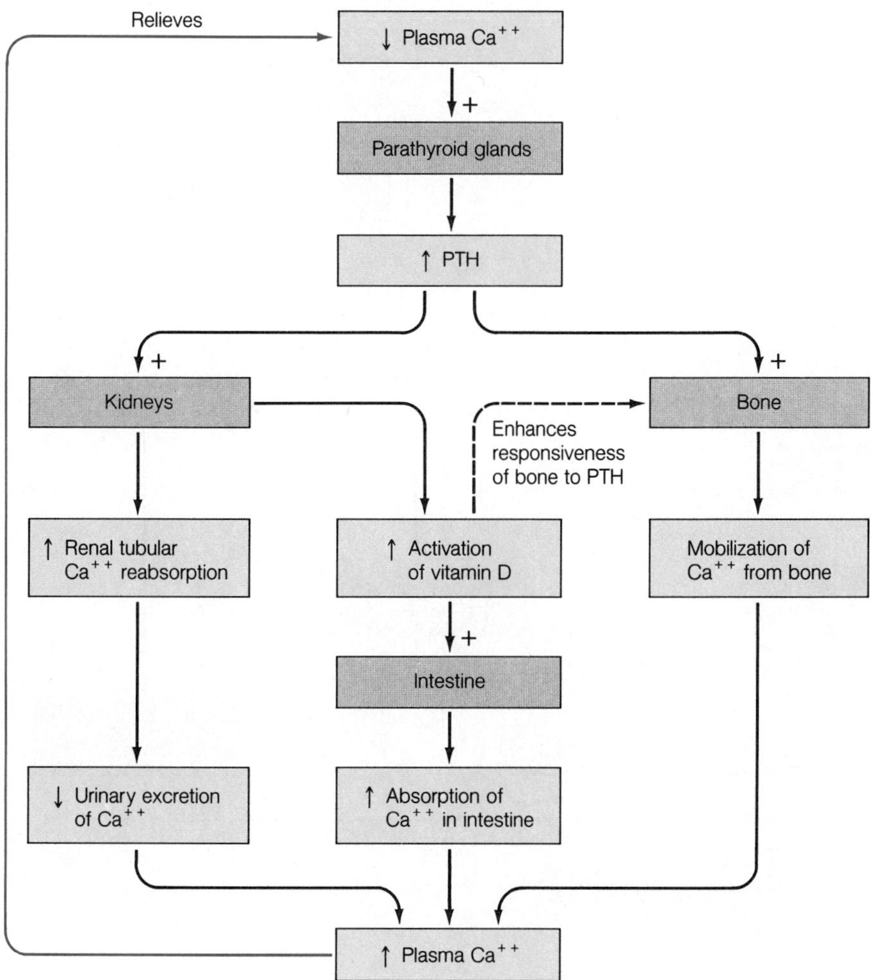

Figure 19–24 Interactions between PTH and Vitamin D in Controlling Plasma Calcium

substantial loss of bone minerals occurs if Ca^{++} intake is reduced for a prolonged period, even though bone is not directly involved in maintaining Ca^{++} input and output in balance.

Phosphate metabolism is controlled by the same mechanisms that regulate calcium metabolism.

Plasma $PO_4^{\equiv}$ concentration is not as tightly controlled as plasma Ca^{++} concentration. Phosphate is regulated directly by vitamin D and indirectly by the plasma Ca^{++}-PTH feedback loop. To illustrate, a fall in plasma $PO_4^{\equiv}$ concentration (**hypophosphatemia**) exerts a two-fold effect to help restore circulating $PO_4^{\equiv}$ level to normal (Fig. 19–25). First, because of the inverse relationship between the $PO_4^{\equiv}$ and Ca^{++} concentrations in the plasma, a fall in plasma $PO_4^{\equiv}$ causes an increase in plasma Ca^{++}, which directly suppresses PTH secretion. In the presence of reduced PTH, $PO_4^{\equiv}$ reabsorption by the kidneys increases, returning plasma $PO_4^{\equiv}$ concentration toward normal. Second, a fall in plasma $PO_4^{\equiv}$ also results in increased activation of vitamin D, which then promotes $PO_4^{\equiv}$ absorption in the intestine. This further contributes to alleviation of the initial hypophosphatemia.

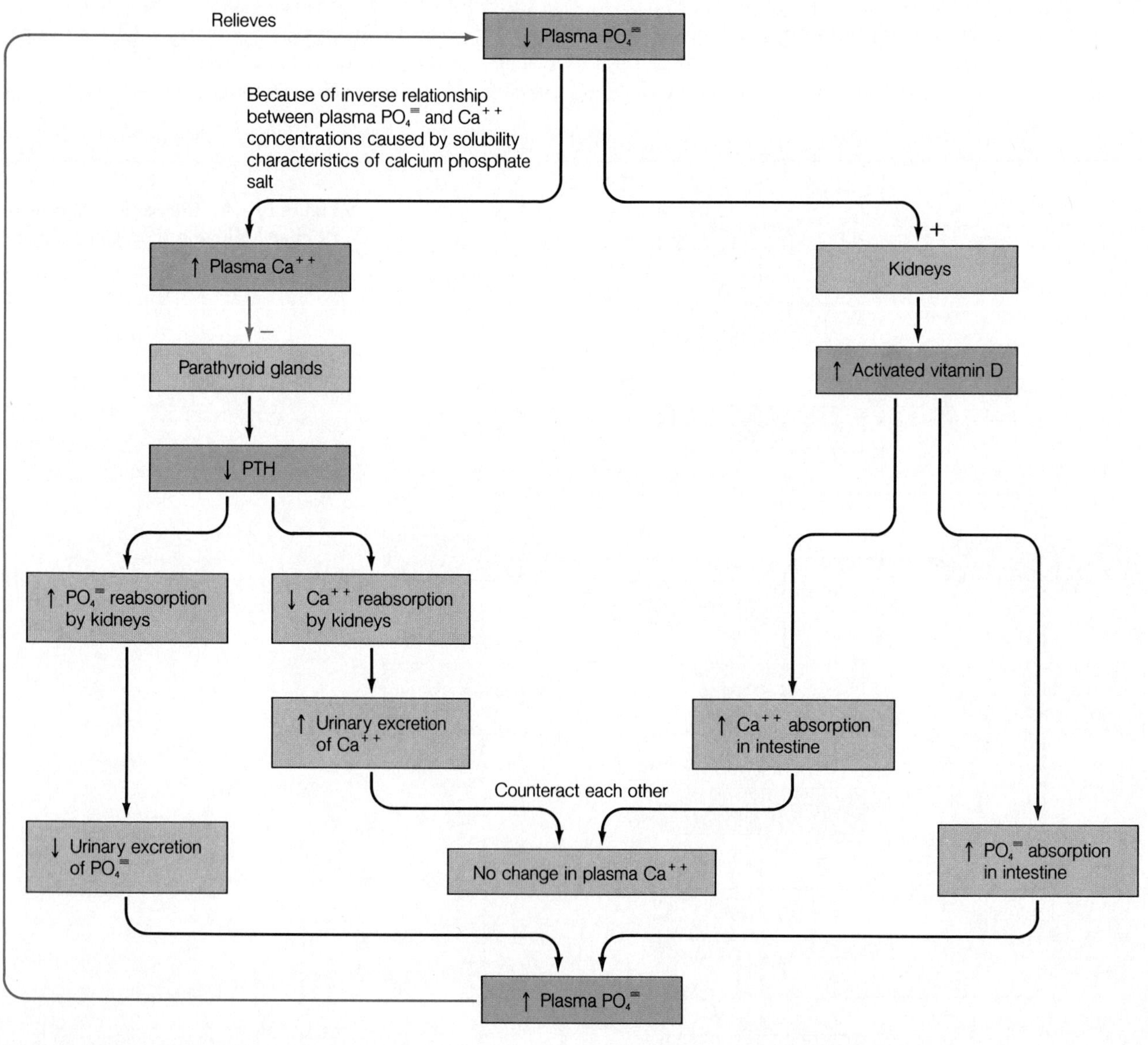

Figure 19–25 Control of Plasma Phosphate

Note that Ca^{++} balance is not compromised by these changes. Although the increase in activated vitamin D stimulates Ca^{++} absorption, the concurrent fall in PTH produces a compensatory increase in urinary Ca^{++} excretion because less of the filtered Ca^{++} is reabsorbed.

Disorders in calcium metabolism may arise from abnormal levels of parathyroid hormone or vitamin D.

Abnormalities in all three endocrine systems involved in Ca^{++} metabolism may occur. However, disorders of calcitonin secretion are extremely rare and cause no obvious abnormalities in Ca^{++} metabolism. The primary disorders that affect Ca^{++} metabolism are too much or too little PTH or a deficiency of vitamin D.

PTH HYPERSECRETION. Excess PTH secretion, or **hyperparathyroidism,** which is usually due to a hypersecreting tumor in one of the parathyroid glands, is characterized by hypercalcemia and hypophosphatemia. The affected individual can be asymptomatic or symptoms can be severe, depending on the magnitude of the problem. The following are among the possible consequences.

1. Hypercalcemia reduces the excitability of muscle and nervous tissue, leading to muscle weakness and neurological disorders, including decreased mentation, poor memory, and depression. Cardiac disturbances may also occur.
2. Excessive mobilization of Ca^{++} and $PO_4^{=}$ from skeletal stores leads to rarefaction of bone, which may result in skeletal deformities and increased incidence of fractures.
3. An increased incidence of formation of Ca^{++} - containing kidney stones occurs because of the excess quantity of Ca^{++} being filtered through the kidneys. These stones may impair renal function. Passage of the stones through the ureters causes extreme pain. Because of these potential multiple consequences, hyperparathyroidism has been coined as a disease of "bones, stones, and abdominal groans."
4. To further account for the "abdominal groans," hypercalcemia can cause gastrointestinal disorders such as peptic ulcers, nausea, and constipation.

PTH HYPOSECRETION. The most common cause of deficient PTH secretion, or **hypoparathyroidism,** used to be inadvertent removal of the parathyroid glands during thyroidectomy. Before their existence was known, these glands were sometimes removed accidentally during surgical excision of the thyroid for treatment of thyroid disease because of their close anatomical relation to the thyroid. If all of the parathyroid tissue was removed, these patients, of course, died, because PTH is essential for life. In fact, it was a puzzle why some patients died soon after thyroid removal even though there were no apparent surgical complications. Now that the location and importance of the parathyroid glands have been discovered, surgeons are careful to leave parathyroid tissue during thyroid removal. Rarely, hyposecretion of the parathyroids glands occurs as a result of primary failure of the parathyroid tissue.

Hypoparathyroidism leads to hypocalcemia and hyperphosphatemia. The symptoms are primarily referable to increased neuromuscular excitability caused by the reduction in the level of free ionized plasma Ca^{++}. In the complete absence of parathyroid hormone, death is imminent because of hypocalcemic spasm of respiratory muscles. With a relative deficiency rather than a complete absence of PTH, milder symptoms of increased neuromuscular excitability are evident. Muscle cramps and twitches derive from spontaneous activity in the motor nerves, while tingling and pins-and-needles sensations result from spontaneous activity in the sensory nerves. Mental changes include irritability and paranoia.

VITAMIN D DEFICIENCY. The major consequence associated with vitamin D deficiency is impaired intestinal absorption of Ca^{++}. In the face of reduced Ca^{++} uptake, PTH maintains the plasma Ca^{++} level at the expense of the bones. As a result, the bone matrix is not properly mineralized because Ca^{++} salts are not available for deposition. The demineralized bones become soft and deformed, bowing to the pressures of weight bearing, especially in children. This condition is known as **rickets** in children and **osteomalacia** in adults.

OSTEOPOROSIS. The rates of bone formation and resorption are about equal throughout most of adult life, so total bone mass usually remains fairly constant during this period. Exceptions include minor hormonally induced fluctuations in bone mass to maintain Ca^{++} homeostasis or mechanically induced adjustments in bone mass in response to changes in compressional weight bearing (for example, taking up exercise or being confined to bed). Such is not the case, however, during the first twenty and last twenty years of an average life span. During the first two decades of life, when growth is occurring, bone deposition exceeds bone resorption under the influence of growth hormone (see p. 656). In contrast, by fifty to sixty years of age, bone resorption often exceeds bone formation. The result is a reduction in bone mass known as **osteoporosis.** This condition is characterized by a diminished laying

A CLOSER LOOK AT EXERCISE PHYSIOLOGY

Osteoporosis: The Bane of Brittle Bones

Osteoporosis, a decrease in bone density resulting from reduced deposition of the bone's organic matrix, is a major health problem in the United States. It is responsible for the greater incidence of bone fractures among women over the age of fifty than among the population at large. Because bone mass is reduced, the bones are more susceptible to fracture in response to a fall, blow, or lifting action that would not normally strain stronger bones. Osteoporosis is the underlying cause for approximately 1.2 million fractures each year, of which 530,000 are vertebral fractures and 227,000 are hip fractures. The cost of rehabilitation is in excess of 6 billion dollars per year. The cost in pain and suffering is not measurable. One half of all American women have spinal pain and deformity by age seventy-five.

There appear to be two types of osteoporosis, which are caused by different mechanisms. Type I osteoporosis affects women soon after menopause and is characterized by vertebral crush fractures or fractures of the arm just above the wrist. It is hypothesized that these fractures occur as a result of the reduction in bone density that accompanies the estrogen deficiency of menopause. Type II osteoporosis occurs in men as well as women, although it affects females twice as often as males. It is characterized by hip fractures as well as fractures at other sites. Because Type II osteoporosis occurs later in life, the decreased ability to absorb Ca^{++} associated with advancing age may play a key role in the development of the condition, although estrogen deficiency probably also contributes, accounting for the higher incidence in women.

Estrogen replacement therapy, Ca^{++} supplementation, and a regular weight-bearing exercise program are among the therapeutic approaches used to minimize or reverse bone loss. Because treatment of osteoporosis is difficult and often less than satisfactory, prevention is by far the best approach to managing this disease. Development of strong bones to begin with before menopause through a good Ca^{++}-rich diet and adequate exercise appear to be the best preventive measures. A large reservoir of bone at midlife may delay the clinical manifestations of osteoporosis in later life. Continued physical activity throughout life appears to retard or prevent bone loss, even in the elderly.

It is well documented that osteoporosis can result from disuse–that is, from reduced mechanical loading of the skeleton. Space travel has clearly shown that lack of gravity results in a decrease in bone density. Studies of athletes, on the other hand, demonstrate that physical activity increases bone density. Within groups of athletes, bone density correlates directly with the load that the bone must bear. If one looks at athletes' femurs (thigh bones), the greatest bone density is found in weight lifters, followed in order by throwers, runners, soccer players, and finally swimmers. In fact, the bone density of swimmers does not differ from that of nonathletic controls. Swimming does not place any strain on bones. The bone density in the playing arm of male tennis players has been found to be as much as 35% greater than in their other arm; female tennis players have been found to have 28% greater density in their playing arm than in their other arm. One study found that very mild activity in nursing-home patients, whose average age was eighty-two years, not only slowed bone loss but even resulted in bone build-up over a thirty-six-month period. Thus exercise is a good defense against osteoporosis.

The exact mechanism responsible for an increase in bone mass as a result of exercise is unknown. According to one proposal, exercise places strain on bone, which causes changes in electrical potential that induce bone formation.

down of organic matrix as a result of reduced osteoblast activity rather than abnormal bone calcification. The underlying cause of osteoporosis is uncertain. Plasma Ca^{++} and $PO_4^{=}$ levels are normal, as are PTH and vitamin D concentrations. Osteoporosis occurs with greatest frequency in postmenopausal women, suggesting that estrogen withdrawal plays a role (see the accompanying boxed feature, A Closer Look at Exercise Physiology).

Chapter in Perspective

A number of peripherally located endocrine organs play key roles in maintaining homeostasis, primarily by means of their regulatory influences over the rate of various metabolic reactions and over electrolyte balance. These endocrine organs all secrete hormones in response to specific stimuli. The hormones, in turn, exert effects that act in negative-feedback fashion to resist the change that induced their secretion, thus maintaining stability in the internal environment.

The thyroid gland secretes two iodine-containing amine hormones: tetraiodothyronine (T_4 or thyroxine) and triiodothyronine (T_3). These thyroid hormones increase the overall metabolic rate and stimulate a number of specific metabolic reactions.

The adrenal cortex secretes three classes of steroid hormones: (1) aldosterone, the primary mineralocorticoid, which is essential for Na^+ and K^+ balance; (2) cortisol, the primary glucocorticoid, which enhances glucose metabolism at the expense of protein and fat metabolism and also plays a key role in adaptation to stress; and (3) adrenal sex hormones, which exert androgen-dependent functions in females.

The adrenal medulla is a modified part of the sympathetic nervous system. The catecholamine hormone that it secretes, epinephrine, generally reinforces activities of the sympathetic system and, in so doing, is important in fight-or-flight responses and in blood pressure regulation.

The endocrine pancreas secretes two peptide hormones, insulin and glucagon. These hormones for the most part exert opposite effects on carbohydrate, fat, and protein metabolism. They are important in shifting metabolic pathways between the absorptive and postabsorptive states to maintain the appropriate blood levels of nutrient molecules.

The parathyroid gland secretes parathyroid hormone (PTH), which increases the plasma concentration of Ca^{++}. This effect is complemented by vitamin D, which promotes intestinal absorption of Ca^{++}. Vitamin D qualifies as a hormone when it is produced by the skin upon adequate exposure to sunlight. Calcitonin secreted by the C cells of the thyroid gland is a hormone that lowers the plasma Ca^{++} level.

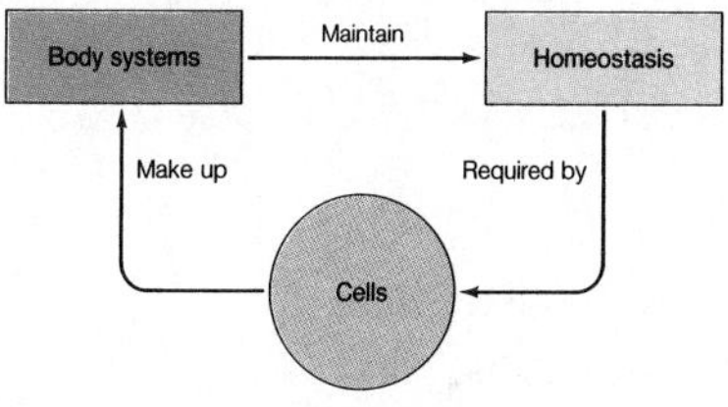

See inside front cover for an expanded version of this model.

Review Exercises

1. Describe the steps of thyroid hormone synthesis.
2. What are the effects of T_3 and T_4? Which is the more potent of the thyroid hormones? What is the source of most circulating T_3?
3. Describe the regulation of thyroid hormone.
4. Discuss the causes and symptoms of both hypothyroidism and hyperthyroidism. Under what circumstances does a goiter occur?
5. What hormones are secreted by each of the three zones of the adrenal cortex? What are the functions and control of each of these hormones?
6. Discuss the causes and symptoms of each type of adrenocortical dysfunction.
7. What is the relationship of the adrenal medulla to the sympathetic nervous system? What are the effects of the adrenomedullary catecholamines? How is epinephrine release controlled?
8. Define stress. Describe the neural and hormonal responses to a stressor.
9. Define intermediary metabolism, anabolism, and catabolism.
10. Distinguish between the absorptive and postabsorptive states with regard to the dispensation of nutrient molecules.
11. List the major cell types of the islets of Langerhans, and indicate the hormonal product of each.
12. Compare the functions and control of insulin secretion with those of glucagon secretion.
13. What are the consequences of diabetes mellitus? Distinguish between Type I and Type II diabetes mellitus.
14. Why must plasma Ca^{++} be closely regulated?
15. Discuss the contributions of parathyroid hormone, calcitonin, and vitamin D to Ca^{++} metabolism. Describe the source and control of each of these hormones.
16. Discuss the major disorders in Ca^{++} metabolism.
17. **A point to ponder:** Why would an infection tend to increase the blood glucose level of a diabetic individual?

CHAPTER 20

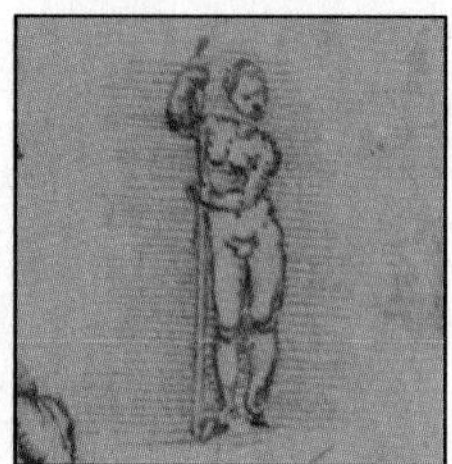

REPRODUCTIVE PHYSIOLOGY

CHAPTER CONTENTS AT A GLANCE

INTRODUCTION *Implicit with our ever-increasing understanding of the mechanisms underlying physiological functions is our growing ability to correct or compensate for defective functions or to manipulate various functions to our advantage. Most advances in the area of biomedical technology have been welcomed as new ways to save, prolong, or enhance human lives. However, one area of new technological control over biological processes has aroused considerable moral, ethical, and legal controversy—namely, our growing power to determine the very existence of new individuals. Reproductive-related technological advances have opened up many new avenues for those desiring to control their potential progeny through artificial means. A brief listing of capabilities already available will indicate how extensive our control over the future generation already is:*

☐ *A variety of methods of* contraception *are available to avoid pregnancy by preventing sperm and egg from joining or, once united, to prevent the fertilized egg from implanting in the uterus where it could develop into another human being.*

☐ *An unwanted pregnancy can be terminated by several different methods of* abortion *that remove the developing individual from its supportive uterine environment.*

☐ *An unborn child can be unobtrusively "viewed" in its mother's womb through* ultrasound *techniques. Furthermore, a sample of its cells can be extracted from the uterus through* amniocentesis *or* placental biopsy *and analyzed to determine the presence or absence of numerous genetic disorders. If defects are discovered, the parents have to make the difficult choice of terminating the pregnancy or knowingly bringing a child into the world who will be physically or mentally impaired.*

☐ *Sperm can be frozen and stored in* sperm banks *where a woman can "shop" for a "father" with particular traits for her future child; the woman can then undergo* artificial insemination *with the chosen sperm and have a normal pregnancy and birth without active participation or even knowledge of the male.*

☐ *Infertile women who fail to ovulate (do not release eggs) can be given* "fertility pills," *which are hormones that promote ovulation.*

☐ *Infertile women who have a mechanical blockage in the pathway where the egg and sperm unite can still have their own child through* in vitro fertilization, *often dubbed the "test tube baby" technique. After hormonally promoting multiple ovulation, the physician collects the eggs through a small incision in the woman's abdomen, then incubates them with sperm donated by the father-to-be. If fertilization takes place in this test-tube environment, the fertilized egg is placed in the mother's uterus, where it develops as would a naturally fertilized egg.*

☐ *It is technically possible, although legally questionable, for a couple unable to have children because of the woman's inability to bear a child to "rent" another woman's uterus to carry their child. The* surrogate mother *can be artificially inseminated with the father's sperm, or an in vitro fertilized egg from the couple can be introduced into the "rented" uterus.*

The reproductive system is essential for survival of the species.

The central theme of this book has been the physiological processes aimed at maintaining homeostasis to insure survival of the individual. We are now going to disembark from this theme to discuss the reproductive system. Normal functioning of the reproductive system is not aimed toward homeostasis and is not necessary for survival of an individual, but it is essential for survival of the species. Only through reproduction can the complex genetic blueprint of each species survive beyond the lives of individual members of the species.

Even though the reproductive system is not essential for survival of an individual, it still plays an important role in a person's life. For example, the manner in which people relate as sexual beings contributes in significant ways to psychosocial behavior, having important influences on how people view themselves and how they interact with others. Reproductive function also has a profound effect on society. The universal organization of societies into family units provides a stable environment that is conducive for perpetuating our species. On the other hand, the population explosion and its resultant drain on dwindling resources has recently led to worldwide concern with the means by which to limit reproduction.

Reproductive capability depends on an intricate relationship among the hypothalamus, anterior pituitary, reproductive organs, and target tissues of the sex hormones. In addition to these basic biological processes, sexual behavior and attitudes are deeply influenced by emotional factors and the sociocultural mores of the society in which the individual lives. We will concentrate on the basic sexual and reproductive functions that are under nervous and hormonal control and will not examine the psychological and social ramifications of sexual behavior.

The reproductive system includes the gonads and reproductive tract.

The male and female **reproductive systems** are designed to enable union of genetic material from the two sexual partners, and the female system is equipped to house and nourish the offspring to the developmental point that it can survive independently in the external environment.

The **primary reproductive organs,** or **gonads,** consist of a pair of **testes** in the male and a pair of **ovaries** in the female. In both sexes, the mature gonads perform the dual function of: (1) producing **gametes** (**gametogenesis**), which are the reproductive or germ cells, those being **spermatozoa** (**sperm**) in the male and **ova** (**eggs**) in the female; and (2) secreting sex hormones, specifically **testosterone** in males and **estrogen** and **progesterone** in females.

In addition to the gonads, the reproductive system in each sex includes a **reproductive tract** encompassing a system of ducts that are specialized to transport or house the gametes after they are produced, plus **accessory sex glands** that empty their supportive secretions into these passageways. In females, the breasts are also considered to be accessory reproductive organs. The externally visible portions of the reproductive system are known as **external genitalia.**

The **secondary sexual characteristics** are the many external characteristics not directly involved in reproduction that distinguish males and females, such as body configuration and hair distribution. Testosterone in the male and estrogen in the female are responsible for the development and maintenance of these characteristics. In some species, the secondary sexual characteristics are of great importance in courting and mating behavior; for example, the rooster's head dressing attracts the female's attention, and the stag's antlers are useful to ward off other males. In humans, the differentiating marks between males and females do serve to attract the opposite sex, but attraction is also strongly influenced by the complexities of human society and cultural behavior.

The essential reproductive functions of the male are: (1) production of sperm (*spermatogenesis*) and (2) delivery of sperm to the female. The sperm-producing organs, the testes, are suspended outside the abdominal cavity in a skin-covered sac, the **scrotum,** which lies within the angle between the legs. The male reproductive system is designed to deliver sperm to the female reproductive tract in a liquid vehicle, **semen,** which is conducive to sperm viability. The major male accessory sex glands, whose secretions provide the bulk of the semen, are the *seminal vesicles*, *prostate gland*, and *bulbourethral glands* (Fig. 20–1). The **penis** is the organ used to deposit semen in the female. Exit for sperm from the testes is through the *epididymis*, *ductus (vas) deferens*, *ejaculatory duct*, and *urethra*, the latter being a canal that runs through the length of the penis.

The female's role in reproduction is more complicated than the male's. The essential female reproductive functions include: (1) the production of ova (*oogenesis*); (2) reception of sperm; (3) transport of the sperm and ovum to a common site for union (*fertilization* or *conception*); (4) maintenance of the developing fetus until it can survive in the outside world (*gestation* or *pregnancy*), including formation of the **placenta,** the organ of exchange between mother and fetus; (5) giving birth to the baby (*parturition*); and (3) nourishing the infant after birth by milk production (*lactation*). The product of fertilization is known as an **embryo** during the first two months of intrauterine development. Beyond this time, it is recognizable as human and is known as a **fetus** during the remainder of gestation.

The ovaries and female reproductive tract lie within the pelvic cavity (Fig. 20–2a and b). The female reproductive tract consists of two **oviducts** (**uterine tubes** or **Fallopian tubes**), which pick up ova upon ovulation and serve as the site for fertilization; the thick-walled hollow **uterus,** which is primarily responsible for maintaining the fetus during its development and expelling it at the end of pregnancy; and the **vagina,** a muscular, expansible tube connecting the uterus to the external environment. The lowest portion of the uterus, the **cervix,** projects into the vagina and contains a single small opening, the **cervical canal.** Sperm are deposited in the vagina by the penis during sexual intercourse. The cervical canal serves as a pathway for sperm through the uterus to the site of fertilization in the oviduct and, when greatly dilated during parturition, as a passageway for delivery of the baby from the uterus.

The **vaginal opening** or **orifice** is located in the **perineal region** between the urethral opening anteriorly and the anal opening posteriorly (Fig. 20–2c). It is partially covered by a thin mucous membrane, the **hymen,** which can be physically disrupted in a variety of ways, including by the first sexual intercourse. The vaginal and urethral openings are surrounded

Figure 20–1 Anatomy of Male Reproductive Organs

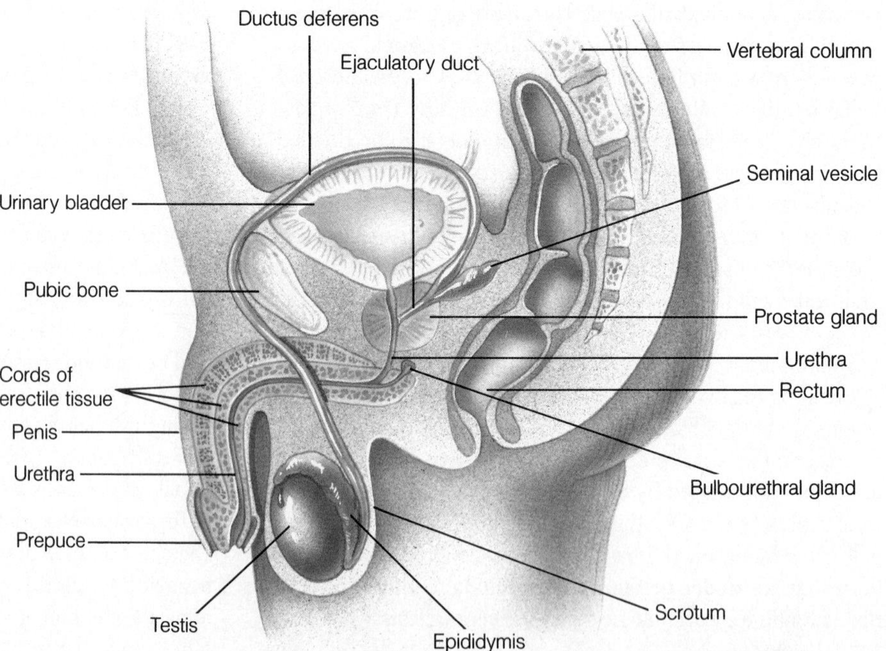

laterally by two pairs of skin folds, the **labia minora** and **labia majora.** The smaller labia minora are located medially to the more prominent labia majora. A small erotic structure composed of tissue identical to the penis, the **clitoris,** lies at the anterior end of the folds of the labia minora. The female external genitalia are collectively referred to as the **vulva.**

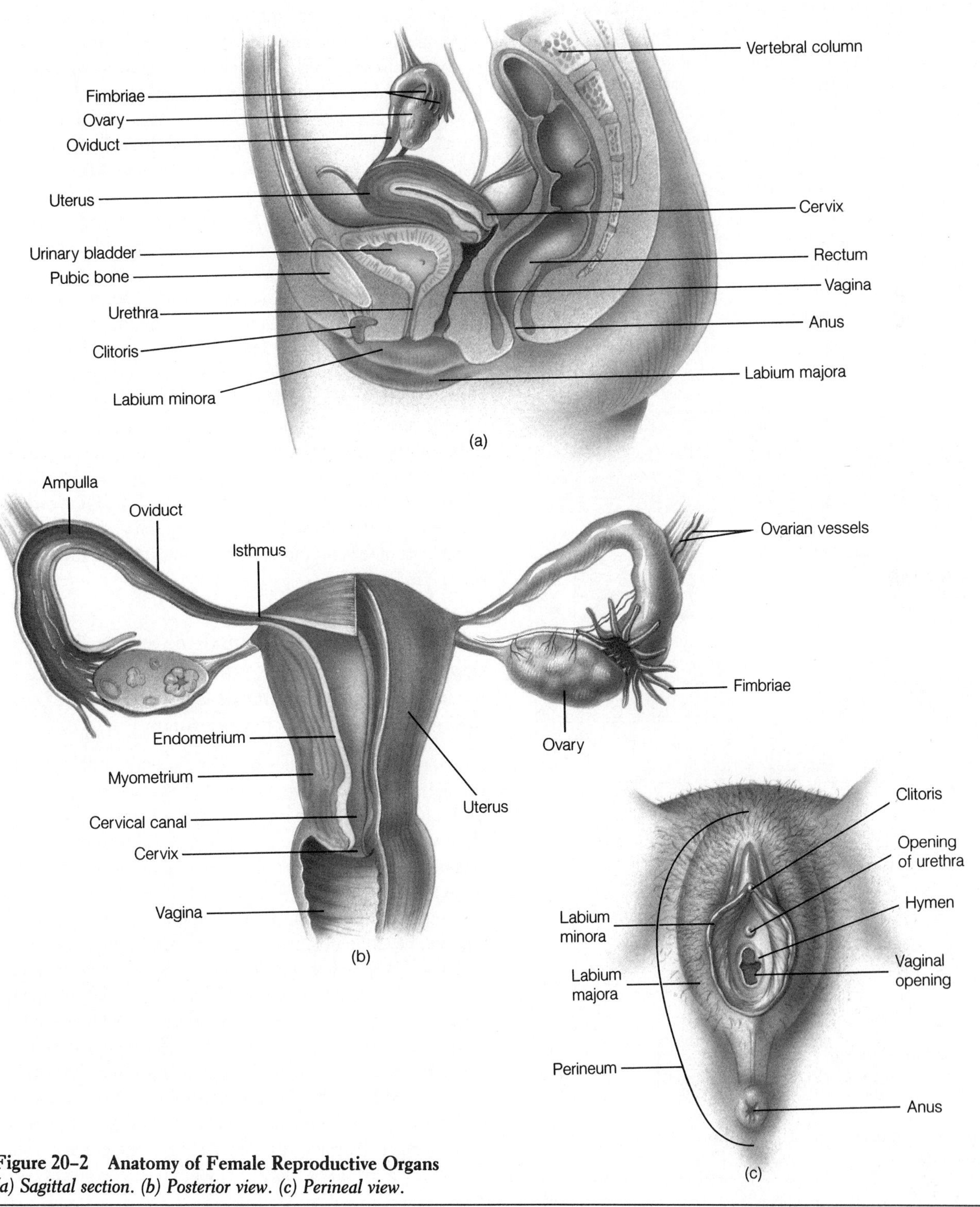

Figure 20–2 Anatomy of Female Reproductive Organs
(a) Sagittal section. (b) Posterior view. (c) Perineal view.

The sex of an individual is determined by the combination of sex chromosomes.

Whether individuals are destined to be males or females is a genetic phenomenon determined by the sex chromosomes they possess. All body (somatic) cells have forty-six chromosomes consisting of twenty-three chromosomal pairs, the *diploid* number of chromosomes. During gametogenesis, the number of chromosomes is reduced to twenty-three, the *haploid* number, as the developing germ cells undergo a meiotic division. The chromosome pairs are separated during meiosis so that each sperm or ovum receives only one member of each chromosome pair (see p. 53). When fertilization takes place, the sperm and ovum fuse to form the start of a new individual with forty-six chromosomes, one member of each chromosomal pair having been inherited from the mother, the other member from the father (Fig. 20–3).

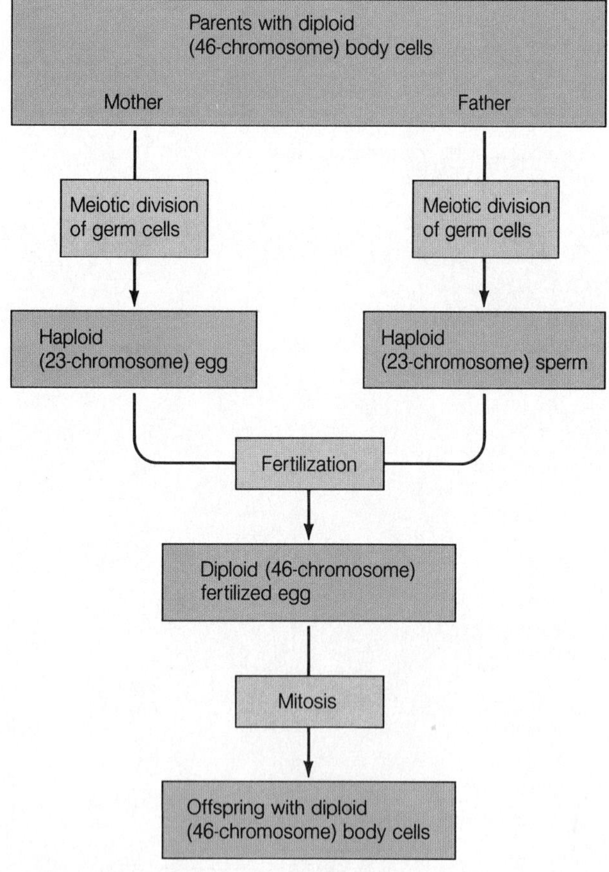

Figure 20–3 Chromosomal Distribution in Sexual Reproduction

Twenty-two of the chromosome pairs are **autosomal chromosomes** that code for general human characteristics as well as for specific traits such as eye color. The remaining pair of chromosomes are the **sex chromosomes,** of which there are two genetically different types—a larger **X chromosome** and a smaller **Y chromosome. Sex determination** depends on the combination of sex chromosomes: **genetic males** have both an X and a Y sex chromosome; **genetic females** have two X sex chromosomes. Thus the genetic difference responsible for all of the anatomical and functional distinctions between males and females is the single Y chromosome. Males have it; females do not.

During the meiotic reduction division of gametogenesis, all chromosome pairs are separated so that each daughter cell contains only one member of each pair, including the sex chromosome pair. When the XY sex chromosome pair separates during sperm formation, half the sperm receive an X chromosome and the other half a Y chromosome. In contrast, during oogenesis, every ovum receives an X chromosome because separation of the XX sex chromosome pair yields only X chromosomes. During fertilization, combination of an X-bearing sperm with an X-bearing ovum produces a genetic female, XX, whereas union of a Y-bearing sperm with an X-bearing ovum results in a genetic male, XY. Thus genetic sex is determined at the time of conception and depends on which type of sex chromosome is contained within the fertilizing sperm.

Sex differentiation along male or female lines depends on the presence or absence of masculinizing determinants during critical periods of embryonic development.

Differences between males and females exist at three levels: genetic, gonadal, and phenotypic (anatomical) sex (Fig. 20–4). **Genetic sex,** which depends on the combination of sex chromosomes at the time of conception, in turn determines **gonadal sex;** that is, whether testes or ovaries develop. The presence or absence of a Y chromosome determines gonadal differentiation. For the first month and a half of gestation, all embryos have the potential to differentiate along either male or female lines, because the developing reproductive tissues of both sexes are identical and indifferent. Gonadal specificity appears during the seventh week of intrauterine life when the indifferent gonadal tissue of a genetic male begins to differentiate into testes under the influence of **testicular determining factor** (**TDF**), the single gene within the Y chromosome that is responsible for sex determination. This gene triggers a chain of reactions that lead to physical development of a male. Testicular determining factor "masculinizes" the gonads (in-

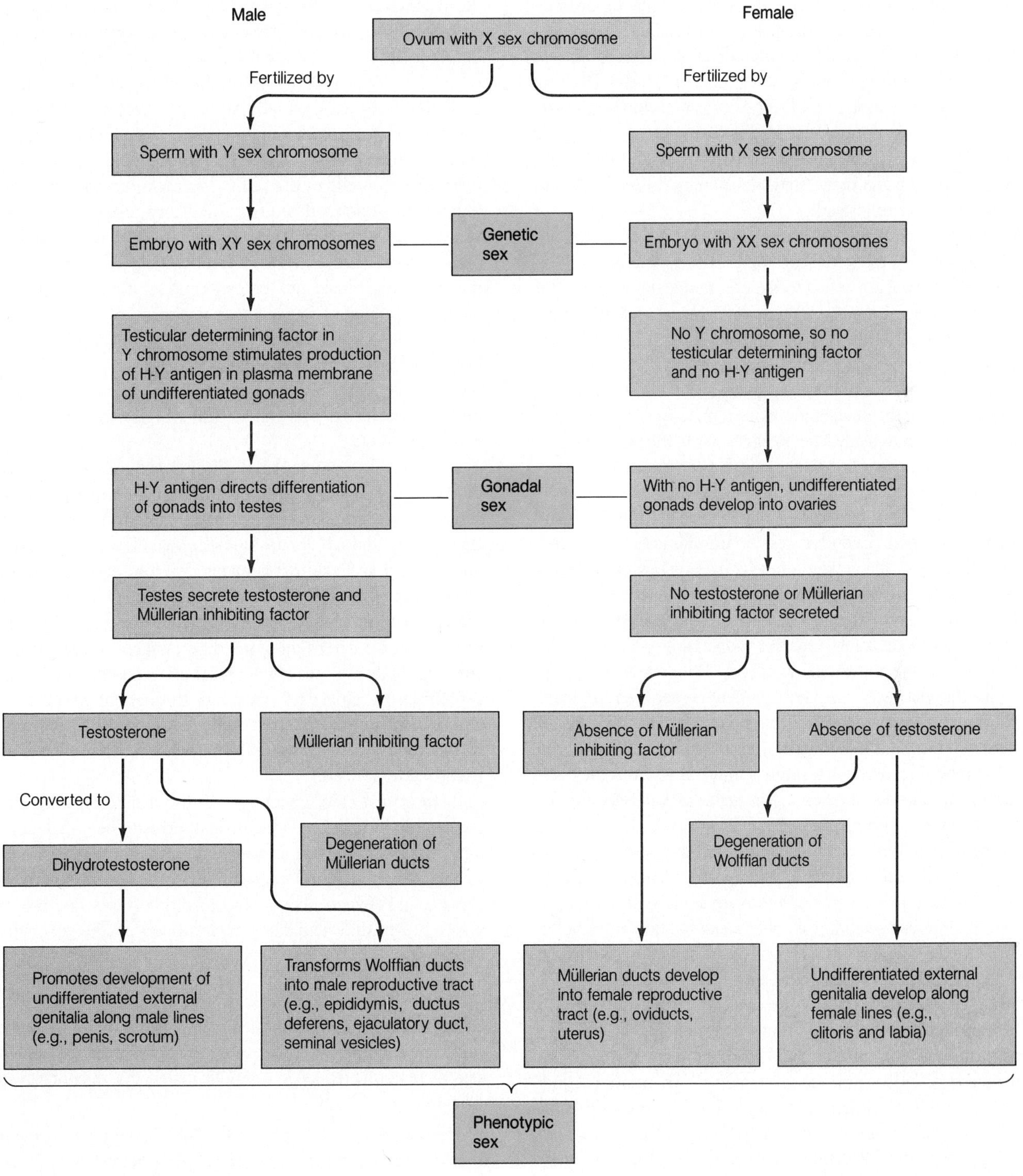

Figure 20–4 Sex Differentiation

duces their development into testes) by stimulating production of **H-Y antigen** by primitive gonadal cells. H-Y antigen, which is a specific plasma-membrane protein found only in males, directs differentiation of the gonads into testes. Because genetic females lack the TDF-bearing Y chromosome and consequently do not produce H-Y antigen, their gonadal cells never receive a signal for testicular formation, so the undifferentiated gonadal tissue starts developing during the ninth week into ovaries instead.

Phenotypic sex, the apparent anatomic sex of an individual, depends on the genetically determined gonadal sex. **Sexual differentiation** refers to the embryonic development of the external genitalia and reproductive tract along either male or female lines. Differentiation into a male-type reproductive system is induced by hormones secreted by the developing testes. The absence of these testicular hormones in female fetuses results in the development of a female-type reproductive system. By ten to twelve weeks of gestation, the sexes can easily be distinguished by the anatomical appearance of the external genitalia.

Male and female external genitalia develop from the same embryonic tissue. In both sexes, the undifferentiated external genitals consist of a genital tubercle, paired urethral folds surrounding a urethral groove, and, more laterally, genital (labioscrotal) swellings (Fig. 20–5). The **genital tubercle** gives rise to exquisitely sensitive erotic tissue—in males the glans penis (the cap at the distal end of the penis) and in females the clitoris. The major distinctions between the glans penis and clitoris are the smaller size of the clitoris and penetration of the glans penis by the urethral opening. The urethra is the tube through which urine is transported from the bladder to the outside, also serving in males as a passageway for exit of semen through the penis to the outside. In males the **urethral folds** fuse around the urethral groove to form the penis, which encircles the urethra. The **genital swellings** similarly fuse to form the scrotum. In females, the urethral folds and genital swellings do not fuse at midline but develop instead into the labia minora and labia majora, respectively. The urethral groove remains open, providing access to the interior through the urethral opening and vaginal orifice.

Although the male and female external genitalia develop from the same undifferentiated embryonic tissue, this is not the case with the reproductive tracts. Two primitive duct systems—the Wolffian ducts and the Müllerian ducts—develop in all embryoes. In males, the reproductive tract develops from the **Wolffian ducts** and the Müllerian ducts degenerate, whereas in females, the **Müllerian ducts** differentiate into the reproductive tract and the Wolffian ducts regress. Since both duct systems are present before sexual differentiation occurs, the early embryo has the potential to develop either a male or female reproductive tract. Development of the reproductive tract along male or female lines is determined by the presence or absence of two hormones secreted by the fetal testes—testosterone and Müllerian-inhibiting factor (Fig. 20–4). A hormone released by the placenta, chorionic gonadotropin, appears to be the stimulus for this early testicular secretion. Testosterone induces development of the Wolffian ducts into the male reproductive tract (epididymis, ductus deferens, ejaculatory duct, and seminal vesicles). This hormone, after being converted into **dihydrotestosterone (DHT)**, is also responsible for differentiating the external genitalia into the penis and scrotum. Meanwhile, Müllerian-inhibiting factor causes regression of the Müllerian ducts. In the absence of testosterone and Müllerian-inhibiting factor in females, the Wolffian ducts regress; the Müllerian ducts develop into the female reproductive tract (oviducts and uterus), and the external genitalia differentiate into the clitoris and labia.

It is important to note that the indifferent embryonic reproductive tissue passively develops into a female structure unless actively acted on by masculinizing factors. In the absence of testosterone and Müllerian-inhibiting factor, a female reproductive tract and external genitalia develop regardless of the genetic sex of the individual. Ovaries do not even need to be present for feminization of the fetal genital tissue. Such a control pattern for determining sex differentiation is appropriate considering that fetuses of both sexes are exposed to high concentrations of female sex hormones throughout gestation. If female sex hormones exerted influence over the development of the reproductive tract and external genitalia, all fetuses would be feminized.

In the usual case, genetic sex and sex differentiation are compatible; that is, a genetic male appears to be a male anatomically and functions as a male, and the same compatibility holds true for females. Occasionally, however, discrepancies occur between genetic and anatomic sexes because of errors in sex differentiation, as illustrated by the following examples:

□ If the testes in a genetic male fail to properly differentiate and secrete hormones, the result is the development of an apparent anatomic female in a genetic male, who, of course, will be sterile.

□ Because testosterone acts on the Wolffian ducts to convert them into a male reproductive tract but the testosterone derivative DHT is responsible for masculinization of the external genitalia, a genetic deficiency of the enzyme that converts testosterone into DHT results in a genetic male with testes and a male reproductive tract but with female external genitalia.

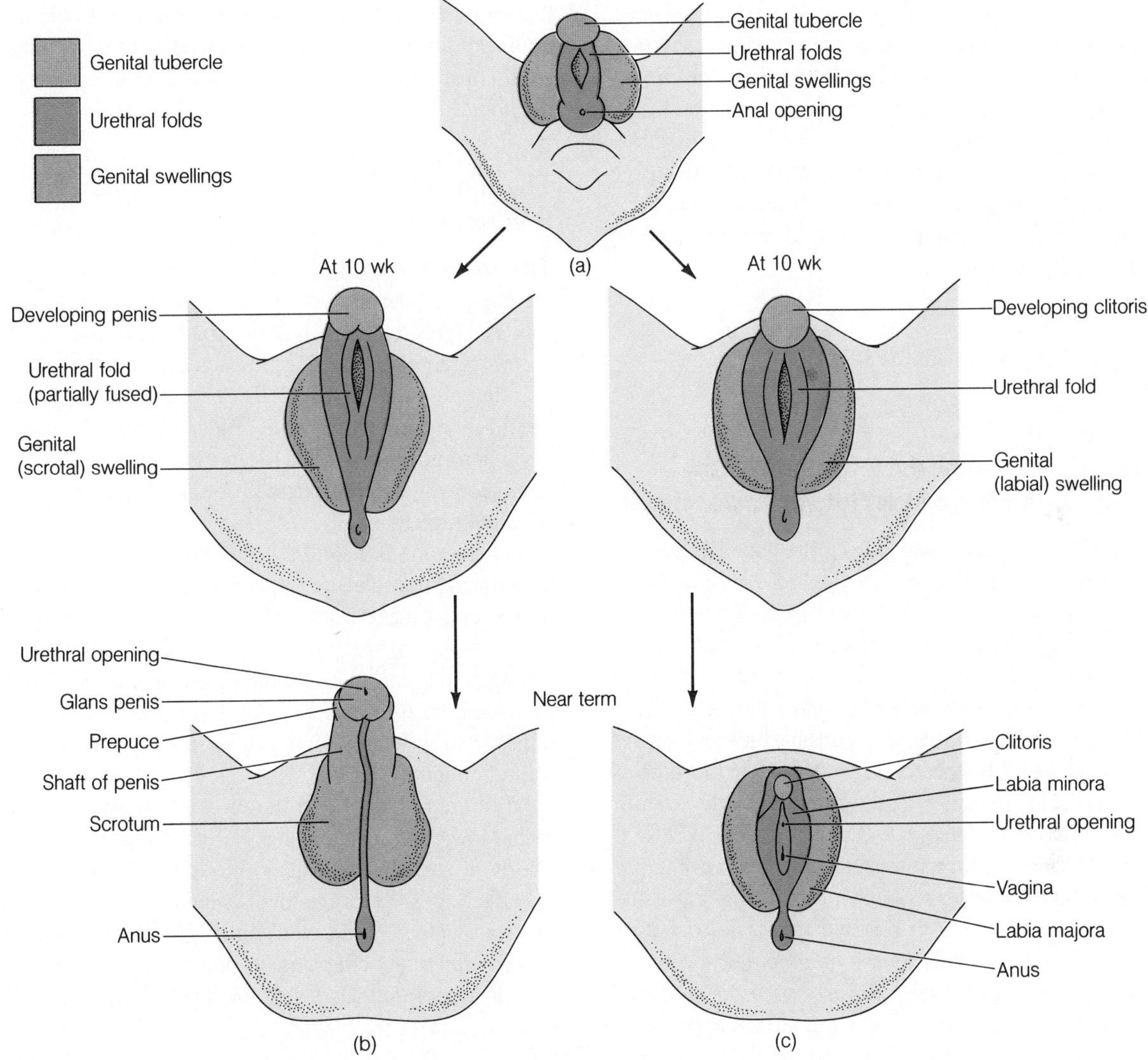

Figure 20–5 Sexual Differentiation of External Genitalia
(a) Undifferentiated stage (7 wk). (b) Male development. (c) Female development.

□ Excess androgen secretion by the adrenal glands (adrenogenital syndrome) in a genetic female during critical developmental stages causes differentiation of a male-type reproductive tract and external genitalia (see p. 675). **Androgens** encompass a group of masculinizing hormones, of which testosterone is the most potent. The adrenal gland normally secretes a weaker androgen, *dehydroepiandrosterone*, in insufficient quantities to masculinize females. However, pathologically excessive secretion of this hormone in a genetically female fetus imposes differentiation of the genitalia along male lines.

Sometimes these discrepancies between genetic sex and apparent sex are not recognized until puberty, at which time the discovery produces a psychologically traumatic gender identity crisis. For example, a masculinized genetic female with ovaries but with male-type external genitalia may be reared as a boy until puberty, when breast enlargement (caused by estrogen secretion by the awakening ovaries) and lack of beard growth (caused by lack of testosterone secretion in the absence of testes) signal an apparent problem. Therefore, it is important to diagnose any problems in sexual differentiation in infancy. Once a sex has been assigned, it can be reinforced, if necessary, with surgical and hormonal treatment so that psychosexual development can proceed as normally as possible. Less dramatic cases of inappropriate sex differentiation often appear as sterility problems.

Male Reproductive Physiology

The scrotal location of the testes provides a cooler environment essential for spermatogenesis.

Embryonically, the testes develop from the gonadal ridge located at the rear of the abdominal cavity. In the last months of fetal life, they begin a slow descent, passing out of the abdominal cavity through the **inguinal canal** into the scrotum, one testis dropping into each pocket of the scrotal sac. As with differentiation of the reproductive tract and external genitalia along male lines, androgen is also responsible for inducing descent of the testes into the scrotum. Although the time is somewhat variable, descent is usually complete by the seventh month of gestation. As a result, descent is complete in 98% of full-term baby boys, but in a substantial percentage of premature male infants, the testes are still within the inguinal canal at birth. In most instances of retained testes, descent does occur naturally before puberty or can be encouraged with administration of testosterone. Rarely, a testis remains undescended into adulthood, a condition known as **cryptorchidism** ("hidden testis"). Following descent of the testes into the scrotum, the opening in the abdominal wall through which the inguinal canal passes closes snuggly around the sperm-carrying duct and blood vessels that traverse between each testis and the abdominal cavity. Incomplete closure or rupture of this opening permits abdominal viscera to slip through, resulting in an **inguinal hernia.**

The temperature within the scrotum averages several degrees Celsius less than normal body (core) temperature. Descent of the testes into this cooler environment is essential, because spermatogenesis is temperature sensitive and cannot occur at normal body temperature. Therefore, a cryptorchid is unable to produce viable sperm.

The relative position of the scrotum in relation to the abdominal cavity can be varied by a spinal reflex mechanism that plays an important role in regulating testicular temperature. Reflex contraction of scrotal muscles upon exposure to a cold environment raises the scrotal sac to bring the testes closer to the warmer abdomen. Conversely, relaxation of the muscles upon exposure to heat permits the scrotal sac to become more pendulous, moving the testes farther from the warm core of the body.

The testicular Leydig cells secrete masculinizing testosterone.

The testes perform the dual function of producing sperm and secreting testosterone. About 80% of the testicular mass consists of highly coiled **seminiferous tubules,** within which spermatogenesis takes place. The endocrine cells that produce testosterone—the **Leydig cells** or **interstitial cells**—are located in the connective tissue (interstitial tissue) between the seminiferous tubules. Thus the portions of the testes that produce sperm and secrete testosterone are structurally and functionally distinct.

Testosterone is a steroid hormone derived from a cholesterol precursor molecule, as are the female sex hormones, estrogen and progesterone. The Leydig cells contain a high concentration of the enzymes required to direct cholesterol through the testosterone-yielding pathway (see p. 635). Once produced, testosterone is secreted into the blood, where it is transported, primarily bound to plasma proteins, to its target sites of action.

Most but not all of testosterone's actions are ultimately directed toward ensuring delivery of sperm to the female. The effects of testosterone can be grouped into five categories: (1) effects on the reproductive system before birth; (2) effects on sex-specific tissues after birth; (3) other reproduction-related effects; (4) effects on secondary sexual characteristics; and (5) nonreproductive actions (Table 20–1).

EFFECTS ON THE REPRODUCTIVE SYSTEM BEFORE BIRTH. Before birth, testosterone secretion by the fetal testes is responsible for masculinizing the reproductive tract and external genitalia and for promoting descent of the testes into the scrotum, as already described. After birth, testosterone secretion ceases, and the testes and remainder of the reproductive system remain small and nonfunctional until puberty.

EFFECTS ON SEX-SPECIFIC TISSUES AFTER BIRTH. **Puberty** refers to the period of arousal and maturation of the previously nonfunctional reproductive system, culminating in attainment of sexual maturity and the ability to reproduce. Its time of onset usually occurs sometime between the ages of ten and fourteen; on the average it begins about two years earlier in females than in males. Usually lasting three to five years, puberty encompasses a complex sequence of endocrine, physical, and behavioral events. **Adolescence** is a broader concept

Table 20-1 Effects of Testosterone

Effects of Testosterone
Effects before Birth
Masculinizes reproductive tract and external genitalia
Promotes descent of testes into scrotum
Effects on Sex-Specific Tissues
Promotes growth and maturation of reproductive system at puberty
Maintains reproductive tract throughout adulthood
Essential for spermatogenesis
Other Reproductive Effects
Development of sex drive at puberty
Control of gonadotropin hormone secretion
Effects on Secondary Sexual Characteristics
Induces male pattern of hair growth (e.g., beard)
Causes voice to deepen because of thickening of vocal cords
Promotes muscle growth responsible for male body configuration
Nonreproductive Actions
Exerts protein anabolic effect
Promotes bone growth at puberty and then closure of epiphyses
Increases secretion of sebaceous glands
May induce aggressive behavior

that refers to the entire transition period between childhood and adulthood, not just sexual maturation.

At puberty, the Leydig cells start secreting testosterone once again, and spermatogenesis is initiated in the seminiferous tubules for the first time. Testosterone is responsible for growth and maturation of the entire male reproductive system. Under the influence of the pubertal surge in testosterone secretion, the testes enlarge and become capable of spermatogenesis, the accessory sex glands enlarge and become secretory, and the penis and scrotum enlarge. Ongoing testosterone secretion is essential for spermatogenesis and for maintaining a mature male reproductive tract throughout adulthood. Following **castration** (surgical removal of the testes) or testicular failure caused by disease, the sex organs regress in size and function.

OTHER REPRODUCTION-RELATED EFFECTS. Testosterone is responsible for development of sexual libido at puberty and helps to maintain the sex drive in the adult male. Stimulation of this behavior by testosterone is important for facilitating delivery of sperm to females. In humans, libido is also influenced by many interacting social and emotional factors. Once libido has developed, testosterone is no longer absolutely required for its maintenance. Castrated males often remain sexually active but at a reduced level.

Also related to reproductive function, testosterone participates in the normal negative-feedback control of gonadotropin hormone secretion by the anterior pituitary, a topic that will be covered more thoroughly later.

EFFECTS ON SECONDARY SEXUAL CHARACTERISTICS. All male secondary sexual characteristics depend on testosterone for their development and maintenance. These nonreproductive male characteristics induced by testosterone include: (1) the male pattern of hair growth (for example, beard, chest hair, axillary and pubic hair, and, in genetically predisposed men, baldness); (2) a deep voice caused by enlargement of the larynx and thickening of the vocal cords; (3) thick skin; and (4) the male body configuration (for example, broad shoulders, heavy arm and leg musculature) as a result of protein deposition. A male castrated before puberty (a **eunuch**) does not mature sexually nor does he develop secondary sexual characteristics.

NONREPRODUCTIVE ACTIONS. Testosterone exerts several important effects not related to reproduction. Androgen has a general protein anabolic (synthesis) effect and promotes bone growth, thus contributing to the more muscular physique of males and to the pubertal growth spurt. Ironically, testosterone not only stimulates bone growth but eventually prevents further growth by sealing the growing ends of the long bones (that is, ossifying or "closing" the epiphyseal plates—see p. 656). Testosterone also stimulates oil secretion by the sebaceous glands. This effect is most striking during the adolescent surge of testosterone secretion, predisposing the young man to develop acne.

In animals, testosterone induces aggressive behavior, but whether or not it influences human behavior other than in the area of sexual behavior is an unresolved issue. Even though some athletes who take testosterone-like anabolic androgenic steroids to gain competitive advantage have been observed to display more aggressive behavior (see p. 240), it is unclear to what extent general behavioral differences between the sexes are hormonally induced or are a result of social conditioning.

Once initiated at puberty, testosterone secretion and spermatogenesis occur continuously throughout the male's life. Testicular efficiency gradually declines after forty-five to fifty years of age, however, even though men in their seventies and beyond may continue to enjoy an active sex life and some even father a child at this late age. The gradual diminution in circulating testosterone levels and in sperm production is not caused by a decrease in stimulation but probably arises instead from degenerative changes associated with aging that occur in the small testicular blood vessels. This gradual decline is often termed "male menopause," although it is not deliberately programmed as is female menopause.

Spermatogenesis yields an abundance of highly specialized, mobile sperm.

About 800 feet (250 meters) of sperm-producing seminiferous tubules are packed within the testes. Two functionally important cell types are present in these tubules: *germ cells*, most of which are in various stages of sperm development, and *Sertoli cells*, which provide crucial support for spermatogenesis. **Spermatogenesis** is a complex process by which relatively undifferentiated primordial germ cells, the **spermatogonia** (each of which contains a diploid complement of forty-six chromosomes), proliferate and are converted into extremely specialized, motile spermatozoa (sperm), each bearing a randomly distributed haploid set of twenty-three chromosomes.

Microscopic examination of a seminiferous tubule reveals layers of germ cells in an anatomical progression of sperm development, starting with the least differentiated in the outer layer and moving inward through various stages of division to the highly differentiated sperm in the lumen, which are ready for exit from the testes (Fig. 20–6). Spermatogenesis takes sixty-four days for development from a spermatogonium to a mature sperm. Up to several hundred million sperm may reach maturity daily. Spermatogenesis encompasses three major stages: *mitotic proliferation*, *meiosis*, and *packaging* (Fig. 20–7).

MITOTIC PROLIFERATION. Spermatogonia located in the outermost layer of the tubule continuously divide mitotically, with all new cells bearing the full complement of forty-six chromosomes that are identical to the parent cell (see p. 52). Such proliferation provides a continual supply of new germ cells. Following mitotic division of a spermatogonium, one of the daughter cells remains at the outer edge of the tubule as an undifferentiated spermatogonium, thus maintaining the germ cell line. The other daughter cell starts moving toward the lumen while undergoing the various steps required to form sperm, which are released into the lumen. In humans, the sperm-forming daughter cell divides mitotically twice more to form four identical **primary spermatocytes.** After the last mitotic division, the primary spermatocytes enter a resting phase during which the chromosomes are duplicated but the doubled strands remain together in preparation for the first meiotic division.

MEIOSIS. Meiosis involves the single duplication of all forty-six chromosomes, followed by two successive divisions so that four haploid daughter cells—each containing twenty-three unpaired chromosomes—are produced from a single diploid germ cell (see p. 53). The first meiotic division yields two **secondary spermatocytes,** each of which contains a half set of doubled chromosomes (chromatids). Each secondary spermatocyte randomly receives some chromosomes from the man's paternal ancestry and some from his maternal ancestry. More than 8 million (2^{23}) different mixtures of the twenty-three paternal and maternal chromosomes are possible. In addition to the reassortment of genes that results from the random distribution of paternal and maternal chromosomes into each daughter cell during this division, further mixing of genetic information occurs as a consequence of chromosomal crossing over, which results in the scrambling of paternal and maternal genetic information within a single chromosome (see p. 56).

No further replication occurs during the second meiotic division. The two secondary spermatocytes rapidly divide to form four **spermatids.** During this division, the twenty-three doubled chromosomes separate and migrate to opposite poles so that each spermatid receives a set of twenty-three single unpaired chromosomes.

No further division takes place beyond this stage of spermatogenesis. Each spermatid is remodeled into a single spermatozoon. Because each sperm-producing spermatogonium mitotically produces four primary spermatocytes and each primary spermatocyte meiotically yields four spermatids (spermatozoa-to-be), the spermatogenic sequence in humans can theoretically produce sixteen spermatozoa each time a spermatogonium initiates this process. Usually, however, some cells are lost at various stages, so the efficiency of productivity is rarely this high.

PACKAGING. Even after meiosis, spermatids still resemble undifferentiated spermatogonia structurally except for the half complement of chromosomes in spermatids. Production of extremely specialized, mobile spermatozoa from spermatids requires extensive remodeling, or packaging, of cellular elements, a process known as **spermiogenesis.** Sperm are essentially "stripped-down" cells in which most of the cytosol and the organelles not needed for the task of delivering the sperm's genetic information to an ovum have been extruded. A **spermatozoon** has four parts (Fig. 20–8): a head, an acrosome, a midpiece, and a tail. The *head* consists primarily of the nu-

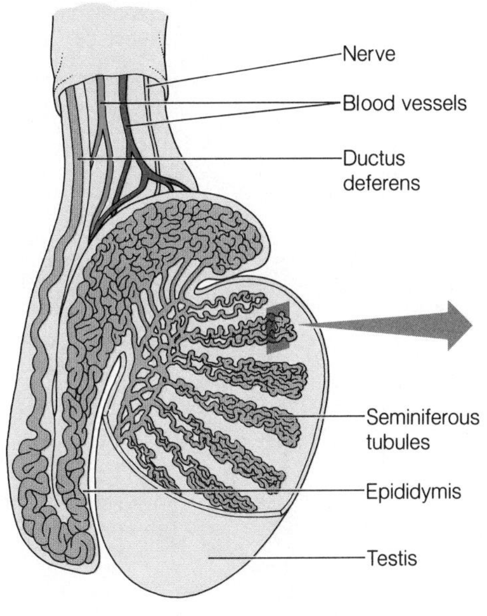

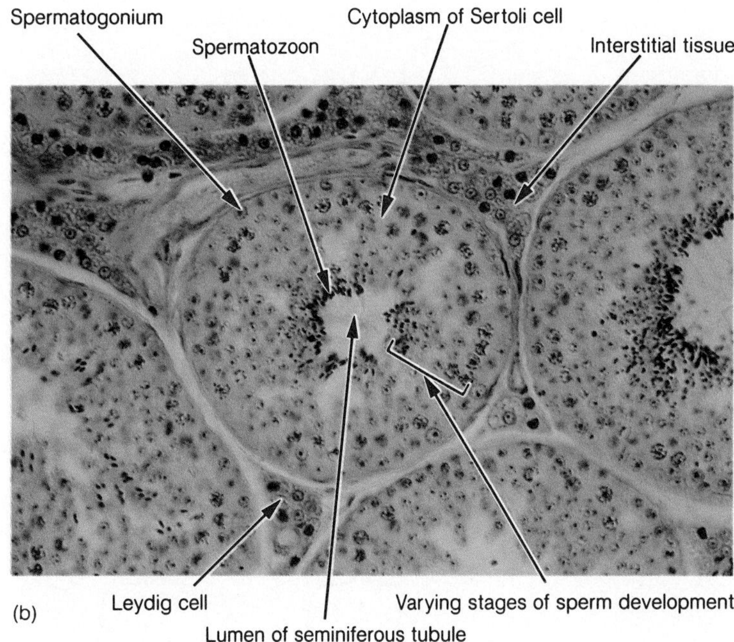

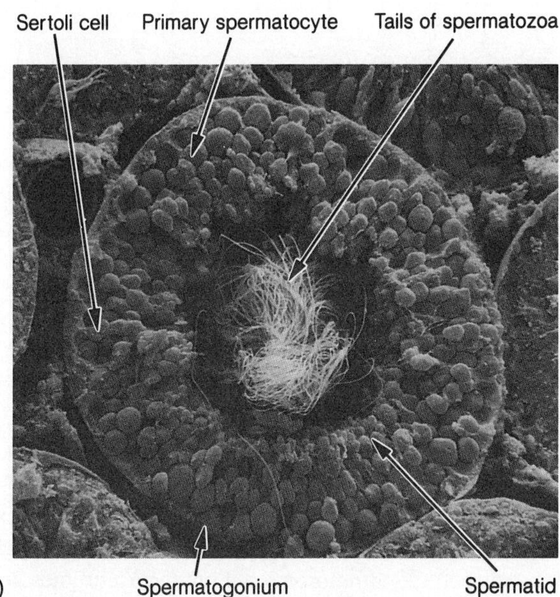

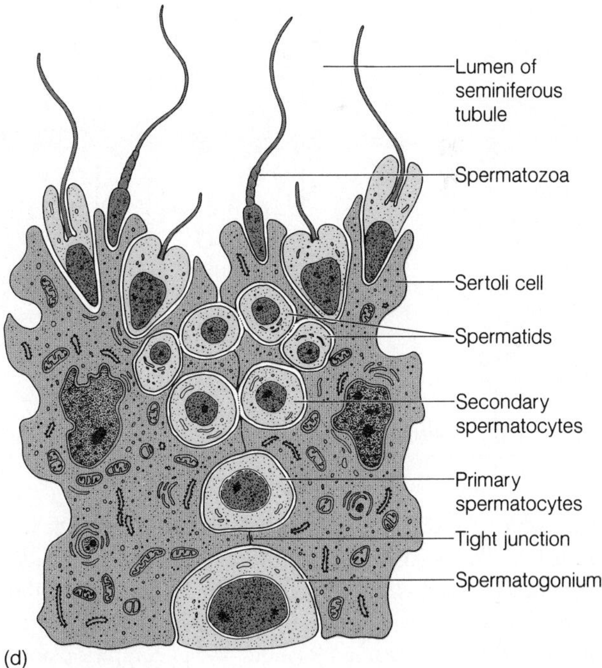

Figure 20–6 Testicular Anatomy Depicting Site of Spermatogenesis *(a) Longitudinal section of testis showing location and arrangement of the seminiferous tubules, the sperm-producing portion of the testis. (b) Light micrograph of cross section of seminiferous tubule. The undifferentiated germ cells (the spermatogonia) lie in the periphery of the tubule, and the differentiated spermatozoa are in the lumen, with the various stages of sperm development in between. (c) Scanning electron micrograph of cross section of seminiferous tubule. (d) Relationship of Sertoli cells to developing sperm cells.*

SOURCE: Part (b) courtesy of Elizabeth R. Walker, Associate Professor, and Dennis O. Overman, Associate Professor, Department of Anatomy, School of Medicine, West Virginia University. Part (c) from *Tissues and Organs: A Text-Atlas of Scanning Electron Microscopy* by Richard G. Kessel and Randy H. Kardon. Copyright ©1979 by W.H. Freeman and Company. Reprinted with permission.

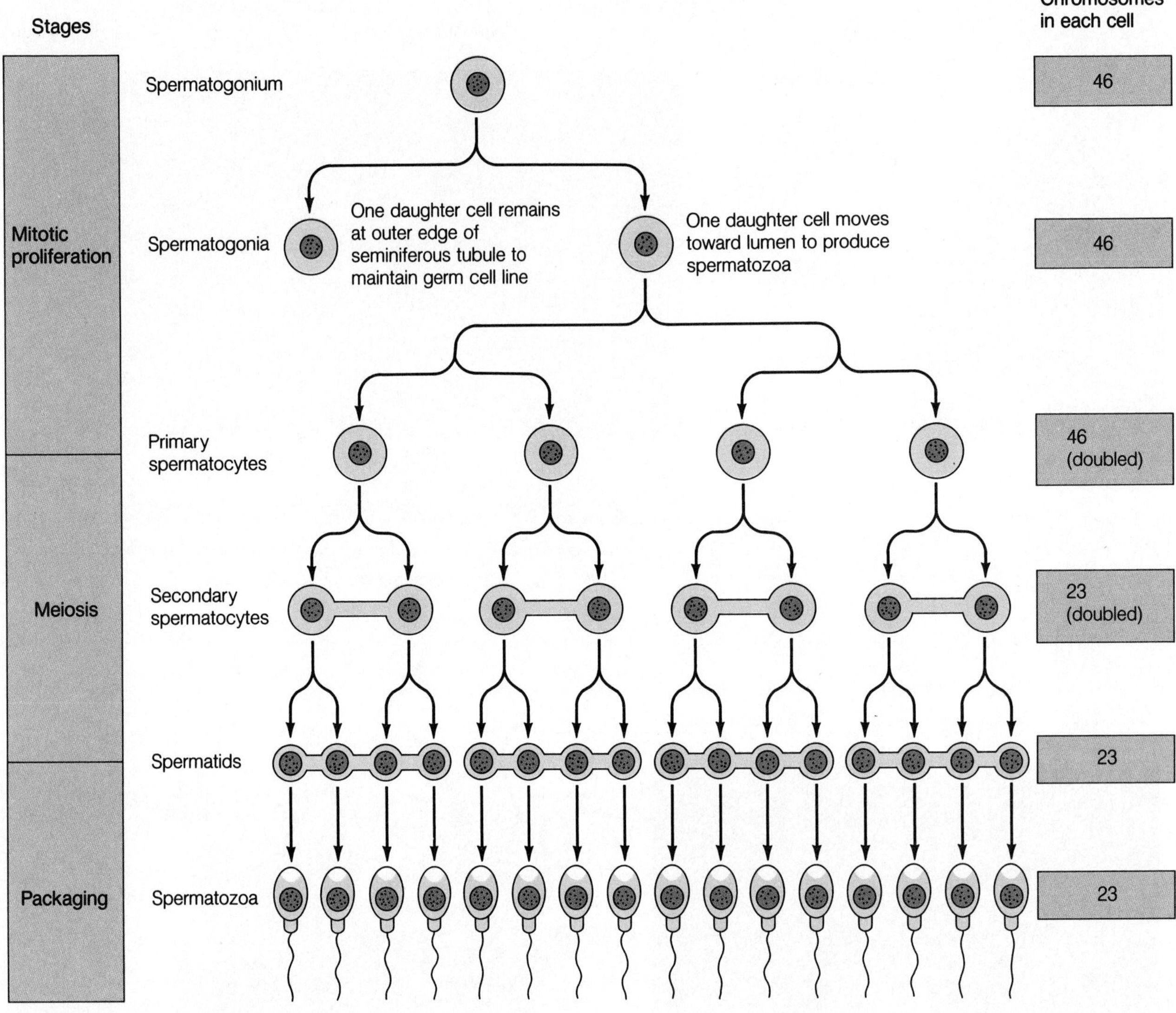

Figure 20–7 Spermatogenesis

cleus, which contains the sperm's complement of genetic information. The **acrosome,** an enzyme-filled vesicle at the tip of the head, is used as an "enzymatic drill" for penetrating the ovum. The acrosome is formed by aggregation of vesicles produced by the endoplasmic reticulum–Golgi complex before these organelles are discarded. Mobility for the spermatozoon is provided by growth of a long, whiplike *tail* out of one of the centriols (see p. 36). Movement of the tail, which occurs as a result of relative sliding of its constituent microtubules (see p. 37), is powered by energy generated by the mitochondria concentrated within the *midpiece* of the sperm.

Until sperm maturation is complete, the developing germ cells arising from a single primary spermatocyte remain joined by cytoplasmic bridges. These connections, which result from incomplete cytoplasmic division, permit cytoplasm to be exchanged between the four developing sperm. This is important, because the X chromosome but not the Y chromosome contains genes that code for cellular products essential for development of the sperm. During meiosis, half the sperm receive an X and the other half a Y chromosome. If it were not for the sharing of cytoplasm so that all the haploid cells are provided with the products coded for by X chromosomes until sperm development is complete, the Y-bearing, male-producing sperm would not be able to develop and survive.

Throughout their development, sperm remain intimately associated with Sertoli cells.

Besides the spermatogonia and developing sperm cells, the seminiferous tubules also house the supportive Sertoli cells. The **Sertoli cells** form a ring that extends from the outer basement membrane to the lumen of the tubule. Each Sertoli cell spans the entire distance from the outer membrane to the fluid-filled lumen (Fig. 20–6d). Adjacent Sertoli cells are joined by tight junctions (see p. 68) at a point slightly beneath the outer membrane. Spermatogonia are tucked between the Sertoli cells at the outer perimeter of the tubule in the spaces between the basement membrane and the tight junctions.

During spermatogenesis, developing sperm cells arising from spermatogonial mitotic activity pass through the tight junctions, which transiently separate to make a path for them, then migrate toward the lumen in intimate association with the adjacent Sertoli cells. The cytoplasm of the Sertoli cells envelops the migrating germ cells, which remain buried within these cytoplasmic recesses throughout their development.

The supportive Sertoli cells perform the following functions essential for spermatogenesis:

1. The tight junctions between adjacent Sertoli cells form a **blood-testes barrier.** Because this barrier prevents blood-borne substances from passing between the cells to gain entry to the tubular lumen, only selected molecules that are able to pass through the Sertoli cells reach the luminal fluid. As a result, the composition of the intratubular fluid varies considerably from that of the blood. The unique composition of this fluid that bathes the germ cells is believed to be critical for the later stages of sperm development. The blood-testes barrier also prevents the antibody-producing cells in the extracellular fluid from reaching the tubular sperm factory, thus preventing the formation of antibodies against the highly differentiated spermatozoa.
2. Since the secluded developing sperm cells do not have direct access to blood-borne nutrients, the Sertoli cells provide nourishment for them.
3. The Sertoli cells have an important phagocytic function. They engulf the cytoplasm extruded from the spermatids during their remodeling and destroy defective germ cells that fail to successfully complete all stages of spermatogenesis.
4. The Sertoli cells secrete into the lumen **seminiferous tubule fluid,** which "flushes" the released sperm from the tubule into the epididymis for storage and further processing.

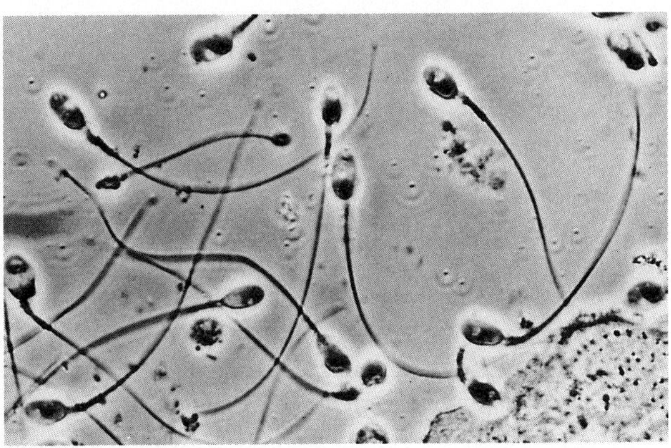

(a)

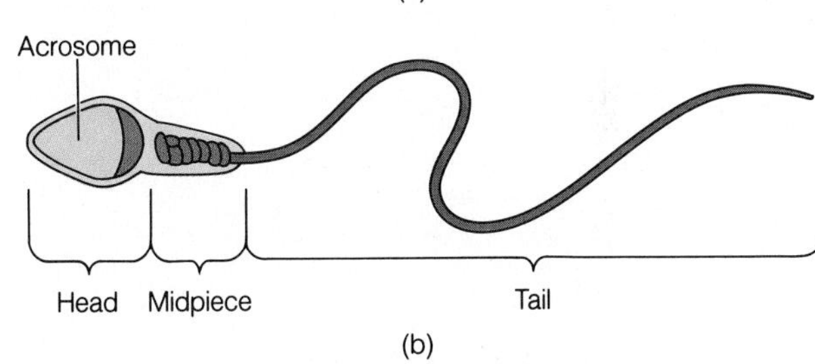

(b)

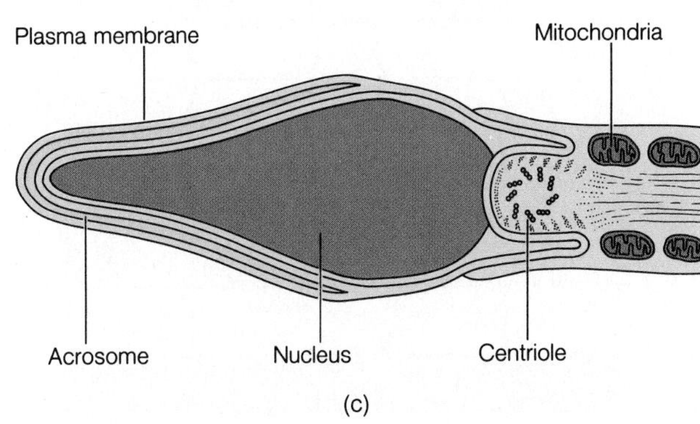

(c)

Figure 20–8 Anatomy of Spermatozoon *(a) Photomicrograph of human spermatozoa. (b) Schematic representation of spermatozoon in frontal view. (c) Enlargement of head portion of spermatozoon in side view.*

5. An important component of this Sertoli secretion is **androgen-binding protein.** As the name implies, this protein binds androgens (that is, testosterone), thus maintaining a very high level of this hormone within the male reproductive tract. This high local concentration of testosterone is essential for sustaining sperm production.
6. The Sertoli cells are the site of action for control of spermatogenesis by both testosterone and follicle-stimulating

hormone (FSH). The Sertoli cells themselves release another hormone, **inhibin,** which acts in negative-feedback fashion to regulate FSH secretion.

The two anterior pituitary gonadotropic hormones, LH and FSH, control testosterone secretion and spermatogenesis.

The testes are controlled by the two gonadotropic hormones secreted by the anterior pituitary, ***luteinizing hormone (LH)*** and ***follicle-stimulating hormone (FSH)***. These hormones act on separate components of the testes (Fig. 20–9). Luteinizing hormone acts on the Leydig cells to regulate testosterone secretion, accounting for its alternative name of ***interstitial cell-stimulating hormone*** (*ICSH*). Follicle-stimulating hormone acts on the seminiferous tubules, specifically the Sertoli cells, to enhance spermatogenesis. Secretion of both LH and FSH from the anterior pituitary is stimulated, in turn, by a single hypothalamic hormone, **gonadotropin-releasing hormone** (**GnRH**) (see p. 649).

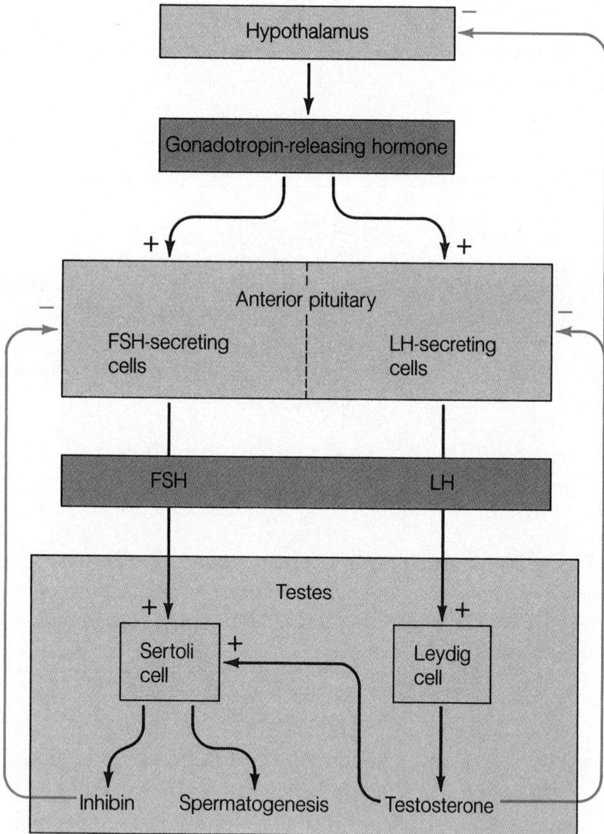

Figure 20–9 Control of Testicular Function

Once every two to three hours, GnRH is released from the hypothalamus in secretory bursts, with no secretion occurring in between. Since GnRH stimulates the gonadotropic hormone-secretory cells in the anterior pituitary, this pulsatile pattern of hypothalamic secretion results in similar episodic bursts in LH and FSH secretion.

Even though GnRH stimulates both LH and FSH secretion, the blood concentrations of these two gonadotropic hormones do not always parallel each other for two reasons. First, LH is removed from the blood more rapidly between the secretory bursts than is the more slowly metabolized FSH, so the pulsatile variations in blood levels of LH are much more pronounced than are those of FSH. Second, two other regulatory factors besides GnRH—testosterone and inhibin—differentially influence the secretory rate of LH and FSH. Testosterone, the product of LH stimulation of the Leydig cells, acts in negative-feedback fashion in two ways to inhibit LH secretion. The predominant negative-feedback effect of testosterone is to decrease the episodes of GnRH release by acting on the hypothalamus, thus indirectly decreasing LH release by the anterior pituitary. Secondly, testosterone acts directly on the anterior pituitary to reduce the responsiveness of the LH-secretory cells to GnRH. The latter action accounts for the fact that testosterone exerts a greater inhibitory effect on LH than on FSH secretion.

The testicular inhibitory signal directed at controlling FSH secretion is believed to be the peptide hormone inhibin, which is secreted by the Sertoli cells. There is strong circumstantial evidence that inhibin acts directly on the anterior pituitary to specifically inhibit FSH secretion. This feedback inhibition of FSH by a Sertoli cell product is appropriate, because FSH stimulates spermatogenesis by acting on the Sertoli cells.

Both FSH and testosterone play critical roles in controlling spermatogenesis, each exerting its effect by means of the Sertoli cell. It appears that FSH is required for spermatid remodeling, whereas testosterone is essential for both mitosis and meiosis of the germ cells. Testosterone concentration in the testes is much higher than in the blood because a substantial portion of this hormone produced locally by the Leydig cells is retained in the luminal fluid complexed with androgen-binding protein secreted by the Sertoli cells. Only this high concentration of testicular testosterone is adequate to sustain sperm production.

Gonadotropin-releasing hormone activity increases at puberty.

Even though the fetal testes secrete testosterone, which directs masculine development of the reproductive system, after birth the testes become quiescent until puberty. During the prepubertal period, LH and FSH are not secreted at adequate levels to stimulate any significant testicular activity. The prepubertal delay in the onset of reproductive capability allows time for adequate physical and psychological maturation for the individual to handle childrearing. (This is especially important in the female, whose body must support the developing fetus.)

The pubertal process is initiated by an increase in GnRH activity some time between eight and twelve years of age. Early in puberty, GnRH pulses occur only at night, causing brief nocturnal increases in LH secretion and, accordingly, testosterone secretion. The duration of episodic GnRH secretion gradually increases as puberty progresses until the adult pattern of GnRH, FSH, LH, and testosterone secretion is established. Under the influence of the rising levels of testosterone during puberty, the physical changes that encompass the secondary sexual characteristics and reproductive maturation become evident.

The low level of GnRH activity during the prepubertal period appears to be due to active inhibition of GnRH release by both hormonal and neural mechanisms. Before puberty, the hypothalamus is extremely sensitive to the negative-feedback actions of testosterone, so the very small amounts of testosterone produced by the prepubertal testes are able to inhibit GnRH release. At puberty, the hypothalamus becomes less sensitive to feedback inhibition by testosterone. Because the low levels of testosterone no longer suppress the hypothalamus, GnRH and gonadotropic hormone levels rise. Also, during the prepubertal period the hypothalamic GnRH cells are subjected to direct neural inhibition, which is similarly removed at puberty.

The factors responsible for removing these inhibitory mechanisms, thus initiating puberty in humans, remain a mystery. Three proposals, none of which is fully satisfactory, have been put forth. One possibility is a preprogrammed, age-related reduction in inhibitory activity. A second suggestion links attainment of a critical body weight or percentage of body fat to increased GnRH release. A final proposal focuses on a potential role for the hormone melatonin, which is secreted by the **pineal gland. Melatonin,** whose secretion decreases during exposure to the light and increases during exposure to the dark, has an antigonadotropic effect in many species. Light striking the eyes inhibits the nerve pathways that are responsible for stimulating melatonin secretion. In many seasonally breeding species, the overall decrease in melatonin secretion in connection with longer days and shorter nights initiates the mating season. Some researchers suggest that an observed reduction in the overall rate of melatonin secretion at puberty in humans—particularly during the night, when the peaks in GnRH secretion first occur—is the trigger for the onset of puberty.

The pineal gland is also believed to have extremely widespread influences besides its possible link with the reproductive system. It may synchronize a gamut of functions (for example, anterior pituitary hormone metabolism, thermoregulation, fat deposition, and immune responses) in relation to the seasons of the year.

The ducts of the reproductive tract store and concentrate sperm as well as increase their motility and fertility.

The remainder of the male reproductive system besides the testes is designed for delivery of sperm to the female reproductive tract. Essentially it consists of: (1) a tortuous pathway of tubes that transport sperm from the testes to the outside of the body; (2) several glands, which contribute secretions that are important to the viability and motility of the sperm; and (3) the penis, which is designed to penetrate and deposit the sperm within the vagina of the female.

A comma-shaped **epididymis** is loosely attached to the posterior surface of each testis (Fig. 20–6a). After sperm are produced in the seminiferous tubules, they are swept into the epididymis as a result of the pressure created by the continual secretion of tubular fluid by the Sertoli cells. The epididymal ducts from each testis converge to form a large, thick-walled, muscular duct called the **ductus deferens.** The ductus deferens from each testis passes up out of the scrotal sac and runs back through the inguinal canal into the abdominal cavity, where it eventually empties into the urethra at the neck of the bladder (Fig. 20–1). The urethra carries sperm out of the penis during ejaculation, the forceful expulsion of semen from the body.

These ducts perform several important functions (Table 20–2). First, the epididymis and ductus deferens serve as a route of exit for sperm from the testes. As they leave the testes, the sperm are incapable of either movement or fertilization. They gain both capabilities during their passage through the epididymis. This maturational process is stimulated by the testosterone retained within the tubular fluid bound to androgen-binding protein. Sperms' capacity to fertilize is enhanced even further by exposure to secretions of the female reproductive tract, a process known as **capacitation.** The epididymis also concentrates the sperm one-hundred-fold by absorbing most of the fluid that enters from the seminiferous tubules. The

Table 20–2 Functions of Components of Male Reproductive System

Testes
Produce sperm
Secrete testosterone
Epididymis and Ductus Deferens
Serve as sperm exit route from testes
Serve as site for maturation of sperm for motility and fertility
Concentrate and store sperm
Seminal Vesicles
Supply fructose to nourish ejaculated sperm
Secrete prostaglandins that stimulate motility within male and female reproductive tracts to help transport sperm
Provide bulk of semen
Provide precursors for clotting of semen
Prostate Gland
Secretes alkaline fluid that neutralizes acidic vaginal secretions
Triggers clotting of semen to keep sperm in vagina during penis withdrawal
Bulbourethral Glands
Secrete mucus for lubrication

maturing sperm are slowly moved through the epididymis into the ductus deferens by rhythmic contractions of the smooth muscle in the walls of these tubes. The ductus deferens serves as an important site for sperm storage. Because the tightly packed sperm are relatively inactive and their metabolic needs are accordingly low, they can be stored in the ductus deferens for many days, even though they have no nutrient blood supply and are nourished only by simple sugars present in the tubular secretions.

A common sterilization procedure in males, a **vasectomy,** involves surgical removal of a small segment of each ductus deferens (alias vas deferens, hence the term *vasectomy*) after it passes from the testis but before it enters the inguinal canal. This blocks the exit of sperm from the testes. The sperm that build up behind the tied-off testicular end of the severed ductus are removed by phagocytosis. Although this procedure blocks sperm exit, it does not interfere with testosterone activity because the Leydig cells secrete testosterone into the blood, not through the ductus deferens. Even though testosterone is essential for spermatogenesis, sperm release is not required for ongoing testosterone secretion. Thus there should be no diminution of testosterone-dependent masculinity or libido following a vasectomy.

The accessory sex glands contribute the bulk of the semen.

Several accessory sex glands—the seminal vesicles and prostate—empty their secretions into the duct system before it joins the urethra (Fig. 20–1). A pair of sac-like **seminal vesicles** empty into the last portion of the two ductus deferens, one on each side. The short segment of duct that passes beyond the entry point of the seminal vesicle to join the urethra constitutes the **ejaculatory duct.** The **prostate** is a large single gland that completely surrounds the ejaculatory ducts and urethra. In a significant number of men, prostatic hypertrophy (enlargement) occurs in middle to older age. Difficulty in urination is often encountered as the enlarging prostate impinges on the portion of the urethra that passes through the prostate. Another pair of accessory sex glands, the **bulbourethral glands,** drain into the urethra after it has passed through the prostate just before it enters the penis. Numerous mucus-secreting glands are also located along the length of the urethra.

During ejaculation, the accessory sex glands contribute secretions that provide support for the continuing viability of the sperm inside the female reproductive tract. These secretions constitute the bulk of the semen, which consists of a mixture of accessory sex gland secretions, sperm, and mucus. Sperm make up only a small percentage of the total ejaculated fluid.

Although the accessory sex gland secretions are not absolutely essential for fertilization, they do make contributions that greatly facilitate the process as follows (Table 20–2):

□ The seminal vesicles: (1) supply fructose, which serves as the primary energy source for ejaculated sperm; (2) secrete prostaglandins, which are believed to stimulate contractions of the smooth muscle in both the male and female reproductive tracts, thereby helping to transport sperm from their storage site in the male to the site of fertilization in the female oviduct; (3) provide more than half the semen, which helps to wash the sperm into the urethra and furthermore dilutes the thick mass of sperm, thus enabling them to develop motility; and (4) secrete fibrinogen, a precursor of fibrin, which forms the meshwork of a clot.

□ The prostate gland: (1) secretes an alkaline fluid that neutralizes the acidic vaginal secretions, an important function because sperm are more viable in a slightly alkaline environment; and (2) provides clotting enzymes and fibrinolysin. The prostatic clotting enzymes act on fibrinogen from the seminal vesicles to produce fibrin, which "clots" the semen. This helps to keep the ejaculated sperm in the female reproductive tract during withdrawal of the penis. Shortly thereafter, the seminal clot is broken down by fibrinolysin, a fibrin-degrading enzyme from the prostate, thus releasing motile sperm within the female tract.

□ During sexual arousal, the bulbourethral glands secrete a mucus-like substance that provides lubrication for sexual intercourse.

Prostaglandins are ubiquitous, locally acting chemical messengers.

Although **prostaglandins** were first identified in the semen and were believed to be of prostate-gland origin (hence their name, even though they are actually secreted into the semen by the seminal vesicles), their production and actions are by no means limited to the reproductive system. These twenty-carbon fatty-acid derivatives are among the most ubiquitous chemical messengers in the body. They are produced in virtually all tissues from arachidonic acid, a fatty-acid constituent of the phospholipids within the plasma membrane. Prostaglandins (and other closely related arachidonic-acid derivatives that are often included for convenience in the category of prostaglandins—namely, ***prostacyclins***, ***thromboxanes***, and ***leukotrienes***)—are among the most biologically active compounds known. Upon appropriate stimulation, arachidonic acid is split from the plasma membrane by a membrane-bound enzyme, then is converted into the appropriate prostaglandin, which acts locally within or near its site of production. After prostaglandins act, they are rapidly inactivated by local enzymes before they gain access to the blood, or, if they do reach the circulatory system, they are swiftly degraded on their first pass through the lungs so that they are not dispersed through the systemic arterial system.

Prostaglandins are designated as belonging to one of three groups—PGA, PGE, or PGF—according to structural variations in the five-carbon ring that they contain at one end (Fig. 20–10). Within each group, prostaglandins are further identified by the number of double bonds present in the two side chains that project from the ring structure (for example, PGE_1 has one double bond and PGE_2 has two double bonds).

Prostaglandins exert a bewildering variety of effects. Not only are slight variations in prostaglandin structure accompanied by profound differences in biological action, but the same prostaglandin molecule may even exert opposite effects in different tissues. Besides enhancing sperm transport in semen, these abundant chemical messengers are known or suspected to exert other actions in the female reproductive system and in the respiratory, urinary, digestive, nervous, and endocrine systems, in addition to having effects on platelet aggregation, fat metabolism, and inflammation (Table 20–3).

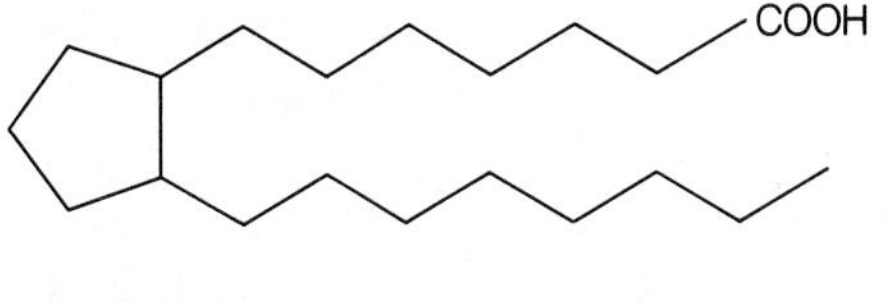

Letter designation (PGA, PGE, PGF) denotes structural variations in the five-carbon ring

Number designation (e.g., PGE_1, PGE_2) denotes number of double bonds present in the two side chains

Figure 20–10 Structure and Nomenclature of Prostaglandins

As prostaglandins' various actions are better understood, new ways of manipulating them therapeutically are becoming available. A classic example is the use of aspirin, which blocks

Table 20–3 Known or Suspected Actions of Prostaglandins

Body System/ Activity	Actions of Prostaglandins
Reproductive system	Promote sperm transport by action on smooth muscle in male and female reproductive tracts
	Important in menstruation
	Play role in ovulation
	Contribute to preparation of maternal portion of placenta
	Contribute to parturition
Respiratory system	Some promote bronchodilation, others bronchoconstriction
Urinary system	Increase renal blood flow
	Increase excretion of water and salt
Digestive system	Inhibit HCl secretion by stomach
	Stimulate intestinal motility
Nervous system	Modulate neurotransmitter release and action
	Act at hypothalamic "thermostat" to increase body temperature
Endocrine system	Enhance cortisol secretion
	Influence tissue responsiveness to hormones in many instances
Circulatory system	Modulate platelet aggregation
Fat metabolism	Inhibit fat breakdown
Defense system	Promote many aspects of inflammation, including fever and development of pain

the conversion of arachidonic acid into prostaglandins, for fever reduction and pain relief. Prostaglandin action is also therapeutically inhibited in the treatment of premenstrual symptoms and menstrual cramping. Furthermore, specific prostaglandins have been medically administered in such diverse situations as inducing labor, treating asthma, and treating gastric ulcers.

Sexual Intercourse between Males and Females

Ultimately, union of male and female gametes in humans requires delivery of semen into the female vagina through the **sex act,** also known as **sexual intercourse, coitus,** or **copulation.**

The male sex act is characterized by erection and ejaculation.

The *male sex act* involves two components: (1) **erection,** or hardening of the normally flaccid penis to permit its entry into the vagina, and (2) **ejaculation,** or forceful expulsion of semen out of the penis (Table 20–4). In addition to these strictly reproductive-related components, the **sexual response cycle** also encompasses broader physiological responses that can be divided into four phases:

1. the *excitement phase*, which includes erection accompanied by testicular vasocongestion and heightened sexual awareness;
2. the *plateau phase*, which is characterized by intensification of these responses, plus more generalized body responses, such as steadily increasing heart rate, blood pressure, respiratory rate, and muscle tension;
3. the *orgasmic phase*, which includes ejaculation as well as other responses that culminate the mounting sexual excitement and are collectively experienced as an intense physical pleasure; and
4. the *resolution phase*, which returns the genitalia and body systems to their prearousal state.

The human sexual response is a multicomponent experience that, in addition to these physiological phenomena, also encompasses emotional, psychological, and sociological factors. We will concentrate only on the physiological aspects of sex.

Erection is not caused by contraction of skeletal muscles within the penis, as might be expected, but is caused instead by engorgement of the penis with blood. The penis consists almost entirely of **erectile tissue** made up of three columns of spongelike vascular spaces extending the length of the organ (Fig. 20–1). In the absence of sexual excitation, the erectile tissues contain little blood, because the arterioles that supply these vascular chambers are constricted. As a result, the penis remains small and flaccid. During sexual arousal, these arterioles reflexly dilate and the erectile tissue starts to fill with blood, causing the penis to enlarge both in length and width and to become more rigid. A larger buildup of blood and further enhancement of erection is achieved by a reduction in venous outflow. The veins that drain the erectile tissue are compressed as a result of engorgement and expansion of the vascular spaces brought about by increased arterial inflow. These local vascular responses—reflexly increased arterial inflow and mechanically reduced venous outflow—transform the penis into a hardened, elongated organ capable of penetrating the vagina.

The erection reflex is a spinal reflex triggered by stimulation of highly sensitive mechanoreceptors located in the glans penis, which caps the tip of the penis. Tactile stimulation of the glans reflexly triggers increased parasympathetic and decreased sympathetic activity to the penile arterioles, the result of which is vasodilation of these arterioles and an ensuing erection (Fig. 20–11). This is the major instance of direct par-

Table 20–4 Components of the Male Sex Act

	Erection	Ejaculation
Definition	Hardening of the normally flaccid penis to permit its entry into the vagina	*Emission phase:* emptying of sperm and accessory sex gland secretions (semen) into urethra *Expulsion phase:* forceful expulsion of semen from penis
Accomplished by	Engorgement of penis erectile tissue with blood as a result of marked parasympathetically induced vasodilation of penile arterioles and mechanical compression of veins	*Emission phase:* sympathetically induced contraction of the smooth muscle in the walls of the ducts and accessory sex glands *Expulsion phase:* motor neuron–induced contraction of skeletal muscles at the base of the penis

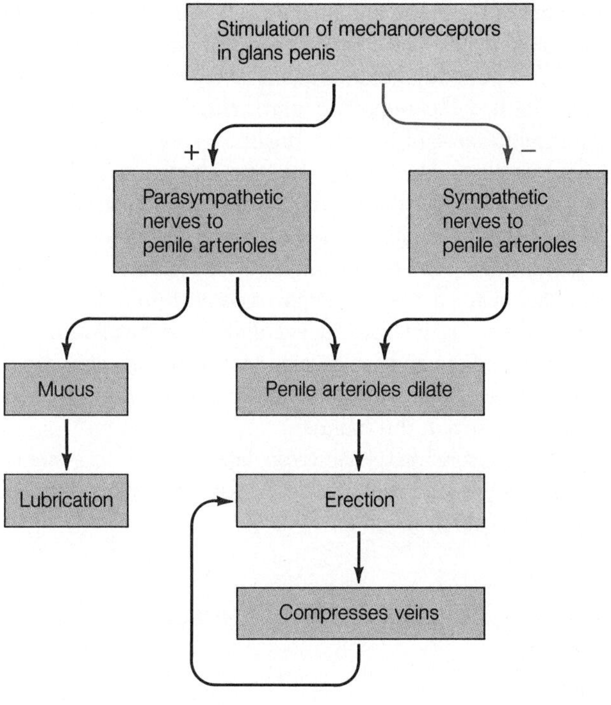

Figure 20–11 Erection Reflex

asympathetic control over blood vessel caliber in the body. Arterioles are typically supplied only by sympathetic nerves, with increased sympathetic activity producing vasoconstriction and decreased sympathetic activity resulting in vasodilation. Concurrent parasympathetic stimulation and sympathetic inhibition of penile arterioles accomplish vasodilation more swiftly and of greater magnitude than is possible elsewhere in the body. Through this efficient means of rapidly increasing blood flow into the penis, complete erection can be accomplished as quickly as five to ten seconds. At the same time, parasympathetic impulses promote secretion of lubricating mucus from the bulbourethral glands and the urethral glands in preparation for coitus.

This basic spinal reflex can either be facilitated or inhibited by higher brain centers through descending pathways that also terminate on the autonomic nerves supplying the penile arterioles. As an example of facilitation, psychic stimuli, such as viewing something sexually exciting, can induce an erection in the complete absence of tactile stimulation of the penis. On the other hand, failure to achieve an erection in spite of appropriate stimulation (**impotence**) may occur as a result of inhibition of the erection reflex by higher brain centers. In fact, impotence is most commonly caused by psychological influences rather than actual physical limitations. A man who becomes overly anxious about his ability to perform the sex act may well be on his way to failure.

The second component of the male sex act is ejaculation. As with erection, ejaculation is accomplished by a spinal reflex. The same types of tactile and psychic stimuli that induce erection cause ejaculation when the level of excitation intensifies to a critical peak. The overall ejaculatory response occurs in two phases. First, sympathetic impulses cause sequential contraction of smooth muscles in the prostate, epididymis, ductus deferens, ejaculatory duct, and seminal vesicles. This contractile activity sequentially delivers first prostatic fluid, then sperm, and finally seminal vesicle fluid (collectively, semen) into the urethra. This phase of the ejaculatory reflex is known as **emission.** During this time, the sphincter at the neck of the bladder is tightly closed to prevent semen from entering the bladder and urine from being expelled along with the ejaculate through the urethra. Second, filling of the urethra with semen triggers nerve impulses that activate a series of skeletal muscles at the base of the penis. Rhythmic contractions of these muscles occur at 0.8 second intervals and increase the pressure within the penis, forcibly expelling the semen through the urethra to the exterior. This is the **expulsion** phase of ejaculation.

The rhythmical contractions that occur during semen expulsion are accompanied by involuntary rhythmic throbbing of pelvic muscles and peak intensity of the overall body responses that were climbing during the earlier phases. Heavy breathing, a heart rate of up to 180 beats per minute, marked generalized skeletal-muscle contraction, and heightened emotions are characteristic. These pelvic and overall systemic responses that culminate the sex act are associated with an intense pleasure characterized by a feeling of release and complete gratification, an experience known as **orgasm.**

During the resolution phase following orgasm, sympathetic vasoconstrictor impulses slow the inflow of blood into the penis so that the erection subsides. A deep relaxation ensues, often accompanied by a feeling of fatigue. Muscle tone returns to its normal state while the cardiovascular and respiratory systems return to their prearousal level of activity. Once ejaculation has occurred, a temporary refractory period of variable duration ensues before sexual stimulation can trigger another erection. Males are therefore unable to experience multiple orgasms within a matter of minutes as females sometimes do.

The volume and sperm content of the ejaculate depend on the length of time between ejaculations. The average volume of semen is 3 ml, ranging from 2.5 to 6 ml, the higher volumes following periods of continence. An average human ejaculate contains about 300–400 million sperm (120 million/ml). Both quantity and quality of the sperm are important determinants of fertility. A man is considered to be clinically infertile if his sperm concentration falls below 20 million/ml of semen. Even though only one spermatozoon actually fertilizes the ovum, large numbers of accompanying sperm are needed to provide sufficient acrosomal enzymes to break down the barriers surrounding the ovum until the victorious sperm penetrates into the ovum's cytoplasm. The quality of sperm also must be taken into account when assessing the fertility potential of a semen sample. The presence of substantial numbers of sperm with abnormal motility or structure, such as sperm with two heads or distorted tails, reduces the chances of fertilization.

The female sexual cycle parallels that of males in many ways.

Both sexes experience the same four phases of the sexual cycle—excitement, plateau, orgasm, and resolution. Furthermore, the physiological mechanisms responsible for orgasm are fundamentally the same in males and females.

The excitement phase in females can be initiated by either psychological or physical stimuli. Tactile stimulation of the clitoris and surrounding perineal area is an especially powerful sexual stimulus. These stimuli trigger spinal reflexes that bring about parasympathetically induced vasodilation of arterioles throughout the vagina and external genitalia. The resultant inflow of blood becomes evident as swelling of the labia and erection of the clitoris. The latter—like its male homologue, the penis—is composed largely of erectile tissue. Vasocongestion of the vaginal capillaries forces fluid out of the vessels into the vaginal lumen. This fluid, which is the first positive indication of sexual arousal, serves as the primary lubricant for intercourse. Additional lubrication is provided by the mucus secretions from the male and by mucus released during sexual arousal from glands located at the outer orifice of the vagina. Also during the excitement phase in the female, the nipples become erect and the breasts enlarge as a result of vasocongestion. In addition, the majority of women show a *sex flush* during this time, which is caused by increased blood flow through the skin.

During the plateau phase, the changes initiated during the excitement phase intensify while systemic responses similar to those in the male occur, such as increased heart rate, blood pressure, respiratory rate, and muscle tension. Further vasocongestion of the outer third of the vagina during this time reduces its inner capacity so that it tightens around the thrusting penis, thus heightening tactile sensation for the male. Simultaneously, the uterus raises upward, lifting the cervix and enlarging the upper two-thirds of the vagina. This ballooning or **tenting effect** creates a space for ejaculate deposition.

If erotic stimulation continues, the sexual response culminates in orgasm as sympathetic impulses trigger rhythmic contractions of the pelvic musculature at 0.8 second intervals, the same as in males. The contractions occur most intensely in the engorged outer third of the vaginal canal; hence this region is often termed the **orgasmic platform.** Systemic responses identical to those of the male orgasm also occur. In fact, the orgasmic experience in females parallels that of males with two exceptions. First, there is no female counterpart to ejaculation. Second, females do not become refractory following an orgasm, so they can respond immediately to continued erotic stimulation and achieve multiple orgasms. If stimulation continues, the sexual intensity only diminishes to the plateau level following orgasm and can quickly be brought to a peak again. Women have been known to achieve as many as twelve successive orgasms in this manner.

During resolution, pelvic vasocongestion and the systemic manifestations gradually subside. As with males, it is a time of great physical relaxation for females.

Female Reproductive Physiology

Complex cycling characterizes female reproductive physiology.

Female reproductive physiology is much more complex than male reproductive physiology. Unlike the continuous sperm production and essentially constant testosterone secretion characteristic of the male, release of ova is intermittent and secretion of female sex hormones displays wide cyclical swings. The tissues influenced by these sex hormones also undergo cyclical changes, the most obvious of which is the monthly menstrual cycle. During each cycle, the female reproductive tract is prepared for the fertilization and implantation of an ovum released from the ovary at ovulation. If fertilization does not occur, the cycle repeats itself. If fertilization does occur, the cycles are interrupted while the female system adapts to nurture and protect the newly conceived human being until it is developed into an individual capable of living outside the maternal environment. Furthermore, the female continues her reproductive duties after birth by producing milk (lactation) for the baby's nourishment. Thus the female reproductive system is characterized by complex cycles that are interrupted only by more complex changes should pregnancy ensue.

The ovaries, as the primary female reproductive organs, perform the dual function of producing ova (oogenesis) and secreting the female sex hormones, estrogen and progesterone. These hormones act together to promote fertilization of the ovum and to prepare the female reproductive system for pregnancy. Estrogen in the female is responsible for many similar functions carried out by testosterone in the male, such as maturation and maintenance of the entire female reproductive system and establishment of female secondary sexual characteristics. In general, the actions of estrogen are important to preconception events. Estrogen is essential for ova maturation and release, development of physical characteristics that are sexually attractive to males, and transport of sperm from the vagina to the site of fertilization in the oviduct. Furthermore, estrogen contributes to breast development in anticipation of lactation. The other ovarian steroid, progesterone, is important in preparing a suitable environment for nourishing a developing embryo/fetus and for contributing to the breasts' ability to produce milk.

As with males, reproductive capability begins at puberty in females, but unlike males, who have reproductive potential throughout life, female reproductive potential ceases during middle-age at menopause.

Chromosome division in oogenesis parallels that in spermatogenesis, but there are major qualitative and quantitative sexual differences in gametogenesis.

Oogenesis contrasts sharply with spermatogenesis in several important aspects, even though the identical steps of chromosome replication and division take place during gamete production in both sexes. The undifferentiated primordial germ cells in the fetal ovaries, the **oogonia** (comparable to the spermatogonia), divide mitotically to give rise to six to seven million oogonia by the fifth month of gestation, at which time mitotic proliferation ceases. During the last part of fetal life, the oogonia begin the early steps of the first meiotic division but do not complete it. Known now as **primary oocytes,** they contain forty-six replicated chromosomes, which are gathered into homologous pairs but which do not separate. The primary oocytes remain in this state of meiotic arrest for years until they are prepared for ovulation.

Before birth, each primary oocyte is surrounded by a single layer of **granulosa cells** to form a **primary follicle.** Oocytes that fail to be incorporated into follicles degenerate, and at birth only about two million primary follicles remain, each containing a single primary oocyte capable of producing a single ovum. No new oocytes or follicles appear after birth; the follicles already present in the ovaries at birth serve as a reservoir from which all ova throughout the reproductive life of a female must arise. Of these follicles, only about four-hundred will mature and release ova. The pool of primary follicles present at birth gives rise to an ongoing trickle of developing follicles. A follicle is destined for one of two fates once it starts to develop: it will reach maturity and ovulate, or it will degenerate to form scar tissue, a process known as **atresia.** Until puberty, all of the follicles that start to develop undergo atresia in the early stages without ever ovulating. Even for the first few years after puberty, many of the cycles are **anovulatory** (that is, no ovum is released). Of the initial pool of follicles, 99.98% never ovulate but instead undergo atresia at some stage in development. By **menopause,** which occurs on average in a woman's early fifties, few, if any, primary follicles remain, having either already ovulated or become atretic. From this point on, the woman's reproductive capacity ceases.

This limited gamete potential, which is already determined at birth in females, is in sharp contrast to the continual process of spermatogenesis in males, who have the potential to produce several hundred million sperm in a single day. Furthermore, there is considerable chromosome wastage in oogenesis as compared with spermatogenesis.

The primary oocyte within a primary follicle is still a diploid cell that contains forty-six chromosomes. Until menopause, a portion of the resting pool of follicles starts developing into **secondary** (or **antral**) **follicles** on a cyclical basis. The number of developing follicles at any time is roughly proportional to the size of the pool, but the mechanisms that determine which follicles in the pool will develop during a given cycle are unknown. Development of a secondary follicle is characterized by growth of the primary oocyte and by expansion and differentiation of the surrounding cell layers. Oocyte enlargement is due to a buildup of cytoplasmic materials that will be needed by the early embryo. Just before ovulation, the primary oocyte, whose nucleus has been in meiotic arrest for years, completes its first meiotic division. This division yields two daughter cells, each receiving a set of twenty-three doubled chromosomes, analogous to the formation of secondary spermatocytes (Fig. 20–12). However, almost all of the cytoplasm remains with one of the daughter cells, now called the **secondary oocyte,** which is destined to become the ovum. The chromosomes of the other daughter cell together with a small share of cytoplasm form the **first polar body.** In this way, the ovum-to-be loses half of its chromosomes to form a haploid gamete but retains all of its nutrient-rich cytoplasm.

It is actually the secondary oocyte and not the mature ovum that is ovulated and fertilized, but common usage refers to the developing female gamete as an ovum even in its primary and secondary oocyte stages. Sperm entry into the secondary oocyte is needed to trigger the second meiotic division. Oocytes that are not fertilized never complete this final division. During this division, a half set of chromosomes along with a thin layer of cytoplasm is extruded as the **second polar body.** The other half set of twenty-three unpaired chromosomes remains behind in what is now the **mature ovum.** These twenty-three maternal chromosomes unite with the twenty-three paternal chromosomes of the penetrating sperm to complete fertilization. If the first polar body has not already degenerated, it too undergoes the second meiotic division at the same time the fertilized secondary oocyte is dividing its chromosomes.

Thus the steps involved in chromosome distribution during oogenesis parallel those of spermatogenesis except that the cytoplasmic distribution and time span for completion sharply differ between these two processes. Just as four haploid spermatids are produced by each primary spermatocyte, four haploid daughter cells are produced by each primary oocyte. In spermatogenesis, each daughter cell is developed into a highly specialized, motile spermatozoon unencumbered by unessential cytoplasm and organelles, its only destiny being to supply half of the genes for a new individual. In oogenesis, however, of the four daughter cells, only the one destined to become the ovum receives cytoplasm. This is important because the ovum, in addition to providing half the genes, provides all of the cytoplasmic components needed to support early development of the fertilized ovum. The large, relatively undifferentiated ovum contains numerous nutrients, organelles, and structural and enzymatic proteins. The three other cytoplasm-scarce daughter cells, the polar bodies, rapidly disintegrate, their chromosomes being wasted.

Note also the considerable difference in time required for completion of spermatogenesis and oogenesis. It takes about two months for development from a spermatogonium to fully remodeled spermatozoa. In contrast, anywhere from eleven years (beginning of ovulatory cycles at onset of puberty) to fifty years (end of ovulation at onset of menopause) is required for development of an oogonium (developed before birth) to a mature ovum. There is considerable speculation that the older age of ova released by women in their late thirties and forties accounts for the higher incidence of genetic abnormalities, such as Down's syndrome, in children borne to women in this age range.

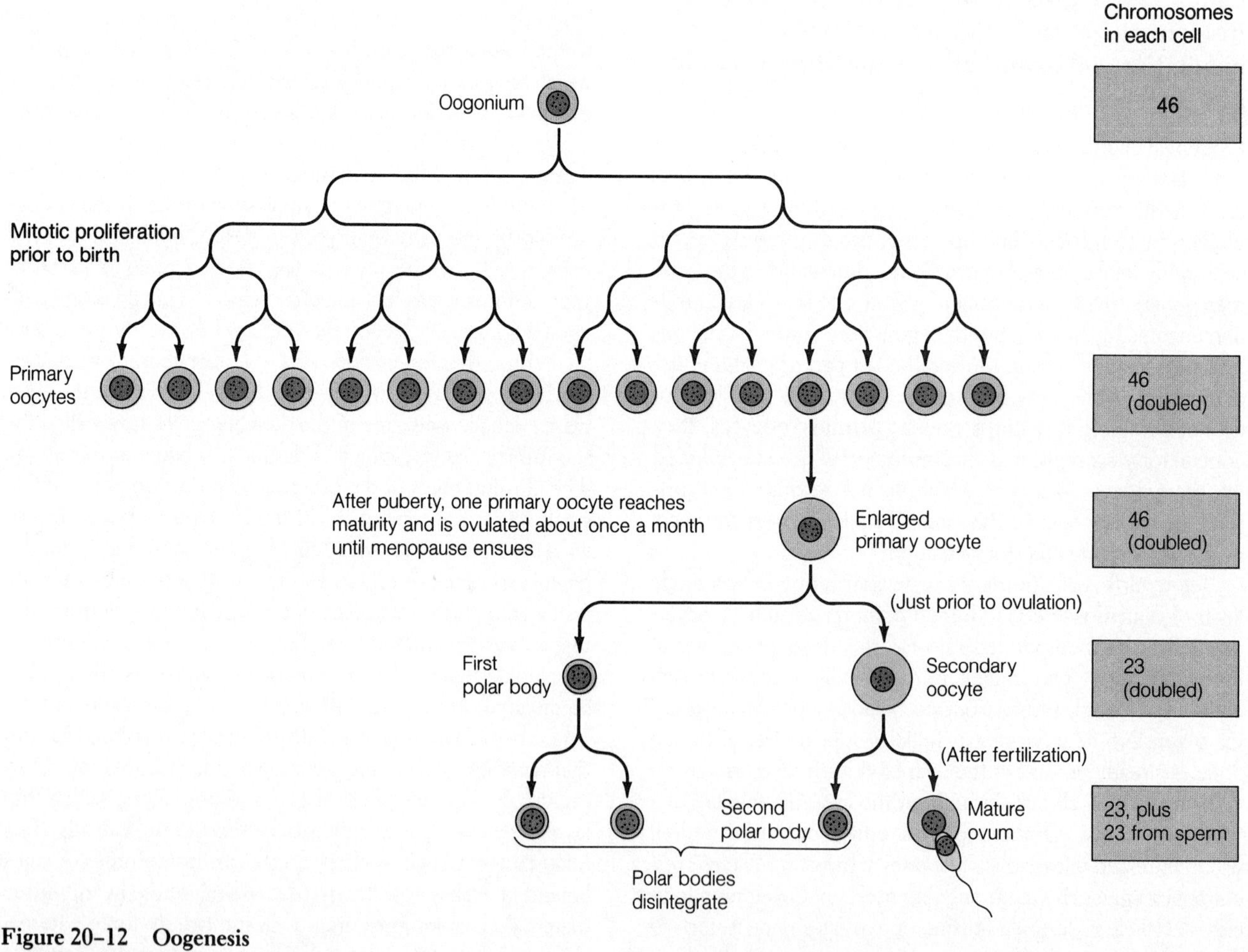

Figure 20–12 Oogenesis

The ovarian cycle consists of alternating follicular and luteal phases.

After the onset of puberty, the ovary constantly alternates between two phases—the **follicular phase,** which is dominated by the presence of maturing follicles, and the **luteal phase,** which is characterized by the presence of the corpus luteum. This cycle is normally interrupted only by pregnancy and is finally terminated by menopause. The average ovarian cycle lasts twenty-eight days, but this varies among women and among cycles in any particular woman. The follicle operates in the first half of the cycle to produce a mature egg ready for ovulation at midcycle. The corpus luteum takes over during the second half of the cycle to prepare the female reproductive tract for pregnancy if fertilization of the released egg occurs.

At any given time throughout the cycle, a portion of the primary follicles are starting to develop. However, only those that do so during the follicular phase, when the hormonal milieu is right to promote their maturation, continue beyond the early stages of development. The others, lacking hormonal support, undergo atresia. During follicular development, as the primary oocyte is synthesizing and storing materials for future use if fertilized, important changes are taking place in the cells surrounding the reactivated oocyte in preparation for the egg's release from the ovary (Fig. 20–13). First, the single layer of granulosa cells in a primary follicle proliferates to form several layers that surround the oocyte. These granulosa cells secrete a thick, gel-like glycoprotein material that covers the oocyte and separates it from the surrounding granulosa cells. This intervening membrane is known as the **zona pellucida.** At the same time, specialized ovarian connective-tissue cells at the edge of the growing follicle proliferate and differentiate to form an outer layer of **thecal cells.** The thecal and granulosa cells, collectively known as **follicular cells,** function as a unit to secrete estrogen. Of the three physiologically important estrogens—estradiol, estrone, and estriol—**estradiol** is the principal ovarian estrogen.

The hormonal environment that exists during the follicular phase promotes enlargement and development of the follicular cells' secretory capacity, converting the primary follicle into a secondary, or antral, follicle capable of estrogen secretion. This stage of follicular development is characterized by the formation of a fluid-filled **antrum** in the midst of the granulosa cells. The follicular fluid originates partially from transudation (passage through capillary pores) of plasma and partially from follicular cell secretions. As the follicular cells start producing estrogen, some of this hormone is secreted into the blood for distribution throughout the body. However, a portion of the estrogen collects in the hormone-rich antral fluid.

The oocyte has reached full size by the time the antrum begins to form. The shift to an antral follicle initiates a period of rapid follicular growth. During this time, the follicle increases in size from a diameter of less than 1 mm to 12 to 16 mm shortly before ovulation. Part of the follicular growth is due to continued proliferation of the granulosa and thecal cells, but most is due to a dramatic expansion of the antrum. As the follicle grows, estrogen is produced in increasing quantities. One of the follicles usually grows more rapidly than the others, developing into a **mature** (**preovulatory, tertiary,** or **Graafian**) **follicle** within about fourteen days after the onset of follicular development. The antrum occupies most of the space in a mature follicle. The oocyte, surrounded by the zona pellucida and a single layer of granulosa cells, is displaced asymmetrically at one side of the growing follicle in a little mound that protrudes into the antrum.

The greatly expanded mature follicle bulges on the ovarian surface, creating a thin area that ruptures to release the oocyte at **ovulation.** Rupture of the follicle is facilitated by the release from the follicular cells of enzymes that digest the connective tissue in the follicular wall. This weakens the bulging wall so that it balloons out even further to the point that it can no longer contain the rapidly expanding follicular contents.

Just before ovulation, the oocyte completes its first meiotic division. The ovum (secondary oocyte), still surrounded by its tightly adhering zona pellucida and granulosa cells (now called the **corona radiata**), is swept out of the ruptured follicle into the abdominal cavity by the leaking antral fluid (Fig. 20–14). The released ovum is quickly drawn into the oviduct, where fertilization may or may not take place.

The other developing follicles that failed to reach maturation and ovulate undergo degeneration, never to be reactivated. Occasionally, two (or perhaps more) follicles reach maturation and ovulate at about the same time. If both are fertilized, **fraternal twins** result. Because fraternal twins arise from separate ova fertilized by separate sperm, they share no more in common than any other two siblings except for the same birth date. **Identical twins,** on the other hand, develop from a single fertilized ovum that completely divides into two separate, genetically identical embryoes at a very early stage in development.

Rupture of the follicle at ovulation signals the end of the follicular phase and ushers in the luteal phase. The ruptured follicle that is left behind in the ovary following release of the ovum undergoes a rapid change. The granulosa and thecal cells remaining in the remnant follicle first collapse into the emptied antral space that has been partially filled by clotted blood. These old follicular cells soon undergo a dramatic structural transformation to form the **corpus luteum,** a process called **luteinization** (Fig. 20–14). The granulosa cells hypertrophy and are converted into very active steroidogenic

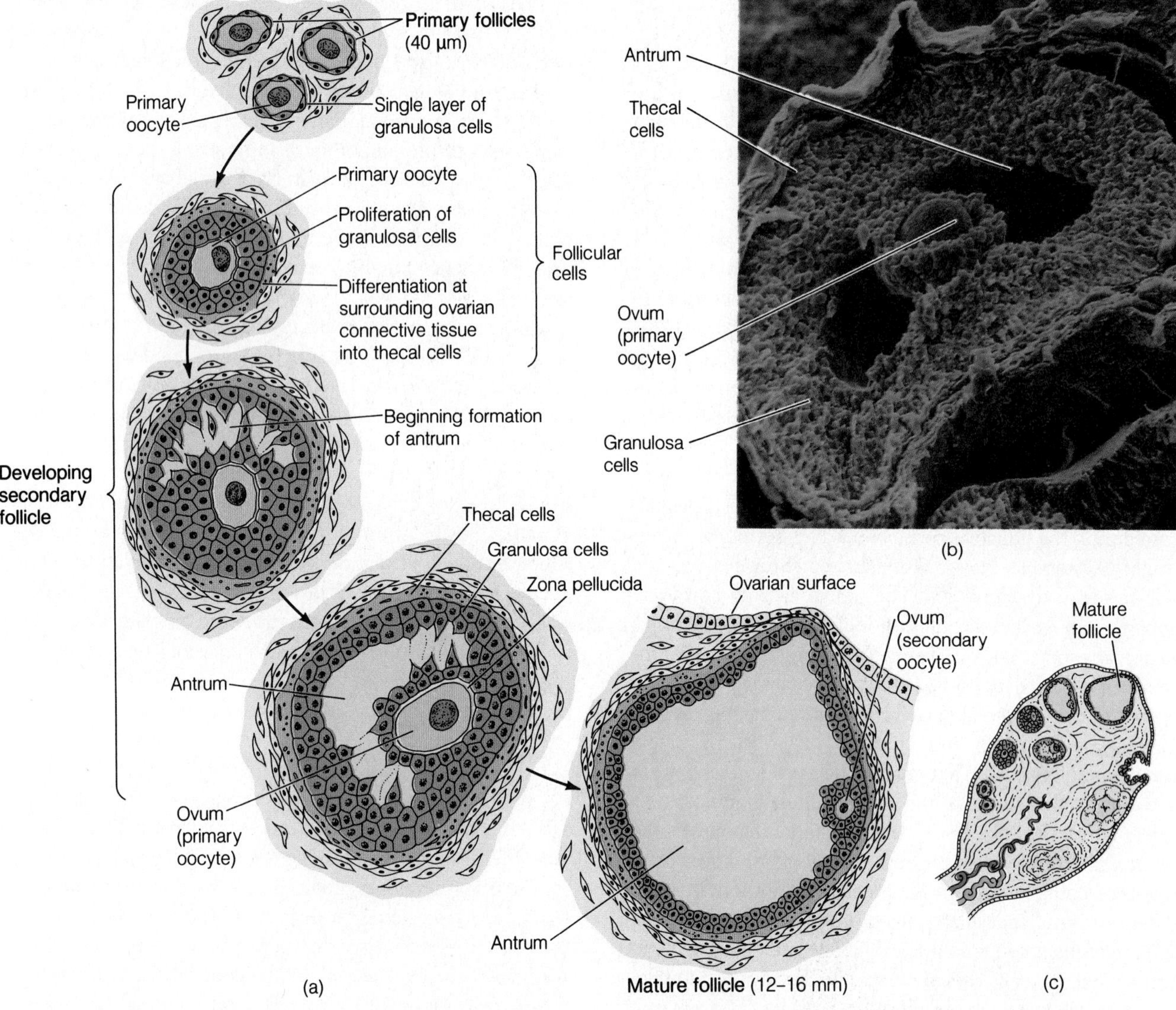

Figure 20–13 Follicular Development *(a) Stages in follicular development from primary follicle through mature follicle. (b) Scanning electron micrograph of developing secondary follicle. (c) Ovary (actual size) showing presence of mature follicle.*

SOURCE: Part (b) courtesy Dr. P. Bagavandoss, Developmental and Reproductive Biology, University of Michigan Medical School.

(steroid hormone–producing) tissue. Abundant storage of cholesterol, the steroid precursor molecule, in lipid droplets within the corpus luteum gives this tissue a yellowish appearance, hence its name (***corpus*** means "body"; ***luteum*** means "yellow"). The corpus luteum becomes highly vascularized as blood vessels from the thecal region invade the luteinizing granulosa. These changes are appropriate for the corpus luteum's function, which is to produce and secrete into the blood abundant quantities of progesterone secretion along with lesser amounts of estrogen. Estrogen secretion in the follicular phase followed by progesterone secretion in the luteal phase is essential for preparing the uterus to be a suitable site for implantation of a fertilized ovum. The corpus luteum becomes fully functional within four days after ovulation, but it continues to increase in size for another four or five days. If the released ovum is not fertilized and does not implant, the corpus luteum regresses within fourteen days after its formation. The luteal cells degenerate and are phagocytized, the vascular supply is withdrawn, and connective tissue rapidly fills in to form a fibrous tissue mass known as the **corpus albicans** ("white body"). The luteal phase is over, and one ovarian cycle is complete. A new wave of follicular development, which begins when regression of the old corpus luteum is completed, signals the onset of a new follicular phase.

If fertilization and implantation do take place, the corpus luteum continues to grow and produce increasing quantities of progesterone and estrogen instead of degenerating. Now called the **corpus luteum of pregnancy,** this ovarian structure persists until the end of pregnancy. It provides the hormones essential for the maintenance of pregnancy until the developing placenta is able to take over this crucial function.

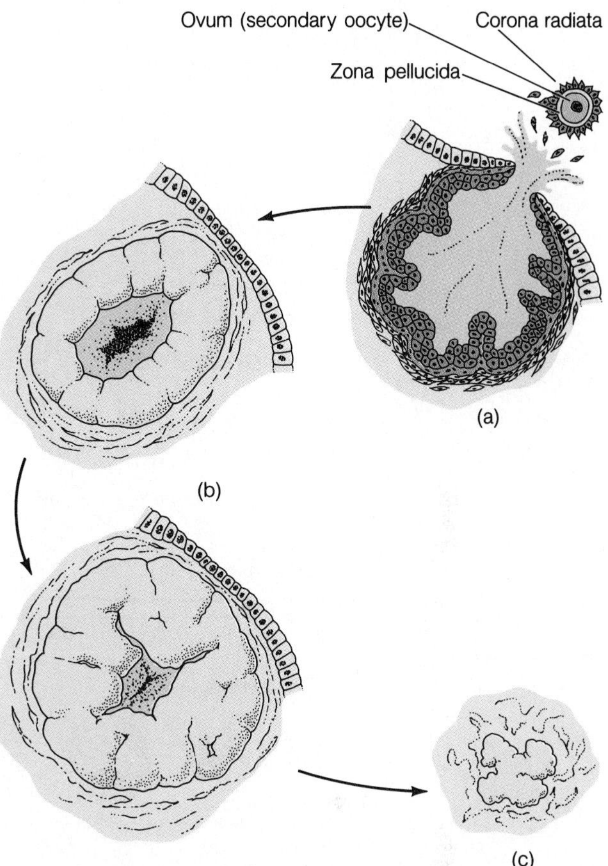

Figure 20–14 Ovulation and Development and Degeneration of Corpus Luteum *(a) Rupture of mature follicle and release of ovum (secondary oocyte) at ovulation. (b) Formation of corpus luteum from old follicular cells following ovulation. (c) Degeneration of corpus luteum if released ovum is not fertilized.*

The ovarian cycle is regulated by complex hormonal interactions among the hypothalamus, anterior pituitary, and ovárian endocrine units.

The ovary has two related endocrine units—the estrogen-secreting follicle during the first half of the cycle, and the corpus luteum, which secretes both progesterone and estrogen during the last half of the cycle. These units are sequentially triggered by complex cyclical hormonal relations among the hypothalamus, anterior pituitary, and these two ovarian endocrine units.

As in the male, gonadal function in the female is directly controlled by the anterior pituitary gonadotropic hormones, follicle-stimulating hormone (FSH) and luteinizing hormone (LH). These hormones, in turn, are regulated by pulsatile episodes of hypothalamic gonadotropin-releasing hormone (GnRH) and feedback actions of gonadal hormones. Differing from the male, however, control of the female gonads is complicated by the cyclic nature of ovarian function. For example, the effects of FSH and LH on the ovaries depend on the stage of the ovarian cycle. Also unlike the male, FSH is not strictly responsible for gametogenesis nor is LH solely responsible for gonadal hormone secretion. We will consider control of follicular function, ovulation, and the corpus luteum separately, using Figure 20–15 as a means of integrating the various concurrent and sequential activities that take place throughout the cycle.

CONTROL OF FOLLICULAR FUNCTION. The factors that initiate follicular development are poorly understood. The early stages of preantral follicular growth and oocyte maturation do not require gonadotropic stimulation. Hormonal support is required, however, for antrum formation, further follicular development, and estrogen secretion. Estrogen, FSH, and LH are all needed. Antrum formation is induced by FSH. Both FSH and estrogen stimulate proliferation of the granulosa cells. Both LH and FSH are required for synthesis and secretion of estrogen by the follicle, but these hormones act on different cells and at different steps in the estrogen production pathway (Fig. 20–16). Both granulos and thecal cells participate in estrogen production. The conversion of cholesterol into estrogen requires a number of sequential steps, the last of which is conversion of androgens into estrogens. Thecal cells readily produce androgens but cannot convert them into estrogens. Granulosa cells, on the other hand, are able to convert androgens into estrogens but are unable to produce androgens in the first place. Luteinizing hormone acts on the thecal cells to stimulate androgen production, whereas FSH acts on the granulosa cells to promote the conversion of thecal androgens (which diffuse into the granulosa cells from the theca cells) into estrogens. Because low basal levels of FSH are sufficient to promote this final conversion to estrogen, the

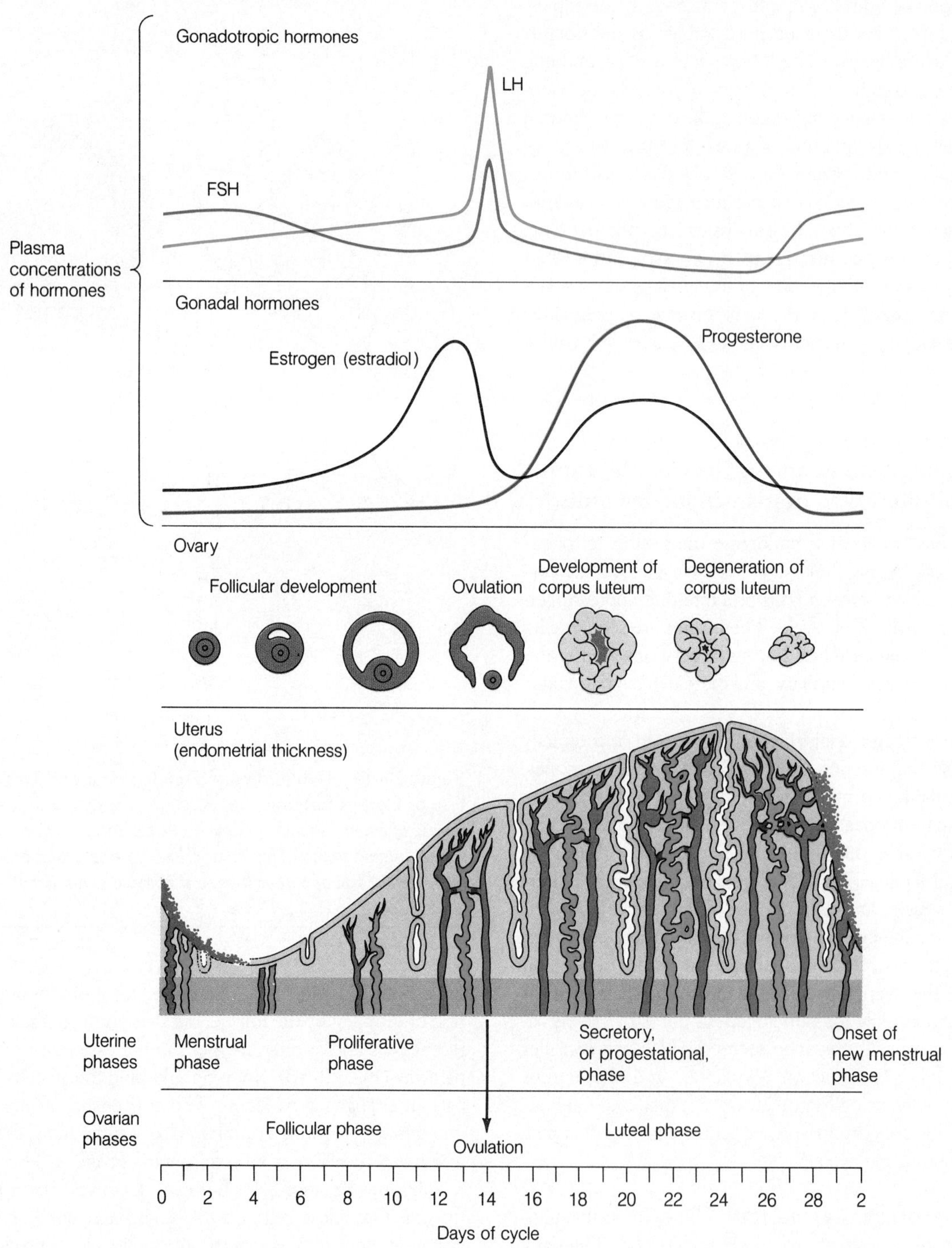

Figure 20–15 Correlation between Hormonal Levels and Cyclical Ovarian and Uterine Changes *See text for an explanation of the various components of this figure.*

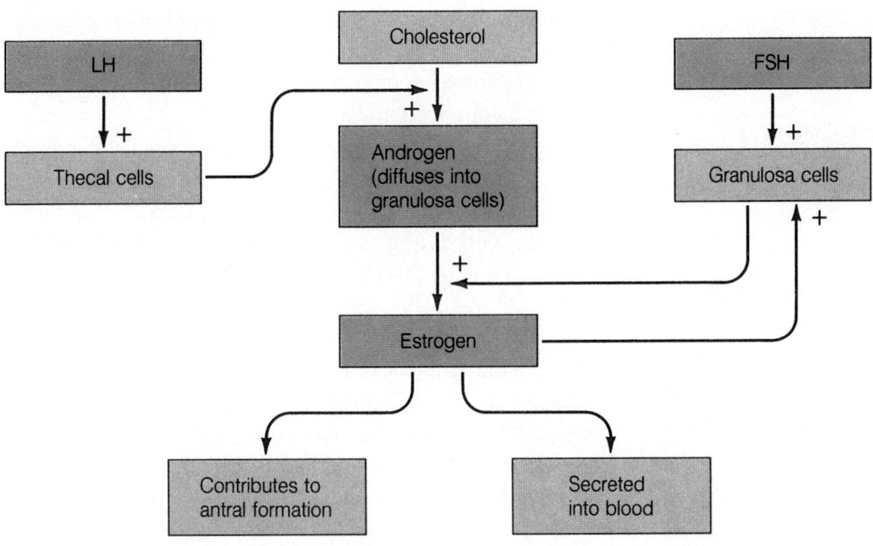

Figure 20–16 Production of Estrogen by Ovarian Follicle

rate of estrogen secretion by the follicle primarily depends on the circulating level of LH, which continues to rise during the follicular phase. Furthermore, as the follicle continues to grow, more estrogen is produced simply because there are more estrogen-producing follicular cells present.

Part of the estrogen produced by the growing follicle is secreted into the blood and is responsible for the steadily increasing plasma estrogen (estradiol) levels during the follicular phase. The remainder of the estrogen remains within the follicle, contributing to the antral fluid and stimulating further granulosa cell proliferation.

The secreted estrogen, in addition to acting on sex-specific tissues such as the uterus, inhibits the hypothalamus and anterior pituitary in negative-feedback fashion (Fig. 20–17). The low, rising levels of estrogen characterizing the follicular phase act directly on the hypothalamus to inhibit GnRH secretion, thus suppressing GnRH-prompted release of FSH and LH from the anterior pituitary. However, estrogen's primary effect is directly on the pituitary itself. Estrogen reduces the sensitivity to GnRH of the cells that produce gonadotropic hormones, especially the FSH-producing cells.

This differential sensitivity of FSH- and LH-producing cells induced by estrogen is at least in part responsible for the fact that the plasma FSH level, unlike the plasma LH concentration, declines during the follicular phase as the estrogen level rises (Fig. 20–15). Another contributing factor to the fall in FSH during the follicular phase may be the secretion of inhibin by the follicle. Inhibin is believed to preferentially inhibit FSH secretion by acting at the anterior pituitary, just as it does in the male. The decline in FSH secretion brings about atresia of all but the one most mature of the developing follicles.

In contrast to FSH, LH secretion continues to slowly rise during the follicular phase despite inhibition of GnRH (and thus indirectly, LH) secretion. This seeming paradox is due to the fact that estrogen alone cannot completely suppress the low-level, ongoing (tonic) LH secretion; both estrogen and progesterone are required to completely inhibit tonic LH secretion. Since progesterone does not appear until the luteal phase of the cycle, the basal level of circulating LH slowly increases during the follicular phase under the incomplete inhibition of estrogen alone.

CONTROL OF OVULATION. Ovulation and subsequent luteinization of the ruptured follicle are triggered by an abrupt, massive increase in LH secretion. This **LH surge** brings about four major changes in the follicle:

1. It halts estrogen synthesis by the follicular cells.
2. It reinitiates meiosis in the oocyte of the developing follicle, apparently by blocking release of an *oocyte maturation-inhibiting substance* produced by granulosa cells. This substance is believed to be responsible for arresting meiosis in the primary oocytes once they are wrapped within granulosa cells in the fetal ovary.
3. It triggers production of specific locally acting prostaglandins, which induce ovulation by promoting vascular changes that cause rapid swelling of the follicle while inducing enzymatic digestion of the follicular wall. Together

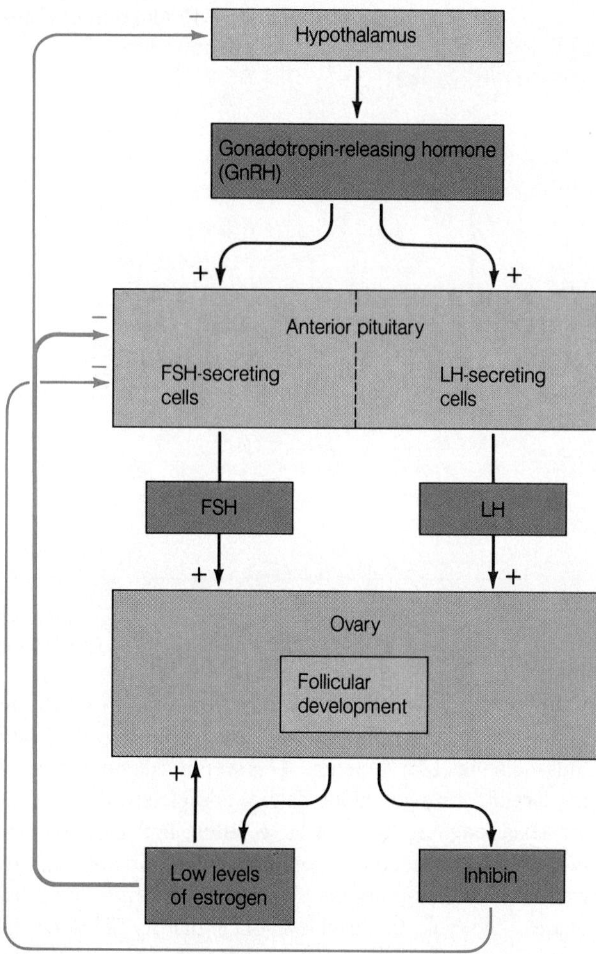

Figure 20–17 Feedback Control of FSH and Tonic LH Secretion during Follicular Phase

these actions lead to rupture of the weakened wall that covers the bulging follicle.

4. It causes differentiation of follicular cells into luteal cells. Because the LH surge triggers both ovulation and luteinization, formation of the corpus luteum automatically follows ovulation. Thus the midcycle burst in LH secretion is a dramatic point in the cycle; it terminates the follicular phase and initiates the luteal phase.

The two different modes of LH secretion—the tonic secretion of LH responsible for promoting ovarian hormone secretion and the LH surge that causes ovulation—not only occur at different times and produce different effects on the ovaries but also are controlled by different mechanisms. Tonic LH secretion is partially suppressed by the inhibitory action of the low, rising levels of estrogen during the follicular phase and is completely suppressed by the increasing levels of progesterone during the luteal phase. Since tonic LH secretion stimulates both estrogen and progesterone secretion, this is a typical negative-feedback control system.

In contrast, the LH surge is triggered by a positive-feedback effect. Whereas the low, rising levels of estrogen early in the follicular phase *inhibit* LH secretion, the high level of estrogen that occurs during peak estrogen secretion late in the follicular phase (Fig. 20–15) *stimulates* LH secretion and initiates the LH surge (Fig. 20–18). Thus LH enhances estrogen production by the follicle, and the resultant peak estrogen concentration stimulates LH secretion. The high plasma concentration of estrogen acts directly on the hypothalamus to increase the frequency of GnRH pulses, in this way increasing both LH and FSH secretion. It also acts directly on the anterior pituitary to specifically increase the sensitivity of LH-secreting cells to GnRH. The latter effect accounts for the much greater surge in LH compared to FSH secretion at midcycle. There is no known role for the modest midcycle surge in FSH that accompanies the pronounced and pivotal LH surge. Because only a mature, preovulatory follicle, not follicles in earlier stages of development, is capable of secreting sufficiently high levels of estrogen to trigger the LH surge, ovulation is not induced until a follicle has reached the proper size and degree of maturation. In a way, then, the follicle lets the hypothalamus know when it is ready to be stimulated to ovulate. The LH surge lasts for only one to two days at midcycle, just before ovulation.

CONTROL OF THE CORPUS LUTEUM. Steroid output by the corpus luteum is controlled by LH alone. Under the influence of LH, the corpus luteum secretes both progesterone and estrogen, with progesterone being its most abundant hormonal product. The plasma progesterone level increases for the first time during the luteal phase. No progesterone is secreted during the follicular phase (except for a slight output of progesterone by the about-to-rupture follicle under the influence of the LH surge). Therefore, the follicular phase is dominated by estrogen and the luteal phase by progesterone (Fig. 20–15).

A transitory drop in the level of circulating estrogen occurs at midcycle as the estrogen-secreting follicle meets its demise at ovulation. The estrogen level climbs again during the luteal phase because of the corpus luteum's activity, although it does not reach the same peak as during the follicular phase. What keeps the moderately high estrogen level during the luteal phase from triggering another LH surge? Progesterone. Even though a high level of estrogen stimulates LH secretion, progesterone, which dominates the luteal phase, powerfully inhibits LH secretion as well as FSH secretion (Fig. 20–19). Inhibition of FSH and LH by progesterone prevents new

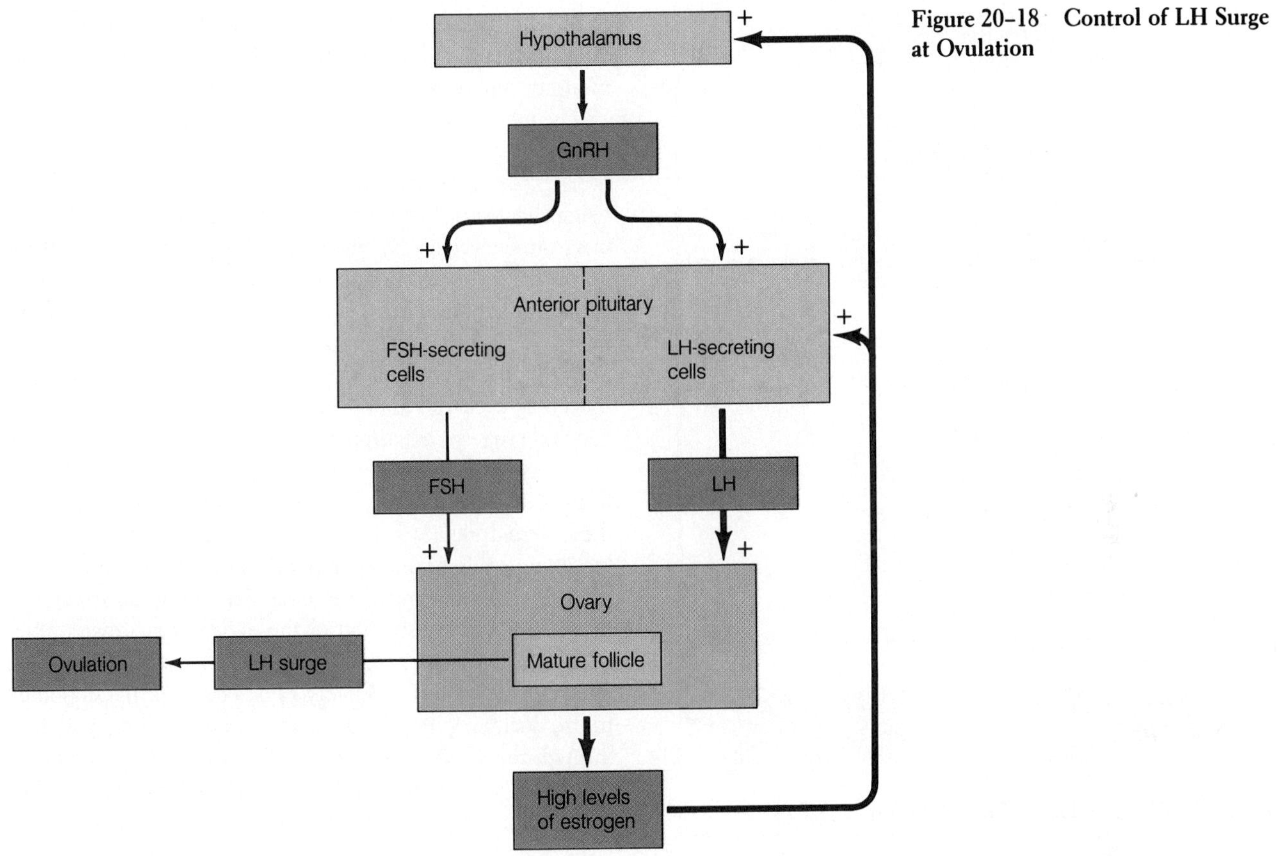

Figure 20–18 Control of LH Surge at Ovulation

follicular maturation and ovulation during the luteal phase. Under progesterone's influence, the reproductive system is gearing up to support the just-released ovum should it be fertilized instead of preparing other ova for release.

The corpus luteum functions for two weeks, then regresses if fertilization does not occur. The mechanisms responsible for degeneration of the corpus luteum are not fully understood. The declining level of circulating LH, driven down by inhibitory actions of progesterone, undoubtedly contributes to the corpus luteum's downfall. Prostaglandins and estrogen released by the luteal cells themselves may play a role. Demise of the corpus luteum terminates the luteal phase and sets the stage for a new follicular phase. As the corpus luteum degenerates, plasma progesterone and estrogen levels fall rapidly because these hormones are no longer being produced. Withdrawal of the inhibitory effects of these hormones on the hypothalamus allows FSH and tonic LH secretion to modestly increase once again. Under the influence of these gonadotropic hormones, another batch of primary follicles are induced to mature as a new follicular phase commences.

The uterine changes that occur during the menstrual cycle reflect hormonal changes during the ovarian cycle.

The fluctuations in circulating levels of estrogen and progesterone that occur during the ovarian cycle induce profound changes in the uterus, giving rise to the **menstrual cycle,** or **uterine cycle.** Because it reflects hormonal changes that occur during the ovarian cycle, the menstrual cycle averages twenty-eight days, as does the ovarian cycle, although there is considerable variation from this mean even in normal adults. This variability is primarily a reflection of differing lengths of the follicular phase; the duration of the luteal phase is fairly constant. The outward manifestation of the cyclical changes that occur in the uterus is the menstrual bleeding that takes place once during each menstrual cycle (that is, once a month). Less obvious changes take place throughout the cycle, however, as the uterus is prepared for implantation should a released ovum be fertilized, then is stripped clean of its prepared lining if implantation does not occur (menstrua-

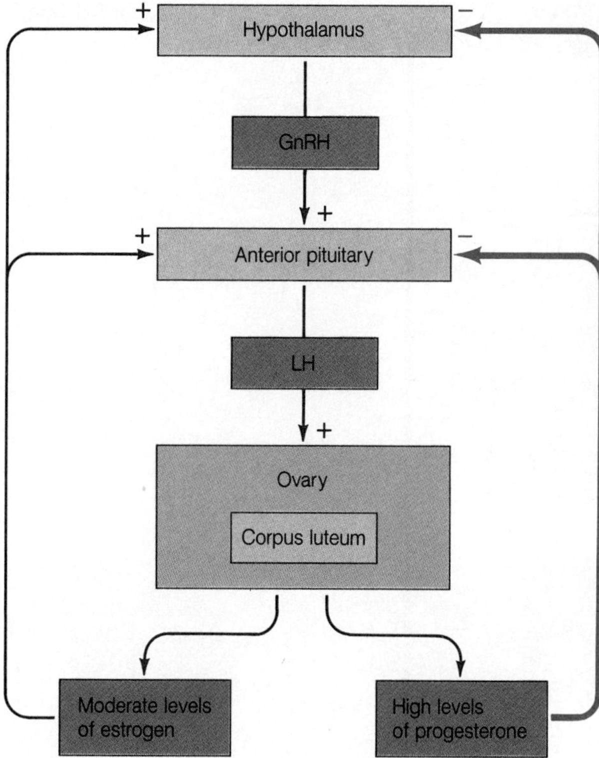

Figure 20–19 Feedback Control during Luteal Phase

tion), only to repair itself and start preparing for the next ovum that will be released during the next cycle.

We will briefly examine the influences of estrogen and progesterone on the uterus and then consider the effects of cyclic fluctuations of these hormones on uterine structure and function. The uterus consists of two layers: the **myometrium,** the outer smooth-muscle layer; and the **endometrium,** the inner lining that contains numerous blood vessels and glands. Estrogen stimulates growth of both the myometrium and endometrium. It also induces the synthesis of progesterone receptors in the endometrium. This means that progesterone is able to exert an effect on the endometrium only after it has been "primed" by estrogen. Progesterone acts on the estrogen-primed endometrium to convert it into a hospitable and nutritious lining suitable for implantation of a fertilized ovum. Under the influence of progesterone, the endometrial connective tissue becomes loose and edematous as a result of an accumulation of electrolytes and water, which facilitates implantation of the fertilized ovum. Progesterone furthermore prepares the endometrium to sustain an early developing embryo by inducing the endometrial glands to secrete and store large quantities of glycogen and by causing tremendous growth of the endometrial blood vessels. Progesterone also reduces the contractility of the uterus to provide a quiet environment for implantation and embryonic growth.

The menstrual cycle consists of three phases: the menstrual phase, the proliferative phase, and the secretory or progestational phase. The **menstrual phase** is the most overt phase, being characterized by discharge of blood and endometrial debris from the vagina. By convention, the first day of menstruation is considered to be the start of a new cycle. It coincides with termination of the ovarian luteal phase and onset of the follicular phase. As the corpus luteum degenerates because fertilization and implantation of the ovum released during the preceding cycle did not take place, circulating levels of estrogen and progesterone drop precipitously. Since the net effect of estrogen and progesterone is preparation of the endometrium for implantation of a fertilized ovum, withdrawal of these steroids deprives the highly vascular, nutrient-rich uterine lining of its hormonal support. The fall in ovarian hormone levels also stimulates release of a uterine prostaglandin that causes vasoconstriction of the endometrial vessels, thus disrupting the blood supply to the endometrium. The subsequent reduction in O_2 delivery causes death of the endometrium, including its blood vessels. The resulting bleeding through the disintegrating vessels flushes the dying endometrial tissue into the uterine lumen. The entire uterine lining sloughs during each menstrual period except for a deep, thin layer of epithelial cells and glands from which the endometrium will regenerate. The same local uterine prostaglandin also stimulates mild rhythmic contractions of the uterine myometrium. These contractions help expel the blood and endometrial debris from the uterine cavity out through the vagina as **menstrual flow.** Excessive uterine contractions caused by the overproduction of prostaglandin are responsible for the menstrual cramps (**dysmenorrhea**) experienced by some women.

The average blood loss during a single menstrual period is 50 to 150 ml. Blood that seeps slowly through the degenerating endometrium clots within the uterine cavity, then is acted on by an anticoagulant, fibrinolysin, which breaks down the fibrin that forms the meshwork of the clot. Therefore, blood in the menstrual flow does not usually clot, because it has already clotted and the clot has been dissolved before it passes out of the vagina. When blood flows rapidly through the leaking vessels, however, it may not be exposed to sufficient fibrinolysin, so blood clots may appear when the menstrual flow is most profuse. In addition to the blood and endometrial debris, large numbers of leukocytes are found in the menstrual flow. These white blood cells play an important defense role in helping the raw endometrium resist infection.

Menstruation typically lasts for about five to seven days after degeneration of the corpus luteum, coinciding in time with the early portion of the ovarian follicular phase (Fig. 20–15). Withdrawal of estrogen and progesterone upon degeneration of the corpus luteum leads simultaneously to sloughing of the endometrium (menstruation) and development of new follicles in the ovary under the influence of rising gonadotropic hormone levels. The drop in gonadal hormone secretion removes inhibitory influences from the hypothalamus and anterior pituitary, so FSH and LH secretion rise and a new follicular phase begins. After five to seven days under the influence of FSH and LH, the newly growing follicles are secreting sufficient quantities of estrogen to induce repair and growth of the endometrium.

Thus menstrual flow ceases and the **proliferative phase** of the uterine cycle begins concurrent with the last portion of the ovarian follicular phase as the endometrium starts to repair itself and proliferate under the influence of estrogen from the newly growing follicles. When the menstrual flow ceases, a thin endometrial layer less than 1 mm thick remains. Estrogen stimulates proliferation of epithelial cells, glands, and blood vessels in the endometrium, increasing this lining to a thickness of 3 to 5 mm. The estrogen-dominant proliferative phase lasts from the end of menstruation to ovulation. Peak estrogen levels trigger the LH surge responsible for ovulation.

Following ovulation, when a new corpus luteum is formed, the uterus enters the **secretory,** or **progestational, phase,** which coincides in time with the ovarian luteal phase. The corpus luteum secretes large amounts of progesterone and estrogen. Progesterone acts on the thickened, estrogen-primed endometrium to convert it to a richly vascularized, glycogen-filled tissue. This period is called either the secretory phase, because the endometrial glands are actively secreting glycogen, or the progestational ("before pregnancy") phase, in reference to the development of a lush endometrial lining capable of supporting an early embryo. If fertilization and implantation do not occur, the corpus luteum degenerates and a new follicular phase and menstruation phase begin once again. (For the effects of exercise, see the accompanying boxed feature, A Closer Look at Exercise Physiology, pp. 742–743.)

Fluctuating estrogen and progesterone levels produce cyclical changes in cervical mucus.

Hormonally induced changes also take place in the cervix during the ovarian cycle. Under the influence of estrogen during the follicular phase, the mucus secreted by the cervix is abundant, clear, and thin. These changes, which are most pronounced as estrogen is at its peak and ovulation is approaching, facilitate passage of sperm through the cervical canal. After ovulation, under the influence of progesterone from the corpus luteum, the mucus becomes thick and sticky, essentially forming a plug across the cervical opening. This constitutes an important defense mechanism by preventing entry from the vagina into the uterus of bacteria that might threaten a pregnancy should conception occur. Sperm also cannot penetrate this thick mucus barrier. Thus the changes in cervical secretion induced by estrogen and progesterone are geared toward facilitating sperm penetration of the cervix at ovulation and preventing entry of vaginal material through the cervix during the time when fertilization might have occurred to protect the conceptus from potential contamination.

Pubertal changes in females are similar to those in males, but menopausal changes are unique to females.

Regular menstrual cycles are absent in both young and aging females, but the reasons for this absence differ in these two age groups.

PUBERTAL CHANGES. The female reproductive system does not become active until puberty. The fetal ovaries do not need to be functional, as are the fetal testes, because feminization of the female reproductive system automatically takes place in the absence of fetal testosterone secretion without the presence of female sex hormones. The female reproductive system remains quiescent from birth until puberty, which occurs at about eleven years of age, because hypothalamic GnRH is actively suppressed by mechanisms that are similar to those in the prepubertal male. As in the male, removal of these inhibitory influences by mechanisms that are not clearly understood is responsible for the onset of puberty.

The resultant secretion of estrogen by the activated ovaries induces growth and maturation of the female reproductive tract as well as development of the female secondary sexual characteristics. Estrogen's prominent action in the latter regard is to promote fat deposition in strategic locations such as the breasts, buttocks, and thighs, giving rise to the typical curvaceous female figure. Enlargement of the breasts at puberty is due primarily to fat deposition in the breast tissue and not to functional development of the mammary glands. Three other pubertal changes in females—growth of axillary and pubic hair, the pubertal growth spurt, and development of libido—are attributable to a spurt in adrenal androgen secretion at puberty, not to estrogen. However, the pubertal rise in estrogen does close the epiphyseal plates, halting further growth in height, similar to the effect of testosterone in males.

MENOPAUSAL CHANGES. The cessation of a woman's menstrual cycles at menopause sometime between the age of forty-five and fifty-five years is attributable to the limited supply of ovarian follicles present at birth. Once this reservoir is de-

A CLOSER LOOK AT EXERCISE PHYSIOLOGY

Menstrual Cycle Irregularity in Athletes

Concurrent with the rise since the 1970s in women's participation in a variety of sports requiring vigorous training regimens has been a growing awareness that many women experience changes in their menstrual cycles as a result of athletic participation. These changes are referred to as *athletic menstrual cycle irregularity* (*AMI*). The menstrual cycle dysfunction can vary in severity from amenorrhea (cessation of menstrual periods) to oligomenorrhea (cycles at irregular or infrequent intervals) to cycles that are normal in length but are anovulatory (no ovulation) or that have a short or inadequate luteal phase.

In early research studies using surveys and questionnaires to determine the prevalence of the problem, the frequency of these sport-related disorders varied from 2% to 51%. In contrast, the rate of occurrence of menstrual cycle dysfunction in females of reproductive age in the general population is 2% to 5%. A major problem of determining the frequency of menstrual cycle irregularity by survey is the questionable accuracy of recall of menstrual periods. Furthermore, without blood tests to determine hormone levels throughout the cycle, a woman would not know whether she was anovulatory or had a shortened luteal phase. Studies in which hormone levels have been determined throughout the menstrual cycle have demonstrated that seemingly normal cycles in athletes frequently have a short luteal phase (less than ten days long with low progesterone levels).

In a study conducted to determine whether strenuous exercise spanning two menstrual cycles would induce menstrual disorders, twenty-eight initially untrained college women with documented ovulation and luteal adequacy served as subjects. The women participated in an eight-week exercise program in which they initially ran four miles per day and progressed to ten miles per day by the fifth week. They were expected to participate daily in three and one-half hours of moderate-intensity sports. Only four women had normal menstrual cycles during the training. Abnormalities that occurred as a result of training included abnormal bleeding, delayed menstrual periods, abnormal luteal function, and loss of LH surge. All women returned to normal cycles within six months after training. The results of this study suggest that the frequency of AMI with strenuous exercise may be much greater than indicated by questionnaire alone. In other

pleted, ovarian cycles, and hence menstrual cycles, cease. Thus the termination of reproductive potential in a middle-aged woman is "preprogrammed" at her own birth. Just as the delay in puberty allows adequate time for females to mature physically and psychologically so that they are better able to handle the stresses of pregnancy and childrearing, menopause may serve as a mechanism to prevent pregnancy in women beyond the time that they could likely rear a child before their own death.

Menopause is preceded by a period of progressive ovarian failure characterized by increasingly irregular cycles, dwindling estrogen levels, and a host of physical and emotional changes. This entire period of transition from sexual maturity to cessation of reproductive capability is known as the **climacteric.** The absence of ovarian estrogen is responsible for the physical postmenopausal changes that occur, such as vaginal dryness that can cause discomfort during sex and gradual atrophy of the genital organs. However, postmenopausal women still have a sex drive because of their adrenal androgens. It is unclear whether the emotional symptoms associated with waning ovarian function, such as depression and irritability, are caused by estrogen withdrawal or are psychological reactions to the implications of menopause.

Males do not experience a similar complete gonadal failure for two reasons. First, their germ cell supply is unlimited because of continuing mitotic activity of the spermatogonia. Second, gonadal hormone secretion in males is not inextricably dependent on gametogenesis, as it is in females. If female sex hormones were produced by separate tissues unrelated to those responsible for gametogenesis, as they are in the male, cessation of estrogen and progesterone secretion would not automatically accompany termination of oogenesis.

studies using low-intensity exercise regimens, AMI was much less frequent.

The mechanisms of AMI are unknown at present, although studies have implicated rapid weight loss, decreased percentage of body fat, dietary insufficiencies, prior menstrual dysfunction, stress, age at onset of training, and the intensity of training as factors that play a role. Epidemiologists have shown that if girls participate in vigorous sports before **menarche** (the first menstrual period), that menarche is delayed. On the average, athletes have their first period when they are about three years older than their nonathletic counterparts. Furthermore, females who participate in sports before menarche seem to have a higher frequency of AMI throughout their athletic careers than those who begin to train after menarche. Hormonal changes that have been found in female athletes include: (1) severely depressed FSH levels, (2) elevated LH levels, (3) low progesterone during the luteal phase, (4) low estrogen levels in the follicular phase, and (5) an FSH/LH environment totally unbalanced as compared to age-matched nonathletic women. The preponderance of evidence indicates that cycles return to normal once vigorous training is stopped.

The major problem associated with athletic amenorrhea is a reduction in bone-mineral density. Studies have shown that the mineral density in the vertebrae of the lower spine of those with athletic amenorrhea is lower than in athletes with normal menstrual cycles and lower than in age-matched nonathletes. However, amenorrheic runners have higher bone-mineral density than amenorrheic nonathletes, presumably because the mechanical stimulus of exercise helps to retard bone loss. Studies have shown that amenorrheic athletes are at higher risk for stress-related fractures compared to athletes with normal menstrual cycles. One study, for example, found stress fractures in six of eleven amenorrheic runners but in only one of six runners with normal menstrual cycles. The mechanism for bone loss is probably the same as is found in postmenopausal osteoporosis—lack of estrogen (see p. 708). The problem is serious enough that an amenorrheic athlete should discuss the possibility of estrogen replacement therapy with her physician.

There may be some positive benefits of athletes' menstrual dysfunction. A recent epidemiologic study to determine if the long-term reproductive and general health of women who had been college athletes differed from that of college nonathletes showed that former athletes had less than half the lifetime occurrence rate of cancers of the reproductive system and half the breast cancer occurrence compared to nonathletes. Because these are hormone-sensitive cancers, the delayed menarche and lower estrogen levels found in women athletes may play a key role in decreasing the risk of cancer of the reproductive system and breast.

The oviduct is the site of fertilization.

You have already learned about the events that take place if fertilization does not occur. Because the primary function of the reproductive system is, of course, reproduction, we will now turn our attention to the sequence of events that ensue when this function is accomplished.

Fertilization, the union of male and female gametes, normally occurs in the **ampulla,** the upper third of the oviduct. This means that both the ovum and sperm must be transported from their gonadal site of production to the ampulla (Fig. 20–20).

OVUM TRANSPORT TO THE OVIDUCT. At ovulation the ovum is released into the peritoneal cavity, but it is quickly picked up by the oviduct. The dilated end of the oviduct cups around the ovary and contains **fimbriae,** fingerlike projections that contract in a sweeping motion to guide the released ovum into the oviduct. Furthermore, the fimbriae are lined by cilia, which are fine hairlike projections that beat in waves toward the interior of the oviduct, further assuring the ovum's passage into the oviduct (see p. 36).

Within the oviduct, the ovum is rapidly propelled by peristaltic contractions and ciliary action to the ampulla. If not fertilized, the ovum begins to disintegrate within twenty-four hours and is subsequently phagocytized by cells that line the reproductive tract.

Fertilization must therefore occur within twenty-four hours after ovulation, when the ovum is still viable. Sperm can survive several days in the female reproductive tract, so sperm deposited within forty-eight hours before or twenty-four hours after ovulation may be able to fertilize the released ovum, al-

Figure 20–20 Sperm and Ovum Transport to Site of Fertilization

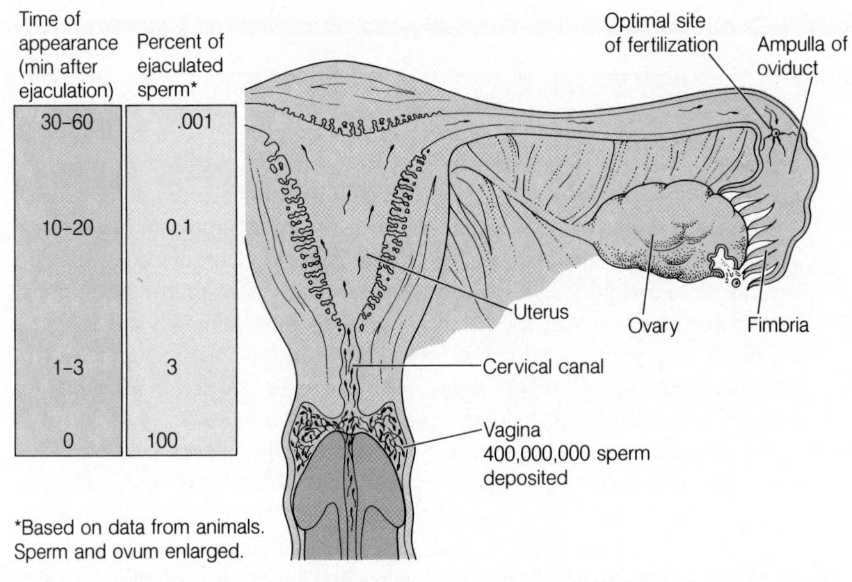

though these times are subject to considerable variation. At any rate, there is a very limited time span in each cycle during which conception can take place (the **fertile period**).

Occasionally an ovum fails to be transported into the oviduct but remains instead in the peritoneal cavity. Rarely, such an ovum gets fertilized, resulting in an **ectopic** (**abdominal**) **pregnancy,** in which the fertilized egg implants in the rich vascular supply to the digestive organs rather than in its usual site in the uterus. If this unusual pregnancy proceeds to term, the baby must be delivered surgically, because the normal vaginal exit is not available to it.

SPERM TRANSPORT TO THE OVIDUCT. Once sperm are deposited in the vagina at ejaculation, they must travel through the cervical canal, through the uterus, and then up to the egg in the upper third of the oviduct. The first sperm arrive in the oviduct within a half hour after ejaculation. Even though sperm are mobile by means of whiplike contractions of their tails, thirty minutes is much too soon for a sperm's own mobility to transport it to the ampulla. To accomplish this formidable journey, sperm need the help of the female reproductive tract. The first hurdle is passage through the cervical canal. The cervical mucus is too thick to permit sperm penetration throughout most of the cycle because of high progesterone or low estrogen levels. Only when estrogen levels are high, as occurs in the presence of a mature follicle about to ovulate, does the cervical mucus become thin and watery enough to permit sperm to penetrate. Sperm migrate up the cervical canal under their own power. The canal remains penetrable for only two or three days during each cycle, around the time of ovulation. Even during this time, of the several hundred million sperm deposited in a single ejaculate, only a few thousand make it to the oviduct (Fig. 20–20). The fact that only a very small percentage of the deposited sperm ever reach their destination is one reason that sperm concentration must be so high ($>$ 20 million/ml of semen) for a man to be fertile. The other reason is that the acrosomal enzymes of many sperm are needed to break down the barriers surrounding the ovum.

Once within the uterus, contractions of the myometrium churn the sperm around in "washing machine" fashion. This action quickly disperses the sperm throughout the uterine cavity. Sperm passing into the oviduct must travel upward against the predominant action of the cilia lining this passageway, most of which beat downward to propel the nonmotile egg toward the uterus. Sperm transport in the oviduct is facilitated by antiperistaltic (upward) contractions of the oviduct smooth muscle, perhaps assisted by tracts of cilia that beat upward in the opposite direction of the main ciliary current. These myometrial and oviductal contractions that facilitate sperm transport are induced by the high estrogen level that exists just prior to ovulation, perhaps aided by seminal prostaglandins.

CAPACITATION. Although sperm are able to arrive at the oviduct within thirty minutes, they must be present within the female tract at least six or seven hours before they are capable of fertilizing the ovum. This is because the sperm must undergo

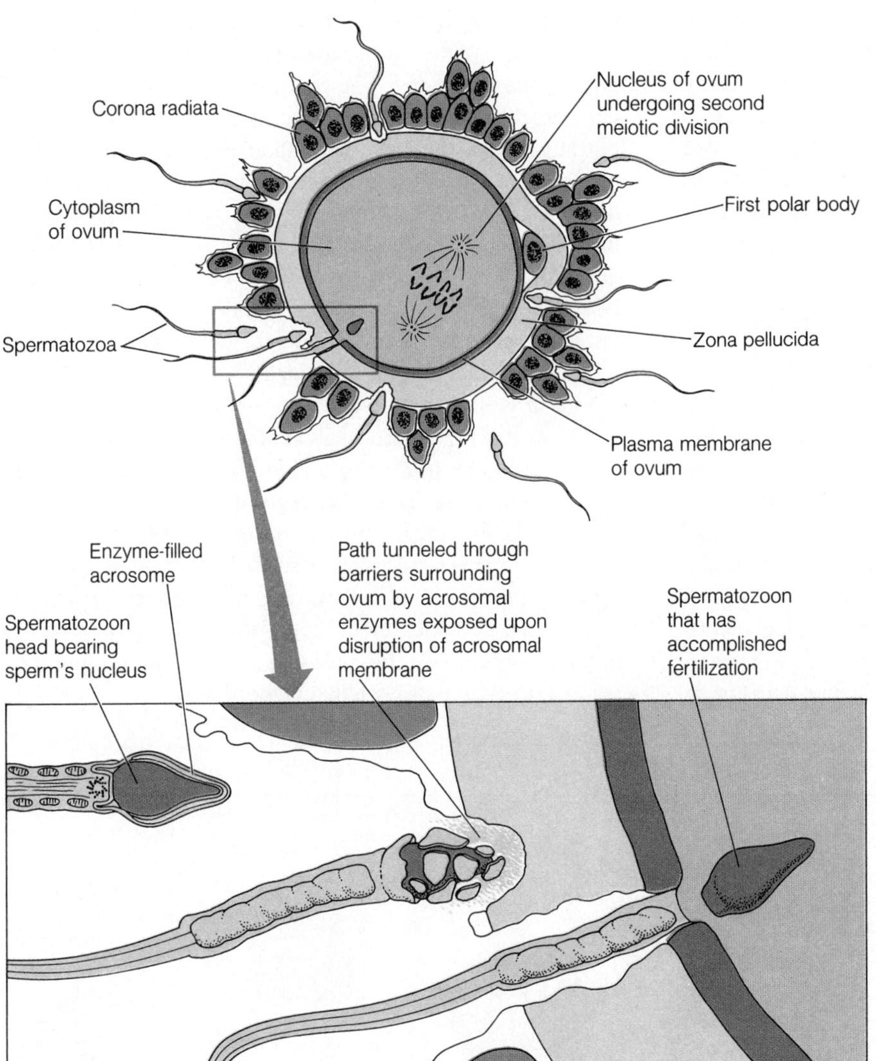

Figure 20–21 Process of Fertilization

a final maturation known as ***capacitation*** within the female reproductive tract before they are fertile. The exact changes responsible for capacitation are unknown.

FERTILIZATION. The tail of the sperm is used to maneuver for final penetration of the ovum. To fertilize an ovum, a sperm must first pass through the corona radiata and zona pellucida surrounding it. The acrosomal enzymes, which are exposed by surface-membrane changes that occur on contact with the corona radiata, enable the sperm to tunnel a path through these protective barriers (Fig. 20–21). Sperm are able to penetrate the zona pellucida only after binding with specific receptor sites on the surface of this layer. (Only sperm of the same species are able to bind to these receptors and pass through.) The first sperm to reach the ovum itself fuses with the plasma membrane of the ovum (actually a secondary oocyte), triggering a chemical change in the ovum's surrounding membrane that makes this outer layer impenetrable to the entry of any more sperm. This phenomenon is known as **block to polyspermy** ("many sperm").

The head of the fused sperm is gradually pulled into the ovum's cytoplasm by a growing cone that engulfs it. The sperm's tail is frequently lost in this process, but it is the head that carries the crucial genetic information anyway. Penetration of the sperm into the cytoplasm triggers the final meiotic division of the secondary oocyte. Within an hour, the sperm and egg nuclei, each bearing a haploid set of chromosomes, fuse. As the paternal and maternal chromosomes mix, a unique diploid cell is formed that will develop through mitotic activity and differentiation into a new individual. In addition to contributing its half of the chromosomes to the fertilized ovum, now called a **zygote,** the victorious sperm also activates ovum enzymes that are essential for the early embryonic developmental program.

The blastocyst implants in the endometrium through the action of its trophoblastic enzymes.

During the first three to four days following fertilization, the zygote remains within the ampulla, because a constriction between the ampulla and the remainder of the oviduct canal prevents further movement of the zygote toward the uterus. The zygote is not idle during this time, however. It is rapidly undergoing a number of mitotic cell divisions to form a solid ball of cells called the **morula** (Fig. 20–22). Meanwhile, the rising levels of progesterone from the newly developing corpus luteum that formed after ovulation stimulate release of glycogen from the endometrium into the reproductive-tract lumen for use as energy by the early embryo. The nutrients stored in the cytoplasm of the ovum can sustain the product of conception for less than a day. The concentration of secreted nutrients increases more rapidly in the small confines of the ampulla than in the uterine lumen. About three to four days after ovulation, progesterone is being produced in sufficient quantities to relax the oviduct constriction, thus permitting the morula to be rapidly propelled into the uterus by oviductal peristaltic contractions and ciliary activity. The temporary delay before passage of the developing embryo into the uterus allows sufficient nutrients to accumulate in the uterine lumen to support the embryo until implantation can take place. If the morula arrives prematurely, it dies.

Occasionally the morula fails to descend into the uterus but instead continues to develop and implant in the lining of the oviduct. This leads to a **tubal pregnancy,** which must be terminated. Such a pregnancy can never succeed, because the oviduct cannot expand as the uterus does to accommodate the growing embryo. The first warning of a tubal pregnancy is pain caused by stretching of the oviduct by the growing embryo. If not removed, the enlarging embryo will rupture the oviduct, causing possibly lethal hemorrhage.

Figure 20–22 Early Stages of Development from Fertilization to Implantation

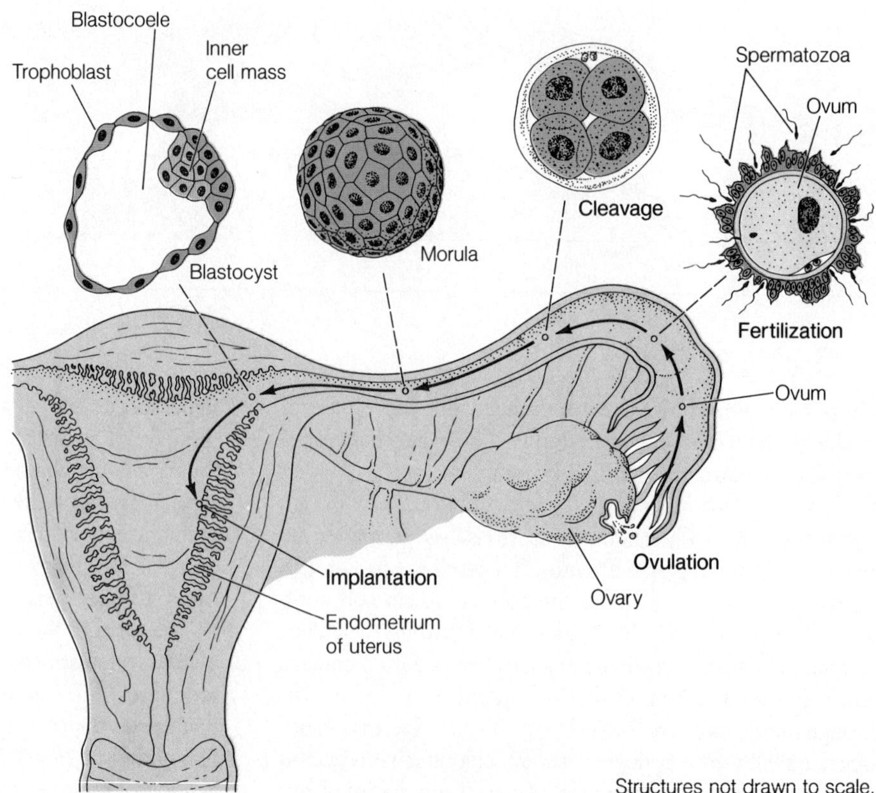

In the usual case, when the morula descends to the uterus, it floats freely within the uterine cavity for another three to four days, living on the endometrial secretions and continuing to divide. During the six to seven days after ovulation, while the developing embryo is in transit in the oviduct and floating in the uterine lumen, the uterine lining is simultaneously being prepared for implantation under the influence of luteal-phase progesterone. During this time, the uterus is in its secretory or progestational phase, storing up glycogen and becoming richly vascularized.

By the time the endometrium is suitable for implantation (about a week after ovulation), the morula has continued to proliferate and differentiate into a **blastocyst** capable of implantation (Figs. 20–22 and 20–23). A blastocyst is a single-layered sphere of cells encircling a fluid-filled cavity with a dense mass of cells grouped together at one side. This dense mass, called the **inner cell mass,** is destined to become the fetus itself. The remainder of the blastocyst will never be incorporated into the fetus but will serve a supportive role during intrauterine life. The thin outermost layer, the **trophoblast,** is responsible for accomplishing implantation, after which it develops into the fetal portion of the placenta. The fluid-filled cavity, the **blastocoele,** will become the **amniotic sac,** which surrounds and cushions the fetus throughout gestation.

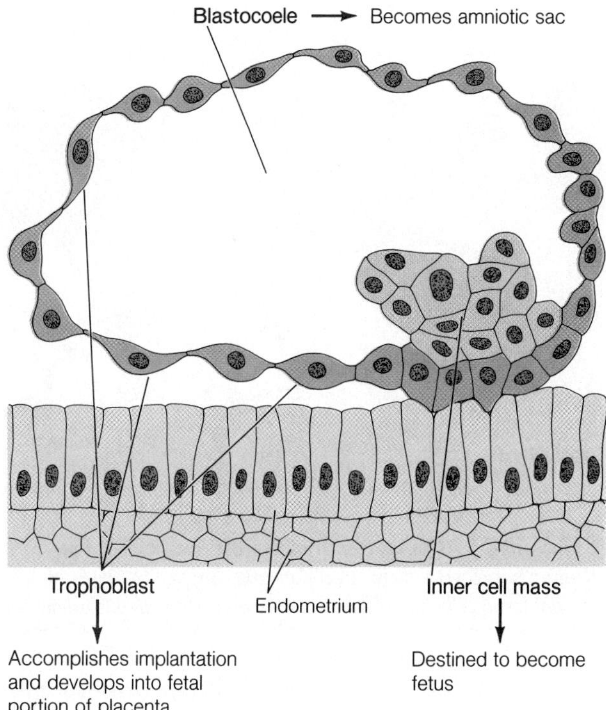

Figure 20–23 Blastocyst Adhering to Endometrial Lining and About to Implant

When the blastocyst is ready to implant, its surface becomes sticky. By this time the endometrium is ready to accept the early embryo. The blastocyst adheres to the uterine lining on its inner-cell-mass side (Fig. 20–24a). **Implantation** begins when the trophoblastic cells overlying the inner cell mass release proteolytic enzymes upon contact with the endometrium. These enzymes digest pathways between the endometrial cells, thus permitting fingerlike cords of trophoblastic cells to penetrate into the depths of the endometrium, where they continue to digest uterine cells (Fig. 20–24b). Through its cannabilistic actions, the trophoblast performs the dual functions of: (1) accomplishing implantation as it carves out a hole in the endometrium for the blastocyst, and (2) making metabolic fuel and raw materials available for the developing embryo as the advancing trophoblastic projections break down the nutrient-rich endometrial tissue.

Stimulated by the invading trophoblast, the endometrial tissue at the site of contact undergoes dramatic changes that enhance its ability to support the implanting embryo. In response to a chemical messenger released by the blastocyst, the underlying endometrial cells secrete prostaglandins, which act to locally increase vascularization, produce edema, and enhance nutrient storage. The endometrial tissue so modified at the implantation site is called the **decidua.** It is into this super-rich decidual tissue that the blastocyst becomes embedded. After the blastocyst burrows into the decidua by means of trophoblastic activity, a layer of endometrial cells covers over the surface of the hole, completely burying the blastocyst within the uterine lining (Fig. 20–24c). The trophoblastic layer continues to digest the surrounding decidual cells, providing energy for the embryo until the placenta develops.

The placenta is the organ of exchange between maternal and fetal blood.

The glycogen stores in the endometrium are only sufficient to nourish the embryo during its first few weeks. To sustain the growing embryo/fetus for the duration of its intrauterine life, the placenta, a specialized organ of exchange between the maternal and fetal blood, is rapidly developed. The placenta is derived from both trophoblastic and decidual tissue.

By day twelve, the embryo is completely embedded in the decidua. By this time the trophoblastic layer is two cell-layers thick and is called the **chorion.** As the chorion continues to release enzymes and expand, it forms an extensive network of cavities within the decidua. These cavities fill with maternal blood as decidual capillary walls are eroded by the expanding chorion (Fig. 20–25). Maternal blood oozing into these

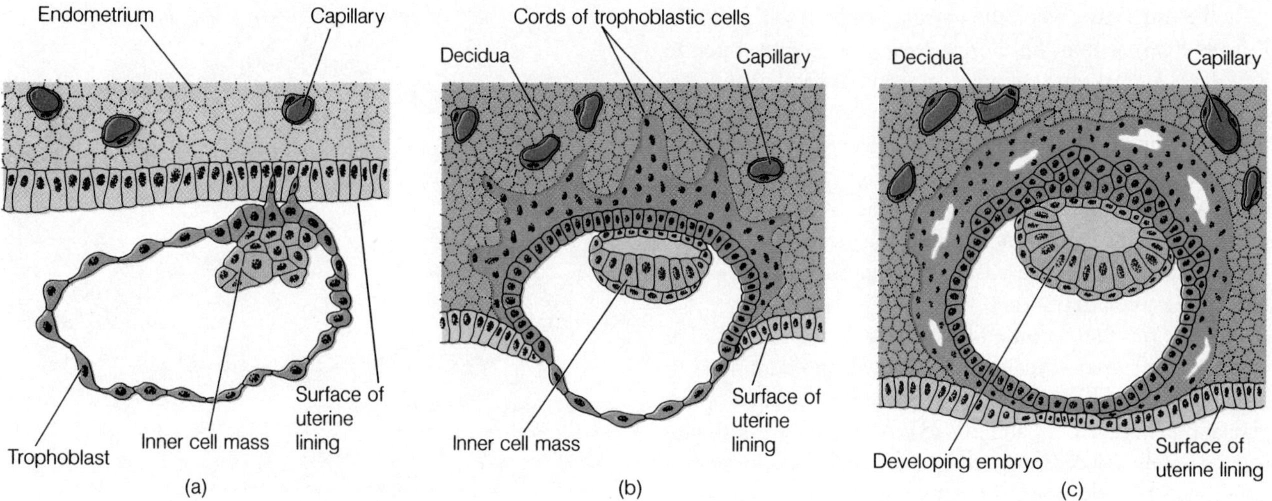

Figure 20–24 Implantation of the Blastocyst *(a) Free-floating blastocyst adheres to endometrial lining. (b) Cords of trophoblastic cells tunnel into endometrium, carving out a hole for the blastocyst. The boundaries between the cells in the advancing trophoblastic tissue disintegrate. (c) When implantation is finished, the blastocyst is completely buried in the endometrium.*

spaces is prevented from clotting by an anticoagulant produced by the chorion. Fingerlike projections of chorionic tissue extend into the pools of maternal blood. Soon the developing embryo sends out capillaries into these chorionic projections to form **placental villi.** Some villi extend completely across the blood-filled spaces to anchor the fetal portion of the placenta to the endometrial tissue, but most simply project into the pool of maternal blood.

Each placental villus contains embryonic (later fetal) capillaries surrounded by a thin layer of chorionic tissue, which separates the embryonic/fetal blood from the maternal blood in the intervillus spaces. There is no actual mingling of maternal and fetal blood, but the barrier between them is extremely thin. To visualize this relationship, think of your hands (the fetal capillary blood vessels) in rubber gloves (the chorionic tissue) immersed in water (the pool of maternal blood). Only the rubber gloves separate your hands from the water. In the same way, only the thin chorionic tissue (plus the capillary wall of the fetal vessels) separates the fetal and maternal blood. It is across this extremely thin barrier that all materials are exchanged between these two bloodstreams. This entire system of interlocking maternal (decidual) and fetal (chorionic) structures makes up the placenta.

The placenta is well-established and operational by five weeks after implantation. By this time, the heart of the developing embryo is pumping blood into the placental villi as well as to the embryonic tissues. Throughout gestation, fetal blood continuously traverses between the placental villi and the circulatory system of the fetus by means of the **umbilical artery** and **umbilical vein,** which are wrapped within the **umbilical cord,** and life-line between the fetus and placenta (Fig. 20–25). The maternal blood within the placenta is continuously replaced as fresh blood enters through the uterine arterioles, percolates through the intervillus spaces, where it exchanges substances with fetal blood in the surrounding villi, then exits through the uterine veins.

During intrauterine life, the placenta performs the functions of the digestive system, the respiratory system, and the kidneys for the "parasitic" fetus. This is not to say that the fetus does not have these organ systems, but rather that they are unable to (and do not need to) function within the uterine environment. Nutrients and O_2 diffuse from the maternal blood across the thin placental barrier into the fetal blood, while CO_2 and other metabolic wastes simultaneously diffuse from the fetal blood into the maternal blood. The nutrients and O_2 brought to the fetus in the maternal blood are acquired by the mother's digestive and respiratory systems, and the CO_2 and wastes transferred into the maternal blood are eliminated by the mother's lungs and kidneys, respectively. Thus the mother's digestive tract, respiratory system, and kidneys serve the fetus's needs as well as her own.

Some substances traverse the placental barrier by special mediated-transport systems in the placental membranes, whereas others move across by simple diffusion. Unfortunately, many drugs, environmental pollutants, other chemical agents, and microorganisms in the mother's bloodstream are able to cross the placental barrier, some of which may be harmful to the developing fetus. Individuals born limbless as a result of exposure to thalidomide, a tranquilizer taken by pregnant women unaware of this drug's devastating effects on the

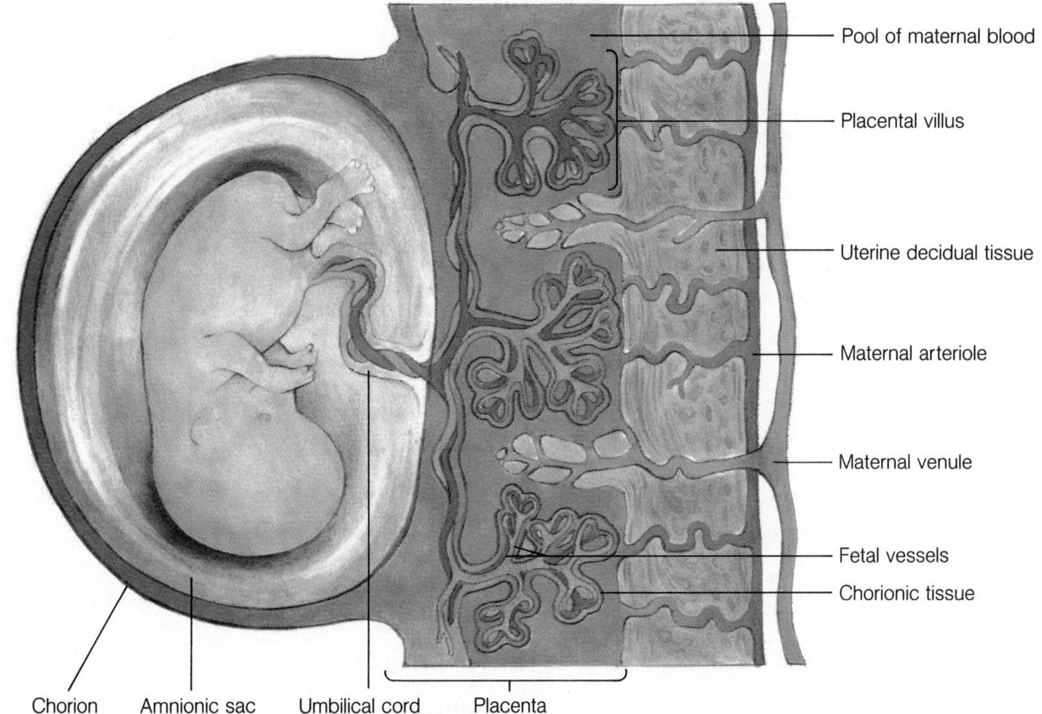

Figure 20–25 Placentation *Schematic representation of interlocking maternal and fetal structures that form the placenta.*

growing fetus, serve as a grim reminder of this fact. Similarly, newborns who have become "addicted" during gestation by their mother's use of an abusive drug such as heroin suffer withdrawal symptoms after birth. Even more common chemical agents such as aspirin, alcohol, and agents in cigarette smoke can reach the fetus and have adverse effects. Likewise, fetuses can acquire AIDS before birth if their mothers are infected with the virus. A pregnant woman should therefore be very cautious about potentially harmful exposures from any source.

In addition to serving as an organ of exchange, the placenta in some way acts as a protective barrier to prevent the embryo from being immunologically rejected by the mother. The embryo is actually a "foreigner" because it is half derived from genetically different paternal chromosomes. Furthermore, the placenta becomes a temporary endocrine organ during pregnancy, a topic to which we now turn our attention.

Hormones secreted by the placenta play a critical role in the maintenance of pregnancy.

The placenta has the remarkable capacity to secrete a number of peptide and steroid hormones essential for the maintenance of pregnancy (Table 20–5). Serving as the major endocrine organ of pregnancy, the placenta is unique among endocrine tissues in two regards. First, it is a transient tissue. Second, secretion of its hormones is not subject to extrinsic control, in contrast to the stringent, often complex mechanisms that regulate the secretion of other hormones. Instead, the type and rate of placental hormone secretion depends primarily on the stage of pregnancy.

One of the first events following implantation is secretion by the developing chorion of **human chorionic gonadotropin (hCG)**, a hormone that prolongs the life span of the corpus luteum. Recall that during the ovarian cycle, the corpus luteum degenerates and the highly prepared, luteal-dependent uterine lining sloughs if fertilization and implantation do not occur. When fertilization does occur, the implanted blastocyst saves itself from being flushed out in menstrual flow by producing hCG. This hormone, which is functionally similar to LH, stimulates and maintains the corpus luteum so that it does not degenerate. Now called the **corpus luteum of pregnancy,** this ovarian endocrine unit grows even larger and produces increasingly greater amounts of estrogen and progesterone for an additional ten weeks until the placenta takes over secretion of these steroid hormones. As a result of the persistence of circulating estrogen and progesterone, the thick, pulpy endometrial tissue is maintained instead

Table 20–5 Placental Hormones

Hormone	Function
Human chorionic gonadotropin	Maintains corpus luteum of pregnancy Stimulates secretion of testosterone by developing testes in XY embryos
Estrogen (also secreted by corpus luteum of pregnancy)	Stimulates growth of myometrium, increasing uterine strength for parturition Helps prepare mammary glands for lactation
Progesterone (also secreted by corpus luteum of pregnancy)	Suppresses uterine contractions to provide quiet environment for fetus Promotes formation of cervical mucus plug to prevent uterine contamination Helps prepare mammary glands for lactation
Human chorionic somatomammotropin	Helps prepare mammary glands for lactation Believed to reduce maternal utilization of glucose so that greater quantities of glucose may be shunted to the fetus
Relaxin (also secreted by corpus luteum of pregnancy)	Softens cervix in preparation for cervical dilation at parturition Loosens connective tissue between pelvic bones in preparation for parturition

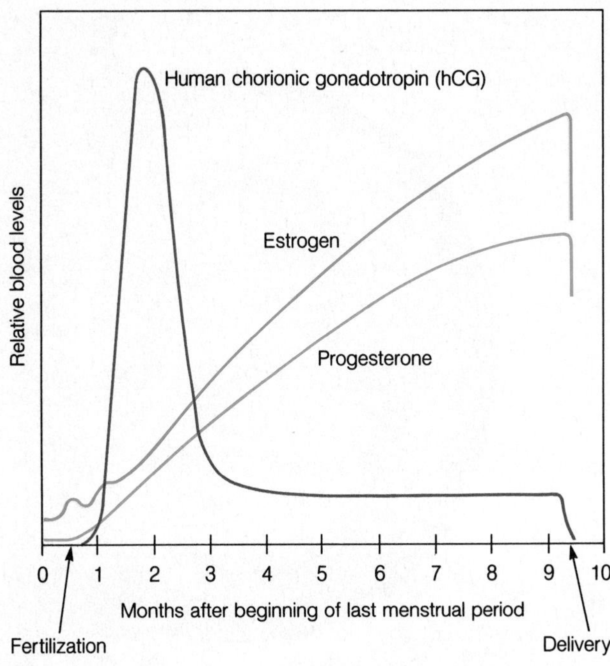

Figure 20–26 Secretion Rates of Placental Hormones

of sloughing. Accordingly, menstruation ceases during pregnancy.

Stimulation by hCG is necessary to maintain the corpus luteum of pregnancy because LH, which maintains the corpus luteum during the normal luteal phase of the uterine cycle, is suppressed through feedback inhibition by the high levels of progesterone. Suppression of the anterior pituitary hormones by the high progesterone levels also precludes further follicular maturation and ovulation for the duration of the pregnancy.

The maintenance of a normal pregnancy depends on high concentrations of progesterone and estrogen. Thus production of hCG is critical during the first trimester to maintain ovarian output of these hormones. In a male fetus, hCG also stimulates the precursor Leydig cells in the fetal testes to secrete testosterone, which masculinizes the developing reproductive tract.

The secretion rate of hCG increases rapidly during early pregnancy to save the corpus luteum from demise. Peak secretion of hCG occurs about sixty days after the end of the last menstrual period (Fig. 20–26). By the tenth week of pregnancy, hCG output declines to a low rate of secretion that is maintained for the duration of gestation. The fall in hCG occurs at a time when the corpus luteum is no longer needed for its steroidal hormone output because the placenta has begun to secrete substantial quantities of estrogen and progesterone. The corpus luteum of pregnancy partially regresses as hCG secretion dwindles, but it is not converted into scar tissue until after delivery of the baby.

Chorionic gonadotropin is eliminated from the body in the urine, where its detection forms the basis of pregnancy diagnosis tests. This placental hormone can be detected in the urine as early as the first month of pregnancy, about two weeks after the first missed menstrual period. This is before the growing embryo can be detected by physical examination, thus permitting early confirmation of pregnancy.

A frequent early clinical sign of pregnancy is **morning sickness,** a daily bout of nausea and vomiting that often occurs in the morning but can take place at any time of day. Since this condition usually appears shortly after implantation and coincides with the time of peak hCG production, it is speculated that this early placental hormone is responsible for

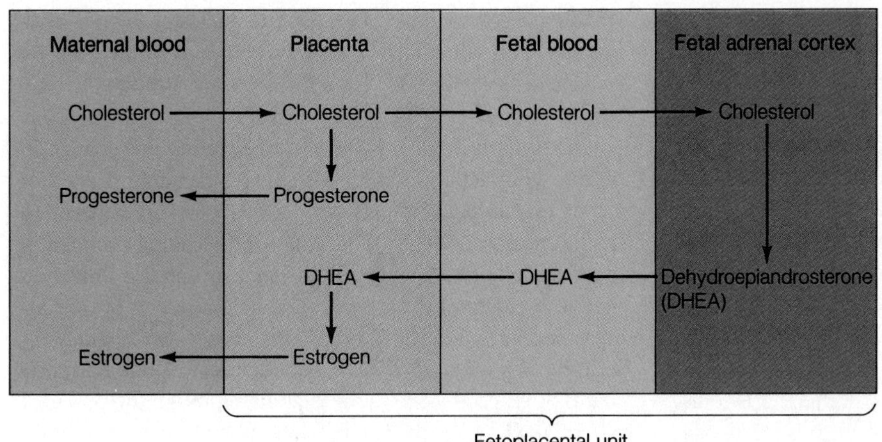

Figure 20–27 Secretion of Estrogen and Progesterone by Placenta

triggering the symptoms, perhaps by acting on the chemoreceptor trigger zone in the vomiting center (see p. 566).

A logical question is why the placenta does not start producing estrogen and progesterone in the first place instead of secreting hCG, which in turn stimulates the corpus luteum to secrete these two critical hormones. The answer lies in the fact that, for different reasons, the placenta is unable to produce sufficient quantities of either estrogen or progesterone in the first trimester of gestation (Fig. 20–27). In the case of estrogen, the placenta does not have all of the enzymes needed for complete synthesis of this hormone. Estrogen synthesis requires a complex interaction between the placenta and the fetus. The placenta is able to convert the androgen hormone produced by the fetal adrenal cortex, dehydroepiandrosterone (DHEA), into estrogen. The placenta is unable to produce estrogen until the fetus has developed to the point that its adrenal cortex is secreting DHEA into the blood. The placenta extracts DHEA from the fetal blood and converts it into estrogen, which it then secretes into the maternal blood. Because both fetal and placental tissues participate in this biosynthetic pathway, estrogen is said to be produced by the **fetoplacental unit.** The primary estrogen synthesized by this unit is **estriol,** in contrast to the main estrogen product of the ovaries, estradiol. Because of this, measurement of estriol levels in the maternal urine can be used clinically to assess the viability of the fetus.

In the case of progesterone, the placenta can synthesize this hormone soon after implantation. Even though the early placenta possesses the enzymes necessary to convert cholesterol extracted from the maternal blood into progesterone, it does not produce much of this hormone because the amount of progesterone produced is proportional to placental weight. The placenta is simply too small in the first ten weeks of pregnancy to produce sufficient quantities of progesterone to maintain the endometrial tissue. The notable increase in circulating progesterone in the last seven months of gestation reflects placental growth during this period.

Placental progesterone performs various roles throughout pregnancy. Its primary function is to prevent miscarriage by suppressing contractions of the uterine myometrium. Progesterone also promotes formation of a mucus plug in the cervical canal by causing the cervical secretions to become very gelatinous. This plug barricades the cervical canal to prevent vaginal contaminants from reaching the uterus. Finally, placental progesterone stimulates the development of milk glands in the breasts in preparation for lactation.

Placental estrogen is thought to stimulate growth of the myometrium, which increases in size throughout pregnancy. The stronger uterine musculature is needed to expel the fetus during labor. Estriol also promotes development of the ducts within the mammary glands through which milk will be ejected during lactation.

Maternal body systems respond to the increased demands of gestation.

The period of **gestation** (**pregnancy**) is about thirty-eight weeks from conception (forty weeks from the end of the last menstrual period). During this time, the embryo/fetus continues to grow and develop to the point of being able to leave its maternal life-support system. Meanwhile, a number of physical changes take place within the mother to accommodate the demands of the pregnancy. The most obvious change is uterine enlargement. The uterus expands and increases in weight more than twenty times, exclusive of its contents. The breasts enlarge and develop the capability of producing milk. Body systems other than the reproductive system also make the needed adjustments. The volume of blood increases by 30%, and the cardiovascular system responds to the increasing demands of the growing placental mass. The weight gain experienced during pregnancy is due only in part to the weight of the baby. The remainder is primarily caused by the increased weight of the uterus, including the placenta, and the increased blood volume. Respiratory activity is increased by about 20% to handle the additional fetal requirements for O_2 utilization and CO_2 removal. Urinary output increases, and the kidneys excrete the additional wastes from the fetus. The increased metabolic demands of the growing fetus results in increased nutritional requirements for the mother. In general, the fetus takes what it needs from the mother, even if this leaves the mother with a nutritional deficit. For example, if the mother does not consume sufficient Ca^{++} in her diet, Ca^{++} will be mobilized from the maternal bones to assure adequate calcification of the fetal bones.

Parturition is accomplished by a positive-feedback cycle.

Parturition (**labor, delivery,** or **birth**) requires: (1) dilation of the cervical canal to accommodate passage of the fetus from the uterus through the vagina and to the outside; and (2) contractions of the uterine myometrium that are sufficiently strong to expel the fetus. During the first two trimesters of gestation, the uterus remains relatively quiescent because of the inhibitory effect of the high levels of progesterone on the uterine musculature. During the last trimester, the uterus becomes progressively more excitable so that mild contractions (**Braxton-Hicks contractions**) are experienced with increasing strength and frequency. Sometimes these contractions become regular enough to be mistaken as the onset of labor, a phenomenon called "false labor."

Several other events take place near the end of pregnancy in preparation for parturition. The cervix begins to soften (or "ripen") as a result of the dissociation of its collagen fibers. This cervical softening is believed to be caused by **relaxin,** a peptide hormone produced by the corpus luteum of pregnancy and by the placenta. Relaxin also "relaxes" the birth canal by loosening the connective tissue between the pelvic bones. Meanwhile, the fetus shifts downward (the baby "drops") and is normally oriented so that the head is in contact with the cervix in preparation for exiting through the birth canal. In a **breech birth,** any part of the body other than the head approaches the birth canal first.

Rhythmic, coordinated contractions, usually painless at first, begin at the onset of true labor. As labor progresses, the contractions occur with increasing frequency and intensity and are accompanied by increasing discomfort. These strong, rhythmical contractions force the fetus against the cervix to dilate it. Then, after having dilated the cervix sufficiently for passage of the fetus, they force the fetus out through the birth canal.

The exact factors responsible for triggering this change in uterine contractility and thus initiating parturition are not clearly established, although endocrine factors are believed to be the most important. According to the leading proposal, a dramatic, progressive increase in the concentration of myometrial receptors for oxytocin, probably induced by the increasing levels of estrogen during pregnancy, is ultimately responsible for initiating labor. **Oxytocin** is a peptide hormone that is produced by the hypothalamus, stored in the posterior pituitary, and released into the blood from the posterior pituitary upon nervous stimulation by the hypothalamus. Oxytocin is a powerful uterine muscle stimulant and is known to play the key role in the progression of labor. However, this hormone was discounted as serving as the trigger to parturition because the circulating levels of oxytocin remain constant prior to the onset of labor. The recent discovery that uterine responsiveness to oxytocin is one-hundred times greater at term than in nonpregnant women (because of the increased concentration of myometrial oxytocin receptors) gives rise to speculation that labor is initiated when the oxytocin receptor concentration reaches a critical threshold level that permits the onset of strong, coordinated contractions in response to ordinary levels of circulating oxytocin.

Once triggered, myometrial contractions progressively increase in frequency, strength, and duration throughout labor until the uterine contents are expelled. At the beginning of labor, contractions lasting thirty seconds or less occur about every twenty-five to thirty minutes; by the end, they last sixty to ninety seconds and occur every two to three minutes.

Myometrial contractions incessantly increase as labor progresses because a positive-feedback cycle involving oxytocin and prostaglandin ensues (Fig. 20–28). Each uterine contraction begins at the top of the uterus and sweeps downward, forcing the fetus toward the cervix. Pressure of the fetus against the cervix accomplishes two things. First, the fetal head pushing against the softened cervix acts as a wedge to dilate the cervical canal. Second, cervical stretch stimulates the release of oxytocin through a neuroendocrine reflex. Stimulation of receptors in the cervix in response to fetal pressure produces a neural signal that travels up the spinal cord to the hypothalamus, which in turn triggers oxytocin release from the posterior pituitary. This additional oxytocin promotes more powerful uterine contractions. As a result, the fetus is pushed more forcefully against the cervix, which stimulates the release of even more oxytocin, and so on. This cycle is reinforced as oxytocin stimulates prostaglandin production by the decidua. As a powerful myometrial stimulant, prostaglandin

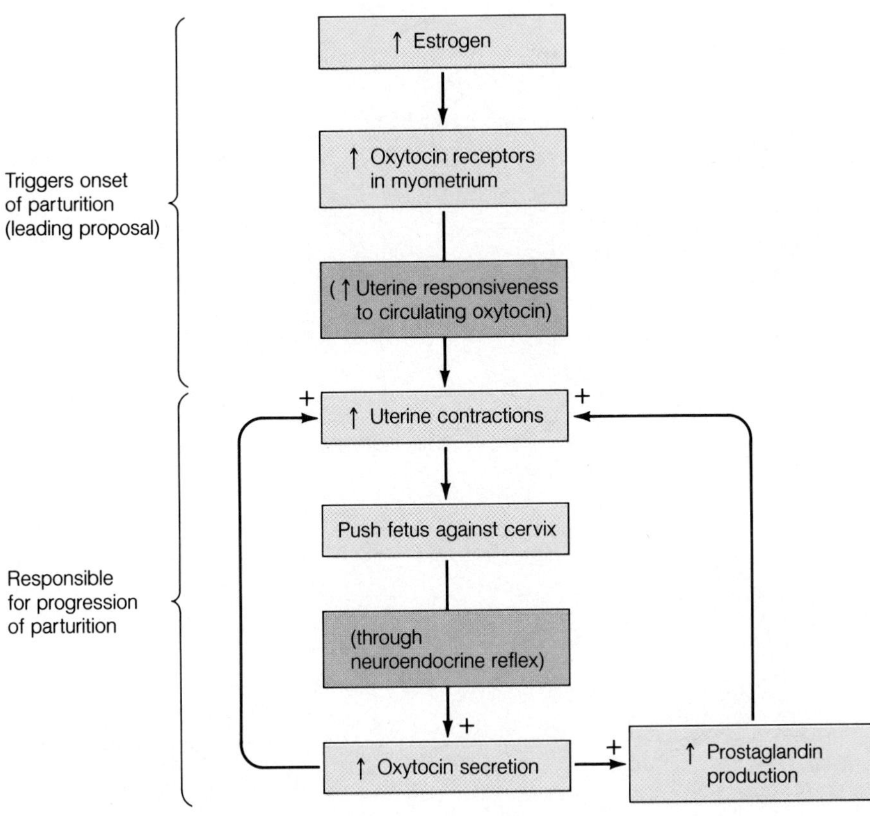

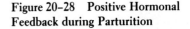

Figure 20–28 Positive Hormonal Feedback during Parturition

further enhances uterine contractions. Oxytocin secretion, prostaglandin production, and uterine contractions continue to increase in positive-feedback fashion throughout labor until the pressure on the cervix is relieved by delivery.

Labor is divided into three stages: (1) cervical dilation; (2) delivery of the baby; and (3) delivery of the placenta (Fig. 20–29). At the onset of labor or sometime during the first stage, the membranes surrounding the amniotic sac or "bag of waters" rupture. As the amniotic fluid escapes out of the vagina, it helps to lubricate the birth canal. During the first stage, the cervix is forced to dilate to accommodate the diameter of the baby's head, usually to a maximum of 10 cm. This stage is the longest, lasting several hours to as long as twenty-four hours in a first pregnancy. If another part of the fetus's body other than the head is oriented against the cervix, it is generally less effective as a wedge than the head. The head has the largest diameter of the baby's body. If the baby approaches the birth canal feet first, the cervix may not be dilated sufficiently by the feet to permit passage of the head. Without medical intervention in such a case, the baby's head would remain stuck behind the too-narrow cervical opening.

The second stage of labor, the actual birth of the baby, begins once cervical dilation is complete. When the infant begins to move through the cervix and vagina, stretch receptors in the vagina activate a neural reflex that triggers contractions of the abdominal wall in synchrony with the uterine contractions. This greatly increases the force pushing the baby through the birth canal. The mother can help deliver the infant by voluntarily contracting the abdominal muscles at this time in unison with each "labor pain" (that is, "push" with each uterine contraction). Stage two is usually much shorter than the first stage, lasting thirty to ninety minutes. The infant is still attached to the placenta by the umbilical cord at birth. The cord is tied and severed, with the stump shriveling up in a few days to form the **umbilicus** (**navel**).

Shortly after delivery of the baby, a second series of uterine contractions causes the placenta to separate from the myometrium and be expelled through the vagina. Delivery of the placenta or **afterbirth** constitutes the third stage of labor. This is typically the shortest stage, being completed within fifteen to thirty minutes after the baby is born. After the placenta is expelled, continued contractions of the myometrium con-

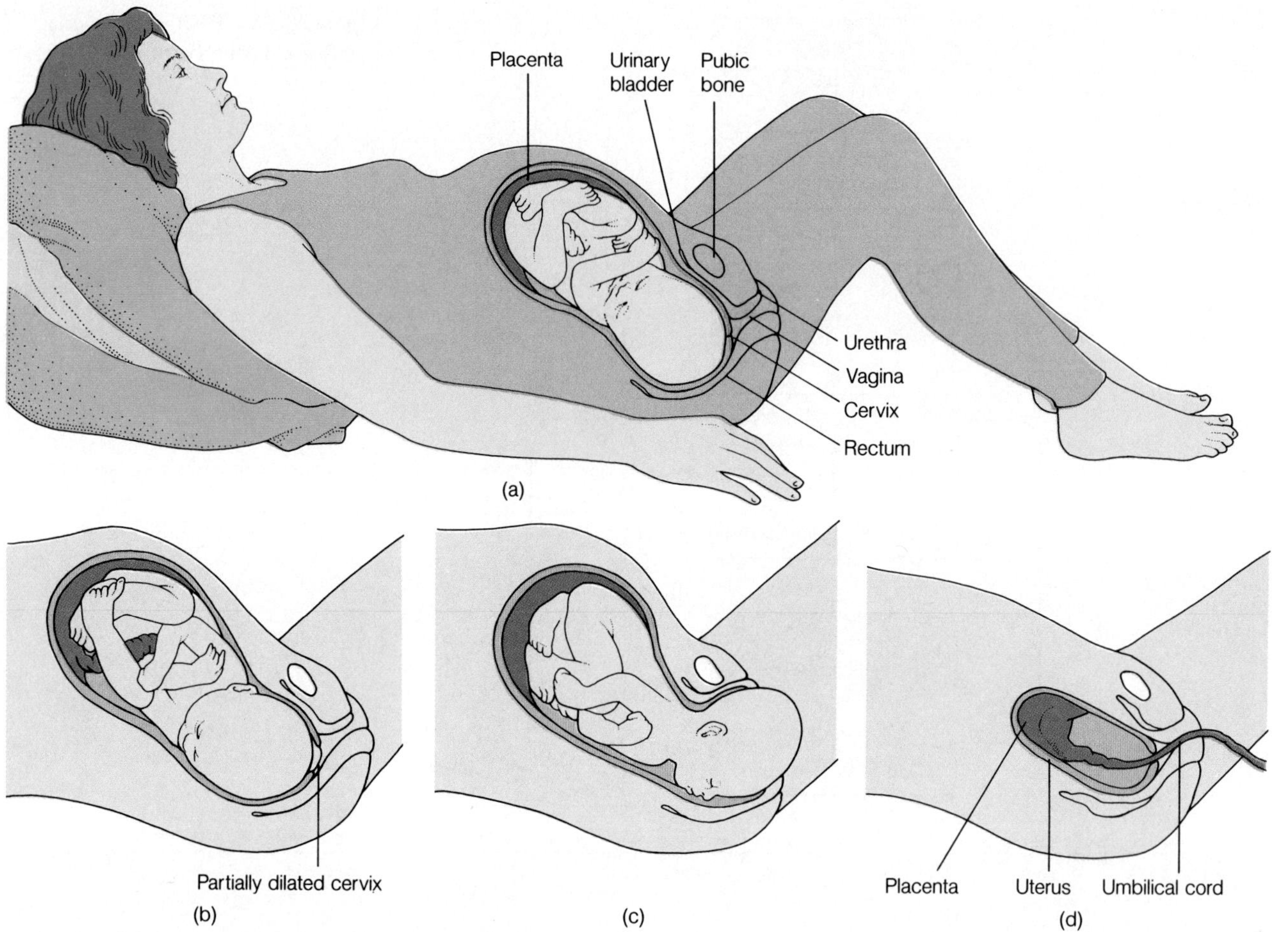

Figure 20–29 Stages of Labor *(a) Position of fetus near the end of pregnancy. (b) First stage of labor: cervical dilation. (c) Second stage of labor: delivery of the baby. (d) Third stage of labor: delivery of the placenta.*

strict the uterine blood vessels supplying the site of placental attachment to prevent hemorrhage.

After delivery, the uterus shrinks to its pregestational size, a process known as **involution** that takes four to six weeks to complete. During involution, the remaining endometrial tissue that was not expelled with the placenta gradually disintegrates and sloughs off, producing a vaginal discharge called **lochia** that continues for three to six weeks following parturition. After this period, the endometrium is restored to its nonpregnant state.

Involution occurs largely because of the precipitous fall in circulating estrogen and progesterone when the placental source of these steroids is lost at delivery. The process is facilitated in mothers who breast-feed their infants because of the oxytocin released in response to suckling. In addition to playing an important role in lactation, this periodic nursing-induced postpartum release of oxytocin promotes myometrial contractions that help maintain uterine muscular tone, thus enhancing involution. Involution is usually complete in about four weeks in nursing mothers but takes about six weeks in those who do not breast-feed.

Lactation requires multiple hormonal inputs.

Milk is essential for survival of the newborn. Accordingly, during gestation the **mammary glands,** or **breasts,** are prepared for **lactation (milk production)**.

The breasts in nonpregnant females are composed mostly of adipose tissue and a rudimentary duct system. The size of the breasts is determined by the amount of adipose tissue, which has nothing to do with the ability to produce milk. Under the hormonal environment present during pregnancy, the mammary glands develop the internal glandular structure and function necessary for milk production. A breast capable of lactating consists of a network of progressively smaller ducts that branch out from the nipple and terminate in lobules (Fig. 20–30). Each *lobule* is made up of a cluster of saclike epithelial-lined *alveoli* that constitute the milk-producing glands. Milk is synthesized by the epithelial cells, then secreted into the alveolar lumen, which is drained by a milk-collecting duct that transports the milk to the surface of the nipple.

During pregnancy, the high concentration of estrogen promotes extensive duct development, whereas the high level of progesterone stimulates abundant alveolar-lobular formation. Elevated concentrations of **prolactin** (an anterior pituitary hormone stimulated by the rising levels of estrogen) and **human chorionic somatomammotropin** (a peptide hormone produced by the placenta) also contribute to mammary gland development by inducing the synthesis of enzymes needed for milk production.

Most of these changes in the breasts occur during the first half of gestation, so the mammary glands are fully capable of producing milk by the middle of pregnancy. However, milk secretion does not occur until parturition. The high estrogen and progesterone concentrations during the last half of pregnancy prevent lactation by blocking prolactin's stimulatory action on milk secretion. Prolactin is the primary stimulant of milk secretion. Thus, even though the high levels of placental steroids induce the development of the milk-producing machinery in the breasts, they prevent these glands from becoming operational until the baby is born and milk is needed. The abrupt decline in estrogen and progesterone that occurs with loss of the placenta at parturition initiates lactation. (The functions of estrogen and progesterone during gestation and lactation as well as throughout the reproductive life of females are summarized in Table 20–6.)

Once milk production begins after delivery, two hormones are critical for maintaining lactation: (1) prolactin, which acts on the alveolar epithelium to promote secretion of milk, and (2) oxytocin, which produces **milk ejection.** The latter refers to the forced expulsion of milk from the lumen of the alveoli out through the ducts. Release of both of these hormones is stimulated by a neuroendocrine reflex triggered by suckling (Fig. 20–31, p. 758). Milk cannot be directly sucked out of the alveolar lumen by the infant. Instead, milk must be actively squeezed out of the alveoli into the ducts and hence toward the nipple by contraction of specialized **myoepithelial cells** that surround each alveoli. Suckling of the breast by the infant stimulates sensory nerve endings in the nipple, the result of which is initiation of action potentials that travel up the spinal cord to the hypothalamus. Thus activated, the hypothalamus triggers a burst of oxytocin release from the posterior pituitary. Oxytocin, in turn, stimulates contraction of the myoepithelial cells in the breasts to bring about milk ejection or *"milk let-down."* Milk let-down continues only as long as the infant continues to nurse. In this way, the milk-ejection reflex assures that milk exits the breasts only when and in the amount needed by the baby. Even though the alveoli may be full of milk, the milk cannot be released without oxytocin. The reflex can become conditioned to stimuli other than suckling, however. For example, the infant's cry can trigger milk let-down,

Table 20–6 Actions of Estrogen and Progesterone

Estrogen

Effects on Sex-Specific Tissues

Essential for egg maturation and release

Stimulates growth and maintenance of entire female reproductive tract

Stimulates granulosa cell proliferation, which leads to follicle maturation

Thins cervical mucus to permit sperm penetration

Enhances transport of sperm to oviduct by stimulating upward contractions of uterus and oviduct

Stimulates growth of endometrium and myometrium

Induces synthesis of progesterone receptors in endometrium and, during gestation, of myometrial receptors for oxytocin

Other Reproductive Effects

Promotes development of secondary sex characteristics

Controls GnRH and gonadotropin secretion

- Low levels inhibit secretion
- High levels responsible for triggering of LH surge

Stimulates duct development in breasts during gestation

Inhibits milk-secreting actions of prolactin during gestation

Nonreproductive Effects

Promotes fat deposition

Closes epiphyseal plates

Reduces blood cholesterol (incidence of atherosclerosis lower in females until after menopause)

Exerts vascular effects (deficiency → "hot flashes" at menopause)

Progesterone

Prepares suitable environment for nourishment of developing embryo/fetus

Promotes formation of thick mucus plug in cervical canal

Inhibits hypothalamic GnRH and gonadotropin secretion

Stimulates alveolar development in breasts during gestation

Inhibits milk-secreting actions of prolactin during gestation

Inhibits uterine contractions during gestation

Figure 20-30 Mammary Gland Anatomy *(a) Internal structure of mammary gland, lateral view. (b) Schematic representation of microscopic structure of an alveolus within the mammary gland.*

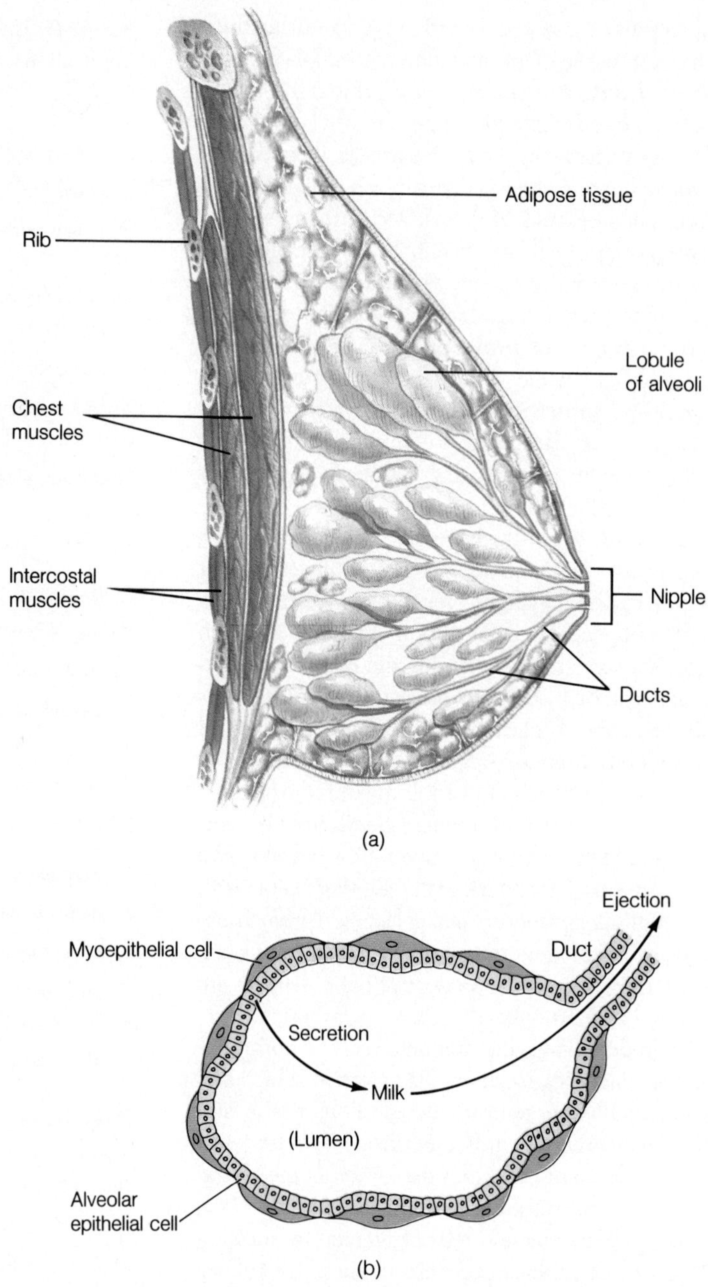

thus causing a spurt of milk to leak from the nipples. On the other hand, psychological stress, acting through the hypothalamus, can easily inhibit milk ejection. For this reason, a positive attitude toward breast-feeding as well as a relaxed environment are essential for successful breast-feeding.

Suckling not only triggers oxytocin release but also stimulates prolactin secretion. Prolactin output by the anterior pituitary is controlled by two hypothalamic secretions: **prolactin-inhibiting hormone** (**PIH**) and **prolactin-releasing factor** (**PRF**). Throughout most of the female's life, PIH is the dominant influence, so prolactin is normally not secreted. During lactation, a burst in prolactin secretion occurs each time the infant suckles. Afferent impulses initiated in the nipple upon suckling are carried by the spinal cord to the hypothalamus. This reflex ultimately leads to prolactin release by the anterior pituitary, although it is unclear whether this outcome is accomplished by inhibition of PIH or stimulation of PRF secretion or both. Prolactin then acts on the alveolar epithelium to promote secretion of milk to replenish the milk lost from the alveoli by milk ejection (Fig. 20–31).

Concurrent stimulation by suckling of both milk ejection and milk production assures that the rate of milk synthesis keeps pace with the baby's needs for milk. The more the infant nurses, the more milk is removed by let-down and the more milk is produced.

In addition to prolactin, which is the most important factor controlling synthesis of milk, at least four other hormones are essential for their permissive role in milk production: these are cortisol, insulin, parathyroid hormone, and growth hormone.

Milk is composed of water, triglyceride fat, the carbohydrate lactose (milk sugar), a number of proteins, vitamins, and the minerals calcium and phosphate. The composition of milk differs during the early part of lactation. Called **colostrum,** the milk for the first five days postpartum contains lower concentrations of fat and lactose but higher concentrations of proteins, most notably lactoferrin and immunoglobulins. These proteins provide early protection against infection until the newborn's own defense system develops. *Lactoferrin* has bactericidal activities, and *immunoglobulins* are antibodies. Furthermore, recent evidence suggests that breast milk may actually stimulate development of the newborn's own immune capabilities. Infants who are bottle-fed on a formula made from cow's milk do not have this protective advantage provided by colostrum and, accordingly, have a higher incidence of colds and other infections than breast-fed babies. Also, the digestive system of a newborn is better equipped to handle human milk than cow milk–derived formula, so bottle-fed babies tend to have more digestive upsets.

Breast-feeding is also advantageous for the mother. Oxytocin release triggered by nursing hastens uterine involution. In addition, suckling suppresses the menstrual cycle by inhibiting LH and FSH secretion, probably through inhibition of GnRH. Lactation, therefore, prevents ovulation and serves as a means of preventing pregnancy. This permits the mother's resources to all be directed toward the newborn instead of being shared with a new embryo. Although not 100% effective as a means of contraception, prolonged nursing of infants in societies in which the young rely on breast milk as a primary energy source for an extended time (two to three years) plays an important role in population control.

When the infant is weaned, two mechanisms contribute to the cessation of milk production. First, prolactin secretion is not stimulated when there is no suckling, thus removing the primary stimulus for continued milk synthesis and secretion. Also because of the lack of suckling, milk let-down does not occur in the absence of oxytocin release. Since milk production does not immediately shut down, milk accumulates in the alveoli, causing engorgement of the breasts. The resultant pressure buildup acts directly on the alveolar epithelial cells to suppress further milk production. Cessation of lactation at weaning therefore results from a lack of suckling-induced stimulation of both prolactin and oxytocin secretion.

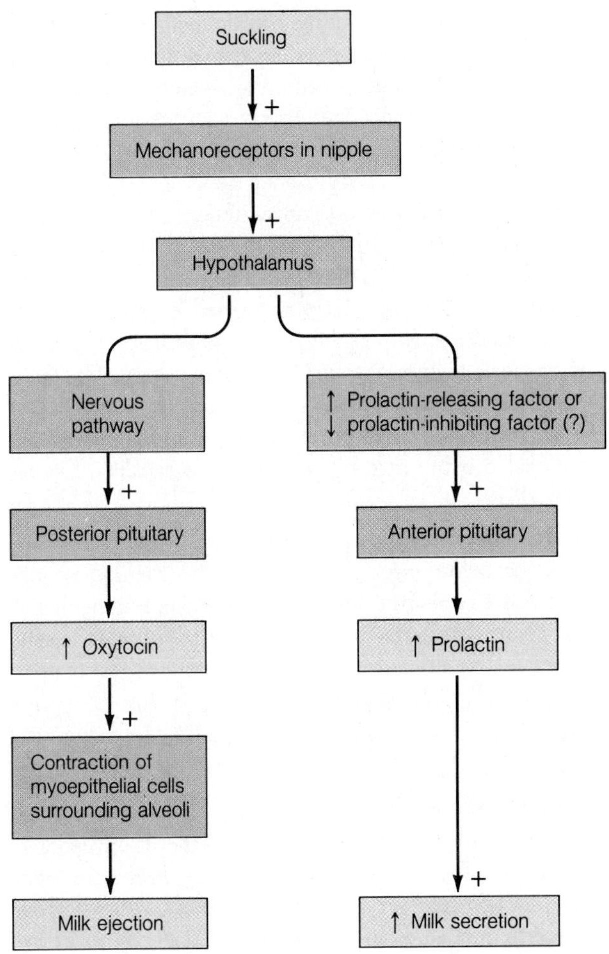

Figure 20–31 Suckling Reflexes

Contraceptive techniques act by blocking sperm transport, ovulation, or implantation.

Couples wishing to engage in sexual intercourse but avoid pregnancy have a number of methods of **contraception** ("against conception") available. These methods, which range in effectiveness (Table 20–7) and ease of use, act by blocking one of three major steps in the reproductive process: sperm transport to the ovum, ovulation, or implantation. The following birth control methods interfere with sperm transport to the ovum.

□ *Natural contraception* or the *rhythm method* of birth control relies on abstinence from intercourse during the woman's fertile period. The woman can predict when ovulation is to occur based on keeping careful records of her menstrual cycles. This technique is only partially effective because of variability in cycles. The time of ovulation can be determined more precisely by recording body temperature each morning before getting up. A slight rise in body temperature occurs about a day after ovulation has taken place. The temperature-rhythm method is not useful in determining when it is safe to engage in intercourse before ovulation, but it can be helpful in determining when it is safe to resume sex after ovulation. This technique is also employed by couples as a method of fertility. The rise in body temperature occurs during the time period that the ovulated ovum may still be viable and capable of being fertilized.

□ *Coitus interruptus* involves withdrawal of the penis from the vagina before ejaculation occurs. This method is only moderately effective, however, because timing is difficult and some sperm may pass out of the urethra prior to ejaculation.

□ *Chemical contraceptives*, such as spermicidal ("sperm-killing") jellies, foams, creams, and suppositories, when inserted into the vagina are toxic to sperm for about an hour following application.

□ *Barrier methods* mechanically prevent sperm transport to the oviduct. The condom is a thin, strong rubber or latex sheath placed over the erect penis prior to ejaculation to prevent sperm from entering the vagina. The diaphragm is a flexible rubber dome that is inserted through the vagina and positioned over the cervix to block sperm entry into the cervical canal. The diaphragm must be fitted by a trained professional and must be left in place for at least six hours but no longer than twenty-four hours after intercourse. Barrier methods are often used in conjunction with spermicidal agents for increased effectiveness.

The most recent combination method is the contraceptive sponge. This nonprescription polyurethane device, once placed in the vagina, acts as a physical deterrant to sperm transport and also releases a spermicidal agent for up to twenty-four hours. The sponge has been implicated as the underlying cause of toxic shock syndrome in a small number of this product's users.

□ *Sterilization*, which involves surgical disruption of either the ductus deferens (vasectomy) in men or the oviduct (tubal ligation) in women, is considered to be a permanent method of preventing sperm and ovum from uniting.

□ A recently patented invention that has not yet been tested in women is an *electrical contraceptive*. It consists of a small, battery-powered cylindrical device that creates an electric field and is designed to be inserted into the cervical canal by a trained professional. The electrical current, which would be imperceptible to the user, has been shown in tests on baboons to immobilize sperm so that they cannot swim through the cervical canal to enter the uterus.

Table 20–7 Average Failure Rate of Various Contraceptive Techniques

Contraceptive Method	Average Failure Rate (Annual Pregnancies/ 100 Women)
None	90
Natural (rhythm) methods	20–30
Coitus interruptus	23
Chemical contraceptives	20
Barrier methods	10–15
Oral contraceptives	2–2.5
Intrauterine device	4

SOURCE: All data (with the exception of chemical contraceptives) for women 20–29 years old taken from Carr, B.R., and Griffin, J.E., in *Williams' Textbook of Endocrinology*, 7th ed. J.D. Wilson and D.W. Foster, editors (Philadelphia: W.B. Saunders Company, 1985).

The second site for blocking the reproductive process is prevention of ovulation. ***Oral contraceptives*** or ***birth-control pills*** prevent ovulation by suppressing gonadotropin secretion. These pills, which contain synthetic estrogen-like and progesterone-like steroids, are taken for three weeks, either in combination or in sequence, and then are withdrawn for one week. These steroids, just as the natural steroids produced during the ovarian cycle, inhibit GnRH and thus FSH and LH secretion. As a result, follicle maturation and ovulation do not take place so conception is impossible. The endometrium responds to the exogenous steroids by thickening and developing secretory capacity, just as it would to the natural hormones. When these synthetic steroids are withdrawn after three weeks, the endometrial lining sloughs and menstruation occurs, as it normally would upon degeneration of the corpus luteum. In addition to blocking ovulation, oral contraceptives also prevent pregnancy by increasing the viscosity of cervical mucus, which makes sperm penetration more difficult, and by decreasing muscular contractions in the female reproductive tract, which reduces sperm transport to the oviduct.

Oral contraceptives are available only by prescription. They have been shown to increase the risk of intravascular clotting, especially in women who also smoke tobacco.

Among the new approaches to contraception are ***long-acting injections*** and ***subcutaneous*** ("under the skin") ***implantation*** of synthetic steroids, which act similarly to oral contraceptives by blocking ovulation and sperm transport but which are effective for three to six months after one treatment.

The third site of interference with the reproductive process, blocking implantation, is most commonly accomplished by insertion of a small ***intrauterine device*** (***IUD***) into the uterus by a physician. The IUD's mechanism of action is not completely understood, although most evidence suggests that the presence of this foreign object in the uterus induces a local inflammatory response that prevents implantation of a fertilized ovum. Although the IUD is a convenient birth control method because it does not require ongoing attention by the user, it is no longer as popular as it once was because of reported complications associated with its use, the most serious of which are pelvic inflammatory disease, permanent infertility, and uterine perforation.

Implantation can also be blocked by the so-called ***morning-after pill***, which is a different type of oral contraceptive than the usual birth-control pill. The high-estrogen content morning-after pill is taken during the early luteal phase within a few days after conception may have occurred. It prevents implantation by inducing premature degeneration of the corpus luteum so that the developing endometrium's hormonal support is withdrawn. Because of side effects, such as nausea and vomiting, and because of the increased risks of cardiovascular disease associated with high doses of estrogen, this contraceptive method is not used on a routine basis. It is beneficial for one-time usage in special circumstances, however, such as for rape victims who may have conceived.

A future birth control technique that holds promise in preventing implantation is oral administration of a ***progesterone antagonist*** during the luteal phase. This drug acts by antagonizing, or blocking, the actions of progesterone on the uterus so that the endometrium is not a suitable environment for implantation. Another technique currently being explored is development of a vaccine that induces the formation of antibodies against human chorionic gonadotropin so that this essential corpus-luteum supporting hormone is not effective should pregnancy occur. Still another possibility under investigation is manipulation of the anterior-pituitary secretion of FSH and LH by GnRH-like drugs.

Chapter in Perspective

The reproductive system is unique in that it is not essential for homeostasis or for survival of the individual, but it is essential for sustaining the thread of life from generation to generation. Reproduction depends on the union of male and female gametes (reproductive cells), each with a half set of chromosomes, to form a new individual with a full, unique set of chromosomes. Unlike the other body systems, which are essentially identical in the two sexes, the reproductive systems of males and females are remarkably different, befitting their different roles in the reproductive process.

The male system is designed to continuously produce huge numbers of motile spermatozoa that are delivered to the female during the sex act. Male gametes must be produced in abundance for two reasons: (1) only a small percentage of them survive the hazardous journey in the female reproductive tract to the site of fertilization; and (2) the cooperative effort of many spermatozoa is required to break down the barriers surrounding the female gamete (ovum or egg) to enable one spermatozoon to penetrate and unite with the ovum.

The female reproductive system undergoes complex changes on a monthly cyclical basis. During the first half of the cycle, a single nonmotile ovum is prepared for release. During the second half, the reproductive system is geared toward preparing a suitable environment for supporting the ovum if fertilization (union with a spermatozoon) occurs. If fertilization does not occur, the prepared supportive environment within the uterus sloughs off, and the cycle starts over again as a new ovum is prepared for release. If fertilization occurs, the female reproductive system adjusts to support growth and development of the new individual until it can survive on its own on the outside.

There are three important parallels in the male and female reproductive systems, even though they differ considerably in structure and function. First, the same set of undifferentiated reproductive tissues in the embryo can develop into either a male or female system, depending on the presence or absence, respectively, of male-determining factors. Second, the same hormones—namely, hypothalamic GnRH and anterior pituitary FSH and LH—control reproductive function in both sexes. In both cases, gonadal steroids and inhibin act in negative-feedback fashion to control hypothalamic and anterior pituitary output. Third, the same events take place in the developing gamete's nucleus during sperm formation and egg formation, despite the fact that males produce millions of sperm in one day, whereas females produce only about four hundred ova in a lifetime.

Reproduction is an appropriate way to end our discussion of "physiology from cells to systems." The single cell resulting from the union of male and female gametes divides mitotically and differentiates into a multicellular individual made up of a number of different organ systems that interact cooperatively to maintain homeostasis (that is, stability in the internal environment). All of the life-supporting homeostatic processes introduced throughout this book begin all over again at the start of a new life.

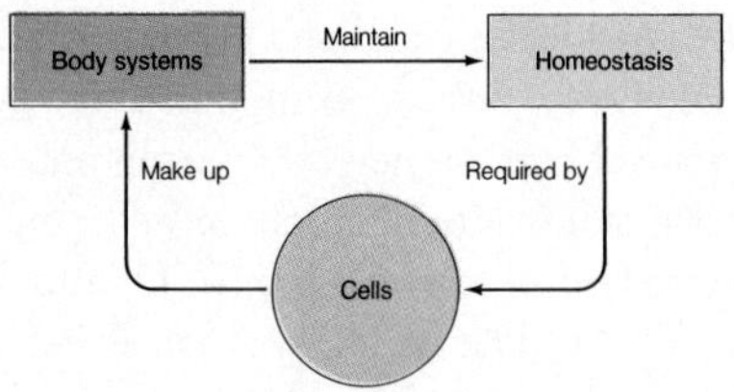

See inside front cover for an expanded version of this model.

Review Exercises

1. Discuss the differences between males and females with regard to genetic, gonadal, and phenotypic sex.
2. Discuss the source and functions of testosterone.
3. Describe the three major stages of spermatogenesis. Discuss the functions of each part of a spermatozoon. What are the roles of Sertoli cells?
4. Discuss the control of testicular function.
5. Describe the functions of each of the male accessory reproductive organs.
6. Compare the sex act in males and females.
7. Compare oogenesis with spermatogenesis.
8. Describe the events of the follicular and luteal phases of the ovarian cycle. Correlate the phases of the uterine cycle with the ovarian cycle.
9. How are the ovum and spermatozoa transported to the site of fertilization? Describe the process of fertilization.
10. Describe the process of implantation and placenta formation. What are the functions of the placenta?
11. What is the role of human chorionic gonadotropin?
12. What is the leading proposal for the mechanism of initiation of parturition?
13. Describe the hormonal factors that play a role in lactation.
14. **A point to ponder:** A promising line of research for a new method of contraception is continuous administration of gonadotropin-releasing hormone-like drugs. How could such drugs act as contraceptives when GnRH is the hypothalamic hormone that triggers the chain of events leading to ovulation? (Hint: The anterior pituitary is "programmed" to respond only to the normal pulsatile pattern of GnRH.)

APPENDIX A

A Review of Chemical Principles

Appendix Contents at a Glance

The chemical nature of all matter makes an understanding of chemistry essential in the study of many disciplines, including human physiology. This appendix contains a brief discussion of some basic chemical concepts that you are encouraged to review as necessary while you study the material in the textbook.

Atoms, Elements, and Compounds

All matter is made up of tiny particles called **atoms**. These particles are too small to be seen individually, even with the most powerful electron microscopes available today. However, the work of generations of scientists has led to an understanding of many characteristics of atoms that help us explain and understand the behavior of matter.

Even though they are tiny, atoms are composed of three smaller particles. **Protons** and **neutrons** are particles of nearly identical mass that make up the nucleus of an atom. Protons carry a positive charge, whereas neutrons have no charge. **Electrons**, the third particle found in atoms, move rapidly around the central nucleus (Fig. A–1). Electrons have a much smaller mass than protons and neutrons and are negatively charged. The charge of a proton exactly matches that of an electron except it is opposite in sign. In all atoms, the number of protons in the nucleus is equal to the number of electrons moving around the nucleus, so their charges balance and the atoms are neutral.

A pure substance that contains only one type of atom is called an **element**. A pure sample of the element carbon contains only carbon atoms even though the atoms might be arranged in a form called diamond or in a form called graphite.

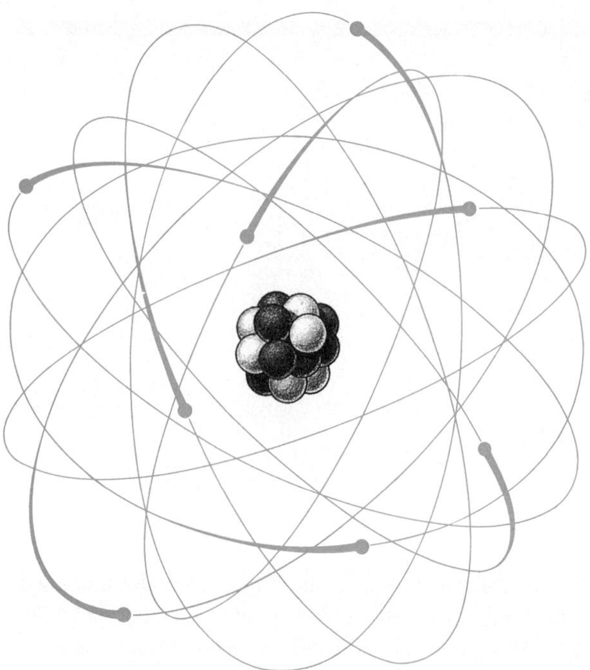

Figure A–1 Electrons Move Rapidly Around a Massive Nucleus

SOURCE: Reprinted by permission from page 61 of *Chemistry: An Introduction* by Michael R. Slabaugh and Spencer L. Seager. Copyright © 1988 by West Publishing Company. All rights reserved.

Pure substances called **compounds** contain more than one type of atom. Pure water, for example, is a compound that contains atoms of hydrogen and atoms of oxygen in a 2:1 ratio, regardless of whether the water is in the form of liquid, solid (ice), or vapor (steam).

When we refer to "type" of atom, just what are we talking about? That is, what makes carbon, hydrogen, and oxygen atoms different? The answer is: the number of protons in the nucleus. Regardless of where they are found, all hydrogen atoms have one proton in the nucleus, all carbon atoms have six, and all oxygen atoms have eight. Of course, these numbers also represent the number of electrons moving around each nucleus because the number of electrons and protons in an atom are equal. This number of protons in the nuclei of the atoms of an element is called the **atomic number** of the element.

As expected, tiny atoms have tiny masses. For example, the actual mass of a hydrogen atom is 1.67×10^{-24} g, that of a carbon atom is 1.99×10^{-23} g, and that of an oxygen atom is 2.66×10^{-23} g. These very small numbers are inconvenient to work with in calculations, so a system of relative masses has been developed. These relative masses simply compare the actual masses of the atoms to each other. Suppose the actual masses of two people were determined to be 45.50 kg and 113.75 kg. Their relative masses are determined by dividing each mass by the smaller mass of the two: $45.50/45.50 = 1.00$ and $113.75/45.50 = 2.50$. Thus, the relative masses of the two people are 1.00 and 2.50, which simply expresses the fact that the mass of the heavier person is 2.50 times that of the other person. The relative masses of atoms are called **atomic masses** or **atomic weights** and are given in atomic mass units (amu). In this system, hydrogen atoms, the least massive of all atoms, have an atomic weight of 1.01 amu. The atomic weight of carbon atoms is 12.01 amu, and that of oxygen atoms is 16.00 amu. Thus, oxygen atoms have a mass about 16 times that of hydrogen atoms. Table A–1 gives the atomic weights and some other characteristics of the elements that are most important physiologically.

Table A–1 Characteristics of Selected Elements

Name	Symbol	Number of Protons	Atomic Number	Atomic Weight (amu)
Hydrogen	H	1	1	1.01
Carbon	C	6	6	12.01
Nitrogen	N	7	7	14.01
Oxygen	O	8	8	16.00
Sodium	Na	11	11	22.99
Magnesium	Mg	12	12	24.31
Phosphorus	P	15	15	30.97
Sulfur	S	16	16	32.06
Chlorine	Cl	17	17	35.45
Potassium	K	19	19	39.10
Calcium	Ca	20	20	40.08

Chemical Bonds

Since all matter is made up of particles called atoms, we must conclude that the atoms are held together some way to form the matter. The forces holding atoms together are called **chemical bonds**. Not all chemical bonds are formed in the same way, but all of them involve the electrons of atoms. It is now understood that the electrons of each atom have energies and other characteristics that allow them to be classified into groupings called **shells**. In general, electrons will belong to the lowest energy shell possible, but the specific shells of an atom have maximum capacities that cannot be exceeded. For example, the first or lowest energy shell of every atom can contain a maximum of only two electrons, while the second or next highest energy shell can contain a maximum of eight electrons. Different atoms have different numbers of electrons in the various shells. Hydrogen atoms have only one electron so it is in the first shell. Helium atoms have two electrons, which are both in the first shell and fill it. Carbon atoms have six electrons, two in the first shell and four in the second shell, while the eight electrons of oxygen are arranged with two in the first shell and six in the second shell.

There is an energy benefit to having filled electron shells. That is, the average energy per electron is lower for the second shell when the shell holds the maximum number of eight electrons instead of any number from one to seven. This energy benefit leads to a general statement about the electronic behavior of atoms: ***atoms tend to undergo processes that result in a filled outermost electron shell.*** Thus, it is the electrons of the outer or higher energy shell that determine the bonding characteristics of an atom.

Consider sodium atoms (Na) and chlorine atoms (Cl). Sodium atoms have eleven electrons, two in the first shell, eight in the second shell, and one in the third shell. Chlorine atoms have seventeen electrons, two in the first shell, eight in the second shell and seven in the third shell. Because the second and third shells are filled by eight electrons, sodium atoms have one electron more than is needed to provide a filled second shell. Chlorine atoms have one less electron than is needed to fill the third shell. Each sodium atom loses an electron to a chlorine atom, leaving each sodium with ten electrons, eight of which are in the second shell, which is now the outer shell occupied by electrons. The acceptance of one electron by each chlorine atom gives each chlorine a total of eighteen electrons, with eight of them in the third or outer shell.

As a result of giving up and accepting electrons, the sodium atoms and chlorine atoms have achieved filled outer shells, but now each atom is unbalanced electrically. While each sodium now has ten electrons, it still has eleven protons in the nucleus and a net electrical charge of $+1$. Similarly, each chlorine now has eighteen electrons, but only seventeen protons. Thus each chlorine has a -1 charge. Charged atoms such as these are called **ions**. Since opposite charges attract, sodium ions (Na^+) and charged chlorine atoms, now called chloride ions (Cl^-), are attracted toward each other. It is this attraction that bonds the ions together in the compound sodium chloride, NaCl, which is common table salt. A sample of sodium chloride actually contains sodium and chloride ions in a three-dimensional geometric arrangement called a crystal lattice. The ions of opposite charge occupy alternate sites within the lattice (Fig. A–2).

Positively charged ions such as Na^+ are called **cations**, whereas negatively charged ions such as Cl^- are called **anions**.

It is not energetically favorable for an atom to give up or accept more than three electrons. In spite of this, carbon atoms with four electrons in their outer shell are known to form compounds. This problem led scientists to propose another bonding mechanism for atoms. Atoms that would have to lose or gain four or more electrons to achieve outer-shell stability usually bond by sharing electrons. Thus, a carbon atom shares its four outer electrons with the four electrons of four hydrogen atoms. This is represented in Equation A–1 where the outer shell electrons are shown as dots around the symbol of each atom.

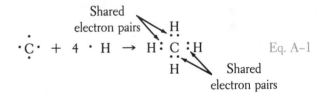

$$\cdot\dot{\underset{\cdot}{C}}\cdot + 4\ \cdot H \rightarrow H:\overset{H}{\underset{H}{\overset{\cdot\cdot}{\underset{\cdot\cdot}{C}}}}:H \qquad \text{Eq. A–1}$$

Shared electron pairs

Shared electron pairs

Each electron shared by two atoms is counted toward the number of electrons needed by each atom to fill the outer shell. Thus, each carbon atom shares four pairs or eight electrons and so has eight in the outer shell. Each hydrogen shares one pair or two electrons and so has a filled outer shell. (Remember, hydrogen atoms need only two electrons to complete their outer shell, which is the first shell.) Atoms that share a pair of electrons are both attracted toward the shared pair. This mutual attraction thus bonds the atoms together in what is called a **covalent bond**. The resulting compound is methane (CH_4), which is a gas made up of individual CH_4 particles called molecules.

Covalent bonds also form between some atoms that are identical. For example, two hydrogen atoms can complete their outer shells by sharing one pair made from the single electron of each atom. This is shown in Equation A–2.

$$H\cdot + \cdot H \rightarrow H:H \qquad \text{Eq. A–2}$$

Figure A–2 Crystal Lattice for Sodium Chloride (Table Salt)

SOURCE: Reprinted by permission from page 113 of *Chemistry: An Introduction* by Michael R. Slabaugh and Spencer L. Seager. Copyright © 1988 by West Publishing Company. All rights reserved.

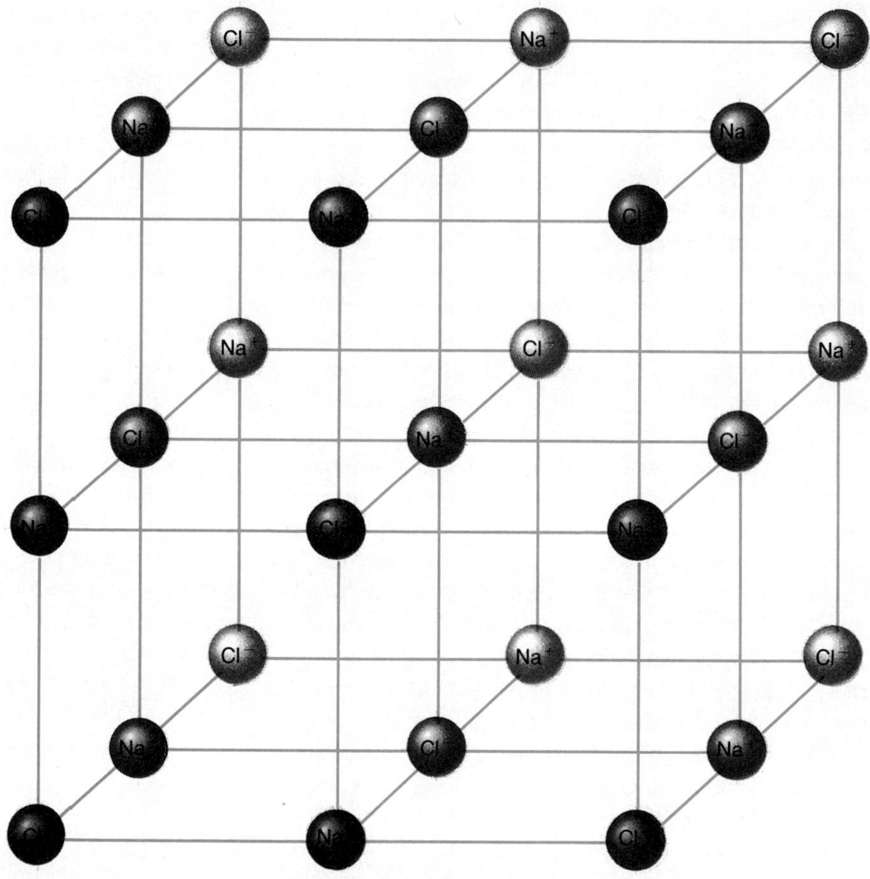

Thus, hydrogen gas consists of individual H_2 molecules. Some other nonmetallic elements also exist as molecules because covalent bonds form between identical atoms. Examples are chlorine (Cl_2), oxygen (O_2), and nitrogen (N_2).

One of the most familiar covalently bonded substances is water (H_2O). The formation of the covalent bonds is represented in Equation A–3.

$$\begin{matrix} H\cdot \\ \\ H\cdot \end{matrix} + \cdot\ddot{\underset{\cdot}{O}}: \rightarrow \begin{matrix} H:\ddot{\underset{\cdot\cdot}{O}}: \\ H \end{matrix} \qquad \text{Eq. A–3}$$

The water molecule is sometimes represented as

$$\begin{matrix} H—O \\ \quad | \\ \quad H \end{matrix}$$

where the nonshared electron pairs are not shown and the covalent bonds or shared pairs are represented by dashes. The water molecule is a good example of a **polar molecule**, or a molecule in which the electrons are not distributed uniformly. Polar molecules result because shared electrons are attracted toward the atoms that share them. When the atoms sharing an electron pair are identical, the electrons are attracted equally by both atoms and so are shared equally; the molecules are **nonpolar**. Examples of molecules containing equally shared electrons are H_2, O_2, and N_2. When the sharing atoms are not identical, the shared pair of electrons is pulled closer to one atom than the other. The side of the molecule to which the electrons are pulled is electrically negative compared to the other side. In water molecules, the oxygen atom pulls shared electrons more strongly than do the hydrogen atoms. Thus, the oxygen side of a water molecule is more negative than the hydrogens. The electron distribution is not uniform, and the water molecule is polar. Atoms of oxygen, nitrogen, phosphorus and chlorine strongly attract electrons when they are bonded to other atoms.

Polar molecules are attracted to other polar molecules. For example, in water there is an attraction between the positive hydrogen ends of some molecules and the negative oxygen ends of others. Hydrogen is not a part of all polar molecules, but when it is covalently bonded to an atom that strongly attracts electrons to form a covalent molecule, the attraction of the positive (hydrogen) end of the polar molecule to the negative end of another polar molecule is called a **hydrogen bond.** Thus, the polar attractions of water molecules to each other is an example of hydrogen bonding.

Chemical Reactions

Processes in which chemical bonds are broken and/or formed are called **chemical reactions**. Reactions are represented by equations in which the reacting substances (**reactants**) are written on the left, the produced substances (**products**) are written on the right, and an arrow points from the reactants to the products. These conventions are illustrated in Equation A-4.

$$\underset{\text{reactants}}{A + B} \rightarrow \underset{\text{products}}{C + D} \quad \text{Eq. A-4}$$

A specific example is the combustion of methane gas, CH_4, given in Equation A-5.

$$CH_4 + 2\,O_2 \rightarrow CO_2 + 2H_2O \quad \text{Eq. A-5}$$

According to this equation, one molecule of methane gas reacts with two molecules of oxygen gas to produce one molecule of carbon dioxide gas and two molecules of water vapor. Coefficients such as the 2 to the left of the O_2 and H_2O are used so that the total number of each type of atom is the same on the left and right side of the equation. Thus, in the reaction above, there is one carbon atom on the left (in one CH_4) and one on the right (in one CO_2). There are four hydrogen atoms on the left (in one CH_4) and four on the right (in two H_2O). There are four oxygen atoms on the left (in two O_2) and four on the right (in one CO_2 and two H_2O). Equations in which the atoms of each type appear in equal numbers on both sides are called **balanced equations**.

Under appropriate conditions, the products of a reaction can be changed back to the reactants. For example, carbon dioxide gas dissolves in and reacts with water to form carbonic acid, H_2CO_3:

$$CO_2 + H_2O \rightarrow H_2CO_3 \quad \text{Eq. A-6}$$

However, the carbonic acid is not very stable and as soon as some is formed, part of it decomposes to give carbon dioxide and water:

$$H_2CO_3 \rightarrow CO_2 + H_2O \quad \text{Eq. A-7}$$

Reactions such as this that go in both directions are called **reversible reactions.** They are usually represented by double arrows pointing in both directions:

$$CO_2 + H_2O \rightleftharpoons H_2CO_3 \quad \text{Eq. A-8}$$

Theoretically, every reaction is reversible. In many reactions, however, conditions create a situation in which a reaction, for all practical purposes, goes only in one direction and is called **irreversible**. For example, in order for the combustion reaction of methane (Eq. A-5) to be reversible, the gaseous CO_2 and water vapor products would have to remain in the vicinity of the reaction site. Otherwise, there is no way they could get together to react. In practice, these gaseous products leave the reaction site and have no chance to recombine. Thus, the ordinary combustion of methane is considered to be irreversible. Some other irreversible reactions are those that take place when an explosion occurs and those involved in the frying of an egg.

The rates (speeds) of chemical reactions are influenced by a number of factors, with one of the most important being catalysts. A **catalyst** is a substance that speeds up a reaction without being used up in the reaction.

Living organisms produce catalysts that are called **enzymes.** These enzymes exert amazing influence on the rates of chemical reactions that take place in the organisms. Reactions that take weeks or even months to occur under normal laboratory conditions take place in seconds under the influence of enzymes in the body. Two of the fastest acting enzymes are ***carbonic anhydrase*** and ***catalase***. Carbonic anhydrase catalyzes the reaction between carbon dioxide and water to form carbonic acid. This reaction is important in the transport of carbon dioxide from tissue cells, where it is produced metabolically, to the lungs, where it is excreted. The reaction is the same one given earlier in Equation A-6. Each molecule of carbonic anhydrase catalyzes the conversion of 36 million CO_2 molecules per minute. Catalase catalyzes the decomposition of hydrogen peroxide (H_2O_2) at a rate of 5.6 million molecules per minute:

$$2H_2O_2 \rightarrow 2H_2O + O_2 \quad \text{Eq. A-9}$$

Hydrogen peroxide is a toxic metabolic product that is formed in cells and that must be detoxified rapidly if the cells are to survive. While most other enzymes act more slowly than these two examples, they are important in essentially every chemical reaction that takes place in living organisms.

Formulas, Equations, and the Mole

Earlier, the concept of atomic weights was discussed. The idea of comparing atomic masses by using relative values is also useful in discussing molecules. Since molecules are made up of atoms, the relative mass of a molecule is simply the sum of the relative masses (atomic weights) of the atoms found in the molecule. The relative masses of molecules are called **molecular masses** or **molecular weights**. The molecular weight of water, H_2O, is thus the sum of the atomic weights of two hydrogen atoms and one oxygen atom, or 1.01 amu + 1.01 amu + 16.00 amu = 18.02 amu.

Not all compounds exist in the form of molecules. Ionically bonded substances such as sodium chloride consist of three-dimensional arrangements of sodium ions (Na^+) and chloride ions (Cl^-) in a 1:1 ratio. The formulas for ionic compounds reflect only the ratio of the ions in the compound and should not be interpreted in terms of molecules. Thus, the formula for sodium chloride, NaCl, indicates that the ions combine in a 1:1 ratio. It is convenient to apply the concept of relative masses to ionic compounds even though they do not exist as molecules. This is done by defining the **formula weight** for such compounds as the sum of the atomic weights of the atoms found in the formula. Thus, the formula weight of magnesium chloride, an ionic compound with the formula $MgCl_2$, is equal to the sum of the atomic weights of one magnesium atom and two chlorine atoms, or 24.31 amu + 35.45 amu + 35.45 amu = 95.21 amu.

As we have seen, chemical reactions can be represented by equations and discussed in terms of numbers of molecules, atoms, and ions reacting with each other. However, it is much more convenient to describe reactions in terms of amounts of reactants and products that can readily be measured, using units such as grams. This is done by using the mole concept. A **mole** of a pure element or compound is the amount of material contained in a sample of the pure substance that has a mass in grams equal to the atomic weight (for elements) or the molecular weight or formula weight (for compounds). Thus, one mole of potassium, K, would be a sample of the element with a mass of 39.10 grams. Similarly, a mole of water, H_2O, would have a mass of 18.02 g, and a mole of sodium chloride, NaCl, would be a sample with a mass of 58.44 grams.

The fact that atomic weights, molecular weights, and formula weights are relative masses leads to an interesting characteristic for moles. One mole of hydrogen atoms has a mass of 1.01 grams, and one mole of oxygen atoms has a mass of 16.00 grams. The ratio of atomic weights for the two is 16.00/1.01, the same as the ratio of the masses of one mole of each element: 16.00 g/1.01 g. This leads to the conclusion that one mole of hydrogen contains exactly the same number of hydrogen atoms as the number of oxygen atoms in one mole of oxygen. Thus, it is possible and sometimes useful to think of a mole as a specific number of particles. The number, called **Avogadro's number**, is equal to 6.02×10^{23}.

SOLUTIONS

Most chemical reactions in the body take place between reactants dissolved to form solutions. **Solutions** are homogeneous mixtures containing a relatively large amount of one substance called the **solvent** and smaller amounts of one or more substances called **solutes**. Salt water, for example, contains mostly water, which is thus the solvent, and a smaller amount of salt, which is the solute. Water is the solvent in most solutions found in the human body.

When ionic solutes are dissolved in water to form solutions, the resulting solution will conduct electricity. This is not true for most covalently bonded solutes. For example, a salt water solution conducts electricity, but a sugar water solution does not. The reason for this behavior is that when salt dissolves in water, the solid lattice of Na^+ and Cl^- is broken down and the individual ions are separated and distributed uniformly throughout the solution. It is the mobile charged ions that conduct electricity through the solution. Solutes that form ions in solution and conduct electricity are called **electrolytes**. Some very polar covalent molecules also behave this way. When sugar dissolves, individual covalently bonded sugar molecules leave the solid and become uniformly distributed throughout the solution. These uncharged molecules cannot conduct a current. Solutes that do not form conductive solutions are called **nonelectrolytes**.

The amount of solute dissolved to form a specific amount of solution can vary. For example, a salt water solution might contain 1 gram of salt in 100 milliliters (ml) of solution, or it could contain 10 grams of salt in 100 ml of solution. Both solutions are salt water solutions, but they have different concentrations of solute. The **concentration** of a solution indicates the relationship between the amount of solute and the amount of solvent or the amount of solution. Concentrations are given in a number of different units.

Concentrations given in terms of **molarity** (abbreviated M) give the number of moles of solute in exactly one liter of solution. Thus, a half molar (0.50 M) solution of NaCl would contain one-half mole or 29.22 g of NaCl in each liter of solution.

Concentrations given in terms of **molality** (abbreviated m) give the number of moles of solute dissolved in 1000 grams (1.00 kg) of solvent. Thus, a half molal (0.50 m) solution of NaCl would contain NaCl and water in the ratio of 29.22 g NaCl to each 1000 g (or 1000 ml) of water.

Note the difference in molarity and molality. Molarity gives the amount of solute in a specific amount of solution, whereas molality gives the amount of solute dissolved in a specific amount of solvent.

When the solute of a solution is an electrolyte, it is sometimes useful to express concentration in a unit that gives information about the amount of ionic charge in the solution. This is done by expressing concentration in terms of **normality** (abbreviated N). The normality of a solution gives the number of equivalents of solute in exactly one liter of solution. An **equivalent** of an electrolyte is the amount that produces one mole of positive (or negative) charges when it dissolves. The number of equivalents of an electrolyte can be calculated by multiplying the number of moles of electrolyte by the total number of positive charges produced when one formula unit of the electrolyte dissolves. Consider sodium chloride (NaCl) and calcium chloride ($CaCl_2$) as examples. The ionization reactions for one formula unit of each solute are:

$$NaCl \rightarrow Na^+ + Cl^- \qquad \text{Eq. A–10}$$

$$MgCl_2 \rightarrow Mg^{++}\ 2Cl^- \qquad \text{Eq. A–11}$$

Thus, one mole of NaCl produces one mole of positive charges (Na^+) and so contains one equivalent:

$$(1 \text{ mole NaCl})(1) = 1 \text{ equivalent}$$

where the number 1 used to multiply the 1 mole of NaCl came from the +1 charge on Na^+.

One mole of $CaCl_2$ produces one mole of Ca^{++}, which is two moles of positive charge. Thus, one mole of $CaCl_2$ contains two equivalents:

$$(1 \text{ mole } CaCl_2)(2) = 2 \text{ equivalents}$$

where the number 2 used in the multiplication came from the +2 charge on Ca^{++}.

If two solutions were made such that one contained one mole of NaCl per liter and the other contained one mole of $CaCl_2$ per liter, the NaCl solution would contain one equivalent of solute per liter and would be one normal (1 N). The $CaCl_2$ solution would contain two equivalents of solute per liter and would be two normal (2 N).

Suspensions and Colloids

Suspensions, like solutions, consist of two or more components, with much more of one component present than the others. In solutions, the component present in largest amount is called the solvent. In suspensions, it is called the **dispersing medium**. In solutions, the components present in smaller amounts are called solutes; in suspensions they are called **dispersed phases**. An important difference between solutions and suspensions is the size of the particles dissolved or dispersed. In solutions, solute particles are present as ions or small molecules. Dispersed phase particles are much larger than ions or small molecules. When the dispersed phase particles are no more than about a hundred times the size of the largest solution solute particles, the suspension is called a **colloid** or **colloidal dispersion**. The dispersed phase particles of colloids generally do not settle out. They are small enough to be kept in suspension by the constant buffeting they receive from the motion of dispersing medium molecules. All dispersed phase particles of colloids carry electrical charges of the same sign. Thus, they repel each other and do not get together to form larger particles that would settle in spite of collisions with dispersing phase molecules. When dispersed phase particles are larger than those in colloids, the dispersed phase will settle out. Such mixtures are usually called suspensions.

The relatively large size of dispersed phase particles in colloids or suspensions causes them to scatter light. As a result, colloids and suspensions appear cloudy. Solutions are clear because the dissolved solute particles are too small to scatter light. In the body, the dispersing medium of most colloids is water, and the colloids are liquids. However, colloids can occur in all three states, depending on the state of the dispersing medium.

Inorganic and Organic Chemicals

It is a common practice to classify chemicals into the two categories of inorganic and organic. The original criterion used to classify chemicals into the categories was the origin of the chemicals. Those that came from living or once-living sources were **organic**, and those that came from other sources were **inorganic**. Today, the basis for classification is the element carbon. Organic chemicals are generally those that contain carbon. All others are classified as inorganic. There are a few carbon-containing chemicals classified as inorganic. The most common are pure carbon in the form of diamond and graphite, carbon dioxide (CO_2), carbon monoxide (CO), carbonates such as limestone ($CaCO_3$), bicarbonates such as baking soda ($NaHCO_3$), and cyanides such as sodium cyanide (NaCN).

The unique ability of carbon atoms to bond to each other to form networks of carbon atoms results in an interesting fact. Even though organic chemicals include those that contain only one specific element, carbon, millions of these compounds have been identified. Some were isolated from natural plant or animal sources, and many have been synthesized in laboratories. Inorganic chemicals include all the other one hundred and eight elements and their compounds, but the number of known inorganic chemicals is estimated to be about two hundred and fifty thousand.

Another result of carbon's ability to bond to itself is the large size of some organic molecules. Molecules classified as organic range in size from methane, CH_4, a small and simple one-carbon-atom molecule, to molecules such as DNA that contain as many as a million carbon atoms. Large molecules such as this are often called **macromolecules**, which include many such as DNA that occur naturally, and many that are synthetically produced. In this latter category are numerous substances that are widely used today, including synthetic textiles (e.g., nylon, dacron, and orlon) and plastics (e.g., lucite, plexiglass, and teflon). Another general name given to these materials is **polymer**, which means "many units," reflecting the fact that these large macromolecules can usually be chemically changed into many smaller molecules. That is, polymeric macromolecules are made by bonding together a large number of smaller molecules.

Acids, Bases, and Salts

Acids, bases, and salts are among the most common and important compounds studied in chemistry. Both inorganic and organic compounds are known that fit these three categories. Until late in the nineteenth century, these substances were classified on the basis of such properties as taste or by the color changes induced in certain dyes. Acids taste sour, bases bitter, and salts salty. Litmus, a dye, is red in the presence of acids and blue in the presence of bases. These and other observations led to the correct conclusions that acids and bases are chemical opposites and that salts are produced when acids and bases react with each other. Today, acids and bases are defined in more precise ways.

In 1887, Swedish chemist Svante Arrhenius proposed a theory in which acids and bases were defined. He said that an **acid** is any substance that will dissociate or break apart when dissolved in water and in the process release a hydrogen ion, H^+. Similarly, **bases** are substances that dissociate when dissolved in water and in the process release a hydroxide ion, OH^-. Hydrogen chloride (HCl) and sodium hydroxide (NaOH) are examples of Arrhenius acids and bases as represented in Equations A–12 and A–13.

$$HCl \rightarrow H^+ + Cl^- \qquad \text{Eq. A–12}$$

$$NaOH \rightarrow Na^+ + OH^- \qquad \text{Eq. A–13}$$

Note that the hydrogen ion is a bare proton, the nucleus of a hydrogen atom. Also note that both HCl and NaOH would behave as electrolytes.

Arrhenius did not know that free hydrogen ions cannot exist in water. It is now believed that they covalently bond to a water molecule to form a hydronium ion as shown in Equation A–14.

$$H^+ + \ddot{O}(H)\text{—}H \rightarrow \left[H\text{—}\ddot{O}(H)\text{—}H \right]^+ \qquad \text{Eq. A–14}$$

In 1923, Johannes Bronsted, in Denmark, and Thomas Lowry, in England, proposed an acid-base theory that took this behavior into account. They defined an acid as any hydrogen-containing substance that donates a proton (hydrogen ion) to another substance, and a base as any substance that accepts a proton. According to these definitions, the acidic behavior of HCl given in Equation A–12 is written as shown in Equation A–15.

$$HCl + H_2O \rightleftarrows H_3O^+ + Cl \qquad \text{Eq. A–15}$$

Note that this reaction is shown to be reversible, and the hydronium ion is represented as H_3O^+.

In Equation A–15, the HCl acts as an acid in the forward (left-to-right) reaction while water acts as a base. In the reverse reaction (right-to-left), the hydronium ion gives up a proton and thus is an acid, while the chloride ion, Cl^-, accepts the proton and so is a base. It is still a common practice to use equations such as A–12 to simplify the representation of the dissociation of an acid even though it is recognized that equations like A–15 are more correct.

At room temperature, **inorganic salts** are crystalline solids that contain the positive ion (cation) of an Arrhenius base such as NaOH and the negative ion (anion) of an acid such as HCl. Salts can be produced by mixing solutions of appropriate acids and bases, allowing a neutralization reaction to occur. In **neutralization reactions**, the acid and base react to form a salt and water. Most salts that form are water soluble and can be recovered by evaporating the water away. Equations A–16 and A–17 are neutralization reactions.

$$HCl + NaOH \rightarrow NaCl + H_2O \qquad \text{Eq. A–16}$$

$$H_2SO_4 + Cu(OH)_2 \rightarrow CuSO_4 + 2H_2O \qquad \text{Eq. A–17}$$

When acids or bases are used as solutes in solutions, the concentrations can be expressed as normalities just as they were earlier for salts. An equivalent of acid is the amount that gives up one mole of H^+ in solution. Thus, one mole of HCl is also one equivalent, but one mole of H_2SO_4 is two equivalents. Bases are described in a similar way, but an equivalent is the amount of base that gives one mole of OH^-.

FUNCTIONAL GROUPS OF ORGANIC MOLECULES

The study of organic compounds is simplified by the concept of functional groups. All organic compounds can be classified according to the functional group or groups they contain. **Functional groups** are specific combinations of atoms that generally behave the same way in reactions, regardless of the number of carbon atoms in the molecule to which they are attached. For example, all aldehydes contain a functional group that contains one carbon atom, one oxygen atom, and one hydrogen atom covalently bonded in a specific way $(-\overset{\overset{\displaystyle O}{\|}}{C}-H)$. The aldehyde group is attached to the rest of the molecule by the covalent bond extending to the left of the carbon atom. Most reactions of aldehydes involve this group, so most aldehyde reactions are the same regardless of the size and nature of the rest of the molecule to which the aldehyde group is attached. Reactions of physiological importance are often between two functional groups or between one functional group and small molecules such as water.

CARBOHYDRATES

Carbohydrates are organic compounds of tremendous biological and commercial importance. They are widely distributed in nature and include such familiar substances as cellulose, starch, and table sugar. Carbohydrates have four important functions in living organisms. They provide energy, they supply carbon atoms for the synthesis of cell components, they serve as a stored form of chemical energy, and they form a part of the structural elements of some cells.

Carbohydrates contain carbon, hydrogen, and oxygen. Their name comes from the fact that most of them contain these three elements in an atomic ratio of one carbon to two hydrogens to one oxygen. This ratio suggests that the general formula is CH_2O and that the compounds are simply carbon hydrates or carbohydrates. It is now known that they are not hydrates of carbon, but the name persists. All carbohydrates have a large number of functional groups per molecule. The most common functional groups in carbohydrates are alcohol (—OH), ketone $(-\overset{\overset{\displaystyle O}{\|}}{C}-)$, aldehyde $(-\overset{\overset{\displaystyle O}{\|}}{C}H)$, or functional groups formed by reactions between pairs of these three.

The simplest carbohydrates are simple sugars, also called **monosaccharides**. As their name indicates, they consist of single (*mono* means "one") units called saccharides. The molecular structure of glucose, an important monosaccharide, is shown in Figure A–3a. In solution, most glucose molecules assume the ring form shown in Figure A–3b. Other common monosaccharides are fructose, galactose, and ribose.

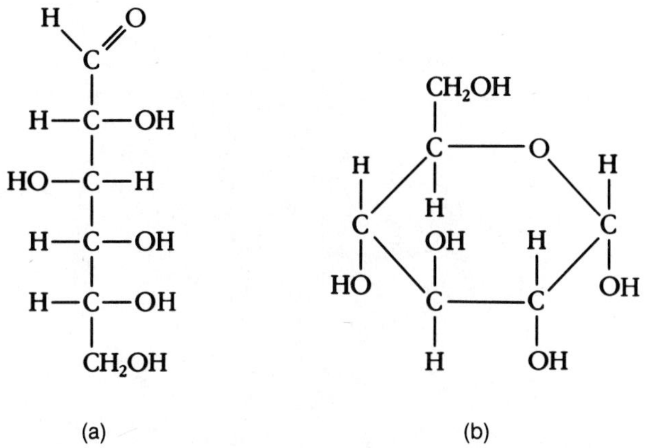

(a) (b)

Figure A–3 Forms of Glucose *(a) Chain. (b) Ring.*

SOURCE: Adapted by permission from page 520 of *Chemistry for Today: General, Organic, and Biochemistry* by Spencer L. Seager and Michael R. Slabaugh. Copyright © 1987 by West Publishing Company. All rights reserved.

Disaccharides (*di* means "two") are sugars formed by a reaction between two monosaccharide molecules. Some common examples of disaccharides are sucrose (common table sugar) and lactose (milk sugar). Sucrose molecules are formed from one glucose and one fructose molecule. Lactose molecules each contain one galactose and one glucose unit.

The large number of functional groups on carbohydrate molecules makes it possible for large numbers of simple carbohydrate molecules to bond together and form long chains and branched networks. These substances are called **polysaccharides** (*poly* means "many"), a name that indicates they contain many saccharide units. Three common polysaccharides that are made up entirely of glucose units are glycogen, starch, and cellulose.

Glycogen is a storage carbohydrate found in animals. It is a highly branched polysaccharide that averages a branch every eight to twelve glucose units. The structure of glycogen is represented in Figure A–4, where each circle represents one glucose unit.

Starch, a storage carbohydrate of plants, consists of two fractions, amylose and amylopectin. Amylose consists of long, essentially unbranched chains of glucose units. Amylopectin is a highly branched network of glucose units averaging twenty-four to thirty glucose units per branch. Thus, it is less highly branched than glycogen. Cellulose, a structural carbohydrate

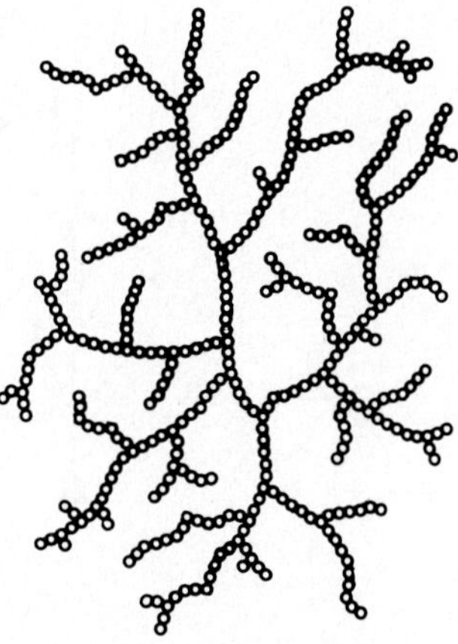

Figure A–4 A Simplified Representation of Glycogen

SOURCE: Adapted by permission from page 532 of *Chemistry for Today: General, Organic, and Biochemistry* by Spencer L. Seager and Michael R. Slabaugh.

of plants, exists in the form of long, unbranched chains of glucose units.

The bonding between glucose units of cellulose is slightly different than the bonding between glucose units of glycogen and starch. Humans have digestive enzymes that catalyze the breaking (hydrolysis) of the glucose-to-glucose bonds in starch but lack the necessary enzymes to hydrolyze cellulose glucose-to-glucose bonds. Thus, starch is a food for humans, but cellulose is not.

Lipids

The group of compounds called lipids is made up of substances with widely different compositions and molecular structures. Unlike carbohydrates, which are classified on the basis of their molecular structure, substances are classified as lipids on the basis of their solubility. **Lipids** are compounds that are insoluble in water but soluble in nonpolar solvents. Thus, lipids are the waxy, greasy, or oily compounds found in plants and animals. Lipids repel water, a useful characteristic of protective wax coatings found on some plants. Fats and oils are energy-rich and have relatively low densities. These properties account for their use as storage forms of energy in plants and animals. Still other lipids occur as structural components, especially in cellular membranes.

Simple lipids contain just two types of components, fatty acids and alcohols. **Fatty acid molecules** consist of a hydrocarbon chain with a carboxylic acid functional group (—COOH) on the end. The hydrocarbon chain can be of variable length, but natural fatty acids always contain an even number of carbon atoms. The hydrocarbon chain can also contain one or more double bonds between carbon atoms. (A double bond is one in which two covalent bonds are formed between the same atoms.) Fatty acids with no double bonds are called **saturated fatty acids**, whereas those with double bonds are called **unsaturated fatty acids**. The more double bonds present, the higher is the degree of unsaturation. The most common alcohol found in simple lipids is **glycerol** (glycerin), a three-carbon alcohol that has three alcohol functional groups (—OH).

Simple lipids called fats and oils are formed by a reaction between the carboxylic acid group of three fatty acids and the three alcohol groups of glycerol. The resulting lipid is called a **triglyceride** or **triacylglycerol**. Such lipids are classified as fats or oils on the basis of their melting points. Fats are solids at room temperature whereas oils are liquids. The melting points depend on the degree of unsaturation in the fatty acids of the triglyceride. The melting point goes down with increasing degree of unsaturation. Thus, oils contain more unsaturated fatty acids than do fats. Examples of the components of fats and oils and a typical triglyceride molecule are shown in Figure A–5.

When triglycerides form, a molecule of water is released as each fatty acid reacts with glycerol. Adipose tissue in the body contains triglycerides. When adipose tissue is used by the body as an energy source, the triglycerides react with water to release free fatty acids into the blood. The fatty acids can be used as an immediate energy source by many organs. In the liver free fatty acids are converted into compounds called **ketone bodies**. Two of the ketone bodies are acids and one is the ketone called acetone.

Complex lipids contain more than two types of components. The different complex lipids usually contain three or more of the following components: glycerol, fatty acids, phosphoric acid, an alcohol other than glycerol, and a carbohydrate. Those that contain phosphoric acid are called **phospholipids**. Figure A–6 contains representations of a few complex lipids that emphasize the components but do not give details of the molecular structures.

Steroids are lipids that have a unique structural feature made up of a fused carbon ring system containing three six-membered rings and a single five-membered ring (Fig. A–7). Different steroids possess this characteristic ring structure but have different functional groups and carbon chains attached.

Cholesterol, a steroidal alcohol, is the most abundant steroid in the human body. It is a component of cell membranes and is used by the body to produce other important steroids that include bile salts, male and female sex hormones, and ad-

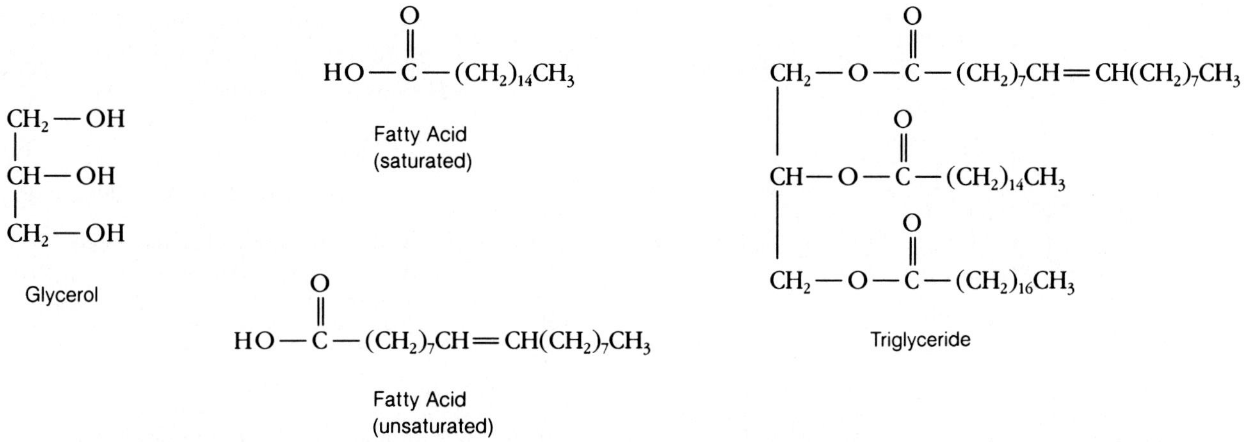

$$\begin{array}{l} CH_2-OH \\ | \\ CH-OH \\ | \\ CH_2-OH \end{array}$$

Glycerol

$$HO-\overset{\overset{\displaystyle O}{\|}}{C}-(CH_2)_{14}CH_3$$

Fatty Acid
(saturated)

$$HO-\overset{\overset{\displaystyle O}{\|}}{C}-(CH_2)_7CH{=}CH(CH_2)_7CH_3$$

Fatty Acid
(unsaturated)

$$\begin{array}{l} CH_2-O-\overset{\overset{\displaystyle O}{\|}}{C}-(CH_2)_7CH{=}CH(CH_2)_7CH_3 \\ | \\ CH-O-\overset{\overset{\displaystyle O}{\|}}{C}-(CH_2)_{14}CH_3 \\ | \\ CH_2-O-\overset{\overset{\displaystyle O}{\|}}{C}-(CH_2)_{16}CH_3 \end{array}$$

Triglyceride

Figure A-5 Triglyceride Components and Structure

SOURCE: Adapted by permission from pages 547 and 548 of *Chemistry for Today: General, Organic, and Biochemistry* by Spencer L. Seager and Michael R. Slabaugh.

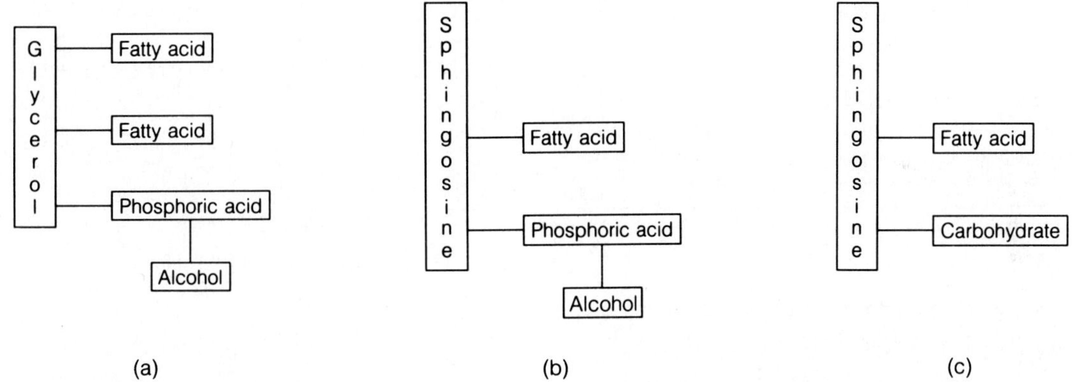

Figure A-6 Examples of Complex Lipids *(a) A phosphoglyceride. (b) A sphingolipid (sphingosine is an alcohol). (c) A glycolipid.*

SOURCE: Adapted by permission from pages 553 and 555 of *Chemistry for Today: General, Organic, and Biochemistry* by Spencer L. Seager and Michael R. Slabaugh.

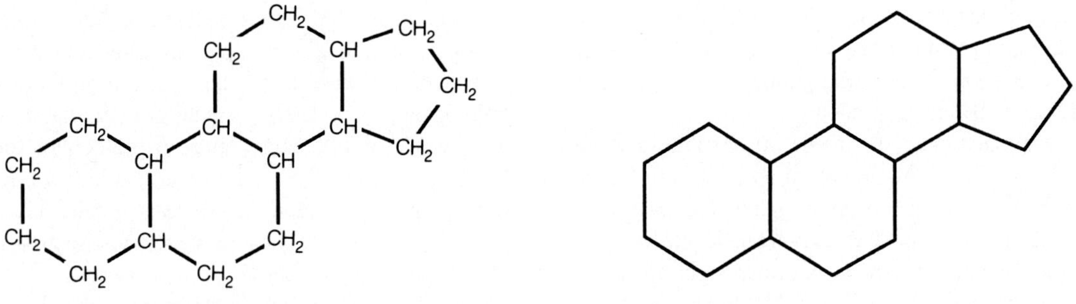

(a) (b)

Figure A-7 The Steroid Ring System *(a) Detailed. (b) Simplified.*

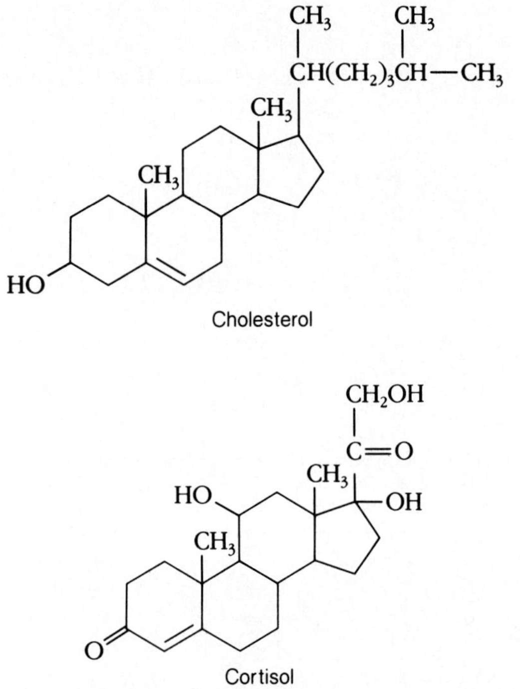

Cholesterol

Cortisol

Figure A–8 Examples of Steroidal Compounds

SOURCE: Adapted by permission from pages 560 and 563 of *Chemistry for Today: General, Organic, and Biochemistry* by Spencer L. Seager and Michael R. Slabaugh. Copyright © 1987 by West Publishing Company. All rights reserved.

renocortical hormones. The structures of cholesterol and cortisol, an important adrenocortical hormone, are given in Figure A–8.

PROTEINS

The name **protein** is derived from the Greek word ***proteios***, which means "of first importance." It is certainly appropriate for these very important biological compounds. Proteins are indispensable components of all living things, where they play crucial roles in all biological processes.

Proteins are macromolecules made up of subunits called **amino acids**. Hundreds of different amino acids, both natural and synthetic, are known, but only twenty are commonly found in natural proteins. Each amino acid molecule has three important parts: an amino functional group (—NH_2), a carboxylic acid functional group (—COOH), and a characteristic side chain or R group. These parts are shown in Figure A-9.

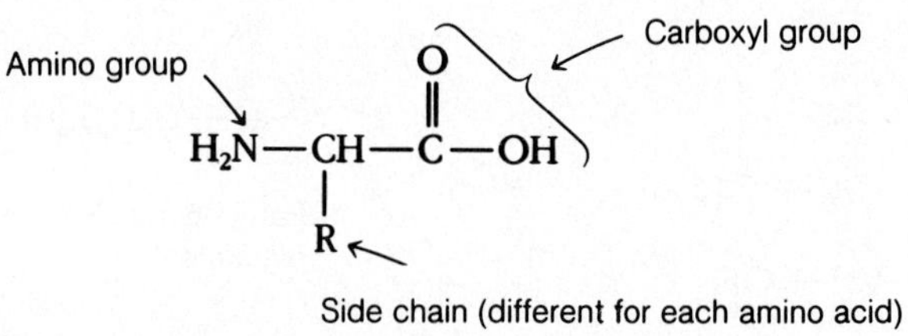

Amino group

Carboxyl group

Side chain (different for each amino acid)

Figure A–9 The General Structure of Amino Acids

SOURCE: Adapted by permission from page 574 of *Chemistry for Today: General, Organic, and Biochemistry* by Spencer L. Seager and Michael R. Slabaugh. Copyright © 1987 by West Publishing Company. All rights reserved.

Amino acids form long chains as a result of reactions between the amino group of one amino acid and the carboxylic acid group of another amino acid. This is illustrated for two amino acids by Equation A–18 in which the carboxylic acid group is shown in an expanded form for clarity.

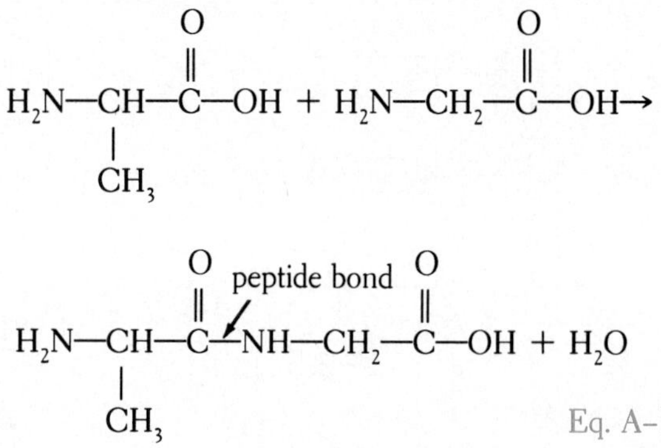

$$H_2N-\underset{\displaystyle CH_3}{\underset{|}{CH}}-\overset{\displaystyle O}{\overset{\|}{C}}-OH + H_2N-CH_2-\overset{\displaystyle O}{\overset{\|}{C}}-OH \rightarrow$$

$$H_2N-\underset{\displaystyle CH_3}{\underset{|}{CH}}-\overset{\displaystyle O}{\overset{\|}{C}}-NH-CH_2-\overset{\displaystyle O}{\overset{\|}{C}}-OH + H_2O \qquad \text{Eq. A–18}$$

peptide bond

Notice that after the two molecules react, the ends of the product still have an amino group and a carboxylic acid group that can react to extend the chain length. The covalent bond formed in the reaction is called a **peptide bond** or peptide linkage.

On a molecular scale, proteins are immense molecules. This can be illustrated by comparing a glucose molecule to a molecule of hemoglobin, a protein. Glucose has a molecular weight of 180 amu and a molecular formula of $C_6H_{12}O_6$. Hemoglobin, a relatively small protein, has a molecular weight of 65,000 amu and a molecular formula of $C_{2952}H_{4664}O_{832}N_{812}S_8Fe_4$. The many atoms in a protein are not arranged in a random way. In fact, proteins have a high degree of structural organization that plays an important role in their behavior in the body.

The first level of protein structure is called the **primary structure**. It is simply the order in which amino acids are bonded together to form the protein chain. It is a common

Thr—Lys—Pro—Thr—Tyr—Phe—Phe—Gly—Arg— · · · · · ·

Figure A–10 A Portion of the Primary Protein Structure of Human Insulin

SOURCE: Adapted by permission from page 589 of *Chemistry for Today: General, Organic, and Biochemistry* by Spencer L. Seager and Michael R. Slabaugh.

practice to represent amino acids by three-letter abbreviations such as Gly for glycine and Arg for arginine. When this practice is followed, the primary structure of a protein is represented as shown in Figure A–10 for a part of the primary structure of human insulin.

The second level of protein structure, called the **secondary structure**, results when hydrogen bonding occurs between the amino hydrogen of one amino acid in the primary chain and the carboxyl oxygen ($-\overset{\overset{\displaystyle O}{\|}}{C}-$) of another amino acid in the same or another chain. When the hydrogen bonding occurs between amino acids in the same chain, the result is to cause the chain to assume a coiled helical shape called the alpha (α) helix, which is by far the most common secondary structure found in natural proteins (Fig. A–11).

The third level of structure in proteins is the **tertiary structure**. It results when functional groups of the side chains of amino acids in the protein chain react with each other. Several different types of interactions are possible as shown in Figure A–12. Tertiary structures can be visualized by letting a length of wire represent the chain of amino acids in the primary structure of a protein. Next, imagine the wire is wound around a pencil to form a helix, which represents the secondary structure. The pencil is removed, and the helical structure is now folded back on itself or carefully wadded into a ball. The folded or spherical structures represent the tertiary structure of a protein.

One of the important functions of proteins is that of enzymes that catalyze the many essential chemical reactions of the body. In addition to catalyzing reactions, proteins can undergo reactions themselves. Two of the most important are hydrolysis and denaturation. Notice that according to Equation A–18, the formation of peptide bonds releases water molecules. Under appropriate conditions, it is possible to reverse such reactions by adding water to the peptide bonds and breaking them. **Hydrolysis** reactions of this type convert large proteins into smaller fragments or even into individual amino acids. **Denaturation** of proteins occurs when the bonds holding a protein chain in its characteristic tertiarty or secondary conformation are broken. When this happens, the protein chain takes on a random, disorganized conformation. Denaturation can result when proteins are subjected to heating, treatment with specific chemicals such as alcohol or heavy metal ions, or extremes of pH. In some instances, denaturation is accompanied by coagulation or precipitation. This is illustrated by the changes that occur in the white of an egg as it is fried.

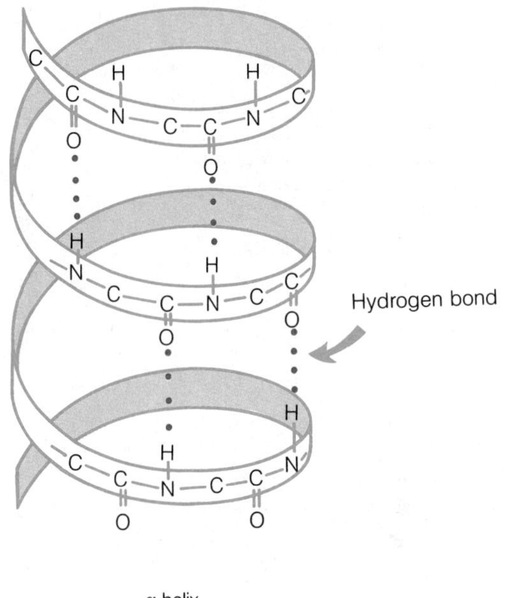

Figure A–11 Secondary Structures of Proteins

SOURCE: Adapted by permission from page 591 of *Chemistry for Today: General, Organic, and Biochemistry* by Spencer L. Seager and Michael R. Slabaugh.

Nucleic Acids

High molecular weight macromolecules called **nucleic acids** are the compounds that enable genetic information to be stored in living cells and passed on to future generations. These important biomolecules are classified into two categories, **ribonucleic acids (RNA)** and **deoxyribonucleic acids (DNA)**. Deoxyribonucleic acids are found primarily in the nuclei of cells and ribonucleic acids are found primarily in the cytoplasm that surrounds cell nuclei.

Both types of nucleic acid are made up of units called **nucleotides** which, in turn, are composed of three simpler components. Each nucleotide contains an organic base, a sugar, and phosphoric acid. The three components are chemically bonded together with the sugar molecule lying between the

Figure A–12 Side Chain Interactions Leading to Tertiary Protein Structure

SOURCE: Adapted by permission from page 595 of *Chemistry for Today: General, Organic, and Biochemistry* by Spencer L. Seager and Michael R. Slabaugh.

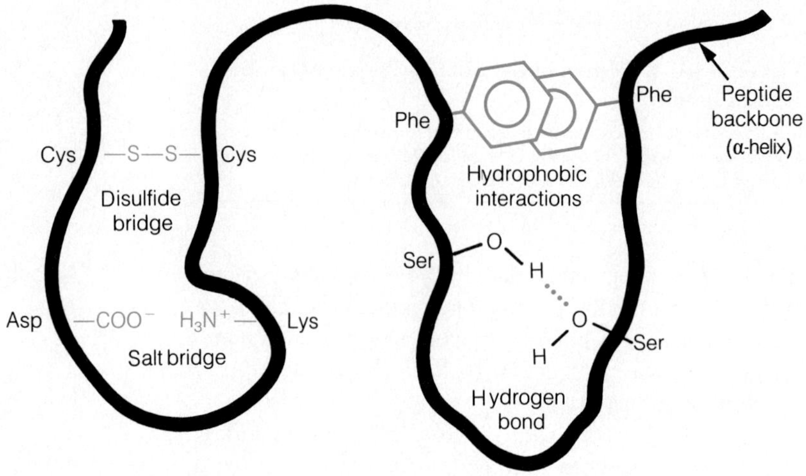

base and the phosphoric acid. In RNA, the sugar is ribose, whereas in DNA it is deoxyribose. When nucleotides bond together to form nucleic acid chains, the bonding is between the phosphoric acid of one nucleotide and the sugar of another. Thus, the resulting nucleic acids consist of chains of alternating phosphoric acid and sugar molecules, with a base molecule extending out of the chain from each sugar molecule.

The chains of nucleic acid assume structural features somewhat like those found in proteins. DNA occurs in the form of two chains that mutually coil around one another to form the well-known double helix. Some RNA occurs in essentially straight chains, while in other types the chain forms specific loops or helices.

High-Energy Biomolecules

Certain molecules in the body store energy that is released during the metabolism of foods and make it available to the parts of the cells where it is needed to do specific cellular work. The primary substance that performs this function is **adenosine triphosphate**, or **ATP**. The structure of this molecule is shown in Figure A–13. Energy is stored in the phosphate bonds of this molecule and is released to the cells when a phosphate bond reacts with water to form adenosine diphosphate (ADP) and phosphoric acid (Eq. A–19).

Figure A–13 The Structure of ATP

SOURCE: Adapted by permission from page 687 of *Chemistry for Today: General, Organic, and Biochemistry* by Spencer L. Seager and Michael R. Slabaugh.

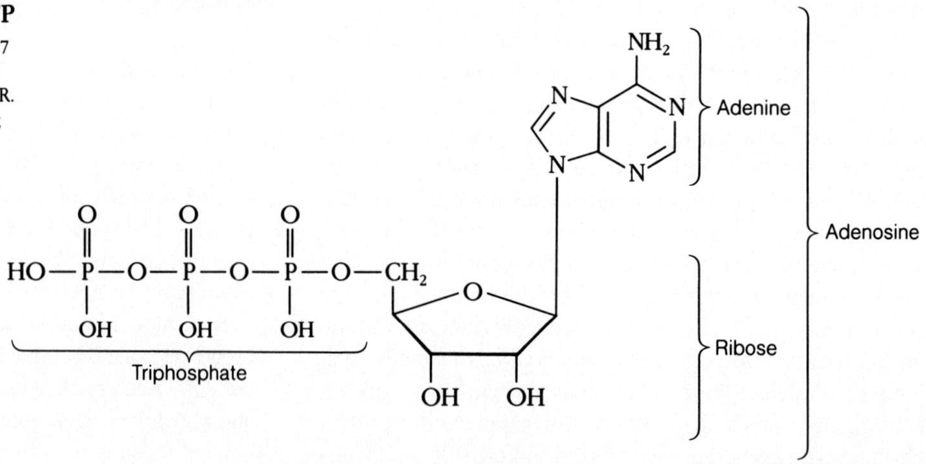

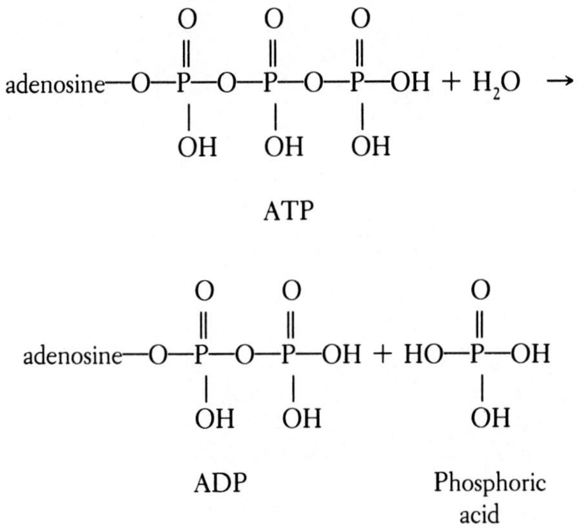

$$\text{adenosine}-O-\overset{\overset{\displaystyle O}{\|}}{\underset{\underset{\displaystyle OH}{|}}{P}}-O-\overset{\overset{\displaystyle O}{\|}}{\underset{\underset{\displaystyle OH}{|}}{P}}-O-\overset{\overset{\displaystyle O}{\|}}{\underset{\underset{\displaystyle OH}{|}}{P}}-OH + H_2O \rightarrow$$

ATP

$$\text{adenosine}-O-\overset{\overset{\displaystyle O}{\|}}{\underset{\underset{\displaystyle OH}{|}}{P}}-O-\overset{\overset{\displaystyle O}{\|}}{\underset{\underset{\displaystyle OH}{|}}{P}}-OH + HO-\overset{\overset{\displaystyle O}{\|}}{\underset{\underset{\displaystyle OH}{|}}{P}}-OH$$

ADP — Phosphoric acid

Eq. A–19

More energy is released when the second phosphate reacts with water in the same way and ADP is converted to adenosine monophosphate (AMP).

Under the influence of an enzyme, AMP can be converted to a cyclic form called **cyclic AMP** or **cAMP**, which affects the activities of a number of enzymes that are involved in important reactions in the body. The formation of cyclic AMP is represented in Equation A–20.

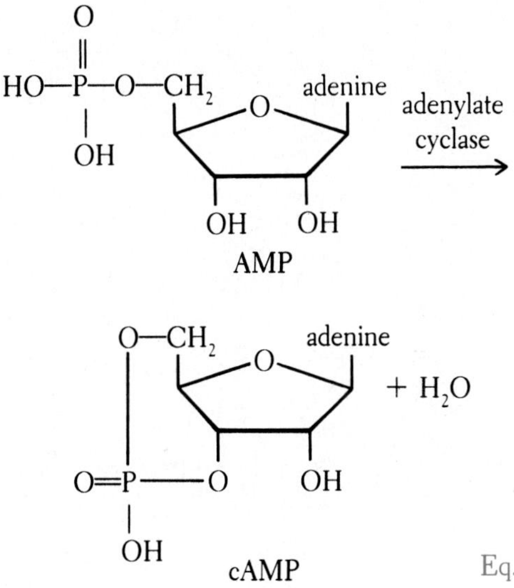

$$HO-\overset{\overset{\displaystyle O}{\|}}{\underset{\underset{\displaystyle OH}{|}}{P}}-O-CH_2-\text{(ribose ring: O, adenine; OH, OH)} \xrightarrow{\text{adenylate cyclase}}$$

AMP

$$O=\underset{\underset{\displaystyle OH}{|}}{P}\text{(–O–}CH_2\text{, –O– ring; adenine; OH)} + H_2O$$

cAMP

Eq. A–20

APPENDIX B

Control, Functions, and Interactions of Immune-Response Effectors

Immune Response Effectors	Description	Controlling Mechanism(s)	Function(s)
Acute-phase proteins[a]	Collection of proteins from the liver	Stimulated by LEM	Exert multitude of wide-ranging effects associated with inflammation, tissue repair, and immune-cell activity
Angry macrophages[b]	Large tissue-bound phagocytic cells	Stimulated by MIF	Display superpower phagocytic ability
Antibodies[a]	Special plasma proteins secreted by plasma cells; same as gamma globulins or immuno-globulins	Production and release stimulated by presence of "matching" antigen, which might be any large, complex molecule that triggers an immune attack against itself	Amplify nonspecific immune responses against the antigen that stimulated their production In some instances, physically hinder antigen by neutralization or agglutination
Autoantibodies[a]	Antibody proteins against self-antigens (own body proteins)	Breakdown of mechanisms for tolerance of self-antigens	Attack one or more of body's own tissues
B lymphocytes (B cells) (see also Plasma cells)[b]	White blood cells	Activated by macrophage-presented antigen; stimulated by helper T cells, IL-l, interferon, B-cell growth factor, and helper T cells	Converted into plasma cells, which produce antibodies Secrete lymphokines
Basophils[b]	White blood cells	Binding of allergen to IgE antibodies coating the basophils	Release histamine; involved in immediate allergic reactions
B-cell growth factor[a]	Lymphokine secreted by helper T cells	Release stimulated by macrophage-presented antigen and IL-l	Enhances antibody secretion by activated B-cell clone
Blocking antibodies[a]	Antibody proteins	Production induced by cancer cells	Protect harmful cancer cells from attack by cytotoxic T cells

[a] Chemical mediator
[b] Effector cell

Appendix B Control, Functions, and Interactions of Immune-Response Effectors (continued)

Immune Response Effectors	Description	Controlling Mechanism(s)	Function(s)
Complement system[a]	Group of inactive plasma proteins that can be sequentially activated	Activated by antibody (classical pathway) and by carbohydrate chains on surface of foreign invaders (alternate pathway); several steps stimulated by kinins	Destroys foreign cells by punching holes in their plasma membranes Facilitates every step of inflammatory process (e.g., various components serve as opsonins, stimulate secretion of histamine, or activate kinins)
Cortisol[a]	Steroid hormone secreted by adrenal cortex	Stimulated by stress	Anti-inflammatory action at pharmacological levels and possibly at physiological levels, the latter of which may prevent stress-activated immune responses from overshooting
Chemotaxins[a]	Local chemical mediators	Released from a variety of sources at site of inflammation	Attract phagocytes to site of inflammation
Cytotoxic T cells[b]	Form of T lymphocytes	Stimulated by interferon, IL-l, and helper T cells; activated by macrophage-presented antigen	Destroy virally infected cells or cancer cells Secrete interferon Mediate delayed hypersensitivity
Endogenous pyrogen (EP)[a]	Chemical mediator released by monocytes/ macrophages; identical to LEM and IL-l	Released in presence of microbes	Causes development of fever
Eosinophil chemotactic factor[a]	Chemical mediator in allergic responses	Released by mast cells or basophils activated by allergen	Attracts eosinophils to site of allergen contact
Eosinophils[b]	White blood cells	Stimulated by eosinophil chemotactic factor and presence of internal parasites	Involved in allergic manifestations Destroy parasitic worms
Fibrinogen (inactive); fibrin (active)[a]	Final plasma protein in clotting cascade	Activated by tissue thromboplastin in injured area and chemical mediators released from microbe-stimulated phagocytes	Forms interstitial fluid clots to wall off injured or invaded area
Fibroblasts[b]	Connective-tissue cells	Start dividing in area vacated by lost tissue cells	Form scar tissue
Gamma globulin (see Antibodies)[a]			
Helper T cells[b]	Form of T lymphocyte	Activated by macrophage-presented antigen; stimulated by IL-l	Enhance activity of B cells, cytotoxic T cells, and suppressor T cells through secretion of B-cell growth factor and T-cell growth factor (IL-2) Activate macrophages

[a] Chemical mediator
[b] Effector cell

Appendix B Control, Functions, and Interactions of Immune-Response Effectors (continued)

Immune Response Effectors	Description	Controlling Mechanism(s)	Function(s)
Histamine[a]	Chemical mediator released from mast cells and basophils	Stimulated by chemical mediators released from microbe-activated phagocytes; complement system; and allergens	Important in inflammation Important in immediate allergic reactions Acts on microcirculation to cause localized vasodilation and increased capillary permeability
Interferon[a]	Family of proteins	Released from cells invaded by virus	Protects noninvaded cells from viral invasion Increases killing ability of cytotoxic T cells, NK cells, and phagocytes Stimulates production of antibodies Slows cell division and suppresses tumor growth; exerts anticancer effect
Interleukin 1 (IL-1)[a]	Chemical mediator released by macrophages; identical to EP and LEM	Released in presence of microbes	Enhances proliferation and differentiation of both B and T lymphocytes
Interleukin 2 (IL-2)[a]	Lymphokine secreted by helper T cells; identical to T-cell growth factor	Released by presence of macrophage-presented antigen; stimulated by IL-1	Augments activity of all T cells
Immunoglobulin (*see* Antibodies)[a]			
Kallikrein[a]	Chemical mediator released from neutrophils	Released in presence of microbes	Activates kinins
Killer (K) cells[b]	Lymphocytelike cells	Stimulated by antibody on surface of target cell	Lyse plasma membrane of target cell
Kininogen (inactive); kinin (active)[a]	Plasma protein	Stimulated by kallikrein released from neutrophils and by complement component	Stimulates complement system Enhances vasodilation and increases capillary permeability in inflammation Activates nearby pain receptors Acts as powerful chemotaxin
Lactoferrin[a]	Chemical mediator secreted by neutrophils	Stimulated by presence of microbes	Binds iron so it is unavailable for essential use by invading bacteria
Leukocyte endogenous mediator (LEM)[a]	Chemical mediator released by monocytes/macrophages; identical to EP and IL-1	Released in presence of microbes	Decreases plasma iron Stimulates granulopoiesis Increases release of acute-phase proteins

Appendix B Control, Functions, and Interactions of Immune-Response Effectors (continued)

Immune Response Effectors	Description	Controlling Mechanism(s)	Function(s)
Lymphocytes[b]	White blood cells–two types: B and T lymphocytes	Stimulated by products specific for each type of cell (see specific cells)	Responsible for specific immune defense
Lymphokines[a]	Chemical mediators, other than antibodies, secreted by lymphocytes (including B- and T-cell growth factor and MIF)	Released by presence of macrophage-presented antigen	Enhance all activity of B and T cells Act as chemotaxins Retain macrophages in inflamed area
Macrophage migration inhibition factor (MIF)[a]	Lymphokine secreted by helper T cells	Released by presence of macrophage-presented antigen	Prevents macrophages from migrating out of inflamed area Increases phagocytic power of macrophages (activates angry macrophages)
Macrophages[b]	Large, tissue-bound phagocytic specialists	Stimulated by presence of microbes; enhanced by interferon; activated by helper T cells; stimulated by antibodies on surface of foreign material; attracted by chemotaxins; linked to target cell by opsonin	Important in nonspecific defense; first line of defense Phagocytize foreign material and cellular debris Secrete chemical mediators that exert a variety of effects (see phagocytic secretions) Clear way for tissue repair by removing debris Secrete interferon Process and present antigen to B and T cells, thereby assisting in antibody formation and T-cell sensitization
Memory cells[b]	B and T cells produced during initial infection but remaining dormant until subsequent invasion by same antigen	Stimulated by subsequent invasion of same antigen	Expand activated clone Launch swifter, more powerful attack on subsequent exposure
Monocytes[b]	White blood cells	Various factors stimulate their derivatives, the macrophages (see Macrophages)	Converted into large, tissue-bound macrophages after leaving the blood
Natural killer (NK) cells[b]	Lymphocytelike cells	Stimulated by interferon	Nonspecifically lyse virus-infected cells and tumor cells on first exposure to them Secrete interferon

[a] Chemical mediator
[b] Effector cell

Appendix B Control, Functions, and Interactions of Immune-Response Effectors (continued)

Immune Response Effectors	Description	Controlling Mechanism(s)	Function(s)
Neutrophils[b]	White blood cells	Stimulated by presence of microbes, by antibodies on surface of foreign material, and by circulating antibodies; attracted by chemotaxins; linked to target cell by opsonins	Important in nonspecific defense Important in inflammatory process; highly mobile phagocytic specialists Release chemicals that exert a variety of effects (see Phagocytic secretions)
Opsonins[a]	Antibodies and one of activated proteins of complement system	Produced in response to factors that stimulate antibody secretion (see Antibodies) and complement activation (see Complement system)	Enhance phagocytosis by linking foreign cell to phagocytic cell
Phagocytic secretions (EP/LEM/IL-1 and others)[a]	Chemical mediators released from neutrophils or macrophages	Released in presence of microbes	Directly kill microbes extracellularly Trigger clotting and anticlotting systems Activitate kinins Induce development of fever Decrease plasma iron Stimulate granulopoiesis Stimulate secretion of acute-phase proteins Enhance proliferation and differentiation of both B and T cells
Plasma cells[b]	Activated form of B lymphocyte	Specific B cells that match invading antigen proliferate and are converted into antibody-secreting plasma cells; stimulated by B-cell growth factor, by helper T cells, and by interferon	Secrete customized antibodies specific to invading antigen
Plasminogen (inactive); plasmin (active)[a]	Final plasma protein in anticlotting cascade	Activated by chemical mediators released from microbe-stimulated phagocytes	Slowly dissolves interstitial fluid clot formed during inflammatory response
Prostaglandin[a]	Fatty acid–derived local chemical mediator	Released in hypothalamus by EP	Raises set point of hypothalamic thermostat to produce fever
Slow reactive substance of anaphylaxis (SRS-A)[a]	Chemical mediator released from mast cells and basophils in allergic reactions	Released by binding of allergen to IgE antibodies coating mast cells and basophils	Induces profound, prolonged constriction of small airways Increases capillary permeability
Suppressor T cells[b]	Form of T lymphocyte	Stimulated by helper T cells	Limit immune reaction in check-and-balance fashion

Appendix B Control, Functions, and Interactions of Immune-Response Effectors (continued)

Immune Response Effectors	Description	Controlling Mechanism(s)	Function(s)
T-cell growth[a] factor	Lymphokine secreted by helper T cells	Release stimulated by macrophage-presented antigen and IL-1	Augments activity of all T cells
Thymosin[a]	Collection of hormones produced by thymus	Unknown	Enhances proliferation of T-cell colonies Enhances immune capabilities of existing T cells
T lymphocytes (T cells) (see Cytotoxic T cells, Helper T cells, and Suppressor T cells)[b]	White blood cells	Activated by macrophage-presented antigen; stimulated by interferon, IL-1, and helper T cells	Responsible for cell-mediated immunity (kills cells through nonphagocytic means) Secrete lymphokines

[a] Chemical mediator
[b] Effector cell

GLOSSARY INDEX

The following key terms are boldfaced and defined on the designated pages.

I N D E X

C R E D I T S

Stock Photographs

Figure 5–1, **p. 117,** Lester V. Bergman & Associates, Inc.; Figure 9–31, **p. 292,** Visuals Unlimited/Sloop-Ober; Figure 11–9, **p. 356,** Copyright Boehringer Ingelheim International GmbH, photo Lennart Nilsson; Figure 12–17, **p. 391,** Copyright Boehringer Ingelheim International GmbH, photo Lennart Nilsson; Figure 12–19, **p. 394,** Copyright Boehringer Ingelheim International GmbH, photo Lennart Nilsson; Figure 12–21, **p. 395,** Copyright Boehringer Ingelheim International GmbH, photo Lennart Nilsson; Figure 18–22, **p. 659,** Lester V. Bergman & Associates, Inc.; Figure 19–5, **p. 669,** Lester V. Bergman & Associates, Inc.; Figure 19–6, **p. 669,** Lester V. Bergman & Associates, Inc.; Figure 20–8a, **p. 723,** Manfred Kage/Peter Arnold, Inc.

Artwork

Wayne Clark Figure 2–2a, **p. 19;** Figure 2–6a, **p. 22;** Figure 2–9b, **p. 25;** Figure 2–10a, **p. 26;** Figure 4–8, **p. 98;** Figure 4–10a, b, **p. 101;** Figure 4–14a, **p. 105;** Figure 4–16 top, **p. 108;** Figure 4–17 top, **p. 111;** Figure 4–18, **p. 112;** Figure 10–14a, **p. 316;** Figure 10–15, **p. 317;** Figure 11–2b, **p. 345;** Figure 11–3, **p. 345;** Figure 11–4, **p. 347;** Figure 12–2, **p. 366;** Figure 13–3, **p. 410;** Figure 13–4, **p. 411;** Figure 13–11, **p. 418;** Figure 14–2, **p. 466;** Figure 14–4, **p. 468;** Figure 14–10, **p. 474;** Figure 14–13, **p. 477;** Figure 20–13 (layout), **p. 734;** Figure 20–14 (layout), **p. 735;** Figure 20–23 (layout), **p. 747;** Figure 20–24 (layout), **p. 748**

Cyndie Clark-Huegel Figure 5–4, **p. 122;** Figure 6–29, **p. 184;** Figure 6–37, **p. 193;** Figure 6–38; **p. 194;** Figure 6-42, **p. 198;** Figure 6–43, **p. 199;** Figure 8–15a, **p. 233;** Figure 8–18, **p. 242;** Figure 8–22, **p. 247;** Figure 9–2, **p. 260;** Figure 9–17, **p. 275;** Figure 10–6a, c, **p. 305;** Figure 10–24, **p. 327;** Figure 10–25a, **p. 328;** Figure 10–28, **p. 330;** Figure 10–30, **p. 332;** Figure 12–1, **p. 364;** Figure 13–2, **p. 409;** Figure 13–20, **p. 430;** Figure 14–1, **p. 464;** Table 16–2, **pp. 550-51;** Figure 16–5, **p. 559;** Figure 16–8, **p. 566;** Figure 17–5, **p. 612;** Figure 18–2, **p. 626;** Figure 19–1, **p. 663;** Figure 20–5 (layout), **p. 717;** Figure 20–6a, d (layout), **p. 721;** Figure 20–8b, c (layout), **p. 723;** Figure 20–20 (layout), **p. 744;** Figure 20–21 (layout), **p. 745;** Figure 20–22 (layout), **p. 746;** Figure 20–25, **p. 749;** Figure 20–29, **p. 754**

Darwen and Vally Hennings Figure 5–6, **p. 124;** Table 5–1, **pp. 126–127;** Figure 5–7, **p. 129;** Figure 5–8, **p. 129;** Figure 5–9, **p. 130;** Figure 5–10, **p. 131;** Figure 5–11, **p. 133;** Figure 5–14, **p. 137;** Figure 5–16, **p. 138;** Figure 5–17, **p .139;** Figure 5–20, **p. 145;** Figure 5–23, **p. 150;** Figure 5–24, **p. 151;** Figure 5–25, **p. 151;** Figure 5–26, **p. 152;** Figure 5–27, **p. 154;** Figure 5–28, **p. 155;** Figure 6–5, **p. 162;** Figure 6–10, **p. 169;** Figure 6–18; **p. 173;** Figure 6–19, **p. 175;** Figure 6–20; **p. 176;** Figure 6-21, **p. 177;** Figure 6–25, **p. 180;** Figure 6–27, **p. 181;** Figure 6–28, **p. 182;** Figure 6–32, **p. 187;** Figure 6–33, **p. 189;** Figure 6–34, **p. 190;** Figure 6–36, **p. 192;** Figure 9–1, **p. 259;** Figure 9–3a, c, **p. 261;** Figure 9–6, **p. 264;** Figure 9–7, **p. 264;** Figure 9–9, **p. 266;** Figure 9–11, **p. 267;** Figure 9–19 bottom, **p. 280;** Figure 9–33, **p. 293;** Figure 9–35, **p. 296;** Figure 10–2, **p. 300;** Figure 10–21a, b, **p. 322;** Figure 13–33, **p. 450;** Figure 13–35, **p. 453;** Figure 14–6, **p. 470;** Figure 16–1, **p. 548;** Figure 16–6, **p. 560;** Figure 16–7, **p. 561;** Figure 16–9, **p. 567;** Figure 16–15; **p. 574;** Figure 16–16; **p. 577;** Figure 16–21, **p. 582;** Figure 16–22, **p. 584;** Figure 18–12, **p. 644;** Figure 19–7, **p. 671**

Sandra McMahon Figure 2–1, **p. 18;** Figure 2–25, **p. 41;** Figure 6–9, **p. 168;** Figure 6–40, **p. 196;** Figure 8–1, **p. 218;** Figure 8–9, **p. 225;** Figure 8–21, **p. 246;** Figure 9–5, **p. 263;** Figure 12–24, **p. 401;** Figure 16–2, **p. 553;** Figure 16–17, **p. 578;** Figure 16–23, **p. 586;** Figure 16–30, **p. 597;** Figure 18–20, **p. 655;** Figure 20–1, **p. 712;** Figure 20–2, **p. 713;** Figure 20–30a, **p. 756**

Rolin Graphics Figure 3–20, **p. 78;** Exercise 3–11, **p. 91;** Figure 4–2, **p. 94;** Figure 4–3, **p. 94;** Figure 4–4, **p. 95;** Figure 4–7, **p. 97;** Figure 4–13, **p. 104;** Figure 4–15, **p. 106;** Figure 4–16 bottom, **p. 108;** Figure 4–17 bottom, **p. 111;** Figure 6–4, **p. 161;** Figure 6–11, **p. 169;** Figure 6–22, **p. 177;** Figure 6–30a, **p. 185;** Figure 6–31, **p. 186;** Figure 6–41, **p. 197;** Figure 8–12, **p. 228;** Figure 8–14, **p. 231;** Figure 8–15b, **p. 233;** Figure 8–16, **p. 234;** Figure 8–19, **p. 243;** Figure 8–25, **p. 253;** Figure 9–10, **p. 267;** Figure 9–12, **p. 268;** Figure 9–15, **p. 272;** Figure 9–16, **p. 273;** Figure 9–18, **p. 276;** Figure 9–19 top, **p. 280;** Figure 9–20, **p. 281;** Figure 9–22a, **p. 284;** Figure 9–23, **p. 286;** Figure 9–25, **p. 288;** Figure 9–27, **p. 289;** Figure 9-29, **p. 291;** Figure 10–5, **p. 304;** Figure 10–6b, **p. 305;** Figure 10–7, **p. 306;** Figure 10–11, **p. 311;** Figure 10-20, **p. 321;** Figure 10–22, **p. 329;** Figure 10–31, **p. 333;** Figure 12–12, **p. 381;** Figure 13–13, **p. 420;** Figure 13–21, **p. 431;** Figure 13–22, **p. 432;** Figure 13–28, **p. 442;** Figure 13–30, **p. 445;** Figure 13–32, **p. 449;** Figure 14–18, **p. 485;** Figure 15–2, **p. 515;** Figure 15–10, **p. 533;** Figure 17–9, **p. 619;** Figure 18–11, **p. 641;** Figure 18–19, **p. 653;** Figure 19–9, **p. 674;** Figure 20-26, **p. 750**

John and Judy Waller Front inside cover endsheets; Figure 1–1, **p. 4;** Figure 1–2, **p. 7;** Figure 1–3, **pp. 8–9;** Figure 1–4, **p. 12;** Figure 2–3, **p. 21;** Figure 2–4, **p. 21;** Figure 2–5, **p. 22;** Figure 2–7, **p. 23;** Figure 2–8, **p. 24;** Figure 2–9a, **p. 25;** Figure 2–11, **p. 27;** Figure 2–12, **p. 28;** Figure 2–13, **p. 30;** Figure 2–14, **p. 31;** Figure 2–15, **p. 32;** Figure 2–17, **p. 36;** Figure 2–18, **p. 36;** Figure 2–19, **p. 37;** Figure 2–20, **p. 37;** Figure 2–21, **p. 38;** Figure 2–22, **p. 38;** Figure 2–23, **p. 39;** Figure

2–26, **p. 43;** Figure 2–27, **p. 44;** Figure 2–28, **p. 45;** Figure 2–29, **p. 47;** Figure 2–30, **p. 48;** Figure 2–31, **p. 49;** Figure 2–32, **p. 50;** Figure 2–33, **p. 51;** Figure 2–34, **p. 52;** Figure 2–35, **pp. 54–55;** Figure 2–37, **p. 56;** Figure 3–2, **p. 60;** Figure 3–3, **p. 61;** Figure 3–4, **p. 63;** Figure 3–5, **p. 65;** Figure 3–6, **p. 66;** Figure 3–7, **p. 67;** Figure 3–8, **p. 68;** Figure 3–9, **p. 69;** Figure 3–10, **p. 69;** Figure 3–11, **p. 70;** Figure 3–12, **p. 71;** Figure 3–13, **p. 73;** Figure 3–14, **p. 73;** Figure 3–15, **p. 73;** Figure 3–16, **p. 74;** Figure 3-17, **p. 74;** Figure 3–18, **p. 75;** Figure 3–19, **p. 77;** Figure 3–21, **p. 80;** Figure 3–22, **p. 81;** Figure 3–23, **p. 82;** Figure 3–24, **p. 85;** Figure 3–25, **p. 87;** Figure 3–26, **p. 87;** Figure 3–27, **p. 88;** Figure 3–28, **p. 89;** Figure 4–1, **p. 93;** Figure 4–5, **p. 96;** Figure 4–6, **p. 96;** Figure 4–9, **p. 99;** Figure 4–10c, **p. 101;** Figure 4–11, **p. 102;** Figure 4–12, **p. 103;** Figure 4–14b, **p. 105;** Figure 5–2, **p. 119;** Figure 5–3, **p. 120;** Figure 5–5, **p. 123;** Figure 5–12, **p. 135;** Figure 5–13, **p. 135;** Figure 5–15, **p. 138;** Figure 5–18; **p. 141;** Figure 5–19, **p. 144;** Figure 5–21, **p. 146;** Figure 5-22, **p. 148;** Figure 6–1, **p. 159;** Figure 6–3, **p. 160;** Figure 6–6, **p. 164;** Figure 6–7, **p. 165;** Figure 6–8, **p. 167;** Figure 6–13, **p. 170;** Figure 6–14, **p. 171;** Figure 6–15, **p. 171;** Figure 6–16, **p. 172;** Figure 6–17, **p. 173;** Figure 6–23; **p. 178;** Figure 6–24; **p. 178;** Figure 6–26, **p. 181;** Figure 6–30b, c, **p. 185;** Figure 6–39, **p. 195;** Figure 7–1, **p. 204;** Figure 7–2, **p. 211;** Figure 7–3, **p. 212;** Figure 8–3, **p. 220;** Figure 8–4, **p. 221;** Figure 8–5, **p. 221;** Figure 8–6, **p. 222;** Figure 8–7, **p. 223;** Figure 8–8, **p. 224;** Figure 8–10, **p. 226;** Figure 8–11, **p. 227;** Figure 8–13, **p. 230;** Figure 8–17, **p. 236;** Figure 8–20, **p. 245;** Figure 8–23, **p. 248;** Figure 8–24, **p. 252;** Figure 8–26, **p. 254;** Figure 8–27, **p. 255;** Figure 9–3b, **p. 261;** Figure 9–4, **p. 262;** Figure 9–8, **p. 265;** Figure 9–13, **p. 269;** Figure 9–14, **p. 271;** Table 9–2, **p. 277;** Figure 9–21, **p. 282;** Figure 9–22b, **p. 284;** Figure 9–24, **p. 288;** Figure 9–26, **p. 289;** Figure 9–28, **p. 290;** Figure 9–30, **p. 291;** Figure 9–32, **p. 293;** Figure 9–34, **p. 295;** Figure 10–1, **p. 299;** Figure 10–3, **p. 302;** Figure 10–4, **p. 303;** Figure 10–8, **p. 307;** Figure 10-9, **p. 308;** Figure 10–10, **p. 309;** Figure 10–12, **p. 313;** Figure 10–13, **p. 315;** Figure 10–14b, **p. 316;** Figure 10–16, **p. 318;** Figure 10–17, **p. 319;** Figure 10–18, **p. 319;** Figure 10–19, **p. 320;** Figure 10–21c, **p. 322;** Figure 10–23, **p. 326;** Figure 10–25b, **p. 328;** Figure 10-26, **p. 329;** Figure 10–27, **p. 330;** Figure 10–29, **p. 331;** Figure 10–32, **p. 333;** Figure 10–33, **p. 334;** Figure 10–34, **p. 338;** Figure 10–35, **p. 339;** Figure 11–1, **p. 344;** Figure 11–5, **p. 349;** Figure 11–7, **p. 352;** Figure 11-8, **p. 355;** Figure 11–10, **p. 357;** Figure 11–11, **p. 358;** Figure 11–12, **p. 359;** Figure 11–13, **p. 360;** Figure 12–3, **p. 368;** Figure 12–4, **pp. 370–371;** Figure 12–5, **p. 372;** Figure 12–7, **p. 375;** Figure 12–9, **p. 377;** Figure 12–10, **p. 378;** Figure 12–11, **p. 380;** Figure 12–13, Figure 382; Figure 12–14, **p. 385;** Figure 12–15, **p. 388;** Figure 12–18, **p. 392;** Figure 12–20, **p. 395;** Figure 12–22, **p. 398;** Figure 12–23, **p. 400;** Figure 12–25, **p. 403;** Figure 13–1, **p. 408;** Figure 13–5, **p. 412;** Figure 13–6, **p. 413;** Figure 13–7, **p. 413;** Figure 13–8, **p. 414;** Figure 13–9, **p. 415;** Figure 13–10, **p. 416;** Figure 13–12, **p. 419;** Figure 13–14, **p. 423;** Figure 13–15, **p. 425;** Figure 13–16, **p. 426;** Figure 13–17, **p. 426;** Figure 13–18, **p. 428;** Figure 13-19, **p. 429;** Figure 13–23, **p. 433;** Figure 13–24, **p. 435;** Figure 13–25, **p. 437;** Figure 13–27, **p. 440;** Figure 13–29, **p. 443;** Figure 13–31, **p. 446;** Figure 13–34, **p. 451;** Figure 13–36, **p. 453;** Figure 13–37, **p. 454;** Figure 13–38, **p. 456;** Figure 14–5, **p. 469;** Table 14–1, **p. 472;** Figure 14–8, **p. 473;** Figure 14–9, **p. 474;** Figure 14–11, **p. 475;** Figure 14–12, **p. 476;** Figure 14–14, **p. 478;** Figure 14–15, **p. 479;** Figure 14–16, **p. 480;** Figure 14–17, **p. 482;** Figure 14–19, **p. 487;** Figure 14–20, **p. 488;** Figure 14–21, **p. 489;** Figure 14–22, **p. 490;** Figure 14–23, **p. 493;** Figure 14–24, **p. 496-97;** Figure 14–25, **p. 498;** Figure 14–26, **p. 499;** Figure 14–27, **p. 500;** Figure 14–28, **p. 502;** Figure 14–29, **p. 507;** Figure 15–1, **p. 511;** Figure 15–3, **p. 521;** Figure 15–4, **p. 522;** Figure 15–5, **p. 523;** Figure 15–6, **p. 525;** Figure 15–7, **p. 526;** Figure 15–8, **p. 527;** Figure 15–9, **p. 530;** Figure 15–11, **p. 534;** Figure 15–12, **p. 535;** Figure 15–13, **p. 536;** Figure 15–14, **p. 537;** Figure 15–15, **p. 538;** Figure 15–16, **p. 539;** Figure 15–17, **p. 543;** Table 16–1, **p. 549;** Figure 16–3, **p. 555;** Figure 16–4, **p. 558;** Figure 16–10, **p. 568;** Figure 16–11, **p. 568;** Table 16–4, **p. 569;** Figure 16–12, **p. 570;** Table 16–5, **p. 571;** Figure 16–13, **p. 572;** Figure 16–14, **p. 573;** Figure 16–18. **p. 579;** Figure 16–19, **p. 580;** Figure 16–20, **p. 581;** Figure 16–25, **p. 589;** Figure 16–26, **p. 590;** Figure 16–27, **p. 592;** Figure 16–28, **p. 593;** Figure 16–29, **p. 595;** Figure 17–1, **p. 604;** Figure 17–3, **p. 608;** Figure 17–4, **p. 611;** Figure 17–6, **p. 615;** Figure 17–7, **p. 617;** Figure 17–8, **p. 618;** Figure 18–1, **p. 624;** Figure 18-3, **p. 627;** Figure 18–4, **p. 634;** Figure 18–5, **p. 635;** Figure 18–6, **p. 637;** Figure 18–7, **p. 638;** Figure 18–8, **p. 638;** Figure 18–9, **p. 639;** Figure 18–10, **p. 640;** Figure 18–13, **p. 645;** Figure 18–14, **p. 647;** Figure 18–15; **p. 649;** Figure 18–16, **p. 650;** Figure 18–17, **p. 651;** Figure 18–18, **p. 652;** Figure 18–21, **p. 657;** Figure 19–2, **p. 664;** Figure 19–3, **p. 667;** Figure 19–4, **p. 668;** Figure 19–8, **p. 673;** Figure 19–10, **p. 676;** Figure 19–11, **p. 678;** Figure 19–12, **p. 680;** Figure 19–13, **p. 682;** Figure 19–14, **p. 685;** Figure 19–15, **p. 686;** Figure 19–16, **p. 689;** Figure 19–17, **p. 691;** Figure 19–18, **p. 694;** Figure 19–19, **p. 697;** Figure 19–20, **p. 698;** Figure 19–21, **p. 702;** Figure 19–22, **p. 703;** Figure 19–23, **p. 704;** Figure 19–24, **p. 705;** Figure 19–25, **p. 706;** Figure 20–3, **p. 714;** Figure 20–4, **p. 715;** Figure 20–5 (final rendering), **p. 717;** Figure 20–6a, d (final rendering), **p. 721;** Figure 20–7, **p. 722;** Figure 20–8b, c (final rendering), **p. 723;** Figure 20–9, **p. 724;** Figure 20–10, **p. 727;** Figure 20–11, **p. 729;** Figure 20–12, **p. 732;** Figure 20–13 (final rendering), **p. 734;** Figure 20–14 (final rendering), **p. 735;** Figure 20–15, **p. 736;** Figure 20–16, **p. 737;** Figure 20–17, **p. 738;** Figure 20-18, **p. 739;** Figure 20–19, **p. 740;** Figure 20–20 (final rendering), **p. 744;** Figure 20–21 (final rendering), **p. 745;** Figure 20–22 (final rendering), **p. 746;** Figure 20–23 (final rendering), **p. 747;** Figure 20–24 (final rendering), **p. 748;** Figure 20–27, **p. 751;** Figure 20–28, **p. 753;** Figure 20–30b, **p. 756;** Figure 20–31, **p. 757**

Anatomical Terms Used to Indicate Direction and Orientation

anterior	situated in front of or in the front part of
posterior	situated behind or toward the rear
ventral	toward the belly or front surface of the body; synonymous with anterior
dorsal	toward the back surface of the body; synonymous with posterior
medial	denoting a position nearer the midline of the body or a body structure
lateral	denoting a position toward the side or farther from the midline of the body or a body structure
superior	toward the head
inferior	away from the head
proximal	closer to a reference point
distal	farther from a reference point
sagittal section	a vertical plane that divides the body or a body structure into right and left sides
longitudinal section	a plane that lies parallel to the length of the body or a body structure
cross section	a plane that runs perpendicular to the length of the body or a body structure
frontal or *coronal section*	a plane parallel to and facing the front part of the body

Word Derivatives Commonly Used in Physiology

a-; an-	absence or lack	*epi-*	above; over	*oto-*	ear
ad-; af-	toward	*erythro-*	red	*para-*	near
adeno-	glandular	*gastr-*	stomach	*pariet-*	wall
angi-	vessel	*-gen; -genic*	produce	*peri-*	around
anti-	against	*gluc-; glyc-*	sweet	*phago-*	eat
archi-	old	*hemo-*	blood	*pod*	footlike
-ase	splitter	*hemi-*	half	*-poiesis*	formation
auto-	self	*hepat-*	liver	*poly-*	many
bi-	two; double	*homeo-*	sameness	*post-*	behind; after
-blast	former	*hyper-*	above; excess	*pre-*	ahead of; before
brady-	slow	*hypo-*	below; deficient	*pro-*	before
cardi-	heart	*inter-*	between	*pseudo-*	false
cephal-	head	*intra-*	within	*pulmon-*	lung
cerebr-	brain	*kal-*	potassium	*rect-*	straight
-cide	kill; destroy	*leuko-*	white	*ren-*	kidney
chondr-	cartilage	*lip-*	fat	*reticul-*	network
contra-	against	*macro-*	large	*retro-*	backward
cost-	rib	*mamm-*	breast	*sacchar-*	sugar
crani-	skull	*mening-*	membrane	*sarc-*	muscle
-crine	secretion	*micro-*	small	*semi-*	half
crypt-	hidden	*mono-*	single	*-some*	body
cutan-	skin	*multi-*	many	*sub-*	under
-cyte	cell	*myo-*	muscle	*supra-*	upon; above
de-	lack of	*natr-*	sodium	*tachy-*	rapid
di-	two; double	*neo-*	new	*therm-*	temperature
dys-	difficult; faulty	*nephr-*	kidney	*-tion*	act or process of
ef-	away from	*neuro-*	nerve	*trans-*	across
-elle	tiny; miniature	*oculo-*	eye	*tri-*	three
encephalo-	brain	*-oid*	resembling	*vaso-*	vessel
endo-	within; inside	*ophthalmo-*	eye	*-uria*	urine
ecto-; exo-; extra-	outside; away from	*oral*	mouth		
-emia	blood	*osteo-*	bone		